Theoretical Global Seismology

Theoretical Global Seismology

F. A. Dahlen

and

Jeroen Tromp

PRINCETON UNIVERSITY PRESS

PRINCETON, NEW JERSEY

Published by Princeton University Press, 41 William Street,
Princeton, New Jersey 08540
In the United Kingdom: Princeton University Press, Chichester, West Sussex

Library of Congress Cataloging-in-Publication Data

Dahlen, F. A., 1942–
Theoretical global seismology / F. A. Dahlen and Jeroen Tromp.
p. cm.
Includes bibliographical references and index.

ISBN: 978-0-691-00124-1

1. Seismology. I. Tromp, Jeroen. II. Title.
QE534.2.D34 1998
551.22—dc21 98-15199

The publisher would like to acknowledge the authors of this volume for providing the camera-ready copy from which this book was printed

The paper used in this publication meets the minimum requirements of ANSI/NISO Z39.48-1992 (R 1997) (*Permanence of Paper*)

http://pup.princeton.edu

Printed in the United States of America

10 9 8 7 6 5 4 3 2 1

10 9 8 7 6 5 4 3 2 1
(Pbk.)

Contents

Part III The Aspherical Earth

Appendixes

Preface

Five years ago, in the summer of 1993, we set out to write a slender monograph tentatively entitled *The Free Oscillations of the Earth*. With each e-mail exchange of draft chapters, our modest ambitions mushroomed; the final result is this book—*Theoretical Global Seismology*—an advanced treatise intended to be read by graduate students and researchers in geophysics and allied fields. Although the present title is more indicative of the scope than the original, the contents nevertheless reflect the book's origins. We devote our attention almost exclusively to the forward problem of computing synthetic seismograms upon a realistic three-dimensional model of the Earth, with a strong emphasis on the normal-mode summation method. Free oscillations have many close associations with surface waves, and we consider them in some detail as well; we give shorter shrift to body waves, and do not discuss seismic instrumentation, data analysis procedures, or geophysical inverse theory at all.

The introductory chapter recounts the history of free-oscillation and surface-wave research, beginning with the earliest theoretical investigation of the oscillations of an elastic sphere in the 1820's, through the first observation of the gravest terrestrial oscillations following the great Chile earthquake of 1960, and concluding with the initial determinations of global upper-mantle heterogeneity using digitally recorded seismograms in the 1980's. The remainder of the text—like ancient Gaul—is divided into three parts. In Part I—*Foundations*—we derive the linearized equations of motion governing both an elastic and anelastic Earth subject to a non-hydrostatic state of initial stress, and show how to express the elastic-gravitational response to an arbitrary earthquake source as a sum of free oscillations or normal modes. We conclude with a discussion of the Rayleigh-Ritz method, which yields a truncated matrix formulation that is identical to the classical theory of the small oscillations of a system with a finite number of degrees of freedom, generalized to account for rotation and anelasticity. In Part II—*The Spherical Earth*—we restrict attention to the case of an Earth model that is non-rotating and spherically symmetric; the toroidal and spheroidal eigenfrequencies and eigenfunctions of such a model can be found essentially exactly by numerical integration of the governing radial differential equations. We show how to calculate synthetic seismograms on a spherical Earth by means of normal-mode summation, and discuss the propagation of Love and Rayleigh surface waves as well as mode-ray duality. These results form the basis for the more general considerations in Part III—*The Aspherical Earth*—where we use perturbation theory to treat the splitting and coupling of the normal-mode multiplets produced by the Earth's rotation, ellipticity and other departures from spherical symmetry, and JWKB theory to describe the propagation of both body waves and surface waves upon a laterally heterogeneous Earth.

The three parts are arranged in order of decreasing "shelf life". The fundamental equations and results obtained in Part I are applicable to a very general Earth model, and should provide the basis for discussions of the elastic-gravitational deformation of the Earth into the foreseeable future. The results pertaining to a spherical Earth in Part II are likewise well established; only relatively minor numerical details are likely to change as the spherically averaged structure of the Earth continues to be refined. The approximate methods of dealing with the Earth's lateral heterogeneity which we discuss in Part III are not as well developed; three-dimensional global tomography is an extremely active research field at the present time, and improvements in the procedures and results described here seem likely in the future. In addition to the fifteen chapters in Parts I through III, there are four mathematical appendixes devoted to vectors and tensors, ordinary and generalized spherical harmonics, and the matrix machinery needed to calculate coupled-mode synthetic seismograms on a rotating, anelastic, laterally heterogeneous Earth.

Variational principles appear in a number of guises, and provide a unifying thread which serves to knit the various chapters together. We enunciate Hamilton's principle for a general elastic Earth model in Chapter 3, and discuss its frequency-domain analogue, Rayleigh's principle, on both a non-rotating and rotating Earth in Chapter 4. We extend Rayleigh's principle to an anelastic Earth in Chapter 6, deduce the equivalent elastic and anelastic matrix principles in Chapter 7, and utilize the orthonormality of the surface spherical harmonics to obtain a purely radial variational principle on a spherically symmetric Earth in Chapter 8. The one-dimensional and three-dimensional versions of Rayleigh's principle provide the basis for the spherical and aspherical perturbation analyses in Chapters 9 and 13. Finally, we develop ray theory for body waves and JWKB theory for surface waves on a smooth laterally heterogeneous Earth using an associated slow variational principle in Chapters 15 and 16.

The subject matter of this book may be described as mathematical in the sense that there is a high proportion of equations to words; however, all of the theoretical considerations are purely formal, with no attempt at rigor whatsoever. We are not finicky about the continuity and differentiability of displacement, strain and stress fields, or the open or closed nature of regions within the Earth, except where it matters to get the physics right. The only mathematical property of the elastic-gravitational operator governing the Earth's free oscillations which is considered to be physically significant is whether or not it is Hermitian. We blithely assume that the normal modes of an elastic Earth model are complete, ignore the presence of a branch cut in developing a mode-sum representation of the response of an anelastic Earth, manipulate infinite matrices without regard for convergence, and seldom worry about the precise nature of the equality in spherical-harmonic and other infinite orthonormal eigenfunction expansions.

Sections denoted by a star* contain more esoteric material which may be omitted upon a first reading. Many of the starred sections deal with the theoretical complications introduced by the Earth's rotation; for example, in analyzing the influence of anelasticity upon the free oscillations, it is necessary to introduce the dual eigenfunctions $\bar{\mathbf{s}}$ associated with the "anti-Earth" having the opposite sense of rotation, as well as the ordinary eigenfunctions $\mathbf{s}$ of the Earth itself. A few unstarred sections make use of these dual eigenfunctions in the interest of brevity and maximum generality; uninterested readers may simply eliminate the overbars, since in the absence of rotation the eigenfunctions and their duals coincide: $\bar{\mathbf{s}} = \mathbf{s}$.

We are deeply indebted to many colleagues for their generous support and assistance during the preparation of this book. First and foremost, we would like to thank Freeman Gilbert for his barrage of encouraging e-mails, filled with valuable commentary upon a variety of topics—ranging from the inherent positivity of the group speed to the application of ray theory to stealth-aircraft detection. We also wish to express our sincere gratitude to Guust Nolet, whose detailed and constructive criticism, particularly of the appendixes, was extremely helpful. An early, incomplete draft was reviewed by Brian Kennett, Guy Masters and Barbara Romanowicz; they suggested a number of improvements which have been incorporated in the final version. Several people graciously complied with our request to read a particular chapter devoted to their field of expertise; we especially wish to thank Henk Marquering and Roel Snieder for their remarks regarding Chapter 11, Li Zhao for his careful review of Chapter 12, and Colin Thomson for his advice on Chapter 15. Appendices B and C on ordinary and generalized spherical harmonics are based in part upon lecture notes by George Backus and John Woodhouse. Further suggestions for improvements and additions were provided by Chris Chapman, Adam Dziewonski, Andy Jackson, Paul Richards, and Philippe Lognonné. Our foray into the early German surface-wave literature was aided by Thomas Meier. Finally, we would like to acknowledge our indebtedness to Miaki Ishii, whose thorough review of the entire manuscript is very much appreciated.

We are grateful to many individuals for helping us to assemble the more than 225 illustrations. The theoretical spectra and seismograms and eigenfunction and Fréchet kernel plots were almost all produced by students in Jeroen Tromp's global seismology courses at Harvard and MIT. We thank John He, Yu Gu, Rishi Jha, Hrafnkell Kárason, Erik Larson, Xian-Feng Liu, Jeff McGuire, Meredith Nettles, Frederik Simons and Mark Taylor for their help in this endeavor. A number of colleagues, including Göran Ekström, Guy Masters, Joe Resovsky, Mike Ritzwoller, Barbara Romanowicz, Genevieve Roult, Peter Shearer, Zheng Wang, Shingo Watada, Ruedi Widmer and Li Zhao, provided us with additional figures; we sincerely thank them all. Most of the cartoons were ably drafted from our slapdash sketches by Dearbhla McHenry and Leslie Hsu; the indispensable Leslie also organized, re-sized, touched-up and unified all of the figures for encapsulation into the final camera-ready copy.

The labor of composing, formatting and typesetting this behemoth of more than 1000 pages and 3800 numbered equations was ameliorated by LaTeX, BibTeX and *MakeIndex*; we benefitted from the expertise of Bob Fischer and Erik Larson. Meredith Nettles and Yu Gu indulged our paranoia by religiously backing up all of the chapter and figure files. It has been a pleasure working with the capable staff at Princeton University Press, particularly Jack Repcheck, who has guided this book to publication from the outset, and Jennifer Slater, who did a splendid job of copy editing.

The awards of a John Simon Guggenheim Memorial Foundation Fellowship to Tony Dahlen and a David and Lucile Packard Foundation Fellowship to Jeroen Tromp are greatly appreciated. In addition, Tony Dahlen would like to express his sincere thanks to Raul Madariaga, Jean-Paul Montagner and Philippe Lognonné for their support and gracious hospitality during his 1993-1994 sabbatical leave at the Institut de Physique du Globe de Paris. A preliminary draft of Part I was completed and the remainder of the book was outlined during this visit. Further financial support was provided by grants from the National Science Foundation to the two authors at Princeton and Harvard.

Finally, we would like to thank Elisabeth, Tracey and Alex for patiently putting up with our seismological gibberish and preoccupation with this project during the past five years. The fact of the matter is that we should be grateful that they put up with us at all.

Princeton and Cambridge
June 1998

Theoretical Global Seismology

Chapter 1

Historical Introduction

After every major earthquake, the Earth rings like a large bell for several days. These free oscillations of the Earth are routinely detected at modern broad-band seismographic stations, which are now distributed globally. The eigenfrequencies and decay rates of the vibrations can be measured and used to constrain the radial and lateral distribution of density, seismic wave speed and anelastic attenuation within the interior. The observed amplitudes and phases can likewise be used to infer the origin times, hypocentral locations, seismic moments, and fault geometries of the earthquakes responsible for the excitation. The analysis of the free oscillations of the Earth and the allied normal-mode methods employed in the determination of the Earth's internal structure and the source mechanisms of earthquakes—the topics considered in this book—constitute one of the cornerstones of quantitative seismology. Excellent reviews which summarize the state of progress at two pivotal points in the development of the field are provided by Stoneley (1961), Lapwood & Usami (1981) and Buland (1981). This introduction contains our own brief historical survey of research on terrestrial free oscillations and the associated propagating surface waves, focusing upon the theoretical and observational advances made prior to 1985. More recent developments are described—with little attention to their historical context—in subsequent chapters of the book.

1.1 Early Theoretical Studies

The theoretical analysis of the Earth's normal modes was initiated over one and one-half centuries ago by the French mathematician Poisson. In a remarkable memoir presented to the Paris Academy of Sciences in August 1828, he developed a general theory of deformation for solid materials, based upon "la considération des actions mutuelles de leurs molécules", and applied it to a large number of special elastostatic and elastodynamic problems, including the determination of the frequencies of the purely radial oscillations of a homogeneous, non-gravitating sphere (Poisson 1829). These investigations, together with the work of his contemporaries Navier and Cauchy, laid the foundations for the modern theory of linear elasticity. The equations of equilibrium and vibration derived by Poisson are now recognized to be incomplete, inasmuch as they characterize the elastic response of an isotropic solid in terms of a single elastic parameter rather than two; the radial-mode eigenfrequencies and eigenfunctions he obtained are, however, correct in the special case that we now refer to as a Poisson solid, which has $\kappa = \frac{5}{3}\mu$, where κ is the incompressibility and μ is the rigidity. Not being a physicist or natural philosopher, Poisson did not seek to estimate or calculate the numerical free periods of radial vibration of the Earth or any man-made spherical objects, but rather expressed his final results in terms of dimensionless ratios.

The first numerical estimate of a vibrational eigenfrequency of the Earth was made by Lord Kelvin in 1863. The prevailing opinion of most geologists and geophysicists at the time was that the Earth was completely molten, except for a thin crust of solid rock. Supporting evidence for this conclusion included the good agreement of the observed ellipticity of figure with the hydrostatic theory of Clairaut, the rapid increase of temperature with depth in mines, and the eruption of lava from active volcanoes. Seeking to challenge this view, Kelvin calculated the fundamental degree-two spheroidal-mode eigenfrequency of the Earth using two different assumptions (Thomson 1863a). For a self-gravitating fluid Earth he found the period of this mode—now designated ${}_0\mathrm{S}_2$—to be 94 minutes, whereas for a solid Earth having the same rigidity as steel he asserted that the period would be approximately 69 minutes. The first value was obtained by means of an exact dynamical analysis for a homogeneous, incompressible ($\kappa = \infty$) fluid ($\mu = 0$) sphere (Thomson 1863b), whereas the second was estimated on the basis of the time required for a shear wave to transit the diameter, using a laboratory value for the rigidity of steel obtained from his brother James in Glasgow. Lacking a means to measure the terrestrial eigenfrequencies, Kelvin devised an ingenious procedure for determining the mean rigidity of the Earth based upon the height of the fortnightly and monthly tides. He noted that the gravitational attraction of the Moon and Sun must raise bodily tides within the solid Earth as well as the tides within the oceans familiar to all seafarers, and pointed out that the observed oceanic tides, which are measured with respect to the deformed seafloor, should be nearly zero on a molten Earth. He determined the elastic-gravitational response of a homogeneous, incompressible solid sphere to an applied tidal potential, and showed that the oceanic tides on an elastic Earth should be reduced relative to their equilibrium value on a rigid Earth by an amount $\eta = (19\mu/2\rho g a)(1 + 19\mu/2\rho g a)^{-1}$, where ρ is the density, a is the radius, and g is the surficial acceleration of gravity. Since Kelvin's analysis was quasi-static, this elastic-Earth reduction factor could not be applied directly to the dominant semi-diurnal and diurnal tides; however, he argued that it should be applicable to the fortnightly and monthly tides, since they are largely devoid of ocean-basin resonance. The available fortnightly and monthly observations were insufficiently accurate for his purpose; accordingly, he persuaded the British Association to establish a Tidal Committee charged with "the evaluation of the long-period tides for the purpose of answering the question of the Earth's rigidity". The harmonic analysis of 66 years of tidal observations from fourteen British, French and Indian ports was undertaken by George Darwin, who published his results in the second edition of the *Treatise on Natural Philosophy* (Thomson & Tait 1883). Averaging the results from all poıts and both tides, Darwin found that $\eta = 0.676 \pm 0.076$, indicating that the tidal-effective rigidity of the Earth is indeed "about equal to that of steel". This celebrated conclusion corroborated Kelvin's 69-minute estimate of the period of the ${}_0S_2$ mode, grounding it upon a measured physical property of the Earth.

An early theoretical investigation of the toroidal modes of a homogeneous sphere was undertaken by Jaerisch (1880); however, the first comprehensive treatment of the free oscillations of a non-gravitating sphere is the classic analysis of Lamb (1882). He distinguished clearly between the spheroidal oscillations, which he called "vibrations of the first class", and the toroidal oscillations, which he called "vibrations of the second class", and concluded that the period of the ${}_0S_2$ mode for a steel sphere the size of the Earth should be 65 minutes in the case $\kappa = \infty$ and 66 minutes in the case $\kappa = \frac{5}{3}\mu$. The good agreement with Kelvin's order-of-magnitude estimate is to some extent coincidental, since Lamb used an improved, slightly higher value for the rigidity μ of steel; the insensitivity to the value of the incompressibility κ is a consequence of the fact that the ellipsoidal deformation is dominated by shear. Lamb conducted his analysis in terms of three-dimensional Cartesian coordinates; however, it was subsequently shown by Chree (1889) that the same results could be obtained much more economically using spherical polar coordinates. Such a spherical-harmonic representation of the elastic-gravitational deformation of the Earth has been employed in the majority of theoretical analyses ever since.

The proximity of the two rigorously derived periods—94 minutes for a fluid sphere whose only restoring force is the mutual gravitation of its parts and 65 minutes for a Poisson-solid sphere devoid of gravitational attraction—is an indication that elasticity and self-gravitation must play roughly equal roles in the ${}_0S_2$ mode of the actual Earth. Bromwich (1898) considered the free oscillations of an incompressible solid sphere, and found that self-gravitation would reduce the period from 65 to 55 minutes. The correct treatment of a compressible, self-gravitating sphere proved to be much more difficult, with many false steps along the way during the next decade. Jeans (1903) was the first person to appreciate and point out the complications, but in order to obtain a self-consistent system of governing equations, he found it necessary to "artificially annul gravitation in the equilibrium configuration, so that this equilibrium configuration may be completely unstressed, and each element of matter be in its normal state". His language illustrates the problem that confounded these early workers: the classical linear theory of elasticity was formulated to deal with deformations away from an unstressed, unstrained equilibrium configuration; however, the total stress within a self-gravitating body such as the Earth was clearly far too great to be related to an infinitesimal strain by Hooke's law. A generalization of the classical theory, which decomposed the total stress into a large *initial stress* balanced by the self-gravitation and an infinitesimal *incremental stress* calculable using Hooke's law, was proposed by Lord Rayleigh (1906):

> "It appears to me that a satisfactory treatment of these problems must start from the condition of the Earth as actually stressed by its self-gravitation and that the difficulties to be faced in following such a course may not be as great as has been supposed.... The conclusion that I draw is that the usual equations may be applied to matter in a state of stress, provided that we allow for altered values of the elasticities."

The elastic-gravitational equations obtained by Rayleigh were not correct, but his perspective upon the problem was highly influential. Love (1907) elaborated upon Rayleigh's idea, and derived an alternative system of equations; however, these too were incorrect because he failed to distinguish between the Eulerian incremental stress at a fixed point in space and the Lagrangian incremental stress experienced by an observer attached to a material particle. He soon realized his error, and not only derived but solved the correct equations in his monumental Adams Prize Essay *Some Problems of Geodynamics* four years later (Love 1911). His argument regarding the stress is extremely clear and well worth repeating:

> "The Earth ought to be regarded as a body in a state of *initial stress*; this initial stress may be regarded as a hydrostatic pressure balancing the self-gravitation of the body in the initial state; the stress in the body, when disturbed, may be taken

> to consist of the initial stress compounded with an *additional stress*; the additional stress may be taken to be connected with the strain, measured from the initial state as unstrained state, by the same formulae as hold in an isotropic elastic body slightly strained from a state of initial stress. The theory, as here described, is ambiguous in the following sense:—The initial stress at a point of the body which is at (x, y, z) in the strained state may be (1) the pressure at (x, y, z) in the initial state, or it may be (2) the pressure in the initial state at that point which is displaced to (x, y, z) when the body is strained. There can be little doubt that the second alternative is the correct one. A small element of the body is moved from one place to another, and during the displacement it suffers compression and distortion. It ought to be regarded as carrying its initial pressure with it, and acquiring an additional state of stress depending upon the compression and distortion."

Love found the period of the ${}_0\mathrm{S}_2$ mode of a self-gravitating Poisson-solid sphere with the rigidity of steel to be "almost exactly 60 minutes". His results are strictly applicable only to a *homogeneous* elastic, self-gravitating sphere; however, the correct dynamical equations and boundary conditions governing a sphere with *radially variable* properties κ, μ, ρ are implicit in his analysis. The general equations do not appear to have been written down explicitly until the work of Hoskins (1920). An apparently independent derivation is given by Jeffreys in the first edition of *The Earth* (1924). Mindful of the checkered history of the subject, he presents his analysis in a separate chapter, which is headed by the celebrated inscription over the Gates of Hell from Dante's *Inferno*, "Lasciate ogni speranza, voi ch'entrate".

Much of the work just summarized was motivated by cosmogonical considerations and a desire to understand the instability mechanisms of massive self-gravitating configurations, with possible applications to the asymmetrical distribution of continents and oceans and the origin by fission of the Moon. Jeans (1927) was the first person to place the normal modes of the Earth in a seismological context; he showed that the superposition of free oscillations or standing waves excited by an earthquake source could alternatively be regarded as a superposition of travelling body and surface waves. His asymptotic relation $\omega p = l + \frac{1}{2}$ between the angular frequency ω and spherical-harmonic degree l of a normal mode of oscillation and the seismological ray parameter p of the corresponding wave remains the basis for discussions of *mode-ray duality* to this day.

The dynamical laws governing the free oscillations of the Earth can alternatively be expressed in the form of a variational principle, known as *Hamilton's principle* in the time domain and as *Rayleigh's principle* in the frequency domain. Such principles can be established either by straightforward manipulation of the linearized conservation laws and constitutive relation, or from first principles by consideration of the kinetic and elastic-gravitational potential energies accompanying an infinitesimal deformation. An early variational calculation of the quasi-static tidal response of an elastic, self-gravitating Earth was attempted by Stoneley (1926b); however, his formulation of Rayleigh's principle, which led him to conclude "that when the expression for the total energy is evaluated, the gravitational terms are of the order of 1/20 of the elastic terms", is incorrect. The first correct variational calculations of the elastic-gravitational eigenfrequencies of a realistic Earth model were made independently and more or less simultaneously by Jobert (1956; 1957; 1961), Pekeris & Jarosch (1958), and Takeuchi (1959). The period of the fundamental toroidal mode ${}_0\mathrm{T}_2$ was found to be 43.5 minutes, whereas that of the ${}_0\mathrm{S}_2$ spheroidal mode was found to be approximately 52 minutes; both values are a few percent smaller than modern determinations due to an intrinsic property of the variational method—it always yields an upper bound upon the eigenfrequency.

The first numerical integration of the radial differential equations describing the elastic-gravitational deformation of a spherically symmetric Earth model was performed by Takeuchi (1950). He derived the single second-order equation governing the toroidal oscillations as well as the three second-order equations governing the spheroidal oscillations, and integrated the latter with the angular frequency $\omega = 0$ in order to determine the degree-two static Love numbers h, k and l of a realistic Earth; the Adams-Williamson relation was used to obtain a consistent second-order differential equation governing the static deformation of the fluid core. Takeuchi's calculation represents a significant achievement, particularly when one recognizes that it was done without the aid of an electronic computer. He obtained values $k = 0.28-0.29$, $h = 0.59-0.61$, $l = 0.07-0.08$ that were in good agreement with a wide variety of geophysical observations, including Darwin's measurement of the fortnightly and monthly tides, the period of the Chandler wobble (Love 1909), and the water-tube tidal tilt measurements of Michelson & Gale (1919).

The modern computational era was inaugurated in a pioneering paper by Alterman, Jarosch & Pekeris (1959). They recast Takeuchi's radial equations into a system of two first-order equations governing the toroidal modes ${}_n\mathrm{T}_l$ and six first-order equations governing the spheroidal modes ${}_n\mathrm{S}_l$, thereby eliminating the dependence upon the radial derivatives of the model parameters, making the results much more amenable to numerical integration. The two first-order equations governing the purely radial modes had been derived earlier by Pekeris & Jarosch (1958). The Runge-Kutta method was used to calculate the eigenfrequencies and eigenfunctions of a number of low-degree free oscillations of the Earth, yielding periods of 44.1 and 53.7 minutes, respectively, for the fundamental modes ${}_0T_2$ and ${}_0S_2$, in excellent agreement with modern determinations. The equations and boundary conditions derived by Pekeris and his colleagues are essentially the same ones that are used in spherical-Earth normal-mode calculations today; much of modern long-period seismology is founded upon their seminal contribution.

A major computational study of the toroidal modes of the Earth was undertaken at approximately the same time by Gilbert & MacDonald (1960); they carried out their calculations using a spherical version of the Thomson-Haskell layer-matrix method in lieu of numerical integration. A large number of fundamental and higher-overtone eigenfrequencies were computed, and Jeans' relation was used to determine the phase and group speeds of the equivalent Love waves. Gilbert subsequently extended the Thomson-Haskell method to the spheroidal modes, forming the basis for the eigenfrequency splitting calculations reported by Backus & Gilbert (1961). Shortly thereafter, he abandoned the method in favor of variable-order, variable-step Runge-Kutta integration, which could be performed much more efficiently. Gilbert and his collaborators went on to dominate the field of normal-mode research during the next two decades; his numerical integration program is the predecessor to `MINEOS` and `OBANI`, two computer codes which are widely used to calculate the Earth's eigenfrequencies and eigenfunctions today.

1.2 Dawn of the Observational Era

The observational era was initiated by Benioff (1958), who reported evidence for a 57-minute oscillation in the Pasadena electromagnetic strainmeter recording of the November 4, 1952 Kamchatka earthquake. No Fourier analysis of the seismogram was performed; the 57-minute period was evidently obtained by visual inspection of a prominent oscillation which terminates after only a few cycles. In retrospect, it is likely that this oscillation is the result of some sort of instrumental malfunctioning; Benioff himself admits that "unfortunately, the noise level... is rather high... so that the measurements are not as reliable as we would like to have them". A modern re-analysis of the original recording by Kanamori (1976) found a number of spectral peaks that could possibly be associated with the Earth's normal modes; however, the amplitudes of these oscillations were too large to have been excited by the $M_0 \approx 4 \times 10^{22}$ N m main shock, calling the results into serious question. Benioff's purported observation was important primarily because it stimulated further development of long-period strainmeters and gravimeters, as well as the ambitious computational programs of theoreticians such as Pekeris and Gilbert.

The largest earthquake in the twentieth century ruptured 1000 kilometers of the Nazca-America plate boundary in Chile on May 22, 1960. From the standpoint of normal-mode seismology, this $M_0 \approx 2 \times 10^{23}$ N m event could not have occurred at a more propitious time: (1) instruments capable of recording the long-period free oscillations had only recently been developed; (2) computers capable of Fourier analyzing lengthy time series had just become generally available; (3) the technology of numerical spectral analysis had just been codified in a widely read book by Blackman & Tukey (1958); and (4) the theoretical eigenfrequencies of the Earth had just been calculated, allowing the identification of peaks in the observed spectra. Barely two months after the earthquake, seismologists from the California Institute of Technology and the University of California at Los Angeles presented spectra at the July 1960 assembly of the International Association of Seismology and Physics of the Earth's Interior (IASPEI) in Helsinki which showed clear evidence for the detection of the Earth's fundamental normal modes. Neither of us is old enough to have been present at this historic IASPEI meeting—indeed, one of us was not even born—but contemporary accounts make it sound like the Woodstock of seismology. The Caltech group recorded the spheroidal oscillations ${}_0S_2$ through ${}_0S_{38}$ and the toroidal oscillations ${}_0T_3$ through ${}_0T_{11}$ on quartz strainmeters located at Ñaña, Peru and Isabella, California and on pendulum seismographs located at Pasadena, whereas the UCLA group recorded the fundamental modes ${}_0S_2$ through ${}_0S_{41}$ on a Lacoste-Romberg tidal gravimeter located at Los Angeles; the absence of any toroidal peaks on the recording from the gravimeter aided in the mode identification. The results of these analyses, together with an analysis of the oscillations recorded by seismologists at the Lamont Geological Observatory of Columbia University on their newly installed quartz strainmeter at Ogdensburg, New Jersey and on pendulum seismographs located at Palisades, New York, were subsequently published in back-to-back papers in the *Journal of Geophysical Research* (Benioff, Press & Smith 1961; Ness, Harrison & Slichter 1961; Alsop, Sutton & Ewing 1961). Analyses of the oscillations recorded on a pendulum located in the Grotta Gigante near Trieste, Italy and on tiltmeters located in Paris, France were published shortly thereafter by Bolt & Marussi (1962) and Connes, Blum, Jobert & Jobert (1962). The measured eigenfrequencies of the Earth were shown by Pekeris, Alterman & Jarosch (1961a) to agree with the theoretical eigenfrequencies to "better than one percent, with a distinct preference shown for the Gutenberg (low velocity) model". The quest to use free-oscillation observations to improve our knowledge of the internal structure of the Earth had begun.

Cable dispatches were received from Caltech and UCLA at the Helsinki meeting reporting that both the ${}_0\mathrm{S}_2$ and ${}_0\mathrm{S}_3$ peaks appeared to be visibly split into two. The source of this splitting was very quickly identified to be the rotation of the Earth, which removes the $2l+1$ degeneracy of the multiplets ${}_n\mathrm{S}_l$ and ${}_n\mathrm{T}_l$. Explanatory analyses of this rotational splitting based upon degenerate Rayleigh-Schrödinger perturbation theory were published by Backus & Gilbert (1961) and Pekeris, Alterman & Jarosch (1961b). Unbeknownst to these workers, the effect of rotation had previously been investigated in an astrophysical context by Cowling & Newing (1949) and Ledoux (1951). To lowest order, the Coriolis force gives rise to $2l+1$ equally spaced singlets with associated eigenfrequency perturbations $\delta\omega_m = m\chi\Omega$, where m is the order of the complex spherical harmonic Y_{lm} characterizing the zeroth-order eigenfunction, and Ω is the angular rate of rotation; the effect is analogous to the Zeeman splitting of atomic spectral lines in a magnetic field. The dimensionless splitting parameter χ is a weak function of the Earth-model parameters κ, μ and ρ for a spheroidal multiplet, whereas it is equal to $[l(l+1)]^{-1}$ for a toroidal multiplet. The observed amplitudes of the singlets depend upon the location, magnitude and geometry of the earthquake and upon the location and polarization of the receiver. The visibly excited peaks excited by the Chile earthquake were shown to correspond to the split angular orders $m = \pm 1$ in the case of the ${}_0\mathrm{S}_2$ multiplet and to $m = \pm 2$ in the case of the ${}_0\mathrm{S}_3$ multiplet.

1.3 Spherical Earth Model Refinement

During the two decades following the 1960 Chile earthquake, most free-oscillation research was directed toward the refinement of the spherically averaged structure of the Earth. In the beginning, the process was iterative: an initial Earth model was used to identify the gravest multiplets, the model was improved by trial-and-error adjustment of κ, μ and ρ, enabling identification of some of the more closely spaced higher-frequency multiplets, and so on. The occurrence of the $M_0 \approx 8 \times 10^{22}$ N m Alaska earthquake on March 28, 1964 supplied additional data for this bootstrap operation (Smith 1966; Slichter 1967). Summarizing the state of progress less than ten years after the free oscillations were first detected, Derr (1969) compiled a catalogue containing 265 measured normal-mode eigenfrequencies; however, he admitted that "only the fundamental spheroidal and torsional modes and a few higher spheroidal modes... can be identified with confidence". A number of high-Q spheroidal overtone modes were observed for the first time by Dratler, Farrell, Block & Gilbert (1971) following the July 31, 1970 deep-focus earthquake in Colombia. Because it is situated on the exponential tail of the associated eigenfunctions, such a deep-focus event does not strongly excite the fundamental modes which normally dominate the response, obscuring the low-amplitude overtones. Attenuation filtering, or elimination of the first several hours of data prior to Fourier transformation, was used to accentuate the high-Q spectral peaks.

Gilbert (1971a) showed that the sum of the first-order eigenfrequency perturbations of a multiplet ${}_n\mathrm{T}_l$ or ${}_n\mathrm{S}_l$ split by the Earth's rotation, ellipticity and lateral heterogeneity is identically zero, $\sum_{j=1}^{2l+1} \delta\omega_j = 0$, as a consequence of the *diagonal sum rule*:

> "We interpret this result as follows. If the Earth is averaged over spherical surfaces centred on its centre of mass the result is an averaged Earth completely free of aspherical perturbations, which we call the terrestrial monopole. The difference between the real Earth and the terrestrial monopole is a strictly aspherical perturbation. It is a consequence of the diagonal sum rule that the averaged frequency for each multiplet, split by first-order perturbations, belongs to the terrestrial monopole. In other words we know that our averaged data belong to an unperturbed spherical Earth model that is, in fact, the spherically averaged Earth."

This provides the justification for using the measured peak frequencies obtained from an ensemble of earthquake sources and receivers as constraints upon the structure of the spherically averaged Earth. Gilbert concluded by pointing out that "although we can hope to obtain good coverage with receivers, we have little, if any, control over the distribution of sources".

The status of the *mode identification problem* was improved dramatically during the next few years by the clever exploitation of data from the global, three-component World-Wide Standard Seismographic Network (WWSSN). The reduced long-period sensitivity of these instruments, which were deployed primarily to monitor underground nuclear explosions, was offset by the availability of large numbers of recordings. Dziewonski & Gilbert (1972; 1973) used 84 WWSSN recordings of the 1964 Alaska earthquake to identify and measure the periods of 249 normal modes. A number of simple discriminants, including polarization and simple histogram analysis as well as attenuation filtering, were used to separate and distinguish multiplets that were closely spaced in the eigenfrequency spectrum. An extremely important advance was made by Mendiguren (1973), who was the first person to utilize the capabilities of the WWSSN network as a global array. He developed a phase-equalization procedure in which spectra were stacked with an appropriate change in sign in order to accentuate the multiplet of interest and reduce the effect of neighboring multiplets. The method, which requires knowledge of the source mechanism of the earthquake, was improved by Gilbert & Dziewonski (1975), who applied it to 213 WWSSN recordings of the 1970 Colombia earthquake and the August 15, 1963 deep-focus event on the Peru-Bolivia border. By combining modes identified in their own and several previous analyses, they were able to compile a standardized data set consisting of 1064 measured eigenfrequencies of the free oscillations of the Earth. Thus, only fifteen years after the first detection in 1960, approximately sixty percent of the Earth's normal modes with periods longer than 80 seconds had been observed and identified.

The rich harvest of high-Q overtones recorded by a single feedback accelerometer following the 1970 Colombia earthquake provided the impetus for the development of the International Deployment of Accelerometers (IDA) network by Agnew, Berger, Buland, Farrell & Gilbert (1976). The low noise level of these instruments permitted the routine detection of normal-mode spectral peaks following earthquakes of relatively modest magnitude:

> "No longer need we wait for truly large and very infrequent earthquakes to add to our observational knowledge of the Earth's free oscillations."

Six stations of the fledgling IDA network recorded the long-period oscillations excited by the $M_0 = 4 \times 10^{21}$ N m Sumbawa, Indonesia earthquake of August 19, 1977, allowing Buland, Berger & Gilbert (1979) to extract all five singlets of the ${}_0S_2$ multiplet and all seven singlets of the ${}_0S_3$ multiplet, using a spherical-harmonic stacking procedure. The measured spacing between the eigenfrequencies compared favorably with updated theoretical calculations of the effects of the Earth's rotation and hydrostatic ellipticity performed by Dahlen & Sailor (1979).

Seismologists engaged in the study of terrestrial free oscillations played a leading role in the development of a powerful new discipline—*geophysical inverse theory*—which enabled the optimal extraction of information regarding the internal structure of the Earth from a finite set of gross Earth data. Models providing a best fit to the data, subject to a variety of imposed constraints, could be obtained by simultaneous adjustment of all the governing parameters (Gilbert 1971b; Jackson 1972; Wiggins 1972); questions critical to model assessment, such as resolution and accuracy, could also be addressed (Backus & Gilbert 1968; 1970). We make no attempt to review this elegant and multi-faceted theory here; excellent and comprehensive summaries are provided by Menke (1984), Tarantola (1987) and Parker (1994). The *Fréchet kernels* relating the perturbation $\delta\omega$ in an eigenfrequency to arbitrary radial perturbations $\delta\kappa$, $\delta\mu$, $\delta\rho$ in the incompressibility, rigidity and density were derived using Rayleigh's principle by Backus & Gilbert (1967). These kernels were used by Gilbert & Dziewonski (1975) to invert their 1064 measured eigenfrequencies, together with the geodetically determined mass and moment of inertia of the Earth; the two resulting spherically symmetric models were whimsically dubbed 1066A and 1066B. A small subset of radial and other modes provided irrefutable evidence for the solidity of the inner core (Dziewonski & Gilbert 1971).

In exhibiting the fit of models 1066A and 1066B to body-wave travel-time data, Gilbert & Dziewonski (1975) added 1.6 seconds to the compressional wave observations and 4.3 seconds to the shear-wave observations. Such a *baseline correction* was commonly thought to be justified by the nature of the iterative procedure used to infer both source origin times and travel times from measured arrival times and by the biased siting of most seismographic stations upon continents. A much more satisfying explanation of the discrepancy was pointed out by Akopyan, Zharkov & Lyubimov (1975; 1976), Randall (1976), and Liu, Anderson & Kanamori (1976): the *physical dispersion* that inevitably accompanies anelastic attenuation renders the Earth's mantle slightly less rigid as "seen" by a typical free oscillation with a period of 200–300 seconds than as "seen" by a teleseismic shear wave with a period of 10–20 seconds. Jeffreys (1958a; 1958b) and Carpenter & Davies (1966) had previously sought unsuccessfully to call attention to this effect; the latter authors investigated whether anelastic dispersion was capable of reconciling the discrepancies between the classical Jeffreys-Bullen and Gutenburg Earth models. Their letter reporting the failure of the attempted reconciliation concludes with an unambiguous warning:

> 'The results do, however, show that the frequency dependence is significant and should at least be considered in any investigation in which attenuation is an accepted part of the Earth model. The differences between the dispersion curves... ignoring or including Q are greater than the standard error obtained in modern studies."

This admonition went unheeded for the next decade, until the frequency dependence of the Earth's elastic parameters was rediscovered in the context of the travel-time baseline correction. The wide dissemination of the results of Akopyan, Zharkov & Lyubimov, Randall, and Liu, Anderson & Kanamori led to a renewed interest in the determination of the bulk and shear quality factors Q_κ and Q_μ within the Earth, based upon the measured decay rates of the free oscillations (Stein & Geller 1978; Sailor & Dziewonski 1978; Geller & Stein 1979; Riedesel, Agnew, Berger & Gilbert 1980).

The Preliminary Reference Earth Model (PREM), which was developed by Dziewonski & Anderson (1981) in response to a request from the International Union of Geodesy and Geophysics for a spherically symmetric model that could be used in geodetic and other geophysical applications, represented the culmination of two decades of progress in measuring and interpreting the free oscillations of the Earth. In addition to being anelastic and therefore dispersive, the PREM model is transversely isotropic, with five rather than two elastic parameters in the uppermost mantle between 24 and 220 km depth. More recent, higher-quality normal-mode eigenfrequency measurements are systematically misfit by the PREM model (Widmer 1991; Masters & Widmer 1995); in addition, a number of its features, such as the conspicuous 220-km discontinuity, are unsupported by body-wave reflectivity analyses (Shearer 1991). Despite these shortcomings, we shall use PREM as the basis for most of the numerical illustrations in this book, for want of a more acceptable model at the time of writing. Subsequent work has led to significantly improved models of the radial anelastic structure of the Earth (Masters & Gilbert 1983; Widmer, Masters & Gilbert 1991; Durek & Ekström 1996). Because decay-rate measurements are intrinsically much noisier than eigenfrequency measurements, and peak widths measured on spectral stacks are biased by splitting, the parameters Q_κ and Q_μ are extremely difficult to constrain.

1.4 Source-Mechanism Determination

Attempts were made by Alterman, Jarosch & Pekeris (1959), Backus & Gilbert (1961), and others to calculate the amplitudes of the oscillations excited by the 1960 Chile earthquake; however, all of these early efforts employed extremely simplified and, in retrospect, unrealistic representations of the source (explosion, point force, single couple, etc.). The response of a spherically symmetric Earth model to a planar fault (double couple) source was first obtained by Saito (1967). His results were subsequently re-derived and extended by Gilbert (1970), who noted that:

> "If one can calculate the excitation of the normal modes of the Earth due to a particular earthquake source, one can use such calculations in an attempt to infer the earthquake mechanism and total moment. Some general results in normal mode theory, due to Rayleigh and Routh about a century ago, make the excitation calculations remarkably simple."

Gilbert and Kostrov (1970) independently introduced the concept of the *seismic moment tensor* **M**, which is capable of representing curved as well as planar faults, and has become the canonical point-source model of an earthquake. The linear dependence of the response upon the elements of $\mathbf{M}$ makes the moment-tensor representation particularly advantageous in source-mechanism studies, as pointed out by Gilbert (1973):

> "Our problem is to determine $\mathbf{M}$ given (displacement) seismograms $\mathbf{u}$ at several locations $\mathbf{r}$ on the surface of the Earth. Since $\mathbf{u}$ is a linear function of $\mathbf{M}$ for any given Earth model, the problem is rather straightforward."

Mendiguren (1973) and Gilbert & Dziewonski (1975) made use of the source-theoretical formulations of Saito and Gilbert, respectively, in carrying out their spherical-Earth stacking analyses of the oscillations excited by the 1970 Colombia and 1963 Peru-Bolivia deep-focus earthquakes. Gilbert & Dziewonski used the observed excitation amplitudes to find the frequency-dependent moment tensors of the two events; an alternative method more suitable for retrieving $\mathbf{M}$ from a sparse network of seismographic stations was subsequently developed by Buland & Gilbert (1976) and Gilbert & Buland (1976).

Dziewonski, Chou & Woodhouse (1981) extended the moment-tensor analysis of Buland & Gilbert to allow for a spatial and temporal shift in the centroid of the source region:

> "We describe in this report an approach that allows us to improve the location parameters and simultaneously to modify the solution for the moment tensor. This, in effect, yields the 'best point source' location, which for earthquakes of finite size need not be the same as the point of initiation of rupture."

The *centroid-moment tensor* or CMT source-mechanism determination procedure developed by these investigators finds the space-time shift $\mathbf{\Delta x}$, Δt of the source centroid and the moment tensor $\mathbf{M}$ by fitting long-period waveforms in the time domain; the kernels relating the response at a given receiver to the unknown source parameters are calculated by normal-mode summation. The method was used in a systematic study of 201 large and moderate earthquakes that occurred in 1981 by Dziewonski & Woodhouse (1983). This marked the beginning of the Harvard centroid-moment tensor project, which now routinely determines the source mechanisms of two to three earthquakes with seismic moments greater than 10^{16}–10^{17} N m every day. The resulting global seismicity catalogue of more than 16,000 CMT solutions extending from 1977 to the present has contributed greatly to our understanding of regional geology and tectonics, and is one of the most valuable legacies of free-oscillation research during the past two decades.

1.5 Surface Waves

The spheroidal and toroidal multiplets ${}_n\mathrm{S}_l$ and ${}_n\mathrm{T}_l$ are equivalent in the limit $n \ll l$ to propagating fundamental and higher-mode Rayleigh and Love surface waves. The study of elastic surface waves has traditionally gone hand-in-hand with the study of the free oscillations of the Earth for this reason. The literature on surface-wave propagation, particularly in plane stratified media, is even more voluminous than that devoted to the elastic-gravitational normal modes; we attempt only an extremely sketchy review here.

The existence of propagating solutions which decay exponentially with depth beneath the surface of a homogeneous elastic half-space was first noted by Rayleigh (1885). The resulting surface waves are non-dispersive. In the case of a Poisson solid, which has $\alpha = \sqrt{3}\beta$, where α and β are the compressional and shear wave speeds, the phase speed of Rayleigh waves is equal to $c = 0.9194\beta$. Rayleigh conducted his investigation at a time when seismology was still in its infancy; however, he concluded with a remarkably prescient comment:

> "It is not improbable that the surface waves here investigated play an important role in earthquakes.... Diverging in two dimensions only, they must acquire at great distance from the source a continually increasing preponderance."

Lamb (1904) extended Rayleigh's results, which were limited to the free propagation of waves, by finding the complete response of a homogeneous half-space to both a line and point force; his classic analysis of what we now call *Lamb's problem* concludes with the first synthetic seismogram ever depicted.

Inasmuch as the first seismometers only measured horizontal motions, the existence of prominent transverse displacements in the "main shock" of an earthquake tremor, following the "primary" and "secondary" arrivals, was one of the earliest established facts of observational seismology. The particle motion of a Rayleigh wave is a retrograde ellipse in the vertical plane passing through the source and receiver, and this initially impeded suggestions that the "main shock" consisted of surface waves. This controversy was resolved by Love (1911), who showed that transverse surface waves could exist in a system consisting of a homogeneous elastic layer overlying a homogeneous half-space. The resulting Love waves are both multi-modal and dispersive, with a phase speed $c_n(\omega)$ that depends upon the overtone number n and the frequency ω. With remarkable foresight, Love argued that surface-wave dispersion was responsible for the observed oscillatory character of the "main shock":

> "The explanation which I wish to suggest is that the oscillations are due to dispersion.... The actual motion may, of course, be analysed into... an aggregate of simple harmonic wave-trains, each travelling with the wave-velocity appropriate to its wave-length."

The dispersion relation for Rayleigh waves in a system consisting of a surficial layer overlying a half-space is considerably more difficult to obtain. Love treated the case in which both the layer and the half-space are incompressible, Bromwich (1898) considered an incompressible fluid layer overlying an elastic half-space, and Stoneley (1926a) extended his analysis to allow for the compressibility of the fluid. The general case of a solid elastic layer overlying an elastic half-space was finally solved by Sezawa (1927) and Stoneley (1928). The non-gravitating version of Rayleigh's variational principle governing propagation in an arbitrarily stratified half-space was first given for Love waves by Meissner (1926) and for Rayleigh waves by Jeffreys (1935).

A number of early investigators, including Angenheister (1921) and Tams (1921), noticed that surface waves propagate at different speeds over oceanic and continental paths. The first systematic study of this phenomenon, which distinguished clearly between waves recorded on the longitudinal and transverse components, and which sought to use Love's theoretical results for the latter to determine the thickness of the oceanic and continental crust, was made by Gutenberg (1924). To measure the speed of Love waves in the period range 10–60 seconds, Gutenberg divided the epicentral distance by the arrival time. Stoneley (1925) pointed out that such measurements corresponded to the theoretical group speed C rather than the phase speed c, as Gutenberg had assumed; the two speeds are related by $C = c + k(dc/dk)$, where k is the wavenumber. Other attempts to constrain crustal structure by comparing measured and theoretical group speeds followed; Stoneley & Tillotson (1928) considered the effect of a two-layer crust—a "granitic" upper layer overlying a "dioritic" lower layer—upon Love-wave propagation. Most of these early efforts were relatively inconclusive due to the scatter in the measurements and the paucity of quantitative theoretical results for pertinent crustal models. The gross thickness and character of the continental crust were inferred primarily by the analysis of Pn, Sn and other body waves from near-focus earthquakes, whereas the structure of the much thinner oceanic crust was ultimately determined by shipboard seismic refraction experiments.

The theoretical understanding of dispersive surface-wave propagation was advanced considerably by the work on underwater sound transmission conducted during World War II by Pekeris (1948) and his collaborators.

Pekeris clarified the role of the group speed in controlling the character of a surface-wave seismogram, and coined the term *Airy phase* to describe the large-amplitude signal associated with a group-speed maximum or minimum. This improved understanding was exploited by seismologists during the post-war years, with the result that reliable group speeds of fundamental Rayleigh and, to a lesser extent, Love waves in both continental and oceanic regions had soon been determined. Ewing & Press (1954) used Benioff's recording of the 1952 Kamchatka earthquake at Pasadena to extend the dispersion curve for Rayleigh waves up to a period of 480 seconds; the multi-orbit wavegroups R6 through R15 could be identified, the latter having undergone more than seven complete circumnavigations of the Earth. The status of post-war surface-wave research was summarized in a major monograph by Ewing, Jardetzky & Press (1957); their description of the dispersive characteristics of mantle Rayleigh waves—"a minimum value of group velocity of 3.5 km/sec at a period of 225 sec, a short-period limit of 3.8 km/sec at 70 sec, and the flattening of the curve for periods greater than 400 sec"—remains accurate to this day. Mantle Love waves on the other hand exhibit a nearly constant group speed of approximately 4.4 kilometers per second in the period range 70–400 seconds, resulting in an impulsive arrival known as a *G wave* after Gutenberg, who first called attention to it (Gutenberg & Richter 1934).

Early measurements of regional surface-wave phase speed employed either a three-station triangulation technique (Press 1956) or a two-station technique which relied upon a suitable geographical alignment of the source and receivers. The two-station method, which was introduced but not applied to the analysis of any earthquake data by Sâto (1955), was used in a classic study of the Canadian shield by Brune & Dorman (1963). The phase speeds of long-period mantle waves were determined by means of a *great-circle* technique, utilizing multiple passages of the same wavegroup through a single station. Sâto (1958) and Nafe & Brune (1960) made the first such measurements for Love and Rayleigh waves, respectively, using G3–G1 and G4–G2 arrivals from the February 1, 1938 New Guinea earthquake, G3–G1 arrivals from the 1952 Kamchatka earthquake, and R5–R3 arrivals from the August 15, 1950 Assam earthquake recorded at Pasadena. Brune, Nafe & Alsop (1961) subsequently showed that these pioneering measurements were slightly too high due to the neglect of the $\pi/2$ *polar phase shift* at the source and antipodal caustics.

A technique for measuring regional surface-wave phase speeds using only minor-arc data was introduced by Brune, Nafe & Oliver (1960); however, this *single-station* method, which requires knowledge of the initial phase of the waves leaving the source, did not become reliable until the double-couple excitation problem was solved, initially by Haskell (1964) and Ben-Menahem & Harkrider (1964), and then in a more useful form by Saito (1967). Recordings of the R1 and G1 minor-arc surface waves from seventeen earthquakes with known focal mechanisms were used in a major study of the structure beneath the eastern Pacific Ocean by Forsyth (1975). The reliability of modern earthquake source-mechanism determinations and the availability of the Harvard CMT catalogue make this the preferred phase-speed measurement method today.

The isolation of higher-mode Love and Rayleigh waves is considerably more difficult than that of fundamental-mode waves due to the near coincidence of the group speeds near 4.4 km/s for periods lower than 100 seconds. The earliest analyses focused upon group-speed measurements of shorter-period crustal-guided phases such as Lg and Rg (Press & Ewing 1952). Nolet (1977) and Cara (1978) used an innovative multi-station stacking technique to measure the phase speeds of higher-mode Rayleigh waves propagating across western Europe and North America, respectively; they obtained results up to overtone number $n = 6$ and phase speed $c = 7.5$ km/s in the period range between 25 and 100 seconds. The Fréchet kernels relating the change δc in the phase speed of a fundamental or higher-mode surface wave to changes $\delta\kappa$, $\delta\mu$, $\delta\rho$ in the model parameters were first obtained using Rayleigh's principle for a stratified half-space by Takeuchi, Dorman & Saito (1964). Successively higher modes of a given period "feel" more deeply into the Earth, thereby offering greater resolution at depth.

The first measurement of the *attenuation* of seismic waves within the Earth was made by Angenheister (1906). He compared the amplitudes of 15–20 second surface waves propagating along the minor and major arcs to Göttingen from five distant earthquakes. It is not clear from his discussion whether the measurements pertain to Love waves or to Rayleigh waves; he refers to the minor-arc and major-arc arrivals as W_1 and W_2, respectively. Expressing the amplitude ratio as $A_2/A_1 = e^{-\gamma d}$, where d is the differential distance in kilometers, he found the attenuation coefficient to be $\gamma = 1.8 \times 10^{-4}$ to $\gamma = 3.4 \times 10^{-4}$ km^{-1}. The corresponding quality factor for 20-second waves is $Q = 150$–300. This pioneering measurement of the Earth's anelastic attenuation is a noteworthy accomplishment, in view of the extremely primitive understanding of surface-wave propagation at the time.

1.6 Lateral Heterogeneity

The observed regional variation in surface-wave dispersion at periods below 100 seconds is the result of the strong contrast in continental and oceanic crustal thickness and other near-surface structural differences; this was clearly recognized by the post-war seismologists who sought to utilize short-period surface waves as a tool for delineating regional variations in crustal structure. The first evidence for lateral variations $\delta\beta$ in the shear-wave speed of the *upper mantle* was provided by Toksöz & Ben-Menahem (1963); they measured the phase speeds of 50–400 second Love waves over six great-circle paths, finding differences in excess of one percent, "much greater than the experimental error". The inverse problem of inferring upper-mantle heterogeneity from phase-speed measurements was investigated by Backus (1964), who pointed out that only the even-degree part of the Earth could be determined from great-circular average data. Regionalization schemes which divided the surface of the Earth into "shield", "oceanic", and "tectonic" provinces were introduced by Toksöz & Anderson (1966) and Kanamori (1970) in an attempt to overcome this ambiguity.

Variations in the apparent central frequencies of fundamental-mode multiplets ${}_0\mathrm{T}_l$ or ${}_0\mathrm{S}_l$ provide an additional constraint upon upper-mantle lateral heterogeneity; such shifts in the location of the spectral peaks are the result of interference among the unresolvably split singlets comprising each multiplet. Jordan (1978) and Dahlen (1979a) used degenerate normal-mode perturbation theory to show that such peak-shift measurements depend only upon the structure underlying the source-receiver great-circle path, in the limit $s_{\mathrm{max}} \ll l$, where s_{max} is the maximum spherical-harmonic degree of the lateral heterogeneity and l is the degree of the multiplet under consideration; normal-mode peak frequencies and great-circular surface-wave phase speeds may be regarded as interchangeable data in this geometrical-optics limit. Silver & Jordan (1981) analyzed 72 accelerograms from the IDA network to obtain 2193 apparent central frequencies for the fundamental spheroidal modes ${}_0\mathrm{S}_5$ through ${}_0\mathrm{S}_{43}$, and sought to interpret these using a more sophisticated regionalization scheme comprised of three oceanic provinces sorted by age and three continental provinces "divided accorded to their generalized tectonic history during the Phanerozoic".

A larger data set consisting of 3934 reliable measurements from 557 IDA recordings was subsequently collected by Masters, Jordan, Silver & Gilbert (1982). Upon plotting their observed peak shifts at the poles of the source-receiver great circles, they discovered a remarkably coherent pattern, which led them to eschew a regionalization approach in favor of a spherical-harmonic representation of the heterogeneity:

> "Free-oscillation data reveal heterogeneity in the Earth's mantle whose geographical pattern is dominated by spherical harmonics of angular degree two... The heterogeneity can be modelled as localized in the transition zone (420–670 km depth) and may be related to a large-scale component of mantle convection."

The variance reduction achieved by their simple degree-two model was a remarkable seventy percent; spherical-harmonic representations of the three-dimensional structure of the Earth have been employed in the majority of global tomographic studies since that time.

Woodhouse & Dziewonski (1984) devised a procedure for the inversion of seismic waveform data in the time domain and applied it to a data set consisting of 2000 IDA and Global Digital Seismographic Network (GDSN) recordings of the mantle waves generated by 53 earthquakes ranging in moment from 4×10^{18} to 3.6×10^{21} N m. The corresponding synthetic seismograms were calculated using a *path-average* or *great-circle approximation*, which accounts for the effect of the Earth's lateral heterogeneity by slight modifications to the eigenfrequency and apparent epicentral distance associated with each multiplet in a conventional mode-summation algorithm; the method assumes that the response depends only upon the laterally averaged structure beneath the source-receiver great-circle path, but it is sensitive to odd as well as even spherical-harmonic degrees because it treats the various arrivals R1, R2,... or G1, G2,... separately. Woodhouse & Dziewonski's model of shear-wave heterogeneity $\delta\beta$ in the upper mantle, "expanded up to degree and order 8 in spherical harmonics, and described by a cubic polynomial in depth for the upper 670 km", represented a significant increase in both lateral and radial resolution. Despite the absence of any a priori tectonic regionalization, their waveform inversion study showed that "shields and ridges are major features in the depth interval 25–250 km".

Global tomographic studies since 1985 have sought not only to improve the resolution and reproducibility of images of the Earth's three-dimensional elastic and anelastic structure, but, more importantly, to address fundamental questions regarding the geodynamical and compositional causes of the heterogeneity. Modern analyses make use of a wide variety of seismological data, including complete spectra of strongly split free-oscillation multiplets, complete waveforms of both body and surface waves, measured phase speeds of R1, R2,... or G1, G2,... waves, measured absolute P, PKP and PKIKP and differential SS–S and ScS–S travel times, and observations of upper-mantle conversions, reflections and reverberations. We do not discuss many of these procedures, nor do we attempt to review or critically appraise the fidelity of the current generation of three-dimensional mantle models, inasmuch as our emphasis is upon the free oscillations of the Earth and related mode-summation and surface-wave ray-theoretical methods of calculating synthetic long-period seismograms. Summary accounts devoted to the broader topic of global seismic tomography are provided by Romanowicz (1991) and Ritzwoller & Lavely (1995).

Part I

Foundations

Chapter 2

Continuum Mechanics

In seismology and geodynamics we regard the Earth as a *continuum*; that is, as a continuous distribution of matter which can interact through both short-range and long-range forces. In this chapter we provide a brief review of the essential principles of continuum mechanics, emphasizing the aspects that are most useful in studying the free oscillations of the Earth and related global seismological phenomena. The mathematical entities that are studied in continuum mechanics are scalar, vector and tensor fields; a synopsis of some of the basic results of multilinear algebra and multivariable calculus that we will use may be found in Appendix A.

Following a discussion of the kinematics of motion and deformation, we summarize the fundamental laws governing the conservation of mass, momentum, angular momentum, and energy. The gravest free oscillations of the Earth and the longest-period surface waves are significantly affected by self-gravitation; in anticipation of this, we also briefly review gravitational potential theory. The four conservation laws and Newton's inverse-square law of attraction must be supplemented by constitutive relations that specify the physical nature of the material before they can be solved. We conclude the chapter by discussing the constitutive relation for a (non-linear) perfectly elastic material.

Our coverage of continuum mechanics, although it is far from complete, is more thorough than that found in most advanced seismology textbooks, including those by Aki & Richards (1980) or Ben-Menahem & Singh (1981). In particular, we introduce the two Piola-Kirchhoff stress tensors in addition to the Cauchy stress, and we derive Lagrangian as well as the more familiar Eulerian forms of all the conservation equations. Our approach to these topics, which seeks to emphasize the physical content, is modelled after the exemplary treatment by Malvern (1969). A more rigorous mathematical discussion, cast in the language of modern differential geometry and functional analysis, is given by Marsden & Hughes (1983). An accessible and authoritative account of potential theory may be found in the treatise by Kellogg (1967).

2.1 Eulerian and Lagrangian Variables

There are two possible ways of describing the motion of a continuum. The first of these, known as the *Lagrangian* description, is the obvious generalization of the standard kinematical description of the motion of one or more point masses, such as the planets orbiting about the Sun. Particles in the material are labelled by their position $\mathbf{x}$ at time $t = 0$, and we denote the position of particle $\mathbf{x}$ at time t by $\mathbf{r}(\mathbf{x}, t)$. Knowledge of $\mathbf{r}(\mathbf{x}, t)$ for all particles $\mathbf{x}$ and all times $t \geq 0$ provides a complete kinematical description of the motion. Since $\mathbf{x}$ is the position of the particle at time $t = 0$ we must have $\mathbf{r}(\mathbf{x}, 0) = \mathbf{x}$. Time derivatives of the motion are defined in precisely the same manner as for isolated point masses; the velocity $\mathbf{u}^{\mathrm{L}}(\mathbf{x}, t)$ of particle $\mathbf{x}$ at time t is, for example, $\mathbf{u}^{\mathrm{L}} = \partial_t \mathbf{r}$.

The second, or *Eulerian*, description, is the description which an idealized, omnipresent weather bureau might seek to provide of the winds in the atmosphere. Attention is focused not upon the motions of individual particles, but rather upon points (such as weather stations) that are fixed in space. We consistently use $\mathbf{r}$, rather than $\mathbf{x}$, to label points that are fixed in space, and we denote the velocity of the particle that is at point $\mathbf{r}$ at time t by $\mathbf{u}^{\mathrm{E}}(\mathbf{r}, t)$. Knowledge of $\mathbf{u}^{\mathrm{E}}(\mathbf{r}, t)$ for all points $\mathbf{r}$ occupied by the continuum and all times $t \geq 0$ provides an alternative, and also complete, description of the motion. It is noteworthy that the two fields $\mathbf{u}^{\mathrm{E}}$ and $\mathbf{u}^{\mathrm{L}}$ are not the same, since the former specifies the velocity as a function of spatial position $\mathbf{r}$, whereas the latter specifies it as a function of particle label $\mathbf{x}$. That is why we use the two superscripts E (for Eulerian) and L (for Lagrangian) to distinguish them.

Every scalar, vector or tensor variable q in a moving continuum has both an Eulerian description $q^{\mathrm{E}}(\mathbf{r}, t)$ and a Lagrangian description $q^{\mathrm{L}}(\mathbf{x}, t)$. If, for example, q is the temperature, then $q^{\mathrm{E}}(\mathbf{r}, t)$ gives the temperature recorded at a fixed point $\mathbf{r}$ in space, whereas $q^{\mathrm{L}}(\mathbf{x}, t)$ gives the temperature recorded by a thermometer attached to a moving particle $\mathbf{x}$. The relation between the two descriptions is

$$q^{\mathrm{E}}(\mathbf{r}(\mathbf{x}, t), t) = q^{\mathrm{L}}(\mathbf{x}, t). \tag{2.1}$$

This result is self-evident, inasmuch as both sides give the value of q recorded by particle $\mathbf{x}$, which is at point $\mathbf{r}$, at time t.

Upon differentiating equation (2.1) with respect to time, using the chain rule, we obtain

$$\partial_t q^{\mathrm{L}} = \partial_t q^{\mathrm{E}} + \mathbf{u}^{\mathrm{E}} \cdot \boldsymbol{\nabla}_{\mathbf{r}} q^{\mathrm{E}} \equiv D_t q^{\mathrm{E}}, \tag{2.2}$$

where $\boldsymbol{\nabla}_{\mathbf{r}}$ denotes the gradient with respect to the fixed spatial position $\mathbf{r}$. Equation (2.2) stipulates that an observer riding on a moving particle $\mathbf{x}$ senses two types of changes in the variable q: in addition to the change $\partial_t q^{\mathrm{E}}$ experienced by a stationary observer, there is a change $\mathbf{u}^{\mathrm{E}} \cdot \boldsymbol{\nabla}_{\mathbf{r}} q^{\mathrm{E}}$ due to the motion of the particle through the spatial gradient $\boldsymbol{\nabla}_{\mathbf{r}} q^{\mathrm{E}}$. The combination of temporal and spatial derivatives $D_t = \partial_t + \mathbf{u}^{\mathrm{E}} \cdot \boldsymbol{\nabla}_{\mathbf{r}}$ is called the *substantial* or *material derivative.*

We shall adhere strictly to the above notational convention throughout the first part of this book, using superscripts E and L to denote, respectively, Eulerian variables q^{E} measured at points $\mathbf{r}$ fixed in space and Lagrangian variables q^{L} measured on moving particles $\mathbf{x}$. The material derivative D_t acts only upon Eulerian quantities q^{E}, in which case $D_t q^{\mathrm{E}}$ denotes the rate of change of $q^{\mathrm{E}}(\mathbf{r}(\mathbf{x}, t), t)$ holding $\mathbf{x}$ fixed. The same rate of change experienced by an observer on a moving particle can be calculated by taking the ordinary partial derivative of the corresponding Lagrangian description, $\partial_t q^{\mathrm{L}}$. The acceleration of the particle at position $\mathbf{r}(\mathbf{x}, t)$ can be written, for example, either as $\mathbf{a}^{\mathrm{L}} = \partial_t \mathbf{u}^{\mathrm{L}} = \partial_t^2 \mathbf{r}$ or as $\mathbf{a}^{\mathrm{E}} = D_t \mathbf{u}^{\mathrm{E}} = \partial_t \mathbf{u}^{\mathrm{E}} + \mathbf{u}^{\mathrm{E}} \cdot \boldsymbol{\nabla}_{\mathbf{r}} \mathbf{u}^{\mathrm{E}}$.

Because an idealized seismometer provides a direct record of the motion $\mathbf{r}(\mathbf{x}, t)$ of the particle $\mathbf{x}$ to which it is attached, it is natural to adopt a Lagrangian viewpoint in seismology. The Cauchy stress tensor, which is the measure of stress that is most familiar to seismologists, is, however, an inherently Eulerian variable. In the usual linearized theory of elasticity, the Eulerian nature of the stress is ignored; this is only permissible if the initial stress in the material is of the same order as the incremental stress, as in most engineering applications. The initial stress at depth within the Earth is large, as noted by Rayleigh (1906) and Love (1911); it is this circumstance that obliges us to distinguish carefully between Eulerian and Lagrangian measures of the stress and other variables in formulating the fundamental laws that govern the elastic-gravitational deformation of the Earth.

2.2 Measures of Deformation

To characterize the deformation, we must either analyze the relative motion at two adjacent points in space, or follow the motion of two adjacent particles. We discuss the various Eulerian and Lagrangian measures of deformation and the relations among them next.

The relative Eulerian velocity $d\mathbf{u}^{\mathrm{E}}$ at two adjacent fixed points in space, $\mathbf{r}$ and $\mathbf{r} + d\mathbf{r}$, is $d\mathbf{u}^{\mathrm{E}} = d\mathbf{r} \cdot \boldsymbol{\nabla}_{\mathbf{r}}\mathbf{u}^{\mathrm{E}}$. We define a tensor $\mathbf{G}^{\mathrm{E}}$ by

$$\mathbf{G}^{\mathrm{E}} = (\boldsymbol{\nabla}_{\mathbf{r}}\mathbf{u}^{\mathrm{E}})^{\mathrm{T}}, \tag{2.3}$$

where the superscript T denotes the transpose, and rewrite this linear relation between $d\mathbf{u}^{\mathrm{E}}$ and the differential distance $d\mathbf{r}$ in the form

$$d\mathbf{u}^{\mathrm{E}} = \mathbf{G}^{\mathrm{E}} \cdot d\mathbf{r}. \tag{2.4}$$

The quantity $\mathbf{G}^{\mathrm{E}}$ is referred to as the Eulerian *deformation-rate tensor*.

If the two particles now located at $\mathbf{r}$ and $\mathbf{r} + d\mathbf{r}$ were initially at $\mathbf{x}$ and $\mathbf{x} + d\mathbf{x}$, then the relative current and initial position vectors $d\mathbf{r}$ and $d\mathbf{x}$ are related by $d\mathbf{r} = d\mathbf{x} \cdot \boldsymbol{\nabla}_{\mathbf{x}}\mathbf{r}$, where $\boldsymbol{\nabla}_{\mathbf{x}}$ denotes the gradient with respect to the particle label $\mathbf{x}$. We again introduce a transposed tensor

$$\mathbf{F} = (\boldsymbol{\nabla}_{\mathbf{x}}\mathbf{r})^{\mathrm{T}}, \tag{2.5}$$

and rewrite this second relation in the form

$$d\mathbf{r} = \mathbf{F} \cdot d\mathbf{x}. \tag{2.6}$$

We shall refer to $\mathbf{F}$ as the *deformation tensor*. Physically impermissible phenomena such as kinking, tearing or inversion are prohibited by the requirement that

$$\det \mathbf{F} \neq 0, \tag{2.7}$$

where $\det \mathbf{F}$ denotes the determinant.

The deformation-rate tensor $\mathbf{G}^{\mathrm{E}}$ and the deformation tensor $\mathbf{F}$ are the two fundamental measures of deformation: $\mathbf{G}^{\mathrm{E}}(\mathbf{r}, t)$ measures the instantaneous rate of deformation in the vicinity of a fixed point $\mathbf{r}$, whereas $\mathbf{F}(\mathbf{x}, t)$ measures the cumulative deformation experienced by a small ball of material surrounding a moving particle $\mathbf{x}$. In more mathematically sophisticated treatments, $\mathbf{F}$ is considered to be a *two-point tensor*, so-called because it relates a vector $d\mathbf{r}$ in the current deformed configuration to a vector $d\mathbf{x}$ in the initial undeformed configuration. The initial vector $d\mathbf{x}$ is related to the current vector $d\mathbf{r}$ by the inverse two-point tensor:

$$d\mathbf{x} = \mathbf{F}^{-1} \cdot d\mathbf{r}. \tag{2.8}$$

The inverse $\mathbf{F}^{-1}$ satisfies $\mathbf{F} \cdot \mathbf{F}^{-1} = \mathbf{F}^{-1} \cdot \mathbf{F} = \mathbf{I}$ where $\mathbf{I}$ is the identity. The existence of $\mathbf{F}^{-1}$ is guaranteed by the restriction (2.7). Strictly speaking, there are two identities $\mathbf{I}$, one in the undeformed configuration and one in the deformed configuration; we shall, however, disregard this and other subtleties that arise from the two-point nature of $\mathbf{F}$ and $\mathbf{F}^{-1}$ in the future.

The two tensors $\mathbf{G}^{\mathrm{E}}$ and $\mathbf{F}$ are related to each other by

$$\partial_t \mathbf{F} = \mathbf{G}^{\mathrm{E}} \cdot \mathbf{F}, \tag{2.9}$$

or, equivalently,

$$\mathbf{G}^{\mathrm{E}} = \partial_t \mathbf{F} \cdot \mathbf{F}^{-1}. \tag{2.10}$$

Equation (2.9) can be verified by taking the time derivative of the definition (2.5), using the chain rule.

We can write the rate-of-deformation tensor $\mathbf{G}^{\mathrm{E}}$ as the sum of a symmetric and an anti-symmetric tensor:

$$\mathbf{G}^{\mathrm{E}} = \mathbf{D}^{\mathrm{E}} + \mathbf{W}^{\mathrm{E}}, \tag{2.11}$$

where

$$\mathbf{D}^{\mathrm{E}} = \tfrac{1}{2}[\mathbf{G}^{\mathrm{E}} + (\mathbf{G}^{\mathrm{E}})^{\mathrm{T}}], \qquad \mathbf{W}^{\mathrm{E}} = \tfrac{1}{2}[\mathbf{G}^{\mathrm{E}} - (\mathbf{G}^{\mathrm{E}})^{\mathrm{T}}]. \tag{2.12}$$

The symmetric part $\mathbf{D}^{\mathrm{E}} = (\mathbf{D}^{\mathrm{E}})^{\mathrm{T}}$ is the Eulerian *strain-rate tensor* whereas the anti-symmetric part $\mathbf{W}^{\mathrm{E}} = -(\mathbf{W}^{\mathrm{E}})^{\mathrm{T}}$ is the rotation-rate or *vorticity tensor*. Upon using equation (2.11) we can decompose the differential velocity $d\mathbf{u}^{\mathrm{E}}$ into a part associated with the local rate of strain and a part associated with the local rate of rotation:

$$d\mathbf{u}^{\mathrm{E}} = \mathbf{D}^{\mathrm{E}} \cdot d\mathbf{r} + \mathbf{W}^{\mathrm{E}} \cdot d\mathbf{r}. \tag{2.13}$$

The curl of the Eulerian velocity, or vector vorticity $\boldsymbol{\nabla}_{\mathbf{r}} \times \mathbf{u}^{\mathrm{E}}$, is related to the vorticity tensor $\mathbf{W}^{\mathrm{E}}$ by

$$\boldsymbol{\nabla}_{\mathbf{r}} \times \mathbf{u}^{\mathrm{E}} = - \wedge \mathbf{W}^{\mathrm{E}}, \tag{2.14}$$

where $\wedge$ denotes the wedge operator defined in Appendix A.1.6. The portion of the Eulerian velocity differential $d\mathbf{u}^{\mathrm{E}}$ associated with $\mathbf{W}^{\mathrm{E}}$ can be written in terms of $\boldsymbol{\nabla}_{\mathbf{r}} \times \mathbf{u}^{\mathrm{E}}$ in the form

$$\mathbf{W}^{\mathrm{E}} \cdot d\mathbf{r} = \tfrac{1}{2}(\boldsymbol{\nabla}_{\mathbf{r}} \times \mathbf{u}^{\mathrm{E}}) \times d\mathbf{r}. \tag{2.15}$$

This justifies the use of the name "vorticity tensor" for $\mathbf{W}^{\mathrm{E}}$, and shows that $\boldsymbol{\nabla}_{\mathbf{r}} \times \mathbf{u}^{\mathrm{E}}$ is twice the instantaneous angular velocity of the material in the vicinity of point $\mathbf{r}$ at time t. The suitability of the name "strain-rate tensor" for $\mathbf{D}^{\mathrm{E}}$ may be seen by considering the rate of change of the squared differential length $\|d\mathbf{r}\|^2$. A straightforward calculation shows that

$$\begin{aligned} \frac{d}{dt}\|d\mathbf{r}\|^2 &= 2(d\mathbf{r} \cdot \partial_t \mathbf{F} \cdot d\mathbf{x}) \\ &= 2(d\mathbf{r} \cdot \partial_t \mathbf{F} \cdot \mathbf{F}^{-1} \cdot d\mathbf{r}) = 2(d\mathbf{r} \cdot \mathbf{G}^{\mathrm{E}} \cdot d\mathbf{r}), \end{aligned} \tag{2.16}$$

where we have used equations (2.6), (2.8) and (2.10). Since the vorticity tensor $\mathbf{W}^{\mathrm{E}}$ is anti-symmetric, it does not contribute to the quadratic product $d\mathbf{r} \cdot \mathbf{G}^{\mathrm{E}} \cdot d\mathbf{r}$; the rate of change of the squared length of the relative position vector $\|d\mathbf{r}\|^2$ is therefore determined only by the strain-rate tensor:

$$\frac{d}{dt}\|d\mathbf{r}\|^2 = 2(d\mathbf{r} \cdot \mathbf{D}^{\mathrm{E}} \cdot d\mathbf{r}). \tag{2.17}$$

The Lagrangian *strain tensor* $\mathbf{E}^{\mathrm{L}}$ is defined in terms of the initial and current squared lengths $\|d\mathbf{x}\|^2$ and $\|d\mathbf{r}\|^2$ by

$$\|d\mathbf{r}\|^2 - \|d\mathbf{x}\|^2 = 2(d\mathbf{x} \cdot \mathbf{E}^{\mathrm{L}} \cdot d\mathbf{x}). \tag{2.18}$$

We can relate $\mathbf{E}^{\mathrm{L}}$ to the deformation tensor $\mathbf{F}$ by noting that

$$\|d\mathbf{r}\|^2 = (\mathbf{F} \cdot d\mathbf{x}) \cdot (\mathbf{F} \cdot d\mathbf{x}) = d\mathbf{x} \cdot (\mathbf{F}^{\mathrm{T}} \cdot \mathbf{F}) \cdot d\mathbf{x}. \tag{2.19}$$

Hence,

$$\mathbf{E}^{\mathrm{L}} = \tfrac{1}{2}(\mathbf{F}^{\mathrm{T}} \cdot \mathbf{F} - \mathbf{I}). \tag{2.20}$$

The tensor $\mathbf{E}^{\mathrm{L}}$ is a measure of the total finite strain acccumulated in the vicinity of particle $\mathbf{x}$ since time $t = 0$. It is clear from equation (2.20) that this cumulative strain is symmetric: $(\mathbf{E}^{\mathrm{L}})^{\mathrm{T}} = \mathbf{E}^{\mathrm{L}}$.

The instantaneous rate of strain experienced by an observer riding upon particle $\mathbf{x}$ is $\partial_t \mathbf{E}^{\mathrm{L}}$. To see how this Lagrangian strain rate is related to the Eulerian strain rate $\mathbf{D}^{\mathrm{E}}$ we consider

$$\frac{d}{dt}\left(\|d\mathbf{r}\|^2 - \|d\mathbf{x}\|^2\right) = 2(d\mathbf{x} \cdot \partial_t \mathbf{E}^{\mathrm{L}} \cdot d\mathbf{x}). \tag{2.21}$$

Since $d\mathbf{x}$ is fixed we may also write, however,

$$\frac{d}{dt}\left(\|d\mathbf{r}\|^2 - \|d\mathbf{x}\|^2\right) = \frac{d}{dt}\|d\mathbf{r}\|^2 = 2[d\mathbf{x} \cdot (\mathbf{F}^{\mathrm{T}} \cdot \mathbf{D}^{\mathrm{E}} \cdot \mathbf{F}) \cdot d\mathbf{x}], \tag{2.22}$$

where we have used equations (2.6) and (2.17). Upon comparing the two results (2.21) and (2.22) we find that

$$\partial_t \mathbf{E}^{\mathrm{L}} = \mathbf{F}^{\mathrm{T}} \cdot \mathbf{D}^{\mathrm{E}} \cdot \mathbf{F}, \tag{2.23}$$

or, equivalently,

$$\mathbf{D}^{\mathrm{E}} = \mathbf{F}^{-\mathrm{T}} \cdot \partial_t \mathbf{E}^{\mathrm{L}} \cdot \mathbf{F}^{-1}. \tag{2.24}$$

The *polar decomposition theorem* asserts that the deformation tensor $\mathbf{F}$ can be written in a unique manner in either of the two forms

$$\mathbf{F} = \mathbf{Q} \cdot \mathbf{R} = \mathbf{L} \cdot \mathbf{Q}. \tag{2.25}$$

The two tensors $\mathbf{R}$ and $\mathbf{L}$ in the decomposition (2.25) are symmetric and positive definite:

$$\mathbf{R}^{\mathrm{T}} = \mathbf{R}, \qquad \mathbf{L}^{\mathrm{T}} = \mathbf{L}, \tag{2.26}$$

whereas $\mathbf{Q}$ is a proper orthogonal tensor:

$$\mathbf{Q}^{\mathrm{T}} \cdot \mathbf{Q} = \mathbf{Q} \cdot \mathbf{Q}^{\mathrm{T}} = \mathbf{I}. \tag{2.27}$$

The symmetric tensors $\mathbf{R}$ and $\mathbf{L}$ are called the *right* and *left stretch tensors*, and the orthogonal tensor $\mathbf{Q}$ is called the *rotation tensor.* The stretch tensors are given explicitly in terms of $\mathbf{F}$ by

$$\mathbf{R} = (\mathbf{F}^{\mathrm{T}} \cdot \mathbf{F})^{1/2}, \qquad \mathbf{L} = (\mathbf{F} \cdot \mathbf{F}^{\mathrm{T}})^{1/2}. \tag{2.28}$$

The right stretch tensor $\mathbf{R}$ and the strain tensor $\mathbf{E}^{\mathrm{L}}$ are closely related; upon comparing equations (2.20) and (2.28) we see that

$$\mathbf{R} = (\mathbf{I} + 2\mathbf{E}^{\mathrm{L}})^{1/2}. \tag{2.29}$$

The physical interpretation of the polar decomposition (2.25) is clear: the deformation of an infinitesimal ball of material surrounding a particle $\mathbf{x}$ can be viewed either as a symmetric stretch $\mathbf{R}$ followed by a rigid rotation $\mathbf{Q}$, or as a finite rotation $\mathbf{Q}$ followed by a symmetric stretch $\mathbf{L}$ (see Figure 2.1). The two stretch tensors are rotated versions of each other, i.e., they are related by an orthogonal transformation: $\mathbf{R} = \mathbf{Q}^{\mathrm{T}} \cdot \mathbf{L} \cdot \mathbf{Q}$.

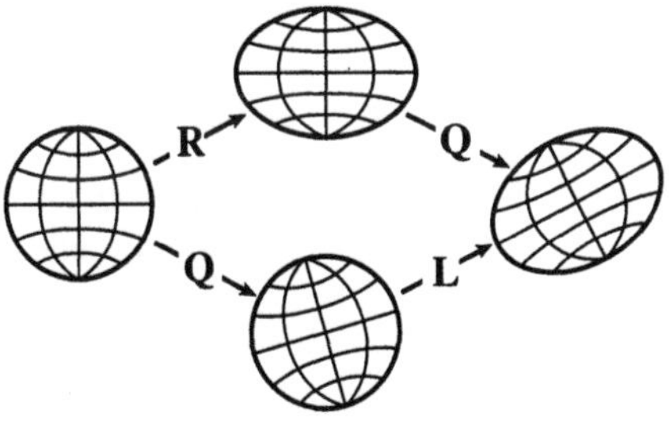

Figure 2.1. Schematic depiction of the polar decomposition of the deformation tensor, $\mathbf{F} = \mathbf{Q}\cdot\mathbf{R} = \mathbf{L}\cdot\mathbf{Q}$. An infinitesimal ball of material surrounding a particle $\mathbf{x}$ is either stretched by an amount $\mathbf{R}$ and then rotated by an amount $\mathbf{Q}$ (*top*) or rotated by an amount $\mathbf{Q}$ and then stretched by an amount $\mathbf{L}$ (*bottom*).

2.3 Volume and Area Changes

To conclude our review of kinematics, we discuss the transformation formulae governing infinitesimal volume and surface elements within a continuum. Consider first an infinitesimal ball of volume dV^0 surrounding particle $\mathbf{x}$ at time $t = 0$. At a later time t, the ball will have been translated to $\mathbf{r}$ and rotated and deformed; we denote its new volume by dV^t. We can calculate the ratio of the deformed to the undeformed volume of this infinitesimal element,

$$J = \frac{dV^t}{dV^0}, \tag{2.30}$$

by noting that it is nothing more than the Jacobian,

$$J = \det \mathbf{F}, \tag{2.31}$$

of the transformation from $\mathbf{x}$ to $\mathbf{r}$ coordinates.

We consider next an infinitesimal oriented surface area element $\hat{\mathbf{n}}^0 d\Sigma^0$ centered upon a particle $\mathbf{x}$ in the undeformed configuration. At time t, this element will have been translated to $\mathbf{r}$ and rotated and stretched to $\hat{\mathbf{n}}^t d\Sigma^t$. We seek the relation between the undeformed and deformed elements $\hat{\mathbf{n}}^0 d\Sigma^0$ and $\hat{\mathbf{n}}^t d\Sigma^t$. To simplify the calculation, we consider two infinitesimal parallelograms with initial edges $d\mathbf{x}$ and $\delta\mathbf{x}$ and current edges $d\mathbf{r}$ and $\delta\mathbf{r}$ (see Figure 2.2). The quantities $\hat{\mathbf{n}}^0 d\Sigma^0$ and $\hat{\mathbf{n}}^t d\Sigma^t$ are in this case given by

$$\hat{\mathbf{n}}^0 d\Sigma^0 = d\mathbf{x} \times \delta\mathbf{x}, \qquad \hat{\mathbf{n}}^t d\Sigma^t = d\mathbf{r} \times \delta\mathbf{r}. \tag{2.32}$$

Introducing a Cartesian axis system $\hat{\mathbf{x}}_1, \hat{\mathbf{x}}_2, \hat{\mathbf{x}}_3$ we rewrite (2.32) using index notation in the form

$$n_i^0 d\Sigma^0 = \varepsilon_{ijk}\, dx_j\, \delta x_k, \qquad n_l^t d\Sigma^t = \varepsilon_{lmn}\, dr_m\, \delta r_n. \tag{2.33}$$

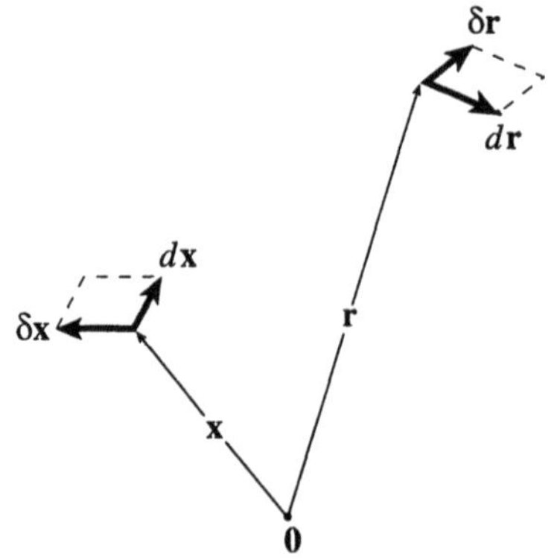

Figure 2.2. At time t the initially undeformed parallelogram $\hat{\mathbf{n}}^0 d\Sigma^0 = d\mathbf{x} \times \delta\mathbf{x}$ at point $\mathbf{x}$ has been transformed into a deformed parallelogram $\hat{\mathbf{n}}^t d\Sigma^t = d\mathbf{r} \times \delta\mathbf{r}$ at point $\mathbf{r}$. The two infinitesimal patches are related by equation (2.37).

Upon substituting $dx_j = F^{-1}_{jm}dr_m$ and $\delta x_k = F^{-1}_{kn}\delta r_n$ into the first of equations (2.33) we obtain

$$n_i^0 d\Sigma^0 = \varepsilon_{ijk}\, F^{-1}_{jm} F^{-1}_{kn}\, dr_m\, \delta r_n. \tag{2.34}$$

Multiplying both sides of (2.34) by F^{-1}_{il} and using the identity

$$J^{-1}\varepsilon_{lmn} = \varepsilon_{ijk}\, F^{-1}_{il} F^{-1}_{jm} F^{-1}_{kn} \tag{2.35}$$

then yields

$$F^{-1}_{il} n_i^0 d\Sigma^0 = J^{-1}\varepsilon_{lmn}\, dr_m\, \delta r_n = J^{-1} n_l^t d\Sigma^t, \tag{2.36}$$

where the last equality follows from the second of equations (2.33). Reverting to an invariant notation, the final result (2.36) takes the form

$$\hat{\mathbf{n}}^t d\Sigma^t = J\, \hat{\mathbf{n}}^0 d\Sigma^0 \cdot \mathbf{F}^{-1}. \tag{2.37}$$

This expression relating $\hat{\mathbf{n}}^t d\Sigma^t$ to $\hat{\mathbf{n}}^0 d\Sigma^0$ is the areal analogue of the volumetric relation $dV^t = J\, dV^0$.

2.4 Reynolds Transport Theorem

Let V^t denote an arbitrary volume which is moving with the material, so that both it and its boundary ∂V^t always consist of the same material particles, as illustrated in Figure 2.3. Let q^{E} be the volumetric density of some extensive mechanical or thermodynamical quantity within the continuum, and consider the rate of change with respect to time t of the total amount of this "q-stuff" within V^t. Both the integrand q^{E} and the domain of integration V^t depend upon t; the total time derivative of the integral over such a *co-moving* volume V^t is

$$\frac{d}{dt}\int_{V^t} q^{\mathrm{E}}\, dV^t = \int_{V^t} \partial_t q^{\mathrm{E}}\, dV^t + \int_{\partial V^t} (\hat{\mathbf{n}}^t \cdot \mathbf{u}^{\mathrm{E}}) q^{\mathrm{E}}\, d\Sigma^t, \tag{2.38}$$

where $\hat{\mathbf{n}}^t$ is the outward unit normal on ∂V^t, and we have affixed a superscript t to the differential volume and surface elements dV^t and $d\Sigma^t$ to serve as a reminder that they are co-moving. The first term in equation (2.38) is due to the local change in the spatial density $\partial_t q^{\mathrm{E}}$, whereas the second term represents the change due to the flux $(\hat{\mathbf{n}}^t \cdot \mathbf{u}^{\mathrm{E}})q^{\mathrm{E}}$ through the moving boundary. Upon applying Gauss' theorem to the surface integral over ∂V^t we obtain

$$\frac{d}{dt}\int_{V^t} q^{\mathrm{E}}\, dV^t = \int_{V^t} [\partial_t q^{\mathrm{E}} + \boldsymbol{\nabla}_{\mathbf{r}} \cdot (\mathbf{u}^{\mathrm{E}} q^{\mathrm{E}})]\, dV^t. \tag{2.39}$$

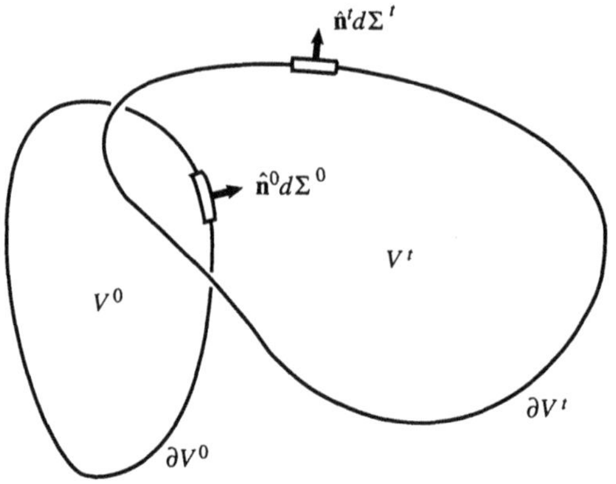

Figure 2.3. The Eulerian conservation laws are obtained by considering a co-moving volume, which always consists of the same material particles. The particles comprising a patch $\hat{\mathbf{n}}^0 d\Sigma^0$ on the boundary ∂V^0 of the initial volume V^0 move to $\hat{\mathbf{n}}^t d\Sigma^t$ on the boundary ∂V^t of the deformed volume V^t.

This result, expressing the time derivative of an integral over a co-moving volume as an integral over the same volume, is known as the *Reynolds transport theorem.* The quantity q^{E} in (2.39) can be a scalar, a vector, or a tensor of any higher order; if q^{E} is not a scalar, it is important to remember that tensor products are non-commutative: $\mathbf{u}^{\mathrm{E}}q^{\mathrm{E}} \neq q^{\mathrm{E}}\mathbf{u}^{\mathrm{E}}$. We will use the Reynolds transport theorem in Section 2.6 to derive the Eulerian forms of the conservation laws for mass, momentum, angular momentum and energy from the corresponding balance principles expressed as integrals over an arbitrary co-moving volume.

2.5 Measures of Stress

Short-range intermolecular forces are represented in continuum mechanics in terms of a *stress tensor.* There are three different ways of measuring or defining the stress, each of which plays a role in the theory governing the elastic-gravitational deformation of the Earth. The most familiar measure is the *Cauchy stress*, which is defined in the following manner. Let $\hat{\mathbf{n}}^t d\Sigma^t$ be an oriented surface-area element centered on a fixed point $\mathbf{r}$ at time t; the side toward which the normal $\hat{\mathbf{n}}^t$ points is referred to as the front or + side of the patch, whereas the other side is referred to as the back or − side (see Figure 2.4). The instantaneous surface force exerted by all of the particles immediately in front of the patch on all of the particles

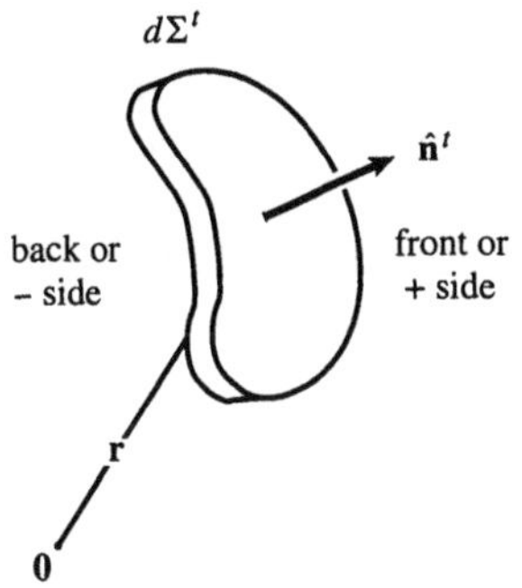

Figure 2.4. Schematic depiction of an oriented surface area element $\hat{\mathbf{n}}^t d\Sigma^t$, showing its front or + and back or − sides. We adhere to this sign convention throughout this book.

immediately in back is denoted by $d\mathbf{f}^{\mathrm{E}}$. The Eulerian Cauchy stress $\mathbf{T}^{\mathrm{E}}$ gives the force $d\mathbf{f}^{\mathrm{E}}$ acting across the patch in the form

$$d\mathbf{f}^{\mathrm{E}} = \hat{\mathbf{n}}^t d\Sigma^t \cdot \mathbf{T}^{\mathrm{E}}. \tag{2.40}$$

The Cauchy stress, like any other variable, has both a Lagrangian and an Eulerian description; the corresponding Lagrangian stress $\mathbf{T}^{\mathrm{L}}$ is defined, as usual, by $\mathbf{T}^{\mathrm{L}}(\mathbf{x}, t) = \mathbf{T}^{\mathrm{E}}(\mathbf{r}(\mathbf{x}, t), t)$. The Eulerian stress $\mathbf{T}^{\mathrm{E}}$ arises naturally in the Eulerian form of the momentum conservation law; the corresponding Lagrangian form cannot, however, be readily expressed in terms of $\mathbf{T}^{\mathrm{L}}$.

The stress measure that does lead to the simplest form of the Lagrangian momentum equation is the so-called *first Piola-Kirchhoff stress.* This quantity, which we shall denote by $\mathbf{T}^{\mathrm{PK}}$, is defined by

$$d\mathbf{f}^{\mathrm{E}} = \hat{\mathbf{n}}^0 d\Sigma^0 \cdot \mathbf{T}^{\mathrm{PK}}. \tag{2.41}$$

Evidently, $\mathbf{T}^{\mathrm{PK}}$ gives the force $d\mathbf{f}^{\mathrm{E}}$ acting across a deformed patch $\hat{\mathbf{n}}^t d\Sigma^t$ at $\mathbf{r}$ in terms of the corresponding undeformed patch $\hat{\mathbf{n}}^0 d\Sigma^0$ at the initial point $\mathbf{x}$. Among other things, the first Piola-Kirchhoff stress $\mathbf{T}^{\mathrm{PK}}$ is a measure of the force per unit undeformed area, whereas both the Eulerian and Lagrangian Cauchy stresses $\mathbf{T}^{\mathrm{E}}$ and $\mathbf{T}^{\mathrm{L}}$ are measures of the force per unit deformed area. We can find the relation between $\mathbf{T}^{\mathrm{PK}}$ and $\mathbf{T}^{\mathrm{L}}$ by using the transformation (2.37) between the deformed and undeformed patches; the result can be expressed in either of the two equivalent forms

$$\mathbf{T}^{\mathrm{PK}} = J\,\mathbf{F}^{-1} \cdot \mathbf{T}^{\mathrm{L}}, \qquad \mathbf{T}^{\mathrm{L}} = J^{-1}\mathbf{F} \cdot \mathbf{T}^{\mathrm{PK}}. \tag{2.42}$$

The surface force vector $d\mathbf{f}^{\mathrm{E}}$ in (2.41) acts upon the displaced point $\mathbf{r}$, whereas the patch vector $\hat{\mathbf{n}}^0 d\Sigma^0$ is affixed to the initial point $\mathbf{x}$; the first Piola-Kirchhoff stress $\mathbf{T}^{\mathrm{PK}}$ is therefore, strictly speaking, a two-point tensor, like the deformation tensor $\mathbf{F}$.

The constitutive relation for a perfectly elastic material is most conveniently expressed in terms of yet a third measure of stress, known as the *second Piola-Kirchhoff stress.* This quantity, which we shall denote by $\mathbf{T}^{\mathrm{SK}}$, gives, instead of the actual force acting upon the deformed element $\hat{\mathbf{n}}^t d\Sigma^t$, a force $d\mathbf{f}^{\mathrm{L}}$ related to $d\mathbf{f}^{\mathrm{E}}$ in the same way that the initial differential $d\mathbf{x}$ is related to the spatial differential $d\mathbf{r}$; that is,

$$d\mathbf{f}^{\mathrm{L}} = \mathbf{F}^{-1} \cdot d\mathbf{f}^{\mathrm{E}}, \tag{2.43}$$

just as $d\mathbf{x} = \mathbf{F}^{-1} \cdot d\mathbf{r}$. Defining $\mathbf{T}^{\mathrm{SK}}$ by

$$d\mathbf{f}^{\mathrm{L}} = \hat{\mathbf{n}}^0 d\Sigma^0 \cdot \mathbf{T}^{\mathrm{SK}}, \tag{2.44}$$

we find that the first and second Piola-Kirchhoff stresses are related by

$$\mathbf{T}^{\mathrm{SK}} = \mathbf{T}^{\mathrm{PK}} \cdot \mathbf{F}^{-\mathrm{T}}, \qquad \mathbf{T}^{\mathrm{PK}} = \mathbf{T}^{\mathrm{SK}} \cdot \mathbf{F}^{\mathrm{T}}. \tag{2.45}$$

Upon comparing with equation (2.42) we obtain the corresponding relation between $\mathbf{T}^{\mathrm{SK}}$ and the Lagrangian Cauchy stress $\mathbf{T}^{\mathrm{L}}$:

$$\mathbf{T}^{\mathrm{SK}} = J\,\mathbf{F}^{-1} \cdot \mathbf{T}^{\mathrm{L}} \cdot \mathbf{F}^{-\mathrm{T}}, \qquad \mathbf{T}^{\mathrm{L}} = J^{-1}\mathbf{F} \cdot \mathbf{T}^{\mathrm{SK}} \cdot \mathbf{F}^{\mathrm{T}}. \tag{2.46}$$

Since the transformed force $d\mathbf{f}^{\mathrm{L}}$ may be considered to act upon the initial position $\mathbf{x}$ rather than upon the displaced position $\mathbf{r}$, the second Piola-Kirchhoff stress $\mathbf{T}^{\mathrm{SK}}$ is an ordinary rather than a two-point tensor. The geometrical relationships between the patches $\hat{\mathbf{n}}^t d\Sigma^t$ and $\hat{\mathbf{n}}^0 d\Sigma^0$ and the forces $d\mathbf{f}^{\mathrm{E}}$ and $d\mathbf{f}^{\mathrm{L}}$ are summarized in Figure 2.5.

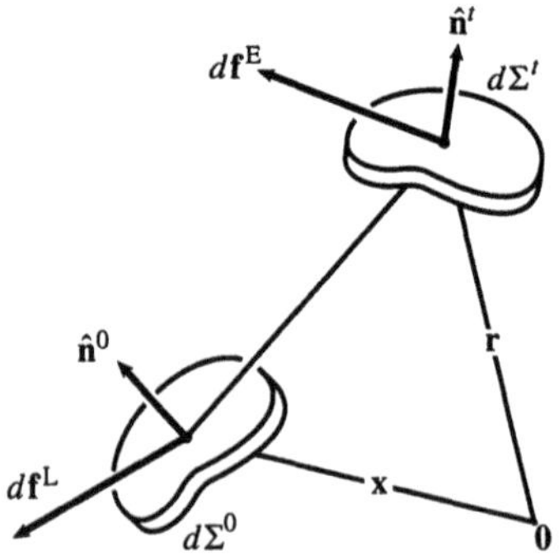

Figure 2.5. The surface forces $d\mathbf{f}^{\mathrm{E}}$ and $d\mathbf{f}^{\mathrm{L}}$ act upon the deformed patch $\hat{\mathbf{n}}^t d\Sigma^t$ at $\mathbf{r}$ and the undeformed patch $\hat{\mathbf{n}}^0 d\Sigma^0$ at $\mathbf{x}$, respectively.

2.6 Eulerian Conservation Laws

Let q^{E} be the density of some extensive "q-stuff" within the continuum, as before. The general form of any local Eulerian conservation law is

$$\partial_t q^{\mathrm{E}} + \boldsymbol{\nabla}_{\mathbf{r}} \cdot \mathbf{k}^{\mathrm{E}} = c^{\mathrm{E}}. \tag{2.47}$$

The quantity $\mathbf{k}^{\mathrm{E}}$ represents the *flux*, whereas c^{E} represents the volumetric *rate of creation* of "q-stuff" at a fixed point $\mathbf{r}$ at time t. The flux $\mathbf{k}^{\mathrm{E}}$ is necessarily a tensor of one order higher than the density q^{E} or the spontaneous creation rate c^{E}. If we integrate equation (2.47) over a *stationary* volume V, we obtain, after an application of Gauss' theorem,

$$\frac{d}{dt}\int_V q^{\mathrm{E}}\, dV = -\int_{\partial V} (\hat{\mathbf{n}} \cdot \mathbf{k}^{\mathrm{E}})\, d\Sigma + \int_V c^{\mathrm{E}}\, dV, \tag{2.48}$$

where ∂V is the boundary of V, and $\hat{\mathbf{n}}$ is the unit outward normal. Equation (2.48) confirms the interpretation of the terms $\mathbf{k}^{\mathrm{E}}$ and c^{E} in the conservation law (2.47). The total rate of change of "q-stuff" within the volume V consists of two terms: a surface integral accounting for the flux $\mathbf{k}^{\mathrm{E}}$ of "q-stuff" through the stationary boundary ∂V, and a volume integral accounting for the pointwise creation of "q-stuff" at a rate c^{E}. The minus sign arises in the surface integral because $\hat{\mathbf{n}}$ is the unit *outward* normal; the rate at which "q-stuff" flows *into* V across the patch $\hat{\mathbf{n}}\, d\Sigma$ is $-(\hat{\mathbf{n}} \cdot \mathbf{k}^{\mathrm{E}})\, d\Sigma$.

2.6.1 Conservation of mass

The simplest law of the form (2.47) is that governing the conservation of mass; we denote the Eulerian mass density by ρ^{E}. Since mass cannot be created or destroyed, the total mass within a co-moving volume of material V^t must be constant:

$$\frac{d}{dt}\int_{V^t} \rho^{\mathrm{E}}\, dV^t = 0. \tag{2.49}$$

Upon using the Reynolds transport theorem (2.39) we can rewrite equation (2.49) in the form

$$\int_{V^t} [\partial_t \rho^{\mathrm{E}} + \boldsymbol{\nabla}_{\mathbf{r}} \cdot (\rho^{\mathrm{E}} \mathbf{u}^{\mathrm{E}})]\, dV^t = 0. \tag{2.50}$$

Equation (2.50) must be valid for every co-moving volume V^t within the continuum; this can only be so if the integrand is identically equal to zero:

$$\partial_t \rho^{\mathrm{E}} + \boldsymbol{\nabla}_{\mathbf{r}} \cdot (\rho^{\mathrm{E}} \mathbf{u}^{\mathrm{E}}) = 0. \tag{2.51}$$

This result, which is the Eulerian form of the law of conservation of mass, is known as the *continuity equation*. Equation (2.51) is of the general form (2.47) with $q^{\mathrm{E}} = \rho^{\mathrm{E}}$, $\mathbf{k}^{\mathrm{E}} = \rho^{\mathrm{E}}\mathbf{u}^{\mathrm{E}}$ and $c^{\mathrm{E}} = 0$.

We can use the continuity equation to obtain an alternative form of the Reynolds transport theorem for an arbitrary extensive "q-stuff". Let χ^{E} be the *specific density* of the quantity under consideration, i.e., the amount per unit mass rather than per unit volume, so that

$$q^{\mathrm{E}} = \rho^{\mathrm{E}}\chi^{\mathrm{E}}. \tag{2.52}$$

Substituting (2.52) into equation (2.39) and using (2.2) and (2.51), we obtain the result

$$\frac{d}{dt}\int_{V^t} \rho^{\mathrm{E}}\chi^{\mathrm{E}}\, dV^t = \int_{V^t} \rho^{\mathrm{E}} D_t\chi^{\mathrm{E}}\, dV^t, \tag{2.53}$$

which we shall use below in deriving the remaining three Eulerian conservation laws. We can likewise use the continuity equation to rewrite the general conservation law (2.47) in terms of the material derivative D_t:

$$\rho^{\mathrm{E}} D_t\chi^{\mathrm{E}} + \boldsymbol{\nabla}_{\mathbf{r}} \cdot \boldsymbol{\kappa}^{\mathrm{E}} = c^{\mathrm{E}}, \tag{2.54}$$

where

$$\boldsymbol{\kappa}^{\mathrm{E}} = \mathbf{k}^{\mathrm{E}} - \mathbf{u}^{\mathrm{E}}q^{\mathrm{E}} = \mathbf{k}^{\mathrm{E}} - \rho^{\mathrm{E}}\mathbf{u}^{\mathrm{E}}\chi^{\mathrm{E}}. \tag{2.55}$$

Physically, the quantity $\boldsymbol{\kappa}^{\mathrm{E}}$ is the flux of "q-stuff" relative to the moving material; the total Eulerian flux $\mathbf{k}^{\mathrm{E}}$ consists of the flux due to *advection* $\mathbf{u}^{\mathrm{E}}q^{\mathrm{E}} = \rho^{\mathrm{E}}\mathbf{u}^{\mathrm{E}}\chi^{\mathrm{E}}$ plus the relative flux $\boldsymbol{\kappa}^{\mathrm{E}}$. The flux of mass is entirely advective: $\chi^{\mathrm{E}} = 1$ and $\boldsymbol{\kappa}^{\mathrm{E}} = \mathbf{0}$, so that (2.54) reduces to a tautology.

It is possible to write the continuity equation (2.51) in terms of the material derivative D_t in the form

$$D_t\rho^{\mathrm{E}} + \rho^{\mathrm{E}}\boldsymbol{\nabla}_{\mathbf{r}} \cdot \mathbf{u}^{\mathrm{E}} = 0. \tag{2.56}$$

Equation (2.56) provides a physical interpretation of the divergence of the Eulerian velocity, or equivalently the trace of the Eulerian strain-rate tensor, $\boldsymbol{\nabla}_{\mathbf{r}} \cdot \mathbf{u}^{\mathrm{E}} = \operatorname{tr}\mathbf{D}^{\mathrm{E}}$. Defining the *specific volume* τ^{E} by

$$\rho^{\mathrm{E}}\tau^{\mathrm{E}} = 1, \tag{2.57}$$

we see that

$$\boldsymbol{\nabla}_{\mathbf{r}} \cdot \mathbf{u}^{\mathrm{E}} = -\frac{D_t\rho^{\mathrm{E}}}{\rho^{\mathrm{E}}} = \frac{D_t\tau^{\mathrm{E}}}{\tau^{\mathrm{E}}}. \tag{2.58}$$

Evidently, $\boldsymbol{\nabla}_{\mathbf{r}} \cdot \mathbf{u}^{\mathrm{E}}$ is the fractional rate of change of the volume per unit mass of the particle that is at point $\mathbf{r}$ at time t. An *incompressible* material is one whose particles have an invariant volume: $\boldsymbol{\nabla}_{\mathbf{r}} \cdot \mathbf{u}^{\mathrm{E}} = 0$.

2.6.2 Conservation of momentum

Newton's second law for a parcel of matter moving in an inertial reference frame is

$$\frac{d}{dt}\int_{V^t} \rho^{\mathrm{E}}\mathbf{u}^{\mathrm{E}}\, dV^t = \boldsymbol{\mathcal{F}}, \tag{2.59}$$

where $\boldsymbol{\mathcal{F}}$ is the total external force exerted on the co-moving volume V^t. The force $\boldsymbol{\mathcal{F}}$ consists of the *short-range surface force* $\boldsymbol{\mathcal{F}}_{\mathrm{s}}$ exerted across the boundary ∂V^t together with the *long-range body force* $\boldsymbol{\mathcal{F}}_{\mathrm{b}}$ exerted on individual particles within V^t:

$$\boldsymbol{\mathcal{F}} = \boldsymbol{\mathcal{F}}_{\mathrm{s}} + \boldsymbol{\mathcal{F}}_{\mathrm{b}}. \tag{2.60}$$

The surface force can be written in terms of the Cauchy stress $\mathbf{T}^{\mathrm{E}}$ acting upon the boundary in the form

$$\boldsymbol{\mathcal{F}}_{\mathrm{s}} = \int_{\partial V^t} (\hat{\mathbf{n}}^t \cdot \mathbf{T}^{\mathrm{E}})\, d\Sigma^t = \int_{V^t} (\boldsymbol{\nabla}_{\mathbf{r}} \cdot \mathbf{T}^{\mathrm{E}})\, dV^t, \tag{2.61}$$

where the second equality follows from Gauss' theorem. The long-range body force $\boldsymbol{\mathcal{F}}_{\mathrm{b}}$ is assumed to be expressible in terms of a specific body-force density $\mathbf{g}^{\mathrm{E}}$ in the form

$$\boldsymbol{\mathcal{F}}_{\mathrm{b}} = \int_{V^t} \rho^{\mathrm{E}}\mathbf{g}^{\mathrm{E}}\, dV^t. \tag{2.62}$$

The prototypical long-range force is of course the attraction of gravity; the quantity $\mathbf{g}^{\mathrm{E}}$ is in that case the *gravitational field* at point $\mathbf{r}$ at time t.

Inserting equations (2.61) and (2.62) into (2.59) and using the Reynolds transport theorem in the form (2.53), we obtain

$$\int_{V^t} [\rho^{\mathrm{E}} D_t \mathbf{u}^{\mathrm{E}} - \boldsymbol{\nabla}_{\mathbf{r}} \cdot \mathbf{T}^{\mathrm{E}} - \rho^{\mathrm{E}}\mathbf{g}^{\mathrm{E}}]\, dV^t = \mathbf{0}. \tag{2.63}$$

Since (2.63) must be valid for every co-moving volume within the continuum, the integrand must vanish identically; the Eulerian version of the momentum conservation law is thus

$$\rho^{\mathrm{E}} D_t \mathbf{u}^{\mathrm{E}} = \boldsymbol{\nabla}_{\mathbf{r}} \cdot \mathbf{T}^{\mathrm{E}} + \rho^{\mathrm{E}}\mathbf{g}^{\mathrm{E}}. \tag{2.64}$$

Equation (2.64) is of the general form (2.54), with $\chi^{\mathrm{E}} = \mathbf{u}^{\mathrm{E}}$, $\boldsymbol{\kappa}^{\mathrm{E}} = -\mathbf{T}^{\mathrm{E}}$ and $c^{\mathrm{E}} = \rho^{\mathrm{E}}\mathbf{g}^{\mathrm{E}}$. The sign convention for the stress tensor was established when we stipulated that $d\mathbf{f}^{\mathrm{E}}$ in equation (2.40) was the force exerted *by* the particles in the *front* of the patch $\hat{\mathbf{n}}^t d\Sigma^t$ *on* the particles in the *back*. With this convention, *minus* the Cauchy stress $-\mathbf{T}^{\mathrm{E}}$ can be interpreted as the flux of momentum relative to the moving material; the flux relative to inertial space is $\mathbf{k}^{\mathrm{E}} = \rho^{\mathrm{E}}\mathbf{u}^{\mathrm{E}}\mathbf{u}^{\mathrm{E}} - \mathbf{T}^{\mathrm{E}}$. For brevity, we refer to the conservation law (2.64) as the *momentum equation.*

2.6.3 Conservation of angular momentum

The rate of change of the angular momentum of a parcel of matter must be equal to the sum of the external torques. Ignoring the possibility of internal angular momentum and surface and body torques associated with any intrinsic spin, we have

$$\frac{d}{dt}\int_{V^t} \rho^{\mathrm{E}}(\mathbf{r}\times\mathbf{u}^{\mathrm{E}})\,dV^t = \boldsymbol{\mathcal{N}}_{\mathrm{s}} + \boldsymbol{\mathcal{N}}_{\mathrm{b}}. \tag{2.65}$$

The quantities $\boldsymbol{\mathcal{N}}_{\mathrm{s}}$ and $\boldsymbol{\mathcal{N}}_{\mathrm{b}}$ are the torques exerted on V^t by the surface and body forces, respectively:

$$\begin{aligned}\boldsymbol{\mathcal{N}}_{\mathrm{s}} &= \int_{\partial V^t} \mathbf{r}\times(\hat{\mathbf{n}}^t\cdot\mathbf{T}^{\mathrm{E}})\,d\Sigma^t = -\int_{\partial V^t}(\hat{\mathbf{n}}^t\cdot\mathbf{T}^{\mathrm{E}})\times\mathbf{r}\,d\Sigma^t \\ &= -\int_{\partial V^t}\hat{\mathbf{n}}\cdot(\mathbf{T}^{\mathrm{E}}\times\mathbf{r})\,d\Sigma^t = -\int_{V^t}\boldsymbol{\nabla}_{\mathbf{r}}\cdot(\mathbf{T}^{\mathrm{E}}\times\mathbf{r})\,dV^t,\end{aligned} \tag{2.66}$$

$$\boldsymbol{\mathcal{N}}_{\mathrm{b}} = \int_{V^t}\rho^{\mathrm{E}}(\mathbf{r}\times\mathbf{g}^{\mathrm{E}})\,dV^t. \tag{2.67}$$

Inserting the representations (2.66) and (2.67) into (2.65) and again using the transport theorem in the form (2.53), we obtain

$$\int_{V^t}[\rho^{\mathrm{E}}D_t(\mathbf{r}\times\mathbf{u}^{\mathrm{E}}) + \boldsymbol{\nabla}_{\mathbf{r}}\cdot(\mathbf{T}^{\mathrm{E}}\times\mathbf{r}) - \rho^{\mathrm{E}}(\mathbf{r}\times\mathbf{g}^{\mathrm{E}})]\,dV^t = \mathbf{0}, \tag{2.68}$$

Once again, the integrand must vanish, since V^t is an arbitrary co-moving volume; the resulting law of conservation of angular momentum is

$$\rho^{\mathrm{E}}D_t(\mathbf{r}\times\mathbf{u}^{\mathrm{E}}) + \boldsymbol{\nabla}_{\mathbf{r}}\cdot(\mathbf{T}^{\mathrm{E}}\times\mathbf{r}) = \rho^{\mathrm{E}}(\mathbf{r}\times\mathbf{g}^{\mathrm{E}}). \tag{2.69}$$

The result (2.69) is of the expected Eulerian form (2.54) with $\chi^{\mathrm{E}} = \mathbf{r}\times\mathbf{u}^{\mathrm{E}}$, $\boldsymbol{\kappa}^{\mathrm{E}} = \mathbf{T}^{\mathrm{E}}\times\mathbf{r}$ and $c^{\mathrm{E}} = \rho^{\mathrm{E}}(\mathbf{r}\times\mathbf{g}^{\mathrm{E}})$; the total flux of angular momentum relative to inertial space is $\mathbf{k}^{\mathrm{E}} = \rho^{\mathrm{E}}\mathbf{u}^{\mathrm{E}}(\mathbf{r}\times\mathbf{u}^{\mathrm{E}}) + \mathbf{T}^{\mathrm{E}}\times\mathbf{r}$.

By combining terms and using the fact that $(D_t\mathbf{r})\times\mathbf{u}^{\mathrm{E}} = \mathbf{u}^{\mathrm{E}}\times\mathbf{u}^{\mathrm{E}} = \mathbf{0}$, we can rewrite equation (2.69) in the form

$$\mathbf{r}\times(\rho^{\mathrm{E}}D_t\mathbf{u}^{\mathrm{E}} - \boldsymbol{\nabla}_{\mathbf{r}}\cdot\mathbf{T}^{\mathrm{E}} - \rho^{\mathrm{E}}\mathbf{g}^{\mathrm{E}}) - \wedge\mathbf{T}^{\mathrm{E}} = \mathbf{0}. \tag{2.70}$$

The expression in parentheses in equation (2.70) is identically zero by virtue of the momentum equation (2.64); hence the angular momentum conservation law reduces to

$$\wedge\mathbf{T}^{\mathrm{E}} = \mathbf{0}. \tag{2.71}$$

Equation (2.71) stipulates that the Cauchy stress $\mathbf{T}^{\mathrm{E}}$ must be *symmetric*:

$$(\mathbf{T}^{\mathrm{E}})^{\mathrm{T}} = \mathbf{T}^{\mathrm{E}}. \tag{2.72}$$

This is the most familiar form of the law of conservation of angular momentum in a spin-free or *non-polar* continuum. The flux of angular momentum can be written in the alternative form $\mathbf{k}^{\mathrm{E}} = \rho^{\mathrm{E}}\mathbf{u}^{\mathrm{E}}(\mathbf{r}\times\mathbf{u}^{\mathrm{E}}) - \mathbf{r}\times\mathbf{T}^{\mathrm{E}}$ by virtue of the symmetry (2.72).

2.6.4 Conservation of energy

A general formulation of the law of conservation of energy must explicitly recognize that heat is a form of energy; this necessitates the introduction of a number of thermodynamical concepts. The macroscopic kinetic energy density $\frac{1}{2}\rho^{\mathrm{E}}(\mathbf{u}^{\mathrm{E}}\cdot\mathbf{u}^{\mathrm{E}})$ does not represent the only form of energy within a continuum; real materials also store and transport thermal kinetic energy and interatomic potential energy. We denote the specific density of this thermodynamic *internal energy* by U^{E}; the total Eulerian energy density is then $\rho^{\mathrm{E}}(\frac{1}{2}\mathbf{u}^{\mathrm{E}}\cdot\mathbf{u}^{\mathrm{E}} + U^{\mathrm{E}})$. The Eulerian flux of internal energy due to advection with the material is $\rho^{\mathrm{E}}U^{\mathrm{E}}\mathbf{u}^{\mathrm{E}}$; in addition, internal energy can be transported relative to the material by the process known as thermal conduction. We denote the Eulerian heat flux vector by $\mathbf{H}^{\mathrm{E}}$; the total flux of heat or internal energy is then $\rho^{\mathrm{E}}U^{\mathrm{E}}\mathbf{u}^{\mathrm{E}} + \mathbf{H}^{\mathrm{E}}$. The final thermodynamic quantity we shall introduce is the specific rate of internal heating h^{E}; this is used in a general formulation to account for microscopic processes such as radioactive decay.

The law of conservation of energy, applied to a co-moving volume V^t, takes the form

$$\begin{aligned}\frac{d}{dt}\int_{V^t} &\rho^{\mathrm{E}}(\tfrac{1}{2}\mathbf{u}^{\mathrm{E}}\cdot\mathbf{u}^{\mathrm{E}} + U^{\mathrm{E}})\,dV^t \\ &= \int_{V^t}\rho^{\mathrm{E}}\mathbf{g}^{\mathrm{E}}\cdot\mathbf{u}^{\mathrm{E}}\,dV^t + \int_{\partial V^t}(\hat{\mathbf{n}}^t\cdot\mathbf{T}^{\mathrm{E}})\cdot\mathbf{u}^{\mathrm{E}}\,d\Sigma^t \\ &\qquad + \int_{V^t}\rho^{\mathrm{E}}h^{\mathrm{E}}\,dV^t - \int_{\partial V^t}(\hat{\mathbf{n}}^t\cdot\mathbf{H}^{\mathrm{E}})\,d\Sigma^t.\end{aligned} \tag{2.73}$$

The quantity on the left of (2.73) is the rate of change of the total energy, kinetic plus internal, within the volume; the four terms on the right account for all the ways in which this energy can be altered. The first two integrals represent the rate at which mechanical energy is added to the volume by the action of body and surface forces, whereas the last two integrals represent the rate at which heat energy is added by internal heating and thermal conduction. The final term appears with a minus sign because $(\hat{\mathbf{n}}^t\cdot\mathbf{H}^{\mathrm{E}})\,d\Sigma^t$ is, by definition, the rate at which heat is conducted from the back of the patch $\hat{\mathbf{n}}^t\,d\Sigma^t$ to the front. Using Gauss' theorem and Reynolds' theorem in the form (2.53), we obtain

$$\int_{V^t} [\rho^{\mathrm{E}} D_t(\tfrac{1}{2}\mathbf{u}^{\mathrm{E}}\cdot\mathbf{u}^{\mathrm{E}} + U^{\mathrm{E}}) \\ + \boldsymbol{\nabla}_{\mathbf{r}}\cdot(\mathbf{H}^{\mathrm{E}} - \mathbf{T}^{\mathrm{E}}\cdot\mathbf{u}^{\mathrm{E}}) - \rho^{\mathrm{E}}(h^{\mathrm{E}} + \mathbf{g}^{\mathrm{E}}\cdot\mathbf{u}^{\mathrm{E}})]\, dV^t = \mathbf{0}. \tag{2.74}$$

Upon setting the integrand in (2.74) equal to zero, extracting a buried copy of the momentum equation as we did in going from (2.69) to (2.70), and making use of the symmetry (2.72), we obtain the Eulerian energy conservation law:

$$\rho^{\mathrm{E}} D_t U^{\mathrm{E}} + \boldsymbol{\nabla}_{\mathbf{r}}\cdot\mathbf{H}^{\mathrm{E}} = \mathbf{T}^{\mathrm{E}}:\mathbf{D}^{\mathrm{E}} + \rho^{\mathrm{E}} h^{\mathrm{E}}. \tag{2.75}$$

The result (2.75), which is known as the *internal-energy equation*, is of the general form (2.54) with $\chi^{\mathrm{E}} = U^{\mathrm{E}}$ and $\boldsymbol{\kappa}^{\mathrm{E}} = \mathbf{H}^{\mathrm{E}}$. The volumetric rate at which internal energy is generated by mechanical processes or "created" by radioactive decay is $c^{\mathrm{E}} = \mathbf{T}^{\mathrm{E}}:\mathbf{D}^{\mathrm{E}} + \rho^{\mathrm{E}} h^{\mathrm{E}}$.

The time and length scales of all seismic phenomena are such that the flow of heat from compressed to dilated regions is negligible during a cycle of oscillation. To see this, we compare the distance $L_{\mathrm{heat}} \approx \sqrt{\kappa T}$ which heat can travel during a time interval T with the distance $L_{\mathrm{wave}} \approx vT$ travelled by a seismic wave. It is evident that $L_{\mathrm{heat}} \ll L_{\mathrm{wave}}$ for oscillations with periods $T \gg \kappa v^{-2}$. Inserting typical upper-mantle values for the thermal diffusivity κ and P-wave speed v, we find that heat flow is significant only for ultra-short period waves: $\kappa v^{-2} \approx 10^{-14}$ seconds. In seismology, we are concerned with waves whose periods are greater than $T \approx 10^{-2}$ seconds; it is therefore an excellent approximation to regard the deformation as *adiabatic*:

$$\mathbf{H}^{\mathrm{E}} = \mathbf{0}. \tag{2.76}$$

Radioactive heating is also thoroughly negligible on seismological time scales; thus we may set $h^{\mathrm{E}} = 0$. With these simplifications, the internal-energy equation in all problems of seismological interest reduces to

$$\rho^{\mathrm{E}} D_t U^{\mathrm{E}} = \mathbf{T}^{\mathrm{E}}:\mathbf{D}^{\mathrm{E}}. \tag{2.77}$$

The mechanical work performed against the stress during a closed cycle of deformation is in general strictly positive,

$$\oint \mathbf{T}^{\mathrm{E}}:\mathbf{D}^{\mathrm{E}}\, dt \geq 0, \tag{2.78}$$

reflecting the inherently dissipative nature of the internal frictional processes within any realistic material. Only in the idealized case of a perfectly elastic continuum is the work *recoverable*, so that equality pertains in equation (2.78).

2.6.5 Boundary conditions

The conservation laws for mass, momentum, angular momentum and energy derived above must be supplemented by continuity conditions at any boundary separating two different types of material. We use the notation $[q]_-^+$ throughout this book to denote the jump in the quantity q in going from the front or $+$ side to the back or $-$ side of a boundary.

At a welded boundary between two solids, the Eulerian velocity must be continuous:

$$[\mathbf{u}^{\mathrm{E}}]_-^+ = \mathbf{0}. \tag{2.79}$$

Tangential slip is allowed at a boundary between a solid and an inviscid fluid; it may also occur on an idealized fault surface separating two solids. At such a slipping boundary, equation (2.79) is replaced by

$$[\hat{\mathbf{n}}^t \cdot \mathbf{u}^{\mathrm{E}}]_-^+ = 0, \tag{2.80}$$

where $\hat{\mathbf{n}}^t$ is the unit normal to the boundary. Equation (2.80) guarantees that there is no separation or interpenetration of the two materials at the boundary.

In addition to the above two kinematical conditions, there is a dynamical requirement that the traction $\hat{\mathbf{n}}^t \cdot \mathbf{T}^{\mathrm{E}}$ must be continuous across both a welded and a slipping boundary:

$$[\hat{\mathbf{n}}^t \cdot \mathbf{T}^{\mathrm{E}}]_-^+ = \mathbf{0}. \tag{2.81}$$

Finally, there can be no shear traction on the boundary between a solid and an inviscid fluid, i.e., $\hat{\mathbf{n}}^t \cdot \mathbf{T}^{\mathrm{E}}$ must be in the direction of the instantaneous normal $\hat{\mathbf{n}}^t$ on a frictionless fluid-solid boundary:

$$\hat{\mathbf{n}}^t \cdot \mathbf{T}^{\mathrm{E}} = \hat{\mathbf{n}}^t(\hat{\mathbf{n}}^t \cdot \mathbf{T}^{\mathrm{E}} \cdot \hat{\mathbf{n}}^t). \tag{2.82}$$

All four conditions (2.79)–(2.82) pertain to the instantaneous position of the boundary, i.e., they must be satisfied at all points $\mathbf{r}$ occupied by the boundary at time t.

2.6.6 Rotating reference frame

Suppose now that the motion is viewed not in an inertial frame, but in a frame that is rotating with a uniform angular velocity $\boldsymbol{\Omega}$ with respect to an inertial frame. The conservation laws for mass and internal energy, as well as the symmetry of the Cauchy stress and the kinematical and dynamical boundary conditions, are all unaffected by this change of reference frame.

The momentum equation (2.64) must, however, be modified to incorporate the apparent *Coriolis* and *centripetal* accelerations:

$$\rho^{\mathrm{E}}[D_t\mathbf{u}^{\mathrm{E}} + 2\boldsymbol{\Omega} \times \mathbf{u}^{\mathrm{E}} + \boldsymbol{\Omega} \times (\boldsymbol{\Omega} \times \mathbf{r})] = \boldsymbol{\nabla}_{\mathbf{r}} \cdot \mathbf{T}^{\mathrm{E}} + \rho^{\mathrm{E}}\mathbf{g}^{\mathrm{E}}. \tag{2.83}$$

It is natural to use a rotating frame to analyze the elastic-gravitational response of the Earth to earthquakes or other forcing phenomena, since seismometers are subject to Coriolis and centripetal accelerations as a result of the Earth's diurnal rotation. To account for rotation, we shall conduct the remainder of the analysis in this and the next chapter using equation (2.83) rather then (2.64).

2.7 Lagrangian Conservation Laws

Each of the four Eulerian conservation laws has an equivalent Lagrangian form, which we consider next. The Eulerian laws were all obtained by first establishing integral balance principles for an arbitrary co-moving volume V^t, and then "erasing" the integral sign. We follow a similar procedure in deriving the Lagrangian laws, first seeking integral balance principles for the *undeformed* volume V^0 in the initial configuration.

2.7.1 Conservation of mass

The mass within a co-moving volume V^t can be written as an integral over the undeformed volume V^0 by transforming from spatial coordinates $\mathbf{r}$ to material coordinates $\mathbf{x}$:

$$\int_{V^t} \rho^{\mathrm{E}}\, dV^t = \int_{V^0} \rho^{\mathrm{L}} J\, dV^0, \tag{2.84}$$

where J is the Jacobian of the transformation. Since mass cannot be created or destroyed, the amount of mass within any volume at time t must be the same as the initial mass within the volume at time $t = 0$:

$$\int_{V^0} \rho^{\mathrm{L}} J\, dV^0 = \int_{V^0} \rho^0\, dV^0, \tag{2.85}$$

where $\rho^0(\mathbf{x}) = \rho^{\mathrm{L}}(\mathbf{x}, 0)$. Equation (2.85) must be valid for every volume V^0 in the initial configuration; this can only be so if the integrands on the left and right are everywhere equal:

$$\rho^{\mathrm{L}} J = \rho^0. \tag{2.86}$$

This result, relating the instantaneous density of a particle ρ^{L} to its initial density ρ^0, is the Lagrangian conservation of mass law.

The Lagrangian specific volume of a particle is defined, as usual, by $\tau^{\mathrm{L}}(\mathbf{x},t)=\tau^{\mathrm{E}}(\mathbf{r}(\mathbf{x},t),t)$, or, equivalently, by $\rho^{\mathrm{L}}\tau^{\mathrm{L}}=1$. Written in terms of τ^{L} rather than ρ^{L}, equation (2.86) takes the form

$$\tau^{\mathrm{L}}=J\tau^{0}, \tag{2.87}$$

where $\tau^{0}(\mathbf{x})=\tau^{\mathrm{L}}(\mathbf{x},0)$. Equation (2.87) is simply the differential volumetric relation $dV^{t}=J\,dV^{0}$, expressed in a different notation.

We can use the Lagrangian conservation of mass law (2.86) to rewrite the transformation formula for an arbitrary "q-stuff", in the same way we used the Eulerian form (2.51) to rewrite the Reynolds transport theorem in the form (2.53). The total amount of "q-stuff" within a co-moving volume V^{t} can be written in terms of an integral over V^{0} in the form

$$\int_{V^{t}}\rho^{\mathrm{E}}\chi^{\mathrm{E}}\,dV^{t}=\int_{V^{0}}\rho^{0}\chi^{\mathrm{L}}\,dV^{0}, \tag{2.88}$$

where $\chi^{\mathrm{L}}(\mathbf{x},t)=\chi^{\mathrm{E}}(\mathbf{r}(\mathbf{x},t),t)$ is the specific density. We will use the result (2.88) below in deriving the Lagrangian forms of the conservation laws for momentum and energy.

2.7.2 Conservation of momentum

Consider the integral of the Eulerian momentum equation (2.83) over an arbitrary co-moving volume V^{t}. Collecting all of the terms proportional to the density ρ^{E} on the left, and using Gauss' theorem, we obtain

$$\begin{aligned}\int_{V^{t}}\rho^{\mathrm{E}}[D_{t}\mathbf{u}^{\mathrm{E}}+2\boldsymbol{\Omega}\times\mathbf{u}^{\mathrm{E}}+\boldsymbol{\Omega}\times(\boldsymbol{\Omega}\times\mathbf{r})-\mathbf{g}^{\mathrm{E}}]\,dV^{t}&\\=\int_{\partial V^{t}}(\hat{\mathbf{n}}^{t}\cdot\mathbf{T}^{\mathrm{E}})\,d\Sigma^{t}.&\end{aligned} \tag{2.89}$$

The quantity on the left can be transformed into an integral over the undeformed volume V^{0} using equation (2.88):

$$\begin{aligned}\int_{V^{t}}\rho^{\mathrm{E}}[D_{t}\mathbf{u}^{\mathrm{E}}+2\boldsymbol{\Omega}\times\mathbf{u}^{\mathrm{E}}+\boldsymbol{\Omega}\times(\boldsymbol{\Omega}\times\mathbf{r})-\mathbf{g}^{\mathrm{E}}]\,dV^{t}&\\=\int_{V^{0}}\rho^{0}[\partial_{t}^{2}\mathbf{r}+2\boldsymbol{\Omega}\times\partial_{t}\mathbf{r}+\boldsymbol{\Omega}\times(\boldsymbol{\Omega}\times\mathbf{r})-\mathbf{g}^{\mathrm{L}}]\,dV^{0},&\end{aligned} \tag{2.90}$$

where $\mathbf{g}^{\mathrm{L}}(\mathbf{x},t)=\mathbf{g}^{\mathrm{E}}(\mathbf{r}(\mathbf{x},t),t)$. The surface integral on the right can likewise be transformed using the definition (2.41) of the first Piola-Kirchhoff stress tensor:

$$\begin{aligned}\int_{\partial V^{t}}(\hat{\mathbf{n}}^{t}\cdot\mathbf{T}^{\mathrm{E}})\,d\Sigma^{t}&=\int_{\partial V^{0}}(\hat{\mathbf{n}}^{0}\cdot\mathbf{T}^{\mathrm{PK}})\,d\Sigma^{0}\\&=\int_{V^{0}}(\boldsymbol{\nabla}_{\mathbf{x}}\cdot\mathbf{T}^{\mathrm{PK}})\,dV^{0},\end{aligned} \tag{2.91}$$

where the second equality follows from an application of Gauss' theorem to the undeformed volume V^0. Combining equations (2.90) and (2.91) we can rewrite the balance principle (2.89) in the form

$$\int_{V^0} \{\rho^0[\partial_t^2\mathbf{r} + 2\boldsymbol{\Omega}\times\partial_t\mathbf{r} + \boldsymbol{\Omega}\times(\boldsymbol{\Omega}\times\mathbf{r})] - \boldsymbol{\nabla}_{\mathbf{x}}\cdot\mathbf{T}^{\mathrm{PK}} - \rho^0\mathbf{g}^{\mathrm{L}}\}\, dV^0 = \mathbf{0}. \tag{2.92}$$

Since the volume V^0 is arbitrary, the integrand in (2.92) must be identically zero; the Lagrangian version of the momentum equation in a uniformly rotating reference frame is therefore

$$\rho^0[\partial_t^2\mathbf{r} + 2\boldsymbol{\Omega}\times\partial_t\mathbf{r} + \boldsymbol{\Omega}\times(\boldsymbol{\Omega}\times\mathbf{r})] = \boldsymbol{\nabla}_{\mathbf{x}}\cdot\mathbf{T}^{\mathrm{PK}} + \rho^0\mathbf{g}^{\mathrm{L}}. \tag{2.93}$$

It is noteworthy that the divergence of the Eulerian Cauchy stress $\boldsymbol{\nabla}_{\mathbf{r}}\cdot\mathbf{T}^{\mathrm{E}}$ in (2.83) is transformed into the divergence of the first Piola-Kirchhoff stress $\boldsymbol{\nabla}_{\mathbf{x}}\cdot\mathbf{T}^{\mathrm{PK}}$ in (2.93); in fact, the original definition (2.41) of $\mathbf{T}^{\mathrm{PK}}$ was motivated by this simple transformation.

It is also possible to write the Lagrangian momentum equation in terms of the second Piola-Kirchhoff stress $\mathbf{T}^{\mathrm{SK}}$ rather than in terms of $\mathbf{T}^{\mathrm{PK}}$; upon substituting the relation (2.45) into (2.93) we obtain

$$\rho^0[\partial_t^2\mathbf{r} + 2\boldsymbol{\Omega}\times\partial_t\mathbf{r} + \boldsymbol{\Omega}\times(\boldsymbol{\Omega}\times\mathbf{r})] = \boldsymbol{\nabla}_{\mathbf{x}}\cdot(\mathbf{T}^{\mathrm{SK}}\cdot\mathbf{F}^{\mathrm{T}}) + \rho^0\mathbf{g}^{\mathrm{L}}. \tag{2.94}$$

The stress-divergence term in equation (2.94) is more complicated than that in (2.93); the angular momentum conservation law is, in contrast, simpler in terms of $\mathbf{T}^{\mathrm{SK}}$ rather than $\mathbf{T}^{\mathrm{PK}}$, as we see next.

2.7.3 Conservation of angular momentum

The first Piola-Kirchhoff stress $\mathbf{T}^{\mathrm{PK}}$ and its transpose $(\mathbf{T}^{\mathrm{PK}})^{\mathrm{T}}$ are related by

$$(\mathbf{T}^{\mathrm{PK}})^{\mathrm{T}} = \mathbf{F}\cdot\mathbf{T}^{\mathrm{PK}}\cdot\mathbf{F}^{-\mathrm{T}}. \tag{2.95}$$

Equation (2.95) can be readily verified using the relation (2.42) together with the symmetry of the Lagrangian Cauchy stress, $(\mathbf{T}^{\mathrm{L}})^{\mathrm{T}} = \mathbf{T}^{\mathrm{L}}$. Using (2.46) instead, we see that the second Piola-Kirchhoff stress $\mathbf{T}^{\mathrm{SK}}$ is symmetric if $\mathbf{T}^{\mathrm{L}}$ is:

$$(\mathbf{T}^{\mathrm{SK}})^{\mathrm{T}} = \mathbf{T}^{\mathrm{SK}}. \tag{2.96}$$

Equations (2.95) and (2.96) are the Lagrangian forms of the law of conservation of angular momentum.

Strictly speaking, we are not permitted to draw any conclusions about the symmetry or asymmetry of $\mathbf{T}^{\mathrm{PK}}$ from the relation (2.95). If we insist on regarding the first Piola-Kirchhoff stress as a two-point tensor, then the concept of symmetry is meaningless, since $\mathbf{T}^{\mathrm{PK}}$ and its transpose $(\mathbf{T}^{\mathrm{PK}})^{\mathrm{T}}$ reside in different spaces. We shall, however, disregard the two-point nature of $\mathbf{T}^{\mathrm{PK}}$ just as we do that of the deformation tensor $\mathbf{F}$; thus we shall say, on the basis of equation (2.95), that *the first Piola-Kirchhoff stress is not symmetric.*

2.7.4 Conservation of energy

Upon integrating the adiabatic internal-energy equation (2.77) over an arbitrary co-moving volume V^t, we obtain

$$\int_{V^t} \rho^{\mathrm{E}} D_t U^{\mathrm{E}}\, dV^t = \int_{V^t} \mathbf{T}^{\mathrm{E}} \!:\! \mathbf{D}^{\mathrm{E}}\, dV^t. \tag{2.97}$$

The left side can be transformed into an integral over the undeformed volume V^0 using (2.88):

$$\int_{V^t} \rho^{\mathrm{E}} D_t U^{\mathrm{E}}\, dV^t = \int_{V^0} \rho^0 \partial_t U^{\mathrm{L}}\, dV^0. \tag{2.98}$$

To transform the right side, we must first express the mechanical rate of internal energy generation $\mathbf{T}^{\mathrm{E}} : \mathbf{D}^{\mathrm{E}}$ in terms of the first and second Piola-Kirchhoff stresses $\mathbf{T}^{\mathrm{PK}}$ and $\mathbf{T}^{\mathrm{SK}}$. Making use of equations (2.10) and (2.42), we find that

$$\begin{aligned} \mathbf{T}^{\mathrm{E}} \!:\! \mathbf{D}^{\mathrm{E}} &= \mathbf{T}^{\mathrm{E}} \!:\! \mathbf{G}^{\mathrm{E}} = \mathrm{tr}\,[J^{-1}(\mathbf{F}\cdot\mathbf{T}^{\mathrm{PK}})\cdot(\partial_t\mathbf{F}\cdot\mathbf{F}^{-1})] \\ &= J^{-1}\,\mathrm{tr}\,(\mathbf{T}^{\mathrm{PK}}\cdot\partial_t\mathbf{F}) = J^{-1}(\mathbf{T}^{\mathrm{PK}} \!:\! \partial_t\mathbf{F}^{\mathrm{T}}). \end{aligned} \tag{2.99}$$

Using (2.24) and (2.46) instead, we obtain the alternative form

$$\begin{aligned} \mathbf{T}^{\mathrm{E}} \!:\! \mathbf{D}^{\mathrm{E}} &= \mathrm{tr}\,[J^{-1}(\mathbf{F}\cdot\mathbf{T}^{\mathrm{SK}}\cdot\mathbf{F}^{\mathrm{T}})\cdot(\mathbf{F}^{-\mathrm{T}}\cdot\partial_t\mathbf{E}^{\mathrm{L}}\cdot\mathbf{F}^{-1})] \\ &= J^{-1}\,\mathrm{tr}\,(\mathbf{T}^{\mathrm{SK}}\cdot\partial_t\mathbf{E}^{\mathrm{L}}) = J^{-1}(\mathbf{T}^{\mathrm{SK}} \!:\! \partial_t\mathbf{E}^{\mathrm{L}}). \end{aligned} \tag{2.100}$$

Upon inserting these two expressions into the left side of (2.97), transforming, and combining with (2.98), we see that

$$\int_{V^0} [\rho^0 \partial_t U^{\mathrm{L}} - \mathbf{T}^{\mathrm{PK}} \!:\! \partial_t \mathbf{F}^{\mathrm{T}}]\, dV^0 = 0, \tag{2.101}$$

$$\int_{V^0} [\rho^0 \partial_t U^{\mathrm{L}} - \mathbf{T}^{\mathrm{SK}} \!:\! \partial_t \mathbf{E}^{\mathrm{L}}]\, dV^0 = 0. \tag{2.102}$$

Since the volumes in both (2.101) and (2.102) are arbitrary, the integrands must vanish; the Lagrangian adiabatic internal energy equation may therefore be written in either of the two equivalent forms

$$\rho^0 \partial_t U^{\mathrm{L}} = \mathbf{T}^{\mathrm{PK}} \!:\! \partial_t \mathbf{F}^{\mathrm{T}} = \mathbf{T}^{\mathrm{SK}} \!:\! \partial_t \mathbf{E}^{\mathrm{L}}. \tag{2.103}$$

The work performed against either $\mathbf{T}^{\mathrm{PK}}$ or $\mathbf{T}^{\mathrm{SK}}$ during a closed cycle of deformation is intrinsically dissipative, i.e.,

$$\oint \mathbf{T}^{\mathrm{PK}} \!:\! \partial_t \mathbf{F}^{\mathrm{T}}\, dt = \oint \mathbf{T}^{\mathrm{SK}} \!:\! \partial_t \mathbf{E}^{\mathrm{L}}\, dt \geq 0, \tag{2.104}$$

where as before the equality pertains only in the case of a perfectly elastic material. We shall use equations (2.103) in our discussion of the constitutive relation for a perfectly elastic material in Section 2.10.

2.7.5 Boundary conditions

Particles that are juxtaposed on either side of a welded boundary at time $t = 0$ must remain juxtaposed; the Lagrangian form of this kinematical boundary condition is simply

$$[\mathbf{r}]_-^+ = \mathbf{0}. \tag{2.105}$$

It is also straightforward to write the dynamical boundary condition (2.81) on a welded boundary in terms of the first Piola-Kirchhoff stress:

$$[\hat{\mathbf{n}}^0 \cdot \mathbf{T}^{\mathrm{PK}}]_-^+ = \mathbf{0}, \tag{2.106}$$

where we have used the equality $\hat{\mathbf{n}}^t \cdot \mathbf{T}^{\mathrm{E}}\, d\Sigma^t = \hat{\mathbf{n}}^0 \cdot \mathbf{T}^{\mathrm{PK}}\, d\Sigma^0$ and the continuity of the differential areas $d\Sigma^t$ and $d\Sigma^0$. Both Lagrangian continuity conditions (2.105) and (2.106) apply on the *undeformed* boundary in the initial configuration, in contrast to the Eulerian conditions (2.79)–(2.82), which apply on the *deformed* boundary in the current configuration. Neither condition is valid on a fluid-solid or a fault boundary, where there is tangential slip; we defer consideration of the Lagrangian boundary conditions on such slipping boundaries until Section 3.4.

2.8 Gravitational Potential Theory

Since gravity is a *conservative* force, we may write the Eulerian gravitational force per unit mass $\mathbf{g}^{\mathrm{E}}$ as the gradient of an *Eulerian gravitational potential* ϕ^{E}:

$$\mathbf{g}^{\mathrm{E}} = -\boldsymbol{\nabla}_{\mathbf{r}} \phi^{\mathrm{E}}. \tag{2.107}$$

The potential ϕ^{E} is everywhere *negative*, by virtue of the minus sign in equation (2.107); we shall adhere to this sign convention throughout this book.

2.8.1 Poisson's equation

Given an instantaneous Eulerian density distribution ρ^{E}, we can find ϕ^{E} by solving *Poisson's equation*

$$\nabla_{\mathbf{r}}^2 \phi^{\mathrm{E}} = 4\pi G \rho^{\mathrm{E}}, \tag{2.108}$$

where G is the gravitational constant. If there are boundaries across which the density ρ^{E} is discontinuous, then (2.108) must be supplemented by the boundary conditions

$$[\phi^{\mathrm{E}}]_-^+ = 0, \tag{2.109}$$

$$[\hat{\mathbf{n}}^t \cdot \boldsymbol{\nabla}_{\mathbf{r}} \phi^{\mathrm{E}}]_-^+ = 0, \tag{2.110}$$

where $\hat{\mathbf{n}}^t$ is the normal to the boundary. Like equations (2.79)–(2.82), the continuity conditions (2.109) and (2.110) must be satisfied on the instantaneous boundary. Note that continuity of the scalars ϕ^{E} and $\hat{\mathbf{n}}^t \cdot \boldsymbol{\nabla}_{\mathbf{r}} \phi^{\mathrm{E}}$ implies continuity of the gravity vector $\mathbf{g}^{\mathrm{E}} = -\boldsymbol{\nabla}_{\mathbf{r}} \phi^{\mathrm{E}}$.

The solution to the boundary-value problem (2.108)–(2.110) is

$$\phi^{\mathrm{E}}(\mathbf{r}, t) = -G \int_{V^t} \frac{\rho^{\mathrm{E}}(\mathbf{r}', t)}{\|\mathbf{r} - \mathbf{r}'\|} \, dV^{t\prime}, \tag{2.111}$$

where the integral is over the matter-filled portion of space at time t, i.e., over all points $\mathbf{r}'$ where $\rho^{\mathrm{E}}(\mathbf{r}', t) > 0$. The gravitational field $\mathbf{g}^{\mathrm{E}}$ obtained by combining (2.107) and (2.111) is

$$\mathbf{g}^{\mathrm{E}}(\mathbf{r}, t) = -G \int_{V^t} \frac{\rho^{\mathrm{E}}(\mathbf{r}', t)\,(\mathbf{r} - \mathbf{r}')}{\|\mathbf{r} - \mathbf{r}'\|^3} \, dV^{t\prime}. \tag{2.112}$$

Equations (2.111) and (2.112) are the mathematical expression of *Newton's inverse square law of attraction.*

Classical gravitational potential theory, as summarized above, is an inherently Eulerian theory; the quantities ϕ^{E} and $\mathbf{g}^{\mathrm{E}}$ are the gravitational potential and gravitational field at a fixed point $\mathbf{r}$ in space. The corresponding Lagrangian variables can, however, readily be found from the relations $\phi^{\mathrm{L}}(\mathbf{x}, t) = \phi^{\mathrm{E}}(\mathbf{r}(\mathbf{x}, t), t)$ and $\mathbf{g}^{\mathrm{L}}(\mathbf{x}, t) = \mathbf{g}^{\mathrm{E}}(\mathbf{r}(\mathbf{x}, t), t)$. Using (2.88) to transform (2.111) and (2.112) into integrals over the corresponding initial volume V^0, we obtain

$$\phi^{\mathrm{L}}(\mathbf{x}, t) = -G \int_{V^0} \frac{\rho^0(\mathbf{x}')}{\|\mathbf{r}(\mathbf{x}, t) - \mathbf{r}(\mathbf{x}', t)\|} \, dV^{0\prime}, \tag{2.113}$$

$$\mathbf{g}^{\mathrm{L}}(\mathbf{x},t) = -G\int_{V^0} \frac{\rho^0(\mathbf{x}')\,[\mathbf{r}(\mathbf{x},t)-\mathbf{r}(\mathbf{x}',t)]}{\|\mathbf{r}(\mathbf{x},t)-\mathbf{r}(\mathbf{x}',t)\|^3}\, dV^{0\prime}, \tag{2.114}$$

where ρ^0 is the initial density. Equation (2.114) provides an explicit formula for the specific body force density $\mathbf{g}^{\mathrm{L}}$ which appears on the right side of the Lagrangian momentum equation (2.93).

2.8.2 Centrifugal potential

The centrifugal force per unit mass $-\boldsymbol{\Omega}\times(\boldsymbol{\Omega}\times\mathbf{r})$ in the Eulerian momentum equation (2.83) can also be written as the gradient of a potential:

$$-\boldsymbol{\Omega}\times(\boldsymbol{\Omega}\times\mathbf{r}) = -\boldsymbol{\nabla}_{\mathbf{r}}\psi. \tag{2.115}$$

The *centrifugal potential* ψ in equation (2.115) is given by

$$\psi(\mathbf{r}) = -\tfrac{1}{2}[\Omega^2 r^2 - (\boldsymbol{\Omega}\cdot\mathbf{r})^2], \tag{2.116}$$

where $\Omega = \|\boldsymbol{\Omega}\|$ and $r = \|\mathbf{r}\|$. Inserting (2.107) and (2.115) we can rewrite the momentum equation (2.83) in the form

$$\rho^{\mathrm{E}}(D_t\mathbf{u}^{\mathrm{E}} + 2\boldsymbol{\Omega}\times\mathbf{u}^{\mathrm{E}}) = \boldsymbol{\nabla}_{\mathbf{r}}\cdot\mathbf{T}^{\mathrm{E}} - \rho^{\mathrm{E}}\boldsymbol{\nabla}_{\mathbf{r}}(\phi^{\mathrm{E}}+\psi). \tag{2.117}$$

This is the form that we shall linearize in Chapter 3.

★2.8.3 Gravitational stress tensor

The Eulerian *gravitational stress tensor*, which is analogous to the Maxwell stress tensor in electromagnetism (Jackson 1962), is defined by

$$\mathbf{N}^{\mathrm{E}} = (8\pi G)^{-1}[(\mathbf{g}^{\mathrm{E}}\cdot\mathbf{g}^{\mathrm{E}})\mathbf{I} - 2\mathbf{g}^{\mathrm{E}}\mathbf{g}^{\mathrm{E}}]. \tag{2.118}$$

It is readily verified that $\boldsymbol{\nabla}_{\mathbf{r}}\cdot\mathbf{N}^{\mathrm{E}} = \rho^{\mathrm{E}}\mathbf{g}^{\mathrm{E}}$. This allows us to write the Eulerian momentum equation in terms of $\mathbf{N}^{\mathrm{E}}$ rather than either $\mathbf{g}^{\mathrm{E}}$ or ϕ^{E}:

$$\rho^{\mathrm{E}}[D_t\mathbf{u}^{\mathrm{E}} + 2\boldsymbol{\Omega}\times\mathbf{u}^{\mathrm{E}} + \boldsymbol{\Omega}\times(\boldsymbol{\Omega}\times\mathbf{r})] = \boldsymbol{\nabla}_{\mathbf{r}}\cdot(\mathbf{T}^{\mathrm{E}}+\mathbf{N}^{\mathrm{E}}). \tag{2.119}$$

If we define the first Piola-Kirchhoff stress $\mathbf{N}^{\mathrm{PK}}$ in terms of the Lagrangian gravitational stress $\mathbf{N}^{\mathrm{L}}(\mathbf{x},t) = \mathbf{N}^{\mathrm{E}}(\mathbf{r}(\mathbf{x},t),t)$ in the usual way,

$$\mathbf{N}^{\mathrm{PK}} = J\,\mathbf{F}^{-1}\cdot\mathbf{N}^{\mathrm{L}}, \qquad \mathbf{N}^{\mathrm{L}} = J^{-1}\mathbf{F}\cdot\mathbf{N}^{\mathrm{PK}}, \tag{2.120}$$

then we can rewrite the Lagrangian form of the momentum equation (2.93) in the corresponding form

$$\rho^0[\partial_t^2\mathbf{r} + 2\boldsymbol{\Omega}\times\partial_t\mathbf{r} + \boldsymbol{\Omega}\times(\boldsymbol{\Omega}\times\mathbf{r})] = \boldsymbol{\nabla}_{\mathbf{x}}\cdot(\mathbf{T}^{\mathrm{PK}}+\mathbf{N}^{\mathrm{PK}}). \tag{2.121}$$

The Eulerian gravitational stress (2.118) is symmetric, $(\mathbf{N}^{\mathrm{E}})^{\mathrm{T}} = \mathbf{N}^{\mathrm{E}}$, but the first Piola-Kirchhoff stress is not; in fact, $(\mathbf{N}^{\mathrm{PK}})^{\mathrm{T}} = \mathbf{F} \cdot \mathbf{N}^{\mathrm{PK}} \cdot \mathbf{F}^{-\mathrm{T}}$. The continuity of the gravitational field $\mathbf{g}^{\mathrm{E}}$ guarantees that the Eulerian gravitational traction acting upon every deformed boundary is continuous: $[\hat{\mathbf{n}}^t \cdot \mathbf{N}^{\mathrm{E}}]_-^+ = \mathbf{0}$. On an undeformed welded boundary, the first Piola-Kirchhoff stress must likewise satisfy $[\hat{\mathbf{n}}^0 \cdot \mathbf{N}^{\mathrm{PK}}]_-^+ = \mathbf{0}$, by virtue of the relation $\hat{\mathbf{n}}^t \, d\Sigma^t \cdot \mathbf{N}^{\mathrm{E}} = \hat{\mathbf{n}}^0 \, d\Sigma^0 \cdot \mathbf{N}^{\mathrm{PK}}$; however, this continuity condition need not apply upon a fluid-solid boundary or a fault boundary, where there is tangential slip.

*2.9 Gravitational Potential Energy

The *gravitational potential energy* $\mathcal{E}_{\mathrm{g}}$ of a mass distribution is the work required to assemble the distribution from matter dispersed at infinity; this work is independent of the details of the assembly process because of the conservative nature of the gravitational force. We can calculate $\mathcal{E}_{\mathrm{g}}$ by considering the assembly to be incremental, in the following manner.

Suppose that most of the mass is already in place, so that the Eulerian density and potential are ρ^{E} and ϕ^{E}. The work required to bring in an infinitesimal additional mass element $\delta\rho^{\mathrm{E}} \, dV^t$ from infinity to $\mathbf{r}$ along any path is $\delta\rho^{\mathrm{E}} \, \phi^{\mathrm{E}} \, dV^t$. The total work needed to augment the density everywhere within V^t by an amount $\delta\rho^{\mathrm{E}}$ is therefore

$$\delta\mathcal{E}_{\mathrm{g}} = \int_{V^t} \delta\rho^{\mathrm{E}} \, \phi^{\mathrm{E}} \, dV^t. \tag{2.122}$$

The change $\delta\phi^{\mathrm{E}}$ in the potential due to the change in the density satisfies $\nabla_{\mathbf{r}}^2(\delta\phi^{\mathrm{E}}) = 4\pi G \, (\delta\rho^{\mathrm{E}})$; thus we may rewrite equation (2.122) in the form

$$\delta\mathcal{E}_{\mathrm{g}} = \frac{1}{4\pi G} \int_{\bigcirc} \nabla_{\mathbf{r}}^2(\delta\phi^{\mathrm{E}}) \, \phi^{\mathrm{E}} \, dV^t. \tag{2.123}$$

The integral may be taken over all of space, which we denote by $\bigcirc$, since $\nabla_{\mathbf{r}}^2(\delta\phi^{\mathrm{E}}) = 0$ wherever there is no mass, i.e., in $\bigcirc - V^t$. Upon applying Gauss' theorem to equation (2.123) we obtain

$$\begin{aligned} \delta\mathcal{E}_{\mathrm{g}} &= -\frac{1}{4\pi G} \int_{\bigcirc} \boldsymbol{\nabla}_{\mathbf{r}}(\delta\phi^{\mathrm{E}}) \cdot \boldsymbol{\nabla}_{\mathbf{r}}\phi^{\mathrm{E}} \, dV^t \\ &= -\frac{1}{8\pi G} \, \delta\left(\int_{\bigcirc} \|\boldsymbol{\nabla}_{\mathbf{r}}\phi^{\mathrm{E}}\|^2 \, dV^t \right). \end{aligned} \tag{2.124}$$

The surface integral contribution in (2.124) vanishes, since $\phi^{\mathrm{E}} \sim r^{-1}$ and $\boldsymbol{\nabla}_{\mathbf{r}}(\delta\phi^{\mathrm{E}}) \sim r^{-2}$, so the integrand $\sim r^{-3}$ whereas the surface area $\sim r^2$, as

$r \to \infty$. Summing up all the contributions $\delta\mathcal{E}_{\mathrm{g}}$, we find the total energy needed to assemble an arbitrary mass distribution from dispersal at infinity:

$$\mathcal{E}_{\mathrm{g}} = -\frac{1}{8\pi G}\int_{\bigcirc} \|\boldsymbol{\nabla}_{\mathbf{r}}\phi^{\mathrm{E}}\|^2\, dV^t. \tag{2.125}$$

The energy $\mathcal{E}_{\mathrm{g}}$ is negative because gravity is an attractive force; this is a characteristic feature of any binding energy.

We can obtain an alternative expression by again applying Gauss' theorem, together with Poisson's equation $\nabla_{\mathbf{r}}^2\phi^{\mathrm{E}} = 4\pi G\rho^{\mathrm{E}}$:

$$\mathcal{E}_{\mathrm{g}} = \tfrac{1}{2}\int_{V^t} \rho^{\mathrm{E}}\phi^{\mathrm{E}}\, dV^t, \tag{2.126}$$

where the integral is again over the instantaneous volume V^t occupied by the matter. The latter form can also be written in terms of the Lagrangian potential ϕ^{L}; using (2.88) we find that

$$\mathcal{E}_{\mathrm{g}} = \tfrac{1}{2}\int_{V^0} \rho^0\phi^{\mathrm{L}}\, dV^0, \tag{2.127}$$

where ρ^0 is the initial density and V^0 is the initial undeformed volume. Upon inserting the representation (2.113) into (2.127), we obtain a final expression for $\mathcal{E}_{\mathrm{g}}$, in the form of a double integral over V^0:

$$\mathcal{E}_{\mathrm{g}} = -\tfrac{1}{2}G\int_{V^0}\int_{V^0} \frac{\rho^0(\mathbf{x})\rho^0(\mathbf{x}')}{\|\mathbf{r}(\mathbf{x},t) - \mathbf{r}(\mathbf{x}',t)\|}\, dV^0\, dV^{0\prime}. \tag{2.128}$$

Equation (2.128) gives the instantaneous gravitational potential energy of a finite continuum of initial density ρ^0, whose particles $\mathbf{x}$ are displaced to points $\mathbf{r}$ at time t; it is valid for an arbitrary piecewise smooth density distribution, subject to arbitrarily large piecewise smooth deformations.

Any potential energy has an arbitrary reference level, and in Chapter 3 we shall find it convenient to redefine $\mathcal{E}_{\mathrm{g}}$ by

$$\begin{aligned}\mathcal{E}_{\mathrm{g}} = &-\tfrac{1}{2}G\int_{V^0}\int_{V^0} \frac{\rho^0(\mathbf{x})\rho^0(\mathbf{x}')}{\|\mathbf{r}(\mathbf{x},t) - \mathbf{r}(\mathbf{x}',t)\|}\, dV^0\, dV^{0\prime} \\ &+ \tfrac{1}{2}G\int_{V^0}\int_{V^0} \frac{\rho^0(\mathbf{x})\rho^0(\mathbf{x}')}{\|\mathbf{x} - \mathbf{x}'\|}\, dV^0\, dV^{0\prime}.\end{aligned} \tag{2.129}$$

This quantity, the work required to assemble the deformed body minus that required to assemble the corresponding undeformed body, is the *change in gravitational potential energy associated with the deformation*. The double integrals in equation (2.129) are evaluated over all particles $\mathbf{x}$ and $\mathbf{x}'$ where $\rho^0(\mathbf{x}) > 0$ and $\rho^0(\mathbf{x}') > 0$.

2.10 Elastic Constitutive Relation

Anelastic dissipation exerts a relatively small effect upon the Earth's free oscillations and equivalent propagating body and surface waves; for this reason, it is useful to consider the idealized case of a *perfectly elastic material.* Every perfectly elastic material has a natural reference configuration to which it will return in the absence of any applied stress; because of this, it is natural to employ a Lagrangian formulation of the constitutive relation. The most general perfectly elastic material is one whose Lagrangian internal energy density U^{L} depends only upon the local instantaneous strain $\mathbf{E}^{\mathrm{L}}$ and the local specific entropy density S^{L}:

$$U^{\mathrm{L}} = U^{\mathrm{L}}(\mathbf{E}^{\mathrm{L}}, S^{\mathrm{L}}). \tag{2.130}$$

The dependence on the strain $\mathbf{E}^{\mathrm{L}}$ guarantees that the constitutive relation (2.130) is consistent with the *principle of material frame indifference.* This principle stipulates that every constitutive relation must be invariant under a Galilean change of reference frame; in the case of a perfectly elastic material, it expresses the obvious physical requirement that there should be no change in the internal energy of a particle due to a rigid rotation. Differentiating the relation (2.130) with respect to time t at fixed $\mathbf{x}$ gives

$$\partial_t U^{\mathrm{L}} = \left(\frac{\partial U^{\mathrm{L}}}{\partial \mathbf{E}^{\mathrm{L}}}\right)_{S^{\mathrm{L}}} : \partial_t \mathbf{E}^{\mathrm{L}} + \left(\frac{\partial U^{\mathrm{L}}}{\partial S^{\mathrm{L}}}\right)_{\mathbf{E}^{\mathrm{L}}} \partial_t S^{\mathrm{L}}. \tag{2.131}$$

In the language of thermodynamics, equation (2.130) is known as a *caloric equation of state*, whereas (2.131) is known as a *Gibbs relation.*

Seismic deformation, in addition to being adiabatic, is to a very good approximation *isentropic*, in the sense

$$\partial_t S^{\mathrm{L}} = D_t S^{\mathrm{E}} = 0. \tag{2.132}$$

Using equation (2.132) to simplify the Gibbs relation (2.131) and multiplying by the initial density ρ^0, we obtain

$$\rho^0 \partial_t U^{\mathrm{L}} = \rho^0 \left(\frac{\partial U^{\mathrm{L}}}{\partial \mathbf{E}^{\mathrm{L}}}\right)_{S^{\mathrm{L}}} : \partial_t \mathbf{E}^{\mathrm{L}}. \tag{2.133}$$

Upon comparing the result (2.133) with the adiabatic form of the internal energy conservation law (2.103), we see that the second Piola-Kirchhoff stress $\mathbf{T}^{\mathrm{SK}}$ within a perfectly elastic material can be written in terms of the internal energy in the form

$$\mathbf{T}^{\mathrm{SK}} = \rho^0 \left(\frac{\partial U^{\mathrm{L}}}{\partial \mathbf{E}^{\mathrm{L}}}\right)_{S^{\mathrm{L}}}. \tag{2.134}$$

The isentropic nature of the deformation enables us to avoid any overt consideration of thermodynamics in seismology. We shall henceforth suppress the dependence upon the entropy density S^{L}, and simply regard U^{L} as an arbitrary function of the strain:

$$U^{\mathrm{L}} = U^{\mathrm{L}}(\mathbf{E}^{\mathrm{L}}). \tag{2.135}$$

It is natural in this case to refer to $\rho^0 U^{\mathrm{L}}$ as the volumetric *elastic strain energy density.* The differential relation (2.134) then stipulates that the second Piola-Kirchhoff stress $\mathbf{T}^{\mathrm{SK}}$ is a function only of the strain:

$$\mathbf{T}^{\mathrm{SK}} = \mathbf{T}^{\mathrm{SK}}(\mathbf{E}^{\mathrm{L}}) = \rho^0 \left(\frac{\partial U^{\mathrm{L}}}{\partial \mathbf{E}^{\mathrm{L}}}\right). \tag{2.136}$$

Equation (2.136) is the fundamental form of the isentropic stress-strain constitutive relation in a general, non-linear elastic material.

Alternatively, we may consider U^{L} to be a function of the right stretch tensor $\mathbf{R} = (\mathbf{I} + 2\mathbf{E}^{\mathrm{L}})^{1/2}$ rather than the strain tensor $\mathbf{E}^{\mathrm{L}}$; the two functions $U^{\mathrm{L}}(\mathbf{E}^{\mathrm{L}})$ and $U^{\mathrm{L}}(\mathbf{R})$ are related by

$$\left(\frac{\partial U^{\mathrm{L}}}{\partial \mathbf{E}^{\mathrm{L}}}\right) = \tfrac{1}{2}\left[\left(\frac{\partial U^{\mathrm{L}}}{\partial \mathbf{R}}\right)\cdot \mathbf{R}^{-1} + \mathbf{R}^{-1}\cdot\left(\frac{\partial U^{\mathrm{L}}}{\partial \mathbf{R}}\right)\right]. \tag{2.137}$$

Using (2.137) we can rewrite the constitutive relation (2.136) in the form

$$\mathbf{T}^{\mathrm{SK}} = \mathbf{T}^{\mathrm{SK}}(\mathbf{R}) = \tfrac{1}{2}\rho^0\left[\left(\frac{\partial U^{\mathrm{L}}}{\partial \mathbf{R}}\right)\cdot \mathbf{R}^{-1} + \mathbf{R}^{-1}\cdot\left(\frac{\partial U^{\mathrm{L}}}{\partial \mathbf{R}}\right)\right]. \tag{2.138}$$

Unlike $\mathbf{T}^{\mathrm{SK}}$, the first Piola-Kirchhoff stress $\mathbf{T}^{\mathrm{PK}}$ and the Lagrangian Cauchy stress $\mathbf{T}^{\mathrm{L}}$ depend upon the entire deformation tensor $\mathbf{F} = \mathbf{Q}\cdot\mathbf{R}$. Regarding U^{L} as a function of the deformation tensor $\mathbf{F}$ rather than $\mathbf{E}^{\mathrm{L}}$, we find from equation (2.20) that

$$\left(\frac{\partial U^{\mathrm{L}}}{\partial \mathbf{F}}\right) = \mathbf{F}\cdot\left(\frac{\partial U^{\mathrm{L}}}{\partial \mathbf{E}^{\mathrm{L}}}\right), \qquad \left(\frac{\partial U^{\mathrm{L}}}{\partial \mathbf{F}^{\mathrm{T}}}\right) = \left(\frac{\partial U^{\mathrm{L}}}{\partial \mathbf{E}^{\mathrm{L}}}\right)\cdot \mathbf{F}^{\mathrm{T}}. \tag{2.139}$$

Because the elastic strain energy must be independent of the rotation tensor $\mathbf{Q}$, the function $U^{\mathrm{L}}(\mathbf{F})$ cannot be specified arbitrarily; from (2.139) we see that it must satisfy the constraint

$$\mathbf{F}\cdot\left(\frac{\partial U^{\mathrm{L}}}{\partial \mathbf{F}^{\mathrm{T}}}\right) = \left(\frac{\partial U^{\mathrm{L}}}{\partial \mathbf{F}}\right)\cdot \mathbf{F}^{\mathrm{T}}. \tag{2.140}$$

Upon inserting (2.136) and (2.139) into equation (2.45), we find that the first Piola-Kirchhoff stress can be written in the form

$$\mathbf{T}^{\mathrm{PK}} = \mathbf{T}^{\mathrm{PK}}(\mathbf{F}) = \rho^0\left(\frac{\partial U^{\mathrm{L}}}{\partial \mathbf{F}^{\mathrm{T}}}\right), \tag{2.141}$$

which is consistent with the rate-of-change relation (2.103). The corresponding result for the isentropic elastic Cauchy stress $\mathbf{T}^{\mathrm{L}}$ can be found using (2.46) and (2.140):

$$\mathbf{T}^{\mathrm{L}} = \mathbf{T}^{\mathrm{L}}(\mathbf{F}) = \rho^0 J^{-1}\mathbf{F}\cdot\left(\frac{\partial U^{\mathrm{L}}}{\partial \mathbf{F}^{\mathrm{T}}}\right) = \rho^0 J^{-1}\left(\frac{\partial U^{\mathrm{L}}}{\partial \mathbf{F}}\right)\cdot\mathbf{F}^{\mathrm{T}}. \tag{2.142}$$

Alternatively, we can use equation (2.137) to rewrite $\mathbf{T}^{\mathrm{PK}}(\mathbf{F})$ in a manner that exhibits its explicit dependence upon the rotation tensor $\mathbf{Q}$:

$$\mathbf{T}^{\mathrm{PK}} = \mathbf{T}^{\mathrm{PK}}(\mathbf{R})\cdot\mathbf{Q}^{\mathrm{T}}, \tag{2.143}$$

where

$$\mathbf{T}^{\mathrm{PK}}(\mathbf{R}) = \tfrac{1}{2}\rho^0\left[\left(\frac{\partial U^{\mathrm{L}}}{\partial \mathbf{R}}\right) + \mathbf{R}^{-1}\cdot\left(\frac{\partial U^{\mathrm{L}}}{\partial \mathbf{R}}\right)\cdot\mathbf{R}\right]. \tag{2.144}$$

Likewise, we may rewrite $\mathbf{T}^{\mathrm{L}}(\mathbf{F})$ in the form

$$\mathbf{T}^{\mathrm{L}} = \mathbf{Q}\cdot\mathbf{T}^{\mathrm{L}}(\mathbf{R})\cdot\mathbf{Q}^{\mathrm{T}}, \tag{2.145}$$

where

$$\mathbf{T}^{\mathrm{L}}(\mathbf{R}) = \tfrac{1}{2}\rho^0 J^{-1}\left[\mathbf{R}\cdot\left(\frac{\partial U^{\mathrm{L}}}{\partial \mathbf{R}}\right) + \left(\frac{\partial U^{\mathrm{L}}}{\partial \mathbf{R}}\right)\cdot\mathbf{R}\right]. \tag{2.146}$$

The physical interpretation of the constitutive relations (2.143)–(2.144) and (2.145)–(2.146) for $\mathbf{T}^{\mathrm{PK}}$ and $\mathbf{T}^{\mathrm{L}}$ is clear: the first Piola-Kirchhoff or Lagrangian Cauchy stress at a particle $\mathbf{x}$ is the stress associated with the stretch $\mathbf{R}$, rotated by the same amount $\mathbf{Q}$ that the ball of material surrounding the particle is. The explicit dependence of the two stresses on $\mathbf{Q}$ is different because of the two-point nature of the tensor $\mathbf{T}^{\mathrm{PK}}$.

The stresses $\mathbf{T}^{\mathrm{SK}}$, $\mathbf{T}^{\mathrm{PK}}$ and $\mathbf{T}^{\mathrm{L}}$ in a so-called *simple anelastic material* do not depend only upon the instantaneous deformation $\mathbf{F}$; rather, they depend upon the *entire past history* of the deformation $\mathbf{F}_{\mathrm{hist}}$. The principle of material frame-indifference imposes a significant restriction on the nature of this dependence; the most general stress-strain relation in such a material may be shown to be of the form

$$\mathbf{T}^{\mathrm{SK}} = \mathbf{T}^{\mathrm{SK}}(\mathbf{R}_{\mathrm{hist}}), \tag{2.147}$$

$$\mathbf{T}^{\mathrm{PK}} = \mathbf{T}^{\mathrm{PK}}(\mathbf{R}_{\mathrm{hist}})\cdot\mathbf{Q}^{\mathrm{T}}, \tag{2.148}$$

$$\mathbf{T}^{\mathrm{L}} = \mathbf{Q}\cdot\mathbf{T}^{\mathrm{L}}(\mathbf{R}_{\mathrm{hist}})\cdot\mathbf{Q}^{\mathrm{T}}. \tag{2.149}$$

There is no dependence upon the history of rigid-body rotation $\mathbf{Q}_{\mathrm{hist}}$; the stress at a particle $\mathbf{x}$ is the stress associated with the stretch history $\mathbf{R}_{\mathrm{hist}}$, rotated by the *current* rotation $\mathbf{Q}$. We shall consider the effect of *linear* anelasticity upon the infinitesimal elastic-gravitational oscillations of the Earth in Chapter 6.

Chapter 3

Equations of Motion

The small magnitudes of seismic deformations permit the formulation of a strictly linear theory of the free oscillations of the Earth; we devote this chapter to a derivation of the linearized equations of motion and boundary conditions governing the infinitesimal elastic-gravitational deformation of an Earth model initially in static equilibrium. We obtain these results in three different ways: (1) by a straightforward, systematic linearization of the exact Eulerian conservation laws and boundary conditions summarized in the previous chapter; (2) by a similar linearization of the exact Lagrangian conservation laws; and (3) by an application of Hamilton's principle, using an action obtained by an independent analysis of the kinetic plus elastic-gravitational energy budget. The results are applicable to a general Earth model, whose density and elastic structure may be specified arbitrarily. The principal complicating factor in all three derivations is the deviatoric initial stress in the solid crust and mantle, which must be accounted for in any self-consistent treatment of an Earth model with lateral density heterogeneities. The analysis presented here elaborates upon and extends the results contained in a number of papers which consider the deformation of a general Earth model, notably Dahlen (1972; 1973), Woodhouse & Dahlen (1978), Valette (1986), and Vermeersen & Vlaar (1991).

3.1 Equilibrium Earth Model

We suppose the Earth to be composed of a number of fluid and solid regions; these regions are separated by non-intersecting, smooth, closed surfaces, termed *interior boundaries*. Some of the interior boundaries, such as the Mohorovičić discontinuity and upper-mantle phase transitions, are welded solid-solid boundaries, whereas others, such as the inner-core boundary, the core-mantle boundary and the seafloor, are frictionless fluid-solid boundaries. We denote the union of the solid regions by $\oplus_S$ and the union of the fluid regions by $\oplus_F$; the entire volume occupied by the model will be denoted by $\oplus = \oplus_S \cup \oplus_F$. The union of all the interior solid-solid boundaries will be denoted by Σ_{SS}, and the union of all the interior fluid-solid boundaries will be denoted by Σ_{FS}. The union of all the boundaries, including the exterior surface $\partial\oplus$, will be denoted by $\Sigma = \partial\oplus \cup \Sigma_{SS} \cup \Sigma_{FS}$. We continue to use $\bigcirc$ to denote all of space, so that $\bigcirc - \oplus$ is the region outside the Earth. The unit outward normal to Σ will be denoted by $\hat{\mathbf{n}}$; we refer to the outside and inside of Σ as the + and − sides, respectively. A schematic cross-section of a general Earth model, summarizing the above notation, is depicted in Figure 3.1; the sizes and shapes of the various regions are not to be taken literally.

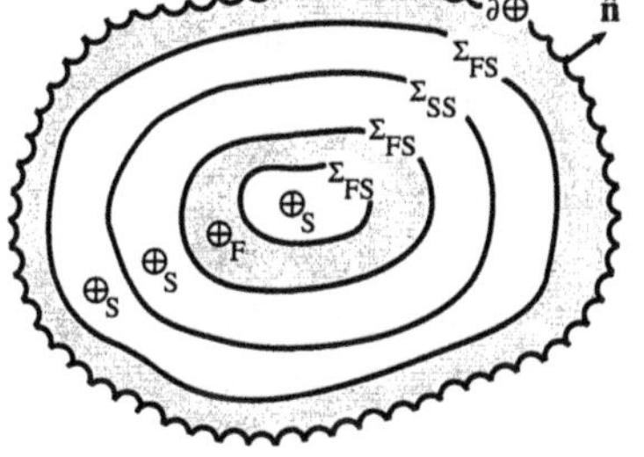

Figure 3.1. Notation used throughout this book to describe the various regions and boundaries within a general Earth model. The solid inner core, mantle and crust (*unshaded*) comprise $\oplus_S$, whereas the fluid outer core and oceans (*shaded*) comprise $\oplus_F$. The unit outward normal to the interior discontinuities Σ_{FS} and Σ_{SS} and the free surface $\partial\oplus$ is denoted by $\hat{\mathbf{n}}$. Notice that all of the boundaries are assumed to be *non-intersecting*; closed fluid-solid "coastlines" are not allowed.

It is possible to extend the theory developed here to a more realistic model with *intersecting* boundaries; however, it complicates the variational and energy-budget analyses unnecessarily. We restrict attention to "onion-like" models of the type shown in the interest of expediency.

Prior to the occurrence of an earthquake, the Earth is presumed to be in a state of mechanical equilibrium, at rest with respect to a set of Cartesian axes $\hat{\mathbf{x}}_1, \hat{\mathbf{x}}_2, \hat{\mathbf{x}}_3$ which are rotating uniformly with diurnal angular velocity $\mathbf{\Omega}$ about an origin $\mathbf{0}$ situated at the center of mass. Points or material particles in $\oplus$ or on Σ will be denoted by their equilibrium position $\mathbf{x}$ in this uniformly rotating frame, and all vectors and tensors will be expressed in terms of their components relative to the rotating axes, whenever index notation is employed. We incorporate the effects of the Coriolis and centrifugal forces associated with the Earth's rotation throughout this chapter because doing so does not lead to any significant additional complications. As a special case, the results are applicable to a non-rotating Earth, which has $\mathbf{\Omega} = \mathbf{0}$.

In Chapter 2 we found it useful to distinguish between the Eulerian and

Lagrangian gradients $\boldsymbol{\nabla}_{\mathbf{r}}$ and $\boldsymbol{\nabla}_{\mathbf{x}}$. In the initial configuration, $\boldsymbol{\nabla}_{\mathbf{r}}$ and $\boldsymbol{\nabla}_{\mathbf{x}}$ are identical, and in the linear equations and boundary conditions, only the gradient with respect to the initial coordinates $\boldsymbol{\nabla}_{\mathbf{x}}$ is required; we shall henceforth drop the subscript $\mathbf{x}$, and use an unadorned $\boldsymbol{\nabla}$ to denote $\boldsymbol{\nabla}_{\mathbf{x}}$. We shall also use dV rather than dV^0 as a pseudonym for an undeformed volume element, and we shall use $d\Sigma$ rather than $d\Sigma^0$ as a pseudonym for an undeformed area element. An oriented patch on the boundaries Σ will thus now be denoted simply by $\hat{\mathbf{n}}\, d\Sigma$, rather than by $\hat{\mathbf{n}}^0 d\Sigma^0$, as we did previously. The purpose of these changes is simply to avoid excessive subscript and superscript proliferation in the equations that follow.

Let ρ^0 denote the initial density distribution within $\oplus$, let ϕ^0 denote the initial gravitational potential, and let

$$\mathbf{g}^0 = -\boldsymbol{\nabla}\phi^0 \tag{3.1}$$

denote the corresponding initial gravitational field. The two quantities ϕ^0 and $\mathbf{g}^0$ are given in terms of ρ^0 by

$$\phi^0 = -G \int_{\oplus} \frac{\rho^{0\prime}}{\|\mathbf{x} - \mathbf{x}'\|}\, dV' \tag{3.2}$$

and

$$\mathbf{g}^0 = -G \int_{\oplus} \frac{\rho^{0\prime}\,(\mathbf{x} - \mathbf{x}')}{\|\mathbf{x} - \mathbf{x}'\|^3}\, dV', \tag{3.3}$$

where the primes denote evaluation at the dummy integration variable $\mathbf{x}'$. The density ρ^0 is assumed to vanish outside the Earth; however, ϕ^0 and $\mathbf{g}^0$ are both non-zero everywhere in $\bigcirc$. The potential ϕ^0 satisfies Poisson's equation

$$\nabla^2\phi^0 = 4\pi G\rho^0 \tag{3.4}$$

within the Earth $\oplus$, together with the continuity conditions

$$[\phi^0]_-^+ = 0, \qquad [\hat{\mathbf{n}}\cdot\boldsymbol{\nabla}\phi^0]_-^+ = 0 \tag{3.5}$$

on the boundaries Σ. In the region $\bigcirc - \oplus$ outside the Earth, the potential is *harmonic*: $\nabla^2\phi^0 = 0$.

In the undeformed configuration, the Cauchy stress and the two Piola-Kirchhoff stresses coincide; we denote this *initial static stress* within the Earth model by $\mathbf{T}^0$. In the fluid regions $\oplus_{\mathrm{F}}$, the initial stress must be hydrostatic: $\mathbf{T}^0 = -p^0\mathbf{I}$, where p^0 is the initial *hydrostatic pressure*. In the solid regions $\oplus_{\mathrm{S}}$, we define the pressure p^0 and the initial *deviatoric stress* $\boldsymbol{\tau}^0$ by

$$p^0 = -\tfrac{1}{3}\,\mathrm{tr}\,\mathbf{T}^0, \qquad \boldsymbol{\tau}^0 = p^0\mathbf{I} + \mathbf{T}^0. \tag{3.6}$$

Equations (3.6) constitute the conventional decomposition of the initial stress $\mathbf{T}^0 = -p^0\mathbf{I} + \boldsymbol{\tau}^0$ into its isotropic and deviatoric parts; the trace of the deviatoric stress vanishes: $\operatorname{tr}\boldsymbol{\tau}^0 = 0$. We use ϖ^0 to denote the negative normal component of the traction on the boundaries Σ, i.e.,

$$\varpi^0 = -(\hat{\mathbf{n}}\cdot\mathbf{T}^0\cdot\hat{\mathbf{n}}). \tag{3.7}$$

On the fluid-solid boundaries Σ_{FS}, we then have $\hat{\mathbf{n}}\cdot\mathbf{T}^0 = -\varpi^0\hat{\mathbf{n}}$, where ϖ^0 is equal to the initial pressure p^0 on the fluid side.

The mechanical equilibrium of the uniformly rotating Earth model is guaranteed by the static momentum equation

$$\boldsymbol{\nabla}\cdot\mathbf{T}^0 = \rho^0\boldsymbol{\nabla}(\phi^0 + \psi), \tag{3.8}$$

where

$$\psi = -\tfrac{1}{2}[\Omega^2x^2 - (\boldsymbol{\Omega}\cdot\mathbf{x})^2] \tag{3.9}$$

is the centrifugal potential. In the fluid regions $\oplus_{\mathrm{F}}$, where the static stress deviator $\boldsymbol{\tau}^0$ vanishes, equation (3.8) reduces to the equation of hydrostatic equilibrium:

$$\boldsymbol{\nabla}p^0 + \rho^0\boldsymbol{\nabla}(\phi^0 + \psi) = 0. \tag{3.10}$$

Equations (3.8) and (3.10) must be satisfied throughout $\oplus_{\mathrm{S}}$ and $\oplus_{\mathrm{F}}$, subject to the traction continuity condition

$$[\hat{\mathbf{n}}\cdot\mathbf{T}^0]_-^+ = \mathbf{0} \tag{3.11}$$

on the boundaries Σ. On the outer free surface of the Earth, the traction must vanish: $\hat{\mathbf{n}}\cdot\mathbf{T}^0 = \mathbf{0}$ on $\partial\oplus$.

3.2 Linear Perturbations

We adopt a Lagrangian description of the motion and write the position vector $\mathbf{r}(\mathbf{x},t)$ in the form

$$\mathbf{r}(\mathbf{x},t) = \mathbf{x} + \mathbf{s}(\mathbf{x},t). \tag{3.12}$$

The quantity $\mathbf{s}$ is the *displacement* of particle $\mathbf{x}$ away from its equilibrium position at time t (see Figure 3.2). To obtain the linearized equations of motion and boundary conditions governing small elastic-gravitational oscillations, we regard $\mathbf{s}$ as a small quantity, and systematically ignore all terms of order $\|\mathbf{s}\|^2$. We discuss some general considerations first, before proceeding to linearize the mass and momentum conservation laws. It is convenient in what follows to define $\mathbf{s}$ to be zero outside the Earth, in $\bigcirc - \oplus$.

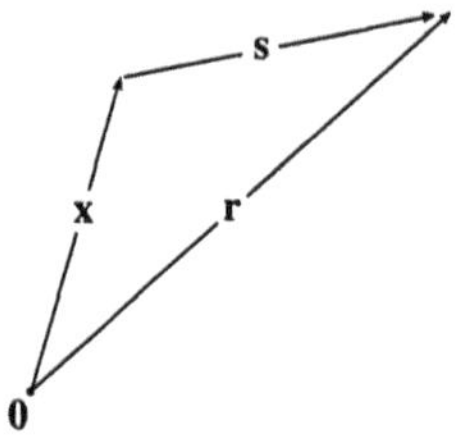

Figure 3.2. The material particle initially located at the point $\mathbf{x}$ moves to the point $\mathbf{r}(\mathbf{x},t) = \mathbf{x}+\mathbf{s}(\mathbf{x},t)$ at time t. In the linearized analysis of deformation developed in the present chapter, the particle *displacement* $\mathbf{s}(\mathbf{x},t)$ is considered to be small.

3.2.1 Eulerian and Lagrangian perturbations

For any physical quantity q, we define the first-order Eulerian and Lagrangian perturbations q^{E1} and q^{L1} by

$$q^{\mathrm{E}}(\mathbf{r},t) = q^0(\mathbf{r}) + q^{\mathrm{E1}}(\mathbf{r},t), \tag{3.13}$$

$$q^{\mathrm{L}}(\mathbf{x},t) = q^0(\mathbf{x}) + q^{\mathrm{L1}}(\mathbf{x},t), \tag{3.14}$$

where q^0 denotes the zeroth-order initial value. If the deformation is small enough to permit linearization, then correct to first order in $\|\mathbf{s}\|$, we may write

$$q^{\mathrm{E1}}(\mathbf{r},t) = q^{\mathrm{E1}}(\mathbf{x},t), \qquad q^{\mathrm{L1}}(\mathbf{x},t) = q^{\mathrm{L1}}(\mathbf{r},t). \tag{3.15}$$

Equations (3.15) show that it is immaterial whether the first-order perturbations q^{E1} and q^{L1} are regarded as functions of $\mathbf{r}$ or $\mathbf{x}$. We shall henceforth regard all zeroth-order and first-order variables as functions of the initial position $\mathbf{x}$; the domain of the resulting linearized equations will then be the volume $\oplus$ of the undeformed Earth. We will supplement these equations with linearized boundary conditions that are valid on the undeformed external and internal boundaries Σ.

Inserting the representations (3.12) and (3.13)–(3.14) into equation (2.1) and ignoring second-order terms, we find that the Lagrangian and Eulerian perturbations q^{L1} and q^{E1} are related by

$$q^{\mathrm{L1}} = q^{\mathrm{E1}} + \mathbf{s}\cdot\boldsymbol{\nabla} q^0. \tag{3.16}$$

Equation (3.16) is in essence a linearized, integrated version of the material derivative relation $D_t = \partial_t + \mathbf{u}^{\mathrm{E}}\cdot\boldsymbol{\nabla}_{\mathbf{r}}$. The physical interpretation is similar: the first-order change q^{L1} experienced by an observer riding on a moving particle consists of the change q^{E1} at a fixed point $\mathbf{x}$ in space, plus the change $\mathbf{s}\cdot\boldsymbol{\nabla} q^0$ due to the displacement $\mathbf{s}$ of the particle through the *initial* spatial gradient $\boldsymbol{\nabla} q^0$. The quantity q may be any physical variable that has a non-zero static value, such as the density, gravity or stress.

3.2.2 Linearized analysis of deformation

Disregarding the two-point nature of the deformation tensor $\mathbf{F}$, we shall write it in the form

$$\mathbf{F} = \mathbf{I} + (\boldsymbol{\nabla}\mathbf{s})^{\mathrm{T}}, \tag{3.17}$$

where $\mathbf{I}$ is the identity. In the case of an infinitesimal deformation, it is convenient to decompose $\mathbf{F}$ into its symmetric and anti-symmetric parts:

$$\mathbf{F} = \mathbf{I} + \boldsymbol{\varepsilon} + \boldsymbol{\omega}, \tag{3.18}$$

where

$$\boldsymbol{\varepsilon} = \tfrac{1}{2}[\boldsymbol{\nabla}\mathbf{s} + (\boldsymbol{\nabla}\mathbf{s})^{\mathrm{T}}], \qquad \boldsymbol{\omega} = -\tfrac{1}{2}[\boldsymbol{\nabla}\mathbf{s} - (\boldsymbol{\nabla}\mathbf{s})^{\mathrm{T}}]. \tag{3.19}$$

The symmetric part $\boldsymbol{\varepsilon} = \boldsymbol{\varepsilon}^{\mathrm{T}}$ is the *infinitesimal strain tensor*, and the anti-symmetric part $\boldsymbol{\omega} = -\boldsymbol{\omega}^{\mathrm{T}}$ is the *infinitesimal rotation tensor*. Correct to first order in $\|\mathbf{s}\|$ we can rewrite (3.18) in the alternative form

$$\mathbf{F} = (\mathbf{I} + \boldsymbol{\omega}) \cdot (\mathbf{I} + \boldsymbol{\varepsilon}) = (\mathbf{I} + \boldsymbol{\varepsilon}) \cdot (\mathbf{I} + \boldsymbol{\omega}). \tag{3.20}$$

Equation (3.20) can be interpreted as the small-deformation version of the *polar decomposition theorem* (2.25). The rotation tensor $\mathbf{Q}$ and the right and left stretch tensors $\mathbf{R}$ and $\mathbf{L}$ are given, correct to first order in $\|\mathbf{s}\|$, by

$$\mathbf{Q} = \mathbf{I} + \boldsymbol{\omega}, \qquad \mathbf{R} = \mathbf{L} = \mathbf{I} + \boldsymbol{\varepsilon}. \tag{3.21}$$

The antisymmetry of $\boldsymbol{\omega}$ guarantees that $\mathbf{Q}$ is orthogonal, correct to first order in $\|\mathbf{s}\|$:

$$(\mathbf{I} + \boldsymbol{\omega})^{\mathrm{T}} \cdot (\mathbf{I} + \boldsymbol{\omega}) = (\mathbf{I} + \boldsymbol{\omega}) \cdot (\mathbf{I} + \boldsymbol{\omega})^{\mathrm{T}} = \mathbf{I}. \tag{3.22}$$

From either equation (2.20) or (2.29) we see that the Lagrangian strain tensor $\mathbf{E}^{\mathrm{L}}$ is given, correct to first order in $\|\mathbf{s}\|$, by

$$\mathbf{E}^{\mathrm{L}} = \boldsymbol{\varepsilon}. \tag{3.23}$$

In view of the results (3.20)–(3.23), we shall henceforth generally drop the adjective "infinitesimal" and simply refer to $\boldsymbol{\varepsilon}$ and $\boldsymbol{\omega}$ as the strain and rotation tensors.

Upon linearizing equations (2.10) and (2.11) we conclude that the Eulerian strain-rate and rotation-rate tensors $\mathbf{D}^{\mathrm{E}}$ and $\mathbf{W}^{\mathrm{E}}$ are given, correct to first order in $\|\mathbf{s}\|$, by

$$\mathbf{D}^{\mathrm{E}} = \partial_t\boldsymbol{\varepsilon}, \qquad \mathbf{W}^{\mathrm{E}} = \partial_t\boldsymbol{\omega}, \tag{3.24}$$

i.e., they are are simply the time derivatives of the strain and rotation tensors. The first-order Eulerian velocity $\mathbf{u}^{\mathrm{E}}$ is likewise simply the time derivative of the displacement:

$$\mathbf{u}^{\mathrm{E}} = \partial_t \mathbf{s}. \tag{3.25}$$

In view of (3.24)–(3.25) it is sometimes asserted that there is no need to distinguish between the Eulerian and Lagrangian viewpoints in the case of an infinitesimal deformation. This is not, however, true for variables such as the density, gravity and stress which have a zeroth-order initial value, as we have seen in Section 3.2.1.

3.2.3 Volume and area perturbations

The Jacobian $J = \det \mathbf{F}$ relating a deformed volume element to the corresponding undeformed element is given, correct to first order in $\|\mathbf{s}\|$, by

$$J = 1 + \operatorname{tr} \boldsymbol{\varepsilon} = 1 + \boldsymbol{\nabla} \cdot \mathbf{s}. \tag{3.26}$$

The inverse of the deformation tensor $\mathbf{F}^{-1}$ is given to the same order by

$$\mathbf{F}^{-1} = \mathbf{I} - (\boldsymbol{\nabla}\mathbf{s})^{\mathrm{T}}. \tag{3.27}$$

Inserting equations (3.26) and (3.27) into (2.37) and ignoring second-order terms, we obtain the linearized relation between a deformed surface area element $\hat{\mathbf{n}}^t d\Sigma^t$ and the corresponding undeformed element $\hat{\mathbf{n}}\, d\Sigma$:

$$\hat{\mathbf{n}}^t d\Sigma^t = (1 + \boldsymbol{\nabla} \cdot \mathbf{s})\, \hat{\mathbf{n}}\, d\Sigma - (\boldsymbol{\nabla}\mathbf{s}) \cdot \hat{\mathbf{n}}\, d\Sigma. \tag{3.28}$$

The terms involving the normal derivative $\partial_n \mathbf{s}$ cancel, so that this result can be rewritten solely in terms of the *surface gradient* $\boldsymbol{\nabla}^{\Sigma} = \boldsymbol{\nabla} - \hat{\mathbf{n}}\partial_n$:

$$\hat{\mathbf{n}}^t d\Sigma^t = (1 + \boldsymbol{\nabla}^{\Sigma} \cdot \mathbf{s})\, \hat{\mathbf{n}}\, d\Sigma - (\boldsymbol{\nabla}^{\Sigma}\mathbf{s}) \cdot \hat{\mathbf{n}}\, d\Sigma. \tag{3.29}$$

The first term in equation (3.29) accounts for the change in the area of the infinitesimal patch, whereas the second term accounts for the deflection of the unit normal. Correct to first order in $\|\mathbf{s}\|$, we may represent these two changes separately:

$$d\Sigma^t = (1 + \boldsymbol{\nabla}^{\Sigma} \cdot \mathbf{s})\, d\Sigma, \tag{3.30}$$

$$\hat{\mathbf{n}}^t = \hat{\mathbf{n}} - (\boldsymbol{\nabla}^{\Sigma}\mathbf{s}) \cdot \hat{\mathbf{n}}. \tag{3.31}$$

Equation (3.30) is the areal analogue of the volumetric relation (3.26); the deformed and undeformed volume elements dV^t and dV are related by

$$dV^t = (1 + \boldsymbol{\nabla} \cdot \mathbf{s})\, dV. \tag{3.32}$$

Equation (3.31) giving the deflection of the unit normal can also be obtained by means of an elementary geometrical argument; together, the two relations (3.30)–(3.31) imply (3.29).

3.2.4 Perturbations in stress

The first-order perturbations in the Eulerian and Lagrangian Cauchy stress are defined by

$$\mathbf{T}^{\mathrm{E}} = \mathbf{T}^{0} + \mathbf{T}^{\mathrm{E1}}, \qquad \mathbf{T}^{\mathrm{L}} = \mathbf{T}^{0} + \mathbf{T}^{\mathrm{L1}}. \tag{3.33}$$

The Lagrangian perturbation $\mathbf{T}^{\mathrm{L1}}$ experienced by an observer riding on a moving particle is related to the Eulerian perturbation $\mathbf{T}^{\mathrm{E1}}$ at a fixed point in space by equation (3.16):

$$\mathbf{T}^{\mathrm{L1}} = \mathbf{T}^{\mathrm{E1}} + \mathbf{s} \cdot \boldsymbol{\nabla}\mathbf{T}^{0}. \tag{3.34}$$

We denote the incremental first and second Piola-Kirchhoff stresses by $\mathbf{T}^{\mathrm{PK1}}$ and $\mathbf{T}^{\mathrm{SK1}}$, respectively. These quantities are defined, in a manner analogous to (3.33), by

$$\mathbf{T}^{\mathrm{PK}} = \mathbf{T}^{0} + \mathbf{T}^{\mathrm{PK1}}, \qquad \mathbf{T}^{\mathrm{SK}} = \mathbf{T}^{0} + \mathbf{T}^{\mathrm{SK1}}. \tag{3.35}$$

Upon inserting (3.26)–(3.27), (3.33) and (3.35) into (2.42) and (2.46) we obtain the linearized equations relating the Piola-Kirchhoff perturbations $\mathbf{T}^{\mathrm{PK1}}$ and $\mathbf{T}^{\mathrm{SK1}}$ to the incremental Lagrangian Cauchy stress $\mathbf{T}^{\mathrm{L1}}$:

$$\mathbf{T}^{\mathrm{PK1}} = \mathbf{T}^{\mathrm{L1}} + \mathbf{T}^{0}(\boldsymbol{\nabla} \cdot \mathbf{s}) - (\boldsymbol{\nabla}\mathbf{s})^{\mathrm{T}} \cdot \mathbf{T}^{0}, \tag{3.36}$$

$$\mathbf{T}^{\mathrm{SK1}} = \mathbf{T}^{\mathrm{L1}} + \mathbf{T}^{0}(\boldsymbol{\nabla} \cdot \mathbf{s}) - (\boldsymbol{\nabla}\mathbf{s})^{\mathrm{T}} \cdot \mathbf{T}^{0} - \mathbf{T}^{0} \cdot \boldsymbol{\nabla}\mathbf{s}. \tag{3.37}$$

The corresponding first-order relation between $\mathbf{T}^{\mathrm{PK1}}$ and $\mathbf{T}^{\mathrm{SK1}}$ can be found either by linearizing (2.45) or by comparing (3.36) and (3.37):

$$\mathbf{T}^{\mathrm{PK1}} = \mathbf{T}^{\mathrm{SK1}} + \mathbf{T}^{0} \cdot \boldsymbol{\nabla}\mathbf{s}. \tag{3.38}$$

It is noteworthy that all three of $\mathbf{T}^{\mathrm{E1}}$, $\mathbf{T}^{\mathrm{L1}}$ and $\mathbf{T}^{\mathrm{SK1}}$ are symmetric, whereas the first incremental Piola-Kirchhoff stress $\mathbf{T}^{\mathrm{PK1}}$ is not; in fact,

$$(\mathbf{T}^{\mathrm{PK1}})^{\mathrm{T}} = \mathbf{T}^{\mathrm{PK1}} - \mathbf{T}^{0} \cdot \boldsymbol{\nabla}\mathbf{s} + (\boldsymbol{\nabla}\mathbf{s})^{\mathrm{T}} \cdot \mathbf{T}^{0}. \tag{3.39}$$

Equation (3.39), which follows from (3.36), is the linearized version of equation (2.95). We discuss the linear elastic constitutive relation, and express the stress perturbations $\mathbf{T}^{\mathrm{L1}}$, $\mathbf{T}^{\mathrm{PK1}}$ and $\mathbf{T}^{\mathrm{SK1}}$ in terms of the displacement gradient $\boldsymbol{\nabla}\mathbf{s}$ in Section 3.6.

3.2.5 Perturbations in gravity

The Eulerian and Lagrangian perturbations to the gravitational potential and field are defined by

$$\phi^{\mathrm{E}} = \phi^0 + \phi^{\mathrm{E1}}, \qquad \mathbf{g}^{\mathrm{E}} = \mathbf{g}^0 + \mathbf{g}^{\mathrm{E1}}, \tag{3.40}$$

and

$$\phi^{\mathrm{L}} = \phi^0 + \phi^{\mathrm{L1}}, \qquad \mathbf{g}^{\mathrm{L}} = \mathbf{g}^0 + \mathbf{g}^{\mathrm{L1}}, \tag{3.41}$$

respectively. These perturbations satisfy the usual first-order relations:

$$\phi^{\mathrm{L1}} = \phi^{\mathrm{E1}} + \mathbf{s}\cdot\boldsymbol{\nabla}\phi^0, \qquad \mathbf{g}^{\mathrm{L1}} = \mathbf{g}^{\mathrm{E1}} + \mathbf{s}\cdot\boldsymbol{\nabla}\mathbf{g}^0. \tag{3.42}$$

The Eulerian field perturbation $\mathbf{g}^{\mathrm{E1}}$ can be written as the gradient of the corresponding potential perturbation:

$$\mathbf{g}^{\mathrm{E1}} = -\boldsymbol{\nabla}\phi^{\mathrm{E1}}. \tag{3.43}$$

The corresponding result for $\mathbf{g}^{\mathrm{L1}}$ is more complicated; from (3.42)–(3.43) we find that

$$\mathbf{g}^{\mathrm{L1}} = -\boldsymbol{\nabla}\phi^{\mathrm{E1}} - \mathbf{s}\cdot\boldsymbol{\nabla}\boldsymbol{\nabla}\phi^0 = -\boldsymbol{\nabla}\phi^{\mathrm{L1}} + \boldsymbol{\nabla}\mathbf{s}\cdot\boldsymbol{\nabla}\phi^0. \tag{3.44}$$

The difference between the two relations (3.43) and (3.44) reflects the inherently Eulerian nature of classical gravitational potential theory. We obtain the linear integral relations between the gravitational perturbations ϕ^{E1}, $\mathbf{g}^{\mathrm{E1}}$, ϕ^{L1} and $\mathbf{g}^{\mathrm{L1}}$ and the displacement $\mathbf{s}$ in Section 3.5.

3.3 Linearized Conservation Laws

The linearized versions of the mass and momentum conservation laws are easily obtained. It is an intrinsic feature of the linear theory that second-order accuracy is required in all energetic considerations; a linearized treatment of the energy conservation law is therefore not adequate. We present a consistent second-order treatment of the elastic-gravitational energy in Sections 3.8 and 3.9.

3.3.1 Linearized continuity equation

The Eulerian and Lagrangian perturbations in density ρ^{E1} and ρ^{L1} are defined by

$$\rho^{\mathrm{E}} = \rho^0 + \rho^{\mathrm{E1}}, \qquad \rho^{\mathrm{L}} = \rho^0 + \rho^{\mathrm{L1}}. \tag{3.45}$$

These two perturbations can be related to the displacement s by linearizing the Eulerian and Lagrangian conservation of mass laws. Upon inserting the decompositions (3.45) into the Eulerian continuity equation (2.51) and integrating with respect to time, we find that

$$\rho^{\mathrm{E1}} = -\boldsymbol{\nabla} \cdot (\rho^0 \mathbf{s}), \tag{3.46}$$

correct to first order in $\|\mathbf{s}\|$. Inserting (3.26) together with (3.45) into the Lagrangian conservation of mass equation (2.86) yields

$$\rho^{\mathrm{L1}} = -\rho^0 (\boldsymbol{\nabla} \cdot \mathbf{s}), \tag{3.47}$$

correct to the same order. It is reassuring that the two perturbations (3.46) and (3.47) are related by

$$\rho^{\mathrm{L1}} = \rho^{\mathrm{E1}} + \mathbf{s} \cdot \boldsymbol{\nabla} \rho^0, \tag{3.48}$$

in agreement with the general relation (3.16).

Equation (3.47) provides a physical interpretation of the divergence of the particle displacement $\boldsymbol{\nabla} \cdot \mathbf{s}$. Defining the perturbation in the Lagrangian specific volume τ^{L1} by

$$\tau^{\mathrm{L}} = \tau^0 + \tau^{\mathrm{L1}}, \tag{3.49}$$

where $\rho^0 \tau^0 = 1$, we see that

$$\boldsymbol{\nabla} \cdot \mathbf{s} = -\frac{\rho^{\mathrm{L1}}}{\rho^0} = \frac{\tau^{\mathrm{L1}}}{\tau^0}. \tag{3.50}$$

Evidently, $\boldsymbol{\nabla} \cdot \mathbf{s}$ is the relative change in the volume per unit mass of particle $\mathbf{x}$ at time t. Equation (3.50) can be regarded as a linearized, integrated version of equation (2.58).

3.3.2 Linearized momentum equation

To obtain the linearized form of the Eulerian momentum equation, we substitute the representations (3.33), (3.40) and (3.45) into the exact relation (2.117). Neglecting terms of second order in $\|\mathbf{s}\|$ and subtracting the static equilibrium condition (3.8) we obtain

$$\rho^0 (\partial_t^2 \mathbf{s} + 2\boldsymbol{\Omega} \times \partial_t \mathbf{s}) = \boldsymbol{\nabla} \cdot \mathbf{T}^{\mathrm{E1}} - \rho^0 \boldsymbol{\nabla} \phi^{\mathrm{E1}} - \rho^{\mathrm{E1}} \boldsymbol{\nabla} (\phi^0 + \psi). \tag{3.51}$$

Strictly speaking, this result is valid at a point $\mathbf{r}$ within the deformed Earth rather than at a point $\mathbf{x}$ within the undeformed Earth, and the gradient is $\boldsymbol{\nabla}_{\mathbf{r}}$ rather than $\boldsymbol{\nabla}_{\mathbf{x}}$. However, correct to first order in $\|\mathbf{s}\|$, this distinction is immaterial, as we have seen in Section 3.2.1. Another way to obtain equation (3.51), which is algebraically slightly more involved, but which yields a result that is explicitly valid in the undeformed Earth, is to substitute into equation (2.117) the alternative relations:

$$\rho^{\mathrm{E}} = \rho^0 + \rho^{\mathrm{E1}} + \mathbf{s}\cdot\boldsymbol{\nabla}\rho^0, \tag{3.52}$$

$$\phi^{\mathrm{E}} = \phi^0 + \phi^{\mathrm{E1}} + \mathbf{s}\cdot\boldsymbol{\nabla}\phi^0, \tag{3.53}$$

$$\mathbf{T}^{\mathrm{E}} = \mathbf{T}^0 + \mathbf{T}^{\mathrm{E1}} + \mathbf{s}\cdot\boldsymbol{\nabla}\mathbf{T}^0, \tag{3.54}$$

$$\boldsymbol{\nabla}_{\mathbf{r}} = \boldsymbol{\nabla} - (\boldsymbol{\nabla}\mathbf{s})\cdot\boldsymbol{\nabla}. \tag{3.55}$$

The additional advective terms $\mathbf{s}\cdot\boldsymbol{\nabla}\rho^0$, $\mathbf{s}\cdot\boldsymbol{\nabla}\phi^0$ and $\mathbf{s}\cdot\boldsymbol{\nabla}\mathbf{T}^0$ appearing in equations (3.52)–(3.54) shift the independent variable from $\mathbf{r}$ to $\mathbf{x}$, and the final relation (3.55) is the corresponding first-order transformation of the spatial gradient $\boldsymbol{\nabla}_{\mathbf{r}}$ into $\boldsymbol{\nabla} = \boldsymbol{\nabla}_{\mathbf{x}}$.

As we shall see in Section 3.6, it is the Lagrangian perturbation in stress $\mathbf{T}^{\mathrm{L1}} = \mathbf{T}^{\mathrm{E1}} + \mathbf{s}\cdot\boldsymbol{\nabla}\mathbf{T}^0$ rather than the Eulerian perturbation $\mathbf{T}^{\mathrm{E1}}$ that is related to $\boldsymbol{\nabla}\mathbf{s}$ by the elastic parameters; for this reason, it is convenient to rewrite (3.51) explicitly in terms of $\mathbf{T}^{\mathrm{L1}}$ in the form

$$\begin{aligned}\rho^0(\partial_t^2\mathbf{s} + 2\boldsymbol{\Omega}\times\partial_t\mathbf{s}) &= \boldsymbol{\nabla}\cdot\mathbf{T}^{\mathrm{L1}}\\ &\quad - \boldsymbol{\nabla}\cdot(\mathbf{s}\cdot\boldsymbol{\nabla}\mathbf{T}^0) - \rho^0\boldsymbol{\nabla}\phi^{\mathrm{E1}} - \rho^{\mathrm{E1}}\boldsymbol{\nabla}(\phi^0+\psi).\end{aligned} \tag{3.56}$$

In the fluid regions $\oplus_{\mathrm{F}}$ equation (3.56) reduces to

$$\begin{aligned}\rho^0(\partial_t^2\mathbf{s} + 2\boldsymbol{\Omega}\times\partial_t\mathbf{s}) &= \boldsymbol{\nabla}\cdot\mathbf{T}^{\mathrm{L1}}\\ &\quad - \boldsymbol{\nabla}[\rho^0\mathbf{s}\cdot\boldsymbol{\nabla}(\phi^0+\psi)] - \rho^0\boldsymbol{\nabla}\phi^{\mathrm{E1}} - \rho^{\mathrm{E1}}\boldsymbol{\nabla}(\phi^0+\psi),\end{aligned} \tag{3.57}$$

where we have used (3.10) to eliminate the initial pressure p^0. The corresponding result in the solid regions $\oplus_{\mathrm{S}}$ is

$$\begin{aligned}\rho^0(\partial_t^2\mathbf{s} + 2\boldsymbol{\Omega}\times\partial_t\mathbf{s}) &= \boldsymbol{\nabla}\cdot\mathbf{T}^{\mathrm{L1}}\\ &\quad - \boldsymbol{\nabla}[\rho^0\mathbf{s}\cdot\boldsymbol{\nabla}(\phi^0+\psi)] - \rho^0\boldsymbol{\nabla}\phi^{\mathrm{E1}} - \rho^{\mathrm{E1}}\boldsymbol{\nabla}(\phi^0+\psi)\\ &\qquad + \boldsymbol{\nabla}[\mathbf{s}\cdot(\boldsymbol{\nabla}\cdot\boldsymbol{\tau}^0)] - \boldsymbol{\nabla}\cdot(\mathbf{s}\cdot\boldsymbol{\nabla}\boldsymbol{\tau}^0),\end{aligned} \tag{3.58}$$

where $\boldsymbol{\tau}^0$ is the initial stress deviator.

The linearized form of the Lagrangian momentum equation is obtained in a similar manner; we substitute (3.35) and (3.41) into equation (2.93) and subtract the static equilibrium condition (3.8). The result, which is exactly valid everywhere within the undeformed Earth $\oplus$, is

$$\rho^0[\partial_t^2\mathbf{s} + 2\boldsymbol{\Omega}\times\partial_t\mathbf{s} + \boldsymbol{\Omega}\times(\boldsymbol{\Omega}\times\mathbf{s})] = \boldsymbol{\nabla}\cdot\mathbf{T}^{\mathrm{PK1}} + \rho^0\mathbf{g}^{\mathrm{L1}}. \tag{3.59}$$

Correct to first order in $\|\mathbf{s}\|$, equation (3.59) is equivalent to

$$\begin{aligned}\rho^0(\partial_t^2\mathbf{s} + 2\boldsymbol{\Omega}\times\partial_t\mathbf{s}) &= \boldsymbol{\nabla}\cdot\mathbf{T}^{\mathrm{PK1}}\\ &\quad - \rho^0\boldsymbol{\nabla}\phi^{\mathrm{E1}} - \rho^0\mathbf{s}\cdot\boldsymbol{\nabla}\boldsymbol{\nabla}(\phi^0+\psi),\end{aligned} \tag{3.60}$$

where we have used (3.44) and the identity $\boldsymbol{\Omega}\times(\boldsymbol{\Omega}\times\mathbf{s})=\mathbf{s}\cdot\boldsymbol{\nabla\nabla}\psi$. For the two versions of the linearized momentum equation (3.56) and (3.60) to be consistent, we must have

$$\begin{aligned}\boldsymbol{\nabla}\cdot\mathbf{T}^{\mathrm{L1}}-\boldsymbol{\nabla}\cdot(\mathbf{s}\cdot\boldsymbol{\nabla}\mathbf{T}^{0})-\rho^{\mathrm{E1}}\boldsymbol{\nabla}(\phi^{0}+\psi)\\ =\boldsymbol{\nabla}\cdot\mathbf{T}^{\mathrm{PK1}}-\rho^{0}\mathbf{s}\cdot\boldsymbol{\nabla\nabla}(\phi^{0}+\psi).\end{aligned}\qquad(3.61)$$

This identity can easily be verified using the relation between the two incremental stresses $\mathbf{T}^{\mathrm{L1}}$ and $\mathbf{T}^{\mathrm{PK1}}$, equation (3.36).

Other forms of the linearized momentum equation can also be obtained; for example, it can be written in terms of the incremental second Piola-Kirchhoff stress $\mathbf{T}^{\mathrm{SK1}}$, using either (3.37) or (3.38):

$$\begin{aligned}\rho^{0}(\partial_t^2\mathbf{s}+2\boldsymbol{\Omega}\times\partial_t\mathbf{s})=\boldsymbol{\nabla}\cdot\mathbf{T}^{\mathrm{SK1}}\\ +\boldsymbol{\nabla}\cdot(\mathbf{T}^{0}\cdot\boldsymbol{\nabla}\mathbf{s})-\rho^{0}\boldsymbol{\nabla}\phi^{\mathrm{E1}}-\rho^{0}\mathbf{s}\cdot\boldsymbol{\nabla\nabla}(\phi^{0}+\psi),\end{aligned}\qquad(3.62)$$

This result, which is the linearized version of equation (2.94), is not commonly employed in global seismology.

3.4 Linearized Boundary Conditions

The linearized equations of motion derived above must be supplemented by linearized kinematic, dynamic and gravitational boundary conditions on the boundaries $\Sigma=\partial\oplus\cup\Sigma_{\mathrm{SS}}\cup\Sigma_{\mathrm{FS}}$.

3.4.1 Kinematic boundary conditions

The kinematic boundary condition that there be no slip upon the welded or solid-solid boundaries Σ_{SS} is obviously

$$[\mathbf{s}]_-^+=\mathbf{0}.\qquad(3.63)$$

On the fluid-solid boundaries Σ_{FS}, tangential slip is allowed; the linearized continuity condition which guarantees that there be no separation or interpenetration is

$$[\hat{\mathbf{n}}\cdot\mathbf{s}]_-^+=0.\qquad(3.64)$$

Equation (3.63) is exact whereas (3.64) is correct only to first order in $\|\mathbf{s}\|$; we consider the second-order tangential slip condition in Section 3.4.4.

3.4.2 Dynamic boundary conditions

The dynamic boundary condition on the welded boundaries Σ_{SS} can be readily obtained by subtracting equation (3.11) from (2.106):

$$[\hat{\mathbf{n}} \cdot \mathbf{T}^{\mathrm{PK1}}]_-^+ = \mathbf{0}. \tag{3.65}$$

The corresponding condition on the outer free surface $\partial\oplus$ is

$$\hat{\mathbf{n}} \cdot \mathbf{T}^{\mathrm{PK1}} = \mathbf{0}. \tag{3.66}$$

Both (3.65) and (3.66) are exact, like the incremental Lagrangian momentum equation (3.59).

The dynamic boundary condition on the fluid-solid boundaries Σ_{FS} requires more consideration. The Lagrangian relation (2.106) is not valid in the presence of tangential slip, since two patches that are initially juxtaposed need not be juxtaposed after deformation, and vice versa. It is necessary to obtain the first-order relations by linearization of the exact Eulerian conditions (2.81)–(2.82). We begin by deriving a continuity condition that is applicable on any slipping interface, including a fault surface, before specializing to the case of a frictionless fluid-solid boundary. Figure 3.3 shows a small portion of a slipping boundary; $\hat{\mathbf{n}}^+ d\Sigma^+$ and $\hat{\mathbf{n}}^- d\Sigma^-$ are two initial elements of surface area centered on particles $\mathbf{x}^+$ and $\mathbf{x}^-$ lying on the front and back sides of the undeformed boundary. Upon deformation, both particles $\mathbf{x}^+$ and $\mathbf{x}^-$ move to the same point $\mathbf{r}$, and the patches $\hat{\mathbf{n}}^+ d\Sigma^+$ and $\hat{\mathbf{n}}^- d\Sigma^-$ merge to form the contiguous element of surface area $\hat{\mathbf{n}}^t d\Sigma^t$ on the deformed boundary. We use superscripts $\pm$ to indicate evaluation at $\mathbf{x}^\pm$;

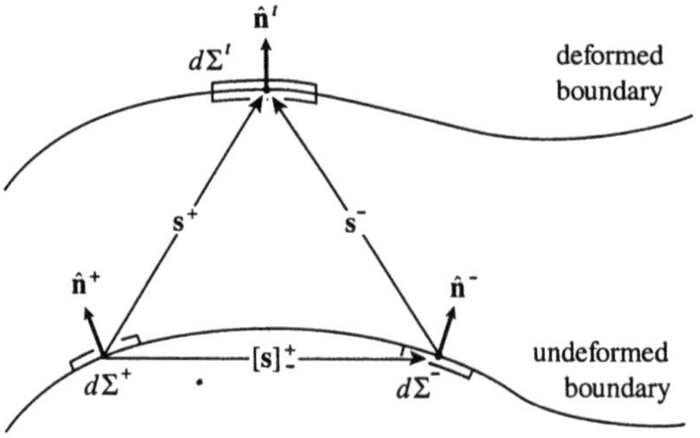

Figure 3.3. Schematic depiction of a small portion of a slipping boundary, both before (*bottom*) and after (*top*) deformation. The initial surface elements $\hat{\mathbf{n}}^+ d\Sigma^+$ and $\hat{\mathbf{n}}^- d\Sigma^-$ merge to form the deformed element $\hat{\mathbf{n}}^t d\Sigma^t$ at time t. The first-order condition that guarantees this is $\mathbf{r} = \mathbf{x}^+ + \mathbf{s}^+ = \mathbf{x}^- + \mathbf{s}^-$.

for example, the displacements of the particles $\mathbf{x}^{\pm}$ will be denoted by $\mathbf{s}^{\pm}$. The general procedure is first to transform the exact relations at $\mathbf{r}$ to $\mathbf{x}^{\pm}$, and then to relate terms at $\mathbf{x}^{+}$ and $\mathbf{x}^{-}$ using

$$\mathbf{x}^{+} - \mathbf{x}^{-} = -[\mathbf{s}]_{-}^{+}. \tag{3.67}$$

Correct to first order in $\|\mathbf{s}\|$, it does not matter precisely where on the undeformed boundary the slip $[\mathbf{s}]_{-}^{+}$ in (3.67) is evaluated.

The exact Eulerian boundary condition at the displaced point $\mathbf{r}$ is

$$[\hat{\mathbf{n}}^{t} d\Sigma^{t} \cdot \mathbf{T}^{\mathrm{E}}]_{-}^{+} = \mathbf{0}, \tag{3.68}$$

where we have inserted the (continuous) differential surface area $d\Sigma^{t}$ into equation (2.81) for convenience. This can be written as a condition on the undeformed boundary using the definition (2.41) of the first Piola-Kirchhoff stress:

$$\hat{\mathbf{n}}^{+} d\Sigma^{+} \cdot (\mathbf{T}^{0+} + \mathbf{T}^{\mathrm{PK1+}}) = \hat{\mathbf{n}}^{-} d\Sigma^{-} \cdot (\mathbf{T}^{0-} + \mathbf{T}^{\mathrm{PK1-}}). \tag{3.69}$$

The infinitesimal surface areas $d\Sigma^{\pm}$ and the unit normals $\hat{\mathbf{n}}^{\pm}$ at the points $\mathbf{x}^{\pm}$ are related by

$$d\Sigma^{+} - d\Sigma^{-} = -(\boldsymbol{\nabla}^{\Sigma} \cdot [\mathbf{s}]_{-}^{+})\, d\Sigma, \tag{3.70}$$

$$\hat{\mathbf{n}}^{+} - \hat{\mathbf{n}}^{-} = (\boldsymbol{\nabla}^{\Sigma} [\mathbf{s}]_{-}^{+}) \cdot \hat{\mathbf{n}}. \tag{3.71}$$

In addition, since the initial traction $\hat{\mathbf{n}} \cdot \mathbf{T}^{0}$ is continuous, the offset quantities $\hat{\mathbf{n}}^{+} \cdot \mathbf{T}^{0+}$ and $\hat{\mathbf{n}}^{-} \cdot \mathbf{T}^{0-}$ are related by

$$\hat{\mathbf{n}}^{+} \cdot \mathbf{T}^{0+} - \hat{\mathbf{n}}^{-} \cdot \mathbf{T}^{0-} = -[\mathbf{s}]_{-}^{+} \cdot \boldsymbol{\nabla}^{\Sigma} (\hat{\mathbf{n}} \cdot \mathbf{T}^{0}). \tag{3.72}$$

Correct to first order in $\|\mathbf{s}\|$, it is immaterial on the right side of equations (3.70)–(3.72) whether $d\Sigma$ be $d\Sigma^{\pm}$, or $\hat{\mathbf{n}}$ be $\hat{\mathbf{n}}^{\pm}$, or $\mathbf{T}^{0}$ be $\mathbf{T}^{0\pm}$. Upon using (3.70) and (3.72) to simplify (3.69) we obtain

$$[\hat{\mathbf{n}} \cdot \mathbf{T}^{\mathrm{PK1}} - \boldsymbol{\nabla}^{\Sigma} \cdot (\mathbf{s}\, \hat{\mathbf{n}} \cdot \mathbf{T}^{0})]_{-}^{+} = \mathbf{0}. \tag{3.73}$$

Equation (3.73) is the linearized condition which guarantees the continuity of traction across any slipping boundary; we shall use this relation to obtain the equivalent-force representation of an idealized earthquake fault in Section 5.3. A welded boundary, which experiences zero slip, can be considered to be a special case of a slipping boundary; if $\mathbf{s}$ is continuous, equation (3.73) reduces to (3.65), as it must.

On the fluid-solid boundaries Σ_{FS}, the initial traction is of the form

$$\hat{\mathbf{n}} \cdot \mathbf{T}^{0} = -\varpi^{0} \hat{\mathbf{n}}, \tag{3.74}$$

where $\varpi^0 = -(\hat{\mathbf{n}}\cdot\mathbf{T}^0\cdot\hat{\mathbf{n}})$ is equal to the initial pressure p^0 on the fluid side. Substitution of (3.74) into (3.73) yields the reduced condition

$$[\hat{\mathbf{n}}\cdot\mathbf{T}^{\mathrm{PK1}} + \boldsymbol{\nabla}^{\Sigma}\cdot(\varpi^0\mathbf{s}\hat{\mathbf{n}})]_-^+ = \mathbf{0}. \tag{3.75}$$

This can be written, after some rearrangement, in the alternative form

$$[\hat{\mathbf{n}}\cdot\mathbf{T}^{\mathrm{PK1}} + \hat{\mathbf{n}}\,\boldsymbol{\nabla}^{\Sigma}\cdot(\varpi^0\mathbf{s}) - \varpi^0(\boldsymbol{\nabla}^{\Sigma}\mathbf{s})\cdot\hat{\mathbf{n}}]_-^+ = \mathbf{0}, \tag{3.76}$$

where we have used the continuity of $\hat{\mathbf{n}}\cdot\mathbf{s}$ and ϖ^0 and the symmetry of the surface curvature tensor: $(\boldsymbol{\nabla}^{\Sigma}\hat{\mathbf{n}})^{\mathrm{T}} = \boldsymbol{\nabla}^{\Sigma}\hat{\mathbf{n}}$. Equation (3.76) is the most useful form of the linearized dynamical continuity condition on Σ_{FS}.

We shall also require a linearized version of the condition (2.82) that the traction at $\mathbf{r}$ must be normal to the deformed fluid-solid boundary:

$$(\hat{\mathbf{n}}^t\cdot\mathbf{T}^{\mathrm{E}})\cdot(\mathbf{I} - \hat{\mathbf{n}}^t\hat{\mathbf{n}}^t) = \mathbf{0}. \tag{3.77}$$

Equation (3.77) can be rewritten as an equation on the undeformed boundary Σ_{FS} in the form

$$\begin{aligned}[-\varpi^{0\pm}&\hat{\mathbf{n}}^{\pm} + \hat{\mathbf{n}}^{\pm}\cdot\mathbf{T}^{\mathrm{PK1}\pm}]\\ &\cdot\{\mathbf{I} - [\hat{\mathbf{n}}^{\pm} - (\boldsymbol{\nabla}^{\Sigma}\mathbf{s}^{\pm})\cdot\hat{\mathbf{n}}][\hat{\mathbf{n}}^{\pm} - (\boldsymbol{\nabla}^{\Sigma}\mathbf{s}^{\pm})\cdot\hat{\mathbf{n}}]\} = \mathbf{0}.\end{aligned} \tag{3.78}$$

Correct to first order in $\|\mathbf{s}\|$, this condition is equivalent to

$$[\hat{\mathbf{n}}\cdot\mathbf{T}^{\mathrm{PK1}} + \hat{\mathbf{n}}\,\boldsymbol{\nabla}^{\Sigma}\cdot(\varpi^0\mathbf{s}) - \varpi^0(\boldsymbol{\nabla}^{\Sigma}\mathbf{s})\cdot\hat{\mathbf{n}}]\cdot(\mathbf{I} - \hat{\mathbf{n}}\hat{\mathbf{n}}) = \mathbf{0}, \tag{3.79}$$

where we have dropped the now immaterial superscripts $\pm$, and added the term $\hat{\mathbf{n}}\,\boldsymbol{\nabla}^{\Sigma}\cdot(\varpi^0\mathbf{s})$ for convenience. Equation (3.79) is the first-order relation that guarantees that there is no shear traction on a fluid-solid boundary. Comparing (3.76) and (3.79), we see that the quantity

$$\mathbf{t}^{\mathrm{PK1}} = \hat{\mathbf{n}}\cdot\mathbf{T}^{\mathrm{PK1}} + \hat{\mathbf{n}}\,\boldsymbol{\nabla}^{\Sigma}\cdot(\varpi^0\mathbf{s}) - \varpi^0(\boldsymbol{\nabla}^{\Sigma}\mathbf{s})\cdot\hat{\mathbf{n}} \tag{3.80}$$

is a continuous, normal vector on Σ_{FS}:

$$[\mathbf{t}^{\mathrm{PK1}}]_-^+ = \mathbf{0}, \tag{3.81}$$

$$\mathbf{t}^{\mathrm{PK1}} = \hat{\mathbf{n}}(\hat{\mathbf{n}}\cdot\mathbf{t}^{\mathrm{PK1}}). \tag{3.82}$$

It is easily verified that equation (3.81) is valid on Σ_{SS} also; frictionless fluid-solid and welded solid-solid boundaries are therefore distinguished only by the normality condition (3.82).

All of the conditions derived above can be expressed in terms of the incremental Lagrangian Cauchy stress $\mathbf{T}^{\mathrm{L1}}$ rather than $\mathbf{T}^{\mathrm{PK1}}$ using the relation (3.36); we shall not, however, have any need for these results. We shall find it convenient to express the dynamical boundary conditions in a *hydrostatic* Earth model in terms of $\mathbf{T}^{\mathrm{L1}}$ in Section 3.11.

3.4.3 Gravitational boundary conditions

The boundary conditions governing the gravitational potential and field perturbations ϕ^{E1} and $\mathbf{g}^{\mathrm{E1}}$ are also obtained by linearizing the exact Eulerian conditions on the deformed boundary. We consider the general case of a boundary that may exhibit tangential slip, referring to Figure 3.3.

The exact continuity condition at the displaced point $\mathbf{r}$ on the deformed boundary,

$$[\phi^{\mathrm{E}}]_-^+ = 0, \tag{3.83}$$

can be written, correct to first order in $\|\mathbf{s}\|$, in the form

$$\phi^{0+} + \mathbf{s}^+ \cdot \boldsymbol{\nabla}\phi^{0+} + \phi^{\mathrm{E1}+} = \phi^{0-} + \mathbf{s}^- \cdot \boldsymbol{\nabla}\phi^{0-} + \phi^{\mathrm{E1}-}, \tag{3.84}$$

where the superscripts $\pm$ denote evaluation at $\mathbf{x}^\pm$. The initial potentials $\phi^{0\pm}$ are related by the first-order expansion

$$\phi^{0+} - \phi^{0-} = -[\mathbf{s}]_-^+ \cdot \boldsymbol{\nabla}^\Sigma\phi^0. \tag{3.85}$$

Combining (3.84) and (3.85) we obtain the simple linearized condition

$$[\phi^{\mathrm{E1}}]_-^+ = 0. \tag{3.86}$$

The other exact condition,

$$[\hat{\mathbf{n}}^t \cdot \mathbf{g}^{\mathrm{E}}]_-^+ = \mathbf{0}, \tag{3.87}$$

can be written, correct to first order in $\|\mathbf{s}\|$, in the form

$$\begin{aligned}[\hat{\mathbf{n}}^+ - (\boldsymbol{\nabla}^\Sigma\mathbf{s}^+)\cdot\hat{\mathbf{n}}] \cdot [\mathbf{g}^{0+} + \mathbf{s}^+ \cdot \boldsymbol{\nabla}\mathbf{g}^{0+} + \mathbf{g}^{\mathrm{E1}+}] \\ = [\hat{\mathbf{n}}^- - (\boldsymbol{\nabla}^\Sigma\mathbf{s}^-)\cdot\hat{\mathbf{n}}] \cdot [\mathbf{g}^{0-} + \mathbf{s}^- \cdot \boldsymbol{\nabla}\mathbf{g}^{0-} + \mathbf{g}^{\mathrm{E1}-}].\end{aligned} \tag{3.88}$$

The quantities $\hat{\mathbf{n}}^\pm \cdot \mathbf{g}^{0\pm}$ are related by an expansion analogous to (3.72):

$$\hat{\mathbf{n}}^+ \cdot \mathbf{g}^{0+} - \hat{\mathbf{n}}^- \cdot \mathbf{g}^{0-} = -[\mathbf{s}]_-^+ \cdot \boldsymbol{\nabla}^\Sigma(\hat{\mathbf{n}} \cdot \mathbf{g}^0). \tag{3.89}$$

Upon combining (3.88) and (3.89) we obtain, after some manipulation,

$$[\hat{\mathbf{n}} \cdot \boldsymbol{\nabla}\phi^{\mathrm{E1}} + 4\pi G\rho^0\hat{\mathbf{n}} \cdot \mathbf{s}]_-^+ = 0. \tag{3.90}$$

In deriving equation (3.90) we have used Poisson's equation (3.4), the continuity of ϕ^0 and its normal derivative $\partial_n\phi^0 = \hat{\mathbf{n}} \cdot \boldsymbol{\nabla}\phi^0$, and the Laplacian formula $\nabla^2 = \boldsymbol{\nabla} \cdot \boldsymbol{\nabla} = \partial_n^2 + (\boldsymbol{\nabla} \cdot \hat{\mathbf{n}})\partial_n + \boldsymbol{\nabla}^\Sigma \cdot \boldsymbol{\nabla}^\Sigma$. Both of the gravitational boundary conditions (3.86) and (3.90) are valid on an idealized fault surface, as well as on all of $\Sigma = \partial\oplus \cup \Sigma_{\mathrm{SS}} \cup \Sigma_{\mathrm{FS}}$.

The complete set of linearized boundary conditions on Σ is summarized for convenience in Table 3.1.

Boundary Type	Linearized Boundary Conditions
$\partial\oplus$: free surface	$\hat{\mathbf{n}}\cdot\mathbf{T}^{\mathrm{PK1}}=\mathbf{0}$
Σ_{SS}: solid-solid	$[\mathbf{s}]_-^+=\mathbf{0}$ $[\hat{\mathbf{n}}\cdot\mathbf{T}^{\mathrm{PK1}}]_-^+=\mathbf{0}$
Σ_{FS}: fluid-solid	$[\hat{\mathbf{n}}\cdot\mathbf{s}]_-^+=0$ $[\mathbf{t}^{\mathrm{PK1}}]_-^+=\hat{\mathbf{n}}[\hat{\mathbf{n}}\cdot\mathbf{t}^{\mathrm{PK1}}]_-^+=\mathbf{0}$
Σ: all boundaries	$[\phi^{\mathrm{E1}}]_-^+=0$ $[\hat{\mathbf{n}}\cdot\boldsymbol{\nabla}\phi^{\mathrm{E1}}+4\pi G\rho^0\,\hat{\mathbf{n}}\cdot\mathbf{s}]_-^+=0$
$\mathbf{t}^{\mathrm{PK1}}=\hat{\mathbf{n}}\cdot\mathbf{T}^{\mathrm{PK1}}+\hat{\mathbf{n}}\,\boldsymbol{\nabla}^{\Sigma}\cdot(\varpi^0\mathbf{s})-\varpi^0(\boldsymbol{\nabla}^{\Sigma}\mathbf{s})\cdot\hat{\mathbf{n}}$	

Table 3.1. Summary of the linearized kinematic, dynamic and gravitational boundary conditions governing a general, non-hydrostatic Earth model.

*3.4.4 Second-order tangential slip condition

In Section 3.9.4 we shall require a *second-order* tangential slip condition to calculate the stored elastic energy in a deformed Earth model having fluid-solid boundaries. The requisite condition can be obtained by expansion of the exact Eulerian continuity condition

$$[\hat{\mathbf{n}}^t\cdot\mathbf{u}^{\mathrm{E}}]_-^+=0, \tag{3.91}$$

keeping terms of order $\|\mathbf{s}\|^2$. We can write equation (3.91) in terms of the corresponding Lagrangian velocities $\partial_t\mathbf{s}^\pm$ at $\mathbf{x}^\pm$ in the form

$$[\hat{\mathbf{n}}^+-(\boldsymbol{\nabla}^{\Sigma}\mathbf{s}^+)\cdot\hat{\mathbf{n}}]\cdot\partial_t\mathbf{s}^+=[\hat{\mathbf{n}}^--(\boldsymbol{\nabla}^{\Sigma}\mathbf{s}^-)\cdot\hat{\mathbf{n}}]\cdot\partial_t\mathbf{s}^-, \tag{3.92}$$

correct to second order in $\|\mathbf{s}\|$. The quantities $\hat{\mathbf{n}}^\pm\cdot\partial_t\mathbf{s}^\pm$ are related by an expression analogous to (3.72):

$$\hat{\mathbf{n}}^+\cdot\partial_t\mathbf{s}^+-\hat{\mathbf{n}}^-\cdot\partial_t\mathbf{s}^-=[\hat{\mathbf{n}}\cdot\partial_t\mathbf{s}-\mathbf{s}\cdot\boldsymbol{\nabla}^{\Sigma}(\hat{\mathbf{n}}\cdot\partial_t\mathbf{s})]_-^+, \tag{3.93}$$

where we have retained the first-order term $[\hat{\mathbf{n}}\cdot\partial_t\mathbf{s}]_-^+$. Upon combining equations (3.92) and (3.93) we find that

$$[\hat{\mathbf{n}}\cdot\partial_t\mathbf{s}-\partial_t\mathbf{s}\cdot\boldsymbol{\nabla}^{\Sigma}(\hat{\mathbf{n}}\cdot\mathbf{s})-\mathbf{s}\cdot\boldsymbol{\nabla}^{\Sigma}(\hat{\mathbf{n}}\cdot\partial_t\mathbf{s})+\partial_t\mathbf{s}\cdot(\boldsymbol{\nabla}^{\Sigma}\hat{\mathbf{n}})\cdot\mathbf{s}]_-^+$$

$$= \partial_t[\hat{\mathbf{n}}\cdot\mathbf{s} - \mathbf{s}\cdot\boldsymbol{\nabla}^{\Sigma}(\hat{\mathbf{n}}\cdot\mathbf{s}) + \tfrac{1}{2}\mathbf{s}\cdot(\boldsymbol{\nabla}^{\Sigma}\hat{\mathbf{n}})\cdot\mathbf{s}]_-^+ = 0, \tag{3.94}$$

where we have used the symmetry $(\boldsymbol{\nabla}^{\Sigma}\hat{\mathbf{n}})^{\mathrm{T}} = \boldsymbol{\nabla}^{\Sigma}\hat{\mathbf{n}}$ to obtain the second equality. The final result is obtained by integration of (3.94), using the initial condition $\mathbf{s}(\mathbf{x},0) = \mathbf{0}$ to eliminate the arbitrary constant:

$$[\hat{\mathbf{n}}\cdot\mathbf{s} - \mathbf{s}\cdot\boldsymbol{\nabla}^{\Sigma}(\hat{\mathbf{n}}\cdot\mathbf{s}) + \tfrac{1}{2}\mathbf{s}\cdot(\boldsymbol{\nabla}^{\Sigma}\hat{\mathbf{n}})\cdot\mathbf{s}]_-^+ = 0. \tag{3.95}$$

Equation (3.95) is the second-order condition that guarantees that there is no separation or interpenetration of the material on either side of the boundary; it is applicable to a solid-solid fault surface as well as to the fluid-solid boundaries Σ_{FS}.

3.5 Linearized Potential Theory

The gravitational perturbations ϕ^{E1}, $\mathbf{g}^{\mathrm{E1}}$, ϕ^{L1} and $\mathbf{g}^{\mathrm{L1}}$ can be related to the displacement $\mathbf{s}$ in two ways. We consider both approaches here and demonstrate their equivalence.

3.5.1 Linearized Poisson's equation

The Eulerian potential perturbation ϕ^{E1} is the solution to the linearized Poisson's equation

$$\nabla^2\phi^{\mathrm{E1}} = 4\pi G\rho^{\mathrm{E1}}, \tag{3.96}$$

subject to the boundary conditions derived above:

$$[\phi^{\mathrm{E1}}]_-^+ = 0, \qquad [\hat{\mathbf{n}}\cdot\boldsymbol{\nabla}\phi^{\mathrm{E1}}]_-^+ = -4\pi G[\rho^0]_-^+(\hat{\mathbf{n}}\cdot\mathbf{s}). \tag{3.97}$$

The perturbed equation (3.96) follows immediately from (2.108), upon subtraction of the initial equation (3.4). Outside of the Earth, where $\rho^{\mathrm{E1}} = 0$, the perturbed potential is harmonic: $\nabla^2\phi^{\mathrm{E1}} = 0$. All of the ingredients in the linearized boundary-value problem (3.96)–(3.97) have a straightforward physical interpretation; in particular, the quantity $-[\rho^0]_-^+(\hat{\mathbf{n}}\cdot\mathbf{s})$ is an *apparent surface mass density* due to the normal displacement of the boundary Σ. The solution ϕ^{E1} is evidently

$$\phi^{\mathrm{E1}} = -G\int_{\oplus}\frac{\rho^{\mathrm{E1}\prime}}{\|\mathbf{x}-\mathbf{x}'\|}\,dV' + G\int_{\Sigma}\frac{[\rho^{0\prime}]_-^+(\hat{\mathbf{n}}'\cdot\mathbf{s}')}{\|\mathbf{x}-\mathbf{x}'\|}\,d\Sigma', \tag{3.98}$$

where the first term accounts for the volumetric density perturbation ρ^{E1} in $\oplus$, and the second term accounts for the apparent surface mass perturbation. Substituting $\rho^{\mathrm{E1}\prime} = -\boldsymbol{\nabla}'\cdot(\rho^{0\prime}\mathbf{s}')$ in equation (3.98) and applying Gauss' theorem, we find that the surface integrals cancel, leaving simply

$$\phi^{\mathrm{E1}} = -G\int_{\oplus}\frac{\rho^{0\prime}\mathbf{s}'\cdot(\mathbf{x}-\mathbf{x}')}{\|\mathbf{x}-\mathbf{x}'\|^3}\,dV'. \tag{3.99}$$

It is necesssary to apply Gauss' theorem to each sub-volume of $\oplus$ separately, and add the results, since the density ρ^0 in the integrand may be discontinuous across Σ. Equation (3.99) is the most convenient analytical representation of the Eulerian potential perturbation ϕ^{E1} as a linear functional of the particle displacement $\mathbf{s}$. The corresponding representation of the incremental Eulerian gravity vector $\mathbf{g}^{\mathrm{E1}} = -\boldsymbol{\nabla}\phi^{\mathrm{E1}}$ can be written in the form

$$\mathbf{g}^{\mathrm{E1}} = G\int_{\oplus} \rho^{0\prime}(\mathbf{s}'\cdot\boldsymbol{\Pi})\,dV', \tag{3.100}$$

where

$$\boldsymbol{\Pi} = \frac{\mathbf{I}}{\|\mathbf{x}-\mathbf{x}'\|^3} - \frac{3(\mathbf{x}-\mathbf{x}')(\mathbf{x}-\mathbf{x}')}{\|\mathbf{x}-\mathbf{x}'\|^5}. \tag{3.101}$$

The Lagrangian perturbations ϕ^{L1} and $\mathbf{g}^{\mathrm{L1}}$ are related to ϕ^{E1} and $\mathbf{g}^{\mathrm{E1}}$ by equations (3.42).

3.5.2 Linearized integral relations

Alternatively, we can obtain ϕ^{E1}, $\mathbf{g}^{\mathrm{E1}}$, ϕ^{L1} and $\mathbf{g}^{\mathrm{L1}}$ by linearizing the exact integral relations (2.111)–(2.114). Considering (2.113) for example, we rewrite it in the form

$$\phi^0 + \phi^{\mathrm{L1}} = -G\int_{\oplus} \frac{\rho^{0\prime}}{\|\mathbf{x}+\mathbf{s}-\mathbf{x}'-\mathbf{s}'\|}\,dV', \tag{3.102}$$

and expand the right side in powers of $\mathbf{s}$ and $\mathbf{s}'$. The zeroth-order term is simply ϕ^0, whereas the first-order term is

$$\phi^{\mathrm{L1}} = -G\int_{\oplus} \frac{\rho^{0\prime}(\mathbf{s}'-\mathbf{s})\cdot(\mathbf{x}-\mathbf{x}')}{\|\mathbf{x}-\mathbf{x}'\|^3}\,dV'. \tag{3.103}$$

By inspection, this equation is precisely $\phi^{\mathrm{L1}} = \phi^{\mathrm{E1}} + \mathbf{s}\cdot\boldsymbol{\nabla}\phi^0$, where ϕ^{E1} is given by (3.99). Expansion of

$$\mathbf{g}^0 + \mathbf{g}^{\mathrm{L1}} = -G\int_{\oplus} \frac{\rho^{0\prime}(\mathbf{x}+\mathbf{s}-\mathbf{x}'-\mathbf{s}')}{\|\mathbf{x}+\mathbf{s}-\mathbf{x}'-\mathbf{s}'\|^3}\,dV' \tag{3.104}$$

likewise leads to

$$\mathbf{g}^{\mathrm{L1}} = G\int_{\oplus} \rho^{0\prime}[(\mathbf{s}'-\mathbf{s})\cdot\boldsymbol{\Pi}]\,dV', \tag{3.105}$$

which is identical to $\mathbf{g}^{\mathrm{L1}} = \mathbf{g}^{\mathrm{E1}} + \mathbf{s}\cdot\boldsymbol{\nabla}\mathbf{g}^0 = -\boldsymbol{\nabla}\phi^{\mathrm{E1}} - \mathbf{s}\cdot\boldsymbol{\nabla}\boldsymbol{\nabla}\phi^0$.

★3.5.3 Incremental gravitational stress tensor

The gravitational stress tensor defined in Section 2.8.3 can be decomposed into an initial stress and a first-order perturbation:

$$\mathbf{N}^{\mathrm{E}} = \mathbf{N}^0 + \mathbf{N}^{\mathrm{E1}}, \qquad \mathbf{N}^{\mathrm{PK}} = \mathbf{N}^0 + \mathbf{N}^{\mathrm{PK1}}, \tag{3.106}$$

where

$$\mathbf{N}^0 = (8\pi G)^{-1}[(\mathbf{g}^0 \cdot \mathbf{g}^0)\mathbf{I} - 2\mathbf{g}^0\mathbf{g}^0]. \tag{3.107}$$

The incremental Eulerian stress is evidently

$$\mathbf{N}^{\mathrm{E1}} = (4\pi G)^{-1}[(\mathbf{g}^0 \cdot \mathbf{g}^{\mathrm{E1}})\mathbf{I} - \mathbf{g}^0\mathbf{g}^{\mathrm{E1}} - \mathbf{g}^{\mathrm{E1}}\mathbf{g}^0], \tag{3.108}$$

whereas the incremental first Piola-Kirchhoff stress is

$$\mathbf{N}^{\mathrm{PK1}} = \mathbf{N}^{\mathrm{E1}} + \mathbf{s} \cdot \boldsymbol{\nabla}\mathbf{N}^0 + \mathbf{N}^0(\boldsymbol{\nabla} \cdot \mathbf{s}) - (\boldsymbol{\nabla}\mathbf{s})^{\mathrm{T}} \cdot \mathbf{N}^0, \tag{3.109}$$

by analogy with equations (3.34) and (3.36). It is readily verified that

$$\boldsymbol{\nabla} \cdot \mathbf{N}^{\mathrm{E1}} = \rho^0\mathbf{g}^{\mathrm{E1}} + \rho^{\mathrm{E1}}\mathbf{g}^0, \qquad \boldsymbol{\nabla} \cdot \mathbf{N}^{\mathrm{PK1}} = \rho^0\mathbf{g}^{\mathrm{L1}}. \tag{3.110}$$

Hence, the linearized momentum equation (3.51) or (3.59) can be written in either of the two alternative forms

$$\begin{aligned} \rho^0[\partial_t^2\mathbf{s} + 2\boldsymbol{\Omega} \times \partial_t\mathbf{s} + \boldsymbol{\Omega} \times (\boldsymbol{\Omega} \times \mathbf{s})] &= \boldsymbol{\nabla} \cdot (\mathbf{T}^{\mathrm{E1}} + \mathbf{N}^{\mathrm{E1}}) \\ &= \boldsymbol{\nabla} \cdot (\mathbf{T}^{\mathrm{PK1}} + \mathbf{N}^{\mathrm{PK1}}). \end{aligned} \tag{3.111}$$

The exact Eulerian continuity condition $[\hat{\mathbf{n}}^t d\Sigma^t \cdot \mathbf{N}^{\mathrm{E}}]_-^+ = \mathbf{0}$ governing the gravitational traction on a deformed boundary can be linearized following the procedure in Section 3.4.2; the resulting first-order relation on the undeformed boundary, analogous to equation (3.73), is

$$[\hat{\mathbf{n}} \cdot \mathbf{N}^{\mathrm{PK1}} - \boldsymbol{\nabla}^\Sigma \cdot (\mathbf{s}\,\hat{\mathbf{n}} \cdot \mathbf{N}^0)]_-^+ = \mathbf{0}. \tag{3.112}$$

Equation (3.112), which must be satisfied at every point on both a welded and a slipping boundary, can also be obtained directly from the potential and gravity boundary conditions (3.86) and (3.90).

3.6 Linearized Elastic Constitutive Relation

All of the results we have obtained so far in this chapter can be regarded as linearized laws of either geometry or physics. To complete these equations, we require linearized constitutive relations between the incremental stresses $\mathbf{T}^{\mathrm{L1}}$, $\mathbf{T}^{\mathrm{PK1}}$ and $\mathbf{T}^{\mathrm{SK1}}$ and the displacement gradient $\boldsymbol{\nabla}\mathbf{s}$. For the time being, we assume that the material comprising the Earth is adiabatic and perfectly elastic, as discussed in Section 2.10. We generalize the results obtained here to the case of a linear anelastic Earth model in Chapter 6.

3.6.1 Elastic strain energy density

An adiabatic, perfectly elastic material is, by definition, one whose volumetric Lagrangian internal energy density $\rho^0 U^{\mathrm{L}}$ depends only upon the local Lagrangian strain tensor $\mathbf{E}^{\mathrm{L}}$. In a consistent linear theory, the energy must be calculated correct to second order in $\|\mathbf{s}\|$; this prohibits the use of the linearized approximation $\mathbf{E}^{\mathrm{L}} = \boldsymbol{\varepsilon}$ in energetic considerations. In specifying $\rho^0 U^{\mathrm{L}}$, we must use the exact relation between $\mathbf{E}^{\mathrm{L}}$ and $\boldsymbol{\nabla}\mathbf{s}$:

$$\mathbf{E}^{\mathrm{L}} = \tfrac{1}{2}[\boldsymbol{\nabla}\mathbf{s} + (\boldsymbol{\nabla}\mathbf{s})^{\mathrm{T}}] + \tfrac{1}{2}(\boldsymbol{\nabla}\mathbf{s})\cdot(\boldsymbol{\nabla}\mathbf{s})^{\mathrm{T}}. \tag{3.113}$$

Invariant notation becomes cumbersome to use in much of what follows, and we shall frequently resort to index notation; expressed in terms of components relative to the Cartesian axes $\hat{\mathbf{x}}_1$, $\hat{\mathbf{x}}_2$, $\hat{\mathbf{x}}_3$, equation (3.113) takes the form

$$E^{\mathrm{L}}_{ij} = \tfrac{1}{2}(\partial_i s_j + \partial_j s_i) + \tfrac{1}{2}\partial_i s_k\, \partial_j s_k. \tag{3.114}$$

Expanding the dependence of the elastic energy density on the strain $\mathbf{E}^{\mathrm{L}}$ to second order in $\|\mathbf{s}\|$, we write $\rho^0 U^{\mathrm{L}}$ in the form

$$\begin{aligned}\rho^0 U^{\mathrm{L}} &= \rho^0 U^0 + \mathbf{T}^0 \!:\! \mathbf{E}^{\mathrm{L}} + \tfrac{1}{2}\mathbf{E}^{\mathrm{L}} \!:\! \boldsymbol{\Xi} \!:\! \mathbf{E}^{\mathrm{L}} \\ &= \rho^0 U^0 + T^0_{ij}\, E^{\mathrm{L}}_{ij} + \tfrac{1}{2}E^{\mathrm{L}}_{ij}\, \Xi_{ijkl}\, E^{\mathrm{L}}_{kl},\end{aligned} \tag{3.115}$$

where $\boldsymbol{\Xi}$ is a fourth-order tensor. The zeroth-order term $\rho^0 U^0$ is the internal energy density in the undeformed reference configuration; the next two terms account for the first-order and second-order dependence of $\rho^0 U^{\mathrm{L}}$ on the strain, respectively. The first-order expansion coefficients in (3.115) are chosen to insure that all three stresses $\mathbf{T}^{\mathrm{L}}$, $\mathbf{T}^{\mathrm{PK}}$ and $\mathbf{T}^{\mathrm{SK}}$ reduce to the initial stress $\mathbf{T}^0$, correct to zeroth order in $\|\mathbf{s}\|$. Without loss of generality, the components Ξ_{ijkl} of the fourth-order tensor $\boldsymbol{\Xi}$ may be assumed to satisfy the symmetry relations

$$\Xi_{ijkl} = \Xi_{jikl} = \Xi_{ijlk} = \Xi_{klij}. \tag{3.116}$$

The second Piola-Kirchhoff stress $\mathbf{T}^{\mathrm{SK}}$ is defined in terms of the energy density $\rho^0 U^{\mathrm{L}}$ by equation (2.136):

$$\mathbf{T}^{\mathrm{SK}} = \mathbf{T}^0 + \mathbf{T}^{\mathrm{SK1}} = \rho^0 \left(\frac{\partial U^{\mathrm{L}}}{\partial \mathbf{E}^{\mathrm{L}}}\right) = \mathbf{T}^0 + \boldsymbol{\Xi} \!:\! \mathbf{E}^{\mathrm{L}}. \tag{3.117}$$

Differentiating the expansion (3.115) twice with respect to $\mathbf{E}^{\mathrm{L}}$, we obtain an explicit formula for the components of the tensor $\boldsymbol{\Xi}$:

$$\Xi_{ijkl} = \rho^0 \left(\frac{\partial U^{\mathrm{L}}}{\partial E^{\mathrm{L}}_{ij}\, \partial E^{\mathrm{L}}_{kl}}\right). \tag{3.118}$$

The equality $\Xi_{ijkl} = \Xi_{klij}$ can be regarded as a consequence of the equality of the mixed partial derivatives of $\rho^0 U^{\mathrm{L}}$. Such relations, which are ubiquitous in thermodynamics, are known as *Maxwell relations.* Once the linear relation (3.117) has been obtained, it is permissible to replace $\mathbf{E}^{\mathrm{L}}$ by the infinitesimal strain tensor $\boldsymbol{\varepsilon}$, and write the constitutive relation for the incremental second Piola-Kirchhoff stress $\mathbf{T}^{\mathrm{SK1}}$, correct to first order in $\|\mathbf{s}\|$, in the form

$$\mathbf{T}^{\mathrm{SK1}} = \boldsymbol{\Xi}:\boldsymbol{\varepsilon}, \tag{3.119}$$

or, equivalently, $T_{ij}^{\mathrm{SK1}} = \Xi_{ijkl}\,\varepsilon_{kl}$. We cannot, however, likewise replace $\mathbf{E}^{\mathrm{L}}$ by $\boldsymbol{\varepsilon}$ in the elastic strain energy density (3.115), because the term $\mathbf{T}^0:\mathbf{E}^{\mathrm{L}}$ is not equal to $\mathbf{T}^0:\boldsymbol{\varepsilon}$, correct to second order in $\|\mathbf{s}\|$.

The linearized constitutive relations for the first Piola-Kirchhoff stress $\mathbf{T}^{\mathrm{PK1}}$ and the incremental Lagrangian Cauchy stress $\mathbf{T}^{\mathrm{L1}}$ can be obtained from (3.119) using the relations (3.36)–(3.38). The results are most conveniently expressed in terms of two new fourth-order tensors $\boldsymbol{\Lambda}$ and $\boldsymbol{\Upsilon}$:

$$\mathbf{T}^{\mathrm{PK1}} = \boldsymbol{\Lambda}:\boldsymbol{\nabla}\mathbf{s}, \tag{3.120}$$

$$\mathbf{T}^{\mathrm{L1}} = \boldsymbol{\Upsilon}:\boldsymbol{\nabla}\mathbf{s}, \tag{3.121}$$

or, equivalently, $T_{ij}^{\mathrm{PK1}} = \Lambda_{ijkl}\,\partial_k s_l$, $T_{ij}^{\mathrm{L1}} = \Upsilon_{ijkl}\,\partial_k s_l$. The components of $\boldsymbol{\Lambda}$ and $\boldsymbol{\Upsilon}$ are defined in terms of those of $\boldsymbol{\Xi}$ by

$$\Lambda_{ijkl} = \Xi_{ijkl} + T_{ik}^0\,\delta_{jl}, \tag{3.122}$$

$$\begin{aligned}\Upsilon_{ijkl} &= \Lambda_{ijkl} + T_{jk}^0\,\delta_{il} - T_{ij}^0\,\delta_{kl}\\ &= \Xi_{ijkl} + T_{ik}^0\,\delta_{jl} + T_{jk}^0\,\delta_{il} - T_{ij}^0\,\delta_{kl},\end{aligned} \tag{3.123}$$

where δ_{ij} is the Kronecker delta. From (3.116) and (3.122)–(3.123) we see that the components Λ_{ijkl} and Υ_{ijkl} satisfy the symmetry relations

$$\Lambda_{ijkl} = \Lambda_{klij}, \tag{3.124}$$

$$\Upsilon_{ijkl} = \Upsilon_{jikl}. \tag{3.125}$$

Upon inserting equation (3.113) into (3.115) and using (3.122), we find that the elastic energy density $\rho^0 U^{\mathrm{L}}$ can be rewritten, correct to second order in $\|\mathbf{s}\|$, in the form

$$\begin{aligned}\rho^0 U^{\mathrm{L}} &= \rho^0 U^0 + \mathbf{T}^0:\boldsymbol{\varepsilon} + \tfrac{1}{2}\boldsymbol{\nabla}\mathbf{s}:\boldsymbol{\Lambda}:\boldsymbol{\nabla}\mathbf{s}\\ &= \rho^0 U^0 + T_{ij}^0\,\varepsilon_{ij} + \tfrac{1}{2}\partial_i s_j\,\Lambda_{ijkl}\,\partial_k s_l.\end{aligned} \tag{3.126}$$

Comparison of (3.120) and (3.126) enables us to write the first Piola-Kirchhoff stress $\mathbf{T}^{\mathrm{PK}}$ in a manner analogous to (3.117):

$$\mathbf{T}^{\mathrm{PK}} = \mathbf{T}^0 + \mathbf{T}^{\mathrm{PK1}} = \rho^0 \left[\frac{\partial U^{\mathrm{L}}}{\partial(\boldsymbol{\nabla}\mathbf{s})} \right]. \tag{3.127}$$

Equation (3.127) is a linearized version of the general relation (2.141). Recalling that $\mathbf{F}^{\mathrm{T}} = \mathbf{I} + \boldsymbol{\nabla}\mathbf{s}$ and making use of the chain rule, we confirm that $\partial U^{\mathrm{L}}/\partial \mathbf{F}^{\mathrm{T}} = \partial U^{\mathrm{L}}/\partial(\boldsymbol{\nabla}\mathbf{s})$. Upon differentiating the result (3.126) twice with respect to the displacement gradient $\boldsymbol{\nabla}\mathbf{s}$, we see that the components of the tensor $\boldsymbol{\Lambda}$ can be written in the form

$$\Lambda_{ijkl} = \rho^0 \left[\frac{\partial U^{\mathrm{L}}}{\partial(\partial_i s_j)\, \partial(\partial_k s_l)} \right]. \tag{3.128}$$

The important symmetry $\Lambda_{ijkl} = \Lambda_{klij}$ is another example of an elastic Maxwell relation.

We could have avoided introducing the second Piola-Kirchhoff stress tensor $\mathbf{T}^{\mathrm{SK}}$ by writing $\rho^0 U^{\mathrm{L}}$ in the form (3.126) rather than (3.115) at the outset. The advantage of the formulation (3.115) is that it is explicitly consistent with the principle of material frame indifference. To guarantee that the representation (3.126) is independent of the rotation tensor $\mathbf{Q}$, we must demand that it satisfy the constraint (2.140). Correct to first order in $\|\mathbf{s}\|$, this leads to the two requirements

$$\Lambda_{jikl} - \Lambda_{ijkl} = T^0_{jk}\,\delta_{il} - T^0_{ik}\,\delta_{jl}, \tag{3.129}$$

$$\Lambda_{ijlk} - \Lambda_{ijkl} = T^0_{il}\delta_{jk} - T^0_{ik}\,\delta_{jl}. \tag{3.130}$$

In the present formulation, these conditions, together with the corresponding results for $\boldsymbol{\Upsilon}$,

$$\Upsilon_{ijlk} - \Upsilon_{ijkl} = T^0_{il}\delta_{jk} - T^0_{ik}\,\delta_{jl} + T^0_{jl}\,\delta_{ik} - T^0_{jk}\,\delta_{il}, \tag{3.131}$$

$$\Upsilon_{klij} - \Upsilon_{ijkl} = T^0_{il}\delta_{jk} - T^0_{jk}\,\delta_{il} + T^0_{ij}\,\delta_{kl} - T^0_{kl}\,\delta_{ij}, \tag{3.132}$$

are an immediate consequence of the relations (3.122)–(3.123) and the symmetries (3.116) of the tensor $\boldsymbol{\Xi}$.

3.6.2 The elastic tensor

In the absence of any initial stress $\mathbf{T}^0$, the classical linear elastic stress-strain relation is

$$\mathbf{T} = \boldsymbol{\Gamma} : \boldsymbol{\varepsilon}, \tag{3.133}$$

where $\mathbf{T}$ can be interpreted as either $\mathbf{T}^{\mathrm{L}}$ or $\mathbf{T}^{\mathrm{E}}$, and where $\mathbf{\Gamma}$ satisfies

$$\Gamma_{ijkl} = \Gamma_{jikl} = \Gamma_{ijlk} = \Gamma_{klij}. \tag{3.134}$$

Equation (3.133) is commonly known as *Hooke's law*; the relations (3.134) are the classical symmetries of a fourth-order *elastic tensor*, with 21 independent coefficients. In the more general case of an incremental stress superimposed upon a zeroth-order initial stress $\mathbf{T}^0$, we shall find it useful to write $\mathbf{\Xi}$, $\mathbf{\Lambda}$ and $\mathbf{\Upsilon}$ in terms of $\mathbf{T}^0$ and a tensor $\mathbf{\Gamma}$ having the elastic symmetries (3.134). The tensor $\mathbf{\Xi}$, according to (3.116), already satisfies these symmetries, so that (3.122) and (3.123) are relations of the desired type; however, $\mathbf{\Xi}$ is not the most satisfactory elastic tensor for our purposes.

In fact, we have considerable latitude in the choice of an elastic tensor; the only requirement is that we satisfy the three sets of symmetries (3.116) and (3.124)–(3.125) as well as the interrelationships (3.122)–(3.123). It is easily verified that all these conditions are satisfied by tensors $\mathbf{\Xi}$, $\mathbf{\Lambda}$ and $\mathbf{\Upsilon}$ of the form

$$\begin{aligned}\Xi_{ijkl} = \Gamma_{ijkl} &+ a(T^0_{ij}\delta_{kl} + T^0_{kl}\delta_{ij}) \\ &+ b(T^0_{ik}\delta_{jl} + T^0_{jk}\delta_{il} + T^0_{il}\delta_{jk} + T^0_{jl}\delta_{ik}),\end{aligned} \tag{3.135}$$

$$\begin{aligned}\Lambda_{ijkl} = \Gamma_{ijkl} &+ a(T^0_{ij}\delta_{kl} + T^0_{kl}\delta_{ij}) \\ &+ (1+b)T^0_{ik}\delta_{jl} + b(T^0_{jk}\delta_{il} + T^0_{il}\delta_{jk} + T^0_{jl}\delta_{ik}),\end{aligned} \tag{3.136}$$

$$\begin{aligned}\Upsilon_{ijkl} = \Gamma_{ijkl} &+ (a-1)T^0_{ij}\delta_{kl} + aT^0_{kl}\delta_{ij} \\ &+ (1+b)(T^0_{ik}\delta_{jl} + T^0_{jk}\delta_{il}) + b(T^0_{il}\delta_{jk} + T^0_{jl}\delta_{ik}),\end{aligned} \tag{3.137}$$

where $\Gamma_{ijkl} = \Gamma_{jikl} = \Gamma_{ijlk} = \Gamma_{klij}$ and the quantities a and b are arbitrary scalars. The two expressions in parentheses multiplying a and b in equation (3.135) are the only two linear combinations of permutations of $\mathbf{T}^0\mathbf{I}$ satisfying the elastic symmetries. Every choice of the scalars a and b defines a possible elastic tensor $\mathbf{\Gamma}$. The most convenient alternative is that adopted by Dahlen (1972):

$$a = -b = \tfrac{1}{2}. \tag{3.138}$$

With this choice, the tensors $\mathbf{\Xi}$, $\mathbf{\Lambda}$ and $\mathbf{\Upsilon}$ are given by

$$\begin{aligned}\Xi_{ijkl} = \Gamma_{ijkl} &+ \tfrac{1}{2}(T^0_{ij}\delta_{kl} + T^0_{kl}\delta_{ij} \\ &- T^0_{ik}\delta_{jl} - T^0_{jk}\delta_{il} - T^0_{il}\delta_{jk} - T^0_{jl}\delta_{ik}),\end{aligned} \tag{3.139}$$

$$\begin{aligned}\Lambda_{ijkl} = \Gamma_{ijkl} &+ \tfrac{1}{2}(T^0_{ij}\delta_{kl} + T^0_{kl}\delta_{ij} \\ &+ T^0_{ik}\delta_{jl} - T^0_{jk}\delta_{il} - T^0_{il}\delta_{jk} - T^0_{jl}\delta_{ik}),\end{aligned} \tag{3.140}$$

$$\Upsilon_{ijkl} = \Gamma_{ijkl} + \tfrac{1}{2}(-T^0_{ij}\delta_{kl} + T^0_{kl}\delta_{ij} + T^0_{ik}\delta_{jl} + T^0_{jk}\delta_{il} - T^0_{il}\delta_{jk} - T^0_{jl}\delta_{ik}). \tag{3.141}$$

Upon inserting equations (3.139)–(3.141) as well as the decompositions $\mathbf{T}^0 = -p^0\mathbf{I}+\boldsymbol{\tau}^0$ and $(\boldsymbol{\nabla}\mathbf{s})^{\mathrm{T}} = \boldsymbol{\varepsilon}+\boldsymbol{\omega}$ into the representations (3.119)–(3.121), we can write the incremental stresses $\mathbf{T}^{\mathrm{SK1}}$, $\mathbf{T}^{\mathrm{PK1}}$ and $\mathbf{T}^{\mathrm{L1}}$ in the form

$$T^{\mathrm{SK1}}_{ij} = \Gamma_{ijkl}\varepsilon_{kl} + p^0(2\varepsilon_{ij} - \varepsilon_{kk}\delta_{ij}) + \tfrac{1}{2}(\tau^0_{ij}\varepsilon_{kk} + \tau^0_{kl}\varepsilon_{kl}\delta_{ij}) - \varepsilon_{ik}\tau^0_{kj} - \tau^0_{ik}\varepsilon_{kj}, \tag{3.142}$$

$$T^{\mathrm{PK1}}_{ij} = \Gamma_{ijkl}\varepsilon_{kl} + p^0(\varepsilon_{ij} + \omega_{ij} - \varepsilon_{kk}\delta_{ij}) + \tfrac{1}{2}(\tau^0_{ij}\varepsilon_{kk} + \tau^0_{kl}\varepsilon_{kl}\delta_{ij}) - \varepsilon_{ik}\tau^0_{kj} - \tau^0_{ik}\omega_{kj}, \tag{3.143}$$

$$T^{\mathrm{L1}}_{ij} = \Gamma_{ijkl}\varepsilon_{kl} + \tfrac{1}{2}(-\tau^0_{ij}\varepsilon_{kk} + \tau^0_{kl}\varepsilon_{kl}\delta_{ij}) + \omega_{ik}\tau^0_{kj} - \tau^0_{ik}\omega_{kj}. \tag{3.144}$$

Equations (3.142), (3.143) and (3.144) are the linearized versions of the general results (2.138), (2.143)–(2.144) and (2.145)–(2.146), respectively. Taken together, these relations constitute the generalization of Hooke's law to the case of a pre-stressed elastic medium; the dependence of $\mathbf{T}^{\mathrm{SK1}}$, $\mathbf{T}^{\mathrm{PK1}}$ and $\mathbf{T}^{\mathrm{L1}}$ upon the initial pressure p^0 and deviatoric stress $\boldsymbol{\tau}^0$ as well as upon the strain and rotation tensors $\boldsymbol{\varepsilon}$ and $\boldsymbol{\omega}$ is exhibited explicitly. It is noteworthy that the incremental Lagrangian Cauchy stress $\mathbf{T}^{\mathrm{L1}}$ does not depend explicitly upon the initial pressure p^0, but only upon the elastic tensor $\boldsymbol{\Gamma}$ and the initial deviatoric stress $\boldsymbol{\tau}^0$. In fact, this is what motivated us to set $a = -b = 1/2$ in equations (3.135)–(3.137); it is the only choice that completely eliminates the explicit dependence of $\mathbf{T}^{\mathrm{L1}}$ upon p^0. The complete Lagrangian Cauchy stress $\mathbf{T}^{\mathrm{L}} = \mathbf{T}^0 + \mathbf{T}^{\mathrm{L1}}$ can be rewritten using invariant notation in the form

$$\mathbf{T}^{\mathrm{L}} = \overbrace{-p^0\mathbf{I} + \boldsymbol{\tau}^0 + \boldsymbol{\omega}\cdot\boldsymbol{\tau}^0 - \boldsymbol{\tau}^0\cdot\boldsymbol{\omega}}^{\text{rotated initial stress}} + \underbrace{\boldsymbol{\Gamma}:\boldsymbol{\varepsilon} - \tfrac{1}{2}\boldsymbol{\tau}^0(\mathrm{tr}\,\boldsymbol{\varepsilon}) + \tfrac{1}{2}(\boldsymbol{\tau}^0:\boldsymbol{\varepsilon})\mathbf{I}}_{\text{strain-dependent stress}}. \tag{3.145}$$

The first four terms in equation (3.145) can be identified as the *rotated initial stress* $\mathbf{Q}^{\mathrm{T}}\cdot\mathbf{T}^0\cdot\mathbf{Q}$, whereas the remaining terms represent the perturbation in stress due to the strain $\boldsymbol{\varepsilon}$.

An *isotropic solid* is, by definition, one whose elastic tensor is of the form

$$\Gamma_{ijkl} = (\kappa - \tfrac{2}{3}\mu)\delta_{ij}\delta_{kl} + \mu(\delta_{ik}\delta_{jl} + \delta_{il}\delta_{jk}), \tag{3.146}$$

where κ is the isentropic *incompressibility* or bulk modulus and μ is the *rigidity* or shear modulus. The approximation (3.146) is most likely to be employed in the case of a *hydrostatic* Earth model, for which $\boldsymbol{\tau}^0 = \mathbf{0}$. The incremental Lagrangian Cauchy stress in such an isotropic, hydrostatic model can be written in the form $\mathbf{T}^{\mathrm{L1}} = -p^{\mathrm{L1}}\mathbf{I}+\boldsymbol{\tau}^{\mathrm{L1}}$, where $p^{\mathrm{L1}} = -\kappa(\boldsymbol{\nabla}\cdot\mathbf{s})$ and $\boldsymbol{\tau}^{\mathrm{L1}} = 2\mu\mathbf{d}$. The quantity $\mathbf{d} = \boldsymbol{\varepsilon} - \frac{1}{3}(\mathrm{tr}\,\boldsymbol{\varepsilon})\mathbf{I}$ is the *deviatoric strain*.

More generally, we can isolate the isotropic part of the elastic tensor $\boldsymbol{\Gamma}$ in either a hydrostatic or a non-hydrostatic Earth model by writing

$$\Gamma_{ijkl} = (\kappa - \tfrac{2}{3}\mu)\delta_{ij}\,\delta_{kl} + \mu(\delta_{ik}\,\delta_{jl} + \delta_{il}\,\delta_{jk}) + \gamma_{ijkl}, \tag{3.147}$$

where $\gamma_{ijkl} = \gamma_{jikl} = \gamma_{ijlk} = \gamma_{klij}$. The incompressibility κ and rigidity μ in the decomposition (3.147) are defined by

$$\kappa = \tfrac{1}{9}\Gamma_{iijj}, \qquad \mu = \tfrac{1}{10}(\Gamma_{ijij} - \tfrac{1}{3}\Gamma_{iijj}), \tag{3.148}$$

in which case

$$\gamma_{iijj} = \gamma_{ijij} = 0. \tag{3.149}$$

The tensor $\Gamma'_{ijkl} = (\kappa - \frac{2}{3}\mu)\delta_{ij}\,\delta_{kl} + \mu(\delta_{ik}\,\delta_{jl} + \delta_{il}\,\delta_{jk})$ is the isotropic tensor that is the best approximation to the elastic tensor in a least-squares sense:

$$(\Gamma_{ijkl} - \Gamma'_{ijkl})(\Gamma_{ijkl} - \Gamma'_{ijkl}) = \text{minimum}. \tag{3.150}$$

The residual $\boldsymbol{\gamma}$ is then the *purely anisotropic part* of the elastic tensor.

In the *fluid* regions of the Earth $\oplus_{\mathrm{F}}$, both the rigidity μ and the anisotropy $\boldsymbol{\gamma}$ are identically zero, and the incremental Lagrangian Cauchy stress is isotropic: $\mathbf{T}^{\mathrm{L1}} = -p^{\mathrm{L1}}\mathbf{I}$, where $p^{\mathrm{L1}} = -\kappa(\boldsymbol{\nabla}\cdot\mathbf{s})$. The corresponding Eulerian stress is $\mathbf{T}^{\mathrm{E1}} = -p^{\mathrm{E1}}\mathbf{I}$, where $p^{\mathrm{E1}} = -\kappa(\boldsymbol{\nabla}\cdot\mathbf{s}) - \mathbf{s}\cdot\boldsymbol{\nabla}p^0$.

★3.6.3 Speed of body-wave propagation

We can gain more insight into the nature of the elastic tensor $\boldsymbol{\Gamma}$ by considering the propagation of localized elastic body waves; this can be done using either of two equivalent methods (Whitham 1974). Following Hadamard, we can consider the propagation of an acceleration jump, that is, a surface across which the particle acceleration suffers a jump discontinuity:

$$\mathbf{a} = [\partial_t^2\mathbf{s}]_-^+. \tag{3.151}$$

Such a jump in the second temporal derivative across a wavefront must be accompanied by an analogous jump in the second spatial derivative:

$$[\boldsymbol{\nabla}\boldsymbol{\nabla}\mathbf{s}]_-^+ = \hat{\mathbf{k}}\hat{\mathbf{k}}\,c^{-2}\mathbf{a}, \tag{3.152}$$

where $\hat{\mathbf{k}}$ is the unit normal to the wavefront, and c is the speed of propagation. Application of the jump operator $[\,\cdot\,]_-^+$ to either form of the linearized momentum equation (3.56) or (3.60) yields

$$\rho^0[\partial_t^2 s_j]_-^+ = \Lambda_{ijkl}\,[\partial_i\partial_k s_l]_-^+ = \Upsilon_{ijkl}\,[\partial_i\partial_k s_l]_-^+, \tag{3.153}$$

where we have used the continuity of ρ^0, $\mathbf{T}^0$, $\boldsymbol{\Gamma}$, $\mathbf{s}$ and $\partial_t\mathbf{s}$. Inserting equations (3.151) and (3.152) into (3.153) leads to the *Christoffel equation* governing the propagation of wavefronts:

$$\mathbf{B}\cdot\mathbf{a} = c^2\,\mathbf{a}, \tag{3.154}$$

where

$$\rho^0 B_{jl} = \Lambda_{ijkl}\,\hat{k}_i\hat{k}_k = \Upsilon_{ijkl}\,\hat{k}_i\hat{k}_k. \tag{3.155}$$

Alternatively, we can consider a JWKB ansatz of the form

$$\mathbf{s} = \mathbf{a}\exp(i\Psi). \tag{3.156}$$

The frequency ω and wavevector $\mathbf{k} = k\hat{\mathbf{k}}$ are defined in terms of the phase Ψ by

$$\omega = -\partial_t\Psi, \qquad \mathbf{k} = \boldsymbol{\nabla}\Psi. \tag{3.157}$$

Substitution of (3.156) into either (3.56) or (3.60) also leads, in the limit of large wavenumber, $k\to\infty$, to the eigenvalue problem (3.154); the quantity c in this case is the *phase speed* of the wave:

$$c = \omega/k. \tag{3.158}$$

In both cases the quantities $\hat{k}_i$ and $\hat{k}_k$ in equation (3.155) denote the corresponding components of the *unit wavevector* or direction of phase propagation $\hat{\mathbf{k}}$. It is noteworthy that there are no rotational or self-gravitational terms in the Christoffel equation (3.154); the latter are absent in a jump discontinuity $[\,\cdot\,]_-^+$ or in the limit $k\to\infty$ by virtue of the fundamentally long-range nature of gravitational attraction.

For every direction of propagation, the Christoffel tensor $\mathbf{B}$ has three eigenvalues c^2 with corresponding eigenvectors $\mathbf{a}$. The eigenvalues are the squared phase speeds of the three independent body-wave types that can propagate in the material, and the eigenvectors describe the associated polarizations. The relations (3.134) and (3.140)–(3.141) guarantee that the Christoffel tensor is always symmetric: $\mathbf{B}^{\mathrm{T}} = \mathbf{B}$. As a result, all three eigenvalues are real, and the eigenvectors may be assumed to be orthogonal. We can write $\mathbf{B}$ explicitly in terms of the incompressibility κ, the rigidity

μ, the anisotropic elastic tensor $\boldsymbol{\gamma}$, and the initial deviatoric stress $\boldsymbol{\tau}^0$ in the form

$$\begin{aligned}\rho^0 B_{jl} = (\kappa + \tfrac{1}{3}\mu)\hat{k}_j\hat{k}_l + \mu\delta_{jl} \\ + \gamma_{ijkl}\,\hat{k}_i\hat{k}_k + \tfrac{1}{2}(\hat{k}_i\tau^0_{ik}\hat{k}_k)\delta_{jl} - \tfrac{1}{2}\tau^0_{jl}.\end{aligned} \tag{3.159}$$

It is noteworthy that there is no explicit dependence of the phase speeds or polarizations upon the initial hydrostatic pressure p^0; once again, only the choice $a = -b = 1/2$ in (3.135)–(3.137) leads to this desirable result. In a physically realizable material, the squared phase speeds c^2, and thus the tensor $\mathbf{B}$, must be non-negative; the condition that guarantees this is

$$\begin{aligned}(\kappa + \tfrac{1}{3}\mu)(\hat{\mathbf{a}}\cdot\hat{\mathbf{k}})^2 + \mu + \hat{\mathbf{a}}\hat{\mathbf{k}}\!:\!\boldsymbol{\gamma}\!:\!\hat{\mathbf{a}}\hat{\mathbf{k}} \\ + \tfrac{1}{2}(\hat{\mathbf{k}}\cdot\boldsymbol{\tau}^0\cdot\hat{\mathbf{k}}) - \tfrac{1}{2}(\hat{\mathbf{a}}\cdot\boldsymbol{\tau}^0\cdot\hat{\mathbf{a}}) \geq 0,\end{aligned} \tag{3.160}$$

for all $\|\hat{\mathbf{k}}\| = \|\hat{\mathbf{a}}\| = 1$. This is a necessary constraint upon the elastic parameters κ, μ, $\boldsymbol{\gamma}$ and the initial deviatoric stress $\boldsymbol{\tau}^0$ in a physically realizable Earth model.

In an isotropic elastic solid medium with a hydrostatic pre-stress, the Christoffel tensor reduces to $\rho^0\mathbf{B} = (\kappa + \frac{1}{3}\mu)\hat{\mathbf{k}}\hat{\mathbf{k}} + \mu\mathbf{I}$. Such a medium can support shear or S waves with transverse polarization $\hat{\mathbf{a}} \perp \hat{\mathbf{k}}$ and speed $\beta = (\mu/\rho^0)^{1/2}$, in addition to compressional or P waves with longitudinal polarization $\hat{\mathbf{a}} \parallel \hat{\mathbf{k}}$ and speed $\alpha = [(\kappa + \frac{4}{3}\mu)/\rho^0]^{1/2}$. More generally, body-wave propagation in a pre-stressed solid medium is *anisotropic*; the phase speeds c and the polarizations $\hat{\mathbf{a}}$ depend upon the direction $\hat{\mathbf{k}}$ of propagation. Both the intrinsic elastic anisotropy $\boldsymbol{\gamma}$ and the anisotropic pre-stress $\boldsymbol{\tau}^0$ contribute to this body-wave anisotropy; if $\|\boldsymbol{\gamma}\| \ll \mu$ and $\|\boldsymbol{\tau}^0\| \ll \mu$, as in the Earth, the independent wave types will be *quasi-P* and *quasi-S waves*. The only waves that can propagate in the fluid regions $\oplus_{\mathrm{F}}$ are P waves with longitudinal polarization $\hat{\mathbf{a}} \parallel \hat{\mathbf{k}}$ and speed $\alpha = (\kappa/\rho^0)^{1/2}$.

Since the phase speeds c depend only upon the direction of the wavevector, and not upon its magnitude k, the propagation of body waves is non-dispersive. The *group velocity* of a wavegroup with wavevector $\mathbf{k}$ is

$$\mathbf{C} = \boldsymbol{\nabla}_{\mathbf{k}}\omega = c\hat{\mathbf{k}} + \boldsymbol{\nabla}_1 c, \tag{3.161}$$

where $\boldsymbol{\nabla}_{\mathbf{k}} = \hat{\mathbf{k}}\partial_k + k^{-1}\boldsymbol{\nabla}_1$. The group velocities $\mathbf{C}$ of the three independent wave types also only depend upon the direction of propagation $\hat{\mathbf{k}}$. Letting $C = \|\mathbf{C}\|$, we see from equation (3.161) that

$$C^2 = c^2 + \|\boldsymbol{\nabla}_1 c\|^2, \tag{3.162}$$

so that the group speed is greater than or equal to the phase speed, $C \geq c$, for every direction $\hat{\mathbf{k}}$.

Name or Description	Linearized Elastic-Gravitational Equation
Continuity equation	$\rho^{\mathrm{E1}} = -\boldsymbol{\nabla}\cdot(\rho^0\mathbf{s})$
Momentum equation	$\rho^0(\partial_t^2\mathbf{s} + 2\boldsymbol{\Omega}\times\partial_t\mathbf{s}) = \boldsymbol{\nabla}\cdot\mathbf{T}^{\mathrm{PK1}} - \rho^0\boldsymbol{\nabla}\phi^{\mathrm{E1}} - \rho^0\mathbf{s}\cdot\boldsymbol{\nabla\nabla}(\phi^0+\psi)$ $\rho^0(\partial_t^2\mathbf{s} + 2\boldsymbol{\Omega}\times\partial_t\mathbf{s}) = \boldsymbol{\nabla}\cdot\mathbf{T}^{\mathrm{L1}} - \boldsymbol{\nabla}\cdot(\mathbf{s}\cdot\boldsymbol{\nabla}\mathbf{T}^0) - \rho^0\boldsymbol{\nabla}\phi^{\mathrm{E1}} - \rho^{\mathrm{E1}}\boldsymbol{\nabla}(\phi^0+\psi)$
Poisson's equation	$\nabla^2\phi^{\mathrm{E1}} = 4\pi G\rho^{\mathrm{E1}}$
Potential perturbation	$\phi^{\mathrm{E1}} = -G\int_\oplus \frac{\rho^{0\prime}\mathbf{s}'\cdot(\mathbf{x}-\mathbf{x}')}{\Vert\mathbf{x}-\mathbf{x}'\Vert^3}\,dV'$
Gravity perturbation	$\mathbf{g}^{\mathrm{E1}} = -\boldsymbol{\nabla}\phi^{\mathrm{E1}} = G\int_\oplus \rho^{0\prime}(\mathbf{s}'\cdot\boldsymbol{\Pi})\,dV'$
Hooke's law	$\mathbf{T}^{\mathrm{PK1}} = \boldsymbol{\Lambda}:\boldsymbol{\nabla}\mathbf{s}$ $\mathbf{T}^{\mathrm{L1}} = \boldsymbol{\Upsilon}:\boldsymbol{\nabla}\mathbf{s}$
Elastic tensor relations	$\Lambda_{ijkl} = \Gamma_{ijkl} + \tfrac{1}{2}(T^0_{ij}\delta_{kl} + T^0_{kl}\delta_{ij} + T^0_{ik}\delta_{jl} - T^0_{jk}\delta_{il} - T^0_{il}\delta_{jk} - T^0_{jl}\delta_{ik})$ $\Upsilon_{ijkl} = \Gamma_{ijkl} + \tfrac{1}{2}(-T^0_{ij}\delta_{kl} + T^0_{kl}\delta_{ij} + T^0_{ik}\delta_{jl} + T^0_{jk}\delta_{il} - T^0_{il}\delta_{jk} - T^0_{jl}\delta_{ik})$
Elastic symmetries	$\Gamma_{ijkl} = \Gamma_{jikl} = \Gamma_{ijlk} = \Gamma_{klij}$ $\Lambda_{ijkl} = \Lambda_{klij}$ $\Upsilon_{ijkl} = \Upsilon_{jikl}$

Table 3.2. Summary of the most important linearized volumetric elastic-gravitational relations governing a general non-hydrostatic Earth.

We conclude this section by summarizing the most important elastic-gravitational equations and constitutive relations in Table 3.2. Both forms of the momentum equation have a mixed Lagrangian-Eulerian character because of the presence of the Lagrangian incremental stresses $\mathbf{T}^{\mathrm{PK1}}$ and $\mathbf{T}^{\mathrm{L1}}$ and the Eulerian density and potential perturbations ρ^{E1} and ϕ^{E1}. The equations involving the first Piola-Kirchhoff stress $\mathbf{T}^{\mathrm{PK1}}$ are more convenient than those involving the Cauchy stress $\mathbf{T}^{\mathrm{L1}}$ for most purposes in a general non-hydrostatic Earth; for this reason, the dynamic boundary conditions summarized in Table 3.1 are expressed in terms of $\mathbf{T}^{\mathrm{PK1}}$ as well. It is not possible to write the equations and boundary conditions governing the free oscillations of a non-hydrostatic Earth in a form that is independent of the initial static stress $\mathbf{T}^0$.

3.7 Hamilton's Principle

The linearized equations and boundary conditions governing the elastic-gravitational deformations of the Earth can be derived from a variational principle. There are two different but equivalent versions of this principle, distinguished by whether or not the Eulerian gravitational potential perturbation ϕ^{E1} and the displacement $\mathbf{s}$ are varied independently. We refer to these as the *displacement* and *displacement-potential* variational principles, respectively. In the present section, we shall simply state the two principles and verify that they yield the governing equations and boundary conditions. Later, in Section 3.10, we shall derive the Lagrangian densities introduced below by means of a first-principles analysis of the energy budget of a deforming Earth.

3.7.1 Displacement variational principle

Let t_1 and t_2 be the starting and ending times of a possible displacement path $\mathbf{s}(\mathbf{x}, t)$. Define the *action* $\mathcal{I}$ for this path by

$$\mathcal{I} = \int_{t_1}^{t_2} \int_{\oplus} L(\mathbf{s}, \partial_t \mathbf{s}, \boldsymbol{\nabla}\mathbf{s})\, dV dt + \int_{t_1}^{t_2} \int_{\Sigma_{\mathrm{FS}}} [L^{\Sigma}(\mathbf{s}, \boldsymbol{\nabla}^{\Sigma}\mathbf{s})]_{-}^{+}\, d\Sigma\, dt, \tag{3.163}$$

where

$$L = \tfrac{1}{2}[\rho^0 \partial_t \mathbf{s} \cdot \partial_t \mathbf{s} - 2\rho^0 \mathbf{s} \cdot \boldsymbol{\Omega} \times \partial_t \mathbf{s} - \boldsymbol{\nabla}\mathbf{s} : \boldsymbol{\Lambda} : \boldsymbol{\nabla}\mathbf{s} - \rho^0 \mathbf{s} \cdot \boldsymbol{\nabla}\phi^{\mathrm{E1}} - \rho^0 \mathbf{s} \cdot \boldsymbol{\nabla}\boldsymbol{\nabla}(\phi^0 + \psi) \cdot \mathbf{s}], \tag{3.164}$$

$$L^{\Sigma}=\tfrac{1}{2}[(\hat{\mathbf{n}}\cdot\mathbf{s})\boldsymbol{\nabla}^{\Sigma}\cdot(\varpi^0\mathbf{s})-\varpi^0\mathbf{s}\cdot(\boldsymbol{\nabla}^{\Sigma}\mathbf{s})\cdot\hat{\mathbf{n}}]. \tag{3.165}$$

The potential perturbation ϕ^{E1} in (3.164) is regarded as a known functional of the displacement $\mathbf{s}$, given explicitly for any $\mathbf{s}$ by equation (3.99). The *volumetric Lagrangian density* L is thus a functional only of $\mathbf{s}$, $\partial_t\mathbf{s}$ and $\boldsymbol{\nabla}\mathbf{s}$, as indicated. The integral in (3.163) involving ϕ^{E1} is given in terms of $\mathbf{s}$ by

$$\int_{\oplus}\rho^0\mathbf{s}\cdot\boldsymbol{\nabla}\phi^{\mathrm{E1}}\,dV=-\int_{\oplus}\int_{\oplus}(\rho^0\mathbf{s}\cdot\boldsymbol{\Pi}\cdot\rho^{0\prime}\mathbf{s}')\,dV\,dV'. \tag{3.166}$$

It is noteworthy that (3.166) is a symmetric quadratic functional of $\mathbf{s}$, by virtue of the symmetry $\boldsymbol{\Pi}(\mathbf{x},\mathbf{x}')=\boldsymbol{\Pi}^{\mathrm{T}}(\mathbf{x}',\mathbf{x})$ of the kernel defined in equation (3.101). The additional *surface Lagrangian density* L^{Σ} is needed in (3.163) to accommodate the extra degrees of freedom associated with the possibility of tangential slip upon the fluid-solid boundaries Σ_{FS}.

Hamilton's principle asserts that the action $\mathcal{I}$ is stationary, correct to first order in an infinitesimal admissible variation of the path $\delta\mathbf{s}$, if and only if $\mathbf{s}$ satisfies the linearized equations of motion and boundary conditions derived above. An *admissible variation* $\delta\mathbf{s}$ is one that vanishes at the starting and ending times t_1 and t_2, and that satisfies $[\delta\mathbf{s}]_-^+=\mathbf{0}$ on Σ_{SS} and $[\hat{\mathbf{n}}\cdot\delta\mathbf{s}]_-^+=0$ on Σ_{FS}. The variation in the action $\delta\mathcal{I}$ due to a variation in the path $\delta\mathbf{s}$ is

$$\begin{aligned}\delta\mathcal{I}=&\int_{t_1}^{t_2}\int_{\oplus}[\delta\mathbf{s}\cdot(\partial_{\mathbf{s}}L)+\partial_t(\delta\mathbf{s})\cdot(\partial_{\partial_t\mathbf{s}}L)+\boldsymbol{\nabla}(\delta\mathbf{s})\!:\!(\partial_{\boldsymbol{\nabla}\mathbf{s}}L)]\,dV dt\\&+\int_{t_1}^{t_2}\int_{\Sigma_{\mathrm{FS}}}[\delta\mathbf{s}\cdot(\partial_{\mathbf{s}}L^{\Sigma})+\boldsymbol{\nabla}^{\Sigma}(\delta\mathbf{s})\!:\!(\partial_{\boldsymbol{\nabla}^{\Sigma}\mathbf{s}}L^{\Sigma})]_-^+\,d\Sigma\,dt,\end{aligned} \tag{3.167}$$

correct to first order in $\|\mathbf{s}\|$. Integrating by parts with respect to time t, and using both the three-dimensional and two-dimensional versions of Gauss' theorem in $\oplus$ and on Σ_{FS}, we can rewrite this in the form

$$\begin{aligned}\delta\mathcal{I}=&\int_{t_1}^{t_2}\int_{\oplus}\delta\mathbf{s}\cdot\Big[\partial_{\mathbf{s}}L-\partial_t(\partial_{\partial_t\mathbf{s}}L)-\boldsymbol{\nabla}\cdot(\partial_{\boldsymbol{\nabla}\mathbf{s}}L)\Big]\,dV dt\\&+\int_{t_1}^{t_2}\int_{\partial\oplus}\delta\mathbf{s}\cdot\Big[\hat{\mathbf{n}}\cdot(\partial_{\boldsymbol{\nabla}\mathbf{s}}L)\Big]\,d\Sigma\,dt\\&-\int_{t_1}^{t_2}\int_{\Sigma_{\mathrm{SS}}}\delta\mathbf{s}\cdot\Big[\hat{\mathbf{n}}\cdot(\partial_{\boldsymbol{\nabla}\mathbf{s}}L)\Big]_-^+\,d\Sigma\,dt\\&+\int_{t_1}^{t_2}\int_{\Sigma_{\mathrm{FS}}}\Big[\delta\mathbf{s}\cdot\Big\{\partial_{\mathbf{s}}L^{\Sigma}-\boldsymbol{\nabla}^{\Sigma}\cdot(\partial_{\boldsymbol{\nabla}^{\Sigma}\mathbf{s}}L^{\Sigma})-\hat{\mathbf{n}}\cdot(\partial_{\boldsymbol{\nabla}\mathbf{s}}L)\Big\}\Big]_-^+\,d\Sigma\,dt.\end{aligned} \tag{3.168}$$

In writing equation (3.168) we have used the fact that $\partial_{\boldsymbol{\nabla}^{\Sigma}\mathbf{s}}L^{\Sigma}$ is a tangent vector: $\hat{\mathbf{n}}\cdot(\partial_{\boldsymbol{\nabla}^{\Sigma}\mathbf{s}}L^{\Sigma})=0$ on Σ_{FS}. Hamilton's principle stipulates that the variation in the action must vanish, $\delta\mathcal{I} = 0$, for an arbitrary admissible variation in the displacement $\boldsymbol{\delta}\mathbf{s}$; this will be true if and only if the following *Euler-Lagrange equation* and associated boundary conditions are satisfied:

$$\partial_{\mathbf{s}} L - \partial_t(\partial_{\partial_t \mathbf{s}} L) - \boldsymbol{\nabla}\cdot(\partial_{\boldsymbol{\nabla}\mathbf{s}} L) = \mathbf{0} \quad \text{in } \oplus, \tag{3.169}$$

$$\hat{\mathbf{n}}\cdot(\partial_{\boldsymbol{\nabla}\mathbf{s}} L) = \mathbf{0} \quad \text{on } \partial\oplus, \tag{3.170}$$

$$[\hat{\mathbf{n}}\cdot(\partial_{\boldsymbol{\nabla}\mathbf{s}} L)]_-^+ = \mathbf{0} \quad \text{on } \Sigma_{\mathrm{SS}}, \tag{3.171}$$

$$\begin{aligned}&[\partial_{\mathbf{s}} L^\Sigma - \boldsymbol{\nabla}^\Sigma\cdot(\partial_{\boldsymbol{\nabla}^\Sigma\mathbf{s}} L^\Sigma) - \hat{\mathbf{n}}\cdot(\partial_{\boldsymbol{\nabla}\mathbf{s}} L)]_-^+ \\ &\quad = \hat{\mathbf{n}}\hat{\mathbf{n}}\cdot[\partial_{\mathbf{s}} L^\Sigma - \boldsymbol{\nabla}^\Sigma\cdot(\partial_{\boldsymbol{\nabla}^\Sigma\mathbf{s}} L^\Sigma) - \hat{\mathbf{n}}\cdot(\partial_{\boldsymbol{\nabla}\mathbf{s}} L)]_-^+ = \mathbf{0} \quad \text{on } \Sigma_{\mathrm{FS}}.\end{aligned} \tag{3.172}$$

The Maxwell relation $\Lambda_{ijkl} = \Lambda_{klij}$ and the symmetry of the curvature tensor, $(\boldsymbol{\nabla}^\Sigma\hat{\mathbf{n}})^{\mathrm{T}} = \boldsymbol{\nabla}^\Sigma\hat{\mathbf{n}}$, suffice to show that

$$\partial_{\boldsymbol{\nabla}\mathbf{s}} L = -\mathbf{T}^{\mathrm{PK1}}, \tag{3.173}$$

$$\partial_{\mathbf{s}} L^\Sigma - \boldsymbol{\nabla}^\Sigma\cdot(\partial_{\boldsymbol{\nabla}^\Sigma\mathbf{s}} L^\Sigma) - \hat{\mathbf{n}}\cdot(\partial_{\boldsymbol{\nabla}\mathbf{s}} L) = \mathbf{t}^{\mathrm{PK1}}. \tag{3.174}$$

Using these results and the quadratic symmetry of the term (3.166), we see that the variational equations (3.169)–(3.172) are equivalent to:

$$\begin{aligned}&\rho^0(\partial_t^2\mathbf{s} + 2\boldsymbol{\Omega}\times\partial_t\mathbf{s}) = \boldsymbol{\nabla}\cdot\mathbf{T}^{\mathrm{PK1}} \\ &\qquad - \rho^0\boldsymbol{\nabla}\phi^{\mathrm{E1}} - \rho^0\mathbf{s}\cdot\boldsymbol{\nabla}\boldsymbol{\nabla}(\phi^0 + \psi) \quad \text{in } \oplus,\end{aligned} \tag{3.175}$$

$$\hat{\mathbf{n}}\cdot\mathbf{T}^{\mathrm{PK1}} = \mathbf{0} \quad \text{on } \partial\oplus, \tag{3.176}$$

$$[\hat{\mathbf{n}}\cdot\mathbf{T}^{\mathrm{PK1}}]_-^+ = \mathbf{0} \quad \text{on } \Sigma_{\mathrm{SS}}, \tag{3.177}$$

$$[\mathbf{t}^{\mathrm{PK1}}]_-^+ = \hat{\mathbf{n}}[\hat{\mathbf{n}}\cdot\mathbf{t}^{\mathrm{PK1}}]_-^+ = \mathbf{0} \quad \text{on } \Sigma_{\mathrm{FS}}. \tag{3.178}$$

Equations (3.175)–(3.178) are precisely the linearized momentum equation (3.60) and the associated dynamical boundary conditions (3.65)–(3.66) and (3.81)–(3.82). The momentum equation is in this case regarded as an *integro-differential equation* for the displacement $\mathbf{s}$ of the form:

$$\begin{aligned}&\rho^0(\partial_t^2\mathbf{s} + 2\boldsymbol{\Omega}\times\partial_t\mathbf{s}) = \boldsymbol{\nabla}\cdot(\boldsymbol{\Lambda}:\boldsymbol{\nabla}\mathbf{s}) \\ &\qquad - \rho^0\mathbf{s}\cdot\boldsymbol{\nabla}\boldsymbol{\nabla}(\phi^0 + \psi) + \rho^0 G\int_\oplus \rho^{0\prime}(\mathbf{s}'\cdot\boldsymbol{\Pi})\, dV'.\end{aligned} \tag{3.179}$$

This integro-differential character of the equation of motion is a consequence of the long-range nature of the gravitational restoring force.

A variation of the form $\boldsymbol{\delta}\mathbf{s} = \varepsilon\mathbf{s}$, where ε is an infinitesimal constant, is certainly admissible, and since the action $\mathcal{I}$ is a quadratic functional of $\mathbf{s}$, its variation in this case is simply $\delta\mathcal{I} = 2\varepsilon\mathcal{I}$. It follows that the value of the action *along the stationary path* is identically zero:

$$\mathcal{I} = 0. \tag{3.180}$$

3.7.2 Displacement-potential variational principle

Alternatively, we may regard the perturbation ϕ^{E1} as the solution to the boundary-value problem:

$$\nabla^2\phi^{\mathrm{E1}} = \begin{cases} -4\pi G\,\boldsymbol{\nabla}\cdot(\rho^0\mathbf{s}) & \text{in } \oplus \\ 0 & \text{in } \bigcirc - \oplus, \end{cases} \tag{3.181}$$

$$[\phi^{\mathrm{E1}}]_-^+ = 0 \quad \text{and} \quad [\hat{\mathbf{n}}\cdot\boldsymbol{\nabla}\phi^{\mathrm{E1}} + 4\pi G\rho^0\hat{\mathbf{n}}\cdot\mathbf{s}]_-^+ = 0 \quad \text{on } \Sigma. \tag{3.182}$$

It is convenient to define an auxiliary vector

$$\boldsymbol{\xi}^{\mathrm{E1}} = (4\pi G)^{-1}\boldsymbol{\nabla}\phi^{\mathrm{E1}} + \rho^0\mathbf{s}, \tag{3.183}$$

where the displacement $\mathbf{s}$ is assumed to vanish outside the Earth, in $\bigcirc - \oplus$. Equations (3.181)–(3.182) can then be condensed into the two equations

$$\boldsymbol{\nabla}\cdot\boldsymbol{\xi}^{\mathrm{E1}} = 0 \quad \text{in } \bigcirc, \tag{3.184}$$

$$[\hat{\mathbf{n}}\cdot\boldsymbol{\xi}^{\mathrm{E1}}]_-^+ = 0 \quad \text{on } \Sigma, \tag{3.185}$$

where we have taken the continuity of ϕ^{E1} for granted. We seek a new variational principle, in which $\mathbf{s}$ and ϕ^{E1} are allowed to vary independently, subject to the constraint that they are related by equations (3.184)–(3.185). An *admissible* potential variation $\delta\phi^{\mathrm{E1}}$ is considered to be one that vanishes at the starting and ending times t_2 and t_2, and that satisfies $[\delta\phi^{\mathrm{E1}}]_-^+ = 0$ on the boundary Σ. To formulate this constrained variational principle, we consider the *modified action*

$$\begin{aligned} \mathcal{I}' = \mathcal{I} &+ \int_{t_1}^{t_2}\int_{\bigcirc} \lambda\,[\nabla^2\phi^{\mathrm{E1}} + 4\pi G\,\boldsymbol{\nabla}\cdot(\rho^0\mathbf{s})]\,dV dt \\ &+ \int_{t_1}^{t_2}\int_{\Sigma} \lambda\,[\hat{\mathbf{n}}\cdot\boldsymbol{\nabla}\phi^{\mathrm{E1}} + 4\pi G\rho^0\hat{\mathbf{n}}\cdot\mathbf{s}]_-^+\,d\Sigma\,dt, \end{aligned} \tag{3.186}$$

where λ is an undetermined *Lagrange multiplier.* The variation of $\mathcal{I}'$ with respect to ϕ^{E1}, holding $\mathbf{s}$ and λ fixed, is

$$\begin{aligned} \delta\mathcal{I}' = &\int_{t_1}^{t_2}\int_{\bigcirc} \delta\phi^{\mathrm{E1}}[\nabla^2\lambda + \tfrac{1}{2}\boldsymbol{\nabla}\cdot(\rho^0\mathbf{s})]\,dV dt \\ &+ \int_{t_1}^{t_2}\int_{\Sigma} \delta\phi^{\mathrm{E1}}[\hat{\mathbf{n}}\cdot\boldsymbol{\nabla}\lambda + \tfrac{1}{2}\rho^0\hat{\mathbf{n}}\cdot\mathbf{s}]_-^+\,d\Sigma\,dt, \end{aligned} \tag{3.187}$$

where we have applied Gauss' theorem twice in succession. A suitable choice for λ, which insures that the variation (3.187) vanishes, is

$$\lambda = (8\pi G)^{-1}\phi^{\mathrm{E1}}. \tag{3.188}$$

Substituting the result (3.188) back into (3.186), we can rewrite the modified action $\mathcal{I}'$ after another application of Gauss' theorem in the form

$$\mathcal{I}' = \int_{t_1}^{t_2} \int_{\bigcirc} L'(\mathbf{s}, \partial_t \mathbf{s}, \boldsymbol{\nabla}\mathbf{s}, \boldsymbol{\nabla}\phi^{\mathrm{E1}})\, dV dt + \int_{t_1}^{t_2} \int_{\Sigma_{\mathrm{FS}}} [L^{\Sigma}(\mathbf{s}, \boldsymbol{\nabla}^{\Sigma}\mathbf{s})]_-^+ \, d\Sigma\, dt, \tag{3.189}$$

where L^{Σ} is unchanged and the *modified Lagrangian density* L' is given by

$$\begin{aligned} L' = \tfrac{1}{2}[\rho^0 \partial_t \mathbf{s} \cdot \partial_t \mathbf{s} - 2\rho^0 \mathbf{s} \cdot \boldsymbol{\Omega} \times \partial_t \mathbf{s} - \boldsymbol{\nabla}\mathbf{s} \mathbin{:} \boldsymbol{\Lambda} \mathbin{:} \boldsymbol{\nabla}\mathbf{s} \\ - 2\rho^0 \mathbf{s} \cdot \boldsymbol{\nabla}\phi^{\mathrm{E1}} - \rho^0 \mathbf{s} \cdot \boldsymbol{\nabla}\boldsymbol{\nabla}(\phi^0 + \psi) \cdot \mathbf{s} \\ - (4\pi G)^{-1} \boldsymbol{\nabla}\phi^{\mathrm{E1}} \cdot \boldsymbol{\nabla}\phi^{\mathrm{E1}}]. \end{aligned} \tag{3.190}$$

The volume integral in equation (3.189) is over all of space $\bigcirc$; however, only the last term involving $\boldsymbol{\nabla}\phi^{\mathrm{E1}} \cdot \boldsymbol{\nabla}\phi^{\mathrm{E1}}$ is non-zero outside of the Earth, in $\bigcirc - \oplus$.

Upon taking the variation of equation (3.189), treating the displacement $\mathbf{s}$ and the potential ϕ^{E1} as independent, we obtain

$$\delta\mathcal{I}' = \delta\mathcal{I} - \int_{t_1}^{t_2} \int_{\bigcirc} \delta\phi^{\mathrm{E1}} [\boldsymbol{\nabla} \cdot (\partial_{\boldsymbol{\nabla}\phi^{\mathrm{E1}}} L')]\, dV - \int_{t_1}^{t_2} \int_{\Sigma} \delta\phi^{\mathrm{E1}} [\hat{\mathbf{n}} \cdot (\partial_{\boldsymbol{\nabla}\phi^{\mathrm{E1}}} L')]_-^+ \, d\Sigma, \tag{3.191}$$

where $\delta\mathcal{I}$ is given in equation (3.168). The term $\delta\mathcal{I}$ arising from the displacement variation $\delta\mathbf{s}$ gives rise to the same linearized equation of motion and dynamical boundary conditions (3.175)–(3.178) as before. The only difference is that the term $\rho^0 \mathbf{s} \cdot \boldsymbol{\nabla}\phi^{\mathrm{E1}}$ is no longer regarded as a quadratic functional of $\mathbf{s}$ given by (3.166); the variation of this term is, however, the same, because of the explicit factor of two in the modified Lagrangian density (3.190). The terms arising from the potential variation $\delta\phi^{\mathrm{E1}}$ vanish if and only if

$$\boldsymbol{\nabla} \cdot (\partial_{\boldsymbol{\nabla}\phi^{\mathrm{E1}}} L') = 0 \quad \text{in } \bigcirc, \tag{3.192}$$

$$[\hat{\mathbf{n}} \cdot (\partial_{\boldsymbol{\nabla}\phi^{\mathrm{E1}}} L')]_-^+ = 0 \quad \text{on } \Sigma. \tag{3.193}$$

It is easily verified that $\partial_{\boldsymbol{\nabla}\phi^{\mathrm{E1}}} L' = -\boldsymbol{\xi}^{\mathrm{E1}}$. Hence, equations (3.192)–(3.193) are equivalent to the perturbed boundary-value problem (3.184)–(3.185).

In conclusion, the complete set of linearized elastic-gravitational equations and boundary conditions can be obtained from the modified Hamilton's principle $\delta\mathcal{I}' = 0$, for arbitrary and independent admissible variations

$\delta\mathbf{s}$ and $\delta\phi^{\mathrm{E1}}$. In this case, the equations are regarded as a coupled system of differential equations for s and ϕ^{E1}, rather than as a single integro-differential equation for s alone, as before. The variation in the modified action $\mathcal{I}'$ for admissible variations of the form $\delta\mathbf{s} = \varepsilon\mathbf{s}$, $\delta\phi^{\mathrm{E1}} = \varepsilon\phi^{\mathrm{E1}}$, where ε is a constant, is $\delta\mathcal{I}' = 2\varepsilon\mathcal{I}'$, so that the stationary value of $\mathcal{I}'$ is also zero:

$$\mathcal{I}' = 0. \tag{3.194}$$

The equality of $\mathcal{I}$ and $\mathcal{I}'$ along the stationary path is guaranteed by the identity

$$\tfrac{1}{2}\int_{\oplus}\rho^0(\mathbf{s}\cdot\boldsymbol{\nabla}\phi^{\mathrm{E1}\prime} + \mathbf{s}'\cdot\boldsymbol{\nabla}\phi^{\mathrm{E1}})\,dV + \frac{1}{4\pi G}\int_{\bigcirc}\boldsymbol{\nabla}\phi^{\mathrm{E1}}\cdot\boldsymbol{\nabla}\phi^{\mathrm{E1}\prime}\,dV = 0, \tag{3.195}$$

which is an elementary consequence of Gauss' theorem for any pair of functions s, ϕ^{E1} and s$'$, $\phi^{\mathrm{E1}\prime}$ satisfying (3.181)–(3.182). The primed and unprimed quantities are identical in the case of interest here; however, we shall require the more general result (3.195) on several occasions in the future.

3.8 Conservation of Energy

The volumetric and surficial *energy densities* E and E^{Σ} are defined in terms of the Lagrangian densities L and L^{Σ} by

$$E = (\partial_{\partial_t\mathbf{s}}L)\cdot\partial_t\mathbf{s} - L, \qquad E^{\Sigma} = -L^{\Sigma}. \tag{3.196}$$

Using equations (3.164) and (3.165) we obtain

$$E = \tfrac{1}{2}[\rho^0\partial_t\mathbf{s}\cdot\partial_t\mathbf{s} + \boldsymbol{\nabla}\mathbf{s}:\boldsymbol{\Lambda}:\boldsymbol{\nabla}\mathbf{s} + \rho^0\mathbf{s}\cdot\boldsymbol{\nabla}\phi^{\mathrm{E1}} + \rho^0\mathbf{s}\cdot\boldsymbol{\nabla}\boldsymbol{\nabla}(\phi^0+\psi)\cdot\mathbf{s}], \tag{3.197}$$

$$E^{\Sigma} = \tfrac{1}{2}[\varpi^0\mathbf{s}\cdot(\boldsymbol{\nabla}^{\Sigma}\mathbf{s})\cdot\hat{\mathbf{n}} - (\hat{\mathbf{n}}\cdot\mathbf{s})\boldsymbol{\nabla}^{\Sigma}\cdot(\varpi^0\mathbf{s})]. \tag{3.198}$$

Upon dotting the linearized momentum equation (3.175) with the velocity $\partial_t\mathbf{s}$, and using the Maxwell relation $\Lambda_{ijkl} = \Lambda_{klij}$, we find that

$$\partial_t E + \boldsymbol{\nabla}\cdot\mathbf{K} = 0, \tag{3.199}$$

where

$$\mathbf{K} = -\mathbf{T}^{\mathrm{PK1}}\cdot\partial_t\mathbf{s} + \tfrac{1}{2}\phi^{\mathrm{E1}}(\partial_t\boldsymbol{\xi}^{\mathrm{E1}}) - \tfrac{1}{2}(\partial_t\phi^{\mathrm{E1}})\boldsymbol{\xi}^{\mathrm{E1}}. \tag{3.200}$$

Equation (3.199) can be interpreted as a *local* conservation law for energy in a general Earth model; the quantity $\mathbf{K}$ is the *energy flux* in the rotating frame. The absence of any Coriolis term $2\rho^0\mathbf{\Omega} \times \partial_t\mathbf{s}$ in the energy density (3.197) is a reflection of the fact that the Coriolis force is always and everywhere orthogonal to the velocity $\partial_t\mathbf{s}$, and thus it can do no work.

Upon integrating equation (3.199) over the Earth model $\oplus$ and applying Gauss' theorem, we obtain

$$\frac{d}{dt}\int_{\oplus} E\,dV + \int_{\Sigma_{\rm FS}} [\hat{\mathbf{n}}\cdot\mathbf{T}^{\rm PK1}\cdot\partial_t\mathbf{s}]_-^+\,d\Sigma. \tag{3.201}$$

The gravitational terms in the normal flux jump $[\hat{\mathbf{n}}\cdot\mathbf{K}]_-^+$ vanish everywhere on Σ by virtue of the continuity conditions (3.182); in addition, the quantity $[\hat{\mathbf{n}}\cdot\mathbf{T}^{\rm PK1}\cdot\partial_t\mathbf{s}]_-^+$ vanishes except on the fluid-solid boundaries $\Sigma_{\rm FS}$, as a consequence of (3.176) and (3.177). Using equations (3.80)–(3.82), we find that

$$\begin{aligned}&\int_{\Sigma_{\rm FS}} [\hat{\mathbf{n}}\cdot\mathbf{T}^{\rm PK1}\cdot\mathbf{v}]_-^+\,d\Sigma\\ &\qquad = \int_{\Sigma_{\rm FS}} [\varpi^0\mathbf{v}\cdot(\boldsymbol{\nabla}^{\Sigma}\mathbf{s})\cdot\hat{\mathbf{n}} - (\hat{\mathbf{n}}\cdot\mathbf{v})\boldsymbol{\nabla}^{\Sigma}\cdot(\varpi^0\mathbf{s})]_-^+\,d\Sigma,\end{aligned} \tag{3.202}$$

for an arbitrary vector field $\mathbf{v}$ satisfying $[\hat{\mathbf{n}}\cdot\mathbf{v}]_-^+ = 0$ on $\Sigma_{\rm FS}$. Applying Gauss' theorem to both terms on the right of (3.202), and using $[\hat{\mathbf{n}}\cdot\mathbf{s}]_-^+ = 0$ and the symmetry $(\boldsymbol{\nabla}^{\Sigma}\hat{\mathbf{n}})^{\rm T} = \boldsymbol{\nabla}^{\Sigma}\hat{\mathbf{n}}$, we obtain an analogous expression with $\mathbf{s}$ and $\mathbf{v}$ interchanged:

$$\begin{aligned}&\int_{\Sigma_{\rm FS}} [\hat{\mathbf{n}}\cdot\mathbf{T}^{\rm PK1}\cdot\mathbf{v}]_-^+\,d\Sigma\\ &\qquad = \int_{\Sigma_{\rm FS}} [\varpi^0\mathbf{s}\cdot(\boldsymbol{\nabla}^{\Sigma}\mathbf{v})\cdot\hat{\mathbf{n}} - (\hat{\mathbf{n}}\cdot\mathbf{s})\boldsymbol{\nabla}^{\Sigma}\cdot(\varpi^0\mathbf{v})]_-^+\,d\Sigma.\end{aligned} \tag{3.203}$$

Upon taking $\mathbf{v} = \partial_t\mathbf{s}$, averaging the two results (3.202) and (3.203), and integrating with respect to time t, we find that

$$\int_{\Sigma_{\rm FS}} [\hat{\mathbf{n}}\cdot\mathbf{T}^{\rm PK1}\cdot\partial_t\mathbf{s}]_-^+\,d\Sigma = \frac{d}{dt}\int_{\Sigma_{\rm FS}} [E^{\Sigma}]_-^+\,d\Sigma, \tag{3.204}$$

where E^{Σ} is given by (3.198). Inserting equation (3.204) into (3.201), we obtain the simple result

$$\frac{d\mathcal{E}}{dt} = 0, \tag{3.205}$$

where

$$\mathcal{E} = \int_{\oplus} E\,dV + \int_{\Sigma_{\rm FS}} [E^{\Sigma}]_-^+\,d\Sigma. \tag{3.206}$$

The relation (3.205) expresses the *global* conservation of energy; the quantity $\mathcal{E}$ can be interpreted as the *total energy*—kinetic plus elastic plus gravitational—of an actively deforming Earth model.

Alternatively, we can define an energy density E' in terms of the modified Lagrangian density L':

$$E' = (\partial_{\partial_t \mathbf{s}} L') \cdot \partial_t \mathbf{s} - L'. \tag{3.207}$$

From (3.190) we find that

$$E' = \tfrac{1}{2}[\rho^0 \partial_t \mathbf{s} \cdot \partial_t \mathbf{s} + \boldsymbol{\nabla}\mathbf{s} : \boldsymbol{\Lambda} : \boldsymbol{\nabla}\mathbf{s} + 2\rho^0 \mathbf{s} \cdot \boldsymbol{\nabla}\phi^{\mathrm{E1}} \\ + \rho^0 \mathbf{s} \cdot \boldsymbol{\nabla}\boldsymbol{\nabla}(\phi^0 + \psi) \cdot \mathbf{s} + (4\pi G)^{-1} \boldsymbol{\nabla}\phi^{\mathrm{E1}} \cdot \boldsymbol{\nabla}\phi^{\mathrm{E1}}]. \tag{3.208}$$

This modified energy density satisfies the conservation law

$$\partial_t E' + \boldsymbol{\nabla} \cdot \mathbf{K}' = 0, \tag{3.209}$$

where

$$\mathbf{K}' = -\mathbf{T}^{\mathrm{PK1}} \cdot \partial_t \mathbf{s} - (\partial_t \phi^{\mathrm{E1}}) \boldsymbol{\xi}^{\mathrm{E1}}. \tag{3.210}$$

Either E or E' can be interpreted as the local energy density; the associated fluxes are then $\mathbf{K}$ and $\mathbf{K}'$, respectively. Upon integrating equation (3.209) over all of space $\bigcirc$, we find, by an argument analogous to that given above, that

$$\frac{d\mathcal{E}}{dt} = 0, \tag{3.211}$$

where

$$\mathcal{E} = \int_{\bigcirc} E' \, dV + \int_{\Sigma_{\mathrm{FS}}} [E^{\Sigma}]_-^+ \, d\Sigma. \tag{3.212}$$

The equivalence of the two expressions (3.206) and (3.212) for the total energy $\mathcal{E}$ is guaranteed by the identity (3.195).

★3.9 Energy Budget

The quantity $\mathcal{E}$ is conserved, and it is composed of integrands such as $\frac{1}{2}\rho^0(\partial_t \mathbf{s} \cdot \partial_t \mathbf{s})$ which have an obvious energetic significance; nevertheless, we have been careful so far to assert only that $\mathcal{E}$ "can be interpreted as" the total energy. In fact, $\mathcal{E}$ is precisely the instantaneous total energy (kinetic plus elastic plus gravitational) of a freely deforming Earth model; to show this, we conduct an analysis of the global energy budget, and calculate the three contributions to the total energy from first principles. As a by-product of this analysis, we provide a physical interpretation of the surface integral over the fluid-solid boundaries Σ_{FS} in equations (3.206) and (3.212).

*3.9.1 Kinetic energy

The kinetic energy of the actively deforming Earth, minus that of the initial uniformly rotating Earth, is

$$\mathcal{E}_{\mathrm{k}} = \tfrac{1}{2}\int_{\oplus} \rho^0 \|\partial_t \mathbf{s} + \boldsymbol{\Omega} \times (\mathbf{x}+\mathbf{s})\|^2 \, dV - \tfrac{1}{2}\int_{\oplus} \rho^0 \|\boldsymbol{\Omega} \times \mathbf{x}\|^2 \, dV. \tag{3.213}$$

Upon making use of the identities

$$(\boldsymbol{\Omega} \times \mathbf{x}) \cdot (\boldsymbol{\Omega} \times \mathbf{s}) = -\mathbf{s} \cdot \boldsymbol{\nabla}\psi, \tag{3.214}$$

$$(\boldsymbol{\Omega} \times \mathbf{s}) \cdot (\boldsymbol{\Omega} \times \mathbf{s}) = -\mathbf{s} \cdot \boldsymbol{\nabla}\boldsymbol{\nabla}\psi \cdot \mathbf{s}, \tag{3.215}$$

we can reduce equation (3.213) to

$$\mathcal{E}_{\mathrm{k}} = \int_{\oplus} \rho^0 [\tfrac{1}{2}\partial_t \mathbf{s} \cdot \partial_t \mathbf{s} - (\mathbf{x}+\mathbf{s}) \cdot (\boldsymbol{\Omega} \times \partial_t \mathbf{s}) - \mathbf{s} \cdot \boldsymbol{\nabla}\psi - \tfrac{1}{2}\mathbf{s} \cdot \boldsymbol{\nabla}\boldsymbol{\nabla}\psi \cdot \mathbf{s}] \, dV. \tag{3.216}$$

In the absence of any external torques, the angular momentum of the Earth must remain constant; the condition that guarantees this is

$$\int_{\oplus} \rho^0 (\mathbf{x}+\mathbf{s}) \times [\partial_t \mathbf{s} + \boldsymbol{\Omega} \times (\mathbf{x}+\mathbf{s})] \, dV - \int_{\oplus} \rho^0 [\mathbf{x} \times (\boldsymbol{\Omega} \times \mathbf{x})] \, dV = \mathbf{0}. \tag{3.217}$$

Dotting equation (3.217) with the rotation vector $\boldsymbol{\Omega}$ and again using the identities (3.214)–(3.215) gives

$$\int_{\oplus} \rho^0 [(\mathbf{x}+\mathbf{s}) \cdot (\boldsymbol{\Omega} \times \partial_t \mathbf{s}) + 2\mathbf{s} \cdot \boldsymbol{\nabla}\psi + \mathbf{s} \cdot \boldsymbol{\nabla}\boldsymbol{\nabla}\psi \cdot \mathbf{s}] \, dV = 0. \tag{3.218}$$

Upon combining the results (3.216) and (3.218), we obtain

$$\mathcal{E}_{\mathrm{k}} = \int_{\oplus} \rho^0 [\tfrac{1}{2}\partial_t \mathbf{s} \cdot \partial_t \mathbf{s} + \mathbf{s} \cdot \boldsymbol{\nabla}\psi + \tfrac{1}{2}\mathbf{s} \cdot \boldsymbol{\nabla}\boldsymbol{\nabla}\psi \cdot \mathbf{s}] \, dV. \tag{3.219}$$

Equation (3.219) is the exact instantaneous kinetic energy of a freely deforming Earth model.

★3.9.2 Elastic energy

The total stored elastic energy in the deformed Earth, relative to that in the initial undeformed Earth, is

$$\mathcal{E}_{\mathrm{e}} = \int_{\oplus} \rho^0 (U^{\mathrm{L}} - U^0)\, dV = \int_{\oplus} [\mathbf{T}^0 : \boldsymbol{\varepsilon} + \tfrac{1}{2} \boldsymbol{\nabla}\mathbf{s} : \boldsymbol{\Lambda} : \boldsymbol{\nabla}\mathbf{s}]\, dV. \tag{3.220}$$

Measuring $\mathcal{E}_{\mathrm{e}}$ relative to the initial internal energy in the Earth is analogous to measuring $\mathcal{E}_{\mathrm{k}}$ relative to the kinetic energy of the uniformly rotating Earth. Physically, $\mathcal{E}_{\mathrm{e}}$ is the elastic energy associated with the deformation.

★3.9.3 Gravitational energy

The gravitational potential energy associated with a deformation $\mathbf{s}$ can be calculated in either of two ways. The simplest method is to note that the *rate of change* of gravitational energy must be equal to the rate of work done *against* gravitational body forces:

$$\frac{d\mathcal{E}_{\mathrm{g}}}{dt} = -\int_{\oplus} \rho^0 \partial_t \mathbf{s} \cdot (\mathbf{g}^0 + \mathbf{g}^{\mathrm{L}1})\, dV. \tag{3.221}$$

Substituting the representations (3.1) and (3.44), and using the quadratic symmetry of the term (3.166), we can write this in the form

$$\frac{d\mathcal{E}_{\mathrm{g}}}{dt} = \frac{d}{dt} \int_{\oplus} \rho^0 [\mathbf{s} \cdot \boldsymbol{\nabla}\phi^0 + \tfrac{1}{2}\mathbf{s} \cdot \boldsymbol{\nabla}\phi^{\mathrm{E}1} + \tfrac{1}{2}\mathbf{s} \cdot \boldsymbol{\nabla}\boldsymbol{\nabla}\phi^0 \cdot \mathbf{s}]\, dV. \tag{3.222}$$

Taking the reference level of gravitational potential energy to be that of the undeformed configuration, as we did in calculating $\mathcal{E}_{\mathrm{k}}$ and $\mathcal{E}_{\mathrm{e}}$, and integrating with respect to time t, we obtain the final result

$$\mathcal{E}_{\mathrm{g}} = \int_{\oplus} \rho^0 [\mathbf{s} \cdot \boldsymbol{\nabla}\phi^0 + \tfrac{1}{2}\mathbf{s} \cdot \boldsymbol{\nabla}\phi^{\mathrm{E}1} + \tfrac{1}{2}\mathbf{s} \cdot \boldsymbol{\nabla}\boldsymbol{\nabla}\phi^0 \cdot \mathbf{s}]\, dV. \tag{3.223}$$

Alternatively, we can make use of the general formula for the energy required to assemble a deformed mass distribution, minus that required to assemble the corresponding undeformed distribution, starting from matter dispersed at infinity. We rewrite this result, equation (2.129), in the form

$$\begin{aligned} \mathcal{E}_{\mathrm{g}} = &-\tfrac{1}{2}G \int_{\oplus}\int_{\oplus} \frac{\rho^0 \rho^{0\prime}}{\|\mathbf{x} + \mathbf{s} - \mathbf{x}' - \mathbf{s}'\|}\, dV\, dV' \\ &+ \tfrac{1}{2}G \int_{\oplus}\int_{\oplus} \frac{\rho^0 \rho^{0\prime}}{\|\mathbf{x} - \mathbf{x}'\|}\, dV\, dV', \end{aligned} \tag{3.224}$$

and expand the first integrand to second order in $\|\mathbf{s}\|$. This procedure gives

$$\begin{aligned}\mathcal{E}_{\mathrm{g}} = G\int_{\oplus}\int_{\oplus}\frac{\rho^0\rho^{0\prime}\mathbf{s}\cdot(\mathbf{x}-\mathbf{x}')}{\|\mathbf{x}-\mathbf{x}'\|^3}\,dV\,dV' \\ - \tfrac{1}{2}G\int_{\oplus}\int_{\oplus}\rho^0\rho^{0\prime}(\mathbf{s}\cdot\boldsymbol{\Pi}\cdot\mathbf{s}')\,dV\,dV' \\ + \tfrac{1}{2}G\int_{\oplus}\int_{\oplus}\rho^0\rho^{0\prime}(\mathbf{s}\cdot\boldsymbol{\Pi}\cdot\mathbf{s})\,dV\,dV',\end{aligned} \tag{3.225}$$

which is just the double integral form of (3.223).

★3.9.4 Total energy

The total energy is the sum of the kinetic, elastic and gravitational energies:

$$\mathcal{E} = \mathcal{E}_{\mathrm{k}} + \mathcal{E}_{\mathrm{e}} + \mathcal{E}_{\mathrm{g}}. \tag{3.226}$$

Each of $\mathcal{E}_{\mathrm{k}}$, $\mathcal{E}_{\mathrm{e}}$ and $\mathcal{E}_{\mathrm{g}}$ has a part that is first order in the infinitesimal displacement $\|\mathbf{s}\|$ and a part that is second order in $\|\mathbf{s}\|$; grouping these accordingly, we obtain

$$\begin{aligned}\mathcal{E} = \int_{\oplus}[\mathbf{T}^0\!:\!\boldsymbol{\varepsilon} + \rho^0\mathbf{s}\cdot\boldsymbol{\nabla}(\phi^0+\psi)]\,dV \\ + \tfrac{1}{2}\int_{\oplus}[\rho^0\partial_t\mathbf{s}\cdot\partial_t\mathbf{s} + \boldsymbol{\nabla}\mathbf{s}\!:\!\boldsymbol{\Lambda}\!:\!\boldsymbol{\nabla}\mathbf{s} \\ + \rho^0\mathbf{s}\cdot\boldsymbol{\nabla}\phi^{\mathrm{E}1} + \rho^0\mathbf{s}\cdot\boldsymbol{\nabla}\boldsymbol{\nabla}(\phi^0+\psi)\cdot\mathbf{s}]\,dV.\end{aligned} \tag{3.227}$$

An application of Gauss' theorem to the first-order term yields

$$\int_{\oplus}[\mathbf{T}^0\!:\!\boldsymbol{\varepsilon} + \rho^0\mathbf{s}\cdot\boldsymbol{\nabla}(\phi^0+\psi)]\,dV = -\int_{\Sigma_{\mathrm{FS}}}[\hat{\mathbf{n}}\cdot\mathbf{T}^0\cdot\mathbf{s}]_-^+\,d\Sigma, \tag{3.228}$$

where we have used the static equilibrium condition (3.8). The surface integral in equation (3.228) is zero on Σ_{SS} and $\partial\oplus$; however, the integral over the fluid-solid boundaries Σ_{FS} *does not vanish*, correct to second order in $\|\mathbf{s}\|$. In fact, invoking the second-order tangential slip condition (3.95), we find that

$$\begin{aligned}\int_{\Sigma_{\mathrm{FS}}}[\hat{\mathbf{n}}\cdot\mathbf{T}^0\cdot\mathbf{s}]_-^+\,d\Sigma &= -\int_{\Sigma_{\mathrm{FS}}}\varpi^0[\hat{\mathbf{n}}\cdot\mathbf{s}]_-^+\,d\Sigma \\ &= -\int_{\Sigma_{\mathrm{FS}}}\varpi^0[\mathbf{s}\cdot\boldsymbol{\nabla}^\Sigma(\hat{\mathbf{n}}\cdot\mathbf{s}) - \tfrac{1}{2}\mathbf{s}\cdot(\boldsymbol{\nabla}^\Sigma\hat{\mathbf{n}})\cdot\mathbf{s}]_-^+\,d\Sigma \\ &= -\tfrac{1}{2}\int_{\Sigma_{\mathrm{FS}}}[\varpi^0\mathbf{s}\cdot(\boldsymbol{\nabla}^\Sigma\mathbf{s})\cdot\hat{\mathbf{n}} - (\hat{\mathbf{n}}\cdot\mathbf{s})\boldsymbol{\nabla}^\Sigma\cdot(\varpi^0\mathbf{s})]_-^+\,d\Sigma,\end{aligned} \tag{3.229}$$

where we have used Gauss' theorem on $\Sigma_{\rm FS}$ to obtain the last equality. Upon combining equations (3.227)–(3.229), we find, finally, that

$$\mathcal{E} = \tfrac{1}{2}\int_{\oplus} [\rho^0 \partial_t \mathbf{s} \cdot \partial_t \mathbf{s} + \boldsymbol{\nabla}\mathbf{s} : \boldsymbol{\Lambda} : \boldsymbol{\nabla}\mathbf{s} + \rho^0 \mathbf{s} \cdot \boldsymbol{\nabla}\phi^{\rm E1} + \rho^0 \mathbf{s} \cdot \boldsymbol{\nabla}\boldsymbol{\nabla}(\phi^0 + \psi) \cdot \mathbf{s}]\, dV + \tfrac{1}{2}\int_{\Sigma_{\rm FS}} [\varpi^0 \mathbf{s} \cdot (\boldsymbol{\nabla}^{\Sigma}\mathbf{s}) \cdot \hat{\mathbf{n}} - (\hat{\mathbf{n}} \cdot \mathbf{s})\boldsymbol{\nabla}^{\Sigma} \cdot (\varpi^0 \mathbf{s})]_-^+\, d\Sigma. \tag{3.230}$$

Comparing with (3.197) and (3.198), we see that this is precisely

$$\mathcal{E} = \int_{\oplus} E\, dV + \int_{\Sigma_{\rm FS}} [E^{\Sigma}]_-^+\, d\Sigma. \tag{3.231}$$

This confirms that $\mathcal{E} = \mathcal{E}_{\rm k} + \mathcal{E}_{\rm e} + \mathcal{E}_{\rm g}$ is the instantaneous total energy of a freely deforming Earth model, and it provides the promised interpretation of the surface integral contribution:

$$\int_{\Sigma_{\rm FS}} [E^{\Sigma}]_-^+\, d\Sigma = -\int_{\Sigma_{\rm FS}} [\hat{\mathbf{n}} \cdot \mathbf{T}^0 \cdot \mathbf{s}]_-^+\, d\Sigma. \tag{3.232}$$

Evidently, this term accounts for the work performed against the initial traction $\hat{\mathbf{n}} \cdot \mathbf{T}^0$ on the fluid-solid boundaries; the first-order work associated with the slip on $\Sigma_{\rm FS}$ vanishes, but the second-order work does not.

★3.9.5 Relative kinetic and potential energy

It is convenient to regard $\mathcal{E}$ as the sum of two new functionals:

$$\mathcal{E} = \mathcal{T} + \mathcal{V}, \tag{3.233}$$

where

$$\mathcal{T} = \tfrac{1}{2}\int_{\oplus} \rho^0 (\partial_t \mathbf{s} \cdot \partial_t \mathbf{s})\, dV, \tag{3.234}$$

$$\mathcal{V} = \tfrac{1}{2}\int_{\oplus} [\boldsymbol{\nabla}\mathbf{s} : \boldsymbol{\Lambda} : \boldsymbol{\nabla}\mathbf{s} + \rho^0 \mathbf{s} \cdot \boldsymbol{\nabla}\phi^{\rm E1} + \rho^0 \mathbf{s} \cdot \boldsymbol{\nabla}\boldsymbol{\nabla}(\phi^0 + \psi) \cdot \mathbf{s}]\, dV + \tfrac{1}{2}\int_{\Sigma_{\rm FS}} [\varpi^0 \mathbf{s} \cdot (\boldsymbol{\nabla}^{\Sigma}\mathbf{s}) \cdot \hat{\mathbf{n}} - (\hat{\mathbf{n}} \cdot \mathbf{s})\boldsymbol{\nabla}^{\Sigma} \cdot (\varpi^0 \mathbf{s})]_-^+\, d\Sigma. \tag{3.235}$$

The quantity $\mathcal{T}$ is the *relative* kinetic energy of motion, as seen by an observer in the uniformly rotating frame; likewise, $\mathcal{V}$ is the *relative* elastic-gravitational potential energy.

As we have seen, the quantity $\mathcal{E}_{\mathrm{g}}$ can be interpreted as the work done against gravitational body forces. In the rotating frame we must also do work against the apparent centrifugal force; by analogy with (3.223) this work is

$$\mathcal{E}_{\psi} = \int_{\oplus} \rho^0 [\mathbf{s} \cdot \boldsymbol{\nabla}\psi + \tfrac{1}{2}\mathbf{s} \cdot \boldsymbol{\nabla}\boldsymbol{\nabla}\psi \cdot \mathbf{s}]\, dV. \tag{3.236}$$

The relative kinetic energy is the total kinetic energy *minus* the work done against the centrifugal force:

$$\mathcal{T} = \mathcal{E}_{\mathrm{k}} - \mathcal{E}_{\psi}, \tag{3.237}$$

whereas the relative potential energy is the total elastic-gravitational energy *plus* the work done against the centrifugal force:

$$\mathcal{V} = \mathcal{E}_{\mathrm{e}} + \mathcal{E}_{\mathrm{g}} + \mathcal{E}_{\psi}. \tag{3.238}$$

The total energy $\mathcal{E}$ can be considered to be either the sum of the total kinetic plus potential energy, or the sum of the relative kinetic and potential energy:

$$\mathcal{E} = \mathcal{E}_{\mathrm{k}} + \mathcal{E}_{\mathrm{e}} + \mathcal{E}_{\mathrm{g}} = \mathcal{T} + \mathcal{V}. \tag{3.239}$$

*3.9.6 Secular stability

In equilibrium, the Earth is at rest in a state of zero relative potential energy. To determine whether or not this equilibrium state is stable, we consider a slight displacement away from equilibrium, and ask whether the system naturally tends to return to the state of zero potential energy or move further away. In the case of a rotating Earth model, it is essential to distinguish between *ordinary stability*, or stability in the absence of any frictional dissipation, and *secular stability*, or stability in the presence of dissipation. Since the condition for secular stability is more restrictive than that for dynamical stability (see Section 4.2.3), and since all real materials exhibit anelastic dissipation, it is secular stability rather than ordinary stability that is limiting for rotating Earth models. We can determine the condition for secular stability of the Earth using a simple energy argument (Lyttleton 1953). We need not formulate a specific phenomenological description for the dissipation, as we shall do in Chapter 6; it is sufficient to note that its effect is to replace the energy conservation law (3.205) or (3.211) by

$$\frac{d\mathcal{E}}{dt} \leq 0. \tag{3.240}$$

The kinetic plus elastic-gravitational energy that is "lost" as a consequence of (3.240) goes into heating up the Earth.

Suppose that, at time $t = 0$, the Earth is at rest in a deformed state, with initial conditions $\mathbf{s}(\mathbf{x}, 0) = \mathbf{s}(\mathbf{x})$, $\partial_t\mathbf{s}(\mathbf{x}, 0) = \mathbf{0}$. The total energy $\mathcal{E}$ in this state consists only of the relative potential energy $\mathcal{V}$ associated with the initial deformation $\mathbf{s}$. Immediately after $t = 0$, the relative kinetic energy $\mathcal{T}$ must increase as a result of the Earth's motion; concurrently, the total energy $\mathcal{E} = \mathcal{T} + \mathcal{V}$ must decrease slightly, as a consequence of (3.240). If the initial potential energy $\mathcal{V}$ is negative, then it must become even more negative, i.e., the Earth will depart from equilibrium at a rate determined by the anelasticity. If, on the other hand, $\mathcal{V}$ is initially positive, it must become less positive; in this case the Earth will return to the state $\mathcal{V} = 0$, at a rate determined by the anelasticity. It follows that the Earth will be stable in the presence of anelastic dissipation, provided that

$$\mathcal{V} \geq 0 \tag{3.241}$$

for every possible initial displacement field $\mathbf{s}$ satisfying $[\mathbf{s}]_-^+ = \mathbf{0}$ on the solid-solid boundaries Σ_{SS} and $[\hat{\mathbf{n}} \cdot \mathbf{s}]_-^+ = 0$ on the fluid-solid boundaries Σ_{FS}. Equation (3.241) is a *necessary and sufficient condition* for the linear secular stability of a general Earth model. Intuitively, the condition is very reasonable; the stored elastic-gravitational potential energy $\mathcal{V}$ must be increased by every admissible deformation $\mathbf{s}$ in order for the Earth to be stable. In other words, the equilibrium configuration must be a local *potential energy minimum*. It is noteworthy that the only effect of rotation is to replace the initial gravitational potential ϕ^0 in the definition (3.235) of $\mathcal{V}$ by $\phi^0 + \psi$. This quantity, the *geopotential*, differs from the purely gravitational potential ϕ^0 by only one part in 300 in the Earth.

★3.10 First-Principles Variational Analysis

At the beginning of this chapter, we promised that we would derive the general elastic-gravitational equations of motion and boundary conditions in three different ways, including "by an application of Hamilton's principle, using an action obtained by an independent analysis of the kinetic plus elastic-gravitational energy budget". So far we have obtained the energy densities from the Lagrangian densities, but not vice versa. To complete the third derivation, we show next how to obtain the Lagrangian densities L and L^{Σ} from the energy densities E and E^{Σ}.

The *first-order momentum density* conjugate to the displacement $\mathbf{s}$ within a rotating Earth is

$$\mathbf{p} = \rho^0(\partial_t\mathbf{s} + \boldsymbol{\Omega} \times \mathbf{s}). \tag{3.242}$$

We regard the volumetric and surficial energy densities E and E^{Σ} in equations (3.230)–(3.231) as *Hamiltonian densities* of the form $H(\mathbf{s}, \mathbf{p}, \boldsymbol{\nabla}\mathbf{s})$ and

$H^{\Sigma}(\mathbf{s}, \boldsymbol{\nabla}^{\Sigma}\mathbf{s})$, where

$$H = \tfrac{1}{2}(\mathbf{p}\cdot\mathbf{p})/\rho^0 - \mathbf{p}\cdot(\boldsymbol{\Omega}\times\mathbf{s}) \\ + \tfrac{1}{2}[\boldsymbol{\nabla}\mathbf{s}:\boldsymbol{\Lambda}:\boldsymbol{\nabla}\mathbf{s} + \rho^0\mathbf{s}\cdot\boldsymbol{\nabla}\phi^{\mathrm{E1}} + \rho^0\mathbf{s}\cdot\boldsymbol{\nabla}\boldsymbol{\nabla}\phi^0\cdot\mathbf{s}], \tag{3.243}$$

$$H^{\Sigma} = \tfrac{1}{2}[\varpi^0\mathbf{s}\cdot(\boldsymbol{\nabla}^{\Sigma}\mathbf{s})\cdot\hat{\mathbf{n}} - (\hat{\mathbf{n}}\cdot\mathbf{s})\boldsymbol{\nabla}^{\Sigma}\cdot(\varpi^0\mathbf{s})]. \tag{3.244}$$

The linearized momentum equation (3.175) can be written in terms of $\mathbf{p}$ in the form

$$\partial_t\mathbf{s} = \partial_{\mathbf{p}}H, \qquad \partial_t\mathbf{p} = -\partial_{\mathbf{s}}H + \boldsymbol{\nabla}\cdot(\partial_{\boldsymbol{\nabla}\mathbf{s}}H), \tag{3.245}$$

where we have used the identity $\boldsymbol{\Omega}\times(\boldsymbol{\Omega}\times\mathbf{s}) = \mathbf{s}\cdot\boldsymbol{\nabla}\boldsymbol{\nabla}\psi$. Equations (3.245) are *Hamilton's equations.* The Lagrangian densities L and L^{Σ} are now defined in terms of H and H^{Σ} by the inverse of the Legendre transformations (3.196):

$$L = (\partial_{\mathbf{p}}H)\cdot\mathbf{p} - H, \qquad L^{\Sigma} = -H^{\Sigma}. \tag{3.246}$$

It is easily verified that equations (3.246) are equivalent to (3.164)–(3.165). The conjugate momentum density (3.242) is given in terms of the volumetric Lagrangian density by $\mathbf{p} = \partial_{\partial_t\mathbf{s}}L$, as expected.

We could have introduced instead of (3.242) the *total momentum density*

$$\mathbf{p}_{\mathrm{tot}} = \rho^0[\partial_t\mathbf{s} + \boldsymbol{\Omega}\times(\mathbf{x}+\mathbf{s})], \tag{3.247}$$

where the second and third terms account for the zeroth-order and first-order effects of the uniform rotation, respectively. The volumetric Hamiltonian density obtained by substituting (3.247) into equation (3.230) is

$$H_{\mathrm{tot}} = \tfrac{1}{2}(\mathbf{p}_{\mathrm{tot}}\cdot\mathbf{p}_{\mathrm{tot}})/\rho^0 - \mathbf{p}_{\mathrm{tot}}\cdot\boldsymbol{\Omega}\times(\mathbf{x}+\mathbf{s}) \\ + \tfrac{1}{2}\rho^0\|\boldsymbol{\Omega}\times\mathbf{x}\|^2 - \rho^0\mathbf{s}\cdot\boldsymbol{\nabla}\psi \\ + \tfrac{1}{2}[\boldsymbol{\nabla}\mathbf{s}:\boldsymbol{\Lambda}:\boldsymbol{\nabla}\mathbf{s} + \rho^0\mathbf{s}\cdot\boldsymbol{\nabla}\phi^{\mathrm{E1}} + \rho^0\mathbf{s}\cdot\boldsymbol{\nabla}\boldsymbol{\nabla}\phi^0\cdot\mathbf{s}]. \tag{3.248}$$

The numerical values of $H_{\mathrm{tot}}(\mathbf{s}, \mathbf{p}_{\mathrm{tot}}, \boldsymbol{\nabla}\mathbf{s})$ and $H(\mathbf{s}, \mathbf{p}, \boldsymbol{\nabla}\mathbf{s})$ are the same; moreover, Hamilton's equations

$$\partial_t\mathbf{s} = \partial_{\mathbf{p}_{\mathrm{tot}}}H_{\mathrm{tot}}, \qquad \partial_t\mathbf{p}_{\mathrm{tot}} = -\partial_{\mathbf{s}}H_{\mathrm{tot}} + \boldsymbol{\nabla}\cdot(\partial_{\boldsymbol{\nabla}\mathbf{s}}H_{\mathrm{tot}}) \tag{3.249}$$

are identical to (3.245). The volumetric Lagrangian density in this case is

$$L_{\mathrm{tot}} = (\partial_{\mathbf{p}_{\mathrm{tot}}}H_{\mathrm{tot}})\cdot\mathbf{p}_{\mathrm{tot}} - H_{\mathrm{tot}} = L + \rho^0\partial_t\mathbf{s}\cdot(\boldsymbol{\Omega}\times\mathbf{x}). \tag{3.250}$$

The total momentum density (3.247) is given by $\mathbf{p}_{\mathrm{tot}} = \partial_{\partial_t\mathbf{s}}L_{\mathrm{tot}}$.

It is the "total-momentum" Lagrangian density L_{tot} rather than the "first-order momentum" density L that arises in a strict "first-principles" variational analysis of the classical action

$$\mathcal{I}_{\mathrm{tot}} = \int_{t_1}^{t_2} [\mathcal{E}_{\mathrm{k}} - (\mathcal{E}_{\mathrm{e}} + \mathcal{E}_{\mathrm{g}})]\, dt. \tag{3.251}$$

Upon inserting equations (3.216), (3.220) and (3.223) into (3.251) we find, after an argument analogous to the one in Section 3.9.4, that

$$\mathcal{I}_{\mathrm{tot}} = \int_{t_1}^{t_2} \int_{\oplus} L_{\mathrm{tot}}(\mathbf{s}, \partial_t\mathbf{s}, \boldsymbol{\nabla}\mathbf{s})\, dV dt + \int_{t_1}^{t_2} \int_{\Sigma_{\mathrm{FS}}} [L^{\Sigma}(\mathbf{s}, \boldsymbol{\nabla}^{\Sigma}\mathbf{s})]_-^+\, d\Sigma\, dt. \tag{3.252}$$

Since $\partial_{\partial_t\mathbf{s}}(L_{\mathrm{tot}} - L) = \rho^0(\boldsymbol{\Omega}\times\mathbf{x})$ is independent of time, the "total-momentum" Euler-Lagrange equation

$$\partial_{\mathbf{s}} L_{\mathrm{tot}} - \partial_t(\partial_{\partial_t\mathbf{s}} L_{\mathrm{tot}}) - \boldsymbol{\nabla}\cdot(\partial_{\boldsymbol{\nabla}\mathbf{s}} L_{\mathrm{tot}}) = \mathbf{0} \tag{3.253}$$

is identical to (3.169). Furthermore, the two Lagrangian densities L_{tot} and L lead to the same (Coriolis-independent) energy density:

$$E = (\partial_{\partial_t\mathbf{s}} L_{\mathrm{tot}})\cdot\partial_t\mathbf{s} - L_{\mathrm{tot}} = (\partial_{\partial_t\mathbf{s}} L)\cdot\partial_t\mathbf{s} - L. \tag{3.254}$$

We have chosen to ground the rotating-Earth analyses in this book upon the quadratic Lagrangian density L rather than the linear-plus-quadratic "first-principles" density L_{tot}, for reasons of simplicity.

3.11 Hydrostatic Earth Model

A general Earth model of the type that we have been considering so far is specified by giving its density ρ^0, initial stress $\mathbf{T}^0$, rotation rate $\boldsymbol{\Omega}$, and elastic tensor $\boldsymbol{\Gamma}$. The first three parameters are related by the static equilibrium condition $\boldsymbol{\nabla}\cdot\mathbf{T}^0 = \rho^0\boldsymbol{\nabla}(\phi^0+\psi)$, together with the boundary conditions $\hat{\mathbf{n}}\cdot\mathbf{T}^0 = \mathbf{0}$ on $\partial\oplus$ and $[\hat{\mathbf{n}}\cdot\mathbf{T}^0]_-^+ = \mathbf{0}$ on Σ_{SS} and Σ_{FS}. For a given ρ^0 and $\boldsymbol{\Omega}$, these equations serve to constrain three of the six independent components of the static stress $\mathbf{T}^0 = -p^0\mathbf{I} + \boldsymbol{\tau}^0$; the remaining three components must be treated as independently specified parameters.

The need to specify $\mathbf{T}^0$, and specifically the initial deviatoric stress $\boldsymbol{\tau}^0$, is obviated if we demand that $\mathbf{T}^0 = -p^0\mathbf{I}$ everywhere in $\oplus$. The equilibrium condition reduces in this case to

$$\boldsymbol{\nabla}p^0 + \rho^0\boldsymbol{\nabla}(\phi^0+\psi) = \mathbf{0}, \tag{3.255}$$

and the boundary conditions reduce to $p^0 = 0$ on $\partial\oplus$ and $[p^0]_-^+ = 0$ on $\Sigma_{\rm FS}$ and $\Sigma_{\rm SS}$, where p^0 is the hydrostatic pressure. We derive the linearized equations of motion and boundary conditions governing such a *hydrostatic Earth model* in this section. The same symbols will be used to denote the Lagrangian and energy densities and associated action and energy integrals on a hydrostatic and a non-hydrostatic Earth, for simplicity.

3.11.1 Applicability of the theory

Upon taking the curl of equation (3.255) we deduce that

$$\boldsymbol{\nabla}\rho^0 \times \boldsymbol{\nabla}(\phi^0 + \psi) = \mathbf{0}. \tag{3.256}$$

Likewise, taking the cross product with $\boldsymbol{\nabla}p^0$, we find that

$$\boldsymbol{\nabla}p^0 \times \boldsymbol{\nabla}(\phi^0 + \psi) = \mathbf{0}. \tag{3.257}$$

These results show that the *level surfaces* of density ρ^0, pressure p^0 and geopotential $\phi^0 + \psi$ *must coincide* in a hydrostatic Earth model. Any boundary surface across which ρ^0 suffers a jump discontinuity, including the outer free surface, must also be a level surface; this means that we must have

$$\boldsymbol{\nabla}^\Sigma\rho^0 = \mathbf{0}, \qquad \boldsymbol{\nabla}^\Sigma(\phi^0 + \psi) = \mathbf{0}, \qquad \boldsymbol{\nabla}^\Sigma p^0 = \mathbf{0} \tag{3.258}$$

on $\Sigma = \partial\oplus \cup \Sigma_{\rm SS} \cup \Sigma_{\rm FS}$ as well. The conditions (3.256)–(3.258) place a very severe restriction upon the allowable density distributions ρ^0 that may be in a state of hydrostatic equilibrium; in fact, every level surface must be an axially symmetric ellipsoid, or, in the limit $\boldsymbol{\Omega} \to \mathbf{0}$, a sphere. To obtain the linearized equations and boundary conditions, it is not sufficient simply to set the deviatoric stress $\boldsymbol{\tau}^0$ equal to zero; as we shall see, we must also systematically utilize the constraints (3.258). The resulting theory is substantially simpler than the general theory; however, it is exactly valid only for a *rotating, ellipsoidal* Earth model or a *non-rotating, spherically symmetric* Earth model.

Any more general, laterally heterogeneous density distribution ρ^0 requires a non-zero initial deviatoric stress $\boldsymbol{\tau}^0$ for its support. To treat such an Earth model exactly, we must specify $\boldsymbol{\tau}^0$ and use the general equations of motion and boundary conditions derived above. At the present time the deviatoric stress within the Earth is not well enough known to be specified everywhere with any precision; because of this, $\boldsymbol{\tau}^0$ is commonly ignored in quantitative global seismology. The procedure for doing this is straightforward: the hydrostatic equations of motion and boundary conditions are simply assumed to apply more generally to a non-hydrostatic Earth model.

We shall refer to this subterfuge as the *quasi-hydrostatic approximation*. From a practical perspective, such an approximation is advantageous, because it rids the theory of any explicit dependence upon the initial stress. A quasi-hydrostatic, laterally heterogeneous Earth model is fully specified by giving its density ρ^0, rotation rate $\mathbf{\Omega}$, and elastic tensor $\mathbf{\Gamma}$. From a mathematical point of view, the best that can be said is that the procedure is self-consistent, and correct to zeroth order in the dimensionless ratio $\|\boldsymbol{\tau}^0\|/\mu$ where μ is the rigidity. The magnitude $\|\boldsymbol{\tau}^0\|$ of the deviatoric stress in the Earth's lithosphere is constrained by laboratory rock-strength measurements to be less than 0.5 GPa, whereas the lithospheric rigidity μ is of order 50 GPa, so that $\|\boldsymbol{\tau}^0\|/\mu \ll 10^{-2}$. The observed lateral heterogeneity of the Earth's rigidity $\delta\mu/\mu$, due to temperature and lithological variations, is several times greater than this; because of this, the quasi-hydrostatic approximation should be a reasonably good one.

3.11.2 Equations of motion and boundary conditions

The equations governing a hydrostatic Earth model are most conveniently expressed in terms of the incremental Lagrangian Cauchy stress $\mathbf{T}^{\mathrm{L1}}$ rather than the incremental Piola-Kirchhoff stress $\mathbf{T}^{\mathrm{PK1}}$. The linearized momentum equation can be written in the form

$$\rho^0(\partial_t^2\mathbf{s} + 2\mathbf{\Omega}\times\partial_t\mathbf{s}) = \boldsymbol{\nabla}\cdot\mathbf{T}^{\mathrm{L1}} - \boldsymbol{\nabla}[\rho^0\mathbf{s}\cdot\boldsymbol{\nabla}(\phi^0+\psi)] - \rho^0\boldsymbol{\nabla}\phi^{\mathrm{E1}} - \rho^{\mathrm{E1}}\boldsymbol{\nabla}(\phi^0+\psi), \tag{3.259}$$

where we have set $\boldsymbol{\tau}^0 = \mathbf{0}$ in equation (3.58). The linearized constitutive relation (3.144) reduces to

$$\mathbf{T}^{\mathrm{L1}} = \mathbf{\Gamma}:\boldsymbol{\varepsilon}, \tag{3.260}$$

where $\mathbf{\Gamma}$ is the elastic tensor, and $\boldsymbol{\varepsilon} = \frac{1}{2}[\boldsymbol{\nabla}\mathbf{s} + (\boldsymbol{\nabla}\mathbf{s})^{\mathrm{T}}]$ is the strain. The Christoffel tensor $\mathbf{B}$ in equation (3.155) reduces to

$$\rho^0 B_{jl} = (\kappa + \tfrac{1}{3}\mu)\hat{k}_j\hat{k}_l + \mu\delta_{jl} + \gamma_{ijkl}\hat{k}_i\hat{k}_k, \tag{3.261}$$

so that the anisotropy of elastic body-wave propagation depends only upon the tensor $\boldsymbol{\gamma}$, as in the classical theory of wave propagation in an unstressed medium.

The elastic energy density (3.115) or (3.126) can be written, correct to second order in $\|\mathbf{s}\|$, in the form

$$\rho^0 U^{\mathrm{L}} = \rho^0 U^0 - p^0(J-1) + \tfrac{1}{2}(\boldsymbol{\varepsilon}:\mathbf{\Gamma}:\boldsymbol{\varepsilon}), \tag{3.262}$$

where

$$J = 1 + \boldsymbol{\nabla}\cdot\mathbf{s} + \tfrac{1}{2}(\boldsymbol{\nabla}\cdot\mathbf{s})^2 - \tfrac{1}{2}(\boldsymbol{\nabla}\mathbf{s}):(\boldsymbol{\nabla}\mathbf{s})^{\mathrm{T}} \tag{3.263}$$

Name or Description	Linearized Elastic-Gravitational Equation
Continuity equation	$\rho^{\mathrm{E1}} = -\boldsymbol{\nabla}\cdot(\rho^0\mathbf{s})$
Momentum equation	$\rho^0(\partial_t^2\mathbf{s} + 2\boldsymbol{\Omega}\times\partial_t\mathbf{s}) = \boldsymbol{\nabla}\cdot\mathbf{T}^{\mathrm{L1}}$ $-\boldsymbol{\nabla}[\rho^0\mathbf{s}\cdot\boldsymbol{\nabla}(\phi^0+\psi)] - \rho^0\boldsymbol{\nabla}\phi^{\mathrm{E1}}$ $-\rho^{\mathrm{E1}}\boldsymbol{\nabla}(\phi^0+\psi)$
Poisson's equation	$\nabla^2\phi^{\mathrm{E1}} = 4\pi G\rho^{\mathrm{E1}}$
Potential perturbation	$\phi^{\mathrm{E1}} = -G\int_\oplus \frac{\rho^{0\prime}\mathbf{s}'\cdot(\mathbf{x}-\mathbf{x}')}{\lVert\mathbf{x}-\mathbf{x}'\rVert^3}\,dV'$
Gravity perturbation	$\mathbf{g}^{\mathrm{E1}} = -\boldsymbol{\nabla}\phi^{\mathrm{E1}} = G\int_\oplus \rho^{0\prime}(\mathbf{s}'\cdot\boldsymbol{\Pi})\,dV'$
Hooke's law	$\mathbf{T}^{\mathrm{L1}} = \boldsymbol{\Gamma}:\boldsymbol{\varepsilon}$
Elastic symmetries	$\Gamma_{ijkl} = \Gamma_{jikl} = \Gamma_{ijlk} = \Gamma_{klij}$

Table 3.3. Linearized elastic-gravitational relations governing a hydrostatic Earth model. There is no explicit dependence upon the initial static stress $\mathbf{T}^0 = -p^0\mathbf{I}$.

is the *exact* Jacobian. The second term on the right in equation (3.262) represents the work done against the isotropic initial stress, whereas the third term is the classical elastic energy density in the absence of any pre-stress. The incremental first Piola-Kirchhoff stress $\mathbf{T}^{\mathrm{PK1}}$ is related to $\mathbf{T}^{\mathrm{L1}}$ in a hydrostatic Earth by

$$\mathbf{T}^{\mathrm{PK1}} = \mathbf{T}^{\mathrm{L1}} - p^0(\boldsymbol{\nabla}\cdot\mathbf{s})\mathbf{I} + p^0(\boldsymbol{\nabla}\mathbf{s})^{\mathrm{T}}. \tag{3.264}$$

The total Piola-Kirchoff stress $\mathbf{T}^0 + \mathbf{T}^{\mathrm{PK1}}$ is the derivative (3.127) of the hydrostatic elastic energy density (3.262), as in the deviatoric case.

The linearized dynamical boundary condition $\hat{\mathbf{n}}\cdot\mathbf{T}^{\mathrm{PK1}} = \mathbf{0}$ on the outer free surface $\partial\oplus$ reduces to

$$\hat{\mathbf{n}}\cdot\mathbf{T}^{\mathrm{L1}} = \mathbf{0}. \tag{3.265}$$

Boundary Type	Linearized Boundary Conditions
$\partial\oplus$: free surface	$\hat{\mathbf{n}}\cdot\mathbf{T}^{\mathrm{L1}}=\mathbf{0}$
Σ_{SS}: solid-solid	$[\mathbf{s}]_-^+=\mathbf{0}$ $[\hat{\mathbf{n}}\cdot\mathbf{T}^{\mathrm{L1}}]_-^+=\mathbf{0}$
Σ_{FS}: fluid-solid	$[\hat{\mathbf{n}}\cdot\mathbf{s}]_-^+=0$ $[\hat{\mathbf{n}}\cdot\mathbf{T}^{\mathrm{L1}}]_-^+=\hat{\mathbf{n}}[\hat{\mathbf{n}}\cdot\mathbf{T}^{\mathrm{L1}}\cdot\hat{\mathbf{n}}]_-^+=\mathbf{0}$
Σ: all boundaries	$[\phi^{\mathrm{E1}}]_-^+=0$ $[\hat{\mathbf{n}}\cdot\boldsymbol{\nabla}\phi^{\mathrm{E1}}+4\pi G\rho^0\,\hat{\mathbf{n}}\cdot\mathbf{s}]_-^+=0$

Table 3.4. Linearized boundary conditions governing a hydrostatic Earth model. Both the equations of motion and the boundary conditions are most conveniently expressed in terms of the incremental Lagrangian Cauchy stress $\mathbf{T}^{\mathrm{L1}}$.

Likewise, on the solid-solid boundaries Σ_{SS}, the condition $[\hat{\mathbf{n}}\cdot\mathbf{T}^{\mathrm{PK1}}]_-^+=\mathbf{0}$ implies that

$$[\hat{\mathbf{n}}\cdot\mathbf{T}^{\mathrm{L1}}]_-^+=\mathbf{0}. \tag{3.266}$$

The auxiliary vector $\mathbf{t}^{\mathrm{PK1}}$ reduces to $\hat{\mathbf{n}}\cdot\mathbf{T}^{\mathrm{L1}}+\hat{\mathbf{n}}(\mathbf{s}\cdot\boldsymbol{\nabla}^{\Sigma}p^0)$; this, however, is simply $\hat{\mathbf{n}}\cdot\mathbf{T}^{\mathrm{L1}}$ by virtue of the isobaric condition (3.258). The dynamic boundary condition on the fluid-solid boundaries Σ_{FS} in a hydrostatic Earth is therefore

$$[\hat{\mathbf{n}}\cdot\mathbf{T}^{\mathrm{L1}}]_-^+=\hat{\mathbf{n}}[\hat{\mathbf{n}}\cdot\mathbf{T}^{\mathrm{L1}}\cdot\hat{\mathbf{n}}]_-^+=\mathbf{0}. \tag{3.267}$$

In conclusion, the traction $\hat{\mathbf{n}}\cdot\mathbf{T}^{\mathrm{L1}}$ is continuous on all of Σ, normal to the boundary on Σ_{FS}, and it vanishes on $\partial\oplus$. The complete set of linearized equations and associated boundary conditions governing a hydrostatic Earth is summarized for convenience in Tables 3.3 and 3.4.

3.11.3 Hamilton's principle

The *action* $\mathcal{I}$ in equations (3.163)–(3.165) reduces to

$$\mathcal{I}=\int_{t_1}^{t_2}\int_{\oplus}L(\mathbf{s},\partial_t\mathbf{s},\boldsymbol{\nabla}\mathbf{s})\,dV dt, \tag{3.268}$$

where

$$L = \tfrac{1}{2}[\rho^0\partial_t\mathbf{s}\cdot\partial_t\mathbf{s} - 2\rho^0\mathbf{s}\cdot\boldsymbol{\Omega}\times\partial_t\mathbf{s} - \boldsymbol{\varepsilon}\!:\!\boldsymbol{\Gamma}\!:\!\boldsymbol{\varepsilon}$$
$$- \rho^0\mathbf{s}\cdot\boldsymbol{\nabla}\phi^{\mathrm{E1}} - \rho^0\mathbf{s}\cdot\boldsymbol{\nabla}\boldsymbol{\nabla}(\phi^0+\psi)\cdot\mathbf{s}$$
$$- \rho^0\boldsymbol{\nabla}(\phi^0+\psi)\cdot(\mathbf{s}\cdot\boldsymbol{\nabla}\mathbf{s} - \mathbf{s}\,\boldsymbol{\nabla}\cdot\mathbf{s})]. \tag{3.269}$$

In obtaining equations (3.268)–(3.269), we have set $\varpi^0 = p^0$, and used Gauss' theorem and the isobaric condition (3.258) to eliminate the surface integral over the fluid-solid boundaries Σ_{FS}. The potential perturbation ϕ^{E1} in (3.269) is regarded as a known functional of the displacement $\mathbf{s}$, given by equation (3.99), so that the hydrostatic Lagrangian density L is a functional only of $\mathbf{s}$, $\partial_t\mathbf{s}$ and $\boldsymbol{\nabla}\mathbf{s}$, as indicated.

The equations of motion and boundary conditions governing a hydrostatic Earth model may be obtained from the displacement variational principle $\delta\mathcal{I} = 0$; an admissible variation $\delta\mathbf{s}$ is one that vanishes at the starting and ending times t_1 and t_2, and that satisfies $[\delta\mathbf{s}]_-^+ = \mathbf{0}$ on Σ_{SS} and $[\hat{\mathbf{n}}\cdot\delta\mathbf{s}]_-^+ = 0$ on Σ_{FS}. The Euler-Lagrange equation and associated continuity conditions are

$$\partial_{\mathbf{s}}L - \partial_t(\partial_{\partial_t\mathbf{s}}L) - \boldsymbol{\nabla}\cdot(\partial_{\boldsymbol{\nabla}\mathbf{s}}L) = \mathbf{0} \quad \text{in } \oplus, \tag{3.270}$$

$$\hat{\mathbf{n}}\cdot(\partial_{\boldsymbol{\nabla}\mathbf{s}}L) = \mathbf{0} \quad \text{on } \partial\oplus, \tag{3.271}$$

$$[\hat{\mathbf{n}}\cdot(\partial_{\boldsymbol{\nabla}\mathbf{s}}L)]_-^+ = \mathbf{0} \quad \text{on } \Sigma_{\mathrm{SS}}, \tag{3.272}$$

$$[\hat{\mathbf{n}}\cdot(\partial_{\boldsymbol{\nabla}\mathbf{s}}L)]_-^+ = \hat{\mathbf{n}}[\hat{\mathbf{n}}\cdot(\partial_{\boldsymbol{\nabla}\mathbf{s}}L)\cdot\hat{\mathbf{n}}]_-^+ = \mathbf{0} \quad \text{on } \Sigma_{\mathrm{FS}}. \tag{3.273}$$

These are precisely the linearized momentum equation (3.259) and the associated boundary conditions (3.265)–(3.267).

The *modified action* $\mathcal{I}'$ in equations (3.189)–(3.190) reduces to

$$\mathcal{I}' = \int_{t_1}^{t_2}\int_{\bigcirc} L'(\mathbf{s},\partial_t\mathbf{s},\boldsymbol{\nabla}\mathbf{s},\boldsymbol{\nabla}\phi^{\mathrm{E1}})\,dV dt, \tag{3.274}$$

where

$$L' = \tfrac{1}{2}[\rho^0\partial_t\mathbf{s}\cdot\partial_t\mathbf{s} - 2\rho^0\mathbf{s}\cdot\boldsymbol{\Omega}\times\partial_t\mathbf{s} - \boldsymbol{\varepsilon}\!:\!\boldsymbol{\Gamma}\!:\!\boldsymbol{\varepsilon}$$
$$- 2\rho^0\mathbf{s}\cdot\boldsymbol{\nabla}\phi^{\mathrm{E1}} - \rho^0\mathbf{s}\cdot\boldsymbol{\nabla}\boldsymbol{\nabla}(\phi^0+\psi)\cdot\mathbf{s}$$
$$- \rho^0\boldsymbol{\nabla}(\phi^0+\psi)\cdot(\mathbf{s}\cdot\boldsymbol{\nabla}\mathbf{s} - \mathbf{s}\,\boldsymbol{\nabla}\cdot\mathbf{s})$$
$$- (4\pi G)^{-1}\boldsymbol{\nabla}\phi^{\mathrm{E1}}\cdot\boldsymbol{\nabla}\phi^{\mathrm{E1}}]. \tag{3.275}$$

Variation of $\mathcal{I}'$ with respect to $\mathbf{s}$ gives equations (3.270)–(3.273), as before, whereas variation with respect to ϕ^{E1} gives the potential boundary-value problem (3.181)–(3.182) in the form

$$\boldsymbol{\nabla}\cdot(\partial_{\boldsymbol{\nabla}\phi^{\mathrm{E1}}}L') = 0 \quad \text{in } \bigcirc, \tag{3.276}$$

$$[\hat{\mathbf{n}}\cdot(\partial_{\boldsymbol{\nabla}\phi^{\mathrm{E1}}}L')]_-^+=0 \quad \text{on } \Sigma. \tag{3.277}$$

An admissible variation $\delta\phi^{\mathrm{E1}}$ is one that vanishes at the starting and ending times t_1 and t_2, and that satisfies $[\delta\phi^{\mathrm{E1}}]_-^+=0$ on Σ. The value of both action integrals along the stationary path is

$$\mathcal{I}=\mathcal{I}'=0, \tag{3.278}$$

by virtue of the equality (3.195).

3.11.4 Conservation of energy

The elastic-gravitational energy densities E and E' are defined in terms of the Lagrangian densities L and L' by

$$E=(\partial_{\partial_t\mathbf{s}}L)\cdot\partial_t\mathbf{s}-L, \qquad E'=(\partial_{\partial_t\mathbf{s}}L')\cdot\partial_t\mathbf{s}-L'. \tag{3.279}$$

From equations (3.269) and (3.275), we find that

$$\begin{aligned}E=\tfrac{1}{2}[&\rho^0\partial_t\mathbf{s}\cdot\partial_t\mathbf{s}+\boldsymbol{\varepsilon}:\boldsymbol{\Gamma}:\boldsymbol{\varepsilon}+\rho^0\mathbf{s}\cdot\boldsymbol{\nabla}\phi^{\mathrm{E1}}\\&+\rho^0\mathbf{s}\cdot\boldsymbol{\nabla}\boldsymbol{\nabla}(\phi^0+\psi)\cdot\mathbf{s}\\&+\rho^0\boldsymbol{\nabla}(\phi^0+\psi)\cdot(\mathbf{s}\cdot\boldsymbol{\nabla}\mathbf{s}-\mathbf{s}\,\boldsymbol{\nabla}\cdot\mathbf{s})],\end{aligned} \tag{3.280}$$

$$\begin{aligned}E'=\tfrac{1}{2}[&\rho^0\partial_t\mathbf{s}\cdot\partial_t\mathbf{s}+\boldsymbol{\varepsilon}:\boldsymbol{\Gamma}:\boldsymbol{\varepsilon}+2\rho^0\mathbf{s}\cdot\boldsymbol{\nabla}\phi^{\mathrm{E1}}\\&+\rho^0\mathbf{s}\cdot\boldsymbol{\nabla}\boldsymbol{\nabla}(\phi^0+\psi)\cdot\mathbf{s}\\&+\rho^0\boldsymbol{\nabla}(\phi^0+\psi)\cdot(\mathbf{s}\cdot\boldsymbol{\nabla}\mathbf{s}-\mathbf{s}\,\boldsymbol{\nabla}\cdot\mathbf{s})\\&+(4\pi G)^{-1}\boldsymbol{\nabla}\phi^{\mathrm{E1}}\cdot\boldsymbol{\nabla}\phi^{\mathrm{E1}}].\end{aligned} \tag{3.281}$$

Upon dotting the linearized momentum equation (3.259) with the velocity $\partial_t\mathbf{s}$, and using the elastic symmetry $\Gamma_{ijkl}=\Gamma_{klij}$ and the hydrostatic condition (3.256), we obtain the local energy conservation laws

$$\partial_t E+\boldsymbol{\nabla}\cdot\mathbf{K}=0, \qquad \partial_t E'+\boldsymbol{\nabla}\cdot\mathbf{K}'=0. \tag{3.282}$$

The elastic-gravitational energy fluxes associated with the two densities E and E' are given by

$$\begin{aligned}\mathbf{K}=-\mathbf{T}^{\mathrm{L1}}\cdot\partial_t\mathbf{s}&-\tfrac{1}{2}\rho^0\boldsymbol{\nabla}(\phi^0+\psi)\cdot[(\partial_t\mathbf{s})\mathbf{s}-\mathbf{s}(\partial_t\mathbf{s})]\\&+\tfrac{1}{2}\phi^{\mathrm{E1}}(\partial_t\boldsymbol{\xi}^{\mathrm{E1}})-\tfrac{1}{2}(\partial_t\phi^{\mathrm{E1}})\boldsymbol{\xi}^{\mathrm{E1}},\end{aligned} \tag{3.283}$$

$$\begin{aligned}\mathbf{K}'=-\mathbf{T}^{\mathrm{L1}}\cdot\partial_t\mathbf{s}&-\tfrac{1}{2}\rho^0\boldsymbol{\nabla}(\phi^0+\psi)\cdot[(\partial_t\mathbf{s})\mathbf{s}-\mathbf{s}(\partial_t\mathbf{s})]\\&-(\partial_t\phi^{\mathrm{E1}})\boldsymbol{\xi}^{\mathrm{E1}},\end{aligned} \tag{3.284}$$

where $\boldsymbol{\xi}^{\mathrm{E1}} = (4\pi G)^{-1}\boldsymbol{\nabla}\phi^{\mathrm{E1}} + \rho^0\mathbf{s}$. Integrating the first of equations (3.282) over the Earth model $\oplus$ or the second over all of space $\bigcirc$ gives the corresponding global energy conservation relation

$$\frac{d\mathcal{E}}{dt} = 0, \tag{3.285}$$

where

$$\mathcal{E} = \int_{\oplus} E\,dV = \int_{\bigcirc} E'\,dV. \tag{3.286}$$

The jump terms $[\hat{\mathbf{n}}\cdot\mathbf{K}]_-^+$ and $[\hat{\mathbf{n}}\cdot\mathbf{K}']_-^+$ vanish on all of Σ by virtue of the boundary conditions (3.182) and (3.265)–(3.267).

The expression (3.286) for the energy in a hydrostatic Earth model can also be obtained from first principles. As in the general case, the total energy is the sum of the instantaneous kinetic energy and the stored elastic and gravitational potential energies:

$$\mathcal{E} = \mathcal{E}_{\mathrm{k}} + \mathcal{E}_{\mathrm{e}} + \mathcal{E}_{\mathrm{g}}. \tag{3.287}$$

The kinetic energy $\mathcal{E}_{\mathrm{k}}$ and gravitational energy $\mathcal{E}_{\mathrm{g}}$ are again given by equations (3.219) and (3.223); the elastic energy in a hydrostatic Earth is, however,

$$\mathcal{E}_{\mathrm{e}} = \int_{\oplus} [-p^0(J-1) + \tfrac{1}{2}\boldsymbol{\varepsilon}:\boldsymbol{\Gamma}:\boldsymbol{\varepsilon}]\,dV. \tag{3.288}$$

Grouping terms and proceeding as in Section 3.9.4, we find that the total energy (3.287) reduces to (3.286); the surface integral arising from the second-order tangential slip condition on Σ_{FS} can be eliminated by an application of Gauss' theorem, together with the isobaric condition (3.258), as in the derivation of the action (3.268)–(3.269).

3.11.5 Relative kinetic and potential energy

The energy of a hydrostatic Earth model can also be regarded as the sum of a relative kinetic and potential energy:

$$\mathcal{E} = \mathcal{T} + \mathcal{V}, \tag{3.289}$$

where

$$\mathcal{T} = \tfrac{1}{2}\int_{\oplus} \rho^0(\partial_t\mathbf{s}\cdot\partial_t\mathbf{s})\,dV, \tag{3.290}$$

and

$$\mathcal{V} = \tfrac{1}{2}\int_{\oplus} [\boldsymbol{\varepsilon}\!:\!\boldsymbol{\Gamma}\!:\!\boldsymbol{\varepsilon} + \rho^0 \mathbf{s}\cdot\boldsymbol{\nabla}\phi^{\mathrm{E1}} + \rho^0\mathbf{s}\cdot\boldsymbol{\nabla\nabla}(\phi^0+\psi)\cdot\mathbf{s}$$
$$+ \rho^0\boldsymbol{\nabla}(\phi^0+\psi)\cdot(\mathbf{s}\cdot\boldsymbol{\nabla}\mathbf{s} - \mathbf{s}\,\boldsymbol{\nabla}\cdot\mathbf{s})]\, dV. \qquad (3.291)$$

A hydrostatic Earth model is secularly stable if and only if the equilibrium state is a local potential energy minimum, so that

$$\mathcal{V} \geq 0, \qquad (3.292)$$

for every possible displacement $\mathbf{s}$ satisfying $[\hat{\mathbf{n}}\cdot\mathbf{s}]_-^+ = 0$ on Σ_{FS}.

It is common to further divide the relative potential energy on a hydrostatic Earth into separate "elastic", "gravitational" and "centrifugal" terms:

$$\mathcal{V} = \mathcal{V}_{\mathrm{e}} + \mathcal{V}_{\mathrm{g}} + \mathcal{V}_{\psi}. \qquad (3.293)$$

The quantity $\mathcal{V}_{\mathrm{e}}$ is simply the classical elastic energy in the absence of any pre-stress:

$$\mathcal{V}_{\mathrm{e}} = \tfrac{1}{2}\int_{\oplus} (\boldsymbol{\varepsilon}\!:\!\boldsymbol{\Gamma}\!:\!\boldsymbol{\varepsilon})\, dV. \qquad (3.294)$$

The remaining terms $\mathcal{V}_{\mathrm{g}}$ and $\mathcal{V}_{\psi}$ express the dependence upon the gravitational and centrifugal potentials, respectively:

$$\mathcal{V}_{\mathrm{g}} = \tfrac{1}{2}\int_{\oplus} [\rho^0\mathbf{s}\cdot\boldsymbol{\nabla}\phi^{\mathrm{E1}} + \rho^0\mathbf{s}\cdot\boldsymbol{\nabla\nabla}\phi^0\cdot\mathbf{s}$$
$$+ \rho^0\boldsymbol{\nabla}\phi^0\cdot(\mathbf{s}\cdot\boldsymbol{\nabla}\mathbf{s} - \mathbf{s}\,\boldsymbol{\nabla}\cdot\mathbf{s})]\, dV, \qquad (3.295)$$

$$\mathcal{V}_{\psi} = \tfrac{1}{2}\int_{\oplus} [\rho^0\mathbf{s}\cdot\boldsymbol{\nabla\nabla}\psi\cdot\mathbf{s} + \rho^0\boldsymbol{\nabla}\psi\cdot(\mathbf{s}\cdot\boldsymbol{\nabla}\mathbf{s} - \mathbf{s}\,\boldsymbol{\nabla}\cdot\mathbf{s})]\, dV. \qquad (3.296)$$

The actual elastic, gravitational and centrifugal energies are $\mathcal{E}_{\mathrm{e}}$, $\mathcal{E}_{\mathrm{g}}$ and $\mathcal{E}_{\psi}$, not $\mathcal{V}_{\mathrm{e}}$, $\mathcal{V}_{\mathrm{g}}$ and $\mathcal{V}_{\psi}$, as we have seen. Nevertheless, the two terms $\mathcal{V}_{\mathrm{e}}$ and $\mathcal{V}_{\mathrm{g}}$ are ubiquitously referred to as the *elastic* and *gravitational potential energies*; we shall adhere to this tradition in the remainder of this book.

Chapter 4

Normal Modes

Because of the time invariance of the properties of the Earth in the uniformly rotating frame, it is natural to seek solutions that are harmonic functions of time:

$$\mathbf{s}(\mathbf{x},t) = \mathbf{s}(\mathbf{x})\exp(i\omega t). \tag{4.1}$$

The quantities ω are referred to as the angular *eigenfrequencies* of the Earth, and the displacement fields $\mathbf{s}(\mathbf{x})$ are referred to as the associated *eigenfunctions*. We devote this chapter to an investigation of the nature of these normal-mode solutions.

Looking for oscillatory solutions of the form (4.1) is equivalent to transforming the equations of motion and boundary conditions to the frequency domain using the relation

$$\mathbf{s}(\mathbf{x},\omega) = \int_{-\infty}^{\infty} \mathbf{s}(\mathbf{x},t)\exp(-i\omega t)\,dt. \tag{4.2}$$

In either case we go from one domain to the other by making the substitution $\partial_t \longleftrightarrow i\omega$. Equation (4.2) is the most natural definition of the Fourier transform in considering the free oscillations of the Earth; it is noteworthy that the sign convention in the exponential differs from the one commonly adopted in the analysis of travelling waves (Aki & Richards 1980).

Much of what we do in this chapter consists of a simple translation of the time-domain results in Chapter 3 into the frequency domain. We use the same symbol for most time-domain and frequency-domain quantities; hopefully, this should not lead to any confusion. For a number of reasons, it is convenient to consider the two cases of a non-rotating Earth and a rotating Earth separately. We take up the simpler non-rotating ($\mathbf{\Omega} = \mathbf{0}$) case first, and relegate the more complicated but parallel case of a rotating elastic Earth model to a separate starred section.

4.1 Non-Rotating Earth Model

The eigenfrequencies ω and eigenfunctions $\mathbf{s}$ of a non-rotating Earth are found by solving

$$-\omega^2\rho^0\mathbf{s} - \boldsymbol{\nabla}\cdot\mathbf{T}^{\mathrm{PK1}} + \rho^0\boldsymbol{\nabla}\phi^{\mathrm{E1}} + \rho^0\mathbf{s}\cdot\boldsymbol{\nabla\nabla}\phi^0 = \mathbf{0} \quad \text{in } \oplus, \tag{4.3}$$

subject to the boundary conditions

$$\hat{\mathbf{n}}\cdot\mathbf{T}^{\mathrm{PK1}} = \mathbf{0} \quad \text{on } \partial\oplus, \tag{4.4}$$

$$[\hat{\mathbf{n}}\cdot\mathbf{T}^{\mathrm{PK1}}]_-^+ = \mathbf{0} \quad \text{on } \Sigma_{\mathrm{SS}}, \tag{4.5}$$

$$[\mathbf{t}^{\mathrm{PK1}}]_-^+ = \hat{\mathbf{n}}[\hat{\mathbf{n}}\cdot\mathbf{t}^{\mathrm{PK1}}]_-^+ = \mathbf{0} \quad \text{on } \Sigma_{\mathrm{FS}}. \tag{4.6}$$

Without loss of generality, we may consider the squared eigenfrequencies ω^2 and the eigenfunctions $\mathbf{s}$ of a non-rotating Earth model to be *real*:

$$(\omega^2)^* = \omega^2, \qquad \mathbf{s}^* = \mathbf{s}, \tag{4.7}$$

where an asterisk denotes the complex conjugate. The dependence of the transformed momentum equation (4.3) on ω^2 indicates that there are two eigenfrequencies associated with every real eigenfunction $\mathbf{s}$; if $\omega^2 > 0$, the eigenfrequencies are real, $\pm\omega$, whereas if $\omega^2 < 0$, they are imaginary, $\pm i|\omega|$. For simplicity, we shall henceforth assume that all of the eigenfrequencies are real; we determine a dynamical stability condition which guarantees this in Section 4.1.5.

4.1.1 Hermitian operator formulation

For the sake of brevity, we rewrite the system of equations (4.3)–(4.6) governing the eigensolutions $\pm\omega$, $\mathbf{s}$ in the symbolic form

$$\mathcal{H}\mathbf{s} = \omega^2\mathbf{s}. \tag{4.8}$$

The symbol $\mathcal{H}$ stands for the *integro-differential operator*

$$\rho^0\mathcal{H}\mathbf{s} = -\boldsymbol{\nabla}\cdot\mathbf{T}^{\mathrm{PK1}} + \rho^0\boldsymbol{\nabla}\phi^{\mathrm{E1}} + \rho^0\mathbf{s}\cdot\boldsymbol{\nabla\nabla}\phi^0 \tag{4.9}$$

in the Earth model $\oplus$, together with the boundary conditions (4.4)–(4.6) on $\Sigma = \partial\oplus \cup \Sigma_{\mathrm{SS}} \cup \Sigma_{\mathrm{FS}}$. The quantities ω^2 and $\mathbf{s}$ can be regarded as the eigenvalue and associated eigenfunction of the linear operator $\mathcal{H}$. For the time being, we consider the Eulerian potential perturbation ϕ^{E1} to be a known functional of $\mathbf{s}$, given by equation (3.99).

We define the *inner product* $\langle \mathbf{s}, \mathbf{s}' \rangle$ of any two piecewise smooth real functions $\mathbf{s}$ and $\mathbf{s}'$ in $\oplus$ by

$$\langle \mathbf{s}, \mathbf{s}' \rangle = \int_{\oplus} \rho^0 \mathbf{s} \cdot \mathbf{s}' \, dV. \tag{4.10}$$

The operator $\mathcal{H}$ is then *Hermitian* or *self-adjoint* with respect to this inner product, i.e.,

$$\langle \mathbf{s}, \mathcal{H}\mathbf{s}' \rangle = \langle \mathcal{H}\mathbf{s}, \mathbf{s}' \rangle = \langle \mathbf{s}', \mathcal{H}\mathbf{s} \rangle. \tag{4.11}$$

It is straightforward to verify equation (4.11); the left and right sides are given explicitly by

$$\langle \mathbf{s}, \mathcal{H}\mathbf{s}' \rangle = \int_{\oplus} \mathbf{s} \cdot [-\boldsymbol{\nabla} \cdot \mathbf{T}^{\mathrm{PK1}\prime} + \rho^0 \boldsymbol{\nabla} \phi^{\mathrm{E1}\prime} + \rho^0 \mathbf{s}' \cdot \boldsymbol{\nabla}\boldsymbol{\nabla} \phi^0] \, dV, \tag{4.12}$$

$$\langle \mathbf{s}', \mathcal{H}\mathbf{s} \rangle = \int_{\oplus} \mathbf{s}' \cdot [-\boldsymbol{\nabla} \cdot \mathbf{T}^{\mathrm{PK1}} + \rho^0 \boldsymbol{\nabla} \phi^{\mathrm{E1}} + \rho^0 \mathbf{s} \cdot \boldsymbol{\nabla}\boldsymbol{\nabla} \phi^0] \, dV, \tag{4.13}$$

where $\mathbf{T}^{\mathrm{PK1}\prime}$ and $\phi^{\mathrm{E1}\prime}$ are the incremental Piola-Kirchhoff stress and Eulerian potential perturbation associated with the primed displacement $\mathbf{s}'$. Making use of Gauss' theorem, and applying the boundary conditions (4.4) and (4.5) on $\partial\oplus$ and Σ_{SS}, we obtain

$$\begin{aligned}\langle \mathbf{s}, \mathcal{H}\mathbf{s}' \rangle = {} & \int_{\oplus} [\boldsymbol{\nabla}\mathbf{s} : \boldsymbol{\Lambda} : \boldsymbol{\nabla}\mathbf{s}' + \rho^0 \mathbf{s} \cdot \boldsymbol{\nabla} \phi^{\mathrm{E1}\prime} + \rho^0 \mathbf{s} \cdot \boldsymbol{\nabla}\boldsymbol{\nabla} \phi^0 \cdot \mathbf{s}'] \, dV \\ & + \int_{\Sigma_{\mathrm{FS}}} [\hat{\mathbf{n}} \cdot \mathbf{T}^{\mathrm{PK1}\prime} \cdot \mathbf{s}]_-^+ \, d\Sigma,\end{aligned} \tag{4.14}$$

$$\begin{aligned}\langle \mathbf{s}', \mathcal{H}\mathbf{s} \rangle = {} & \int_{\oplus} [\boldsymbol{\nabla}\mathbf{s}' : \boldsymbol{\Lambda} : \boldsymbol{\nabla}\mathbf{s} + \rho^0 \mathbf{s}' \cdot \boldsymbol{\nabla} \phi^{\mathrm{E1}} + \rho^0 \mathbf{s}' \cdot \boldsymbol{\nabla}\boldsymbol{\nabla} \phi^0 \cdot \mathbf{s}] \, dV \\ & + \int_{\Sigma_{\mathrm{FS}}} [\hat{\mathbf{n}} \cdot \mathbf{T}^{\mathrm{PK1}} \cdot \mathbf{s}']_-^+ \, d\Sigma.\end{aligned} \tag{4.15}$$

The volume integrals on the right of (4.14) and (4.15) are equal by virtue of the Maxwell relation $\Lambda_{ijkl} = \Lambda_{klij}$ and the gravitational identity

$$\begin{aligned}\int_{\oplus} \rho^0 \mathbf{s} \cdot \boldsymbol{\nabla} \phi^{\mathrm{E1}\prime} \, dV &= \int_{\oplus} \rho^0 \mathbf{s}' \cdot \boldsymbol{\nabla} \phi^{\mathrm{E1}} \, dV \\ &= -\int_{\oplus} \int_{\oplus} (\rho^0 \mathbf{s} \cdot \boldsymbol{\Pi} \cdot \rho^{0\prime} \mathbf{s}') \, dV \, dV'.\end{aligned} \tag{4.16}$$

The symmetric kernel $\mathbf{\Pi}(\mathbf{x},\mathbf{x}') = \mathbf{\Pi}^{\mathrm{T}}(\mathbf{x}',\mathbf{x})$ is given by equation (3.101). The surface integrals over Σ_{FS} are also equal; from (3.202) and (3.203) we find that

$$\begin{aligned}\int_{\Sigma_{\mathrm{FS}}} [\hat{\mathbf{n}}\cdot\mathbf{T}^{\mathrm{PK1}'}\cdot\mathbf{s}]_-^+\,d\Sigma &= \int_{\Sigma_{\mathrm{FS}}} [\hat{\mathbf{n}}\cdot\mathbf{T}^{\mathrm{PK1}}\cdot\mathbf{s}']_-^+\,d\Sigma \\ &= \tfrac{1}{2}\int_{\Sigma_{\mathrm{FS}}} [\varpi^0\mathbf{s}\cdot(\boldsymbol{\nabla}^\Sigma\mathbf{s}')\cdot\hat{\mathbf{n}} + \varpi^0\mathbf{s}'\cdot(\boldsymbol{\nabla}^\Sigma\mathbf{s})\cdot\hat{\mathbf{n}} \\ &\qquad - (\hat{\mathbf{n}}\cdot\mathbf{s})\boldsymbol{\nabla}^\Sigma\cdot(\varpi^0\mathbf{s}') - (\hat{\mathbf{n}}\cdot\mathbf{s}')\boldsymbol{\nabla}^\Sigma\cdot(\varpi^0\mathbf{s})]_-^+\,d\Sigma.\end{aligned} \tag{4.17}$$

It is noteworthy that the detailed manipulations needed to establish the Hermitian character of the elastic-gravitational operator $\mathcal{H}$ are identical to those used to establish the law of conservation of energy, $d\mathcal{E}/dt = 0$, in Section 3.8. This is an illustration of a general principle—that physical systems governed by Hermitian operators are energy-conserving.

4.1.2 Orthonormality

Taking the inner product of $\mathcal{H}\mathbf{s} = \omega^2\mathbf{s}$ with $\mathbf{s}'$ yields

$$\omega^2\langle\mathbf{s}',\mathbf{s}\rangle = \langle\mathbf{s}',\mathcal{H}\mathbf{s}\rangle, \tag{4.18}$$

whereas taking the inner product of $\mathcal{H}\mathbf{s}' = \omega'^2\mathbf{s}'$ with $\mathbf{s}$ yields

$$\omega'^2\langle\mathbf{s},\mathbf{s}'\rangle = \langle\mathbf{s},\mathcal{H}\mathbf{s}'\rangle. \tag{4.19}$$

Upon subtracting equation (4.19) from (4.18) and using the Hermitian symmetry (4.11), we find that two eigenfunctions $\mathbf{s}$ and $\mathbf{s}'$ associated with distinct positive eigenfrequencies are *orthogonal* in the sense

$$\langle\mathbf{s},\mathbf{s}'\rangle = 0 \quad \text{if} \quad \omega \neq \omega'. \tag{4.20}$$

Because of this orthogonality, every eigensolution $\pm\omega$, $\mathbf{s}$ is referred to as a *normal mode.*

If $\mathbf{s}$ is an eigenfunction associated with $\pm\omega$, then so is $c\mathbf{s}$, where c is an arbitrary constant. To determine $|c|$ we impose the *normalization condition*

$$\langle\mathbf{s},\mathbf{s}\rangle = 1. \tag{4.21}$$

This specifies the eigenfunctions $\mathbf{s}$ completely—apart from an immaterial overall sign. Both rotation and linear anelasticity act to modify the orthonormality relations, as we shall see in Sections 4.2.1, 6.2.1 and 6.3.1. In all cases, we require these more complicated relations to agree with equations (4.20)–(4.21) in the appropriate limit.

4.1.3 Rayleigh's principle

Every self-adjoint eigenvalue problem of the form $\mathcal{H}\mathbf{s} = \omega^2\mathbf{s}$ is associated with a variational principle, known as *Rayleigh's principle.* We regard the right side of the equation

$$\omega^2 = \frac{\langle \mathbf{s}, \mathcal{H}\mathbf{s}\rangle}{\langle \mathbf{s}, \mathbf{s}\rangle} \tag{4.22}$$

as a functional which assigns a scalar ω^2 to every possible displacement field $\mathbf{s}$. Rayleigh's principle asserts that this functional is stationary for an arbitrary variation $\delta\mathbf{s}$ if and only if $\mathbf{s}$ is an eigenfunction of $\mathcal{H}$ with associated squared eigenfrequency ω^2. To verify this we note that, correct to first order in $\|\delta\mathbf{s}\|$,

$$\begin{aligned}\delta\omega^2 &= \frac{\langle \delta\mathbf{s}, \mathcal{H}\mathbf{s}\rangle + \langle \mathbf{s}, \mathcal{H}\delta\mathbf{s}\rangle - \omega^2\langle \delta\mathbf{s}, \mathbf{s}\rangle - \omega^2\langle \mathbf{s}, \delta\mathbf{s}\rangle}{\langle \mathbf{s}, \mathbf{s}\rangle} \\ &= \frac{2\langle \delta\mathbf{s}, \mathcal{H}\mathbf{s} - \omega^2\mathbf{s}\rangle}{\langle \mathbf{s}, \mathbf{s}\rangle},\end{aligned} \tag{4.23}$$

where we have used the self-adjointness (4.11) of the operator $\mathcal{H}$. It is clear from (4.23) that the variation $\delta\omega^2$ vanishes for an arbitrary $\delta\mathbf{s}$ if and only if ω^2 and $\mathbf{s}$ satisfy $\mathcal{H}\mathbf{s} = \omega^2\mathbf{s}$. This establishes Rayleigh's principle; the ratio of the two quantities $\langle \mathbf{s}, \mathcal{H}\mathbf{s}\rangle$ and $\langle \mathbf{s}, \mathbf{s}\rangle$ is known as the *Rayleigh quotient.*

Alternatively and equivalently, we may consider the quantity

$$\mathcal{I} = \tfrac{1}{2}\omega^2\langle \mathbf{s}, \mathbf{s}\rangle - \tfrac{1}{2}\langle \mathbf{s}, \mathcal{H}\mathbf{s}\rangle \tag{4.24}$$

rather than the squared eigenfrequency ω^2 to be the stationary functional. We regard $\mathcal{I}$ as a quadratic functional of the displacement field $\mathbf{s}$, for a fixed value of ω^2; correct to first order in $\|\delta\mathbf{s}\|$ we may then write

$$\delta\mathcal{I} = \langle \delta\mathbf{s}, \omega^2\mathbf{s} - \mathcal{H}\mathbf{s}\rangle, \tag{4.25}$$

where we have again used the self-adjointness of $\mathcal{H}$. Clearly, $\delta\mathcal{I}$ vanishes for an arbitrary $\delta\mathbf{s}$ if and only if $\delta\omega^2$ does; the two variations are related by $\delta\omega^2 = -2\langle \mathbf{s}, \mathbf{s}\rangle^{-1}\delta\mathcal{I}$. The stationarity of the squared eigenfrequencies ω^2 has an appealing physical significance; however, the quadratic dependence of $\mathcal{I}$ upon the trial eigenfunction $\mathbf{s}$ makes it easier to manipulate in future applications. In addition, the stationarity of $\mathcal{I}$ can be more readily generalized to the case of a rotating and/or anelastic Earth, as we shall see in Sections 4.2.3, 6.2.2 and 6.3.2.

The above "proof" of Rayleigh's principle for an arbitrary self-adjoint eigenvalue problem $\mathcal{H}\mathbf{s} = \omega^2\mathbf{s}$ is simply schematic; in any particular application, we must take account of the boundary conditions that are associated with the operator $\mathcal{H}$. It is also necessary to impose a restriction upon the allowable, or *admissible*, variations $\delta\mathbf{s}$; in the present case, an admissible variation is one that satisfies $[\delta\mathbf{s}]_-^+ = \mathbf{0}$ on the solid-solid boundaries Σ_{SS} and $[\hat{\mathbf{n}} \cdot \delta\mathbf{s}]_-^+ = 0$ on the fluid-solid boundaries Σ_{FS}. To obtain a more precise statement of the *displacement version* of Rayleigh's principle, it is convenient to rewrite the action $\mathcal{I}$ in the form

$$\mathcal{I} = \tfrac{1}{2}(\omega^2\mathcal{T} - \mathcal{V}), \tag{4.26}$$

where

$$\mathcal{T} = \int_{\oplus} \rho^0(\mathbf{s} \cdot \mathbf{s})\, dV, \tag{4.27}$$

$$\begin{aligned}\mathcal{V} = &\int_{\oplus} [\boldsymbol{\nabla}\mathbf{s} : \boldsymbol{\Lambda} : \boldsymbol{\nabla}\mathbf{s} + \rho^0\mathbf{s} \cdot \boldsymbol{\nabla}\phi^{\mathrm{E1}} + \rho^0\mathbf{s} \cdot \boldsymbol{\nabla}\boldsymbol{\nabla}\phi^0 \cdot \mathbf{s}]\, dV \\ &+ \int_{\Sigma_{\mathrm{FS}}} [\varpi^0\mathbf{s} \cdot (\boldsymbol{\nabla}^{\Sigma}\mathbf{s}) \cdot \hat{\mathbf{n}} - (\hat{\mathbf{n}} \cdot \mathbf{s})\boldsymbol{\nabla}^{\Sigma} \cdot (\varpi^0\mathbf{s})]_-^+\, d\Sigma.\end{aligned} \tag{4.28}$$

The quantities $\mathcal{T}$ and $\mathcal{V}$ defined by (4.27)–(4.28) are quadratic functionals of the displacement $\mathbf{s}$. For obvious reasons, we refer to these as the *kinetic energy* and *elastic-gravitational potential energy* functionals, respectively. Upon making use of both the three-dimensional and two-dimensional versions of Gauss' theorem in $\oplus$ and on Σ_{FS}, as we did in verifying Hamilton's principle in Section 3.7.1, we find that the variation (4.25) of $\mathcal{I}$ can be written in the form

$$\begin{aligned}\delta\mathcal{I} = &\int_{\oplus} \delta\mathbf{s} \cdot [\omega^2\rho^0\mathbf{s} + \boldsymbol{\nabla} \cdot \mathbf{T}^{\mathrm{PK1}} - \rho^0\boldsymbol{\nabla}\phi^{\mathrm{E1}} - \rho^0\mathbf{s} \cdot \boldsymbol{\nabla}\boldsymbol{\nabla}\phi^0]\, dV \\ &- \int_{\partial\oplus} \delta\mathbf{s} \cdot (\hat{\mathbf{n}} \cdot \mathbf{T}^{\mathrm{PK1}})\, d\Sigma \\ &+ \int_{\Sigma_{\mathrm{SS}}} \delta\mathbf{s} \cdot [\hat{\mathbf{n}} \cdot \mathbf{T}^{\mathrm{PK1}}]_-^+\, d\Sigma \\ &+ \int_{\Sigma_{\mathrm{FS}}} [\delta\mathbf{s} \cdot \mathbf{t}^{\mathrm{PK1}}]_-^+\, d\Sigma.\end{aligned} \tag{4.29}$$

Equation (4.29) shows that $\delta\mathcal{I}$ vanishes for an arbitrary admissible variation $\delta\mathbf{s}$ if and only if ω^2 and $\mathbf{s}$ satisfy the normal-mode equation (4.3) and the associated boundary conditions (4.4)–(4.6). In the context of the Earth's normal modes, the frequency-domain quantity $\mathcal{I} = \frac{1}{2}(\omega^2\mathcal{T} - \mathcal{V})$ is often referred to as the *action*.

Thus far we have only considered variations of the displacement eigenfunction $\mathbf{s}$, regarding the Eulerian potential perturbation ϕ^{E1} as a known functional of s given by equation (3.99). Just as with Hamilton's principle, however, there is also a *displacement-potential version* of Rayleigh's principle, in which the quantities s and ϕ^{E1} are varied independently. The stationary functional in this case is the *modified action*

$$\mathcal{I}' = \tfrac{1}{2}(\omega^2\mathcal{T} - \mathcal{V}'), \tag{4.30}$$

where

$$\begin{aligned}\mathcal{V}' = \int_{\bigcirc} [\boldsymbol{\nabla}\mathbf{s}:\boldsymbol{\Lambda}:\boldsymbol{\nabla}\mathbf{s} + 2\rho^0\mathbf{s}\cdot\boldsymbol{\nabla}\phi^{\mathrm{E1}} \\ + \rho^0\mathbf{s}\cdot\boldsymbol{\nabla\nabla}\phi^0\cdot\mathbf{s} + (4\pi G)^{-1}\boldsymbol{\nabla}\phi^{\mathrm{E1}}\cdot\boldsymbol{\nabla}\phi^{\mathrm{E1}}]\,dV \\ + \int_{\Sigma_{\mathrm{FS}}} [\varpi^0\mathbf{s}\cdot(\boldsymbol{\nabla}^{\Sigma}\mathbf{s})\cdot\hat{\mathbf{n}} - (\hat{\mathbf{n}}\cdot\mathbf{s})\boldsymbol{\nabla}^{\Sigma}\cdot(\varpi^0\mathbf{s})]_-^+\,d\Sigma.\end{aligned} \tag{4.31}$$

An *admissible* potential variation $\delta\phi^{\mathrm{E1}}$ is one that satisfies $[\delta\phi^{\mathrm{E1}}]_-^+ = 0$ on the boundary Σ. Correct to first order in $\|\delta\mathbf{s}\|$ and $\delta\phi^{\mathrm{E1}}$, the variation of the modified action is

$$\delta\mathcal{I}' = \delta\mathcal{I} + \int_{\bigcirc} \delta\phi^{\mathrm{E1}}(\boldsymbol{\nabla}\cdot\boldsymbol{\xi}^{\mathrm{E1}})\,dV + \int_{\Sigma} \delta\phi^{\mathrm{E1}}[\hat{\mathbf{n}}\cdot\boldsymbol{\xi}^{\mathrm{E1}}]_-^+\,d\Sigma, \tag{4.32}$$

where $\boldsymbol{\xi}^{\mathrm{E1}} = (4\pi G)^{-1}\boldsymbol{\nabla}\phi^{\mathrm{E1}} + \rho^0\mathbf{s}$. The first term in (4.32), given by (4.29), vanishes for an arbitrary admissible variation $\delta\mathbf{s}$ if and only if ω^2, s and ϕ^{E1} satisfy the momentum equation (4.3) and dynamical boundary conditions (4.4)–(4.6) as before, whereas the remaining terms vanish for an arbitrary admissible variation $\delta\phi^{\mathrm{E1}}$ if and only if ϕ^{E1} is related to s by the gravitational boundary-value problem

$$\boldsymbol{\nabla}\cdot\boldsymbol{\xi}^{\mathrm{E1}} = 0 \quad \text{in } \bigcirc, \qquad [\hat{\mathbf{n}}\cdot\boldsymbol{\xi}^{\mathrm{E1}}]_-^+ = 0 \quad \text{on } \Sigma. \tag{4.33}$$

This establishes the displacement-potential variational principle. The two potential energy functionals $\mathcal{V}$ and $\mathcal{V}'$ and the two actions $\mathcal{I}$ and $\mathcal{I}'$ are equal for any eigensolution ω^2, s, ϕ^{E1} by virtue of the identity (3.195). The stationary value of both actions is

$$\mathcal{I} = \mathcal{I}' = 0, \tag{4.34}$$

by virtue of equations (4.22) and (4.24).

4.1.4 Lagrangian and energy density

We shall find it convenient in later portions of this book to rewrite the frequency-domain actions in the explicit form

$$\mathcal{I} = \int_{\oplus} L(\mathbf{s},\boldsymbol{\nabla}\mathbf{s})\,dV + \int_{\Sigma_{\mathrm{FS}}} [L^{\Sigma}(\mathbf{s},\boldsymbol{\nabla}^{\Sigma}\mathbf{s})]_-^+\,d\Sigma, \tag{4.35}$$

$$\mathcal{I}' = \int_{\bigcirc} L'(\mathbf{s}, \boldsymbol{\nabla}\mathbf{s}, \boldsymbol{\nabla}\phi^{\mathrm{E1}})\, dV + \int_{\Sigma_{\mathrm{FS}}} [L^{\Sigma}(\mathbf{s}, \boldsymbol{\nabla}^{\Sigma}\mathbf{s})]_{-}^{+}\, d\Sigma, \tag{4.36}$$

where

$$L = \tfrac{1}{2}[\omega^2\rho^0\mathbf{s}\cdot\mathbf{s} - \boldsymbol{\nabla}\mathbf{s}{:}\boldsymbol{\Lambda}{:}\boldsymbol{\nabla}\mathbf{s} - \rho^0\mathbf{s}\cdot\boldsymbol{\nabla}\phi^{\mathrm{E1}} - \rho^0\mathbf{s}\cdot\boldsymbol{\nabla}\boldsymbol{\nabla}\phi^0\cdot\mathbf{s}], \tag{4.37}$$

$$L' = \tfrac{1}{2}[\omega^2\rho^0\mathbf{s}\cdot\mathbf{s} - \boldsymbol{\nabla}\mathbf{s}{:}\boldsymbol{\Lambda}{:}\boldsymbol{\nabla}\mathbf{s} - 2\rho^0\mathbf{s}\cdot\boldsymbol{\nabla}\phi^{\mathrm{E1}} - \rho^0\mathbf{s}\cdot\boldsymbol{\nabla}\boldsymbol{\nabla}\phi^0\cdot\mathbf{s} - (4\pi G)^{-1}\boldsymbol{\nabla}\phi^{\mathrm{E1}}\cdot\boldsymbol{\nabla}\phi^{\mathrm{E1}}], \tag{4.38}$$

$$L^{\Sigma} = \tfrac{1}{2}[(\hat{\mathbf{n}}\cdot\mathbf{s})\boldsymbol{\nabla}^{\Sigma}\cdot(\varpi^0\mathbf{s}) - \varpi^0\mathbf{s}\cdot(\boldsymbol{\nabla}^{\Sigma}\mathbf{s})\cdot\hat{\mathbf{n}}]. \tag{4.39}$$

We shall refer to L, L' and L^{Σ} as the volumetric and surficial *Lagrangian densities*. The corresponding frequency-domain *energy densities*, defined by analogy with equations (3.196) and (3.207), are

$$E = \omega\partial_\omega L - L, \qquad E' = \omega\partial_\omega L' - L', \qquad E^{\Sigma} = -L^{\Sigma}, \tag{4.40}$$

or, equivalently,

$$E = \tfrac{1}{2}[\omega^2\rho^0\mathbf{s}\cdot\mathbf{s} + \boldsymbol{\nabla}\mathbf{s}{:}\boldsymbol{\Lambda}{:}\boldsymbol{\nabla}\mathbf{s} + \rho^0\mathbf{s}\cdot\boldsymbol{\nabla}\phi^{\mathrm{E1}} + \rho^0\mathbf{s}\cdot\boldsymbol{\nabla}\boldsymbol{\nabla}\phi^0\cdot\mathbf{s}], \tag{4.41}$$

$$E' = \tfrac{1}{2}[\omega^2\rho^0\mathbf{s}\cdot\mathbf{s} + \boldsymbol{\nabla}\mathbf{s}{:}\boldsymbol{\Lambda}{:}\boldsymbol{\nabla}\mathbf{s} + 2\rho^0\mathbf{s}\cdot\boldsymbol{\nabla}\phi^{\mathrm{E1}} + \rho^0\mathbf{s}\cdot\boldsymbol{\nabla}\boldsymbol{\nabla}\phi^0\cdot\mathbf{s} + (4\pi G)^{-1}\boldsymbol{\nabla}\phi^{\mathrm{E1}}\cdot\boldsymbol{\nabla}\phi^{\mathrm{E1}}], \tag{4.42}$$

$$E^{\Sigma} = \tfrac{1}{2}[\varpi^0\mathbf{s}\cdot(\boldsymbol{\nabla}^{\Sigma}\mathbf{s})\cdot\hat{\mathbf{n}} - (\hat{\mathbf{n}}\cdot\mathbf{s})\boldsymbol{\nabla}^{\Sigma}\cdot(\varpi^0\mathbf{s})]. \tag{4.43}$$

The integrated energy of a mode,

$$\mathcal{E} = \int_{\oplus} E\, dV + \int_{\Sigma_{\mathrm{FS}}} [E^{\Sigma}]_{-}^{+}\, d\Sigma = \int_{\bigcirc} E'\, dV + \int_{\Sigma_{\mathrm{FS}}} [E^{\Sigma}]_{-}^{+}\, d\Sigma, \tag{4.44}$$

can be written in terms of the kinetic and potential energy quadratic functionals in the form

$$\mathcal{E} = \tfrac{1}{2}(\omega^2\mathcal{T} + \mathcal{V}) = \tfrac{1}{2}(\omega^2\mathcal{T} + \mathcal{V}'). \tag{4.45}$$

The result (4.34) expresses the *equipartition* between the kinetic and elastic-gravitational potential energies of a mode:

$$\omega^2\mathcal{T} = \mathcal{V} = \mathcal{V}'. \tag{4.46}$$

The total energy of an oscillation is therefore simply twice the kinetic energy: $\mathcal{E} = \omega^2\mathcal{T}$.

4.1.5 Dynamical stability

As we have seen, the reality of the Rayleigh quotient guarantees that the eigenfrequencies are either purely real, $\pm\omega$, or purely imaginary, $\pm i|\omega|$. Any imaginary eigenfrequencies $\pm i|\omega|$ are indicative of instability, with the associated initial disturbance growing exponentially like $\exp(|\omega|t)$. This is referred to as *ordinary instability*, since it occurs even in the absence of any infinitesimal anelasticity. An Earth model will be dynamically stable in this ordinary sense if and only if every squared eigenfrequency $\omega^2 = \mathcal{V}/\mathcal{T}$ is non-negative. The kinetic energy functional $\mathcal{T}$ is inherently positive; it follows that a necessary and sufficient condition for ordinary stability is that

$$\mathcal{V} \geq 0, \tag{4.47}$$

for all piecewise smooth displacements $\mathbf{s}$ satisfying $[\mathbf{s}]_-^+ = \mathbf{0}$ on the solid-solid boundaries $\Sigma_{\rm SS}$ and $[\hat{\mathbf{n}} \cdot \mathbf{s}]_-^+ = 0$ on the fluid-solid boundaries $\Sigma_{\rm FS}$. This is precisely the condition for secular stability which we obtained in Section 3.9.6; the conditions for linear stability in the presence or absence of friction coincide on a non-rotating Earth model.

★4.1.6 Rigid-body and geostrophic modes

There is a class of "modes" whose associated eigenfrequencies $\pm\omega$ are identically zero; the associated eigenspace consists of all the time-independent displacement fields $\mathbf{s}$ that do not alter the elastic-gravitational potential energy $\mathcal{V}$ of the Earth. We refer to these as the trivial modes, or for brevity, the *trivials*. On a non-rotating Earth, these modes are so inconsequential that they are hardly worth mentioning; however, their counterparts on a rotating Earth play a troublesome role in the analysis of the response to an imposed force, so it is useful to discuss them now. The non-rotating trivials consist of two types of modes: rigid-body modes of the whole Earth, and geostrophic modes confined to the fluid regions $\oplus_{\rm F}$.

The *rigid-body modes* have real displacement eigenfunctions of the form

$$\mathbf{s} = \mathbf{X} + \mathbf{Q} \cdot \mathbf{x}, \tag{4.48}$$

where $\mathbf{X}$ is a constant vector, and $\mathbf{Q}$ is a constant orthogonal tensor, satisfying $\mathbf{Q}^{\rm T} \cdot \mathbf{Q} = \mathbf{Q} \cdot \mathbf{Q}^{\rm T} = \mathbf{I}$. Such displacements, consisting of a rigid *translation* $\mathbf{X}$ and *rotation* $\mathbf{Q}$, clearly have no effect on the elastic-gravitational energy of the Earth. It is easily verified that the frequency-domain momentum equation (4.3) is satisfied; the associated changes in the Eulerian density and gravitational potential are $\rho^{\rm E1} = -\mathbf{s} \cdot \boldsymbol{\nabla}\rho^0$ and $\phi^{\rm E1} = -\mathbf{s} \cdot \boldsymbol{\nabla}\phi^0$.

In general, there are six such modes, each with an associated squared eigenfrequency $\omega^2 = 0$, accounting for the six rigid-body degrees of freedom of the Earth. In certain circumstances, this catalogue must be amended in an obvious way; for example, if the boundaries of the fluid outer core are spherically or ellipsoidally symmetric, then there will be separate rigid-body rotational modes $\mathbf{Q} \cdot \mathbf{x}$ for the crust and mantle and the solid inner core.

In the fluid regions of the Earth $\oplus_{\rm F}$, equation (4.3) reduces to

$$-\omega^2 \rho^0 \mathbf{s} + \boldsymbol{\nabla} p^{\rm E1} + \rho^0 \boldsymbol{\nabla} \phi^{\rm E1} + \rho^{\rm E1} \boldsymbol{\nabla} \phi^0 = \mathbf{0}, \tag{4.49}$$

where $\rho^{\rm E1} = -\boldsymbol{\nabla} \cdot (\rho^0 \mathbf{s})$ and $p^{\rm E1} = -\kappa(\boldsymbol{\nabla} \cdot \mathbf{s}) - \mathbf{s} \cdot \boldsymbol{\nabla} p^0$. A *geostrophic mode* has a displacement field of the form

$$\mathbf{s} = \mathbf{0} \quad \text{in } \oplus_{\rm S}, \tag{4.50}$$

$$\hat{\mathbf{n}} \cdot \mathbf{s} = 0 \quad \text{on } \Sigma_{\rm FS}, \tag{4.51}$$

$$\mathbf{s} = (1/\rho^0)(\hat{\boldsymbol{\gamma}}^0 \times \boldsymbol{\nabla} \chi) \quad \text{in } \oplus_{\rm F}. \tag{4.52}$$

The unit vector $\hat{\boldsymbol{\gamma}}^0$ is the normal to the level surfaces of ρ^0, ϕ^0 and p^0 in $\oplus_{\rm F}$, and χ is an arbitrary scalar. The elastic-gravitational potential energy of the Earth is unaffected by any such displacement; the static momentum equation (4.49) is satisfied because the associated Eulerian perturbations in density $\rho^{\rm E1}$, gravitational potential $\phi^{\rm E1}$ and pressure $p^{\rm E1}$ are zero everywhere throughout the Earth. There is an infinite-dimensional family of such geostrophic modes in $\oplus_{\rm F}$, each with an associated squared eigenfrequency $\omega^2 = 0$. Any steady flow that moves along the common level surfaces ($\hat{\boldsymbol{\gamma}}^0 \cdot \mathbf{s} = 0$) and preserves the density at every spatial point is a member of the geostrophic eigenspace.

Within any regions of $\oplus_{\rm F}$ that are neutrally stratified, the family of geostrophic flows is even larger; the *Adams-Williamson condition* which characterizes a state of neutral stratification in a general Earth model is $\boldsymbol{\nabla} p^0 = (\kappa/\rho^0)\boldsymbol{\nabla}\rho^0$. Wherever this relation is satisfied, the representation (4.52) can be replaced by

$$\mathbf{s} = (1/\rho^0)(\boldsymbol{\nabla} \times \boldsymbol{\chi}), \tag{4.53}$$

where $\boldsymbol{\chi}$ is an arbitrary vector.

4.1.7 Green tensor

The response of the Earth to an earthquake, or any other seismic source which excites its free oscillations and the equivalent travelling body and surface waves, can be conveniently expressed in terms of the second-order

Green tensor or *impulse response* $\mathbf{G}(\mathbf{x},\mathbf{x}';t)$. By definition, the quantity $G_{pq}(\mathbf{x},\mathbf{x}';t)$ is the $\hat{\mathbf{x}}_p$ component of the displacement response at $\mathbf{x}$, t to a unit impulsive force in the $\hat{\mathbf{x}}_q$ direction acting at $\mathbf{x}'$, 0. Alternatively and equivalently, we can characterize $\mathbf{G}$ as the solution to the homogeneous equation

$$\rho^0(\partial_t^2\mathbf{G} + \mathcal{H}\mathbf{G}) = \mathbf{0}, \tag{4.54}$$

subject to the inhomogeneous initial conditions

$$\mathbf{G}(\mathbf{x},\mathbf{x}';0) = \mathbf{0}, \qquad \partial_t\mathbf{G}(\mathbf{x},\mathbf{x}';0) = (1/\rho^0)\mathbf{I}\,\delta(\mathbf{x}-\mathbf{x}'). \tag{4.55}$$

To solve the initial-value problem (4.54)–(4.55), we label the eigenfrequencies $\pm\,\omega_k$ and the associated eigenfunctions $\mathbf{s}_k$ with an index k. The orthonormality relations (4.20)–(4.21) can then be written in the form $\langle\mathbf{s}_k,\mathbf{s}_{k'}\rangle = \delta_{kk'}$, or, equivalently,

$$\int_\oplus \rho^0\mathbf{s}_k\cdot\mathbf{s}_{k'}\,dV = \delta_{kk'}. \tag{4.56}$$

We assume that the normalized eigenfunctions constitute a *complete orthonormal basis*, and write the impulse response $\mathbf{G}$ as a linear combination of real free oscillations, of the form

$$\mathbf{G}(\mathbf{x},\mathbf{x}',t) = \sum_k \mathbf{s}_k(\mathbf{x})[\mathbf{a}_k(\mathbf{x}')\cos\omega_k t + \mathbf{b}_k(\mathbf{x}')\sin\omega_k t], \tag{4.57}$$

where the sum is over all of the non-negative eigenfrequencies, $\omega_k \geq 0$. Such a normal-mode sum satisfies equation (4.54) by inspection; the initial conditions (4.55) will be satisfied provided that

$$\sum_k \mathbf{s}_k\mathbf{a}_k = \mathbf{0}, \qquad \sum_k \omega_k\mathbf{s}_k\mathbf{b}_k = (1/\rho^0)\mathbf{I}\,\delta(\mathbf{x}-\mathbf{x}'). \tag{4.58}$$

The coefficients $\mathbf{a}_k$ and $\mathbf{b}_k$ are easily obtained by taking the inner product of (4.58) with $\mathbf{s}_k$, and using the orthonormality relation (4.56):

$$\mathbf{a}_k = \mathbf{0}, \qquad \mathbf{b}_k = \omega_k^{-1}\mathbf{s}_k(\mathbf{x}'). \tag{4.59}$$

The Green tensor of a non-rotating Earth is therefore given in terms of the normal-mode eigenfrequencies and eigenfunctions by

$$\mathbf{G}(\mathbf{x},\mathbf{x}';t) = \sum_k \omega_k^{-1}\mathbf{s}_k(\mathbf{x})\mathbf{s}_k(\mathbf{x}')\sin\omega_k t. \tag{4.60}$$

The result (4.60) is valid for all times $t \geq 0$; obviously, $\mathbf{G} = \mathbf{0}$ for $t < 0$.

The imposition of a unit impulsive force causes every mode to begin oscillating like $\sin\omega_k t$. Since the eigenfunctions $\mathbf{s}_k$ are real, the phase of every oscillation is the same $\pm\pi$ throughout the Earth. This is characteristic of a *standing wave*. It is noteworthy that $\mathbf{G}$ is symmetric in the sense

$$\mathbf{G}(\mathbf{x},\mathbf{x}';t) = \mathbf{G}^{\mathrm{T}}(\mathbf{x}',\mathbf{x};t). \tag{4.61}$$

Equation (4.61) expresses the *principle of seismic reciprocity*. Loosely speaking, this principle states that the source and receiver can be interchanged; note that the orientations as well as the locations must be swapped, as depicted schematically in Figure 4.1.

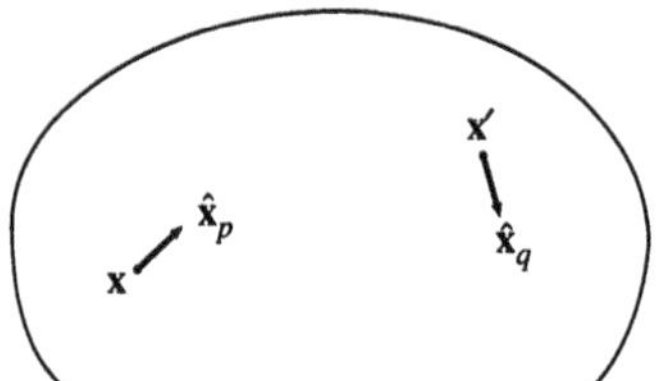

Figure 4.1. Seismic reciprocity on a non-rotating Earth: the $\hat{\mathbf{x}}_p$ component of the response at $\mathbf{x}$ to a point force acting in the $\hat{\mathbf{x}}_q$ direction at $\mathbf{x}'$ is equal to the $\hat{\mathbf{x}}_q$ component of the response at $\mathbf{x}'$ to a point force acting in the $\hat{\mathbf{x}}_p$ direction at $\mathbf{x}$. The components of the Green tensor or unit-impulse response satisfy $G_{pq}(\mathbf{x},\mathbf{x}';t) = G_{qp}(\mathbf{x}',\mathbf{x};t)$.

The contribution of each of the Earth's trivial modes to the Green tensor $\mathbf{G}$ grows linearly in time, since $\omega_k^{-1}\sin\omega_k t \to t$ as $\omega_k \to 0$. Such a linear growth is not indicative of instability, because the associated particle velocities $\partial_t\mathbf{G}$ remain bounded. In any case, none of the trivials can be excited by an earthquake source. The rigid-body modes are unexcited because no indigenous source can exert a net force or torque upon the Earth, and the geostrophic modes are unexcited because they are confined to the fluid regions of the Earth, whereas an earthquake must be situated in the solid crust or mantle. We shall refer to the quantity $\mathbf{G}$ with the trivials excised from the sum (4.60) as the *seismic Green tensor.*

4.1.8 Response to a transient force

In Chapter 5 we shall show that any indigenous source, such as an earthquake, can always be represented by an equivalent body force density $\mathbf{f}$ acting in $\oplus$ and an equivalent surface force density $\mathbf{t}$ acting upon $\partial\oplus$. The displacement $\mathbf{s}$ produced by any such source can be written as a convolution of the impulse response $\mathbf{G}$ with the entire past history of the equivalent forces $\mathbf{f}$ and $\mathbf{t}$:

$$\mathbf{s}(\mathbf{x},t) = \int_{-\infty}^{t}\int_{\oplus} \mathbf{G}(\mathbf{x},\mathbf{x}';t-t')\cdot\mathbf{f}(\mathbf{x}',t')\,dV'\,dt' \tag{4.62}$$
$$+ \int_{-\infty}^{t}\int_{\partial\oplus} \mathbf{G}(\mathbf{x},\mathbf{x}';t-t')\cdot\mathbf{t}(\mathbf{x}',t')\,d\Sigma'\,dt'.$$

This result, which embodies the principles of superposition and causality, should be self-evident; for doubters who demand a derivation, we provide one (on a rotating Earth) in Section 5.3.

We can write $\mathbf{s}$ as a sum of normal modes by inserting the representation (4.60) into (4.62); this gives

$$\mathbf{s}(\mathbf{x},t) = \sum_k \omega_k^{-1}\mathbf{s}_k(\mathbf{x}) \int_{-\infty}^{t} A_k(t') \sin\omega_k(t-t')\, dt', \tag{4.63}$$

where

$$A_k(t) = \int_{\oplus} \mathbf{f}(\mathbf{x},t)\cdot\mathbf{s}_k(\mathbf{x})\, dV + \int_{\partial\oplus} \mathbf{t}(\mathbf{x},t)\cdot\mathbf{s}_k(\mathbf{x})\, d\Sigma. \tag{4.64}$$

Integrating by parts with respect to time, we obtain the equivalent result

$$\mathbf{s}(\mathbf{x},t) = \sum_k \omega_k^{-2}\mathbf{s}_k(\mathbf{x}) \int_{-\infty}^{t} \partial_{t'} A_k(t')[1-\cos\omega_k(t-t')]\, dt', \tag{4.65}$$

where

$$\partial_t A_k(t) = \int_{\oplus} \partial_t\mathbf{f}(\mathbf{x},t)\cdot\mathbf{s}_k(\mathbf{x})\, dV + \int_{\partial\oplus} \partial_t\mathbf{t}(\mathbf{x},t)\cdot\mathbf{s}_k(\mathbf{x})\, d\Sigma. \tag{4.66}$$

Equation (4.65) expresses the response $\mathbf{s}$ as a superposition of Heaviside or step-function responses $\omega_k^{-2}[1-\cos\omega_k(t-t')]$, whereas (4.63) expresses it as a superposition of Dirac or impulse responses $\omega_k^{-1}\sin\omega_k(t-t')$.

The equivalent forces $\mathbf{f}$ and $\mathbf{t}$ associated with an earthquake are zero prior to some origin time t_0, and they attain constant static values $\mathbf{f}_{\mathrm{f}}$ and $\mathbf{t}_{\mathrm{f}}$ after a time t_{f}, as illustrated in Figure 4.2. In the case of a fault source, the

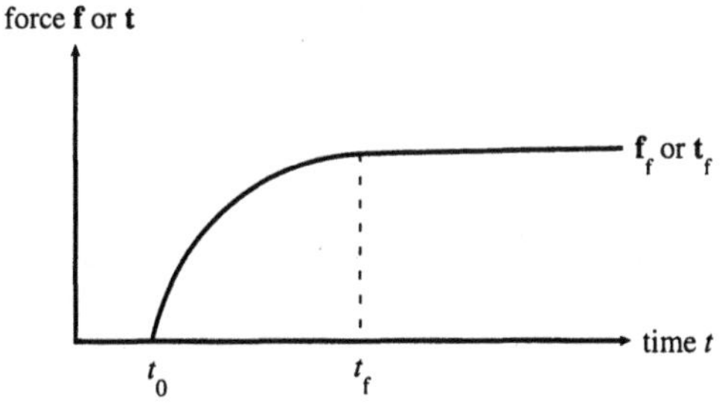

Figure 4.2. Schematic time dependence of a transient equivalent force $\mathbf{f}(\mathbf{x},t)$ or $\mathbf{t}(\mathbf{x},t)$, which initiates at time $t=t_0$ and attains a static constant value $\mathbf{f}_{\mathrm{f}}(\mathbf{x})$ or $\mathbf{t}_{\mathrm{f}}(\mathbf{x})$ for times $t \geq t_{\mathrm{f}}$.

time interval $t_{\mathrm{f}} - t_0$ represents the duration of faulting, and the quantities $\mathbf{f}_{\mathrm{f}}$ and $\mathbf{t}_{\mathrm{f}}$ are associated with the final static slip on the fault, as we shall see in Chapter 5. The response to such a *transient source* takes an especially simple form for times $t \geq t_{\mathrm{f}}$, after the temporal variation of the equivalent forces has ceased. From equation (4.65) we find that

$$\mathbf{s}(\mathbf{x},t) = \sum_k \omega_k^{-2}(a_k^{\mathrm{f}} - a_k \cos\omega_k t - b_k \sin\omega_k t)\,\mathbf{s}_k(\mathbf{x}), \quad t \geq t_{\mathrm{f}}, \tag{4.67}$$

where

$$a_k^{\mathrm{f}} = \int_{\oplus} \mathbf{f}_{\mathrm{f}}(\mathbf{x}) \cdot \mathbf{s}_k(\mathbf{x})\,dV + \int_{\partial\oplus} \mathbf{t}_{\mathrm{f}}(\mathbf{x}) \cdot \mathbf{s}_k(\mathbf{x})\,d\Sigma, \tag{4.68}$$

$$\begin{aligned} a_k = &\int_{t_0}^{t_{\mathrm{f}}} \int_{\oplus} \partial_t \mathbf{f}(\mathbf{x},t) \cdot \mathbf{s}_k(\mathbf{x}) \cos\omega_k t\,dV\,dt \\ &+ \int_{t_0}^{t_{\mathrm{f}}} \int_{\partial\oplus} \partial_t \mathbf{t}(\mathbf{x},t) \cdot \mathbf{s}_k(\mathbf{x}) \cos\omega_k t\,d\Sigma\,dt, \end{aligned} \tag{4.69}$$

$$\begin{aligned} b_k = &\int_{t_0}^{t_{\mathrm{f}}} \int_{\oplus} \partial_t \mathbf{f}(\mathbf{x},t) \cdot \mathbf{s}_k(\mathbf{x}) \sin\omega_k t\,dV\,dt \\ &+ \int_{t_0}^{t_{\mathrm{f}}} \int_{\partial\oplus} \partial_t \mathbf{t}(\mathbf{x},t) \cdot \mathbf{s}_k(\mathbf{x}) \sin\omega_k t\,d\Sigma\,dt. \end{aligned} \tag{4.70}$$

The physical interpretation of the result (4.67) is clear. The oscillatory terms $a_k \cos\omega_k t$ and $b_k \sin\omega_k t$ represent the *free oscillations* of the Earth excited by the earthquake. The inevitable presence of anelasticity within the Earth will cause these oscillations to decay with time; the resulting time-independent displacement in the limit $t \to \infty$ is in that case

$$\mathbf{s}_{\mathrm{f}}(\mathbf{x}) = \sum_k \omega_k^{-2} a_k^{\mathrm{f}} \mathbf{s}_k(\mathbf{x}). \tag{4.71}$$

The quantity $\mathbf{s}_{\mathrm{f}}$ represents the *static response* of the Earth to the final static values of the equivalent forces $\mathbf{f}_{\mathrm{f}}$ and $\mathbf{t}_{\mathrm{f}}$. In the case of a fault source, the offset of one wall relative to the other produces a permanent deformation at every point $\mathbf{x}$ within the Earth.

The particle *acceleration* $\mathbf{a} = \partial_t^2 \mathbf{s}$ subsequent to the cessation of faulting is given by

$$\mathbf{a}(\mathbf{x},t) = \sum_k (a_k \cos\omega_k t + b_k \sin\omega_k t)\,\mathbf{s}_k(\mathbf{x}), \quad t \geq t_{\mathrm{f}}. \tag{4.72}$$

We shall generally express the response of the Earth to an earthquake as an accelerogram, because it simplifies a number of theoretical results in later sections. In fact, an idealized accelerometer responds to changes in the Earth's gravitational field in addition to the acceleration of the instrument housing. We show how to account for these effects in Section 4.4.

*4.2 Rotating Earth Model

On a rotating Earth, the frequency-domain momentum equation (4.3) is replaced by

$$-\omega^2\rho^0\mathbf{s} + 2i\omega\rho^0\boldsymbol{\Omega}\times\mathbf{s} - \boldsymbol{\nabla}\cdot\mathbf{T}^{\mathrm{PK1}} + \rho^0\boldsymbol{\nabla}\phi^{\mathrm{E1}} + \rho^0\mathbf{s}\cdot\boldsymbol{\nabla}\boldsymbol{\nabla}(\phi^0+\psi) = \mathbf{0}, \tag{4.73}$$

where ψ is the centrifugal potential. The presence of the Coriolis force term $2i\omega\rho^0\boldsymbol{\Omega}\times\mathbf{s}$ prevents us from regarding (4.73) as a real equation, as we did with (4.3). The eigenfunctions $\mathbf{s}$ of a rotating Earth model are *inherently complex.* We shall continue to assume that the eigenfrequencies ω are real (we show below that this is guaranteed for any secularly stable Earth model). Upon taking the complex conjugate of (4.73), we observe that $-\omega$, $\mathbf{s}^*$ is an eigensolution if and only if ω, $\mathbf{s}$ is. A particular normal mode of a rotating Earth is characterized by the pair of complex conjugate eigensolutions ω, $\mathbf{s}$ and $-\omega$, $\mathbf{s}^*$.

The eigenfrequencies and eigenfunctions of the *anti-Earth*, with a reversed sense of rotation, are also of interest. Noting that equation (4.73) is invariant under the simultaneous transformations $\boldsymbol{\Omega}\to-\boldsymbol{\Omega}$ and $\omega\to-\omega$, we see that ω, $\mathbf{s}^*$ is a mode of the anti-Earth if and only if ω, $\mathbf{s}$ is a mode of the actual Earth. The eigenfrequencies ω thus do not depend on the sense of the Earth's rotation.

*4.2.1 Orthonormality

We redefine the integro-differential operator $\mathcal{H}$ on a rotating Earth so that it incorporates the centrifugal potential ψ:

$$\rho^0\mathcal{H}\mathbf{s} = -\boldsymbol{\nabla}\cdot\mathbf{T}^{\mathrm{PK1}} + \rho^0\boldsymbol{\nabla}\phi^{\mathrm{E1}} + \rho^0\mathbf{s}\cdot\boldsymbol{\nabla}\boldsymbol{\nabla}(\phi^0+\psi). \tag{4.74}$$

The frequency-domain momentum equation (4.73) together with the boundary conditions (4.4)–(4.6) can then be written symbolically in the form

$$\mathcal{H}\mathbf{s} + 2i\omega\boldsymbol{\Omega}\times\mathbf{s} = \omega^2\mathbf{s}. \tag{4.75}$$

Equation (4.75) is a non-standard eigenvalue problem, because the eigenfrequency ω occurs linearly in the Coriolis term $2i\omega\boldsymbol{\Omega}\times\mathbf{s}$ as well as quadratically in the inertial term $\omega^2\mathbf{s}$.

We define the *inner product* of any two complex functions $\mathbf{s}$ and $\mathbf{s}'$ in a rotating Earth $\oplus$ by

$$\langle \mathbf{s}, \mathbf{s}' \rangle = \int_{\oplus} \rho^0 \mathbf{s}^* \cdot \mathbf{s}' \, dV. \tag{4.76}$$

It is easily verified that both operators $\mathcal{H}$ and $i\boldsymbol{\Omega}\times$ appearing in (4.75) are *Hermitian* or *self-adjoint* with respect to the complex inner product (4.76):

$$\langle \mathbf{s}, \mathcal{H}\mathbf{s}' \rangle = \langle \mathcal{H}\mathbf{s}, \mathbf{s}' \rangle = \langle \mathbf{s}', \mathcal{H}\mathbf{s} \rangle^*, \tag{4.77}$$

$$\langle \mathbf{s}, i\boldsymbol{\Omega} \times \mathbf{s}' \rangle = \langle i\boldsymbol{\Omega} \times \mathbf{s}, \mathbf{s}' \rangle = \langle \mathbf{s}', i\boldsymbol{\Omega} \times \mathbf{s} \rangle^*. \tag{4.78}$$

Equation (4.77) can be established by a straightforward extension of the argument used to verify (4.11) on a non-rotating Earth, whereas (4.78) is an elementary triple product identity. The Hermitian character of equation (4.75) should come as no surprise—energy is conserved on either a rotating or a non-rotating Earth model, as we have seen in Section 3.8.

Taking the inner product of $\mathcal{H}\mathbf{s} + 2i\omega\boldsymbol{\Omega} \times \mathbf{s} = \omega^2\mathbf{s}$ with $\mathbf{s}'$ yields

$$\omega^2 \langle \mathbf{s}', \mathbf{s} \rangle - 2\omega \langle \mathbf{s}', i\boldsymbol{\Omega} \times \mathbf{s} \rangle - \langle \mathbf{s}', \mathcal{H}\mathbf{s} \rangle = 0, \tag{4.79}$$

whereas taking the inner product of $\mathcal{H}\mathbf{s}' + 2i\omega'\boldsymbol{\Omega} \times \mathbf{s}' = \omega'^2\mathbf{s}'$ with $\mathbf{s}$ yields

$$\omega'^2 \langle \mathbf{s}, \mathbf{s}' \rangle - 2\omega' \langle \mathbf{s}, i\boldsymbol{\Omega} \times \mathbf{s}' \rangle - \langle \mathbf{s}, \mathcal{H}\mathbf{s}' \rangle = 0. \tag{4.80}$$

Upon subtracting equation (4.79) from (4.80) and using the Hermitian symmetries (4.77)–(4.78), we obtain the result

$$\langle \mathbf{s}, \mathbf{s}' \rangle - 2(\omega + \omega')^{-1} \langle \mathbf{s}, i\boldsymbol{\Omega} \times \mathbf{s}' \rangle = 0 \quad \text{if} \quad \omega \neq \omega', \tag{4.81}$$

where we have divided by the sum of the eigenfrequencies $\omega + \omega'$ for convenience. Equation (4.81), which is the analogue of (4.20), is the expression of normal-mode *orthogonality* on a rotating elastic Earth. The complex eigenfunctions $\mathbf{s}$ and $\mathbf{s}'$ associated with distinct positive eigenfrequencies $\omega \neq \omega'$ are not orthogonal in the ordinary sense $\langle \mathbf{s}, \mathbf{s}' \rangle = 0$; instead, the orthogonality condition explicitly involves the Coriolis term. We *normalize* the eigenfunctions by requiring that

$$\langle \mathbf{s}, \mathbf{s} \rangle - \omega^{-1} \langle \mathbf{s}, i\boldsymbol{\Omega} \times \mathbf{s} \rangle = 1. \tag{4.82}$$

With this choice, the orthonormality relations (4.81)–(4.82) reduce to the previous results (4.20)–(4.21) in the limit of no rotation, $\boldsymbol{\Omega} = \mathbf{0}$.

*4.2.2 Reduction to a standard eigenvalue problem

Following Dyson & Schutz (1979) and Wahr (1981a), we can convert the non-standard eigenvalue problem (4.75) into an ordinary eigenvalue problem at the expense of doubling the dimension of the eigenfunctions. We associate a *six-dimensional* vector

$$\mathbf{z} = \begin{pmatrix} \mathbf{s} \\ \omega\mathbf{s} \end{pmatrix} \tag{4.83}$$

with every three-dimensional eigenfunction $\mathbf{s}$ and eigenfrequency ω. It is then easily shown that (4.75) is equivalent to

$$\mathcal{K}\mathbf{z} = \omega\mathbf{z}, \tag{4.84}$$

where

$$\mathcal{K} = \begin{pmatrix} 0 & 1 \\ \mathcal{H} & 2i\boldsymbol{\Omega}\times \end{pmatrix}. \tag{4.85}$$

The quantities ω and $\mathbf{z}$ can therefore be regarded as the eigenfrequency and associated eigenfunction of the six-dimensional linear operator $\mathcal{K}$.

We define the dot or scalar product $\mathbf{z}\cdot\mathbf{z}'$ of any two six-vectors

$$\mathbf{z} = \begin{pmatrix} \mathbf{s} \\ \mathbf{v} \end{pmatrix}, \qquad \mathbf{z}' = \begin{pmatrix} \mathbf{s}' \\ \mathbf{v}' \end{pmatrix} \tag{4.86}$$

by $\mathbf{z}\cdot\mathbf{z}' = \mathbf{s}\cdot\mathbf{s}' + \mathbf{v}\cdot\mathbf{v}'$, and we define their *inner product* $\langle\langle\mathbf{z},\mathbf{z}'\rangle\rangle$ by

$$\langle\langle\mathbf{z},\mathbf{z}'\rangle\rangle = \int_{\oplus} \mathbf{z}^*\cdot\mathcal{P}\mathbf{z}'\,dV, \tag{4.87}$$

where

$$\mathcal{P} = \rho^0\begin{pmatrix} \mathcal{H} & 0 \\ 0 & 1 \end{pmatrix}. \tag{4.88}$$

Written out explicitly in terms of the associated three-vectors, this six-dimensional inner product is $\langle\langle\mathbf{z},\mathbf{z}'\rangle\rangle = \langle\mathbf{s},\mathcal{H}\mathbf{s}'\rangle + \langle\mathbf{v},\mathbf{v}'\rangle$. If $\mathbf{z}$ and $\mathbf{z}'$ are both six-dimensional eigenfunctions, so that $\mathbf{v} = \omega\mathbf{s}$ and $\mathbf{v}' = \omega'\mathbf{s}'$, then $\langle\langle\mathbf{z},\mathbf{z}'\rangle\rangle = \langle\mathbf{s},\mathcal{H}\mathbf{s}'\rangle + \omega\omega'\langle\mathbf{s},\mathbf{s}'\rangle$.

The operator $\mathcal{K}$ is formally *self-adjoint* or *Hermitian* with respect to the six-dimensional inner product (4.87), i.e.,

$$\langle\langle\mathbf{z},\mathcal{K}\mathbf{z}'\rangle\rangle = \langle\langle\mathcal{K}\mathbf{z},\mathbf{z}'\rangle\rangle = \langle\langle\mathbf{z}',\mathcal{K}\mathbf{z}\rangle\rangle^* \tag{4.89}$$

for any two six-vectors $\mathbf{z}$ and $\mathbf{z}'$. This can be demonstrated by means of a straightforward calculation; the quantities $\langle\langle \mathbf{z}, \mathcal{K}\mathbf{z}'\rangle\rangle$ and $\langle\langle \mathbf{z}', \mathcal{K}\mathbf{z}\rangle\rangle$ are given explicitly by

$$\langle\langle \mathbf{z}, \mathcal{K}\mathbf{z}'\rangle\rangle = \int_\oplus \rho^0 \left[\begin{pmatrix} \mathbf{s} & \mathbf{v} \end{pmatrix}^* \cdot \begin{pmatrix} 0 & \mathcal{H} \\ \mathcal{H} & 2i\boldsymbol{\Omega}\times \end{pmatrix} \begin{pmatrix} \mathbf{s}' \\ \mathbf{v}' \end{pmatrix} \right] dV, \tag{4.90}$$

$$\langle\langle \mathbf{z}', \mathcal{K}\mathbf{z}\rangle\rangle = \int_\oplus \rho^0 \left[\begin{pmatrix} \mathbf{s}' & \mathbf{v}' \end{pmatrix}^* \cdot \begin{pmatrix} 0 & \mathcal{H} \\ \mathcal{H} & 2i\boldsymbol{\Omega}\times \end{pmatrix} \begin{pmatrix} \mathbf{s} \\ \mathbf{v} \end{pmatrix} \right] dV, \tag{4.91}$$

where we have used the operator identity

$$\mathcal{P}\mathcal{K} = \rho^0 \begin{pmatrix} \mathcal{H} & 0 \\ 0 & 1 \end{pmatrix} \begin{pmatrix} 0 & 1 \\ \mathcal{H} & 2i\boldsymbol{\Omega}\times \end{pmatrix} = \rho^0 \begin{pmatrix} 0 & \mathcal{H} \\ \mathcal{H} & 2i\boldsymbol{\Omega}\times \end{pmatrix}. \tag{4.92}$$

Upon performing the indicated operations in equations (4.90) and (4.91), and rewriting the results in terms of the three-dimensional inner product, we obtain

$$\langle\langle \mathbf{z}, \mathcal{K}\mathbf{z}'\rangle\rangle = \langle \mathbf{s}, \mathcal{H}\mathbf{v}'\rangle + \langle \mathbf{v}, \mathcal{H}\mathbf{s}'\rangle + 2\langle \mathbf{v}, i\boldsymbol{\Omega}\times\mathbf{v}'\rangle, \tag{4.93}$$

$$\langle\langle \mathbf{z}', \mathcal{K}\mathbf{z}\rangle\rangle = \langle \mathbf{s}', \mathcal{H}\mathbf{v}\rangle + \langle \mathbf{v}', \mathcal{H}\mathbf{s}\rangle + 2\langle \mathbf{v}', i\boldsymbol{\Omega}\times\mathbf{v}\rangle. \tag{4.94}$$

Equation (4.93) is equal to the complex conjugate of (4.94) by virtue of the Hermitian symmetries (4.77)–(4.78); this establishes the six-dimensional Hermiticity relation (4.89). Fundamentally, it is the Hermitian character of $\mathcal{H}$ and $i\boldsymbol{\Omega}\times$ that is responsible for the self-adjointness of the operator $\mathcal{K}$.

Upon taking the six-dimensional inner product of $\mathcal{K}\mathbf{z} = \omega\mathbf{z}$ with $\mathbf{z}'$ as well as that of $\mathcal{K}\mathbf{z}' = \omega'\mathbf{z}'$ with $\mathbf{z}$, and subtracting the results, we obtain the six-dimensional orthogonality relation

$$\langle\langle \mathbf{z}, \mathbf{z}'\rangle\rangle = 0 \quad \text{if} \quad \omega \neq \omega'. \tag{4.95}$$

It is readily verified that (4.95) is equivalent to the three-dimensional orthogonality relation (4.81); using equation (4.79) to eliminate $\langle \mathbf{s}, \mathcal{H}\mathbf{s}'\rangle$, we find that $\langle\langle \mathbf{z}, \mathbf{z}'\rangle\rangle = \omega(\omega+\omega')[\langle \mathbf{s}, \mathbf{s}'\rangle - 2(\omega+\omega')^{-1}\langle \mathbf{s}, i\boldsymbol{\Omega}\times\mathbf{s}'\rangle]$. The six-dimensional normalization condition equivalent to (4.82) is

$$\langle\langle \mathbf{z}, \mathbf{z}\rangle\rangle = 2\omega^2. \tag{4.96}$$

The unorthodox orthonormality relations involving the Coriolis force on a rotating Earth are therefore seen to be ordinary orthonormality relations in the space of six-dimensional eigenfunctions.

The above argument, though elegant, contains a subtle flaw: the relation (4.87) does not define a valid inner product in the space of all piecewise smooth six-vectors $\mathbf{z}$ in the Earth model $\oplus$, since there is a class of trivial modes whose six-dimensional norm $\langle\langle \mathbf{z}, \mathbf{z} \rangle\rangle$ vanishes. The remedy for this problem is straightforward—we simply excise the trivials and other related vectors that also fail to satisfy the constraint $\langle\langle \mathbf{z}, \mathbf{z} \rangle\rangle > 0$ from the space under consideration; the technical details of this surgical procedure may be found in Wahr (1981a). We enumerate the trivials, and briefly describe what must be done to account for them in the normal-mode excitation problem, in Sections 4.2.5 and 4.2.7–4.2.8.

*4.2.3 Rayleigh's principle

Rayleigh's principle can be readily generalized to the case of a rotating Earth model. Upon taking the six-dimensional inner product of $\mathcal{K}\mathbf{z} = \omega\mathbf{z}$ with $\mathbf{z}$, we obtain the *Rayleigh quotient*

$$\omega = \frac{\langle\langle \mathbf{z}, \mathcal{K}\mathbf{z} \rangle\rangle}{\langle\langle \mathbf{z}, \mathbf{z} \rangle\rangle}. \tag{4.97}$$

We regard the right side of equation (4.97) as a functional that assigns a scalar ω to every non-trivial six-vector $\mathbf{z}$. Rayleigh's principle states that this functional is stationary for an arbitrary variation $\boldsymbol{\delta}\mathbf{z}$ if and only if $\mathbf{z}$ is an eigenfunction of the operator $\mathcal{K}$ with associated eigenfrequency ω. To verify this, we note that, correct to first order in $\|\boldsymbol{\delta}\mathbf{z}\|$,

$$\begin{aligned} \delta\omega &= \frac{\langle\langle \boldsymbol{\delta}\mathbf{z}, \mathcal{K}\mathbf{z} \rangle\rangle + \langle\langle \mathbf{z}, \mathcal{K}\boldsymbol{\delta}\mathbf{z} \rangle\rangle - \omega\langle\langle \boldsymbol{\delta}\mathbf{z}, \mathbf{z} \rangle\rangle - \omega\langle\langle \mathbf{z}, \boldsymbol{\delta}\mathbf{z} \rangle\rangle}{\langle\langle \mathbf{z}, \mathbf{z} \rangle\rangle} \\ &= \frac{2\,\mathrm{Re}\,\langle\langle \boldsymbol{\delta}\mathbf{z}, \mathcal{K}\mathbf{z} - \omega\mathbf{z} \rangle\rangle}{\langle\langle \mathbf{z}, \mathbf{z} \rangle\rangle}, \end{aligned} \tag{4.98}$$

where we have used the self-adjointness of the operator $\mathcal{K}$. It is clear from the result (4.98) that $\delta\omega$ vanishes for an arbitrary $\boldsymbol{\delta}\mathbf{z}$ if and only if ω and $\mathbf{z}$ satisfy $\mathcal{K}\mathbf{z} = \omega\mathbf{z}$. This establishes the six-dimensional version of Rayleigh's principle.

Alternatively and equivalently, we can regard the *action*

$$\mathcal{I} = \tfrac{1}{2}\langle\langle \mathbf{z}, \mathbf{z} \rangle\rangle - \tfrac{1}{2}\omega^{-1}\langle\langle \mathbf{z}, \mathcal{K}\mathbf{z} \rangle\rangle \tag{4.99}$$

rather than the eigenfrequency ω as the stationary functional. The variation of $\mathcal{I}$ is given, correct to first order in $\|\boldsymbol{\delta}\mathbf{z}\|$, by

$$\delta\mathcal{I} = \omega^{-1}\mathrm{Re}\,\langle\langle \boldsymbol{\delta}\mathbf{z}, \omega\mathbf{z} - \mathcal{K}\mathbf{z} \rangle\rangle, \tag{4.100}$$

where we have again used the self-adjointness of $\mathcal{K}$. Evidently, the two variations $\delta\omega$ and $\delta\mathcal{I}$ are related by $\delta\omega = -2\omega\langle\langle \mathbf{z}, \mathbf{z} \rangle\rangle^{-1}\delta\mathcal{I}$, so that the Rayleigh quotient ω and the action $\mathcal{I}$ share the same stationary points $\mathbf{z}$.

The action $\mathcal{I}$ can be rewritten in terms of the three-dimensional eigenfunctions $\mathbf{s}$ in the form

$$\mathcal{I} = \tfrac{1}{2}\omega^2\langle \mathbf{s}, \mathbf{s}\rangle - \omega\langle \mathbf{s}, i\boldsymbol{\Omega}\times\mathbf{s}\rangle - \tfrac{1}{2}\langle \mathbf{s}, \mathcal{H}\mathbf{s}\rangle. \tag{4.101}$$

A factor of ω^{-1} was introduced into the six-dimensional definition (4.99) expressly in order to obtain the result (4.101), which is the natural generalization of (4.24). The variation of $\mathcal{I}$ due to an infinitesimal variation of the displacement field $\delta\mathbf{s}$ is

$$\delta\mathcal{I} = \mathrm{Re}\,\langle \delta\mathbf{s}, \omega^2\mathbf{s} - 2i\omega\boldsymbol{\Omega}\times\mathbf{s} - \mathcal{H}\mathbf{s}\rangle, \tag{4.102}$$

where we have used the self-adjointness (4.77)–(4.78) of the operators $\mathcal{H}$ and $i\boldsymbol{\Omega}\times$. It is clear from equation (4.102) that $\delta\mathcal{I}$ vanishes for an arbitrary variation $\delta\mathbf{s}$ if and only if $\mathcal{H}\mathbf{s} + 2i\omega\boldsymbol{\Omega}\times\mathbf{s} = \omega^2\mathbf{s}$; this is an alternative, three-dimensional statement of Rayleigh's principle.

As in the non-rotating case, the above schematic "proof" is too cavalier about the boundary conditions associated with the operator $\mathcal{H}$ and the admissibility conditions that must be imposed on the variations $\delta\mathbf{s}$. To obtain a more precise formulation, we rewrite the three-dimensional action (4.101) in the form

$$\mathcal{I} = \tfrac{1}{2}(\omega^2\mathcal{T} - 2\omega\mathcal{W} - \mathcal{V}), \tag{4.103}$$

where

$$\mathcal{T} = \int_{\oplus} \rho^0\mathbf{s}^*\cdot\mathbf{s}\,dV, \tag{4.104}$$

$$\mathcal{W} = \int_{\oplus} \rho^0\mathbf{s}^*\cdot(i\boldsymbol{\Omega}\times\mathbf{s})\,dV, \tag{4.105}$$

$$\begin{aligned}\mathcal{V} = \int_{\oplus} &[\boldsymbol{\nabla}\mathbf{s}^*{:}\boldsymbol{\Lambda}{:}\boldsymbol{\nabla}\mathbf{s} + \tfrac{1}{2}\rho^0(\mathbf{s}^*\cdot\boldsymbol{\nabla}\phi^{\mathrm{E1}} + \mathbf{s}\cdot\boldsymbol{\nabla}\phi^{\mathrm{E1}*}) \\ &+ \rho^0\mathbf{s}^*\cdot\boldsymbol{\nabla}\boldsymbol{\nabla}(\phi^0+\psi)\cdot\mathbf{s}]\,dV \\ + \tfrac{1}{2}\int_{\Sigma_{\mathrm{FS}}} &[\varpi^0\mathbf{s}^*\cdot(\boldsymbol{\nabla}^{\Sigma}\mathbf{s})\cdot\hat{\mathbf{n}} + \varpi^0\mathbf{s}\cdot(\boldsymbol{\nabla}^{\Sigma}\mathbf{s}^*)\cdot\hat{\mathbf{n}} \\ &- (\hat{\mathbf{n}}\cdot\mathbf{s}^*)\boldsymbol{\nabla}^{\Sigma}\cdot(\varpi^0\mathbf{s}) - (\hat{\mathbf{n}}\cdot\mathbf{s})\boldsymbol{\nabla}^{\Sigma}\cdot(\varpi^0\mathbf{s}^*)]_-^+\,d\Sigma.\end{aligned} \tag{4.106}$$

We refer to the quantities $\mathcal{T}$, $\mathcal{W}$ and $\mathcal{V}$ as the *kinetic energy*, *Coriolis* and *potential energy functionals*, respectively. The kinetic and potential energy functionals are defined in a manner analogous to (4.27) and (4.28) on a non-rotating Earth, except that we allow for the complexity of the eigenfunctions $\mathbf{s}$, and we replace the gravitational potential ϕ^0 by the geopotential

$\phi^0 + \psi$. All three functionals (4.104)–(4.106) are real quadratic functions of the argument s, and so therefore is the action $\mathcal{I}$.

The *displacement version* of Rayleigh's principle asserts that $\mathcal{I}$ is stationary with respect to arbitrary admissible variations $\delta\mathbf{s}$ if and only if s is an eigenfunction with associated eigenfrequency ω. Upon making use of both the three-dimensional and two-dimensional versions of Gauss' theorem, as we did in Section 4.1.3, we obtain

$$\begin{aligned}\delta\mathcal{I} = \mathrm{Re}\int_{\oplus} \delta\mathbf{s}^* \cdot [\omega^2\rho^0\mathbf{s} - 2i\omega\rho^0\boldsymbol{\Omega}\times\mathbf{s} + \boldsymbol{\nabla}\cdot\mathbf{T}^{\mathrm{PK1}} \\ - \rho^0\boldsymbol{\nabla}\phi^{\mathrm{E1}} - \rho^0\mathbf{s}\cdot\boldsymbol{\nabla}\boldsymbol{\nabla}(\phi^0+\psi)]\,dV \\ - \mathrm{Re}\int_{\partial\oplus} \delta\mathbf{s}^*\cdot(\hat{\mathbf{n}}\cdot\mathbf{T}^{\mathrm{PK1}})\,d\Sigma \\ + \mathrm{Re}\int_{\Sigma_{\mathrm{SS}}} \delta\mathbf{s}^*\cdot[\hat{\mathbf{n}}\cdot\mathbf{T}^{\mathrm{PK1}}]_-^+\,d\Sigma \\ + \mathrm{Re}\int_{\Sigma_{\mathrm{FS}}} [\delta\mathbf{s}^*\cdot\mathbf{t}^{\mathrm{PK1}}]_-^+\,d\Sigma.\end{aligned} \tag{4.107}$$

Equation (4.107) shows that $\delta\mathcal{I} = 0$ for an arbitrary admissible variation $\delta\mathbf{s}$ if and only if the eigenfrequency ω and associated eigenfunction s satisfy the rotating normal-mode equation (4.73) and the dynamical boundary conditions (4.4)–(4.6), as asserted.

There is, of course, a *displacement-potential version* of Rayleigh's principle on a rotating Earth as well. We define the modified potential energy functional $\mathcal{V}'$ in a manner analogous to (4.31), allowing for the complexity of the eigenfunctions and replacing ϕ^0 by $\phi^0 + \psi$:

$$\begin{aligned}\mathcal{V}' = \int_{\bigcirc} [\boldsymbol{\nabla}\mathbf{s}^*:\boldsymbol{\Lambda}:\boldsymbol{\nabla}\mathbf{s} + \rho^0(\mathbf{s}^*\cdot\boldsymbol{\nabla}\phi^{\mathrm{E1}} + \mathbf{s}\cdot\boldsymbol{\nabla}\phi^{\mathrm{E1}*}) \\ + \rho^0\mathbf{s}^*\cdot\boldsymbol{\nabla}\boldsymbol{\nabla}(\phi^0+\psi)\cdot\mathbf{s} + (4\pi G)^{-1}\boldsymbol{\nabla}\phi^{\mathrm{E1}*}\cdot\boldsymbol{\nabla}\phi^{\mathrm{E1}}]\,dV \\ + \tfrac{1}{2}\int_{\Sigma_{\mathrm{FS}}} [\varpi^0\mathbf{s}^*\cdot(\boldsymbol{\nabla}^{\Sigma}\mathbf{s})\cdot\hat{\mathbf{n}} + \varpi^0\mathbf{s}\cdot(\boldsymbol{\nabla}^{\Sigma}\mathbf{s}^*)\cdot\hat{\mathbf{n}} \\ - (\hat{\mathbf{n}}\cdot\mathbf{s}^*)\boldsymbol{\nabla}^{\Sigma}\cdot(\varpi^0\mathbf{s}) - (\hat{\mathbf{n}}\cdot\mathbf{s})\boldsymbol{\nabla}^{\Sigma}\cdot(\varpi^0\mathbf{s}^*)]_-^+\,d\Sigma.\end{aligned} \tag{4.108}$$

The corresponding modified action

$$\mathcal{I}' = \tfrac{1}{2}(\omega^2\mathcal{T} - 2\omega\mathcal{W} - \mathcal{V}') \tag{4.109}$$

is then stationary with respect to arbitrary and independent admissible variations $\delta\mathbf{s}$ and $\delta\phi^{\mathrm{E1}}$ if and only if ω, s, ϕ^{E1} is an eigensolution. The two potential energy functionals $\mathcal{V}$ and $\mathcal{V}'$ and the two actions $\mathcal{I}$ and $\mathcal{I}'$ are equal for any eigensolution by virtue of the identity (3.195). The stationary value of the actions is

$$\mathcal{I} = \mathcal{I}' = 0, \tag{4.110}$$

by virtue of equations (4.97) and (4.99). The result (4.110) can also be obtained by dotting the three-dimensional momentum equation (4.73) with $\mathbf{s}^*$ and integrating over the Earth model $\oplus$ or, equivalently, by equating the primed and unprimed eigensolutions in equations (4.79) and (4.80).

The stationarity of $\mathcal{I}$ and $\mathcal{I}'$ can also be deduced directly from Hamilton's principle by considering a time-domain displacement of the form

$$\mathbf{s}(\mathbf{x}, t) = \tfrac{1}{2}[\mathbf{s}(\mathbf{x}) \exp(i\omega t) + \mathbf{s}^*(\mathbf{x}) \exp(-i\omega t)], \tag{4.111}$$

for a fixed frequency ω. If the time interval $t_2 - t_1$ is an integral number of cycles of the oscillation, then the time-domain actions (3.163) and (3.189) are constants times (4.103) and (4.109), so that Rayleigh's principle is a special case of Hamilton's principle. The distinction between the instantaneous relative kinetic and elastic-gravitational potential energies in equations (3.234)–(3.235) and the time-independent functionals $\mathcal{T}$ and $\mathcal{V}$ in equations (4.104) and (4.106) is noteworthy. Physically, the functionals $\omega^2\mathcal{T}$ and $\mathcal{V}$ are *four times the average* relative kinetic and potential energies during an oscillatory cycle of the form (4.111).

★4.2.4 Dynamical stability

Regarding the energy-balance relation $\omega^2\mathcal{T} - 2\omega\mathcal{W} - \mathcal{V} = 0$ as a quadratic equation for ω, we see that the eigenfrequency associated with an eigenfunction $\mathbf{s}$ must be one of the two solutions

$$\omega = \frac{\mathcal{W} \pm \sqrt{\mathcal{W}^2 + \mathcal{T}\mathcal{V}}}{\mathcal{T}}. \tag{4.112}$$

Since $\mathcal{W}^2 \geq 0$ and $\mathcal{T} > 0$, the eigenfrequencies will all be real, and thus the Earth will be dynamically stable, provided that

$$\mathcal{V} \geq 0 \tag{4.113}$$

for all piecewise smooth functions $\mathbf{s}$ in $\oplus$. Equation (4.113) is the condition for secular stability obtained in Section 3.9.6, and we see that secular stability implies dynamical stability. In this case, however, $\mathcal{V} \geq 0$ is not a necessary as well as a sufficient condition; it is only necessary that the discriminant $\mathcal{W}^2 + \mathcal{T}\mathcal{V}$ in equation (4.112) be non-negative, and this may be so even if $\mathcal{V} < 0$, i.e., even if the elastic-gravitational potential energy of the initial configuration is a local maximum rather than a local minimum.

A classical example of such a dynamically stable but secularly unstable configuration is a quasi-rigid Earth model rotating about its axis of least principal inertia. In the absence of any frictional dissipation, a slight perturbation will excite a stable Eulerian free nutation or Chandler wobble. If, however, there is any anelasticity whatsoever, the amplitude of this oscillation will grow at a rate determined by the dissipation, and the Earth will reorient itself until the rotation axis coincides with the axis of greatest principal inertia. Uniform rotation about the axis of greatest principal inertia is the only secularly stable configuration of an Earth model of fixed angular momentum; a slight perturbation of this state will excite a decaying Chandler wobble. We shall henceforth assume that the Earth, prior to an earthquake, is in such a secularly stable state, so that $\mathcal{V} \geq 0$ for all possible elastic-gravitational deformations.

⋆4.2.5 Rigid-body and geostrophic modes

The trivial modes of a rotating Earth consist of the eigensolutions whose six-dimensional norm $\langle\langle \mathbf{z}, \mathbf{z}' \rangle\rangle$ vanishes; the corresponding condition expressed in terms of the three-dimensional kinetic and potential energy functionals is $\mathcal{V} + \omega^2 \mathcal{T} = 0$. As on a non-rotating Earth, there are two types of trivials: rigid-body modes of the whole Earth, and geostrophic modes confined to the fluid regions $\oplus_{\mathrm{F}}$.

The rigid-body modes of a rotating Earth are more complicated than those of a non-rotating Earth; fundamentally, however, they still account for the six degrees of freedom of rigid-body motion. To enumerate these modes, it is convenient to adopt a Cartesian coordinate system which has $\hat{\mathbf{z}}$ aligned along the rotation axis, $\mathbf{\Omega} = \Omega\hat{\mathbf{z}}$, so that $\hat{\mathbf{x}}$ and $\hat{\mathbf{y}}$ are equatorial. The eigenfrequency and un-normalized eigenfunction associated with the *axial translational mode* are of the form

$$\omega = 0, \qquad \mathbf{s} = \hat{\mathbf{z}}, \tag{4.114}$$

whereas those associated with the *axial spin mode* are of the form

$$\omega = 0, \qquad \mathbf{s} = \hat{\mathbf{z}} \times \mathbf{x}. \tag{4.115}$$

Equations (4.114) and (4.115) correspond, respectively, to a rigid translation along the rotation axis and a rigid rotation about the rotation axis. The two *equatorial translational modes* have associated eigensolutions of the form

$$\omega = \pm\Omega, \qquad \mathbf{s} = \hat{\mathbf{x}} \pm i\hat{\mathbf{y}}. \tag{4.116}$$

In this case, the motion consists of a constant equatorial translation in an inertial frame; this fixed displacement is aliased into an apparent diurnal motion in the rotating frame. The final rigid-body mode that can be specified analytically is the *tilt-over mode*, whose eigensolution is of the form

$$\omega = \pm\Omega, \qquad \mathbf{s} = (\hat{\mathbf{x}} \pm i\hat{\mathbf{y}}) \times \mathbf{x}. \tag{4.117}$$

This mode corresponds to a fixed equatorial rotation, or tilt of the Earth relative to the rotation axis, in an inertial frame; it too is aliased into a diurnal motion in the reference frame of a terrestrial observer. All of the displacement fields (4.114)–(4.117) are strain-free, $\boldsymbol{\varepsilon} = \mathbf{0}$, so that the associated Eulerian perturbations in the density and gravitational potential are purely advective: $\rho^{\mathrm{E1}} = -\mathbf{s}\cdot\boldsymbol{\nabla}\rho^0$ and $\phi^{\mathrm{E1}} = -\mathbf{s}\cdot\boldsymbol{\nabla}\phi^0$. The elastic-gravitational operator (4.74) reduces to $\rho^0\mathcal{H}\mathbf{s} = \rho^0\mathbf{s}\cdot\boldsymbol{\nabla\nabla}\psi - \rho^0\boldsymbol{\nabla}\psi\cdot\boldsymbol{\nabla}\mathbf{s}$, so that equation (4.75) is satisfied. The discriminant $\mathcal{W}^2 + \mathcal{T}\mathcal{V}$ vanishes for both of the equatorial translational modes, so that they account for both equatorial translational degrees of freedom. The tilt-over mode, on the other hand, accounts for only one of the equatorial rotational degrees of freedom; the other equatorial rotational mode is the Chandler wobble (Smith & Dahlen 1981). The first four rigid-body modes (4.114)–(4.116) all have $\langle\langle\mathbf{z},\mathbf{z}\rangle\rangle = 0$; hence, they must be removed from the space of allowable six-vectors prior to employing (4.87) as an inner product. The tilt-over mode (4.117) is not a trivial, since it has $\langle\langle\mathbf{z},\mathbf{z}\rangle\rangle = 2C\Omega^2$, where C is the polar moment of inertia of the Earth.

The geostrophic modes are identical in character to those of a non-rotating Earth; the associated displacement eigenfunctions $\mathbf{s}$ are of the form (4.50)–(4.52), with associated eigenfrequency $\omega = 0$. The only difference is that the level surfaces of ρ^0 and p^0 coincide with the level surfaces of the geopotential $\phi^0 + \psi$ rather than those of ϕ^0 alone. Within any regions of $\oplus_{\mathrm{F}}$ that are neutrally stratified, the family of geostrophic flows is even larger, as discussed in Section 4.1.6. In general, the geostrophic eigenspace consists of all the steady flows in $\oplus_{\mathrm{F}}$ that do not alter the elastic-gravitational potential energy $\mathcal{V}$ of the Earth.

★4.2.6 Green tensor

The Green tensor or impulse response $\mathbf{G}(\mathbf{x},\mathbf{x}';t)$ of a rotating Earth satisfies the homogeneous equation

$$\rho^0(\partial_t^2\mathbf{G} + 2i\boldsymbol{\Omega}\times\partial_t\mathbf{G} + \mathcal{H}\mathbf{G}) = \mathbf{0}, \tag{4.118}$$

together with the initial conditions

$$\mathbf{G}(\mathbf{x},\mathbf{x}';0) = \mathbf{0}, \qquad \partial_t\mathbf{G}(\mathbf{x},\mathbf{x}';0) = (1/\rho^0)\mathbf{I}\,\delta(\mathbf{x}-\mathbf{x}'). \tag{4.119}$$

To solve (4.118)–(4.119) we assume that the eigenfunctions $\mathbf{s}_k$ and their complex conjugates $\mathbf{s}_k^*$ form a complete set, and write $\mathbf{G}$ in the form

$$\mathbf{G}(\mathbf{x}, \mathbf{x}'; t) = \mathrm{Re} \sum_k \mathbf{s}_k(\mathbf{x}) \mathbf{c}_k(\mathbf{x}') \exp(i\omega_k t). \tag{4.120}$$

Such an expansion will satisfy the initial conditions (4.119) provided that

$$\sum_k (\mathbf{s}_k \mathbf{c}_k + \mathbf{s}_k^* \mathbf{c}_k^*) = \mathbf{0}, \tag{4.121}$$

$$\sum_k (\omega_k \mathbf{s}_k \mathbf{c}_k - \omega_k \mathbf{s}_k^* \mathbf{c}_k^*) = -(2i/\rho^0)\mathbf{I}\, \delta(\mathbf{x} - \mathbf{x}'). \tag{4.122}$$

We can write (4.121)–(4.122) as a single equation in the space of six-dimensional eigenfunctions:

$$\sum_k (\mathbf{z}_k \mathbf{c}_k + \mathbf{z}_k^* \mathbf{c}_k^*) = \begin{pmatrix} \mathbf{0} \\ -(2i/\rho^0)\mathbf{I}\, \delta(\mathbf{x} - \mathbf{x}') \end{pmatrix}, \tag{4.123}$$

where

$$\mathbf{z}_k = \begin{pmatrix} \mathbf{s}_k \\ \omega_k \mathbf{s}_k \end{pmatrix}, \qquad \mathbf{z}_k^* = \begin{pmatrix} \mathbf{s}_k^* \\ -\omega_k \mathbf{s}_k^* \end{pmatrix}. \tag{4.124}$$

Upon taking the six-dimensional inner product of (4.123) with $\mathbf{z}_k$ and $\mathbf{z}_k^*$, and using the orthonormality relation $\langle\langle \mathbf{z}_k, \mathbf{z}_{k'} \rangle\rangle = 2\omega_k^2\, \delta_{kk'}$, or, equivalently,

$$\int_\oplus \rho^0 \mathbf{s}_k^* \cdot \mathbf{s}_{k'}\, dV - 2(\omega_k + \omega_{k'})^{-1} \int_\oplus \rho^0 \mathbf{s}_k^* \cdot (i\boldsymbol{\Omega} \times \mathbf{s}_{k'})\, dV = \delta_{kk'}, \tag{4.125}$$

we obtain

$$\mathbf{c}_k = (i\omega_k)^{-1} \mathbf{s}_k^*(\mathbf{x}). \tag{4.126}$$

The Green tensor of a rotating Earth is therefore given by

$$\mathbf{G}(\mathbf{x}, \mathbf{x}'; t) = \mathrm{Re} \sum_k (i\omega_k)^{-1} \mathbf{s}_k(\mathbf{x}) \mathbf{s}_k^*(\mathbf{x}') \exp(i\omega_k t). \tag{4.127}$$

Equations (4.125) and (4.127) reduce to the corresponding results (4.56) and (4.60) in the non-rotating limit $\boldsymbol{\Omega} \rightarrow \mathbf{0}$, as expected.

Due to the complexity of the eigenfunctions $\mathbf{s}_k$, the imposition of a unit impulsive force does not cause every normal mode to begin oscillating with the same phase. Instead, each term in the mode sum (4.127) has the character of an eastward or westward *travelling wave*. As a consequence of this, the Green tensor on a rotating Earth does not satisfy the principle of source-receiver reciprocity: $\mathbf{G}(\mathbf{x},\mathbf{x}';t) \neq \mathbf{G}^{\mathrm{T}}(\mathbf{x}',\mathbf{x};t)$. We can, however, obtain a related but more complicated result by considering the impulse response of the anti-Earth with a reversed sense of rotation, $\boldsymbol{\Omega} \rightarrow -\boldsymbol{\Omega}$. We denote this *anti-Green tensor* by $\overline{\mathbf{G}}(\mathbf{x},\mathbf{x}';t)$; to find it, we must solve

$$\rho^0(\partial_t^2\overline{\mathbf{G}} - 2i\boldsymbol{\Omega} \times \partial_t\overline{\mathbf{G}} + \mathcal{H}\overline{\mathbf{G}}) = \mathbf{0}, \tag{4.128}$$

subject to the initial conditions

$$\overline{\mathbf{G}}(\mathbf{x},\mathbf{x}';0) = \mathbf{0}, \qquad \partial_t\overline{\mathbf{G}}(\mathbf{x},\mathbf{x}';0) = (1/\rho^0)\mathbf{I}\,\delta(\mathbf{x}-\mathbf{x}'). \tag{4.129}$$

The eigensolutions of the anti-Earth are ω_k, $\mathbf{s}_k^*$ if those of the real Earth are ω_k, $\mathbf{s}_k$; hence,

$$\overline{\mathbf{G}}(\mathbf{x},\mathbf{x}';t) = \mathrm{Re}\sum_k (i\omega_k)^{-1}\mathbf{s}_k^*(\mathbf{x})\mathbf{s}_k(\mathbf{x}')\exp(i\omega_k t). \tag{4.130}$$

Upon comparing (4.127) and (4.130) we see that

$$\mathbf{G}(\mathbf{x},\mathbf{x}';t) = \overline{\mathbf{G}}^{\mathrm{T}}(\mathbf{x}',\mathbf{x};t). \tag{4.131}$$

The physical interpretation of the result (4.131) is straightforward. The lack of reciprocity on a rotating Earth is a consequence of the *Doppler effect*: eastward-travelling and westward-travelling waves have different speeds of propagation, as viewed by a terrestrial observer. To obtain the same response upon interchanging the source and receiver, we must also reverse the sense of rotation in order to account for this; we refer to (4.131) as the principle of *generalized reciprocity* (see Figure 4.3).

The trivial modes must be excised from the basis prior to taking the six-dimensional inner product of (4.123) with $\mathbf{z}_k$ and $\mathbf{z}_k^*$; the sum in equation (4.127) is thus over all of the non-trivial modes with positive eigenfrequencies, $\omega_k > 0$. The tilt-over mode is not a trivial so that it should, strictly speaking, be included; however, it cannot be excited by an indigenous source such as an earthquake, which exerts no net torque upon the Earth. We refer to $\mathbf{G}$ with the tilt-over mode as well as all the trivials excised from the mode-sum as the *seismic Green tensor*. The Green tensor on a rotating Earth was first obtained using the above method by Wahr (1981a); a less elegant three-dimensional derivation is given by Dahlen & Smith (1975) and Dahlen (1977; 1978; 1980b). It is noteworthy that the result (4.127) is valid even in the presence of accidental eigenfrequency degeneracy; this will not be the case for the mode-sum Green tensors on an anelastic Earth, which we obtain in Sections 6.2.3 and 6.3.3. Although the

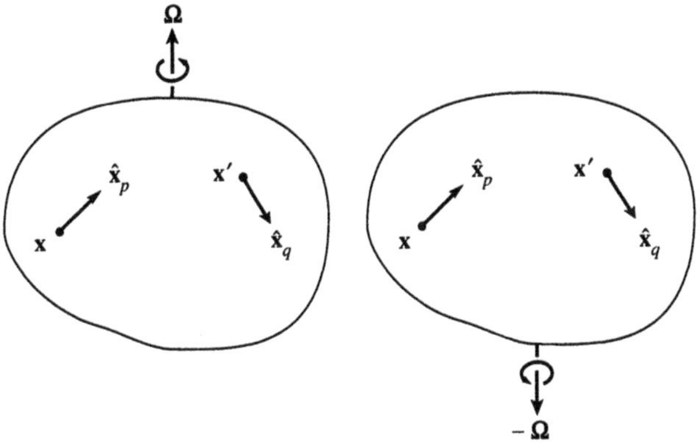

Figure 4.3. Seismic reciprocity on a rotating Earth: the $\hat{\mathbf{x}}_p$ component of the response at $\mathbf{x}$ to a point force acting in the $\hat{\mathbf{x}}_q$ direction at a point $\mathbf{x}'$ in the Earth (*left*) is equal to the $\hat{\mathbf{x}}_q$ component of the response at $\mathbf{x}'$ to a point force acting in the $\hat{\mathbf{x}}_p$ direction at a point $\mathbf{x}$ in the anti-Earth (*right*). The components of the Green and anti-Green tensors are related by $G_{pq}(\mathbf{x},\mathbf{x}';t) = \overline{G}_{qp}(\mathbf{x}',\mathbf{x};t)$.

tilt-over mode can be ignored in seismological applications, it exerts a significant effect upon the nutational response of the Earth to the externally applied luni-solar tidal torque; an adjacent non-rigid-body mode, known as the free core nutation or nearly diurnal free wobble, plays an even more important role in this regard (Smith 1977; Wahr 1981c).

*4.2.7 Response to a transient force

The complete response to an applied body force density $\mathbf{f}$ and surface force density $\mathbf{t}$ is of the form

$$\begin{aligned}\mathbf{s}(\mathbf{x},t) = \int_{-\infty}^{t}\int_{\oplus} \mathbf{G}(\mathbf{x},\mathbf{x}';t-t')\cdot\mathbf{f}(\mathbf{x}',t')\,dV'\,dt' \\ + \int_{-\infty}^{t}\int_{\partial\oplus} \mathbf{G}(\mathbf{x},\mathbf{x}';t-t')\cdot\mathbf{t}(\mathbf{x}',t')\,d\Sigma'\,dt' \\ + \mathbf{s}_{\mathrm{triv}}(\mathbf{x},t),\end{aligned} \tag{4.132}$$

where $\mathbf{s}_{\mathrm{triv}}$ is a linear combination of the trivials, which must still be determined. Ignoring the trivial contribution for the moment, and inserting the Green tensor (4.127) into (4.132), we obtain the mode-sum representation of the seismic response:

$$\mathbf{s}(\mathbf{x},t) = \mathrm{Re}\sum_k (i\omega_k)^{-1}\mathbf{s}_k(\mathbf{x}) \int_{-\infty}^{t} C_k(t') \exp i\omega_k(t-t')\, dt', \tag{4.133}$$

where

$$C_k(t) = \int_\oplus \mathbf{f}(\mathbf{x},t)\cdot\mathbf{s}_k^*(\mathbf{x})\, dV + \int_{\partial\oplus} \mathbf{t}(\mathbf{x},t)\cdot\mathbf{s}_k^*(\mathbf{x})\, d\Sigma. \tag{4.134}$$

Upon integrating by parts with respect to time, we obtain the equivalent result, analogous to equation (4.65):

$$\mathbf{s}(\mathbf{x},t) = \mathrm{Re}\sum_k \omega_k^{-2}\mathbf{s}_k(\mathbf{x}) \int_{-\infty}^{t} \partial_{t'} C_k(t')[1 - \exp i\omega_k(t-t')]\, dt', \tag{4.135}$$

where

$$\partial_t C_k(t) = \int_\oplus \partial_t\mathbf{f}(\mathbf{x},t)\cdot\mathbf{s}_k^*(\mathbf{x})\, dV + \int_{\partial\oplus} \partial_t\mathbf{t}(\mathbf{x},t)\cdot\mathbf{s}_k^*(\mathbf{x})\, d\Sigma. \tag{4.136}$$

As on a non-rotating Earth, we are especially interested in the response to the *transient forces* equivalent to an earthquake, which are zero prior to some origin time t_0, and which attain the constant values $\mathbf{f}_\mathrm{f}$ and $\mathbf{t}_\mathrm{f}$ after a time t_f. The displacement produced by such a force, subsequent to the cessation of faulting, is of the form

$$\mathbf{s}(\mathbf{x},t) = \mathrm{Re}\sum_k \omega_k^{-2}[c_k^\mathrm{f} - c_k \exp(i\omega_k t)]\mathbf{s}_k(\mathbf{x}), \quad t \geq t_\mathrm{f}, \tag{4.137}$$

where

$$c_k^\mathrm{f} = \int_\oplus \mathbf{f}_\mathrm{f}(\mathbf{x})\cdot\mathbf{s}_k^*(\mathbf{x})\, dV + \int_{\partial\oplus} \mathbf{t}_\mathrm{f}(\mathbf{x})\cdot\mathbf{s}_k^*(\mathbf{x})\, d\Sigma, \tag{4.138}$$

$$\begin{aligned} c_k = {} & \int_{t_0}^{t_\mathrm{f}} \int_\oplus \partial_t\mathbf{f}(\mathbf{x},t)\cdot\mathbf{s}_k^*(\mathbf{x}) \exp(-i\omega_k t)\, dV\, dt \\ & + \int_{t_0}^{t_\mathrm{f}} \int_{\partial\oplus} \partial_t\mathbf{t}(\mathbf{x},t)\cdot\mathbf{s}_k^*(\mathbf{x}) \exp(-i\omega_k t)\, d\Sigma\, dt. \end{aligned} \tag{4.139}$$

The *free oscillations* $c_k \exp(i\omega_k t)$ in equation (4.137) decay with time due to the anelasticity of the Earth; the final *static displacement* in the long-time limit $t \to \infty$ is

$$\mathbf{s}_\mathrm{f}(\mathbf{x}) = \mathrm{Re}\sum_k \omega_k^{-2} c_k^\mathrm{f}\mathbf{s}_k(\mathbf{x}), \tag{4.140}$$

whereas the particle *acceleration* subsequent to the cessation of faulting is

$$\mathbf{a}(\mathbf{x},t) = \mathrm{Re}\sum_k c_k \exp(i\omega_k t)\,\mathbf{s}_k(\mathbf{x}), \quad t \geq t_{\mathrm{f}}. \tag{4.141}$$

Equations (4.137)–(4.141) reduce to (4.67)–(4.72) upon setting $c_k = a_k - ib_k$ and $c_k^{\mathrm{f}} = a_k^{\mathrm{f}} - ib_k^{\mathrm{f}}$. The sum (4.141) constitutes a complete description of the motion that can be recorded at times $t \geq t_{\mathrm{f}}$ by a seismometer; we turn our attention next to the trivial response $\mathbf{s}_{\mathrm{triv}}$ in equation (4.132).

*4.2.8 Change in rotation rate

Only one trivial—the axial spin mode—can be excited by an earthquake in the crust or mantle; the geostrophic modes cannot be excited because they are confined to the fluid regions of the Earth, whereas the translational modes can only be excited by a source that exerts a net force on the Earth. The axial spin mode is excited despite the absence of any net torque on the Earth, because the static deformation $\mathbf{s}_{\mathrm{f}}$ gives rise to perturbation in the principal moment of inertia C, which must be offset by a change in the rate of rotation Ω. The infinitesimal spin response produced by an arbitrary, time-dependent force density $\mathbf{f}$, $\mathbf{t}$ has been determined by Wahr (1981a). We simply quote his result here:

$$\mathbf{s}_{\mathrm{triv}}(\mathbf{x},t) = c_{\mathrm{triv}}(t)(\hat{\mathbf{z}} \times \mathbf{x}), \tag{4.142}$$

where

$$c_{\mathrm{triv}}(t) = 2(C\Omega)^{-1}\mathrm{Re}\sum_k \omega_k^{-2} \int_\oplus \rho^0 \mathbf{s}_k \cdot \boldsymbol{\nabla}\psi\, dV \times \int_{-\infty}^{t} C_k(t')\, dt'. \tag{4.143}$$

In the case of a transient force, the time dependence in equation (4.143) reduces to

$$\int_{-\infty}^{t} C_k(t')\, dt' = \int_{t_0}^{t_{\mathrm{f}}} C_k(t')\, dt' + c_k^{\mathrm{f}}(t - t_{\mathrm{f}}). \tag{4.144}$$

The result (4.143) has been derived under the assumption that the displacement $\mathbf{s}_{\mathrm{triv}}$ can be linearized, and equation (4.144) is quite obviously valid only for short times after t_{f}. The term proportional to $t - t_{\mathrm{f}}$ can, however, be interpreted as an infinitesimal *change in the rotation rate of the Earth*, by an amount

$$\delta\Omega = 2(C\Omega)^{-1}\mathrm{Re}\sum_k \omega_k^{-2} c_k^{\mathrm{f}} \int_\oplus \rho^0 \mathbf{s}_k \cdot \boldsymbol{\nabla}\psi\, dV. \tag{4.145}$$

The seismic displacement (4.137) can be regarded as the displacement measured by an observer situated in a reference frame rotating with the new angular velocity $\Omega + \delta\Omega$. This is consistent with the interpretation that the quantity $\mathbf{s}_{\mathrm{f}}$ represents the infinitesimal static deformation of the Earth.

The result (4.145) can also be derived from first principles, by appealing to the law of conservation of angular momentum, which stipulates that the change in rotation rate $\delta\Omega$ is related to the infinitesimal change in the principal moment of inertia δC by

$$C\,\delta\Omega + \delta C\,\Omega = 0. \tag{4.146}$$

The first-order perturbation in the inertia tensor $\mathbf{C} = \int_{\oplus} \rho^0[(\mathbf{x}\cdot\mathbf{x})\mathbf{I} - \mathbf{x}\mathbf{x}]\,dV$ associated with the static redistribution of mass is

$$\boldsymbol{\delta}\mathbf{C} = \int_{\oplus} \rho^0[2(\mathbf{x}\cdot\mathbf{s}_{\mathrm{f}})\mathbf{I} - \mathbf{x}\mathbf{s}_{\mathrm{f}} - \mathbf{s}_{\mathrm{f}}\mathbf{x}]\,dV. \tag{4.147}$$

Setting $\delta C = \hat{\mathbf{z}}\cdot\boldsymbol{\delta}\mathbf{C}\cdot\hat{\mathbf{z}}$ and substituting into (4.146), we obtain

$$\delta\Omega = 2(C\Omega)^{-1}\int_{\oplus} \rho^0 \mathbf{s}_{\mathrm{f}}\cdot\boldsymbol{\nabla}\psi\,dV, \tag{4.148}$$

which agrees with equation (4.145).

4.3 Hydrostatic Earth Model

All of the results obtained in this chapter are applicable to an Earth model having a hydrostatic initial stress, with very little alteration. We summarize the necessary modifications here, restricting attention to the non-rotating case for the sake of brevity; the corresponding results on a rotating, hydrostatic Earth may be readily deduced by analogy.

The eigenfrequencies ω and eigenfunctions $\mathbf{s}$ of a non-rotating, hydrostatic Earth are found by solving

$$\begin{aligned} &-\omega^2\rho^0\mathbf{s} - \boldsymbol{\nabla}\cdot\mathbf{T}^{\mathrm{L1}} \\ &\qquad + \boldsymbol{\nabla}(\rho^0\mathbf{s}\cdot\boldsymbol{\nabla}\phi^0) + \rho^0\boldsymbol{\nabla}\phi^{\mathrm{E1}} + \rho^{\mathrm{E1}}\boldsymbol{\nabla}\phi^0 = \mathbf{0}, \end{aligned} \tag{4.149}$$

subject to the boundary conditions

$$\hat{\mathbf{n}}\cdot\mathbf{T}^{\mathrm{L1}} = \mathbf{0} \quad \text{on } \partial\oplus, \tag{4.150}$$

$$[\hat{\mathbf{n}}\cdot\mathbf{T}^{\mathrm{L1}}]_-^+ = \mathbf{0} \quad \text{on } \Sigma_{\mathrm{SS}}, \tag{4.151}$$

$$[\hat{\mathbf{n}}\cdot\mathbf{T}^{\mathrm{L1}}]_-^+ = \hat{\mathbf{n}}[\hat{\mathbf{n}}\cdot\mathbf{T}^{\mathrm{L1}}\cdot\hat{\mathbf{n}}]_-^+ = \mathbf{0} \quad \text{on } \Sigma_{\mathrm{FS}}. \tag{4.152}$$

The incremental Lagrangian Cauchy stress $\mathbf{T}^{\mathrm{L1}}$ and the Eulerian density perturbation ρ^{E1} are given by $\mathbf{T}^{\mathrm{L1}} = \boldsymbol{\Gamma}:\boldsymbol{\varepsilon}$ and $\rho^{\mathrm{E1}} = -\boldsymbol{\nabla}\cdot(\rho^0\mathbf{s})$.

4.3.1 Hermiticity and orthonormality

We can write the normal-mode eigenvalue problem (4.149)–(4.152) in the symbolic form

$$\mathcal{H}\mathbf{s} = \omega^2\mathbf{s}, \tag{4.153}$$

where

$$\rho^0\mathcal{H}\mathbf{s} = -\boldsymbol{\nabla}\cdot\mathbf{T}^{\mathrm{L1}} + \boldsymbol{\nabla}(\rho^0\mathbf{s}\cdot\boldsymbol{\nabla}\phi^0) + \rho^0\boldsymbol{\nabla}\phi^{\mathrm{E1}} + \rho^{\mathrm{E1}}\boldsymbol{\nabla}\phi^0. \tag{4.154}$$

An application of Gauss' theorem together with the boundary conditions (4.150)–(4.151) shows that

$$\begin{aligned}\langle\mathbf{s},\mathcal{H}\mathbf{s}'\rangle = \int_\oplus [&\boldsymbol{\varepsilon}\!:\!\boldsymbol{\Gamma}\!:\!\boldsymbol{\varepsilon}' + \rho^0\mathbf{s}\cdot\boldsymbol{\nabla}\phi^{\mathrm{E1}\prime} + \rho^0\mathbf{s}\cdot\boldsymbol{\nabla}\boldsymbol{\nabla}\phi^0\cdot\mathbf{s}' \\ &+ \rho^0\boldsymbol{\nabla}\phi^0\cdot(\mathbf{s}\cdot\boldsymbol{\nabla}\mathbf{s}' - \mathbf{s}\boldsymbol{\nabla}\cdot\mathbf{s}')]\,dV,\end{aligned} \tag{4.155}$$

$$\begin{aligned}\langle\mathbf{s}',\mathcal{H}\mathbf{s}\rangle = \int_\oplus [&\boldsymbol{\varepsilon}'\!:\!\boldsymbol{\Gamma}\!:\!\boldsymbol{\varepsilon} + \rho^0\mathbf{s}'\cdot\boldsymbol{\nabla}\phi^{\mathrm{E1}} + \rho^0\mathbf{s}'\cdot\boldsymbol{\nabla}\boldsymbol{\nabla}\phi^0\cdot\mathbf{s} \\ &+ \rho^0\boldsymbol{\nabla}\phi^0\cdot(\mathbf{s}'\cdot\boldsymbol{\nabla}\mathbf{s} - \mathbf{s}'\boldsymbol{\nabla}\cdot\mathbf{s})]\,dV.\end{aligned} \tag{4.156}$$

Equations (4.155) and (4.156) are equal by virtue of the Maxwell relation $\Gamma_{ijkl} = \Gamma_{klij}$, the gravitational symmetry (4.16), and the identity

$$\begin{aligned}&\int_\oplus \rho^0\boldsymbol{\nabla}\phi^0\cdot(\mathbf{s}\cdot\boldsymbol{\nabla}\mathbf{s}' - \mathbf{s}\boldsymbol{\nabla}\cdot\mathbf{s}')]\,dV \\ &\qquad = \int_\oplus \rho^0\boldsymbol{\nabla}\phi^0\cdot(\mathbf{s}'\cdot\boldsymbol{\nabla}\mathbf{s} - \mathbf{s}'\boldsymbol{\nabla}\cdot\mathbf{s})]\,dV.\end{aligned} \tag{4.157}$$

It follows that the operator $\mathcal{H}$ is Hermitian with respect to the inner product (4.10):

$$\langle\mathbf{s},\mathcal{H}\mathbf{s}'\rangle = \langle\mathcal{H}\mathbf{s},\mathbf{s}'\rangle = \langle\mathbf{s}',\mathcal{H}\mathbf{s}\rangle. \tag{4.158}$$

In obtaining the result (4.157) we have used the volumetric and surficial hydrostatic conditions $\boldsymbol{\nabla}\rho^0\times\boldsymbol{\nabla}\phi^0 = \mathbf{0}$ in $\oplus$ and $\boldsymbol{\nabla}^\Sigma\phi^0 = \mathbf{0}$ on Σ. The Hermitian symmetry (4.158) guarantees that the normal-mode orthogonality relation (4.20) is applicable on a non-rotating, hydrostatic Earth model; we continue to normalize the eigenfunctions by $\langle\mathbf{s},\mathbf{s}\rangle = 1$.

4.3.2 Rayleigh's principle

The potential energy functional (4.28) and the modified potential energy functional (4.31) reduce to

$$\mathcal{V} = \int_{\oplus} [\boldsymbol{\varepsilon}\!:\!\boldsymbol{\Gamma}\!:\!\boldsymbol{\varepsilon} + \rho^0\mathbf{s}\cdot\boldsymbol{\nabla}\phi^{\mathrm{E1}} + \rho^0\mathbf{s}\cdot\boldsymbol{\nabla}\boldsymbol{\nabla}\phi^0\cdot\mathbf{s} \\ + \rho^0\boldsymbol{\nabla}\phi^0\cdot(\mathbf{s}\cdot\boldsymbol{\nabla}\mathbf{s} - \mathbf{s}\boldsymbol{\nabla}\cdot\mathbf{s})]\, dV, \tag{4.159}$$

$$\mathcal{V}' = \int_{\bigcirc} [\boldsymbol{\varepsilon}\!:\!\boldsymbol{\Gamma}\!:\!\boldsymbol{\varepsilon} + 2\rho^0\mathbf{s}\cdot\boldsymbol{\nabla}\phi^{\mathrm{E1}} + \rho^0\mathbf{s}\cdot\boldsymbol{\nabla}\boldsymbol{\nabla}\phi^0\cdot\mathbf{s} \\ + \rho^0\boldsymbol{\nabla}\phi^0\cdot(\mathbf{s}\cdot\boldsymbol{\nabla}\mathbf{s} - \mathbf{s}\boldsymbol{\nabla}\cdot\mathbf{s}) + (4\pi G)^{-1}\boldsymbol{\nabla}\phi^{\mathrm{E1}}\cdot\boldsymbol{\nabla}\phi^{\mathrm{E1}}]\, dV. \tag{4.160}$$

The various versions of Rayleigh's principle all remain valid with the above substitutions. The action $\mathcal{I} = \frac{1}{2}(\omega^2\mathcal{T} - \mathcal{V})$ is stationary for an arbitrary admissible variation $\delta\mathbf{s}$ if and only if ω and $\mathbf{s}$ satisfy the frequency-domain momentum equation and boundary conditions (4.149)–(4.152), whereas the modified action $\mathcal{I}' = \frac{1}{2}(\omega^2\mathcal{T} - \mathcal{V}')$ is stationary for arbitrary and independent admissible variations $\delta\mathbf{s}$ and $\delta\phi^{\mathrm{E1}}$ if and only if $\mathbf{s}$ and ϕ^{E1} satisfy the potential boundary-value problem (4.33) as well. The unprimed and primed potential energy functionals and actions are equal by virtue of the identity (3.195). The stationary value of both of the hydrostatic actions is $\mathcal{I} = \mathcal{I}' = 0$ for every eigensolution $\mathbf{s}$, ϕ^{E1}.

4.3.3 Lagrangian and energy density

We can write the actions on a non-rotating, hydrostatic Earth in the form

$$\mathcal{I} = \int_{\oplus} L(\mathbf{s}, \boldsymbol{\nabla}\mathbf{s})\, dV, \tag{4.161}$$

$$\mathcal{I}' = \int_{\bigcirc} L'(\mathbf{s}, \boldsymbol{\nabla}\mathbf{s}, \boldsymbol{\nabla}\phi^{\mathrm{E1}})\, dV, \tag{4.162}$$

where

$$L = \tfrac{1}{2}[\omega^2\rho^0\mathbf{s}\cdot\mathbf{s} - \boldsymbol{\varepsilon}\!:\!\boldsymbol{\Gamma}\!:\!\boldsymbol{\varepsilon} - \rho^0\mathbf{s}\cdot\boldsymbol{\nabla}\phi^{\mathrm{E1}} \\ - \rho^0\mathbf{s}\cdot\boldsymbol{\nabla}\boldsymbol{\nabla}\phi^0\cdot\mathbf{s} - \rho^0\boldsymbol{\nabla}\phi^0\cdot(\mathbf{s}\cdot\boldsymbol{\nabla}\mathbf{s} - \mathbf{s}\boldsymbol{\nabla}\cdot\mathbf{s})], \tag{4.163}$$

$$L' = \tfrac{1}{2}[\omega^2\rho^0\mathbf{s}\cdot\mathbf{s} - \boldsymbol{\varepsilon}\!:\!\boldsymbol{\Gamma}\!:\!\boldsymbol{\varepsilon} - 2\rho^0\mathbf{s}\cdot\boldsymbol{\nabla}\phi^{\mathrm{E1}} \\ - \rho^0\mathbf{s}\cdot\boldsymbol{\nabla}\boldsymbol{\nabla}\phi^0\cdot\mathbf{s} - \rho^0\boldsymbol{\nabla}\phi^0\cdot(\mathbf{s}\cdot\boldsymbol{\nabla}\mathbf{s} - \mathbf{s}\boldsymbol{\nabla}\cdot\mathbf{s}) \\ - (4\pi G)^{-1}\boldsymbol{\nabla}\phi^{\mathrm{E1}}\cdot\boldsymbol{\nabla}\phi^{\mathrm{E1}}]. \tag{4.164}$$

The corresponding frequency-domain energy densities $E = \omega\partial_\omega L - L$ and $E' = \omega\partial_\omega L' - L'$ are

$$E = \tfrac{1}{2}[\omega^2\rho^0\mathbf{s}\cdot\mathbf{s} + \boldsymbol{\varepsilon}\!:\!\boldsymbol{\Gamma}\!:\!\boldsymbol{\varepsilon} + \rho^0\mathbf{s}\cdot\boldsymbol{\nabla}\phi^{\mathrm{E1}} + \rho^0\mathbf{s}\cdot\boldsymbol{\nabla}\boldsymbol{\nabla}\phi^0\cdot\mathbf{s} + \rho^0\boldsymbol{\nabla}\phi^0\cdot(\mathbf{s}\cdot\boldsymbol{\nabla}\mathbf{s} - \mathbf{s}\boldsymbol{\nabla}\cdot\mathbf{s})], \tag{4.165}$$

$$E' = \tfrac{1}{2}[\omega^2\rho^0\mathbf{s}\cdot\mathbf{s} + \boldsymbol{\varepsilon}\!:\!\boldsymbol{\Gamma}\!:\!\boldsymbol{\varepsilon} + 2\rho^0\mathbf{s}\cdot\boldsymbol{\nabla}\phi^{\mathrm{E1}} + \rho^0\mathbf{s}\cdot\boldsymbol{\nabla}\boldsymbol{\nabla}\phi^0\cdot\mathbf{s} + \rho^0\boldsymbol{\nabla}\phi^0\cdot(\mathbf{s}\cdot\boldsymbol{\nabla}\mathbf{s} - \mathbf{s}\boldsymbol{\nabla}\cdot\mathbf{s}) + (4\pi G)^{-1}\boldsymbol{\nabla}\phi^{\mathrm{E1}}\cdot\boldsymbol{\nabla}\phi^{\mathrm{E1}}]. \tag{4.166}$$

The volume integral of the energy density $\mathcal{E} = \frac{1}{2}(\omega^2\mathcal{T} + \mathcal{V}) = \frac{1}{2}(\omega^2\mathcal{T} + \mathcal{V}')$ is four times the average kinetic plus potential energy present in the Earth during a cycle of free oscillation of the form $\mathbf{s}(\mathbf{x})\cos\omega t$ or $\mathbf{s}(\mathbf{x})\sin\omega t$.

4.3.4 Elastic and gravitational energy

As in the time domain, we can decompose the potential energy of a mode on a non-rotating, hydrostatic Earth into separate elastic and gravitational energies:

$$\mathcal{V} = \mathcal{V}_{\mathrm{e}} + \mathcal{V}_{\mathrm{g}}, \tag{4.167}$$

where

$$\mathcal{V}_{\mathrm{e}} = \int_{\oplus} (\boldsymbol{\varepsilon}\!:\!\boldsymbol{\Gamma}\!:\!\boldsymbol{\varepsilon})\,dV, \tag{4.168}$$

$$\mathcal{V}_{\mathrm{g}} = \int_{\oplus} [\rho^0\mathbf{s}\cdot\boldsymbol{\nabla}\phi^{\mathrm{E1}} + \rho^0\mathbf{s}\cdot\boldsymbol{\nabla}\boldsymbol{\nabla}\phi^0\cdot\mathbf{s} + \rho^0\boldsymbol{\nabla}\phi^0\cdot(\mathbf{s}\cdot\boldsymbol{\nabla}\mathbf{s} - \mathbf{s}\boldsymbol{\nabla}\cdot\mathbf{s})]\,dV. \tag{4.169}$$

With the adopted normalization, $\mathcal{T} = 1$, the energy equipartition relation $\omega^2\mathcal{T} = \mathcal{V}$ reduces to

$$f_{\mathrm{e}} + f_{\mathrm{g}} = 1, \tag{4.170}$$

where $f_{\mathrm{e}} = \omega^{-2}\mathcal{V}_{\mathrm{e}}$ and $f_{\mathrm{g}} = \omega^{-2}\mathcal{V}_{\mathrm{g}}$ are the *fractional elastic* and *gravitational energies*. The elastic energy content of a non-trivial mode is necessarily non-negative, $\mathcal{V}_{\mathrm{e}} \geq 0$, whereas the gravitational energy may be either positive or negative, depending upon whether gravity acts as a stabilizing or destabilizing influence.

4.3.5 Non-gravitating limit

Physically, we expect long-range gravitational restoring forces to exert a negligible influence upon short-period, short-wavelength oscillations. To determine a criterion for the neglect of gravity, we compare the relative magnitudes of the terms in equation (4.149):

$$\begin{aligned}&\text{inertial:}\quad -\,\omega^2\rho^0\mathbf{s},\\ &\text{elastic:}\quad -\,\boldsymbol{\nabla}\cdot\mathbf{T}^{\mathrm{L1}},\\ &\text{gravitational:}\quad \boldsymbol{\nabla}(\rho^0\mathbf{s}\cdot\boldsymbol{\nabla}\phi^0)+\rho^0\boldsymbol{\nabla}\phi^{\mathrm{E1}}+\rho^{\mathrm{E1}}\boldsymbol{\nabla}\phi^0.\end{aligned}$$

Using $\rho^{\mathrm{E1}} = -\boldsymbol{\nabla}\cdot(\rho^0\mathbf{s})$ and $\nabla^2\phi^{\mathrm{E1}} = 4\pi G\rho^{\mathrm{E1}}$ to estimate the magnitudes of the incremental density and gravitational potential, we find that

$$\|-\,\omega^2\rho^0\mathbf{s}\| \approx \|-\,\boldsymbol{\nabla}\cdot\mathbf{T}^{\mathrm{L1}}\| \approx \omega^2(\rho^0 S), \tag{4.171}$$

$$\|\boldsymbol{\nabla}(\rho^0\mathbf{s}\cdot\boldsymbol{\nabla}\phi^0)+\rho^0\boldsymbol{\nabla}\phi^{\mathrm{E1}}+\rho^{\mathrm{E1}}\boldsymbol{\nabla}\phi^0\| \approx 4\pi G\rho^0(\rho^0 S), \tag{4.172}$$

where $S \approx \|\mathbf{s}\|$ is the magnitude of the particle displacement. Comparing (4.171) and (4.172) we see that the effects of gravity are comparable to those of inertia and elasticity only for oscillations of very low frequency,

$$\omega \approx (4\pi G\rho^0)^{1/2}. \tag{4.173}$$

Setting $\rho^0 \approx 5500\,\mathrm{kg/m^3}$, the mean density of the Earth, we find that this corresponds to a period $2\pi/\omega \approx 3000$ seconds. Not coincidentally, this is very nearly equal to the period of the ${}_0\mathrm{S}_2$ football mode, an oscillation that is strongly influenced by gravity (see Sections 1.1 and 8.8.10). In practice, gravity is routinely ignored in body-wave, regional-wave and exploration seismology, at periods shorter than about 30 seconds.

Wave propagation is governed in the non-gravitating approximation by the classical elastodynamic equation

$$-\omega^2\rho^0\mathbf{s} - \boldsymbol{\nabla}\cdot\mathbf{T}^{\mathrm{L1}} = \mathbf{0} \quad\text{where}\quad \mathbf{T}^{\mathrm{L1}} = \boldsymbol{\Gamma}:\boldsymbol{\varepsilon}. \tag{4.174}$$

The classical Lagrangian and energy densities corresponding to (4.174) are

$$L = \tfrac{1}{2}(\omega^2\rho^0\mathbf{s}\cdot\mathbf{s} - \boldsymbol{\varepsilon}:\boldsymbol{\Gamma}:\boldsymbol{\varepsilon}), \qquad E = \tfrac{1}{2}(\omega^2\rho^0\mathbf{s}\cdot\mathbf{s} + \boldsymbol{\varepsilon}:\boldsymbol{\Gamma}:\boldsymbol{\varepsilon}). \tag{4.175}$$

The potential energy of an oscillation in this approximation is, of course, purely elastic; a non-gravitating fluid or solid elastic body is *inherently stable*, since $\mathcal{V} = \mathcal{V}_{\mathrm{e}} \geq 0$.

It is also possible to ignore the first-order perturbation in the gravitational potential ϕ^{E1}, but to retain the initial potential ϕ^0. This reduces (4.149) to a purely differential eigenvalue problem:

$$-\omega^2\rho^0\mathbf{s} - \boldsymbol{\nabla}\cdot\mathbf{T}^{\mathrm{L1}} + \boldsymbol{\nabla}(\rho^0\mathbf{s}\cdot\boldsymbol{\nabla}\phi^0) + \rho^{\mathrm{E1}}\boldsymbol{\nabla}\phi^0 = \mathbf{0}. \tag{4.176}$$

The acoustic version of equation (4.176), with $\mathbf{T}^{\mathrm{L1}} = -\kappa(\boldsymbol{\nabla}\cdot\mathbf{s})\mathbf{I}$, is referred to in the astrophysical literature as the *Cowling approximation*; it permits the qualitative features of the predominantly gravitational ($f_{\mathrm{g}} \gg f_{\mathrm{e}}$) as well as the predominantly acoustic ($f_{\mathrm{e}} \gg f_{\mathrm{g}}$) oscillations of the Sun and other stars to be investigated without having to solve an integro-differential equation (Cowling 1941; Cox 1980; Unno, Osaki, Ando, Saio & Shibahashi 1989). All of the observed oscillations of the Earth are predominantly elastic in character; for this reason, the Cowling approximation has received little if any use in terrestrial seismology.

*4.4 Response of an Idealized Seismometer

Before we can employ the mode-sum representations (4.72) and (4.141) of the acceleration response $\mathbf{a}(\mathbf{x},t)$ in the analysis of seismological data, we must account for the response of the recording instrument. Following Gilbert (1980) and Wahr (1981b), we model a three-component seismometer as a sensing element or mass with three degrees of freedom, contained within a housing attached to a moving particle $\mathbf{x}$. We account for the effect of the Earth's rotation $\boldsymbol{\Omega}$ in the discussion that follows, since it does not present any significant additional complications.

Let $\boldsymbol{\nu}$ denote the displacement of the mass with respect to the instrument housing; its position relative to the uniformly rotating frame is then $\mathbf{r} = \mathbf{x} + \mathbf{s} + \boldsymbol{\nu}$. The equation of motion of the sensing element is

$$\partial_t^2\mathbf{r} + 2\boldsymbol{\Omega}\times\partial_t\mathbf{r} + \boldsymbol{\Omega}\times(\boldsymbol{\Omega}\times\mathbf{r}) = \mathbf{F} + \mathbf{g}^{\mathrm{L}}, \tag{4.177}$$

where $\mathbf{g}^{\mathrm{L}}$ is the total gravitational force exerted on the particle $\mathbf{x}$ by the Earth, and $\mathbf{F}$ is the electro-mechanical restoring force applied by the instrument. Rewritten with the terms depending upon the relative displacement $\boldsymbol{\nu}$ on the left and the remaining terms on the right, equation (4.177) is

$$\begin{aligned}\partial_t^2\boldsymbol{\nu} + 2\boldsymbol{\Omega}\times\partial_t\boldsymbol{\nu} + \boldsymbol{\Omega}\times(\boldsymbol{\Omega}\times\boldsymbol{\nu}) &= \mathbf{F} - \boldsymbol{\nabla}(\phi^0+\psi)\\ &\quad - \partial_t^2\mathbf{s} - 2\boldsymbol{\Omega}\times\partial_t\mathbf{s} - \boldsymbol{\nabla}\phi^{\mathrm{E1}} - \mathbf{s}\cdot\boldsymbol{\nabla}\boldsymbol{\nabla}(\phi^0+\psi),\end{aligned} \tag{4.178}$$

where we have made use of the first-order relation (3.44) as well as the centrifugal identities $\boldsymbol{\Omega}\times(\boldsymbol{\Omega}\times\mathbf{x}) = \boldsymbol{\nabla}\psi$ and $\boldsymbol{\Omega}\times(\boldsymbol{\Omega}\times\mathbf{s}) = \mathbf{s}\cdot\boldsymbol{\nabla}\boldsymbol{\nabla}\psi$. In a modern force-balance seismometer, the restoring force $\mathbf{F}$ is continually adjusted in such a way that the position of the sensing element relative to the housing remains invariant; the equilibrium force in the absence of any ground motion is $\mathbf{F}^0 = \boldsymbol{\nabla}(\phi^0+\psi)$, whereas the feedback force needed to keep $\boldsymbol{\nu} = \mathbf{0}$ is $\mathbf{F} = \boldsymbol{\nabla}(\phi^0+\psi) + \partial_t^2\mathbf{s} + 2\boldsymbol{\Omega}\times\partial_t\mathbf{s} + \boldsymbol{\nabla}\phi^{\mathrm{E1}} + \mathbf{s}\cdot\boldsymbol{\nabla}\boldsymbol{\nabla}(\phi^0+\psi)$. The recorded signal is proportional to the quantity

$$A = \hat{\boldsymbol{\nu}}\cdot\mathbf{F} - \hat{\boldsymbol{\nu}}^0\cdot\mathbf{F}^0 = (\hat{\boldsymbol{\nu}} - \hat{\boldsymbol{\nu}}^0)\cdot\boldsymbol{\nabla}(\phi^0+\psi)$$

$$+\hat{\boldsymbol{\nu}}^0\cdot[\partial_t^2\mathbf{s}+2\boldsymbol{\Omega}\times\partial_t\mathbf{s}+\boldsymbol{\nabla}\phi^{\mathrm{E}1}+\mathbf{s}\cdot\boldsymbol{\nabla\nabla}(\phi^0+\psi)], \tag{4.179}$$

where $\hat{\boldsymbol{\nu}}$ and $\hat{\boldsymbol{\nu}}^0$ are the instantaneous and equilibrium polarization directions of the instrument, respectively. We have ignored terms of second order in $\mathbf{s}$ in writing the second equality.

A *vertical seismometer* has $\hat{\boldsymbol{\nu}}^0=-\hat{\boldsymbol{\gamma}}^0$ and $\hat{\boldsymbol{\nu}}-\hat{\boldsymbol{\nu}}^0=-\boldsymbol{\nabla}_{\perp}\mathbf{s}\cdot\hat{\boldsymbol{\nu}}^0$, where $\boldsymbol{\gamma}^0=-\boldsymbol{\nabla}(\phi^0+\psi)$ is the unperturbed gravitational-plus-centripetal acceleration, and $\boldsymbol{\nabla}_{\perp}=\boldsymbol{\nabla}-\hat{\boldsymbol{\gamma}}^0(\hat{\boldsymbol{\gamma}}^0\cdot\boldsymbol{\nabla})$ is the gradient in the direction perpendicular to the polarization. Upon inserting these relations into (4.179) and making use of the orthogonality $\hat{\boldsymbol{\gamma}}^0\cdot\boldsymbol{\nabla}_{\perp}=0$ and the differential symmetry $(\boldsymbol{\nabla}_{\perp}\hat{\boldsymbol{\nu}}^0)^{\mathrm{T}}=\boldsymbol{\nabla}_{\perp}\hat{\boldsymbol{\nu}}^0$, we find that

$$A_{\mathrm{vert}}=\hat{\boldsymbol{\nu}}^0\cdot(\partial_t^2\mathbf{s}+2\boldsymbol{\Omega}\times\partial_t\mathbf{s}+\boldsymbol{\nabla}\phi^{\mathrm{E}1})+\mathbf{s}\cdot\boldsymbol{\nabla}\gamma^0, \tag{4.180}$$

where $\gamma^0=\|\boldsymbol{\gamma}^0\|$. We see from equation (4.180) that a vertically polarized instrument responds to the Coriolis acceleration $2\boldsymbol{\Omega}\times\partial_t\mathbf{s}$, the perturbation in gravitational potential $\boldsymbol{\nabla}\phi^{\mathrm{E}1}$, and the free-air change in gravity $\mathbf{s}\cdot\boldsymbol{\nabla}\gamma^0$, in addition to the particle acceleration $\partial_t^2\mathbf{s}$. A *horizontal seismometer* has $\hat{\boldsymbol{\nu}}^0\cdot\hat{\boldsymbol{\gamma}}^0=0$ and $\hat{\boldsymbol{\nu}}-\hat{\boldsymbol{\nu}}^0=\hat{\boldsymbol{\nu}}^0\cdot\boldsymbol{\nabla}\mathbf{s}$; equation (4.179) reduces in this case to

$$A_{\mathrm{horiz}}=\hat{\boldsymbol{\nu}}^0\cdot[\partial_t^2\mathbf{s}+2\boldsymbol{\Omega}\times\partial_t\mathbf{s}+\boldsymbol{\nabla}\phi^{\mathrm{E}1}-\boldsymbol{\nabla}(\boldsymbol{\gamma}^0\cdot\mathbf{s})]. \tag{4.181}$$

Such an instrument responds to the Coriolis acceleration $2\boldsymbol{\Omega}\times\partial_t\mathbf{s}$, the perturbation in gravitational potential $\boldsymbol{\nabla}\phi^{\mathrm{E}1}$, and the tilt of the ground surface $\boldsymbol{\nabla}(\boldsymbol{\gamma}^0\cdot\mathbf{s})$, in addition to the particle acceleration $\partial_t^2\mathbf{s}$.

In conclusion, a vertically or horizontally polarized accelerometer does not simply sense the acceleration of the instrument housing $\mathbf{a}=\partial_t^2\mathbf{s}$; instead it senses a *modified acceleration*

$$\begin{aligned}\mathbf{A}&=\partial_t^2\mathbf{s}+2\boldsymbol{\Omega}\times\partial_t\mathbf{s}+\boldsymbol{\nabla}\phi^{\mathrm{E}1}-\hat{\boldsymbol{\nu}}^0\mathbf{s}\cdot\boldsymbol{\nabla}(\hat{\boldsymbol{\nu}}^0\cdot\boldsymbol{\gamma}^0)\\&\quad-\hat{\boldsymbol{\nu}}^0[\hat{\boldsymbol{\nu}}^0-\hat{\boldsymbol{\gamma}}^0(\hat{\boldsymbol{\nu}}^0\cdot\hat{\boldsymbol{\gamma}}^0)]\cdot\boldsymbol{\nabla}(\boldsymbol{\gamma}^0\cdot\mathbf{s}),\end{aligned} \tag{4.182}$$

where $\hat{\boldsymbol{\nu}}^0$ is the component being measured. We can account for these self-gravitational effects on a non-rotating Earth by substituting for the eigenfunction $\mathbf{s}_k(\mathbf{x})$ in equation (4.72):

$$\begin{aligned}\mathbf{s}_k\rightarrow\mathbf{s}_k-\omega_k^{-2}\{&\boldsymbol{\nabla}\phi_k^{\mathrm{E}1}-\hat{\boldsymbol{\nu}}^0\mathbf{s}_k\cdot\boldsymbol{\nabla}(\hat{\boldsymbol{\nu}}^0\cdot\mathbf{g}^0)\\&-\hat{\boldsymbol{\nu}}^0[\hat{\boldsymbol{\nu}}^0-\hat{\mathbf{g}}^0(\hat{\boldsymbol{\nu}}^0\cdot\hat{\mathbf{g}}^0)]\cdot\boldsymbol{\nabla}(\mathbf{g}^0\cdot\mathbf{s}_k)\}.\end{aligned} \tag{4.183}$$

On a rotating Earth we must account for the effects of the Coriolis and centrifugal forces as well, by substituting in equation (4.141):

$$\begin{aligned}\mathbf{s}_k\rightarrow\mathbf{s}_k-2i\omega_k^{-1}\boldsymbol{\Omega}\times\mathbf{s}_k-\omega_k^{-2}\{&\boldsymbol{\nabla}\phi_k^{\mathrm{E}1}-\hat{\boldsymbol{\nu}}^0\mathbf{s}_k\cdot\boldsymbol{\nabla}(\hat{\boldsymbol{\nu}}^0\cdot\boldsymbol{\gamma}^0)\\&-\hat{\boldsymbol{\nu}}^0[\hat{\boldsymbol{\nu}}^0-\hat{\boldsymbol{\gamma}}^0(\hat{\boldsymbol{\nu}}^0\cdot\hat{\boldsymbol{\gamma}}^0)]\cdot\boldsymbol{\nabla}(\boldsymbol{\gamma}^0\cdot\mathbf{s}_k)\}.\end{aligned} \tag{4.184}$$

We shall continue to use the simple equations (4.72) and (4.141) to represent the acceleration response to a transient body and surface force, with the understanding that the results should be modified in accordance with (4.183) and (4.184) in any actual application. The additional effects are quite significant for some of the low-frequency spheroidal oscillations, as we shall see in Section 10.4.

Chapter 5

Seismic Source Representation

In the previous chapter, we showed how to calculate the normal-mode response of the Earth to an applied body force density $\mathbf{f}$ and surface force density $\mathbf{t}$. In this chapter, we describe the *equivalent-force representation* of an earthquake fault source. The equivalent body and surface force distribution is the unique combination of $\mathbf{f}$ and $\mathbf{t}$ that—if applied in the absence of the faulting—produces exactly the same motion $\mathbf{s}$ everywhere. We obtain this representation of an earthquake source in two different ways, first, by using the concept of the stress glut, introduced by Backus & Mulcahy (1976a; 1976b), and, second, by a straightforward extension of an earlier classical argument due to Burridge & Knopoff (1964). The first method is more elegant, and it shows that *any indigenous seismic source* can be represented by an equivalent body and surface force distribution; the second method is more restrictive, since attention is restricted to a fault source at the outset.

The equivalent-force representation of an idealized fault source is one of the theoretical cornerstones of modern global seismology; other derivations of this fundamental result may be found in a number of advanced textbooks and specialized monographs, including Aki & Richards (1980), Ben-Menahem & Singh (1981), Kennett (1983) and Kostrov & Das (1988). Our derivation is more general than these standard elastodynamic treatments, since it accounts for self-gravitation, and the presence of a non-hydrostatic initial stress. We also allow for the rotation of the Earth, since it complicates matters very little, and does not affect the representation of the seismic source.

It is convenient to introduce one minor change in the notation employed previously; in this chapter, we shall refrain from using the symbol Σ to denote the union of interfaces $\partial\oplus \cup \Sigma_{\mathrm{SS}} \cup \Sigma_{\mathrm{FS}}$. Rather, in the case of an earthquake fault source, we shall now use Σ^t to denote the *instantaneous fault surface* at time t.

5.1 Stress Glut

For the sake of brevity, we shall restrict our analysis at the outset to the case of an Earth model that is everywhere solid and smooth. This is appropriate, since our goal is to determine the equivalent force densities $\mathbf{f}$ and $\mathbf{t}$, and their form depends only upon the conditions within the source region. At the end of this section, we shall briefly discuss the relatively minor complications introduced by the presence of the interior solid-solid and fluid-solid discontinuities within the Earth.

The linearized elastic-gravitational equation of motion and dynamical free-surface boundary condition governing the deformation of a smooth, solid Earth model are

$$\rho^0(\partial_t^2\mathbf{s} + 2\boldsymbol{\Omega} \times \partial_t\mathbf{s}) - \boldsymbol{\nabla} \cdot \mathbf{T}^{\mathrm{PK1}} \\ + \rho^0\boldsymbol{\nabla}\phi^{\mathrm{E1}} + \rho^0\mathbf{s} \cdot \boldsymbol{\nabla\nabla}(\phi^0 + \psi) = \mathbf{0} \quad \text{in } \oplus, \tag{5.1}$$

$$\hat{\mathbf{n}} \cdot \mathbf{T}^{\mathrm{PK1}} = \mathbf{0} \quad \text{on } \partial\oplus. \tag{5.2}$$

The quantity ϕ^{E1} is the Eulerian perturbation in the gravitational potential, given by

$$\phi^{\mathrm{E1}} = -G\int_{\oplus} \frac{\rho^{0\prime}\mathbf{s}' \cdot (\mathbf{x} - \mathbf{x}')}{\|\mathbf{x} - \mathbf{x}'\|^3}\,dV', \tag{5.3}$$

whereas $\mathbf{T}^{\mathrm{PK1}}$ is the incremental first Piola-Kirchhoff stress, given by

$$\mathbf{T}^{\mathrm{PK1}} = \boldsymbol{\Lambda}\!:\!\boldsymbol{\nabla}\mathbf{s}. \tag{5.4}$$

An *indigenous source* is any phenomenon occurring within or upon the surface of the Earth that does not involve forces exerted by any other bodies. Slip on a fault, no matter how complicated, is certainly an indigenous source, as is a sudden phase change, whereas a comet or meteor strike is not. If the Earth is initially at rest, and no external forces act, the unique solution to the homogeneous system of equations (5.1)–(5.4) is eternal seismological quiescence: $\mathbf{s} = \mathbf{0}$ for all times $t \geq 0$. The existence of earthquakes is manifest evidence that $\mathbf{s} \neq \mathbf{0}$; hence, one of the four equations (5.1)–(5.4) must break down. The momentum equation (5.1), the free-surface boundary condition (5.2) and the Newtonian gravitational relation (5.3) are all linearized laws of physics, which must be satisfied by any infinitesimal displacement $\mathbf{s}$; only Hooke's empirical "law" (5.4) is not. Earthquakes and other indigenous sources can therefore be considered to be the result of a *localized, transient failure of the linearized elastic constitutive relation*; this elementary but insightful observation was first articulated by Backus & Mulcahy (1976a; 1976b).

We regard the linearized equations and boundary conditions (5.1)–(5.3) as correct for the physical or *true stress* $\mathbf{T}^{\rm PK1}_{\rm true}$, whereas equation (5.4) is simply the definition of the Hooke or *model stress* $\mathbf{T}^{\rm PK1}_{\rm model}$. The *stress glut* $\mathbf{S}$ is defined to be the difference between the model stress and the true stress:

$$\mathbf{S} = \mathbf{T}^{\rm PK1}_{\rm model} - \mathbf{T}^{\rm PK1}_{\rm true}. \tag{5.5}$$

The *equivalent body and surface force densities* $\mathbf{f}$ and $\mathbf{t}$ are then defined in terms of $\mathbf{S}$ by

$$\mathbf{f} = -\boldsymbol{\nabla}\cdot\mathbf{S} \quad \text{in } \oplus, \qquad \mathbf{t} = \hat{\mathbf{n}}\cdot\mathbf{S} \quad \text{on } \partial\oplus. \tag{5.6}$$

The governing equations (5.1)–(5.2) can be rewritten in terms of these quantities in the form

$$\begin{aligned}\rho^0(\partial_t^2\mathbf{s} + 2\boldsymbol{\Omega}\times\partial_t\mathbf{s}) - \boldsymbol{\nabla}\cdot\mathbf{T}^{\rm PK1}&\\ + \rho^0\boldsymbol{\nabla}\phi^{\rm E1} + \rho^0\mathbf{s}\cdot\boldsymbol{\nabla}\boldsymbol{\nabla}(\phi^0+\psi) = \mathbf{f} &\quad \text{in } \oplus,\end{aligned} \tag{5.7}$$

$$\hat{\mathbf{n}}\cdot\mathbf{T}^{\rm PK1} = \mathbf{t} \quad \text{on } \partial\oplus. \tag{5.8}$$

The quantity $\mathbf{T}^{\rm PK1}$ in (5.7)–(5.8) is the incremental elastic stress $\mathbf{T}^{\rm PK1}_{\rm model}$, given by equation (5.4). The initial-value problem for the non-homogeneous system of equations (5.7)–(5.8) has a unique solution $\mathbf{s}\neq\mathbf{0}$; the equivalent forces $\mathbf{f}$ and $\mathbf{t}$ arising from the breakdown of Hooke's law act to excite the Earth's free oscillations and equivalent propagating waves. The above argument shows conclusively that any indigenous source can be represented by an equivalent body and surface force density of the form (5.6).

The incremental first Piola-Kirchhoff stress $\mathbf{T}^{\rm PK1}$ is related to the incremental Lagrangian Cauchy stress $\mathbf{T}^{\rm L1}$ by

$$\mathbf{T}^{\rm PK1} = \mathbf{T}^{\rm L1} + \mathbf{T}^0(\boldsymbol{\nabla}\cdot\mathbf{s}) - (\boldsymbol{\nabla}\mathbf{s})^{\rm T}\cdot\mathbf{T}^0, \tag{5.9}$$

where $\mathbf{T}^0$ is the initial static stress. We regard equation (5.9) as a linearized geometrical relation which is valid for both the true stresses $\mathbf{T}^{\rm PK1}_{\rm true}$ and $\mathbf{T}^{\rm L1}_{\rm true}$ and the model stresses $\mathbf{T}^{\rm PK1}_{\rm model}$ and $\mathbf{T}^{\rm L1}_{\rm model}$. With this proviso, the stress glut $\mathbf{S}$ may alternatively be considered to be the difference between the model and true Cauchy stresses:

$$\mathbf{S} = \mathbf{T}^{\rm L1}_{\rm model} - \mathbf{T}^{\rm L1}_{\rm true}. \tag{5.10}$$

It follows from equation (5.10) that $\mathbf{S}$ is symmetric:

$$\mathbf{S}^{\mathrm{T}} = \mathbf{S}. \tag{5.11}$$

We shall employ the representation (5.10) rather than (5.5) in deriving the stress glut of an idealized earthquake fault source in Section 5.2, because the true Cauchy stress $\mathbf{T}^{\mathrm{L1}}_{\mathrm{true}}$ can be easily specified in terms of the displacement gradient $\boldsymbol{\nabla}\mathbf{s}$ using elementary physical considerations.

In general, the breakdown of Hooke's law will be spatially localized within some source region S^t. The superscript serves as a reminder of the time-dependent nature of the failure process; typically, the source will nucleate at a point and spread catastrophically to adjacent points thereafter. The instantaneous source volume may either be embedded entirely within the Earth, or its surface ∂S^t may intersect the free surface, as depicted in Figure 5.1. The stress glut vanishes everywhere outside of the instantaneous failure region, including on its buried boundary:

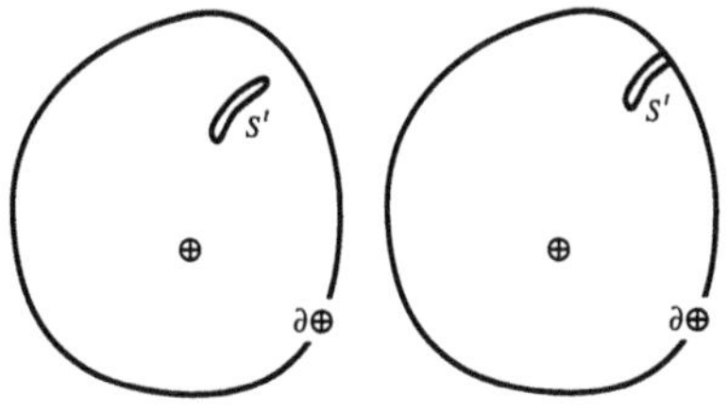

Figure 5.1. Schematic representation of an instantaneous failure volume S^t within a smooth, everywhere solid Earth model. (*Left*) A completely buried source, with $\partial S^t \cap \partial\oplus = 0$. (*Right*) An outcropping source, with $\partial S^t \cap \partial\oplus \neq 0$.

$$\mathbf{S} = \mathbf{0} \quad \text{in } \oplus - S^t \text{ and on } \partial S^t - \partial S^t \cap \partial\oplus. \tag{5.12}$$

Note, however, that $\mathbf{S}$ may be non-zero on $\partial S^t \cap \partial\oplus$; for example, the slip on a shallow fault may extend up to the free surface.

The *total force* exerted upon the Earth by the equivalent body and surface force densities $\mathbf{f}$ and $\mathbf{t}$ is

$$\mathcal{F} = \int_{S^t} \mathbf{f}\, dV + \int_{\partial S^t \cap \partial\oplus} \mathbf{t}\, d\Sigma, \tag{5.13}$$

where we have made use of the restrictions that $\mathbf{f} = \mathbf{0}$ in $\oplus - S^t$ and $\mathbf{t} = \mathbf{0}$ on $\partial\oplus - \partial S^t \cap \partial\oplus$ in expressing the domains of integration. Upon substituting the representations (5.6) into (5.13), we find after an application of Gauss' theorem that

$$\mathcal{F} = -\int_{\partial S^t - \partial S^t \cap \partial\oplus} (\hat{\mathbf{n}} \cdot \mathbf{S})\, d\Sigma. \tag{5.14}$$

The *total torque* exerted upon the Earth,

$$\mathcal{N} = \int_{S^t} (\mathbf{x} \times \mathbf{f})\, dV + \int_{\partial S^t \cap \partial\oplus} (\mathbf{x} \times \mathbf{t})\, d\Sigma, \tag{5.15}$$

can likewise be written in terms of the stress glut in the form

$$\boldsymbol{\mathcal{N}} = \int_{S^t} (\wedge \mathbf{S})\, dV - \int_{\partial S^t - \partial S^t \cap \partial \oplus} \mathbf{x} \times (\hat{\mathbf{n}} \cdot \mathbf{S})\, d\Sigma, \tag{5.16}$$

where $\wedge$ is the wedge operator defined in Appendix A.1.6. The surface integrals over the buried source boundary $\partial S^t - \partial S^t \cap \partial \oplus$ in equations (5.14) and (5.16) vanish by virtue of the condition (5.12), whereas the volume integral involving the quantity $\wedge \mathbf{S}$ vanishes as a consequence of the symmetry (5.11). In conclusion, the equivalent force densities associated with an indigenous source do not exert a net force or torque upon the Earth:

$$\boldsymbol{\mathcal{F}} = \mathbf{0}, \qquad \boldsymbol{\mathcal{N}} = \mathbf{0}. \tag{5.17}$$

This result is to be expected, since any event that occurs entirely within or upon the surface of the Earth cannot alter its linear or angular momentum.

The solution to the inhomogeneous excitation problem (5.7)–(5.8) can be expressed as a sum of normal modes:

$$\mathbf{s}(\mathbf{x}, t) = \mathrm{Re} \sum_k (i\omega_k)^{-1} \mathbf{s}_k(\mathbf{x}) \int_{-\infty}^{t} C_k(t') \exp i\omega_k(t - t')\, dt', \tag{5.18}$$

where

$$C_k(t) = \int_{\oplus} \mathbf{f}(\mathbf{x}, t) \cdot \mathbf{s}_k^*(\mathbf{x})\, dV + \int_{\partial\oplus} \mathbf{t}(\mathbf{x}, t) \cdot \mathbf{s}_k^*(\mathbf{x})\, d\Sigma, \tag{5.19}$$

as we have seen in Chapter 4. Upon inserting the equivalent-force representations (5.6) and making use of Gauss' theorem, we can rewrite equation (5.19) in terms of the stress glut in the form

$$C_k(t) = \int_{S^t} \mathbf{S}(\mathbf{x}, t) : \boldsymbol{\varepsilon}_k^*(\mathbf{x})\, dV, \tag{5.20}$$

where $\boldsymbol{\varepsilon}_k = \frac{1}{2}[\boldsymbol{\nabla}\mathbf{s}_k + (\boldsymbol{\nabla}\mathbf{s}_k)^{\mathrm{T}}]$ is the strain associated with the displacement eigenfunction $\mathbf{s}_k$. The acceleration response (4.141) to a transient indigenous source that begins at t_0 and attains a steady failure state at a later time t_{f} is likewise given by

$$\mathbf{a}(\mathbf{x}, t) = \mathrm{Re} \sum_k c_k \exp(i\omega_k t) \mathbf{s}_k(\mathbf{x}), \quad t \geq t_{\mathrm{f}}, \tag{5.21}$$

where

$$c_k = \int_{t_0}^{t_{\mathrm{f}}} \int_{S^t} \partial_t \mathbf{S}(\mathbf{x}, t) : \boldsymbol{\varepsilon}_k^*(\mathbf{x}) \exp(-i\omega_k t)\, dV\, dt. \tag{5.22}$$

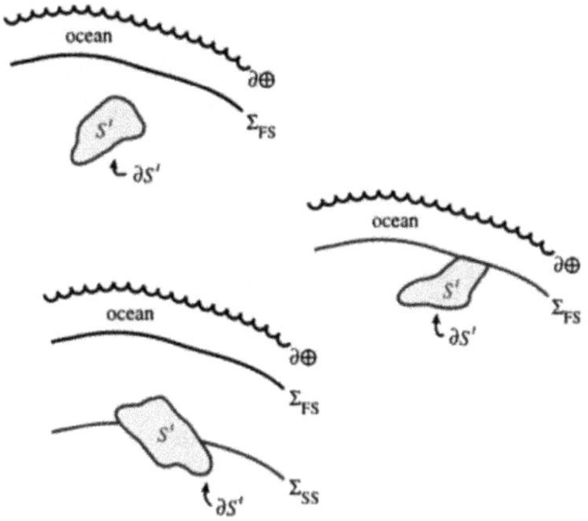

Figure 5.2. Three examples of a source volume S^t within a more general piecewise continuous Earth model. (*Top left*) A completely buried source, away from any internal discontinuities, has $\mathbf{t} = \mathbf{0}$. (*Bottom left*) A source spanning an interior solid-solid discontinuity has $\mathbf{t} = -[\hat{\mathbf{n}} \cdot \mathbf{S}]_-^+$ on Σ_{SS}. (*Middle right*) A source within the solid Earth but abutting the seafloor has $\mathbf{t} = -[\hat{\mathbf{n}} \cdot \mathbf{S}]_-^+$ on Σ_{FS}.

Equations (5.20) and (5.22) are valid regardless of whether or not the Earth is rotating; the asterisk on $\boldsymbol{\varepsilon}_k^*$ is unnecessary in the absence of rotation.

The above results have been obtained under the assumption that the Earth is everywhere solid and smooth; however, they are applicable more generally to any Earth model composed of a number of piecewise smooth fluid and solid regions. Earthquakes are confined to the solid crust and upper mantle, so that the stress glut $\mathbf{S}$ vanishes in the fluid core and the oceans $\oplus_F$. In addition to the external equivalent surface force density $\mathbf{t} = \hat{\mathbf{n}} \cdot \mathbf{S}$ on $\partial\oplus$, it is necessary to consider internal equivalent surface force densities $\mathbf{t} = -[\hat{\mathbf{n}} \cdot \mathbf{S}]_-^+$ on Σ_{SS} and Σ_{FS} whenever the source region S^t spans a solid-solid interface or abuts the seafloor, as shown in Figure 5.2. Gauss' theorem must be applied to each smooth region of the Earth separately and the results combined in order to obtain the normal-mode excitation formulae (5.20) and (5.22).

5.2 Earthquake Fault Source

Earthquakes are generally considered to be the consequence of slip on a fault within the Earth. We calculate the stress glut $\mathbf{S}$ and the corresponding body and surface force densities $\mathbf{f}$ and $\mathbf{t}$ for such an earthquake fault source in this section. We begin with an informal qualitative discussion which clarifies the origin of the name stress "glut", and then present a more formal treatment for an infinitesimally thin fault using the theory of distributions.

5.2.1 The essential idea

To illustrate the essential idea, we consider a concrete example—an earthquake on a vertical strike-slip fault such as the San Andreas Fault in California. We visualize the fault as consisting of a thin gouge or failure zone which experiences intense deformation during the earthquake, surrounded by country rock which behaves in a linear elastic manner. Let x be measured in the direction of right-lateral slip, and let y be measured away from the fault in the direction of the normal. The co-seismic slip s_x and the corresponding shear strain $\varepsilon_{xy} = \varepsilon_{yx} = \frac{1}{2}(\partial_y s_x)$ are depicted schematically in Figure 5.3. The model shear stress is calculated within both the fault zone and the country rock using Hooke's law, $T_{xy}^{\text{model}} = T_{yx}^{\text{model}} = \mu(\partial_y s_x)$, even though the strain within the gouge is far too large for this classical elastic constitutive relation to be valid. We have assumed that the crust is isotropic, and dropped the incremental Lagrangian superscript L1 to avoid clutter; the rigidity μ is assumed to be approximately the same within the fault zone as within the country rock. The true shear stress $T_{xy}^{\text{true}} = T_{yx}^{\text{true}}$, by continuity, cannot be too different within the gouge zone than within the surrounding elastic country rock. We depict the variation of the two stresses T_{xy}^{model} and T_{xy}^{true} using solid and dashed curves, respectively, in Figure 5.3. The difference $S_{xy} = T_{xy}^{\text{model}} - T_{xy}^{\text{true}}$ is non-zero only within the failure zone, as shown. There is an excess or "glut" of model stress within the fault zone due to the large anelastic strain—this is the reason for the terminology.

The model stress increases whereas the true stress remains essentially unchanged as the failure zone becomes narrower; the stress glut of a narrow vertical strike-slip fault is therefore $S_{xy} = S_{yx} \approx \mu(\partial_y s_x)$, to a good approximation. In the idealized case of an infinitesimally thin fault, the displacement s_x tends to a step function, and the stress glut tends to a Dirac delta function:

$$S_{xy} = S_{yx} = \mu \Delta s\, \delta(y), \tag{5.23}$$

where $\Delta s = [s_x]_-^+$ is the slip on the fault. In Section 5.2.3 we shall calculate the stress glut for an ideal, infinitesimally thin fault, taking into account both elastic anisotropy and the presence of a non-hydrostatic initial stress. As we shall see, the result agrees with the above heuristic analysis in the case of an isotropic, hydrostatic Earth model.

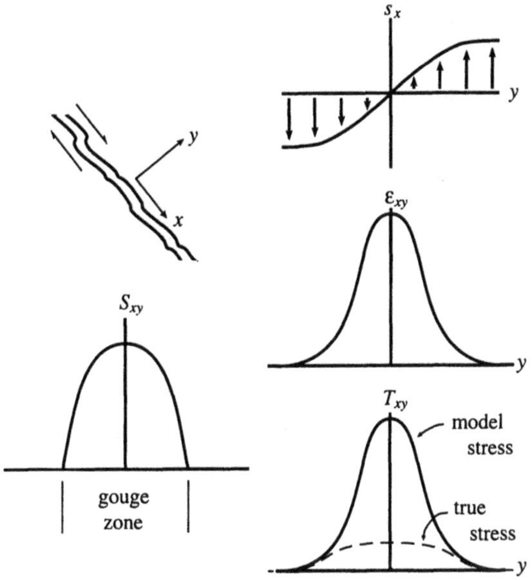

Figure 5.3. (*Top left*) Schematic map view of a right-lateral vertical strike-slip fault, showing the horizontal coordinates x and y. Shading depicts the idealized gouge zone within which Hooke's law fails. (*Top right*) Displacement s_x versus perpendicular distance y within the gouge zone. (*Middle right*) Corresponding shear strain ε_{xy}. (*Bottom right*) Model and shear stresses T_{xy}. (*Bottom left*) The stress glut $S_{xy} = T_{xy}^{\text{model}} - T_{xy}^{\text{true}}$ is zero outside of the gouge zone.

5.2.2 Distribution theory

Before considering a general ideal fault, it is useful to review some simple notions from the theory of distributions. By definition, a *distribution* is a continuous linear functional on a space of smooth test functions. We denote the scalar that a distribution f assigns to a test function φ by $\langle f, \varphi \rangle$. In the case of an Earth model that is everywhere solid and smooth, the test functions are required to vanish in empty space $\bigcirc - \oplus$; more generally, we demand that they vanish outside of the solid portion of the Earth containing the source. We shall refer to the domain of integration as $\oplus$ and to its boundary as $\partial\oplus$ in what follows, with the understanding that these regions must be modified if there are any fluid-solid discontinuities. In an Earth model with a fluid outer core and ocean, the test functions are assumed to be non-zero throughout the crust and mantle, going smoothly to zero on the core-mantle boundary and the seafloor.

Every ordinary function f can be regarded as a distribution, provided that we define

$$\langle f, \varphi\rangle = \int_{\oplus} f\varphi \, dV, \tag{5.24}$$

where $\oplus$ is to be interpreted in the generalized sense described above. Every distribution that is simply an ordinary function regarded as a distribution in the sense (5.24) is said to be *regular*, whereas all other distributions are said to be *singular*. When it is important to distinguish between functions and distributions, we shall denote the regular distribution associated with the function f by $\mathcal{D}f$. By analogy, we shall frequently write the scalar assigned to φ by a singular distribution f in the form (5.24) as well; in this case, the "integration" is purely symbolic.

For an ordinary differentiable function f, we can write

$$\int_{\oplus} (\boldsymbol{\nabla} f)\varphi \, dV = -\int_{\oplus} f(\boldsymbol{\nabla}\varphi) \, dV, \tag{5.25}$$

where we have applied Gauss' theorem and used the fact that φ vanishes on the integration boundary $\partial\oplus$. The gradient of a singular distribution f is defined, by analogy with the result (5.25), in the form

$$\langle \boldsymbol{\nabla} f, \varphi\rangle = -\langle f, \boldsymbol{\nabla}\varphi\rangle. \tag{5.26}$$

If the test functions φ are smooth enough, then every distribution f can be differentiated any number of times in this sense.

The most familiar example of a singular distribution is the *Dirac delta distribution* δ_0, defined by

$$\langle \delta_0, \varphi\rangle = \int_{\oplus} \delta_0 \varphi \, dV = \varphi(\mathbf{x}_0). \tag{5.27}$$

Perhaps a more familiar notation for (5.27), which we shall also employ, is

$$\int_{\oplus} \delta(\mathbf{x} - \mathbf{x}_0) \, \varphi(\mathbf{x}) \, dV = \varphi(\mathbf{x}_0). \tag{5.28}$$

The gradient of the delta distribution $\boldsymbol{\nabla}\delta_0$ is given in accordance with the definition (5.26) by

$$\begin{aligned} \langle \boldsymbol{\nabla}\delta_0, \varphi\rangle &= \int_{\oplus} \boldsymbol{\nabla}\delta(\mathbf{x} - \mathbf{x}_0) \, \varphi(\mathbf{x}) \, dV \\ &= -\int_{\oplus} \delta(\mathbf{x} - \mathbf{x}_0) \, \boldsymbol{\nabla}\varphi(\mathbf{x}) \, dV = -\boldsymbol{\nabla}\varphi(\mathbf{x}_0). \end{aligned} \tag{5.29}$$

In common parlance $\delta(\mathbf{x} - \mathbf{x}_0)$ is often referred to loosely as the Dirac delta "function".

If w is an ordinary function defined on a surface Σ situated within $\oplus$, we define the singular distribution $w\delta_\Sigma$ by

$$\langle w\delta_\Sigma, \varphi\rangle = \int_\oplus (w\delta_\Sigma)\varphi \, dV = \int_\Sigma w\varphi \, d\Sigma. \tag{5.30}$$

We shall also write this surface Dirac distribution using a more suggestive notation:

$$w\delta_\Sigma(\mathbf{x}) = \int_\Sigma w(\mathbf{x}') \, \delta(\mathbf{x} - \mathbf{x}') \, d\Sigma'. \tag{5.31}$$

We can think of $w\delta_\Sigma$ as a weighted "distribution" of Dirac delta functions on the surface Σ, in the same way that we regard $\sum_k w_k\delta(\mathbf{x} - \mathbf{x}_k)$ as a weighted discrete "distribution". The value of $w\delta_\Sigma$ is zero everywhere except on Σ, just as the value of $\sum_k w_k\delta(\mathbf{x} - \mathbf{x}_k)$ is zero everywhere except at the points $\mathbf{x}_k$ (the intuitively obvious concept of the pointwise value of a singular distribution can be rigorously defined by considering the support of the test functions φ). A weighted "distribution" of Dirac gradients

$$w\boldsymbol{\nabla}\delta_\Sigma(\mathbf{x}) = \int_\Sigma w(\mathbf{x}') \, \boldsymbol{\nabla}\delta(\mathbf{x} - \mathbf{x}') \, d\Sigma' \tag{5.32}$$

has the replication property

$$\langle w\boldsymbol{\nabla}\delta_\Sigma, \varphi\rangle = -\int_\Sigma w(\boldsymbol{\nabla}\varphi) \, d\Sigma, \tag{5.33}$$

for any test function φ.

Suppose now that f is a function which is smooth everywhere within the domain $\oplus$, except upon an embedded surface Σ, where it suffers a jump discontinuity $[f]_-^+$. The gradient of f does not exist as an ordinary function within $\oplus$, notably on Σ. We can, however, calculate the gradient of the associated regular distribution $\mathcal{D}f$. To find $\boldsymbol{\nabla}(\mathcal{D}f)$, we consider the scalar that it assigns to an arbitrary test function:

$$\langle \boldsymbol{\nabla}(\mathcal{D}f), \varphi\rangle = -\langle \mathcal{D}f, \boldsymbol{\nabla}\varphi\rangle. \tag{5.34}$$

We can regard the quantity on the right side of (5.34) as the limiting value of an ordinary integral over a punctured volume $\oplus - \oplus_\varepsilon$ that excludes the discontinuity:

$$\langle \boldsymbol{\nabla}(\mathcal{D}f), \varphi\rangle = -\lim_{\varepsilon\to 0} \int_{\oplus - \oplus_\varepsilon} f(\boldsymbol{\nabla}\varphi) \, dV. \tag{5.35}$$

The interior surface $\partial\oplus_\varepsilon$ of the integration volume $\oplus - \oplus_\varepsilon$ completely surrounds Σ, and in the limit $\varepsilon \to 0$ collapses onto it, as depicted in Figure 5.4.

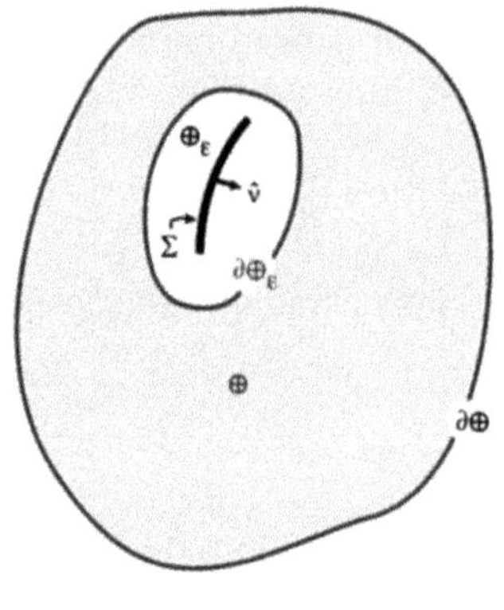

Figure 5.4. Shading depicts the integration volume $\oplus - \oplus_\varepsilon$ in equation (5.35). In the limit $\varepsilon \to 0$ the interior boundary $\partial\oplus_\varepsilon$ collapses onto the embedded discontinuity surface Σ.

Gauss' theorem may be applied to the integral (5.35) prior to taking the limit. Remembering that φ vanishes on $\partial\oplus$, we obtain

$$\langle \boldsymbol{\nabla}(\mathcal{D}f), \varphi\rangle = \int_\oplus (\boldsymbol{\nabla} f)\varphi\, dV + \int_\Sigma \hat{\boldsymbol{\nu}}[f]_-^+ \varphi\, d\Sigma, \tag{5.36}$$

where $\hat{\boldsymbol{\nu}}$ is the unit normal to the surface Σ. The function $\boldsymbol{\nabla} f$ in the volume integral on the right side of equation (5.36) will in general be discontinuous across the surface Σ; its value *on* Σ is irrelevant from the point of view of distribution theory. Replacing $\boldsymbol{\nabla} f$ by the associated regular distribution $\mathcal{D}(\boldsymbol{\nabla} f)$, and using the definition (5.30), we find that

$$\langle \boldsymbol{\nabla}(\mathcal{D}f), \varphi\rangle = \langle \mathcal{D}(\boldsymbol{\nabla} f) + \hat{\boldsymbol{\nu}}[f]_-^+ \delta_\Sigma, \varphi\rangle. \tag{5.37}$$

Equation (5.37) shows that $\boldsymbol{\nabla}(\mathcal{D}f)$ and $\mathcal{D}(\boldsymbol{\nabla} f) + \hat{\boldsymbol{\nu}}[f]_-^+ \delta_\Sigma$ assign the same scalar to every test function φ; they must therefore be the same singular distribution:

$$\boldsymbol{\nabla}(\mathcal{D}f) = \mathcal{D}(\boldsymbol{\nabla} f) + \hat{\boldsymbol{\nu}}[f]_-^+ \delta_\Sigma. \tag{5.38}$$

Loosely speaking, the result (5.38) states that the gradient of f consists of the "ordinary" gradient $\boldsymbol{\nabla} f$ everywhere except upon the surface Σ, where there is an additional delta-function contribution arising from the jump discontinuity $[f]_-^+$. The analogous result in one rather than three dimensions may be more familiar:

$$d(\mathcal{D}f)/dx = \mathcal{D}(df/dx) + [f]_-^+ \delta_0. \tag{5.39}$$

The function f in equation (5.39) is smooth everywhere except at $x = x_0$, where it suffers a jump discontinuity $[f]_-^+$. Its derivative $d(\mathcal{D}f)/dx$ has both an "ordinary" contribution $\mathcal{D}(df/dx)$ and a singular contribution $[f]_-^+\,\delta_0$.

5.2.3 Ideal fault

An *ideal fault* is a surface Σ^t embedded within $\oplus$, across which there is a tangential slip discontinuity $\boldsymbol{\Delta}\mathbf{s} = [\mathbf{s}]_-^+$. The walls of the fault are not allowed to separate or interpenetrate; hence, we must have

$$\hat{\boldsymbol{\nu}} \cdot \boldsymbol{\Delta}\mathbf{s} = 0 \quad \text{on } \Sigma^t. \tag{5.40}$$

In addition, the slip $\boldsymbol{\Delta}\mathbf{s}$ must vanish on the instantaneous edge $\partial\Sigma^t$ of the fault, except where that edge may intersect the solid surface of the Earth or the seafloor. The superscripts on Σ^t and $\partial\Sigma^t$ indicate that the region that has experienced failure will in general depend on time, as the slip nucleates at a point and expands and propagates laterally. The linear elastic constitutive relation $\mathbf{T}^{\mathrm{L1}} = \boldsymbol{\Upsilon} : \boldsymbol{\nabla}\mathbf{s}$ is assumed to be valid everywhere within the Earth except upon the surface Σ^t. The breakdown of Hooke's law is thus considered to be confined to the infinitesimally thin fault.

The true physical stress $\mathbf{T}^{\mathrm{L1}}_{\mathrm{true}}$ will in general be discontinuous across the fault, but the associated distribution $\mathcal{D}(\mathbf{T}^{\mathrm{L1}}_{\mathrm{true}})$ is regular:

$$\mathcal{D}(\mathbf{T}^{\mathrm{L1}}_{\mathrm{true}}) = \boldsymbol{\Upsilon} : \mathcal{D}(\boldsymbol{\nabla}\mathbf{s}). \tag{5.41}$$

The Hooke or model stress is, on the other hand, a singular distribution, given by

$$\mathbf{T}^{\mathrm{L1}}_{\mathrm{model}} = \boldsymbol{\Upsilon} : \boldsymbol{\nabla}(\mathcal{D}\mathbf{s}). \tag{5.42}$$

The stress glut, or difference $\mathbf{T}^{\mathrm{L1}}_{\mathrm{model}} - \mathcal{D}(\mathbf{T}^{\mathrm{L1}}_{\mathrm{true}})$ between the model and true stresses, is evidently

$$\mathbf{S} = \boldsymbol{\Upsilon} : [\boldsymbol{\nabla}(\mathcal{D}\mathbf{s}) - \mathcal{D}(\boldsymbol{\nabla}\mathbf{s})] = (\boldsymbol{\Upsilon} : \hat{\boldsymbol{\nu}}\boldsymbol{\Delta}\mathbf{s})\delta_{\Sigma^t}, \tag{5.43}$$

where we have used the distributional derivative relation (5.38). It is convenient to define the *stress-glut density* on the fault surface by

$$\mathbf{m} = \boldsymbol{\Upsilon} : \hat{\boldsymbol{\nu}}\boldsymbol{\Delta}\mathbf{s}. \tag{5.44}$$

Using this shorthand notation, we may rewrite equation (5.43) in the form

$$\mathbf{S} = \mathbf{m}\delta_{\Sigma^t}. \tag{5.45}$$

The stress glut (5.45) of an ideal fault is a singular distribution that is identically equal to zero everywhere except on the fault surface Σ^t, where the failure of Hooke's law occurs. The quantity $\mathbf{m}$, which is also known as the *surface moment-density tensor*, is the stress glut per unit area on each infinitesimal patch of the slipping fault. The fourth-order tensor $\boldsymbol{\Upsilon}$ is related to the elastic tensor $\boldsymbol{\Gamma}$ in a general, anisotropic Earth model by $\Upsilon_{ijkl} = \Gamma_{ijkl} + \frac{1}{2}(-T^0_{ij}\delta_{kl} + T^0_{kl}\delta_{ij} + T^0_{ik}\delta_{jl} + T^0_{jk}\delta_{il} - T^0_{il}\delta_{jk} - T^0_{jl}\delta_{ik})$. The product of a discontinuous function and a Dirac delta distribution is not defined; hence, it is necessary to stipulate that $\boldsymbol{\Upsilon}$ must be continuous across the fault surface:

$$[\boldsymbol{\Upsilon}]^+_- = \mathbf{0} \quad \text{on } \Sigma^t. \tag{5.46}$$

The result (5.45) is valid if the fault intersects the seafloor or cuts across a solid-solid discontinuity; however, it does not apply to a fault that *coincides* with a part of Σ_{SS}, because of this restriction.

The equivalent body and surface force densities found using (5.6) are

$$\mathbf{f} = -\mathbf{m}\cdot\boldsymbol{\nabla}\delta_{\Sigma^t} \quad \text{in } \oplus, \qquad \mathbf{t} = (\hat{\mathbf{n}}\cdot\mathbf{m})\,\delta_{\Sigma^t} \quad \text{on } \partial\oplus. \tag{5.47}$$

The body-force density $\mathbf{f}$ vanishes everywhere except upon the instantaneous fault surface Σ^t, whereas the surface-force density $\mathbf{t}$ vanishes everywhere except upon the visible rupture trace $\partial\Sigma^t \cap \partial\oplus$. It is easily verified that the net force and torque exerted upon the Earth by the equivalent forces (5.47) satisfy the constraints (5.17). The torque $\boldsymbol{\mathcal{N}}$ vanishes by virtue of the symmetry of the surface moment-density tensor: $\mathbf{m}^{\mathrm{T}} = \mathbf{m}$. The same symmetry guarantees that $\mathbf{m}$ can always be written in diagonal form:

$$\mathbf{m} = m_+\hat{\mathbf{e}}_+\hat{\mathbf{e}}_+ + m_0\hat{\mathbf{e}}_0\hat{\mathbf{e}}_0 + m_-\hat{\mathbf{e}}_-\hat{\mathbf{e}}_-, \tag{5.48}$$

where $m_+ \geq m_0 \geq m_-$ are the eigenvalues and $\hat{\mathbf{e}}_+$, $\hat{\mathbf{e}}_0$, $\hat{\mathbf{e}}_-$ are the associated normalized eigenvectors. The equivalent body force density $\mathbf{f}$ can be considered to be a distribution of mutually perpendicular *linear vector dipoles* on Σ^t of strengths m_+, m_0, m_-, as shown on the left of Figure 5.5. An individual dipole is "outward pointing" if the associated eigenvalue is positive and "inward pointing" if it is negative.

In the case of an isotropic Earth model with a hydrostatic initial stress, the tensor $\boldsymbol{\Upsilon}$ is given by $\Upsilon_{ijkl} = \Gamma_{ijkl} = (\kappa - \frac{2}{3}\mu)\delta_{ij}\delta_{kl} + \mu(\delta_{ik}\delta_{jl} + \delta_{il}\delta_{jk})$, where κ is the isentropic incompressibility and μ is the rigidity. The surface moment-density tensor (5.44) reduces in this case to

$$\mathbf{m} = \mu\Delta s(\hat{\boldsymbol{\nu}}\hat{\boldsymbol{\sigma}} + \hat{\boldsymbol{\sigma}}\hat{\boldsymbol{\nu}}), \tag{5.49}$$

where Δs is the magnitude of the slip vector and $\hat{\boldsymbol{\sigma}}$ is the slip direction, i.e., $\boldsymbol{\Delta}\mathbf{s} = \Delta s\,\hat{\boldsymbol{\sigma}}$. The equivalent body force density $\mathbf{f}$ in an isotropic Earth model can be regarded as a distribution of *double couples* on the fault

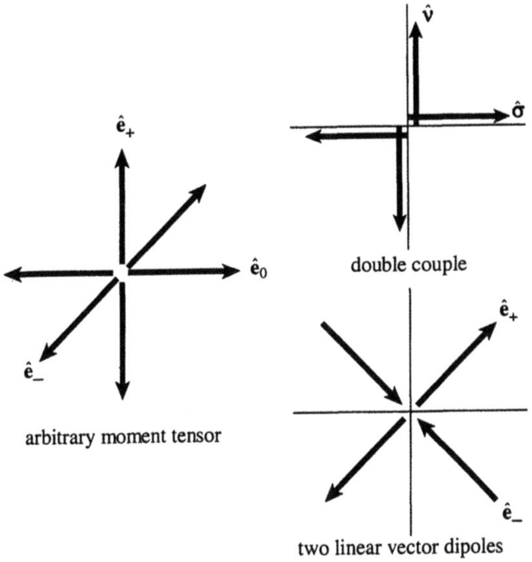

Figure 5.5. (*Left*) Equivalent body-force density **f** associated with an arbitrary moment-density tensor **m**. It is obvious that such a distribution of linear vector dipoles exerts no net force or torque upon the Earth. (*Right*) In an isotropic Earth with a hydrostatic initial stress, **f** may be visualized either as a distribution of opposed couples with moment $\mu\Delta s$, called a double couple (*top*), or as a distribution of mutually orthogonal linear vector dipoles oriented at 45° to the fault surface (*bottom*).

surface Σ^t; this well-known representation has been exploited in numerous seismological applications. Alternatively, we can write **m** in diagonal form,

$$\mathbf{m} = \mu\Delta s(\hat{\mathbf{e}}_+\hat{\mathbf{e}}_+ - \hat{\mathbf{e}}_-\hat{\mathbf{e}}_-), \tag{5.50}$$

where $\hat{\mathbf{e}}_\pm = (\hat{\boldsymbol{\nu}} \pm \hat{\boldsymbol{\sigma}})/\sqrt{2}$, and view **f** as a distribution of equal and opposite linear vector dipoles, as shown on the right of Figure 5.5. The representation (5.49) agrees with the previously derived result (5.23) for a vertical strike-slip fault, as expected.

The displacement produced by an ideal fault source can be written in terms of the surface moment-tensor density in the form

$$\mathbf{s}(\mathbf{x}, t) = \mathrm{Re}\sum_k (i\omega_k)^{-1}\mathbf{s}_k(\mathbf{x}) \int_{-\infty}^{t} C_k(t') \exp i\omega_k(t - t')\, dt', \tag{5.51}$$

where

$$C_k(t) = \int_{\Sigma^t} \mathbf{m}(\mathbf{x}, t) : \boldsymbol{\varepsilon}_k^*(\mathbf{x})\, d\Sigma. \tag{5.52}$$

The acceleration response after the faulting has ceased is likewise given by

$$\mathbf{a}(\mathbf{x}, t) = \mathrm{Re} \sum_k c_k \exp(i\omega_k t)\mathbf{s}_k(\mathbf{x}), \quad t \geq t_\mathrm{f}, \tag{5.53}$$

where

$$c_k = \int_{t_0}^{t_\mathrm{f}} \int_{\Sigma^t} \partial_t \mathbf{m}(\mathbf{x}, t) : \boldsymbol{\varepsilon}_k^*(\mathbf{x}) \exp(-i\omega_k t)\, d\Sigma\, dt. \tag{5.54}$$

The above results follow immediately upon substituting (5.45) into the general formulae (5.18)–(5.22).

The above derivation admittedly has a certain air of magic about it; once the mathematical underpinnings have been established, the equivalent-force representation (5.47) is obtained so effortlessly that it seems to materialize out of thin air. We have tried to keep our discussion as physical as possible; however, it must be admitted that the stress glut is not, upon first encounter, an easy concept to grasp physically. In particular, it is difficult to conceive of a "stress-glut meter" that could be designed and constructed by an experimentalist. Because of this, and because of the importance of the result (5.47), we rederive it using a more conventional approach before discussing it further.

*5.3 Burridge-Knopoff Method

To find the displacement $\mathbf{s}$ produced by an imposed body force density $\mathbf{f}$ and surface traction $\mathbf{t}$, we are required to solve the system of equations

$$\begin{aligned} &\rho^0(\partial_t^2 \mathbf{s} + 2\boldsymbol{\Omega} \times \partial_t \mathbf{s}) - \boldsymbol{\nabla} \cdot \mathbf{T}^{\mathrm{PK1}} \\ &\qquad + \rho^0 \boldsymbol{\nabla} \phi^{\mathrm{E1}} + \rho^0 \mathbf{s} \cdot \boldsymbol{\nabla}\boldsymbol{\nabla}(\phi^0 + \psi) = \mathbf{f} \quad \text{in } \oplus, \end{aligned} \tag{5.55}$$

$$\hat{\mathbf{n}} \cdot \mathbf{T}^{\mathrm{PK1}} = \mathbf{t} \quad \text{on } \partial\oplus, \tag{5.56}$$

$$[\hat{\mathbf{n}} \cdot \mathbf{T}^{\mathrm{PK1}}]_-^+ = \mathbf{0} \quad \text{on } \Sigma_{\mathrm{SS}}, \tag{5.57}$$

$$[\mathbf{t}^{\mathrm{PK1}}]_-^+ = \hat{\mathbf{n}}[\hat{\mathbf{n}} \cdot \mathbf{t}^{\mathrm{PK1}}]_-^+ = \mathbf{0} \quad \text{on } \Sigma_{\mathrm{FS}}. \tag{5.58}$$

The response $\overline{\mathbf{s}}$ of the anti-Earth, with a reversed sense of rotation $\boldsymbol{\Omega} \to -\boldsymbol{\Omega}$, to a different set of force densities $\overline{\mathbf{f}}$, $\overline{\mathbf{t}}$ can likewise be found by solving

$$\begin{aligned} &\rho^0(\partial_t^2 \overline{\mathbf{s}} - 2\boldsymbol{\Omega} \times \partial_t \overline{\mathbf{s}}) - \boldsymbol{\nabla} \cdot \overline{\mathbf{T}}^{\mathrm{PK1}} \\ &\qquad + \rho^0 \boldsymbol{\nabla} \overline{\phi}^{\mathrm{E1}} + \rho^0 \overline{\mathbf{s}} \cdot \boldsymbol{\nabla}\boldsymbol{\nabla}(\phi^0 + \psi) = \overline{\mathbf{f}} \quad \text{in } \oplus, \end{aligned} \tag{5.59}$$

$$\hat{\mathbf{n}}\cdot\overline{\mathbf{T}}^{\mathrm{PK1}}=\overline{\mathbf{t}}\quad\text{on }\partial\oplus, \tag{5.60}$$

$$[\hat{\mathbf{n}}\cdot\overline{\mathbf{T}}^{\mathrm{PK1}}]_-^+=\mathbf{0}\quad\text{on }\Sigma_{\mathrm{SS}}, \tag{5.61}$$

$$[\overline{\mathbf{t}}^{\mathrm{PK1}}]_-^+=\hat{\mathbf{n}}[\hat{\mathbf{n}}\cdot\overline{\mathbf{t}}^{\mathrm{PK1}}]_-^+=\mathbf{0}\quad\text{on }\Sigma_{\mathrm{FS}}. \tag{5.62}$$

We take the dot product of the momentum equation (5.55) with $\overline{\mathbf{s}}$ and the dot product of the anti-momentum equation (5.59) with $\mathbf{s}$, integrate over the Earth model $\oplus$, and apply Gauss' theorem. The surface integrals over $\partial\oplus\cup\Sigma_{\mathrm{SS}}\cup\Sigma_{\mathrm{FS}}$ involving the jumps $[\hat{\mathbf{n}}\cdot\mathbf{T}^{\mathrm{PK1}}\cdot\overline{\mathbf{s}}]_-^+$ and $[\hat{\mathbf{n}}\cdot\overline{\mathbf{T}}^{\mathrm{PK1}}\cdot\mathbf{s}]_-^+$ as well as the Hermitian terms in the volume integrals cancel, as in previous calculations, leaving

$$\begin{aligned}\int_\oplus[\rho^0\overline{\mathbf{s}}\cdot\partial_t^2\mathbf{s}+2\rho^0\overline{\mathbf{s}}\cdot(\boldsymbol{\Omega}\times\partial_t\mathbf{s})-\overline{\mathbf{s}}\cdot\mathbf{f}]\,dV-\int_{\partial\oplus}\overline{\mathbf{s}}\cdot\mathbf{t}\,d\Sigma\\ =\int_\oplus[\rho^0\mathbf{s}\cdot\partial_t^2\overline{\mathbf{s}}-2\rho^0\mathbf{s}\cdot(\boldsymbol{\Omega}\times\partial_t\overline{\mathbf{s}})-\mathbf{s}\cdot\overline{\mathbf{f}}]\,dV-\int_{\partial\oplus}\mathbf{s}\cdot\overline{\mathbf{t}}\,d\Sigma.\end{aligned} \tag{5.63}$$

Equation (5.63) is true even if the unbarred and barred quantities are evaluated at different times; it is convenient to suppose that $\mathbf{f}$, $\mathbf{t}$, $\mathbf{s}$ are evaluated at time t, whereas $\overline{\mathbf{f}}$, $\overline{\mathbf{t}}$, $\overline{\mathbf{s}}$ are evaluated at time $-t$. Upon integrating by parts with respect to time over the interval $-\infty\leq t\leq\infty$, we then obtain

$$\begin{aligned}&\int_{-\infty}^{\infty}\int_\oplus\mathbf{s}(\mathbf{x},t)\cdot\overline{\mathbf{f}}(\mathbf{x},-t)\,dV\,dt\\ &\quad+\int_{-\infty}^{\infty}\int_{\partial\oplus}\mathbf{s}(\mathbf{x},t)\cdot\overline{\mathbf{t}}(\mathbf{x},-t)\,d\Sigma\,dt\\ &\qquad=\int_{-\infty}^{\infty}\int_\oplus\overline{\mathbf{s}}(\mathbf{x},-t)\cdot\mathbf{f}(\mathbf{x},t)\,dV\,dt\\ &\qquad\quad+\int_{-\infty}^{\infty}\int_{\partial\oplus}\overline{\mathbf{s}}(\mathbf{x},-t)\cdot\mathbf{t}(\mathbf{x},t)\,d\Sigma\,dt.\end{aligned} \tag{5.64}$$

The forces $\mathbf{f}$, $\mathbf{t}$ and $\overline{\mathbf{f}}$, $\overline{\mathbf{t}}$ are assumed either to have been turned on gradually or to have begun at some finite time in the past, so that the contributions from $\pm\infty$ vanish by causality. Equation (5.64) is the generalization of the classical *Betti reciprocal relation* to the case of a rotating, self-gravitating Earth. The dependence upon the anti-Earth as well as upon the Earth itself is noteworthy.

Suppose now that both of the applied surface tractions are identically zero, $\mathbf{t}=\overline{\mathbf{t}}=\mathbf{0}$, and that both forces are impulsive in space and time:

$$\mathbf{f}(\mathbf{x},t)=\hat{\mathbf{x}}_q\,\delta(\mathbf{x}-\mathbf{x}')\,\delta(t-t'), \tag{5.65}$$

$$\overline{\mathbf{f}}(\mathbf{x}',t') = \hat{\mathbf{x}}_p\,\delta(\mathbf{x}'-\mathbf{x})\,\delta(t'+t). \tag{5.66}$$

The corresponding responses can in that case be written in terms of the Green and anti-Green tensors $\mathbf{G}$ and $\overline{\mathbf{G}}$ in the form

$$s_p(\mathbf{x},t) = G_{pq}(\mathbf{x},\mathbf{x}';t-t'), \tag{5.67}$$

$$\overline{s}_q(\mathbf{x}',t') = \overline{G}_{qp}(\mathbf{x}',\mathbf{x};t'+t). \tag{5.68}$$

Betti's relation (5.64) reduces to $s_p(\mathbf{x},t) = \overline{s}_q(\mathbf{x}',-t')$ or, equivalently,

$$\mathbf{G}(\mathbf{x},\mathbf{x}';t-t') = \overline{\mathbf{G}}^{\mathrm{T}}(\mathbf{x}',\mathbf{x};t-t'). \tag{5.69}$$

Equation (5.69) is the *generalized source-receiver reciprocity principle* which we established earlier in Section 4.2.6.

Upon taking $\overline{\mathbf{t}}$ to be zero and $\overline{\mathbf{f}}$ to be impulsive, of the form (5.66), but letting $\mathbf{f}$ and $\mathbf{t}$ be arbitrary, we obtain

$$\mathbf{s}(\mathbf{x},t) = \int_{-\infty}^{t}\int_{\oplus} \overline{\mathbf{G}}^{\mathrm{T}}(\mathbf{x}',\mathbf{x};t-t')\cdot\mathbf{f}(\mathbf{x}',t')\,dV'\,dt' + \int_{-\infty}^{t}\int_{\partial\oplus} \overline{\mathbf{G}}^{\mathrm{T}}(\mathbf{x}',\mathbf{x};t-t')\cdot\mathbf{t}(\mathbf{x}',t')\,d\Sigma'\,dt', \tag{5.70}$$

where we have made the identification (5.68) and used the fact that the anti-Green tensor $\overline{\mathbf{G}}(\mathbf{x},\mathbf{x}';t-t')$ vanishes for $t<t'$. We can eliminate any dependence upon the anti-Earth from the representation (5.70) by making use of the reciprocity relation (5.69):

$$\mathbf{s}(\mathbf{x},t) = \int_{-\infty}^{t}\int_{\oplus} \mathbf{G}(\mathbf{x},\mathbf{x}';t-t')\cdot\mathbf{f}(\mathbf{x}',t')\,dV'\,dt' + \int_{-\infty}^{t}\int_{\partial\oplus} \mathbf{G}(\mathbf{x},\mathbf{x}';t-t')\cdot\mathbf{t}(\mathbf{x}',t')\,d\Sigma'\,dt'. \tag{5.71}$$

Equation (5.71) gives the response $\mathbf{s}$ to the applied forces $\mathbf{f}$ and $\mathbf{t}$ in terms of a convolution with the Green tensor $\mathbf{G}$. We used this intuitively obvious result to express the response as a normal-mode sum in Section 4.2.7.

Suppose next that the applied surface tractions $\mathbf{t}$ and $\overline{\mathbf{t}}$ are zero, but that there is a fault surface Σ^t within the Earth across which the displacements $\mathbf{s}$ and $\overline{\mathbf{s}}$ may be discontinuous. Upon repeating the argument we used to find equation (5.64), we obtain in this case

$$\int_{-\infty}^{\infty}\int_{\oplus} \mathbf{s}(\mathbf{x},t)\cdot\overline{\mathbf{f}}(\mathbf{x},-t)\,dV\,dt - \int_{-\infty}^{\infty}\int_{\Sigma^t} [\hat{\boldsymbol{\nu}}(\mathbf{x})\cdot\overline{\mathbf{T}}^{\mathrm{PK1}}(\mathbf{x},-t)\cdot\mathbf{s}(\mathbf{x},t)]_-^+\,d\Sigma\,dt =$$

$$\int_{-\infty}^{\infty}\int_{\oplus} \bar{\mathbf{s}}(\mathbf{x},-t)\cdot\mathbf{f}(\mathbf{x},t)\,dV\,dt$$
$$-\int_{-\infty}^{\infty}\int_{\Sigma^t}[\hat{\boldsymbol{\nu}}(\mathbf{x})\cdot\mathbf{T}^{\mathrm{PK1}}(\mathbf{x},t)\cdot\bar{\mathbf{s}}(\mathbf{x},-t)]_-^+\,d\Sigma\,dt, \tag{5.72}$$

which is a generalization of Betti's reciprocal relation. The surface integrals in equation (5.72) do not vanish, unlike the corresponding integrals over the fluid-solid boundary Σ_{FS}, because the fault surface Σ^t is not frictionless. It is unnecessary to suppose that the fault is situated entirely within a smooth region of the crust or upper mantle in order for the result (5.72) to be valid. If Σ^t intersects the solid free surface $\partial\oplus$ or the seafloor Σ_{FS}, or if it cuts across a solid-solid discontinuity Σ_{SS}, then we are obliged to integrate as shown in Figure 5.6 in applying Gauss' theorem.

To apply the result (5.72) we again take $\bar{\mathbf{f}}$ to be a unit impulsive force of the form (5.66), and we take $\mathbf{f}$ to be zero; we regard the displacement $\mathbf{s}$ as the response to the faulting. Upon using the generalized reciprocity relation (5.69) and the Maxwell relation $\Lambda_{ijkl}=\Lambda_{klij}$ to write the displacement $\bar{\mathbf{s}}$ and associated stress $\overline{\mathbf{T}}^{\mathrm{PK1}}$ in the anti-Earth in terms of the Green tensor $\mathbf{G}$, we obtain

$$s_p(\mathbf{x},t)=\int_{-\infty}^{t}\int_{\Sigma^t}[\partial_i' G_{pj}(\mathbf{x},\mathbf{x}';t-t')\Lambda_{ijkl}(\mathbf{x}')\nu_k(\mathbf{x}')s_l(\mathbf{x}',t')$$
$$-\,G_{pj}(\mathbf{x},\mathbf{x}';t-t')\nu_i(\mathbf{x}')T_{ij}^{\mathrm{PK1}}(\mathbf{x}',t')]_-^+\,d\Sigma'\,dt', \tag{5.73}$$

where ∂_i' denotes the partial derivative with respect to x_i'. The incremental traction $\hat{\boldsymbol{\nu}}\cdot\mathbf{T}^{\mathrm{PK1}}$ satisfies the linearized dynamical boundary condition (3.73), which we rewrite in index notation here for convenience:

$$[\nu_i T_{ij}^{\mathrm{PK1}}]_-^+=[\partial_i^{\Sigma}(s_i\nu_k T_{kj}^0)]_-^+, \tag{5.74}$$

where $\partial_i^{\Sigma}=\partial_i-\nu_i(\nu_j\partial_j)$. Using the result (5.74) and applying the two-dimensional version of Gauss' theorem (A.77)–(A.78), we can rewrite the second term on the right of (5.73) in the form

$$\int_{\Sigma^t}[G_{pj}\nu_i T_{ij}^{\mathrm{PK1}}]_-^+\,d\Sigma'=\int_{\partial\Sigma_{\mathrm{c}}^t}G_{pj}[\nu_k T_{kj}^0 b_i s_i]_-^+\,dL'$$
$$-\int_{\Sigma^t}\partial_i' G_{pj}[\nu_k T_{kj}^0 s_i]_-^+\,d\Sigma', \tag{5.75}$$

where $\hat{\mathbf{b}}$ is the unit normal tangent to Σ^t and pointing out of Σ^t on $\partial\Sigma^t$. The line integral in equation (5.75) is only over the buried part of the fault boundary where the integration surface depicted in Figure 5.6 folds back upon itself; the slip $\Delta\mathbf{s}$ goes to zero on this instantaneous rupture

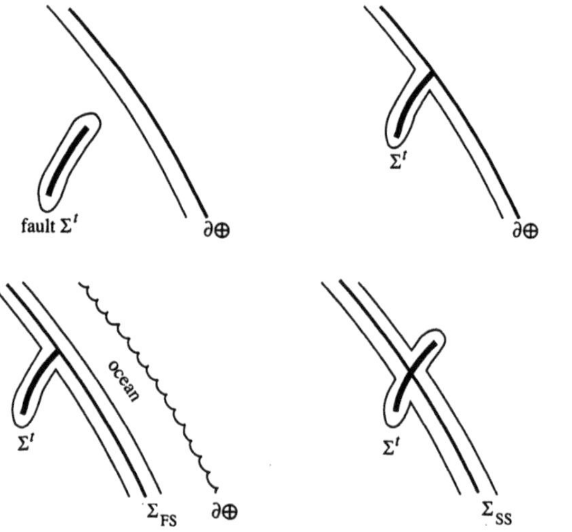

Figure 5.6. Integration schemes used in the derivation of the generalized Betti reciprocal relation (5.72). Thin lines depict the surfaces of integration which must be considered in applying Gauss' theorem; in each case, these must be imagined to deform so that the integration volume is the entire Earth $\oplus$. (*Upper left*) If the fault surface is entirely embedded within a smooth solid portion of $\oplus$, then the integration surface consists of the outer free surface $\partial\oplus$ together with both sides of Σ^t. (*Upper right*) If the fault intersects the solid free surface, then the integration surface must envelop Σ^t in a "one-sided" fashion; in this case, there is no contribution to the line integral in equation (5.75) from the exposed fault trace $\partial\Sigma^t \cap \partial\oplus$. (*Lower right*) If the fault cuts across a solid-solid discontinuity, then the integration surface must envelop both Σ^t and $\Sigma_{\rm SS}$. (*Lower left*) In the case of a submarine fault that intersects the seafloor, there is no contribution to the line integral in (5.75) from $\partial\Sigma^t \cap \Sigma_{\rm FS}$.

front or *crack tip* $\partial\Sigma_{\rm c}^t$, so that this contribution vanishes. Upon combining the remaining surface integrals in (5.73) and (5.75), and using the elastic tensor relation (3.123), we can rewrite the response to a prescribed slip distribution $\mathbf{\Delta s}$ upon Σ^t in the compact form

$$s_p(\mathbf{x},t) = \int_{-\infty}^{t} \int_{\Sigma^t} [\partial_i' G_{pj}(\mathbf{x},\mathbf{x}';t-t') m_{ij}(\mathbf{x}',t')]_-^+ \, d\Sigma' \, dt', \qquad (5.76)$$

where $\mathbf{m} = \boldsymbol{\Upsilon} : \hat{\boldsymbol{\nu}}\mathbf{\Delta s}$ is the surface moment-density tensor. In obtaining the result (5.76), we have assumed that the initial static stress is continuous across the fault, $[\mathbf{T}^0]_-^+ = \mathbf{0}$, and made use of the kinematic boundary condition $\hat{\boldsymbol{\nu}} \cdot \mathbf{\Delta s} = 0$. We have also assumed that the Green tensor and its gradient are continuous: $[\mathbf{G}]_-^+ = \mathbf{0}$ and $[\boldsymbol{\nabla}'\mathbf{G}]_-^+ = \mathbf{0}$. The fault surface cannot coincide with a portion of $\Sigma_{\rm SS}$ because of the latter restriction. The representation (5.76) of the displacement $\mathbf{s}$ in terms of $\mathbf{m}$ is sometimes referred to as the *Volterra representation theorem*.

To determine the equivalent body and surface force densities $\mathbf{f}$ and $\mathbf{t}$, we rewrite the gradient of the Green tensor in the form

$$\begin{aligned}\partial_i' G_{pj}(\mathbf{x},\mathbf{x}';t-t') &= \int_\oplus \delta(\mathbf{x}''-\mathbf{x}')\,\partial_i'' G_{pj}(\mathbf{x},\mathbf{x}'';t-t')\,dV'' \\ &= \int_{\partial\oplus} \delta(\mathbf{x}''-\mathbf{x}')\,n_i(\mathbf{x}'')G_{pj}(\mathbf{x},\mathbf{x}'';t-t')\,d\Sigma'' \\ &\qquad - \int_\oplus \partial_i''\delta(\mathbf{x}''-\mathbf{x}')\,G_{pj}(\mathbf{x},\mathbf{x}'';t-t')\,dV'',\end{aligned} \tag{5.77}$$

where $\delta(\mathbf{x}''-\mathbf{x}')$ is the Dirac delta distribution. Substituting (5.77) into the Volterra representation (5.76), we obtain

$$\begin{aligned}\mathbf{s}(\mathbf{x},t) &= \int_{-\infty}^{t}\int_\oplus \mathbf{G}(\mathbf{x},\mathbf{x}';t-t')\cdot\mathbf{f}(\mathbf{x}',t')\,dV'\,dt' \\ &\qquad + \int_{-\infty}^{t}\int_{\partial\oplus} \mathbf{G}(\mathbf{x},\mathbf{x}';t-t')\cdot\mathbf{t}(\mathbf{x}',t')\,d\Sigma'\,dt',\end{aligned} \tag{5.78}$$

where

$$\mathbf{f}(\mathbf{x},t) = -\int_{\Sigma^t} \mathbf{m}(\mathbf{x}',t)\cdot\boldsymbol{\nabla}\delta(\mathbf{x}-\mathbf{x}')\,d\Sigma' \quad \text{in } \oplus, \tag{5.79}$$

$$\mathbf{t}(\mathbf{x},t) = \hat{\mathbf{n}}(\mathbf{x})\cdot\int_{\Sigma^t} \mathbf{m}(\mathbf{x}',t)\,\delta(\mathbf{x}-\mathbf{x}')\,d\Sigma' \quad \text{on } \partial\oplus. \tag{5.80}$$

Equations (5.79)–(5.80) are the same as (5.47), rewritten using an alternative notation; both results give the equivalent force densities $\mathbf{f}$ and $\mathbf{t}$ for an ideal fault. The equivalent body and surface force densities for a fault in a self-gravitating Earth with a non-hydrostatic initial stress were first derived in the above manner by Dahlen (1972).

5.4 Point-Source Approximation

In normal-mode and surface-wave seismology, we are generally concerned with periods that are much longer than the duration of the source, and with wavelengths that are much longer than the source dimensions. It is possible in this case to simplify the above results substantially. We describe a number of point-source approximations that are commonly employed, considering a general indigenous source with a prescribed stress-glut tensor $\mathbf{S}$ in addition to an ideal fault source with a dynamic slip vector $\boldsymbol{\Delta}\mathbf{s}$.

5.4.1 Moment tensor

The complex excitation amplitude c_k of a normal mode is an integral (5.22) over the source duration $t_0 \leq t \leq t_{\mathrm{f}}$ and the source volume S^t. In the limit of long wavelengths and long periods, we can replace the quantity $\boldsymbol{\varepsilon}_k^*(\mathbf{x})\exp(-i\omega_k t)$ in this integral by a constant:

$$\boldsymbol{\varepsilon}_k^*(\mathbf{x})\exp(-i\omega_k t) = \boldsymbol{\varepsilon}_k^*(\mathbf{x}_{\mathrm{s}})\exp(-i\omega_k t_{\mathrm{s}}). \tag{5.81}$$

The amplitude c_k then reduces to $c_k = \mathbf{M}\!:\!\boldsymbol{\varepsilon}_k^*(\mathbf{x}_{\mathrm{s}})\exp(-i\omega_k t_{\mathrm{s}})$, where

$$\mathbf{M} = \int_{t_0}^{t_{\mathrm{f}}}\int_{S^t} \partial_t \mathbf{S}\, dV\, dt. \tag{5.82}$$

Noting that $\mathbf{S}$ vanishes in $S^{\mathrm{f}} - S^t$, where S^{f} is the source volume at time t_{f}, and interchanging the order of integration, we can rewrite equation (5.82) in the form

$$\mathbf{M} = \int_{S^{\mathrm{f}}} \mathbf{S}_{\mathrm{f}}\, dV, \tag{5.83}$$

where $\mathbf{S}_{\mathrm{f}}$ denotes the final static value of the stress glut. The quantity $\mathbf{M}$ is known as the *moment tensor* of the source.

The glut rate in this lowest-order point-source approximation is a Dirac distribution in space-time:

$$\partial_t \mathbf{S} = \mathbf{M}\,\delta(\mathbf{x} - \mathbf{x}_{\mathrm{s}})\,\delta(t - t_{\mathrm{s}}). \tag{5.84}$$

The associated equivalent body and surface force densities (5.6) are

$$\mathbf{f} = -\mathbf{M}\cdot\boldsymbol{\nabla}\delta(\mathbf{x} - \mathbf{x}_{\mathrm{s}})\,H(t - t_{\mathrm{s}}), \qquad \mathbf{t} = \mathbf{0}, \tag{5.85}$$

where $H(t - t_{\mathrm{s}})$ is the Heaviside step function. The quantities $\mathbf{x}_{\mathrm{s}}$ and t_{s} are the fiducial *hypocentral location* and *origin time*. The displacement response (5.18) to such an impulsive moment-tensor source is

$$\mathbf{s}(\mathbf{x}, t) = \mathrm{Re}\sum_k \omega_k^{-2}\mathbf{M}\!:\!\boldsymbol{\varepsilon}_k^*(\mathbf{x}_{\mathrm{s}})\,\mathbf{s}_k(\mathbf{x})[1 - \exp i\omega_k(t - t_{\mathrm{s}})], \tag{5.86}$$

where the asterisk on the strain is unnecessary in the absence of rotation. The associated acceleration (5.21) is

$$\mathbf{a}(\mathbf{x}, t) = \mathrm{Re}\sum_k \mathbf{M}\!:\!\boldsymbol{\varepsilon}_k^*(\mathbf{x}_{\mathrm{s}})\,\mathbf{s}_k(\mathbf{x})\exp i\omega_k(t - t_{\mathrm{s}}). \tag{5.87}$$

The results (5.86)–(5.87), which have been used as the basis of numerous source-mechanism determinations, were first obtained by Gilbert (1970); he mistakenly considered $\mathbf{M}$ to be the integral of the stress drop rather than the stress glut.

The moment tensor of an ideal fault is given by

$$\mathbf{M} = \int_{\Sigma^{\mathrm{f}}} \boldsymbol{\Upsilon} : \hat{\boldsymbol{\nu}} \boldsymbol{\Delta}\mathbf{s}_{\mathrm{f}} \, d\Sigma, \tag{5.88}$$

where Σ^{f} is the final fault surface and $\boldsymbol{\Delta}\mathbf{s}_{\mathrm{f}}$ is the final static fault slip. In the case of an isotropic, hydrostatic Earth model, this result reduces to

$$\mathbf{M} = \int_{\Sigma^{\mathrm{f}}} \mu \Delta s_{\mathrm{f}} (\hat{\boldsymbol{\nu}}\hat{\boldsymbol{\sigma}} + \hat{\boldsymbol{\sigma}}\hat{\boldsymbol{\nu}}) \, d\Sigma, \tag{5.89}$$

where $\Delta s_{\mathrm{f}} = \|\boldsymbol{\Delta}\mathbf{s}_{\mathrm{f}}\|$. For a planar fault with uni-directional slip, the moment tensor (5.89) is simply

$$\mathbf{M} = M_0(\hat{\boldsymbol{\nu}}\hat{\boldsymbol{\sigma}} + \hat{\boldsymbol{\sigma}}\hat{\boldsymbol{\nu}}) \quad \text{where} \quad M_0 = \int_{\Sigma^{\mathrm{f}}} \mu \Delta s_{\mathrm{f}} \, d\Sigma. \tag{5.90}$$

The quantity M_0 is the *scalar seismic moment*, which has served as the standard measure of the size of an earthquake ever since its introduction into quantitative seismology by Aki (1966). The equivalent body force (5.85) associated with the moment tensor (5.90) is a classical *double-couple* point source. As is well known, it is impossible to distinguish the fault plane with normal $\hat{\boldsymbol{\nu}}$ from the *auxiliary plane* with normal $\hat{\boldsymbol{\sigma}}$ in this approximation.

More generally, the scalar moment of an arbitrary moment-tensor source $\mathbf{M}_0$ can be defined by

$$M_0 = \tfrac{1}{\sqrt{2}}(\mathbf{M} : \mathbf{M})^{1/2}. \tag{5.91}$$

The factor of $1/\sqrt{2}$ guarantees that equation (5.91) agrees with Aki's classical definition in the case of a double-couple source. We can write the moment tensor of an arbitrary source, by analogy with (5.90), in the form

$$\mathbf{M} = \sqrt{2} M_0 \hat{\mathbf{M}}. \tag{5.92}$$

The quantity $\hat{\mathbf{M}}$, which satisfies the normalization relation $\hat{\mathbf{M}} : \hat{\mathbf{M}} = 1$, is referred to as the *unit source-mechanism tensor*.

5.4.2 Centroid-moment tensor

The constant $\varepsilon_k^*(\mathbf{x}_{\mathrm{s}}) \exp(-i\omega_k t_{\mathrm{s}})$ in equation (5.81) is simply the zeroth-order term in a Taylor series expansion of the quantity $\varepsilon_k^*(\mathbf{x}) \exp(-i\omega_k t)$.

We can obtain an improved approximation by taking the next term of this expansion into account as well:

$$\begin{aligned}\boldsymbol{\varepsilon}_k^*(\mathbf{x})\exp(-i\omega_k t) &= \boldsymbol{\varepsilon}_k^*(\mathbf{x}_\mathrm{s})\exp(-i\omega_k t_\mathrm{s})\\ &\quad + (\mathbf{x}-\mathbf{x}_\mathrm{s})\cdot\boldsymbol{\nabla}\boldsymbol{\varepsilon}_k^*(\mathbf{x}_\mathrm{s})\exp(-i\omega_k t_\mathrm{s})\\ &\qquad - i\omega_k(t-t_\mathrm{s})\boldsymbol{\varepsilon}_k^*(\mathbf{x}_\mathrm{s})\exp(-i\omega_k t_\mathrm{s}).\end{aligned} \tag{5.93}$$

The excitation amplitude of a mode in this case becomes

$$\begin{aligned}c_k &= \mathbf{M}:\boldsymbol{\varepsilon}_k^*(\mathbf{x}_\mathrm{s})\exp(-i\omega_k t_\mathrm{s}) + \mathbf{D}\,\vdots\,\boldsymbol{\nabla}\boldsymbol{\varepsilon}_k^*(\mathbf{x}_\mathrm{s})\exp(-i\omega_k t_\mathrm{s})\\ &\quad - i\omega_k\mathbf{H}:\boldsymbol{\varepsilon}_k^*(\mathbf{x}_\mathrm{s})\exp(-i\omega_k t_\mathrm{s}),\end{aligned} \tag{5.94}$$

where $\mathbf{M}$ is given by equation (5.83) as before, and where

$$\mathbf{D} = \int_{t_0}^{t_\mathrm{f}}\int_{S^t}(\mathbf{x}-\mathbf{x}_\mathrm{s})\partial_t\mathbf{S}\,dV\,dt, \tag{5.95}$$

$$\mathbf{H} = \int_{t_0}^{t_\mathrm{f}}\int_{S^t}(t-t_\mathrm{s})\partial_t\mathbf{S}\,dV\,dt. \tag{5.96}$$

The quantities $\mathbf{D}$ and $\mathbf{H}$ are the *first spatial* and *temporal moments* of the glut rate $\partial_t\mathbf{S}$, just as the moment tensor $\mathbf{M}$ is its zeroth moment. The moments are measured with respect to the fiducial source location $\mathbf{x}_\mathrm{s}$ and origin time t_s.

Following Backus (1977a; 1977b) we define the *centroid location* $\mathbf{x}_\mathrm{c}$ and the *centroid time* t_c to be the values of $\mathbf{x}_\mathrm{s}$ and t_s for which $\mathbf{D}$ and $\mathbf{H}$ are minimized in a least-squares sense. Denoting the moments with respect to this space-time centroid by $\mathbf{D}_\mathrm{c}$ and $\mathbf{H}_\mathrm{c}$, we demand that

$$\mathbf{D}_\mathrm{c}\,\vdots\,\mathbf{D}_\mathrm{c} = (\mathbf{D}-\boldsymbol{\Delta}\mathbf{x}\,\mathbf{M})\,\vdots\,(\mathbf{D}-\boldsymbol{\Delta}\mathbf{x}\,\mathbf{M}) = \text{minimum}, \tag{5.97}$$

$$\mathbf{H}_\mathrm{c}:\mathbf{H}_\mathrm{c} = (\mathbf{H}-\Delta t\,\mathbf{M}):(\mathbf{H}-\Delta t\,\mathbf{M}) = \text{minimum}, \tag{5.98}$$

where

$$\boldsymbol{\Delta}\mathbf{x} = \mathbf{x}_\mathrm{c}-\mathbf{x}_\mathrm{s}, \qquad \Delta t = t_\mathrm{c}-t_\mathrm{s}. \tag{5.99}$$

The deviations $\boldsymbol{\Delta}\mathbf{x}$, Δt of the centroid away from the fiducial origin $\mathbf{x}_\mathrm{s}$, t_s are given by

$$\boldsymbol{\Delta}\mathbf{x} = \frac{\mathbf{D}:\mathbf{M}}{\mathbf{M}:\mathbf{M}}, \qquad \Delta t = \frac{\mathbf{H}:\mathbf{M}}{\mathbf{M}:\mathbf{M}}. \tag{5.100}$$

The minimum tensors

$$\mathbf{D}_\mathrm{c} = \int_{t_0}^{t_\mathrm{f}}\int_{S^t}(\mathbf{x}-\mathbf{x}_\mathrm{c})\partial_t\mathbf{S}\,dV\,dt, \tag{5.101}$$

$$\mathbf{H}_c = \int_{t_0}^{t_f} \int_{S^t} (t - t_c)\partial_t \mathbf{S}\, dV\, dt \tag{5.102}$$

must satisfy the constraints $\mathbf{D}_c : \mathbf{M} = \mathbf{0}$ and $\mathbf{H}_c : \mathbf{M} = 0$.

We can provide a physical interpretation of the centroid $\mathbf{x}_c$, t_c of an arbitrary source by defining a *normalized scalar glut-rate density*:

$$\dot{m} = \tfrac{1}{2} M_0^{-2} (\mathbf{M} : \partial_t \mathbf{S}), \tag{5.103}$$

where M_0 is the scalar moment given in equation (5.91). The space-time integral of this scalar density over the source is unity:

$$\int_{t_0}^{t_f} \int_{S^t} \dot{m}\, dV\, dt = 1. \tag{5.104}$$

The centroid shift (5.100) can be rewritten in terms of $\dot{m}$ in the form

$$\boldsymbol{\Delta}\mathbf{x} = \int_{t_0}^{t_f} \int_{S^t} (\mathbf{x} - \mathbf{x}_s)\dot{m}\, dV\, dt, \tag{5.105}$$

$$\Delta t = \int_{t_0}^{t_f} \int_{S^t} (t - t_s)\dot{m}\, dV\, dt, \tag{5.106}$$

or, equivalently,

$$\mathbf{x}_c = \int_{t_0}^{t_f} \int_{S^t} \mathbf{x}\, \dot{m}\, dV\, dt, \qquad t_c = \int_{t_0}^{t_f} \int_{S^t} t\, \dot{m}\, dV\, dt. \tag{5.107}$$

We see from equation (5.107) that $\mathbf{x}_c, t_c$ can be interpreted as the space-time centroid or hypocenter of the normalized glut rate; we can regard $\dot{m}$ as a sort of scalar source "charge" density.

For a planar fault with uni-directional slip, situated in an isotropic Earth with a hydrostatic initial stress, the normalized glut-rate density (5.103) is

$$\dot{m} = M_0^{-1} \mu \partial_t \Delta s\, \delta_{\Sigma^t}, \tag{5.108}$$

and the hypocentral coordinates (5.107) are

$$\mathbf{x}_c = \frac{1}{M_0} \int_{\Sigma^f} \mathbf{x}\, \mu \Delta s_f\, d\Sigma, \tag{5.109}$$

$$t_c = \frac{1}{M_0} \int_{t_0}^{t_f} \int_{\Sigma^t} t\, \mu \partial_t \Delta s\, d\Sigma\, dt, \tag{5.110}$$

where the moment M_0 is given by (5.90). Equation (5.109) shows that the spatial centroid $\mathbf{x}_c$ is situated on the planar fault surface, as is obvious from physical considerations.

In practical applications of the above formalism, the moment tensors $\mathbf{D}_c$ and $\mathbf{H}_c$ with respect to the centroid $\mathbf{x}_c$, t_c are commonly ignored. This is justified for an earthquake on a planar fault with uni-directional slip in an isotropic Earth with a hydrostatic initial stress. For such a source, in fact, $\mathbf{D}_c = \mathbf{0}$ and $\mathbf{H}_c = \mathbf{0}$. If we adopt this simplification, the acceleration response (5.21) reduces to

$$\begin{aligned}\mathbf{a}(\mathbf{x},t) &= \mathrm{Re}\sum_k (1 - i\omega_k\Delta t)\mathbf{M}:\boldsymbol{\varepsilon}_k^*(\mathbf{x}_s)\,\mathbf{s}_k(\mathbf{x})\exp i\omega_k(t-t_s)\\ &\quad + \mathrm{Re}\sum_k \boldsymbol{\Delta}\mathbf{x}\,\mathbf{M}:\boldsymbol{\nabla}\boldsymbol{\varepsilon}_k^*(\mathbf{x}_s)\,\mathbf{s}_k(\mathbf{x})\exp i\omega_k(t-t_s). \qquad (5.111)\end{aligned}$$

A seismic source is represented in this approximation by its centroid shift $\boldsymbol{\Delta}\mathbf{x}$, Δt with respect to the fiducial origin $\mathbf{x}_s$, t_s in addition to its moment tensor $\mathbf{M}$. If $\mathbf{x}_s$ and t_s have been determined from the measured arrival times of high-frequency body waves, as is usually the case, then they are estimates of the location and time of the initiation of rupture; the centroid location $\mathbf{x}_c$ and time t_c will generally be different. For this reason, it is desirable to allow for a centroid shift $\boldsymbol{\Delta}\mathbf{x}$, Δt by using (5.111) rather than (5.87) in source-mechanism determinations, even if the initial estimates $\mathbf{x}_s$, t_s are thought to be accurate. The total number of parameters that must be determined in a *centroid-moment tensor* or *CMT solution* based on (5.111) is ten (Dziewonski, Chou & Woodhouse 1981).

5.4.3 Deviatoric and double-couple sources

Every symmetric moment tensor can be decomposed into its isotropic and deviatoric parts:

$$\mathbf{M} = \tfrac{1}{3}(\mathrm{tr}\,\mathbf{M})\mathbf{I} + \boldsymbol{\mathcal{M}}, \qquad (5.112)$$

where $\mathrm{tr}\,\boldsymbol{\mathcal{M}} = 0$. An ideal fault source in an isotropic Earth with a hydrostatic initial stress has no isotropic part,

$$\mathrm{tr}\,\mathbf{M} = 0, \qquad (5.113)$$

by virtue of the restriction $\hat{\boldsymbol{\nu}}\cdot\hat{\boldsymbol{\sigma}} = 0$. Because of this, and because attempts to detect the isotropic parts of deep-focus and other sources have not been successful (Kawakatsu 1991; 1996; Hara, Kuge & Kawakatsu 1995; 1996; Okal 1996), it is common to impose the linear constraint (5.113) in source-mechanism determinations. This reduces the number of unknown CMT parameters in equation (5.111) to nine.

To compare with near-field geodetic observations and for other reasons, we frequently wish to find a double-couple moment tensor of the form (5.90). The stipulation that the determinant must vanish,

$$\det \mathbf{M} = 0, \tag{5.114}$$

in conjunction with (5.113), guarantees that $\mathbf{M}$ has the sought-after form; however, this constraint is non-linear and difficult to apply in practice. It is customary instead to find the double-couple moment tensor $\mathbf{M}_{\mathrm{bfdc}}$ that provides the best approximation to the deviatoric moment tensor $\boldsymbol{\mathcal{M}}$ in a least-squares sense:

$$(\mathbf{M}_{\mathrm{bfdc}} - \boldsymbol{\mathcal{M}}):(\mathbf{M}_{\mathrm{bfdc}} - \boldsymbol{\mathcal{M}}) = \text{minimum}. \tag{5.115}$$

The solution to the minimization problem (5.115) may be readily determined by first transforming to a system of principal axes; we write the diagonalized deviatoric moment tensor as

$$\boldsymbol{\mathcal{M}} = \begin{pmatrix} \mathcal{M} & 0 & 0 \\ 0 & -\mathcal{M} - \mathcal{M}' & 0 \\ 0 & 0 & \mathcal{M}' \end{pmatrix}, \tag{5.116}$$

where $|\mathcal{M}| \geq |\mathcal{M} + \mathcal{M}'| \geq |\mathcal{M}'|$, and decompose it in the form

$$\boldsymbol{\mathcal{M}} = \mathbf{M}_{\mathrm{bfdc}} + \mathbf{M}_{\mathrm{clvd}}, \tag{5.117}$$

where

$$\mathbf{M}_{\mathrm{bfdc}} = \begin{pmatrix} \mathcal{M} + \frac{1}{2}\mathcal{M}' & 0 & 0 \\ 0 & -\mathcal{M} - \frac{1}{2}\mathcal{M}' & 0 \\ 0 & 0 & 0 \end{pmatrix}, \tag{5.118}$$

$$\mathbf{M}_{\mathrm{clvd}} = \begin{pmatrix} -\frac{1}{2}\mathcal{M}' & 0 & 0 \\ 0 & -\frac{1}{2}\mathcal{M}' & 0 \\ 0 & 0 & \mathcal{M}' \end{pmatrix}. \tag{5.119}$$

The tensor $\mathbf{M}_{\mathrm{bfdc}}$ is then the *best-fitting double couple*; the traceless residual $\mathbf{M}_{\mathrm{clvd}}$ is a so-called *compensated linear vector dipole* (Knopoff & Randall 1970). Note that if $\mathcal{M}$ is the greatest positive eigenvalue, then $-\mathcal{M} - \mathcal{M}'$ is the greatest negative eigenvalue, whereas if $\mathcal{M}$ is the greatest negative eigenvalue, then $-\mathcal{M} - \mathcal{M}'$ is the greatest positive eigenvalue; in other words, $\mathcal{M}'$ is always the intermediate eigenvalue. An alternative decomposition which is occasionally employed in lieu of equation (5.117) is

$$\boldsymbol{\mathcal{M}} = \mathbf{M}_{\mathrm{maj}} + \mathbf{M}_{\mathrm{min}}, \tag{5.120}$$

where

$$\mathbf{M}_{\mathrm{maj}} = \begin{pmatrix} \mathcal{M} & 0 & 0 \\ 0 & -\mathcal{M} & 0 \\ 0 & 0 & 0 \end{pmatrix}, \tag{5.121}$$

$$\mathbf{M}_{\mathrm{min}} = \begin{pmatrix} 0 & 0 & 0 \\ 0 & -\mathcal{M}' & 0 \\ 0 & 0 & \mathcal{M}' \end{pmatrix}. \tag{5.122}$$

The tensors $\mathbf{M}_{\mathrm{maj}}$ and $\mathbf{M}_{\mathrm{min}}$ are referred to as the *major* and *minor* double couples, respectively.

A convenient measure of the extent to which a deviatoric source $\boldsymbol{\mathcal{M}}$ departs from a double couple is provided by the parameter $\varepsilon = \mathcal{M}'/|\mathcal{M}|$. In general, this ratio lies in the range $|\varepsilon| \leq 1/2$, where $\varepsilon = 0$ corresponds to a double couple and $|\varepsilon| = 1/2$ corresponds to a pure compensated linear vector dipole. The moment tensor of a planar fault with uni-directional slip can differ from a double couple as a consequence of the Earth's anisotropy and deviatoric initial stress; however, these effects, which are accounted for in the general result (5.88), are likely to be small. Multi-directional slip on a curved or segmented fault surface is a more likely cause of any significant deviations from a double-couple mechanism (Ekström 1994; Kuge & Lay 1994). The candidate geometries are, however, more limited than one might suspect, since constancy of any one of the three vectors $\hat{\boldsymbol{\nu}}$, $\hat{\boldsymbol{\sigma}}$ or $\hat{\boldsymbol{\nu}} \times \hat{\boldsymbol{\sigma}}$ reduces the moment tensor (5.89) of a general fault in an isotropic Earth to a double couple (Frohlich 1990). Fewer than four percent of the more than 16,000 mechanisms in the Harvard CMT catalogue have $|\varepsilon| \geq 0.3$ (Ekström 1994).

5.4.4 Beachballs

The locus of points $\mathbf{x}_s + \hat{\mathbf{p}}_s$ surrounding an earthquake hypocenter $\mathbf{x}_s$ is referred to as the *focal sphere*. The unit vectors $\hat{\mathbf{p}}_s$ specify the takeoff directions of the seismic rays leaving the source. We show in Sections 12.5.5 and 15.7.2 that the far-field amplitude of the outgoing compressional waves is proportional to the scalar product $\hat{\mathbf{p}}_s \cdot \hat{\mathbf{M}} \cdot \hat{\mathbf{p}}_s$. For historical reasons, it is customary to depict earthquake focal mechanisms $\hat{\mathbf{M}}$ by plotting either a stereographic or equal-area projection of this *P-wave radiation pattern* on the lower focal hemisphere; regions in which $-1 \leq \hat{\mathbf{p}}_s \cdot \hat{\mathbf{M}} \cdot \hat{\mathbf{p}}_s < 0$ and $0 < \hat{\mathbf{p}}_s \cdot \hat{\mathbf{M}} \cdot \hat{\mathbf{p}}_s \leq 1$ are filled and unfilled, respectively, to obtain the traditional black-and-white *beachball representation*. If we express the unit moment tensor in diagonal form,

$$\hat{\mathbf{M}} = \hat{M}_+ \hat{\mathbf{e}}_+ \hat{\mathbf{e}}_+ + \hat{M}_0 \hat{\mathbf{e}}_0 \hat{\mathbf{e}}_0 + \hat{M}_- \hat{\mathbf{e}}_- \hat{\mathbf{e}}_-, \tag{5.123}$$

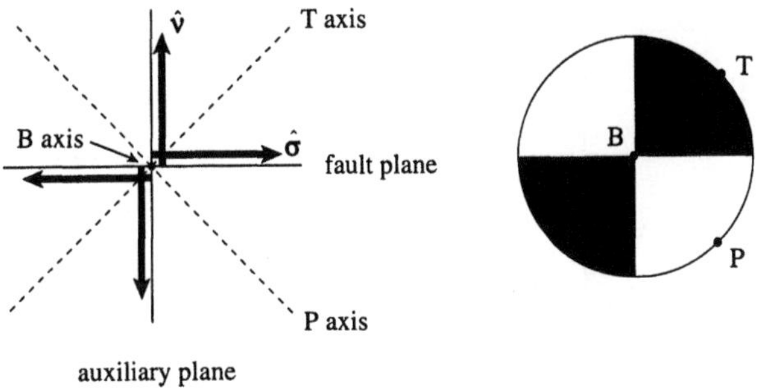

Figure 5.7. (*Left*) Map view of the static force $\mathbf{f}_{\mathrm{f}} = -M_0(\hat{\boldsymbol{\nu}}\hat{\boldsymbol{\sigma}} + \hat{\boldsymbol{\sigma}}\hat{\boldsymbol{\nu}}) \cdot \boldsymbol{\nabla}\delta(\mathbf{x} - \mathbf{x}_{\mathrm{s}})$ equivalent to a right-lateral strike-slip fault. (*Right*) Corresponding beachball with the projections of the T, B and P axes $\hat{\mathbf{e}}_+ = (\hat{\boldsymbol{\nu}} + \hat{\boldsymbol{\sigma}})/\sqrt{2}$, $\hat{\mathbf{e}}_0 = \hat{\boldsymbol{\nu}} \times \hat{\boldsymbol{\sigma}}$ and $\hat{\mathbf{e}}_- = (\hat{\boldsymbol{\nu}} - \hat{\boldsymbol{\sigma}})/\sqrt{2}$ indicated. The intermediate or B axis of such a double-couple source lies at the intersection of the two orthogonal P-wave nodal planes. Note that it is impossible to distinguish the illustrated case of right-lateral slip on an east-west striking fault from left-lateral slip on a north-south striking fault; the point-source P-wave radiation patterns are the same.

where $\hat{M}_+ \geq \hat{M}_0 \geq \hat{M}_-$ and $\hat{M}_+\hat{M}_+ + \hat{M}_0\hat{M}_0 + \hat{M}_-\hat{M}_- = 1$, then the three mutually perpendicular eigenvectors $\hat{\mathbf{e}}_+$, $\hat{\mathbf{e}}_0$ and $\hat{\mathbf{e}}_-$ are referred to as the T, B and P axes of the source. The T and P axes lie in the middle of the black and white quadrants of the focal sphere, respectively; mnemonically, we may think of the outward and inward first motions as resulting from perpendicular states of tension and compression at the source. Figure 5.7 illustrates the beachball and eigenvector axes $\hat{\mathbf{e}}_+ = (\hat{\boldsymbol{\nu}} + \hat{\boldsymbol{\sigma}})/\sqrt{2}$, $\hat{\mathbf{e}}_0 = \hat{\boldsymbol{\nu}} \times \hat{\boldsymbol{\sigma}}$ and $\hat{\mathbf{e}}_- = (\hat{\boldsymbol{\nu}} - \hat{\boldsymbol{\sigma}})/\sqrt{2}$ associated with a right-lateral strike-slip fault having a unit normal $\hat{\boldsymbol{\nu}}$ and slip vector $\hat{\boldsymbol{\sigma}}$.

It is conventional in global seismology to express the moment tensor of an earthquake by giving its spherical polar components at the hypocenter:

$$\mathbf{M} = \begin{pmatrix} M_{rr} & M_{r\theta} & M_{r\phi} \\ M_{\theta r} & M_{\theta\theta} & M_{\theta\phi} \\ M_{\phi r} & M_{\phi\theta} & M_{\phi\phi} \end{pmatrix}, \tag{5.124}$$

where $M_{r\theta} = M_{\theta r}$, $M_{r\phi} = M_{\phi r}$ and $M_{\theta\phi} = M_{\phi\theta}$. The radial, colatitudinal and longitudinal unit vectors point up, south and east, respectively. An east-west striking, right-lateral strike-slip fault such as the one illustrated in Figure 5.7 has $\hat{M}_{\theta\phi} = \hat{M}_{\phi\theta} = -1/\sqrt{2}$ and all other components equal to zero. A pictorial glossary of this and other elementary double-couple and non-double-couple focal mechanisms $\hat{\mathbf{M}}$ is given in Table 5.1.

Moment Tensor	Beachball	Moment Tensor	Beachball
$\frac{1}{\sqrt{3}}\begin{pmatrix} 1 & 0 & 0 \\ 0 & 1 & 0 \\ 0 & 0 & 1 \end{pmatrix}$		$-\frac{1}{\sqrt{3}}\begin{pmatrix} 1 & 0 & 0 \\ 0 & 1 & 0 \\ 0 & 0 & 1 \end{pmatrix}$	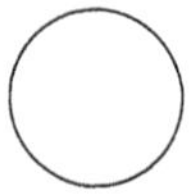
$-\frac{1}{\sqrt{2}}\begin{pmatrix} 0 & 0 & 0 \\ 0 & 0 & 1 \\ 0 & 1 & 0 \end{pmatrix}$		$\frac{1}{\sqrt{2}}\begin{pmatrix} 0 & 0 & 0 \\ 0 & 1 & 0 \\ 0 & 0 & -1 \end{pmatrix}$	
$\frac{1}{\sqrt{2}}\begin{pmatrix} 0 & 1 & 0 \\ 1 & 0 & 0 \\ 0 & 0 & 0 \end{pmatrix}$		$\frac{1}{\sqrt{2}}\begin{pmatrix} 0 & 0 & 1 \\ 0 & 0 & 0 \\ 1 & 0 & 0 \end{pmatrix}$	
$\frac{1}{\sqrt{2}}\begin{pmatrix} 1 & 0 & 0 \\ 0 & -1 & 0 \\ 0 & 0 & 0 \end{pmatrix}$		$\frac{1}{\sqrt{2}}\begin{pmatrix} 1 & 0 & 0 \\ 0 & 0 & 0 \\ 0 & 0 & -1 \end{pmatrix}$	
$\frac{1}{\sqrt{6}}\begin{pmatrix} 1 & 0 & 0 \\ 0 & 1 & 0 \\ 0 & 0 & -2 \end{pmatrix}$		$\frac{1}{\sqrt{6}}\begin{pmatrix} 1 & 0 & 0 \\ 0 & -2 & 0 \\ 0 & 0 & 1 \end{pmatrix}$	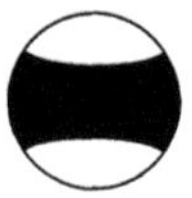
$\frac{1}{\sqrt{6}}\begin{pmatrix} -2 & 0 & 0 \\ 0 & 1 & 0 \\ 0 & 0 & 1 \end{pmatrix}$		$-\frac{1}{\sqrt{6}}\begin{pmatrix} -2 & 0 & 0 \\ 0 & 1 & 0 \\ 0 & 0 & 1 \end{pmatrix}$	

Table 5.1. Selected unit moment tensors $\hat{\mathbf{M}}$ and their associated beachballs. The components $\hat{M}_{rr}, \hat{M}_{\theta\theta}, \ldots, \hat{M}_{\theta\phi}$ are arranged in accordance with the convention (5.124). The two sources in the top row are a pure explosion $\hat{\mathbf{M}} = \mathbf{I}/\sqrt{3}$, with an entirely black beachball, and a pure implosion $\hat{\mathbf{M}} = -\mathbf{I}/\sqrt{3}$, with an entirely white beachball. The next three rows show a number of double couples, corresponding to vertical strike-slip faults (*second from top*), vertical dip-slip faults (*third from top*) and 45°-dip thrust faults (*fourth from top*). The sources in the fifth and sixth rows are pure compensated linear vector dipoles; the lowermost right entry is an idealized "eyeball" or "fried-egg" mechanism analogous to those in Figure 5.9. With the exception of the explosion and implosion, all of these sources are purely deviatoric: $\mathrm{tr}\,\hat{\mathbf{M}} = 0$.

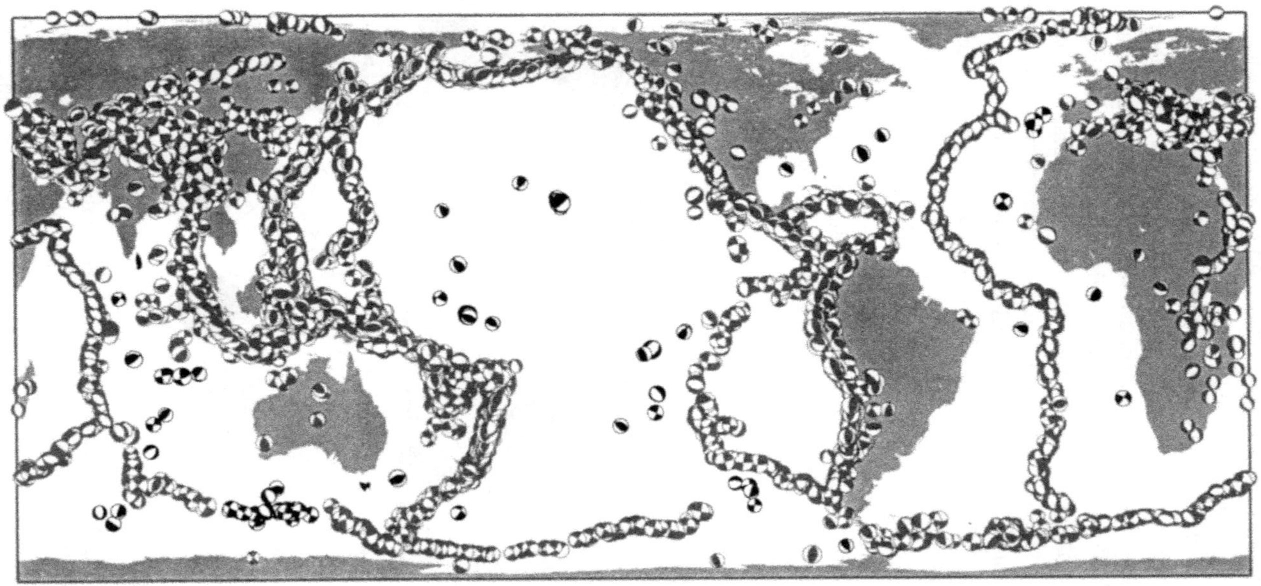

Figure 5.8. Epicentral locations and source mechanisms of 10,219 earthquakes in the Harvard CMT catalogue, with focal depths less than 50 km, during the period 1976–1997. The size of each beachball is proportional to the logarithm of the seismic moment M_0. The world map is a cylindrical equal-area projection, with landmasses shaded. (Courtesy of E. Larson.)

Most earthquakes in the Harvard CMT and other focal-mechanism catalogues are shallow-focus plate-boundary events, for which the fault plane and auxiliary plane of the best-fitting double couple may be readily distinguished. A map-view representation of these events is illustrated in Figure 5.8; the consistent occurrence of strike-slip mechanisms along transform faults and shallow-angle thrust mechanisms along subduction zones is one of the most striking manifestations of plate tectonics. The tangential components of the slip vectors of these earthquakes provide an important constraint upon the Euler vectors describing the current motions of the plates (Minster & Jordan 1978; De Mets, Gordon, Argus & Stein 1990). Figure 5.9 shows an unusual but extremely interesting suite of non-double-couple earthquakes in the vicinity of the Bárdarbunga Volcano in Iceland; such "eyeball" or "fried-egg" focal mechanisms have been detected in a number of other active volcanic settings as well. These earthquakes are thought to be the result of slip on a curved, outward-dipping, caldera-

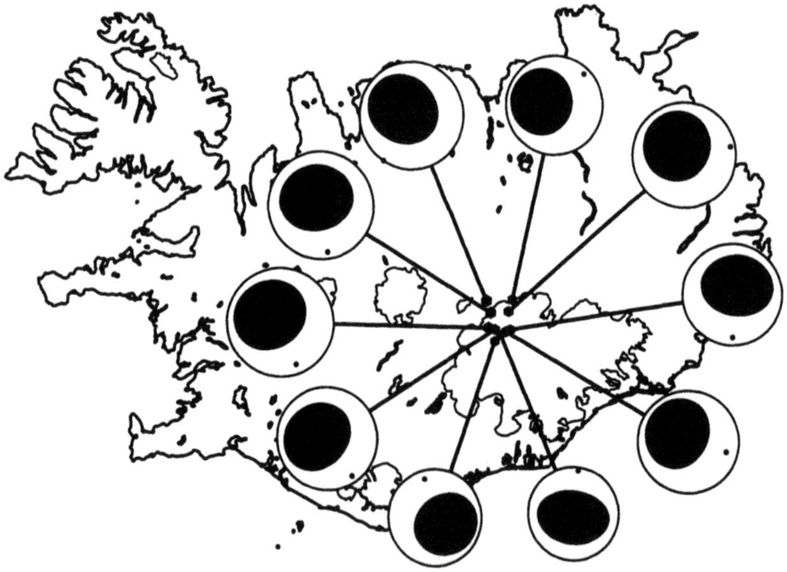

Figure 5.9. Harvard CMT mechanisms of ten earthquakes (1976–1996) with epicenters (*at ends of attached lines*) near the sub-glacial Bárdarbunga Volcano in Iceland. All of these "eyeball" or "fried-egg" events are nearly pure compensated linear vector dipoles with quasi-vertical T axes, indicative of simultaneous vertical extension and horizontal compression. The "mote" in each "eyeball" is the P axis. (Courtesy of M. Nettles & G. Ekström.)

bounding ring fault, in response to the deflation of an underlying deep-seated magma chamber (Ekström 1994).

5.4.5 Source time function

At higher frequencies, particularly in the analysis of teleseismic body waves, it is necessary to account for the finite character of the source. In the simplest approximation, the centroid shift $\boldsymbol{\Delta}\mathbf{x}$ and finite spatial extent are ignored; the finite temporal history is modelled by generalizing the impulsive glut rate in equation (5.84):

$$\partial_t \mathbf{S} = \dot{\mathbf{M}}(t)\,\delta(\mathbf{x} - \mathbf{x}_{\mathrm{s}}). \tag{5.125}$$

The quantity $\dot{M}(t)$ is the *moment-rate tensor*, or volume integral of the instantaneous glut rate:

$$\dot{\mathbf{M}}(t) = \int_{S^t} \partial_t \mathbf{S}\, dV. \tag{5.126}$$

The normal-mode response (5.21) to such a time-dependent point source is

$$\mathbf{a}(\mathbf{x}, t) = \mathrm{Re} \sum_k \mathbf{M}(\omega_k) : \boldsymbol{\varepsilon}_k^*(\mathbf{x}_{\mathrm{s}})\, \mathbf{s}_k(\mathbf{x}) \exp i\omega_k(t - t_{\mathrm{s}}), \tag{5.127}$$

where

$$\mathbf{M}(\omega) = \exp(i\omega t_{\mathrm{s}}) \int_{t_0}^{t_{\mathrm{f}}} \dot{\mathbf{M}}(t) \exp(-i\omega t)\, dt. \tag{5.128}$$

Equation (5.127) is identical to the Dirac-delta glut response (5.87), with $\mathbf{M}$ replaced by a *frequency-dependent moment tensor* (5.128). It is noteworthy that $\mathbf{M}(\omega)$ is not the Fourier transform of a time-dependent moment tensor $\mathbf{M}(t)$; rather, it is a time-shift factor $\exp(i\omega t_{\mathrm{s}})$ times the Fourier transform of the moment-rate tensor $\dot{\mathbf{M}}(t)$.

In applying (5.127) it is usually assumed that the source is *synchronous*, in the sense that all the components of $\dot{\mathbf{M}}(t)$ exhibit the same time dependence. It is convenient to write the moment-rate tensor of such a synchronous source in the form

$$\dot{\mathbf{M}}(t) = \sqrt{2} M_0 \hat{\mathbf{M}}\, \dot{m}(t), \tag{5.129}$$

where M_0 and $\hat{\mathbf{M}}$ are the time-independent scalar moment and unit mechanism tensor defined in equations (5.91)–(5.92). The quantity $\dot{m}(t)$ is a normalized *source time function* satisfying

$$\int_{t_0}^{t_{\mathrm{f}}} \dot{m}(t)\, dt = 1. \tag{5.130}$$

The normalized glut-rate density (5.103) is given in this approximation by $\dot{m}(\mathbf{x},t) = \dot{m}(t)\,\delta(\mathbf{x}-\mathbf{x}_{\mathrm{s}})$. A planar fault with uni-directional slip in a hydrostatic Earth with a non-deviatoric initial stress is certainly synchronous; its source time function is a weighted average of the slip speed:

$$\dot{m}(t) = \frac{1}{M_0}\int_{\Sigma^t} \mu\partial_t\Delta s\, d\Sigma. \tag{5.131}$$

In observational analyses of teleseismic body-wave pulses, the source time function $\dot{m}(t)$ of large earthquakes is commonly expressed as a sequence of boxcars or overlapping trapezoids or isosceles triangles (Langston 1981; Kikuchi & Kanamori 1982; Nábělek 1985; Ekström 1989).

The frequency-dependent moment tensor of a synchronous point source is $\mathbf{M}(\omega) = \sqrt{2}M_0\hat{\mathbf{M}}\, m(\omega)$, where

$$m(\omega) = \exp(i\omega t_{\mathrm{s}})\int_{t_0}^{t_{\mathrm{f}}} \dot{m}(t)\exp(-i\omega t)\, dt. \tag{5.132}$$

In the limit of low frequencies, $\omega \to 0$, the transform (5.132) can be approximated by a truncated Taylor series:

$$m(\omega) = 1 - i\omega(\Delta t) - \tfrac{1}{2}\omega^2(\Delta t)^2 - \tfrac{1}{6}\omega^2\tau_{\mathrm{h}}^2, \tag{5.133}$$

where

$$\Delta t = t_{\mathrm{c}} - t_{\mathrm{s}} = \int_{t_0}^{t_{\mathrm{f}}} (t - t_{\mathrm{s}})\dot{m}(t)\, dt, \tag{5.134}$$

$$\tau_{\mathrm{h}}^2 = 3\int_{t_0}^{t_{\mathrm{f}}} (t - t_{\mathrm{c}})^2\dot{m}(t)\, dt. \tag{5.135}$$

The quantity Δt is the time shift of the centroid t_{c} of $\dot{m}(t)$ with respect to the fiducial origin time t_{s}, as illustrated in Figure 5.10, whereas τ_{h} is a measure of the temporal *half-duration* of the source. The factor of three in the definition (5.135) makes τ_{h} exactly equal to $\frac{1}{2}(t_{\mathrm{f}} - t_0)$ in the case of a boxcar source time function,

$$\dot{m}(t) = (t_{\mathrm{f}} - t_0)^{-1}[H(t - t_0) - H(t - t_{\mathrm{f}})]. \tag{5.136}$$

The routine Harvard CMT processing makes use of such a centered boxcar with $t_{\mathrm{c}} = \frac{1}{2}(t_0 + t_{\mathrm{f}})$ and a nominal half-duration $\tau_{\mathrm{h}} = 2.4\times 10^{-6} M_0^{1/3}$, where τ_{h} and M_0 are measured in s and N m, respectively (G. Ekström, personal communication 1996). Silver & Jordan (1982; 1983) and Ihmlé & Jordan (1994; 1995) employ an expansion analogous to (5.133) in their analysis of so-called "slow" earthquakes, with τ_{h} replaced by a "characteristic" source duration $\tau_{\mathrm{c}} = (4/3)^{1/2}\tau_{\mathrm{h}}$.

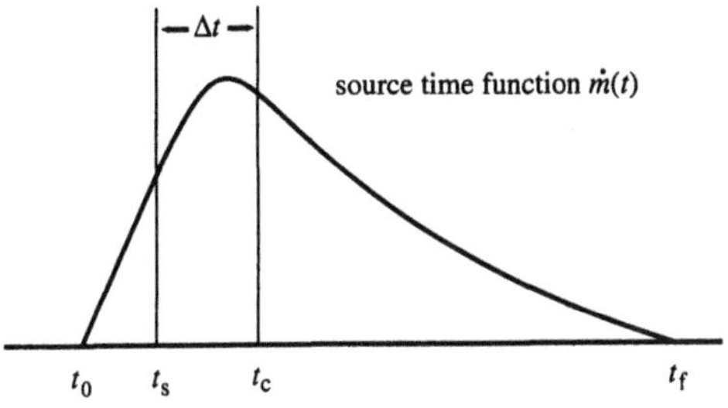

Figure 5.10. Schematic depiction of an earthquake source time function $\dot{m}(t)$ which commences at t_0 and ends at t_f. The centroid time $t_c = \int_{t_0}^{t_f} t\,\dot{m}(t)\,dt$ is shifted with respect to the fiducial origin time t_s by an amount Δt.

A consistent seismic source representation should account for finiteness in both the spatial and temporal coordinates to the same order. Estimates of the source "half-duration" obtained by applying the above formalism need to be interpreted with some caution for this reason; in reality τ_h is only an apparent half-duration, which measures the spatial as well as the temporal extent of the earthquake rupture. Backus (1977a; 1977b) carries out the polynomial moment expansion of the glut-rate tensor $\partial_t\mathbf{S}$ to second degree, and shows how to relate the resulting spatio-temporal moments to the duration, dimensions and directivity of the source. The number of parameters that must be determined in such a general second-degree analysis is, however, daunting. To reduce this number to something more manageable, Bukchin (1995) considers a *spatially extended synchronous source* of the form $\partial_t\mathbf{S} = \sqrt{2}M_0\hat{\mathbf{M}}\,\dot{m}(\mathbf{x},t)$. Such a source is characterized by the four first moments and ten second moments of the normalized scalar glut-rate density $\dot{m}(\mathbf{x},t)$, in addition to the moment tensor $\mathbf{M} = \sqrt{2}M_0\hat{\mathbf{M}}$.

We shall ignore the spatial and temporal centroid shifts $\mathbf{\Delta x}$ and Δt, as well as the spatially extended nature of earthquakes, throughout the subsequent chapters of this book in the interest of simplicity. In addition, we shall henceforth take the fiducial origin time to be zero:

$$t_s = 0. \tag{5.137}$$

Generally, we shall employ an impulsive moment-tensor source with a glut rate $\partial_t\mathbf{S} = \mathbf{M}\,\delta(\mathbf{x}-\mathbf{x}_s)\,\delta(t)$ and an associated equivalent force density $\mathbf{f} = -\mathbf{M}\cdot\boldsymbol{\nabla}\delta(\mathbf{x}-\mathbf{x}_s)\,H(t)$. We shall write the response to such a source in the form (5.86)–(5.87), with t_s set to zero in accordance with the convention (5.137). In any mode-sum expression for the particle acceleration $\mathbf{a}(\mathbf{x},t)$ or $\mathbf{a}(\mathbf{x},\omega)$, it is straightforward to allow for a finite-duration source by substituting $\mathbf{M} \to \mathbf{M}(\omega_k) = \int_{t_0}^{t_f} \dot{\mathbf{M}}(t) \exp(-i\omega_k t)\, dt$, where $\dot{\mathbf{M}}(t)$ is a time-dependent moment-rate tensor and ω_k is the eigenfrequency of interest. In our analysis of body-wave propagation within an elastic and anelastic Earth in Chapters 12 and 15, we shall consider a synchronous time-dependent source $\dot{\mathbf{M}}(t) = \sqrt{2} M_0 \hat{\mathbf{M}}\, \dot{m}(t)$.

★5.4.6 Problematical sources

There are two situations in which the point-source moment-tensor representation leads to difficulties in the determination of earthquake source mechanisms; the first is a source that is located in the vicinity of the solid free surface or the seafloor, and the second is a source that spans a solid-solid discontinuity Σ_{SS}. We discuss these two problematical cases briefly in this section, restricting attention to the case of an isotropic Earth model with a hydrostatic initial stress.

The response to a moment-tensor source on a rotating Earth model is, as we have seen, proportional to the product $\mathbf{M}:\boldsymbol{\varepsilon}^*$, where we have dropped the mode identification label on the strain eigenfunction $\boldsymbol{\varepsilon}_k$ for simplicity. The dynamical boundary condition on the solid free surface, $\hat{\mathbf{n}} \cdot \mathbf{T}^{\mathrm{L1}} = \mathbf{0}$, implies that

$$\varepsilon_{xz} = \varepsilon_{yz} = 0, \tag{5.138}$$

$$(\kappa + \tfrac{4}{3}\mu)\varepsilon_{zz} + (\kappa - \tfrac{2}{3}\mu)(\varepsilon_{xx} + \varepsilon_{yy}) = 0, \tag{5.139}$$

where we have adopted a local Cartesian coordinate system in which the $\hat{\mathbf{z}}$ axis is normal to the boundary. The contribution to $\mathbf{M}:\boldsymbol{\varepsilon}^*$ from M_{xz} and M_{yz} vanishes as the source location $\mathbf{x}_s$ approaches the solid free surface or the seafloor, by virtue of the shear-strain conditions (5.138). As a result, these vertical dip-slip components of the moment tensor $\mathbf{M}$ are not well constrained for shallow-focus earthquake sources. The condition (5.139) likewise implies that there will be a negligible contribution to the excitation amplitude $\mathbf{M}:\boldsymbol{\varepsilon}^*$ from any near-surface source having

$$M_{zz} = \left(\frac{\kappa + \frac{4}{3}\mu}{\kappa - \frac{2}{3}\mu}\right) M_{xx} = \left(\frac{\kappa + \frac{4}{3}\mu}{\kappa - \frac{2}{3}\mu}\right) M_{yy}. \tag{5.140}$$

This latter indeterminacy involving the two diagonal components M_{zz} and $M_{xx} + M_{yy}$ can be eliminated by adopting the constraint (5.113) that the moment tensor be traceless: $M_{xx} + M_{yy} + M_{zz} = 0$. However, there is no such easy remedy for the off-diagonal components M_{xz} and M_{yz}. A common although obviously arbitrary expedient is simply to set M_{xz} and M_{yz} equal to zero in shallow earthquake source-mechanism determinations; Ekström

& Dziewonski (1985) show that this has little effect upon the remaining well-constrained components M_{xy}, $M_{xx} - M_{yy}$ and $M_{zz} = -M_{xx} - M_{yy}$. A shallow sub-horizontal landslide, such as that which accompanied the 1980 Mt. St. Helens eruption, is a particularly pronounced example of this type of problematical event. It has been shown, both observationally and theoretically, that such sources can be alternatively represented by a horizontal surface point force (Kanamori & Given 1982; Okal 1990; Dahlen 1993).

Suppose next that the final source region is divided into two subregions, $S^{\mathrm{f}} = S^+ \cup S^-$, by a portion of the surface Σ_{SS}, across which the elastic moduli κ and μ are discontinuous, as shown in Figure 5.11. Such a source spanning a solid-solid discontinuity has been considered by Woodhouse (1981a); we summarize his conclusions here. The product $\mathbf{M}:\boldsymbol{\varepsilon}^*$ is replaced in this case by $\mathbf{M}^+ : \boldsymbol{\varepsilon}^{+*} + \mathbf{M}^- : \boldsymbol{\varepsilon}^{-*}$, where $\boldsymbol{\varepsilon}^{\pm}$ are the strains on either side of the boundary Σ_{SS}, and $\mathbf{M}^{\pm}$ are the corresponding partial moment tensors,

$$\mathbf{M}^{\pm} = \int_{S^{\pm}} \mathbf{S}_{\mathrm{f}}\, dV. \tag{5.141}$$

The continuity of the displacement eigenfunction, $[\mathbf{s}]_-^+ = \mathbf{0}$, guarantees that the tangential strains are continuous,

$$[\varepsilon_{xx}]_-^+ = [\varepsilon_{yy}]_-^+ = [\varepsilon_{xy}]_-^+, \tag{5.142}$$

and the continuity of the traction, $[\hat{\mathbf{n}} \cdot \mathbf{T}^{\mathrm{L1}}]_-^+ = \mathbf{0}$, guarantees that

$$[\mu\varepsilon_{xz}]_-^+ = [\mu\varepsilon_{yz}]_-^+ = 0, \tag{5.143}$$

$$[(\kappa + \tfrac{4}{3}\mu)\varepsilon_{zz} + (\kappa - \tfrac{2}{3}\mu)(\varepsilon_{xx} + \varepsilon_{yy})]_-^+ = 0, \tag{5.144}$$

where we have again adopted a local Cartesian coordinate system in which $\hat{\mathbf{z}}$ is normal to the boundary. We can use the conditions (5.142)–(5.144) to

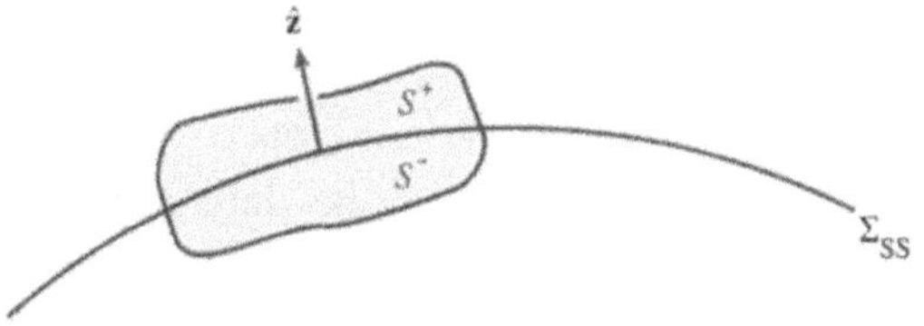

Figure 5.11. Schematic depiction of an earthquake source volume $S^{\mathrm{f}} = S^+ \cup S^-$ spanning a solid-solid discontinuity Σ_{SS}. The local unit normal $\hat{\mathbf{z}}$ points toward the + side of the discontinuity, as usual.

rewrite the excitation solely in terms of the strain $\boldsymbol{\varepsilon}^+$ or $\boldsymbol{\varepsilon}^-$ on one side of the discontinuity or the other:

$$\mathbf{M}^+ : \boldsymbol{\varepsilon}^{+*} + \mathbf{M}^- : \boldsymbol{\varepsilon}^{-*} = \boldsymbol{\mathcal{M}}^+ : \boldsymbol{\varepsilon}^{+*} = \boldsymbol{\mathcal{M}}^- : \boldsymbol{\varepsilon}^{-*}. \tag{5.145}$$

The quantities $\boldsymbol{\mathcal{M}}^+$ and $\boldsymbol{\mathcal{M}}^-$ in the two alternative representations (5.145) are given explicitly by

$$\mathcal{M}_{xx}^{\pm} = M_{xx}^{+} + M_{xx}^{-} + a_{\pm} M_{zz}^{\mp}, \tag{5.146}$$

$$\mathcal{M}_{yy}^{\pm} = M_{yy}^{+} + M_{yy}^{-} + a_{\pm} M_{zz}^{\mp}, \tag{5.147}$$

$$\mathcal{M}_{zz}^{\pm} = M_{zz}^{\pm} + b_{\pm} M_{zz}^{\mp}, \tag{5.148}$$

$$\mathcal{M}_{xz}^{\pm} = M_{xz}^{\pm} + c_{\pm} M_{xz}^{\mp}, \tag{5.149}$$

$$\mathcal{M}_{yz}^{\pm} = M_{yz}^{\pm} + c_{\pm} M_{yz}^{\mp}, \tag{5.150}$$

$$\mathcal{M}_{xy}^{\pm} = M_{xy}^{+} + M_{xy}^{-}, \tag{5.151}$$

where

$$a_{\pm} = \frac{(\kappa_{\pm} - \frac{2}{3}\mu_{\pm}) - (\kappa_{\mp} - \frac{2}{3}\mu_{\mp})}{\kappa_{\mp} + \frac{4}{3}\mu_{\mp}}, \tag{5.152}$$

$$b_{\pm} = \frac{\kappa_{\pm} + \frac{4}{3}\mu_{\pm}}{\kappa_{\mp} + \frac{4}{3}\mu_{\mp}}, \qquad c_{\pm} = \frac{\mu_{\pm}}{\mu_{\mp}}. \tag{5.153}$$

Evidently, the true source is equivalent to a moment-tensor source $\boldsymbol{\mathcal{M}}^+$ placed on the + side of Σ_{SS} or to a source $\boldsymbol{\mathcal{M}}^-$ placed on the − side of Σ_{SS}. The two apparent moment tensors $\boldsymbol{\mathcal{M}}^+$ and $\boldsymbol{\mathcal{M}}^-$ are the source mechanisms that would be obtained from inversion of seismic data under the assumption that the earthquake is situated on the + or − side of Σ_{SS}, respectively. Neither of these sources represents the true moment tensor, which is given by

$$\mathbf{M} = \mathbf{M}^+ + \mathbf{M}^-. \tag{5.154}$$

In fact, the true moment tensor cannot be determined without making further physical assumptions about the source. One possible assumption is that $\mathbf{M}^+$ and $\mathbf{M}^-$ have the same orientation, in which case

$$\mathbf{M}^+ = \gamma \mathbf{M}, \qquad \mathbf{M}^- = (1-\gamma)\mathbf{M}, \tag{5.155}$$

where $0 \le \gamma \le 1$ is a parameter representing the fraction of the source lying on the + side of the discontinuity. An example of a source satisfying (5.155) is a planar fault with uni-directional slip which cuts across the discontinuity. The true moment tensor $\mathbf{M}$ can be readily determined in terms of either $\boldsymbol{\mathcal{M}}^{+}$ or $\boldsymbol{\mathcal{M}}^{-}$ if we adopt this assumption; however, the result depends upon the proportionality parameter γ. Generally, there is insufficient information for this parameter to be independently specified, and it is simply assumed that $\gamma = 0$ or $\gamma = 1$, i.e., the source is presumed to lie entirely on one side of the discontinuity or the other.

Neither of the above two difficulties is fundamental to the equivalent-force representation of a seismic source; they are artifacts of the point-source approximation. Both the equivalent body-force density $\mathbf{f}$ and surface traction $\mathbf{t}$ are well defined for a finite source that intersects the solid free surface or the seafloor or that spans a solid-solid discontinuity Σ_{SS}. The only fundamentally problematical source, for which $\mathbf{f}$ and $\mathbf{t}$ are undefined, is an ideal fault that *coincides with* a portion of Σ_{SS}.

★5.5 Earthquake Energy Balance

We conclude this chapter with a discussion of the balance of energy in earthquake faulting. For simplicity we restrict consideration to the case of a *non-rotating Earth*; the complications introduced by rotation will be briefly mentioned in Section 5.5.4.

★5.5.1 Net energy release

Since earthquakes occur spontaneously, they must act to release energy that is stored in the Earth prior to the faulting. The total amount of energy released is independent of the details of the temporal history of the faulting; it only depends on the final static equilibrium state of the Earth, after the faulting has ceased and all the normal modes of oscillation have decayed. We take the zero or reference level of both elastic and gravitational potential energy to be that of the initial equilibrium state prior to the earthquake; the *released energy* $\mathcal{E}_{\mathrm{r}}$, which is an inherently positive quantity, is then *minus* the static elastic-gravitational energy of the final state:

$$\mathcal{E}_{\mathrm{r}} = -\mathcal{E}_{\mathrm{e}} - \mathcal{E}_{\mathrm{g}}, \tag{5.156}$$

where

$$\mathcal{E}_{\mathrm{e}} = \int_{\oplus} [\mathbf{T}^0 \!:\! \boldsymbol{\varepsilon}_{\mathrm{f}} + \tfrac{1}{2}\boldsymbol{\nabla}\mathbf{s}_{\mathrm{f}} \!:\! \boldsymbol{\Lambda} \!:\! \boldsymbol{\nabla}\mathbf{s}_{\mathrm{f}}]\, dV, \tag{5.157}$$

$$\mathcal{E}_{\mathrm{g}} = \int_{\oplus} \rho^0 [\mathbf{s}_{\mathrm{f}} \cdot \boldsymbol{\nabla}\phi^0 + \tfrac{1}{2}\mathbf{s}_{\mathrm{f}} \cdot \boldsymbol{\nabla}\phi_{\mathrm{f}}^{\mathrm{E1}} + \tfrac{1}{2}\mathbf{s}_{\mathrm{f}} \cdot \boldsymbol{\nabla}\boldsymbol{\nabla}\phi^0 \cdot \mathbf{s}_{\mathrm{f}}]\, dV. \tag{5.158}$$

The quantity $\phi_{\rm f}^{\rm E1}$ in (5.158) is the final static Eulerian gravitational potential perturbation.

We can rewrite $\mathcal{E}_{\rm r}$ as an integral over the final fault surface $\Sigma^{\rm f}$ by applying Gauss' theorem to equations (5.156)–(5.158), taking into account the discontinuity $\mathbf{\Delta}\mathbf{s}_{\rm f} = [\mathbf{s}_{\rm f}]_-^+$. Upon making use of the second-order tangential slip condition $[\hat{\mathbf{n}}\cdot\mathbf{s}_{\rm f} - \mathbf{s}_{\rm f}\cdot\boldsymbol{\nabla}^\Sigma(\hat{\mathbf{n}}\cdot\mathbf{s}_{\rm f}) + \frac{1}{2}\mathbf{s}_{\rm f}\cdot(\boldsymbol{\nabla}^\Sigma\hat{\mathbf{n}})\cdot\mathbf{s}_{\rm f}]_-^+ = 0$ on the fluid-solid discontinuities $\Sigma_{\rm FS}$, as in Section 3.9.4, we obtain

$$\begin{aligned}\mathcal{E}_{\rm r} = &\int_\oplus \mathbf{s}_{\rm f}\cdot[\boldsymbol{\nabla}\cdot\mathbf{T}^0 - \rho^0\boldsymbol{\nabla}\phi^0]\,dV\\ &+ \tfrac{1}{2}\int_\oplus \mathbf{s}_{\rm f}\cdot[\boldsymbol{\nabla}\cdot\mathbf{T}_{\rm f}^{\rm PK1} - \rho^0\mathbf{s}_{\rm f}\cdot\phi_{\rm f}^{\rm E1} - \rho^0\mathbf{s}_{\rm f}\cdot\boldsymbol{\nabla}\boldsymbol{\nabla}\phi^0]\,dV\\ &- \tfrac{1}{2}\int_{\partial\oplus}[\hat{\mathbf{n}}\cdot(2\mathbf{T}^0 + \mathbf{T}_{\rm f}^{\rm PK1})\cdot\mathbf{s}_{\rm f}]\,d\Sigma\\ &+ \tfrac{1}{2}\int_{\Sigma_{\rm SS}}[\hat{\mathbf{n}}\cdot(2\mathbf{T}^0 + \mathbf{T}_{\rm f}^{\rm PK1})\cdot\mathbf{s}_{\rm f}]_-^+\,d\Sigma\\ &+ \tfrac{1}{2}\int_{\Sigma_{\rm FS}}[(2\hat{\mathbf{n}}\cdot\mathbf{T}^0 + \mathbf{t}_{\rm f}^{\rm PK1})\cdot\mathbf{s}_{\rm f}]_-^+\,d\Sigma\\ &+ \tfrac{1}{2}\int_{\Sigma^{\rm f}}[\hat{\boldsymbol{\nu}}\cdot(2\mathbf{T}^0 + \mathbf{T}_{\rm f}^{\rm PK1})\cdot\mathbf{s}_{\rm f}]_-^+\,d\Sigma,\end{aligned} \tag{5.159}$$

where $\mathbf{T}_{\rm f}^{\rm PK1} = \boldsymbol{\Lambda}\!:\!\boldsymbol{\nabla}\mathbf{s}_{\rm f}$ and $\mathbf{t}_{\rm f}^{\rm PK1} = \hat{\mathbf{n}}\cdot\mathbf{T}_{\rm f}^{\rm PK1} + \hat{\mathbf{n}}\,\boldsymbol{\nabla}^\Sigma\cdot(\varpi^0\mathbf{s}_{\rm f}) - \varpi^0(\boldsymbol{\nabla}^\Sigma\mathbf{s}_{\rm f})\cdot\hat{\mathbf{n}}$. The two volume integrals in equation (5.159) vanish as a consequence of the static elastic-gravitational equilibrium conditions $\boldsymbol{\nabla}\cdot\mathbf{T}^0 - \rho^0\boldsymbol{\nabla}\phi^0 = \mathbf{0}$ and $\boldsymbol{\nabla}\cdot\mathbf{T}_{\rm f}^{\rm PK1} - \rho^0\mathbf{s}_{\rm f}\cdot\boldsymbol{\nabla}\phi_{\rm f}^{\rm E1} - \rho^0\mathbf{s}_{\rm f}\cdot\boldsymbol{\nabla}\boldsymbol{\nabla}\phi^0 = \mathbf{0}$ in the initial and final Earth models, respectively. The surface integrals over the free surface and internal discontinuities vanish by virtue of the continuity of the initial traction, $[\hat{\mathbf{n}}\cdot\mathbf{T}^0]_-^+ = \mathbf{0}$, together with the associated boundary conditions governing the static displacement: $\hat{\mathbf{n}}\cdot\mathbf{T}_{\rm f}^{\rm PK1} = \mathbf{0}$ on $\partial\oplus$, $[\hat{\mathbf{n}}\cdot\mathbf{T}_{\rm f}^{\rm PK1}]_-^+ = \mathbf{0}$ on $\Sigma_{\rm SS}$, and $[\mathbf{t}_{\rm f}^{\rm PK1}]_-^+ = \hat{\mathbf{n}}[\hat{\mathbf{n}}\cdot\mathbf{t}_{\rm f}^{\rm PK1}]_-^+ = \mathbf{0}$ on $\Sigma_{\rm FS}$. The total elastic-gravitational energy released by an ideal fault is therefore simply

$$\mathcal{E}_{\rm r} = \tfrac{1}{2}\int_{\Sigma^{\rm f}}[\hat{\boldsymbol{\nu}}\cdot(2\mathbf{T}^0 + \mathbf{T}_{\rm f}^{\rm PK1})\cdot\mathbf{s}_{\rm f}]_-^+\,d\Sigma. \tag{5.160}$$

The vector $\frac{1}{2}\hat{\boldsymbol{\nu}}\cdot(2\mathbf{T}^0 + \mathbf{T}_{\rm f}^{\rm PK1})$ is the average of the initial and final traction acting upon either side of the fault surface $\Sigma^{\rm f}$.

Equation (5.160) bears a strong resemblance to the classical expression for the energy released by an ideal fault, first used by Reid (1910) in his celebrated investigation of the 1906 San Francisco earthquake. The usual derivation of this result ignores self-gravitation, and makes the unrealistic assumption that the initial static stress is the result of an infinitesimal initial strain away from some natural unstrained, unstressed state (Steketee 1958; Savage 1969). The present point of view, which accounts for the Earth's gravity and makes no assumptions about the magnitude or origin of the initial stress $\mathbf{T}^0$, is much more satisfactory. As is well known, the explicit dependence of $\mathcal{E}_{\mathrm{r}}$ upon $\mathbf{T}^0$ makes it impossible to measure the energy release of an earthquake using only seismological methods.

*5.5.2 Dissipation of released energy

The ultimate fate of all the energy that is released by an earthquake is to be dissipated somewhere within the volume $\oplus$ of the Earth. Three regions of energy dissipation may be distinguished. First, energy will be dissipated during faulting $t_0 \leq t \leq t_{\mathrm{f}}$ in heating on the walls of the instantaneous fault surface Σ^t, where work must generally be performed against the frictional traction which acts there to resist slip. Second, energy will also be dissipated during this interval either in overcoming material cohesion and thereby creating fresh fault surface area, or simply in additional heating, along the instantaneously expanding edge or rupture front $\partial\Sigma_{\mathrm{c}}^t$ of the fault. Third, and finally, the remainder of the released energy must be the energy of oscillation of the normal modes of the Earth excited by the earthquake; this energy will be dissipated throughout the Earth by the infinitesimal anelasticity which causes the modes to decay. Bodily dissipation will commence at the onset of faulting, i.e., at $t = t_0$, and it will continue until every normal mode of oscillation has completely decayed. We examine now the extent to which the released energy $\mathcal{E}_{\mathrm{r}}$ is partitioned among these three dissipation mechanisms.

The instantaneous rate of change with time of the kinetic plus elastic-gravitational potential energy of the Earth is given by

$$\dot{\mathcal{E}} = \dot{\mathcal{E}}_{\mathrm{k}} + \dot{\mathcal{E}}_{\mathrm{e}} + \dot{\mathcal{E}}_{\mathrm{g}}, \tag{5.161}$$

where

$$\dot{\mathcal{E}}_{\mathrm{k}} = \frac{d}{dt}\int_{\oplus} [\tfrac{1}{2}\rho^0 \partial_t \mathbf{s} \cdot \partial_t \mathbf{s}]\, dV, \tag{5.162}$$

$$\dot{\mathcal{E}}_{\mathrm{e}} = \frac{d}{dt}\int_{\oplus} [\mathbf{T}^0 \!:\! \boldsymbol{\varepsilon} + \tfrac{1}{2}\boldsymbol{\nabla}\mathbf{s} \!:\! \boldsymbol{\Lambda} \!:\! \boldsymbol{\nabla}\mathbf{s}]\, dV, \tag{5.163}$$

$$\dot{\mathcal{E}}_{\mathrm{g}} = \frac{d}{dt}\int_{\oplus} \rho^0[\mathbf{s} \cdot \boldsymbol{\nabla}\phi^0 + \tfrac{1}{2}\mathbf{s} \cdot \boldsymbol{\nabla}\phi^{\mathrm{E1}} + \tfrac{1}{2}\mathbf{s} \cdot \boldsymbol{\nabla}\boldsymbol{\nabla}\phi^0 \cdot \mathbf{s}]\, dV. \tag{5.164}$$

To evaluate the time derivatives during the time interval $t_0 \leq t \leq t_{\mathrm{f}}$ when active faulting is occurring, we must devote special attention to the propagating crack tip $\partial\Sigma_{\mathrm{c}}^t$. Let ε denote the distance measured away from this

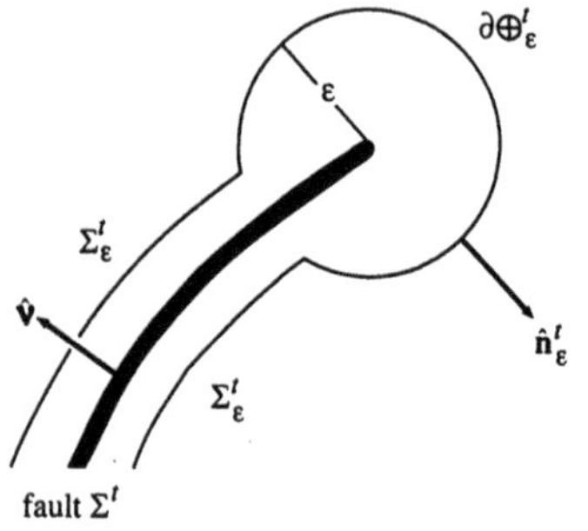

Figure 5.12. Blow-up of the surface of integration surrounding the tip of a propagating crack. The "pancakes" Σ^t_ε adjacent to the walls of the fault Σ^t expand as the fixed-radius torus $\partial\oplus^t_\varepsilon$ moves with the crack tip.

curve. The macroscopic manifestation of either a non-zero material cohesion or rupture-front heating will be $\varepsilon^{-1/2}$ singularities in both $\partial_t\mathbf{s}$ and $\boldsymbol{\nabla}\mathbf{s}$, provided that the only catastrophic breakdown of Hooke's law occurs on the fault surface Σ^t, as we assume. To handle these singularities, we regard each of the volume integrals in (5.162)–(5.164) as the limit over a punctured volume which envelops the fault surface Σ^t, so that only a torus surrounding $\partial\Sigma^t_{\rm c}$ remains. Let $\partial\oplus^t_\varepsilon$ denote the surface of this torus, let $\hat{\mathbf{n}}^t_\varepsilon$ be its unit outward normal, and let Σ^t_ε be the portion of the fault surface lying just inside of Σ^t, so that the surface enclosing the fault consists of $\partial\oplus^t_\varepsilon$ and both sides of Σ^t_ε, as shown in Figure 5.12. The torus is assumed to move with $\partial\Sigma^t_{\rm c}$, maintaining a time-independent radius ε; the inward normal velocity of the surface $\partial\oplus^t_\varepsilon$ relative to the material is then $-\hat{\mathbf{n}}^t_\varepsilon\cdot\mathbf{v}$, where $\mathbf{v}$ is the crack-tip or *rupture velocity*. Upon applying the standard rule for differentiating an integral over a moving volume, as well as Gauss' theorem, we can write the rate of change of the total energy $\dot{\mathcal{E}}$ in the form

$$\begin{aligned}
\dot{\mathcal{E}} = \int_{\oplus} & \partial_t\mathbf{s}\cdot[\rho^0\boldsymbol{\nabla}\phi^0 - \boldsymbol{\nabla}\cdot\mathbf{T}^0]\,dV \\
& + \int_{\oplus}\partial_t\mathbf{s}\cdot[\rho^0\partial_t^2\mathbf{s} - \boldsymbol{\nabla}\cdot\mathbf{T}^{\rm PK1} + \rho^0\mathbf{s}\cdot\boldsymbol{\nabla}\phi^{\rm E1} + \rho^0\mathbf{s}\cdot\boldsymbol{\nabla}\boldsymbol{\nabla}\phi^0]\,dV \\
& + \int_{\partial\oplus}[\hat{\mathbf{n}}\cdot(\mathbf{T}^0 + \mathbf{T}^{\rm PK1})\cdot\partial_t\mathbf{s}]\,d\Sigma \\
& \quad - \int_{\Sigma_{\rm SS}}[\hat{\mathbf{n}}\cdot(\mathbf{T}^0 + \mathbf{T}^{\rm PK1})\cdot\partial_t\mathbf{s}]_-^+\,d\Sigma \\
& \quad\quad - \int_{\Sigma_{\rm FS}}[(\hat{\mathbf{n}}\cdot\mathbf{T}^0 + \mathbf{t}^{\rm PK1})\cdot\partial_t\mathbf{s}]_-^+\,d\Sigma
\end{aligned}$$

$$- \lim_{\varepsilon \to 0} \int_{\Sigma_\varepsilon^t} [\hat{\boldsymbol{\nu}} \cdot (\mathbf{T}^0 + \mathbf{T}^{\mathrm{PK1}}) \cdot \partial_t \mathbf{s}]_-^+ \, d\Sigma$$

$$- \lim_{\varepsilon \to 0} \int_{\partial \oplus_\varepsilon^t} [e(\hat{\mathbf{n}}_\varepsilon^t \cdot \mathbf{v}) + \hat{\mathbf{n}}_\varepsilon^t \cdot \mathbf{T}^{\mathrm{PK1}} \cdot \partial_t \mathbf{s}] \, d\Sigma, \qquad (5.165)$$

where $e = \frac{1}{2}(\rho^0 \partial_t \mathbf{s} \cdot \partial_t \mathbf{s} + \boldsymbol{\nabla}\mathbf{s} : \boldsymbol{\Lambda} : \boldsymbol{\nabla}\mathbf{s})$. In obtaining the result (5.165), we have made use of the Maxwell relation $\Lambda_{ijkl} = \Lambda_{klij}$, the symmetry (4.16), and the second-order tangential slip condition (3.95) on the fluid-solid boundaries Σ_{FS}. The volume integrals over $\oplus$ vanish as a consequence of the static equilibrium condition $\rho^0 \boldsymbol{\nabla}\phi^0 - \boldsymbol{\nabla} \cdot \mathbf{T}^0 = \mathbf{0}$ and the momentum equation $\rho^0 \partial_t^2 \mathbf{s} - \boldsymbol{\nabla} \cdot \mathbf{T}^{\mathrm{PK1}} + \rho^0 \mathbf{s} \cdot \boldsymbol{\nabla}\phi^{\mathrm{E1}} + \rho^0 \mathbf{s} \cdot \boldsymbol{\nabla}\boldsymbol{\nabla}\phi^0 = \mathbf{0}$, whereas the surface integrals over $\partial\oplus$, Σ_{SS} and Σ_{FS} vanish by virtue of the continuity of the initial traction, $[\hat{\mathbf{n}} \cdot \mathbf{T}^0]_-^+ = \mathbf{0}$, and the boundary conditions (3.176)–(3.178). Equation (5.165) therefore reduces to

$$\dot{\mathcal{E}} = -\dot{\mathcal{E}}_{\mathrm{w}} - \dot{\mathcal{E}}_{\mathrm{c}}, \qquad (5.166)$$

where

$$\dot{\mathcal{E}}_{\mathrm{w}} = \lim_{\varepsilon \to 0} \int_{\Sigma_\varepsilon^t} [\hat{\boldsymbol{\nu}} \cdot (\mathbf{T}^0 + \mathbf{T}^{\mathrm{PK1}}) \cdot \partial_t \mathbf{s}]_-^+ \, d\Sigma, \qquad (5.167)$$

$$\dot{\mathcal{E}}_{\mathrm{c}} = \lim_{\varepsilon \to 0} \int_{\partial \oplus_\varepsilon^t} [e(\hat{\mathbf{n}}_\varepsilon^t \cdot \mathbf{v}) + \hat{\mathbf{n}}_\varepsilon^t \cdot \mathbf{T}^{\mathrm{PK1}} \cdot \partial_t \mathbf{s}] \, d\Sigma. \qquad (5.168)$$

The first term $\dot{\mathcal{E}}_{\mathrm{w}}$ in equation (5.166) is the instantaneous rate of frictional energy dissipation on the *walls* of the fault surface Σ^t, whereas the second term $\dot{\mathcal{E}}_{\mathrm{c}}$ represents the flux of total energy into the propagating crack tip $\partial\Sigma_{\mathrm{c}}^t$. Both of these quantities are inherently positive, reflecting the irreversible nature of the faulting process; the flux $\dot{\mathcal{E}}_{\mathrm{c}}$ provides a macroscopic measure of the rate at which energy must be supplied to create fresh fault surface at the rupture front. There are two contributions to this flux, each of which has a simple physical interpretation; the first is an advective contribution due to the motion of the rupture front, whereas the second arises from the work being performed by the traction exerted upon $\partial\oplus_\varepsilon^t$. Other terms multiplying $\hat{\mathbf{n}}_\varepsilon^t \cdot \mathbf{v}$ also arise, but they do not contribute in the limit $\varepsilon \to 0$, provided that $\partial_t \mathbf{s}$ and $\boldsymbol{\nabla}\mathbf{s}$ are no more singular than $\varepsilon^{-1/2}$, as we have assumed.

In fracture mechanics, it is customary to measure the rate of energy flow into the crack tip *per unit distance* of crack advance along the fault surface Σ^t rather than per unit time. The resulting quantity—the so-called dynamical *energy release rate*—is defined by $G = v^{-1}\dot{\mathcal{E}}_{\mathrm{c}}$, where $v = \|\mathbf{v}\|$ is the rupture speed. Freund (1990) gives an extensive discussion of the mechanical significance of G, and its role in the formulation of dynamical fracture criteria.

★5.5.3 Seismic energy

We can obtain a precise statement of the above-mentioned tripartite partition of the energy released by an earthquake by integrating the rate of change of energy (5.166) with respect to time from the onset of faulting at $t = t_0$ to $t = \infty$. Thus far the anelasticity of the Earth, which is responsible for the decay of the oscillations subsequent to the earthquake, has not been explicitly introduced. We now assume that its only effect is to add an infinitesimal negative quantity to the sum $-\dot{\mathcal{E}}_{\mathrm{w}} - \dot{\mathcal{E}}_{\mathrm{c}}$ on the right, representing the rate of bodily dissipation. The left side of the integrated equation is the total change in energy, or minus the static energy release; the energy-balance relation can therefore be written in the form

$$\mathcal{E}_{\mathrm{r}} = \mathcal{E}_{\mathrm{s}} + \int_{t_0}^{t_{\mathrm{f}}} \dot{\mathcal{E}}_{\mathrm{w}}\, dt + \int_{t_0}^{t_{\mathrm{f}}} \dot{\mathcal{E}}_{\mathrm{c}}\, dt, \tag{5.169}$$

where $\mathcal{E}_{\mathrm{s}}$ is the total amount of energy dissipated against bodily anelasticity during the interval $t_0 \leq t \leq \infty$. The quantity $\mathcal{E}_{\mathrm{s}}$ is usually referred to as the *seismic energy*.

The rate of energy dissipation on the walls of the fault can be rewritten in the form

$$\dot{\mathcal{E}}_{\mathrm{w}} = \frac{d}{dt}\left(\lim_{\varepsilon\to 0}\int_{\Sigma_\varepsilon^t} [\hat{\boldsymbol{\nu}}\cdot(\mathbf{T}^0 + \mathbf{T}^{\mathrm{PK1}})\cdot\mathbf{s}]_-^+\, d\Sigma\right) - \lim_{\varepsilon\to 0}\int_{\Sigma_\varepsilon^t} [\hat{\boldsymbol{\nu}}\cdot\partial_t\mathbf{T}^{\mathrm{PK1}}\cdot\mathbf{s}]_-^+\, d\Sigma, \tag{5.170}$$

where we have used the fact that $\boldsymbol{\Delta}\mathbf{s} = \mathbf{0}$ on $\partial\Sigma_{\mathrm{c}}^t$. Upon inserting the identity (5.170) into the energy-balance equation (5.169), and making use of the expression (5.160) for the energy release $\mathcal{E}_{\mathrm{r}}$, we obtain the result

$$\mathcal{E}_{\mathrm{s}} = -\tfrac{1}{2}\int_{\Sigma_{\mathrm{f}}} [\hat{\boldsymbol{\nu}}\cdot\mathbf{T}_{\mathrm{f}}^{\mathrm{PK1}}\cdot\mathbf{s}_{\mathrm{f}}]_-^+\, d\Sigma + \int_{t_0}^{\infty}\lim_{\varepsilon\to 0}\int_{\Sigma_\varepsilon^t} [\hat{\boldsymbol{\nu}}\cdot\partial_t\mathbf{T}^{\mathrm{PK1}}\cdot\mathbf{s}]_-^+\, d\Sigma - \int_{t_0}^{t_{\mathrm{f}}} \dot{\mathcal{E}}_{\mathrm{c}}\, dt. \tag{5.171}$$

Equation (5.171) expresses the seismic energy dissipated against bodily friction throughout the Earth in terms only of quantities on the fault surface Σ^t and rupture front $\partial\Sigma_{\mathrm{c}}^t$. The first term depends only upon the static displacement $\mathbf{s}_{\mathrm{f}}$ and incremental traction $\hat{\boldsymbol{\nu}}\cdot\mathbf{T}_{\mathrm{f}}^{\mathrm{PK1}}$ on the final fault Σ_{f}, but the second term depends upon the entire history of the displacement and incremental traction on the expanding fault surface Σ^t. The derivative $\partial_t\mathbf{T}^{\mathrm{PK1}}$ in the second term of (5.171) need not vanish for $t \geq t_{\mathrm{f}}$, even though the slip $\mathbf{\Delta s}$ has attained its final static value; note that we replaced the upper limit of integration by infinity prior to using (5.170) to eliminate $\mathcal{E}_{\mathrm{r}}$ from equation (5.169). The seismic energy, by definition, must depend quadratically upon the infinitesimal displacement $\mathbf{s}$; this is obviously true of each of the terms in the representation (5.171). Kostrov (1974) and Kostrov & Das (1988) derive an expression for the seismic energy $\mathcal{E}_{\mathrm{s}}$ by considering the radiant flux of energy through a sphere surrounding a fault in an infinite elastic medium; they ignore self-gravitation and assume that the initial static stress $\mathbf{T}^0$ is infinitesimal. Equation (5.171) extends their result to the realistic case of a finite, self-gravitating Earth model.

The quantity $\mathcal{E}_{\mathrm{s}}$ includes the energy dissipated by bodily friction during the interval $t_0 \leq t \leq t_{\mathrm{f}}$ that faulting is occurring. During that interval the acceleration cannot be expressed as a sum

$$\mathbf{a}(\mathbf{x}, t) = \sum_k (a_k \cos \omega_k t + b_k \sin \omega_k t)\, \mathbf{s}_k(\mathbf{x}) \tag{5.172}$$

of free oscillations. For that reason, a simple expression for $\mathcal{E}_{\mathrm{s}}$ in terms of the observable excitation amplitudes a_k and b_k cannot be obtained. Let us consider instead only that part of the seismic energy that is dissipated during the free decay after the faulting has terminated; we denote this *modified seismic energy* by $\mathcal{E}_{\mathrm{s}}'$. Presumably $\mathcal{E}_{\mathrm{s}}' \approx \mathcal{E}_{\mathrm{s}}$, since the duration of faulting is very short compared to the decay time of the oscillations; in any case, we must always have $\mathcal{E}_{\mathrm{s}}' \leq \mathcal{E}_{\mathrm{s}}$. After the faulting has ceased, bodily friction is the only active mechanism of dissipation; integration of equation (5.166) from $t = t_{\mathrm{f}}$ to $t = \infty$ yields

$$\mathcal{E}_{\mathrm{s}}' = \mathcal{E}(t_{\mathrm{f}}) - \mathcal{E}(\infty), \tag{5.173}$$

since both $\dot{\mathcal{E}}_{\mathrm{w}}$ and $\dot{\mathcal{E}}_{\mathrm{c}}$ are equal to zero. The energy $\mathcal{E}(t_{\mathrm{f}})$ is the total kinetic-plus-potential energy of oscillation immediately after the cessation of faulting, whereas $\mathcal{E}(\infty)$ is the remaining elastic-gravitational potential energy after all the modes have decayed. This energy difference is transformed into heat by the dissipation; the equality (5.173) is therefore an obvious consequence of the conservation of energy. Upon making use of the Maxwell relation $\Lambda_{ijkl} = \Lambda_{klij}$ and the symmetry (4.16), we can write the modified seismic energy in the form

$$\begin{aligned} \mathcal{E}_{\mathrm{s}}' = \tfrac{1}{2} \int_{\oplus} [\rho^0 \partial_t \mathbf{s} \cdot \partial_t \mathbf{s} + \boldsymbol{\nabla}(\mathbf{s} - \mathbf{s}_{\mathrm{f}}) : \boldsymbol{\Lambda} : \boldsymbol{\nabla}(\mathbf{s} - \mathbf{s}_{\mathrm{f}}) \\ + \rho^0 (\mathbf{s} - \mathbf{s}_{\mathrm{f}}) \cdot \boldsymbol{\nabla}(\phi^{\mathrm{E1}} - \phi^{\mathrm{E1}}_{\mathrm{f}}) \\ + \rho^0 (\mathbf{s} - \mathbf{s}_{\mathrm{f}}) \cdot \boldsymbol{\nabla\nabla}\phi^0 \cdot (\mathbf{s} - \mathbf{s}_{\mathrm{f}})]\, dV, \end{aligned} \tag{5.174}$$

where the unsubscripted quantities $\mathbf{s}$ and ϕ^{E1} are evaluated at the cessation of faulting $t = t_{\mathrm{f}}$, and the subscripted quantities $\mathbf{s}_{\mathrm{f}}$ and $\phi^{\mathrm{E1}}_{\mathrm{f}}$ are evaluated at $t = \infty$, as usual. Inserting the representation (5.172) into (5.174) and making use of the orthonormality relation (4.56), we obtain

$$\mathcal{E}_{\mathrm{s}}' = \tfrac{1}{2}\sum_k (a_k^2 + b_k^2). \tag{5.175}$$

This simple expression relating the modified seismic energy $\mathcal{E}_{\mathrm{s}}'$ to the excitation amplitudes a_k and b_k of the Earth's normal modes is due to McCowan & Dziewonski (1977). It is clear from the derivation that (5.175) is valid for any transient indigenous source, not only for an ideal fault. The only requirement is that the motion be free, i.e., a slowly decaying version of equation (5.172), during the interval $t_{\mathrm{f}} \leq t \leq \infty$.

★5.5.4 Discussion

The balance of energy on a rotating Earth is complicated by the change in rotation rate $\delta\Omega$, which occurs in response to the static deformation $\mathbf{s}_{\mathrm{f}}$. In addition to the changes in elastic and gravitational potential energy, given by equations (5.157) and (5.158), there is a permanent change in the kinetic energy of a rotating Earth:

$$\mathcal{E}_{\mathrm{k}} = \tfrac{1}{2}(C + \delta C)(\Omega + \delta\Omega)^2 - \tfrac{1}{2}C\Omega^2 = \tfrac{1}{2}C\Omega^2(\delta\Omega/\Omega), \tag{5.176}$$

where we have used the exact law of conservation of angular momentum, $(C + \delta C)(\Omega + \delta\Omega) = C\Omega$, to obtain the final equality. The change in rotational kinetic energy $\mathcal{E}_{\mathrm{k}}$ can alternatively be interpreted as the work performed against the apparent centrifugal force associated with the Earth's rotation; the sum of the two changes $\mathcal{E}_{\mathrm{g}} + \mathcal{E}_{\mathrm{k}}$ is the total work that is performed against both real and apparent body forces. The total energy released by an earthquake on a rotating Earth, $\mathcal{E}_{\mathrm{r}} = -\mathcal{E}_{\mathrm{e}} - \mathcal{E}_{\mathrm{g}} - \mathcal{E}_{\mathrm{k}}$, can be reduced by an argument analogous to that in Section 5.5.1 to the same integral over the final fault surface Σ^{f}:

$$\mathcal{E}_{\mathrm{r}} = \tfrac{1}{2}\int_{\Sigma^{\mathrm{f}}} [\hat{\boldsymbol{\nu}} \cdot (2\mathbf{T}^0 + \mathbf{T}_{\mathrm{f}}^{\mathrm{PK1}}) \cdot \mathbf{s}_{\mathrm{f}}]_-^+ \, d\Sigma. \tag{5.177}$$

The tripartite energy-balance relation (5.169) and the expression (5.171) giving the seismic energy $\mathcal{E}_{\mathrm{s}}$ in terms only of quantities on the instantaneous fault surface Σ^t and rupture front $\partial\Sigma_{\mathrm{c}}^t$ are also valid on a rotating Earth. The modified seismic energy $\mathcal{E}_{\mathrm{s}}'$ is given in terms of the complex excitation amplitudes c_k in equation (4.141) by

$$\mathcal{E}_{\mathrm{s}}' = \tfrac{1}{2}\sum_k c_k^* c_k. \tag{5.178}$$

It is remarkable that neither of the two results (5.177)–(5.178) has any explicit dependence upon either the gravitational constant G or the angular rate of rotation $\Omega = \|\boldsymbol{\Omega}\|$.

To gain a better understanding of the balance of energy released during faulting, we consider a specific example—the great Chile earthquake of May 22, 1960. This event, whose seismic moment, $M_0 \approx 2 \times 10^{23}$ N m, is the largest ever measured, gave rise to a coseismic decrease in the length of day of only a few microseconds (Chao & Gross 1995). Such a change, corresponding to $\delta\Omega/\Omega \approx 10^{-10}$, is negligible in comparison to the observed changes in the rotation rate of the Earth, which are of order $\delta\Omega/\Omega \approx 10^{-8}$, due to a wide variety of geophysical phenomena other than earthquakes. Nevertheless, since the reservoir of the Earth's rotational kinetic energy is large, $\frac{1}{2}C\Omega^2 = 2 \times 10^{29}$ J, the change in energy due to the earthquake is substantial: $\mathcal{E}_{\mathrm{k}} \approx 2 \times 10^{19}$ J. Surprisingly, this change in rotational kinetic energy exceeds our best estimate of the total energy released by the Chile earthquake: $\mathcal{E}_{\mathrm{r}} \approx 1 \times 10^{19}$ J, according to the empirical Gutenburg-Richter formula $\mathcal{E}_{\mathrm{r}} = 0.5 \times 10^{-4}\, M_0$ (Kanamori 1977). It is clear that there is a serious discrepancy between the result that $\mathcal{E}_{\mathrm{k}} \approx 2 \times \mathcal{E}_{\mathrm{r}}$ and the strong intuitive notion that the rotation of the Earth must play a completely negligible role in the mechanics of faulting! This apparent paradox was first pointed out by Dahlen (1977), who showed how it can be resolved.

Since $\mathcal{E}_{\mathrm{g}} + \mathcal{E}_{\mathrm{k}}$ is the total work performed against gravitational and centrifugal body forces, and the former are about 300 times stronger than the latter in the Earth, we expect the magnitude of the gravitational energy change $\mathcal{E}_{\mathrm{g}}$ to be roughly 300 times that of $\mathcal{E}_{\mathrm{k}}$. Numerical calculations confirm this expectation; for the 1960 Chile earthquake, Chao, Gross & Dong (1995) find that $\mathcal{E}_{\mathrm{g}} \approx -600 \times \mathcal{E}_{\mathrm{r}}$. Such a decrease in the gravitational energy, corresponding to a net compression of the Earth, is typical of shallow subduction-zone earthquakes. Evidently, the magnitude of the elastic energy change $\mathcal{E}_{\mathrm{e}}$ accompanying the Chile earthquake must also be about 600 times larger than $\mathcal{E}_{\mathrm{r}}$, and the two changes $\mathcal{E}_{\mathrm{e}}$ and $\mathcal{E}_{\mathrm{g}} + \mathcal{E}_{\mathrm{k}}$ must be very nearly equal and opposite. This fine balance of the elastic and gravitational-centrifugal energy changes is a universal feature of other earthquakes as well. The individual energy changes $\mathcal{E}_{\mathrm{e}}$ and $\mathcal{E}_{\mathrm{g}} + \mathcal{E}_{\mathrm{k}}$ may be of either sign; however, it is generally true that $|\mathcal{E}_{\mathrm{e}}| \approx |\mathcal{E}_{\mathrm{g}} + \mathcal{E}_{\mathrm{k}}| \gg \mathcal{E}_{\mathrm{r}}$. The dominant contribution to the elastic energy change $\mathcal{E}_{\mathrm{e}}$ is the work performed against the initial hydrostatic pressure p^0, and it can be shown that the total energy release (5.177) is explicitly independent of p^0. This makes it clear that the balance between $\mathcal{E}_{\mathrm{e}}$ and $\mathcal{E}_{\mathrm{g}} + \mathcal{E}_{\mathrm{k}}$ is in essence a balance between the work performed against body forces and that concomitantly performed against the initial hydrostatic pressure. Such a balance is not surprising, inasmuch as the origin of the initial hydrostatic pressure p^0 in the Earth may be attributed directly to the body forces. The deviatoric stress $\boldsymbol{\tau}^0$, which is tectonic in origin, and which is fundamentally responsible for the occurrence of earthquakes, is significantly smaller than the pressure p^0 except very near the surface of the Earth; for this reason, the energy $\mathcal{E}_{\mathrm{r}}$ released during earthquake faulting is in general much smaller than its constituent parts $-\mathcal{E}_{\mathrm{e}}$ and $-(\mathcal{E}_{\mathrm{g}}+\mathcal{E}_{\mathrm{k}})$. It is incorrect to regard the classical expression for the energy release $\mathcal{E}_{\mathrm{r}}$ in a non-gravitating medium as the change in elastic energy, and to account separately for the change in gravitational energy in the case that permanent elevation changes have occurred in the vicinity of the focus. This ad hoc procedure, which might be tempting on physical grounds, would amount to counting the relatively enormous change in gravitational energy twice.

Chapter 6

Anelasticity and Attenuation

We have in several instances in this book recognized that anelasticity causes the normal modes of oscillation of the Earth to decay following their excitation by an earthquake. So far, however, we have assumed that the anelasticity is infinitesimal, so that its only effect is to replace the elastic-gravitational eigenfrequencies ω by $\omega + i0$, without altering the associated eigenfunctions $\mathbf{s}$. In reality, the anelasticity of the Earth is slight, but not infinitesimal; a quantitative model of seismic attenuation is formulated in this chapter. We shall see that even in the weak-attenuation approximation, it is not possible to treat the elastic and anelastic responses independently; instead, they must be considered to be two intimately related aspects of the Earth's rheology.

We begin by reviewing the mathematical theory of linear anelasticity, with an emphasis on those aspects that are most pertinent to global seismology. Our coverage is not as complete as that provided by a number of materials-science monographs devoted to the topic, notably Zener (1948), Gross (1953) and Nowick & Berry (1972). We adopt a strictly macroscopic, phenomenological point of view, and make no attempt to inquire into the microscopic solid-state mechanisms responsible for seismic attenuation. Speculations regarding the likely mechanisms of anelasticity within the Earth may be found in Minster (1980), Karato & Spetzler (1990) and Anderson (1991). A review of relevant high-pressure, high-temperature laboratory creep and attenuation experiments is given by Jackson (1993).

We shall use the symbol $\nu = \omega + i\gamma$ systematically throughout this chapter and the rest of this book to denote a complex angular frequency of oscillation, with real part ω and imaginary part γ. We shall always use ω to denote a *real* frequency, as we have done so far. If $\nu = \omega + i\gamma$ is an angular eigenfrequency of one of the Earth's normal modes, the imaginary part $\gamma > 0$ is the mode's *decay rate*.

6.1 Linear Isotropic Anelasticity

The linear elastic constitutive relation which we have employed so far stipulates that the incremental stress $\mathbf{T}^{\mathrm{L1}}$ or $\mathbf{T}^{\mathrm{PK1}}$ depends only upon the local *instantaneous* strain $\boldsymbol{\varepsilon}$ or deformation $\boldsymbol{\nabla}\mathbf{s}$. For an *isotropic* material with a *hydrostatic* initial stress, the elastic constitutive relation is

$$\mathbf{T}^{\mathrm{L1}} = \kappa\theta\mathbf{I} + 2\mu\mathbf{d}, \tag{6.1}$$

where κ and μ are the isentropic incompressibility and rigidity, and $\theta = \mathrm{tr}\,\boldsymbol{\varepsilon}$ and $\mathbf{d} = \boldsymbol{\varepsilon} - \frac{1}{3}(\mathrm{tr}\,\boldsymbol{\varepsilon})\mathbf{I}$ are the scalar isotropic and the deviatoric strain. In an isotropic *anelastic* material, the stress experienced by a particle is assumed to depend linearly upon its *entire past history* of strain. Equation (6.1) is generalized in that case to

$$\mathbf{T}^{\mathrm{L1}}(t) = \int_{-\infty}^{t} \kappa(t-t')\partial_{t'}\theta(t')\,dt'\,\mathbf{I} + \int_{-\infty}^{t} 2\mu(t-t')\partial_{t'}\mathbf{d}(t')\,dt', \tag{6.2}$$

where the dependence on the particle $\mathbf{x}$ is regarded as understood. The lower limit of integration $-\infty$ reflects the dependence of $\mathbf{T}^{\mathrm{L1}}$ upon the *entire* past history of the deformation, whereas the upper limit t insures agreement with the *principle of causality*—there is no dependence upon the deformation in the future. Equation (6.2) is known as the *Boltzmann superposition principle*; the representation in terms of the strain rate $\partial_t\boldsymbol{\varepsilon}$ rather than the strain $\boldsymbol{\varepsilon}$ is conventional.

6.1.1 Creep and stress relaxation functions

Following Nowick & Berry (1972) we shall rewrite the anelastic constitutive relation (6.2) using an abbreviated scalar notation:

$$\sigma(t) = \int_{-\infty}^{t} M(t-t')\dot{\varepsilon}(t')\,dt'. \tag{6.3}$$

The quantity $\varepsilon(t)$ is a pseudonym for either of the strain variables θ or $\mathbf{d}$, whereas $\sigma(t)$ stands for the complementary isotropic or deviatoric stress

Variable	Isotropic	Deviatoric
Strain ε	$\theta = \mathrm{tr}\,\boldsymbol{\varepsilon}$	$\mathbf{d} = \boldsymbol{\varepsilon} - \frac{1}{3}(\mathrm{tr}\,\boldsymbol{\varepsilon})\mathbf{I}$
Stress σ	$-p^{\mathrm{L1}} = \frac{1}{3}\mathrm{tr}\,\mathbf{T}^{\mathrm{L1}}$	$\boldsymbol{\tau}^{\mathrm{L1}} = \mathbf{T}^{\mathrm{L1}} - \frac{1}{3}(\mathrm{tr}\,\mathbf{T}^{\mathrm{L1}})\mathbf{I}$
Modulus M	κ	μ

Table 6.1. Summary of scalar notation used in the theory of linear isotropic anelasticity. The quantities $\sigma(t)$ and $\varepsilon(t)$ are complementary stress and strain variables.

$-p^{\mathrm{L1}} = \frac{1}{3}\mathrm{tr}\,\mathbf{T}^{\mathrm{L1}}$ or $\boldsymbol{\tau}^{\mathrm{L1}} = \mathbf{T}^{\mathrm{L1}} - \frac{1}{3}(\mathrm{tr}\,\mathbf{T}^{\mathrm{L1}})\mathbf{I}$. The quantity $M(t)$ denotes, in the first instance, the time-dependent incompressibility κ, and in the second instance, the time-dependent rigidity μ. The dot denotes differentiation with respect to time. The correspondence between the scalar and tensor notations is summarized in Table 6.1.

We may alternatively consider the strain to be a functional of the entire past history of the stress, rather than vice versa:

$$\varepsilon(t) = \int_{-\infty}^{t} J(t - t')\dot{\sigma}(t')\,dt'. \tag{6.4}$$

Physically, the two quantities $M(t)$ and $J(t)$ are the stress and strain responses to a unit step $H(t)$ in the complementary variable. The unit step stress response $M(t)$ is known as the *stress relaxation function*, whereas the corresponding strain response $J(t)$ is known as the *creep function*.

The values of $M(t)$ and $J(t)$ at $t = 0$ and $t = \infty$ are denoted by special symbols, namely,

$$M(0) = M_{\mathrm{u}}, \qquad M(\infty) = M_{\mathrm{r}}, \tag{6.5}$$

$$J(0) = J_{\mathrm{u}}, \qquad J(\infty) = J_{\mathrm{r}}. \tag{6.6}$$

The quantities M_{u} and M_{r} are referred to as the *unrelaxed and relaxed moduli*, respectively, whereas J_{u} and J_{r} are the corresponding *unrelaxed and relaxed compliances*. The unrelaxed modulus and compliance M_{u} and J_{u}, which describe the *instantaneous elastic response*, and the relaxed modulus and compliance M_{r} and J_{r}, which describe the *long-term equilibrium response*, must be reciprocals of each other, i.e.,

$$M_{\mathrm{u}} = 1/J_{\mathrm{u}}, \qquad M_{\mathrm{r}} = 1/J_{\mathrm{r}}. \tag{6.7}$$

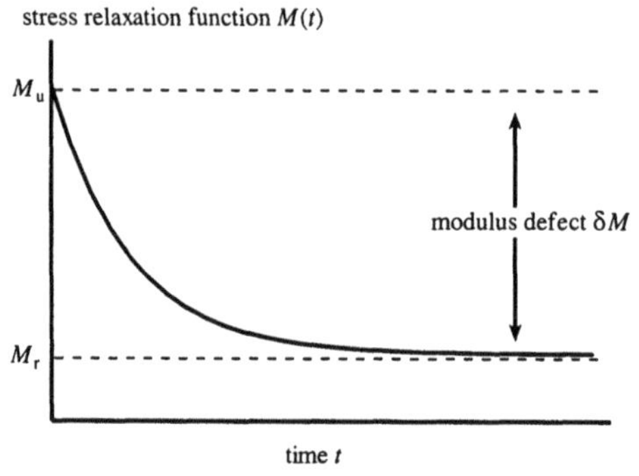

Figure 6.1. The stress relaxation function $M(t)$ of an anelastic material decreases monotonically from its unrelaxed value $M_{\rm u}$ at $t = 0$ to its relaxed value $M_{\rm r}$ at $t = \infty$.

The stress $M(t)$ required to maintain a step strain $\varepsilon(t) = H(t)$ relaxes with time from $M_{\rm u}$ at $t = 0$ to $M_{\rm r}$ at $t = \infty$, as illustrated in Figure 6.1. The imposition of a unit step stress $\sigma(t) = H(t)$ likewise causes $J(t)$ to creep from $J_{\rm u}$ at $t = 0$ to $J_{\rm r}$ at $t = \infty$, as shown in Figure 6.2. This is the reason for the nomenclature. The differences

$$\delta M = M_{\rm u} - M_{\rm r}, \qquad \delta J = J_{\rm r} - J_{\rm u} \tag{6.8}$$

are known as the *modulus defect* and the *relaxation of the compliance*, respectively.

We have taken it for granted in the above discussion that there is both a non-zero instantaneous equilibrium response at time $t = 0$ and a finite equilibrium response at time $t = \infty$, so that

$$0 < M_{\rm r} \leq M_{\rm u} < \infty, \qquad 0 < J_{\rm u} \leq J_{\rm r} < \infty. \tag{6.9}$$

These two assumptions are not strictly necessary, as we shall see in Sections 6.1.4 and 6.1.11, but they simplify much of the ensuing theoretical development. Materials satisfying the constraints (6.9) are referred to as *anelastic*, to distinguish them from the more general *viscoelastic* materials, which may have $M_{\rm r} = 1/J_{\rm r} = 0$ or $M_{\rm u} = 1/J_{\rm u} = \infty$.

Knowledge of the stress relaxation function $M(t)$ for all positive times $0 \leq t \leq \infty$ implies knowledge of the creep function $J(t)$ and vice versa. An implicit relation between the two functions can be obtained by setting $\sigma(t) = H(t)$ in equation (6.3). Noting that $\varepsilon(t) = J(t)$ undergoes a jump

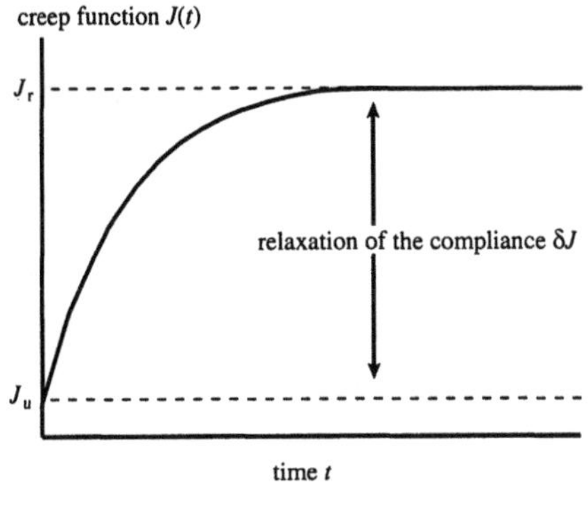

Figure 6.2. The creep function $J(t)$ increases monotonically from its unrelaxed value J_u at $t=0$ to its relaxed value J_r at $t=\infty$.

to J_u at $t=0$, we obtain

$$1 = J_\mathrm{u}M(t) + \int_0^t M(t-t')\dot{J}(t')\,dt'. \tag{6.10}$$

Alternatively, upon either setting $\varepsilon(t) = H(t)$ in equation (6.4) or integrating (6.10) by parts, we find

$$1 = M_\mathrm{u}J(t) + \int_0^t J(t-t')\dot{M}(t')\,dt'. \tag{6.11}$$

Another useful relation which can be iterated to find $J(t)$ in terms of $M(t)$ or vice versa is

$$t = \int_0^t J(t')M(t-t')\,dt' = \int_0^t M(t')J(t-t')\,dt'. \tag{6.12}$$

The equivalence of equations (6.10)–(6.11) and (6.12) can be demonstrated by differentiating the latter with respect to time. Substituting the elementary relation $M(t-t') = M(t) - [M(t) - M(t-t')]$ into (6.10) and carrying out the integration over the term involving $M(t)$ yields

$$M(t)J(t) = 1 + \int_0^t [M(t) - M(t-t')]\dot{J}(t')\,dt'. \tag{6.13}$$

The two functions $M(t)$ and $J(t)$ are experimentally observed to be monotonic; the integrand in (6.13) is therefore strictly negative, so that

$$M(t)J(t) \leq 1. \tag{6.14}$$

We already know that equality holds in (6.14) for $t=0$ and $t=\infty$; we shall see in Section 6.1.9 that there is approximate equality even more generally.

6.1.2 Harmonic variations

So far we have been considering the transient response to a unit step in stress or strain. We turn next to the case of a harmonic variation in stress and strain, of the form

$$\sigma(t) = \mathrm{Re}\,[\sigma(\nu)\exp(i\nu t)], \qquad \varepsilon(t) = \mathrm{Re}\,[\varepsilon(\nu)\exp(i\nu t)]. \tag{6.15}$$

The linearity of the response guarantees that $\sigma(t)$ will be of the form (6.15) if $\varepsilon(t)$ is, and vice versa. Substituting the representations (6.15) into the Boltzmann superposition principle (6.3)–(6.4), we find that the complex amplitudes $\sigma(\nu)$ and $\varepsilon(\nu)$ are related by

$$\sigma(\nu) = M(\nu)\varepsilon(\nu), \qquad \varepsilon(\nu) = J(\nu)\sigma(\nu), \tag{6.16}$$

where

$$M(\nu) = i\nu \int_0^\infty M(t)\exp(-i\nu t)\,dt, \tag{6.17}$$

$$J(\nu) = i\nu \int_0^\infty J(t)\exp(-i\nu t)\,dt. \tag{6.18}$$

The *complex modulus* $M(\nu)$ and the *complex compliance* $J(\nu)$ are evidently related to each other by

$$M(\nu)J(\nu) = 1. \tag{6.19}$$

Both $M(\nu)$ and $J(\nu)$ are *analytic* functions satisfying

$$M(-\nu^*) = M^*(\nu), \qquad J(-\nu^*) = J^*(\nu) \tag{6.20}$$

in the *lower half* of the complex frequency plane, $\mathrm{Im}\,\nu \leq 0$. At very low frequencies, the stress and strain are related by the relaxed modulus and compliance:

$$\lim_{\nu\to 0} M(\nu) = \lim_{\nu\to 0} 1/J(\nu) = M_{\mathrm{r}} = 1/J_{\mathrm{r}}. \tag{6.21}$$

Conversely, at very high frequencies, they are related by the unrelaxed modulus and compliance:

$$\lim_{\nu\to\infty} M(\nu) = \lim_{\nu\to\infty} 1/J(\nu) = M_{\mathrm{u}} = 1/J_{\mathrm{u}}. \tag{6.22}$$

It is conventional to decompose the complex modulus and compliance on the real frequency axis into real and imaginary parts:

$$M(\omega) = M_1(\omega) + iM_2(\omega), \qquad J(\omega) = J_1(\omega) - iJ_2(\omega). \tag{6.23}$$

The quantities $M_1(\omega)$, $M_2(\omega)$, $J_1(\omega)$ and $J_2(\omega)$ can be written explicitly in terms of $M(t)$ and $J(t)$ in the form

$$M_1(\omega) = \omega \int_0^\infty M(t) \sin \omega t \, dt, \tag{6.24}$$

$$M_2(\omega) = \omega \int_0^\infty M(t) \cos \omega t \, dt, \tag{6.25}$$

$$J_1(\omega) = \omega \int_0^\infty J(t) \sin \omega t \, dt, \tag{6.26}$$

$$J_2(\omega) = -\omega \int_0^\infty J(t) \cos \omega t \, dt. \tag{6.27}$$

The real modulus and compliance are both even functions of frequency, whereas the imaginary modulus and compliance are odd:

$$M_1(-\omega) = M_1(\omega), \qquad M_2(-\omega) = -M_2(\omega), \tag{6.28}$$

$$J_1(-\omega) = J_1(\omega), \qquad J_2(-\omega) = -J_2(\omega). \tag{6.29}$$

The choice of signs in (6.23) guarantees that all four functions $M_1(\omega)$, $M_2(\omega)$, $J_1(\omega)$ and $J_2(\omega)$ are positive for $\omega > 0$.

6.1.3 Springs and dashpots

Many simple anelastic and viscoelastic materials may be modelled or visualized as "tinker-toy" networks of linear elastic springs and viscous dashpots. The elements comprising the network may be combined in either series or parallel; the rules for analyzing such lumped-parameter mechanical analogues are simple:

1. Each spring has a constitutive relation of the form $\sigma = M\varepsilon$ or $\varepsilon = J\sigma$, where M is the elastic modulus and $J = 1/M$ is the compliance.

2. Each dashpot has a constitutive relation of the form $\sigma = \eta\dot{\varepsilon}$, where η is the viscosity.

3. Whenever two elements are combined in series, the stresses are equal, whereas the strains are additive: $\sigma = \sigma_1 = \sigma_2$ and $\varepsilon = \varepsilon_1 + \varepsilon_2$.

4. Whenever two elements are combined in parallel, the stresses are additive, whereas the strains are equal: $\sigma = \sigma_1 + \sigma_2$ and $\varepsilon = \varepsilon_1 = \varepsilon_2$.

Every conglomeration of linear springs and dashpots connected in series or parallel corresponds to a viscoelastic material with a linear *differential constitutive relation* of the form

$$a_0\sigma + a_1\dot{\sigma} + a_2\ddot{\sigma} + \cdots = b_0\varepsilon + b_1\dot{\varepsilon} + b_2\ddot{\varepsilon} + \cdots, \tag{6.30}$$

where $a_0, a_1, a_2, \ldots$ and $b_0, b_1, b_2, \ldots$ are real constants. The associated complex modulus and compliance are rational polynomials:

$$M(\nu) = \frac{1}{J(\nu)} = \frac{b_0 + b_1(i\nu) + b_2(i\nu)^2 + \cdots}{a_0 + a_1(i\nu) + a_2(i\nu)^2 + \cdots}. \tag{6.31}$$

The coefficients $a_0, a_1, a_2, \ldots$ and $b_0, b_1, b_2, \ldots$ can be determined by applying the above rules to a sketch or graph of the network, in the same way that Kirchhoff's laws can be used to find the aggregate properties of a lumped-parameter electrical circuit. Bland (1960) gives a systematic formulation of the theory of linear viscoelasticity based upon this approach.

6.1.4 Maxwell and Kelvin-Voigt solids

The simplest such tinker-toy materials are a *Maxwell solid*, consisting of a spring and dashpot in series, and a *Kelvin-Voigt* solid, consisting of a spring and dashpot in parallel. Labelling the constituent elements as in Figure 6.3 and adding or equating the stresses and strains as indicated, we find the differential constitutive relations of these two single-spring-plus-dashpot solids to be

$$\underbrace{\sigma + \tau_\sigma\dot{\sigma} = \tau_\sigma M_{\mathrm{u}}\dot{\varepsilon},}_{\text{Maxwell}} \qquad \underbrace{\sigma = M_{\mathrm{r}}(\varepsilon + \tau_\varepsilon\dot{\varepsilon})}_{\text{Kelvin-Voigt}}. \tag{6.32}$$

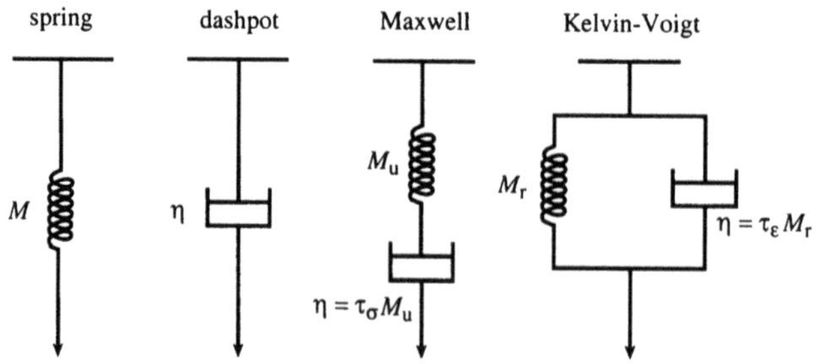

Figure 6.3. (*Left to right*) Schematic depiction of a spring, dashpot, Maxwell and Kelvin-Voigt solid. Each of the networks is anchored at the top. The stress relaxation and creep functions $M(t)$ and $J(t)$ are the responses to a unit step strain and stress applied at the arrow.

Property	Maxwell Solid	Kelvin-Voigt Solid
$M(t)$	$M_{\mathrm{u}}\exp(-t/\tau_\sigma)$	$M_{\mathrm{r}}[\tau_\varepsilon\delta(t)+H(t)]$
$J(t)$	$J_{\mathrm{u}}[H(t)+t/\tau_\sigma]$	$J_{\mathrm{r}}[1-\exp(-t/\tau_\varepsilon)]$
$M(\nu)$	$M_{\mathrm{u}}(i\nu\tau_\sigma)(1+i\nu\tau_\sigma)^{-1}$	$M_{\mathrm{r}}(1+i\nu\tau_\varepsilon)$
$J(\nu)$	$J_{\mathrm{u}}(i\nu\tau_\sigma)^{-1}(1+i\nu\tau_\sigma)$	$J_{\mathrm{r}}(1+i\nu\tau_\varepsilon)^{-1}$

Table 6.2. Time-domain (*top*) and frequency-domain (*bottom*) properties of the two single-spring-plus-dashpot solids.

The Maxwell and Kelvin-Voigt stress relaxation function $M(t)$ and creep function $J(t)$ may be found by setting $\varepsilon = H(t)$ and $\sigma = H(t)$, respectively, in equations (6.32). These quantities, together with the complex moduli and compliances $M(\nu)$ and $J(\nu)$, are summarized in Table 6.2. The quantities $\tau_\sigma = \eta/M_{\mathrm{u}}$ and $\tau_\varepsilon = \eta/M_{\mathrm{r}}$ are referred to as the stress and strain *relaxation times*; the exponentially decaying character of $M_{\mathrm{u}}\exp(-t/\tau_\sigma)$ and the exponentially growing character of $J_{\mathrm{r}}[1-\exp(-t/\tau_\varepsilon)]$ are the source of this terminology. Neither of these two-parameter materials is anelastic in the sense (6.9): a Maxwell solid creeps forever under the application of a step stress, $J_{\mathrm{r}} = 1/M_{\mathrm{r}} = \infty$, whereas a Kelvin-Voigt solid has a singular instantaneous elastic response, $M_{\mathrm{u}} = 1/J_{\mathrm{u}} = \infty$.

6.1.5 Standard linear solid

The simplest example of a material which does satisfy all of the anelastic constraints is a so-called *standard linear solid*. It may be visualized either as a Maxwell solid and a spring connected in parallel or as a Kelvin-Voigt solid and a spring connected in series; see Figure 6.4. Upon applying the stress and strain additivity and equality rules, we find that the differential constitutive relation for either of these equivalent three-parameter networks may be written in the form

$$\dot{\sigma} + \tau_\sigma^{-1}\sigma = M_{\mathrm{u}}(\dot{\varepsilon} + \tau_\varepsilon^{-1}\varepsilon), \tag{6.33}$$

where

$$\delta M/M_{\mathrm{u}} = \delta J/J_{\mathrm{r}} = 1 - \tau_\sigma/\tau_\varepsilon. \tag{6.34}$$

The stress relaxation and creep functions corresponding to equation (6.33) are both exponentials, but with different decay and growth times:

$$M(t) = M_{\rm u} - \delta M[1 - \exp(-t/\tau_\sigma)], \tag{6.35}$$

$$J(t) = J_{\rm u} + \delta J[1 - \exp(-t/\tau_\varepsilon)]. \tag{6.36}$$

We see from equation (6.34) that the stress and strain relaxation times must satisfy $\tau_\sigma \leq \tau_\varepsilon$.

The complex modulus and compliance of a standard linear solid are found from either (6.17)–(6.18) or a frequency-domain analysis of Figure 6.4 to be

$$M(\nu) = M_{\rm u} - \delta M(1 + i\nu\tau_\sigma)^{-1}, \tag{6.37}$$

$$J(\nu) = J_{\rm u} + \delta J(1 + i\nu\tau_\varepsilon)^{-1}. \tag{6.38}$$

It is noteworthy that equations (6.37) and (6.38) have imaginary simple poles at $\nu = i/\tau_\sigma$ and $\nu = i/\tau_\varepsilon$, respectively. This is illustrative of a general anelastic phenomenon—that the analytic continuations of $M(\nu)$ and $J(\nu)$ exhibit singularities in the *upper half* of the complex frequency plane, $\mathrm{Im}\,\nu > 0$.

The real and imaginary moduli and compliances on the real frequency axis are given by

$$M_1(\omega) = M_{\rm u} - \frac{\delta M}{1 + \omega^2\tau_\sigma^2}, \qquad M_2(\omega) = \frac{\omega\tau_\sigma\,\delta M}{1 + \omega^2\tau_\sigma^2}, \tag{6.39}$$

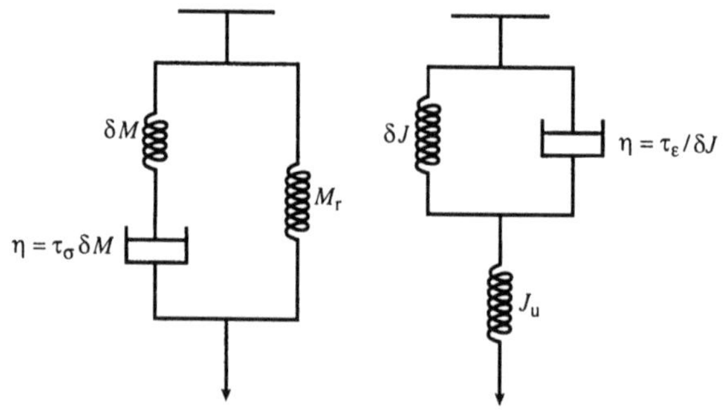

Figure 6.4. Equivalent three-element representations of a standard linear solid. (*Left*) A Maxwell solid and a spring connected in parallel. (*Right*) A Kelvin-Voigt solid and a spring connected in series.

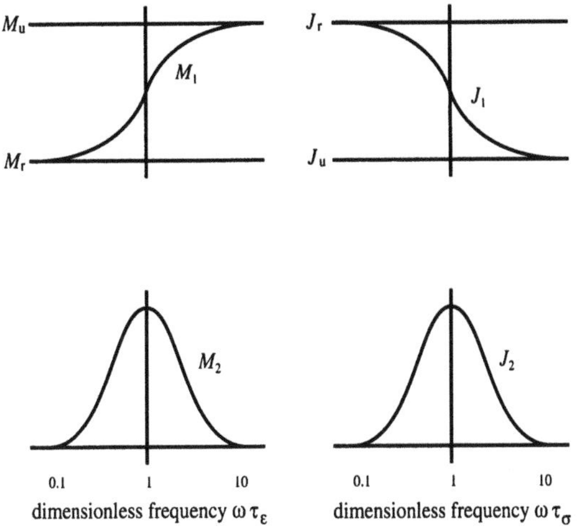

Figure 6.5. (*Left*) Frequency variation of the real and complex moduli $M_1(\omega)$ and $M_2(\omega)$ of a standard linear solid. (*Right*) Corresponding variation of the real and complex compliances $J_1(\omega)$ and $J_2(\omega)$. Note the logarithmic frequency scale.

$$J_1(\omega) = J_u + \frac{\delta J}{1+\omega^2\tau_\varepsilon^2}, \qquad J_2(\omega) = \frac{\omega\tau_\varepsilon\,\delta J}{1+\omega^2\tau_\varepsilon^2}. \tag{6.40}$$

Both $M_2(\omega)$ and $J_2(\omega)$ have the form of a so-called *Debye relaxation peak* centered upon $\omega\tau = 1$, where τ denotes either τ_σ or τ_ε. Outside of the frequency band $1/10 \leq \omega\tau \leq 10$, the imaginary modulus and compliance are relatively insignificant, as shown in Figure 6.5. The real modulus $M_1(\omega)$ increases from M_r to M_u and the corresponding compliance $J_1(\omega)$ decreases from J_r to J_u over the same band $1/10 \leq \omega\tau \leq 10$. The *dispersion* within this anelastic band is said to be *normal*.

6 1.6 Energy dissipation and Q

The stress $\sigma(t) = \mathrm{Re}\,[\sigma(\omega)\exp(i\omega t)]$ and the strain $\varepsilon(t) = \mathrm{Re}\,[\varepsilon(\omega)\exp(i\omega t)]$ follow an elliptical hysteresis loop during a harmonic cycle of forced oscillation at real frequency ω, as illustrated in Figure 6.6. The work done against internal friction in a unit volume during a full cycle of oscillation is the area of the hysteresis loop, which is easily shown to be

$$\oint \dot{E}\,dt = \oint \sigma\dot{\varepsilon}\,dt = \pi M_2(\omega)\,|\varepsilon(\omega)|^2 = \pi J_2(\omega)\,|\sigma(\omega)|^2. \tag{6.41}$$

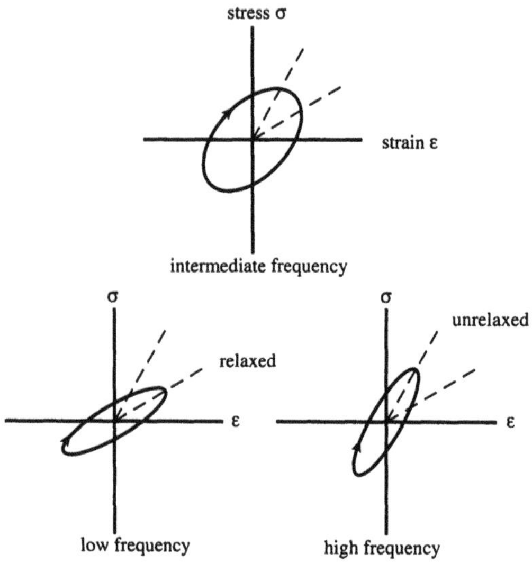

Figure 6.6. Trajectory of stress versus strain during a harmonic cycle of forced oscillation. Dashed lines represent relaxed and unrelaxed perfectly elastic behavior in each case. (*Top*) At intermediate frequencies the stress-strain hysteresis loop is quasi-circular. (*Bottom*) At low and high frequencies the loop is highly elliptical; the response is either very nearly relaxed, $\sigma \approx M_{\mathrm{r}}\varepsilon$, or very nearly unrelaxed, $\sigma \approx M_{\mathrm{u}}\varepsilon$, respectively.

The imaginary modulus $M_2(\omega)$ and compliance $J_2(\omega)$ thus provide a measure of the volumetric rate of energy dissipation due to the anelasticity; the dissipated energy appears as heat within the material.

The intrinsic *quality factor* $Q(\omega)$ of an anelastic material is defined, following O'Connell & Budiansky (1978), by

$$Q(\omega) = \frac{M_1(\omega)}{M_2(\omega)} = \frac{J_1(\omega)}{J_2(\omega)}. \tag{6.42}$$

Upon writing the complex modulus and compliance in the polar form

$$M(\omega) = |M(\omega)| \exp[i\phi(\omega)], \qquad J(\omega) = |J(\omega)| \exp[-i\phi(\omega)], \tag{6.43}$$

we see that the inverse quality factor is

$$Q^{-1}(\omega) = \tan\phi(\omega). \tag{6.44}$$

The angle $\phi(\omega)$ is the *phase lag* of the stress behind the strain in the harmonic hysteresis loop.

The *stored* elastic energy, in contrast to the *dissipated* energy, is an ambiguous concept in a general anelastic or viscoelastic material; however, it is easily defined for a Maxwell, Kelvin-Voigt or standard linear solid, as well as all other materials whose rheology can be replicated by a network of linear springs and dashpots. Bland (1960) shows that the density of elastic energy stored in the springs of such a network during a cycle of harmonic oscillation at real frequency ω is

$$\begin{aligned} E &= \tfrac{1}{4}M_1(\omega)\,|\varepsilon(\omega)|^2 \\ &\qquad + \tfrac{1}{4}\mathrm{Re}\,\{[M(\omega) - \omega\partial_\omega M(\omega)]\varepsilon^2(\omega)\exp(2i\omega t)\} \\ &= \tfrac{1}{4}J_1(\omega)\,|\sigma(\omega)|^2 \\ &\qquad + \tfrac{1}{4}\mathrm{Re}\,\{[J(\omega) + \omega\partial_\omega J(\omega)]\sigma^2(\omega)\exp(2i\omega t)\}. \end{aligned} \tag{6.45}$$

As in the purely elastic case, the instantaneous energy density E consists of a constant part in addition to a term proportional to $\exp(2i\omega t)$, which averages to zero over a full cycle of the oscillation. Because of this, the intrinsic quality factor can be given an energetic interpretation—the quantity $4\pi Q^{-1}(\omega)$ is the *fractional average energy dissipated per cycle*:

$$4\pi Q^{-1} = \frac{1}{\langle E\rangle}\oint \dot{E}\,dt, \tag{6.46}$$

where

$$\langle E\rangle = \frac{\omega}{2\pi}\oint E\,dt. \tag{6.47}$$

As noted by O'Connell & Budiansky (1978), "it is curious ... but incontrovertible" that the stored energy in an anelastic material depends not only upon the values of the complex modulus or compliance at the frequency of interest, but also upon their derivatives $\partial_\omega M(\omega)$ and $\partial_\omega J(\omega)$ with respect to frequency. This precludes any simple interpretation of the quality factor in terms of the *maximum* stored energy instead. The interpretation (6.46) is, as we have noted, only valid for a material that can be modelled by a network of linear elastic springs and dashpots. Moreover, it pertains only to a purely sinusoidal or cosinusoidal variation at real frequency ω; such a harmonic variation is commonly employed to measure $Q(\omega)$ using a resonance apparatus in the laboratory, but it never occurs naturally in the Earth. The free oscillations excited by an earthquake oscillate like $\exp(i\omega t)$ but they also decay with time like $\exp(-\gamma t)$ subsequent to the cessation of faulting. The slow time scale of the decay relative to that of the oscillation, $\gamma \ll \omega$, permits a meaningful measurement of the Earth's intrinsic attenuation, as we shall see in Section 6.1.13.

Upon inserting equations (6.39)–(6.40) into the definition (6.42) and rearranging terms, we find that the inverse quality factor of a standard linear solid has the form of a Debye peak centered upon the geometric mean of the stress and strain relaxation times:

$$Q^{-1}(\omega) = \frac{\delta M}{\sqrt{M_{\mathrm{u}} M_{\mathrm{r}}}} \left(\frac{\omega \overline{\tau}}{1 + \omega^2 \overline{\tau}^2} \right), \tag{6.48}$$

where $\overline{\tau} = \sqrt{\tau_\sigma \tau_\varepsilon}$. The dissipation within such a material is negligible outside of the anelastic frequency band $1/10 \leq \omega\overline{\tau} \leq 10$.

★6.1.7 Kramers-Kronig relations

The causal nature of the response in the time domain gives rise to an important relation between the real and imaginary moduli $M_1(\omega)$ and $M_2(\omega)$ in the frequency domain. To derive this relation, we consider a complex function $f(\nu)$ with the following properties:

1. $f(\nu)$ is analytic in the lower half plane, $\mathrm{Im}\,\nu \leq 0$,
2. $f(\nu) \to 0$ as $\nu \to \infty$.

The first property together with Cauchy's theorem assures us that this function satisfies

$$\oint_C \frac{f(\nu)}{\nu - \omega}\, d\nu = 0, \tag{6.49}$$

where ω is an arbitrary real number, and C is the contour shown in Figure 6.7. The contribution from the large semi-circular arc vanishes in the

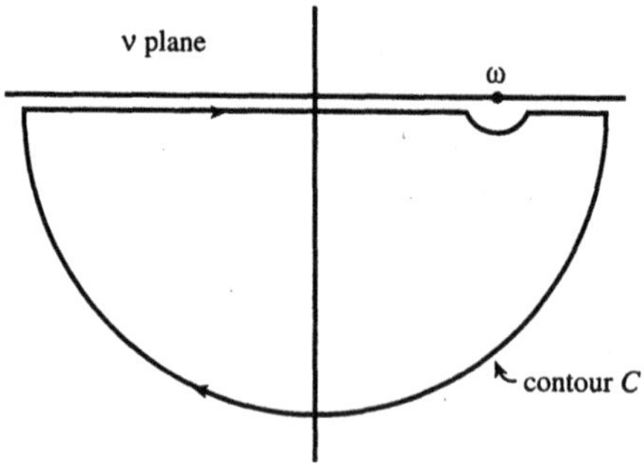

Figure 6.7. Integration contour C used in the derivation of the Kramers-Kronig relations (6.54)–(6.57).

limit $\nu \to \infty$ by virtue of the second property. We may thus rewrite equation (6.49) in the form

$$\lim_{\varepsilon\to 0}\int_{C_\varepsilon}\frac{f(\nu)}{\nu-\omega}\,d\nu+\lim_{\varepsilon\to 0}\left(\int_{-\infty}^{-\varepsilon}+\int_{\varepsilon}^{\infty}\right)\frac{f(\omega')}{\omega'-\omega}\,d\omega'=0, \tag{6.50}$$

where C_ε is the small semi-circular arc that excludes the point ω from the interior, and ω' is a real dummy integration variable. The first integral on the right of (6.50) can be evaluated by substituting $\nu=\omega+\varepsilon e^{i\theta}$:

$$\lim_{\varepsilon\to 0}\int_{C_\varepsilon}\frac{f(\nu)}{\nu-\omega}\,d\nu=\lim_{\varepsilon\to 0} i\int_{\pi}^{0} f(\omega+\varepsilon e^{i\theta})\,d\theta=-i\pi f(\omega). \tag{6.51}$$

The second term defines the *Cauchy principal value* of the otherwise improper integral, which we henceforth designate by a special symbol:

$$⨍_{-\infty}^{\infty}\frac{f(\omega')}{\omega'-\omega}\,d\omega'=\lim_{\varepsilon\to 0}\left(\int_{-\infty}^{-\varepsilon}+\int_{\varepsilon}^{\infty}\right)\frac{f(\omega')}{\omega'-\omega}\,d\omega'. \tag{6.52}$$

Combining equations (6.51) and (6.52), we see that $f(\omega)$ must satisfy

$$f(\omega)=\frac{1}{i\pi}⨍_{-\infty}^{\infty}\frac{f(\omega')}{\omega'-\omega}\,d\omega' \tag{6.53}$$

everywhere on the real axis.

Both $M(\nu)-M_{\rm u}$ and $J(\nu)-J_{\rm u}$ have the two properties required of $f(\nu)$ for any anelastic solid satisfying the constraints (6.9). Upon substituting $f(\omega)=M(\omega)-M_{\rm u}$ and $f(\omega)=J(\omega)-J_{\rm u}$ into equation (6.53) and equating the real and imaginary parts, we obtain

$$M_1(\omega)-M_{\rm u}=\frac{1}{\pi}⨍_{-\infty}^{\infty}\frac{M_2(\omega')}{\omega'-\omega}\,d\omega', \tag{6.54}$$

$$M_2(\omega)=-\frac{1}{\pi}⨍_{-\infty}^{\infty}\frac{M_1(\omega')-M_{\rm u}}{\omega'-\omega}\,d\omega', \tag{6.55}$$

$$J_1(\omega)-J_{\rm u}=\frac{1}{\pi}⨍_{-\infty}^{\infty}\frac{J_2(\omega')}{\omega'-\omega}\,d\omega', \tag{6.56}$$

$$J_2(\omega)=-\frac{1}{\pi}⨍_{-\infty}^{\infty}\frac{J_1(\omega')-J_{\rm u}}{\omega'-\omega}\,d\omega'. \tag{6.57}$$

The quantities $M_1(\omega)-M_{\rm u}$ and $M_2(\omega)$ are *Hilbert transforms* of each other, as are the quantities $J_1(\omega)-J_{\rm u}$ and $J_2(\omega)$. Equations (6.54)–(6.55) and (6.56)–(6.57), which were first introduced in a study of dielectric relaxation, are known as the *Kramers-Kronig relations*. It is noteworthy that the modulus relations (6.54)–(6.55) are valid even in the absence of a finite equilibrium response, i.e., even if $M_{\mathrm{r}} = 0$, as long as $M_{\mathrm{u}} < \infty$. The compliance relations (6.56)–(6.57) are likewise valid even in the absence of an instantaneous elastic response, i.e., even if $J_{\mathrm{r}} = 0$, as long as $J_{\mathrm{u}} < \infty$. Both pairs of relations are simultaneously valid only for an anelastic solid satisfying the constraints (6.9).

By using the even and odd symmetries (6.28) and (6.29), we can rewrite the Kramers-Kronig relations in the alternative form

$$M_1(\omega) - M_{\mathrm{u}} = \frac{2}{\pi} \int_0^{\infty} \frac{\omega' M_2(\omega')}{(\omega' + \omega)(\omega' - \omega)}\, d\omega', \tag{6.58}$$

$$M_2(\omega) = -\frac{2}{\pi} \int_0^{\infty} \frac{\omega' [M_1(\omega') - M_{\mathrm{u}}]}{(\omega' + \omega)(\omega' - \omega)}\, d\omega', \tag{6.59}$$

$$J_1(\omega) - J_{\mathrm{u}} = \frac{2}{\pi} \int_0^{\infty} \frac{\omega' J_2(\omega')}{(\omega' + \omega)(\omega' - \omega)}\, d\omega', \tag{6.60}$$

$$J_2(\omega) = -\frac{2}{\pi} \int_0^{\infty} \frac{\omega' [J_1(\omega') - J_{\mathrm{u}}]}{(\omega' + \omega)(\omega' - \omega)}\, d\omega'. \tag{6.61}$$

We see from (6.58)–(6.61) that knowledge of $M_2(\omega)$ and $J_2(\omega)$ for all positive frequencies $0 < \omega < \infty$ implies knowledge of $M_1(\omega) - M_{\mathrm{u}}$ and $J_1(\omega) - J_{\mathrm{u}}$, and vice versa.

6.1.8 Relaxation and retardation spectrum

The intrinsic quality factor $Q(\omega)$ in the Earth is observed to be roughly independent of frequency within the band from 0.3 mHz to 1 Hz. Because of this, a constant-Q material is of particular interest in global seismology. A useful phenomenological model of such a material can be constructed by considering a superposition of standard linear solids with a continuous spectrum of relaxation times.

Generalizing equations (6.35)–(6.36), we write the stress relaxation and creep functions in the form

$$M(t) = M_{\mathrm{u}} - \int_0^{\infty} \tau^{-1} Y(\tau)[1 - \exp(-t/\tau)]\, d\tau, \tag{6.62}$$

$$J(t) = J_{\mathrm{u}} + \int_0^{\infty} \tau^{-1} X(\tau)[1 - \exp(-t/\tau)]\, d\tau. \tag{6.63}$$

The complex modulus and compliance (6.37)–(6.38) are likewise generalized to

$$M(\nu) = M_{\mathrm{u}} - \int_0^{\infty} \tau^{-1} Y(\tau)(1 + i\nu\tau)^{-1}\, d\tau, \tag{6.64}$$

$$J(\nu) = J_{\rm u} + \int_0^\infty \tau^{-1} X(\tau)(1 + i\nu\tau)^{-1}\, d\tau. \tag{6.65}$$

For real frequencies ω, the real and imaginary parts of (6.64)–(6.65) are given by

$$M_1(\omega) = M_{\rm u} - \int_0^\infty \frac{Y(\tau)}{1+\omega^2\tau^2}\,\frac{d\tau}{\tau}, \tag{6.66}$$

$$M_2(\omega) = \int_0^\infty Y(\tau)\,\frac{\omega\tau}{1+\omega^2\tau^2}\,\frac{d\tau}{\tau}, \tag{6.67}$$

$$J_1(\omega) = J_{\rm u} + \int_0^\infty \frac{X(\tau)}{1+\omega^2\tau^2}\,\frac{d\tau}{\tau}, \tag{6.68}$$

$$J_2(\omega) = \int_0^\infty X(\tau)\,\frac{\omega\tau}{1+\omega^2\tau^2}\,\frac{d\tau}{\tau}. \tag{6.69}$$

The quantities $Y(\tau)$ and $X(\tau)$, which are called the *relaxation and retardation spectra*, respectively, satisfy

$$\delta M = \int_0^\infty \tau^{-1} Y(\tau)\, d\tau, \qquad \delta J = \int_0^\infty \tau^{-1} X(\tau)\, d\tau. \tag{6.70}$$

Evidently, $\tau^{-1}Y(\tau)$ is a measure of the contribution to the modulus defect δM and $\tau^{-1}X(\tau)$ is a measure of the contribution to the relaxation of the compliance δJ from anelastic processes in the material with relaxation times between τ and $\tau + d\tau$. The "extra" factor of τ^{-1} in the integrands is conventional. Materials whose anelastic properties are described by equations of the form (6.66)–(6.69) are referred to as *absorption-band solids*.

6.1.9 Approximate relations

The Kramers-Kronig relations (6.58)–(6.61) prohibit the existence of any exact local relationship between the real and imaginary moduli $M_1(\omega)$ and $M_2(\omega)$ or the real and imaginary compliances $J_1(\omega)$ and $J_2(\omega)$. There is, however, a simple approximate relationship which is valid whenever the relaxation and retardation spectra $Y(\tau)$ and $X(\tau)$ are slowly varying over a broad range of relaxation times.

As we have seen, the Debye peak function $\omega\tau/(1+\omega^2\tau^2)$ is negligible outside of the two-decade range $1/10 \le \omega\tau \le 10$. If the relaxation spectrum $Y(\tau)$ does not vary significantly within this range, we can take it outside of the integral in equation (6.67); this leads to the approximation

$$M_2(\omega) \approx \omega Y(1/\omega) \int_0^\infty \frac{d\tau}{1+\omega^2\tau^2} = \tfrac{1}{2}\pi Y(1/\omega). \tag{6.71}$$

The partial derivative of the real modulus (6.66) with respect to frequency can likewise be approximated by

$$\begin{aligned}\frac{\partial M_1(\omega)}{\partial\omega} &= 2\omega\int_0^\infty \frac{\tau Y(\tau)}{(1+\omega^2\tau^2)^2}\,d\tau \\ &\approx 2Y(1/\omega)\int_0^\infty \frac{\tau\,d\tau}{(1+\omega^2\tau^2)^2} = \omega^{-1}Y(1/\omega).\end{aligned} \tag{6.72}$$

The function $1/(1+\omega^2\tau^2)$ in equation (6.66) has a steplike character, decreasing from one to zero in the same range $1/10 \leq \omega\tau \leq 10$. Approximating this steplike function by an abrupt step at $\omega\tau = 1$ yields

$$M_1(\omega) \approx M_{\mathrm{u}} - \int_0^{1/\omega} \tau^{-1}Y(\tau)\,d\tau. \tag{6.73}$$

Differentiation of (6.73) reproduces (6.72) so that these two approximations are self-consistent. Comparison of the two results (6.71) and (6.72) leads to the desired relation between the real and imaginary moduli:

$$\frac{\partial M_1(\omega)}{\partial\omega} \approx \frac{2M_2(\omega)}{\pi\omega}, \tag{6.74}$$

or, equivalently,

$$\frac{\partial \ln M_1(\omega)}{\partial \ln\omega} \approx \frac{2}{\pi Q(\omega)}, \tag{6.75}$$

where we have used the definition (6.42). The approximation (6.75) relates the logarithmic dispersion of the real modulus $\partial \ln M_1(\omega)/\partial \ln\omega$ to the quality factor $Q(\omega)$ at the same frequency ω. This is a useful approximation for any $Q(\omega)$ that does not vary strongly with frequency.

Upon integrating the result (6.75) we obtain

$$\frac{M_1(\omega)}{M_1(\omega_0)} \approx 1 + \frac{2}{\pi}\int_{\omega_0}^{\omega} \frac{d\omega'}{\omega' Q(\omega')}, \tag{6.76}$$

where it has been assumed that ω and ω_0 are positive, and that $Q(\omega) \gg 1$. An analogous result obviously pertains to the real compliance, but with a minus sign, namely,

$$\frac{J_1(\omega)}{J_1(\omega_0)} \approx 1 - \frac{2}{\pi}\int_{\omega_0}^{\omega} \frac{d\omega'}{\omega' Q(\omega')}. \tag{6.77}$$

A *constant-Q material* having $Q(\omega) \approx Q$ in the interval between ω and ω_0 exhibits *logarithmic dispersion* of the form

$$\frac{M_1(\omega)}{M_1(\omega_0)} \approx 1 + \frac{2}{\pi Q}\ln\left(\frac{\omega}{\omega_0}\right), \tag{6.78}$$

$$\frac{J_1(\omega)}{J_1(\omega_0)} \approx 1 - \frac{2}{\pi Q} \ln\left(\frac{\omega}{\omega_0}\right). \tag{6.79}$$

Upon comparing equations (6.76) and (6.77), we see that the real modulus and compliance are approximate inverses in the case $Q(\omega) \gg 1$, i.e.,

$$M_1(\omega)J_1(\omega) \approx 1. \tag{6.80}$$

Equation (6.80) is an approximation to the exact modulus-compliance relation $M_1(\omega)J_1(\omega) + M_2(\omega)J_2(\omega) = 1$, which follows from (6.19).

Similar approximations can be developed for the stress relaxation function $M(t)$ and the creep function $J(t)$. The function $1 - \exp(-t/\tau)$ in equation (6.62) is also a steplike function, which increases from zero to one over the range $1/10 \leq t/\tau \leq 10$. Replacing this by an abrupt step at $t/\tau = 1$, we obtain the result

$$M(t) \approx M_{\mathrm{u}} - \int_0^t \tau^{-1} Y(\tau)\, d\tau. \tag{6.81}$$

Comparing this with equation (6.73), we see that there is a simple relation between the stress relaxation function $M(t)$ and the real modulus $M_1(\omega)$:

$$M(t) \approx M_1(\omega)\big|_{\omega=1/t}. \tag{6.82}$$

The creep function $J(t)$ and the real compliance $J_1(\omega)$ likewise satisfy

$$J(t) \approx J_1(\omega)\big|_{\omega=1/t}. \tag{6.83}$$

A final relation of interest can be obtained by combining the three approximations (6.80) and (6.82)–(6.83), namely,

$$M(t)J(t) \approx 1. \tag{6.84}$$

It is interesting to compare this with the exact result (6.14).

⋆6.1.10 Constant-Q absorption-band model

Following Liu, Anderson & Kanamori (1976) and Kanamori & Anderson (1977), we can construct a constant-Q model within a specified frequency band $1/\tau_{\mathrm{M}} \ll \omega \ll 1/\tau_{\mathrm{m}}$ by choosing a boxcar for either the relaxation or retardation spectrum:

$$Y(\tau) = \begin{cases} \delta M/\ln(\tau_{\mathrm{M}}/\tau_{\mathrm{m}}) & \text{if } \tau_{\mathrm{m}} \leq \tau \leq \tau_{\mathrm{M}} \\ 0 & \text{otherwise,} \end{cases} \tag{6.85}$$

$$X(\tau) = \begin{cases} \delta J/\ln(\tau_{\mathrm{M}}/\tau_{\mathrm{m}}) & \text{if } \tau_{\mathrm{m}} \leq \tau \leq \tau_{\mathrm{M}} \\ 0 & \text{otherwise.} \end{cases} \tag{6.86}$$

Upon inserting (6.85)–(6.86) into equations (6.64)–(6.65) and evaluating the integrals, we obtain

$$M(\nu) = M_{\rm r} + \left[\frac{\delta M}{\ln(\tau_{\rm M}/\tau_{\rm m})}\right] \ln\left(\frac{i\nu\tau_{\rm M}+1}{i\nu\tau_{\rm m}+1}\right), \tag{6.87}$$

$$J(\nu) = J_{\rm r} - \left[\frac{\delta J}{\ln(\tau_{\rm M}/\tau_{\rm m})}\right] \ln\left(\frac{i\nu\tau_{\rm M}+1}{i\nu\tau_{\rm m}+1}\right), \tag{6.88}$$

where $M_{\rm r} = M_{\rm u} - \delta M$ and $J_{\rm r} = J_{\rm u} + \delta J$ as usual. The complex modulus (6.87) and compliance (6.88) are analytic everywhere except on the positive imaginary axis, where they exhibit *logarithmic branch cuts* lying between the points $\nu = i/\tau_{\rm M}$ and $\nu = i/\tau_{\rm m}$. For real frequencies ω, the real and imaginary parts of (6.87) and (6.88) are given by

$$M_1(\omega) = M_{\rm r} + \tfrac{1}{2}\left[\frac{\delta M}{\ln(\tau_{\rm M}/\tau_{\rm m})}\right] \ln\left(\frac{1+\omega^2\tau_{\rm M}^2}{1+\omega^2\tau_{\rm m}^2}\right), \tag{6.89}$$

$$M_2(\omega) = \left[\frac{\delta M}{\ln(\tau_{\rm M}/\tau_{\rm m})}\right] \arctan\left[\frac{\omega(\tau_{\rm M}-\tau_{\rm m})}{1+\omega^2\tau_{\rm m}\tau_{\rm M}}\right] \tag{6.90}$$

and

$$J_1(\omega) = J_{\rm r} - \tfrac{1}{2}\left[\frac{\delta J}{\ln(\tau_{\rm M}/\tau_{\rm m})}\right] \ln\left(\frac{1+\omega^2\tau_{\rm M}^2}{1+\omega^2\tau_{\rm m}^2}\right), \tag{6.91}$$

$$J_2(\omega) = \left[\frac{\delta J}{\ln(\tau_{\rm M}/\tau_{\rm m})}\right] \arctan\left[\frac{\omega(\tau_{\rm M}-\tau_{\rm m})}{1+\omega^2\tau_{\rm m}\tau_{\rm M}}\right]. \tag{6.92}$$

The two four-parameter prescriptions (6.89)–(6.90) and (6.91)–(6.92) are equivalent only in the case of a wide-band, weak-attenuation medium, satisfying

$$\tau_{\rm m} \ll \tau_{\rm M}, \qquad \delta M \ll M_{\rm u}, M_{\rm r}, \qquad \delta J \ll J_{\rm u}, J_{\rm r}. \tag{6.93}$$

The quality factor $Q(\omega) = M_1(\omega)/M_2(\omega) = J_1(\omega)/J_2(\omega)$ of such a medium is essentially independent of frequency within the band $1/\tau_{\rm M} \ll \omega \ll 1/\tau_{\rm m}$, as illustrated in Figure 6.8. Its value within this *constant-Q absorption band* is given by any of the equivalent relations

$$\frac{2}{\pi Q}\ln\left(\frac{\tau_{\rm M}}{\tau_{\rm m}}\right) \approx \frac{\delta M}{M_{\rm u}} \approx \frac{\delta M}{M_{\rm r}} \approx \frac{\delta J}{J_{\rm u}} \approx \frac{\delta J}{M_{\rm r}}. \tag{6.94}$$

The conditions (6.93) guarantee that $Q \gg 1$. The real modulus (6.89) and compliance (6.91) are well approximated within the constant-Q band by

$$M_1(\omega) \approx M_{\rm u}\left[1 + \frac{2}{\pi Q}\ln(\omega\tau_{\rm m})\right] \approx M_{\rm r}\left[1 + \frac{2}{\pi Q}\ln(\omega\tau_{\rm M})\right], \tag{6.95}$$

$$J_1(\omega) \approx J_u\left[1 - \frac{2}{\pi Q}\ln(\omega\tau_m)\right] \approx J_r\left[1 - \frac{2}{\pi Q}\ln(\omega\tau_M)\right]. \tag{6.96}$$

It is noteworthy that $M_1(\omega)J_1(\omega) \approx 1$, in accordance with equation (6.80). The stress relaxation function $M(t)$ and creep function $J(t)$ corresponding to the modulus (6.87) and compliance (6.88) may be determined by inserting the spectra (6.85)–(6.86) into the representations (6.62)–(6.63). This leads to a pair of exponential integrals that can be approximated by

$$M(t) \approx M_u\left[1 - \frac{2}{\pi Q}\ln\left(\frac{t}{\tau_m}\right)\right] \approx M_r\left[1 - \frac{2}{\pi Q}\ln\left(\frac{t}{\tau_M}\right)\right], \tag{6.97}$$

$$J(t) \approx J_u\left[1 + \frac{2}{\pi Q}\ln\left(\frac{t}{\tau_m}\right)\right] \approx J_r\left[1 + \frac{2}{\pi Q}\ln\left(\frac{t}{\tau_M}\right)\right], \tag{6.98}$$

in the interval $\tau_m \ll t \ll \tau_M$. The results (6.97)–(6.98) are consistent with the approximations (6.82)–(6.84), as expected. Logarithmic behavior of the form (6.98) was noted in early laboratory observations of rock deformation by Lomnitz (1956; 1957), and is known in geophysical circles as *Lomnitz's law of creep*.

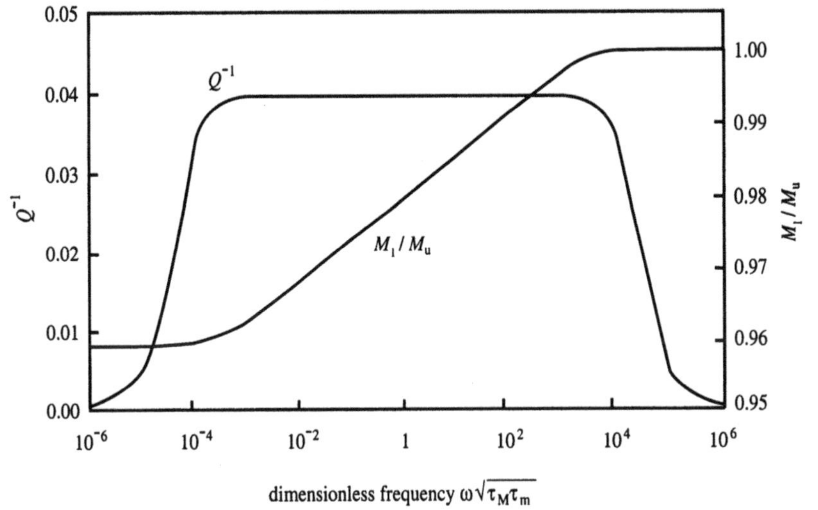

Figure 6.8. Frequency variation of the dimensionless real modulus $M_1(\omega)/M_u$ and inverse quality factor $Q^{-1}(\omega)$ of a typical absorption-band solid. The ratio of the upper and lower band edges in this case is $\tau_M/\tau_m = 10^7$, and the quality factor within the absorption band is $Q = 250$; the resulting dimensionless modulus defect $\delta M/M_u$ is 4.1 percent.

★6.1.11 Strictly constant Q model

No anelastic material can have $Q(\omega)$ exactly constant for all frequencies $0 < \omega < \infty$; however, it is easy find such a material if we relax the anelastic constraints (6.9). Following Kjartansson (1979), we consider the creep function

$$J(t) = \frac{(\omega_0 t)^q}{M_0\,\Gamma(1+q)} H(t), \tag{6.99}$$

where ω_0, M_0 and q are positive constants, and $\Gamma(1+q)$ denotes the gamma function. The complex compliance corresponding to (6.99) is easily found from (6.18) to be

$$J(\nu) = \frac{1}{M_0}\left(\frac{i\nu}{\omega_0}\right)^{-q}. \tag{6.100}$$

The complex modulus must be the reciprocal of (6.100), namely,

$$M(\nu) = M_0\left(\frac{i\nu}{\omega_0}\right)^{q}. \tag{6.101}$$

Upon inverting the Fourier transform (6.101) to find the stress relaxation function, we obtain

$$M(t) = \frac{M_0(\omega_0 t)^{-q}}{\Gamma(1-q)} H(t). \tag{6.102}$$

A viscoelastic material characterized by the four relations (6.99)–(6.102) has $J_{\mathrm{u}} = 1/M_{\mathrm{u}} = 0$ and $J_{\mathrm{r}} = 1/M_{\mathrm{r}} = \infty$; hence, it has neither an instantaneous elastic response nor a long-term equilibrium response. The constant M_0 is seen to be the value of the absolute modulus $|M(\omega)|$ at the real fiducial frequency $\omega = \omega_0$.

The real and imaginary parts of the complex modulus (6.101) and compliance (6.100) are

$$M_1(\omega) = M_0\left|\frac{\omega}{\omega_0}\right|^{q} \cos\left(\tfrac{1}{2}\pi q \operatorname{sgn}\omega\right), \tag{6.103}$$

$$M_2(\omega) = M_0\left|\frac{\omega}{\omega_0}\right|^{q} \sin\left(\tfrac{1}{2}\pi q \operatorname{sgn}\omega\right), \tag{6.104}$$

$$J_1(\omega) = \frac{1}{M_0}\left|\frac{\omega}{\omega_0}\right|^{-q} \cos\left(\tfrac{1}{2}\pi q \operatorname{sgn}\omega\right), \tag{6.105}$$

$$J_2(\omega) = \frac{1}{M_0}\left|\frac{\omega}{\omega_0}\right|^{-q} \sin\left(\tfrac{1}{2}\pi q \operatorname{sgn}\omega\right), \tag{6.106}$$

where ω is presumed to be real, and $\operatorname{sgn}\omega$ is the sign function:

$$\operatorname{sgn}\omega = \begin{cases} +1 & \text{if } \omega > 0 \\ -1 & \text{if } \omega < 0. \end{cases} \tag{6.107}$$

As advertised, the quality factor $Q(\omega) = M_1(\omega)/M_2(\omega) = J_1(\omega)/J_2(\omega)$ is strictly independent of frequency:

$$Q(\omega) = Q \operatorname{sgn}\omega \quad \text{where} \quad Q^{-1} = \tan(\tfrac{1}{2}\pi q). \tag{6.108}$$

The discontinuity at $\omega = 0$ guarantees that $Q(\omega)$ has the requisite odd symmetry $Q(-\omega) = -Q(\omega)$. The dispersion relation for real positive frequencies, $\omega > 0$, is

$$\frac{M_1(\omega)}{M_1(\omega_0)} = \left(\frac{\omega}{\omega_0}\right)^q \approx 1 + \frac{2}{\pi Q}\ln\left(\frac{\omega}{\omega_0}\right), \tag{6.109}$$

$$\frac{J_1(\omega)}{J_1(\omega_0)} = \left(\frac{\omega}{\omega_0}\right)^{-q} \approx 1 - \frac{2}{\pi Q}\ln\left(\frac{\omega}{\omega_0}\right), \tag{6.110}$$

where the final approximation in both cases is valid for $Q \gg 1$.

The agreement among the three independent expressions (6.78)–(6.79), (6.95)–(6.96) and (6.109)–(6.110), despite the very different assumptions used to derive them, is indicative of the robustness of this fundamental result. The logarithmic nature of the dispersion is insensitive to the details of the anelastic behavior outside of the band between the two frequencies being compared; the only important requirement is that $Q(\omega)$ be independent of frequency within the band.

*6.1.12 Power-law Q model

Following Anderson & Minster (1979), we can allow for a weak frequency dependence of $Q(\omega)$ by considering a relaxation or retardation spectrum of the form

$$Y(\tau) = \begin{cases} \alpha\,(\delta M)\tau^{\alpha}/(\tau_{\mathrm{M}}^{\alpha} - \tau_{\mathrm{m}}^{\alpha}) & \text{if } \tau_{\mathrm{m}} \le \tau \le \tau_{\mathrm{M}} \\ 0 & \text{otherwise,} \end{cases} \tag{6.111}$$

$$X(\tau) = \begin{cases} \alpha\,(\delta J)\tau^{\alpha}/(\tau_{\mathrm{M}}^{\alpha} - \tau_{\mathrm{m}}^{\alpha}) & \text{if } \tau_{\mathrm{m}} \le \tau \le \tau_{\mathrm{M}} \\ 0 & \text{otherwise,} \end{cases} \tag{6.112}$$

where $0 < \alpha \ll 1$. We insert the spectra (6.111)–(6.112) into equations (6.66)–(6.69) and use the approximations (6.93) to evaluate the integrals in the absorption band $1/\tau_{\mathrm{M}} \ll \omega \ll 1/\tau_{\mathrm{m}}$. The resulting quality factor $Q(\omega) = M_1(\omega)/M_2(\omega) = J_1(\omega)/J_2(\omega)$ exhibits a power-law dependence upon frequency:

$$Q(\omega) \approx Q_0(\omega/\omega_0)^\alpha, \tag{6.113}$$

where $Q_0 = Q(\omega_0)$. The real modulus and compliance associated with an attenuation law of the form (6.113) are

$$\frac{M_1(\omega)}{M_1(\omega_0)} \approx 1 + \frac{2}{\alpha\pi Q_0}\left[1 - \left(\frac{\omega}{\omega_0}\right)^{-\alpha}\right], \tag{6.114}$$

$$\frac{J_1(\omega)}{J_1(\omega_0)} \approx 1 - \frac{2}{\alpha\pi Q_0}\left[1 - \left(\frac{\omega}{\omega_0}\right)^{-\alpha}\right]. \tag{6.115}$$

Equations (6.114)–(6.115) reduce to the logarithmic constant-Q result in the frequency-independent limit $\alpha \to 0$, and they agree with the general dispersion law (6.76)–(6.77) in the case of a weak frequency dependence of the form (6.113), as expected.

★6.1.13 Behavior near the real frequency axis

A freely decaying normal mode "sees" the anelastic modulus and compliance of the Earth at its complex frequency $\nu = \omega + i\gamma$. Because the decay time is much longer than the natural period of oscillation, it is of interest to investigate the behavior of $M(\nu)$ and $J(\nu)$ in the vicinity of the real positive frequency axis. Following O'Connell & Budiansky (1978), we write the modulus in terms of its real and imaginary parts in the form

$$\begin{aligned} M(\omega + i\gamma) &= F(\omega,\gamma) + iG(\omega,\gamma) \\ &\approx F(\omega,0) + \gamma\partial_\gamma F(\omega,0) + i[G(\omega,0) + \gamma\partial_\gamma G(\omega,0)], \end{aligned} \tag{6.116}$$

where the approximation is valid for $\gamma \ll \omega$. Assuming that $M(\omega + i\gamma)$ is analytic in the vicinity of the real axis, we can employ the Cauchy-Riemann equations to rewrite (6.116) in the form

$$\begin{aligned} M(\omega + i\gamma) &\approx F(\omega,0) - \gamma\partial_\omega G(\omega,0) + i[G(\omega,0) + \gamma\partial_\omega F(\omega,0)] \\ &= M_1(\omega) - \gamma\partial_\omega M_2(\omega) + i[M_2(\omega) + \gamma\partial_\omega M_1(\omega)]. \end{aligned} \tag{6.117}$$

Upon substituting the definition (6.42) and the approximation (6.74) into equation (6.117) we obtain

$$\begin{aligned} M(\omega + i\gamma) \approx\ & M_1(\omega)\{1 + Q^{-2}(\omega)[\gamma\partial_\omega Q(\omega) - 2\gamma/\pi\omega]\} \\ & + iM_1(\omega)Q^{-1}(\omega)(1 + 2\gamma/\pi\omega), \end{aligned} \tag{6.118}$$

which is correct to second order in Q^{-1} and γ/ω. The compliance is given to the same order in these quantities by

$$J(\omega + i\gamma) \approx J_1(\omega)\{1 + Q^{-2}(\omega)[\gamma\partial_\omega Q(\omega) - 2\gamma/\pi\omega]\} \\ - iJ_1(\omega)Q^{-1}(\omega)(1 + 2\gamma/\pi\omega). \tag{6.119}$$

Even more simply, we can ignore the second-order terms in (6.118)–(6.119) and write, correct to first order in Q^{-1} and γ/ω:

$$M(\omega + i\gamma) \approx M_1(\omega)[1 + iQ^{-1}(\omega)], \tag{6.120}$$

$$J(\omega + i\gamma) \approx J_1(\omega)[1 - iQ^{-1}(\omega)]. \tag{6.121}$$

Since $Q(\omega) \gg 1$ and, thus, $\gamma \ll \omega$ for the Earth, the latter two approximations are generally considered adequate for seismological purposes. The results (6.120)–(6.121) simply state that every normal mode "sees" the anelasticity of the Earth at its real frequency of oscillation ω. This is in accord with elementary physical intuition.

6.1.14 Bulk and shear quality factors

In summary, the dissipation of seismic energy can be accounted for in an isotropic, hydrostatic Earth model by replacing the elastic incompressibility κ and rigidity μ by complex, frequency-dependent, anelastic parameters. The frequency-domain constitutive relation equivalent to equation (6.2) in such an Earth is

$$\mathbf{T}^{\mathrm{L1}}(\omega) = \kappa(\omega)[1 + iQ_\kappa^{-1}(\omega)]\theta(\omega)\mathbf{I} \\ + 2\mu(\omega)[1 + iQ_\mu^{-1}(\omega)]\mathbf{d}(\omega), \tag{6.122}$$

where we have dropped the subscript 1 on the real moduli $\kappa(\omega)$ and $\mu(\omega)$ to achieve a cleaner notation. The quantities $Q_\kappa(\omega)$ and $Q_\mu(\omega)$ are referred to as the *bulk and shear quality factors*, respectively.

More generally, we shall write the incremental Lagrangian Cauchy stress in an arbitrary hydrostatic Earth model in the form

$$\mathbf{T}^{\mathrm{L1}}(\omega) = \mathbf{\Gamma}(\omega) : \boldsymbol{\varepsilon}(\omega), \tag{6.123}$$

where

$$\Gamma_{ijkl}(\omega) = \{\kappa(\omega)[1 + iQ_\kappa^{-1}(\omega)] - \tfrac{2}{3}\mu(\omega)[1 + iQ_\mu^{-1}(\omega)]\}\,\delta_{ij}\,\delta_{kl} \\ + \mu(\omega)[1 + iQ_\mu^{-1}(\omega)]\,(\delta_{ik}\,\delta_{jl} + \delta_{il}\,\delta_{jk}) + \gamma_{ijkl}, \tag{6.124}$$

and we shall write the incremental Piola-Kirchhoff stress in an arbitrary non-hydrostatic Earth model in the form

$$\mathbf{T}^{\mathrm{PK1}}(\omega) = \mathbf{\Lambda}(\omega) : \boldsymbol{\nabla}\mathbf{s}(\omega), \tag{6.125}$$

where

$$\Lambda_{ijkl}(\omega) = \{\kappa(\omega)[1 + iQ_\kappa^{-1}(\omega)] - \tfrac{2}{3}\mu(\omega)[1 + iQ_\mu^{-1}(\omega)]\}\,\delta_{ij}\,\delta_{kl}$$
$$+ \mu(\omega)[1 + iQ_\mu^{-1}(\omega)]\,(\delta_{ik}\,\delta_{jl} + \delta_{il}\,\delta_{jk}) + \gamma_{ijkl} \qquad (6.126)$$
$$+ \tfrac{1}{2}(T_{ij}^0\,\delta_{kl} + T_{kl}^0\,\delta_{ij} + T_{ik}^0\,\delta_{jl} - T_{jk}^0\,\delta_{il} - T_{il}^0\delta_{jk} - T_{jl}^0\,\delta_{ik}).$$

The anisotropic tensor γ in (6.124) and (6.126) is regarded as purely elastic; the only complex dispersive terms in $\mathbf{\Gamma}(\omega)$ and $\mathbf{\Lambda}(\omega)$ are thus the incompressibility $\kappa(\omega)[1 + iQ_\kappa^{-1}(\omega)]$ and rigidity $\mu(\omega)[1 + iQ_\mu^{-1}(\omega)]$. Equations (6.123) and (6.125) are consistent with the principle of material frame indifference (2.148)–(2.149), as of course they must be. A fully anisotropic theory of anelasticity, which replaces γ by a complex frequency-dependent tensor $\gamma(\omega)$, could easily be formulated; however, there is no seismic evidence that such a theory is required at the present time. Both the Earth's isotropic anelasticity, embodied in $Q_\kappa^{-1}(\omega)$ and $Q_\mu^{-1}(\omega)$, and its elastic anisotropy, embodied in the real tensor γ, are slight; even carefully controlled experiments provide rather limited information regarding the distribution of these properties. Anisotropic anelasticity—although undoubtedly present in any polycrystalline solid—is "doubly slight". For this reason it is common to regard the anelasticity of the Earth as isotropic.

To a first approximation, the bulk and shear quality factors in equations (6.124) and (6.126) can be regarded as frequency-independent:

$$Q_\kappa(\omega) \approx Q_\kappa, \qquad Q_\mu(\omega) \approx Q_\mu, \qquad (6.127)$$

for positive frequencies, $\omega > 0$. The associated dispersion is in that case logarithmic:

$$\frac{\kappa(\omega)}{\kappa(\omega_0)} \approx 1 + \frac{2}{\pi Q_\kappa} \ln\left(\frac{\omega}{\omega_0}\right), \qquad (6.128)$$

$$\frac{\mu(\omega)}{\mu(\omega_0)} \approx 1 + \frac{2}{\pi Q_\mu} \ln\left(\frac{\omega}{\omega_0}\right). \qquad (6.129)$$

We shall employ the constant-Q logarithmic dispersion law (6.128)–(6.129) throughout most of the rest of this book, for simplicity. Any weak frequency dependence of the bulk or shear quality factors within the frequency band of interest in global seismology can be taken into account using the more general approximations (6.76) or (6.114) if desired.

6.2 Non-Rotating Anelastic Earth

We show in this section and the next that the results in Chapter 4 can be generalized to account for anelasticity. There is no need in these purely formal considerations to restrict attention to an isotropic constant-Q Earth, or to make the approximation that a mode "sees" the anelasticity at its real frequency of oscillation ω. In fact, it is notationally simpler to consider a general anelastic Earth model governed by a complex, frequency-dependent constitutive relation of the form

$$\mathbf{T}^{\mathrm{PK1}}(\nu) = \mathbf{\Lambda}(\nu) : \boldsymbol{\nabla}\mathbf{s}(\nu) \quad \text{where} \quad \mathbf{\Lambda}(-\nu^*) = \mathbf{\Lambda}^*(\nu). \tag{6.130}$$

We shall specialize to the case of actual interest only occasionally. It is again convenient to treat the cases of a non-rotating Earth and a rotating Earth separately; we take up the simpler non-rotating ($\mathbf{\Omega} = \mathbf{0}$) case first.

6.2.1 Duality and biorthonormality

The complex eigenfrequencies ν and associated eigenfunctions $\mathbf{s}$ of a non-rotating anelastic Earth are found by solving the abstract operator equation

$$\mathcal{H}(\nu)\mathbf{s} = \nu^2\mathbf{s}, \tag{6.131}$$

where $\mathcal{H}(\nu)$ stands for the integro-differential operator

$$\rho^0\mathcal{H}(\nu)\mathbf{s} = -\boldsymbol{\nabla}\cdot[\mathbf{\Lambda}(\nu) : \boldsymbol{\nabla}\mathbf{s}] + \rho^0\boldsymbol{\nabla}\phi^{\mathrm{E1}} + \rho^0\mathbf{s}\cdot\boldsymbol{\nabla}\boldsymbol{\nabla}\phi^0 \tag{6.132}$$

in the Earth model $\oplus$, together with the associated boundary conditions on $\partial\oplus$, Σ_{SS} and Σ_{FS}. The anelastic-gravitational operator (6.132) has the dynamical symmetry

$$\mathcal{H}(-\nu^*) = \mathcal{H}^*(\nu), \tag{6.133}$$

where $\rho^0\mathcal{H}^*(\nu)\mathbf{s} = -\boldsymbol{\nabla}\cdot[\mathbf{\Lambda}^*(\nu) : \boldsymbol{\nabla}\mathbf{s}] + \rho^0\boldsymbol{\nabla}\phi^{\mathrm{E1}} + \rho^0\mathbf{s}\cdot\boldsymbol{\nabla}\boldsymbol{\nabla}\phi^0$. Upon taking the complex conjugate of equation (6.131) and using the result (6.133), we observe that $-\nu^*$, $\mathbf{s}^*$ is an eigensolution of a non-rotating anelastic Earth if and only if ν, $\mathbf{s}$ is.

The *inner product* of any two piecewise smooth complex functions $\mathbf{s}$ and $\mathbf{s}'$ in $\oplus$ is defined by

$$\langle\mathbf{s}, \mathbf{s}'\rangle = \int_{\oplus} \rho^0\mathbf{s}^*\cdot\mathbf{s}'\, dV. \tag{6.134}$$

By an argument analogous to that on an elastic, non-rotating Earth in Section 4.1.1, we can use the thermodynamic symmetry $\Lambda_{ijkl}(\nu) = \Lambda_{klij}(\nu)$ to show that

$$\langle\mathbf{s}, \mathcal{H}(\nu)\mathbf{s}'\rangle = \langle\mathcal{H}^*(\nu)\mathbf{s}, \mathbf{s}'\rangle = \langle\mathbf{s}', \mathcal{H}^*(\nu)\mathbf{s}\rangle^*. \tag{6.135}$$

The operator $\mathcal{H}^*(\nu)$ is thus the *Hermitian adjoint* of the operator $\mathcal{H}(\nu)$. We can consider $\mathbf{s}^*$ to be either an eigenfunction of the original operator

$\mathcal{H}(-\nu^*)$ with associated eigenfrequency $-\nu^*$, or an eigenfunction of the adjoint operator $\mathcal{H}^*(\nu)$ with associated eigenfrequency ν. It is generally true that an operator and its adjoint have complex conjugate spectra; in the present case, the relationship between the associated eigenfunctions is particularly simple—they are also complex conjugates of each other.

The *quality factor* Q of a normal mode of oscillation is defined in terms of its real positive eigenfrequency $\omega = \operatorname{Re}\nu$ and its decay rate $\gamma = \operatorname{Im}\nu$ by

$$Q^{-1} = 2\gamma/\omega. \tag{6.136}$$

Upon taking the imaginary part of the inner product of $\mathcal{H}(\nu)\mathbf{s} = \nu^2\mathbf{s}$ with the eigenfunction $\mathbf{s}$, we find that

$$2(\operatorname{Re}\nu)(\operatorname{Im}\nu)\int_\oplus \rho^0\mathbf{s}^*\cdot\mathbf{s}\,dV = \int_\oplus \boldsymbol{\nabla}\mathbf{s}^*\!:\operatorname{Im}\boldsymbol{\Lambda}(\nu)\!:\boldsymbol{\nabla}\mathbf{s}\,dV. \tag{6.137}$$

In the case of an anelastically isotropic Earth with frequency-independent intrinsic quality factors Q_κ and Q_μ, this relation reduces to

$$Q^{-1} = \frac{\displaystyle\int_\oplus [\kappa(\omega)Q_\kappa^{-1}(\boldsymbol{\nabla}\cdot\mathbf{s}^*)(\boldsymbol{\nabla}\cdot\mathbf{s}) + 2\mu(\omega)Q_\mu^{-1}(\mathbf{d}^*\!:\mathbf{d})]\,dV}{\displaystyle\omega^2\int_\oplus \rho^0\mathbf{s}^*\cdot\mathbf{s}\,dV}, \tag{6.138}$$

where we have used the weak-attenuation approximation (6.120) for simplicity. The positivity of the bulk and shear quality factors Q_κ and Q_μ implies that $Q > 0$, which guarantees that all of the eigenfrequencies $\nu = \omega + i\gamma$ lie in the upper half of the complex frequency plane. This confirms the physically obvious result that the normal modes must *decay* with time as a result of the anelastic dissipation.

In addition to the inner product (6.134), it is convenient to define a second bilinear product between any two complex functions in $\oplus$, obtained by eliminating the asterisk:

$$[\mathbf{s},\mathbf{s}'] = \int_\oplus \rho^0\mathbf{s}\cdot\mathbf{s}'\,dV. \tag{6.139}$$

Equation (6.135) can be rewritten in the form

$$[\mathbf{s},\mathcal{H}(\nu)\mathbf{s}'] = [\mathcal{H}(\nu)\mathbf{s},\mathbf{s}'] = [\mathbf{s}',\mathcal{H}(\nu)\mathbf{s}], \tag{6.140}$$

so that $\mathcal{H}(\nu)$ is *symmetric* with respect to this new product. We shall refer to (6.139) as a *duality product*; the reason for this terminology will be clarified in Section 6.3.1. Taking the duality product of $\mathcal{H}(\nu)\mathbf{s} = \nu^2\mathbf{s}$ with $\mathbf{s}'$ yields

$$\nu^2[\mathbf{s}',\mathbf{s}] = [\mathbf{s}',\mathcal{H}(\nu)\mathbf{s}], \tag{6.141}$$

whereas taking the duality product of $\mathcal{H}(\nu')\mathbf{s}' = \nu'^2\mathbf{s}'$ with $\mathbf{s}$ yields

$$\nu'^2[\mathbf{s}, \mathbf{s}'] = [\mathbf{s}, \mathcal{H}(\nu')\mathbf{s}']. \tag{6.142}$$

Upon subtracting equation (6.142) from (6.141) and using the symmetry (6.140), we obtain the result

$$[\mathbf{s}, \mathbf{s}'] - (\nu^2 - \nu'^2)^{-1}[\mathbf{s}, \{\mathcal{H}(\nu) - \mathcal{H}(\nu')\}\mathbf{s}'] = 0 \quad \text{if } \nu \neq \nu'. \tag{6.143}$$

Equation (6.143) is the anelastic generalization of the normal-mode orthogonality condition (4.20).

In the present case, it is imprecise to say that two eigenfunctions $\mathbf{s}$ and $\mathbf{s}'$ associated with distinct eigenfrequencies ν and ν' are orthogonal in the sense (6.143), because $[\mathbf{s}, \mathbf{s}']$ is not a valid inner product in a complex space. In fact (6.143) is an expression of the *biorthogonality* between the eigenfunctions $\mathbf{s}^*$ and $\mathbf{s}'$ associated with $-\nu^*$ and ν'. Any eigenfunction $\mathbf{s}$ and its complex conjugate $\mathbf{s}^*$ have distinct eigenfrequencies $\nu \neq -\nu^*$, and the biorthogonality (6.143) in that case reduces to (6.137). The quantity on the left of equation (6.143) is well defined in the limit $\nu' \to \nu$. It is convenient to normalize the eigenfunctions so that, in that limit,

$$[\mathbf{s}, \mathbf{s}] - \tfrac{1}{2}\nu^{-1}[\mathbf{s}, \partial_\nu\mathcal{H}(\nu)\mathbf{s}] = 1. \tag{6.144}$$

This has the virtue that it reduces to the normalization (4.21) adopted on a non-rotating elastic Earth in the non-dispersive case $\partial_\nu\mathcal{H}(\nu) = 0$.

6.2.2 Rayleigh's principle

The various forms of Rayleigh's principle enunciated in Section 4.1.3 all remain valid, provided that the inner product (4.10) is replaced by the duality product (6.139). Generalizing equation (4.24), we define the *action* in an anelastic Earth by

$$\mathcal{I} = \tfrac{1}{2}\nu^2[\mathbf{s}, \mathbf{s}] - \tfrac{1}{2}[\mathbf{s}, \mathcal{H}(\nu)\mathbf{s}]. \tag{6.145}$$

Rayleigh's principle states that this complex functional is stationary for an arbitrary admissible variation $\delta\mathbf{s}$ if and only if $\mathbf{s}$ is an eigenfunction of the anelastic Earth with associated eigenfrequency ν. To verify this, we note that, correct to first order in $\|\delta\mathbf{s}\|$,

$$\delta\mathcal{I} = \mathrm{Re}\,[\delta\mathbf{s}, \nu^2\mathbf{s} - \mathcal{H}(\nu)\mathbf{s}], \tag{6.146}$$

where we have used the symmetry (6.140) of the operator $\mathcal{H}(\nu)$. Clearly, $\delta\mathcal{I}$ vanishes for an arbitrary $\delta\mathbf{s}$ if and only if $\mathcal{H}(\nu)\mathbf{s} = \nu^2\mathbf{s}$.

Written out explicitly, the action (6.145) is

$$\mathcal{I} = \int_{\oplus} L(\mathbf{s}, \boldsymbol{\nabla}\mathbf{s})\, dV + \int_{\Sigma_{\rm FS}} L^{\Sigma}(\mathbf{s}, \boldsymbol{\nabla}^{\Sigma}\mathbf{s})\, d\Sigma, \tag{6.147}$$

where

$$L = \tfrac{1}{2}[\nu^2\rho^0\mathbf{s}\cdot\mathbf{s} - \boldsymbol{\nabla}\mathbf{s}{:}\boldsymbol{\Lambda}(\nu){:}\boldsymbol{\nabla}\mathbf{s} - \rho^0\mathbf{s}\cdot\boldsymbol{\nabla}\phi^{\rm E1} - \rho^0\mathbf{s}\cdot\boldsymbol{\nabla}\boldsymbol{\nabla}\phi^0\cdot\mathbf{s}], \tag{6.148}$$

$$L^{\Sigma} = \tfrac{1}{2}[(\hat{\mathbf{n}}\cdot\mathbf{s})\boldsymbol{\nabla}^{\Sigma}\cdot(\varpi^0\mathbf{s}) - \varpi^0(\mathbf{s}\cdot\boldsymbol{\nabla}^{\Sigma}\mathbf{s}\cdot\hat{\mathbf{n}})]. \tag{6.149}$$

The Eulerian potential perturbation $\phi^{\rm E1}$ in equation (6.148) is regarded as a known functional of $\mathbf{s}$, given by equation (3.99), so that $\delta\mathcal{I} = 0$ is a displacement variational principle.

As on a non-rotating elastic Earth, there is also a displacement-potential principle in which the quantities $\mathbf{s}$ and $\phi^{\rm E1}$ are varied independently; the modified action for this principle is

$$\mathcal{I}' = \int_{\bigcirc} L'(\mathbf{s}, \boldsymbol{\nabla}\mathbf{s}, \boldsymbol{\nabla}\phi^{\rm E1})\, dV + \int_{\Sigma_{\rm FS}} L^{\Sigma}(\mathbf{s}, \boldsymbol{\nabla}^{\Sigma}\mathbf{s})\, d\Sigma, \tag{6.150}$$

where

$$L' = \tfrac{1}{2}[\nu^2\rho^0\mathbf{s}\cdot\mathbf{s} - \boldsymbol{\nabla}\mathbf{s}{:}\boldsymbol{\Lambda}(\nu){:}\boldsymbol{\nabla}\mathbf{s} - 2\rho^0\mathbf{s}\cdot\boldsymbol{\nabla}\phi^{\rm E1} - \rho^0\mathbf{s}\cdot\boldsymbol{\nabla}\boldsymbol{\nabla}\phi^0\cdot\mathbf{s} - (4\pi G)^{-1}\boldsymbol{\nabla}\phi^{\rm E1}\cdot\boldsymbol{\nabla}\phi^{\rm E1}]. \tag{6.151}$$

The stationary value of both of the anelastic actions is $\mathcal{I} = \mathcal{I}' = 0$ for every anelastic eigensolution ν, $\mathbf{s}$, $\phi^{\rm E1}$.

6.2.3 Green tensor

In Section 4.1.7 we determined the Green tensor $\mathbf{G}(\mathbf{x}, \mathbf{x}'; t)$ of a non-rotating elastic Earth directly in the time domain. On an anelastic Earth it is more convenient to consider the Fourier transform of the Green tensor,

$$\mathbf{G}(\mathbf{x}, \mathbf{x}'; \nu) = \int_0^{\infty} \mathbf{G}(\mathbf{x}, \mathbf{x}'; t)\exp(-i\nu t)\, dt, \tag{6.152}$$

in the complex frequency domain. The corresponding inverse transform is

$$\mathbf{G}(\mathbf{x}, \mathbf{x}'; t) = \frac{1}{2\pi}\int_{-\infty}^{\infty} \mathbf{G}(\mathbf{x}, \mathbf{x}'; \nu)\exp(i\nu t)\, d\nu, \tag{6.153}$$

where the contour of integration runs along the real frequency axis as shown in Figure 6.9. To find the frequency-domain Green tensor $\mathbf{G}(\mathbf{x}, \mathbf{x}'; \nu)$ we must solve

$$\rho^0[-\nu^2\mathbf{G} + \mathcal{H}(\nu)\mathbf{G}] = \mathbf{I}\,\delta(\mathbf{x} - \mathbf{x}'). \tag{6.154}$$

Upon taking the complex conjugate of equation (6.154) and using the symmetry (6.133) we see that

$$\mathbf{G}(\mathbf{x}, \mathbf{x}'; -\nu^*) = \mathbf{G}^*(\mathbf{x}, \mathbf{x}'; \nu). \tag{6.155}$$

It therefore suffices to find $\mathbf{G}(\mathbf{x}, \mathbf{x}'; \nu)$ in the right half plane, $\mathrm{Re}\,\nu > 0$; its value in the left half plane, $\mathrm{Re}\,\nu < 0$, may be found using equation (6.155).

We label the right half-plane eigenfrequencies ν_k and the associated eigenfunctions $\mathbf{s}_k$ with an index k; the corresponding left half-plane eigenfrequencies and eigenfunctions are then $-\nu_k^*$ and $\mathbf{s}_k^*$. The biorthogonality relation (6.143) for the right half-plane eigenfunctions with discrete eigenfrequencies $\nu_k \neq \nu_{k'}$ can be written explicitly in the form

$$\int_\oplus \rho^0 \mathbf{s}_k \cdot \mathbf{s}_{k'}\, dV - (\nu_k^2 - \nu_{k'}^2)^{-1} \int_\oplus \boldsymbol{\varepsilon}_k : [\boldsymbol{\Gamma}(\nu_k) - \boldsymbol{\Gamma}(\nu_{k'})] : \boldsymbol{\varepsilon}_{k'}\, dV = 0. \tag{6.156}$$

The anelastic normalization condition (6.144) is likewise

$$\int_\oplus \rho^0 \mathbf{s}_k \cdot \mathbf{s}_k\, dV - \tfrac{1}{2}\nu_k^{-1} \int_\oplus \boldsymbol{\varepsilon}_k : \partial_\nu \boldsymbol{\Gamma}(\nu_k) : \boldsymbol{\varepsilon}_k\, dV = 1. \tag{6.157}$$

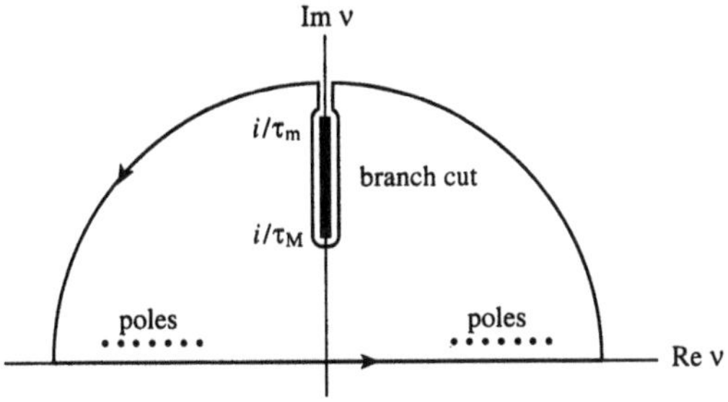

Figure 6.9. Contour of integration used in the evaluation of the inverse Fourier transform (6.153). For $t < 0$ the contour is closed in the lower half-plane, where there are no poles or other singularities, so that $\mathbf{G}(\mathbf{x}, \mathbf{x}'; t) = \mathbf{0}$. For $t > 0$ the contour is closed in the upper half-plane, as shown. The residues of the simple poles at the complex eigenfrequencies ν_k and $-\nu_k^*$ are accounted for, but any discontinuity in the integrand across the logarithmic branch cut $i/\tau_{\mathrm{M}} \leq \nu \leq i/\tau_{\mathrm{m}}$ is ignored. This leads to the simple result (6.164).

The elastic tensor $\mathbf{\Gamma}(\nu)$ appears in (6.156)–(6.157) rather than $\mathbf{\Lambda}(\nu)$, since the difference $\mathbf{\Lambda}(\nu) - \mathbf{\Gamma}(\nu)$ depends only upon the initial deviatoric stress $\boldsymbol{\tau}^0$, so it is independent of the frequency ν. We seek a solution to the inhomogeneous equation (6.154) of the form

$$\mathbf{G}(\mathbf{x}, \mathbf{x}'; \nu) = \sum_k \mathbf{s}_k(\mathbf{x})\mathbf{c}_k(\mathbf{x}'; \nu), \quad \operatorname{Re}\nu > 0. \tag{6.158}$$

To obtain a formal normal-mode sum for the Green tensor $\mathbf{G}(\mathbf{x}, \mathbf{x}'; t)$ in the time domain, we make two approximations:

1. We ignore the possibility of *eigenfrequency degeneracy*, and assume that all of the right half-plane eigenfrequencies are distinct.

2. We also ignore the effect of the *logarithmic branch cut* or any other singularities exhibited by $\mathbf{\Lambda}(\nu)$ in the upper half of the complex frequency plane, $\operatorname{Im}\nu \geq 0$. The only singularities of $\mathbf{G}(\mathbf{x}, \mathbf{x}'; \nu)$ accounted for are thus the *simple poles* at the complex eigenfrequencies ν_k and $-\nu_k^*$.

We evaluate the integral (6.153) using the method of residues, closing the contour in the upper half plane for $t > 0$, to insure that the contribution from the arc at infinity vanishes. Upon invoking the above two approximations we obtain

$$\mathbf{G}(\mathbf{x}, \mathbf{x}'; t) = 2\operatorname{Re}\sum_k \mathbf{s}_k(\mathbf{x}) \left[\lim_{\nu\to\nu_k} i(\nu - \nu_k)\mathbf{c}_k(\mathbf{x}'; \nu)\right] \exp(i\nu_k t)\,. \tag{6.159}$$

The quantity in brackets in equation (6.159) is i times the residue of the simple pole at $\nu = \nu_k$; the corresponding poles at $\nu = -\nu_k^*$ have been accounted for using the symmetry relation (6.155). To find the residue, we insert the expansion (6.158) into equation (6.154), take the duality product with $\mathbf{s}_k$, and rearrange terms to obtain

$$\begin{aligned}\mathbf{c}_k(\mathbf{x}'; \nu) = {} & \frac{\mathbf{s}_k(\mathbf{x}')}{[\mathbf{s}_k, \mathcal{H}(\nu)\mathbf{s}_k - \nu^2\mathbf{s}_k]} \\ & - \sum_{k'\neq k} \frac{[\mathbf{s}_k, \mathcal{H}(\nu)\mathbf{s}_{k'} - \nu^2\mathbf{s}_{k'}]}{[\mathbf{s}_k, \mathcal{H}(\nu)\mathbf{s}_k - \nu^2\mathbf{s}_k]}\,\mathbf{c}_{k'}(\mathbf{x}'; \nu).\end{aligned} \tag{6.160}$$

Upon adding the term $\mathcal{H}(\nu_k)\mathbf{s}_k - \nu_k^2\mathbf{s}_k = \mathbf{0}$ to the denominators to facilitate the passage to the limit, we find, after some manipulation,

$$\begin{aligned}\lim_{\nu\to\nu_k} i(\nu - \nu_k)\mathbf{c}_k(\mathbf{x}'; \nu) = {} & \frac{i\mathbf{s}_k(\mathbf{x}')}{[\mathbf{s}_k, \partial_\nu\mathcal{H}(\nu_k)\mathbf{s}_k] - 2\nu_k[\mathbf{s}_k, \mathbf{s}_k]} \\ & - \sum_{k'\neq k} \frac{i[\mathbf{s}_k, \mathcal{H}(\nu_k)\mathbf{s}_{k'} - \nu_k^2\mathbf{s}_{k'}]}{[\mathbf{s}_k, \partial_\nu\mathcal{H}(\nu_k)\mathbf{s}_k] - 2\nu_k[\mathbf{s}_k, \mathbf{s}_k]}\,\mathbf{c}_{k'}(\mathbf{x}'; \nu_k).\end{aligned} \tag{6.161}$$

The sum over $k' \neq k$ in (6.161) vanishes by virtue of the symmetry (6.140) of the operator $\mathcal{H}(\nu)$:

$$[\mathbf{s}_k, \mathcal{H}(\nu_k)\mathbf{s}_{k'} - \nu_k^2\mathbf{s}_{k'}] = [\mathbf{s}_{k'}, \mathcal{H}(\nu_k)\mathbf{s}_k - \nu_k^2\mathbf{s}_k] = 0. \tag{6.162}$$

As a result the expansion coefficient in (6.159) reduces to

$$\lim_{\nu\to\nu_k} i(\nu - \nu_k)\mathbf{c}_k(\mathbf{x}';\nu) = \tfrac{1}{2}(i\nu_k)^{-1}\mathbf{s}_k(\mathbf{x}'), \tag{6.163}$$

where we have used the normalization condition (6.157). The Green tensor is therefore given in terms of the complex eigenfrequencies and eigenfunctions by

$$\mathbf{G}(\mathbf{x},\mathbf{x}';t) = \mathrm{Re}\sum_k (i\nu_k)^{-1}\mathbf{s}_k(\mathbf{x})\mathbf{s}_k(\mathbf{x}')\exp(i\nu_k t). \tag{6.164}$$

The trivial modes of a non-rotating anelastic Earth are the same as those of the corresponding non-rotating elastic Earth, since the rigid-body modes have no associated deformation, and the geostrophic modes are confined to the fluid regions $\oplus_{\mathrm{F}}$, where the dissipation is negligible. We eliminate the trivials from the normal-mode sum (6.164), since they cannot be excited by an earthquake in the solid crust or mantle, and refer to the resulting sum over only the seismic modes as the *seismic Green tensor*. It is noteworthy that (6.164) satisfies

$$\mathbf{G}(\mathbf{x},\mathbf{x}';t) = \mathbf{G}^{\mathrm{T}}(\mathbf{x}',\mathbf{x};t), \tag{6.165}$$

so that the principle of *source-receiver reciprocity* is maintained upon a non-rotating anelastic Earth.

*6.3 Rotating Anelastic Earth

The general case of a rotating anelastic Earth was first treated correctly by Lognonné (1991). We summarize his findings in this section.

*6.3.1 Duality and biorthonormality

The complex eigenfrequencies ν and associated eigenfunctions $\mathbf{s}$ of a rotating anelastic Earth are found by solving the abstract operator equation

$$\mathcal{H}(\nu)\mathbf{s} + 2i\nu\boldsymbol{\Omega}\times\mathbf{s} = \nu^2\mathbf{s}. \tag{6.166}$$

The symbol $\mathcal{H}(\nu)$ in this case stands for the centrifugally modified integro-differential operator

$$\rho^0\mathcal{H}(\nu)\mathbf{s} = -\boldsymbol{\nabla}\cdot[\boldsymbol{\Lambda}(\nu):\boldsymbol{\nabla}\mathbf{s}] + \rho^0\boldsymbol{\nabla}\phi^{\mathrm{E1}} + \rho^0\mathbf{s}\cdot\boldsymbol{\nabla\nabla}(\phi^0+\psi), \tag{6.167}$$

together with the associated boundary conditions on $\partial\oplus$, $\Sigma_{\rm SS}$ and $\Sigma_{\rm FS}$. The symmetry $\mathbf{\Lambda}(-\nu^*) = \mathbf{\Lambda}^*(\nu)$ of the elastic tensor guarantees that

$$\mathcal{H}(-\nu^*) = \mathcal{H}^*(\nu), \tag{6.168}$$

where $\rho^0\mathcal{H}^*(\nu)\mathbf{s} = -\boldsymbol{\nabla}\cdot[\mathbf{\Lambda}^*(\nu):\boldsymbol{\nabla}\mathbf{s}] + \rho^0\boldsymbol{\nabla}\phi^{\rm E1} + \rho^0\mathbf{s}\cdot\boldsymbol{\nabla}\boldsymbol{\nabla}(\phi^0+\psi)$. Upon taking the complex conjugate of equation (6.166) and using (6.168), we observe that $-\nu^*$, $\mathbf{s}^*$ is an eigensolution if and only if ν, $\mathbf{s}$ is. The operator $\mathcal{H}^*(\nu)$ is the *adjoint* of $\mathcal{H}(\nu)$ with respect to the inner product (6.134), as on a non-rotating anelastic Earth.

The eigenfrequencies and eigenfunctions of the anti-Earth with a reversed sense of rotation, $\boldsymbol{\Omega}\rightarrow-\boldsymbol{\Omega}$, are also of interest. The Earth and the anti-Earth have the *same complex eigenfrequencies* ν, but different associated eigenfunctions, which we shall denote by $\mathbf{s}$ and $\overline{\mathbf{s}}$, respectively. We refer to the eigenfunctions $\overline{\mathbf{s}}$ of the anti-Earth as the *dual eigenfunctions*; to find them, we must solve

$$\mathcal{H}(\nu)\overline{\mathbf{s}} - 2i\nu\boldsymbol{\Omega}\times\overline{\mathbf{s}} = \nu^2\overline{\mathbf{s}}, \tag{6.169}$$

in addition to equation (6.166). The symmetry (6.168) guarantees that $-\nu^*$, $\overline{\mathbf{s}}^*$ is an eigensolution of (6.169) if ν, $\overline{\mathbf{s}}$ is.

Upon taking the imaginary part of the inner product of equation (6.167) with $\mathbf{s}$, and using the approximations (6.120) and (6.126), we obtain

$$Q^{-1} = \frac{\displaystyle\int_\oplus [\kappa(\omega)Q_\kappa^{-1}(\boldsymbol{\nabla}\cdot\mathbf{s}^*)(\boldsymbol{\nabla}\cdot\mathbf{s}) + 2\mu(\omega)Q_\mu^{-1}(\mathbf{d}^*:\mathbf{d})]\,dV}{\omega^2\left[\displaystyle\int_\oplus \rho^0\mathbf{s}^*\cdot\mathbf{s}\,dV - \omega^{-1}\int_\oplus \rho^0\mathbf{s}^*\cdot(i\boldsymbol{\Omega}\times\mathbf{s})\,dV\right]}, \tag{6.170}$$

which generalizes the result (6.138) found on a non-rotating, anelastically isotropic, constant-Q Earth model. The quantity in brackets in the denominator is the normalization integral (4.125), which we set equal to unity on a rotating *elastic* Earth. It is necessarily positive for every seismic mode, which has $\omega \ll \Omega$; this, together with the positivity of the intrinsic quality factors Q_κ and Q_μ, guarantees that $Q > 0$, so that all of the seismic eigenfrequencies must lie in the upper half of the complex ν plane.

The *duality product* of an unprimed dual eigenfunction $\overline{\mathbf{s}}$ and a primed eigenfunction $\mathbf{s}'$ is defined by

$$[\overline{\mathbf{s}},\mathbf{s}'] = \int_\oplus \rho^0\overline{\mathbf{s}}\cdot\mathbf{s}'\,dV. \tag{6.171}$$

As on a non-rotating Earth, the elastic-gravitational operator $\mathcal{H}(\nu)$ is symmetric with respect to this duality product:

$$[\overline{\mathbf{s}},\mathcal{H}(\nu)\mathbf{s}'] = [\mathcal{H}(\nu)\overline{\mathbf{s}},\mathbf{s}'] = [\mathbf{s}',\mathcal{H}(\nu)\overline{\mathbf{s}}]. \tag{6.172}$$

The Coriolis operator is, on the other hand, *anti-symmetric*:

$$[\bar{\mathbf{s}}, i\boldsymbol{\Omega}\times\mathbf{s}'] = -[i\boldsymbol{\Omega}\times\bar{\mathbf{s}}, \mathbf{s}'] = -[\mathbf{s}', i\boldsymbol{\Omega}\times\bar{\mathbf{s}}]. \tag{6.173}$$

Symmetric operators must always have identical spectra, and the two equations (6.172) and (6.173) are the reason that the eigenfrequencies ν do not depend upon the sense of rotation; the anti-symmetry (6.173) is annulled by the transformation $\boldsymbol{\Omega}\to-\boldsymbol{\Omega}$.

Taking the duality product of $\mathcal{H}(\nu)\bar{\mathbf{s}} - 2i\nu\,\boldsymbol{\Omega}\times\bar{\mathbf{s}} = \nu^2\bar{\mathbf{s}}$ with $\mathbf{s}'$ yields

$$\nu^2[\mathbf{s}',\bar{\mathbf{s}}] + 2\nu[\mathbf{s}', i\boldsymbol{\Omega}\times\bar{\mathbf{s}}] - [\mathbf{s}',\mathcal{H}(\nu)\bar{\mathbf{s}}] = 0, \tag{6.174}$$

whereas taking the duality product of $\mathcal{H}(\nu')\mathbf{s}' + 2i\nu'\boldsymbol{\Omega}\times\mathbf{s}' = \nu'^2\mathbf{s}'$ with $\bar{\mathbf{s}}$ yields

$$\nu'^2[\bar{\mathbf{s}},\mathbf{s}'] - 2\nu'[\bar{\mathbf{s}}, i\boldsymbol{\Omega}\times\mathbf{s}'] - [\bar{\mathbf{s}},\mathcal{H}(\nu')\mathbf{s}'] = 0. \tag{6.175}$$

Upon subtracting equation (6.175) from (6.174) and making use of the two results (6.172) and (6.173), we obtain

$$\begin{aligned}&[\bar{\mathbf{s}},\mathbf{s}'] - 2(\nu+\nu')^{-1}[\bar{\mathbf{s}}, i\boldsymbol{\Omega}\times\mathbf{s}'] \\ &\quad - (\nu^2-\nu'^2)^{-1}[\bar{\mathbf{s}}, \{\mathcal{H}(\nu)-\mathcal{H}(\nu')\}\mathbf{s}'] = 0 \quad \text{if } \nu\neq\nu'.\end{aligned} \tag{6.176}$$

Equation (6.176) is the generalization of equation (4.81) on a rotating elastic Earth and of equation (6.143) on a non-rotating anelastic Earth; the dual eigenfunctions $\bar{\mathbf{s}}$ and eigenfunctions $\mathbf{s}'$ associated with distinct eigenfrequencies $\nu\neq\nu'$ are *biorthogonal* in this sense. By analogy with (4.82) and (6.144), we normalize the right half-plane eigenfunctions $\mathbf{s}$ and their duals $\bar{\mathbf{s}}$ such that

$$[\bar{\mathbf{s}},\mathbf{s}] - \nu^{-1}[\bar{\mathbf{s}}, i\boldsymbol{\Omega}\times\mathbf{s}] - \tfrac{1}{2}\nu^{-1}[\bar{\mathbf{s}}, \partial_\nu\mathcal{H}(\nu)\mathbf{s}] = 1. \tag{6.177}$$

It is now evident why we referred to (6.139) in Section 6.2.1 as a *duality product*—on a rotating Earth, the first slot accepts dual eigenfunctions $\bar{\mathbf{s}}$ of the anti-Earth, whereas the second slot accepts eigenfunctions $\mathbf{s}$ of the actual Earth. In general, these two spaces are physically distinct; only in the non-rotating limit $\boldsymbol{\Omega}\to\mathbf{0}$ can they be confused. It is common in dealing with non-Hermitian eigenvalue problems to have to solve a dual eigenvalue problem involving the adjoint operator as well. In the present instance, the adjoint of the elastic-gravitational operator $\mathcal{H}^*(\nu)$ satisfies the dynamical symmetry (6.168), and the Coriolis operator $i\boldsymbol{\Omega}\times$ is self-adjoint; because of this, the dual eigensolutions ν, $\bar{\mathbf{s}}$ have an appealing physical interpretation—they represent the normal modes of the anti-Earth just as the primal eigensolutions ν, $\mathbf{s}$ represent the normal modes of the actual Earth.

★6.3.2 Rayleigh's principle

The equations (6.166) and (6.169) governing the eigensolutions and dual eigensolutions of a rotating anelastic Earth and anti-Earth can be obtained from a straightforward extension of Rayleigh's principle. Generalizing (4.101), we define the *action* by

$$\mathcal{I} = \tfrac{1}{2}\nu^2[\bar{\mathbf{s}}, \mathbf{s}] - \nu[\bar{\mathbf{s}}, i\boldsymbol{\Omega}\times\mathbf{s}] - [\bar{\mathbf{s}}, \mathcal{H}(\nu)\mathbf{s}]. \tag{6.178}$$

We regard $\mathcal{I}$ as a *bilinear functional* of the two complex fields $\mathbf{s}$ and $\bar{\mathbf{s}}$, for a fixed value of the parameter ν. The variation of this functional, correct to first order in $\|\delta\mathbf{s}\|$ and $\|\delta\bar{\mathbf{s}}\|$, is

$$\begin{aligned}\delta\mathcal{I} = {} & \tfrac{1}{2}[\delta\bar{\mathbf{s}}, \nu^2\mathbf{s} - 2i\nu\boldsymbol{\Omega}\times\mathbf{s} - \mathcal{H}(\nu)\mathbf{s}] \\ & + \tfrac{1}{2}[\delta\mathbf{s}, \nu^2\bar{\mathbf{s}} + 2i\nu\boldsymbol{\Omega}\times\bar{\mathbf{s}} - \mathcal{H}(\nu)\bar{\mathbf{s}}],\end{aligned} \tag{6.179}$$

where we have used equations (6.172) and (6.173). It is evident that $\delta\mathcal{I}$ vanishes for arbitrary and independent variations $\delta\mathbf{s}$ and $\delta\bar{\mathbf{s}}$ if and only if $\mathbf{s}$ is an eigenfunction of the Earth and $\bar{\mathbf{s}}$ is the corresponding dual eigenfunction of the anti-Earth, with associated eigenfrequency ν. The variational procedure leads naturally to both sets of governing equations $\mathcal{H}(\nu)\mathbf{s} + 2i\nu\boldsymbol{\Omega}\times\mathbf{s} = \nu^2\mathbf{s}$ and $\mathcal{H}(\nu)\bar{\mathbf{s}} - 2i\nu\boldsymbol{\Omega}\times\bar{\mathbf{s}} = \nu^2\bar{\mathbf{s}}$.

Written out explicitly, the action (6.178) governing the displacement version of Rayleigh's principle on a rotating anelastic Earth is

$$\mathcal{I} = \int_{\oplus} L(\mathbf{s}, \boldsymbol{\nabla}\mathbf{s}\,;\,\bar{\mathbf{s}}, \boldsymbol{\nabla}\bar{\mathbf{s}})\,dV + \int_{\Sigma_{\mathrm{FS}}} L^{\Sigma}(\mathbf{s}, \boldsymbol{\nabla}^{\Sigma}\mathbf{s}\,;\,\bar{\mathbf{s}}, \boldsymbol{\nabla}^{\Sigma}\bar{\mathbf{s}})\,d\Sigma, \tag{6.180}$$

where

$$\begin{aligned}L = {} & \tfrac{1}{2}[\nu^2\rho^0\bar{\mathbf{s}}\cdot\mathbf{s} - 2\nu\bar{\mathbf{s}}\cdot(i\boldsymbol{\Omega}\times\mathbf{s}) - \boldsymbol{\nabla}\bar{\mathbf{s}}\!:\!\boldsymbol{\Lambda}(\nu)\!:\!\boldsymbol{\nabla}\mathbf{s} \\ & - \tfrac{1}{2}\rho^0(\bar{\mathbf{s}}\cdot\boldsymbol{\nabla}\phi^{\mathrm{E1}} + \mathbf{s}\cdot\boldsymbol{\nabla}\bar{\phi}^{\mathrm{E1}}) - \rho^0\bar{\mathbf{s}}\cdot\boldsymbol{\nabla}\boldsymbol{\nabla}(\phi^0 + \psi)\cdot\mathbf{s}],\end{aligned} \tag{6.181}$$

$$\begin{aligned}L^{\Sigma} = {} & \tfrac{1}{4}[(\hat{\mathbf{n}}\cdot\bar{\mathbf{s}})\boldsymbol{\nabla}^{\Sigma}\cdot(\varpi^0\mathbf{s}) + (\hat{\mathbf{n}}\cdot\mathbf{s})\boldsymbol{\nabla}^{\Sigma}\cdot(\varpi^0\bar{\mathbf{s}}) \\ & - \varpi^0(\bar{\mathbf{s}}\cdot\boldsymbol{\nabla}^{\Sigma}\mathbf{s}\cdot\hat{\mathbf{n}}) - \varpi^0(\mathbf{s}\cdot\boldsymbol{\nabla}^{\Sigma}\bar{\mathbf{s}}\cdot\hat{\mathbf{n}})].\end{aligned} \tag{6.182}$$

The modified action governing the associated displacement-potential variational principle is

$$\begin{aligned}\mathcal{I}' = {} & \int_{\bigcirc} L'(\mathbf{s}, \boldsymbol{\nabla}\mathbf{s}, \boldsymbol{\nabla}\phi^{\mathrm{E1}}; \bar{\mathbf{s}}, \boldsymbol{\nabla}\bar{\mathbf{s}}, \boldsymbol{\nabla}\bar{\phi}^{\mathrm{E1}})\,dV \\ & + \int_{\Sigma_{\mathrm{FS}}} L^{\Sigma}(\mathbf{s}, \boldsymbol{\nabla}^{\Sigma}\mathbf{s}\,;\,\bar{\mathbf{s}}, \boldsymbol{\nabla}^{\Sigma}\bar{\mathbf{s}})\,d\Sigma,\end{aligned} \tag{6.183}$$

where

$$L' = \tfrac{1}{2}[\nu^2\rho^0\overline{\mathbf{s}}\cdot\mathbf{s} - 2\nu\overline{\mathbf{s}}\cdot(i\mathbf{\Omega}\times\mathbf{s}) - \boldsymbol{\nabla}\overline{\mathbf{s}}\!:\!\mathbf{\Lambda}(\nu)\!:\!\boldsymbol{\nabla}\mathbf{s}$$
$$- \rho^0(\overline{\mathbf{s}}\cdot\boldsymbol{\nabla}\phi^{\mathrm{E1}} + \mathbf{s}\cdot\boldsymbol{\nabla}\overline{\phi}^{\mathrm{E1}}) - \rho^0\overline{\mathbf{s}}\cdot\boldsymbol{\nabla\nabla}(\phi^0+\psi)\cdot\mathbf{s}],$$
$$- (4\pi G)^{-1}\boldsymbol{\nabla}\overline{\phi}^{\mathrm{E1}}\cdot\boldsymbol{\nabla}\phi^{\mathrm{E1}}. \qquad (6.184)$$

The salient feature that distinguishes Rayleigh's principle in this, the most general case, is the dependence of the actions $\mathcal{I}$ and $\mathcal{I}'$ upon the dual eigenfunctions $\overline{\mathbf{s}}$ and $\overline{\phi}^{\mathrm{E1}}$ as well as the eigenfunctions $\mathbf{s}$ and ϕ^{E1}. The variation with respect to $\overline{\mathbf{s}}$ and $\overline{\phi}^{\mathrm{E1}}$ yields the frequency-domain equations and associated boundary conditions governing the actual Earth, whereas that with respect to $\mathbf{s}$ and ϕ^{E1} yields the corresponding equations and boundary conditions governing the anti-Earth. The common value of the two actions is $\mathcal{I}=\mathcal{I}'=0$ at each of the stationary points $\mathbf{s}$, ϕ^{E1} and $\overline{\mathbf{s}}$, $\overline{\phi}^{\mathrm{E1}}$.

*6.3.3 Green tensor

To find the frequency-domain Green tensor $\mathbf{G}(\mathbf{x},\mathbf{x}';\nu)$, we must solve the rotating version of equation (6.154):

$$\rho^0[-\nu^2\mathbf{G} + 2i\nu\mathbf{\Omega}\times\mathbf{G} + \mathcal{H}(\nu)\mathbf{G}] = \mathbf{I}\,\delta(\mathbf{x}-\mathbf{x}'). \qquad (6.185)$$

We label the right half-plane eigenfrequencies ν_k and the associated eigenfunctions and dual eigenfunctions $\mathbf{s}_k$, $\overline{\mathbf{s}}_k$ with an index k. The corresponding left half-plane eigenfrequencies and associated eigenfunctions and their duals are then $-\nu_k^*$ and $\mathbf{s}_k^*$, $\overline{\mathbf{s}}_k^*$. The biorthogonality relation (6.176) for the right half-plane eigenfunctions and dual eigenfunctions with discrete eigenfrequencies $\nu_k \neq \nu_{k'}$ can be written explicitly in the form

$$\int_\oplus \rho^0\overline{\mathbf{s}}_k\cdot\mathbf{s}_{k'}\,dV - 2(\nu_k+\nu_{k'})^{-1}\int_\oplus \rho^0\overline{\mathbf{s}}_k\cdot(i\mathbf{\Omega}\times\mathbf{s}_{k'})\,dV$$
$$- (\nu_k^2-\nu_{k'}^2)^{-1}\int_\oplus \overline{\boldsymbol{\varepsilon}}_k\!:\![\mathbf{\Gamma}(\nu_k)-\mathbf{\Gamma}(\nu_{k'})]\!:\!\boldsymbol{\varepsilon}_{k'}\,dV = 0. \qquad (6.186)$$

The normalization condition (6.177) on a rotating anelastic Earth is likewise

$$\int_\oplus \rho^0\overline{\mathbf{s}}_k\cdot\mathbf{s}_k\,dV - \nu_k^{-1}\int_\oplus \rho^0\overline{\mathbf{s}}_k\cdot(i\mathbf{\Omega}\times\mathbf{s}_k)\,dV$$
$$- \tfrac{1}{2}\nu_k^{-1}\int_\oplus \overline{\boldsymbol{\varepsilon}}_k\!:\!\partial_\nu\mathbf{\Gamma}(\nu_k)\!:\!\boldsymbol{\varepsilon}_k\,dV = 1. \qquad (6.187)$$

Proceeding as we did in in Section 6.2.3, we seek a solution to (6.185) in the form (6.158). The same naive application of Cauchy's theorem yields an analogous result for the Green tensor in the time domain:

$$\mathbf{G}(\mathbf{x},\mathbf{x}';t) = 2\,\mathrm{Re}\sum_k \mathbf{s}_k(\mathbf{x})\left[\lim_{\nu\to\nu_k} i(\nu-\nu_k)\mathbf{c}_k(\mathbf{x}';\nu)\right]\exp(i\nu_k t), \tag{6.188}$$

where

$$\mathbf{c}_k(\mathbf{x}';\nu) = \frac{\overline{\mathbf{s}}_k(\mathbf{x}')}{[\overline{\mathbf{s}}_k, \mathcal{H}(\nu)\mathbf{s}_k + 2i\nu\boldsymbol{\Omega}\times\mathbf{s}_k - \nu^2\mathbf{s}_k]} \tag{6.189}$$
$$- \sum_{k'\neq k} \frac{[\overline{\mathbf{s}}_k, \mathcal{H}(\nu)\mathbf{s}_{k'} + 2i\nu\boldsymbol{\Omega}\times\mathbf{s}_{k'} - \nu^2\mathbf{s}_{k'}]}{[\overline{\mathbf{s}}_k, \mathcal{H}(\nu)\mathbf{s}_k + 2i\nu\boldsymbol{\Omega}\times\mathbf{s}_k - \nu^2\mathbf{s}_k]}\,\mathbf{c}_{k'}(\mathbf{x}';\nu).$$

The sum over $k'\neq k$ once again vanishes in the limit $\nu\to\nu_k$, by virtue of the symmetry (6.172) and the anti-symmetry (6.173). Using the normalization condition (6.187) to simplify the first term in (6.189), we obtain

$$\lim_{\nu\to\nu_k} i(\nu-\nu_k)\mathbf{c}_k(\mathbf{x}';\nu) = \tfrac{1}{2}(i\nu_k)^{-1}\overline{\mathbf{s}}_k(\mathbf{x}'). \tag{6.190}$$

The Green tensor on a rotating anelastic Earth is therefore given by

$$\mathbf{G}(\mathbf{x},\mathbf{x}';t) = \mathrm{Re}\sum_k (i\nu_k)^{-1}\mathbf{s}_k(\mathbf{x})\overline{\mathbf{s}}_k(\mathbf{x}')\exp(i\nu_k t). \tag{6.191}$$

As on a rotating elastic Earth, we exclude the tilt-over mode as well as the trivials from the expansion (6.191), and refer to the sum over only the seismic modes as the *seismic Green tensor*.

To find the frequency-domain seismic Green tensor $\overline{\mathbf{G}}(\mathbf{x},\mathbf{x}';\nu)$ on the anti-Earth, we must solve

$$\rho^0[-\nu^2\overline{\mathbf{G}} - 2i\nu\boldsymbol{\Omega}\times\overline{\mathbf{G}} + \mathcal{H}(\nu)\overline{\mathbf{G}}] = \mathbf{I}\,\delta(\mathbf{x}-\mathbf{x}'), \tag{6.192}$$

rather than (6.185). Introducing an expansion in terms of the dual eigenfunctions $\overline{\mathbf{s}}_k$ rather than the eigenfunctions $\mathbf{s}_k$, and applying Cauchy's theorem as above, we find that

$$\overline{\mathbf{G}}(\mathbf{x},\mathbf{x}';t) = \mathrm{Re}\sum_k (i\nu_k)^{-1}\overline{\mathbf{s}}_k(\mathbf{x})\mathbf{s}_k(\mathbf{x}')\exp(i\nu_k t). \tag{6.193}$$

Upon comparing (6.191) and (6.193), we see that the Green tensor and anti-Green tensor satisfy the same principle of *generalized reciprocity* as on an elastic Earth:

$$\mathbf{G}(\mathbf{x},\mathbf{x}';t) = \overline{\mathbf{G}}^{\mathrm{T}}(\mathbf{x}',\mathbf{x};t). \tag{6.194}$$

The physical interpretation of equation (6.194) in terms of the Doppler effect is the same. The general conclusion is that neither reciprocity on a non-rotating Earth nor generalized reciprocity on a rotating Earth is affected by anelastic attenuation.

We summarize and compare the four expressions for the seismic Green tensor $\mathbf{G}(\mathbf{x},\mathbf{x}';t)$, with and without rotation and anelasticity, in Table 6.3. The associated normalization conditions are collected for convenience in Table 6.4. Both the dual eigenfunctions $\overline{\mathbf{s}}_k$ and the eigenfunctions $\mathbf{s}_k$ are required to express the response in the most general case of a rotating anelastic Earth model. In the absence of rotation, the eigenfunctions and the duals coincide: $\overline{\mathbf{s}}_k = \mathbf{s}_k$. The general results (6.186)–(6.187) and (6.191) reduce in that case to (6.156)–(6.157) and (6.164), as expected. In the absence of anelasticity, the eigenfrequencies are real, $\nu_k = \omega_k$, and the eigenfunctions and their duals are complex conjugates: $\overline{\mathbf{s}}_k = \mathbf{s}_k^*$. Equations (6.186)–(6.187) and (6.191) reduce in that case to (4.125) and (4.127). Finally, in the absence of both rotation and anelasticity, the eigenfunctions $\mathbf{s}_k$ as well as the eigenfrequencies ω_k are real, and the duals are simply the eigenfunctions themselves: $\overline{\mathbf{s}}_k = \mathbf{s}_k$. In that simplest case, equations (6.186)–(6.187) and (6.191) reduce to the corresponding results (4.56) and (4.60) on a non-rotating elastic Earth, as of course they must. Unlike the corresponding results (4.60) and (4.127) on an elastic Earth, the seismic

Earth Model	Time-Domain Green Tensor
Non-Rotating Elastic	$\mathbf{G}(\mathbf{x},\mathbf{x}';t) = \sum_k \omega_k^{-1}\mathbf{s}_k(\mathbf{x})\mathbf{s}_k(\mathbf{x}')\sin\omega_k t$
Rotating Elastic	$\mathbf{G}(\mathbf{x},\mathbf{x}';t) = \mathrm{Re}\sum_k (i\omega_k)^{-1}\mathbf{s}_k(\mathbf{x})\mathbf{s}_k^*(\mathbf{x}')\exp(i\nu_k t)$
Non-Rotating Anelastic	$\mathbf{G}(\mathbf{x},\mathbf{x}';t) = \mathrm{Re}\sum_k (i\nu_k)^{-1}\mathbf{s}_k(\mathbf{x})\mathbf{s}_k(\mathbf{x}')\exp(i\nu_k t)$
Rotating Anelastic	$\mathbf{G}(\mathbf{x},\mathbf{x}';t) = \mathrm{Re}\sum_k (i\nu_k)^{-1}\mathbf{s}_k(\mathbf{x})\overline{\mathbf{s}}_k(\mathbf{x}')\exp(i\nu_k t)$

Table 6.3. Seismic Green tensor on an Earth model in the presence and absence of both rotation and anelasticity. Each of the sums is over all of the seismic normal modes with associated real or complex eigenfrequencies ω_k or $\nu_k = \omega_k + i\gamma_k$ in the right half-plane.

Green tensors (6.164) and (6.191) on an anelastic Earth are not exact. As noted above, we have assumed implicitly that the eigenfrequencies ν_k are distinct—the quantity in brackets in equations (6.159) and (6.188) is the residue only for a simple pole—and we have ignored the contribution from the integral along both sides of the logarithmic branch cut in Figure 6.7. Degeneracies that arise from a geometrical symmetry, such as the $2l+1$ degeneracy of a non-rotating, spherically symmetric Earth or the doublet degeneracy of a slowly rotating, ellipsoidal Earth, do not affect the results. A basis of orthonormal eigenfunctions $\mathbf{s}_k$ or biorthonormal eigenfunctions and dual eigenfunctions $\mathbf{s}_k$, $\overline{\mathbf{s}}_k$ still exists in both of these cases, as we discuss in Section 9.9. Only *accidental degeneracies* can alter the form of the Green tensor; they give rise to *secular* solutions of the form $t^n \exp(i\nu_k t)$, where n is the number of missing eigenfunctions associated with the degenerate eigenfrequency ν_k. A simple example illustrating this so-called *defective* case for a constant-Q system with two mechanical degrees of freedom is discussed by Tromp & Dahlen (1990b). We ignore the possibility that the elastic-gravitational operator $\mathcal{H}(\nu)$ of the Earth is defective, on the grounds that accidental degeneracy in systems with an infinite number of degrees of freedom is exceedingly rare.

Earth Model	Eigenfunction Normalization
Non-Rotating Elastic	$\int_\oplus \rho^0 \mathbf{s}_k \cdot \mathbf{s}_k \, dV = 1$
Rotating Elastic	$\int_\oplus \rho^0 \mathbf{s}_k^* \cdot \mathbf{s}_k \, dV - \omega_k^{-1} \int_\oplus \rho^0 \mathbf{s}_k^* \cdot (i\mathbf{\Omega} \times \mathbf{s}_k) \, dV = 1$
Non-Rotating Anelastic	$\int_\oplus \rho^0 \mathbf{s}_k \cdot \mathbf{s}_k \, dV - \frac{1}{2}\nu_k^{-1} \int_\oplus \boldsymbol{\varepsilon}_k : \partial_\nu \mathbf{\Gamma}(\nu_k) : \boldsymbol{\varepsilon}_k \, dV = 1$
Rotating Anelastic	$\int_\oplus \rho^0 \overline{\mathbf{s}}_k \cdot \mathbf{s}_k \, dV - \nu_k^{-1} \int_\oplus \rho^0 \overline{\mathbf{s}}_k \cdot (i\mathbf{\Omega} \times \mathbf{s}_k) \, dV$ $-\frac{1}{2}\nu_k^{-1} \int_\oplus \overline{\boldsymbol{\varepsilon}}_k : \partial_\nu \mathbf{\Gamma}(\nu_k) : \boldsymbol{\varepsilon}_k \, dV = 1$

Table 6.4. Eigenfunction normalization implicit in the representation of the four Green tensors $\mathbf{G}(\mathbf{x}, \mathbf{x}'; t)$.

6.4 Hydrostatic Anelastic Earth

All of the results in Sections 6.2 and 6.3 can be extended to the case of a hydrostatic Earth model with very little modification. On a non-rotating hydrostatic Earth the anelastic-gravitational operator $\rho^0\mathcal{H}(\nu)$ is given by

$$\begin{aligned}\rho^0\mathcal{H}(\nu)\mathbf{s} = -\boldsymbol{\nabla}[\boldsymbol{\Gamma}(\nu)\!:\!\boldsymbol{\varepsilon}] + \boldsymbol{\nabla}(\rho^0\mathbf{s}\cdot\boldsymbol{\nabla}\phi^0)\\ + \rho^0\boldsymbol{\nabla}\phi^{\mathrm{E1}} + \rho^{\mathrm{E1}}\boldsymbol{\nabla}\phi^0.\end{aligned} \tag{6.195}$$

It is easily verified that (6.195) is symmetric with respect to the duality product (6.139):

$$[\mathbf{s}, \mathcal{H}(\nu)\mathbf{s}'] = [\mathcal{H}(\nu)\mathbf{s}, \mathbf{s}'] = [\mathbf{s}', \mathcal{H}(\nu)\mathbf{s}]. \tag{6.196}$$

The result (6.196) guarantees that the biorthogonality and normalization conditions (6.156)–(6.157) are valid on a non-rotating hydrostatic Earth model. Since those conditions are the only essential ingredients in the derivation of the seismic Green tensor (6.164), it is clear that it remains valid as well.

The various forms of Rayleigh's principle are also applicable on a hydrostatic Earth, with $\mathcal{H}(\nu)$ given by equation (6.195). The action (6.147) and modified action (6.150) on a non-rotating hydrostatic Earth reduce to

$$\mathcal{I} = \int_\oplus L(\mathbf{s}, \boldsymbol{\nabla}\mathbf{s})\,dV, \qquad \mathcal{I}' = \int_\bigcirc L'(\mathbf{s}, \boldsymbol{\nabla}\mathbf{s}, \boldsymbol{\nabla}\phi^{\mathrm{E1}})\,dV, \tag{6.197}$$

where

$$\begin{aligned}L = \tfrac{1}{2}[\nu^2\rho^0\mathbf{s}\cdot\mathbf{s} - \boldsymbol{\varepsilon}\!:\!\boldsymbol{\Gamma}(\nu)\!:\!\boldsymbol{\varepsilon} - \rho^0\mathbf{s}\cdot\boldsymbol{\nabla}\phi^{\mathrm{E1}}\\ - \rho^0\mathbf{s}\cdot\boldsymbol{\nabla}\boldsymbol{\nabla}\phi^0\mathbf{s} - \rho^0\boldsymbol{\nabla}\phi^0\cdot(\mathbf{s}\cdot\boldsymbol{\nabla}\mathbf{s} - \mathbf{s}\boldsymbol{\nabla}\cdot\mathbf{s})],\end{aligned} \tag{6.198}$$

$$\begin{aligned}L' = \tfrac{1}{2}[\nu^2\rho^0\mathbf{s}\cdot\mathbf{s} - \boldsymbol{\varepsilon}\!:\!\boldsymbol{\Gamma}(\nu)\!:\!\boldsymbol{\varepsilon} - 2\rho^0\mathbf{s}\cdot\boldsymbol{\nabla}\phi^{\mathrm{E1}}\\ - \rho^0\mathbf{s}\cdot\boldsymbol{\nabla}\boldsymbol{\nabla}\phi^0\mathbf{s} - \rho^0\boldsymbol{\nabla}\phi^0\cdot(\mathbf{s}\cdot\boldsymbol{\nabla}\mathbf{s} - \mathbf{s}\boldsymbol{\nabla}\cdot\mathbf{s})\\ - (4\pi G)^{-1}\boldsymbol{\nabla}\phi^{\mathrm{E1}}\cdot\boldsymbol{\nabla}\phi^{\mathrm{E1}}].\end{aligned} \tag{6.199}$$

The stationary value of both of the hydrostatic actions is $\mathcal{I} = \mathcal{I}' = 0$ for any eigensolution ν, $\mathbf{s}$, ϕ^{E1}. Similar remarks apply to the case of a rotating hydrostatic Earth model.

6.5 Response to a Moment-Tensor Source

All of the considerations regarding the body-force density $\mathbf{f}$ and surface-force density $\mathbf{t}$ equivalent to an ideal fault source obtained in Chapter 5 pertain to an anelastic Earth as well, without any modification. It is simply necessary to replace the Hookean elastic constitutive relation (5.4) by (6.130) or the equivalent convolution in the time domain:

$$\mathbf{T}^{\mathrm{PK1}}(\mathbf{x},t) = \int_{-\infty}^{t} \mathbf{\Lambda}(\mathbf{x},t-t'):\partial_{t'}\boldsymbol{\nabla}\mathbf{s}(\mathbf{x},t')\,dt'. \tag{6.200}$$

An indigenous source then corresponds to a breakdown of this linear anelastic "law" relating the model stress to the entire past history of the deformation rate; the stress glut $\mathbf{S} = \mathbf{T}^{\mathrm{PK1}}_{\mathrm{model}} - \mathbf{T}^{\mathrm{PK1}}_{\mathrm{true}} = \mathbf{T}^{\mathrm{L1}}_{\mathrm{model}} - \mathbf{T}^{\mathrm{L1}}_{\mathrm{true}}$ is a measure of the extent of this breakdown. In the instantaneous point-source approximation, the moment tensor $\mathbf{M}$ is the integral over the source volume S^{f} of the final static stress glut $\mathbf{S}_{\mathrm{f}}$, just as on an elastic Earth.

The displacement response to an impulsive moment-tensor source on a rotating anelastic Earth is

$$\mathbf{s}(\mathbf{x},t) = \mathrm{Re}\sum_k \nu_k^{-2}\mathbf{M}:\overline{\boldsymbol{\varepsilon}}(\mathbf{x}_{\mathrm{s}})\mathbf{s}_k(\mathbf{x})[1-\exp(i\nu_k t)], \tag{6.201}$$

where we have set the fiducial origin time t_{s} equal to zero in accordance with the convention (5.137). The constant term in (6.201) represents the final static displacement:

$$\mathbf{s}_{\mathrm{f}}(\mathbf{x}) = \mathrm{Re}\sum_k \nu_k^{-2}\mathbf{M}:\overline{\boldsymbol{\varepsilon}}(\mathbf{x}_{\mathrm{s}})\mathbf{s}_k(\mathbf{x}). \tag{6.202}$$

The acceleration subsequent to the cessation of faulting is

$$\mathbf{a}(\mathbf{x},t) = \mathrm{Re}\sum_k \mathbf{M}:\overline{\boldsymbol{\varepsilon}}(\mathbf{x}_{\mathrm{s}})\mathbf{s}_k(\mathbf{x})\exp(i\nu_k t). \tag{6.203}$$

This can be rewritten in terms of the real positive eigenfrequencies $\omega_k > 0$ and the mode quality factors Q_k in the form

$$\begin{aligned}\mathbf{a}(\mathbf{x},t) = \sum_k\; & \mathrm{Re}\,[\mathbf{M}:\overline{\boldsymbol{\varepsilon}}(\mathbf{x}_{\mathrm{s}})\mathbf{s}_k(\mathbf{x})]\cos\omega_k t\exp(-\omega_k t/2Q_k) \\ & - \mathrm{Im}\,[\mathbf{M}:\overline{\boldsymbol{\varepsilon}}(\mathbf{x}_{\mathrm{s}})\mathbf{s}_k(\mathbf{x})]\sin\omega_k t\exp(-\omega_k t/2Q_k).\end{aligned} \tag{6.204}$$

Each mode oscillates with time like a linear combination of $\cos\omega_k t$ and $\sin\omega_k t$, and decays with time like $\exp(-\omega_k t/2Q_k)$. The anelasticity of the Earth is slight, so that $Q_k \gg 1$ for every mode in the sum (6.204).

The Fourier transform of the acceleration response is a weighted sum of Lorentzian resonance peaks centered on the positive and negative real eigenfrequencies $\pm\omega_k$:

$$\mathbf{a}(\mathbf{x},\omega) = \tfrac{1}{2}\sum_k \left[\frac{\mathbf{M}:\overline{\boldsymbol{\varepsilon}}_k(\mathbf{x}_{\mathrm{s}})\mathbf{s}_k(\mathbf{x})}{\gamma_k + i(\omega-\omega_k)} + \frac{\mathbf{M}:\overline{\boldsymbol{\varepsilon}}_k^*(\mathbf{x}_{\mathrm{s}})\mathbf{s}_k^*(\mathbf{x})}{\gamma_k + i(\omega+\omega_k)}\right]. \tag{6.205}$$

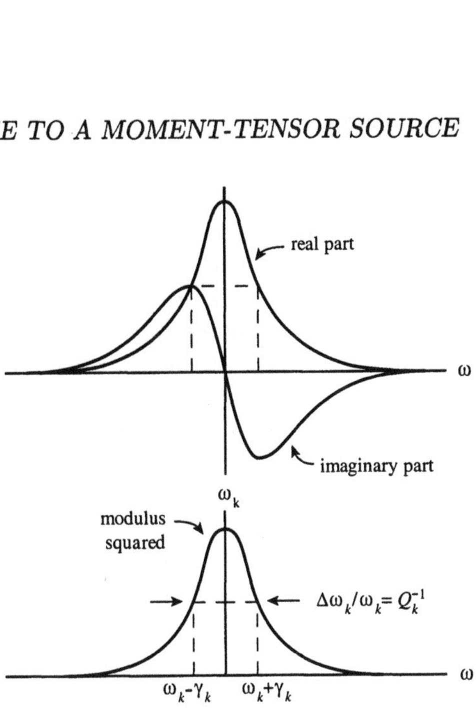

Figure 6.10. (*Top*) Real and imaginary parts of the unit Lorentzian resonance peak $\eta_k(\omega) = \frac{1}{2}[\gamma_k + i(\omega - \omega_k)]^{-1}$. (*Bottom*) Unit power spectrum $|\eta_k(\omega)|^2$.

The positive and negative peaks overlap very little in the case $Q_k \gg 1$, so that for positive frequencies, $\omega > 0$, we can replace (6.205) to a very good approximation by

$$\mathbf{a}(\mathbf{x}, \omega) = \sum_k \mathbf{M}:\overline{\boldsymbol{\varepsilon}}_k(\mathbf{x}_s)\mathbf{s}_k(\mathbf{x})\eta_k(\omega), \tag{6.206}$$

where

$$\eta_k(\omega) = \tfrac{1}{2}[\gamma_k + i(\omega - \omega_k)]^{-1}. \tag{6.207}$$

The characteristic features of the *unit Lorentzian* $\eta_k(\omega)$ are illustrated in Figure 6.10. The maximum of the corresponding unit power spectrum $|\eta_k(\omega)|^2$ occurs at $\omega = \omega_k$, and the half-power points are at $\omega = \omega_k \pm \gamma_k$. The width $\Delta\omega_k$ of each resonance peak at its half-power points is therefore related to the quality factor Q_k of the associated mode by

$$\Delta\omega_k/\omega_k = Q_k^{-1}. \tag{6.208}$$

In accurate quantitative studies, it is advisable to eschew the approximation (6.206) and use equation (6.205) as it stands. In that case, the maximum of each peak is displaced away from $\omega = \omega_k$ by a small amount, of order Q_k^{-2}, and the half-power width $\Delta\omega_k$ differs from the familiar result (6.208) by an amount of the same order.

Equations (6.201)–(6.204) and (6.205)–(6.207) constitute the complete solution to the *forward problem* of normal-mode seismology—given a model of the Earth and a moment-tensor source $\mathbf{M}$ situated at $\mathbf{x}_s$, they enable us to calculate the long-period response at a seismic station $\mathbf{x}$ in either the time or the frequency domain. The time-domain acceleration $\mathbf{a}(\mathbf{x}, t)$ is a weighted sum of decaying sinusoids and cosinusoids, whereas the corresponding frequency-domain response $\mathbf{a}(\mathbf{x}, \omega)$ is a weighted sum of dissipatively broadened resonance peaks. The weights or complex excitation amplitudes of each of the modes depend upon the size and orientation $\mathbf{M}$ and hypocentral location $\mathbf{x}_s$ of the source. All of the results in this section may be applied to a non-rotating anelastic Earth by simply eliminating the overbars upon the dual strains: $\bar{\boldsymbol{\varepsilon}}_k = \boldsymbol{\varepsilon}_k$.

Chapter 7

Rayleigh-Ritz Method

Thus far we have regarded the normal-mode eigenfrequencies and eigenfunctions of the Earth as the solutions to an integro-differential boundary-value problem. The *Rayleigh-Ritz method*, which we discuss in this chapter, provides a means of calculating ω, $\mathbf{s}$ or ν, $\mathbf{s}$ using only linear algebra, and not calculus. The method is extremely simple—we represent each eigenfunction $\mathbf{s}$ as a linear combination of real *trial or basis functions*,

$$\mathbf{s} = \sum_k q_k \mathbf{s}_k, \tag{7.1}$$

and solve for the expansion coefficients q_k. The basis functions $\mathbf{s}_k$ are, for the moment, regarded as completely arbitrary; the only restriction we impose—in addition to being real—is that they must be continuous everywhere within $\oplus$, except on the fluid-solid boundary Σ_{FS}, where they must satisfy $[\hat{\mathbf{n}} \cdot \mathbf{s}_k]_-^+ = 0$. In Chapter 13 we shall investigate the effect of the Earth's slow rotation and slight lateral heterogeneity by populating the basis set with the eigenfunctions of a non-rotating spherical Earth model.

We use lower-case and upper-case sans serif letters, respectively, to denote the column vectors and matrices that arise in the variational analysis. The vectors and matrices are finite-dimensional and the resulting eigensolutions are only approximate if the number of basis functions $\mathbf{s}_k$ in the expansion (7.1) is finite, as in any practical application. More generally, we can consider the basis set to be infinite-dimensional and complete, in which case the results can be regarded as exact. We shall refer to the column vectors as $\infty \times 1$ and the matrices as $\infty \times \infty$ in what follows, with the understanding that they must be truncated in practice. The truncated results can be regarded as an extension of the classical theory of small oscillations of a system with a finite number of degrees of freedom (Rayleigh 1877;

Goldstein 1980), generalized to account for rotation and linear anelasticity in a completely self-consistent manner. Most of the material in this chapter is simply a transcription of previously obtained results into the language of vectors and matrices; for that reason, we shall be terse.

7.1 Non-Rotating Elastic Earth

Upon inserting the expansion (7.1) into either equations (4.26) or (4.30), we obtain a purely algebraic form for the action governing a non-rotating elastic Earth:

$$\mathcal{I} = \tfrac{1}{2}\mathsf{q}^{\mathrm{T}}(\omega^2\mathsf{T} - \mathsf{V})\mathsf{q}, \tag{7.2}$$

where

$$\mathsf{q} = \begin{pmatrix} \vdots \\ q_k \\ \vdots \end{pmatrix} \tag{7.3}$$

is the real $\infty \times 1$ column vector of unknown coefficients, and the superscript T denotes the transpose. The elements of the $\infty \times \infty$ *kinetic and potential energy matrices*

$$\mathsf{T} = \begin{pmatrix} & \vdots & \\ \cdots & T_{kk'} & \cdots \\ & \vdots & \end{pmatrix}, \qquad \mathsf{V} = \begin{pmatrix} & \vdots & \\ \cdots & V_{kk'} & \cdots \\ & \vdots & \end{pmatrix} \tag{7.4}$$

are given by

$$T_{kk'} = \int_{\oplus} \rho^0 \mathbf{s}_k \cdot \mathbf{s}_{k'}\, dV, \tag{7.5}$$

$$\begin{aligned} V_{kk'} = {} & \int_{\oplus} [\boldsymbol{\nabla}\mathbf{s}_k : \boldsymbol{\Lambda} : \boldsymbol{\nabla}\mathbf{s}_{k'} + \tfrac{1}{2}\rho^0(\mathbf{s}_k \cdot \boldsymbol{\nabla}\phi_{k'}^{\mathrm{E1}} + \mathbf{s}_{k'} \cdot \boldsymbol{\nabla}\phi_k^{\mathrm{E1}}) \\ & \qquad + \rho^0 \mathbf{s}_k \cdot \boldsymbol{\nabla}\boldsymbol{\nabla}\phi^0 \cdot \mathbf{s}_{k'}]\, dV \\ & + \tfrac{1}{2}\int_{\Sigma_{\mathrm{FS}}} [\varpi^0 \mathbf{s}_k \cdot (\boldsymbol{\nabla}^{\Sigma}\mathbf{s}_{k'}) \cdot \hat{\mathbf{n}} + \varpi^0 \mathbf{s}_{k'} \cdot (\boldsymbol{\nabla}^{\Sigma}\mathbf{s}_k) \cdot \hat{\mathbf{n}} \\ & \qquad - (\hat{\mathbf{n}} \cdot \mathbf{s}_k)\boldsymbol{\nabla}^{\Sigma} \cdot (\varpi^0 \mathbf{s}_{k'}) - (\hat{\mathbf{n}} \cdot \mathbf{s}_{k'})\boldsymbol{\nabla}^{\Sigma} \cdot (\varpi^0 \mathbf{s}_k)]_-^+\, d\Sigma. \end{aligned} \tag{7.6}$$

The potential basis functions ϕ_k^{E1} are defined in terms of the displacement basis functions $\mathbf{s}_k$ by the integral relation

$$\phi_k^{\mathrm{E1}} = -G \int_{\oplus} \frac{\rho^{0\prime}\mathbf{s}_k' \cdot (\mathbf{x} - \mathbf{x}')}{\|\mathbf{x} - \mathbf{x}'\|^3}\, dV', \tag{7.7}$$

or, equivalently, by the boundary-value problem

$$\nabla^2 \phi_k^{\mathrm{E1}} = -4\pi G\, \boldsymbol{\nabla} \cdot (\rho^0 \mathbf{s}_k) \quad \text{in } \oplus, \tag{7.8}$$

$$\nabla^2 \phi_k^{\mathrm{E1}} = 0 \quad \text{in } \bigcirc - \oplus, \tag{7.9}$$

$$[\phi_k^{\mathrm{E1}}]_-^+ = 0 \quad \text{on } \Sigma, \tag{7.10}$$

$$[\hat{\mathbf{n}} \cdot \boldsymbol{\nabla}\phi_k^{\mathrm{E1}} + 4\pi G \rho^0 \hat{\mathbf{n}} \cdot \mathbf{s}_k]_-^+ = 0 \quad \text{on } \Sigma. \tag{7.11}$$

The displacement and displacement-potential versions of Rayleigh's principle lead to the same algebraic variational principle for the expansion-coefficient vector q, by virtue of the identity (3.195). Both T and V are real and symmetric:

$$\mathsf{T}^{\mathrm{T}} = \mathsf{T}, \quad \mathsf{V}^{\mathrm{T}} = \mathsf{V}. \tag{7.12}$$

In addition, the kinetic energy matrix is positive definite, and the potential energy matrix is positive semi-definite for any dynamically stable Earth model: $\mathsf{q}^{\mathrm{T}}\mathsf{T}\mathsf{q} > 0$ and $\mathsf{q}^{\mathrm{T}}\mathsf{V}\mathsf{q} \geq 0$ for all $\mathsf{q} \neq 0$.

The variation of the action $\mathcal{I}$, for a fixed value of the eigenfrequency, is $\delta\mathcal{I} = \delta\mathsf{q}^{\mathrm{T}}(\omega^2\mathsf{T} - \mathsf{V})\mathsf{q}$, where we have used the symmetries (7.12). This vanishes for an arbitrary variation $\delta\mathsf{q}$ if and only if

$$\mathsf{V}\mathsf{q} = \omega^2 \mathsf{T}\mathsf{q}. \tag{7.13}$$

Equation (7.13) is a generalized algebraic eigenvalue problem which can be solved to find the eigenfrequencies ω and associated eigenvectors q of the Earth model. The eigenfrequencies are the roots of the *secular equation*

$$\det(\mathsf{V} - \omega^2\mathsf{T}) = 0. \tag{7.14}$$

We could alternatively have regarded the *Rayleigh quotient*

$$\omega^2 = \frac{\mathsf{q}^{\mathrm{T}}\mathsf{V}\mathsf{q}}{\mathsf{q}^{\mathrm{T}}\mathsf{T}\mathsf{q}} \tag{7.15}$$

rather than the action $\mathcal{I}$ as the stationary functional. The two variations are related by $\delta\omega^2 = -2(\mathsf{q}^{\mathrm{T}}\mathsf{T}\mathsf{q})^{-1}\delta\mathcal{I}$, so that both $\delta\omega^2 = 0$ and $\delta\mathcal{I} = 0$ lead to the same equation (7.13). The stationary value of the action is $\mathcal{I} = 0$ at each of the eigensolutions ω^2, q by virtue of equation (7.15).

Since the kinetic energy matrix T is symmetric and positive definite, its inverse T^{-1} exists. This enables us to rewrite (7.13) as an ordinary algebraic eigenvalue problem of the form

$$(\mathsf{T}^{-1}\mathsf{V})\mathsf{q} = \omega^2\mathsf{q}. \tag{7.16}$$

The matrix $\mathsf{T}^{-1}\mathsf{V}$ in equation (7.16) is not in general symmetric since $(\mathsf{T}^{-1}\mathsf{V})^{\mathrm{T}} = \mathsf{V}\mathsf{T}^{-1} \neq \mathsf{T}^{-1}\mathsf{V}$. It is, however, self-adjoint with respect to the *kinetic-energy inner product*

$$\langle \mathsf{q}, \mathsf{q}' \rangle = \mathsf{q}^{\mathrm{T}}\mathsf{T}\mathsf{q}'. \tag{7.17}$$

The Hermiticity relation

$$\langle \mathsf{q}, \mathsf{T}^{-1}\mathsf{V}\mathsf{q}' \rangle = \langle \mathsf{T}^{-1}\mathsf{V}\mathsf{q}, \mathsf{q}' \rangle = \langle \mathsf{q}', \mathsf{T}^{-1}\mathsf{V}\mathsf{q} \rangle \tag{7.18}$$

is an elementary consequence of the symmetry of the potential energy matrix V. It follows that eigenvectors q and q' associated with distinct eigenfrequencies $\omega \neq \omega'$ are *orthogonal* in the sense

$$\langle \mathsf{q}, \mathsf{q}' \rangle = \mathsf{q}^{\mathrm{T}}\mathsf{T}\mathsf{q}' = 0 \quad \text{if } \omega \neq \omega'. \tag{7.19}$$

If we *normalize* the eigenvectors q such that

$$\langle \mathsf{q}, \mathsf{q} \rangle = \mathsf{q}^{\mathrm{T}}\mathsf{T}\mathsf{q} = 1, \tag{7.20}$$

then the two conditions (7.19)–(7.20) together can be rewritten in the form

$$\mathsf{Q}^{\mathrm{T}}\mathsf{T}\mathsf{Q} = \mathsf{I}, \tag{7.21}$$

where Q is the $\infty \times \infty$ matrix whose columns are the eigenvectors q, and I is the $\infty \times \infty$ identity. Equation (7.21) is the matrix version of the eigenfunction orthonormality relations (4.20)–(4.21) on a non-rotating elastic Earth.

The generalized eigenvalue problem (7.13) can be rewritten in the form

$$\mathsf{V}\mathsf{Q} = \mathsf{T}\mathsf{Q}\Omega^2, \tag{7.22}$$

where $\Omega = \mathrm{diag}\,[\cdots\ \omega\ \cdots]$ denotes the diagonal matrix of eigenfrequencies. Multiplying both sides of equation (7.22) on the left by Q^{T} and using (7.21), we obtain

$$\mathsf{Q}^{\mathrm{T}}\mathsf{V}\mathsf{Q} = \Omega^2. \tag{7.23}$$

Equations (7.21) and (7.23) show that the matrices T and V can be *simultaneously diagonalized* by the *congruent transformation* Q. Furthermore, these two results together imply that

$$\mathsf{Q}^{-1}(\mathsf{T}^{-1}\mathsf{V})\mathsf{Q} = \Omega^2, \tag{7.24}$$

so that Q is also the *similarity transformation* which diagonalizes $\mathsf{T}^{-1}\mathsf{V}$. It is noteworthy that Q is not an orthogonal matrix, $\mathsf{Q}^{-1} \neq \mathsf{Q}^{\mathrm{T}}$, so that none of the above transformations are rigid rotations in the space of $\infty \times 1$ column vectors q (Horn & Johnson 1985).

The $\infty\times\infty$ *Green matrix* $\mathsf{G}(t)$ is the solution to the initial-value problem

$$\mathsf{T}\ddot{\mathsf{G}} + \mathsf{V}\mathsf{G} = 0, \tag{7.25}$$

$$\mathsf{G}(0) = 0, \qquad \dot{\mathsf{G}}(0) = \mathsf{T}^{-1}, \tag{7.26}$$

where a dot denotes differentiation with respect to time. We seek a solution to equations (7.25)–(7.26) that is a sum of normal modes:

$$\mathsf{G}(t) = \mathsf{Q}\cos(\Omega t)\,\mathsf{A} + \mathsf{Q}\sin(\Omega t)\,\mathsf{B}, \tag{7.27}$$

where A and B are real $\infty\times\infty$ matrices of unknown coefficients. Upon multiplying both sides of the initial conditions

$$\mathsf{Q}\mathsf{A} = 0, \qquad \mathsf{Q}\Omega\mathsf{B} = \mathsf{T}^{-1} \tag{7.28}$$

on the left by $\mathsf{Q}^{\mathrm{T}}\mathsf{T}$ and invoking the orthonormality relation (7.21), we find that these coefficient matrices are

$$\mathsf{A} = 0, \qquad \mathsf{B} = \Omega^{-1}\mathsf{Q}^{\mathrm{T}}. \tag{7.29}$$

The Green matrix $\mathsf{G}(t)$ is therefore given in terms of the eigenfrequencies Ω and eigenvectors Q by

$$\mathsf{G}(t) = \mathsf{Q}\Omega^{-1}\sin(\Omega t)\,\mathsf{Q}^{\mathrm{T}}. \tag{7.30}$$

Equation (7.30) is valid for $t \geq 0$; obviously, $\mathsf{G}(t) = 0$ for $t < 0$.

The Green matrix can also be expressed directly in terms of the kinetic and potential energy matrices T and V without solving for the eigenfrequencies and eigenvectors. Following Woodhouse (1983) we define a matrix X by

$$\mathsf{X}^2 = \mathsf{T}^{-1}\mathsf{V}, \tag{7.31}$$

and write the right side of (7.30) as a power-series expansion:

$$\begin{aligned}\mathsf{Q}\Omega^{-1}\sin(\Omega t)\,\mathsf{Q}^{\mathrm{T}} &= (\mathsf{Q}\mathsf{Q}^{\mathrm{T}})t - \frac{1}{3!}(\mathsf{Q}\Omega^2\mathsf{Q}^{\mathrm{T}})t^3 + \frac{1}{5!}(\mathsf{Q}\Omega^4\mathsf{Q}^{\mathrm{T}})t^5 - \cdots \\ &= \left[t - \frac{1}{3!}\mathsf{X}^2t^3 + \frac{1}{5!}\mathsf{X}^4t^5 - \cdots\right]\mathsf{Q}\mathsf{Q}^{\mathrm{T}} \\ &= \mathsf{X}^{-1}\sin(\mathsf{X}t)\mathsf{Q}\mathsf{Q}^{\mathrm{T}} = \mathsf{X}^{-1}\sin(\mathsf{X}t)\mathsf{T}^{-1},\end{aligned} \tag{7.32}$$

where we have used (7.24) and (7.21) to obtain the second and fourth equalities, respectively. The Green matrix on a non-rotating elastic Earth can accordingly be written in the form

$$\mathsf{G}(t) = \mathsf{X}^{-1}\sin(\mathsf{X}t)\mathsf{T}^{-1}. \tag{7.33}$$

The frequency-domain Green matrix $\mathsf{G}(\omega)$ corresponding to the two time-domain expressions (7.30) and (7.33) is

$$\mathsf{G}(\omega) = \mathsf{Q}(\Omega^2 - \omega^2\mathsf{I})^{-1}\mathsf{Q}^{\mathrm{T}} = (\mathsf{V} - \omega^2\mathsf{T})^{-1}, \tag{7.34}$$

where we have used (7.21) and (7.23) to obtain the second equality. *Source-receiver reciprocity* is guaranteed by the symmetry $\mathsf{G}^{\mathrm{T}} = \mathsf{G}$, where G denotes either the time-domain or frequency-domain impulse response.

★7.2 Rotating Elastic Earth

The action obtained by inserting the expansion (7.1) into either (4.103) or (4.109) on a rotating elastic Earth is

$$\mathcal{I} = \tfrac{1}{2}\mathsf{q}^{\mathrm{H}}(\omega^2\mathsf{T} - 2\omega\mathsf{W} - \mathsf{V})\mathsf{q}, \tag{7.35}$$

where the column vector of unknown coefficients q is now complex, and the superscript H denotes the Hermitian or complex-conjugate transpose. The elements of the kinetic and potential energy matrices T and V are the same as on a non-rotating Earth, with the gravitational potential ϕ^0 replaced by the geopotential $\phi^0 + \psi$:

$$\begin{aligned} V_{kk'} = \int_{\oplus} [&\boldsymbol{\nabla}\mathbf{s}_k : \boldsymbol{\Lambda} : \boldsymbol{\nabla}\mathbf{s}_{k'} + \tfrac{1}{2}\rho^0(\mathbf{s}_k \cdot \boldsymbol{\nabla}\phi_{k'}^{\mathrm{E1}} + \mathbf{s}_{k'} \cdot \boldsymbol{\nabla}\phi_k^{\mathrm{E1}}) \\ &+ \rho^0\mathbf{s}_k \cdot \boldsymbol{\nabla}\boldsymbol{\nabla}(\phi^0 + \psi) \cdot \mathbf{s}_{k'}]\, dV \\ + \tfrac{1}{2}\int_{\Sigma_{\mathrm{FS}}} &[\varpi^0\mathbf{s}_k \cdot (\boldsymbol{\nabla}^{\Sigma}\mathbf{s}_{k'}) \cdot \hat{\mathbf{n}} + \varpi^0\mathbf{s}_{k'} \cdot (\boldsymbol{\nabla}^{\Sigma}\mathbf{s}_k) \cdot \hat{\mathbf{n}} \\ &- (\hat{\mathbf{n}} \cdot \mathbf{s}_k)\boldsymbol{\nabla}^{\Sigma} \cdot (\varpi^0\mathbf{s}_{k'}) - (\hat{\mathbf{n}} \cdot \mathbf{s}_{k'})\boldsymbol{\nabla}^{\Sigma} \cdot (\varpi^0\mathbf{s}_k)]_-^+\, d\Sigma. \end{aligned} \tag{7.36}$$

The elements of the $\infty \times \infty$ *Coriolis matrix*

$$\mathsf{W} = \begin{pmatrix} & \vdots & \\ \cdots & W_{kk'} & \cdots \\ & \vdots & \end{pmatrix} \tag{7.37}$$

are given by

$$W_{kk'} = \int_{\oplus} \rho^0\mathbf{s}_k \cdot (i\boldsymbol{\Omega} \times \mathbf{s}_{k'})\, dV. \tag{7.38}$$

Both T and V are real and symmetric as before; however, W is imaginary and anti-symmetric. As a result, all three matrices are Hermitian:

$$\mathsf{T}^{\mathrm{H}} = \mathsf{T}, \quad \mathsf{V}^{\mathrm{H}} = \mathsf{V}, \quad \mathsf{W}^{\mathrm{H}} = \mathsf{W}. \tag{7.39}$$

In addition, the kinetic energy matrix T is positive definite, and the potential energy matrix V is positive semi-definite for any secularly stable Earth model.

The variation of the action (7.35) is $\delta\mathcal{I} = \mathrm{Re}\,[\delta\mathsf{q}^{\mathrm{H}}(\omega^2\mathsf{T} - 2\omega\mathsf{W} - \mathsf{V})\mathsf{q}]$, where we have used the symmetries (7.39). Clearly, $\delta\mathcal{I}$ vanishes for an arbitrary variation $\delta\mathsf{q}$ if and only if q is an eigenvector of the rotating Earth with associated eigenfrequency ω:

$$(\mathsf{V} + 2\omega\mathsf{W} - \omega^2\mathsf{T})\mathsf{q} = 0. \tag{7.40}$$

Upon taking the complex conjugate of (7.40), we see that $-\omega$, q^* is an eigensolution if and only if ω, q is. Furthermore, the combination ω, q^* is an eigensolution of the anti-Earth with a reversed sense of rotation, $\mathsf{W} \rightarrow -\mathsf{W}$, if and only if ω, q is an eigensolution of the actual Earth. The eigenfrequencies, which are the roots of the *secular equation*

$$\det(\mathsf{V} \pm 2\omega\mathsf{W} - \omega^2\mathsf{T}) = 0, \tag{7.41}$$

are independent of the sense of the Earth's rotation. The stationary value of the action is $\mathcal{I} = 0$ at each of the eigensolutions ω, q.

The normal-mode orthogonality relation which is obtained by subtracting $\mathsf{q}'^{\mathrm{H}}(\mathsf{V} + 2\omega\mathsf{W} - \omega^2\mathsf{T})\mathsf{q}$ from $\mathsf{q}^{\mathrm{H}}(\mathsf{V} + 2\omega'\mathsf{W} - \omega'^2\mathsf{T})\mathsf{q}'$ is

$$\mathsf{q}^{\mathrm{H}}\mathsf{T}\mathsf{q}' - 2(\omega + \omega')^{-1}\mathsf{q}^{\mathrm{H}}\mathsf{W}\mathsf{q}' = 0 \quad \text{if } \omega \neq \omega'. \tag{7.42}$$

We normalize the eigenvectors q by

$$\mathsf{q}^{\mathrm{H}}\mathsf{T}\mathsf{q} - \omega^{-1}\mathsf{q}^{\mathrm{H}}\mathsf{W}\mathsf{q} = 1. \tag{7.43}$$

Equations (7.42)–(7.43) are the matrix version of the three-dimensional eigenvector orthonormality relations (4.81)–(4.82).

The non-standard generalized algebraic eigenvalue problem (7.40) can be converted into an ordinary eigenvalue problem at the expense of doubling its size. We define the $2\infty \times 1$ column vector

$$\mathsf{z} = \begin{pmatrix} \mathsf{q} \\ \omega\mathsf{q} \end{pmatrix} \tag{7.44}$$

and the $2\infty \times 2\infty$ matrices

$$\mathsf{K} = \begin{pmatrix} 0 & \mathsf{I} \\ \mathsf{V} & 2\mathsf{W} \end{pmatrix}, \qquad \mathsf{M} = \begin{pmatrix} \mathsf{I} & 0 \\ 0 & \mathsf{T} \end{pmatrix}. \tag{7.45}$$

It is then easily shown that (7.40) is equivalent to

$$\mathsf{K}\mathsf{z} = \omega\mathsf{M}\mathsf{z}. \tag{7.46}$$

The augmented kinetic energy matrix M is Hermitian and positive definite, so its inverse M^{-1} exists; hence, equation (7.46) can be rewritten in the form

$$\mathsf{M}^{-1}\mathsf{K}\mathbf{z} = \omega\mathbf{z}. \tag{7.47}$$

The matrix $\mathsf{M}^{-1}\mathsf{K}$ is self-adjoint with respect to the augmented energy inner product

$$\langle\langle \mathbf{z}, \mathbf{z}' \rangle\rangle = \mathbf{z}^{\mathrm{H}}\mathsf{P}\mathbf{z}', \tag{7.48}$$

where

$$\mathsf{P} = \begin{pmatrix} \mathsf{V} & 0 \\ 0 & \mathsf{T} \end{pmatrix}. \tag{7.49}$$

The Hermiticity relation

$$\langle\langle \mathbf{z}, \mathsf{M}^{-1}\mathsf{K}\mathbf{z}' \rangle\rangle = \langle\langle \mathsf{M}^{-1}\mathsf{K}\mathbf{z}, \mathbf{z}' \rangle\rangle = \langle\langle \mathbf{z}', \mathsf{M}^{-1}\mathsf{K}\mathbf{z} \rangle\rangle^{*} \tag{7.50}$$

is a straightforward consequence of the $2\infty \times 2\infty$ Hermitian symmetry $\mathsf{P}\mathsf{M}^{-1}\mathsf{K} = (\mathsf{P}\mathsf{M}^{-1}\mathsf{K})^{\mathrm{H}}$. It follows from equation (7.50) that the $2\infty \times 1$ eigenvectors $\mathbf{z}$ and $\mathbf{z}'$ associated with distinct eigenfrequencies $\omega \neq \omega'$ are orthogonal in the sense

$$\langle\langle \mathbf{z}, \mathbf{z}' \rangle\rangle = \mathbf{z}^{\mathrm{H}}\mathsf{P}\mathbf{z}' = 0 \quad \text{if } \omega \neq \omega'. \tag{7.51}$$

We normalize the eigenvectors $\mathbf{z}$ such that

$$\langle\langle \mathbf{z}, \mathbf{z} \rangle\rangle = \mathbf{z}^{\mathrm{H}}\mathsf{P}\mathbf{z} = 2\omega^2. \tag{7.52}$$

Equations (7.51)–(7.52) are then equivalent to the six-dimensional orthonormality relations (4.95)–(4.96), as well as to the $\infty \times 1$ eigenvector relations (7.42)–(7.43).

We can express the above results in a manner analogous to (7.21)–(7.24), provided we account for the eigenvectors

$$\mathbf{z}^{*} = \begin{pmatrix} \mathbf{q}^{*} \\ -\omega\mathbf{q}^{*} \end{pmatrix} \tag{7.53}$$

associated with the negative eigenfrequencies $-\omega$ explicitly. Let Z be the $2\infty \times 2\infty$ matrix whose columns consist of all of the eigenvectors $\mathbf{z}$ followed by all of the eigenvectors $\mathbf{z}^{*}$, and let $\Sigma = \mathrm{diag}\,[\cdots \omega \cdots -\omega \cdots]$ be the associated diagonal matrix of eigenfrequencies. The orthonormality relations (7.51)–(7.52) can be rewritten using this notation in the succinct form

$$\mathsf{Z}^{\mathrm{H}}\mathsf{P}\mathsf{Z} = 2\Sigma^2. \tag{7.54}$$

Multiplying both sides of the generalized eigenvalue problem

$$\mathsf{KZ} = \mathsf{MZ\Sigma} \tag{7.55}$$

on the left by $\mathsf{Z}^{\mathrm{H}}\mathsf{PM}^{-1}$, and invoking (7.54), we also obtain

$$\mathsf{Z}^{\mathrm{H}}(\mathsf{PM}^{-1}\mathsf{K})\mathsf{Z} = 2\Sigma^3. \tag{7.56}$$

Equations (7.54) and (7.56) reveal the algebraic structure of the elastic rotating eigenproblem clearly: the two Hermitian matrices

$$\mathsf{P} = \begin{pmatrix} \mathsf{V} & 0 \\ 0 & \mathsf{T} \end{pmatrix} \quad \text{and} \quad \mathsf{PM}^{-1}\mathsf{K} = \begin{pmatrix} 0 & \mathsf{V} \\ \mathsf{V} & 2\mathsf{W} \end{pmatrix} \tag{7.57}$$

are simultaneously diagonalized by the congruent transformation Z. In addition, Z is the similarity transformation that diagonalizes

$$\mathsf{M}^{-1}\mathsf{K} = \begin{pmatrix} 0 & \mathsf{I} \\ \mathsf{T}^{-1}\mathsf{V} & 2\mathsf{T}^{-1}\mathsf{W} \end{pmatrix}, \tag{7.58}$$

since

$$\mathsf{Z}^{-1}(\mathsf{M}^{-1}\mathsf{K})\mathsf{Z} = \Sigma. \tag{7.59}$$

Fundamentally, the Hermitian structure of the $2\infty \times 2\infty$ eigenvalue problem on a rotating elastic Earth is a consequence of the $\infty \times \infty$ symmetries (7.39).

The Green matrix on a rotating elastic Earth satisfies

$$\mathsf{T}\ddot{\mathsf{G}} - 2i\mathsf{W}\dot{\mathsf{G}} + \mathsf{V}\mathsf{G} = 0, \tag{7.60}$$

or, equivalently,

$$\frac{d}{dt}\begin{pmatrix} \mathsf{G} \\ \dot{\mathsf{G}} \end{pmatrix} = \begin{pmatrix} 0 & \mathsf{I} \\ \mathsf{T}^{-1}\mathsf{V} & -2i\mathsf{T}^{-1}\mathsf{W} \end{pmatrix}\begin{pmatrix} \mathsf{G} \\ \dot{\mathsf{G}} \end{pmatrix}, \tag{7.61}$$

subject to the initial conditions

$$\begin{pmatrix} \mathsf{G}(0) \\ \dot{\mathsf{G}}(0) \end{pmatrix} = \begin{pmatrix} 0 \\ \mathsf{T}^{-1} \end{pmatrix}. \tag{7.62}$$

The matrix $i\mathsf{W}$ is real, so that (7.61)–(7.62) is a real initial-value problem. We consider a solution consisting of a sum over the $\pm\Omega$ eigensolutions of the form

$$\mathsf{G}(t) = \mathrm{Re}\,[\mathsf{Q}\exp(i\Omega t)\,\mathsf{C}], \tag{7.63}$$

and use the initial conditions

$$\begin{pmatrix} \mathsf{QC} + \mathsf{Q}^*\mathsf{C}^* \\ \mathsf{Q\Omega C} - \mathsf{Q}^*\Omega\mathsf{C}^* \end{pmatrix} = \begin{pmatrix} 0 \\ -2i\mathsf{T}^{-1} \end{pmatrix} \tag{7.64}$$

or, equivalently,

$$\mathsf{Z}\begin{pmatrix} \mathsf{C} \\ \mathsf{C}^* \end{pmatrix} = \begin{pmatrix} 0 \\ -2i\mathsf{T}^{-1} \end{pmatrix} \tag{7.65}$$

to solve for the complex coefficient matrix C. Multiplying both sides of equation (7.65) on the left by $\mathsf{Z}^{\mathrm{H}}\mathsf{P}$ and making use of the orthonormality relation (7.54), we obtain

$$\mathsf{C} = (i\Omega)^{-1}\mathsf{Q}^{\mathrm{H}}. \tag{7.66}$$

The Green matrix is therefore given in terms of Ω and Q by

$$\mathsf{G}(t) = \mathrm{Re}\,[\mathsf{Q}(i\Omega)^{-1}\exp(i\Omega t)\,\mathsf{Q}^{\mathrm{H}}]. \tag{7.67}$$

The corresponding *anti-Green matrix* on an Earth with the sense of rotation reversed is

$$\overline{\mathsf{G}}(t) = \mathrm{Re}\,[\mathsf{Q}^*(i\Omega)^{-1}\exp(i\Omega t)\,\mathsf{Q}^{\mathrm{T}}], \tag{7.68}$$

where Q^* is the matrix whose columns are the anti-eigenvectors q^*.

The Fourier transform of the Green matrix can be written in either of the two forms

$$\begin{aligned}\mathsf{G}(\omega) &= \tfrac{1}{2}\mathsf{Q}\Omega^{-1}(\Omega - \omega\mathsf{I})^{-1}\mathsf{Q}^{\mathrm{H}} + \tfrac{1}{2}\mathsf{Q}^*\Omega^{-1}(\Omega + \omega\mathsf{I})^{-1}\mathsf{Q}^{\mathrm{T}} \\ &= (\mathsf{V} + 2\omega\mathsf{W} - \omega^2\mathsf{T})^{-1},\end{aligned} \tag{7.69}$$

whereas the Fourier transform of the anti-Green matrix is

$$\begin{aligned}\overline{\mathsf{G}}(\omega) &= \tfrac{1}{2}\mathsf{Q}^*\Omega^{-1}(\Omega - \omega\mathsf{I})^{-1}\mathsf{Q}^{\mathrm{T}} + \tfrac{1}{2}\mathsf{Q}\Omega^{-1}(\Omega + \omega\mathsf{I})^{-1}\mathsf{Q}^{\mathrm{H}} \\ &= (\mathsf{V} - 2\omega\mathsf{W} - \omega^2\mathsf{T})^{-1}.\end{aligned} \tag{7.70}$$

The principle of *generalized source-receiver reciprocity* is guaranteed by the matrix symmetry relation $\mathsf{G} = \overline{\mathsf{G}}^{\mathrm{T}}$.

7.3 Non-Rotating Anelastic Earth

The action obtained by inserting the expansion (7.1) into either equation (6.147) or (6.150) on a non-rotating anelastic Earth can be written in the form

$$\mathcal{I} = \tfrac{1}{2}\mathsf{q}^{\mathrm{T}}[\nu^2\mathsf{T} - \mathsf{V}(\nu)]\mathsf{q}. \tag{7.71}$$

The elements of the $\infty \times \infty$ kinetic and potential energy matrices T and $\mathsf{V}(\nu)$ are again given by by equations (7.5) and (7.6), except that $\mathbf{\Lambda}$ is now replaced by the complex, frequency-dependent tensor $\mathbf{\Lambda}(\nu)$:

$$\begin{aligned} V_{kk'}(\nu) = \int_{\oplus} [\boldsymbol{\nabla}\mathbf{s}_k : \mathbf{\Lambda}(\nu) : \boldsymbol{\nabla}\mathbf{s}_{k'} + \tfrac{1}{2}\rho^0(\mathbf{s}_k \cdot \boldsymbol{\nabla}\phi^{\mathrm{E1}}_{k'} + \mathbf{s}_{k'} \cdot \boldsymbol{\nabla}\phi^{\mathrm{E1}}_{k}) \\ + \rho^0 \mathbf{s}_k \cdot \boldsymbol{\nabla\nabla}\phi^0 \cdot \mathbf{s}_{k'}]\, dV \qquad (7.72) \\ + \tfrac{1}{2}\int_{\Sigma_{\mathrm{FS}}} [\varpi^0 \mathbf{s}_k \cdot (\boldsymbol{\nabla}^{\Sigma}\mathbf{s}_{k'}) \cdot \hat{\mathbf{n}} + \varpi^0 \mathbf{s}_{k'} \cdot (\boldsymbol{\nabla}^{\Sigma}\mathbf{s}_{k}) \cdot \hat{\mathbf{n}} \\ - (\hat{\mathbf{n}} \cdot \mathbf{s}_k)\boldsymbol{\nabla}^{\Sigma} \cdot (\varpi^0 \mathbf{s}_{k'}) - (\hat{\mathbf{n}} \cdot \mathbf{s}_{k'})\boldsymbol{\nabla}^{\Sigma} \cdot (\varpi^0 \mathbf{s}_{k})]_-^+ \, d\Sigma. \end{aligned}$$

The matrix $\mathsf{V}(\nu)$ is now complex as well as symmetric:

$$\mathsf{V}^{\mathrm{T}}(\nu) = \mathsf{V}(\nu). \qquad (7.73)$$

The variation of the action, $\delta\mathcal{I} = \mathrm{Re}\,\{\delta\mathsf{q}^{\mathrm{T}}[\nu^2\mathsf{T} - \mathsf{V}(\nu)]\mathsf{q}\}$, vanishes for an arbitrary $\delta\mathsf{q}$ if and only if

$$\mathsf{V}(\nu)\mathsf{q} = \nu^2\mathsf{T}\mathsf{q}. \qquad (7.74)$$

The eigenfrequencies ν and associated eigenvectors q obtained by solving (7.74) are *complex*; the eigenfrequencies are the roots of the secular equation

$$\det\,[\mathsf{V}(\nu) - \nu^2\mathsf{T}] = 0. \qquad (7.75)$$

The symmetry $\mathbf{\Lambda}(-\nu^*) = \mathbf{\Lambda}^*(\nu)$ of the anelasticity tensor guarantees that $\mathsf{V}(-\nu^*) = \mathsf{V}^*(\nu)$, so that $-\nu^*$, q^* is an eigensolution if ν, q is. The value of the action at each of the stationary points is $\mathcal{I} = 0$.

The normal-mode orthogonality relation which is obtained by subtracting $\mathsf{q}'^{\mathrm{T}}[\mathsf{V}(\nu) - \nu^2\mathsf{T}]\mathsf{q}$ from $\mathsf{q}^{\mathrm{T}}[\mathsf{V}(\nu') - \nu'^2\mathsf{T}]\mathsf{q}'$ is

$$\mathsf{q}^{\mathrm{T}}\mathsf{T}\mathsf{q}' - (\nu^2 - \nu'^2)^{-1}\mathsf{q}^{\mathrm{T}}[\mathsf{V}(\nu) - \mathsf{V}(\nu')]\mathsf{q}' = 0 \quad \text{if } \nu \neq \nu'. \qquad (7.76)$$

We normalize the complex eigenvectors q by

$$\mathsf{q}^{\mathrm{T}}\mathsf{T}\mathsf{q} - \tfrac{1}{2}\nu^{-1}\mathsf{q}^{\mathrm{T}}\partial_\nu\mathsf{V}(\nu)\mathsf{q} = 1. \qquad (7.77)$$

Equations (7.76)–(7.77) are the matrix version of the non-rotating, anelastic orthonormality relations (6.143)–(6.144).

The $\infty \times \infty$ Green matrix on an anelastic non-rotating Earth is

$$\mathsf{G}(t) = \mathrm{Re}\,[\mathsf{Q}(i\mathsf{N})^{-1}\exp(i\mathsf{N}t)\,\mathsf{Q}^{\mathrm{T}}], \qquad (7.78)$$

where $\mathsf{N} = \mathrm{diag}\,[\cdots\,\nu\,\cdots]$ is the diagonal matrix of complex eigenfrequencies. The corresponding impulse response in the frequency domain is

$$\begin{aligned} \mathsf{G}(\nu) &= \tfrac{1}{2}\mathsf{Q}\mathsf{N}^{-1}(\mathsf{N} - \nu\mathsf{I})^{-1}\mathsf{Q}^{\mathrm{T}} + \tfrac{1}{2}\mathsf{Q}^*\mathsf{N}^{*-1}(\mathsf{N}^* + \nu\mathsf{I})^{-1}\mathsf{Q}^{\mathrm{H}} \\ &= [\mathsf{V}(\nu) - \nu^2\mathsf{T}]^{-1}. \end{aligned} \qquad (7.79)$$

The time-domain Green matrix $\mathsf{G}(t)$ can be obtained by an application of the residue theorem analogous to that used to find $\mathbf{G}(\mathbf{x},\mathbf{x}';t)$ in Section 6.2.3. Eigenfrequency degeneracy and the effect of the logarithmic branch cut along the positive imaginary axis must again be ignored, so that the result (7.78) is equivalent to equation (6.164). Fundamentally, it is the symmetry (7.73) of the potential energy matrix $\mathsf{V}(\nu)$ that is responsible for the simplicity of the response $\mathsf{G}(t)$. Source-receiver reciprocity is guaranteed by the symmetry of the Green matrix: $\mathsf{G}^{\mathrm{T}} = \mathsf{G}$.

★7.4 Rotating Anelastic Earth

On a rotating anelastic Earth, we must expand the *dual eigenfunctions* $\overline{\mathbf{s}}$ of the anti-Earth as well as the eigenfunctions $\mathbf{s}$ of the actual Earth:

$$\mathbf{s} = \sum_k q_k \mathbf{s}_k, \qquad \overline{\mathbf{s}} = \sum_k \overline{q}_k \mathbf{s}_k. \tag{7.80}$$

Note that the real basis functions $\mathbf{s}_k$ in the two expansions are identical; only the complex coefficients q_k and $\overline{q}_k$ are different. Inserting (7.80) into either equation (6.180) or (6.183) yields the algebraic action:

$$\mathcal{I} = \tfrac{1}{2}\overline{\mathsf{q}}^{\mathrm{T}}[\nu^2\mathsf{T} - 2\nu\mathsf{W} - \mathsf{V}(\nu)]\mathsf{q}, \tag{7.81}$$

where

$$\mathsf{q} = \begin{pmatrix} \vdots \\ q_k \\ \vdots \end{pmatrix}, \qquad \overline{\mathsf{q}} = \begin{pmatrix} \vdots \\ \overline{q}_k \\ \vdots \end{pmatrix}. \tag{7.82}$$

The elements of the potential energy matrix $\mathsf{V}(\nu)$ now include the effects of both anelasticity $\mathbf{\Lambda}(\nu)$ and the centrifugal potential ψ:

$$\begin{aligned} V_{kk'}(\nu) = \int_{\oplus} [&\nabla\mathbf{s}_k : \mathbf{\Lambda}(\nu) : \nabla\mathbf{s}_{k'} + \tfrac{1}{2}\rho^0(\mathbf{s}_k\cdot\nabla\phi_{k'}^{\mathrm{E1}} + \mathbf{s}_{k'}\cdot\nabla\phi_k^{\mathrm{E1}}) \\ &+ \rho^0\mathbf{s}_k\cdot\nabla\nabla(\phi^0+\psi)\cdot\mathbf{s}_{k'}]\,dV \\ + \tfrac{1}{2}\int_{\Sigma_{\mathrm{FS}}} &[\varpi^0\mathbf{s}_k\cdot(\nabla^\Sigma\mathbf{s}_{k'})\cdot\hat{\mathbf{n}} + \varpi^0\mathbf{s}_{k'}\cdot(\nabla^\Sigma\mathbf{s}_k)\cdot\hat{\mathbf{n}} \\ &- (\hat{\mathbf{n}}\cdot\mathbf{s}_k)\nabla^\Sigma\cdot(\varpi^0\mathbf{s}_{k'}) - (\hat{\mathbf{n}}\cdot\mathbf{s}_{k'})\nabla^\Sigma\cdot(\varpi^0\mathbf{s}_k)]_-^+\,d\Sigma. \end{aligned} \tag{7.83}$$

The variation of the action (7.81) is given by

$$\begin{aligned} \delta\mathcal{I} = \tfrac{1}{2}\delta\overline{\mathsf{q}}^{\mathrm{T}}&[\nu^2\mathsf{T} - 2\nu\mathsf{W} - \mathsf{V}(\nu)]\mathsf{q} \\ &+ \tfrac{1}{2}\delta\mathsf{q}^{\mathrm{T}}[\nu^2\mathsf{T} + 2\nu\mathsf{W} - \mathsf{V}(\nu)]\overline{\mathsf{q}}, \end{aligned} \tag{7.84}$$

where we have used the symmetry of the kinetic and potential energy matrices, $\mathsf{T}^{\mathrm{T}} = \mathsf{T}$ and $\mathsf{V}^{\mathrm{T}}(\nu) = \mathsf{V}(\nu)$, in addition to the anti-symmetry of the Coriolis matrix:

$$\mathsf{W}^{\mathrm{T}} = -\mathsf{W}. \tag{7.85}$$

Evidently, $\delta\mathcal{I}$ vanishes for arbitrary and independent variations $\delta\mathsf{q}$ and $\delta\overline{\mathsf{q}}$ if and only if q and $\overline{\mathsf{q}}$ are an eigenvector and dual eigenvector with associated eigenfrequency ν:

$$[\mathsf{V}(\nu) + 2\nu\mathsf{W} - \nu^2\mathsf{T}]\mathsf{q} = 0, \tag{7.86}$$

$$[\mathsf{V}(\nu) - 2\nu\mathsf{W} - \nu^2\mathsf{T}]\overline{\mathsf{q}} = 0. \tag{7.87}$$

The eigenfrequencies ν are the roots of the secular equation

$$\det[\mathsf{V}(\nu) \pm 2\nu\mathsf{W} - \nu^2\mathsf{T}] = 0. \tag{7.88}$$

The transposed matrices $\mathsf{V}(\nu) \pm 2\nu\mathsf{W} - \nu^2\mathsf{T}$ have the same determinant; as a result, the eigenfrequencies do not depend on the sense of the Earth's rotation. The complex frequency symmetry $\mathsf{V}(-\nu^*) = \mathsf{V}^*(\nu)$ guarantees that $-\nu^*$, q^*, $\overline{\mathsf{q}}^*$ are an eigensolution and dual eigensolution if ν, q, $\overline{\mathsf{q}}$ are. The value of the action at each of the stationary points ν, q, $\overline{\mathsf{q}}$ is $\mathcal{I} = 0$. The primal–dual eigenvector *biorthonormality* relations analogous to equations (6.186)–(6.187) are

$$\begin{aligned}&\overline{\mathsf{q}}^{\mathrm{T}}\mathsf{T}\mathsf{q}' - 2(\nu+\nu')^{-1}\overline{\mathsf{q}}^{\mathrm{T}}\mathsf{W}\mathsf{q}' \\ &\quad - (\nu^2 - \nu'^2)^{-1}\overline{\mathsf{q}}^{\mathrm{T}}[\mathsf{V}(\nu) - \mathsf{V}(\nu')]\mathsf{q}' = 0 \quad \text{if } \nu \neq \nu',\end{aligned} \tag{7.89}$$

and

$$\overline{\mathsf{q}}^{\mathrm{T}}\mathsf{T}\mathsf{q} - \nu^{-1}\overline{\mathsf{q}}^{\mathrm{T}}\mathsf{W}\mathsf{q} - \tfrac{1}{2}\nu^{-1}\overline{\mathsf{q}}^{\mathrm{T}}\partial_\nu\mathsf{V}(\nu)\mathsf{q} = 1. \tag{7.90}$$

The Green matrices on the Earth and anti-Earth are given by

$$\mathsf{G}(t) = \mathrm{Re}\,[\mathsf{Q}(i\mathsf{N})^{-1}\exp(i\mathsf{N}t)\,\overline{\mathsf{Q}}^{\mathrm{T}}], \tag{7.91}$$

$$\overline{\mathsf{G}}(t) = \mathrm{Re}\,[\overline{\mathsf{Q}}(i\mathsf{N})^{-1}\exp(i\mathsf{N}t)\,\mathsf{Q}^{\mathrm{T}}], \tag{7.92}$$

where $\overline{\mathsf{Q}}$ is the $\infty \times \infty$ matrix whose columns are the dual eigenvectors $\overline{\mathsf{q}}$. The corresponding results in the frequency domain are

$$\begin{aligned}\mathsf{G}(\nu) &= \tfrac{1}{2}\mathsf{Q}\mathsf{N}^{-1}(\mathsf{N} - \nu\mathsf{I})^{-1}\overline{\mathsf{Q}}^{\mathrm{T}} + \tfrac{1}{2}\mathsf{Q}^*\mathsf{N}^{*-1}(\mathsf{N}^* + \nu\mathsf{I})^{-1}\overline{\mathsf{Q}}^{\mathrm{H}} \\ &= [\mathsf{V}(\nu) + 2\nu\mathsf{W} - \nu^2\mathsf{T}]^{-1},\end{aligned} \tag{7.93}$$

$$\begin{aligned}\overline{\mathsf{G}}(\nu) &= \tfrac{1}{2}\overline{\mathsf{Q}}\mathsf{N}^{-1}(\mathsf{N} - \nu\mathsf{I})^{-1}\mathsf{Q}^{\mathrm{T}} + \tfrac{1}{2}\overline{\mathsf{Q}}^*\mathsf{N}^{*-1}(\mathsf{N}^* + \nu\mathsf{I})^{-1}\mathsf{Q}^{\mathrm{H}} \\ &= [\mathsf{V}(\nu) - 2\nu\mathsf{W} - \nu^2\mathsf{T}]^{-1}.\end{aligned} \tag{7.94}$$

Generalized reciprocity is again guaranteed by the relation $\mathsf{G} = \overline{\mathsf{G}}^{\mathrm{T}}$.

7.5 Hydrostatic Earth

All of the results obtained in Sections 7.1 through 7.4 are obviously applicable on a hydrostatic Earth model as well. The elements of the $\infty \times \infty$ potential energy matrix V on a non-rotating, elastic Earth reduce to

$$V_{kk'} = \int_{\oplus} [\boldsymbol{\varepsilon}_k : \boldsymbol{\Gamma} : \boldsymbol{\varepsilon}_{k'} + \tfrac{1}{2}\rho^0(\mathbf{s}_k \cdot \boldsymbol{\nabla}\phi_{k'}^{\mathrm{E1}} + \mathbf{s}_{k'} \cdot \boldsymbol{\nabla}\phi_k^{\mathrm{E1}})$$
$$+ \tfrac{1}{2}\rho^0\boldsymbol{\nabla}\phi^0 \cdot (\mathbf{s}_k \cdot \boldsymbol{\nabla}\mathbf{s}_{k'} + \mathbf{s}_{k'} \cdot \boldsymbol{\nabla}\mathbf{s}_k - \mathbf{s}_k\boldsymbol{\nabla} \cdot \mathbf{s}_{k'} - \mathbf{s}_{k'}\boldsymbol{\nabla} \cdot \mathbf{s}_k)$$
$$+ \rho^0\mathbf{s}_k \cdot \boldsymbol{\nabla}\boldsymbol{\nabla}\phi^0 \cdot \mathbf{s}_{k'}]\, dV. \qquad (7.95)$$

The hydrostatic action $\mathcal{I} = \frac{1}{2}\mathsf{q}^{\mathrm{T}}(\omega^2\mathsf{T} - \mathsf{V})\mathsf{q}$ is the result of substituting the Rayleigh-Ritz eigenfunction expansion (7.1) into either of the two integral representations (4.161) or (4.162); the generalized algebraic eigenvalue equation $\mathsf{Vq} = \omega^2\mathsf{Tq}$ follows from Rayleigh's variational principle $\delta\mathcal{I} = 0$. Rotation and anelasticity are accounted for by the modifications $\phi^0 \to \phi^0 + \psi$ and $\boldsymbol{\Gamma} \to \boldsymbol{\Gamma}(\nu)$, as before.

★7.6 Effect of a Small Perturbation

Suppose now that we have an initial model of the Earth characterized by the kinetic energy, Coriolis, and anelastic potential energy matrices T, W, and $\mathsf{V}(\nu)$. If the properties of the Earth are modified slightly, these matrices will be perturbed:

$$\mathsf{T} \to \mathsf{T} + \delta\mathsf{T}, \qquad \mathsf{W} \to \mathsf{W} + \delta\mathsf{W}, \qquad \mathsf{V}(\nu) \to \mathsf{V}(\nu) + \delta\mathsf{V}(\nu). \qquad (7.96)$$

We seek to determine the resulting perturbed Green matrix:

$$\mathsf{G}(\nu) \to \mathsf{G}(\nu) + \delta\mathsf{G}(\nu). \qquad (7.97)$$

Upon subtracting the unperturbed relation

$$[\mathsf{V}(\nu) + 2\nu\mathsf{W} - \nu^2\mathsf{T}]\mathsf{G}(\nu) = \mathsf{I} \qquad (7.98)$$

from the corresponding perturbed relation

$$[\mathsf{V}(\nu) + \delta\mathsf{V}(\nu) + 2\nu(\mathsf{W} + \delta\mathsf{W})$$
$$- \nu^2(\mathsf{T} + \delta\mathsf{T})][\mathsf{G}(\nu) + \delta\mathsf{G}(\nu)] = \mathsf{I}, \qquad (7.99)$$

we obtain

$$[\mathsf{V}(\nu) + 2\nu\mathsf{W} - \nu^2\mathsf{T}]\delta\mathsf{G}(\nu)$$
$$= [\nu^2\delta\mathsf{T} - 2\nu\delta\mathsf{W} - \delta\mathsf{V}(\nu)][\mathsf{G}(\nu) + \delta\mathsf{G}(\nu)]. \qquad (7.100)$$

Note that the right side of equation (7.100) depends upon the complete response $\mathsf{G}(\nu) + \delta\mathsf{G}(\nu)$; in the jargon of quantum mechanics, this result is known as the *Lippmann-Schwinger equation* (Schiff 1968).

In the lowest-order *Born approximation*, the perturbation $\delta\mathsf{G}(\nu)$ on the right of the Lippmann-Schwinger equation is ignored. This leads to the result

$$\delta\mathsf{G}(\nu) = [\mathsf{V}(\nu) + 2\nu\mathsf{W} - \nu^2\mathsf{T}]^{-1}[\nu^2\delta\mathsf{T} - 2\nu\delta\mathsf{W} - \delta\mathsf{V}(\nu)] \\ [\mathsf{V}(\nu) + 2\nu\mathsf{W} - \nu^2\mathsf{T}]^{-1}. \tag{7.101}$$

More generally, we can express the perturbation $\delta\mathsf{G}(\nu)$ as an infinite *Born series* of the form

$$\delta\mathsf{G}(\nu) = \delta\mathsf{G}^{(1)}(\nu) + \delta\mathsf{G}^{(2)}(\nu) + \cdots, \tag{7.102}$$

where the superscripts $(1), (2), \ldots$ denote the order of the dependence upon $\delta\mathsf{T}$, $\delta\mathsf{W}$ and $\delta\mathsf{V}(\nu)$. The complete sequence of perturbations in (7.102) can be found by solving the Lippmann-Schwinger equation iteratively. The first iterate is the Born approximation (7.101), which we rewrite in the form

$$\delta\mathsf{G}^{(1)}(\nu) = \mathsf{F}(\nu)\mathsf{G}(\nu), \tag{7.103}$$

where $\mathsf{F}(\nu)$ is a convenient auxiliary matrix defined by

$$\mathsf{F}(\nu) = [\mathsf{V}(\nu) + 2\nu\mathsf{W} - \nu^2\mathsf{T}]^{-1}[\nu^2\delta\mathsf{T} - 2\nu\delta\mathsf{W} - \delta\mathsf{V}(\nu)]. \tag{7.104}$$

The second-order Born approximation, which is obtained by substituting the first-order result (7.103) into the right side of the Lippmann-Schwinger equation, is

$$\delta\mathsf{G}^{(2)}(\nu) = \mathsf{F}^2(\nu)\mathsf{G}(\nu). \tag{7.105}$$

Each succeeding term in the series (7.102) incorporates an additional factor of $\mathsf{F}(\nu)$, so that the total response is given by

$$\mathsf{G}(\nu) + \delta\mathsf{G}^{(1)}(\nu) + \delta\mathsf{G}^{(2)}(\nu) + \cdots \\ = [\mathsf{I} + \mathsf{F}(\nu) + \mathsf{F}^2(\nu) + \cdots]\mathsf{G}(\nu). \tag{7.106}$$

Formally, the geometric series in equation (7.106) can be evaluated,

$$\mathsf{I} + \mathsf{F}(\nu) + \mathsf{F}^2(\nu) + \cdots = [\mathsf{I} - \mathsf{F}(\nu)]^{-1}, \tag{7.107}$$

yielding the final perturbed Green matrix,

$$\mathsf{G}(\nu) + \delta\mathsf{G}(\nu) \\ = [\mathsf{V}(\nu) + \delta\mathsf{V}(\nu) + 2\nu(\mathsf{W} + \delta\mathsf{W}) - \nu^2(\mathsf{T} + \delta\mathsf{T})]^{-1}. \tag{7.108}$$

The infinite Born matrix representation of the normal-mode response of the Earth was first summed in the above manner by Tromp & Dahlen (1990a). Of course, the result (7.108) can be obtained much more directly by simply inverting equation (7.99).

If the initial model is elastic, we can replace the unperturbed potential energy matrix by $\mathsf{V}(\nu) \rightarrow \mathsf{V}$, and if it is non-rotating, we can make the substitutions $\mathsf{W} \rightarrow 0$ and $\delta\mathsf{W} \rightarrow \mathsf{W}$. The lowest-order Born approximation (7.101) reduces in that case to

$$\delta\mathsf{G}(\nu) = [\mathsf{V} - \nu^2\mathsf{T}]^{-1}[\nu^2\delta\mathsf{T} - 2\nu\mathsf{W} - \delta\mathsf{V}(\nu)][\mathsf{V} - \nu^2\mathsf{T}]^{-1}, \tag{7.109}$$

and the complete response (7.108) becomes

$$\mathsf{G}(\nu) + \delta\mathsf{G}(\nu) = [\mathsf{V} + \delta\mathsf{V}(\nu) + 2\nu\mathsf{W} - \nu^2(\mathsf{T} + \delta\mathsf{T})]^{-1}. \tag{7.110}$$

In equations (7.109) and (7.110) the rotation and anelasticity of the Earth are both considered to be small perturbations. This is a good approximation for all of the seismic modes, which have $\omega \gg \Omega$.

7.7 Response to a Moment-Tensor Source

To express the acceleration response to a step-function moment-tensor source, it is convenient to define the $\infty \times 1$ *receiver and source vectors*:

$$\mathsf{r} = \begin{pmatrix} \vdots \\ \hat{\boldsymbol{\nu}} \cdot \mathbf{s}_k(\mathbf{x}) \\ \vdots \end{pmatrix}, \qquad \mathsf{s} = \begin{pmatrix} \vdots \\ \mathbf{M} : \boldsymbol{\varepsilon}_k(\mathbf{x}_\mathrm{s}) \\ \vdots \end{pmatrix}, \tag{7.111}$$

where $\hat{\boldsymbol{\nu}}$ is the polarization of the accelerometer and $\boldsymbol{\varepsilon}_k = \frac{1}{2}[\boldsymbol{\nabla}\mathbf{s}_k + (\boldsymbol{\nabla}\mathbf{s}_k)^\mathrm{T}]$ is the strain associated with the displacement eigenfunction $\mathbf{s}_k$. We denote the $\hat{\boldsymbol{\nu}}$ component of the acceleration by

$$a(t) = \hat{\boldsymbol{\nu}} \cdot \mathbf{a}(\mathbf{x}, t). \tag{7.112}$$

This scalar accelerogram is given in terms of the time derivative $\dot{\mathsf{G}}(t)$ of the Green matrix $\mathsf{G}(t)$ by

$$a(t) = \mathsf{r}^\mathrm{T}\dot{\mathsf{G}}(t)\,\mathsf{s}, \tag{7.113}$$

where we have set the fiducial origin time $t_\mathrm{s} = 0$, in accordance with the convention (5.137). In the most general case of a rotating anelastic Earth, we can rewrite equation (7.113) in the form

$$a(t) = \mathrm{Re}\,[\mathsf{r}'^\mathrm{T}\exp(i\mathsf{N}t)\,\mathsf{s}'], \tag{7.114}$$

where r' and s' are *transformed receiver and source vectors* given by

$$\mathsf{r}' = \mathsf{Q}^{\mathrm{T}}\mathsf{r}, \qquad \mathsf{s}' = \overline{\mathsf{Q}}^{\mathrm{T}}\mathsf{s}. \tag{7.115}$$

The result (7.114)–(7.115) is equivalent to equation (6.203), as long as the basis functions $\mathbf{s}_k$ are complete. On a non-rotating anelastic Earth, the dual eigenvector matrix $\overline{\mathsf{Q}}$ in equation (7.115) is replaced by Q.

The Fourier transform of the $\hat{\boldsymbol{\nu}}$ component of the acceleration is given in terms of the frequency-domain Green matrix $\mathsf{G}(\omega)$ by

$$a(\omega) = i\omega\mathsf{r}^{\mathrm{T}}\mathsf{G}(\omega)\mathsf{s}. \tag{7.116}$$

This can be written, for $\omega > 0$, in a form analogous to the result (6.206):

$$a(\omega) = \tfrac{1}{2}i\mathsf{r}'^{\mathrm{T}}(\mathsf{N} - \omega\mathsf{I})^{-1}\mathsf{s}', \tag{7.117}$$

where we have ignored the contribution from the negative-frequency peaks as before. The quantity $\frac{1}{2}i(\mathsf{N} - \omega\mathsf{I})^{-1}$ is a diagonal matrix of unit Lorentzian resonance peaks of the form $\eta_k(\omega) = \frac{1}{2}i(\omega_k + i\gamma_k - \omega)^{-1}$ centered upon the real positive eigenfrequencies.

Alternatively, we can write the frequency-domain acceleration response explicitly in terms of the matrices $\mathsf{V}(\omega)$, W and T and the unperturbed receiver and source vectors r and s in the form

$$a(\omega) = i\omega\mathsf{r}^{\mathrm{T}}[\mathsf{V}(\omega) + 2\omega\mathsf{W} - \omega^2\mathsf{T}]^{-1}\mathsf{s}. \tag{7.118}$$

The result (7.118), which is exact for all frequencies $-\infty \leq \omega \leq \infty$, can be made the basis of a *direct solution method* of calculating synthetic accelerograms without solving for the eigenfrequencies N and associated eigenvectors and anti-eigenvectors Q and $\overline{\mathsf{Q}}$. To synthesize a suite of accelerograms at a number of receivers $\hat{\boldsymbol{\nu}}$, $\mathbf{x}$ produced by a given earthquake $\mathbf{M}$, $\mathbf{x}_s$ we first solve

$$[\mathsf{V}(\omega) + 2\omega\mathsf{W} - \omega^2\mathsf{T}]\mathsf{d}(\omega) = \mathsf{s} \tag{7.119}$$

for the *source response vector* $\mathsf{d}(\omega)$. The acceleration at each receiver is then given by the scalar product

$$a(\omega) = i\omega\mathsf{r}^{\mathrm{T}}\mathsf{d}(\omega). \tag{7.120}$$

If, on the other hand, we seek the response at a single receiver to a suite of sources, it is more convenient to solve

$$[\mathsf{V}(\omega) - 2\omega\mathsf{W} - \omega^2\mathsf{T}]\mathsf{e}(\omega) = \mathsf{r} \tag{7.121}$$

for the *receiver response vector* $\mathsf{e}(\omega)$. The acceleration excited by each earthquake is then

$$a(\omega) = i\omega\mathsf{e}^{\mathrm{T}}(\omega)\mathsf{s}. \tag{7.122}$$

The difference in sign of the term $\pm 2\omega \mathsf{W}$ in equations (7.119) and (7.121) arising from the anti-symmetry $\mathsf{W}^{\mathrm{T}} = -\mathsf{W}$ is noteworthy; on a non-rotating Earth, this difference is obviously immaterial, since $\mathsf{W} = \mathsf{0}$. We discuss the practical implementation of these direct solution techniques in Chapter 13.

Part II

The Spherical Earth

Chapter 8

Spheroidal and Toroidal Oscillations

With this chapter we commence the study of the free oscillations of a spherically symmetric, non-rotating Earth model. We begin by considering an Earth model with an isotropic stress-strain relation, and subsequently generalize to a transversely isotropic Earth model. By exploiting the separability of the normal-mode eigenvalue problem, it is possible to convert the governing vector and tensor equations into an equivalent system of radial scalar equations, which can be solved by numerical integration; this enables the eigenfrequencies ω and associated eigenfunctions $\mathbf{s}$ of either an isotropic or transversely isotropic Earth model to be found essentially exactly. Every spherically symmetric, non-rotating Earth model supports two independent types of free oscillations—*spheroidal* oscillations which alter the external shape of the Earth and *toroidal* oscillations which do not. This separation into two types of normal modes is the standing-wave analogue of the independent propagation of Rayleigh and Love surface waves and P-SV and SH polarized body waves, as we shall see in Chapters 11 and 12.

8.1 Change in Notation

Up until now in this book, we have carefully distinguished between Eulerian and Lagrangian dynamical variables by using superscripts E and L, respectively. In addition, we have consistently labelled parameters that pertain to the equilibrium Earth model by a superscript 0 and incremental parameters that describe small perturbations away from the equilibrium configuration by a superscript 1. Now that the fundamental results which are applicable

Variable Name or Description	Old Notation	New Notation
Initial density	ρ^0	ρ
Initial gravitational potential	ϕ^0	Φ
Centrifugal potential†	ψ	ψ
Initial hydrostatic pressure	p^0	p
Initial deviatoric stress	$\boldsymbol{\tau}^0$	$\boldsymbol{\tau}$
Negative normal traction	ϖ^0	ϖ
Isentropic incompressibility†	κ	κ
Rigidity†	μ	μ
Compressional-wave speed†	α	α
Shear-wave speed†	β	β
Fourth-order elastic tensor†	$\mathbf{\Gamma}$	$\mathbf{\Gamma}$
Auxiliary elastic tensor†	$\mathbf{\Lambda}$	$\mathbf{\Lambda}$
Anisotropic elastic tensor†	$\boldsymbol{\gamma}$	$\boldsymbol{\gamma}$
Displacement†	$\mathbf{s}$	$\mathbf{s}$
Strain†	$\boldsymbol{\varepsilon}$	$\boldsymbol{\varepsilon}$
Deviatoric strain†	$\mathbf{d}$	$\mathbf{d}$
Eulerian potential perturbation	ϕ^{E1}	ϕ
Auxiliary gravity vector	$\boldsymbol{\xi}^{\mathrm{E1}}$	$\boldsymbol{\xi}$
Incremental Lagrangian Cauchy stress	$\mathbf{T}^{\mathrm{L1}}$	$\mathbf{T}$
Incremental first Piola-Kirchhoff stress	$\mathbf{T}^{\mathrm{PK1}}$	$\tilde{\mathbf{T}}$
Auxiliary traction vector	$\mathbf{t}^{\mathrm{PK1}}$	$\tilde{\mathbf{t}}$

Table 8.1. Correspondence between the notation employed in Part I and that employed in Parts II and III. Unsuperscripted symbols for the quantities designated by a dagger† are unchanged. We shall henceforth denote the total initial stress by $-p\mathbf{I}+\boldsymbol{\tau}$ and the Eulerian density perturbation by $-\boldsymbol{\nabla}\cdot(\rho\mathbf{s})$ in order to avoid introducing new symbols for these quantities.

to a general Earth model have been established, it is advantageous to discard this clear but cluttered notation, and replace it by a less elaborate one. The correspondence between the new "naked" variables, which we shall employ throughout the rest of this book, and the old superscripted variables, which we used for clarity in Part I, is summarized in Table 8.1. Several of the tabulated quantities, notably $\boldsymbol{\tau}$, ϖ, $\tilde{\mathbf{T}}$ and $\tilde{\mathbf{t}}$, do not arise in our discussion of a spherical Earth in Part II; however, they are included here because they will be needed when we consider a slightly aspherical Earth in Part III. One unsatisfactory feature of the new streamlined notation is worth mentioning: we use an unadorned ϕ to denote both the Eulerian gravitational potential perturbation ϕ^{E1} and the longitude in a system of spherical polar coordinates r, θ, ϕ. We do not believe that this will cause any confusion; the meaning should always be clear from the context.

8.2 SNREI Earth Model

A SNREI Earth model is one that is spherically symmetric, non-rotating, perfectly elastic and isotropic. The final qualifier "isotropic" has a double meaning—the initial stress is isotropic, so that there is no deviatoric stress, $\boldsymbol{\tau} = \mathbf{0}$, and the fourth-order elastic tensor $\boldsymbol{\Gamma}$ is isotropic, of the form

$$\Gamma_{ijkl} = (\kappa - \tfrac{2}{3}\mu)\delta_{ij}\delta_{kl} + \mu(\delta_{ik}\delta_{jl} + \delta_{il}\delta_{jk}). \tag{8.1}$$

A general SNREI Earth model is completely characterized by specifying its density ρ, isentropic incompressibility κ and rigidity μ as functions of the radial distance r from the center. Alternatively, we may specify the radial variation of the compressional-wave speed $\alpha = [(\kappa+\frac{4}{3}\mu)/\rho]^{1/2}$ and the shear-wave speed $\beta = (\mu/\rho)^{1/2}$ rather than the incompressibility and rigidity. Figure 8.1 shows the isotropic version of the Preliminary Reference Earth Model, henceforth referred to as PREM (Dziewonski & Anderson 1981); the speeds α and β are those "seen" by a P or S wave with a frequency of one Hz. The canonical version of the PREM model is transversely isotropic as well as anelastic; we consider the effects of these two complications in Sections 8.9 and 9.7, respectively.

As in Part I we denote the union of the various solid regions of the Earth by $\oplus_{\mathrm{S}}$, the union of the various fluid regions by $\oplus_{\mathrm{F}}$, and the entire volume of the Earth by $\oplus = \oplus_{\mathrm{S}} \cup \oplus_{\mathrm{F}}$. All of space will still be denoted by $\bigcirc$, so that $\bigcirc - \oplus$ is the region outside of the Earth. For a typical SNREI Earth model such as PREM, the domain $\oplus_{\mathrm{S}}$ consists of the solid inner core and the mantle, whereas $\oplus_{\mathrm{F}}$ consists of the fluid outer core and the ocean (we use the term "mantle" somewhat loosely in what follows to refer to the mantle plus the overlying crust). We also continue to denote the union of

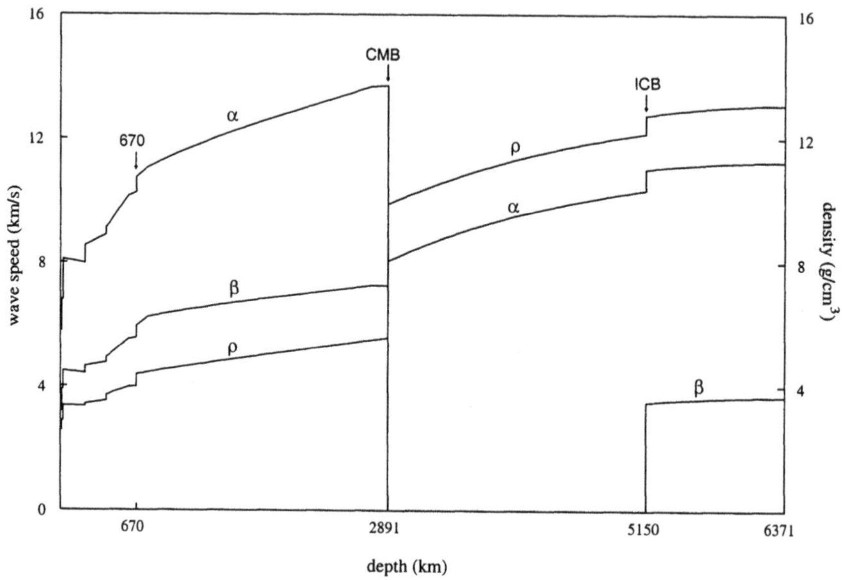

Figure 8.1. Compressional-wave speed α, shear-wave speed β and density ρ in the "equivalent" isotropic Preliminary Reference Earth Model (PREM). The locations of the inner-core boundary (ICB) and the core-mantle boundary (CMB) are marked. The model is capped by a 3 km thick uniform oceanic layer. There are also a number of solid-solid boundaries, including a Mohorovičić discontinuity at a depth of 24.4 km and upper-mantle discontinuities at depths of 220, 400 and 670 km. More recent analyses have questioned the existence of a global 220 km discontinuity, and refined the locations of the 400 km and 670 km discontinuities.

all the internal solid-solid discontinuities by $\Sigma_{\rm SS}$, the union of all the fluid-solid discontinuities by $\Sigma_{\rm FS}$, and the union of all the boundaries, including the exterior free surface, by $\Sigma = \partial\oplus \cup \Sigma_{\rm SS} \cup \Sigma_{\rm FS}$. The outward normal $\hat{\mathbf{n}}$ to the boundary Σ in a spherically symmetric Earth model is the unit radial vector $\hat{\mathbf{r}}$; we continue to refer to the outside and inside of Σ as the $+$ and the $-$ sides of the boundary, respectively.

The radii of all the solid-solid discontinuities will be collectively denoted by $d_{\rm SS}$, whereas the radii of all the fluid-solid discontinuities will be denoted by $d_{\rm FS}$. The radii of the free surface, the core-mantle boundary, the inner-core boundary and the seafloor will be denoted by a, b, c and s, respectively. The ensemble of all discontinuity radii, including the external boundary, will be denoted by $d = a \cup d_{\rm SS} \cup d_{\rm FS}$. For convenience in what follows, we define the model parameters ρ, κ, μ, α and β of a SNREI Earth model to be zero in the region $r > a$ outside of the Earth. We shall henceforth use a dot, $\dot{q}$, to denote the derivative dq/dr of any function q that depends only upon radius.

8.2.1 Gravity and hydrostatic pressure

The gravitational field $\mathbf{g} = -\boldsymbol{\nabla}\Phi$ within a SNREI Earth points radially downward:

$$\mathbf{g} = -g\hat{\mathbf{r}} \quad \text{where} \quad g = \dot{\Phi}. \tag{8.2}$$

The scalar acceleration of gravity $g = \|\mathbf{g}\|$ satisfies the first-order differential relation

$$\dot{g} + 2r^{-1}g = 4\pi G\rho, \tag{8.3}$$

where G is the Newtonian gravitational constant. Equation (8.3) can be readily integrated to yield an explicit formula for g and the associated gravitational potential Φ in terms of the density distribution ρ:

$$g(r) = \frac{4\pi G}{r^2}\int_0^r \rho'\, r'^2 dr', \qquad \Phi(r) = -\frac{4\pi G}{r}\int_0^r \rho'\, r'^2 dr', \tag{8.4}$$

where a prime denotes evaluation at the radial integration variable r'. As is well known, there is no contribution from the spherical shell of material outside of the observation point, and the sphere of material inside attracts like an equivalent point mass. The external gravitational acceleration and potential are $g(r) = GM/r^2$ and $\Phi(r) = -GM/r$, where $M = 4\pi\int_0^a \rho\, r^2 dr$ is the total mass of the Earth.

The mechanical equilibrium of a SNREI Earth model is guaranteed by the hydrostatic balance equation

$$\dot{p} + \rho g = 0, \tag{8.5}$$

where p is the initial hydrostatic pressure. This equation can also be integrated with the result

$$p(r) = \int_r^a \rho' g'\, dr', \tag{8.6}$$

where we have used the free-surface boundary condition $p(a) = 0$. The radial variation of the gravity g and hydrostatic pressure p within the PREM model are shown in Figure 8.2.

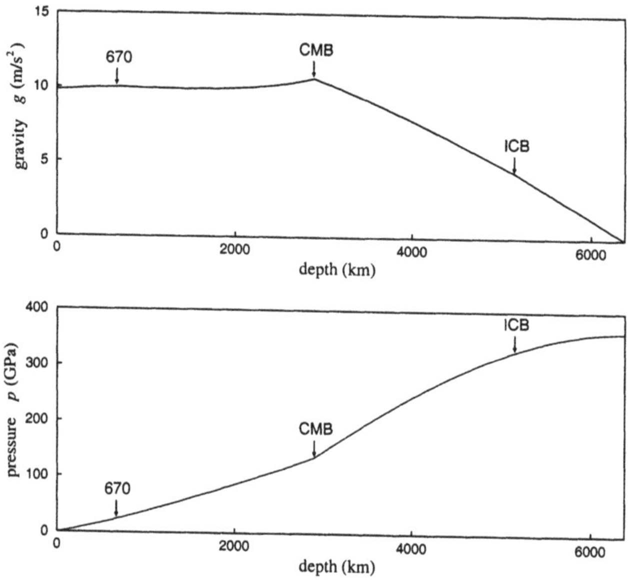

Figure 8.2. Variation of the acceleration of gravity g and hydrostatic pressure p with depth in the Preliminary Reference Earth Model. Gravity (*top*) is roughly uniform (9.8–10.1 m/s^2) throughout the mantle. The pressure (*bottom*) increases monotonically to a maximum value of 364 GPa at the center of the Earth.

★8.2.2 Brunt-Väisälä frequency

An auxiliary parameter that is of interest, particularly in the fluid regions of the Earth $\oplus_{\mathrm{FS}}$, is the *Brunt-Väisälä frequency* $N(r)$, defined by

$$N^2 = -\frac{\dot{\rho} g}{\rho} - \frac{\rho g^2}{\kappa}. \tag{8.7}$$

The physical significance of this quantity can be understood by considering the virtual displacement of a small parcel of fluid (Eckart 1960; Tolstoy 1973). For definiteness, it is convenient to regard the parcel as surrounded by a membrane that is flaccid, so that the interior and ambient pressures are always identical, but perfectly insulating, so that density changes within the displaced fluid occur adiabatically. If the parcel is displaced upward by an infinitesimal amount ξ, it experiences a change in pressure $\delta p = -\rho g \xi$ and an adiabatic change in density $\delta\rho_{\mathrm{ad}} = \rho\kappa^{-1}\delta p = -\rho^2 g \kappa^{-1}\xi$. The ambient density of the surrounding fluid has changed by a different amount

$\delta\rho_{\mathrm{am}} = \dot{\rho}\xi$; as a result, the displaced parcel is acted upon by a buoyancy force per unit volume equal to $(\delta\rho_{\mathrm{am}} - \delta\rho_{\mathrm{ad}})g = -\rho N^2\xi$. Equating this to the inertial force per unit volume, we find the equation of motion of such an adiabatic fluid parcel:

$$\frac{d^2\xi}{dt^2} + N^2\xi = 0. \tag{8.8}$$

If N^2 is positive the parcel will oscillate sinusoidally about its initial equilibrium position with angular frequency N; if, on the other hand, N^2 is negative it will depart from its initial position exponentially. The density stratification in the fluid is gravitationally stable in the first case whereas it is gravitationally unstable in the second; for this reason N is sometimes referred to as the stability frequency.

A related measure of the extent to which the density gradient $\dot{\rho}$ departs from the *Adams-Williamson relation* $\dot{\rho} = -\rho^2 g/\kappa$ is the dimensionless stratification parameter $\eta_{\mathrm{B}}(r)$ of Bullen (1963), defined by $N^2 = \rho g^2(\eta_{\mathrm{B}} - 1)/\kappa$. Bullen's parameter can be rewritten in terms of the pressure derivative of the incompressibility in the form

$$\eta_{\mathrm{B}} = \frac{d\kappa}{dp} + \frac{1}{g}\frac{d}{dr}\left(\frac{\kappa}{\rho}\right). \tag{8.9}$$

A neutrally stable fluid region is characterized by the equivalent conditions $N^2 = 0$ and $\eta_{\mathrm{B}} = 1$.

A typical SNREI Earth model such as PREM, which has been obtained by fitting free-oscillation and body-wave travel-time data for ρ, α and β, has a number of alternating regions where $N^2 > 0$ and $N^2 < 0$ in the fluid outer core. The physical significance of these stable and unstable regions is moot, however, since neither N^2 nor η_{B} is very well constrained; the available seismological evidence is consistent with the hypothesis that the fluid core is everywhere neutrally stratified (Masters 1979). The compressible, constant-density oceanic layer of PREM is also, strictly speaking, unstable; in this case the problem is a practical one—resolving the actual stable stratification of the oceans would require a much finer discretization than is necessary for any global seismological applications. The cataloguing of the possible free oscillations is significantly simplified in a SNREI Earth model which has $N^2 = 0$ everywhere in $\oplus_{\mathrm{F}}$, as we shall see in Sections 8.8.2 and 8.8.11.

8.3 Equations of Motion

The linearized equations and boundary equations governing the free oscillations of a SNREI Earth model can be obtained from the equations for a non-rotating, hydrostatic Earth model, discussed in Sections 3.11 and 4.3. The frequency-domain momentum equation (4.149), rewritten in our new streamlined notation, is

$$-\omega^2\rho\mathbf{s} - \boldsymbol{\nabla}\cdot\mathbf{T} + (4\pi G\rho^2 s_r)\hat{\mathbf{r}} + \rho\boldsymbol{\nabla}\phi \\ + \rho g[\boldsymbol{\nabla}s_r - (\boldsymbol{\nabla}\cdot\mathbf{s} + 2r^{-1}s_r)\hat{\mathbf{r}}] = \mathbf{0}, \tag{8.10}$$

where $s_r = \hat{\mathbf{r}}\cdot\mathbf{s}$. We have used equation (8.3) to eliminate $\dot{g}$ and consolidate the gravitational terms. The incremental Cauchy stress $\mathbf{T}$ is given by the isotropic constitutive relation

$$\mathbf{T} = \kappa(\boldsymbol{\nabla}\cdot\mathbf{s})\mathbf{I} + 2\mu\mathbf{d}, \tag{8.11}$$

where $\mathbf{d} = \frac{1}{2}[\boldsymbol{\nabla}\mathbf{s} + (\boldsymbol{\nabla}\mathbf{s})^{\mathrm{T}}] - \frac{1}{3}(\boldsymbol{\nabla}\cdot\mathbf{s})\mathbf{I}$ is the deviatoric strain. Upon making use of the kinematic identity

$$\hat{\mathbf{r}}\cdot\boldsymbol{\varepsilon} = \partial_r\mathbf{s} + \tfrac{1}{2}\hat{\mathbf{r}}\times(\boldsymbol{\nabla}\times\mathbf{s}), \tag{8.12}$$

we can rewrite equation (8.10) entirely in terms of the displacement $\mathbf{s}$ and the incremental gravitational potential ϕ in the form

$$-\omega^2\rho\mathbf{s} - (\kappa + \tfrac{1}{3}\mu)\boldsymbol{\nabla}(\boldsymbol{\nabla}\cdot\mathbf{s}) - \mu\nabla^2\mathbf{s} - (\dot{\kappa} - \tfrac{2}{3}\dot{\mu})(\boldsymbol{\nabla}\cdot\mathbf{s})\hat{\mathbf{r}} \\ - 2\dot{\mu}[\partial_r\mathbf{s} + \tfrac{1}{2}\hat{\mathbf{r}}\times(\boldsymbol{\nabla}\times\mathbf{s})] + (4\pi G\rho^2 s_r)\hat{\mathbf{r}} + \rho\boldsymbol{\nabla}\phi \\ + \rho g[\boldsymbol{\nabla}s_r - (\boldsymbol{\nabla}\cdot\mathbf{s} + 2r^{-1}s_r)\hat{\mathbf{r}}] = \mathbf{0}. \tag{8.13}$$

The vector Laplacian is defined by $\nabla^2\mathbf{s} = \boldsymbol{\nabla}(\boldsymbol{\nabla}\cdot\mathbf{s}) - \boldsymbol{\nabla}\times(\boldsymbol{\nabla}\times\mathbf{s})$, as usual.

The kinematic boundary conditions require the displacement to be continuous everywhere except on the fluid-solid boundaries, where tangential slip is allowed, i.e., $[\mathbf{s}]_-^+ = \mathbf{0}$ on Σ_{SS} and $[s_r]_-^+ = 0$ on Σ_{FS}. The corresponding dynamical boundary conditions (4.150)–(4.152) are

$$\hat{\mathbf{r}}\cdot\mathbf{T} = \mathbf{0} \quad \text{on } \partial\oplus, \tag{8.14}$$

$$[\hat{\mathbf{r}}\cdot\mathbf{T}]_-^+ = \mathbf{0} \quad \text{on } \Sigma_{\mathrm{SS}}, \tag{8.15}$$

$$[\hat{\mathbf{r}}\cdot\mathbf{T}]_-^+ = \hat{\mathbf{r}}[\hat{\mathbf{r}}\cdot\mathbf{T}\cdot\hat{\mathbf{r}}]_-^+ = \mathbf{0} \quad \text{on } \Sigma_{\mathrm{FS}}. \tag{8.16}$$

Equation (8.12) can also be used to obtain a simple expression for the incremental traction on any spherical surface, namely,

$$\hat{\mathbf{r}}\cdot\mathbf{T} = (\kappa - \tfrac{2}{3}\mu)(\boldsymbol{\nabla}\cdot\mathbf{s})\hat{\mathbf{r}} + 2\mu(\partial_r\mathbf{s}) + \mu\hat{\mathbf{r}}\times(\boldsymbol{\nabla}\times\mathbf{s}). \tag{8.17}$$

In the fluid regions of the Earth the rigidity μ is identically equal to zero; as a result, the incremental stress is hydrostatic and the associated traction is purely radial: $\mathbf{T} = \kappa(\boldsymbol{\nabla}\cdot\mathbf{s})\mathbf{I}$ and $\hat{\mathbf{r}}\cdot\mathbf{T} = \kappa(\boldsymbol{\nabla}\cdot\mathbf{s})\hat{\mathbf{r}}$.

The Eulerian perturbation ϕ in the gravitational potential is determined by the incremental version of Poisson's equation, which in a spherically symmetric Earth takes the form

$$\nabla^2\phi = -4\pi G(\rho\boldsymbol{\nabla}\cdot\mathbf{s} + \dot{\rho}s_r). \tag{8.18}$$

The potential perturbation must be continuous everywhere, including on the boundaries Σ, where $[\phi]_-^+ = 0$. In addition, we must have

$$[\dot{\phi} + 4\pi G\rho s_r]_-^+ = 0 \quad \text{on } \Sigma. \tag{8.19}$$

The second term in (8.19) accounts for the apparent surface mass associated with the radial displacement s_r of any density discontinuities $[\rho]_-^+$. To find the normal-mode eigensolutions ω, $\mathbf{s}$, ϕ of a SNREI Earth model, we must solve equation (8.13) in $\oplus$ and equation (8.18) in all of space $\bigcirc$, subject to the boundary conditions (8.14)–(8.16) and (8.19) on Σ. The incremental potential ϕ is given explicitly in terms of the displacement $\mathbf{s}$ by

$$\begin{aligned}\phi &= G\int_{\oplus}\frac{(\rho'\boldsymbol{\nabla}'\cdot\mathbf{s}' + \dot{\rho}'s_r')}{\|\mathbf{x}-\mathbf{x}'\|}\,dV' + G\int_{\Sigma}\frac{[\rho']_-^+ s_r'}{\|\mathbf{x}-\mathbf{x}'\|}\,d\Sigma' \\ &= -G\int_{\oplus}\frac{\rho'\mathbf{s}'\cdot(\mathbf{x}-\mathbf{x}')}{\|\mathbf{x}-\mathbf{x}'\|^3}\,dV',\end{aligned} \tag{8.20}$$

where a prime denotes evaluation at the dummy integration variable $\mathbf{x}'$, and the second equality follows from Gauss' theorem.

8.4 Rayleigh's Principle

The linearized elastic-gravitational equations of motion and boundary conditions governing a SNREI Earth may be deduced from either a displacement or displacement-potential version of Rayleigh's variational principle. The action in the first instance is of the form (4.161):

$$\mathcal{I} = \int_{\oplus} L(\mathbf{s}, \boldsymbol{\nabla}\mathbf{s})\,dV. \tag{8.21}$$

The hydrostatic Lagrangian density (4.163) can be reduced with the aid of Poisson's equation (8.3) to

$$\begin{aligned}L = \tfrac{1}{2}[&\omega^2\rho\mathbf{s}\cdot\mathbf{s} - \kappa(\boldsymbol{\nabla}\cdot\mathbf{s})^2 - 2\mu(\mathbf{d}:\mathbf{d}) - 4\pi G\rho^2 s_r^2 \\ &- \rho\mathbf{s}\cdot\boldsymbol{\nabla}\phi - \rho g(\mathbf{s}\cdot\boldsymbol{\nabla}s_r - s_r\boldsymbol{\nabla}\cdot\mathbf{s} - 2r^{-1}s_r^2)].\end{aligned} \tag{8.22}$$

The potential perturbation ϕ in (8.22) is regarded as a known functional of the displacement $\mathbf{s}$, given by equation (8.20). The variation $\delta\mathcal{I}$ vanishes for an arbitrary admissible variation $\boldsymbol{\delta}\mathbf{s}$ if and only if ω and $\mathbf{s}$ satisfy the frequency-domain momentum equation (8.13) and the dynamical boundary conditions (8.14)–(8.16). The corresponding modified action (4.162) is

$$\mathcal{I}' = \int_{\bigcirc} L'(\mathbf{s}, \boldsymbol{\nabla}\mathbf{s}, \boldsymbol{\nabla}\phi)\, dV, \tag{8.23}$$

where

$$\begin{aligned} L' = \tfrac{1}{2}[&\omega^2\rho\mathbf{s}\cdot\mathbf{s} - \kappa(\boldsymbol{\nabla}\cdot\mathbf{s})^2 - 2\mu(\mathbf{d}\!:\!\mathbf{d}) - 4\pi G\rho^2 s_r^2 \\ &- 2\rho\mathbf{s}\cdot\boldsymbol{\nabla}\phi - \rho g(\mathbf{s}\cdot\boldsymbol{\nabla}s_r - s_r\boldsymbol{\nabla}\cdot\mathbf{s} - 2r^{-1}s_r^2) \\ &\quad - (4\pi G)^{-1}\boldsymbol{\nabla}\phi\cdot\boldsymbol{\nabla}\phi]. \end{aligned} \tag{8.24}$$

The variation $\delta\mathcal{I}'$ vanishes for arbitrary and independent admissible variations $\boldsymbol{\delta}\mathbf{s}$ and $\delta\phi$ if and only if ϕ and $\mathbf{s}$ are related by equations (8.18) and (8.19) as well. An admissible displacement variation $\boldsymbol{\delta}\mathbf{s}$ is one that satisfies $[\boldsymbol{\delta}\mathbf{s}]_-^+ = \mathbf{0}$ on Σ_{SS} and $[\delta s_r]_-^+ = 0$ on Σ_{FS}, whereas an admissible potential variation $\delta\phi$ is one that satisfies $[\delta\phi]_-^+ = 0$ on all boundaries Σ. The stationary value of both actions is

$$\mathcal{I} = \mathcal{I}' = 0 \tag{8.25}$$

at every eigensolution ω, $\mathbf{s}$, ϕ. The result (8.25) can be verified by integrating the dot product of $\mathbf{s}$ with equation (8.10) and the product of $(4\pi G)^{-1}\phi$ with equation (8.18) over the Earth $\oplus$ and all of space $\bigcirc$, respectively.

8.5 Energy Budget and Stability

The frequency-domain energy density $E = \omega\partial_\omega L - L$ in a SNREI Earth model is given by

$$\begin{aligned} E = \tfrac{1}{2}[&\omega^2\rho\mathbf{s}\cdot\mathbf{s} + \kappa(\boldsymbol{\nabla}\cdot\mathbf{s})^2 + 2\mu(\mathbf{d}\!:\!\mathbf{d}) + 4\pi G\rho^2 s_r^2 \\ &+ \rho\mathbf{s}\cdot\boldsymbol{\nabla}\phi + \rho g(\mathbf{s}\cdot\boldsymbol{\nabla}s_r - s_r\boldsymbol{\nabla}\cdot\mathbf{s} - 2r^{-1}s_r^2)]. \end{aligned} \tag{8.26}$$

The total integrated energy of a normal mode of oscillation is a sum of kinetic and elastic-gravitational potential energies:

$$\mathcal{E} = \tfrac{1}{2}(\omega^2\mathcal{T} + \mathcal{V}), \tag{8.27}$$

where

$$\mathcal{T} = \int_{\oplus} \rho\mathbf{s}\cdot\mathbf{s}\, dV. \tag{8.28}$$

Equation (8.25) can be interpreted as an energy *equipartition* relation, $\omega^2\mathcal{T} = \mathcal{V}$, so that the total energy of an oscillation is simply twice the kinetic energy: $\mathcal{E} = \omega^2\mathcal{T}$. The potential energy can be further decomposed into separate elastic *compressional*, elastic *shear* and *gravitational energies*:

$$\mathcal{V} = \mathcal{V}_\kappa + \mathcal{V}_\mu + \mathcal{V}_{\mathrm{g}}, \tag{8.29}$$

where

$$\mathcal{V}_\kappa = \int_\oplus \kappa(\boldsymbol{\nabla}\cdot\mathbf{s})^2\, dV, \tag{8.30}$$

$$\mathcal{V}_\mu = \int_\oplus 2\mu(\mathbf{d}\!:\!\mathbf{d})\, dV, \tag{8.31}$$

$$\begin{aligned}\mathcal{V}_{\mathrm{g}} = \int_\oplus \rho[4\pi G\rho s_r^2 + \mathbf{s}\cdot\boldsymbol{\nabla}\phi \\ + g(\mathbf{s}\cdot\boldsymbol{\nabla}s_r - s_r\boldsymbol{\nabla}\cdot\mathbf{s} - 2r^{-1}s_r^2)]\, dV.\end{aligned} \tag{8.32}$$

The extent to which a spheroidal normal mode subdivides its potential energy among $\mathcal{V}_\kappa$, $\mathcal{V}_\mu$ and $\mathcal{V}_{\mathrm{g}}$ is one of its principal distinguishing characteristics, as we shall see in Section 8.8.10.

A SNREI Earth model is dynamically stable only if the potential energy functional (8.29) is non-negative, $\mathcal{V}_\kappa + \mathcal{V}_\mu + \mathcal{V}_{\mathrm{g}} \geq 0$, for every piecewise smooth displacement $\mathbf{s}$ satisfying $[s_r]_-^+ = 0$ on $r = d_{\mathrm{FS}}$, as discussed in Section 4.1.5. The compressional energy $\mathcal{V}_\kappa$ and the shear energy $\mathcal{V}_\mu$ are inherently non-negative; in contrast, the gravitational energy $\mathcal{V}_{\mathrm{g}}$ may be of either sign, so that gravity may exert either a stabilizing or destabilizing influence, depending upon the shape of the oscillation. Upon applying Gauss' theorem and rearranging terms, we can rewrite the potential energy in terms of the Brunt-Väisälä frequency N defined in equation (8.7):

$$\begin{aligned}\mathcal{V} = \int_\oplus [\kappa(\boldsymbol{\nabla}\cdot\mathbf{s} - \kappa^{-1}\rho g s_r)^2 + 2\mu(\mathbf{d}\!:\!\mathbf{d}) + \rho N^2 s_r^2]\, dV \\ - \frac{1}{4\pi G}\int_\bigcirc \boldsymbol{\nabla}\phi\cdot\boldsymbol{\nabla}\phi\, dV - \int_\Sigma [\rho]_-^+ g s_r^2\, d\Sigma.\end{aligned} \tag{8.33}$$

Equation (8.33) shows that a SNREI Earth model will be *locally* stable against *all short-wavelength* deformations—those for which the perturbation ϕ in the gravitational potential is negligible—provided that $N^2 \geq 0$ in $0 \leq r \leq a$ and $[\rho]_-^+ \leq 0$ on $r = d$. In fact, these two conditions can be shown to render a model *globally* stable against *all non-radial* perturbations. The proof of this global stability result is too lengthy to reproduce here; it uses a "tricky" transformation of the term involving $\boldsymbol{\nabla}\phi\cdot\boldsymbol{\nabla}\phi$ together with "standard" inequalities to verify that $\mathcal{V} \geq 0$ for all deformations satisfying the constraint $\int_\Omega \phi\, d\Omega = 0$ (Aly & Pérez 1992). In the case of a fluid ($\mu = 0$) sphere the physically plausible conditions $N^2 \geq 0$ and $[\rho]_-^+ \leq 0$ are necessary as well as sufficient for non-radial stability; this result is known in astrophysics as the *Antonov-Lebovitz theorem* (Binney & Tremaine 1987).

All of the relations (8.10)–(8.33) in this and the preceding two sections are simplified in the *non-gravitating limit*, obtained by setting the gravitational constant G, the acceleration of gravity g and the potential perturbation ϕ all equal to zero. The momentum equation, boundary conditions, and associated Lagrangian and energy densities are then those governing a classical isotropic elastic body. Such a body is inherently stable, since the potential energy is entirely elastic: $\mathcal{V} = \mathcal{V}_\kappa + \mathcal{V}_\mu \geq 0$.

8.6 Radial Scalar Equations

To compute the eigenfrequencies and eigenfunctions of a SNREI Earth model, we need to convert the linearized equations of motion and associated boundary conditions into an equivalent system of coupled scalar equations. We shall accomplish this important task using three different methods. The first is a straightforward brute-force approach, which makes use only of the classical spherical-harmonic representation of scalar and vector fields, reviewed in Appendix B. The second method requires less tedious calculation but a more elaborate theoretical foundation; it is based upon the generalized spherical-harmonic representation of an arbitrary tensor field, developed in Appendix C. The third and final method makes use of Rayleigh's principle; the first step in this case is to find equivalent scalar expressions for the actions $\mathcal{I}$ and $\mathcal{I}'$. Readers who are unfamiliar with the geometric and algebraic properties of either ordinary or generalized spherical harmonics may wish to consult Appendices B and C before proceeding further.

8.6.1 Approach 1: Vector spherical harmonics

We employ a system of spherical polar coordinates r, θ, ϕ with its origin at the center of the SNREI Earth, and seek separable eigensolutions of equations (8.13)–(8.16) and (8.18)–(8.19) of the form

$$\mathbf{s} = U\mathbf{P}_{lm} + V\mathbf{B}_{lm} + W\mathbf{C}_{lm}, \qquad \phi = P\mathcal{Y}_{lm}, \tag{8.34}$$

where the *radial eigenfunctions* $U(r)$, $V(r)$, $W(r)$ and $P(r)$ are functions only of the radius. The real scalar and vector spherical harmonics $\mathcal{Y}_{lm}$ and

$\mathbf{P}_{lm}$, $\mathbf{B}_{lm}$, $\mathbf{C}_{lm}$ of degree $0 \leq l \leq \infty$ and order $-l \leq m \leq l$ are defined by

$$\mathcal{Y}_{lm}(\theta,\phi) = \left(\frac{2l+1}{4\pi}\right)^{1/2} \frac{1}{2^l l!} \left[\frac{(l-|m|)!}{(l+|m|)!}\right]^{1/2}$$
$$\times (\sin\theta)^{|m|} \left(\frac{1}{\sin\theta}\frac{d}{d\theta}\right)^{l+|m|} (\sin\theta)^{2l}$$
$$\times \begin{cases} \sqrt{2}\cos m\phi & \text{if } -l \leq m < 0 \\ 1 & \text{if } m = 0 \\ \sqrt{2}\sin m\phi & \text{if } 0 < m \leq l \end{cases} \tag{8.35}$$

and

$$\mathbf{P}_{lm}(\theta,\phi) = \hat{\mathbf{r}}\,\mathcal{Y}_{lm}(\theta,\phi), \qquad \mathbf{B}_{lm}(\theta,\phi) = k^{-1}\boldsymbol{\nabla}_1\mathcal{Y}_{lm}(\theta,\phi),$$
$$\mathbf{C}_{lm}(\theta,\phi) = -k^{-1}(\hat{\mathbf{r}}\times\boldsymbol{\nabla}_1)\mathcal{Y}_{lm}(\theta,\phi). \tag{8.36}$$

The dimensionless tangent-vector operators $\boldsymbol{\nabla}_1 = \hat{\boldsymbol{\theta}}\partial_\theta + \hat{\boldsymbol{\phi}}(\sin\theta)^{-1}\partial_\phi$ and $\hat{\mathbf{r}}\times\boldsymbol{\nabla}_1 = -\hat{\boldsymbol{\theta}}(\sin\theta)^{-1}\partial_\phi + \hat{\boldsymbol{\phi}}\partial_\theta$ in (8.36) are the surface gradient and curl on the unit sphere Ω, and

$$k = \sqrt{l(l+1)}. \tag{8.37}$$

We will see in Chapters 11 and 12 that the quantity k is the asymptotic angular *wavenumber* of the travelling body or surface waves associated with a normal mode of oscillation in the limit $l \to \infty$. In the present more general context, however, equation (8.37) is simply a definition that is valid for all angular degrees $0 \leq l \leq \infty$.

It is demonstrated in Appendix B.12 that the quantities $\boldsymbol{\nabla}\cdot\mathbf{s}$, $\boldsymbol{\nabla}\times\mathbf{s}$, $\boldsymbol{\nabla}(\boldsymbol{\nabla}\cdot\mathbf{s})$, $\boldsymbol{\nabla}\times\boldsymbol{\nabla}\times\mathbf{s}$ and $\nabla^2\mathbf{s} = \boldsymbol{\nabla}(\boldsymbol{\nabla}\cdot\mathbf{s}) - \boldsymbol{\nabla}\times(\boldsymbol{\nabla}\times\mathbf{s})$ can be expressed in terms of the radial eigenfunctions U, V and W in the form

$$\boldsymbol{\nabla}\cdot\mathbf{s} = [\dot{U} + r^{-1}(2U - kV)]\mathcal{Y}_{lm}, \tag{8.38}$$

$$\boldsymbol{\nabla}\times\mathbf{s} = (kr^{-1}W)\mathbf{P}_{lm} + (\dot{W} + r^{-1}W)\mathbf{B}_{lm}$$
$$- [\dot{V} + r^{-1}(V - kU)]\mathbf{C}_{lm}, \tag{8.39}$$

$$\boldsymbol{\nabla}(\boldsymbol{\nabla}\cdot\mathbf{s}) = [\ddot{U} + r^{-1}(2\dot{U} - k\dot{V}) - r^{-2}(2U - kV)]\mathbf{P}_{lm}$$
$$+ kr^{-1}[\dot{U} + r^{-1}(2U - kV)]\mathbf{B}_{lm}, \tag{8.40}$$

$$\boldsymbol{\nabla}\times\boldsymbol{\nabla}\times\mathbf{s} = -kr^{-1}[\dot{V} + r^{-1}(V - kU)]\mathbf{P}_{lm}$$
$$- (\ddot{V} + 2r^{-1}\dot{V} - kr^{-1}\dot{U})\mathbf{B}_{lm}$$
$$- (\ddot{W} + 2r^{-1}\dot{W} - k^2r^{-2}W)\mathbf{C}_{lm}, \tag{8.41}$$

$$\begin{aligned}\nabla^2\mathbf{s} = [\ddot{U} + 2r^{-1}\dot{U} - 2r^{-2}U + kr^{-2}(2V - kU)]\mathbf{P}_{lm}\\ + [\ddot{V} + 2r^{-1}\dot{V} + kr^{-2}(2U - kV)]\mathbf{B}_{lm}\\ + (\ddot{W} + 2r^{-1}\dot{W} - k^2r^{-2}W)\mathbf{C}_{lm}.\end{aligned} \tag{8.42}$$

Upon substituting the expansions (8.34) and (8.38)–(8.42) into the linearized equation of motion (8.13) and collecting terms dependent upon the harmonics $\mathbf{P}_{lm}$, $\mathbf{B}_{lm}$ and $\mathbf{C}_{lm}$, we obtain three second-order ordinary differential equations:

$$\begin{aligned}r^{-2}\frac{d}{dr}[r^2(\kappa + \tfrac{4}{3}\mu)\dot{U} + (\kappa - \tfrac{2}{3}\mu)r(2U - kV)]\\ + r^{-1}[(\kappa + \tfrac{4}{3}\mu)\dot{U} + (\kappa - \tfrac{2}{3}\mu)r^{-1}(2U - kV)]\\ - 3\kappa r^{-1}(\dot{U} + 2r^{-1}U - kr^{-1}V)\\ - k\mu r^{-1}(\dot{V} - r^{-1}V + kr^{-1}U) + \omega^2\rho U\\ - \rho[\dot{P} + (4\pi G\rho - 4gr^{-1})U + kgr^{-1}V] = 0,\end{aligned} \tag{8.43}$$

$$\begin{aligned}r^{-2}\frac{d}{dr}[\mu r^2(\dot{V} - r^{-1}V + kr^{-1}U)] + \mu r^{-1}(\dot{V} - r^{-1}V + kr^{-1}U)\\ + k(\kappa - \tfrac{2}{3}\mu)r^{-1}\dot{U} + k(\kappa + \tfrac{1}{3}\mu)r^{-2}(2U - kV)\\ + [\omega^2\rho - (k^2 - 2)\mu r^{-2}]V - k\rho r^{-1}(P + gU) = 0,\end{aligned} \tag{8.44}$$

$$\begin{aligned}r^{-2}\frac{d}{dr}[\mu r^2(\dot{W} - r^{-1}W)] + \mu r^{-1}(\dot{W} - r^{-1}W)\\ + [\omega^2\rho - (k^2 - 2)\mu r^{-2}]W = 0.\end{aligned} \tag{8.45}$$

The kinematic continuity conditions upon the solid-solid and fluid-solid boundaries $\Sigma_{\rm SS}$ and $\Sigma_{\rm FS}$ can be expressed in terms of the radial eigenfunctions in the form $[U]_-^+ = [V]_-^+ = [W]_-^+ = 0$ on $r = d_{\rm SS}$ and $[U]_-^+ = 0$ on $r = d_{\rm FS}$.

The traction (8.17) exerted upon any spherical surface is given in terms of the displacement scalars U, V and W by

$$\hat{\mathbf{r}}\cdot\mathbf{T} = R\mathbf{P}_{lm} + S\mathbf{B}_{lm} + T\mathbf{C}_{lm}, \tag{8.46}$$

where

$$R = (\kappa + \tfrac{4}{3}\mu)\dot{U} + (\kappa - \tfrac{2}{3}\mu)r^{-1}(2U - kV), \tag{8.47}$$

$$S = \mu(\dot{V} - r^{-1}V + kr^{-1}U), \tag{8.48}$$

$$T = \mu(\dot{W} - r^{-1}W). \tag{8.49}$$

The dynamical boundary conditions (8.14)–(8.16) expressing the continuity of traction on the various boundaries imply that

$$R = S = T = 0 \quad \text{on } r = a, \tag{8.50}$$

$$[R]_-^+ = [S]_-^+ = [T]_-^+ = 0 \quad \text{on } r = d_{\rm SS}, \tag{8.51}$$

$$[R]_-^+ = S = T = 0 \quad \text{on } r = d_{\rm FS}. \tag{8.52}$$

The shear tractions S and T are zero throughout the fluid regions $\oplus_{\rm F}$ by virtue of the vanishing of the rigidity, $\mu = 0$. Consequently, the two quantities $\dot{V} - r^{-1}V + kr^{-1}U$ and $\dot{W} - r^{-1}W$ must vanish on the solid side of $\Sigma_{\rm FS}$.

Poisson's equation (8.18) is equivalent to the second-order ordinary differential equation

$$\begin{aligned}&\ddot{P} + 2r^{-1}\dot{P} - k^2r^{-2}P \\ &\qquad = -4\pi G\dot{\rho}U - 4\pi G\rho[\dot{U} + r^{-1}(2U - kV)].\end{aligned} \tag{8.53}$$

The associated gravitational boundary conditions are $[P]_-^+ = 0$ and

$$[\dot{P} + 4\pi G\rho U]_-^+ = 0 \quad \text{on } r = d. \tag{8.54}$$

The solution to the boundary-value problem (8.53)–(8.54) is readily shown by substitution to be

$$\begin{aligned}P(r) = -\frac{4\pi G}{2l+1}\bigg\{&r^{-l-1}\int_0^r \rho'[lU' + kV']r'^{\,l+1}\,dr' \\ &+ r^l\int_r^a \rho'[-(l+1)U' + kV']r'^{\,-l}\,dr'\bigg\} \\ &\qquad\text{in } 0 \leq r \leq a,\end{aligned} \tag{8.55}$$

$$\begin{aligned}P(r) = -\frac{4\pi G}{2l+1}\bigg\{&r^{-l-1}\int_0^a \rho'[lU' + kV']r'^{\,l+1}\,dr'\bigg\} \\ &\qquad\text{in } a \leq r \leq \infty.\end{aligned} \tag{8.56}$$

Equations (8.55) and (8.56), which express the Eulerian potential perturbation P explicitly in terms of U and V, are the scalar analogue of equation (8.20). The results (8.43)–(8.56) were first obtained in the above brute-force manner by Pekeris & Jarosch (1958). Because the authors of this classic study defined the analogue of $\mathbf{B}_{lm}$ and $\mathbf{C}_{lm}$ in equations (8.36) differently, their tangential scalars V and W are smaller by a factor of k than ours. A number of other seminal free-oscillation analyses, including Backus & Gilbert (1967) and Woodhouse (1980), also use this alternative vector spherical-harmonic convention.

8.6.2 Decoupling and degeneracy

Inspection of equations (8.43)–(8.56) enables us to make two important observations. First, the scalar equations and boundary conditions that determine the radial eigenfunctions U, V and P are completely decoupled from those that determine W. As a result, a SNREI Earth model has two distinct types of normal modes—*spheroidal* modes with displacements of the form $U\mathbf{P}_{lm} + V\mathbf{B}_{lm}$ and *toroidal* modes with displacements of the form $W\mathbf{C}_{lm}$. The spheroidal oscillations alter the external shape and internal density of the Earth; hence, they are accompanied by perturbations $P\mathcal{Y}_{lm}$ in the gravitational potential. The toroidal oscillations, in contrast, have purely tangential displacements and zero divergence; hence, they leave the shape and the radial density distribution ρ of the Earth unaffected. The radial equations and boundary conditions governing the toroidal modes are the same on either a gravitating or non-gravitating Earth. Second, we note that none of the scalar relations governing the radial eigenfunctions U, V, P or W exhibit any dependence upon the azimuthal order m. Because of this, every spheroidal or toroidal eigenfrequency ω is *degenerate*, with an associated $(2l+1)$-dimensional eigenspace spanned by the real surface spherical harmonics $\mathcal{Y}_{l\,-l}, \ldots, \mathcal{Y}_{l0}, \ldots, \mathcal{Y}_{ll}$. This $2l+1$ degeneracy is an expected mathematical consequence of the spherical symmetry of the model.

Whenever it is necessary to distinguish between SNREI-Earth eigenfrequencies and eigenfunctions, we shall do so using index notation such as ${}_n\omega_l^{\mathrm{S}}$, ${}_n\omega_l^{\mathrm{T}}$ and ${}_nU_l$, ${}_nV_l$, ${}_nW_l$. We have introduced the prefix *overtone number* $n = 0, 1, 2, \ldots$ in anticipation of the fact that for a given value of the spherical-harmonic degree l there will be an infinite number of spheroidal and toroidal modes with eigenfrequencies ${}_n\omega_l^{\mathrm{S}}$ and ${}_n\omega_l^{\mathrm{T}}$ which tend to infinity in the limit $n \to \infty$. The $2l+1$ oscillations associated with a given eigenfrequency ${}_n\omega_l^{\mathrm{S}}$ or ${}_n\omega_l^{\mathrm{T}}$ are referred to as a *multiplet*, designated by ${}_n\mathrm{S}_l$ for spheroidal modes and by ${}_n\mathrm{T}_l$ for mantle toroidal modes. Each spheroidal eigenfunction ${}_nU_l\mathbf{P}_{lm} + {}_nV_l\mathbf{B}_{lm}$ within a multiplet ${}_n\mathrm{S}_l$ and each toroidal eigenfunction ${}_nW_l\mathbf{C}_{lm}$ within a multiplet ${}_n\mathrm{T}_l$ is referred to as a *singlet*. Every different choice of the polar axis ($\theta = 0$) and zero meridian ($\phi = 0$) gives rise to different spherical harmonics $\mathcal{Y}_{l\,-l}, \ldots, \mathcal{Y}_{l0}, \ldots, \mathcal{Y}_{ll}$ and therefore a different basis of $2l+1$ singlets spanning each multiplet. Any departure of the Earth model away from spherical symmetry removes the eigenfrequency degeneracy and causes the multiplets ${}_n\mathrm{S}_l$ and ${}_n\mathrm{T}_l$ to split and couple, as we shall discuss in Chapters 13–14. As a general rule, we shall affix only as many of the identifying superscripts S, T and subscripts n, l, m to the eigenfrequencies ω and eigenfunctions U, V, W as is necessary for clarity. Normally, we shall omit these multiplet identifiers, as in equations (8.34), to avoid notational clutter.

⋆8.6.3 Approach 2: Generalized spherical harmonics

Vectors and tensors in the generalized spherical-harmonic representation are expressed in terms of the so-called *canonical basis*

$$\hat{\mathbf{e}}_- = \tfrac{1}{\sqrt{2}}(\hat{\boldsymbol{\theta}} - i\hat{\boldsymbol{\phi}}), \qquad \hat{\mathbf{e}}_0 = \hat{\mathbf{r}}, \qquad \hat{\mathbf{e}}_+ = -\tfrac{1}{\sqrt{2}}(\hat{\boldsymbol{\theta}} + i\hat{\boldsymbol{\phi}}), \tag{8.57}$$

rather than the conventional real basis $\hat{\mathbf{r}}$, $\hat{\boldsymbol{\theta}}$, $\hat{\boldsymbol{\phi}}$. Greek indices $\alpha, \beta, \gamma, \ldots$ that range over the values $\{-, 0, +\}$ are used to designate components with respect to the vectors (8.57) in what follows. The displacement and gravitational potential perturbations (8.34) are rewritten in the form

$$\mathbf{s} = s^- Y_{lm}^{-1}\hat{\mathbf{e}}_- + s^0 Y_{lm}^0 \hat{\mathbf{e}}_0 + s^+ Y_{lm}^1 \hat{\mathbf{e}}_+, \qquad \phi = P Y_{lm}^0, \tag{8.58}$$

where the canonical components s^α depend only upon the radius. The strain tensor $\boldsymbol{\varepsilon}$ can be represented in an analogous manner, namely,

$$\begin{aligned}\boldsymbol{\varepsilon} = {} & \varepsilon^{--} Y_{lm}^{-2}\hat{\mathbf{e}}_-\hat{\mathbf{e}}_- + \varepsilon^{0-} Y_{lm}^{-1}(\hat{\mathbf{e}}_0\hat{\mathbf{e}}_- + \hat{\mathbf{e}}_-\hat{\mathbf{e}}_0) \\ & + (\varepsilon^{00} + 2\varepsilon^{-+}) Y_{lm}^0 \hat{\mathbf{e}}_0\hat{\mathbf{e}}_0 + \varepsilon^{0+} Y_{lm}^1 (\hat{\mathbf{e}}_0\hat{\mathbf{e}}_+ + \hat{\mathbf{e}}_+\hat{\mathbf{e}}_0) \\ & \quad + \varepsilon^{++} Y_{lm}^2 \hat{\mathbf{e}}_+\hat{\mathbf{e}}_+ .\end{aligned} \tag{8.59}$$

The symmetry $\boldsymbol{\varepsilon} = \boldsymbol{\varepsilon}^{\mathrm{T}}$ guarantees that $\varepsilon^{\alpha\beta} = \varepsilon^{\beta\alpha}$. The angular dependence of $\mathbf{s}$, ϕ and $\boldsymbol{\varepsilon}$ is embedded in the complex *generalized spherical harmonics* Y_{lm}^N of degree $0 \leq l \leq \infty$, order $-l \leq m \leq l$ and upper index $-l \leq N \leq l$, which are defined by

$$\begin{aligned} Y_{lm}^N(\theta, \phi) = {} & \left(\frac{2l+1}{4\pi}\right)^{1/2} \left[\frac{1}{(l+N)!(l-N)!}\right]^{1/2} \left[\frac{(l-m)!}{(l+m)!}\right]^{1/2} \\ & \times 2^{l+m} (\sin\tfrac{1}{2}\theta)^{m-N} (\cos\tfrac{1}{2}\theta)^{m+N} \\ & \times \left(\frac{1}{\sin\theta}\frac{d}{d\theta}\right)^{l+m} \left[(\sin\tfrac{1}{2}\theta)^{2l+2N}(\cos\tfrac{1}{2}\theta)^{2l-2N}\right] \\ & \times \exp(im\phi). \end{aligned} \tag{8.60}$$

Note that only Y_{lm}^0, $Y_{lm}^{\pm 1}$ and $Y_{lm}^{\pm 2}$ are required to represent the scalar, vector and second-order tensor fields (8.58)–(8.59).

The six independent canonical components of the strain $\varepsilon^{\alpha\beta}$ are related to the three components of the displacement s^α by

$$\varepsilon^{00} = \dot{s}^0, \qquad \varepsilon^{\pm\pm} = \Omega_l^2 r^{-1} s^{\pm}, \tag{8.61}$$

$$\varepsilon^{0\pm} = \varepsilon^{\pm 0} = \tfrac{1}{2}[\dot{s}^{\pm} - r^{-1}(s^{\pm} - \Omega_l^0 s^0)], \tag{8.62}$$

$$\varepsilon^{\pm\mp} = \tfrac{1}{2}\Omega_l^0 r^{-1}(s^- + s^+) - r^{-1}s^0, \tag{8.63}$$

where

$$\Omega_l^0 = \sqrt{\tfrac{1}{2}l(l+1)}, \qquad \Omega_l^2 = \sqrt{\tfrac{1}{2}(l-1)(l+2)}. \tag{8.64}$$

The trace of the strain tensor, or alternatively the divergence of the displacement, is given by

$$\boldsymbol{\nabla}\cdot\mathbf{s} = [\dot{s}^0 + 2r^{-1}s^0 - \Omega_l^0 r^{-1}(s^- + s^+)]Y_{lm}^0. \tag{8.65}$$

The gradient of the potential perturbation is

$$\boldsymbol{\nabla}\phi = \Omega_l^0 r^{-1}PY_{lm}^{-1}\hat{\mathbf{e}}_- + \dot{P}Y_{lm}^0\hat{\mathbf{e}}_0 + \Omega_l^0 r^{-1}PY_{lm}^1\hat{\mathbf{e}}_+. \tag{8.66}$$

Upon inserting equations (8.59), (8.61)–(8.63) and (8.65) into the isotropic, elastic constitutive relation (8.11), we obtain the generalized spherical-harmonic representation of the incremental stress:

$$\begin{aligned}\mathbf{T} = T^{--}Y_{lm}^{-2}\hat{\mathbf{e}}_-\hat{\mathbf{e}}_- + T^{0-}Y_{lm}^{-1}(\hat{\mathbf{e}}_0\hat{\mathbf{e}}_- + \hat{\mathbf{e}}_-\hat{\mathbf{e}}_0)\\ + (T^{00} + 2T^{-+})Y_{lm}^0\hat{\mathbf{e}}_0\hat{\mathbf{e}}_0 + T^{0+}Y_{lm}^1(\hat{\mathbf{e}}_0\hat{\mathbf{e}}_+ + \hat{\mathbf{e}}_+\hat{\mathbf{e}}_0)\\ + T^{++}Y_{lm}^2\hat{\mathbf{e}}_+\hat{\mathbf{e}}_+,\end{aligned} \tag{8.67}$$

where

$$T^{00} = (\kappa + \tfrac{4}{3}\mu)\dot{s}^0 + (\kappa - \tfrac{2}{3}\mu)r^{-1}[2s^0 - \Omega_l^0(s^- + s^+)], \tag{8.68}$$

$$T^{\pm\pm} = 2\Omega_l^2\mu r^{-1}s^{\pm}, \tag{8.69}$$

$$T^{0\pm} = T^{\pm 0} = \mu[\dot{s}^{\pm} - r^{-1}(s^{\pm} - \Omega_l^0 s^0)], \tag{8.70}$$

$$T^{\pm\mp} = -(\kappa - \tfrac{2}{3}\mu)\dot{s}^0 - (\kappa + \tfrac{1}{3}\mu)r^{-1}[2s^0 - \Omega_l^0(s^- + s^+)]. \tag{8.71}$$

The traction vector and the divergence of the stress tensor can be written in terms of the canonical components $T^{\alpha\beta} = T^{\beta\alpha}$ in the form

$$\hat{\mathbf{r}}\cdot\mathbf{T} = T^{0-}Y_{lm}^{-1}\hat{\mathbf{e}}_- + T^{00}Y_{lm}^0\hat{\mathbf{e}}_0 + T^{0+}Y_{lm}^1\hat{\mathbf{e}}_+, \tag{8.72}$$

$$\begin{aligned}\boldsymbol{\nabla}\cdot\mathbf{T} = [\dot{T}^{0-} + r^{-1}(3T^{0-} - \Omega_l^0 T^{-+} - \Omega_l^2 T^{--})]Y_{lm}^{-1}\hat{\mathbf{e}}_-\\ + [\dot{T}^{00} + 2r^{-1}(T^{00} + T^{-+}) - \Omega_l^0 r^{-1}(T^{0-} + T^{0+})]Y_{lm}^0\hat{\mathbf{e}}_0\\ + [\dot{T}^{0+} + r^{-1}(3T^{0+} - \Omega_l^0 T^{-+} - \Omega_l^2 T^{++})]Y_{lm}^1\hat{\mathbf{e}}_+.\end{aligned} \tag{8.73}$$

Putting the above results together, we find that the equation of motion (8.10) is equivalent to the three scalar equations:

$$\begin{aligned}-\omega^2\rho s^0 - \dot{T}^{00} - 2r^{-1}T^{00} - 2r^{-1}T^{-+} + \Omega_l^0 r^{-1}(T^{0-} + T^{0+})\\ + \rho[\dot{P} + (4\pi G\rho - 4r^{-1}g)s^0 + \Omega_l^0 g r^{-1}(s^- + s^+)] = 0,\end{aligned} \tag{8.74}$$

$$-\omega^2\rho s^{\pm} - \dot{T}^{0\pm} - 3r^{-1}T^{0\pm} + r^{-1}(\Omega_l^0 T^{-+} + \Omega_l^2 T^{\pm\pm}) \\ + \Omega_l^0 \rho r^{-1}(P + gs^0) = 0. \tag{8.75}$$

The kinematic continuity conditions can be written in terms of the canonical components s^0 and $s^{\pm}$ in the form $[s^0]_-^+ = [s^{\pm}]_-^+ = 0$ on $r = d_{\mathrm{SS}}$ and $[s^0]_-^+ = 0$ on $r = d_{\mathrm{FS}}$. The dynamical boundary conditions (8.14)–(8.16) take the form $T^{00} = T^{0\pm} = 0$ on $r = a$, $[T^{00}]_-^+ = [T^{0\pm}]_-^+ = 0$ on $r = d_{\mathrm{SS}}$, and $[T^{00}]_-^+ = T^{0\pm} = 0$ on $r = d_{\mathrm{FS}}$. Poisson's equation (8.18) is equivalent to the canonical relation

$$\ddot{P} + 2r^{-1}\dot{P} - l(l+1)r^{-2}P \\ = -4\pi G\dot{\rho}s^0 - 4\pi G\rho[\dot{s}^0 + 2r^{-1}s^0 - \Omega_l^0 r^{-1}(s^- + s^+)]. \tag{8.76}$$

The latter must be solved subject to the boundary conditions $[P]_-^+ = 0$ and $[\dot{P} + 4\pi G\rho s^0]_-^+ = 0$ on $r = d$.

Upon adding and subtracting the two versions of equation (8.75) we find that the sum $s^- + s^+$ and the difference $s^- - s^+$ satisfy

$$-\omega^2\rho(s^- + s^+) - (\dot{T}^{0-} + \dot{T}^{0+}) - 3r^{-1}(T^{0-} + T^{0+}) \\ + r^{-1}[2\Omega_l^0 T^{-+} + \Omega_l^2(T^{--} + T^{++})] \\ + 2\Omega_l^0 \rho r^{-1}(P + gs^0) = 0, \tag{8.77}$$

$$-\omega^2\rho(s^- - s^+) - (\dot{T}^{0-} - \dot{T}^{0+}) - 3r^{-1}(T^{0-} - T^{0+}) \\ + \Omega_l^2 r^{-1}(T^{--} - T^{++}) = 0, \tag{8.78}$$

where

$$T^{0-} + T^{0+} = \mu[(\dot{s}^- + \dot{s}^+) - r^{-1}(s^- + s^+) - 2\Omega_l^0 r^{-1}s^0], \tag{8.79}$$

$$T^{--} + T^{++} = 2\Omega_l^2 \mu r^{-1}(s^- + s^+), \tag{8.80}$$

$$T^{0-} - T^{0+} = \mu[(\dot{s}^- - \dot{s}^+) - r^{-1}(s^- - s^+)], \tag{8.81}$$

$$T^{--} - T^{++} = 2\Omega_l^2 \mu r^{-1}(s^- - s^+). \tag{8.82}$$

The existence of independent spheroidal and toroidal free oscillations is manifested in the present approach by the decoupling of the relations that determine s^0, $s^- + s^+$ and P from those that determine $s^- - s^+$. In fact, equations (8.74) and (8.76)–(8.78) are identical to (8.43)–(8.45) and (8.53), by virtue of the transformation relations

$$U = s^0, \qquad V = \tfrac{1}{\sqrt{2}}(s^- + s^+), \qquad W = \tfrac{i}{\sqrt{2}}(s^- - s^+), \tag{8.83}$$

$$R = T^{00}, \quad S = \tfrac{1}{\sqrt{2}}(T^{0-} + T^{0+}), \quad T = \tfrac{i}{\sqrt{2}}(T^{0-} - T^{0+}). \tag{8.84}$$

Generally, the relation between the real and canonical expansion coefficients U, V, W, R, S, T and $s^0, s^{\pm}, T^{00}, T^{0\pm}$ is more complicated than (8.83)–(8.84); the simplicity in the present instance is due to the lack of any dependence upon the order m.

The advantage of the generalized spherical-harmonic representation is that it lessens the labor of manipulating vector and, especially, higher-order tensor quantities in spherical coordinates. We shall employ the more familiar vector spherical-harmonic representations (8.34) and (8.46) of the displacement $\mathbf{s}$ and traction $\hat{\mathbf{r}}\cdot\mathbf{T}$ for pedagogical purposes throughout the rest of this book, resorting to generalized spherical harmonics only when necessary for intricate calculations.

8.6.4 Approach 3: Rayleigh's principle

Our final approach to deriving the radial scalar equations and boundary conditions that govern the free oscillations of a SNREI Earth model is based upon Rayleigh's variational principle. We substitute the classical representation (8.34) of the displacement $\mathbf{s}$ into the three-dimensional action integral (8.21) and perform the integration over the unit sphere Ω; this integration only involves real surface spherical harmonics $\mathcal{Y}_{lm}$ and their tangential derivatives. Upon making use of the scalar and vector orthonormality relations

$$\int_\Omega \mathbf{P}_{lm}\cdot\mathbf{P}_{l'm'}\,d\Omega = \int_\Omega \mathbf{B}_{lm}\cdot\mathbf{B}_{l'm'}\,d\Omega$$
$$= \int_\Omega \mathbf{C}_{lm}\cdot\mathbf{C}_{l'm'}\,d\Omega = \int_\Omega \mathcal{Y}_{lm}\mathcal{Y}_{l'm'}\,d\Omega = \delta_{ll'}\delta_{mm'}, \qquad (8.85)$$

together with the tensor integral relations summarized in Table B.3, we find that $\mathcal{I}$ decomposes naturally into separate spheroidal and toroidal radial action integrals:

$$\mathcal{I} = \mathcal{I}_{\mathrm{S}} + \mathcal{I}_{\mathrm{T}}, \qquad (8.86)$$

where

$$\mathcal{I}_{\mathrm{S}} = \int_0^a L_{\mathrm{S}}(U,\dot{U},V,\dot{V})\,r^2 dr, \qquad (8.87)$$

$$\mathcal{I}_{\mathrm{T}} = \int_0^a L_{\mathrm{T}}(W,\dot{W})\,r^2 dr. \qquad (8.88)$$

The *radial spheroidal Lagrangian density* L_{S} and the *radial toroidal Lagrangian density* L_{T} are

$$L_{\mathrm{S}} = \tfrac{1}{2}[\omega^2\rho(U^2+V^2) - \kappa(\dot{U} + 2r^{-1}U - kr^{-1}V)^2$$

$$- \tfrac{1}{3}\mu(2\dot{U} - 2r^{-1}U + kr^{-1}V)^2 - \mu(\dot{V} - r^{-1}V + kr^{-1}U)^2$$
$$- (k^2 - 2)\mu r^{-2}V^2 - \rho(U\dot{P} + kr^{-1}VP)$$
$$- 4\pi G\rho^2 U^2 + 2\rho g r^{-1}U(2U - kV)], \tag{8.89}$$

$$L_{\mathrm{T}} = \tfrac{1}{2}[\omega^2\rho W^2 - \mu(\dot{W} - r^{-1}W)^2 - (k^2 - 2)\mu r^{-2}W^2]. \tag{8.90}$$

The potential eigenfunction P in equation (8.89) is regarded as a known functional of the spheroidal displacement eigenfunctions U and V, given by the radial integral (8.55).

The variations of the spheroidal and toroidal actions are given, correct to first order in δU, δV and δW, by

$$\delta\mathcal{I}_{\mathrm{S}} = \int_0^a \delta U\left[\partial_U L_{\mathrm{S}} - r^{-2}\frac{d}{dr}(r^2\partial_{\dot{U}}L_{\mathrm{S}})\right]r^2dr$$
$$+ \int_0^a \delta V\left[\partial_V L_{\mathrm{S}} - r^{-2}\frac{d}{dr}(r^2\partial_{\dot{V}}L_{\mathrm{S}})\right]r^2dr$$
$$- \sum_d d^2\left[\delta U(\partial_{\dot{U}}L_{\mathrm{S}}) + \delta V(\partial_{\dot{V}}L_{\mathrm{S}})\right]_-^+, \tag{8.91}$$

$$\delta\mathcal{I}_{\mathrm{T}} = \int_0^a \delta W\left[\partial_W L_{\mathrm{T}} - r^{-2}\frac{d}{dr}(r^2\partial_{\dot{W}}L_{\mathrm{T}})\right]r^2dr$$
$$- \sum_d d^2\left[\delta W(\partial_{\dot{W}}L_{\mathrm{T}})\right]_-^+, \tag{8.92}$$

where the summation is over all of the discontinuities, including the outer free surface. Rayleigh's principle stipulates that $\delta\mathcal{I}_{\mathrm{S}} = 0$ and $\delta\mathcal{I}_{\mathrm{T}} = 0$ for arbitrary independent variations δU, δV and δW satisfying the admissibility constraints $[\delta U]_-^+ = [\delta V]_-^+ = [\delta W]_-^+ = 0$ on $r = d_{\mathrm{SS}}$ and $[\delta U]_-^+ = 0$ on $r = d_{\mathrm{FS}}$. This will be so if and only if the displacement scalars U, V and W satisfy the *radial Euler-Lagrange equations*

$$\partial_U L_{\mathrm{S}} - r^{-2}\frac{d}{dr}(r^2\partial_{\dot{U}}L_{\mathrm{S}}) = 0, \tag{8.93}$$

$$\partial_V L_{\mathrm{S}} - r^{-2}\frac{d}{dr}(r^2\partial_{\dot{V}}L_{\mathrm{S}}) = 0, \tag{8.94}$$

$$\partial_W L_{\mathrm{T}} - r^{-2}\frac{d}{dr}(r^2\partial_{\dot{W}}L_{\mathrm{T}}) = 0 \tag{8.95}$$

in the Earth model, together with the associated boundary conditions

$$\partial_{\dot{U}}L_{\mathrm{S}} = \partial_{\dot{V}}L_{\mathrm{S}} = \partial_{\dot{W}}L_{\mathrm{T}} = 0 \quad \text{on } r = a, \tag{8.96}$$

$$[\partial_{\dot{U}} L_{\mathrm{S}}]_-^+ = [\partial_{\dot{V}} L_{\mathrm{S}}]_-^+ = [\partial_{\dot{W}} L_{\mathrm{T}}]_-^+ = 0 \quad \text{on } r = d_{\mathrm{SS}}, \tag{8.97}$$

$$[\partial_{\dot{U}} L_{\mathrm{S}}]_-^+ = \partial_{\dot{V}} L_{\mathrm{S}} = \partial_{\dot{W}} L_{\mathrm{T}} = 0 \quad \text{on } r = d_{\mathrm{FS}}. \tag{8.98}$$

The partial derivatives with respect to the quantities $\dot{U}$, $\dot{V}$ and $\dot{W}$ are precisely the three negative traction scalars (8.47)–(8.49):

$$\partial_{\dot{U}} L_{\mathrm{S}} = -R, \qquad \partial_{\dot{V}} L_{\mathrm{S}} = -S, \qquad \partial_{\dot{W}} L_{\mathrm{S}} = -T. \tag{8.99}$$

Using these results together with the quadratic nature of the terms involving P and $\dot{P}$ in the spheroidal Lagrangian density (8.89), it may be readily shown that equations (8.93)–(8.98) are equivalent to (8.43)–(8.45) and (8.50)–(8.52).

We may alternatively insert the representation (8.34) of $\mathbf{s}$ and ϕ into equation (8.23) to obtain the *modified spheroidal action integral*:

$$\mathcal{I}'_{\mathrm{S}} = \int_0^\infty L'_{\mathrm{S}}(U, \dot{U}, V, \dot{V}, P, \dot{P})\, r^2 dr \tag{8.100}$$

where

$$\begin{aligned} L'_{\mathrm{S}} = \tfrac{1}{2}[&\omega^2\rho(U^2 + V^2) - \kappa(\dot{U} + 2r^{-1}U - kr^{-1}V)^2 \\ &- \tfrac{1}{3}\mu(2\dot{U} - 2r^{-1}U + kr^{-1}V)^2 - \mu(\dot{V} - r^{-1}V + kr^{-1}U)^2 \\ &- (k^2 - 2)\mu r^{-2}V^2 - 2\rho(U\dot{P} + kr^{-1}VP) \\ &- 4\pi G\rho^2 U^2 + 2\rho g r^{-1} U(2U - kV) \\ &- (4\pi G)^{-1}(\dot{P}^2 + k^2 r^{-2} P^2)]. \end{aligned} \tag{8.101}$$

There is no need to distinguish between the action and the modified action for the toroidal modes, since they are unaccompanied by a perturbation P in the gravitational potential. The upper limit of integration in equation (8.100) is infinity; however, only the final term involving $\dot{P}^2 + k^2 r^{-2} P^2$ is non-zero in $a \le r \le \infty$. The variation of the modified action $\mathcal{I}'_{\mathrm{S}}$ with respect to U and V, treating P as an independent variable, gives rise to the spheroidal momentum equations (8.43)–(8.44) together with the dynamical boundary conditions governing R and S, as before, whereas the variation with respect to P leads to Poisson's equation (8.53) and the associated boundary condition (8.54), expressed in the form

$$\partial_P L'_{\mathrm{S}} - r^{-2}\frac{d}{dr}(r^2 \partial_{\dot{P}} L'_{\mathrm{S}}) = 0, \tag{8.102}$$

$$[\partial_{\dot{P}} L'_{\mathrm{S}}]_-^+ = 0 \quad \text{on } r = d. \tag{8.103}$$

It is easily verified that $\partial_{\dot{P}} L'_{\rm S} = -(4\pi G)^{-1}\dot{P} - \rho U$ in $0 \leq r \leq \infty$. The stationary value of both of the spheroidal actions and the toroidal action is

$$\mathcal{I}_{\rm S} = \mathcal{I}'_{\rm S} = 0, \qquad \mathcal{I}_{\rm T} = 0, \tag{8.104}$$

at every eigensolution U, V, P or W. The result (8.104), which follows from equation (8.25), can also be obtained by integrating the products of U, V, W with equations (8.43)–(8.45) and the product of P with (8.53) over $0 \leq r \leq a$ and $0 \leq r \leq \infty$, respectively. The scalar analogue of the three-dimensional identity (3.195) is

$$\tfrac{1}{2}\int_0^a \rho(U\dot{P}' + k'r^{-1}VP' + U'\dot{P} + kr^{-1}V'P)\, r^2 dr + \frac{1}{4\pi G}\int_0^\infty (\dot{P}\dot{P}' + kk'r^{-2}PP')\, r^2 dr = 0, \tag{8.105}$$

for any two spheroidal displacement and incremental potential fields U, V, P and U', V', P' satisfying equations (8.53)–(8.54).

8.6.5 Orthonormality

The displacement eigenfunctions of a SNREI Earth model must satisfy the general orthonormality relation

$$\int_\oplus \rho\, \mathbf{s}_k \cdot \mathbf{s}_{k'}\, dV = \delta_{kk'}, \tag{8.106}$$

where k is a pseudonym for the quadripartite identifier $\{n, l, m; \text{S or T}\}$. The case $k = k'$ specifies our adopted normalization. It is easily verified that spheroidal and toroidal eigenfunctions of different degree or order as well as spheroidal-toroidal pairs of eigenfunctions are orthogonal, by virtue of the spherical-harmonic relations (8.85). The spheroidal and toroidal radial eigenfunctions *of the same degree l* must be orthonormal in the sense

$$\int_0^a \rho({}_nU_l\, {}_{n'}U_l + {}_nV_l\, {}_{n'}V_l)\, r^2 dr = \delta_{nn'}, \tag{8.107}$$

$$\int_0^a \rho({}_nW_l\, {}_{n'}W_l)\, r^2 dr = \delta_{nn'}. \tag{8.108}$$

The scalar orthonormality relations (8.107)–(8.108), which are an immediate consequence of (8.106), can also be obtained directly from the governing radial equations and boundary conditions (8.43)–(8.54). We proceed to discuss the two independent classes of free oscillations separately in Sections 8.7 and 8.8, concentrating upon the simpler toroidal modes first.

8.7 Toroidal Oscillations

The toroidal oscillations of a SNREI Earth model have tangential displacement and traction vectors of the form

$$\mathbf{s} = W\mathbf{C}_{lm}, \qquad \hat{\mathbf{r}}\cdot\mathbf{T} = T\mathbf{C}_{lm}, \tag{8.109}$$

where $T = \mu(\dot{W} - r^{-1}W)$ and $\mathbf{C}_{lm} = k^{-1}[\hat{\boldsymbol{\theta}}(\sin\theta)^{-1}\partial_\phi - \hat{\boldsymbol{\phi}}\,\partial_\theta]\mathcal{Y}_{lm}$. The eigenfrequencies ω and associated radial eigenfunctions W and T depend upon the overtone number n and the spherical-harmonic degree l, but they are independent of the azimuthal order m. There are no toroidal modes of angular degree $l = 0$ because $\mathbf{C}_{00} = \mathbf{0}$.

8.7.1 Toroidal energy

The kinetic plus potential energy of a toroidal oscillation can be written as a sum of radial integrals:

$$\mathcal{E} = \tfrac{1}{2}(\omega^2\mathcal{T} + \mathcal{V}_\mu), \tag{8.110}$$

where

$$\mathcal{T} = \int_0^a \rho W^2\, r^2 dr, \tag{8.111}$$

$$\mathcal{V}_\mu = \int_0^a \mu[(\dot{W} - r^{-1}W)^2 + (k^2 - 2)r^{-2}W^2]\, r^2 dr. \tag{8.112}$$

Equations (8.111)–(8.112) are the result of substituting the toroidal displacement (8.109) into the volumetric integral representations (8.28) and (8.30)–(8.32); alternatively, we can define $\mathcal{E}$ in terms of the radial toroidal Lagrangian density L_{T} by

$$\mathcal{E} = \int_0^a (\omega\partial_\omega L - L)\, r^2 dr. \tag{8.113}$$

The potential energy of a toroidal mode is stored entirely in elastic shear; there is no compressional or gravitational energy: $\mathcal{V}_\kappa = \mathcal{V}_{\mathrm{g}} = 0$. Toroidal oscillations are never unstable: $\omega^2 = \mathcal{T}^{-1}\mathcal{V}_\mu \geq 0$, since the shear energy $\mathcal{V}_\mu$ is non-negative for any piecewise continuous displacement W in $0 \leq r \leq a$.

*8.7.2 Trivial modes

We begin by considering the trivial toroidal "modes", which have $\omega = 0$ and associated displacements s that do not alter the elastic shear energy $\mathcal{V}_\mu$ of the Earth. The first class of trivials consists of the geostrophic modes, which have $W = 0$ in the solid inner core and mantle $\oplus_{\rm S}$, and W equal to an arbitrary function of radius in the fluid outer core and ocean $\oplus_{\rm F}$. Since $\mu = 0$ in $\oplus_{\rm F}$, these modes have zero traction, $T = 0$, throughout the entire Earth model $0 \leq r \leq a$. This infinite-dimensional eigenspace of zero-frequency modes accommodates the possibility of rearranging the fluid outer core and ocean on spherical shells without affecting the elastic-gravitational energy.

The solid regions of the Earth also support zero-frequency trivials of degree $l = 1$, corresponding to rigid-body rotations. The unnormalized radial eigenfunctions in this case are of the form $W = r$ in the inner core or mantle and $W = 0$ elsewhere; the associated traction T again vanishes everywhere in $0 \leq r \leq a$. The toroidal vector spherical harmonic of degree one and order zero is $\mathbf{C}_{10} = (3/8\pi)^{1/2}\sin\theta\,\hat{\boldsymbol{\phi}}$, so that the unnormalized displacement of the $m = 0$ singlet is $\mathbf{s}_{10} = W\mathbf{C}_{10} = (3/8\pi)^{1/2}r\sin\theta\,\hat{\boldsymbol{\phi}}$. By inspection, this represents a rigid rotation of the solid inner core or mantle about the $\hat{\mathbf{z}}$ axis. It is likewise easily demonstrated that the two vectors $\mathbf{s}_{1\,-1} = W\mathbf{C}_{1\,-1}$ and $\mathbf{s}_{11} = W\mathbf{C}_{11}$ correspond to rigid rotations about the $\hat{\mathbf{x}}$ and $\hat{\mathbf{y}}$ axes, respectively.

8.7.3 First-order radial equations

The second-order differential equation (8.45) that governs the toroidal oscillations can be rewritten as a system of coupled first-order equations for the quantities W and T in the form:

$$\dot{W} = r^{-1}W + \mu^{-1}T, \tag{8.114}$$

$$\dot{T} = [-\omega^2\rho + (k^2 - 2)\mu r^{-2}]W - 3r^{-1}T. \tag{8.115}$$

Both the displacement and the traction must be continuous across every solid-solid discontinuity: $[W]_-^+ = 0$ and $[T]_-^+ = 0$ on $r = d_{\rm SS}$. Tangential slip is allowed on the fluid-solid boundaries, but the traction must vanish there and on the outer free surface:

$$T = 0 \quad \text{on } r = d_{\rm FS} \text{ and } r = a. \tag{8.116}$$

It is noteworthy that equations (8.114)–(8.116) contain only the model parameters μ and ρ, and not their radial derivatives $\dot{\mu}$ and $\dot{\rho}$; this makes the system ideally suited for numerical integration, particularly in the case of a gridded model specified at a finite number of knots in the range $0 \leq r \leq a$. There is no dependence upon the incompressibility κ because of the pure-shear nature of a toroidal deformation.

★8.7.4 Homogeneous sphere

The simplest example of a SNREI Earth model that exhibits toroidal free oscillations is a homogeneous solid sphere, with radius a and shear-wave speed $\beta = (\mu/\rho)^{1/2}$. The second-order differential equation (8.45) governing the oscillations of such a glass marble or metallic ball-bearing reduces to

$$\ddot{W} + 2r^{-1}\dot{W} + [\omega^2\beta^{-2} - l(l+1)r^{-2}]W = 0. \tag{8.117}$$

The solution to (8.117) that is regular at the origin $r = 0$ is the *spherical Bessel function* of degree l:

$$W = j_l(\omega r/\beta) = \frac{(\omega r/\beta)^l}{2^{l+1}\,l!}\int_0^\pi (\sin\zeta)^{2l+1}\cos[(\omega r/\beta)\cos\zeta]\,d\zeta. \tag{8.118}$$

The traction $T = \mu(\dot{W} - r^{-1}W)$ associated with the displacement (8.118) is

$$T = \mu r^{-1}[(l-1)j_l(\omega r/\beta) - (\omega r/\beta)j_{l+1}(\omega r/\beta)], \tag{8.119}$$

so that the free-surface boundary condition $T(a) = 0$ reduces to

$$(l-1)j_l(\omega a/\beta) - (\omega a/\beta)j_{l+1}(\omega a/\beta) = 0. \tag{8.120}$$

The eigenfrequency ${}_n\omega_l$ of the nth overtone mode corresponds to the nth root of equation (8.120).

To illustrate the above results, let us estimate the lowest three toroidal-mode eigenfrequencies of angular degree $l = 1$ of the Earth's solid inner core. From (8.120) we deduce that the eigenfrequencies ${}_n\omega_1$ are determined by the trigonometric relation

$$\tan(\omega a/\beta) = \frac{3(\omega a/\beta)}{3 - (\omega a/\beta)^2}. \tag{8.121}$$

The first root of equation (8.121) is ${}_0\omega_1 = 0$, corresponding to the rigid-body rotational triplet discussed in Section 8.7.2. The next three non-trivial roots are ${}_1\omega_1 = 5.76(\beta/a)$, ${}_2\omega_1 = 9.10(\beta/a)$ and ${}_3\omega_1 = 12.32(\beta/a)$, corresponding to the fundamental mode and first two overtones, respectively. The Earth's inner core has a radius $a = 1221$ km and a relatively uniform shear-wave speed $\beta \approx 3.5$ km/s; the frequencies of a homogeneous sphere with these properties are ${}_1\omega_1 = 16.5\times10^{-3}$ rad/s, ${}_2\omega_1 = 26.1\times10^{-3}$ rad/s, and ${}_3\omega_1 = 35.3\times10^{-3}$ rad/s, corresponding to periods of 380, 241 and 178 seconds, respectively. For the isotropic PREM model the exact periods calculated using numerical integration are 379.4, 239.1 and 176.1 seconds, quite close to the results obtained in the homogeneous approximation.

8.7.5 Numerical integration

Determination of the non-trivial eigenfrequencies ω and associated eigenfunctions W, T of an Earth model with realistic radial variations of density ρ and rigidity μ requires numerical integration. The solid inner core and the mantle exhibit independent toroidal oscillations, which are decoupled by the presence of the fluid outer core. To calculate the mantle-mode eigensolutions, we start with inhomogeneous initial values $W(b) = 1$, $T(b) = 0$ at the core-mantle boundary $r = b$, and integrate the system of first-order equations (8.114)–(8.115) up to the seafloor $r = s$. It is noteworthy that both of the independent variables W and T are continuous everywhere in $b \leq r \leq s$, including across any solid-solid discontinuities $r = d_{\mathrm{SS}}$. In general, for a given trial eigenfrequency ω, the value of the traction $T(s)$ at the seafloor will be non-zero; a root-finding procedure is required to determine the eigenfrequencies ω and associated eigenfunctions W, T for each fixed angular degree l. If the model does not have an oceanic layer, then $s = a$, and the integration is continued up to the free surface. The arbitrariness implicit in the initial choice of $W(b)$ is determined by the normalization (8.108).

It is customary to denote the toroidal modes *of the mantle* by ${}_n\mathrm{T}_l$; the overtone index $n = 0, 1, \ldots, \infty$ designates the order of the eigenfrequencies:

$$ {}_0\omega_l < {}_1\omega_l < \cdots < {}_\infty\omega_l = \infty. \tag{8.122}$$

The ensemble of modes ${}_n\mathrm{T}_0, {}_n\mathrm{T}_1, \ldots, {}_n\mathrm{T}_\infty$ with the same overtone number n is referred to as a *dispersion branch*. The lowest-frequency multiplet ${}_0\mathrm{T}_l$ of every degree l is, by definition, the *fundamental mode*, the next multiplet ${}_1\mathrm{T}_l$ is the first *overtone*, and so on. The system of first-order equations (8.114)–(8.115) together with the boundary condition (8.116) is a classical Sturm-Liouville eigenvalue problem. It is a fundamental property of such systems that the number of radial nodes in the eigenfunction ${}_nW_l$ is equal to the overtone number n; this is known technically as Sturm's oscillation theorem. All of the toroidal modes along the fundamental branch have no nodes in the mantle $b \leq r \leq s$, and all of the modes along the nth overtone branch have n nodes, for this reason. For $l = 1$ the rigid-body mantle triplet is conventionally labelled ${}_0\mathrm{T}_1$, so that the lowest-frequency non-trivial mode is the first overtone ${}_1\mathrm{T}_1$, with one radial node.

In the case of the toroidal modes of the solid inner core, the lower boundary condition is replaced by the requirement that the radial eigenfunctions W, T must be regular at the origin $r = 0$. In practice, a small portion of the Earth $0 \leq r \leq \varepsilon$ is regarded as homogeneous, and the analytical solutions (8.118)–(8.119) for a homogeneous sphere are employed as starting values $W(\varepsilon)$, $T(\varepsilon)$ in the vicinity of the center. The integration is continued up to the inner-core boundary $r = c$, at which point the traction $T(c)$ is inspected to determine the eigenfrequencies ω. The inner-core toroidal oscillations of a SNREI Earth are of little intrinsic interest, because they can neither be excited by earthquakes in the crust or upper mantle nor observed by seismometers on the surface. In fact, these modes are generally considered to be so obscure that seismologists have not agreed upon a nomenclature analogous to ${}_n\mathrm{T}_l$ to describe them. We shall refer to them as ${}_n\mathrm{C}_l$, where $n = 0, 1, 2, \ldots$ is the overtone number. Integration of the second-order system of equations (8.114)–(8.115) in the direction of increasing radius is a numerically stable procedure which can be accomplished using standard algorithms. A variable-order, variable-step Runge-Kutta technique is used to control the accuracy of the integration in the widely available programs `MINEOS` and `OBANI` developed collaboratively by Freeman Gilbert, Guy Masters and John Woodhouse. For many modes, the radial eigenfunctions W and T decay quasi-exponentially beneath the turning point of the equivalent SH body waves; in these cases, it is possible to save time by using the evanescent JWKB representation discussed in Section 12.3.4 to start the integration at a higher level. Sturm's theo-

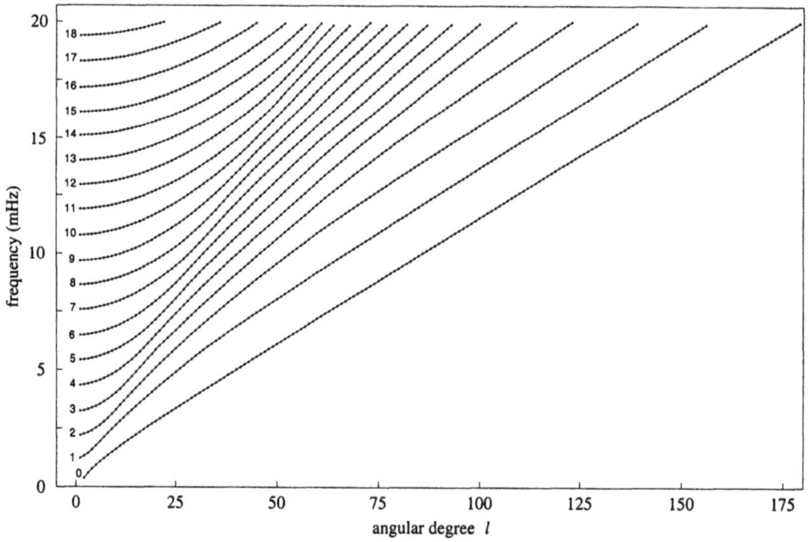

Figure 8.3. Dispersion diagram showing the mantle toroidal oscillations ${}_n\mathrm{T}_l$ of the isotropic PREM model. All of the observable eigenfrequencies ${}_n\omega_l^{\mathrm{T}}/2\pi$ below 20 mHz are plotted versus the spherical-harmonic degree l. Modes having the same overtone number n are connected by straight lines.

rem makes it possible to calculate the number of eigenfrequencies that lie between two trial values of ω by counting the radial nodes of the trial eigenfunctions W. This enables all the roots of $T(s)$ or $T(c)$ to be bracketed by upper and lower bounds; once such bounds have been established, it is easy to converge upon the eigenfrequencies by means of an infallible scheme based upon bisection. The energy equipartition relation $\omega^2\mathcal{T} = \mathcal{V}_\mu$ can be used as a check upon the final accuracy of the computation.

8.7.6 Toroidal-mode menagerie

We conclude this section with a brief qualitative survey of the principal features of the Earth's toroidal oscillations. Figure 8.3 shows the eigenfrequencies ${}_n\omega_l$ of the observable toroidal modes of the mantle, plotted versus angular degree l, for the isotropic Preliminary Reference Earth Model. The modes along each dispersion branch $n = 0, 1, 2, \ldots$ are connected by straight lines; a plot of this type is referred to as a *dispersion diagram*. Any SNREI Earth model obtained by fitting post-1975 free-oscillation and travel-time data will have a dispersion diagram that is indistinguishable from that of PREM, at the level of resolution shown. Models are discriminated on the basis of frequency residuals measured in μHz rather than mHz.

The radial variation of the fundamental-mode displacement eigenfunc-

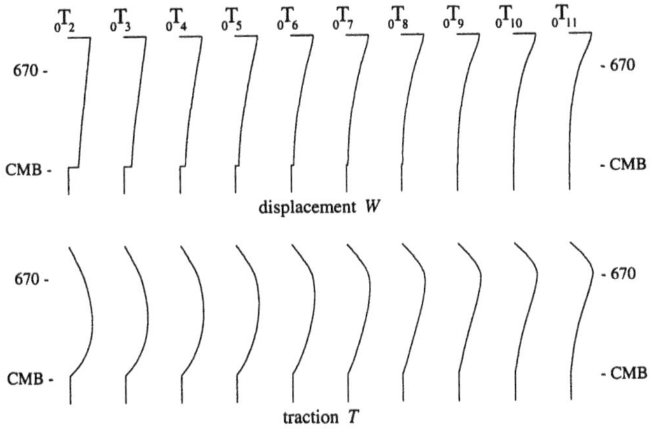

Figure 8.4. Displacement ${}_0W_l$ (*top*) and associated traction ${}_0T_l$ (*bottom*) of the first ten fundamental toroidal modes. Vertical axis is depth beneath the surface of the Earth; the locations of the 670 km discontinuity and the core-mantle boundary (CMB) are indicated.

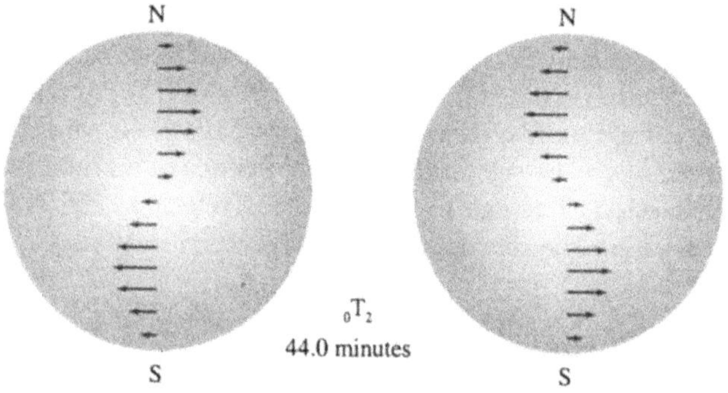

Figure 8.5. Surficial displacement pattern of the $_0\mathrm{T}_2$ toroidal mode. Two extremes of the oscillation cycle of the $m = 0$ singlet are shown. The North Pole ($\theta = 0$) and the South Pole ($\theta = \pi$) are denoted by N and S, respectively.

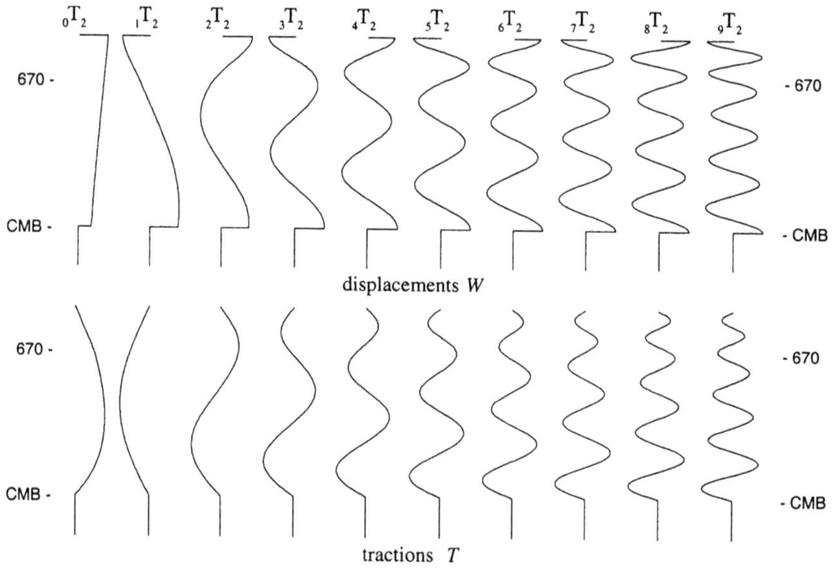

Figure 8.6. Displacement $_nW_2$ (*top*) and associated traction $_nT_2$ (*bottom*) of the first ten degree-two toroidal modes. Vertical axis is depth beneath the surface of the Earth; the locations of the 670 km discontinuity and the core-mantle boundary (CMB) are indicated.

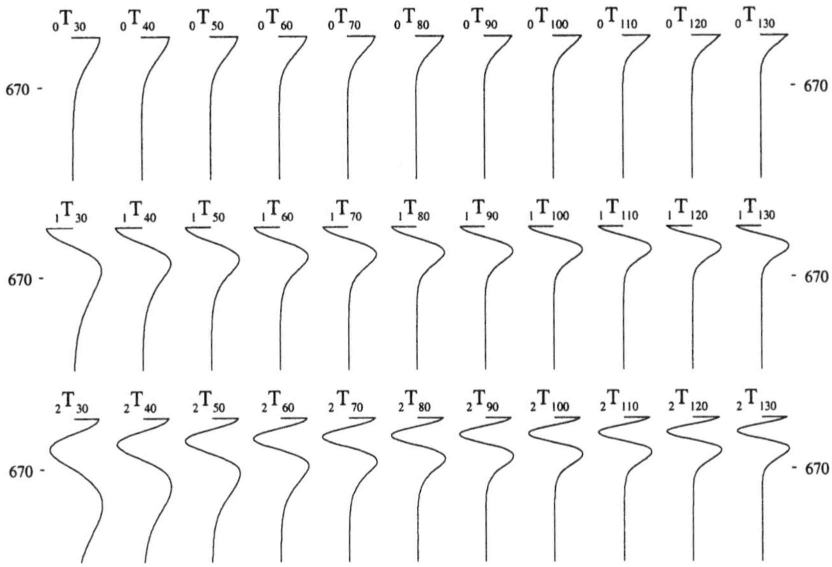

Figure 8.7. Eigenfunctions ${}_0W_l$, ${}_1W_l$ and ${}_2W_l$ along the fundamental (*top row*), first-overtone (*middle row*) and second-overtone (*bottom row*) branches of the Love wave-equivalent modes. Vertical axis extends from the surface of the Earth to 1500 km depth; the location of the 670 km discontinuity is indicated.

tions ${}_0W_2, \ldots, {}_0W_{11}$ and associated tractions ${}_0T_2, \ldots, {}_0T_{11}$ is illustrated in Figure 8.4. The gravest toroidal mode is the ${}_0\mathrm{T}_2$ quintuplet, which has a degenerate period of 44.0 minutes. The first unambiguous detection of this oscillation did not occur until after the large ($M_0 = 2 \times 10^{21}$ N m) strike-slip earthquake on the Macquarie Rise in 1989 (Widmer, Zürn & Masters 1992). The $m = 0$ singlet has an associated vector displacement of the form $\mathbf{s} = -(15/8\pi)^{1/2} W(r) \sin\theta \cos\theta\, \hat{\boldsymbol{\phi}}$, which corresponds to a shearing or twisting of the Earth, such that the northern hemisphere moves clockwise when the southern hemisphere moves counterclockwise, and vice versa, as illustrated in Figure 8.5. The displacements ${}_0W_2, \ldots, {}_9W_2$ and tractions ${}_0T_2, \ldots, {}_9T_2$ of the fundamental and next nine overtones of degree $l = 2$ are depicted in Figure 8.6. It is evident that ${}_nW_2$ has n radial nodes within the mantle, in accordance with Sturm's theorem, and that ${}_nT_2$ vanishes at both the core-mantle boundary $r = b$ and the seafloor $r = s$; this is the defining property of a mantle eigensolution.

Figure 8.7 depicts the eigenfunctions ${}_0W_l$, ${}_1W_l$ and ${}_2W_l$ of a number of higher-frequency toroidal modes along the fundamental and first two

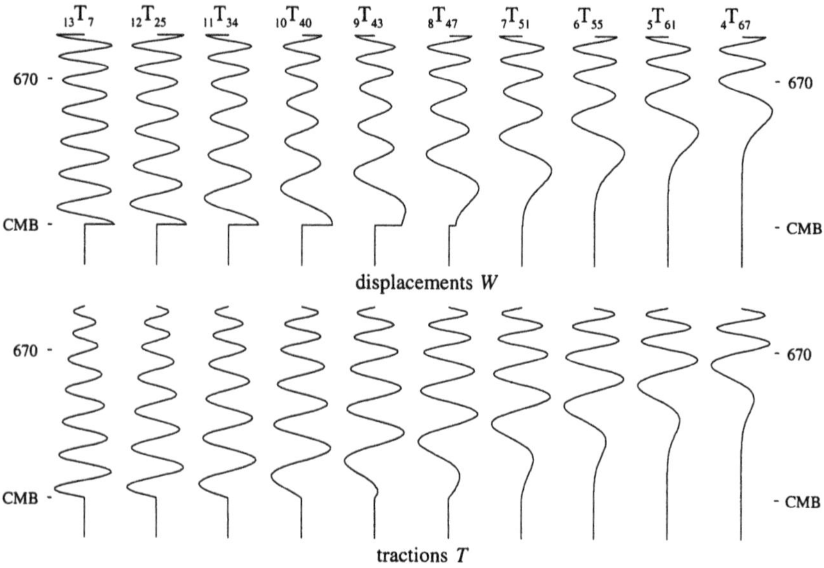

Figure 8.8. Displacement ${}_nW_l$ (*top*) and associated traction ${}_nT_l$ (*bottom*) along a constant-frequency line at ${}_n\omega_l/2\pi \approx 14$ mHz, showing the transition between $\mathrm{ScS_{SH}}$-equivalent and turning SH-equivalent modes at $n \approx l/4$. Vertical axis is depth beneath the surface of the Earth; the locations of the 670 km discontinuity and the core-mantle boundary (CMB) are indicated.

overtone branches. As the angular degree l and the frequency ${}_n\omega_l$ increase along a branch of fixed overtone number n, the displacement ${}_nW_l$ penetrates less and less deeply into the mantle. Since the lower mantle participates very little in the oscillation, it is obvious that its properties cannot have a strong influence upon the eigenfrequency; we will conduct a more quantitative examination of this depth "sensitivity" in Chapter 9. These modes ${}_0\mathrm{T}_l$, ${}_1\mathrm{T}_l$, ${}_2\mathrm{T}_l$, etc. correspond in the limit $l \gg 1$ to fundamental and higher-overtone *Love surface waves* or, equivalently, to constructively interfering SH body waves that turn in the upper mantle and are reflected beneath the seafloor.

Figure 8.8 shows the radial variation of ${}_nW_l$ and ${}_nT_l$ for a number of modes having approximately the same eigenfrequency ${}_n\omega_l/2\pi \approx 14$ mHz. The lowest-overtone modes, whose eigenfrequencies lie near the abscissa in the dispersion diagram, have oscillatory displacements of roughly uniform amplitude throughout the entire mantle, and correspond to $\mathrm{ScS_{SH}}$ body waves that are multiply reflected at both the core-mantle boundary and the seafloor. The spacing ${}_{n+1}\omega_l - {}_n\omega_l$ between adjacent $l \approx 1$ toroidal-mode eigenfrequencies is approximately equal to $2\pi/T_{\mathrm{ScS}}$, where T_{ScS} is the two-way travel time of a vertically propagating $\mathrm{ScS}_{\mathrm{SH}}$ wave. The slope or asymptotic group speed $d\omega/dk \approx {}_n\omega_{l+1} - {}_n\omega_l$ of the $\mathrm{ScS}_{\mathrm{SH}}$-equivalent dispersion curves approaches zero in the vicinity of the abscissa. The transition between the reflected $\mathrm{ScS}_{\mathrm{SH}}$-equivalent and turning SH-equivalent modes occurs along the slant line $\omega/k \approx 2 \times 10^{-3}$ rad/s, or equivalently $n \approx l/4$, where each of the dispersion curves has a single inflection point. Toroidal modes ${}_n\mathrm{T}_l$ for which $n \gg l/4$ have $d^2\omega/dk^2 > 0$ and are $\mathrm{ScS}_{\mathrm{SH}}$-equivalent, whereas those for which $n \ll l/4$ have $d^2\omega/dk^2 < 0$ and are SH-equivalent. We undertake a more systematic JWKB analysis of the asymptotic correspondence between the high-frequency toroidal oscillations of the mantle and propagating $\mathrm{ScS}_{\mathrm{SH}}$ and SH body waves in Chapter 12.

8.8 Spheroidal Oscillations

The spheroidal oscillations of a SNREI Earth model have associated eigenfunctions of the form

$$\mathbf{s} = U\mathbf{P}_{lm} + V\mathbf{B}_{lm}, \qquad \hat{\mathbf{r}}\cdot\mathbf{T} = R\mathbf{P}_{lm} + S\mathbf{B}_{lm}, \tag{8.123}$$

where $\mathbf{P}_{lm} = \hat{\mathbf{r}}\mathcal{Y}_{lm}$ and $\mathbf{B}_{lm} = k^{-1}[\hat{\boldsymbol{\theta}}\partial_\theta + \hat{\boldsymbol{\phi}}(\sin\theta)^{-1}\partial_\phi]\mathcal{Y}_{lm}$. The radial and tangential traction scalars R, S are related to the corresponding displacement scalars U, V by $R = (\kappa + \frac{4}{3}\mu)\dot{U} + (\kappa - \frac{2}{3}\mu)r^{-1}(2U - kV)$, $S = \mu(\dot{V} - r^{-1}V + kr^{-1}U)$. The associated perturbation in the gravitational potential is $\phi = P\mathcal{Y}_{lm}$, where P is given explicitly in terms of U and V by equation (8.55). It is convenient to define an auxiliary function

$$B = \dot{P} + 4\pi G\rho U + (l+1)r^{-1}P, \tag{8.124}$$

which vanishes at the surface of the Earth $r = a$ by virtue of the boundary condition (8.54). The eigenfrequencies ω and the six radial scalars U, V, R, S, P and B depend upon the overtone number n and the angular degree l, but they are independent of the azimuthal order m. Spheroidal oscillations of angular degree $l = 0$ have purely radial displacement and traction vectors, because the tangential vector spherical harmonic $\mathbf{B}_{00}$ is identically equal to zero; these are referred to as the *radial modes*.

8.8.1 Spheroidal energy

The kinetic plus elastic-gravitational potential energy of a spheroidal oscillation, obtained by substituting the representation (8.123) into equations (8.28) and (8.30)–(8.32), is

$$\mathcal{E} = \tfrac{1}{2}(\omega^2\mathcal{T} + \mathcal{V}_\kappa + \mathcal{V}_\mu + \mathcal{V}_{\mathrm{g}}), \tag{8.125}$$

where

$$\mathcal{T} = \int_0^a \rho(U^2 + V^2)\, r^2 dr, \tag{8.126}$$

$$\mathcal{V}_\kappa = \int_0^a \kappa(\dot{U} + 2r^{-1}U - kr^{-1}V)^2\, r^2 dr, \tag{8.127}$$

$$\begin{aligned}\mathcal{V}_\mu = \int_0^a & \mu[\tfrac{1}{3}(2\dot{U} - 2r^{-1}U + kr^{-1}V)^2 \\ & + (\dot{V} - r^{-1}V + kr^{-1}U)^2 + (k^2 - 2)r^{-2}V^2]\, r^2 dr,\end{aligned} \tag{8.128}$$

$$\begin{aligned}\mathcal{V}_{\mathrm{g}} = \int_0^a & \rho[U\dot{P} + kr^{-1}VP + 4\pi G\rho U^2 \\ & - 2gr^{-1}U(2U - kV)]\, r^2 dr.\end{aligned} \tag{8.129}$$

The energy $\mathcal{E}$ can also be expressed in terms of the radial spheroidal Lagrangian density L_{S} in the form (8.113). The Rayleigh quotient of a spheroidal mode is

$$\omega^2 = \frac{\mathcal{V}_\kappa + \mathcal{V}_\mu + \mathcal{V}_{\mathrm{g}}}{\mathcal{T}}. \tag{8.130}$$

With the normalization adopted here, $\mathcal{T} = 1$, the energy equipartition relation (8.130) reduces to

$$f_\kappa + f_\mu + f_{\mathrm{g}} = 1, \tag{8.131}$$

where we have defined the three dimensionless quantities

$$f_\kappa = \omega^{-2}\mathcal{V}_\kappa, \qquad f_\mu = \omega^{-2}\mathcal{V}_\mu, \qquad f_{\mathrm{g}} = \omega^{-2}\mathcal{V}_{\mathrm{g}}. \tag{8.132}$$

Evidently, f_κ, f_μ and f_{g} can be interpreted as the *fractional compressional, shear* and *gravitational energies* of a non-trivial mode, with eigenfrequency $\omega \neq 0$. The term "fractional" must be given a broader meaning than usual in this context, since the sign of the gravitational potential energy $\mathcal{V}_{\mathrm{g}}$ may be either positive or negative. A toroidal mode is devoid of compressional and gravitational energy; all of its energy is stored in shear, so that $f_\mu = 1$.

*8.8.2 Trivial modes

We have seen in Section 8.7.2 that the trivial toroidal multiplet ${}_0\mathrm{T}_1$ corresponds to the rigid-body rotations of the inner core and mantle. There is likewise a trivial spheroidal triplet called ${}_0\mathrm{S}_1$ which accommodates the three rigid-body translational degrees of freedom. This zero-frequency $l = 1$ "mode" has unnormalized displacement eigenfunctions $U = 1$ and $V = \sqrt{2}$; the associated traction vanishes, $R = 0$ and $S = 0$, and the perturbation in the gravitational potential is purely advective: $P = -g$ so that $B = 0$. The unnormalized vector displacement ${}_0\mathbf{s}_{10} = \mathbf{P}_{10} + \sqrt{2}\,\mathbf{B}_{10}$ of the $m = 0$ singlet is

$$ {}_0\mathbf{s}_{10} = (3/4\pi)^{1/2}(\hat{\mathbf{r}}\cos\theta - \hat{\boldsymbol{\theta}}\sin\theta) = (3/4\pi)^{1/2}\hat{\mathbf{z}}, \tag{8.133} $$

which represents a rigid translation of the Earth along the $\hat{\mathbf{z}}$ axis. Similarly, the two vectors ${}_0\mathbf{s}_{1\,-1} = \mathbf{P}_{1\,-1} + \sqrt{2}\,\mathbf{B}_{1\,-1}$ and ${}_0\mathbf{s}_{11} = \mathbf{P}_{11} + \sqrt{2}\,\mathbf{B}_{11}$ correspond to translations along the $\hat{\mathbf{x}}$ and $\hat{\mathbf{y}}$ axes, respectively. Note that it is not possible for a portion of the Earth such as the solid inner core to translate entirely independently of the remainder.

There are no geostrophic spheroidal modes unless the squared Brunt-Väisälä frequency vanishes, $N^2 = 0$, in all or a portion of the fluid outer core and ocean $\oplus_\mathrm{F}$. In that case, there are zero-frequency modes with adiabatic displacements confined to the neutrally stratified fluid regions. Any displacement satisfying $\boldsymbol{\nabla}\cdot(\rho\mathbf{s}) = 0$, or equivalently

$$ V = (k\rho r)^{-1}\frac{d}{dr}(\rho r^2 U), \tag{8.134} $$

wherever $N^2 = 0$, is a geostrophic spheroidal-mode eigenfunction, with associated eigenfrequency $\omega = 0$. The radial displacement U in (8.134) may be chosen arbitrarily, subject to the constraint that $U = 0$ outside of the neutrally stratified regions; if $N^2 = 0$ everywhere in $\oplus_\mathrm{F}$, then U must vanish on the fluid-solid boundaries $r = d_\mathrm{FS}$. The associated traction is purely gravitational, $R = \rho g U$ and $S = 0$, and the potential perturbation vanishes: $P = 0$ so that $B = 4\pi G\rho U$. There is an infinite family of such geostrophic spheroidal modes for every angular degree $l > 0$; these modes owe their existence to the fact that it is possible to overturn any neutrally stratified regions of $\oplus_\mathrm{F}$ without altering the elastic-gravitational potential energy of the Earth.

8.8.3 First-order radial equations

The three second-order differential equations (8.43)–(8.44) and (8.53) governing the spheroidal modes can be written as a system of six coupled first-order equations for the quantities U, V, P, R, S and B in the form:

$$\begin{aligned}\dot{U} &= -2(\kappa + \tfrac{4}{3}\mu)^{-1}(\kappa - \tfrac{2}{3}\mu)r^{-1}U \\ &\quad + k(\kappa + \tfrac{4}{3}\mu)^{-1}(\kappa - \tfrac{2}{3}\mu)r^{-1}V + (\kappa + \tfrac{4}{3}\mu)^{-1}R,\end{aligned} \tag{8.135}$$

$$\dot{V} = -kr^{-1}U + r^{-1}V + \mu^{-1}S, \tag{8.136}$$

$$\dot{P} = -4\pi G\rho U - (l+1)r^{-1}P + B, \tag{8.137}$$

$$\begin{aligned}\dot{R} &= [-\omega^2\rho - 4\rho g r^{-1} + 12\kappa\mu(\kappa + \tfrac{4}{3}\mu)^{-1}r^{-2}]U \\ &\quad + [k\rho g r^{-1} - 6k\kappa\mu(\kappa + \tfrac{4}{3}\mu)^{-1}r^{-2}]V \\ &\qquad - 4\mu(\kappa + \tfrac{4}{3}\mu)^{-1}r^{-1}R + kr^{-1}S \\ &\qquad\quad - (l+1)\rho r^{-1}P + \rho B,\end{aligned} \tag{8.138}$$

$$\begin{aligned}\dot{S} &= [k\rho g r^{-1} - 6k\kappa\mu(\kappa + \tfrac{4}{3}\mu)^{-1}r^{-2}]U \\ &\quad - [\omega^2\rho + 2\mu r^{-2} - 4k^2\mu(\kappa + \tfrac{1}{3}\mu)(\kappa + \tfrac{4}{3}\mu)^{-1}r^{-2}]V \\ &\qquad - k(\kappa - \tfrac{2}{3}\mu)(\kappa + \tfrac{4}{3}\mu)^{-1}r^{-1}R - 3r^{-1}S + k\rho r^{-1}P,\end{aligned} \tag{8.139}$$

$$\dot{B} = -4\pi G(l+1)\rho r^{-1}U + 4\pi Gk\rho r^{-1}V + (l-1)r^{-1}B. \tag{8.140}$$

All of the dependent variables in (8.135)–(8.140) are continuous everywhere in $0 \le r \le a$, except for the tangential displacement V, which can exhibit jump discontinuities at the fluid-solid boundaries: $[V]^+_- = 0$ on $r = d_{\mathrm{SS}}$ and $[U]^+_- = 0$, $[P]^+_- = 0$, $[R]^+_- = 0$, $[S]^+_- = 0$, $[B]^+_- = 0$ on $r = d_{\mathrm{SS}}$ and $r = d_{\mathrm{FS}}$. The quantity B in equation (8.124) was introduced expressly to make the boundary conditions on the Earth's surface homogeneous:

$$R = S = 0 \quad \text{and} \quad B = 0 \quad \text{on } r = a. \tag{8.141}$$

In addition, the shear traction must vanish on the slipping interfaces:

$$S = 0 \quad \text{on } r = d_{\mathrm{FS}}. \tag{8.142}$$

The system of equations (8.135)–(8.140) is independent of the radial model derivatives $\dot{\kappa}$, $\dot{\mu}$ and $\dot{\rho}$; this makes it suitable for numerical integration.

8.8.4 Fluid regions

In the fluid regions of the Earth the rigidity μ is equal to zero, so the shear traction and its radial derivative vanish: $S = 0$ and $\dot{S} = 0$ in $\oplus_{\mathrm{F}}$.

Equation (8.139) reduces in this case to an algebraic equation expressing the tangential displacement V in terms of U, P and R:

$$V = \frac{kr^{-1}(\rho gU + \rho P - R)}{\omega^2\rho}. \tag{8.143}$$

We can use this result to eliminate V from the remaining four equations governing the radial eigenfunctions U, P, R and B:

$$\dot{U} = (\omega^{-2}k^2gr^{-2} - 2r^{-1})U + (\kappa^{-1} - \omega^{-2}k^2\rho^{-1}r^{-2})R + \omega^{-2}k^2r^{-2}P, \tag{8.144}$$

$$\dot{P} = -4\pi G\rho U - (l+1)r^{-1}P + B, \tag{8.145}$$

$$\dot{R} = (-\omega^2\rho - 4\rho gr^{-1} + \omega^{-2}k^2\rho g^2r^{-2})U - \omega^{-2}k^2gr^{-2}R + [\omega^{-2}k^2\rho gr^{-2} - (l+1)\rho r^{-1}]P + \rho B, \tag{8.146}$$

$$\dot{B} = 4\pi G\rho[\omega^{-2}k^2gr^{-2} - (l+1)r^{-1}]U - 4\pi G\omega^{-2}k^2r^{-2}R + 4\pi G\omega^{-2}k^2\rho r^{-2}P + (l-1)r^{-1}B. \tag{8.147}$$

The lack of any dependence upon $\dot{\kappa}$ and $\dot{\rho}$ makes equations (8.143)–(8.147) the preferred system of linearized fluid-dynamical equations in terrestrial normal-mode seismology. Astrophysicists and helioseismologists who study the adiabatic oscillations of the Sun and other stars generally solve an equivalent system of four equations, in which the Eulerian pressure perturbation $p^{\mathrm{E1}} = -R + \rho gU$ is employed as one of the dependent variables, and the model is specified in terms of the density ρ, the sound speed $\alpha = (\kappa/\rho)^{1/2}$ and the Brunt-Väisälä or buoyancy frequency N (Cox 1980; Unno, Osaki, Ando, Saio & Shibahashi 1989).

8.8.5 Radial oscillations

The radial modes, which have $V = 0$ and $S = 0$, require separate consideration. The associated perturbation in the gravitational potential P is given in terms of the radial displacement U by

$$P(r) = \begin{cases} 4\pi G \displaystyle\int_r^a \rho' U'\, dr' & \text{in } 0 \leq r \leq a \\ 0 & \text{in } r \geq a. \end{cases} \tag{8.148}$$

This is a special case of equations (8.55)–(8.56). There is no perturbation in the external potential because the Earth model remains spherically symmetric and there is no change in the total mass. Upon making use of the explicit representation (8.148), we can reduce the four first-order equations governing U, R, P and B to a system of two first-order equations for the radial displacement U and traction R:

$$\dot{U} = -2(\kappa + \tfrac{4}{3}\mu)^{-1}(\kappa - \tfrac{2}{3}\mu)r^{-1}U + (\kappa + \tfrac{4}{3}\mu)^{-1}R, \tag{8.149}$$

$$\begin{aligned}\dot{R} = [-\omega^2\rho - 4\rho g r^{-1} + 12\kappa\mu(\kappa + \tfrac{4}{3}\mu)^{-1}r^{-2}]U \\ - 4\mu(\kappa + \tfrac{4}{3}\mu)^{-1}r^{-1}R.\end{aligned} \tag{8.150}$$

Equations (8.149)–(8.150), together with the associated kinematic and dynamical boundary conditions $[U]_-^+ = 0$, $[R]_-^+ = 0$ on $r = d$ and

$$R = 0 \quad \text{on } r = a, \tag{8.151}$$

constitute a Sturm-Liouville eigenvalue problem. The number of radial nodes of the displacement eigenfunction ${}_nU_0$ is, as a result, equal to the overtone number n.

The action (8.87) for the radial modes reduces to

$$\mathcal{I}_{\mathrm{R}} = \int_0^a L_{\mathrm{R}}(U, \dot{U})\, r^2 dr, \tag{8.152}$$

where

$$\begin{aligned}L_{\mathrm{R}} = \tfrac{1}{2}[\omega^2\rho U^2 - \kappa(\dot{U} + 2r^{-1}U)^2 \\ - \tfrac{4}{3}\mu(\dot{U} - r^{-1}U)^2 + 4\rho g r^{-1}U^2].\end{aligned} \tag{8.153}$$

Rayleigh's principle asserts that $\delta\mathcal{I}_{\mathrm{R}} = 0$ for an arbitrary admissible variation δU if and only if U is a radial-mode eigenfunction with associated eigenfrequency ω. The Euler-Lagrange equation

$$\partial_U L_{\mathrm{R}} - r^{-2}\frac{d}{dr}(r^2\partial_{\dot{U}}L_{\mathrm{R}}) = 0 \tag{8.154}$$

and associated variational boundary conditions $[\partial_{\dot{U}}L_{\mathrm{R}}]_-^+ = 0$ on $r = d$ and $\partial_{\dot{U}}L_{\mathrm{R}} = 0$ on $r = a$ are equivalent to the Sturm-Liouville eigenvalue problem (8.149)–(8.151).

The compressional, shear and gravitational energies (8.127)–(8.129) of a radial mode reduce to

$$\mathcal{V}_\kappa = \int_0^a \kappa(\dot{U} + 2r^{-1}U)^2\, r^2 dr, \tag{8.155}$$

$$\mathcal{V}_\mu = \tfrac{4}{3}\int_0^a \mu(\dot{U} - r^{-1}U)^2\, r^2 dr, \tag{8.156}$$

$$\mathcal{V}_{\rm g} = -4\int_0^a \rho(gr^{-1}U^2)\, r^2 dr. \tag{8.157}$$

The negativity of the gravitational energy $\mathcal{V}_{\rm g}$ is indicative of the *destabilizing effect of self-gravitation* on the radial modes. The physical reason for this is obvious—any spherical shell that moves closer to the center of the sphere experiences an increased gravitational force. If the restoring forces provided by the incompressibility and rigidity are insufficient, so that $\mathcal{V}_\kappa + \mathcal{V}_\mu + \mathcal{V}_{\rm g} \leq 0$, the configuration will suffer a *gravitational collapse*.

★8.8.6 Neglect of self-gravitation

For oscillations with periods shorter than about thirty seconds, the Earth's self-gravitation plays a negligible role as a restoring force compared to its elastic incompressibility and rigidity. The six governing first-order differential equations in the solid regions of the Earth $\oplus_{\rm S}$ can then be replaced by a system of four equations, obtained by setting the gravitational constant G and acceleration g, as well as the gravitational perturbations P and B, equal to zero in (8.135)–(8.140):

$$\begin{aligned}\dot{U} &= -2(\kappa + \tfrac{4}{3}\mu)^{-1}(\kappa - \tfrac{2}{3}\mu)r^{-1}U \\ &\quad + k(\kappa + \tfrac{4}{3}\mu)^{-1}(\kappa - \tfrac{2}{3}\mu)r^{-1}V + (\kappa + \tfrac{4}{3}\mu)^{-1}R,\end{aligned} \tag{8.158}$$

$$\dot{V} = -kr^{-1}U + r^{-1}V + \mu^{-1}S, \tag{8.159}$$

$$\begin{aligned}\dot{R} &= [-\omega^2\rho + 12\kappa\mu(\kappa + \tfrac{4}{3}\mu)^{-1}r^{-2}]U \\ &\quad - 6k\kappa\mu(\kappa + \tfrac{4}{3}\mu)^{-1}r^{-2}V \\ &\qquad - 4\mu(\kappa + \tfrac{4}{3}\mu)^{-1}r^{-1}R + kr^{-1}S,\end{aligned} \tag{8.160}$$

$$\begin{aligned}\dot{S} &= -6k\kappa\mu(\kappa + \tfrac{4}{3}\mu)^{-1}r^{-2}U \\ &\quad - [\omega^2\rho + 2\mu r^{-2} - 4k^2\mu(\kappa + \tfrac{1}{3}\mu)(\kappa + \tfrac{4}{3}\mu)^{-1}r^{-2}]V \\ &\qquad - k(\kappa - \tfrac{2}{3}\mu)(\kappa + \tfrac{4}{3}\mu)^{-1}r^{-1}R - 3r^{-1}S.\end{aligned} \tag{8.161}$$

In the fluid regions of the Earth $\oplus_{\rm F}$, the tangential displacement (8.143) is given by $V = -(kr^{-1}R)/(\omega^2\rho)$, and we obtain a second-order system of differential equations for the radial displacement and traction U and R:

$$\dot{U} = -2r^{-1}U + (\kappa^{-1} - \omega^{-2}k^2\rho^{-1}r^{-2})R, \tag{8.162}$$

$$\dot{R} = -\omega^2\rho U. \tag{8.163}$$

Equations (8.158)–(8.161) and (8.162)–(8.163) must be solved subject to the boundary conditions that $R = S = 0$ on $r = a$. This non-gravitating approximation is rarely employed in numerical applications anymore, because the advantage of having to integrate only a fourth-order rather than a sixth-order system in $\oplus_{\rm S}$ and only a second-order rather than a fourth-order system in $\oplus_{\rm F}$ has been obviated by the power of modern computers. We shall use these results to find quasi-analytical JWKB eigenfrequencies and eigenfunctions that are valid in the limit $\omega \to \infty$ in Section 12.3.

Oscillations that owe their existence to gravity, such as those we shall discuss in Section 8.8.11, have been eliminated in equations (8.158)–(8.161) and (8.162)–(8.163). Short-wavelength predominantly gravitational modes can be investigated using a less restrictive approximation, in which P, B and G are ignored, but the initial acceleration of gravity g is retained. This results in a slightly more complicated system of four equations governing U, V, R and S in $\oplus_{\rm S}$ and two equations governing U and R in $\oplus_{\rm F}$. The generalizations of (8.162)–(8.163) in this *Cowling approximation* are

$$\dot{U} = (\omega^{-2}k^2gr^{-2} - 2r^{-1})U + (\kappa^{-1} - \omega^{-2}k^2\rho^{-1}r^{-2})R, \tag{8.164}$$

$$\dot{R} = (-\omega^2\rho - 4\rho gr^{-1} + \omega^{-2}k^2\rho g^2r^{-2})U - \omega^{-2}k^2gr^{-2}R. \tag{8.165}$$

Equations (8.164)–(8.165) were used prior to the advent of the computational era to investigate the predominantly gravitational ($f_{\rm g} \gg f_\kappa$) and predominantly acoustic ($f_\kappa \gg f_{\rm g}$) oscillations of the Sun and other stars, as we noted in Section 4.3.5.

★8.8.7 Homogeneous sphere: radial oscillations

The radial oscillations of a self-gravitating solid sphere of radius a, with a homogeneous density ρ, compressional-wave speed $\alpha = [(\kappa + \frac{4}{3}\mu)/\rho]^{1/2}$ and shear-wave speed $\beta = (\mu/\rho)^{1/2}$, may be readily calculated. The acceleration of gravity within such a homogeneous sphere varies linearly with radius: $g = \frac{4}{3}\pi G\rho r$ in $0 \leq r \leq a$. Equations (8.149) and (8.150) can be combined to yield the second-order ordinary differential equation

$$\ddot{U} + 2r^{-1}\dot{U} + (\gamma^2 - 2r^{-2})U = 0, \tag{8.166}$$

where $\gamma^2 = (\omega^2 + \frac{16}{3}\pi G\rho)\alpha^{-2}$. We recognize the result (8.166) as the spherical Bessel equation of degree $l = 1$, with solution

$$U = j_1(\gamma r). \tag{8.167}$$

The radial traction $R = (\kappa + \frac{4}{3}\mu)(\dot{U} + 2r^{-1}U) - 4\mu r^{-1}U$ associated with the displacement (8.167) is

$$R = (\kappa + \tfrac{4}{3}\mu)\gamma\, j_0(\gamma r) - 4\mu r^{-1}j_1(\gamma r). \tag{8.168}$$

The boundary condition $R(a) = 0$ on the free surface implies that

$$\cot(\gamma a) = (\gamma a)^{-1} - \tfrac{1}{4}\mu^{-1}(\kappa + \tfrac{4}{3}\mu)\gamma a. \tag{8.169}$$

The radial-mode eigenfrequencies ${}_n\omega_0$ of a homogeneous sphere are determined in terms of the roots ${}_n\gamma_0$ of (8.169) by

$$\omega^2 = \gamma^2\alpha^2 - \tfrac{16}{3}\pi G\rho. \tag{8.170}$$

For a fluid ($\mu = 0$) sphere the roots are simply ${}_n\gamma_0 = (n+1)\pi/a$. In the case of a Poisson solid ($\kappa = \frac{5}{3}\mu$) they are slightly smaller: ${}_0\gamma_0 = 0.82\,\pi/a$ and ${}_1\gamma_0 = 1.98\,\pi/a$. The eigenfrequencies of a *non-gravitating* ($G = 0$) Poisson solid sphere satisfy

$$\cot(\omega a/\alpha) = (\omega a/\alpha)^{-1} - \tfrac{3}{4}(\omega a/\alpha). \tag{8.171}$$

The result (8.171) is venerable—its derivation by Poisson (1829) marks the beginning of the theoretical analysis of the free oscillations of the Earth, as we noted in Chapter 1.

The destabilizing effect of self-gravitation is manifested in the negative final term $-\frac{16}{3}\pi G\rho$ in equation (8.170). The least stable oscillation is the fundamental mode ${}_0\mathrm{S}_0$, whose squared eigenfrequency ${}_0\omega_0^2$ is negative if the incompressibility and rigidity are small enough or the density and radius are large enough so that

$$\alpha < \alpha_{\mathrm{crit}} = {}_0\gamma_0^{-1}(\tfrac{16}{3}\pi G\rho)^{1/2}. \tag{8.172}$$

For a homogeneous fluid sphere having the same radius ($a = 696,000$ km) and mean density ($\rho = 1408$ kg/m^3) as the Sun, the critical sound-wave speed needed to prevent gravitational collapse is $\alpha_{\mathrm{crit}} = 278$ km/s. For a homogeneous Poisson solid sphere having the same radius ($a = 6371$ km) and mean density ($\rho = 5514$ kg/m^3) as the Earth, the critical wave speeds are $\alpha_{\mathrm{crit}} = \sqrt{3}\,\beta_{\mathrm{crit}} = 6.18$ km/s. The actual sound-wave speed within the solar interior increases from nearly zero at the surface to approximately 500 km/s at the center, and the actual compressional-wave speed in the Earth's mantle increases from 8.1 km/s beneath the Mohorovičić discontinuity to 13.7 km/s above the core-mantle boundary. The stratification of α, β and ρ within the two bodies acts to stabilize them against radial deformations more strongly than the above simple examples indicate.

★8.8.8 Homogeneous sphere: non-radial oscillations

The non-radial free oscillations of a self-gravitating homogeneous solid sphere are discussed in detail by Love (1911), Pekeris & Jarosch (1958) and Takeuchi & Saito (1972); we simply quote their results here. Two of the three linearly independent regular solutions to the radial eigenfunction equations (8.135)–(8.140) can be expressed in terms of the spherical Bessel functions of degree l and $l+1$ in the form

$$U = l\xi r^{-1} j_l(\gamma r) - \zeta\gamma\, j_{l+1}(\gamma r), \tag{8.173}$$

$$V = k\xi r^{-1} j_l(\gamma r) + k\gamma\, j_{l+1}(\gamma r), \tag{8.174}$$

$$P = -4\pi G\rho\zeta\, j_l(\gamma r), \tag{8.175}$$

$$\begin{aligned} R = &-[(\kappa + \tfrac{4}{3}\mu)\zeta\gamma^2 + 2l(l-1)\mu\xi r^{-2}] j_l(\gamma r) \\ &- 2\mu(2\zeta + k^2)\gamma r^{-1} j_{l+1}(\gamma r), \end{aligned} \tag{8.176}$$

$$\begin{aligned} S = &\, k\mu[\gamma^2 + 2(l-1)\xi r^{-2}] j_l(\gamma r) \\ &- 2k\mu(\zeta + 1)\gamma r^{-1} j_{l+1}(\gamma r), \end{aligned} \tag{8.177}$$

$$B = -4\pi G\rho r^{-1}[k^2 - (l+1)\zeta] j_l(\gamma r), \tag{8.178}$$

where

$$\begin{aligned} \gamma^2 = &\frac{\omega^2}{2\beta^2} + \frac{\omega^2 + \frac{16}{3}\pi G\rho}{2\alpha^2} \\ &\pm \frac{1}{2}\left[\left(\frac{\omega^2}{\beta^2} - \frac{\omega^2 + \frac{16}{3}\pi G\rho}{\alpha^2}\right)^2 + \left(\frac{8\pi Gk\rho}{3\alpha\beta}\right)^2\right]^{1/2}, \end{aligned} \tag{8.179}$$

$$\zeta = \tfrac{3}{4}(\pi G\rho)^{-1}\beta^2(\gamma^2 - \omega^2/\beta^2), \qquad \xi = \zeta - (l+1). \tag{8.180}$$

Note that there are two solutions of the form (8.173)–(8.178) because of the $\pm$ sign in the definition (8.179). The third linearly independent regular solution is given by the algebraic expressions

$$U = lr^{l-1}, \qquad V = kr^{l-1}, \qquad P = (\omega^2 - \tfrac{4}{3}\pi G\rho l)r^l, \tag{8.181}$$

$$R = 2l(l-1)\mu r^{l-2}, \qquad S = 2k(l-1)\mu r^{l-2}, \tag{8.182}$$

$$B = [(2l+1)\omega^2 - \tfrac{8}{3}\pi Gl(l-1)\rho]r^{l-1}. \tag{8.183}$$

Equations (8.181)–(8.183) are a generalization of the rigid-body eigensolution discussed in Section 8.8.2; the boundary conditions $R = 0$, $S = 0$, $B = 0$ are satisfied by this solution alone only in the case $l = 1$, $\omega = 0$.

The non-trivial eigenfrequencies $\omega \neq 0$ of a homogeneous solid sphere are determined by the roots of the determinant equation

$$\det \begin{vmatrix} R_1(a) & R_2(a) & R_3(a) \\ S_1(a) & S_2(a) & S_3(a) \\ B_1(a) & B_2(a) & B_3(a) \end{vmatrix} = 0, \tag{8.184}$$

where the subscripts 1, 2 and 3 denote the three linearly independent solutions. The above analysis assumes that both roots γ^2 of (8.179) are positive; this condition is demonstrably valid for any Poisson solid ($\kappa = \frac{5}{3}\mu$) sphere that is gravitationally stable. It can furthermore be demonstrated that the gravitational stability of such a sphere against radial deformations guarantees its complete stability, against non-radial deformations as well.

The gravest free oscillation of a stable homogeneous sphere is the *football mode* ${}_0\mathrm{S}_2$, so-called because the deformed shape alternates between that of an American football (or lemon if you prefer) and a pumpkin. For a Poisson solid sphere the same mass and size of the Earth, and having "a rigidity equal to that of steel", which Love (1911) took to be $\mu = 8.19 \times 10^{10}$ Pa, the period of this fundamental mode is "almost exactly 60 minutes". The corresponding period for the ${}_0\mathrm{S}_2$ mode of the isotropic PREM model, obtained by numerical integration, is 53.9 minutes.

In the case of a homogeneous fluid sphere, there is just a single radial wavenumber γ, rather than the two values defined by (8.179):

$$\gamma^2 = \frac{\omega^2 + \frac{16}{3}\pi G\rho}{\alpha^2} - \left(\frac{4\pi Gk\rho}{3\omega\alpha}\right)^2, \tag{8.185}$$

$$\zeta = -\tfrac{3}{4}(\pi G\rho)^{-1}\omega^2, \qquad \xi = \zeta - (l+1). \tag{8.186}$$

As a result there are two rather than three linearly independent solutions; the two-by-two determinant equation

$$\det \begin{vmatrix} R_1(a) & R_2(a) \\ B_1(a) & B_2(a) \end{vmatrix} = 0 \tag{8.187}$$

reduces to $j_l(\gamma a) = 0$. Every root ${}_n\gamma_l a$ of this spherical Bessel function specifies both a purely real eigenfrequency (${}_n\omega_l^2 > 0$) and a purely imaginary eigenfrequency (${}_n\omega_l^2 < 0$) through the quadratic relation (8.185). The real eigenfrequencies correspond to the stable, predominantly acoustic modes of seismological interest, whereas the imaginary eigenfrequencies reflect the inherent gravitational instability of a homogeneous compressible fluid sphere, which has $N^2 = -\rho g^2/\kappa < 0$ everywhere. There are no such unstable gravity modes of degree $l = 0$; the only radial oscillations of a homogeneous fluid sphere are the gravitationally modified acoustic modes, whose eigenfrequencies are given by equation (8.169).

The only stable homogeneous fluid sphere is one that is *incompressible*, with $\kappa = \infty$. Such a sphere has no radial modes and no overtones; for every degree $l \geq 1$ there is a single fundamental oscillation whose eigenfrequency is given by the classical formula of Lord Kelvin (Thomson 1863b):

$$\omega^2 = \frac{\frac{8}{3}\pi l(l-1)G\rho}{2l+1}. \tag{8.188}$$

The special case $l = 1$ corresponds to the rigid-body translational triplet, with $\omega = 0$. The gravest non-trivial oscillation is the football mode ${}_0\mathrm{S}_2$, which has a period of 47.7 or 94.3 minutes for a sphere having the same mean density as the Sun or the Earth, respectively.

8.8.9 Numerical integration

Numerical integration is needed to calculate the spheroidal normal modes of a realistic SNREI Earth model with a solid inner core, fluid outer core, solid mantle and crust, and fluid ocean. The eigenfrequencies ω and associated radial eigenfunctions U, V, P, R, S and B of the non-radial oscillations can be determined by commencing the numerical integration just above the center of the Earth model, using initial values based upon the exact solutions for a homogeneous sphere (8.173)–(8.178) and (8.181)–(8.183). Because there are three regular linearly independent starting solutions, we must integrate the system of six first-order equations (8.135)–(8.140) three times from the starting radius $r = \varepsilon$ to the inner-core boundary $r = c$. At the boundary there will be two linearly independent combinations of these solutions that satisfy the boundary condition $S(c) = 0$. The continuous quantities $U(c)$, $R(c)$, $P(c)$ and $B(c)$ corresponding to these two linear combinations serve as initial values for the integration of the four first-order equations (8.144)–(8.147) through the fluid outer core to the core-mantle boundary $r = b$. The values of $U(b)$, $P(b)$, $R(b)$ and $B(b)$ together with the initializations $V(b) = 0$, $S(b) = 0$ comprise two linearly independent solutions at the base of the mantle; a third solution, which allows for the possibility of tangential slip at the core-mantle boundary, is $U(b) = 0$, $V(b) = 1$, $P(b) = 0$, $R(b) = 0$, $S(b) = 0$, $B(b) = 0$. These three solutions are integrated through the mantle and overlying crust using equations (8.135)–(8.140); the continuity conditions $[U]_-^+ = 0$, $[V]_-^+ = 0$, $[P]_-^+ = 0$, $[R]_-^+ = 0$, $[S]_-^+ = 0$ and $[B]_-^+ = 0$ are imposed at any intervening solid-solid interfaces $r = d_{\mathrm{SS}}$. If the Earth model does not have an ocean, there will be three linearly independent solutions, which we label by subscripts 1, 2 and 3, at the free surface $r = a$. The eigenfrequencies are then the roots of the secular determinant equation (8.184); once a particular ${}_n\omega_l$ has been determined, the unique linear combination corresponding to the associated radial eigenfunctions ${}_nU_l$, ${}_nV_l$, ${}_nP_l$, ${}_nR_l$, ${}_nS_l$ and ${}_nB_l$ is easily found. If there is an ocean, there will be two linearly independent solutions that satisfy the boundary condition $S(s) = 0$ at the seafloor. These solutions are integrated up to the free surface $r = a$ using (8.144)–(8.147), and the eigenfrequencies ${}_n\omega_l$ are then found using (8.187). A separate integration and root search of the determinant must be performed for every value of the angular degree l. The eigenfrequencies are arranged in numerical order, starting with the gravest, and the overtone numbers $n = 0, 1, 2, \ldots$ are assigned accordingly. As with the toroidal modes, it is frequently possible to save time by starting the integration at a higher level, taking care that it is situated sufficiently far beneath the turning point of the equivalent P-SV body waves. The radial modes are calculated separately by integrating the two first-order equations (8.149)–(8.150), starting at $r = \varepsilon$ with the analytical solution for a homogeneous sphere (8.167)–(8.168), and finding the roots ${}_n\omega_0$ of $R(a)$.

Practical numerical integration algorithms such as those implemented in `MINEOS` and `OBANI` differ from the above description in two important respects. First, the linear systems of differential equations (8.135)–(8.140) and (8.144)–(8.147) are stiff, so that direct integration can lead to grossly unstable results, even at moderately high frequencies. Starting solutions that are linearly independent at $r = \varepsilon$ become more and more dependent with increasing radius, so that on a computer with finite precision, they become indistinguishable numerically from a set of dependent solutions. The determinants (8.184) and (8.187) are in that case very badly behaved functions of the trial eigenfrequency ω. This difficulty is overcome by solving instead for the 2×2 and 3×3 minors of the 2×4 and 3×6 solid and fluid solution matrices; the resulting twenty-dimensional and six-dimensional minor vectors in the fluid and solid, respectively, uniquely characterize the linearly independent solution space, and are independent of the starting basis, to within a constant multiplying factor. The systems of first-order ordinary differential equations satisfied by these vectors in $\oplus_S$ and $\oplus_F$, and the associated continuity conditions at the fluid-solid boundaries $r = d_{FS}$ and boundary conditions at the free surface $r = a$, are derived and discussed by Gilbert & Backus (1966; 1969) and Woodhouse (1988). The latter reference also describes how to reconstruct the primitive eigenfunctions U, V, P, R, S and B from the minor vectors. For the toroidal and radial modes, which require the integration of a single 1×2 solution vector for W and T or U and R, the minor formalism is degenerate.

The second problem is guaranteeing that all of the eigenfrequencies and eigenfunctions have been found. This is particularly problematical for the non-radial spheroidal modes, whose spectrum contains many closely and irregularly spaced eigenfrequencies, as we shall see in Section 8.8.10. A

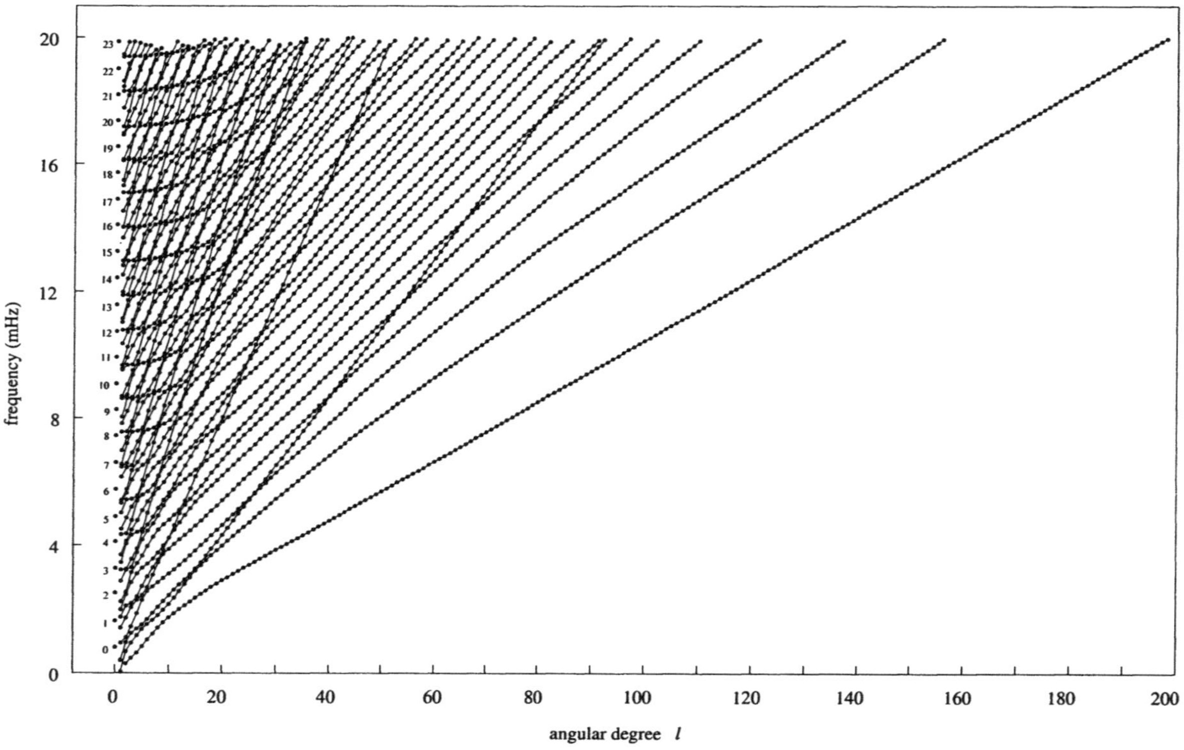

Figure 8.9. Dispersion diagram showing the spheroidal oscillations of the isotropic PREM model. All of the predominantly elastic eigenfrequencies ${}_n\omega_l^{\mathrm{S}}/2\pi$ below 20 mHz are plotted versus the spherical-harmonic degree l. Small numbers on left label the radial-mode ($n = 0$) eigenfrequencies. See Figure 8.10 for greater detail.

straightforward exhaustive search requires an extremely small step in frequency to be certain of locating all the roots of the secular determinant; moreover, it is impossible to specify a priori what a suitable frequency step might be. To overcome this difficulty, Woodhouse (1988) has established a generalization of Sturm's theorem which enables the number of eigenfrequencies between two bounding values ω_{min} and ω_{max} to be determined by counting the zero-crossings of a scalar product of the minor vectors with just two integrations. This remarkable result allows all of the spheroidal-mode eigenfrequencies to be efficiently bracketed, so that they can be refined by bisection without fear of overlooking any.

8.8.10 Spheroidal-mode menagerie

Figure 8.9 shows the dispersion diagram for the predominantly elastic spheroidal oscillations of the isotropic PREM Earth model. The spacing between the eigenfrequencies is highly irregular, as noted above; the solid lines connecting each overtone branch ${}_n\omega_0 - {}_n\omega_1 - {}_n\omega_2 \cdots$ exhibit a characteristic "terraced" appearance, as a consequence of this irregularity. A magnified version of the near-origin region of the spheroidal-mode dispersion diagram is depicted in Figure 8.10; the horizontal scale has been stretched (and truncated at $l = 60$) so that the osculation of the branches can be discerned more clearly. The terracing is most pronounced for the high-frequency, high-overtone modes situated in the upper left corner of the diagram.

The regularly spaced modes ${}_0\mathrm{S}_l$, ${}_1\mathrm{S}_l$, ${}_2\mathrm{S}_l$, etc. on the right of Figure 8.9 correspond in the limit $l \gg 1$ to fundamental and higher-mode *Rayleigh surface waves*, or equivalently to constructively interfering multiply reflected P and SV body waves that turn in the upper mantle. Figure 8.11 shows the radial variation of ${}_0U_l$, ${}_1U_l$, ${}_2U_l$ and ${}_0V_l$, ${}_1V_l$, ${}_2V_l$ for a number of these Rayleigh-wave-equivalent modes. The depth to which a mode ${}_0\mathrm{S}_l$, ${}_1\mathrm{S}_l$, ${}_2\mathrm{S}_l$ penetrates or "feels" into the mantle decreases as the angular degree l increases along the fundamental and each overtone branch, just as for the analogous Love-wave-equivalent toroidal modes ${}_0\mathrm{T}_l$, ${}_1\mathrm{T}_l$, ${}_2\mathrm{T}_l$ discussed in Section 8.7.6. The radial displacement ${}_0U_l$ of a fundamental Rayleigh-wave-equivalent mode has no radial nodes whereas the tangential displacement ${}_0V_l$ has a single radial node. The sign difference between ${}_0U_l$ and ${}_0V_l$ at the Earth's surface $r = a$ is indicative of the characteristic *prograde particle motion* of a fundamental Rayleigh wave, as we shall see in Section 11.4.

The irregularly spaced modes of high frequency and high overtone number correspond to steeply propagating P and SV body waves which interact with the solid inner and fluid outer core boundaries. There are three distinct families of these modes, each of which exhibits a fairly regular pattern of eigenfrequencies ${}_n\omega_l$; the terracing of the fixed-n dispersion curves is

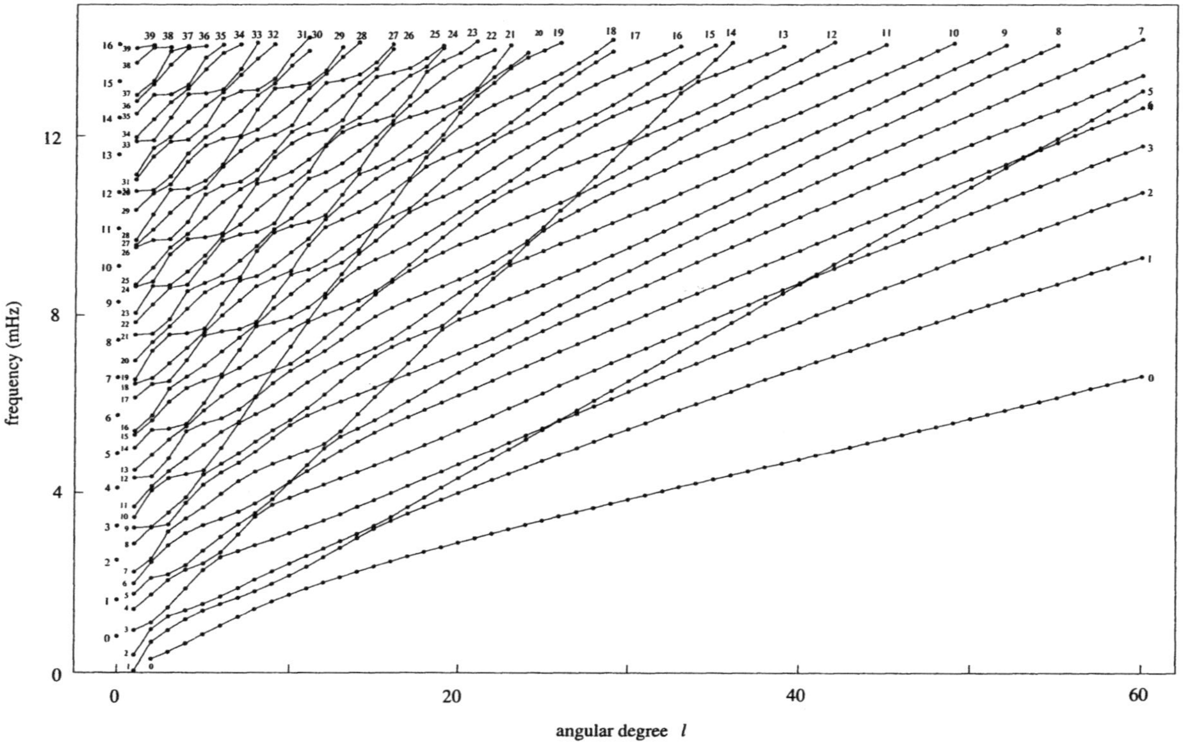

Figure 8.10. Detailed view of the lower left corner of the spheroidal-mode dispersion diagram in Figure 8.9. Leftmost numbers 0 through 16 label the radial-mode ($n = 0$) eigenfrequencies; repeated numbers 0 through 39 label the low-frequency and high-frequency ends of the various overtone branches.

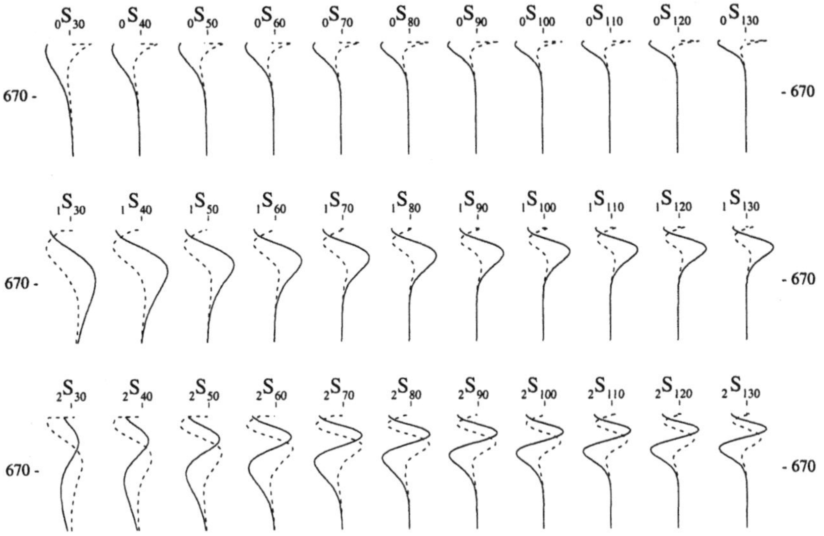

Figure 8.11. Eigenfunctions ${}_0U_l$, ${}_1U_l$, ${}_2U_l$ (*solid line*) and ${}_0V_l$, ${}_1V_l$, ${}_2V_l$ (*dotted line*) along the fundamental (*top row*), first-overtone (*middle row*) and second-overtone (*bottom row*) branches of the Rayleigh-wave-equivalent modes. Vertical axis extends from the free surface to 1200 km depth; the location of the 670 km discontinuity is indicated.

due to the "avoided crossings" of these three mode types. The constituent curves can be discerned by placing one's eye near the plane of Figure 8.9 and viewing it at a grazing angle; for clarity, we also plot the three families of eigenfrequencies separately in Figure 8.12. The first group of modes, which have asymptotic group speed $d\omega/dk \approx 0$ in the vicinity of the abscissa, correspond to constructively interfering $\mathrm{ScS_{SV}}$ body waves in the mantle; the second group, which have intermediate group speed $d\omega/dk$, correspond to constructively interfering PKIKP waves throughout the whole Earth; and the third and final group, which have the highest group speed $d\omega/dk$, correspond to constructively interfering $\mathrm{J_{SV}}$ waves confined to the solid inner core. The $\mathrm{ScS_{SV}}$-equivalent and $\mathrm{J_{SV}}$-equivalent modes are the spheroidal analogues of the $\mathrm{ScS_{SH}}$-equivalent mantle toroidal modes ${}_n\mathrm{T}_l$ and the $\mathrm{J_{SH}}$-equivalent inner-core toroidal modes ${}_n\mathrm{C}_l$, respectively. Comparison of Figures 8.3 and 8.12 reveals that the $\mathrm{ScS_{SV}}$ and $\mathrm{ScS_{SH}}$ dispersion curves are nearly coincident; the spacing between adjacent $l \approx 1$ eigenfrequencies is in both cases approximately equal to $2\pi/T_{\mathrm{ScS}}$, where T_{ScS} is the two-way vertical shear-wave travel time. The $\mathrm{J_{SV}}$ and $\mathrm{J_{SH}}$

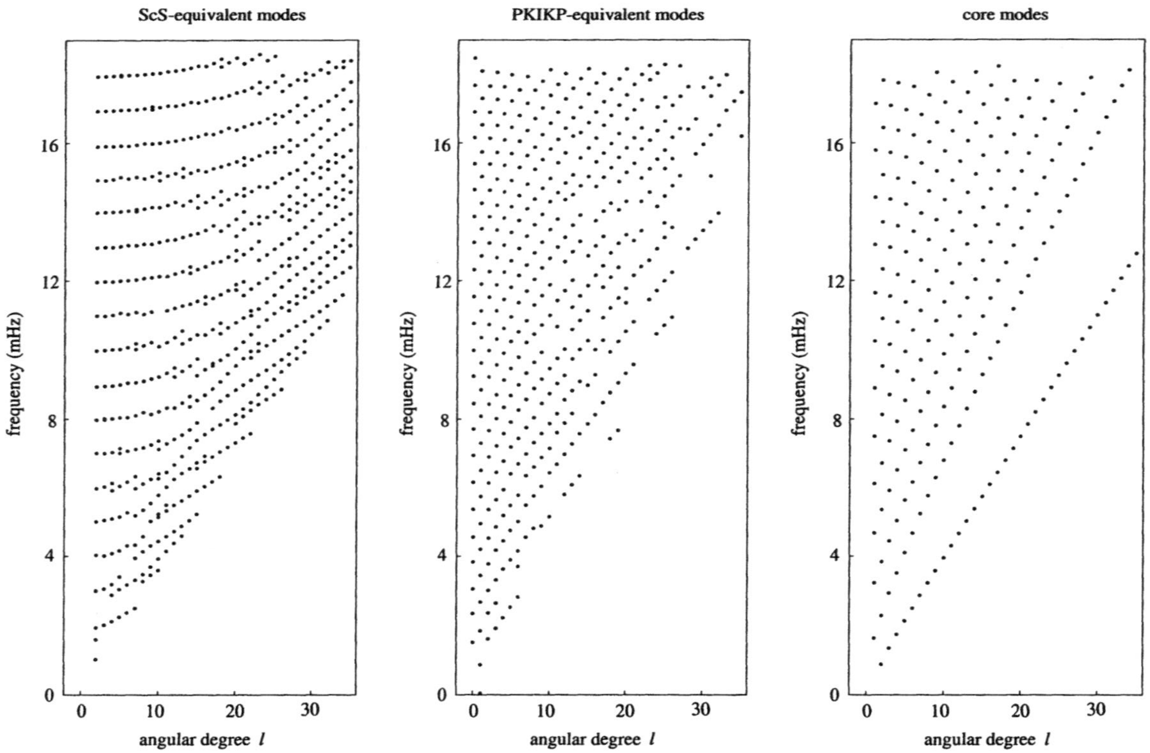

Figure 8.12. Subdivision of the high-frequency, high-overtone spheroidal oscillations into mantle $\mathrm{ScS_{SV}}$ modes (*left*), PKIKP modes (*middle*) and inner-core $\mathrm{J_{SV}}$ modes (*right*). The $\mathrm{ScS_{SV}}$ modes have $f_\mu > 0.5$ and $f_{\mu,\,\mathrm{inner\ core}} \ll f_{\mu,\,\mathrm{mantle}}$; the PKIKP modes have $f_\kappa > 0.5$; the $\mathrm{J_{SV}}$ modes have $f_\mu > 0.5$ and $f_{\mu,\,\mathrm{inner\ core}} \gg f_{\mu,\,\mathrm{mantle}}$.

modes have the same asymptotic group speed $d\omega/dk$ and the same $l \approx 1$ spacing, namely $2\pi/T_{\mathrm{J}}$, where T_{J} is the time required for a $\mathrm{J_{SV}}$ or $\mathrm{J_{SH}}$ shear wave to traverse the inner core; unlike the $\mathrm{ScS_{SV}}$ and $\mathrm{ScS_{SH}}$ modes, however, they are staggered or interleaved, as illustrated in Figure 8.13. Each of the PKIKP-equivalent dispersion curves is terminated at degree zero by a radial mode ${}_n\mathrm{S}_0$. The spacing ${}_{n+1}\omega_0 - {}_n\omega_0$ between adjacent radial-mode eigenfrequencies is approximately equal to $2\pi/T_{\mathrm{PKIKP}}$, where T_{PKIKP} is the transit time of a PKIKP wave that passes through the Earth's center. The subdivision of the high-frequency, high-overtone spheroidal modes into quasi-independent $\mathrm{ScS_{SV}}$, $\mathrm{J_{SV}}$ and PKIKP families can be understood in terms of the JWKB mode-ray duality analysis which we undertake in Chapter 12.

In a sense, the regularity of the toroidal-mode dispersion diagram in Figure 8.3 is illusory: if both the mantle modes ${}_n\mathrm{T}_l$ and the inner-core modes ${}_n\mathrm{C}_l$ were displayed together with the overtone numbers n re-assigned accordingly, then the "whole-Earth" toroidal-mode dispersion curves would exhibit terracing and osculation just as the spheroidal-mode curves do. The mantle and inner-core toroidal oscillations are fully decoupled by the intervening presence of the fluid outer core; hence, they may be considered independently. The three families of high-frequency, high-overtone spheroidal modes are, in contrast, only quasi-independent; the determinants (8.184) and (8.187) indiscriminately exhibit all three types of roots; for this reason, they *must* be displayed on a common spheroidal dispersion diagram.

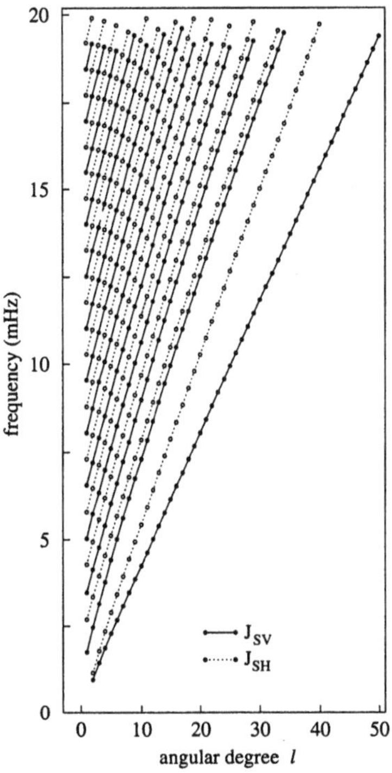

Figure 8.13. Eigenfrequencies of the $\mathrm{J_{SV}}$ and $\mathrm{J_{SH}}$ inner-core modes. The interleaving of the two sets of dispersion curves can be attributed to the fact that the horizontal component of a $\mathrm{J_{SV}}$ wave exhibits a reversal in sign upon turning near the center of the Earth, whereas that of a $\mathrm{J_{SH}}$ wave does not.

The radial and tangential displacement eigenfunctions U and V of a number of degree-two spheroidal modes ${}_n\mathrm{S}_2$ are illustrated in Figure 8.14.

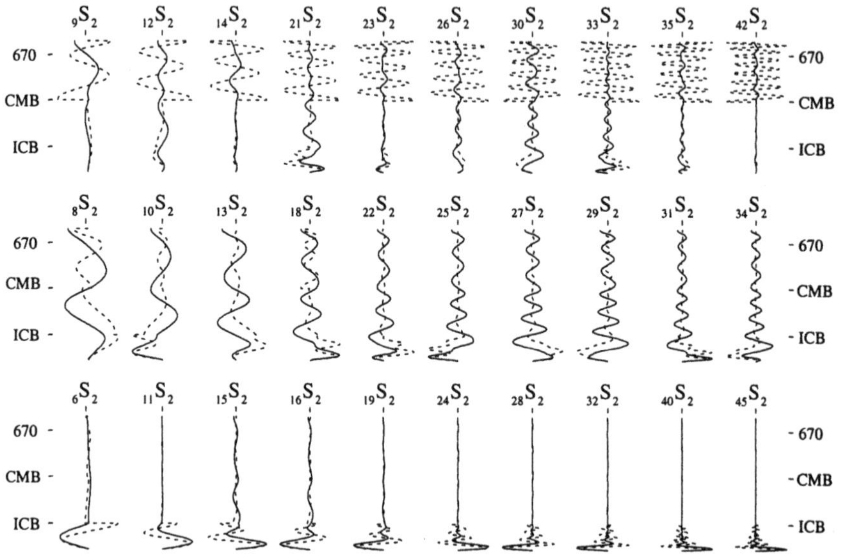

Figure 8.14. Eigenfunctions ${}_nU_2$ (*solid line*) and ${}_nV_2$ (*dashed line*) of a number of $l = 2$ spheroidal modes sorted according to type: $\mathrm{ScS_{SV}}$ modes (*top row*), PKIKP modes (*middle row*) and $\mathrm{J_{SV}}$ modes (*bottom row*). Overtone number n and frequency ${}_n\omega_l$ increase to the right. Vertical axis extends from the free surface to the center of the Earth; the locations of the 670 km discontinuity, the core-mantle boundary (CMB) and the inner-core boundary (ICB) are indicated.

The $\mathrm{ScS_{SV}}$-equivalent modes have large, predominantly tangential displacements in the mantle, the PKIKP-equivalent modes have significant, predominantly radial displacements throughout the whole Earth, and the $\mathrm{J_{SV}}$-equivalent modes have negligible displacements outside of the solid inner core; the only significant radial displacements of the $\mathrm{J_{SV}}$ modes occur in the vicinity of the center of the Earth, $r = 0$. All of these features are exhibited most clearly by the high-frequency, high-overtone modes such as ${}_{42}\mathrm{S}_2$, ${}_{34}\mathrm{S}_2$ and ${}_{45}\mathrm{S}_2$ on the right. Some of the lower-frequency modes are less pure *hybrids*; for example, the three $\mathrm{ScS_{SV}}$ modes ${}_{21}\mathrm{S}_2$, ${}_{30}\mathrm{S}_2$ and ${}_{33}\mathrm{S}_2$ and the two $\mathrm{J_{SV}}$ modes ${}_{15}\mathrm{S}_2$ and ${}_{16}\mathrm{S}_2$ all exhibit secondary PKIKP characteristics (non-negligible radial displacements throughout the whole Earth). The toroidal displacement eigenfunctions W of the decoupled $\mathrm{ScS_{SH}}$ and $\mathrm{J_{SH}}$ modes vanish identically outside of the mantle and inner core; the high-frequency $\mathrm{ScS_{SV}}$ and $\mathrm{J_{SV}}$ eigenfunctions, though predominantly confined to the mantle and inner core, respectively, are non-zero everywhere.

The eigenfrequencies ${}_n\omega_l$ lying along the two distinctly straight lines

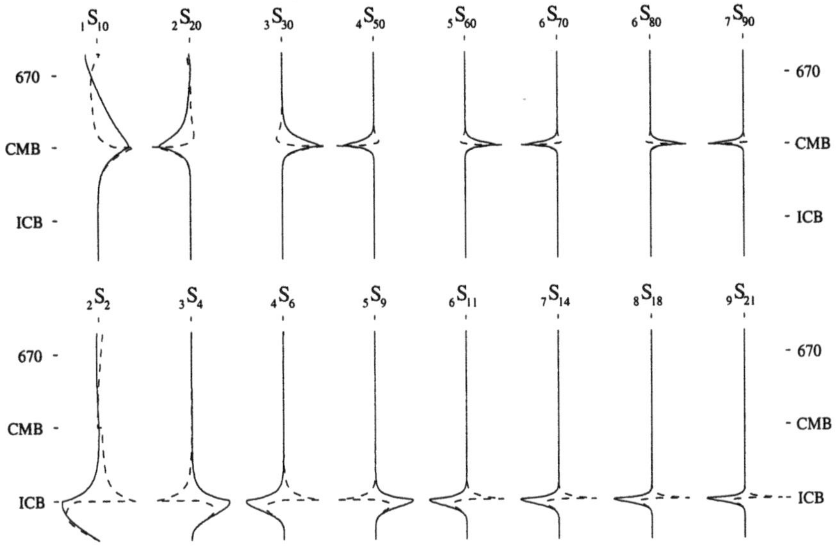

Figure 8.15. Eigenfunctions ${}_nU_l$ (*solid line*) and ${}_nV_l$ (*dashed line*) of a number of core-mantle boundary (CMB) Stoneley modes (*top row*) and inner-core boundary (ICB) Stoneley modes (*bottom row*). Vertical axis extends from the free surface to the center of the Earth.

with slopes of approximately 0.2 mHz and 0.4 mHz per degree l in Figure 8.9 are associated with the so-called *Stoneley modes* trapped at the core-mantle boundary and inner-core boundary, respectively. Classical Stoneley waves at a fluid-solid interface are non-dispersive (Stoneley 1924; Scholte 1947); this is why the eigenfrequencies are so well aligned. Figure 8.15 shows the radial eigenfunctions ${}_nU_l$ and ${}_nV_l$ for a number of core-mantle boundary and inner-core boundary Stoneley modes. The radial displacement exhibits a quasi-exponential decrease away from a maximum at the boundary, whereas the tangential displacement has a node within the solid; the displacements become increasingly confined to the vicinity of the boundary as the frequency increases. The eigenfunctions along the first, second and third J_{SV}-equivalent dispersion branches are similar in character to the inner-core Stoneley eigenfunctions, except that the radial displacements ${}_nU_l$ have one, two and three inner-core nodes, respectively. In effect, the inner-core Stoneley modes constitute the "fundamental" J_{SV} branch, and the first, second and third J_{SV} branches are ICB Stoneley-mode "overtones". Neither the J_{SV} inner-core oscillations nor the high-frequency Stoneley oscillations are of much practical importance, because they cannot be excited

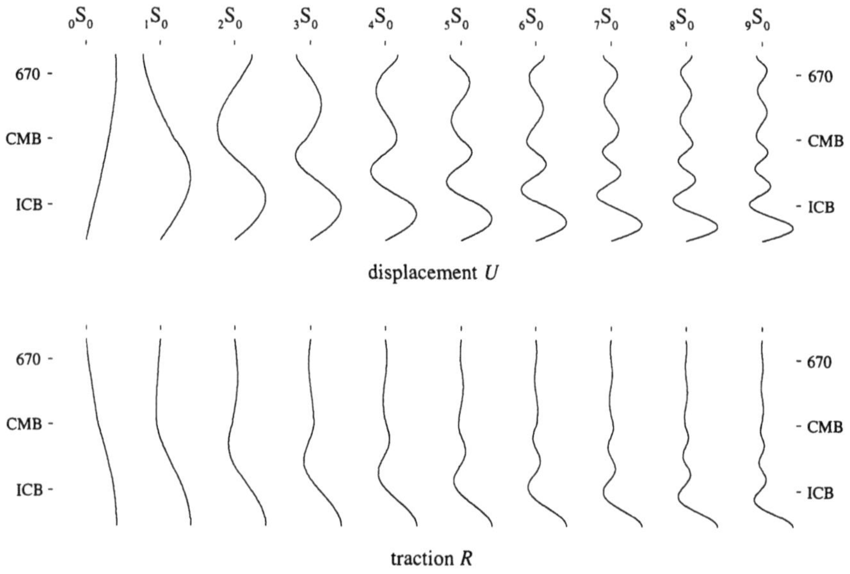

Figure 8.16. Displacement $_nU_0$ (*top*) and associated traction $_nR_0$ (*bottom*) of the first ten radial modes. Vertical axis extends from the free surface to the center of the Earth; the locations of the 670 km discontinuity, the core-mantle boundary (CMB) and the inner-core boundary (ICB) are indicated.

significantly by an earthquake in the crust or upper mantle, or observed by a seismometer situated on the Earth's surface.

The displacement eigenfunctions $_0U_0, \ldots, _9U_0$ and associated tractions $_0R_0, \ldots, _9R_0$ of the first ten radial oscillations are shown in Figure 8.16; the number of radial displacement nodes is equal to the overtone number n, in accordance with Sturm's theorem. The radial "wavelength" of the oscillations is that of the equivalent radially propagating PKIKP waves. The large displacements near the center of the Earth are due to an asymptotic r^{-1} divergence there (see Section 12.3.5). The fundamental radial mode $_0S_0$, with an isotropic PREM period of 20.5 minutes, has no nodes and corresponds to a roughly uniform compression and dilatation, or "breathing" in and out of the whole Earth, as illustrated in Figure 8.17.

The gravest observed free oscillation of the Earth is the celebrated $_0S_2$ football mode, which has an isotropic PREM period of 53.9 minutes. The $m = 0$ singlet has an axially symmetric vector displacement field of the form $\mathbf{s} = \frac{1}{4}(5/\pi)^{1/2}\, {}_0U_2(r)(3\cos^2\theta - 1)\hat{\mathbf{r}} - \frac{3}{2}(5/6\pi)^{1/2}\, {}_0V_2(r)\sin\theta\cos\theta\,\hat{\boldsymbol{\theta}}$; the surface of the Earth alternately assumes the shape of a prolate and

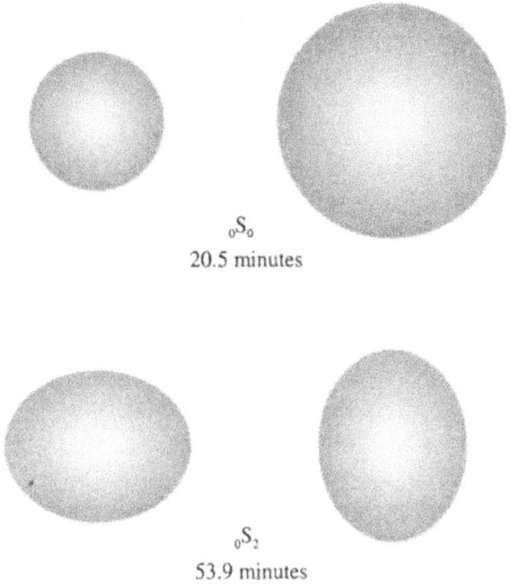

Figure 8.17. Schematic depiction of the fundamental radial mode $_0\mathrm{S}_0$ (*top*) and the $m = 0$ singlet of the football mode $_0\mathrm{S}_2$ (*bottom*). Two extremes of the oscillation cycle are shown.

oblate ellipsoid, as illustrated in Figure 8.17. The radial and tangential eigenfunctions $_0U_2$, $_0V_2$ and associated tractions $_0R_2$, $_0S_2$ are displayed in Figure 8.18. It is evident that the entire Earth partakes in the motion.

The only graver oscillation in Figure 8.9 is the Slichter or *translational inner-core mode* $_1\mathrm{S}_1$, which has a theoretical period of 325 minutes or approximately five and one-half hours! This mode, whose existence was first pointed out by Slichter (1961), consists essentially of a rigid translation of the solid inner core relative to the fluid outer core and mantle; the displacements $_1U_1$, $_1V_1$ and associated tractions $_1R_1$, $_1S_1$ are shown in Figure 8.19. Both the radial and tangential displacements are approximately constant ($_1V_1 \approx \sqrt{2}\,_1U_1$) within the inner core, $0 \leq r \leq c$; the motion within the outer core $c \leq r \leq b$ represents the "return flow" of the fluid, which must move out of the way to enable the inner core to be displaced. The $m = -1$, $m = 1$ and $m = 0$ singlets correspond to translations along the x, y and z axes, respectively. The eigenfrequency $_1\omega_1$ of the Slichter mode is a strong function of the density jump at the inner-core boundary, inasmuch as the principal restoring force is the negative buoyancy of the solid inner core (Smith 1976). In principle, a translation of the solid inner core could be

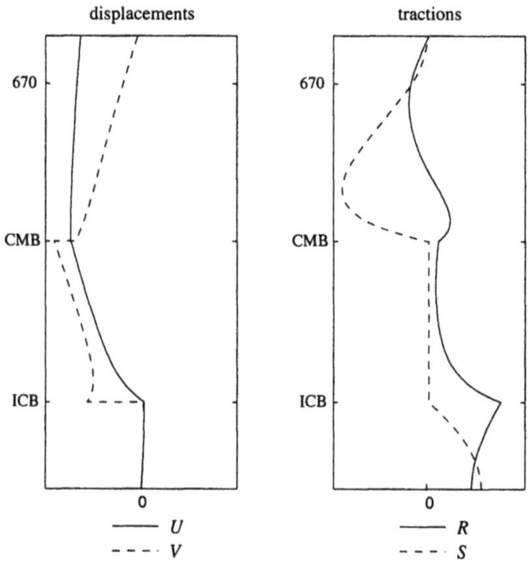

Figure 8.18. Radial variation of the displacement eigenfunctions ${}_0U_2$, ${}_0V_2$ (*left*) and associated tractions ${}_0R_2$, ${}_0S_2$ (*right*) of the football mode ${}_0\mathrm{S}_2$. Vertical axis extends from the free surface to the center of the Earth; the locations of the 670 km discontinuity, the core-mantle boundary (CMB) and the inner-core boundary (ICB) are indicated.

observed on a gravimeter at the Earth's surface; so far, however, it has never been unambiguously detected. Both the inner-core and core-mantle boundary Stoneley modes and the Slichter mode represent a computational challenge for numerical codes, such as `MINEOS` and `OBANI`, because the free surface $r = a$, where the boundary conditions (8.141) must be imposed, is essentially a node.

The fractional compressional, shear and gravitational potential energies of a number of representative spheroidal modes are listed in Table 8.2. Note that in every case $f_\kappa + f_\mu + f_{\mathrm{g}} = 1$. The two modes which have the largest fractional gravitational energies are the Slichter mode ${}_1\mathrm{S}_1$ and the football mode ${}_0\mathrm{S}_2$; in both cases $\mathcal{V}_{\mathrm{g}}$ is positive, indicating that the effect of gravity is to restore the deformed Earth to its equilibrium spherical configuration, as is obvious from physical considerations. The negative gravitational energies of the radial oscillations ${}_n\mathrm{S}_0$ are indicative of the destabilizing influence of self-gravitation upon these modes. The extremely low shear-energy content of the fundamental radial mode ${}_0\mathrm{S}_0$, which consists of almost a pure

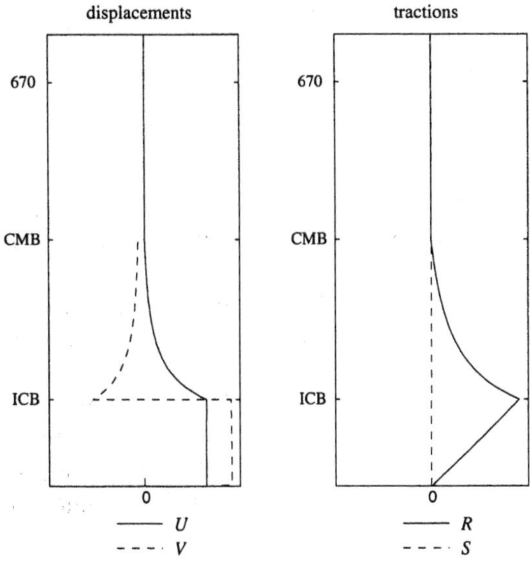

Figure 8.19. Radial variation of the displacement eigenfunctions ${}_1U_1$, ${}_1V_1$ (*left*) and associated tractions ${}_1R_1$, ${}_1S_1$ (*right*) of the Slichter or translational inner-core mode ${}_1\mathrm{S}_1$. Vertical axis extends from the free surface to the center of the Earth; the locations of the 670 km discontinuity, the core-mantle boundary (CMB) and the inner-core boundary (ICB) are indicated.

compression and dilatation of the Earth, is also noteworthy. The mantle $\mathrm{ScS_{SV}}$ and the inner-core $\mathrm{J_{SV}}$ modes store almost all of their energy in shear, whereas the PKIKP modes store the bulk of their energy in compression. A portion—between twenty and thirty percent—of the PKIKP energy is shear, reflecting the mixed compressional-shear nature of the P and I waves in the mantle and solid inner core. All of the modes trapped at boundaries, including the short-period Rayleigh-wave-equivalent modes and the inner-core and core-mantle boundary Stoneley modes, are shear-dominated, with only a small fraction—ten to twenty percent—of their energy in compression. This is a characteristic feature of elastic interface waves.

It is conventional to include all of the predominantly elastic modes, including the inner-core $\mathrm{J_{SV}}$ and Stoneley modes, in assigning the identifying labels ${}_n\mathrm{S}_l$. An annoying feature of the resulting nomenclature is that it is Earth-model dependent. To see this, consider a PKIKP-equivalent mode ${}_n\mathrm{S}_l$ and a $\mathrm{J_{SV}}$-equivalent mode ${}_{n+1}\mathrm{S}_l$ whose eigenfrequencies for a given

Mode	mHz	f_κ	f_μ	f_g	Description or Name
${}_0S_0$	0.8143	1.30	0.03	−0.33	fundamental radial
${}_1S_0$	1.6313	0.95	0.16	−0.11	radial overtone
${}_2S_0$	2.5105	0.88	0.17	−0.05	radial overtone
${}_0S_2$	0.3093	0.12	0.55	0.33	football mode
${}_0S_3$	0.4686	0.16	0.65	0.19	pear-shaped
${}_0S_{10}$	1.7265	0.23	0.81	−0.04	fundamental Rayleigh
${}_0S_{20}$	2.8784	0.20	0.81	−0.01	fundamental Rayleigh
${}_0S_{30}$	3.8155	0.16	0.85	−0.01	fundamental Rayleigh
${}_0S_{40}$	4.7101	0.14	0.86	−0.00	fundamental Rayleigh
${}_{10}S_6$	4.9142	0.03	1.01	−0.04	inner-core J_{SV}
${}_{16}S_3$	6.2225	0.02	1.02	−0.04	inner-core J_{SV}
${}_{24}S_8$	10.8312	0.04	0.99	−0.03	inner-core J_{SV}
${}_{14}S_3$	5.4075	0.03	0.98	−0.02	mantle ScS_{SV}
${}_{17}S_6$	7.5806	0.10	0.92	−0.02	mantle ScS_{SV}
${}_{24}S_5$	9.6478	0.02	0.99	−0.02	mantle ScS_{SV}
${}_{11}S_5$	5.0744	0.72	0.30	−0.02	PKIKP
${}_{16}S_6$	7.1537	0.75	0.26	−0.01	PKIKP
${}_{27}S_2$	9.8653	0.80	0.21	−0.01	PKIKP
${}_1S_{10}$	2.1484	0.14	0.80	0.06	CMB Stoneley
${}_2S_{18}$	3.8745	0.20	0.77	0.03	CMB Stoneley
${}_3S_4$	1.8333	0.08	0.93	−0.01	ICB Stoneley
${}_6S_{11}$	4.5350	0.06	0.94	−0.00	ICB Stoneley
${}_1S_1$	0.0513	0.64	0.00	0.36	Slichter mode

Table 8.2. Frequency $\omega/2\pi$ (mHz) and fractional compressional, shear and gravitational potential energies of selected spheroidal modes. The tabulated values are for the isotropic PREM Earth model; the acronyms CMB and ICB denote the core-mantle boundary and inner-core boundary, respectively.

Earth model are near an avoided crossing so that they are very nearly equal. Suppose now that the shear-wave speed β within the solid inner core is reduced by a small amount while keeping the compressional-wave speed α everywhere the same; the frequency ${}_{n+1}\omega_l$ of the J_{SV} mode will be decreased slightly but the frequency ${}_n\omega_l$ of the PKIKP mode will remain essentially unchanged. If the change in β is great enough, the modification can cause the two modes to exchange identities, so that ${}_n S_l$ becomes the inner-core J_{SV} mode and ${}_{n+1}S_l$ becomes the PKIKP mode. For example, the observed PKIKP mode with a SNREI frequency of 4.04 mHz is ${}_{10}S_2$ for Earth model 1066A, which has a mean inner-core shear-wave speed $\beta \approx 3.6$ km/s, whereas it is ${}_{11}S_2$ for model 1066B, which has $\beta \approx 3.5$ km/s (Gilbert & Dziewonski 1975). Other instances of such quasi-degenerate PKIKP and J_{SV} modes include ${}_6S_2$–${}_7S_2$, ${}_5S_{10}$–${}_6S_{10}$ and ${}_{11}S_8$–${}_{12}S_8$.

*8.8.11 Tsunami and core gravity modes

In addition to the predominantly elastic modes discussed above, a realistic SNREI Earth model such as PREM has two types of predominantly gravitational spheroidal oscillations, neither of which is displayed in Figure 8.9 or accounted for in assigning the identifying labels ${}_nS_l$. The first of these "ex-

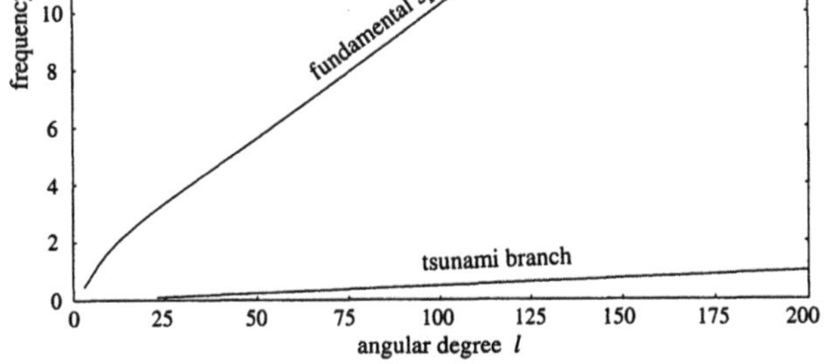

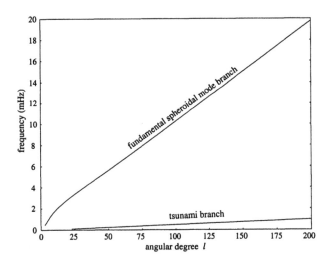

Figure 8.20. Dispersion diagram comparing the tsunami and fundamental spheroidal-mode branches for a modified PREM model with a 4 km deep oceanic layer. For $l \leq 200$ the tsunami modes are non-dispersive to a very good approximation.

otic" spheroidal oscillations are oceanic surface-gravity or *tsunami modes*. Figure 8.20 compares the tsunami-mode eigenfrequencies with those situated along the ${}_0\mathrm{S}_l$ mode branch; it is evident that the designation of ${}_0\mathrm{S}_l$ as the "fundamental" spheroidal-mode branch of the Earth is something of a misnomer! The displacement eigenfunctions of the tsunami modes are predominantly confined to the homogeneous ocean, as illustrated in Figure 8.21. At a period of 1505 seconds (angular degree $l = 135$) the radial eigenfunction U within the ocean is roughly ten times smaller than the tangential eigenfunction V—the water sloshes back and forth with very little free-surface deformation. Furthermore, V is nearly constant within the ocean, whereas U decreases roughly linearly from a maximum at the surface to almost zero at the seafloor. These are the characteristic features of a non-dispersive shallow-water surface-gravity wave (Lamb 1932; Lighthill

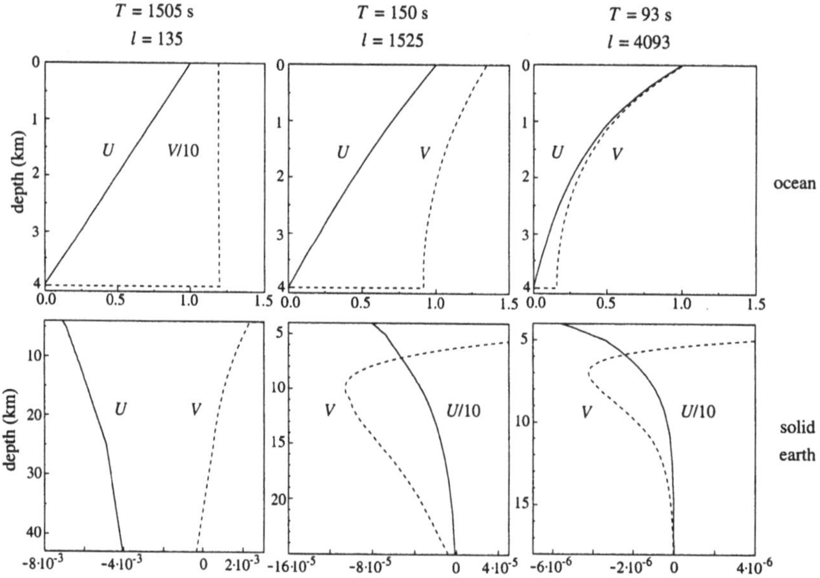

Figure 8.21. Tsunami-mode displacement eigenfunctions U (*solid line*) and V (*dashed line*) at periods of 1505, 150 and 93 seconds, corresponding to angular degrees $l = 135$, $l = 1525$ and $l = 4093$, respectively. (*Top row*) Eigenfunctions within the 4 km oceanic layer. (*Bottom row*) Magnified view of the solid-Earth eigenfunctions. Vertical axis in all cases is depth $z = a - r$ beneath the sea surface; note the differing depth scales in the solid. The eigenfunctions are normalized for plotting such that $U = 1$ at $z = 0$ for all periods.

1978). The corresponding displacements U and V within the underlying solid Earth are more than two orders of magnitude smaller than the fluid displacements; a 1505-second tsunami is able to "feel" only about 50 km into the solid Earth. At a period of 150 seconds ($l \approx 1500$) the radial and tangential displacements are of comparable magnitude within the ocean; the associated solid-Earth displacements are almost four orders of magnitude smaller, and "feel" to a depth of only about 20 km beneath the seafloor. Finally, at a period of 93 seconds ($l \approx 4000$) U and V are nearly identical within the ocean; both exhibit a quasi-exponential $\exp(-kz/a)$ decay with depth z, which is characteristic of a deep-water surface-gravity wave. The solid-Earth displacements accompanying such a 90-second tsunami are thoroughly negligible 10 km beneath the seafloor. The transition between shallow-water and deep-water behavior occurs at a period between 50 and 100 seconds; tsunami modes with periods shorter than this cannot be excited by earthquakes, because they are oblivious to the presence of the underlying solid Earth. The dominant observed period of tsunamis is for this reason between a few hundred and a few thousand seconds. Even at long periods it is evident that virtually the entire solid Earth is a node of the deformation; this is, of course, why tsunamis are only excited by large shallow-focus submarine earthquakes. The potential energy $\mathcal{V}_\kappa + \mathcal{V}_\mu + \mathcal{V}_{\mathrm{g}}$ of all of the tsunami modes is more than 95 percent gravitational. The application of the SNREI-Earth normal-mode formalism to the study of tsunami generation and propagation was pioneered by Ward (1980); an asymptotic analysis of the dispersion along the tsunami-mode branch has been conducted by Okal (1982).

Fluid-core gravity modes, also known as core *undertones*, constitute the second type of "exotic" gravitational oscillation. The theoretical eigenfrequencies of these as yet unobserved modes depend sensitively upon the radial distribution of the Brunt-Väisälä frequency N in the fluid outer core. If $N^2 > 0$ throughout the entire outer core, $c \leq r \leq b$, there are an infinite number of undertone modes with real eigenfrequencies in the range $0 \leq \omega \leq N_{\mathrm{max}}$; on the other hand, if $N^2 < 0$ throughout $c \leq r \leq b$, there are an infinite number of unstable undertone modes with purely imaginary eigenfrequencies ($\omega^2 < 0$). Models that are constrained to fit seismic eigenfrequency observations typically have alternating regions in the core where $N^2 > 0$ and $N^2 < 0$; in that case, there are both $\omega^2 > 0$ and $\omega^2 < 0$ core modes, with associated eigenfunctions that are predominantly confined to the stably and unstably stratified regions, respectively. The details of these undertone oscillations for any particular best-fitting model, such as PREM, are of very little interest because of the poorly constrained character of N within the core. There are no non-trivial gravity modes if the fluid core is neutrally stratified (and there is no surficial ocean) so that

$N = 0$ everywhere in $\oplus_{\rm F}$. Such a neutrally stratified model has instead an infinite-dimensional geostrophic eigenspace, with an associated degenerate eigenfrequency $\omega = 0$, as discussed in Section 8.8.2.

By judiciously combining equations (8.143)–(8.147) it is possible to derive the relation

$$\omega^2 \frac{d}{dr}(rV) = \omega^2 kU - kN^2(\rho g)^{-1}R, \tag{8.189}$$

which must be satisfied by every mode in $\oplus_{\rm F}$. In any neutrally stratified region of the fluid outer core, equation (8.189) is valid for the trivial modes by virtue of the fact that $\omega = 0$. For a non-trivial mode, with $\omega \neq 0$, equation (8.189) reduces to $\dot{V} + r^{-1}V - kr^{-1}U = 0$ wherever $N = 0$. This simple relation must be satisfied by every non-trivial spheroidal oscillation, including the predominantly elastic oscillations ${}_n\mathrm{S}_l$, in an Earth model with a neutrally stratified fluid outer core.

*8.8.12 Atmospheric modes

The presence of the Earth's atmosphere can, in principle, be accounted for by continuing the numerical integration of the four fluid-dynamical equations (8.143)–(8.147) above the surface oceanic layer, subject to the continuity conditions $[U]_-^+ = 0$, $[P]_-^+ = 0$, $[R]_-^+ = 0$, $[B]_-^+ = 0$ at the air-sea interface $r = a$. In the limit $r \to \infty$ (in practice at an elevation $r - a \approx 200$ km) it is necessary to impose a frequency-dependent boundary condition which stipulates that there is an outward radiation of acoustic-gravitational energy into the exosphere, where the density ρ and incompressibility κ are both infinitesimally small (Watada 1995; Lognonné, Clévédé & Kanamori 1998). Earth models with an overlying atmosphere have—in addition to the modes already discussed—a rich spectrum of predominantly gravitational modes governed by the atmospheric Brunt-Väisälä frequency N and predominantly acoustic modes governed by the atmospheric sound speed $\alpha = (\kappa/\rho)^{1/2}$. The latter are, to a very good approximation, non-dispersive, with eigenfrequencies ω that are essentially independent of the angular degree in the range $0 \leq l \leq 100$, as illustrated in Figure 8.22. The frequencies and quality factors of the fundamental and first seven overtone acoustic modes are summarized in Table 8.3. The attenuation is almost entirely due to the outward radiation; the modes have a tiny exponential tail down into the underlying fluid-solid Earth; however, the effect of intrinsic attenuation within the mantle is insignificant. Only the fundamental mode and first overtone have $Q > 10$; the low quality factors of the higher overtones indicate that they are imperfectly trapped within the atmosphere.

The fundamental seismic mode branch ${}_0\mathrm{S}_l$ crosses the first two atmospheric mode branches at $l = 28-29$ and $l = 36-37$, respectively. The

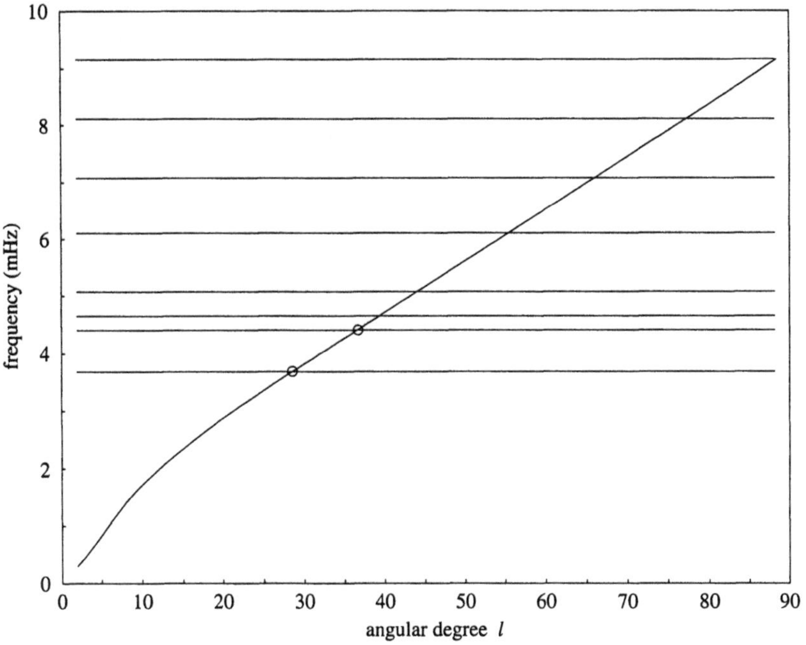

Figure 8.22. Dispersion diagram showing the real eigenfrequencies $\omega/2\pi$ of the fundamental and first seven overtone acoustic modes. The negligible group speed, $d\omega/dk \approx 0$, indicates that the propagation of the associated atmospheric acoustic waves is very nearly non-dispersive. The crossing curve is the dispersion relation of the fundamental spheroidal mantle modes ${}_0\mathrm{S}_l$. Small circles mark the first two crossover points at angular degrees $l = 28$–29 ($\omega/2\pi = 3.68$ mHz) and $l = 36$–37 ($\omega/2\pi = 4.40$ mHz).

proximity of the eigenfrequencies gives rise to an enhanced coupling between the fluid-solid Earth and the atmosphere; the atmospheric energy content of the affected modes ${}_0\mathrm{S}_{28}$–${}_0\mathrm{S}_{29}$ and ${}_0\mathrm{S}_{36}$–${}_0\mathrm{S}_{37}$ can be as high as 0.04 percent. Fundamental spheroidal oscillations with frequencies $\omega/2\pi \approx 3.68$ mHz and $\omega/2\pi \approx 4.40$ mHz (periods $T \approx 272$ s and $T \approx 227$ s) can be excited by strong atmospheric sources for this reason. Such quasi-monochromatic oscillations were first detected following the 1982 El Chichón, Mexico and 1991 Mount Pinatubo, Philippines volcanic eruptions (Kanamori & Mori 1992; Widmer & Zürn 1992; Zürn & Widmer 1996). Atmospheric modes with angular degrees $l \approx 28-29$ and $l \approx 36-37$ can likewise be excited by sources situated within the solid Earth. This provides an explanation for the microbarographic and ionospheric disturbances that have occasionally

Frequency (mHz)	3.68	4.40	4.65	5.07	6.10	7.07	8.11	9.16
Quality Factor	117	21	3	9	6	7	7	8

Table 8.3. Theoretical frequency $\omega/2\pi$ and quality factor Q of the predominantly acoustic atmospheric modes in the non-dispersive range $0 \leq l \leq 100$ (Lognonné, Clévédé & Kanamori 1998). The density ρ and sound speed α above $a = 6371$ km are those of the U.S. Standard Atmosphere; the underlying fluid-solid Earth model is PREM.

been reported following large earthquakes (Mikumo 1968; Yuen, Weaver, Suzuki & Furumoto 1969).

Throughout the rest of the book, we shall continue to assume, for the sake of simplicity, that the region $\bigcirc - \oplus$ outside of the free surface $\partial\oplus$ of the fluid-solid Earth $\oplus$ is a vacuum.

*8.9 Transversely Isotropic Earth Model

A spherically symmetric Earth model is, by definition, one that is invariant under all rigid rotations about its center. The most general fourth-order elastic tensor $\mathbf{\Gamma}$ with this property is not isotropic, but rather *transversely isotropic*, with a radial symmetry axis (Love 1927; Stoneley 1949):

$$\begin{aligned}\mathbf{\Gamma} = C\,\hat{\mathbf{r}}\hat{\mathbf{r}}\hat{\mathbf{r}}\hat{\mathbf{r}} &+ A(\hat{\boldsymbol{\theta}}\hat{\boldsymbol{\theta}}\hat{\boldsymbol{\theta}}\hat{\boldsymbol{\theta}} + \hat{\boldsymbol{\phi}}\hat{\boldsymbol{\phi}}\hat{\boldsymbol{\phi}}\hat{\boldsymbol{\phi}}) \\ &+ F(\hat{\mathbf{r}}\hat{\mathbf{r}}\hat{\boldsymbol{\theta}}\hat{\boldsymbol{\theta}} + \hat{\boldsymbol{\theta}}\hat{\boldsymbol{\theta}}\hat{\mathbf{r}}\hat{\mathbf{r}} + \hat{\mathbf{r}}\hat{\mathbf{r}}\hat{\boldsymbol{\phi}}\hat{\boldsymbol{\phi}} + \hat{\boldsymbol{\phi}}\hat{\boldsymbol{\phi}}\hat{\mathbf{r}}\hat{\mathbf{r}}) \\ &+ (A - 2N)(\hat{\boldsymbol{\theta}}\hat{\boldsymbol{\theta}}\hat{\boldsymbol{\phi}}\hat{\boldsymbol{\phi}} + \hat{\boldsymbol{\phi}}\hat{\boldsymbol{\phi}}\hat{\boldsymbol{\theta}}\hat{\boldsymbol{\theta}}) \\ &+ N(\hat{\boldsymbol{\theta}}\hat{\boldsymbol{\phi}}\hat{\boldsymbol{\theta}}\hat{\boldsymbol{\phi}} + \hat{\boldsymbol{\phi}}\hat{\boldsymbol{\theta}}\hat{\boldsymbol{\phi}}\hat{\boldsymbol{\theta}} + \hat{\boldsymbol{\theta}}\hat{\boldsymbol{\phi}}\hat{\boldsymbol{\phi}}\hat{\boldsymbol{\theta}} + \hat{\boldsymbol{\phi}}\hat{\boldsymbol{\theta}}\hat{\boldsymbol{\theta}}\hat{\boldsymbol{\phi}}) \\ &+ L(\hat{\mathbf{r}}\hat{\boldsymbol{\theta}}\hat{\mathbf{r}}\hat{\boldsymbol{\theta}} + \hat{\boldsymbol{\theta}}\hat{\mathbf{r}}\hat{\boldsymbol{\theta}}\hat{\mathbf{r}} + \hat{\mathbf{r}}\hat{\boldsymbol{\theta}}\hat{\boldsymbol{\theta}}\hat{\mathbf{r}} + \hat{\boldsymbol{\theta}}\hat{\mathbf{r}}\hat{\mathbf{r}}\hat{\boldsymbol{\theta}} \\ &+ \hat{\mathbf{r}}\hat{\boldsymbol{\phi}}\hat{\mathbf{r}}\hat{\boldsymbol{\phi}} + \hat{\boldsymbol{\phi}}\hat{\mathbf{r}}\hat{\boldsymbol{\phi}}\hat{\mathbf{r}} + \hat{\mathbf{r}}\hat{\boldsymbol{\phi}}\hat{\boldsymbol{\phi}}\hat{\mathbf{r}} + \hat{\boldsymbol{\phi}}\hat{\mathbf{r}}\hat{\mathbf{r}}\hat{\boldsymbol{\phi}}).\end{aligned} \tag{8.190}$$

A general spherically symmetric, transversely isotropic Earth model can be characterized by specifying the radial variation of the density ρ and the five elastic parameters C, A, L, N and F. The speeds of localized elastic body waves within the mantle and solid inner core of such an Earth model are azimuthally invariant, but they depend upon the polarization and the angle of inclination between the wavevector and the local radial axis. Instead of the parameters C, A, L, N and F, we can specify the speeds of vertically and horizontally propagating compressional waves $\alpha_{\mathrm{v}} = (C/\rho)^{1/2}$, $\alpha_{\mathrm{h}} = (A/\rho)^{1/2}$ and the speeds of vertically and horizontally propagating SH-polarized shear waves $\beta_{\mathrm{v}} = (L/\rho)^{1/2}$, $\beta_{\mathrm{h}} = (N/\rho)^{1/2}$, together with the dimensionless parameter $\eta = F/(A-2L)$. Vertically propagating shear waves in a transversely isotropic Earth model do not exhibit any splitting; all polarizations propagate with speed β_{v}. Both vertically and horizontally propagating SV-polarized waves have the same wave speed, so that β_{v} can be interpreted either as the speed of a vertically propagating SH wave or as the speed of a horizontally propagating SV wave. The incompressibility and rigidity of a transversely isotropic Earth model are defined by

$$\kappa = \tfrac{1}{9}(C + 4A - 4N + 4F), \tag{8.191}$$

$$\mu = \tfrac{1}{15}(C + A + 6L + 5N - 2F). \tag{8.192}$$

These are the parameters of an "equivalent" isotropic model whose elastic tensor $\mathbf{\Gamma}'$ is the best approximation to $\mathbf{\Gamma}$ in the least-squares sense (3.150); equations (8.191)–(8.192) are a special case of (3.148). In the limiting case of zero anisotropy $C = A = \kappa + \frac{4}{3}\mu$, $L = N = \mu$ and $F = \kappa - \frac{2}{3}\mu$; as a result $\alpha_{\mathrm{v}} = \alpha_{\mathrm{h}} = \alpha$, $\beta_{\mathrm{v}} = \beta_{\mathrm{h}} = \beta$, and $\eta = 1$. The fluid regions of the Earth must be isotropic, with $C = A = F = \kappa$ and $L = N = 0$. We note finally that a completely general spherically symmetric Earth model may also have a transversely isotropic initial stress tensor, of the form $-p_{\mathrm{v}}\hat{\mathbf{r}}\hat{\mathbf{r}} - p_{\mathrm{h}}(\hat{\boldsymbol{\theta}}\hat{\boldsymbol{\theta}} + \hat{\boldsymbol{\phi}}\hat{\boldsymbol{\phi}})$; we shall ignore that eventuality and assume that $p_{\mathrm{v}} = p_{\mathrm{h}} = p$. Figure 8.23 shows the radial variation of α_{v}, α_{h}, β_{v}, β_{h} and η in the transversely isotropic version of the Preliminary Reference Earth Model (Dziewonski & Anderson 1981). Only the uppermost mantle of the model, between 24.4 and 220 km depth, is anisotropic; vertically propagating P and S waves in this region travel two to four percent more slowly than horizontally propagating ones do.

The frequency-domain momentum equation (8.10) and the dynamical boundary conditions (8.14)–(8.16) are valid for any spherically symmetric Earth model; the only difference is that the constitutive relation (8.11) must be replaced by $\mathbf{T} = \mathbf{\Gamma}:\boldsymbol{\varepsilon}$, where $\mathbf{\Gamma}$ is given by equation (8.190). The displacement and displacement-potential versions of Rayleigh's principle are also both valid on a transversely isotropic Earth model, provided that the elastic energy density $\kappa(\boldsymbol{\nabla}\cdot\mathbf{s})^2 + 2\mu(\mathbf{d}:\mathbf{d})$ is replaced by $\boldsymbol{\varepsilon}:\mathbf{\Gamma}:\boldsymbol{\varepsilon}$ in equations (8.22) and (8.24). We can convert the governing vector and tensor equations into an equivalent system of radial scalar equations using any of the three approaches outlined in Section 8.6. Regardless of how we do the conversion, we find that there is a natural decomposition into decoupled spheroidal and toroidal eigenvalue problems, just as in the case of a SNREI Earth. We consider these two classes of oscillations separately in Sections 8.9.1 and 8.9.2, starting, as before, with the simpler toroidal modes first.

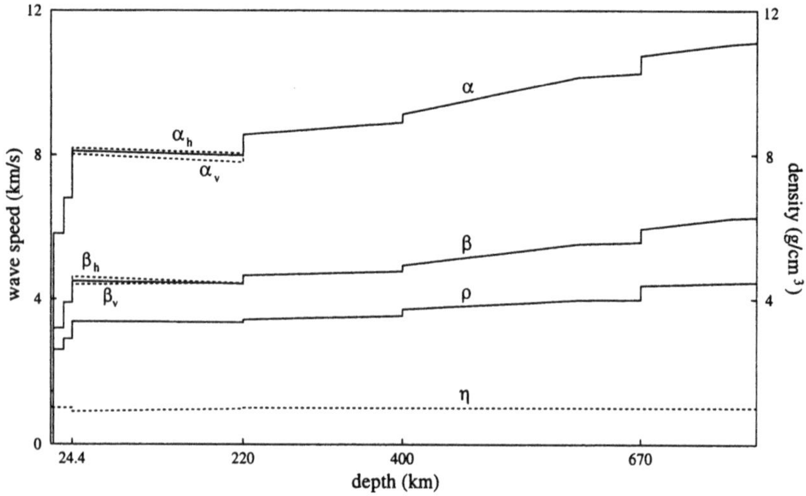

Figure 8.23. Dashed lines show speeds α_v, α_h and β_v, β_h of vertically and horizontally propagating compressional and shear waves and anisotropic parameter η in the upper mantle of the Preliminary Reference Earth Model (PREM). Solid lines show density ρ and compressional and shear wave speeds α and β of the "equivalent" isotropic model for comparison. All of the speeds are those "seen" by waves with a frequency $\omega/2\pi = 1$ Hz.

⋆8.9.1 Toroidal oscillations

The toroidal oscillations of a transversely isotropic Earth model have displacement vectors of the form $\mathbf{s} = W\mathbf{C}_{lm}$ and associated traction vectors of the form $\hat{\mathbf{r}} \cdot \mathbf{T} = T\mathbf{C}_{lm}$, where

$$T = L(\dot{W} - r^{-1}W). \tag{8.193}$$

The radial action integral $\mathcal{I}_\mathrm{T}$ governing these modes is of the form (8.88), where

$$L_\mathrm{T} = \tfrac{1}{2}[\omega^2\rho W^2 - L(\dot{W} - r^{-1}W)^2 - (k^2 - 2)Nr^{-2}W^2]. \tag{8.194}$$

Neither the toroidal Lagrangian density L_T nor the spheroidal and radial Lagrangian densities L_S, L_S' and L_R which we introduce below should be confused with the elastic parameter L denoted by the same symbol. The meaning should always be clear from the context, even when the subscripts T, S and R are dropped from the densities, as we do in Section 9.4. The

Euler-Lagrange equation (8.95) is equivalent to the system of first-order ordinary differential equations:

$$\dot{W} = r^{-1}W + L^{-1}T, \tag{8.195}$$

$$\dot{T} = [-\omega^2\rho + (k^2 - 2)Nr^{-2}]W - 3r^{-1}T. \tag{8.196}$$

These equations must be solved subject to the boundary conditions that $T = 0$ on $r = d_{\mathrm{FS}}$ and $r = a$. The elastic potential energy of a toroidal mode is stored entirely in shear; the total kinetic plus potential energy is $\mathcal{E} = \frac{1}{2}(\omega^2\mathcal{T} + \mathcal{V}_{\mathrm{e}})$, where

$$\mathcal{V}_{\mathrm{e}} = \int_0^a [L(\dot{W} - r^{-1}W)^2 + (k^2 - 2)Nr^{-2}W^2]\, r^2 dr. \tag{8.197}$$

It is noteworthy that the toroidal normal modes depend only upon the density ρ and the two elastic parameters L and N, or, equivalently, only upon ρ and the two SH-wave velocities β_{v} and β_{h}. The toroidal modes of a transversely isotropic Earth are always stable, $\omega^2 = \mathcal{V}_{\mathrm{e}}/\mathcal{T} \geq 0$, by virtue of the elastic constraints $L \geq 0$ and $N \geq 0$.

*8.9.2 Spheroidal oscillations

The spheroidal oscillations have displacement and traction vectors of the form $\mathbf{s} = U\mathbf{P}_{lm} + V\mathbf{B}_{lm}$ and $\hat{\mathbf{r}} \cdot \mathbf{T} = R\mathbf{P}_{lm} + S\mathbf{B}_{lm}$, where

$$R = C\dot{U} + Fr^{-1}(2U - kV), \tag{8.198}$$

$$S = L(\dot{V} - r^{-1}V + kr^{-1}U). \tag{8.199}$$

The displacement and displacement-potential radial action integrals $\mathcal{I}_{\mathrm{S}}$ and $\mathcal{I}'_{\mathrm{S}}$ are of the form (8.87) and (8.100), where

$$\begin{aligned} L_{\mathrm{S}} = \tfrac{1}{2}[&\omega^2\rho(U^2 + V^2) - C\dot{U}^2 - 2Fr^{-1}\dot{U}(2U - kV) \\ &- (A - N)r^{-2}(2U - kV)^2 - L(\dot{V} - r^{-1}V + kr^{-1}U)^2 \\ &\quad - (k^2 - 2)Nr^{-2}V^2 - \rho(U\dot{P} + kr^{-1}VP) \\ &\qquad - 4\pi G\rho^2U^2 + 2\rho g r^{-1}U(2U - kV)], \end{aligned} \tag{8.200}$$

$$\begin{aligned} L'_{\mathrm{S}} = \tfrac{1}{2}[&\omega^2\rho(U^2 + V^2) - C\dot{U}^2 - 2Fr^{-1}\dot{U}(2U - kV) \\ &- (A - N)r^{-2}(2U - kV)^2 - L(\dot{V} - r^{-1}V + kr^{-1}U)^2 \\ &\quad - (k^2 - 2)Nr^{-2}V^2 - 2\rho(U\dot{P} + kr^{-1}VP) \\ &\qquad - 4\pi G\rho^2U^2 + 2\rho g r^{-1}U(2U - kV) \\ &\qquad\quad - (4\pi G)^{-1}(\dot{P}^2 + k^2r^2P^2)]. \end{aligned} \tag{8.201}$$

The Euler-Lagrange equations (8.93)–(8.94) and (8.102) are equivalent to the system of six first-order equations:

$$\dot{U} = -2C^{-1}Fr^{-1}U + kC^{-1}Fr^{-1}V + C^{-1}R, \tag{8.202}$$

$$\dot{V} = -kr^{-1}U + r^{-1}V + L^{-1}S, \tag{8.203}$$

$$\dot{P} = -4\pi G\rho U - (l+1)r^{-1}P + B, \tag{8.204}$$

$$\begin{aligned}\dot{R} = {} & [-\omega^2\rho - 4\rho g r^{-1} + 4(A - N - C^{-1}F^2)r^{-2}]U \\ & + [k\rho g r^{-1} - 2k(A - N - C^{-1}F^2)r^{-2}]V \\ & - 2(1 - C^{-1}F)r^{-1}R + kr^{-1}S \\ & - (l+1)\rho r^{-1}P + \rho B,\end{aligned} \tag{8.205}$$

$$\begin{aligned}\dot{S} = {} & [k\rho g r^{-1} - 2k(A - N - C^{-1}F^2)r^{-2}]U \\ & - [\omega^2\rho + 2Nr^{-2} - k^2(A - C^{-1}F^2)r^{-2}]V \\ & - kC^{-1}Fr^{-1}R - 3r^{-1}S + k\rho r^{-1}P,\end{aligned} \tag{8.206}$$

$$\dot{B} = -4\pi G(l+1)\rho r^{-1}U + 4\pi Gk\rho r^{-1}V + (l-1)r^{-1}B. \tag{8.207}$$

These equations must be solved subject to the boundary conditions that $R = 0$ and $B = 0$ on $r = a$ and $S = 0$ on $r = a$ and $r = d_{\rm FS}$. The kinetic plus elastic-gravitational potential energy of a spheroidal oscillation is $\mathcal{E} = \frac{1}{2}(\omega^2\mathcal{T} + \mathcal{V}_{\rm e} + \mathcal{V}_{\rm g})$, where

$$\begin{aligned}\mathcal{V}_{\rm e} = \int_0^a & [C\dot{U}^2 + 2Fr^{-1}\dot{U}(2U - kV) + (A - N)r^{-2}(2U - kV)^2 \\ & + L(\dot{V} - r^{-1}V + kr^{-1}U)^2 + (k^2 - 2)Nr^{-2}V^2]\, r^2 dr,\end{aligned} \tag{8.208}$$

and $\mathcal{V}_{\rm g}$ is given by equation (8.129) as before. The spheroidal modes depend upon the density ρ and all five of the elastic parameters C, A, L, N and F. The governing sixth-order system (8.202)–(8.207) was first derived by Backus (1967).

*8.9.3 Radial oscillations

The radial action $\mathcal{I}_{\rm R}$ governing the radial modes of a transversely isotropic Earth model is of the form (8.152), where

$$L_{\rm R} = \tfrac{1}{2}[\omega^2\rho U^2 - C\dot{U}^2 - 4Fr^{-1}\dot{U}U - 4(A - N)r^{-2}U^2]. \tag{8.209}$$

The Euler-Lagrange equation (8.154) is equivalent to the pair of first-order equations

$$\dot{U} = -2C^{-1}Fr^{-1}U + C^{-1}R, \tag{8.210}$$

$$\begin{aligned}\dot{R} = [-\omega^2\rho + 4(A - N - C^{-1}F^2) - 4\rho g r^{-1}]U \\ - 2(1 - C^{-1}F)r^{-1}R,\end{aligned} \tag{8.211}$$

which must be solved subject to the boundary condition that $R = 0$ on $r = a$. The energy of a radial mode is $\mathcal{E} = \frac{1}{2}(\omega^2\mathcal{T} + \mathcal{V}_{\mathrm{e}} + \mathcal{V}_{\mathrm{g}})$, where

$$\mathcal{V}_{\mathrm{e}} = \int_0^a [C\dot{U}^2 + 4Fr^{-1}\dot{U}U + 4(A - N)r^{-2}U^2]\, r^2 dr, \tag{8.212}$$

and $\mathcal{V}_{\mathrm{g}}$ is given by (8.157). The radial oscillations depend upon ρ and the four parameters C, A, N and F, but they are independent of the fifth elastic parameter L.

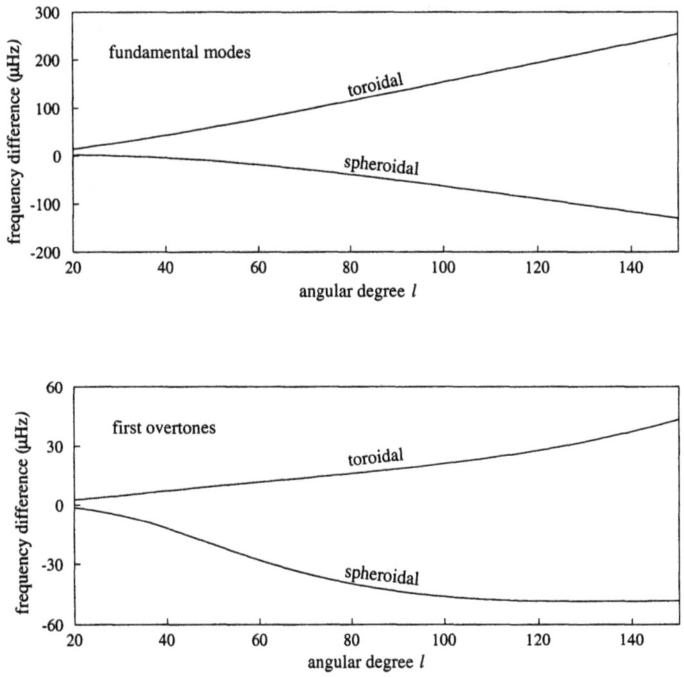

Figure 8.24. Difference $\omega_{\mathrm{TI}} - \omega_{\mathrm{EI}}$ between the fundamental-mode (*top*) and first-overtone (*bottom*) eigenfrequencies of the transversely isotropic (TI) and "equivalent" isotropic (EI) Preliminary Reference Earth Model.

★8.9.4 Effect upon the eigenfrequencies

In conclusion, the elastic-gravitational eigenfrequencies and eigenfunctions of a transversely isotropic Earth model can be calculated by making extremely modest changes in the formalism developed for a SNREI Earth. These changes have been incorporated into both the `MINEOS` and `OBANI` codes, enabling them to deal with either isotropic or transversely isotropic Earth models. Figure 8.24 illustrates the influence of the transverse isotropy of the PREM model upon the eigenfrequencies of the fundamental and first-overtone dispersion branches ${}_0\mathrm{S}_l$, ${}_0\mathrm{T}_l$ and ${}_1\mathrm{S}_l$, ${}_1\mathrm{T}_l$. The quantity plotted is the difference $\omega_{\mathrm{TI}} - \omega_{\mathrm{EI}}$ between the eigenfrequency of the transversely isotropic model and that of the "equivalent" isotropic model. The fundamental toroidal eigenfrequencies ${}_0\omega_l^{\mathrm{T}}$, which depend much more strongly upon β_{h} than upon β_{v}, are increased by 0.5–1.5 percent, whereas the fundamental spheroidal eigenfrequencies ${}_0\omega_l^{\mathrm{S}}$, which are practically independent of β_{h}, are decreased by 0.3–1 percent by the anisotropy. There is a similar but much weaker effect upon the eigenfrequencies ${}_1\omega_l^{\mathrm{T}}$ and ${}_1\omega_l^{\mathrm{S}}$ along the first-overtone branch. The transverse isotropy of the PREM model was introduced originally in an attempt to reconcile the measured eigenfrequencies of fundamental Rayleigh-wave-equivalent and Love-wave-equivalent modes ${}_0\mathrm{S}_l$ and ${}_0\mathrm{T}_l$ of degree l greater than 40–50. More recent analyses have reduced the magnitude of this discrepancy (Widmer 1991; Ekström, Tromp and Larson 1997), and it is unclear at the present time how much if any upper-mantle anisotropy is required to fit the globally averaged fundamental-mode data.

Chapter 9

Elastic and Anelastic Perturbations

In this chapter we determine the effect of a slight spherically symmetric perturbation upon the normal-mode eigenfrequencies of a SNREI or transversely isotropic Earth model. Using Rayleigh's principle, we obtain an explicit formula for the first-order perturbation $\delta\omega$ in the eigenfrequency of a spheroidal or toroidal oscillation in terms of the unperturbed radial eigenfunctions. By considering a complex perturbation of the isotropic elastic parameters, we derive an analogous first-order formula for the inverse quality factor Q^{-1} of a free oscillation due to the presence of bodily anelasticity. These results provide the basis for the inversion of the measured eigenfrequencies and decay rates of the Earth's free oscillations for the spherically averaged elastic and anelastic structure within the interior.

9.1 Spherical Perturbation

Let us suppose that we know the eigenfrequencies ω and associated radial eigenfunctions U, V, P or W of an initial, unperturbed spherically symmetric, non-rotating Earth model characterized by its density ρ and associated gravitational acceleration g, together with its elastic parameters, either κ and μ in the case of an isotropic model or C, A, L, N and F in the case of a transversely isotropic model. The model parameters may exhibit jump discontinuities at a number of internal solid-solid or fluid-solid and external radii $d = a \cup d_{\mathrm{SS}} \cup d_{\mathrm{FS}}$. We wish to calculate the effect of an infinitesimal spherically symmetric perturbation, which is specified in the following manner:

(1) the density is everywhere perturbed, $\rho \to \rho + \delta\rho$;

(2) and so is the gravitational acceleration, $g \to g + \delta g$;

(3) the elastic parameters are perturbed, either $\kappa \to \kappa + \delta\kappa$, $\mu \to \mu + \delta\mu$ or $C \to C+\delta C$, $A \to A+\delta A$, $L \to L+\delta L$, $N \to N+\delta N$, $F \to F+\delta F$;

(4) finally, the radii of the discontinuities are displaced, $d \to d + \delta d$, and the perturbed density, gravitational acceleration and elastic parameters are all redefined by extrapolation to be smooth on either side of the unperturbed boundary.

The perturbation in the gravitational acceleration is related to the perturbations $\delta\rho$ and δd by

$$\delta g(r) = 4\pi G r^{-2}\left\{\int_0^r \delta\rho' \, r'^2 dr' - \sum_{d<r} \delta d \, d^2 [\rho]_-^+ \right\}, \tag{9.1}$$

where the first term accounts for the underlying density perturbations, and the second term accounts for the displacements of any discontinuities with radii d smaller than r. It is unnecessary in what follows to specify either the initial hydrostatic pressure p or its perturbation δp.

9.2 Application of Rayleigh's Principle

As a result of these changes in the Earth model, the eigenfrequency of every mode is perturbed, $\omega \to \omega + \delta\omega$, and so are the associated eigenfunctions, $U \to U + \delta U$, $V \to V + \delta V$, $P \to P + \delta P$ or $W \to W + \delta W$. Rayleigh's principle enables us to calculate the change $\delta\omega$ in terms of the unperturbed eigenfunctions U, V, P or W without solving simultaneously for the perturbations δU, δV, δP or δW. For a unified treatment of the spheroidal, toroidal and radial oscillations, we rewrite the associated displacement and displacement-potential actions $\mathcal{I}_{\rm S}$, $\mathcal{I}_{\rm T}$, $\mathcal{I}_{\rm R}$ and $\mathcal{I}'_{\rm S}$ using a generic notation:

$$\mathcal{I} = \int_0^\infty L(X, \dot{X}; \omega, \oplus)\, r^2 dr, \tag{9.2}$$

where X stands for U and V in the case of the spheroidal Lagrangian density (8.89) or (8.200), for W in the case of the toroidal Lagrangian density (8.90) or (8.194), for U in the case of the radial Lagrangian density (8.153) or (8.209), and for U, V and P in the case of the modified spheroidal Lagrangian density (8.101) or (8.201). To accommodate the latter case we write the upper limit of integration in (9.2) as ∞, and define U, V and W to be zero in the region $r > a$ outside of the Earth model. The additional arguments in the Lagrangian density L express the dependence upon the eigenfrequency ω and the Earth-model parameters; the symbol $\oplus$ stands for κ, μ, ρ and g in the case of an isotropic Earth model and for C, A, L, N, F, ρ and g in the case of a transversely isotropic Earth model. The generic version of Rayleigh's principle asserts that the action $\mathcal{I}$ is stationary for an arbitrary admissible variation δX if and only if X satisfies the Euler-Lagrange equation

$$\partial_X L - r^{-2}\frac{d}{dr}(r^2 \partial_{\dot{X}} L) = 0, \tag{9.3}$$

together with the boundary conditions $[\partial_{\dot{X}} L]_-^+ = 0$ wherever X is continuous and $\partial_{\dot{X}} L = 0$ wherever it is not. The latter case accounts for the possibility of tangential slip on the fluid-solid boundaries $r = d_{\mathrm{FS}}$ and allows for non-zero displacement on the free surface $r = a$.

The stationary value of the action is $\mathcal{I} = 0$, whether it be evaluated at ω and X in the unperturbed Earth model $\oplus$ with discontinuities at $r = d$, or at $\omega + \delta\omega$ and $X + \delta X$ in the perturbed Earth model $\oplus + \delta\oplus$ with discontinuities at $r = d + \delta d$. The *total variation* of $\mathcal{I}$ with respect to all of its arguments, including ω, $\oplus$ and the radii d of the discontinuities, is given, correct to first order in the various perturbations, by

$$\begin{aligned} \delta\mathcal{I}_{\mathrm{total}} &= \int_0^\infty [\delta X(\partial_X L) + \delta\dot{X}(\partial_{\dot{X}} L)]\, r^2 dr \\ &+ \int_0^\infty [\delta\omega(\partial_\omega L) + \delta\oplus(\partial_\oplus L)]\, r^2 dr - \sum_d \delta d\, d^2 [L]_-^+ = 0, \end{aligned} \tag{9.4}$$

where the sum over discontinuities accounts for the alterations in the composite region of integration due to the shift in the boundaries. Upon integrating the term involving $\delta\dot{X}$ by parts, as in the derivation of Rayleigh's principle, we can rewrite equation (9.4) in the form

$$\begin{aligned} \delta\mathcal{I}_{\mathrm{total}} &= \int_0^\infty \delta X \left[\partial_X L - r^{-2}\frac{d}{dr}(r^2\partial_{\dot{X}} L)\right] r^2 dr \\ &+ \int_0^\infty [\delta\omega(\partial_\omega L) + \delta\oplus(\partial_\oplus L)]\, r^2 dr \\ &\quad - \sum_d d^2[\delta X(\partial_{\dot{X}} L) + \delta d\, L]_-^+ = 0. \end{aligned} \tag{9.5}$$

The first integral in (9.5) vanishes by virtue of the Euler-Lagrange equation (9.3). In the present instance, however, the sum involving δX does *not* vanish, because the perturbation δd of the discontinuity radii renders the variation δX *inadmissible*. The first-order equation guaranteeing continuity of the perturbed eigenfunction $X + \delta X$ across the perturbed boundary is $[\delta X]_-^+ = -\delta d\,[\dot{X}]_-^+$. This condition need not prevail for the tangential displacements V, W on $r = d_{\mathrm{FS}}$ nor for U, V, W on $r = a$; however, the quantity $\partial_{\dot{X}}L$ vanishes in these cases, so we must have

$$[\delta X\,(\partial_{\dot{X}}L)]_-^+ = -\delta d\,[\dot{X}\,(\partial_{\dot{X}}L)]_-^+ \tag{9.6}$$

on all boundaries $r = d$. Upon inserting the relation (9.6) into (9.5) and rearranging terms, we obtain

$$\delta\omega \int_0^\infty (\partial_\omega L)\, r^2 dr = -\int_0^\infty \delta\oplus(\partial_\oplus L)\, r^2 dr + \sum_d \delta d\, d^2 [L - \dot{X}(\partial_{\dot{X}}L)]_-^+. \tag{9.7}$$

This is the desired result, which enables us to calculate the first-order perturbation in an eigenfrequency $\delta\omega$ due to an infinitesimal change in the properties $\delta\oplus$ and the radii of the discontinuities δd of a spherically symmetric, non-rotating Earth model.

Two properties of the radial action (9.2) have been exploited in the above derivation: first, Rayleigh's principle, which stipulates that $\mathcal{I}$ is stationary for arbitrary admissible variations δX in the absence of any perturbations to the Earth model and, second, the energy equipartition relation, which guarantees that the total variation $\delta\mathcal{I}_{\mathrm{total}}$—or difference between $\mathcal{I}$ evaluated at $\omega + \delta\omega$, $X + \delta X$, $\oplus + \delta\oplus$, $d + \delta d$ and at ω, X, $\oplus$, d—must be zero. The recipe for calculating $\delta\omega$ in terms of $\delta\oplus$ and the unperturbed eigenfunction X in the absence of any boundary perturbations is classical, having been noted for a system with a finite number of degrees of freedom by Rayleigh (1877). Subsequently, Backus & Gilbert (1967) presented a formula of the form (9.7) which was supposed to account for the effect of δd on the free oscillations of the Earth, but which omitted the term $-\dot{X}(\partial_{\dot{X}}L)$. The correct result (9.7), which allows for the non-admissible nature of the eigenfunction perturbation δX, was first obtained by Woodhouse (1976). We apply equation (9.7) to a SNREI Earth model and a transversely isotropic Earth model in Sections 9.3 and 9.4, respectively.

9.3 SNREI-to-SNREI Perturbation

Upon substituting the spheroidal, toroidal, radial and modified spheroidal Lagrangian densities (8.89)–(8.90), (8.153) and (8.101) governing a SNREI Earth into the generic result (9.7), we obtain

$$\delta\omega \int_0^a (\partial_\omega L_{\mathrm{S}})\, r^2 dr$$

$$= -\int_0^a [\delta\kappa(\partial_\kappa L_{\rm S}) + \delta\mu(\partial_\mu L_{\rm S}) + \delta\rho(\partial_\rho L_{\rm S}) + \delta g(\partial_g L_{\rm S})]\, r^2 dr$$
$$+ \sum_d \delta d\, d^2 [L_{\rm S} - \dot{U}(\partial_{\dot{U}} L_{\rm S}) - \dot{V}(\partial_{\dot{V}} L_{\rm S})]_-^+, \tag{9.8}$$

$$\delta\omega \int_0^a (\partial_\omega L_{\rm T})\, r^2 dr = -\int_0^a [\delta\mu(\partial_\mu L_{\rm T}) + \delta\rho(\partial_\rho L_{\rm T})]\, r^2 dr$$
$$+ \sum_d \delta d\, d^2 [L_{\rm T} - \dot{W}(\partial_{\dot{W}} L_{\rm T})]_-^+, \tag{9.9}$$

$$\delta\omega \int_0^a (\partial_\omega L_{\rm R})\, r^2 dr$$
$$= -\int_0^a [\delta\kappa(\partial_\kappa L_{\rm R}) + \delta\mu(\partial_\mu L_{\rm R}) + \delta\rho(\partial_\rho L_{\rm R}) + \delta g(\partial_g L_{\rm R})]\, r^2 dr$$
$$+ \sum_d \delta d\, d^2 [L_{\rm R} - \dot{U}(\partial_{\dot{U}} L_{\rm R})]_-^+, \tag{9.10}$$

$$\delta\omega \int_0^\infty (\partial_\omega L'_{\rm S})\, r^2 dr \tag{9.11}$$
$$= -\int_0^\infty [\delta\kappa(\partial_\kappa L'_{\rm S}) + \delta\mu(\partial_\mu L'_{\rm S}) + \delta\rho(\partial_\rho L'_{\rm S}) + \delta g(\partial_g L'_{\rm S})]\, r^2 dr$$
$$+ \sum_d \delta d\, d^2 [L'_{\rm S} - \dot{U}(\partial_{\dot{U}} L'_{\rm S}) - \dot{V}(\partial_{\dot{V}} L'_{\rm S}) - \dot{P}(\partial_{\dot{P}} L'_{\rm S})]_-^+,$$

where we have replaced the upper limit of integration by the radius of the Earth a whenever appropriate. For a toroidal mode of the mantle the lower and upper limits in (9.9) can be replaced by b and s, respectively, whereas for a toroidal mode of the inner core the upper limit can be replaced by c; the sums are likewise only over the discontinuities d within the mantle and inner core, respectively. With the normalization (8.107)–(8.108) adopted in this book, the left sides of (9.8)–(9.11) all reduce simply to $\omega\,\delta\omega$. Upon carrying out the indicated differentiations on the right, we find that the perturbation in the eigenfrequency of either a spheroidal or toroidal or radial free oscillation of a SNREI Earth model can be expressed in terms of the four perturbations $\delta\kappa$, $\delta\mu$, $\delta\rho$ and δd in the form

$$\delta\omega = \int_0^a (\delta\kappa\, K_\kappa + \delta\mu\, K_\mu + \delta\rho\, K_\rho)\, dr + \sum_d \delta d\, [K_d]_-^+, \tag{9.12}$$

where

$$2\omega K_\kappa = (r\dot{U} + 2U - kV)^2, \tag{9.13}$$

$$\begin{aligned}2\omega K_\mu &= \tfrac{1}{3}(2r\dot{U} - 2U + kV)^2 \\ &\quad + (r\dot{V} - V + kU)^2 + (r\dot{W} - W)^2 \\ &\qquad + (k^2 - 2)(V^2 + W^2),\end{aligned} \tag{9.14}$$

$$\begin{aligned}2\omega K_\rho &= -\omega^2 r^2(U^2 + V^2 + W^2) + 8\pi G\rho r^2 U^2 \\ &\quad + 2r^2(U\dot{P} + kr^{-1}VP) - 2grU(2U - kV) \\ &\qquad - 8\pi G r^2 \int_r^a \rho' U'(2U' - kV')\, r'^{-1} dr',\end{aligned} \tag{9.15}$$

$$\begin{aligned}2\omega K_d &= -\kappa(2\omega K_\kappa) - \mu(2\omega K_\mu) - \rho(2\omega K_\rho) \\ &\quad + 2\kappa r\dot{U}(r\dot{U} + 2U - kV) + \tfrac{4}{3}\mu r\dot{U}(2r\dot{U} - 2U + kV) \\ &\qquad + 2\mu r\dot{V}(r\dot{V} - V + kU) + 2\mu r\dot{W}(r\dot{W} - W).\end{aligned} \tag{9.16}$$

For the sake of brevity we have consolidated the results for all three types of modes in equations (9.13)–(9.16); for spheroidal modes we set W equal to zero, for toroidal modes we set U, V and P equal to zero, and for radial modes we set V and W equal to zero. Integration by parts based upon the representation (9.1) has been used to eliminate the dependence upon the gravitational perturbation δg:

$$\begin{aligned}&\int_0^a \delta g\,[2\rho r^{-1}U(2U - kV)]\, r^2 dr \\ &\quad = 8\pi G \int_0^a \delta\rho \int_r^a \rho' U'(2U' - kV')\, r'^{-1} dr'\, r^2 dr \\ &\quad - 8\pi G \sum_d \delta d\, d^2\, [\rho]_-^+ \int_r^a \rho' U'(2U' - kV')\, r'^{-1} dr'.\end{aligned} \tag{9.17}$$

Both (9.8) and (9.11) lead to identical results, as expected; it is necessary to account for the density dependence of P, as expressed in equation (8.55), when evaluating the derivative of the unmodified spheroidal action $\partial_\rho L_S$. The so-called *Fréchet kernels* K_κ, K_μ, K_ρ and K_d provide a direct measure of the sensitivity of a normal mode to spherical perturbations in the incompressibility $\delta\kappa$, rigidity $\delta\mu$, density $\delta\rho$ and discontinuity radii δd.

The eigenfrequencies of many modes depend primarily upon either the compressional-wave speed α or the shear-wave speed β, with only a weak dependence upon the density ρ. To exhibit this sensitivity explicitly, it is convenient to express the perturbation $\delta\omega$ in terms of $\delta\alpha$, $\delta\beta$ and $\delta\rho$ rather than $\delta\kappa$, $\delta\mu$ and $\delta\rho$, using the first-order relations

$$\delta\kappa = \delta\rho(\alpha^2 - \tfrac{4}{3}\beta^2) + 2\rho(\alpha\,\delta\alpha - \tfrac{4}{3}\beta\,\delta\beta), \tag{9.18}$$

$$\delta\mu = \delta\rho\,\beta^2 + 2\rho\beta\,\delta\beta. \tag{9.19}$$

Upon inserting (9.18)–(9.19) into equation (9.12) we obtain

$$\delta\omega = \int_0^a (\delta\alpha\, K_\alpha + \delta\beta\, K_\beta + \delta\rho\, K'_\rho)\, dr + \sum_d \delta d\, [K_d]_-^+, \tag{9.20}$$

where

$$K_\alpha = 2\rho\alpha K_\kappa, \tag{9.21}$$

$$K_\beta = 2\rho\beta(K_\mu - \tfrac{4}{3}K_\kappa), \tag{9.22}$$

$$K'_\rho = (\alpha^2 - \tfrac{4}{3}\beta^2)K_\kappa + \beta^2 K_\mu + K_\rho. \tag{9.23}$$

The representation (9.20) in terms of the Fréchet kernels K_α, K_β, K'_ρ and K_d is obviously preferable to (9.12) whenever body-wave travel-time data as well as normal-mode eigenfrequency data are taken into account in the refinement of SNREI Earth reference models.

Loosely speaking, equations (9.12) and (9.20) enable us to calculate the first-order *partial derivatives* of ω with respect to the depth-dependent properties of the Earth model. We may write, for instance,

$$\left(\frac{\partial\omega}{\partial\alpha}\right)_{\beta,\rho,d} = K_\alpha, \qquad \left(\frac{\partial\omega}{\partial\beta}\right)_{\alpha,\rho,d} = K_\beta, \tag{9.24}$$

$$\left(\frac{\partial\omega}{\partial\rho}\right)_{\alpha,\beta,d} = K'_\rho, \qquad \left(\frac{\partial\omega}{\partial d}\right)_{\alpha,\beta,\rho} = K_d, \tag{9.25}$$

where the subscripts specify the variables that are held fixed during the differentiation. The partial derivatives with respect to the incompressibility, rigidity and density are defined analogously in terms of the original Fréchet kernels K_κ, K_μ and K_ρ.

Upon making use of the kinetic-potential energy equipartition relation $\omega^2\mathcal{T} = \mathcal{V}_\kappa + \mathcal{V}_\mu + \mathcal{V}_{\mathrm{g}}$, we find that the partial derivative with respect to the density ρ, holding the seismic wave speeds α and β fixed, satisfies

$$2\omega \int_0^a \rho \left(\frac{\partial\omega}{\partial\rho}\right)_{\alpha,\beta,d} dr = \int_0^a [4\pi G\rho^2 U^2 + \rho(U\dot{P} + kr^{-1}VP)]\, r^2 dr$$
$$- 8\pi G \int_0^a \int_r^a \rho' U'(2U' - kV')\, r'^{-1} dr'\, r^2 dr. \tag{9.26}$$

The right side of equation (9.26) is identically zero for a toroidal mode. Furthermore—because every term is proportional to either G or P—it is negligible for any high-frequency spheroidal mode that is insignificantly influenced by self-gravitation. It follows that the eigenfrequency ω of such a purely or predominantly elastic mode is insensitive to a constant relative perturbation $\delta\rho/\rho$. High-frequency normal-mode data help to determine the radial *stratification* of the spherical-Earth density; however, the overall *magnitude* of ρ is constrained only by the low-frequency gravitationally sensitive spheroidal modes. We have arrived at this conclusion using first-order perturbation theory; however, it is easy to see that the result itself is even more general. The equations (8.114)–(8.115) and (8.158)–(8.161) governing a toroidal or gravitationally insensitive spheroidal mode are invariant under the transformation

$$\rho \rightarrow c\rho, \qquad \kappa \rightarrow c\kappa, \qquad \mu \rightarrow c\mu, \tag{9.27}$$

where c is a constant. The eigenfrequencies of any such mode are therefore invariant under the same transformation; the normalized eigenfunctions are altered to $U \rightarrow c^{-1/2}U$, $V \rightarrow c^{-1/2}V$, $W \rightarrow c^{-1/2}W$ and $R \rightarrow c^{1/2}R$, $S \rightarrow c^{1/2}S$, $T \rightarrow c^{1/2}T$. It is well known that the period of a simple pendulum is independent of the attached mass, since any change in the mass affects the inertial and restoring forces equally. In the same way, the transformation (9.27) has no effect upon the eigenfrequencies, because it maintains the proportionality between the inertial and elastic restoring forces upon every volume element within a non-gravitating Earth model; this elementary observation was first enunciated by Nolet (1976).

Finally, we note that if the gravitational constant G is treated as a variable parameter (i.e., a part of $\oplus$) in equation (9.7), then

$$G\left(\frac{\partial\omega}{\partial G}\right)_{\alpha,\beta,\rho,d} = \int_0^a \rho\left(\frac{\partial\omega}{\partial\rho}\right)_{\alpha,\beta,G,d} dr. \tag{9.28}$$

Equation (9.28) implies that it is impossible to distinguish a slight increase or decrease in $\delta G/G$ from a slight constant decrease or increase in $\delta\rho/\rho$. This first-order result has its own exact generalization: the eigenfrequencies of a self-gravitating Earth model are invariant under the transformation

$$\rho \rightarrow c\rho, \qquad \kappa \rightarrow c\kappa, \qquad \mu \rightarrow c\mu, \qquad G \rightarrow c^{-1}G. \tag{9.29}$$

The non-gravitating displacement and traction transformation relations are supplemented by $P \rightarrow c^{-1/2}P$. The invariance under a parameter change of the form (9.29) precludes the possibility of using low-frequency normal-mode data to constrain the planetary-scale value of the gravitational constant. Furthermore, any constraints that such data place upon the density distribution ρ depend upon the independently measured (laboratory-scale) value of G. This situation is reminiscent of our inability to determine the mass M of the Earth using either astrometric or spatial geodetic data; only the product GM can be determined. The famous laboratory measurement of G by Cavendish (1798) is often referred to as "weighing the Earth".

*9.4 Transversely Isotropic Perturbation

More generally, we can make use of equation (9.7) to determine the perturbation $\delta\omega$ in an eigenfrequency of a transversely isotropic Earth model due to spherically symmetric perturbations δC, δA, δL, δN, δF, $\delta\rho$ and δd in its properties. Upon making use of the transversely isotropic Lagrangian densities (8.194), (8.200)–(8.201) and (8.209), we obtain results analogous to equations (9.8)–(9.11), but with $\delta\kappa(\partial_\kappa L)$ and $\delta\mu(\partial_\mu L)$ replaced by $\delta C(\partial_C L)$, $\delta A(\partial_A L)$, $\delta L(\partial_L L)$, $\delta N(\partial_N L)$ and $\delta F(\partial_F L)$. The final formula for the eigenfrequency perturbation $\delta\omega$ of either a spheroidal or toroidal or radial free oscillation can be written in the form

$$\delta\omega = \int_0^a (\delta C\, K_C + \delta A\, K_A + \delta L\, K_L + \delta N\, K_N + \delta F\, K_F + \delta\rho\, K_\rho)\, dr + \sum_d \delta d\, [K_d]_-^+, \tag{9.30}$$

where

$$2\omega K_C = r^2\dot{U}^2, \qquad 2\omega K_A = (2U - kV)^2, \tag{9.31}$$

$$2\omega K_L = (r\dot{V} - V + kU)^2 + (r\dot{W} - W)^2, \tag{9.32}$$

$$2\omega K_N = -(2U - kV)^2 + (k^2 - 2)(V^2 + W^2), \tag{9.33}$$

$$2\omega K_F = 2r\dot{U}(2U - kV), \tag{9.34}$$

$$\begin{aligned} 2\omega K_d = {} & C(2\omega K_C) - A(2\omega K_A) - L(2\omega K_L) - N(2\omega K_N) \\ & - \rho(2\omega K_\rho) + 2Lr\dot{V}(r\dot{V} - V + kU) + 2Lr\dot{W}(r\dot{W} - W). \end{aligned} \tag{9.35}$$

The quantities K_C, K_A, K_L, K_N and K_F are the Fréchet kernels describing the sensitivity of the eigenfrequency ω to the transversely isotropic elastic parameters C, A, L, N and F; it is noteworthy that the discontinuity kernel K_d is independent of the fifth parameter F.

Alternatively, we can use the first-order relations

$$\delta C = \delta\rho\, \alpha_{\mathrm{v}}^2 + 2\rho\alpha_{\mathrm{v}}\, \delta\alpha_{\mathrm{v}}, \qquad \delta A = \delta\rho\, \alpha_{\mathrm{h}}^2 + 2\rho\alpha_{\mathrm{h}}\, \delta\alpha_{\mathrm{h}}, \tag{9.36}$$

$$\delta L = \delta\rho\, \beta_{\mathrm{v}}^2 + 2\rho\beta_{\mathrm{v}}\, \delta\beta_{\mathrm{v}}, \qquad \delta N = \delta\rho\, \beta_{\mathrm{h}}^2 + 2\rho\beta_{\mathrm{h}}\, \delta\beta_{\mathrm{h}}, \tag{9.37}$$

$$\delta F = \delta\eta(A - 2L) + \eta(\delta A - 2\delta L) \tag{9.38}$$

to express the eigenfrequency perturbation $\delta\omega$ in terms of the perturbations in the vertical and horizontal wave speeds $\delta\alpha_{\rm v}$, $\delta\alpha_{\rm h}$, $\delta\beta_{\rm v}$, $\delta\beta_{\rm h}$, the dimensionless parameter $\delta\eta$ and the density $\delta\rho$:

$$\delta\omega = \int_0^a (\delta\alpha_{\rm v} K_{\alpha_{\rm v}} + \delta\alpha_{\rm h} K_{\alpha_{\rm h}} + \delta\beta_{\rm v} K_{\beta_{\rm v}} + \delta\beta_{\rm h} K_{\beta_{\rm h}} + \delta\eta K_\eta + \delta\rho K'_\rho)\, dr + \sum_d \delta d\, [K_d]_-^+, \tag{9.39}$$

where

$$K_{\alpha_{\rm v}} = 2\rho\alpha_{\rm v} K_C, \qquad K_{\alpha_{\rm h}} = 2\rho\alpha_{\rm h}(K_A + \eta K_F), \tag{9.40}$$

$$K_{\beta_{\rm v}} = 2\rho\beta_{\rm v}(K_L - 2\eta K_F), \qquad K_{\beta_{\rm h}} = 2\rho\beta_{\rm h} K_N, \tag{9.41}$$

$$K_\eta = \rho(\alpha_{\rm h}^2 - 2\beta_{\rm v}^2)K_F, \tag{9.42}$$

$$K'_\rho = \alpha_{\rm v}^2 K_C + \alpha_{\rm h}^2(K_A + \eta K_F) + \beta_{\rm v}^2(K_L - 2\eta K_F) + \beta_{\rm h}^2 K_N + K_\rho. \tag{9.43}$$

Equation (9.39) is the generalization of the isotropic result (9.20) just as (9.30) is the generalization of (9.12).

★9.5 An Alternative Derivation

We can also obtain the perturbation $\delta\omega$ in an eigenfrequency of a spherical Earth model using a straightforward brute-force approach, which eschews Rayleigh's principle. We illustrate this for the simple case of a mantle toroidal mode of a transversely isotropic Earth in this section. The second-order ordinary differential equation governing such an oscillation can be written in the form

$$r^{-2}\frac{d}{dr}(r^2 T) + r^{-1}T + [\omega^2\rho - (k^2 - 2)Nr^{-2}]W = 0, \tag{9.44}$$

where $T = L(\dot{W} - r^{-1}W)$. The first-order perturbation of this equation is

$$r^{-2}\frac{d}{dr}(r^2\delta T) + r^{-1}\delta T + [\omega^2\rho - (k^2 - 2)Nr^{-2}]\delta W + [2\omega\,\delta\omega\,\rho + \omega^2\delta\rho - (k^2 - 2)\delta N r^{-2}]W = 0, \tag{9.45}$$

where $\delta T = \delta L(\dot{W} - r^{-1}W) + L(\delta\dot{W} - r^{-1}\delta W)$. Our objective is to find the eigenfrequency perturbation $\delta\omega$ without having to solve simultaneously for the perturbations in the eigenfunctions δW and δT. We can accomplish this by subtracting $r^2\delta W$ times equation (9.44) from $r^2 W$ times equation (9.45), and integrating by parts from the core-mantle boundary $r = b$ to the seafloor $r = s$; the integrals involving δW and δT cancel, leaving the result

$$2\omega\,\delta\omega \int_b^s \rho W^2 r^2 dr = \int_b^s [\delta L(r\dot{W} - W)^2 + \delta N(k^2 - 2)W^2 + \delta\rho(-\omega^2 r^2 W^2)]\,dr + \sum_{b\leq d\leq s} \delta d\,[W\,\delta T - \delta W\,T]_-^+. \tag{9.46}$$

To eliminate δW and δT from the sum over discontinuities, we employ the first-order condition guaranteeing that the product $(W + \delta W)(T + \delta T)$ is continuous across the perturbed boundary $r = d + \delta d$:

$$[W\,\delta T - \delta W\,T]_-^+ = -\delta d\,[W\dot{T} - \dot{W}T]_-^+. \tag{9.47}$$

Upon substituting equation (9.47) into (9.46) and making use of the normalization condition (8.108), we obtain

$$2\omega\,\delta\omega = \int_b^s [\delta L(r\dot{W} - W)^2 + \delta N(k^2 - 2)W^2 + \delta\rho(-\omega^2 r^2 W^2)]\,dr + \sum_{b\leq d\leq s} \delta d\,[Lr\dot{W}(r\dot{W} - W) - (k^2 - 2)NW^2 + \omega^2\rho r^2 W^2]_-^+,$$

which is equivalent to the previous result (9.30). The corresponding formulae for a spheroidal or radial mode can be derived in an analogous manner, by manipulating the perturbed and unperturbed radial equations governing these oscillations.

9.6 Rogues' Gallery of Fréchet Kernels

The sensitivity of a mode to either isotropic perturbations $\delta\alpha$, $\delta\beta$, $\delta\rho$, δd or transversely isotropic perturbations $\delta\alpha_{\mathrm{v}}$, $\delta\alpha_{\mathrm{h}}$, $\delta\beta_{\mathrm{v}}$, $\delta\beta_{\mathrm{h}}$, $\delta\eta$, $\delta\rho$, δd in the Earth model varies in a distinctive and well-understood manner from one type of oscillation to another, as we illustrate graphically in this section. We limit ourselves here to a qualitative discussion of the features that are most readily apparent upon an eyeball inspection of the plotted Fréchet kernels. Most of the plots and the accompanying discussion are for the case of a SNREI-to-SNREI perturbation; we conclude by examining the effect of a transversely isotropic perturbation upon the surface-wave equivalent modes. The unperturbed model is in every case the isotropic or transversely isotropic version of PREM. A reminder regarding our convention may be appropriate: since the differential element in equation (9.12) is dr rather than r^2dr, the Fréchet kernels K_κ, K_μ and K_ρ express the effect of radially rather than volumetrically weighted perturbations $\delta\kappa$, $\delta\mu$ and $\delta\rho$. Similar remarks apply to equations (9.20), (9.30) and (9.39). A more quantitative asymptotic analysis of the Fréchet kernels of a SNREI Earth model is presented in Section 12.4.3.

Figure 9.1 shows the kernels K_β and K'_ρ for the first ten toroidal modes of degree $l = 2$. These modes are sensitive to changes in both the shear-wave speed β and the density ρ throughout the entire mantle; the radial "wavelength" of the oscillations is roughly twice that of the equivalent monochromatic $\mathrm{ScS_{SH}}$ wave. An increase in β always increases the eigenfrequency, since $K_\beta \geq 0$ for every toroidal mode. This is in accord with elementary physical intuition: if waves travel faster within the Earth, then the pitch of the oscillations formed by the constructive interference of these waves must rise. An increase in the density may, on the other hand, have either a positive or a negative effect, depending upon the depth. This is consistent with the requirement (9.26) that $\int_b^s \rho K'_\rho\, dr = 0$ for every toroidal mode. A uniform relative perturbation $\delta\rho/\rho$ has no first-order effect whatsoever, as we have seen; the effect of a long-wavelength perturbation (longer than the oscillations in K'_ρ) is slight. Bear in mind that these results pertain to a change in the density with the shear-wave speed held fixed; if the

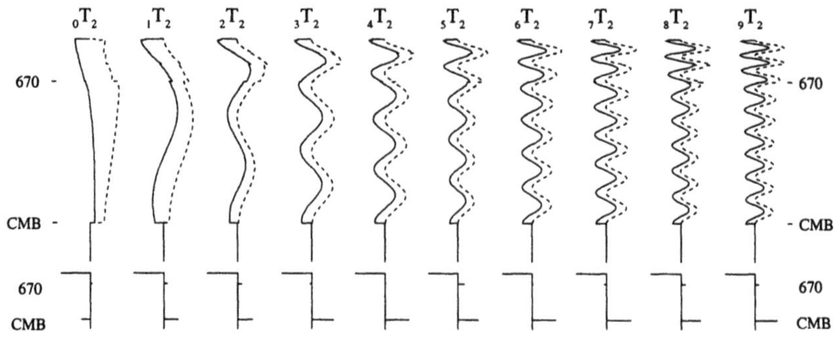

Figure 9.1. Fréchet kernels K_β (*dashed line*) and K'_ρ (*solid line*) for the degree-two toroidal modes ${}_0\mathrm{T}_2, \ldots, {}_9\mathrm{T}_2$ (see Figure 8.6 for displacements W and tractions T). Vertical axis is depth beneath the surface of the Earth; the locations of the 670 km discontinuity and the core-mantle boundary (CMB) are indicated. Lower diagrams show the sensitivity to perturbations in the location of these three major boundaries. Left-pointing and right-pointing "limbs" are associated with negative and positive kernels $[K_s]_-^+$, $[K_{670}]_-^+$, $[K_b]_-^+$, respectively. Each graph is scaled independently, so that the maximum values of the kernels are the same.

rigidity μ is held fixed instead, then an increase in ρ always decreases the eigenfrequency of a toroidal mode, since the pertinent kernel in that case is everywhere non-positive: $K_\rho \leq 0$. Such a perturbation increases the inertia of every volume element within the Earth, but leaves the restoring force unchanged. The "espalier" diagrams at the bottom of Figure 9.1 depict the kernels $[K_s]_-^+$, $[K_{670}]_-^+$ and $[K_b]_-^+$ which express the effects of a perturbation in the location of the seafloor, the 670 km discontinuity and the core-mantle boundary. An increase $\delta s > 0$ in the elevation of the seafloor lowers all of the toroidal eigenfrequencies, whereas an increase $\delta b > 0$ in the radius of the core-mantle boundary raises the eigenfrequencies of all but the ${}_0\mathrm{T}_2$ fundamental mode. These results too have a simple physical explanation: the two perturbations respectively expand and contract the dimensions of the mantle within which the equivalent $\mathrm{ScS}_{\mathrm{SH}}$ waves are propagating. The decrease $[K_b]_-^+ < 0$ in the eigenfrequency of the ${}_0\mathrm{T}_2$ mode cannot be predicted by means of such ray-mode duality arguments, which are strictly valid only in the limit $\omega \to \infty$.

Figure 9.2 shows the Fréchet kernels K_β, K'_ρ and $[K_s]_-^+$, $[K_{670}]_-^+$, $[K_b]_-^+$ for a number of toroidal modes having approximately the same eigenfrequency, $\omega/2\pi \approx 14$ mHz. Many of the characteristic features are again apparent; note, in particular, the dominant and inherently positive ($K_\beta > 0$) sensitivity to perturbations in the shear-wave speed and the zero-mean ($\int_b^s \rho K'_\rho \, dr = 0$) oscillatory dependence upon the density. The most notable new feature is the transition at $n \approx l/4$ between reflected $\mathrm{ScS}_{\mathrm{SH}}$-equivalent

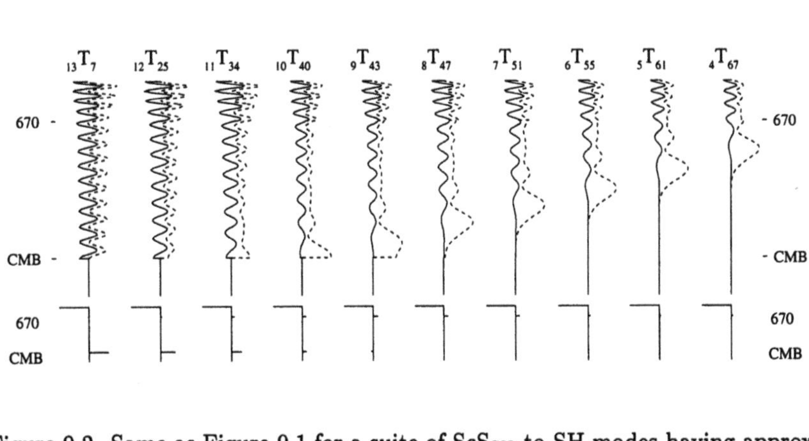

Figure 9.2. Same as Figure 9.1 for a suite of $\mathrm{ScS}_{\mathrm{SH}}$-to-SH modes having approximately the same eigenfrequency $\omega/2\pi \approx 14$ mHz (see Figure 8.8 for displacements W and tractions T). Dashed and solid lines lines denote K_β and K'_ρ, respectively. Mode ${}_4\mathrm{T}_{67}$ on the far right is in essence a fourth-overtone Love wave, which "feels" down to a depth of about 1000 km.

modes which "feel" the core-mantle boundary and turning SH-equivalent modes which do not. The shear-wave kernel K_β for the SH-equivalent modes exhibits a maximum in the vicinity of the turning radius, where the constructively interfering waves spend most of their time. Perturbations $\delta\beta$ and $\delta\rho$ beneath the turning radius and changes δb in the location of the core-mantle boundary have a negligible effect upon the eigenfrequencies of the SH-equivalent modes.

Figure 9.3 displays the Fréchet kernels K_β and K'_ρ for two suites of toroidal modes having approximately the same turning radii. The associated eigenfrequencies lie along two straight lines in the toroidal-mode dispersion diagram, as shown in Figure 9.4. All of the modes with $n \approx l/20$ exhibit a maximum in K_β at a depth $h \approx 900$ km, whereas those with $n \approx l/10$ exhibit a maximum in K_β at a depth $h \approx 1800$ km. This illustrates a general principle of mode-ray duality, which we shall explore more fully in Chapter 12: high-frequency toroidal or spheroidal modes having the same phase speed $\omega/k \approx \omega/(l+\frac{1}{2})$ are composed of constructively interfering SH or P-SV body waves having the same ray parameter $p = (l+\frac{1}{2})/\omega$.

The Fréchet kernels K_α, K_β, K'_ρ and $[K_s]^+_-$, $[K_{670}]^+_-$, $[K_b]^+_-$, $[K_c]^+_-$ for

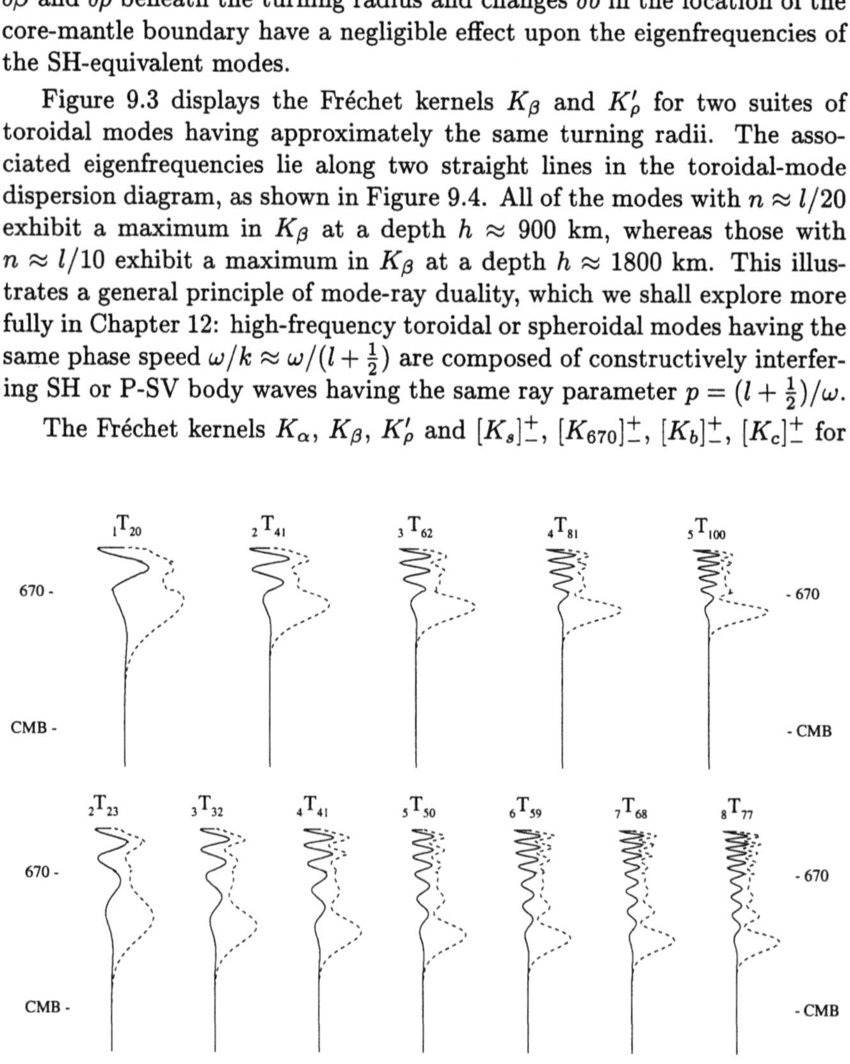

Figure 9.3. Fréchet kernels K_β (*dashed line*) and K'_ρ (*solid line*) for a number of toroidal modes ${}_nT_l$ with $n \approx l/20$ (*top*) and $n \approx l/10$ (*bottom*). The associated eigenfrequencies ${}_n\omega_l^{\mathrm{T}}$ lie along the two heavy straight lines in Figure 9.4.

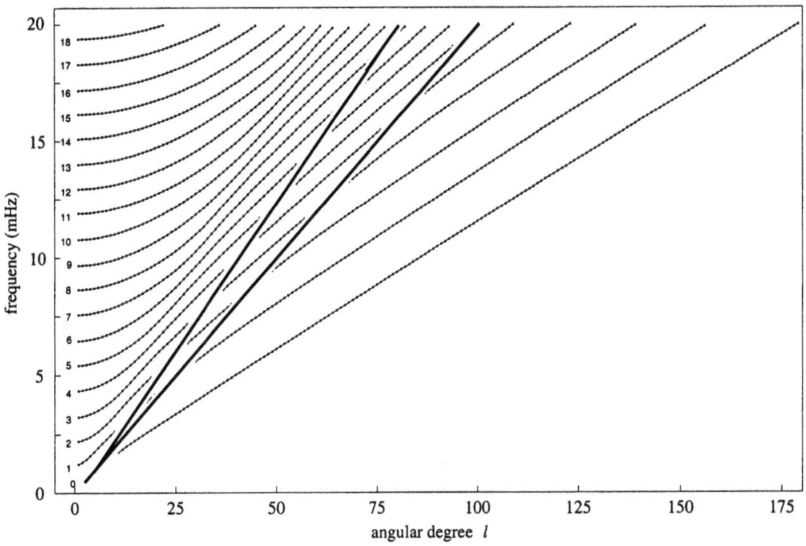

Figure 9.4. Toroidal-mode dispersion diagram showing the eigenfrequencies of the modes displayed in Figure 9.3. The shallower and steeper heavy solid lines identify the top ($n \approx l/20$) and bottom ($n \approx l/10$) suites of modes, respectively.

the fundamental spheroidal modes ${}_0S_2$ through ${}_0S_9$ and the first nine radial modes ${}_nS_0$ are depicted in Figures 9.5 and 9.6. The eigenfrequency of the ${}_0S_2$ "football" mode, the gravest observed free oscillation of the Earth, depends upon α, β and ρ throughout the entire Earth; however, its most pronounced sensitivity is to the shear-wave speed at the base of the lower mantle. The peak in the K_β sensitivity kernel shifts up to approximately 1500 km depth as we move up the fundamental-mode branch to mode ${}_0S_9$, which is equivalent to a very-long-period (634 s) Rayleigh wave. The fundamental radial mode ${}_0S_0$ is sensitive to both wave speeds as well as the density throughout the whole Earth, just as the "football" mode is; however, the higher-overtone radial modes such as ${}_9S_0$, which are equivalent to radially propagating PKIKP waves, are primarily sensitive to the compressional-wave speed α, as expected. The "wavelength" of the kernel oscillations is roughly twice that of a monochromatic PKIKP wave. An increase in the radius of the core-mantle boundary decreases the (fast) mantle ray path length and increases the (slow) core ray path length; this leads to an increase in the PKIKP transit time, and thus to a decrease in the eigenfrequencies of these and other PKIKP-equivalent modes.

Figure 9.7 compares the volumetric Fréchet kernels K_α, K_β, K'_ρ for a

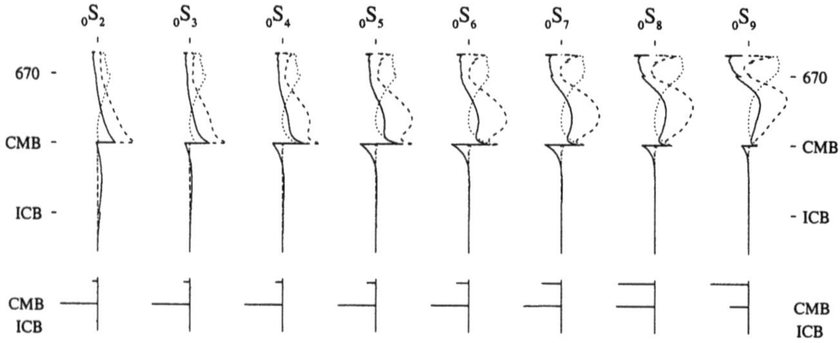

Figure 9.5. Fréchet kernels K_α (*dotted line*), K_β (*dashed line*) and K'_ρ (*solid line*) for the fundamental spheroidal modes ${}_0S_2, \ldots, {}_0S_9$ (see Figure 8.18 for displacements U, V and tractions R, S of the football mode). Vertical axis is depth beneath the surface of the Earth; the locations of the 670 km discontinuity, the core-mantle boundary (CMB) and the inner-core boundary (ICB) are indicated. Lower diagrams show the sensitivity to perturbations in the location of the seafloor and these three boundaries. Left-pointing and right-pointing "limbs" are associated with negative and positive kernels $[K_s]_-^+$, $[K_{670}]_-^+$, $[K_b]_-^+$, $[K_c]_-^+$ (*top to bottom*). Each graph is scaled independently, so that the maximum values of the kernels are the same.

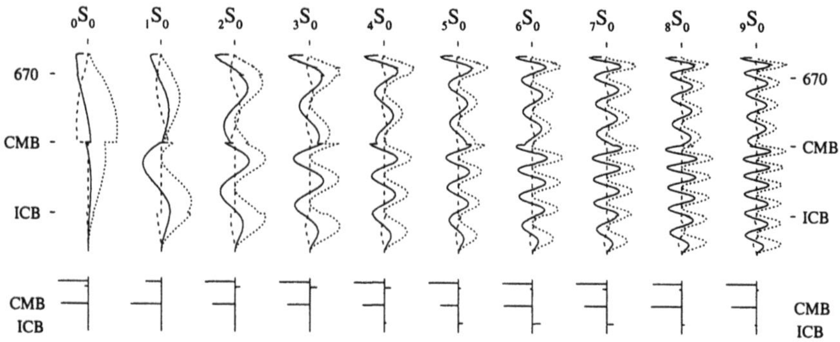

Figure 9.6. Same as Figure 9.5 for the fundamental and higher-overtone radial modes ${}_0S_0, \ldots, {}_9S_0$ (see Figure 8.16 for displacements U and tractions R). Dotted, dashed, and solid lines denote K_α, K_β and K'_ρ, respectively.

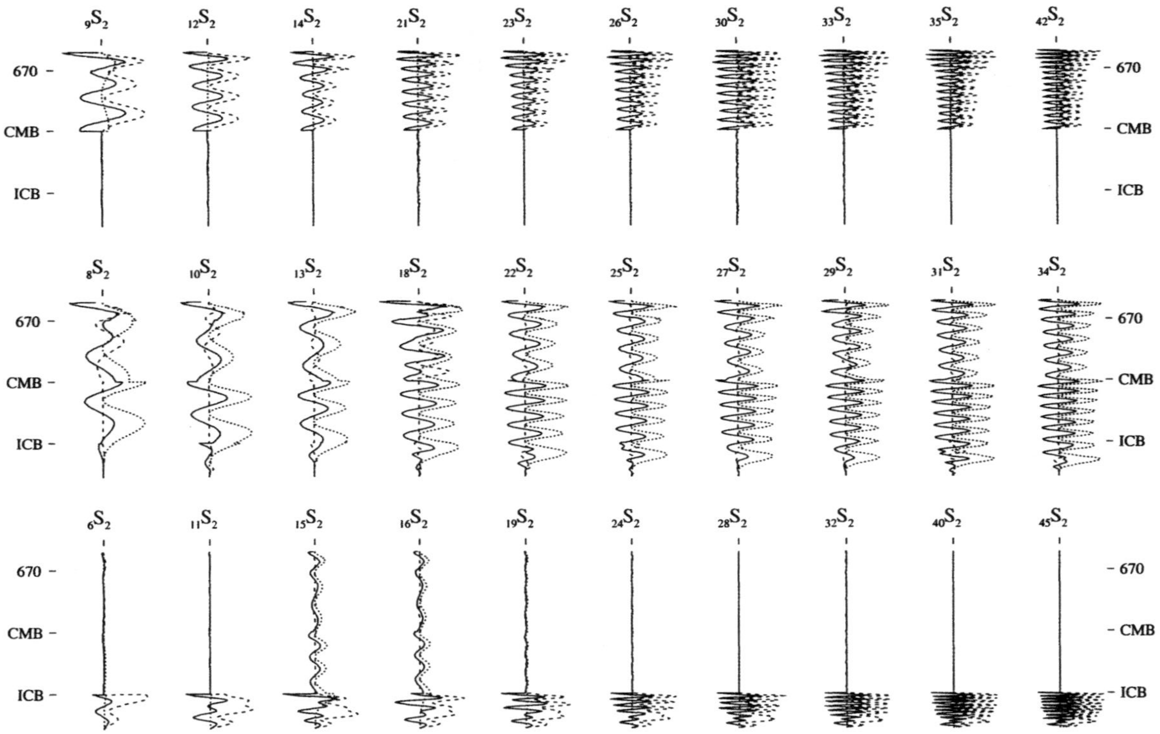

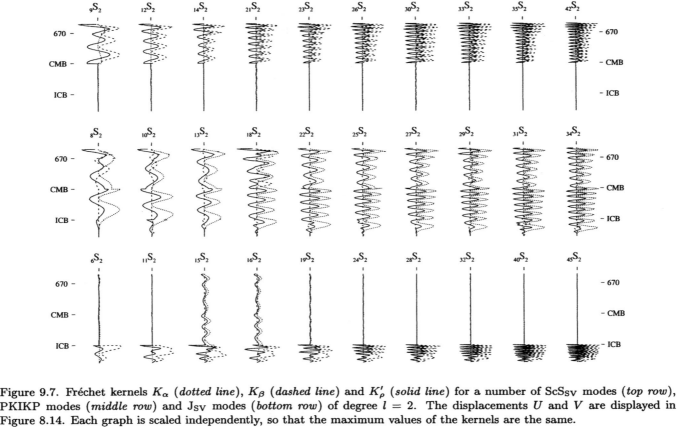

Figure 9.7. Fréchet kernels K_α (*dotted line*), K_β (*dashed line*) and K'_ρ (*solid line*) for a number of ScS_{SV} modes (*top row*), PKIKP modes (*middle row*) and J_{SV} modes (*bottom row*) of degree $l = 2$. The displacements U and V are displayed in Figure 8.14. Each graph is scaled independently, so that the maximum values of the kernels are the same.

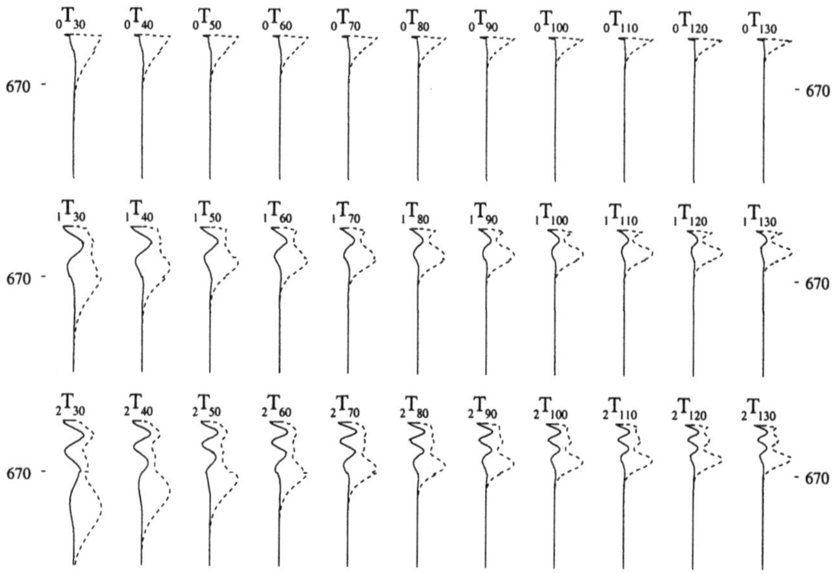

Figure 9.8. Variation of the Fréchet kernels K_β (*dashed line*) and K'_ρ (*solid line*) along the fundamental (*top row*), first-overtone (*middle row*) and second-overtone (*bottom row*) branches of the Love-wave-equivalent modes. Vertical axis extends from the free surface to 1500 km depth; the location of the 670 km discontinuity is indicated. Displacements W are shown in Figure 8.7. Each graph is scaled independently, so that the maximum values of K_β are all the same; in fact, the fundamental-mode kernels are about three times as large as the overtone kernels.

number of degree-two spheroidal modes, sorted according to type. The $\mathrm{ScS_{SV}}$-equivalent modes are the spheroidal analogue of the $\mathrm{ScS_{SH}}$ toroidal modes in Figure 9.1; they are predominantly sensitive to the shear-wave speed β in the mantle. The PKIKP-equivalent modes have properties similar to those of the radial modes in Figure 9.6; the characteristic variation in the "wavelength" of the compressional-wave sensitivity kernels K_α upon going from the lower mantle ($\alpha = 13.7$ km/s) to the fluid outer core ($\alpha = 8.1$ km/s) is evident, particularly for the higher-frequency modes such as ${}_{31}\mathrm{S}_2$ and ${}_{34}\mathrm{S}_2$. Finally, the $\mathrm{J_{SV}}$-equivalent modes are predominantly sensitive to the shear-wave speed β in the solid inner core, as expected. The density kernels for all three types of modes are large but oscillatory, with roughly equal positive and negative values, in accordance with the constraint $\int_0^a \rho K'_\rho\, dr = 0$, so that the effect of a long-wavelength perturbation in density $\delta\rho$ is slight.

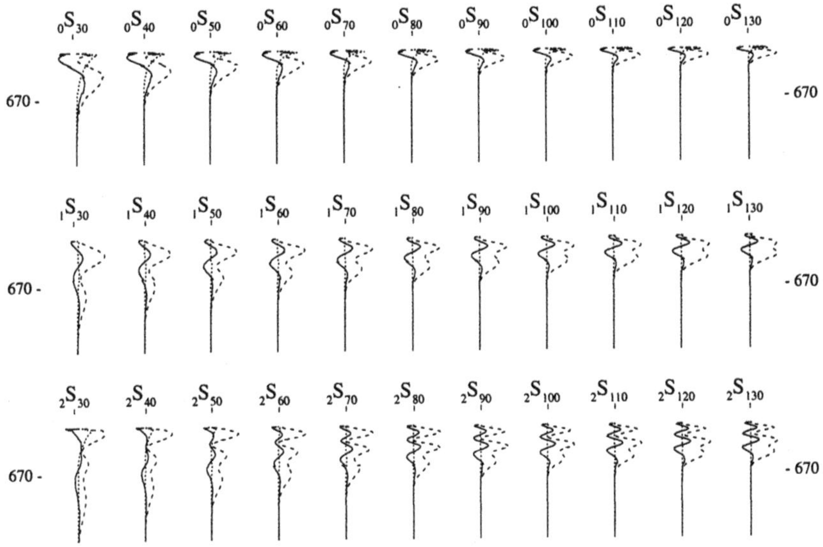

Figure 9.9. Variation of the Fréchet kernels K_α (*dotted line*), K_β (*dashed line*) and K'_ρ (*solid line*) along the fundamental (*top row*), first-overtone (*middle row*) and second-overtone (*bottom row*) branches of the Rayleigh-wave-equivalent modes. Vertical axis extends from the free surface to 1500 km depth; the location of the 670 km discontinuity is indicated. Displacements U and V are shown in Figure 8.11. Each graph is scaled independently, so that the maximum values of K_β are all the same; in fact, the fundamental-mode kernels are about three times as large as the overtone kernels.

The variation of the isotropic Fréchet kernels K_α, K_β and K'_ρ along the fundamental and first two overtone surface-wave dispersion branches is illustrated in Figures 9.8 and 9.9. It is evident that both the Love and Rayleigh modes are predominantly sensitive to variations in the shear-wave speed β in the upper mantle. The fundamental Rayleigh modes ${}_0\mathrm{S}_l$ exhibit a weak dependence upon the compressional-wave speed α; however, its influence upon the higher modes ${}_1\mathrm{S}_l$ and ${}_2\mathrm{S}_l$ is virtually negligible. The shear-wave kernel K_β for the fundamental Love modes ${}_0\mathrm{T}_l$ attains a maximum at a depth of approximately $0.1-0.2\,\lambda$, where $\lambda = 2\pi a/k$ is the wavelength of the equivalent travelling waves. The corresponding maximum sensitivity for the fundamental Rayleigh modes is significantly deeper, approximately 0.3–0.4 λ. The higher modes ${}_1\mathrm{T}_l$, ${}_2\mathrm{T}_l$ and ${}_1\mathrm{S}_l$, ${}_2\mathrm{S}_l$ are much more sensitive to deep-seated variations $\delta\beta$ than the fundamental modes are. Thus, for example, mode ${}_0\mathrm{S}_{100}$, with a period of 97 seconds, has very limited sensi-

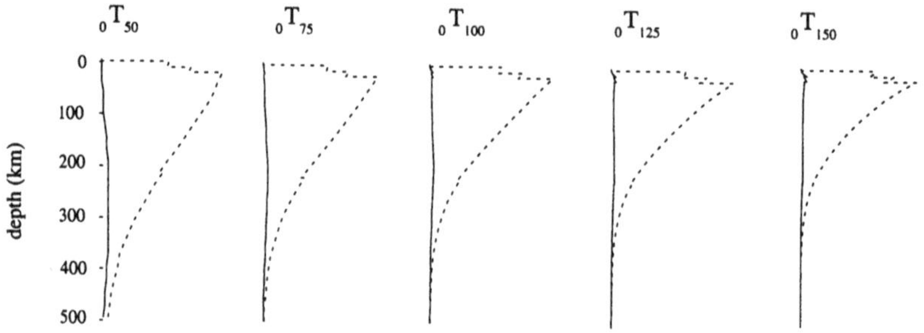

Figure 9.10. Variation of the transversely isotropic Fréchet kernels K_{β_v} (*solid line*) and K_{β_h} (*dashed line*) along the fundamental toroidal-mode branch. Each graph is scaled independently, so that the maximum values of K_{β_h} are all the same.

tivity beneath 250 km, whereas the first and second overtone modes $_1S_{68}$ and $_2S_{56}$ of approximately the same period "feel" down well beneath the 670 km discontinuity. We relate the perturbation δc in the phase speed of a

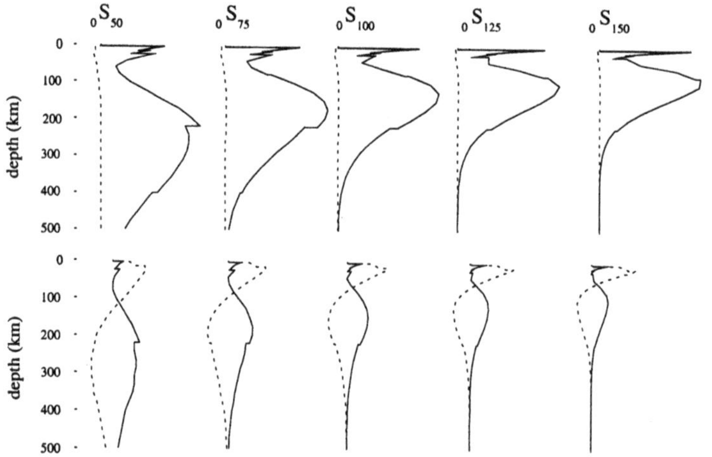

Figure 9.11. Variation of the transversely isotropic Fréchet kernels along the fundamental spheroidal-mode branch. Top row shows shear-wave speed kernels K_{β_v} (*solid line*) and K_{β_h} (*dashed line*). Bottom row shows compressional-wave kernels K_{α_v} (*solid line*) and K_{α_h} (*dashed line*). Each graph is scaled independently, so that the maximum values of K_{β_v} are all the same.

monochromatic surface wave to the eigenfrequency perturbation $\delta\omega$ of the equivalent free oscillation in Section 11.8.

Figures 9.10 and 9.11 show the radial variation of the Fréchet kernels K_{β_v}, K_{β_h} and K_{α_v}, K_{α_h} along the ${}_0\mathrm{T}_l$ and ${}_0\mathrm{S}_l$ branches of the transversely isotropic version of PREM. The fundamental toroidal eigenfrequencies are almost completely insensitive to perturbations in the vertically propagating shear-wave speed β_v; the fundamental spheroidal eigenfrequencies are, on the other hand, primarily sensitive to β_v and essentially independent of the horizontally propagating shear-wave speed β_h. This nearly complete "decoupling" of β_h and β_v is what enables a transversely isotropic upper-mantle model to fit otherwise irreconcilable Love and Rayleigh fundamental-mode eigenfrequency measurements. The ${}_0\mathrm{S}_l$ modes also exhibit a weak sensitivity to upper-mantle perturbations in the two compressional-wave speeds α_v and α_h.

9.7 Anelasticity and Attenuation

As we have seen in Chapter 6, isotropic anelasticity can be accounted for by replacing the quantities $\delta\kappa$ and $\delta\mu$ by complex frequency-dependent perturbations. Changing the notation slightly, we shall henceforth denote the incompressibility and rigidity of an unperturbed SNREI Earth model by κ_0 and μ_0, respectively. The subscript zero indicates that these are deemed to be an appropriate description of the elastic properties of the Earth at a reference or fiducial frequency ω_0; the reference frequency of the PREM model is one Hz, i.e., $\omega_0 = 2\pi$ rad/s. We consider an infinitesimal anelastic perturbation of the form

$$\kappa_0 \to \kappa_0 + \delta\kappa(\omega) + i\kappa_0 Q_\kappa^{-1}, \tag{9.48}$$

$$\mu_0 \to \mu_0 + \delta\mu(\omega) + i\mu_0 Q_\mu^{-1}, \tag{9.49}$$

where Q_κ and Q_μ are the bulk and shear quality factors, which are assumed to be frequency independent. The real incompressibility and rigidity perturbations "seen" by a normal mode whose positive unperturbed eigenfrequency is ω are

$$\delta\kappa(\omega) = \delta\kappa_0 + \tfrac{2}{\pi}\kappa_0 Q_\kappa^{-1}\ln(\omega/\omega_0), \tag{9.50}$$

$$\delta\mu(\omega) = \delta\mu_0 + \tfrac{2}{\pi}\mu_0 Q_\mu^{-1}\ln(\omega/\omega_0), \tag{9.51}$$

where the first term is the perturbation at the reference frequency ω_0, and the second term accounts for the anelastic dispersion.

The complex perturbation in the positive eigenfrequency of an oscillation will now be written in the form

$$\omega \to \omega + \delta\omega_0 + \delta\omega_{\rm d} + i\gamma, \tag{9.52}$$

where $\delta\omega_0$ represents the effect of the perturbation at the reference frequency, and $\delta\omega_{\rm d}$ represents the additional effect of the logarithmic dispersion. The quantity γ is the decay rate, which is related to the quality factor Q of a mode by

$$\gamma = \tfrac{1}{2}\omega Q^{-1}. \tag{9.53}$$

To find the three perturbations $\delta\omega_0$, $\delta\omega_{\rm d}$ and γ we substitute (9.48)–(9.53) into equation (9.12) and isolate the relevant terms accordingly. The inverse quality factor obtained by equating the imaginary parts is

$$Q^{-1} = 2\omega^{-1} \int_0^a [(\kappa_0 K_\kappa)Q_\kappa^{-1} + (\mu_0 K_\mu)Q_\mu^{-1}]\, dr. \tag{9.54}$$

If our objective is simply to calculate the effect of the infinitesimal anelasticity upon the SNREI eigenfrequency ω, we set $\delta\kappa_0$ and $\delta\mu_0$ equal to zero, in which case $\delta\omega_0 = 0$ and

$$\delta\omega_{\rm d} = \tfrac{1}{\pi}\omega Q^{-1} \ln(\omega/\omega_0). \tag{9.55}$$

More generally, we may regard $\delta\kappa_0$ and $\delta\mu_0$, together with $\delta\rho$ and δd, as specified perturbations of the reference SNREI Earth model. This results in an additional perturbation of the real eigenfrequency, given by equation (9.12) with $\delta\omega$, $\delta\kappa$ and $\delta\mu$ replaced by $\delta\omega_0$, $\delta\kappa_0$ and $\delta\mu_0$. The dispersive correction (9.55) is negative for any free oscillation whose unperturbed eigenfrequency is less than the fiducial frequency, $\omega < \omega_0$, and positive for any oscillation with $\omega > \omega_0$. This is physically reasonable since such a lower-frequency or higher-frequency mode "sees" an Earth that is more or less compliant than the reference SNREI Earth model.

A transversely isotropic elastic Earth model with elastic parameters C_0, A_0, L_0, N_0 and F_0 is rendered isotropically anelastic by replacing its "equivalent" incompressibility $\kappa_0 = \frac{1}{9}(C_0 + 4A_0 - 4N_0 + 4F_0)$ and rigidity $\mu_0 = \frac{1}{15}(C_0 + A_0 + 6L_0 + 5N_0 - 2F_0)$ by complex frequency-dependent parameters, as in equations (9.48)–(9.51). The five "purely" anisotropic parameters $C_0' = C_0 - \kappa_0 - \frac{4}{3}\mu_0$, $A_0' = A_0 - \kappa_0 - \frac{4}{3}\mu_0$, $L_0' = L_0 - \mu_0$, $N_0' = N_0 - \mu_0$ and $F_0' = F_0 - \kappa_0 + \frac{2}{3}\mu_0$ are regarded as real and frequency independent. The inverse quality factor Q^{-1} of a normal mode is still given by equation (9.54); the only difference is that the displacement eigenfunctions U, V and W are those of the transversely isotropic reference model. Any perturbations in the structure of the Earth at the reference frequency

ω_0 give rise to an additional real eigenfrequency perturbation given by equation (9.30), with $\delta\omega$, δC, δA, δL, δN and δF replaced by $\delta\omega_0$, δC_0, δA_0, δL_0, δN_0 and δF_0.

In summary, a spherically symmetric, non-rotating, anelastic, isotropic (SNRAI) Earth model is completely characterized by five positive functions of radius—κ_0, μ_0, ρ, Q_κ and Q_μ—and a transversely isotropic model is characterized by eight functions of radius—C_0, A_0, L_0, N_0, F_0, ρ, Q_κ and Q_μ—where the subscript zero signifies that these are appropriate at a reference or fiducial frequency ω_0. Equation (9.54) provides the theoretical basis for using the observed decay rates γ or quality factors Q of the Earth's free oscillations to invert for the radial variation of the intrinsic attenuation within the Earth. This inverse problem for the reciprocal bulk and shear quality factors Q_κ^{-1} and Q_μ^{-1} is notable for its linearity. The anelastic Fréchet kernels have a straightforward physical interpretation: $2\omega^{-1}r^{-2}(\kappa_0 K_\kappa)$ and $2\omega^{-1}r^{-2}(\mu_0 K_\mu)$ are the fractional compressional and shear energy densities of each mode (the factors of r^{-2} stem from our convention that dr rather than r^2dr is the differential element in any Fréchet kernel relation). In the special case that the intrinsic quality factors are only weakly dependent upon radius, equation (9.54) reduces to

$$Q^{-1} \approx f_\kappa Q_\kappa^{-1} + f_\mu Q_\mu^{-1}, \tag{9.56}$$

where $f_\kappa = 2\omega^{-1}\int_0^a(\kappa_0 K_\kappa)\,dr$ and $f_\mu = 2\omega^{-1}\int_0^a(\mu_0 K_\mu)\,dr$ are the net fractional energies. It is evident that bulk attenuation is only likely to have a discernible effect upon those modes which have a significant fraction of compressional energy, i.e., the radial and other PKIKP-equivalent spheroidal modes. The toroidal modes are devoid of compression, so that their damping is dependent only upon Q_μ; if the shear attenuation is nearly independent of radius, then every toroidal mode will have nearly the same quality factor: $Q \approx Q_\mu$.

Two strategies are available in inverting the observed eigenfrequencies ω of the Earth for the best-fitting spherically symmetric elastic structure κ_0, μ_0 and ρ or, more generally, C_0, A_0, L_0, N_0, F_0 and ρ. Either one can calculate the unperturbed eigenfrequencies of the perfectly elastic, non-dispersive reference Earth model, as described above, and "correct" the residuals $\omega_{\mathrm{meas}} - \omega_{\mathrm{calc}}$ for dispersion by subtracting $\delta\omega_d$ prior to inversion, or one can account for the effects of dispersion directly while solving the radial scalar equations for ω and U, V, W. The latter is the preferred procedure in most applications; both `MINEOS` and `OBANI` incorporate physical dispersion, so that every normal mode "sees" the real isotropic elastic parameters $\kappa_0[1 + \frac{2}{\pi}Q_\kappa^{-1}\ln(\omega/\omega_0)]$ and $\mu_0[1 + \frac{2}{\pi}Q_\mu^{-1}\ln(\omega/\omega_0)]$ at its unperturbed frequency of oscillation ω. Note that no matter how it is calculated, the correction for dispersion depends upon the attenuation model Q_κ, Q_μ.

It is advisable to choose the reference frequency ω_0 near the center of the normal-mode frequency band, in order to minimize the effect of uncertainties in these parameters. The one Hz reference frequency of PREM is not ideal in this regard, since it results in dispersive corrections $\delta\omega_{\mathrm{d}}$ for many of the fundamental ${}_0\mathrm{S}_l$ and ${}_0\mathrm{T}_l$ modes that are more than ten times their observational errors (Widmer 1991).

In some applications it may be more convenient to parameterize the anelasticity in terms of the compressional and shear wave speeds rather than the incompressibility and rigidity. In this case the reference Earth model is characterized by isotropic speeds $\alpha_0 = [(\kappa_0 + \frac{4}{3}\mu_0)/\rho]^{1/2}$ and $\beta_0 = (\mu_0/\rho)^{1/2}$, and we consider complex perturbations of the form

$$\alpha_0 \to \alpha_0 + \delta\alpha(\omega) + \tfrac{1}{2} i\alpha_0 Q_\alpha^{-1}, \tag{9.57}$$

$$\beta_0 \to \beta_0 + \delta\beta(\omega) + \tfrac{1}{2} i\beta_0 Q_\beta^{-1}. \tag{9.58}$$

The P-wave and S-wave quality factors Q_α and Q_β are related to the bulk and shear quality factors Q_κ and Q_μ by

$$Q_\alpha^{-1} = (1 - \tfrac{4}{3}\beta_0^2/\alpha_0^2) Q_\kappa^{-1} + \tfrac{4}{3}(\beta_0^2/\alpha_0^2) Q_\mu^{-1}, \tag{9.59}$$

$$Q_\beta^{-1} = Q_\mu^{-1}. \tag{9.60}$$

The real perturbations in wave speed at the reference frequency ω_0 are given by the analogue of (9.50)–(9.51):

$$\delta\alpha(\omega) = \delta\alpha_0 + \tfrac{1}{\pi}\alpha_0 Q_\alpha^{-1} \ln(\omega/\omega_0), \tag{9.61}$$

$$\delta\beta(\omega) = \delta\beta_0 + \tfrac{1}{\pi}\beta_0 Q_\beta^{-1} \ln(\omega/\omega_0). \tag{9.62}$$

The perturbation $\delta\omega_0$ is given in terms of the wave-speed perturbations at the reference frequency ω_0 by equation (9.20), with $\delta\omega$, $\delta\alpha$ and $\delta\beta$ replaced by $\delta\omega_0$, $\delta\alpha_0$ and $\delta\beta_0$, whereas the inverse quality factor Q^{-1} of a free oscillation can be rewritten in terms of Q_α^{-1} and Q_β^{-1} in the form

$$Q^{-1} = \omega^{-1} \int_0^a [(\alpha_0 K_\alpha) Q_\alpha^{-1} + (\beta_0 K_\beta) Q_\beta^{-1}]\, dr, \tag{9.63}$$

where $K_\alpha = 2\rho\alpha_0 K_\kappa$ and $K_\beta = 2\rho\beta_0(K_\mu - \frac{4}{3}K_\kappa)$. Equation (9.63) can be used to invert for the radial variation of the P-wave and S-wave quality factors Q_α and Q_β in the same way that (9.54) is used to invert for the bulk and shear quality factors Q_κ and Q_μ.

9.8 Q Kernels, Measurements and Models

The anelastic Fréchet kernels $\mu_0 K_\mu$ for a representative collection of toroidal modes are depicted in Figure 9.12. The top row shows the first ten modes ${}_n\mathrm{T}_2$ of degree two, and the bottom row shows a number of $\mathrm{ScS_{SH}}$-to-SH transitional modes having nearly the same eigenfrequency. The sensitivity to variations in Q_μ^{-1} exhibits a maximum in the vicinity of the turning point of the equivalent SH waves; the SH-equivalent modes in the bottom row have turning points in the lower mantle, between the core-mantle boundary and 670 km depth. The kernels $\kappa_0 K_\kappa$ and $\mu_0 K_\mu$ of a number of spheroidal oscillations of degree $l = 2$ are plotted in Figure 9.13. As expected, the $\mathrm{ScS_{SV}}$-equivalent modes are predominantly sensitive to Q_μ^{-1} in the mantle, the PKIKP modes are sensitive to both Q_κ^{-1} and Q_μ^{-1} throughout the whole Earth, and the $\mathrm{J_{SV}}$ modes are sensitive to Q_μ^{-1} within the solid inner core. Finally, in Figures 9.14 and 9.15 we show the variations in $\kappa_0 K_\kappa$ and $\mu_0 K_\mu$ along the fundamental and two gravest overtone branches ${}_0\mathrm{T}_l$, ${}_1\mathrm{T}_l$, ${}_2\mathrm{T}_l$ and ${}_0\mathrm{S}_l$, ${}_1\mathrm{S}_l$, ${}_2\mathrm{S}_l$. The damping of these surface-wave-equivalent modes

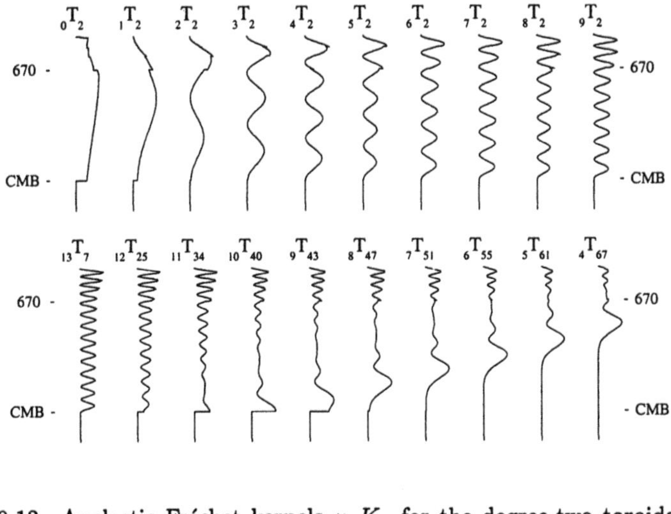

Figure 9.12. Anelastic Fréchet kernels $\mu_0 K_\mu$ for the degree-two toroidal modes ${}_0\mathrm{T}_2, \ldots, {}_9\mathrm{T}_2$ (*top row*) and a suite of $\mathrm{ScS_{SH}}$-to-SH modes having approximately the same eigenfrequency $\omega/2\pi \approx 14$ mHz (*bottom row*). Vertical axis is depth beneath the surface of the Earth; the locations of the 670 km discontinuity and the core-mantle boundary (CMB) are indicated. The constant factor $2\omega^{-1}$ in equation (9.54) is irrelevant, since each mode is scaled independently; the maximum values of $\mu_0 K_\mu$ are all the same. See Figures 9.1 and 9.2 for elastic Fréchet kernels K_β and K'_ρ.

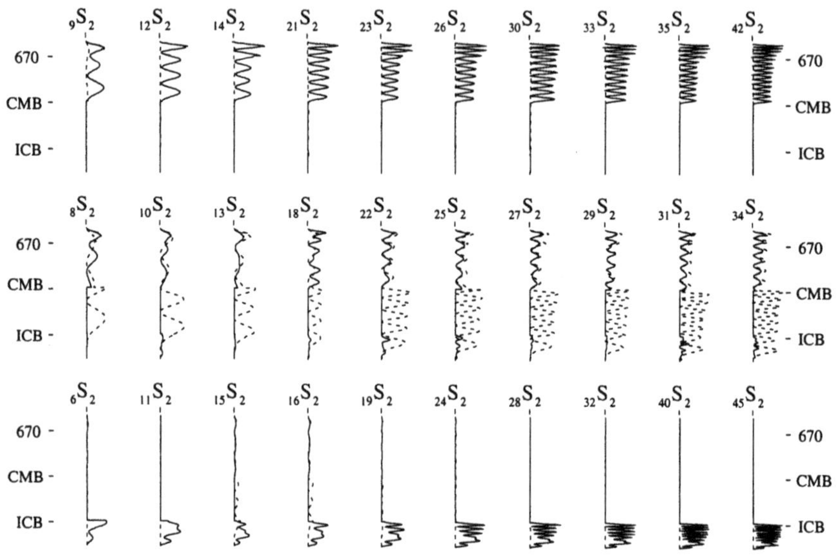

Figure 9.13. Anelastic Fréchet kernels $\kappa_0 K_\kappa$ (*dashed line*) and $\mu_0 K_\mu$ (*solid line*) for a number of ScS_{SV} modes (*top row*), PKIKP modes (*middle row*) and J_{SV} modes (*bottom row*) of degree $l = 2$. Vertical axis is depth beneath the surface of the Earth; the locations of the 670 km discontinuity, the core-mantle boundary (CMB) and the inner-core boundary (ICB) are indicated. See Figure 9.7 for elastic Fréchet kernels K_α, K_β and K'_ρ.

is strongly dominated by shear attenuation Q_μ^{-1} in the upper mantle. Low-frequency Love and Rayleigh modes "feel" the Earth's anelasticity more deeply than high-frequency modes do, and overtones "feel" more deeply than the fundamental mode. As already noted, the Q_κ^{-1} and Q_μ^{-1} sensitivity kernels of all these modes may alternatively be viewed as plots of the compressional and shear energy density—bear in mind that they are radially rather than volumetrically weighted.

The decay rate γ of a normal mode of oscillation can be measured by fitting a straight line to the observed log-amplitude variation after an earthquake, using a straightforward time-lapse method. Alternatively, it is possible to measure the frequency ω, quality factor Q, and complex excitation amplitude of an isolated mode simultaneously by fitting a Lorentzian to an observed resonance peak in the frequency domain. Regardless of which method is used, the compilation of a high-quality, unbiased Q data set is difficult for two reasons. The first is a purely statistical effect: least-squares estimates of Q are inherently more uncertain than those of ω, because deter-

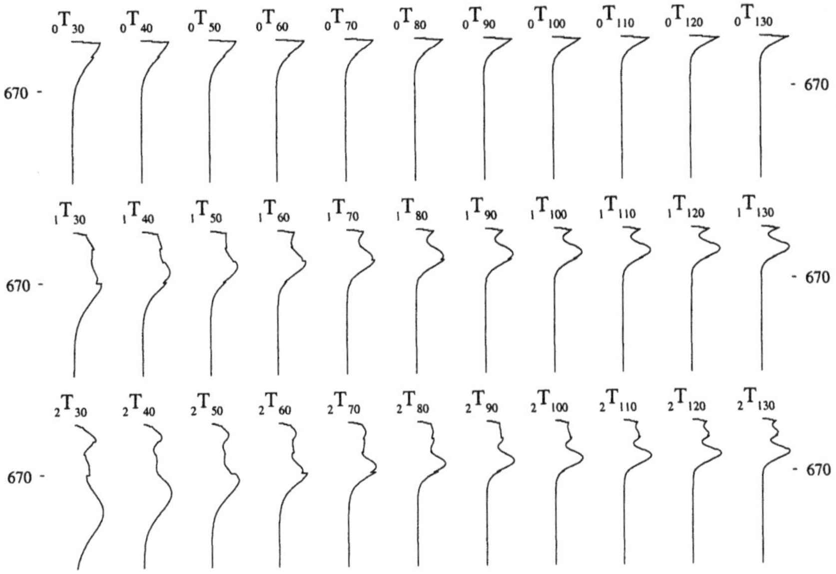

Figure 9.14. Variation of the anelastic Fréchet kernels $\mu_0 K_\mu$ along the fundamental (*top row*) first-overtone (*middle row*) and second-overtone (*bottom row*) branches of the Love-wave-equivalent modes. Vertical axis extends from the free surface to 1500 km depth; the location of the 670 km discontinuity is indicated. See Figure 9.8 for elastic Fréchet kernels K_β and K'_ρ.

mining the latter is tantamount to counting zero crossings, whereas determining the former is tantamount to counting e^{-1} decay times. The relative standard deviation of the decay rate $\gamma = \frac{1}{2}\omega Q^{-1}$ is $2Q$ times larger than the relative standard deviation in frequency; since typical values of Q lie between 10^2 and 10^3, attenuation measurements will always be 200 to 2000 times less precise than eigenfrequency measurements (Dahlen 1982). Second, and more importantly, attenuation data are contaminated by splitting due to the Earth's rotation, ellipticity and lateral heterogeneity; the superposition of $2l + 1$ closely spaced singlets within each multiplet ${}_n\mathrm{S}_l$ or ${}_n\mathrm{T}_l$ gives rise to beating in the time domain and peak shifts and distortion in the frequency domain. Unedited single-station attenuation measurements fail to exhibit any coherent geographical pattern (Smith & Masters 1989a) and measurements made on spectral stacks are biased toward low Q by 20–40% (Widmer 1991) as a consequence of this splitting. The attenuation of the ${}_0\mathrm{S}_l$ and ${}_0\mathrm{T}_l$ modes can alternatively be studied by measuring the rate of decay with *distance* of the equivalent fundamental Rayleigh and

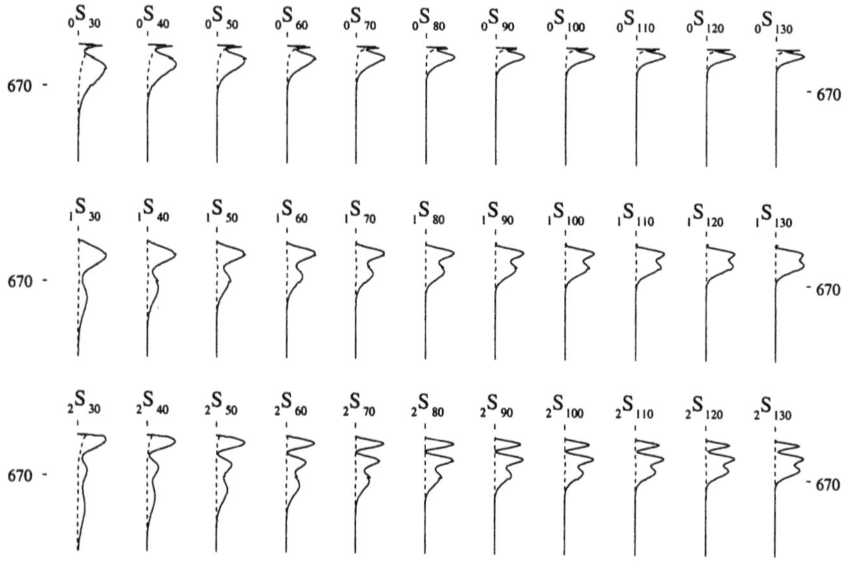

Figure 9.15. Variation of the anelastic Fréchet kernels $\kappa_0 K_\kappa$ (*dashed line*) and $\mu_0 K_\mu$ (*solid line*) along the fundamental (*top row*) first-overtone (*middle row*) and second-overtone (*bottom row*) branches of the Rayleigh-wave-equivalent modes. Vertical axis extends from the free surface to 1500 km depth; the location of the 670 km discontinuity is indicated. See Figure 9.9 for elastic Fréchet kernels K_α, K_β and K'_ρ.

Love waves (see Section 11.4). Such travelling-wave measurements are also corrupted by lateral heterogeneity, which gives rise to geometrical amplitude variations due to the focusing and defocusing of the ray tubes on the surface of the Earth, as we discuss in Chapter 16.

The radial modes ${}_n\mathrm{S}_0$ constitute a notable exception; since their eigenfrequencies are non-degenerate, they do not exhibit any splitting. Furthermore, the oscillations are in phase and of equal amplitude at all locations on the surface of the Earth; the data from a number of stations can therefore be stacked without any knowledge of the hypocentral position or source mechanism. Figure 9.16 shows a number of recordings of the ${}_4\mathrm{S}_0$ mode following the June 9, 1994 deep-focus earthquake in Bolivia; Durek & Ekström (1995) used this data to measure the quality factor of this as well as five other radial overtone modes. The fundamental radial mode ${}_0\mathrm{S}_0$ was not well excited by the Bolivian event; however, its attenuation and that of the first radial overtone ${}_1\mathrm{S}_0$ were measured by Riedesel, Agnew, Berger & Gilbert

(1980) following the August 17, 1977 earthquake in Sumbawa, Indonesia. The slow decay of the fundamental radial mode ($Q \approx 5700$) required the analysis of three months of continuous seismic data! The reason for this extraordinarily high Q is the low shear energy content ($f_\mu = 0.03$) of ${}_0S_0$. The radial overtones have fractional shear energies that are five to six times larger than this (see Table 8.2), so their quality factors are correspondingly lower; for example, $Q \approx 2000$ for ${}_1S_0$ whereas $Q \approx 1200$ for ${}_4S_0$. The long duration of the radial oscillations enables their eigenfrequencies to be measured quite precisely; indeed, the frequencies 0.814664 ± 0.000004 mHz and 1.63151 ± 0.00003 mHz of the ${}_0S_0$ and ${}_1S_0$ modes are among the best determined of all geophysical constants.

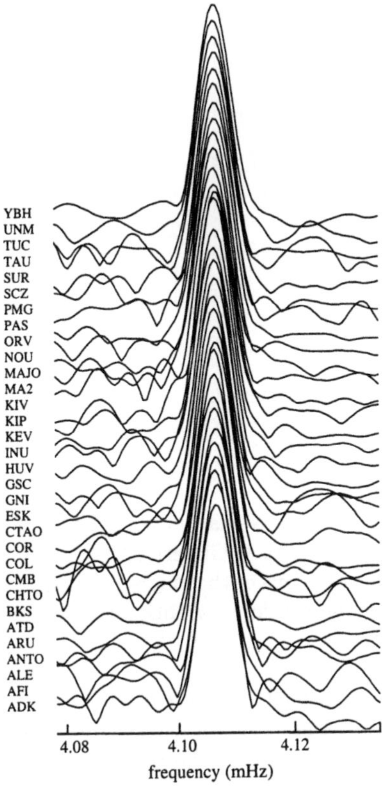

Figure 9.16. Single-station spectra in the vicinity of the ${}_4S_0$ resonance peak following the deep-focus Bolivia earthquake of June 9, 1994 (Durek & Ekström 1995). The quantity plotted in each case is the absolute Fourier transform of a Hann-tapered recording of duration 50–80 hours; station symbols are shown on the left. The fluctuations on either side of the main peak are due to noise. (Courtesy of G. Ekström.)

The results of two recent spherical attenuation studies are displayed in Figure 9.17. Widmer, Masters & Gilbert (1991) obtained model QM1 by inverting 146 radial, spheroidal and toroidal quality factors Q obtained principally by fitting single-station resonance peaks in the frequency band between 0.3 and 6 milliHertz. Durek & Ekström (1996) used essentially the same non-radial overtone data—mostly high-Q PKIKP-equivalent oscillations with visibly split spectra—but replaced the fundamental-mode peak fits with travelling-wave measurements "corrected" for focusing-defocusing effects to obtain model QL6. The discrepancies between the two models are indicative of the current level of imprecision in our knowledge of Q_κ and Q_μ within the Earth, due to observational uncertainties, the limited resolution of the data set and—perhaps most

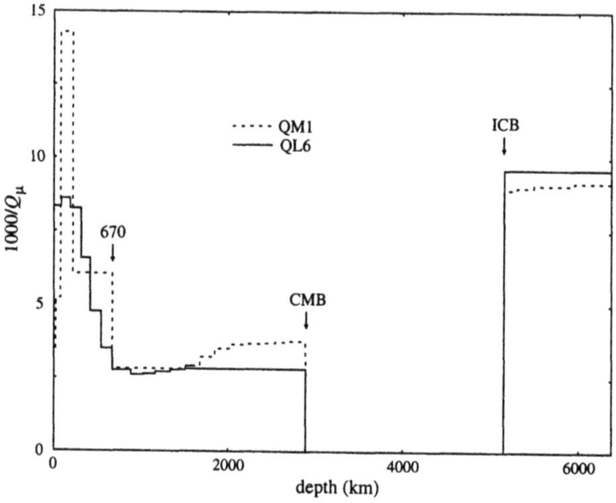

Figure 9.17. Variation with depth of $1000\,Q_\mu^{-1}$: (*dashed line*) model QM1 (Widmer, Masters & Gilbert 1991); (*solid line*) model QL6 (Durek & Ekström 1996). The locations of the inner-core boundary (ICB), the core-mantle boundary (CMB) and the 670 km discontinuity are indicated.

importantly—disagreement regarding the relative reliability of standing-wave and travelling-wave measurements (Durek & Ekström 1997). The average value of the shear quality factor in the mantle is well established: $\overline{Q}_\mu = 250 \pm 5$ in model QM1 whereas $\overline{Q}_\mu = 253 \pm 5$ in model QL6. As a result of the physical dispersion that accompanies this shear attenuation, the mantle is 0.6–0.8 percent more rigid at the frequency of a teleseismic shear wave, say 50–100 mHz, than in the middle of the normal-mode band, at a frequency of about 4 mHz. This provides a natural explanation for the observed 4–5 second baseline discrepancy between observed shear-wave travel times and the times predicted by Earth models that fit normal-mode eigenfrequency measurements, as pointed out by Akopyan, Zharkov & Lyubimov (1975; 1976), Randall (1976) and Liu, Anderson & Kanamori (1976).

Shear attenuation due to microscopic mechanisms such as dislocation motion and grain boundary sliding is much more significant than bulk attenuation in all crystalline materials, so that $Q_\kappa \gg Q_\mu$ everywhere within the Earth (Minster 1980; Karato & Spetzler 1990). A slight amount of bulk attenuation must, however, be present to account for the observed decay rates of the radial and other PKIKP-equivalent modes. Model QM1 has $Q_\kappa = 2920$ in the upper mantle and $Q_\kappa = 12,000$ in the fluid outer core; model QL6 has slightly higher bulk attenuation ($Q_\kappa = 943$) confined to the upper mantle. Durek & Ekström (1995) have used their improved radial-mode Q measurements after the Bolivian earthquake to refine the distribution of bulk attenuation within the mantle; their optimal model has $Q_\kappa = 213$ within the asthenosphere (80–220 km depth) and $Q_\kappa = 27,700$ elsewhere. A small amount of bulk dissipation is an expected consequence of the polycrystalline nature of the mantle—the application of a macroscopic isotropic stress to such an aggregate results in microscopic shear stresses within the crystals due to the local mismatch of the elastic moduli of neighboring grains. Heinz, Jeanloz & O'Connell (1982) show that $Q_\kappa/Q_\mu \approx 50-100$ for typical mantle mineral assemblages, as a result of this mechanism. Any asthenospheric bulk dissipation significantly in excess of this—if it can be confirmed—may be indicative of partial melting. Very little reliable attenuation data is available for modes that sample the solid inner core, so that neither Q_κ nor Q_μ are well determined in that region.

Careful comparisons between normal-mode and body-wave attenuation measurements provide subtle but credible evidence for a slight decrease of Q_μ with frequency within the seismic band. The theoretical quality factor

$$Q_{\mathrm{ScS}} = \frac{\displaystyle\int_b^a \beta^{-1}\, dr}{\displaystyle\int_b^a \beta^{-1} Q_\beta^{-1}\, dr} \tag{9.64}$$

of a radially propagating $\mathrm{ScS_{SH}}$ or $\mathrm{ScS_{SV}}$ wave is $Q_{\mathrm{ScS}} = 236 \pm 5$ for model QM1 and $Q_{\mathrm{ScS}} = 233 \pm 5$ for model QL6; these well-constrained values pertain to a hypothetical ultralow-frequency (4 mHz) wave. Direct measurements of the attenuation of intermediate-frequency (40 mHz) multiply reflected ScS waves yield values in the range $Q_{\mathrm{ScS}} = 170 \pm 34$ (Sipkin & Jordan 1980) to $Q_{\mathrm{ScS}} = 207 \pm 25$ (Revenaugh & Jordan 1991). The discrepancy between these two determinations and the large uncertainties reflect the difficulty in determining a reliable global average in the face of strong regional variations and uneven geographical coverage; taken at face value, however, these results imply a 10–20% decrease in mantle shear attenuation over one decade in frequency. The routine detection of much shorter-period ScS arrivals suggests that Q_μ must increase again in the vicinity of one Hz. Sipkin & Jordan (1979) show that this high-frequency increase is consistent with the presence of an absorption band-edge situated at $\tau_{\mathrm{m}} \approx 0.5$ seconds.

There have been a number of attempts to study the behavior of Q_μ below the seismic frequency band by exploiting the Chandler wobble, which samples the Earth's anelasticity at a period of 435 days. These studies generally assume a power-law dependence of the form $Q_\mu \sim \omega^\alpha$ as in equations (6.113)–(6.114), so that a single parameter—the exponent α—is being estimated. The first such study was conducted by Jeffreys (1958a; 1958b);

he used the observed damping rate of the Chandler wobble together with the qualitative observation that teleseismic S waveforms are not strongly dispersed to conclude that "$\alpha = 0.17$ nearly". Other investigators have sought to refine this estimate, using the period as well as the damping rate of the Chandler wobble, together with more recent seismic information. Anderson & Minster (1979) conclude that "the data can accommodate α from about 0.2 to 0.4", whereas Smith & Dahlen (1981) determined after a lengthy analysis that $0.04 \leq \alpha \leq 0.19$. Dickman (1988; 1993), on the other hand, finds that the evidence is compatible with frequency-independent anelasticity, i.e., $\alpha \approx 0$. The discrepancy between the last two conclusions is due to the difficulty of calculating the theoretical Chandler wobble period of an elastic Earth model that is demonstrably not in hydrostatic equilibrium. Gravimetric or geodetic observations of the solid-Earth tides and associated nutations and monthly and fortnightly variations in the length of day can be used to constrain the behavior of Q_μ within the four-decade gap between the seismic band and the Chandler wobble. In the past, tidal anelasticity investigations were hampered by the inability to account for poorly known atmospheric and oceanic effects (Zschau 1978; 1986; Wahr & Bergen 1986; Lambeck 1988). This situation has, however, recently been ameliorated by direct measurement of open-ocean tides using satellite radar altimetry (Ray, Eanes & Chao 1996; Baker, Curtis & Dodson 1996).

★9.9 Exact Anelasticity

In the previous section we used Rayleigh's principle to calculate the complex perturbation $\delta\omega_d + i\gamma$ in an eigenfrequency ω of a SNREI Earth model κ_0, μ_0, ρ or a transversely isotropic Earth model C_0, A_0, L_0, N_0, F_0, ρ due to an isotropic anelastic perturbation

$$\kappa_0 \rightarrow \kappa_0[1 + iQ_\kappa^{-1} + \tfrac{2}{\pi}Q_\kappa^{-1}\ln(\omega/\omega_0)], \tag{9.65}$$

$$\mu_0 \rightarrow \mu_0[1 + iQ_\mu^{-1} + \tfrac{2}{\pi}Q_\mu^{-1}\ln(\omega/\omega_0)]. \tag{9.66}$$

The commonly employed result (9.53)–(9.55) is valid only to first order in the anelasticity; moreover, the corresponding perturbations to the radial eigenfunctions U, V and W have been ignored. In fact, both the complex eigenfrequencies and associated eigenfunctions of a spherically symmetric anelastic Earth model can be calculated essentially exactly with a little more effort. A straightforward method of doing this is to solve the governing first-order ordinary differential equations and associated boundary conditions directly with κ and μ replaced by (9.65)–(9.66). Yuen & Peltier (1982) adopted this approach in an analytical calculation of the complex eigenfrequencies of a homogeneous anelastic solid Earth, and Buland, Yuen, Konstanty & Widmer (1985) used it subsequently in a numerical investigation of the effect of anelasticity upon the toroidal modes of a more realistic radially stratified Earth model. Finding all of the complex roots of a determinant equation such as (8.184) or (8.187) numerically would be extremely difficult; for this reason the direct method has never been applied to the spheroidal modes. An alternative approach is to consider the anelastic coupling between the oscillations of an elastic Earth model using a simple variant of the Rayleigh-Ritz algorithm discussed in Chapter 7. We briefly review this numerically feasible method here, following the description given by Tromp & Dahlen (1990b).

The spheroidal and toroidal eigenfunctions of the anelastically perturbed Earth model can be written in the form

$$\mathbf{s}^{\mathrm{S}} = \mathcal{U}\mathbf{P}_{lm} + \mathcal{V}\mathbf{B}_{lm}, \qquad \mathbf{s}^{\mathrm{T}} = \mathcal{W}\mathbf{C}_{lm}, \tag{9.67}$$

where $\mathbf{P}_{lm}$, $\mathbf{B}_{lm}$ and $\mathbf{C}_{lm}$ are the real vector spherical harmonics (8.36). The basic idea is to expand the complex radial eigenfunctions $\mathcal{U}$, $\mathcal{V}$ and $\mathcal{W}$ in terms of the real eigenfunctions U, V and W of the unperturbed model, as in equation (7.1):

$$\mathcal{U} = \sum_n q_n^{\mathrm{S}} U_n, \qquad \mathcal{V} = \sum_n q_n^{\mathrm{S}} V_n, \qquad \mathcal{W} = \sum_n q_n^{\mathrm{T}} W_n, \tag{9.68}$$

where we seek to determine the complex coefficients q_n^{S} and q_n^{T}. There is no spheroidal-toroidal coupling, nor is there any coupling between modes of the same type but differing angular degree l or order m, as a consequence of the spherical symmetry of the perturbation (9.65)–(9.66). The only coupling is that between toroidal modes ${}_n\mathrm{T}_l - {}_{n'}\mathrm{T}_l$ or spheroidal modes ${}_n\mathrm{S}_l - {}_{n'}\mathrm{S}_l$ of the same type and degree l. The summation index in equation (9.68) is thus just the overtone number n. For simplicity, we have adopted the shorthand notation U_n, V_n and W_n for ${}_nU_l$, ${}_nV_l$ and ${}_nW_l$; likewise, we shall abbreviate the unperturbed eigenfrequencies ${}_n\omega_l$ by ω_n.

We insert the representation (9.67)–(9.68) into the linearized equation of motion (8.13), dot the result with an arbitrary elastic eigenfunction $U_n\mathbf{P}_{lm} + V_n\mathbf{B}_{lm}$ or $W_n\mathbf{C}_{lm}$, and integrate by parts over the Earth model $\oplus$, making use of the boundary conditions (8.14)–(8.16). This leads to a pair of non-linear $\infty \times \infty$ algebraic eigenvalue equations of the form

$$[\Omega^2 + i\mathsf{A} + \tfrac{2}{\pi}\ln(\omega/\omega_0)\mathsf{A}]\mathsf{q} = (\omega + i\gamma)^2\mathsf{q}, \tag{9.69}$$

where q is the column vector of unknown expansion coefficients q_n^{S} or q_n^{T}, Ω is the diagonal matrix of unperturbed eigenfrequencies ω_n^{S} or ω_n^{T}, and A

is the anelastic potential-energy perturbation matrix, with elements

$$A^{\mathrm{S}}_{nn'} = \int_0^a \Big\{\kappa_0 Q_\kappa^{-1}(r\dot{U}_n + 2U_n - kV_n)(r\dot{U}_{n'} + 2U_{n'} - kV_{n'}) \\ + \mu_0 Q_\mu^{-1}\big[\tfrac{1}{3}(2r\dot{U}_n - 2U_n + kV_n)(2r\dot{U}_{n'} - 2U_{n'} + kV_{n'}) \\ + (r\dot{V}_n - V_n + kU_n)(r\dot{V}_{n'} - V_{n'} + kU_{n'}) \\ + (k^2 - 2)V_n V_{n'}\big]\Big\}\, dr, \tag{9.70}$$

$$A^{\mathrm{T}}_{nn'} = \int_0^a \mu_0 Q_\mu^{-1}\big[(r\dot{W}_n - W_n)(r\dot{W}_{n'} - W_{n'}) \\ + (k^2 - 2)W_n W_{n'}\big]\, dr. \tag{9.71}$$

The complex eigenfrequencies $\omega + i\gamma$ and associated radial eigenfunctions $\mathcal{U}$, $\mathcal{V}$ and $\mathcal{W}$ of a spherical anelastic Earth can be obtained by iteratively solving a truncated version of equation (9.69). The dimension of the matrix A is just the number of overtones considered in the analysis, so that the method is computationally straightforward.

The symmetries $A^{\mathrm{S}}_{n'n} = A^{\mathrm{S}}_{nn'}$ and $A^{\mathrm{T}}_{n'n} = A^{\mathrm{T}}_{nn'}$ guarantee that the complex spheroidal and toroidal eigenfunctions are biorthogonal in the sense

$$\int_b^a \rho(\mathcal{U}_n\mathcal{U}_{n'} + \mathcal{V}_n\mathcal{V}_{n'})\, r^2 dr - \frac{2}{\pi}\frac{\ln(\omega_n/\omega_{n'})}{\omega_n^2 - \omega_{n'}^2} \\ \times \int_0^a \Big\{\kappa_0 Q_\kappa^{-1}(r\dot{\mathcal{U}}_n + 2\mathcal{U}_n - k\mathcal{V}_n)(r\dot{\mathcal{U}}_{n'} + 2\mathcal{U}_{n'} - k\mathcal{V}_{n'}) \\ + \mu_0 Q_\mu^{-1}\big[\tfrac{1}{3}(2r\dot{\mathcal{U}}_n - 2\mathcal{U}_n + k\mathcal{V}_n)(2r\dot{\mathcal{U}}_{n'} - 2\mathcal{U}_{n'} + k\mathcal{V}_{n'}) \\ + (r\dot{\mathcal{V}}_n - \mathcal{V}_n + k\mathcal{U}_n)(r\dot{\mathcal{V}}_{n'} - \mathcal{V}_{n'} + k\mathcal{U}_{n'}) \\ + (k^2 - 2)\mathcal{V}_n\mathcal{V}_{n'}\big]\Big\}\, dr = 0 \quad \text{if } \omega_n \neq \omega_{n'}, \tag{9.72}$$

$$\int_0^a \rho\mathcal{W}_n\mathcal{W}_{n'}\, r^2 dr - \frac{2}{\pi}\frac{\ln(\omega_n/\omega_{n'})}{\omega_n^2 - \omega_{n'}^2} \\ \times \int_0^a \mu_0 Q_\mu^{-1}\big[(r\dot{\mathcal{W}}_n - \mathcal{W}_n)(r\dot{\mathcal{W}}_{n'} - \mathcal{W}_{n'}) \\ + (k^2 - 2)\mathcal{W}_n\mathcal{W}_{n'}\big]\, dr = 0 \quad \text{if } \omega_n \neq \omega_{n'}. \tag{9.73}$$

The associated normalization conditions in the limiting case $\omega_{n'} \to \omega_n$ are

$$\int_0^a \rho(\mathcal{U}_n^2 + \mathcal{V}_n^2)\, r^2 dr - \frac{1}{\pi}\omega_n^{-2} \int_0^a \Big\{\kappa_0 Q_\kappa^{-1}(r\dot{\mathcal{U}}_n + 2\mathcal{U}_n - k\mathcal{V}_n)^2 \\ + \mu_0 Q_\mu^{-1}\big[\tfrac{1}{3}(2r\dot{\mathcal{U}}_n - 2\mathcal{U}_n + k\mathcal{V}_n)^2 + (r\dot{\mathcal{V}}_n - \mathcal{V}_n + k\mathcal{U}_n)^2 \\ + (k^2 - 2)\mathcal{V}_n^2\big]\Big\}\, dr = 1, \tag{9.74}$$

$$\int_0^a \rho\mathcal{W}_n^2\, r^2 dr - \frac{1}{\pi}\omega_n^{-2}\int_0^a \mu_0 Q_\mu^{-1}\big[(r\dot{\mathcal{W}}_n - \mathcal{W}_n)^2 + (k^2-2)\mathcal{W}_n^2\big]\, dr = 1. \tag{9.75}$$

As with the unperturbed eigenfunctions, we have abbreviated ${}_n\mathcal{U}_l$, ${}_n\mathcal{V}_l$ and ${}_n\mathcal{W}_l$ by $\mathcal{U}_n$, $\mathcal{V}_n$ and $\mathcal{W}_n$. Equations (9.72)–(9.75) are a special case of the general biorthonormality relations (6.143)–(6.144) on a non-rotating anelastic Earth.

Tromp & Dahlen (1990b) show that the decay rates $\gamma = \frac{1}{2}\omega Q^{-1}$ obtained by solving the eigenvalue problem (9.69) are generally very well approximated by the first-order formula for Q^{-1} obtained using Rayleigh's principle. This justifies the continued usage of the approximation (9.54) in normal-mode attenuation inversion studies. The effects of spherical anelastic coupling are negligible for the toroidal modes ${}_n\mathrm{T}_l$ due to the regular spacing of the unperturbed eigenfrequencies. They are strongest for spheroidal modes ${}_n\mathrm{S}_l$ with nearly degenerate unperturbed eigenfrequencies, lying near the avoided crossings of the PKIKP-equivalent, $\mathrm{ScS_{SV}}$-equivalent and $\mathrm{J_{SV}}$-equivalent dispersion branches. For a few strongly coupled modes, the dispersive eigenfrequency shifts $\omega - \omega_n$ obtained by solving equation (9.69) can differ from the corresponding first-order shifts (9.55) by one to two μHz, which is of the same order as the precision of many eigenfrequency measurements. In addition, the complex eigenfunctions $\mathcal{U}_n$ and $\mathcal{V}_n$ of some strongly coupled modes can differ substantially from the corresponding unperturbed real eigenfunctions U_n and V_n. This could have a significant effect on the theoretical phase and amplitude of the associated oscillations following an earthquake. Fortunately, the modes with the largest anomalies are predominantly $\mathrm{J_{SV}}$-equivalent inner-core modes which are difficult to either excite or observe.

We conclude this chapter by calling attention to the necessity of using the *real* vector spherical harmonics $\mathbf{P}_{lm} = \hat{\mathbf{r}}\mathcal{Y}_{lm}$, $\mathbf{B}_{lm} = k^{-1}\boldsymbol{\nabla}_1\mathcal{Y}_{lm}$ and $\mathbf{C}_{lm} = -k^{-1}(\hat{\mathbf{r}}\times\boldsymbol{\nabla}_1\mathcal{Y}_{lm})$ in the representations (9.67). The vector biorthonormality relations (6.143)–(6.144) would *not* reduce to the scalar relations (9.72)–(9.75) if we sought instead to express $\mathbf{s}^{\mathrm{S}}$ and $\mathbf{s}^{\mathrm{T}}$ in terms of the complex vector harmonics $\hat{\mathbf{r}}Y_{lm}$, $k^{-1}\boldsymbol{\nabla}_1 Y_{lm}$ and $-k^{-1}(\hat{\mathbf{r}}\times\boldsymbol{\nabla}_1 Y_{lm})$. The only complexity of the anelastic eigenfunctions $\mathbf{s}^{\mathrm{S}}$ and $\mathbf{s}^{\mathrm{T}}$ must be that inherent in the radial scalars $\mathcal{U}$, $\mathcal{V}$ and $\mathcal{W}$; this is essential in normal-mode excitation calculations upon a non-rotating, spherical anelastic Earth, as we shall see in Section 10.2. The rotation of the Earth gives rise to geographical as well as radial complexity; in fact, correct to zeroth order in $\Omega = \|\boldsymbol{\Omega}\|$, the eigenfunctions of a rotating elastic or anelastic Earth are identical to those of the corresponding non-rotating Earth, with $\mathcal{Y}_{lm}$ replaced by Y_{lm} (see Section 14.2.1). We summarize the nature of the exact

Spherical Earth Model	Exact or Zeroth-Order Displacement Eigenfunction
Non-Rotating Elastic	$\mathbf{s} = U\hat{\mathbf{r}}\mathcal{Y}_{lm} + k^{-1}V\,\boldsymbol{\nabla}_1\mathcal{Y}_{lm} - k^{-1}W(\hat{\mathbf{r}} \times \boldsymbol{\nabla}_1\mathcal{Y}_{lm})$
Rotating Elastic	$\mathbf{s} = U\hat{\mathbf{r}}Y_{lm} + k^{-1}V\,\boldsymbol{\nabla}_1 Y_{lm} - k^{-1}W(\hat{\mathbf{r}} \times \boldsymbol{\nabla}_1 Y_{lm})$
Non-Rotating Anelastic	$\mathbf{s} = \mathcal{U}\hat{\mathbf{r}}\mathcal{Y}_{lm} + k^{-1}\mathcal{V}\,\boldsymbol{\nabla}_1\mathcal{Y}_{lm} - k^{-1}\mathcal{W}(\hat{\mathbf{r}} \times \boldsymbol{\nabla}_1\mathcal{Y}_{lm})$
Rotating Anelastic	$\mathbf{s} = \mathcal{U}\hat{\mathbf{r}}Y_{lm} + k^{-1}\mathcal{V}\,\boldsymbol{\nabla}_1 Y_{lm} - k^{-1}\mathcal{W}(\hat{\mathbf{r}} \times \boldsymbol{\nabla}_1 Y_{lm})$

Table 9.1. Displacement eigenfunctions of a spherically symmetric Earth model in the presence and absence of rotation and anelasticity. The scalars U, V, W and the spherical harmonics $\mathcal{Y}_{lm}$ are real, whereas $\mathcal{U}$, $\mathcal{V}$, $\mathcal{W}$ and Y_{lm} are complex. Toroidal eigenfunctions have $U = V = 0$ and $\mathcal{U} = \mathcal{V} = 0$, whereas spheroidal eigenfunctions have $W = 0$ and $\mathcal{W} = 0$. The eigenfunctions of a non-rotating spherical Earth model are exact, whereas those of a rotating spherical Earth model are only correct to zeroth order in the angular rate of rotation, $\Omega = \|\boldsymbol{\Omega}\|$.

and zeroth-order eigenfunctions in the presence and absence of rotation and anelasticity in Table 9.1.

Chapter 10

Synthetic Seismograms

Now that we know how to calculate the eigenfrequencies and associated eigenfunctions of a spherically symmetric, non-rotating Earth model, it is a straightforward matter to synthesize long-period seismograms and spectra by means of normal-mode summation. The response of an elastic or anelastic spherical Earth is a superposition of all of the non-trivial spheroidal and toroidal singlet eigenfunctions ${}_n\mathbf{s}^{\mathrm{S}}_{lm}$ and ${}_n\mathbf{s}^{\mathrm{T}}_{lm}$. In this chapter we develop an operator formalism which enables the sum over the spherical-harmonic order m to be performed explicitly in an ordinary geographic (as opposed to epicentral) coordinate system. We use a prime, as in $\mathbf{x}'$, and a subscript s, as in $\mathbf{x}_{\mathrm{s}}$, interchangeably to denote the source location, conducting the initial geometrical analysis in terms of the former. The Earth's anelasticity is fully accounted for in the final formulae for the Green tensor $\mathbf{G}(\mathbf{x}, \mathbf{x}'; t)$ and the displacement response $\mathbf{s}(\mathbf{x}, t)$ to a moment-tensor source $\mathbf{M}$ situated at $\mathbf{x}_{\mathrm{s}}$, since doing so does not significantly complicate the development.

10.1 Source-Receiver Geometry

Let $\hat{\mathbf{r}}$, $\hat{\boldsymbol{\theta}}$, $\hat{\boldsymbol{\phi}}$ be the triad of mutually perpendicular unit vectors in the direction of increasing radius r, colatitude θ, and longitude ϕ at the location of a receiver $\mathbf{x}$, and let $\hat{\mathbf{r}}'$, $\hat{\boldsymbol{\theta}}'$, $\hat{\boldsymbol{\phi}}'$ be the corresponding triad of vectors in the direction of increasing r', θ', ϕ' at the location of a seismic point source $\mathbf{x}'$. The response of a spherical Earth is most readily expressed in terms of the receiver and source radii r and r', and a pair of angular epicentral coordinates which we shall denote by Θ and Φ. The quantity Θ is the *angular epicentral distance* between the source and receiver, given by

$$\cos\Theta = \hat{\mathbf{r}} \cdot \hat{\mathbf{r}}' = \cos\theta\cos\theta' + \sin\theta\sin\theta'\cos(\phi - \phi'), \tag{10.1}$$

whereas Φ is the *azimuth* to the receiver, measured in a *counterclockwise* sense *from due south* at the source. For a fixed source location $\mathbf{x}'$, we may regard r, Θ, Φ as an epicentral system of spherical polar coordinates specifying the position of the receiver. In most quantitative seismological discussions, the epicentral distance is called Δ rather than Θ; in this book, we use Δ to denote the angular distance travelled by a *multi-orbit* surface or body wave (see Sections 11.3–11.5 and 12.5).

It is convenient to define two unit vectors tangent to the great circle that passes through the projections onto the unit sphere of the receiver and the source:

$$\hat{\mathbf{\Theta}} = \boldsymbol{\nabla}_1 \Theta, \qquad \hat{\mathbf{\Theta}}' = -\boldsymbol{\nabla}_1' \Theta, \tag{10.2}$$

where $\boldsymbol{\nabla}_1 = \hat{\boldsymbol{\theta}}\partial_\theta + \hat{\boldsymbol{\phi}}(\sin\theta)^{-1}\partial_\phi$ and $\boldsymbol{\nabla}_1' = \hat{\boldsymbol{\theta}}'\partial_{\theta'} + \hat{\boldsymbol{\phi}}'(\sin\theta')^{-1}\partial_{\phi'}$ are the surface gradients with respect to the receiver and source coordinates, respectively. Upon using equation (10.1) to evaluate (10.2) we obtain the explicit expressions

$$\begin{aligned}\hat{\mathbf{\Theta}} = (\sin\Theta)^{-1}\{\hat{\boldsymbol{\theta}}\,[\sin\theta\cos\theta' - \cos\theta\sin\theta'\cos(\phi-\phi')] \\ + \hat{\boldsymbol{\phi}}\,\sin\theta'\sin(\phi-\phi')\},\end{aligned} \tag{10.3}$$

$$\begin{aligned}\hat{\mathbf{\Theta}}' = (\sin\Theta)^{-1}\{\hat{\boldsymbol{\theta}}'\,[-\cos\theta\sin\theta' + \sin\theta\cos\theta'\cos(\phi-\phi')] \\ + \hat{\boldsymbol{\phi}}'\,\sin\theta\sin(\phi-\phi')\}.\end{aligned} \tag{10.4}$$

The unit vectors $\hat{\mathbf{r}}$ and $\hat{\mathbf{\Theta}}$ at the receiver are related to the corresponding vectors $\hat{\mathbf{r}}'$ and $\hat{\mathbf{\Theta}}'$ at the source by a rotation about the pole of the source-receiver great circle:

$$\hat{\mathbf{r}} = \hat{\mathbf{r}}'\cos\Theta + \hat{\mathbf{\Theta}}'\sin\Theta, \qquad \hat{\mathbf{\Theta}} = -\hat{\mathbf{r}}'\sin\Theta + \hat{\mathbf{\Theta}}'\cos\Theta. \tag{10.5}$$

To obtain a complete triad of mutually perpendicular epicentral unit vectors at the receiver and source, we define

$$\hat{\mathbf{\Phi}} = \hat{\mathbf{r}} \times \hat{\mathbf{\Theta}}, \qquad \hat{\mathbf{\Phi}}' = \hat{\mathbf{r}}' \times \hat{\mathbf{\Theta}}'. \tag{10.6}$$

Equations (10.5) ensure that $\hat{\mathbf{\Phi}} = \hat{\mathbf{\Phi}}'$; physically, this third vector is the unit normal to the source-receiver great-circle plane. The source vectors $\hat{\mathbf{\Theta}}'$ and $\hat{\mathbf{\Phi}}'$ are simply related to the azimuth Φ by

$$\hat{\mathbf{\Theta}}' = \hat{\boldsymbol{\theta}}'\cos\Phi + \hat{\boldsymbol{\phi}}'\sin\Phi, \qquad \hat{\mathbf{\Phi}}' = -\hat{\boldsymbol{\theta}}'\sin\Phi + \hat{\boldsymbol{\phi}}'\cos\Phi. \tag{10.7}$$

We could likewise define $\hat{\mathbf{\Theta}}$ and $\hat{\mathbf{\Phi}}$ in terms of the back-azimuth to the source measured at the receiver, but this will not be necessary in the development that follows. A sketch summarizing our coordinate conventions may be found in Figure 10.1.

The directions $\hat{\mathbf{r}}$, $\hat{\mathbf{\Theta}}$, $\hat{\mathbf{\Phi}}$ correspond to the *radial*, *longitudinal* and *transverse* components, respectively, of a seismograph situated at the point $\mathbf{x}$. This terminology for the receiver polarization is natural in studies of the free oscillations of the Earth, since it preserves the usual geometrical meaning of the the word "radial" in spherical polar coordinates. In everyday seismological parlance $\hat{\mathbf{\Theta}}$ is commonly referred to as the "radial" direction and $\hat{\mathbf{r}}$ is referred to as the "vertical" direction. We shall refrain from using these informal terms, to avoid any possibility of confusion.

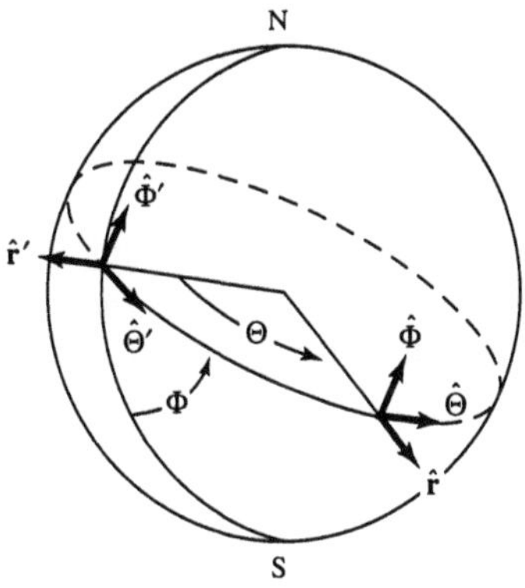

Figure 10.1. Schematic diagram illustrating the coordinate convention used here. The receiver is situated at $\mathbf{x} = (r, \theta, \phi)$ and the source is at $\mathbf{x}' = (r', \theta', \phi')$. For simplicity the two radii r and r' are shown equal in the diagram. Note that both $\hat{\mathbf{\Theta}}$ and $\hat{\mathbf{\Theta}}'$ point in the direction of wave propagation along the minor arc between the source and receiver.

Using the definitions (10.3)–(10.6) together with the geometrical identities (A.119)–(A.121) it is straightforward to show that

$$\boldsymbol{\nabla}_1\hat{\mathbf{r}} = \hat{\mathbf{\Theta}}\hat{\mathbf{\Theta}} + \hat{\mathbf{\Phi}}\hat{\mathbf{\Phi}}, \tag{10.8}$$

$$\boldsymbol{\nabla}_1\hat{\mathbf{\Theta}} = -\hat{\mathbf{\Theta}}\hat{\mathbf{r}} + \hat{\mathbf{\Phi}}\hat{\mathbf{\Phi}}\cot\Theta, \tag{10.9}$$

$$\boldsymbol{\nabla}_1\hat{\mathbf{\Phi}} = -\hat{\mathbf{\Phi}}\hat{\mathbf{r}} - \hat{\mathbf{\Phi}}\hat{\mathbf{\Theta}}\cot\Theta, \tag{10.10}$$

$$\boldsymbol{\nabla}_1'\hat{\mathbf{r}}' = \hat{\mathbf{\Theta}}'\hat{\mathbf{\Theta}}' + \hat{\mathbf{\Phi}}'\hat{\mathbf{\Phi}}', \tag{10.11}$$

$$\boldsymbol{\nabla}_1'\hat{\mathbf{\Theta}}' = -\hat{\mathbf{\Theta}}'\hat{\mathbf{r}}' - \hat{\mathbf{\Phi}}'\hat{\mathbf{\Phi}}'\cot\Theta, \tag{10.12}$$

$$\boldsymbol{\nabla}_1'\hat{\mathbf{\Phi}}' = -\hat{\mathbf{\Phi}}'\hat{\mathbf{r}}' + \hat{\mathbf{\Phi}}'\hat{\mathbf{\Theta}}'\cot\Theta, \tag{10.13}$$

$$\boldsymbol{\nabla}_1\hat{\mathbf{r}}' = \mathbf{0}, \tag{10.14}$$

$$\boldsymbol{\nabla}_1\hat{\mathbf{\Theta}}' = \hat{\mathbf{\Phi}}\hat{\mathbf{\Phi}}'(\sin\Theta)^{-1}, \tag{10.15}$$

$$\boldsymbol{\nabla}_1\hat{\mathbf{\Phi}}' = -\hat{\mathbf{\Phi}}\hat{\mathbf{\Theta}}'(\sin\Theta)^{-1}, \tag{10.16}$$

$$\boldsymbol{\nabla}_1'\hat{\mathbf{r}} = \mathbf{0}, \tag{10.17}$$

$$\boldsymbol{\nabla}_1'\hat{\boldsymbol{\Theta}} = -\hat{\boldsymbol{\Phi}}'\hat{\boldsymbol{\Phi}}(\sin\Theta)^{-1}, \tag{10.18}$$

$$\boldsymbol{\nabla}_1'\hat{\boldsymbol{\Phi}} = \hat{\boldsymbol{\Phi}}'\hat{\boldsymbol{\Theta}}(\sin\Theta)^{-1}. \tag{10.19}$$

Equations (10.15)–(10.16) account for the perturbation in the tangent vectors $\hat{\boldsymbol{\Theta}}', \hat{\boldsymbol{\Phi}}'$ at the source due to a change in the coordinates θ, ϕ of the receiver, whereas (10.18)–(10.19) account for the analogous effect upon the receiver vectors $\hat{\boldsymbol{\Theta}}$, $\hat{\boldsymbol{\Phi}}$ due to a change in the coordinates θ', ϕ' of the source.

We may write the surface gradient operator $\boldsymbol{\nabla}_1$ at the receiver in terms of the epicentral distance Θ and azimuth Φ in the form:

$$\boldsymbol{\nabla}_1 = (\boldsymbol{\nabla}_1\Theta)\partial_\Theta + (\boldsymbol{\nabla}_1\Phi)\partial_\Phi. \tag{10.20}$$

Upon making use of equations (10.7) and (10.15) we obtain the result

$$\boldsymbol{\nabla}_1\Phi = (\sin\Theta)^{-1}\hat{\boldsymbol{\Phi}}. \tag{10.21}$$

Substituting (10.2) and (10.21) into equation (10.20) we deduce that

$$\boldsymbol{\nabla}_1 = \hat{\boldsymbol{\Theta}}\partial_\Theta + (\sin\Theta)^{-1}\hat{\boldsymbol{\Phi}}\partial_\Phi. \tag{10.22}$$

Equation (10.22) could alternatively have been written down from first principles, inasmuch as r, Θ, Φ comprise an epicentral system of spherical polar coordinates.

10.2 Green Tensor

We saw in Section 6.2.3 that the time-domain Green tensor of a non-rotating, anelastic Earth model may be expressed in terms of its complex eigenfrequencies $\nu_k = \omega_k + i\gamma_k$ and eigenfunctions $\mathbf{s}_k$ in the form

$$\mathbf{G}(\mathbf{x}, \mathbf{x}'; t) = \mathrm{Re}\sum_k (i\nu_k)^{-1}\mathbf{s}_k(\mathbf{x})\mathbf{s}_k(\mathbf{x}')\exp(i\nu_k t). \tag{10.23}$$

In the case of a spherically symmetric Earth, the index k denotes the quadruplet $\{n, l, m, \mathrm{S} \text{ or } \mathrm{T}\}$, where n is the radial overtone number, l is the spherical-harmonic degree, and m is the order. The spheroidal and toroidal eigenfrequencies ${}_n\nu_l^{\mathrm{S}} = {}_n\omega_{lm}^{\mathrm{S}} + i_n\gamma_{lm}^{\mathrm{S}}$ and ${}_n\nu_l^{\mathrm{T}} = {}_n\omega_{lm}^{\mathrm{T}} + i_n\gamma_{lm}^{\mathrm{T}}$ are independent of the order m and thus $2l+1$ degenerate. The associated singlet eigenfunctions ${}_n\mathbf{s}_{lm}^{\mathrm{S}}$ and ${}_n\mathbf{s}_{lm}^{\mathrm{T}}$ may be written in terms of the real surface spherical harmonics (8.35) in the form

$$\mathbf{s}_k(\mathbf{x}) = {}_n\boldsymbol{\mathcal{D}}_l(r, \theta, \phi)\mathcal{Y}_{lm}(\theta, \phi), \tag{10.24}$$

where, for each fixed overtone number n and angular degree l,

$$\mathcal{D} = \mathcal{U}\hat{\mathbf{r}} + k^{-1}\mathcal{V}\boldsymbol{\nabla}_1 - k^{-1}\mathcal{W}(\hat{\mathbf{r}} \times \boldsymbol{\nabla}_1). \tag{10.25}$$

The exact *displacement operator* $\mathcal{D}$, which is also independent of m, is complex because the radial eigenfunctions $\mathcal{U}, \mathcal{V}$ and $\mathcal{W}$ of a spherical anelastic Earth are complex, as discussed in Section 9.9. For brevity, we employ a single symbol $\mathcal{D}$ in what follows, rather than defining separate spheroidal and toroidal operators $\mathcal{D}^{\mathrm{S}}$ and $\mathcal{D}^{\mathrm{T}}$. By definition, $\mathcal{U}$ and $\mathcal{V}$ are zero for a toroidal mode whereas $\mathcal{W}$ is zero for a spheroidal mode in the definition (10.25). The analogue of (10.24)–(10.25) at the source is

$$\mathbf{s}_k(\mathbf{x}') = {}_n\mathcal{D}_l{}'(r', \theta', \phi')\mathcal{Y}_{lm}(\theta', \phi') \tag{10.26}$$

where

$$\mathcal{D}' = \mathcal{U}'\hat{\mathbf{r}}' + k^{-1}\mathcal{V}'\boldsymbol{\nabla}_1' - k^{-1}\mathcal{W}'(\hat{\mathbf{r}}' \times \boldsymbol{\nabla}_1'). \tag{10.27}$$

The primes on the radial eigenfunctions $\mathcal{U}'$, $\mathcal{V}'$ and $\mathcal{W}'$ signify that they are evaluated at the radius r'.

Upon substituting the representations (10.24) and (10.26) into (10.23), we can rewrite the Green tensor on a spherical anelastic Earth as

$$\mathbf{G}(\mathbf{x}, \mathbf{x}'; t) = \mathrm{Re} \sum_{n=0}^{\infty} \sum_{l=0}^{\infty} (i_n\nu_l)^{-1} {}_n\mathcal{G}_l(\mathbf{x}, \mathbf{x}') \exp(i_n\nu_l t). \tag{10.28}$$

The time-independent tensor ${}_n\mathcal{G}_l(\mathbf{x}, \mathbf{x}')$ in equation (10.28) is given by

$$\mathcal{G} = \mathcal{D}\mathcal{D}' \sum_{m=-l}^{l} \mathcal{Y}_{lm}(\theta, \phi)\mathcal{Y}_{lm}(\theta', \phi'). \tag{10.29}$$

The summation over m in equation (10.29) can be performed using the addition theorem for surface spherical harmonics (B.74), with the result

$$\mathcal{G} = \left(\frac{2l+1}{4\pi}\right) \mathcal{D}\mathcal{D}' P_l(\cos\Theta). \tag{10.30}$$

It is evident from the form of (10.30) that

$$\mathcal{G}(\mathbf{x}, \mathbf{x}') = \mathcal{G}^{\mathrm{T}}(\mathbf{x}', \mathbf{x}), \tag{10.31}$$

so that equation (10.28) satisfies the source-receiver reciprocity relation $\mathbf{G}(\mathbf{x}, \mathbf{x}'; t) = \mathbf{G}^{\mathrm{T}}(\mathbf{x}', \mathbf{x}; t)$, as it must. The summation in equation (10.28) is over all of the non-trivial spheroidal and toroidal multiplets ${}_n\mathrm{S}_l$ and ${}_n\mathrm{T}_l$.

To obtain a form of the tensor $\mathcal{G}$ suitable for numerical evaluation, we must determine the effect of the operators $\mathcal{D}$ and $\mathcal{D}'$ upon the Legendre polynomial $P_l(\cos\Theta)$. Using the results (10.8)–(10.19) and the identities

$$\boldsymbol{\nabla}_1 P_{l0} = -P_{l1}\hat{\boldsymbol{\Theta}}, \qquad \boldsymbol{\nabla}_1 P_{l1} = \tfrac{1}{2}(k^2 P_{l0} - P_{l2})\hat{\boldsymbol{\Theta}}, \tag{10.32}$$

$$\boldsymbol{\nabla}_1' P_{l0} = P_{l1}\hat{\boldsymbol{\Theta}}', \qquad \boldsymbol{\nabla}_1' P_{l1} = -\tfrac{1}{2}(k^2 P_{l0} - P_{l2})\hat{\boldsymbol{\Theta}}', \tag{10.33}$$

where $P_{lm}(\cos\Theta)$ is the associated Legendre function of order m, it is straightforward to demonstrate that

$$\begin{aligned}\mathcal{G} = \left(\frac{2l+1}{4\pi}\right)\Big[&\mathcal{U}\mathcal{U}'\hat{\mathbf{r}}\hat{\mathbf{r}}' P_{l0} + k^{-1}(\mathcal{U}\mathcal{V}'\hat{\mathbf{r}}\hat{\boldsymbol{\Theta}}' - \mathcal{V}\mathcal{U}'\hat{\boldsymbol{\Theta}}\hat{\mathbf{r}}')P_{l1} \\ &+ \tfrac{1}{2}k^{-2}(\mathcal{V}\mathcal{V}'\hat{\boldsymbol{\Theta}}\hat{\boldsymbol{\Theta}}' + \mathcal{W}\mathcal{W}'\hat{\boldsymbol{\Phi}}\hat{\boldsymbol{\Phi}}')(k^2 P_{l0} - P_{l2}) \\ &\quad + k^{-2}(\mathcal{V}\mathcal{V}'\hat{\boldsymbol{\Phi}}\hat{\boldsymbol{\Phi}}' + \mathcal{W}\mathcal{W}'\hat{\boldsymbol{\Theta}}\hat{\boldsymbol{\Theta}}')(\sin\Theta)^{-1}P_{l1}\Big].\end{aligned} \tag{10.34}$$

All of the terms involving the product of a spheroidal and a toroidal eigenfunction vanish because $\mathcal{U}$ and $\mathcal{V}$ are zero whenever $\mathcal{W}'$ is non-zero, and vice versa. In verifying that equation (10.34) satisfies the symmetry (10.31), it must be remembered that the tangent vectors $\hat{\boldsymbol{\Theta}}$, $\hat{\boldsymbol{\Phi}}$ and $\hat{\boldsymbol{\Theta}}'$, $\hat{\boldsymbol{\Phi}}'$ change sign whenever $\mathbf{x}$ and $\mathbf{x}'$ are interchanged.

In closing, we would like to call attention to the necessity of using real surface spherical harmonics $\mathcal{Y}_{lm}$ in the representations (10.24) and (10.26). If we had naively employed the complex harmonics Y_{lm} instead, we would have been led to a non-rotationally-invariant sum $\sum_m Y_{lm}(\theta,\phi)Y_{lm}(\theta',\phi')$ in equation (10.29). The only complexity of the anelastic eigenfunctions $\mathbf{s}_k$ must be that present in the radial scalars $\mathcal{U}, \mathcal{V}$ and $\mathcal{W}$; this is dictated by the biorthonormality relations (6.143)–(6.144), as noted in Section 9.9. For further comments on the nature of the mode-sum Green tensor $\mathbf{G}$ in the presence and absence of anelasticity, see Section 6.3.3.

10.3 Moment-Tensor Response

The displacement response of a non-rotating, anelastic Earth to a step-function moment-tensor source $\mathbf{M}$ situated at $\mathbf{x}_s$ is given by

$$\mathbf{s}(\mathbf{x},t) = \mathrm{Re}\sum_k \nu_k^{-2}\mathbf{M}\!:\!\boldsymbol{\varepsilon}_k(\mathbf{x}_s)\,\mathbf{s}_k(\mathbf{x})\,[1-\exp(i\nu_k t)], \tag{10.35}$$

where $\boldsymbol{\varepsilon}_k = \frac{1}{2}[\boldsymbol{\nabla}\mathbf{s}_k + (\boldsymbol{\nabla}\mathbf{s}_k)^{\mathrm{T}}]$ is the strain. Upon inserting the representations (10.24) and (10.26) of the eigenfunctions $\mathbf{s}_k(\mathbf{x})$ and $\mathbf{s}_k(\mathbf{x}_s)$ on a spherically symmetric Earth, and applying the spherical-harmonic addition theorem as in Section 10.2, we can rewrite the result (10.35) in the form

$$\mathbf{s}(\mathbf{x},t) = \mathrm{Re}\sum_{n=0}^{\infty}\sum_{l=0}^{\infty} {}_n\nu_l^{-2}\, {}_n\boldsymbol{\mathcal{A}}_l(\mathbf{x})\,[1-\exp(i_n\nu_l t)]. \tag{10.36}$$

The complex amplitude of excitation ${}_n\boldsymbol{\mathcal{A}}_l(\mathbf{x})$ is given by an expression analogous to (10.30):

$$\boldsymbol{\mathcal{A}} = \left(\frac{2l+1}{4\pi}\right)\boldsymbol{\mathcal{D}}[(\mathbf{M}\!:\!\boldsymbol{\mathcal{E}}_{\mathrm{s}})\,P_l(\cos\Theta)]. \tag{10.37}$$

The tensor strain operator $\boldsymbol{\mathcal{E}}_{\mathrm{s}}$ is related to the vector operator $\boldsymbol{\mathcal{D}}_{\mathrm{s}}$ by

$$\boldsymbol{\mathcal{E}}_{\mathrm{s}} = \tfrac{1}{2}[\boldsymbol{\nabla}_{\mathrm{s}}\boldsymbol{\mathcal{D}}_{\mathrm{s}} + (\boldsymbol{\nabla}_{\mathrm{s}}\boldsymbol{\mathcal{D}}_{\mathrm{s}})^{\mathrm{T}}]. \tag{10.38}$$

The subscript s in (10.37)–(10.38) signifies evaluation at the source $\mathbf{x}_{\mathrm{s}}$, as usual. Upon making use of the results (10.8)–(10.19) and (10.32)–(10.33) we find that

$$\begin{aligned}\boldsymbol{\mathcal{E}}_{\mathrm{s}}P_l &= \dot{\mathcal{U}}_{\mathrm{s}}\hat{\mathbf{r}}_{\mathrm{s}}\hat{\mathbf{r}}_{\mathrm{s}}P_{l0} + r_{\mathrm{s}}^{-1}(\mathcal{U}_{\mathrm{s}} - \tfrac{1}{2}k\mathcal{V}_{\mathrm{s}})(\hat{\boldsymbol{\Theta}}_{\mathrm{s}}\hat{\boldsymbol{\Theta}}_{\mathrm{s}} + \hat{\boldsymbol{\Phi}}_{\mathrm{s}}\hat{\boldsymbol{\Phi}}_{\mathrm{s}})P_{l0}\\ &\quad + \tfrac{1}{2}k^{-1}(\dot{\mathcal{V}}_{\mathrm{s}} - r_{\mathrm{s}}^{-1}\mathcal{V}_{\mathrm{s}} + kr_{\mathrm{s}}^{-1}\mathcal{U}_{\mathrm{s}})(\hat{\mathbf{r}}_{\mathrm{s}}\hat{\boldsymbol{\Theta}}_{\mathrm{s}} + \hat{\boldsymbol{\Theta}}_{\mathrm{s}}\hat{\mathbf{r}}_{\mathrm{s}})P_{l1}\\ &\quad - \tfrac{1}{2}k^{-1}(\dot{\mathcal{W}}_{\mathrm{s}} - r_{\mathrm{s}}^{-1}\mathcal{W}_{\mathrm{s}})(\hat{\mathbf{r}}_{\mathrm{s}}\hat{\boldsymbol{\Phi}}_{\mathrm{s}} + \hat{\boldsymbol{\Phi}}_{\mathrm{s}}\hat{\mathbf{r}}_{\mathrm{s}})P_{l1}\\ &\quad + \tfrac{1}{2}k^{-1}r_{\mathrm{s}}^{-1}\mathcal{V}_{\mathrm{s}}(\hat{\boldsymbol{\Theta}}_{\mathrm{s}}\hat{\boldsymbol{\Theta}}_{\mathrm{s}} - \hat{\boldsymbol{\Phi}}_{\mathrm{s}}\hat{\boldsymbol{\Phi}}_{\mathrm{s}})P_{l2}\\ &\quad - \tfrac{1}{2}k^{-1}r_{\mathrm{s}}^{-1}\mathcal{W}_{\mathrm{s}}(\hat{\boldsymbol{\Theta}}_{\mathrm{s}}\hat{\boldsymbol{\Phi}}_{\mathrm{s}} + \hat{\boldsymbol{\Phi}}_{\mathrm{s}}\hat{\boldsymbol{\Theta}}_{\mathrm{s}})P_{l2}.\end{aligned} \tag{10.39}$$

For convenience, we denote the contraction of the moment tensor $\mathbf{M}$ with the tensor (10.39) by

$$\mathcal{A}(\Theta,\Phi) = (\mathbf{M}\!:\!\boldsymbol{\mathcal{E}}_{\mathrm{s}})P_l(\cos\Theta). \tag{10.40}$$

Upon expanding the moment tensor in the form

$$\begin{aligned}\mathbf{M} &= M_{rr}\hat{\mathbf{r}}_{\mathrm{s}}\hat{\mathbf{r}}_{\mathrm{s}} + M_{\theta\theta}\hat{\boldsymbol{\theta}}_{\mathrm{s}}\hat{\boldsymbol{\theta}}_{\mathrm{s}} + M_{\phi\phi}\hat{\boldsymbol{\phi}}_{\mathrm{s}}\hat{\boldsymbol{\phi}}_{\mathrm{s}}\\ &\quad + 2M_{r\theta}\hat{\mathbf{r}}_{\mathrm{s}}\hat{\boldsymbol{\theta}}_{\mathrm{s}} + 2M_{r\phi}\hat{\mathbf{r}}_{\mathrm{s}}\hat{\boldsymbol{\phi}}_{\mathrm{s}} + 2M_{\theta\phi}\hat{\boldsymbol{\theta}}_{\mathrm{s}}\hat{\boldsymbol{\phi}}_{\mathrm{s}},\end{aligned} \tag{10.41}$$

and making use of equations (10.7), we find that

$$\mathcal{A}(\Theta,\Phi) = \sum_{m=0}^{2} P_{lm}(\cos\Theta)(\mathcal{A}_m\cos m\Phi + \mathcal{B}_m\sin m\Phi), \tag{10.42}$$

where

$$\mathcal{A}_0 = M_{rr}\dot{\mathcal{U}}_{\mathrm{s}} + (M_{\theta\theta} + M_{\phi\phi})r_{\mathrm{s}}^{-1}(\mathcal{U}_{\mathrm{s}} - \tfrac{1}{2}k\mathcal{V}_{\mathrm{s}}), \tag{10.43}$$

$$\mathcal{B}_0 = 0, \tag{10.44}$$

$$\mathcal{A}_1 = k^{-1}[M_{r\theta}(\dot{\mathcal{V}}_s - r_s^{-1}\mathcal{V}_s + kr_s^{-1}\mathcal{U}_s) - M_{r\phi}(\dot{\mathcal{W}}_s - r_s^{-1}\mathcal{W}_s)], \tag{10.45}$$

$$\mathcal{B}_1 = k^{-1}[M_{r\phi}(\dot{\mathcal{V}}_s - r_s^{-1}\mathcal{V}_s + kr_s^{-1}\mathcal{U}_s) + M_{r\theta}(\dot{\mathcal{W}}_s - r_s^{-1}\mathcal{W}_s)], \tag{10.46}$$

$$\mathcal{A}_2 = k^{-1}r_s^{-1}[\tfrac{1}{2}(M_{\theta\theta} - M_{\phi\phi})\mathcal{V}_s - M_{\theta\phi}\mathcal{W}_s], \tag{10.47}$$

$$\mathcal{B}_2 = k^{-1}r_s^{-1}[M_{\theta\phi}\mathcal{V}_s + \tfrac{1}{2}(M_{\theta\theta} - M_{\phi\phi})\mathcal{W}_s]. \tag{10.48}$$

Note that $M_{rr}, M_{\theta\theta}, \ldots, M_{\theta\phi}$ are the spherical polar components of $\mathbf{M}$ *at the source.* A more consistent notation might be $M_{r_s r_s}, M_{\theta_s\theta_s}, \ldots, M_{\theta_s\phi_s}$; we adhere to the convention enunciated in Section 5.4.4, and eliminate the unwieldy subscripts-upon-subscripts in the interest of simplicity. The vector amplitude (10.37) is given in terms of $\mathcal{A}$ by

$$\boldsymbol{\mathcal{A}}(\mathbf{x}) = \left(\frac{2l+1}{4\pi}\right)\boldsymbol{\mathcal{D}}(r,\Theta,\Phi)\mathcal{A}(\Theta,\Phi). \tag{10.49}$$

The explicit epicentral representation of the displacement operator (10.25) is

$$\boldsymbol{\mathcal{D}} = \hat{\mathbf{r}}\mathcal{U} + \hat{\boldsymbol{\Theta}}\,k^{-1}[\mathcal{V}\partial_\Theta + \mathcal{W}(\sin\Theta)^{-1}\partial_\Phi] + \hat{\boldsymbol{\Phi}}\,k^{-1}[\mathcal{V}(\sin\Theta)^{-1}\partial_\Phi - \mathcal{W}\partial_\Theta]. \tag{10.50}$$

The scalar field $\mathcal{A}$ is a surface spherical harmonic of degree l and orders $-2 \le m \le 2$ in the epicentral coordinates Θ and Φ. The evaluation of the derivatives ∂_Θ and ∂_Φ in (10.50) is, as a result, straightforward.

In practical calculations of synthetic seismograms on a spherical Earth, the effect of anelasticity upon the radial eigenfunctions is generally ignored, and only its effect upon the eigenfrequencies ${}_n\nu_l = {}_n\omega_l + i\,{}_n\gamma_l$ is retained. In this approximation the displacement response (10.36) is

$$\mathbf{s}(\mathbf{x},t) = \sum_{n=0}^{\infty}\sum_{l=0}^{\infty}\left(\frac{1}{{}_n\omega_l^2 + {}_n\gamma_l{}^2}\right){}_n\mathbf{A}_l(\mathbf{x}) \times\left\{\left(\frac{{}_n\omega_l^2 - {}_n\gamma_l{}^2}{{}_n\omega_l^2 + {}_n\gamma_l{}^2}\right)[1 - \cos({}_n\omega_l t)\exp(-{}_n\gamma_l t)] - \left(\frac{2\,{}_n\omega_l\,{}_n\gamma_l}{{}_n\omega_l^2 + {}_n\gamma_l{}^2}\right)\sin({}_n\omega_l t)\exp(-{}_n\gamma_l t)\right\}, \tag{10.51}$$

where, for each fixed overtone number n and angular degree l,

$$\mathbf{A}(\mathbf{x}) = \left(\frac{2l+1}{4\pi}\right)\mathbf{D}(r,\Theta,\Phi)A(\Theta,\Phi). \tag{10.52}$$

The real scalar function A is obtained by substituting the real radial eigenfunctions $\mathcal{U}_s \to U_s$, $\mathcal{V}_s \to V_s$ and $\mathcal{W}_s \to W_s$ into equations (10.43)–(10.48):

$$A(\Theta, \Phi) = \sum_{m=0}^{2} P_{lm}(\cos\Theta)(A_m \cos m\Phi + B_m \sin m\Phi), \tag{10.53}$$

where

$$A_0 = M_{rr}\dot{U}_s + (M_{\theta\theta} + M_{\phi\phi})r_s^{-1}(U_s - \tfrac{1}{2}kV_s), \tag{10.54}$$

$$B_0 = 0, \tag{10.55}$$

$$\begin{aligned} A_1 = k^{-1}[&M_{r\theta}(\dot{V}_s - r_s^{-1}V_s + kr_s^{-1}U_s) \\ &- M_{r\phi}(\dot{W}_s - r_s^{-1}W_s)], \end{aligned} \tag{10.56}$$

$$\begin{aligned} B_1 = k^{-1}[&M_{r\phi}(\dot{V}_s - r_s^{-1}V_s + kr_s^{-1}U_s) \\ &+ M_{r\theta}(\dot{W}_s - r_s^{-1}W_s)], \end{aligned} \tag{10.57}$$

$$A_2 = k^{-1}r_s^{-1}[\tfrac{1}{2}(M_{\theta\theta} - M_{\phi\phi})V_s - M_{\theta\phi}W_s], \tag{10.58}$$

$$B_2 = k^{-1}r_s^{-1}[M_{\theta\phi}V_s + \tfrac{1}{2}(M_{\theta\theta} - M_{\phi\phi})W_s]. \tag{10.59}$$

Likewise, the displacement operator $\mathbf{D}$ is the real analogue of (10.50):

$$\begin{aligned} \mathbf{D} = \hat{\mathbf{r}}\,U &+ \hat{\boldsymbol{\Theta}}\,k^{-1}[V\partial_\Theta + W(\sin\Theta)^{-1}\partial_\Phi] \\ &+ \hat{\boldsymbol{\Phi}}\,k^{-1}[V(\sin\Theta)^{-1}\partial_\Phi - W\partial_\Theta]. \end{aligned} \tag{10.60}$$

In any case of practical interest, the receiver is situated upon the Earth's surface, so that the eigenfunctions U, V and W in equation (10.60) are evaluated at $r = a$.

The decay rate ${}_n\gamma_l$ of each multiplet in equation (10.51) is calculated using Rayleigh's principle, as outlined in Section 9.7. The effect of anelastic dispersion upon the eigenfrequencies ${}_n\omega_l$ is accounted for by allowing each mode to "see" the appropriate incompressibility and rigidity, either by means of equation (9.55) or directly in a spherical-Earth normal-mode code such as `MINEOS` or `OBANI`. Since ${}_n\gamma_l \ll {}_n\omega_l$ we may generally replace the cumbersome expression (10.51) by

$$\mathbf{s}(\mathbf{x}, t) \approx \sum_{n=0}^{\infty}\sum_{l=0}^{\infty} {}_n\omega_l^{-2}\,{}_n\mathbf{A}_l(\mathbf{x})\,[1 - \cos({}_n\omega_l t)\,\exp(-{}_n\gamma_l t)]. \tag{10.61}$$

The constant term in both equations (10.51) and (10.61) is the final static displacement of the spherically symmetric Earth after all of the oscillations have decayed:

$$\mathbf{s}_{\rm f}(\mathbf{x}) = \sum_{n=0}^{\infty}\sum_{l=0}^{\infty}({}_n\omega_l^2 + {}_n\gamma_l{}^2)^{-2}({}_n\omega_l^2 - {}_n\gamma_l{}^2)\,{}_n\mathbf{A}_l(\mathbf{x})$$
$$\approx \sum_{n=0}^{\infty}\sum_{l=0}^{\infty}{}_n\omega_l^{-2}\,{}_n\mathbf{A}_l(\mathbf{x}). \tag{10.62}$$

The acceleration found by differentiating (10.51) twice with respect to time is relatively simple even without making the high-Q approximation:

$$\mathbf{a}(\mathbf{x},t) = \sum_{n=0}^{\infty}\sum_{l=0}^{\infty}{}_n\mathbf{A}_l(\mathbf{x})\,\cos({}_n\omega_l t)\,\exp(-{}_n\gamma_l t). \tag{10.63}$$

The reality of the excitation amplitude ${}_n\mathbf{A}_l(\mathbf{x})$ guarantees that the phase of every oscillation ${}_n\mathrm{S}_l$ or ${}_n\mathrm{T}_l$ is the same $\pm\pi$ at every point $\mathbf{x}$ throughout the Earth. In the absence of anelasticity, this mode-sum representation of the spherical-Earth response to a step-function moment-tensor source is *exact*.

For positive frequencies, $\omega > 0$, the Fourier transform of the acceleration (10.63) is shown in Section 6.5 to be well approximated by

$$\mathbf{a}(\mathbf{x},\omega) = \sum_{n=0}^{\infty}\sum_{l=0}^{\infty}{}_n\mathbf{A}_l(\mathbf{x})\,{}_n\eta_l(\omega), \tag{10.64}$$

where

$${}_n\eta_l(\omega) = \tfrac{1}{2}[{}_n\gamma_l + i(\omega - {}_n\omega_l)]^{-1}. \tag{10.65}$$

The spectral response (10.64)–(10.65), which is the frequency-domain analogue of (10.63), is a weighted sum of Lorentzian resonance peaks centered upon the real eigenfrequencies ${}_n\omega_l$. The real amplitude of each peak depends upon the location $\mathbf{x}$ and polarization $\hat{\mathbf{r}}$, $\hat{\mathbf{\Theta}}$, $\hat{\mathbf{\Phi}}$ of the receiver, as well as upon the location $\mathbf{x}_{\rm s}$ and moment tensor $\mathbf{M}$ of the source. In any practical numerical implementation of equations (10.51) or (10.61)–(10.64) it is obviously necessary to truncate the infinite sums over overtone number n and angular degree l. Incorporation of all spheroidal and oscillations below a cutoff frequency $\omega_{\rm max}$ yields the complete band-limited response in either the time or frequency domain. It is, of course, this guaranteed completeness that is the distinguishing characteristic of the mode-summation method.

The character of the radial excitation amplitude $A(\Theta,\Phi)$ is illustrated in Figure 10.2. An M_{rr} or $M_{\theta\theta} + M_{\phi\phi}$ source has no azimuthal nodes in the range $0 \leq \Phi \leq 2\pi$ and l longitudinal nodes in the range $0 < \Theta < \pi$; an

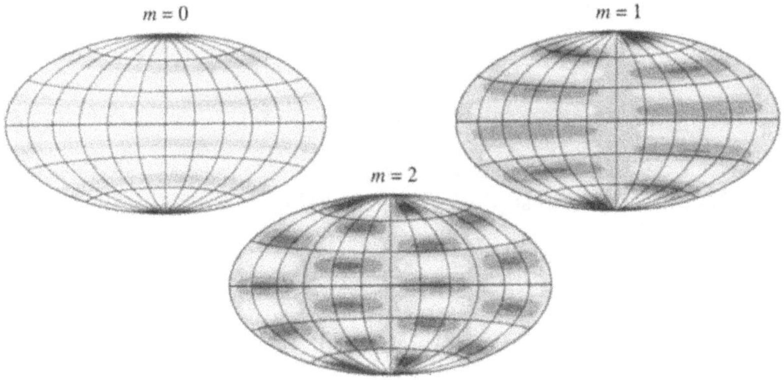

Figure 10.2. Schematic illustration of the scalar response $A(\Theta, \Phi)$ to three "canonical" moment-tensor sources situated at the North Pole. (*Upper left*) M_{rr} or $M_{\theta\theta} + M_{\phi\phi}$ source. (*Upper right*) $M_{r\theta}$ or $M_{r\phi}$ (vertical dip-slip) source. (*Lower center*) $M_{\theta\theta} - M_{\phi\phi}$ or $M_{\theta\phi}$ (vertical strike-slip) source. These exhibit isotropic, dipolar and quadrupolar azimuthal amplitude patterns, respectively. Any high-degree mode such as this ($l = 10$) has many more longitudinal than azimuthal nodes. Map is an Aitoff equal-area projection with a number of colatitude and longitude lines superimposed. In the case of a spheroidal oscillation, one should visualize the dark patches as going up whenever the light patches are going down, and vice versa.

$M_{r\theta}$ or $M_{r\phi}$ (vertical dip-slip) source has two azimuthal nodes in the range $0 \leq \Phi \leq 2\pi$ and $l-1$ longitudinal nodes in the range $0 < \Theta < \pi$; finally, an $M_{\theta\theta} - M_{\phi\phi}$ or $M_{\theta\phi}$ (vertical strike-slip) source has four azimuthal nodes in the range $0 \leq \Phi \leq 2\pi$ and $l-2$ longitudinal nodes in the range $0 < \Theta < \pi$. A general source $M_{rr}, M_{\theta\theta}, M_{\phi\phi}, M_{r\theta}, M_{r\phi}, M_{\theta\phi}$ exhibits a mixture of these three "canonical" patterns. The toroidal oscillations obviously have no radial or $\hat{\mathbf{r}}$ component; note, however, that they do have both a longitudinal $\hat{\mathbf{\Theta}}$ and a transverse $\hat{\mathbf{\Phi}}$ component. The spheroidal oscillations of a spherical Earth may be observed on all three components of a seismometer. For modes of angular degree $l \gg 2$, the longitudinal derivative ∂_Θ in (10.60) will be significantly greater than $(\sin\Theta)^{-1}\partial_\Phi$, except near the source $\Theta = 0$ and its antipode $\Theta = \pi$. For that reason, high-degree toroidal oscillations will generally be most prominent on the transverse component, whereas high-degree spheroidal oscillations will be most prominent on the radial and longitudinal components.

The determination of shallow-focus earthquake source mechanisms is problematical, as we have noted in Section 5.4.6. The free-surface boundary conditions (8.50) specify that

$$\dot{V}_s - r_s^{-1}V_s + kr_s^{-1}U_s \to 0, \qquad \dot{W}_s - r_s^{-1}W_s \to 0, \tag{10.66}$$

$$(\kappa_s + \tfrac{4}{3}\mu_s)\dot{U}_s + (\kappa_s - \tfrac{2}{3}\mu_s)r_s^{-1}(2U_s - kV_s) \to 0, \tag{10.67}$$

as the source depth $h = a - r_s$ tends to zero. For this reason, a shallow-focus $M_{r\theta}$ or $M_{r\phi}$ (vertical dip-slip) source or a shallow source having

$$M_{rr} = \left(\frac{\kappa_s + \frac{4}{3}\mu_s}{\kappa_s - \frac{2}{3}\mu_s}\right) M_{\theta\theta} = \left(\frac{\kappa_s + \frac{4}{3}\mu_s}{\kappa_s - \frac{2}{3}\mu_s}\right) M_{\phi\phi} \tag{10.68}$$

do not excite any spheroidal or toroidal oscillations. The indeterminacy of the elements M_{rr} and $M_{\theta\theta} + M_{\phi\phi}$ is generally eliminated by imposing the constraint that the source has no isotropic component:

$$M_{rr} + M_{\theta\theta} + M_{\phi\phi} = 0. \tag{10.69}$$

Equation (10.54) reduces in this case to $A_0 = M_{rr}[\dot{U}_s - r_s^{-1}(U_s - \frac{1}{2}kV_s)]$.

★10.4 Seismometer Response

As we have seen in Section 4.4 an accelerometer responds to changes in the Earth's gravitational field in addition to the acceleration $\partial_t^2\mathbf{s}$ of the instrument housing. These effects can be accounted for by specializing (4.183) to the case of a spherically symmetric Earth. We restrict attention to the customary case of a seismometer situated upon the surface of the Earth, $r = a$, and make use of the external gravity gradient formula $\dot{g} = -2a^{-1}g$. The spheroidal eigenfunctions in equation (10.60) must then be replaced by $U \to U + \omega^{-2}[2a^{-1}gU + (l+1)a^{-1}P]$ and $V \to V - \omega^{-2}(ka^{-1}gU + ka^{-1}P)$. The toroidal oscillations are, of course, unaffected. Labelling the gravitationally modified spheroidal eigenfunctions with a star, we replace $\mathbf{D}$ in equation (10.52) by a modified displacement operator:

$$\begin{aligned}\mathbf{D}_\star = \hat{\mathbf{r}}\,U_\star + \hat{\mathbf{\Theta}}\,k^{-1}[V_\star\partial_\Theta + W(\sin\Theta)^{-1}\partial_\Phi] \\ + \hat{\mathbf{\Phi}}\,k^{-1}[V_\star(\sin\Theta)^{-1}\partial_\Phi - W\partial_\Theta],\end{aligned} \tag{10.70}$$

where

$$\begin{aligned}U_\star = U + U_{\mathrm{free}} + U_{\mathrm{pot}}, \qquad U_{\mathrm{free}} = 2\omega^{-2}ga^{-1}U, \\ U_{\mathrm{pot}} = (l+1)\omega^{-2}a^{-1}P,\end{aligned} \tag{10.71}$$

$$\begin{aligned}V_\star = V + V_{\mathrm{tilt}} + V_{\mathrm{pot}}, \qquad V_{\mathrm{tilt}} = -k\omega^{-2}ga^{-1}U, \\ V_{\mathrm{pot}} = -k\omega^{-2}a^{-1}P.\end{aligned} \tag{10.72}$$

Mode	mHz	$\frac{U}{U_\star}$	$\frac{U_{\rm free}}{U_\star}$	$\frac{U_{\rm pot}}{U_\star}$	$\frac{V}{V_\star}$	$\frac{V_{\rm tilt}}{V_\star}$	$\frac{V_{\rm pot}}{V_\star}$
${}_0S_2$	0.3093	0.812	0.664	−0.476	−0.121	2.148	−1.027
${}_0S_3$	0.4686	0.870	0.310	−0.180	0.502	0.701	−0.203
${}_0S_4$	0.6471	0.914	0.171	−0.084	0.672	0.408	−0.080
${}_0S_5$	0.8404	0.941	0.104	−0.045	0.760	0.280	−0.040
${}_0S_6$	1.0382	0.957	0.069	−0.027	0.810	0.214	−0.024
${}_0S_7$	1.2318	0.968	0.050	−0.018	0.839	0.177	−0.016
${}_0S_8$	1.4135	0.975	0.038	−0.013	0.855	0.156	−0.012
${}_0S_9$	1.5783	0.979	0.031	−0.010	0.865	0.145	−0.009
${}_0S_{10}$	1.7265	0.983	0.026	−0.008	0.870	0.138	−0.008
${}_1S_1$	0.0513	0.032	0.960	0.008	−0.028	1.019	0.009
${}_1S_2$	0.6799	1.032	0.175	−0.207	1.009	−0.041	0.032
${}_1S_3$	0.9398	0.991	0.088	−0.078	1.034	−0.061	0.027
${}_1S_4$	1.1729	0.985	0.056	−0.041	1.061	−0.086	0.025
${}_1S_5$	1.3703	0.983	0.041	−0.024	1.141	−0.176	0.034
${}_1S_6$	1.5220	0.982	0.033	−0.016	—	—	—
${}_1S_7$	1.6555	0.984	0.028	−0.012	0.694	0.342	−0.036
${}_1S_8$	1.7993	0.986	0.024	−0.009	0.770	0.252	−0.022
${}_2S_3$	1.2422	0.949	0.048	0.003	0.963	0.036	0.001
${}_3S_1$	0.9439	0.919	0.081	0.000	0.953	0.047	0.000
${}_3S_2$	1.1062	0.938	0.060	0.003	0.959	0.040	0.001
${}_3S_8$	2.8196	0.988	0.010	0.002	0.990	0.010	0.001

Table 10.1. Relative magnitudes of the inertial, free-air, tilt and potential-perturbation contributions to the accelerometer response operator $\mathbf{D}_\star$ for some representative spheroidal oscillations. The tabulated values are for the transversely isotropic PREM model. Second column lists the associated eigenfrequency $\omega/2\pi$ in mHz. Mode ${}_1S_6$ is unlikely to be observed on a horizontally polarized instrument inasmuch as it has $V \ll U$.

The terms $U_{\rm free}$ and $V_{\rm tilt}$ account for the free-air change in gravity due to the radial displacement of the seismometer and the indistinguishability of ground tilt and horizontal acceleration, respectively, whereas the terms $U_{\rm pot}$ and $V_{\rm pot}$ account for the perturbation P in the gravitational potential due to the redistribution of the Earth's mass. The relative magnitudes of the various terms in (10.71)–(10.72) are given for a number of spheroidal oscillations in Table 10.1. For all high-frequency modes $\omega^2 \gg ga^{-1}$ and $P \ll gU$, so that $U_\star \approx U$ and $V_\star \approx V$. Free-air effects U_{free} and tilt V_{tilt} dominate the response of the Slichter mode ${}_1\mathrm{S}_1$, and contribute significantly to the response of the low-degree fundamental modes such as ${}_0\mathrm{S}_2$. The gravitational potential perturbation $U_{\mathrm{pot}}, V_{\mathrm{pot}}$ is also important for the low-degree fundamental modes. Tilt effects V_{tilt} are surprisingly non-negligible even for intermediate-frequency ${}_0\mathrm{S}_l$ and ${}_1\mathrm{S}_l$ modes. In general, these self-gravitational corrections are sufficiently easily incorporated that it is advisable to make routine use of $\mathbf{D}_\star$ rather than $\mathbf{D}$ when computing synthetic seismograms and spectra.

10.5 Wiggly Lines—At Last!

Finally—after a scant 376 pages—we have assembled the tools needed to compute synthetic seismograms by means of mode summation upon a spherically symmetric, non-rotating, slightly anelastic Earth. We present in this section a representative sample of acceleration spectra, time-domain accelerograms and displacement seismograms, and describe their salient features. We do not elaborate at length upon every wiggle in every record, but instead rely upon the adage that "one picture is worth a thousand words". Our intent is simply to illustrate the richness and completeness of mode-sum synthetics, by means of a few visual examples. We shall supplement this impressionistic account with a more extended discussion of Love and Rayleigh surface waves and P-SV body waves in Chapters 11 and 12.

10.5.1 Computational details

A step-by-step description of our customary procedure for computing synthetic accelerograms and spectra is given below. We use a unit vector $\hat{\boldsymbol{\nu}}$ to denote the *polarization* of the accelerometer—generally either radial ($\hat{\boldsymbol{\nu}} = \hat{\mathbf{r}}$), longitudinal ($\hat{\boldsymbol{\nu}} = \hat{\boldsymbol{\Theta}}$) or transverse ($\hat{\boldsymbol{\nu}} = \hat{\boldsymbol{\Phi}}$).

1. The time-domain response $\hat{\boldsymbol{\nu}} \cdot \mathbf{a}(\mathbf{x}, t)$ to an impulsive moment-rate tensor $\dot{\mathbf{M}}(t) = \sqrt{2}M_0\hat{\mathbf{M}}\,\delta(t)$ is calculated at equally spaced times $t = n\Delta t$, $n = 0, 1, 2, \ldots$, using equation (10.63). The quantity Δt is the chosen digitization interval.

2. The response to a finite-duration source $\dot{\mathbf{M}}(t) = \sqrt{2}M_0\hat{\mathbf{M}}\,\dot{m}(t)$ can be found by convolving (10.63) with the source time function $\dot{m}(t)$.

In practice, it is simpler to replace the moment tensor $\mathbf{M}$ in equations (10.54)–(10.59) by $\sqrt{2}M_0\hat{\mathbf{M}}\,m(\omega)$ where ω is either ${}_n\omega_l^{\mathrm{T}}$ or ${}_n\omega_l^{\mathrm{S}}$ and $m(\omega)$ is the Fourier transform of $\dot{m}(t)$. To reduce "ringing" and other acausal artifacts caused by the abrupt truncation of the mode sum, we generally employ a symmetric boxcar source time function $\dot{m}(t) = (2\tau_{\mathrm{h}})^{-1}[H(t+\tau_{\mathrm{h}}) - H(t-\tau_{\mathrm{h}})]$ with a half-duration $\tau_{\mathrm{h}} = 10.5$ s. The Fourier transform in that case is a "sinc" function $m(\omega) = (\omega\tau_{\mathrm{h}})^{-1}\sin(\omega\tau_{\mathrm{h}})$. Note the normalization: $\int_{-\tau_{\mathrm{h}}}^{\tau_{\mathrm{h}}} \dot{m}(t)\,dt = 1$.

3. Free-air, tilt and potential-perturbation effects upon the accelerometer are accounted for in all cases using the modified displacement operator $\mathbf{D}_\star$ given in equation (10.70).

4. The spectrum $\hat{\boldsymbol{\nu}}\cdot\mathbf{a}(\mathbf{x},\omega)$ is calculated by means of a time-to-frequency discrete Fourier transformation, rather than by a direct frequency-domain implementation of equation (10.64). The desired portion of the time-domain record $\hat{\boldsymbol{\nu}}\cdot\mathbf{a}(\mathbf{x},t)$ is isolated and multiplied by a Hann or cosine-bell taper in order to reduce the effects of spectral leakage and peak distortion due to mode-mode interference (Harris 1978; Dahlen 1982). If the windowed segment consists of N samples at times $t = n\Delta t$, $n = 0, 1, \ldots, N-1$ then the Hann-taper weights are $w_n = w(n\Delta t) = \frac{1}{2}\{1-\cos[2\pi n/(N-1)]\}$. The tapered time series is padded with zeros after $t = (N-1)\Delta t$ prior to fast Fourier transformation. This has the effect of interpolating the spectrum between the Fourier frequencies $\omega_n = (2\pi n)/(N\Delta t)$, $n = -N/2, \ldots, N/2$.

Unless stated otherwise, our synthetic seismograms and spectra are computed by summing over all mantle toroidal modes ${}_n\mathrm{T}_l$ and all spheroidal modes ${}_n\mathrm{S}_l$ with periods greater than or equal to eight seconds. This amounts to more than 60,000 toroidal and more than 100,000 spheroidal oscillations. The spherical Earth model used to construct the eigenfrequency-eigenfunction catalogue is transversely isotropic PREM. For display purposes we generally plot the absolute value $|\hat{\boldsymbol{\nu}}\cdot\mathbf{a}(\mathbf{x},\omega)|$, which is referred to as the *amplitude spectrum*. It is also possible to plot the squared absolute value $|\hat{\boldsymbol{\nu}}\cdot\mathbf{a}(\mathbf{x},\omega)|^2$, which is known as the *power spectrum*.

10.5.2 Spectra

We begin by considering the spectral or frequency-domain response to an hypothetical shallow-focus thrust event; the strike and dip of the fault plane are W45°N and 45°NE, respectively; the depth of the moment-tensor point source is 33 km. The transverse-component amplitude spectrum $|\hat{\boldsymbol{\Phi}}\cdot\mathbf{a}(\mathbf{x},\omega)|$ at an epicentral distance $\Theta = 60°$ due east of such an event is shown in

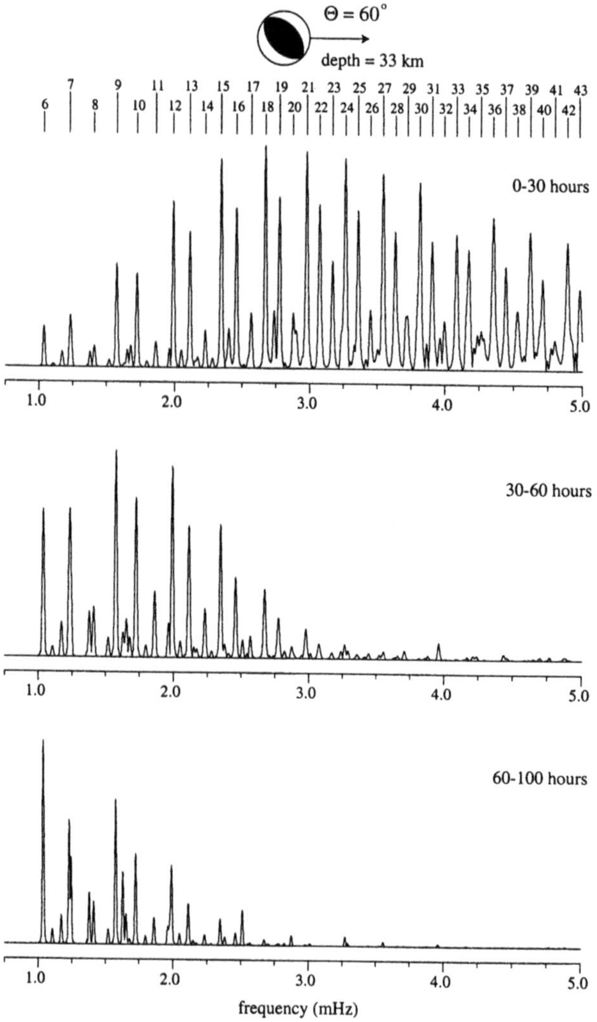

Figure 10.5. Synthetic radial-component acceleration spectra for an hypothetical shallow-focus thrust fault on the PREM Earth model. (*Top*) t = 0–30 hours. (*Middle*) t = 30–60 hours. (*Bottom*) t = 60–100 hours. The maximum amplitudes of the attenuation-filtered spectra are smaller than the maximum amplitude of the unfiltered spectrum, by factors of 0.093 and 0.027, respectively. The earthquake source mechanism and receiver azimuth are depicted schematically at the very top. Vertical lines denote the locations of the fundamental spheroidal modes ${}_0S_6$ through ${}_0S_{43}$. See Figure 10.3 for corresponding transverse component.

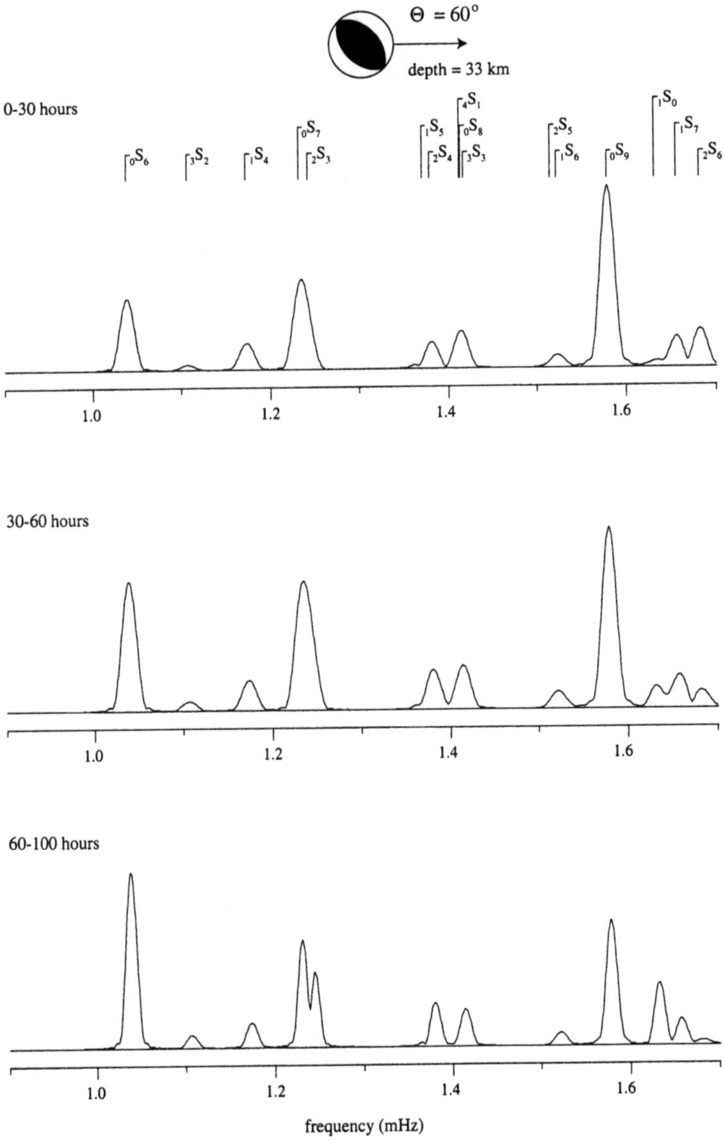

Figure 10.6. Blow-up of the low-frequency portion of Figure 10.5. The maximum amplitudes of the attenuation-filtered spectra are smaller than the maximum amplitude of the unfiltered spectrum, by factors of 0.200 and 0.059, respectively. A number of spheroidal modes visible in the spectra are identified along the top. See Figure 10.4 for corresponding transverse component.

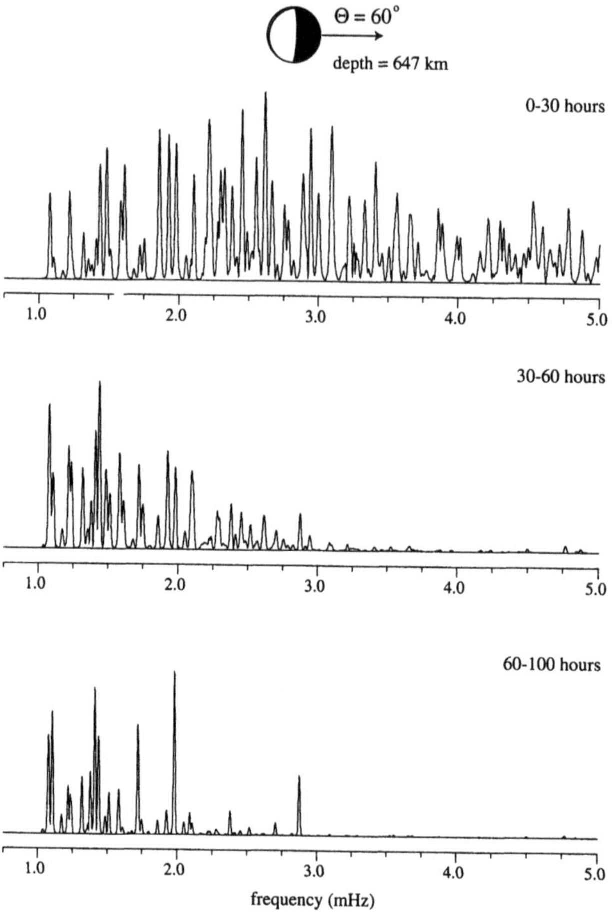

Figure 10.7. Synthetic transverse-component acceleration spectra for the great Bolivia deep-focus earthquake of June 9, 1994. (*Top*) $t = 0$–30 hours. (*Middle*) $t = 30$–60 hours. (*Bottom*) $t = 60$–100 hours. The maximum amplitudes of the attenuation-filtered spectra are smaller than the maximum amplitude of the unfiltered spectrum, by factors of 0.089 and 0.022, respectively. The beachball with attached arrow at the very top shows the relation of the receiver azimuth to the earthquake focal mechanism; this diagram has been rotated by approximately 90° for convenience of display. The correctly oriented mechanism of the Bolivia earthquake is shown in Figure 10.19; the receiver is situated to the south. See Figure 10.9 for corresponding radial component.

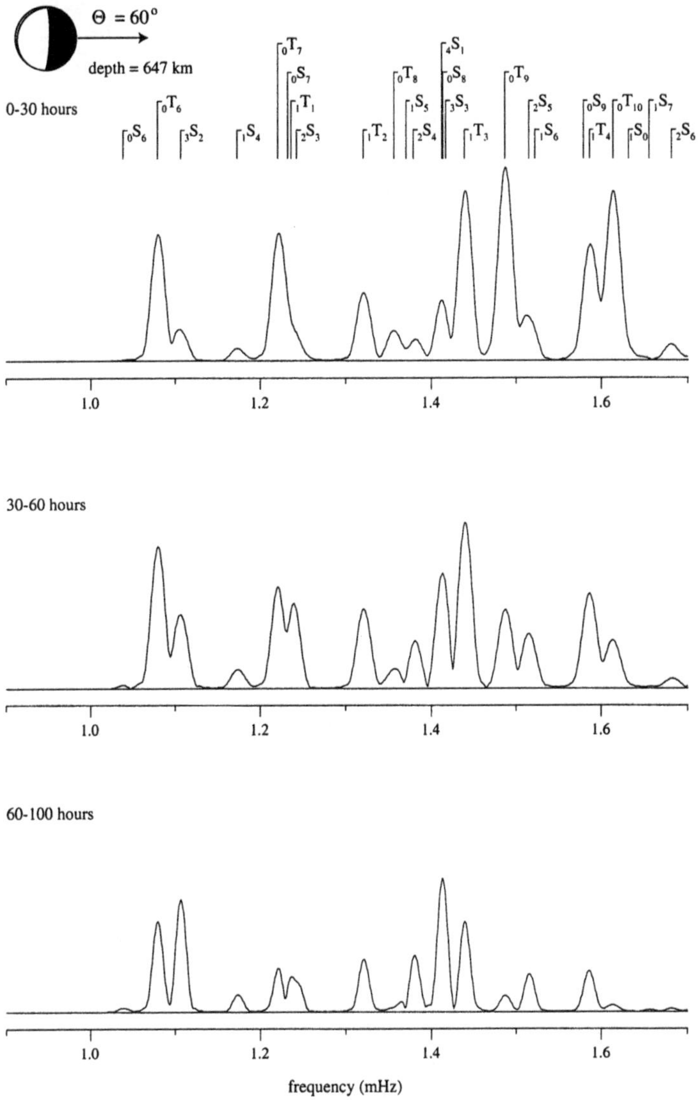

Figure 10.8. Blow-up of the low-frequency portion of Figure 10.7. The maximum amplitudes of the attenuation-filtered spectra are smaller than the maximum amplitude of the unfiltered spectrum, by factors of 0.128 and 0.028, respectively. A number of toroidal and spheroidal oscillations visible in the spectra are identified along the top.

Figure 10.3. The uppermost spectrum was obtained by isolating and Hann tapering the mode-sum accelerogram $\hat{\mathbf{\Phi}} \cdot \mathbf{a}(\mathbf{x}, t)$ between the origin time $t = 0$ and $t = 30$ hours. The most prominent peaks in the 1–5 mHz band correspond to the fundamental toroidal oscillations ${}_0\mathrm{T}_5$ through ${}_0\mathrm{T}_{40}$. The theoretical eigenfrequencies of these strongly excited modes are indicated along the top; the maximum spectral amplitude occurs at 3.11 mHz—mode ${}_0\mathrm{T}_{23}$. The middle spectrum was obtained by isolating and Hann tapering the accelerogram between $t = 30$ hours and $t = 60$ hours. This simple process of discarding data and re-calculating the spectrum is known as *attenuation filtering*. Modes with relatively low values of Q—in this case the higher-frequency fundamental modes—tend to be eliminated from the filtered spectrum; the highest remaining peak is mode ${}_0\mathrm{T}_7$. Even fewer high-Q modes remain in the lowermost spectrum, which was obtained by isolating and Hann tapering the transverse-component accelerogram between $t = 60$ hours and $t = 100$ hours; note that this is from two and one-half to four days after the event.

In Figure 10.4 we zoom in on the low-frequency portion of Figure 10.3, between 0.9 and 1.7 mHz, in order to investigate the effects of the attenuation filtering further. The toroidal and spheroidal oscillations of interest in this frequency range are labeled along the top. Naive ray-theoretical considerations (which are strictly valid only in the limit $\omega \to \infty$) suggest that we should never expect to see spheroidal modes on the transverse component of an accelerometer. In fact, we see that a number of long-period, high-Q spheroidal modes actually dominate the 60–100 hour spectrum! The largest peak is the overtone ${}_2\mathrm{S}_3$, which can be seen emerging on the high-frequency shoulder of the fundamental toroidal mode ${}_0\mathrm{T}_7$ thirty hours after the earthquake. The modes ${}_0\mathrm{S}_6$, ${}_3\mathrm{S}_2$, ${}_1\mathrm{S}_4$ and ${}_2\mathrm{S}_4$ are all also prominent after a lapse time of sixty hours.

Figure 10.5 shows the radial-component spectra $|\hat{\mathbf{r}} \cdot \mathbf{a}(\mathbf{x}, \omega)|$ for the same source-receiver combination. The dominant peaks in the 0–30 hour spectrum are the fundamental spheroidal oscillations ${}_0\mathrm{S}_6$ through ${}_0\mathrm{S}_{43}$, labelled along the top. The maximum amplitude of ${}_0\mathrm{S}_{18}$ is about one and one-half times larger than that of ${}_0\mathrm{T}_{23}$ in the unfiltered transverse spectrum. The higher-frequency, low-Q fundamental modes tend once again to be eliminated upon discarding the first thirty or sixty hours of data. A blow-up of the unfiltered and filtered spectra between 0.9 and 1.7 mHz is shown in Figure 10.6. It is noteworthy that many of the low-frequency spheroidal modes present in the filtered transverse spectra also dominate the filtered radial spectra. One prominent new peak is the first radial overtone mode ${}_1\mathrm{S}_0$, which by definition cannot be observed on a horizontal accelerometer.

As a second example, we consider the earthquake which occurred at a depth of 647 km beneath Bolivia on June 9, 1994. This $M_0 = 2.4 \times 10^{21}$ Nm

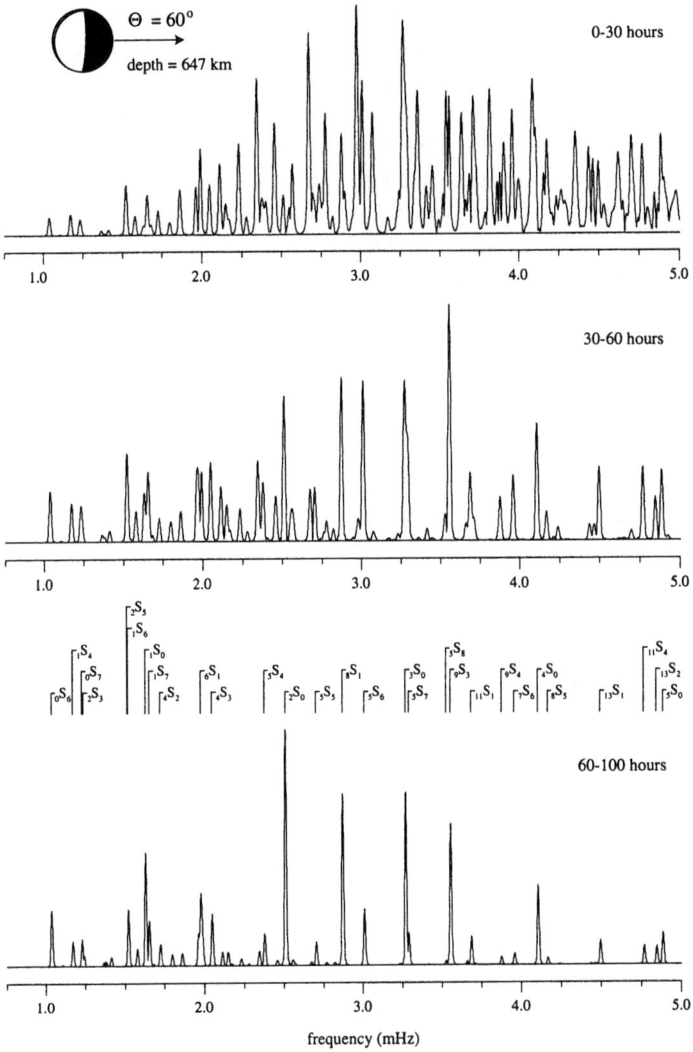

Figure 10.9. Synthetic radial-component acceleration spectra for the great Bolivia deep-focus earthquake of June 9, 1994. (*Top*) t = 0–30 hours. (*Middle*) t = 30–60 hours. (*Bottom*) t = 60–100 hours. The maximum amplitudes of the attenuation-filtered spectra are smaller than the maximum amplitude of the unfiltered spectrum, by factors of 0.127 and 0.049, respectively. The earthquake source mechanism and receiver azimuth are depicted (in a rotated view) at the very top. A number of high-Q spheroidal oscillations visible in the 60–100 hour spectra are identified. See Figure 10.7 for corresponding transverse component.

event within the subducting Nazca Plate is the largest deep-focus earthquake ever recorded by modern seismological instruments. We present a number of synthetic acceleration spectra at an hypothetical station located in the compressive quadrant of the P-wave focal mechanism, at an angular epicentral distance $\Theta = 60^\circ$. The transverse-component spectra $|\hat{\mathbf{\Phi}}\cdot\mathbf{a}(\mathbf{x},\omega)|$ of the same three Hann-tapered time intervals 0–30 hours, 30–60 hours and 60–100 hours are shown in Figure 10.7. As before, a number of low-frequency, high-Q spheroidal modes are prominent in the filtered spectra; these are identified, together with the intermingled toroidal modes, in the blowup of the 0.9–1.7 mHz band in Figure 10.8.

The synthetic radial-component spectra $|\hat{\mathbf{r}}\cdot\mathbf{a}(\mathbf{x},\omega)|$ for the Bolivia event are shown in Figure 10.9. The contrast with the shallow event in Figure 10.5 is striking: many more intermediate-frequency and high-frequency spectral peaks remain visible after the first thirty to sixty hours of data have been discarded. The remaining peaks in the 60–100 hour filtered spectrum are identified above the lowermost panel. They are all high-Q PKIKP-equivalent spheroidal modes, including the first five radial overtones ${}_1\mathrm{S}_0 - {}_5\mathrm{S}_0$, as well as other oscillations such as ${}_2\mathrm{S}_3$, ${}_8S_1$, ${}_9\mathrm{S}_3$, ${}_{11}\mathrm{S}_1$, ${}_8\mathrm{S}_5$, ${}_{13}\mathrm{S}_1$, ${}_{11}\mathrm{S}_4$ and ${}_{13}\mathrm{S}_2$. The latter are extremely interesting oscillations that are sensitive to the properties of the deepest interior of the Earth. In Section 14.2.8 we shall see that these multiplets are split much more than expected as a result of well-characterized perturbations, including the Earth's rotation, hydrostatic ellipticity, and large-scale mantle heterogeneity. This so-called *anomalous splitting* provided the first evidence for the anisotropy of the inner core (Woodhouse, Giardini & Li 1986; Tromp 1993).

10.5.3 Seismograms

In Figure 10.10 we show a number of transverse-component synthetic accelerograms $\hat{\mathbf{\Phi}}\cdot\mathbf{a}(\mathbf{x},t)$. The hypothetical source in this example is a vertical strike-slip fault situated at a depth of 33 km. The stations are located at epicentral distances $\Theta = 30^\circ$, $\Theta = 60^\circ$ and $\Theta = 90^\circ$ along the extension of the right-lateral fault, as shown along the top. The geometry is such that only the toroidal modes ${}_n\mathrm{T}_l$ are excited. The top three records have been bandpass filtered between 20 and 80 seconds in order to accentuate the direct and reflected body-wave arrivals. A number of prominent phases such as SH, $\mathrm{SS_{SH}}$, $\mathrm{SSS_{SH}}$ and $\mathrm{SKKS_{SH}}$ may be readily identified. The bottom three records have been bandpass filtered between 50 and 250 seconds in order to display the long-period fundamental Love or G waves.

Figure 10.11 depicts the radial and longitudinal accelerograms $\hat{\mathbf{r}}\cdot\mathbf{a}(\mathbf{x},t)$ and $\hat{\mathbf{\Theta}}\cdot\mathbf{a}(\mathbf{x},t)$ at the same three epicentral distances $\Theta = 30^\circ$, $\Theta = 60^\circ$ and $\Theta = 90^\circ$. The source is once again a vertical strike-slip fault at a depth of 33 km; however, the strike has been rotated by 45°, so that now only the spheroidal modes ${}_nS_l$ are excited. The uppermost six records have been filtered between 20 and 80 seconds to highlight the P-SV body waves; it is noteworthy that the long-period shear-wave arrivals such as SV and ScS_{SV} are more prominent than the compressional-wave arrivals such as P and PcP. The lowermost six traces show the same accelerograms filtered to emphasize the fundamental-mode Rayleigh waves. The dispersed character of the radial and longitudinal wavetrains is evident: the 50–100 second waves arrive noticeably earlier than the 200–250 second "mantle" Rayleigh waves. We discuss the dispersion of both Rayleigh and Love waves in greater detail in Section 11.6.

In Figure 10.12 we illustrate the effect of source depth upon the first two transverse-component body-wave arrivals. The vertical strike-slip source

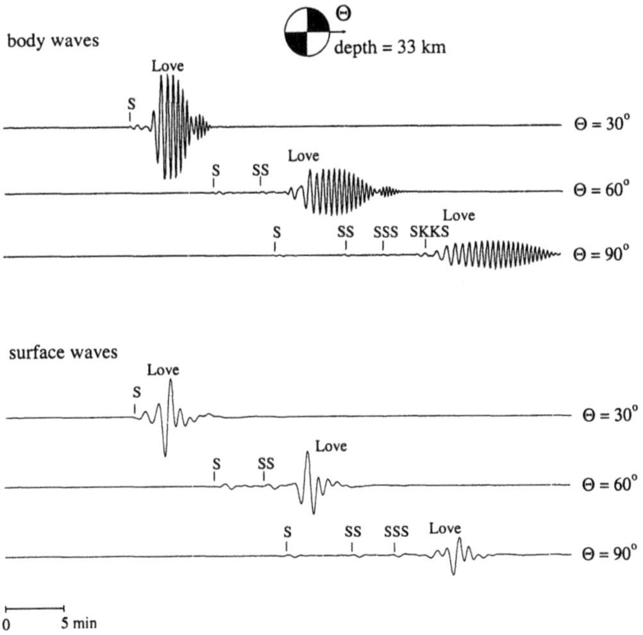

Figure 10.10. Transverse-component synthetic accelerograms at epicentral distances $\Theta = 30°$, $\Theta = 60°$ and $\Theta = 90°$ due east of an hypothetical shallow-focus right-lateral strike-slip fault. The earthquake source mechanism and receiver azimuth are depicted schematically at the very top. (*Top*) Bandpass filtered between 20 and 80 seconds to highlight the SH body waves. (*Bottom*) Bandpass filtered between 50 and 250 seconds to highlight the Love surface waves.

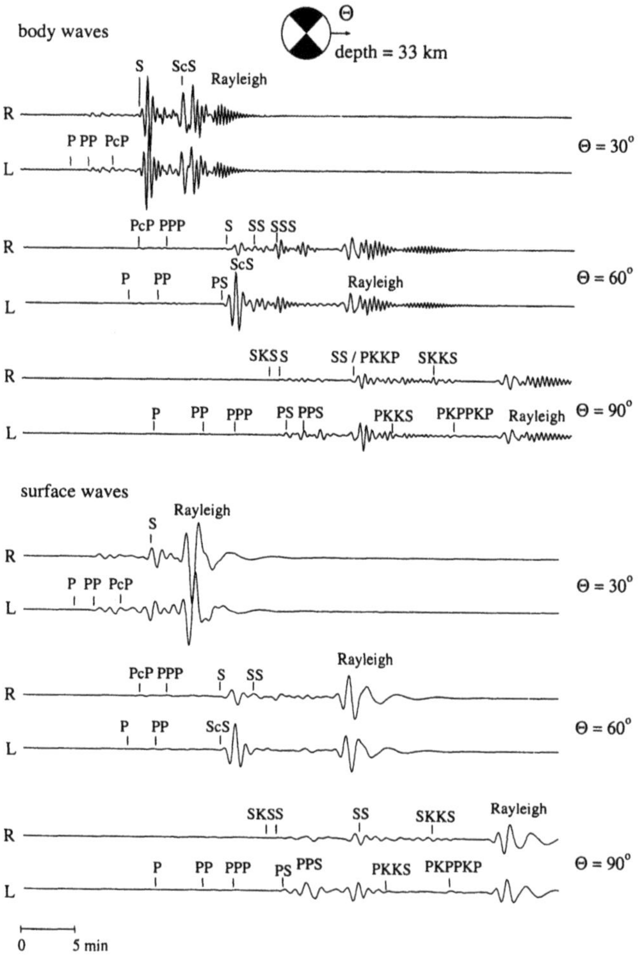

Figure 10.11. Radial (R) and longitudinal (L) synthetic accelerograms at epicentral distances $\Theta = 30°$, $\Theta = 60°$ and $\Theta = 90°$ due east of an hypothetical shallow-focus right-lateral strike-slip fault. The earthquake source mechanism and receiver azimuth are depicted schematically at the very top. (*Top*) Bandpass filtered between 20 and 80 seconds to highlight the P and SV body-wave arrivals. (*Bottom*) Bandpass filtered between 50 and 250 seconds to highlight the Rayleigh surface waves.

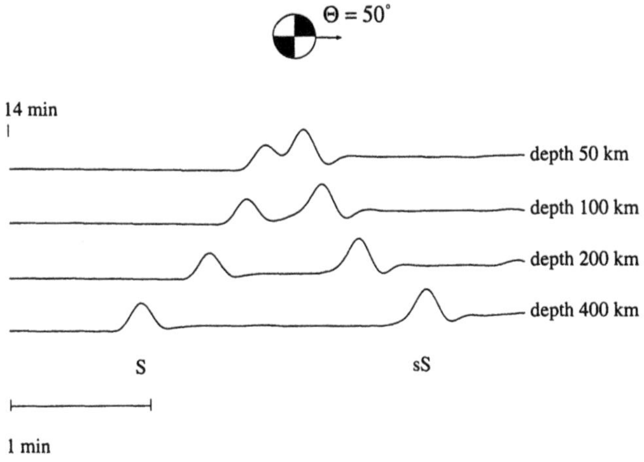

Figure 10.12. Transverse-component displacement seismograms at an epicentral distance $\Theta = 50°$ due east of an hypothetical right-lateral strike-slip fault. The earthquake source mechanism and receiver azimuth are depicted schematically at the top. The surface-reflected phase $\mathrm{sS_{SH}}$ "moves out" from the direct phase SH as the source depth increases from 50 km (*top*) to 400 km (*bottom*).

mechanism and receiver azimuth are chosen so that only the toroidal modes ${}_n\mathrm{T}_l$ are excited; the epicentral distance is $\Theta = 50°$. As expected, the direct SH wave arrives earlier, whereas the surface-reflected phase $\mathrm{sS_{SH}}$ arrives later as the depth of the event is increased from 50 to 400 km. The particle displacement $\hat{\mathbf{\Phi}} \cdot \mathbf{s}(\mathbf{x}, t)$ rather than the acceleration $\hat{\mathbf{\Phi}} \cdot \mathbf{a}(\mathbf{x}, t)$ has been plotted, and an isosceles triangle with a half-duration $\tau_{\mathrm{h}} = 10.5$ s has been used as the source time function $\dot{m}(t)$ rather than a boxcar, in order to clarify this "depth-phase moveout". Figure 10.13 shows the analogous effect of source depth upon the radial and longitudinal P-SV body waves. The station is located at an epicentral distance $\Theta = 90°$ from a strike-slip event, oriented so that only the spheroidal modes ${}_n\mathrm{S}_l$ modes are excited. As the depth of the event is increased from 100 to 650 km, the pP–P and pPP–PP time intervals both grow. In practice, the observed differences in arrival time between these various direct and surface-reflected phases provide an important constraint upon the depths of earthquakes.

Figures 10.14 and 10.15 illustrate the effect of source depth upon the excitation of the fundamental-mode surface waves. The source-receiver geometries are the same as in Figures 10.12 and 10.13, so that only Love waves are excited in the first instance and only Rayleigh waves in the second. The characteristic retrograde particle motion of the Rayleigh waves is evident

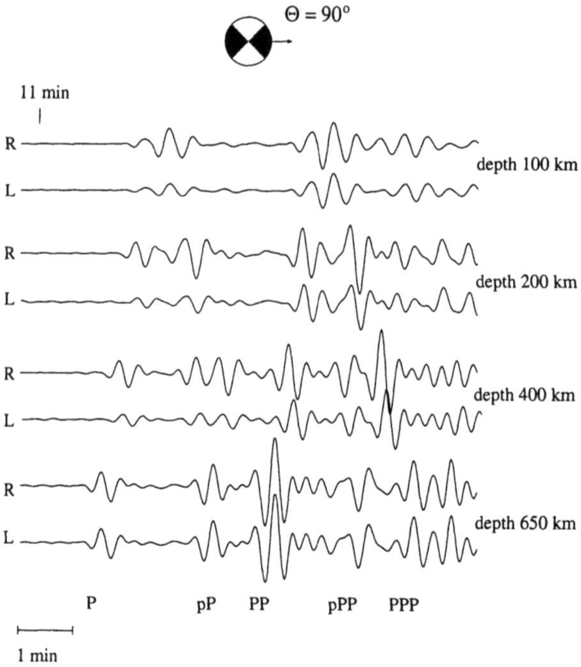

Figure 10.13. Radial (R) and longitudinal (L) synthetic accelerograms at an epicentral distance $\Theta = 90^\circ$ due east of an hypothetical right-lateral strike-slip fault. The earthquake source mechanism and receiver azimuth are depicted schematically at the top. The surface-reflected "depth phases" pP and pPP "move out" from the direct phases P and PP as the source depth increases from 100 km (*top*) to 650 km (*bottom*).

upon comparing the radial and longitudinal components. Both types of surface waves decrease in amplitude as the depth of the strike-slip event increases from 10 to 400 km. A deep-focus earthquake radiates relatively weak fundamental-mode waves because the source is situated on the quasi-exponential tail—i.e., essentially on a node—of the associated ${}_0\mathrm{T}_l$ and ${}_0\mathrm{S}_l$ eigenfunctions.

Figure 10.16 explores the effect of a vertical strike-slip earthquake source mechanism upon the radiation pattern of the SH and P-SV body waves. The position of each set of three-component accelerograms relative to the central beachball is indicative of the azimuth of the corresponding station relative to the event. The epicentral distance to all eight three-component stations is $\Theta = 60^\circ$; the synthetic records have been bandpass filtered

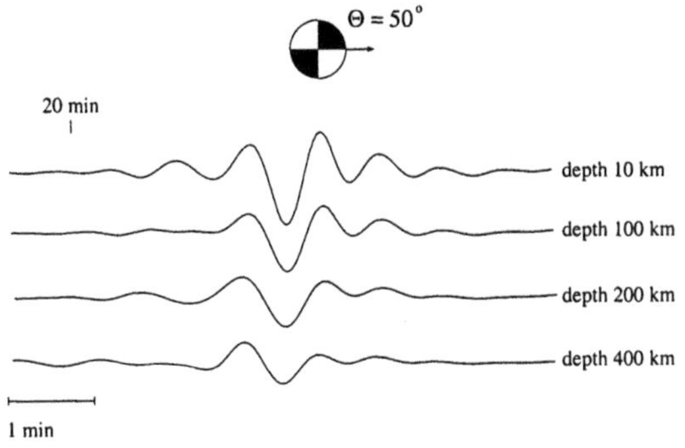

Figure 10.14. Transverse-component synthetic accelerograms at an epicentral distance $\Theta = 50^\circ$ due east of an hypothetical right-lateral strike-slip fault, showing the reduction in Love-wave amplitude as the focal depth increases from 10 km (*top*) to 400 km (*bottom*).

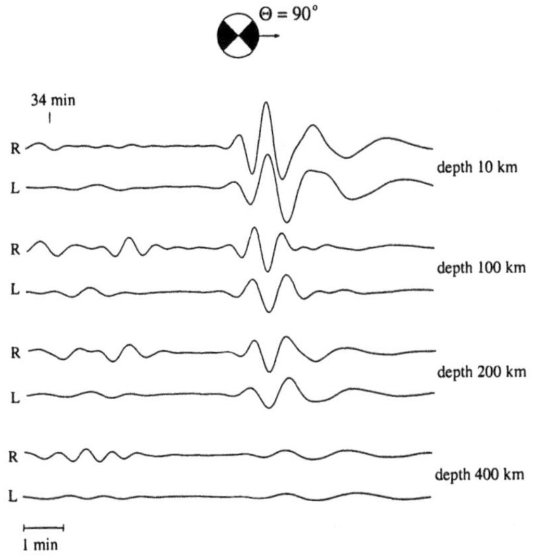

Figure 10.15. Radial (R) and longitudinal (L) synthetic accelerograms at an epicentral distance $\Theta = 90^\circ$ due east of an hypothetical right-lateral strike-slip fault, showing the reduction in Rayleigh-wave amplitude as the focal depth increases from 10 km (*top*) to 400 km (*bottom*).

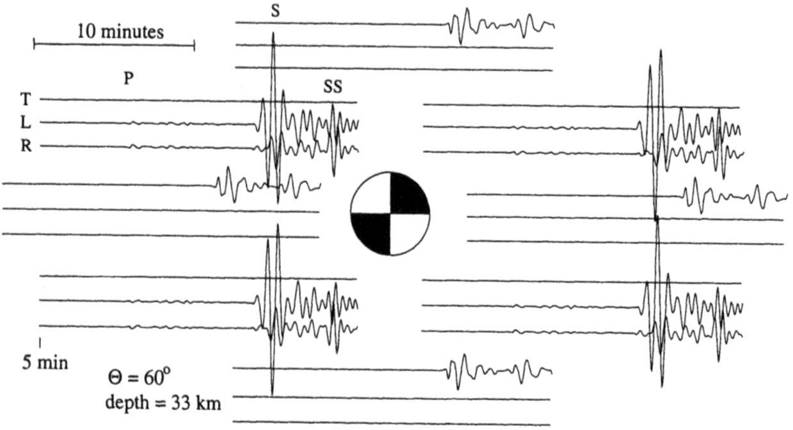

Figure 10.16. Illustration of the effect of the body-wave radiation pattern upon synthetic accelerograms. The source is a right-lateral strike-slip fault situated at a focal depth $h = 33$ km; black and white areas on the central beachball correspond to compressional and dilatational P-wave quadrants on the lower focal hemisphere, respectively. The eight three-component seismographic stations are situated (*clockwise from top*) to the north, northeast, east, southeast, south, southwest, west and northwest, all at an angular distance $\Theta = 60°$ from the epicenter. The transverse (T), longitudinal (L) and radial (R) components and the P-wave, S-wave and SS-wave arrivals are labelled on the northwest-station recordings.

between 20 and 80 seconds to enhance the body waves. There are no visible arrivals upon the radial and longitudinal components at the north, east, south and west stations in Figure 10.16, because both the P and SV waves exhibit nodes along the fault plane and auxiliary plane. The anti-nodal character of the SH-wave radiation pattern at these four stations results, however, in clear SH and SS_{SH} arrivals on the transverse components. The northeast, southeast, southwest and northwest stations exhibit large anti-nodal P, SV and SS_{SV} arrivals on the radial and longitudinal components, and no SH arrivals on the transverse components. Figure 10.17 shows the radial, longitudinal and transverse accelerograms at the same eight stations due to a north-south striking, 45° dipping thrust fault. The most notable features in this case are the predominantly longitudinally polarized shear waves in the two along-strike directions.

Following the September 2, 1992 earthquake in Nicaragua, Kanamori (1993) observed a curious long-period waveform between the P and S pulses

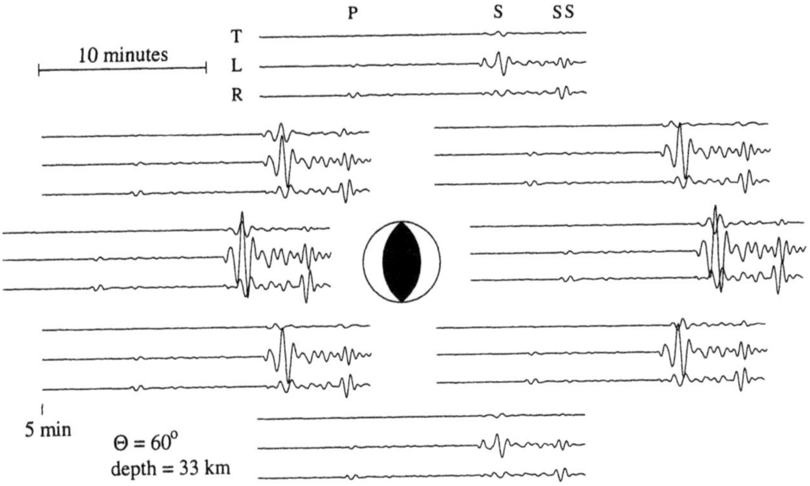

Figure 10.17. Same as Figure 10.16 except for a shallow-focus ($h = 33$ km) thrust fault. The transverse (T), longitudinal (L) and radial (R) components and the P-wave, S-wave and SS-wave arrivals are labelled on the synthetic accelerograms at the station situated to the north. All of the records have been bandpass filtered between 20 and 80 seconds to accentuate the predominantly P-SV body waves.

at a number of broad-band seismic stations. He attributed this hitherto unnoticed teleseismic arrival to a whispering-gallery effect caused by the constructive interference of far-field body waves, and accordingly dubbed it the *W phase*. The Nicaragua event, which gave rise to a large tsunami, was characterized by anomalously slow rupture on the subduction boundary between the Cocos and North American plates (Kanamori & Kikuchi 1993). In Figure 10.18 we illustrate that the observed radial displacement at Pasadena is reasonably well matched by a synthetic normal-mode seismogram computed for a source having a half-duration $\tau_{\mathrm{h}} = 40$ seconds. Vidale, Goes & Richards (1995) have pointed out that the W phase may be regarded as a spherical-Earth manifestation of the intermediate-field and near-field energy that propagates at speeds between α and β and attenuates like (distance)$^{-2}$ and (distance)$^{-3}$ rather than like (distance)$^{-1}$ in an infinite homogeneous medium (Aki & Richards 1980). This unradiated energy is not amenable to analysis by the ray-theoretical methods which we develop in Chapter 15; however, it is fully accounted for in the spherical-Earth mode sum (10.51).

Our final—and favorite—example shows the synthetic particle displacement $\mathbf{s}(\mathbf{x}, t)$ at a near-field site following the great June 9, 1994 Bolivia

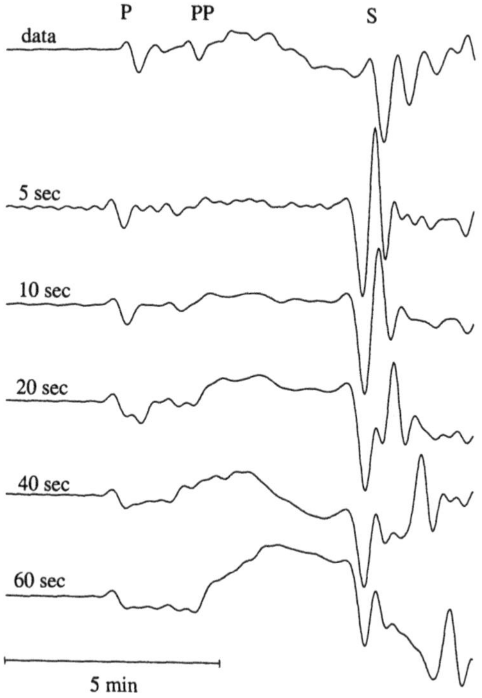

Figure 10.18. (*Top*) Radial-component displacement seismogram $\hat{\mathbf{r}} \cdot \mathbf{s}(\mathbf{x}, t)$ recorded at the broad-band station PAS in Pasadena, California (epicentral distance $\Theta = 36.4°$) following the September 2, 1992 Nicaragua tsunamigenic earthquake. In addition to the indicated P, PP and S arrivals, there is a characteristic long-period W phase which arrives in the 5–6 minute time interval between the P and S waves. (*Bottom*) Five synthetic radial-component seismograms obtained by normal-mode summation; the half-durations $\tau_{\rm h} = 5$–60 s of the boxcar source time functions $\dot{m}(t) = (2\tau_{\rm h})^{-1}[H(t + \tau_{\rm h}) - H(t - \tau_{\rm h})]$ are indicated in each case. All seismograms have been Butterworth filtered between 1 and 30 mHz.

deep-focus earthquake (Ekström 1995). The three-component seismograph is situated at station ST4 of the BANJO array, 647 km above and approximately the same distance south of the hypocenter (Jiao, Wallace, Beck, Silver & Zandt 1995). The radial, north-south and east-west seismograms have been computed by convolving an isosceles triangle of half-duration $\tau_{\rm h} = 20$ s with the approximate mode-sum impulse response (10.61). Such a source time function $\dot{m}(t)$ is in good agreement with a number of detailed teleseismic studies of the rupture history of this well-characterized event

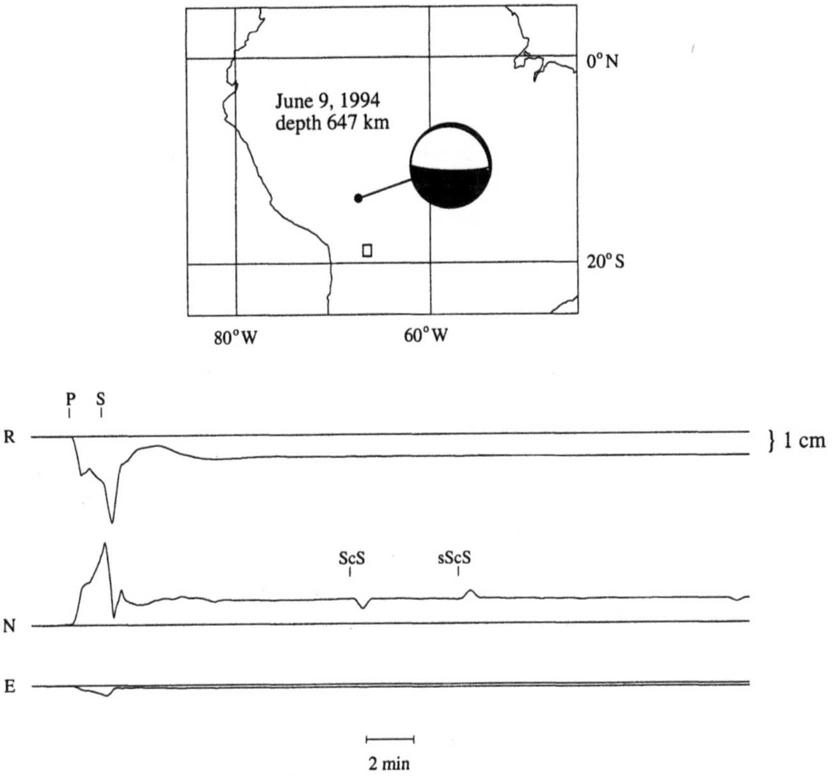

Figure 10.19. (*Top*) Dot with attached beachball shows location and focal mechanism of the great deep-focus Bolivia earthquake of June 9, 1994. (*Bottom*) Near-field synthetic displacement seismograms at station ST4 of the Banjo broad-band array, denoted by the open square in map. The radial (R), north-south (N) and east-west (E) components as well as the P, S, ScS and sScS body-wave arrivals are labelled. The first motion of the P wave is downward and toward the north, whereas that of the S wave is downward and toward the south; the station was permanently displaced approximately one centimeter downward and toward the north by the faulting. The ramp between P and S is a manifestation of the same intermediate-field and near-field energy that gives rise to the teleseismic W phase, particularly after slow, shallow-focus earthquakes such as the one in Nicaragua (see Figure 10.18).

(Kikuchi & Kanamori 1994; Ihmlé & Jordan 1995). The resulting synthetic displacement records, shown in Figure 10.19, are remarkable: they exhibit a ramp between the arrivals of the far-field P and S waves, corresponding to the expected intermediate-field and near-field contributions to the point-source moment-tensor response (Aki & Richards 1980; Vidale, Goes & Richards 1995. In addition, there is a clear permanent offset of the receiver—approximately one centimeter—downward and toward the north; this represents the final static displacement (10.62) due to the static slip on the fault. Later-arriving far-field waves such as $\mathrm{ScS_{SV}}$ and $\mathrm{sScS_{SV}}$ on the north-south component do not affect this static displacement $\mathbf{s}_{\mathrm{f}}(\mathbf{x})$.

10.6 Stacking and Stripping

Suppose we wish to isolate a particular toroidal or spheroidal multiplet from a collection of long-period seismograms recorded at a number of seismic stations distributed all around the globe. This can be done by combining or *stacking* the spectra associated with a single large earthquake in a manner that enhances the characteristics of the target mode while simultaneously reducing the noise. To describe this procedure, we ignore the effect of anelasticity upon the normal-mode eigenfunctions, and utilize the spectral approximation (10.64)–(10.65) for simplicity. Upon making use of the vector spherical-harmonic orthogonality relations (8.85), it is easily shown that the real vector amplitudes (10.52) satisfy

$$\int_\Omega {}_n\mathbf{A}_l(\mathbf{x})\cdot {}_{n'}\mathbf{A}_{l'}(\mathbf{x})\,d\Omega = D_{nn'}\delta_{ll'}, \tag{10.73}$$

where the integration is performed over the epicentral angular coordinates Θ, Φ, and where

$$\begin{aligned} D_{nn'} &= (UU' + VV' + WW') \\ &\quad\times [A_0A_0' + \tfrac{1}{2}k^2(A_1A_1' + B_1B_1') \\ &\qquad + k^2(k^2-2)(A_2A_2' + B_2B_2')]. \end{aligned} \tag{10.74}$$

For brevity we have omitted the mode-identification subscripts n, n' and l on the right side of equation (10.74); unprimed quantities depend upon the overtone number n, whereas primed quantities depend upon the overtone number n'. Both the primed and unprimed eigenfunctions are evaluated at the radius of the receiver, assumed to be $r = a$.

The ideal stack for the multiplet ${}_n\mathrm{S}_l$ or ${}_n\mathrm{T}_l$ is a weighted integral over the surface of the Earth:

$${}_n\Sigma_l(\omega) = \int_\Omega {}_n\mathbf{A}_l(\mathbf{x})\cdot\mathbf{a}(\mathbf{x},\omega)\,d\Omega. \tag{10.75}$$

Using the result (10.73), we find that

$$\mathsf{\Sigma}(\omega) = \mathsf{D}\mathsf{\eta}(\omega), \tag{10.76}$$

where

$$\boldsymbol{\Sigma} = \begin{pmatrix} \vdots \\ {}_n\Sigma_l \\ \vdots \end{pmatrix}, \qquad \boldsymbol{\eta} = \begin{pmatrix} \vdots \\ {}_n\eta_l \\ \vdots \end{pmatrix}, \tag{10.77}$$

$$\mathsf{D} = \begin{pmatrix} & \vdots & \\ \cdots & D_{nn'} & \cdots \\ & \vdots & \end{pmatrix}. \tag{10.78}$$

We can think of (10.75) as a phase-equalization process in which the complex spectra $\mathbf{a}(\mathbf{x}, \omega)$ are first aligned and then summed or stacked to enhance the signal of a target multiplet ${}_n\mathrm{S}_l$ or ${}_n\mathrm{T}_l$. All of the multiplets whose type (spheroidal or toroidal) and angular degree l differ from those of the target are eliminated by this procedure; only modes of the same type and degree but different overtone number n are retained.

In practice we can never evaluate the integral in equation (10.75) exactly, because our funding agencies have not seen fit to provide us with a luxurious carpet of three-component seismometers covering the surface of the Earth. We must be content with a finite collection of spectra obtained from a global network of receivers situated at points $\mathbf{x}_j$, $j = 1, 2, \ldots, J$. The stack (10.75) is in that case approximated by the sum

$$\Sigma(\omega) = 4\pi J^{-1} \sum_{j=1}^{J} \mathbf{A}(\mathbf{x}_j) \cdot \mathbf{a}(\mathbf{x}_j, \omega), \tag{10.79}$$

where we have dropped the subscripts n and l for simplicity. Data from any number of different events $i = 1, 2, \ldots, I$ can easily be incorporated into the stack by replacing equation (10.79) by

$$\Sigma(\omega) = 16\pi^2 I^{-1} J^{-1} \sum_{i=1}^{I} \sum_{j=1}^{J} \mathbf{A}_i(\mathbf{x}_j) \cdot \mathbf{a}_i(\mathbf{x}_j, \omega). \tag{10.80}$$

The quantity $\mathbf{A}_i(\mathbf{x}_j)$ is the theoretical amplitude pattern and $\mathbf{a}_i(\mathbf{x}_j, \omega)$ is the observed acceleration for the ith earthquake source. We have assigned the same differential area element $d\Omega = 4\pi I^{-1}$ or $d\Omega = 4\pi J^{-1}$ to each earthquake $\mathbf{x}_i$ or seismometer $\mathbf{x}_j$ in (10.79)–(10.80) for simplicity; more generally, one may wish to introduce a weighting intended to account for the non-uniform distribution of sources and receivers. The incompleteness of the geographical coverage as well as other practical problems such as

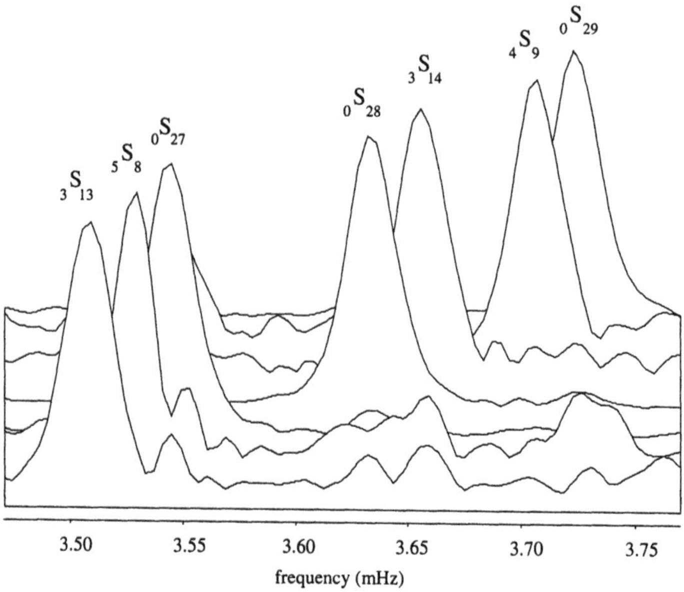

Figure 10.20. Results of multiplet stripping in a small frequency band between 3.50 and 3.75 mHz. The target multiplets are (*front to back*) $_3S_{13}$, $_5S_8$, $_0S_{27}$, $_0S_{28}$, $_3S_{14}$, $_4S_9$ and $_0S_{29}$. The quantity plotted in each case is the stripped amplitude spectrum $|_n\eta_l(\omega)|$. (Courtesy of R. Widmer.)

missing or noisy components destroys the theoretical block-diagonal character of the matrix D in (10.76); as a result, multiplets of different type, or different angular degree, or both, may be present in the non-ideal stacks $_n\Sigma_l(\omega)$. The rotation, hydrostatic ellipticity and lateral heterogeneity of the Earth, which are neglected in (10.73)–(10.74), also contribute to this interference. Despite these difficulties, stacking is a useful procedure inasmuch as the target multiplet and overtones of the same type and degree are usually enhanced relative to the interfering modes.

In general, the well-excited overtones of fixed type and degree l are well separated in frequency, with rare exceptions near PKIKP–ScS$_{\rm SV}$ avoided crossings in the case of the spheroidal modes. There is therefore little difficulty in identifying the peaks associated with $_n\mathrm{S}_l$ and $_n\mathrm{T}_l$ in the stacks $_n\Sigma_l(\omega)$. An additional step, known as *stripping*, often leads to improved results. Since D is a known, symmetric, positive-definite matrix it is possible to invert equation (10.76) to recover the individual resonance peaks:

$$\eta(\omega) = \mathsf{D}^{-1}\Sigma(\omega). \tag{10.81}$$

In practical applications of equation (10.81) the generalized inverse D^{-g} obtained by singular-value decomposition of D is generally employed in place of D^{-1} in order to suppress numerical instabilities. Each resulting entry ${}_n\eta_l(\omega)$ in the column vector $\eta(\omega)$ is known as a *multiplet strip*. The degenerate eigenfrequencies ${}_n\omega_l^{\mathrm{S}}$ and ${}_n\omega_l^{\mathrm{T}}$ can be measured by least-squares fitting a Lorentzian (10.65) to each strip. Careful attention must be paid to a host of subtle details in order to implement stacking and stripping successfully, and minimize the bias (Widmer 1991). Both methods require a knowledge of the locations $\mathbf{x}_s$ and moment tensors $\mathbf{M}$ of every source.

Mendiguren (1973) was the first seismologist to apply the principle of stacking to a global network of seismic stations. He used $\operatorname{sgn}\mathbf{A}(\mathbf{x}_j)$ rather than $\mathbf{A}(\mathbf{x}_j)$ as a weight in (10.79), i.e., he simply reversed the polarity of records whose amplitude of excitation was negative before summing. The seismograms employed in this pioneering study were manually digitized World-Wide Standard Seismographic Network recordings of the July 31, 1970 Colombia deep-focus earthquake. The stacking and stripping procedure outlined above was then developed by Gilbert & Dziewonski (1975); they measured the degenerate eigenfrequencies of more than 800 of the Earth's free oscillations, and used these together with previously available data to construct two SNREI Earth models—1066A and 1066B—that were compatible with the observations. The transversely isotropic and anelastic PREM model was subsequently fit to essentially the same data set of approximately 1000 normal-mode eigenfrequencies, in addition to 500 summary travel-time measurements and 100 multiplet quality factors (Dziewonski & Anderson 1981).

More recent efforts have focused upon the correction of occasional mode misidentifications and, more importantly, the reduction of the eigenfrequency measurement error by the analysis of many more digital recordings. The high quality of modern digital strips is illustrated nicely in Figure 10.20. The overtone modes ${}_3\mathrm{S}_{13}$, ${}_5\mathrm{S}_8$, ${}_3\mathrm{S}_{14}$ and ${}_4\mathrm{S}_9$ are beautifully separated from the much more strongly excited fundamental modes ${}_0\mathrm{S}_{27}$, ${}_0\mathrm{S}_{28}$ and ${}_0\mathrm{S}_{29}$ in the narrow frequency band between 3.50 and 3.75 mHz. Figures 10.21 and 10.22 show that all of the fundamental spheroidal and toroidal modes in the frequency range 1.5–4.0 mHz can be successfully stripped; each peak has been plotted on a separate shifted frequency scale $\omega - \omega_{\mathrm{PREM}}$ relative to PREM. The distinctive offsets in the observed residual dispersion curves are a manifestation of the rotation of the Earth; the Coriolis coupling between spheroidal and toroidal multiplets gives rise to a "repulsion" of the strips associated with closely spaced pairs having adjacent angular degrees such as ${}_0\mathrm{S}_{11}$–${}_0\mathrm{T}_{12}$ and ${}_0\mathrm{S}_{19}$–${}_0\mathrm{T}_{20}$. This effect is well understood theoretically, as we shall see in Section 14.3.2, so that the resulting bias can be removed. A modern high-quality data set consisting of more than 600 Coriolis-corrected

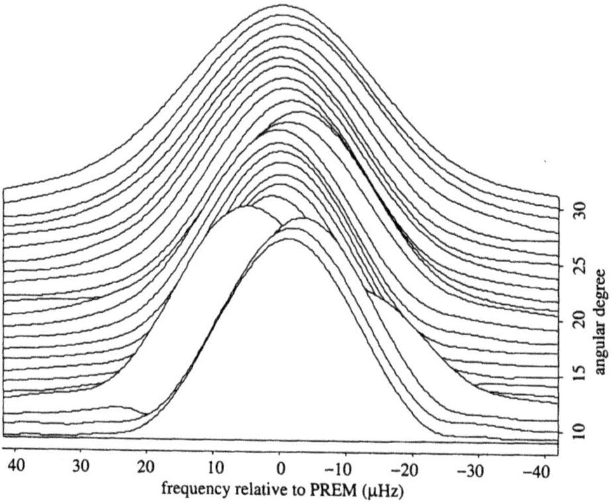

Figure 10.21. Strips $|_n\eta_l(\omega - \omega_{\mathrm{PREM}})|$ of the fundamental spheroidal modes ${}_0\mathrm{S}_8$ (*front*) through ${}_0\mathrm{S}_{30}$ (*back*). Note that the ordinate has been inverted so that $\omega > \omega_{\mathrm{PREM}}$ on the left and $\omega < \omega_{\mathrm{PREM}}$ on the right. (Courtesy of R. Widmer.)

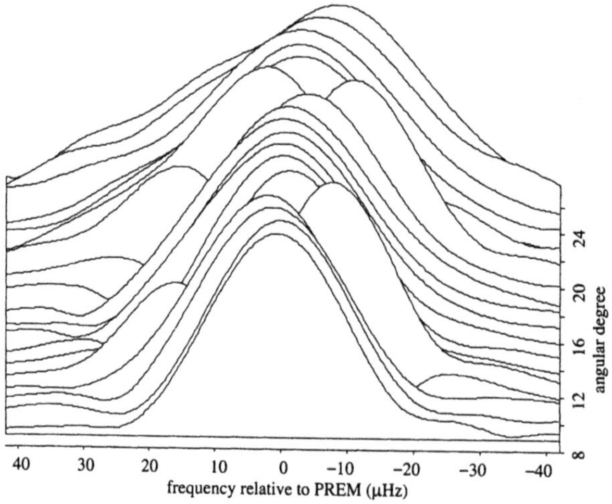

Figure 10.22. Strips $|_n\eta_l(\omega - \omega_{\mathrm{PREM}})|$ of the fundamental toroidal modes ${}_0\mathrm{T}_8$ (*front*) through ${}_0\mathrm{T}_{26}$ (*back*). (Courtesy of R. Widmer.)

eigenfrequencies has been compiled by Masters & Widmer (1995). These measured modes are shown in Figure 10.23. Most of the measurements have been made on multiplet strips involving several thousand recordings.

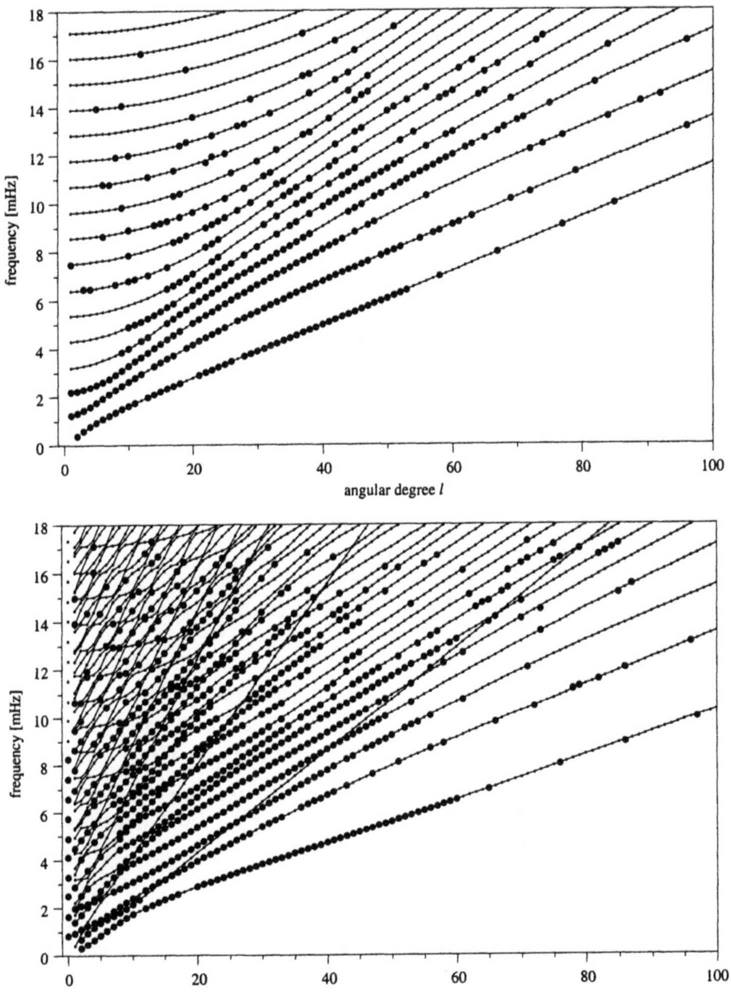

Figure 10.23. Large dots in toroidal-mode (*top*) and spheroidal-mode (*bottom*) dispersion diagrams indicate those degenerate eigenfrequencies that are considered to be "reliably determined" at this time. Many of these high-precision data are not adequately fit by PREM. (Courtesy of R. Widmer.)

★10.7 Alternatives to Mode Summation

There are two alternatives to calculating synthetic spherical-Earth seismograms by normal-mode summation, which we describe briefly in this section. Both methods solve for the response to a moment-tensor source directly in the frequency domain, and compute the time-domain displacement $\mathbf{s}(\mathbf{x}, t)$ or acceleration $\mathbf{a}(\mathbf{x}, t)$ using a numerical inverse Fourier transform. The advantage of such a direct approach is that the spacing between frequency samples can be tailored to the application at hand; the numerical effort is proportional to the number of frequencies and thus to the length of the seismograms that must be synthesized. This may be useful in certain high-frequency applications where short seismograms—perhaps only a single body-wave arrival—are required. In the normal-mode method the spacing between adjacent eigenfrequencies is obviously inalterable; the most laborious step in the procedure is the initial construction of a complete catalogue of eigenfrequencies, multiplet quality factors and eigenfunctions. Given such a catalogue, the calculation of a suite of short mode-sum seismograms takes very little less time than that of a suite of long ones.

The starting point of the first direct method is the non-homogeneous momentum equation

$$\begin{aligned}-\omega^2\rho\mathbf{s} - \boldsymbol{\nabla}\cdot\mathbf{T} + (4\pi G\rho^2 s_r)\hat{\mathbf{r}} + \rho\boldsymbol{\nabla}\phi \\ + \rho g[\boldsymbol{\nabla}s_r - (\boldsymbol{\nabla}\cdot\mathbf{s} + 2r^{-1}s_r)\hat{\mathbf{r}}] \\ = -[\pi\delta(\omega) + (i\omega)^{-1}]\,\mathbf{M}\!:\!\boldsymbol{\nabla}\delta(\mathbf{x} - \mathbf{x}_{\mathrm{s}}).\end{aligned} \tag{10.82}$$

The left side of (10.82) consists of all the terms on the left side of the corresponding homogeneous equation (8.10). The right side is the equivalent force $\mathbf{f}(\mathbf{x}, \omega)$ associated with the earthquake. The quantity $\pi\delta(\omega) + (i\omega)^{-1}$ is the Fourier transform of the Heaviside step function $H(t)$; a more general source time history may be allowed for by employing a frequency-dependent moment tensor $\mathbf{M}(\omega)$ if desired. Upon expanding both the displacement $\mathbf{s}(\mathbf{x}, \omega)$ and the force $\mathbf{f}(\mathbf{x}, \omega)$ in vector spherical harmonics $\mathbf{P}_{lm}$, $\mathbf{B}_{lm}$ and $\mathbf{C}_{lm}$, we obtain three second-order ordinary differential equations analogous to (8.43)–(8.45) governing the toroidal, spheroidal and radial modes. Abbreviating the corresponding homogeneous systems of first-order equations (8.114)–(8.115), (8.135)–(8.140), (8.144)–(8.147) and (8.149)–(8.150) by $\dot{\mathsf{y}} = \mathsf{A}\mathsf{y}$, we are required to solve a number of non-homogeneous systems of the form

$$\dot{\mathsf{y}} = \mathsf{A}\mathsf{y} + \mathsf{z}_1\delta(r - r_{\mathrm{s}}) + \mathsf{z}_2\dot{\delta}(r - r_{\mathrm{s}}), \tag{10.83}$$

where the vectors $\mathsf{z}_1(\omega)$ and $\mathsf{z}_2(\omega)$ depend upon the geometry $\mathbf{M}$ and angular location $\theta_{\mathrm{s}}, \phi_{\mathrm{s}}$ of the source. The presence of the non-homogeneous term removes the $2l+1$ degeneracy of $\mathsf{y}(r,\omega)$; placing the source at the pole ($\theta_s \to 0$) reduces the computational burden substantially, since $\mathbf{z}_1(\omega)$ and $\mathbf{z}_2(\omega)$ are then non-zero only for orders $m=0$, $m=\pm 1$ and $m=\pm 2$. The solution to equation (10.83) is of the form $\mathsf{y}(r,\omega) = \mathsf{v}(r,\omega) + \mathsf{z}_2(\omega)\delta(r-r_s)$ where $\dot{\mathsf{v}} = \mathsf{A}\mathsf{v} + (\mathsf{z}_1 + \mathsf{A}\mathsf{z}_2)\delta(r-r_s)$. The vector $\mathsf{v}(r,\omega)$ can be found by numerical integration; the inhomogeneous term gives rise to a jump discontinuity $\mathsf{z}_1(\omega) + \mathsf{A}(r_s,\omega)\mathsf{z}_2(\omega)$ at the source depth $r=r_s$. The boundary conditions (8.116), (8.141)–(8.142) and (8.151) on the free surface $r=a$ and any fluid-solid discontinuities $r=d_{\mathrm{FS}}$ are identical to those governing the toroidal, spheroidal and radial normal modes. Numerical instabilities arising from the stiffness of the system (10.83) can be overcome as in the homogeneous case by solving for the minors of the solution matrices instead of the matrices themselves. Anelasticity can be accounted for by using complex, frequency-dependent elastic parameters $\kappa_0[1+iQ_\kappa^{-1}+\frac{2}{\pi}Q_\kappa^{-1}\ln(\omega/\omega_0)]$ and $\mu_0[1+iQ_\mu^{-1}+\frac{2}{\pi}Q_\mu^{-1}\ln(\omega/\omega_0)]$ in the coefficient matrices $\mathsf{A}(r,\omega)$. A numerical implementation of this *direct radial integration* method is described by Friederich & Dalkolmo (1995). They discuss strategies for minimizing aliasing artifacts, and present comparisons with mode-sum seismograms to establish the validity of their results.

The second alternative is a *direct algebraic* approach, based upon the development outlined in Section 7.7. Choosing basis functions proportional to the vector spherical harmonics $\mathbf{P}_{lm}$, $\mathbf{B}_{lm}$ and $\mathbf{C}_{lm}$ decomposes the kinetic and potential energy matrices T and $\mathsf{V}(\omega)$ into independent toroidal, spheroidal or radial matrices for each angular degree l. The toroidal matrix elements are given, for example, by

$$T_{kk'} = \int_b^s \rho X_k X_{k'}\, r^2 dr, \tag{10.84}$$

$$V_{kk'}(\omega) = \int_b^s \mu_0[1+iQ_\mu^{-1}+\tfrac{2}{\pi}Q_\mu^{-1}\ln(\omega/\omega_0)] \\ \times [(r\dot{X}_k - X_k)(r\dot{X}_{k'} - X_{k'}) + (l-1)(l+2)X_k X_{k'}]\, dr, \tag{10.85}$$

where $X_k(r)$ are the radial basis functions, which are still at our disposal. Local triangles or linear splines are a popular choice, because they are simple and lead to sparse matrices:

$$X_k(r) = \begin{cases} (r-r_{k-1})/(r_k-r_{k-1}) & \text{if } r_{k-1}\le r\le r_k \\ (r_{k+1}-r)/(r_{k+1}-r_k) & \text{if } r_k\le r\le r_{k+1} \\ 0 & \text{otherwise,} \end{cases} \tag{10.86}$$

where $b = r_1 < r_2 < \cdots < r_K = s$. The first and second lines of (10.86) are ignored at the end points $k=1$ and $k=K$, respectively. Placing the source at the pole ($\theta_s \to 0$) is once again advantageous, since it limits the source vector s to elements of order $|m| \leq 2$. Using (B.98) to transform the limiting relations (D.22)–(D.27) we find the non-zero toroidal elements:

$$s_{k\,\pm 1} = \begin{cases} -\left(\frac{2l+1}{8\pi}\right)^{1/2} M_{r\theta}(\dot{X}_k - r^{-1}X_k)_{r=r_s} \\ \left(\frac{2l+1}{8\pi}\right)^{1/2} M_{r\phi}(\dot{X}_k - r^{-1}X_k)_{r=r_s}, \end{cases} \tag{10.87}$$

$$s_{k\,\pm 2} = \begin{cases} \frac{1}{2}\left(\frac{2l+1}{8\pi}\right)^{1/2} \sqrt{(l-1)(l+2)}\,(M_{\theta\theta} - M_{\phi\phi})(r^{-1}X_k)_{r=r_s} \\ -\left(\frac{2l+1}{8\pi}\right)^{1/2} \sqrt{(l-1)(l+2)}\,M_{\theta\phi}(r^{-1}X_k)_{r=r_s}. \end{cases}$$

The corresponding elements of the toroidal receiver vector r are

$$r_{k\,\pm 1} = X_k(r)\,\hat{\boldsymbol{\nu}} \cdot \mathbf{C}_{l\,\pm 1}(\theta, \phi), \tag{10.88}$$

$$r_{k\,\pm 2} = X_k(r)\,\hat{\boldsymbol{\nu}} \cdot \mathbf{C}_{l\,\pm 2}(\theta, \phi). \tag{10.89}$$

To find the degree l and order $m = \pm 1, \pm 2$ response, we solve the linear frequency-dependent equation

$$[\mathsf{V}(\omega) - \omega^2 \mathsf{T}]\mathsf{d}(\omega) = \mathsf{s}. \tag{10.90}$$

The acceleration $a(\omega) = \hat{\boldsymbol{\nu}} \cdot \mathbf{a}(\mathbf{x}, \omega)$ is then given in terms of the toroidal response vector $\mathsf{d}(\omega)$ by

$$a(\omega) = i\omega \mathsf{r}^{\mathrm{T}} \mathsf{d}(\omega). \tag{10.91}$$

Numerical results obtained using a perfectly elastic, complex spherical-harmonic version of equations (10.84)–(10.91) are presented by Cummins, Geller, Hatori & Takeuchi (1994). They describe a number of minor modifications which improve the accuracy for a given knot spacing. The corresponding spheroidal formulation is more complicated, because of the need to account for the different number of degrees of freedom in the solid inner core and mantle $\oplus_{\mathrm{F}}$ and the fluid outer core and ocean $\oplus_{\mathrm{S}}$, as well as the possibility of slip upon the fluid-solid boundaries d_{FS} (Cummins, Geller & Takeuchi 1994; Takeuchi, Geller & Cummins 1996). The validity of the spherical-Earth direct-solution code is demonstrated by Cummins (1997); he presents an impressive wiggle-for-wiggle reproduction of the radial near-field seismogram in Figure 10.19.

Chapter 11

Love and Rayleigh Waves

Thus far in this book, we have expressed the response of the Earth to an earthquake source as a superposition of orthonormal free oscillations, or *standing waves*. Alternatively, it is possible to decompose the response into a sum of *travelling waves*. The conversion of a standing-wave representation into a travelling-wave representation on a sphere is a classical analytical procedure, which is applicable to acoustic or electromagnetic waves as well as to elastic waves. The mathematical basis of the procedure is the so-called *Watson transformation*, which we introduce in Section 11.1. Our subsequent seismological application of this transformation in Sections 11.2 and 11.4 extends the analyses of Gilbert (1976a), Dahlen (1979a) and Snieder & Nolet (1987), and provides a unified treatment that accounts for both body waves and surface waves.

Following the transformation of the mode-sum response into a travelling-wave representation that is uniformly valid upon a slightly anelastic spherical Earth, we devote the remainder of the present chapter to Love and Rayleigh surface waves, which are equivalent to the free-oscillation multiplets ${}_n\mathrm{T}_l$ and ${}_n\mathrm{S}_l$ with $n \ll l/4$. The geometrical spreading of these trapped waves is confined to two rather than three dimensions; because of this, and because they are strongly excited by shallow-focus earthquakes, fundamental-mode surface waves are the largest-amplitude arrivals on most seismograms. Long-period waves on a given overtone branch $n = 0, 1, 2, \ldots$ "feel" the elasticity and density of the Earth more deeply than short-period waves do; as a result, Love and Rayleigh wave propagation is *dispersive*. In practice, surface-wave dispersion is regionally variable because of geographical differences in crustal and upper-mantle structure; much of the present analysis is directly applicable to a laterally heterogeneous Earth in the context of the JWKB approximation, as we shall see in Chapter 16.

11.1 Watson Transformation

The first step in the conversion procedure is to express the sum over angular degree l as an integral over wavenumber k. This is accomplished by means of the so-called *Watson transformation*, which states that

$$\sum_{l=0}^{\infty} f(l+\tfrac{1}{2}) = \tfrac{1}{2}\int_C f(k)e^{-ik\pi}(\cos k\pi)^{-1}dk. \tag{11.1}$$

Equation (11.1) is an identity which is valid for any function $f(k)$ that is analytic in the vicinity of the real k axis; the integration is along the closed contour C in the complex k plane shown in Figure 11.1. The integrand has simple poles at the positive half-integer values $k = 1/2,\, 3/2,\, 5/2, \ldots$, so that the result is easily verified by evaluating the contour integral using the residue theorem.

A more useful variant of the Watson transformation for our purposes is the *Poisson sum formula*, which may be readily obtained from equation (11.1), as we show next. On the portion C^- of the contour situated in the lower half of the k plane, we can write

$$(\cos k\pi)^{-1} = \frac{2e^{-ik\pi}}{1+e^{-2ik\pi}} = -2\sum_{s=1}^{\infty}(-1)^s e^{-i(2s-1)k\pi}. \tag{11.2}$$

Since $\operatorname{Im} k < 0$ on C^- the summation over s converges. By a similar argument we can write, on the portion C^+ of the contour in the upper half

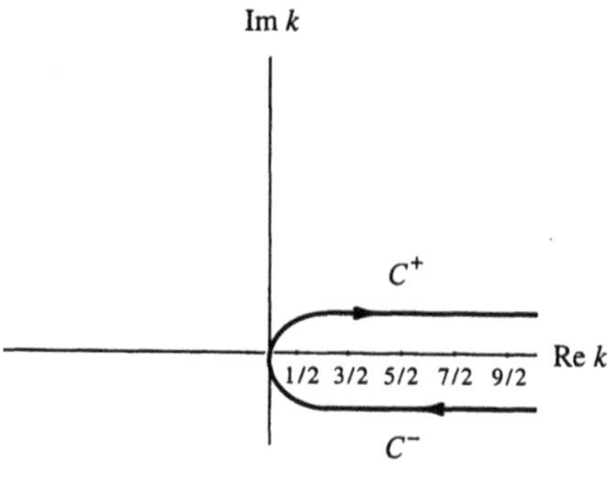

Figure 11.1. Schematic diagram of the complex wavenumber plane showing the Watson contour $C = C_- + C_+$. The residue of the simple pole at $k = l + 1/2$ is $i\pi^{-1}f(l+1/2)$.

plane,

$$(\cos k\pi)^{-1} = \frac{2e^{ik\pi}}{1+e^{2ik\pi}} = 2\sum_{s=-\infty}^{0}(-1)^s e^{-i(2s-1)k\pi}. \tag{11.3}$$

Upon combining equations (11.2) and (11.3) we can rewrite the transformation (11.1) in the form

$$\sum_{l=0}^{\infty} f(l+\tfrac{1}{2}) = \sum_{s=-\infty}^{\infty}(-1)^s \int_0^{\infty} f(k)e^{-2isk\pi}\,dk, \tag{11.4}$$

where it must be remembered that in the limit $s \to \infty$ the path of integration runs from the origin to $\infty - i0$ *below* the real k axis, whereas in the limit $s \to -\infty$ it runs from the origin to $\infty + i0$ *above* the axis. The precise location of the path for finite values of the index s is unimportant. We will use the identity (11.4) in the next three sections to obtain a travelling-wave representation of the spherical-Earth Green tensor and the acceleration response to an earthquake.

11.2 Travelling-Wave Decomposition

Changing the notation slightly from that employed in Section 10.2, we write the Fourier transform of the spherical-Earth Green tensor

$$\mathbf{G}(\mathbf{x},\mathbf{x}';\omega) = \int_0^{\infty} \mathbf{G}(\mathbf{x},\mathbf{x}';t)e^{-i\omega t}\,dt \tag{11.5}$$

in the form

$$\mathbf{G} = \frac{1}{2\pi}\sum_{n=0}^{\infty}\sum_{l=0}^{\infty}(l+\tfrac{1}{2})\,{}_n\mathbf{G}_l\,P_l(\cos\Theta), \tag{11.6}$$

where

$${}_n\mathbf{G}_l = {}_n\mathbf{D}_l\left[\frac{\frac{1}{2}(i\,{}_n\omega_l)^{-1}}{{}_n\gamma_l + i(\omega - {}_n\omega_l)} + \frac{\frac{1}{2}(-i\,{}_n\omega_l)^{-1}}{{}_n\gamma_l + i(\omega + {}_n\omega_l)}\right]{}_n\mathbf{D}_l'. \tag{11.7}$$

Equation (11.7) is correct to first order in the intrinsic reciprocal quality factors Q_κ^{-1} and Q_μ^{-1}. The complex eigenfrequencies ${}_n\nu_l = {}_n\omega_l + i\,{}_n\gamma_l$ have been approximated by ${}_n\omega_l$ except in the Lorentzians $\frac{1}{2}[{}_n\gamma_l + i(\omega \pm {}_n\omega_l)]^{-1}$. In addition, we have ignored the effect of anelasticity upon the radial eigenfunctions, and replaced the complex receiver and source operators $\mathcal{D}$ and $\mathcal{D}'$ in equations (10.28)–(10.30) by $\mathbf{D} = U\hat{\mathbf{r}} + k^{-1}V\boldsymbol{\nabla}_1 - k^{-1}W(\hat{\mathbf{r}}\times\boldsymbol{\nabla}_1)$ and $\mathbf{D}' = U'\hat{\mathbf{r}}' + k^{-1}V'\boldsymbol{\nabla}_1' - k^{-1}W'(\hat{\mathbf{r}}'\times\boldsymbol{\nabla}_1')$.

Upon using the Poisson sum formula (11.4) to convert the summation over angular degree l in (11.6) to an integral over the wavenumber k, we obtain the representation

$$\mathbf{G} = \frac{1}{2\pi}\sum_{n=0}^{\infty}\sum_{s=-\infty}^{\infty}(-1)^s\int_0^{\infty}\mathbf{G}_n(k)\,P_{k-\frac{1}{2}}(\cos\Theta)\,e^{-2isk\pi}\,kdk. \qquad (11.8)$$

The double-differential operator $\mathbf{G}_n(k)$ in (11.8) is defined for $\mathrm{Re}\,k > 0$ by analytic continuation between the integer values of angular degree l:

$$\mathbf{G}_n(k) = {}_n\mathbf{G}_l \quad \text{at the points} \quad k = \sqrt{l(l+1)}. \qquad (11.9)$$

The radial eigenfunction equations (8.43)–(8.45) and (8.53) and the associated boundary conditions (8.50)–(8.52) and (8.54), together with their counterparts on a transversely isotropic Earth, are invariant under the transformations $k \to -k$ and $V_n(k) \to -V_n(-k)$, $W_n(k) \to -W_n(-k)$; we may use the resulting reflection symmetry of the differential operators $\mathbf{D}_n(k)$ and $\mathbf{D}'_n(k)$ and the associated eigenfrequencies $\omega_n(k)$ and decay rates $\gamma_n(k)$ to extend the definition (11.9) of $\mathbf{G}_n(k)$ to the left half of the complex k plane:

$$\mathbf{G}_n(-k) = \mathbf{G}_n(k). \qquad (11.10)$$

The relation (11.10) stipulates that the operator $\mathbf{G}_n(k)$ is an even function of the complex wavenumber k.

Equation (11.8) is still a standing-wave representation of the Green tensor $\mathbf{G}$ because of the presence of the Legendre function $P_{k-1/2}(\cos\Theta)$. To facilitate the transformation into a travelling-wave representation, we make use of the standard decomposition (B.133) into Legendre functions of the first and second kinds:

$$P_{k-\frac{1}{2}}(\cos\Theta) = Q^{(1)}_{k-\frac{1}{2}}(\cos\Theta) + Q^{(2)}_{k-\frac{1}{2}}(\cos\Theta), \qquad (11.11)$$

where $Q^{(1,2)}_{k-1/2}$ correspond to waves that are propagating in the directions of increasing and decreasing Θ, respectively. Upon inserting equation (11.11) into (11.8) and rearranging the sum from $s = -\infty$ to $s = \infty$ so that it is a sum over the positive odd and even integers, we obtain

$$\begin{aligned}\mathbf{G} = \frac{1}{2\pi}\sum_{n=0}^{\infty}\Bigg\{ &\sum_{s=1,3,5,\ldots}^{\infty}(-1)^{(s-1)/2}\int_0^{\infty}\mathbf{G}_n(k)\,Q^{(1)}_{k-\frac{1}{2}}(\cos\Theta) \\ &\qquad\times\left[e^{-i(s-1)k\pi} - e^{i(s+1)k\pi}\right]kdk \\ &+ \sum_{s=2,4,6,\ldots}^{\infty}(-1)^{s/2}\int_0^{\infty}\mathbf{G}_n(k)\,Q^{(2)}_{k-\frac{1}{2}}(\cos\Theta)\end{aligned}$$

$$\times \left[e^{-isk\pi} - e^{i(s-2)k\pi}\right] kdk \Bigg\}. \qquad (11.12)$$

The integrals involving the exponentials $\exp[-i(s-1)k\pi]$ and $\exp[-isk\pi]$ in (11.12) are evaluated just below the real k axis, whereas those involving $\exp[i(s+1)k\pi]$ and $\exp[i(s-2)k\pi]$ are evaluated just above. The last step in obtaining a travelling-wave Green tensor is to make the substitution $k \rightarrow -k$ in each of the upper-half-plane terms, in order to obtain a sum of integrals that extend from $k = -\infty$ to $k = \infty$. Equations (B.138), which we reproduce here for convenience, relate the travelling-wave Legendre functions with negative values of k to those with positive values:

$$Q^{(1,2)}_{-k-\frac{1}{2}}(\cos\Theta) = e^{\pm 2ik\pi} Q^{(1,2)}_{k-\frac{1}{2}}(\cos\Theta) \\ + e^{\pm ik\pi} \tan k\pi P_{k-\frac{1}{2}}(-\cos\Theta). \qquad (11.13)$$

Using these relations, together with the symmetry (11.10), the representation (11.12) can be rewritten in the form

$$\mathbf{G} = \frac{1}{2\pi} \sum_{n=0}^{\infty} \Bigg[\sum_{s=1,3,5,\ldots}^{\infty} (-1)^{(s-1)/2} \int_{-\infty}^{\infty} \mathbf{G}_n(k)\, Q^{(1)}_{k-\frac{1}{2}}(\cos\Theta) \\ \times e^{-i(s-1)k\pi}\, kdk \;+ \sum_{s=2,4,6,\ldots}^{\infty} (-1)^{s/2} \\ \times \int_{-\infty}^{\infty} \mathbf{G}_n(k)\, Q^{(2)}_{k-\frac{1}{2}}(\cos\Theta) e^{-isk\pi}\, kdk \Bigg], \qquad (11.14)$$

where we have used the fact that the integrals involving $P_{k-1/2}(-\cos\Theta)$ cancel. The substitution $k \rightarrow -k$ shifts the $s \rightarrow \infty$ path of integration from the first quadrant into the third quadrant, so that the contour now runs just beneath the real wavenumber axis, as shown in Figure 11.2.

Equation (11.14) is the most convenient representation of the travelling-wave Green tensor $\mathbf{G}(\mathbf{x}, \mathbf{x}'; \omega)$ on a spherical Earth. It is applicable to all types of waves, including SH and P-SV body waves as well as Love and Rayleigh surface waves. The first two terms $s = 1, 2$ correspond to waves that propagate from the source at $\mathbf{x}'$ to the receiver at $\mathbf{x}$ along the minor arc and the major arc, respectively, whereas the remaining terms $s = 3, 5, \ldots$ and $s = 4, 6, \ldots$ correspond to multi-orbit waves that encircle the Earth one or more times before arriving. The two travelling-wave Legendre functions $Q^{(1,2)}_{k-1/2}(\cos\Theta)$ diverge like $\ln\Theta$ in the vicinity of the epicenter $\Theta = 0$ and like $\ln(\pi - \Theta)$ in the vicinity of the antipode $\Theta = \pi$, respectively. The infinite sum over all of the odd-plus-even travelling-wave arrivals is nevertheless regular at all points $\mathbf{x}$ within the Earth; this is obvious, inasmuch

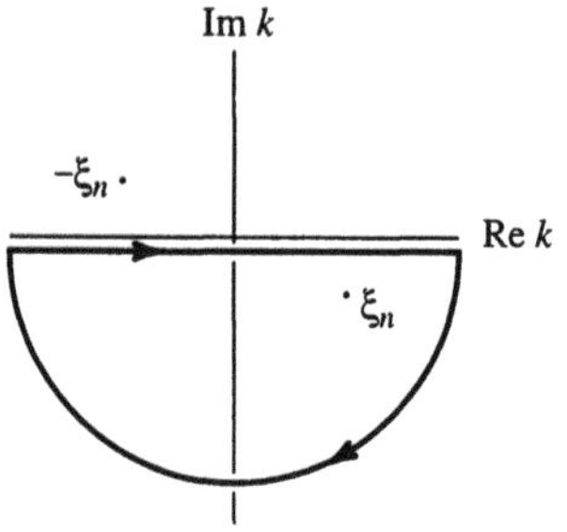

Figure 11.2. After the substitution $k \to -k$ the path of integration in equation (11.14) runs from $-\infty - i0$ to $\infty - i0$. The integrand has two surface-wave poles lying in the fourth and second quadrants, respectively. Closure of the contour in $\operatorname{Im} k < 0$ encircles $\xi_n(\omega)$ but not $-\xi_n(\omega)$.

as it is equivalent to the standing-wave sum (11.6)–(11.7). We shall use the uniformly valid result (11.14) to find the surface-wave Green tensor in the next section. Further consideration of the body-wave response will be deferred until Section 12.5.

11.3 Surface-Wave Green Tensor

The fundamental and higher-overtone Love and Rayleigh surface waves all propagate independently of each other on a spherically symmetric Earth. To obtain the Green tensor corresponding to each dispersion branch, we evaluate the integral over wavenumber k in (11.14) for each fixed overtone number n and arrival number s by a straightforward application of the residue theorem. For a given angular frequency ω, the quantity $\mathbf{G}_n(k)$ has two simple poles $k = \pm\xi_n(\omega)$ in the complex wavenumber plane. For concreteness, we shall suppose in what follows that $\omega > 0$. The poles then stem from the presence of the positive-frequency Lorentzian in equation (11.7), and are given implicitly by

$$\gamma_n(\xi_n) + i[\omega - \omega_n(\xi_n)] = 0. \tag{11.15}$$

In the limit of weak attenuation, we can obtain an explicit formula for $\xi_n(\omega)$ by expanding the eigenfrequency and decay rate about the point $k = k_n(\omega)$, where $k_n(\omega)$ is the real wavenumber of waves of frequency ω on the unperturbed elastic Earth:

$$\begin{aligned} \omega_n(\xi_n) &= \omega_n(k_n) + C_n(\xi_n - k_n) + \cdots \\ &= \omega + C_n(\xi_n - k_n) + \cdots, \end{aligned} \tag{11.16}$$

$$\gamma_n(\xi_n) = \gamma_n(k_n) + \cdots, \tag{11.17}$$

where the ellipses denote higher-order terms that will be ignored. For simplicity, and because it is the case of greatest seismological interest, we shall assume that the angular eigenfrequency ω_n is a *monotonically increasing* function of the wavenumber $k > 0$ along each dispersion branch. The positive slope of the nth dispersion curve,

$$C_n(\omega) = \left(\frac{d\omega_n}{dk}\right)_{k=k_n(\omega)}, \tag{11.18}$$

is then the angular *group speed* of the associated surface waves, measured in radians per second on the unit sphere. Upon inserting the approximations (11.16)–(11.17) into (11.15) we find, correct to first order in the anelasticity, that

$$\xi_n = k_n - i\gamma_n/C_n = k_n - i\omega/2C_nQ_n, \tag{11.19}$$

where $Q_n(\omega)$ is the temporal quality factor. The conditions $C_n > 0$ and $Q_n > 0$ ensure that the surface-wave poles $\xi_n(\omega)$ and $-\xi_n(\omega)$ are situated just below and above the axis, in the fourth and second quadrants of the wavenumber plane, respectively.

The asymptotic behavior of the travelling-wave Legendre functions in the limit $k \to \infty$ and the presence of the exponential terms $\exp[-i(s-1)k\pi]$ and $\exp(-isk\pi)$ require us to close the contour in the lower k plane in order to avoid a contribution from the arc at infinity when evaluating the integral in (11.14) by means of the residue theorem. As a result, the only pole that we encircle is (11.19) in the fourth quadrant, as shown in Figure 11.2. The surface-wave Green tensor $\mathbf{G} = \mathbf{G}_{\text{Love}} + \mathbf{G}_{\text{Rayleigh}}$ obtained by evaluation of the residue is

$$\begin{aligned} \mathbf{G} = \tfrac{1}{2}\sum_{n=0}^{\infty}(c_nC_n)^{-1}\Big\{ &\sum_{s=1,3,5,\ldots}^{\infty} \exp i[(s-2)\pi/2 - (s-1)\xi_n\pi] \\ &\qquad \times \mathbf{D}_n\mathbf{D}'_n Q^{(1)}_{\xi_n-\frac{1}{2}}(\cos\Theta) \\ + &\sum_{s=2,4,6,\ldots}^{\infty} \exp i[(s-1)\pi/2 - s\xi_n\pi] \\ &\qquad \times \mathbf{D}_n\mathbf{D}'_n Q^{(2)}_{\xi_n-\frac{1}{2}}(\cos\Theta)\Big\}. \end{aligned} \tag{11.20}$$

The quantity c_n is the angular surface-wave *phase speed*, given by

$$c_n(\omega) = \omega/k_n(\omega), \tag{11.21}$$

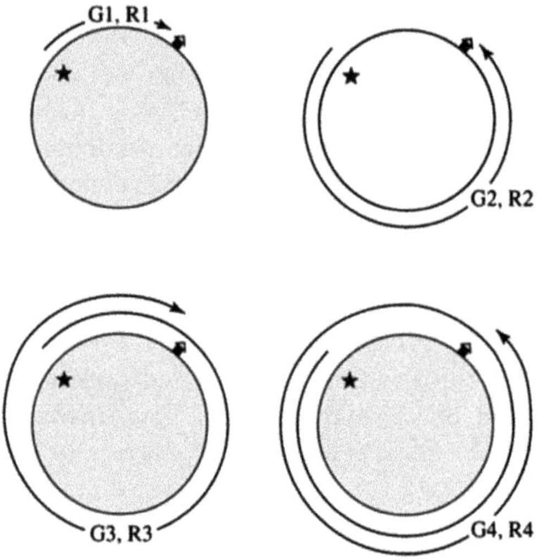

Figure 11.3. Nomenclature for multi-orbit surface waves. The source and seismic station are symbolized by a star and a doghouse. The letters G and R refer to Love and Rayleigh waves, respectively; the numbers denote the order of arrival at the receiver.

and the differential operators $\mathbf{D}_n = U_n\hat{\mathbf{r}} + k_n^{-1}V_n\boldsymbol{\nabla}_1 - k_n^{-1}W_n(\hat{\mathbf{r}} \times \boldsymbol{\nabla}_1)$ and $\mathbf{D}'_n = U'_n\hat{\mathbf{r}}' + k_n^{-1}V'_n\boldsymbol{\nabla}'_1 - k_n^{-1}W'_n(\hat{\mathbf{r}}' \times \boldsymbol{\nabla}'_1)$ are now regarded as functions of frequency. The representation (11.20) is *exact* on a perfectly elastic Earth, and it is correct to first order in the inverse surface-wave quality factor Q_n^{-1} on an anelastic Earth. Every term represents an exponentially decaying wave that departs from the source at $\Theta = 0$; exponentially growing waves associated with the second-quadrant pole $-\xi_n(\omega)$ have been eliminated. The odd arrivals $s = 1, 3, 5, \ldots$ correspond to the multi-orbit Love and Rayleigh wavegroups G1, G3, G5, ... and R1, R3, R5, ..., whereas the even arrivals $s = 2, 4, 6, \ldots$ correspond to G2, G4, G6, ... and R2, R4, R6, ..., as shown in Figure 11.3.

Away from the epicenter $\Theta = 0$ and its antipode $\Theta = \pi$, and in the limit of large real wavenumber, $k_n \gg 1$, and slight anelasticity, $Q_n \gg 1$, we can simplify $\mathbf{G}(\mathbf{x}, \mathbf{x}'; \omega)$ by making use of the asymptotic approximation to the travelling-wave Legendre functions:

$$Q^{(1,2)}_{\xi_n - \frac{1}{2}}(\cos\Theta) \approx (2\pi k_n \sin\Theta)^{-1/2} \times \exp[\mp i(k_n\Theta - \pi/4) \mp \omega\Theta/2C_nQ_n]. \tag{11.22}$$

The resulting *far-field Green tensor* can be written as a double sum over *modes* or dispersion branches $n = 0, 1, 2, \ldots$ and *surface-wave orbits* or sequential arrivals $s = 1, 2, 3, \ldots$ in the form

$$\mathbf{G}(\mathbf{x}, \mathbf{x}'; \omega) = \sum_{\text{modes}} \sum_{\text{rays}} (cC)^{-1} (8\pi k |\sin\Delta|)^{-1/2}$$
$$\times [\hat{\mathbf{r}}U - i\hat{\mathbf{k}}V + i(\hat{\mathbf{r}} \times \hat{\mathbf{k}})W][\hat{\mathbf{r}}'U' + i\hat{\mathbf{k}}'V' - i(\hat{\mathbf{r}}' \times \hat{\mathbf{k}}')W']$$
$$\times \exp i\,[-k\Delta + (s-1)\pi/2 - \pi/4] \exp(-\omega\Delta/2CQ), \tag{11.23}$$

where we have dropped the numerical indices and identifying subscripts for simplicity. The quantity Δ, which is the total angular distance traversed by a given arrival, is given explicitly by

$$\Delta = \begin{cases} \Theta + (s-1)\pi, & s \text{ odd} \\ s\pi - \Theta, & s \text{ even.} \end{cases} \tag{11.24}$$

For G1 and R1 waves, $\Delta = \Theta$; for G2 and R2 waves, $\Delta = 2\pi - \Theta$; for G3 and R3 waves, $\Delta = \Theta + 2\pi$; etc. The quantities $\hat{\mathbf{k}}$ and $\hat{\mathbf{k}}'$ are *unit wavevectors* that point in the direction of propagation at $\mathbf{x}$ and the ray takeoff direction at $\mathbf{x}'$, respectively, i.e.,

$$\hat{\mathbf{k}} = \begin{cases} \hat{\mathbf{\Theta}}, & s \text{ odd} \\ -\hat{\mathbf{\Theta}}, & s \text{ even,} \end{cases} \qquad \hat{\mathbf{k}}' = \begin{cases} \hat{\mathbf{\Theta}}', & s \text{ odd} \\ -\hat{\mathbf{\Theta}}', & s \text{ even,} \end{cases} \tag{11.25}$$

$$\hat{\mathbf{r}} \times \hat{\mathbf{k}} = \begin{cases} \hat{\mathbf{\Phi}}, & s \text{ odd} \\ -\hat{\mathbf{\Phi}}, & s \text{ even,} \end{cases} \qquad \hat{\mathbf{r}}' \times \hat{\mathbf{k}}' = \begin{cases} \hat{\mathbf{\Phi}}', & s \text{ odd} \\ -\hat{\mathbf{\Phi}}', & s \text{ even.} \end{cases} \tag{11.26}$$

Physically, we can regard $\hat{\mathbf{r}}$, $\hat{\mathbf{k}}$ and $\hat{\mathbf{r}} \times \hat{\mathbf{k}}$ as the radial, longitudinal and transverse polarizations of surface waves which propagate around the Earth in either a clockwise or counterclockwise sense, as illustrated in Figure 11.4. It is noteworthy that the far-field Green tensor (11.23) is consistent with the principle of source-receiver reciprocity, $\mathbf{G}(\mathbf{x}, \mathbf{x}'; \omega) = \mathbf{G}^{\mathrm{T}}(\mathbf{x}', \mathbf{x}; \omega)$, by virtue of the reversal in the direction of propagation, $\hat{\mathbf{k}} \to -\hat{\mathbf{k}}$ and $\hat{\mathbf{k}}' \to -\hat{\mathbf{k}}'$, upon interchange of $\mathbf{x}$ and $\mathbf{x}'$.

As already noted, the results in this section are valid only for positive angular frequencies, $\omega > 0$. The corresponding results for $\omega < 0$ can be obtained by a similar application of the residue theorem to the surface-wave pole in the third quadrant, which is associated with the negative-frequency Lorentzian: $\gamma_n(\xi_n) + i[\omega + \omega_n(\xi_n)] = 0$. Alternatively, and more simply, we can make use of the general relation $\mathbf{G}(\mathbf{x}, \mathbf{x}'; -\omega) = \mathbf{G}^*(\mathbf{x}, \mathbf{x}'; \omega)$, which is a consequence of the reality of the time-domain response $\mathbf{G}(\mathbf{x}, \mathbf{x}'; t)$. The $\exp(-ik\Delta)$ rather than $\exp(ik\Delta)$ dependence of $\mathbf{G}(\mathbf{x}, \mathbf{x}'; \omega)$ can be

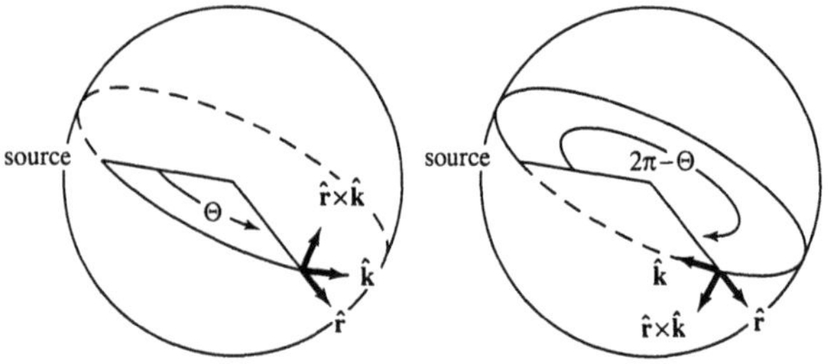

Figure 11.4. Schematic diagram of a G1 or R1 surface-wave ray path traversing the minor arc of length Θ (*left*) and a G2 or R2 ray path traversing the major arc of length $2\pi - \Theta$ (*right*). Heavy arrows show the radial, longitudinal and transverse unit polarization vectors $\hat{\mathbf{r}}$, $\hat{\mathbf{k}}$ and $\hat{\mathbf{r}} \times \hat{\mathbf{k}}$ at the receiver.

traced back to the Fourier-transform sign convention which we adopted in equations (4.2) and (11.5). Waves propagating away from the source, are of the form $\exp i(\omega t - k\Delta)$ rather than $\exp i(k\Delta - \omega t)$.

In certain highly unusual circumstances, fundamental and higher-mode Rayleigh waves on a self-gravitating Earth may have an eigenfrequency ω_n that is a *decreasing* rather than an increasing function of wavenumber $k > 0$ (Gilbert 1967). The angular group speed is in that case defined by $C_n = -(d\omega/dk)_{k=k_n}$ rather than by (11.18), and the surface-wave poles ξ_n and $-\xi_n$ are situated in the third and first quadrants rather than the fourth and second. Only the third-quadrant pole contributes upon closure of the integration contour in the lower k plane; the resulting far-field Green tensor is identical to (11.23), with the phase factor $\exp(-ik\Delta)$ replaced by $\exp(ik\Delta)$. Every monochromatic wave is then of the form $\exp i(\omega t + k\Delta)$; that is, individual crests and troughs propagate toward the source rather than away from it. The inherent positivity of the group speed, $C_n > 0$, guarantees that the associated wave energy always propagates away from a broad-band seismic source, regardless of the nature of the dispersion. We shall continue to restrict attention to "normally" dispersed waves having $d\omega/dk > 0$ throughout the remainder of this book.

11.4 Moment-Tensor Response

The standing-wave representation of the acceleration response $\mathbf{a}(\mathbf{x}, \omega)$ to a step-function, moment-tensor source is given by

$$\mathbf{a} = \frac{1}{2\pi} \sum_{n=0}^{\infty} \sum_{l=0}^{\infty} (l + \tfrac{1}{2}) \left[\frac{\frac{1}{2}}{{}_n\gamma_l + i(\omega - {}_n\omega_l)} + \frac{\frac{1}{2}}{{}_n\gamma_l + i(\omega + {}_n\omega_l)} \right]$$

$$\times\, {}_n\mathbf{D}_l \sum_{m=0}^{2} P_{lm}(\cos\Theta)(A_m \cos m\Phi + B_m \sin m\Phi), \tag{11.27}$$

where we have retained the negative-frequency Lorentzian dropped in equations (10.64)–(10.65) in the interest of greater generality. The corresponding travelling-wave representation can be obtained by a straightforward generalization of the argument used above—we apply the Poisson sum formula (11.4), decompose the associated Legendre functions in the form

$$P_{k-\frac{1}{2}\,m}(\cos\Theta) = Q^{(1)}_{k-\frac{1}{2}\,m}(\cos\Theta) + Q^{(2)}_{k-\frac{1}{2}\,m}(\cos\Theta), \tag{11.28}$$

and deform the contour so that it runs from $-\infty - i0$ to $\infty - i0$ just beneath the real wavenumber axis. This gives rise to the expression

$$\begin{aligned}\mathbf{a} = \frac{1}{2\pi}\sum_{n=0}^{\infty}\Bigg\{ &\sum_{s=1,3,5,\ldots}^{\infty} (-1)^{(s-1)/2} \int_{-\infty}^{\infty} a_n(k)\,\mathbf{D}_n \sum_{m=0}^{2} Q^{(1)}_{k-\frac{1}{2}\,m}(\cos\Theta) \\ &\quad\times (A_m \cos m\Phi + B_m \sin m\Phi)\, e^{-i(s-1)k\pi}\, kdk \\ + &\sum_{s=2,4,6,\ldots}^{\infty} (-1)^{s/2} \int_{-\infty}^{\infty} a_n(k)\,\mathbf{D}_n \sum_{m=0}^{2} Q^{(2)}_{k-\frac{1}{2}\,m}(\cos\Theta) \\ &\quad\times (A_m \cos m\Phi + B_m \sin m\Phi)\, e^{-isk\pi}\, kdk\Bigg\},\end{aligned} \tag{11.29}$$

where we have let

$$a_n(k) = \frac{\frac{1}{2}}{\gamma_n(k) + i[\omega - \omega_n(k)]} + \frac{\frac{1}{2}}{\gamma_n(k) + i[\omega + \omega_n(k)]}. \tag{11.30}$$

This result, like equation (11.14), is applicable to both body waves and surface waves; it can be rendered exact even on an anelastic Earth by replacing the real operator $\mathbf{D}_n$ by $\boldsymbol{\mathcal{D}}_n$ and the real coefficients A_m and B_m by $\mathcal{A}_m$ and $\mathcal{B}_m$.

To obtain the surface-wave response $\mathbf{a} = \mathbf{a}_{\text{Love}} + \mathbf{a}_{\text{Rayleigh}}$, we evaluate the wavenumber integral in (11.29) for each fixed dispersion branch n and each odd or even wavegroup arrival s using the method of residues. This leads to the representation, analogous to equation (11.20),

$$\begin{aligned}\mathbf{a} = \tfrac{1}{2} i\omega \sum_{n=0}^{\infty} (c_n C_n)^{-1} \Bigg\{ &\sum_{s=1,3,5,\ldots}^{\infty} \exp i[(s-2)\pi/2 - (s-1)\xi_n\pi] \\ &\times \mathbf{D}_n \sum_{m=0}^{2} Q^{(1)}_{\xi_n-\frac{1}{2}\,m}(\cos\Theta)(A_m \cos m\Phi + B_m \sin m\Phi)\end{aligned}$$

$$+ \sum_{s=2,4,6,\ldots}^{\infty} \exp i[(s-1)\pi/2 - s\xi_n\pi] \tag{11.31}$$
$$\times \mathbf{D}_n \sum_{m=0}^{2} Q^{(2)}_{\xi_n-\frac{1}{2}m}(\cos\Theta)(A_m \cos m\Phi + B_m \sin m\Phi)\bigg\}.$$

Equation (11.31) is uniformly valid everywhere within the Earth; away from the epicenter and its antipode, we can simplify this result by using the $0 \leq m \leq 2$ generalization of the asymptotic representation (11.22):

$$Q^{(1,2)}_{\xi_n-\frac{1}{2}m}(\cos\Theta) \approx (-k_n)^m(2\pi k_n \sin\Theta)^{-1/2}$$
$$\times \exp[\mp i(k_n\Theta + m\pi/2 - \pi/4) \mp \omega\Theta/2C_nQ_n]. \tag{11.32}$$

The far-field response to an earthquake obtained in this manner can be written as a sum over surface-wave modes and rays, analogous to the far-field Green tensor (11.23):

$$\mathbf{a}(\mathbf{x},\omega) = \sum_{\text{modes}} \sum_{\text{rays}} (cC)^{-1}(8\pi k|\sin\Delta|)^{-1/2}[\hat{\mathbf{r}}U - i\hat{\mathbf{k}}V + i(\hat{\mathbf{r}}\times\hat{\mathbf{k}})W]$$
$$\times R(\Phi)\exp i\,[-k\Delta + (s-1)\pi/2]\exp(-\omega\Delta/2CQ), \tag{11.33}$$

where

$$R(\Phi) = \omega\{[M_{rr}\dot{U}_{\mathrm{s}} + (M_{\theta\theta} + M_{\phi\phi})r_{\mathrm{s}}^{-1}(U_{\mathrm{s}} - \tfrac{1}{2}kV_{\mathrm{s}})]e^{i\pi/4}$$
$$+ (-1)^s(\dot{V}_{\mathrm{s}} - r_{\mathrm{s}}^{-1}V_{\mathrm{s}} + kr_{\mathrm{s}}^{-1}U_{\mathrm{s}})(M_{r\phi}\sin\Phi + M_{r\theta}\cos\Phi)e^{-i\pi/4}$$
$$- kr_{\mathrm{s}}^{-1}V_{\mathrm{s}}[M_{\theta\phi}\sin 2\Phi + \tfrac{1}{2}(M_{\theta\theta} - M_{\phi\phi})\cos 2\Phi]e^{i\pi/4}$$
$$+ (-1)^s(\dot{W}_{\mathrm{s}} - r_{\mathrm{s}}^{-1}W_{\mathrm{s}})(M_{r\theta}\sin\Phi - M_{r\phi}\cos\Phi)e^{-i\pi/4}$$
$$- kr_{\mathrm{s}}^{-1}W_{\mathrm{s}}[\tfrac{1}{2}(M_{\theta\theta} - M_{\phi\phi})\sin 2\Phi - M_{\theta\phi}\cos 2\Phi]e^{i\pi/4}\}. \tag{11.34}$$

The roman subscripts on U_{s}, V_{s} and W_{s} in equation (11.34) denote evaluation at the source radius r_{s}; they should not be confused with the italic wavegroup-arrival index s. Both the general and far-field results (11.31) and (11.33) are valid for positive frequencies, $\omega > 0$; the corresponding expressions for negative frequencies, $\omega < 0$, can be obtained using the symmetry $\mathbf{a}(\mathbf{x},-\omega) = \mathbf{a}^*(\mathbf{x},\omega)$.

Every term in the representation (11.33) has a straightforward physical interpretation. The oscillatory factor $\exp(-ik\Delta) = \exp(-i\omega\Delta/c)$ represents the phase delay due to propagation at angular speed c through the angular distance Δ along a great-circular surface-wave ray. The vector $\hat{\mathbf{r}}U - i\hat{\mathbf{k}}V + i(\hat{\mathbf{r}}\times\hat{\mathbf{k}})W$ describes the *polarization* of the surface wave upon arrival at the receiver; in the long-wavelength limit $k \gg 1$, a Love wave exhibits purely transverse particle motion $i(\hat{\mathbf{r}}\times\hat{\mathbf{k}})W$, whereas a Rayleigh wave exhibits motion that is radial and longitudinal. The factor of i in the Rayleigh-wave polarization $\hat{\mathbf{r}}U - i\hat{\mathbf{k}}V$ is an indication that the particle motion is elliptical—retrograde at depths where $\operatorname{sgn} U \neq \operatorname{sgn} V$ and prograde at depths where $\operatorname{sgn} U = \operatorname{sgn} V$. The amplitude factor $|\sin\Delta|^{-1/2}$ accounts for the *geometrical spreading* of monochromatic surface waves on a sphere. The differential angular width of an infinitesimal tube of surface-wave rays upon a sphere varies as the absolute sine of the epicentral distance: $dw \propto |\sin\Delta|$. In the absence of anelastic dissipation, the wave energy within a ray tube is conserved; the wave amplitude A, which is proportional to the square root of the energy, therefore varies as

$$\frac{A_2}{A_1} = \left(\frac{dw_2}{dw_1}\right)^{-1/2} = \left|\frac{\sin\Delta_2}{\sin\Delta_1}\right|^{-1/2}, \tag{11.35}$$

where the subscripts 1 and 2 refer to two sequential points along the ray tube. The exponential decay $\exp(-\omega\Delta/2CQ)$ accounts for the additional *anelastic attenuation* of the waves due to the presence of friction within the Earth; the factor of C rather than c in the denominator reflects the fact that the wave energy—which is what is being dissipated—travels with the group speed. The factor $\exp[i(s-1)\pi/2]$ is the so-called "polar" phase shift, whose importance was first pointed out in this context by Brune, Nafe & Alsop (1961). In fact, this frequency-independent $\pi/2$ phase advance upon every passage through the antipode and the source is an example of a more general phenomenon—it is a *caustic phase shift*. If we consider the differential ray-tube width to be a smoothly varying, positive or negative function of the total propagation distance, $dw \propto \sin\Delta$, then we can associate this phase shift with the sign change, $\operatorname{sgn} dw_2 \neq \operatorname{sgn} dw_1$, which accompanies every passage through a caustic:

$$\left(\frac{dw_2}{dw_1}\right)^{-1/2} = \left(\frac{\sin\Delta_2}{\sin\Delta_1}\right)^{-1/2} = \left|\frac{\sin\Delta_2}{\sin\Delta_1}\right|^{-1/2} \exp(i\pi/2). \tag{11.36}$$

Finally, the frequency-dependent quantity $R(\Phi)$ given in equation (11.34) represents the complex *Love or Rayleigh radiation pattern*. For a known source location $\mathbf{x}_s$ and moment tensor $\mathbf{M}$, equations (11.33) and (11.34) provide a basis for the measurement and interpretation of surface-wave phase speeds c and quality factors Q. Alternatively, if the phase speed c and quality factor Q are regarded as known, we can use these results to retrieve the earthquake depth and source mechanism $\mathbf{M}$ from the spectra of G1, G2, G3,... or R1, R2, R3,... surface waves. Free-air, tilt and potential-perturbation effects are small for all but very long-period surface waves; however, they are extremely easy to account for—it is simply necessary to replace the polarization vector in (11.33) by its gravitationally modified counterpart: $\hat{\mathbf{r}}U - i\hat{\mathbf{k}}V + i(\hat{\mathbf{r}}\times\hat{\mathbf{k}})W \rightarrow \hat{\mathbf{r}}U_\star - i\hat{\mathbf{k}}V_\star + i(\hat{\mathbf{r}}\times\hat{\mathbf{k}})W$.

The result (11.33)–(11.34) can alternatively be obtained by starting with the far-field surface-wave Green tensor (11.23) and applying the general formula

$$\mathbf{a}(\mathbf{x},\omega) = i\omega\mathbf{M}\!:\!\boldsymbol{\nabla}_{\mathrm{s}}\mathbf{G}^{\mathrm{T}}(\mathbf{x},\mathbf{x}_{\mathrm{s}};\omega). \tag{11.37}$$

To lowest order, the gradient $\boldsymbol{\nabla}_{\mathrm{s}} = \hat{\mathbf{r}}_{\mathrm{s}}\partial_{r_s} + r_{\mathrm{s}}^{-1}\boldsymbol{\nabla}_{1\mathrm{s}}$ with respect to the source coordinates $\mathbf{x}_{\mathrm{s}}$ acts only upon the oscillatory term $\exp(-ik\Delta)$ and the source polarization vector $\hat{\mathbf{r}}_{\mathrm{s}}U_{\mathrm{s}} + i\hat{\mathbf{k}}_{\mathrm{s}}V_{\mathrm{s}} - i(\hat{\mathbf{r}}_{\mathrm{s}}\times\hat{\mathbf{k}}_{\mathrm{s}})W_{\mathrm{s}}$. The radiation pattern (11.34) can be rewritten using invariant notation in the form

$$R(\Phi) = i\omega(\mathbf{M}\!:\!\mathbf{E}_{\mathrm{s}}^{*})\exp(-i\pi/4). \tag{11.38}$$

The complex symmetric tensor

$$\begin{aligned}\mathbf{E}_{\mathrm{s}} = {}& \dot{U}_{\mathrm{s}}\hat{\mathbf{r}}_{\mathrm{s}}\hat{\mathbf{r}}_{\mathrm{s}} + r_{\mathrm{s}}^{-1}(U_{\mathrm{s}} - kV_{\mathrm{s}})\hat{\mathbf{k}}_{\mathrm{s}}\hat{\mathbf{k}}_{\mathrm{s}} + r_{\mathrm{s}}^{-1}U_{\mathrm{s}}(\hat{\mathbf{r}}_{\mathrm{s}}\times\hat{\mathbf{k}}_{\mathrm{s}})(\hat{\mathbf{r}}_{\mathrm{s}}\times\hat{\mathbf{k}}_{\mathrm{s}}) \\ & -\tfrac{1}{2}i(\dot{V}_{\mathrm{s}} - r_{\mathrm{s}}^{-1}V_{\mathrm{s}} + kr_{\mathrm{s}}^{-1}U_{\mathrm{s}})(\hat{\mathbf{r}}_{\mathrm{s}}\hat{\mathbf{k}}_{\mathrm{s}} + \hat{\mathbf{k}}_{\mathrm{s}}\hat{\mathbf{r}}_{\mathrm{s}}) \\ & \quad + \tfrac{1}{2}i(\dot{W}_{\mathrm{s}} - r_{\mathrm{s}}^{-1}W_{\mathrm{s}})[\hat{\mathbf{r}}_{\mathrm{s}}(\hat{\mathbf{r}}_{\mathrm{s}}\times\hat{\mathbf{k}}_{\mathrm{s}}) + (\hat{\mathbf{r}}_{\mathrm{s}}\times\hat{\mathbf{k}}_{\mathrm{s}})\hat{\mathbf{r}}_{\mathrm{s}}] \\ & \qquad + \tfrac{1}{2}kr_{\mathrm{s}}^{-1}W_{\mathrm{s}}[\hat{\mathbf{k}}_{\mathrm{s}}(\hat{\mathbf{r}}_{\mathrm{s}}\times\hat{\mathbf{k}}_{\mathrm{s}}) + (\hat{\mathbf{r}}_{\mathrm{s}}\times\hat{\mathbf{k}}_{\mathrm{s}})\hat{\mathbf{k}}_{\mathrm{s}}]\end{aligned} \tag{11.39}$$

is the *surface-wave strain* at the source. We use the JWKB approximation to generalize the far-field impulse and moment-tensor responses (11.23) and (11.33)–(11.34) to a smooth laterally heterogeneous Earth model in Section 16.5.

11.5 Stationary-Phase Approximation

A distinguishing feature of Love and Rayleigh wave propagation is the dependence of the phase speed $c = \omega/k$ upon the wavenumber k and the frequency ω. Since waves having different wavenumbers and frequencies propagate at different velocities, the signals emitted by a localized source arrive at different times—for this reason, the propagation is said to be *dispersive*. In this section we review the well-known kinematic and energetic significance of the group speed $C = d\omega/dk$. If we consider the phase speed c to be a function of the wavenumber k, then the group and phase speeds are related by

$$C = d(ck)/dk = c + k(dc/dk). \tag{11.40}$$

If, on the other hand, we regard c as a function of the frequency ω, we have

$$C = \frac{c}{1 - (\omega/c)(dc/d\omega)}. \tag{11.41}$$

None of the results presented here are unique to seismic surface waves; indeed, they are applicable to dispersive waves in any linear medium. More systematic and authoritative treatments of linear dispersive wave propagation are provided by Whitham (1974) and Lighthill (1978). A more rigorous exposition of the *method of stationary phase*, which is the mathematical cornerstone of the analysis, may be found in Bender & Orszag (1978).

Every mode branch $n = 0, 1, 2, \ldots$ and multi-orbit arrival $s = 1, 2, 3, \ldots$ in the response (11.33) is subjected to an independent kinematic analysis; the double summation over modes and surface-wave rays will henceforth be regarded as understood. Each term in the sum may be written in the abbreviated form

$$\mathbf{a}(\mathbf{x}, \omega) = \mathbf{A}(\mathbf{x}, \omega) \exp[-ik(\omega)\Delta], \tag{11.42}$$

where

$$\begin{aligned} \mathbf{A} = \omega(cC)^{-1}(8\pi k|\sin\Delta|)^{-1/2}[\hat{\mathbf{r}}U - i\hat{\mathbf{k}}V + i(\hat{\mathbf{r}} \times \hat{\mathbf{k}})W] \\ \times R(\Phi) \exp[i(s-1)\pi/2] \exp(-\omega\Delta/2CQ). \end{aligned} \tag{11.43}$$

Note that both anelastic damping $\exp(-\omega\Delta/2CQ)$ and geometrical spreading $|\sin\Delta|^{-1/2}$ have been incorporated into the amplitude factor $\mathbf{A}(\mathbf{x}, \omega)$. This is permissible as long as the anelasticity is slight; in practice, amplitude variations due to these two effects are of the same order of magnitude. The single-mode, single-ray time-domain response can be obtained by evaluating the inverse Fourier transform of (11.42):

$$\mathbf{a}(\mathbf{x}, t) = \frac{1}{\pi} \operatorname{Re} \int_0^\infty \mathbf{A}(\mathbf{x}, \omega) \exp[i\omega t - ik(\omega)\Delta]\, d\omega, \tag{11.44}$$

where we have made use of the identity $\mathbf{a}(\mathbf{x}, -\omega) = \mathbf{a}^*(\mathbf{x}, \omega)$ to obtain a one-sided integral. Upon defining the quantity

$$\Psi(\omega) = \omega - k(\omega)\Delta/t, \tag{11.45}$$

we can rewrite (11.44) in a form that is suitable for the application of the method of stationary phase:

$$\mathbf{a} = \frac{1}{\pi} \operatorname{Re} \int_0^\infty \mathbf{A}(\omega) \exp[it\Psi(\omega)]\, d\omega, \tag{11.46}$$

where the inessential arguments have been eliminated. We seek to evaluate the acceleration (11.46) asymptotically in the limit $t \to \infty$ *for a fixed value of the parameter* Δ/t; the resulting response will be that viewed by an observer moving at a fixed speed. The essential idea is that at large times the integrand is highly oscillatory, leading to a high degree of cancellation, except where the stationarity of the phase $\Psi(\omega)$ allows mutual reinforcement of waves with neighboring frequencies.

Let ω_0 denote the angular frequency (or possibly frequencies) at which the phase of the integrand in (11.46) is stationary:

$$\Psi'(\omega_0) = 0. \tag{11.47}$$

Throughout the rest of this section we use a prime to denote differentiation with respect to frequency, $d/d\omega$, and a subscript 0 to denote evaluation at the stationary angular frequency, as in $C_0 = C(\omega_0)$. Equation (11.47) is equivalent to the result

$$\Delta/t = C_0, \tag{11.48}$$

so that the observer referred to above is moving with the stationary group speed C_0. In the limit $t \to \infty$, the acceleration $\mathbf{a}$ is entirely determined by the frequencies in the immediate vicinity of the stationary point ω_0. In this vicinity we can approximate the amplitude and phase in (11.46) by the zeroth-order and second-order Taylor expansions

$$\mathbf{A}(\omega) \approx \mathbf{A}_0, \qquad \Psi(\omega) \approx \Psi_0 + \tfrac{1}{2}(\omega - \omega_0)^2\Psi_0''. \tag{11.49}$$

It is noteworthy that there is no phase term proportional to $\omega - \omega_0$, inasmuch as $\Psi_0' = 0$. Upon making use of the approximations (11.49) we obtain

$$\mathbf{a} \approx \frac{1}{\pi}\,\mathrm{Re}\left\{\mathbf{A}_0 \exp[i(\omega_0 t - k_0\Delta)] \times \int_{-\infty}^{\infty} \exp\left[\tfrac{1}{2}i\Psi_0''(\omega - \omega_0)^2 t\right] d\omega\right\}, \tag{11.50}$$

where we have used the the same mutual cancellation argument in reverse to extend the limits of integration from the narrow interval around ω_0 where (11.49) is valid to the full frequency axis $-\infty \leq \omega \leq \infty$. By rotating the path of integration through $\pm\pi/4$, the remaining integral in (11.50) may be evaluated analytically:

$$\int_{-\infty}^{\infty} \exp\left[\tfrac{1}{2}i\Psi_0''(\omega - \omega_0)^2 t\right] d\omega = \left(\frac{2\pi C_0}{|C_0'|t}\right)^{1/2} \exp(\tfrac{1}{4}i\pi\,\mathrm{sgn}\,C_0'), \tag{11.51}$$

where we have made use of the fact that $\Psi_0'' = C_0'/C_0$. As a result the response $\mathbf{a}(\mathbf{x}, t)$ reduces to

$$\mathbf{a} \approx \mathrm{Re}\left[\boldsymbol{\mathcal{A}} \exp i\,(\omega_0 t - k_0\Delta)\right], \tag{11.52}$$

where

$$\boldsymbol{\mathcal{A}} = \mathbf{A}_0 \left(\frac{2C_0}{\pi|C_0'|t}\right)^{1/2} \exp(\tfrac{1}{4}i\pi\,\mathrm{sgn}\,C_0'). \tag{11.53}$$

The requirement $t \to \infty$ guarantees that the waves left the source in the distant past, so that they are well dispersed. The asymptotic signal at an angular epicentral distance Δ and time t then appears to consist of a *single wave* of variable frequency ω_0, wavenumber k_0 and amplitude $\mathcal{A}$. If there is more than one stationary frequency ω_0, the response is a sum of terms of the form (11.52)–(11.53).

The stationarity condition (11.48) stipulates that $\Delta = C_0 t$ is the locus at time t of waves of frequency ω_0 and wavenumber k_0; an observer moving away from the source at the speed C_0 is always surrounded by a *group* of waves of this frequency and wavenumber—this is the reason for the nomenclature. The individual waves within the group are always changing, since they propagate with the phase speed c_0. Whether the new crests and troughs enter the group at the rear and leave in the front or vice versa depends upon whether $c_0 > C_0$ or $c_0 < C_0$. From equation (11.41) we see that the phase speed exceeds the group speed if $c_0' < 0$, whereas the group speed exceeds the phase speed if $c_0' > 0$. An individual wave crest or trough must either accelerate or decelerate with time, since its frequency ω, wavenumber k and phase speed c keep changing. At a fixed epicentral distance Δ the condition (11.48) specifies the instantaneous frequency ω_0 of a surface-wave seismogram at time t; on the other hand, at a fixed time t it specifies the local wavenumber k_0 of a "snapshot" of the wavetrain at distance Δ.

The dependence of the absolute amplitude $A = |\mathcal{A}|$ upon time and distance along a ray can be understood by considering the energy of an infinitesimal wavegroup bounded on its sides by the walls of the ray tube and on its back and front by the two group lines corresponding to the frequencies ω_0 and $\omega_0 + d\omega_0$. The differential surface area of the group patch at any point along the ray is $d\Sigma = |\sin\Delta|\, d\Delta\, d\Phi = C_0' t\, |\sin\Delta|\, d\omega_0\, d\Phi$, where we have compared $C(\omega_0) = \Delta/t$ with $C(\omega_0 + d\omega_0) = (\Delta + d\Delta)/t$ to obtain the second equality. The energy contained within the patch is a constant in the absence of anelastic attenuation; it follows that the amplitudes at two sequential points upon a perfectly elastic Earth are related by

$$\frac{A_2}{A_1} = \left(\frac{d\Sigma_2}{d\Sigma_1}\right)^{-1/2} = \left|\frac{t_2 \sin\Delta_2}{t_1 \sin\Delta_1}\right|^{-1/2}. \tag{11.54}$$

The group lines in back and front of the patch diverge at a rate proportional to t as a result of the dispersion; hence, the amplitude A of a wavegroup decreases like $t^{-1/2}$. This dispersive $t^{-1/2}$ decay is superimposed upon the geometrical $|\sin\Delta|^{-1/2}$ and anelastic $\exp(-\omega\Delta/2CQ)$ amplitude variations of a monochromatic wave. The agreement of the elastic amplitude ratio (11.54) with (11.52)–(11.53) verifies that *the energy propagates with the group speed*. The above argument is based only upon the premise that the energy is proportional to the amplitude squared; we give a more precise definition of the energy of a travelling surface wave in Chapter 16, where we extend the theory developed here to the case of a laterally heterogeneous Earth model.

The asymptotic result (11.52)–(11.53) is invalid in the vicinity of group speed maxima or minima, where $C_0' = 0$, and the waves associated with the two stationary points on either side interfere. This interference gives rise to a particularly high-amplitude surface-wave signal, which is known as an *Airy phase* in honor of Airy's celebrated analysis of the mathematically related problem of the diffraction of light at a caustic. An analytical expression which captures the principal features of the interference can be obtained by expanding the phase $\Psi(\omega)$ up to order $(\omega-\omega_0)^3$ before evaluating the response (Ben-Menahem & Singh 1981); however, this higher-order result is of little practical utility. In quantitative applications, it is preferable to eschew the stationary-phase approximation entirely, and either use the representation (11.42)–(11.43) directly in the frequency domain, or evaluate the inverse Fourier transform (11.44) numerically. Alternatively, it is possible to calculate the surface-wave response $\mathbf{a}(\mathbf{x}, t)$ by summation of the equivalent normal modes ${}_n\mathrm{T}_l$ and ${}_n\mathrm{S}_l$. The value of the approximate result (11.52)–(11.53) is the physical insight it gives into the nature of dispersive wave propagation.

11.6 Dispersion Relation and Group Speed

Any equation expressing the angular frequency ω of a travelling wave in terms of its wavenumber k is referred to as a *dispersion relation*. There is a separate such relation governing the independent propagation along every fundamental ($n = 0$) or higher-overtone ($n = 1, 2, \ldots$) Love or Rayleigh wave branch. Following Jeffreys (1961) we introduce a convenient new notation for describing the dispersion of surface waves on a spherical Earth, and use it in conjunction with Rayleigh's principle to derive a useful expression for the group speed $C = d\omega/dk$. Most of the material in this section is simply a recasting of results obtained in Chapter 8 in terms of the continuous wavenumber $0 \leq k \leq \infty$ rather than the discrete angular degree $l = 0, 1, 2, \ldots$.

11.6.1 Love waves

The radial action integral (8.88) governing the propagation of Love waves may be rewritten in the form

$$\mathcal{I} = \tfrac{1}{2}(\omega^2 I_1 - k^2 I_2 - I_3), \tag{11.55}$$

where

$$I_1 = \int_0^a \rho W^2 \, r^2 dr, \tag{11.56}$$

$$I_2 = \int_0^a \mu W^2 \, dr, \tag{11.57}$$

$$I_3 = \int_0^a \mu[(\dot{W} - r^{-1}W)^2 - 2r^{-2}W^2] \, r^2 dr. \tag{11.58}$$

In practice, the integrations in (11.56)–(11.58) can be confined to the mantle $b \leq r \leq s$, since we have no interest in surface-wave propagation within the solid inner core. We may regard the energy equipartition relation $\mathcal{I} = 0$ as the dispersion relation for Love waves:

$$\omega^2 I_1 - k^2 I_2 - I_3 = 0. \tag{11.59}$$

Equation (11.59) is an implicit relation between the frequency ω and the wavenumber k because of the dependence of the radial eigenfunction W upon those variables. The kinetic-plus-potential energy (8.110) of a Love wave can also be expressed in terms of the integrals (11.56)–(11.58):

$$\mathcal{E} = \tfrac{1}{2}(\omega^2 I_1 + k^2 I_2 + I_3). \tag{11.60}$$

Rayleigh's principle stipulates that the action (11.55) is stationary with respect to perturbations $W \to W + \delta W$ if and only if W is a Love-wave eigenfunction with associated angular frequency ω and wavenumber k:

$$\omega^2 \delta I_1 - k^2 \delta I_2 - \delta I_3 = 0, \tag{11.61}$$

where

$$\delta I_1 = 2 \int_0^a \rho W \, \delta W \, r^2 dr, \tag{11.62}$$

$$\delta I_2 = 2 \int_0^a \mu W \, \delta W \, dr, \tag{11.63}$$

$$\delta I_3 = 2 \int_0^a \mu[(\dot{W} - r^{-1}W)(\delta\dot{W} - r^{-1}\delta W) - 2r^{-2}W \, \delta W] \, r^2 dr. \tag{11.64}$$

By subtracting the dispersion relations for two waves with associated frequencies ω and $\omega + \delta\omega$, wavenumbers k and $k + \delta k$ and associated radial eigenfunctions W and $W + \delta W$, we find, on the other hand, that

$$2\omega \, \delta\omega \, I_1 - \omega^2 \delta I_1 - 2k \, \delta k \, I_2 - k^2 \delta I_2 - \delta I_3 = 0. \tag{11.65}$$

Equation (11.65) can alternatively be obtained by taking the total variation of (11.59) with respect to all three variables ω, k and W. The terms involving the eigenfunction perturbation δW cancel by virtue of (11.61), leaving a relation between $\delta\omega$ and δk:

$$2\omega\,\delta\omega\,I_1 = 2k\,\delta k\,I_2. \tag{11.66}$$

This result can be rearranged to yield an *exact* expression for the group speed $C = \delta\omega/\delta k$ in terms of the radial integrals (11.56)–(11.58):

$$C = \frac{I_2}{cI_1}, \tag{11.67}$$

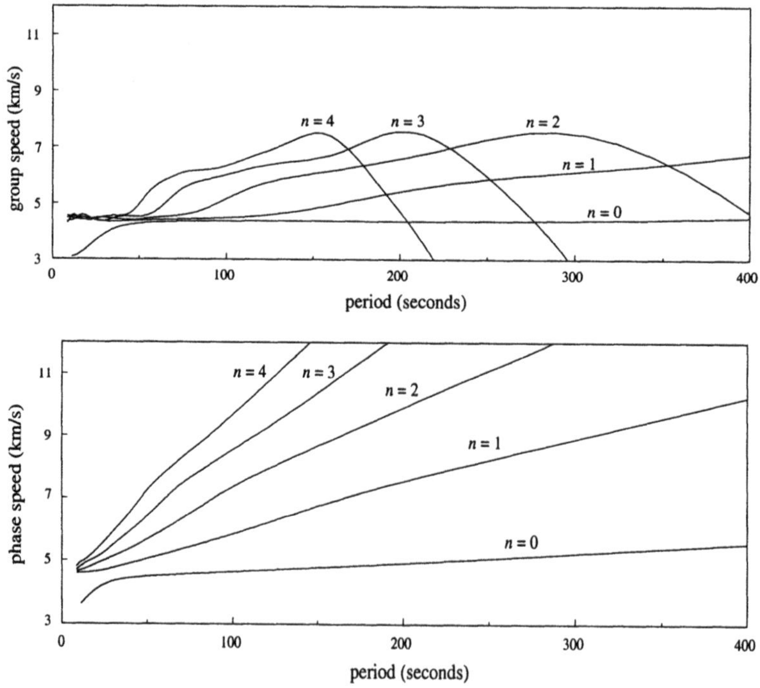

Figure 11.5. Group speed aC (*top*) and phase speed ac (*bottom*) of Love waves. Integers $n = 0$ and $n = 1$–4 denote the fundamental and first four overtone branches, respectively. At periods shorter than $2\pi/\omega \approx 40$ seconds, the dispersion illustrated here is unlike that anywhere on the Earth because of the unusual hybrid crustal structure of the spherically averaged PREM model—seventy percent oceanic and thirty percent continental.

where the quantity $c = \omega/k$ is the Love-wave phase speed. The advantage of the relation (11.67) is that it provides a recipe for calculating C without having to resort to numerical differentiation.

Figure 11.5 shows the variation of Love-wave group speed aC and phase speed ac on the surface of the Earth, $r = a$, as a function of period $2\pi/\omega$ along the fundamental ($n = 0$) and first four overtone ($n = 1$–4) dispersion branches. Fundamental Love waves with periods greater than about 40 seconds propagate with a relatively constant group speed, $aC \approx 4.4$ km/s. This accounts for the relatively non-dispersed or impulsive character of the G wave, as we shall see in Section 11.7. It takes approximately two and one-half hours for a long-period G wavegroup to orbit the Earth once. The decrease in group speed of the higher modes at long periods is a manifestation of the presence of the Earth's fluid core; the associated $n = 2, 3, 4$ group arrivals are $\mathrm{ScS_{SH}}$-equivalent, and therefore not true surface waves.

11.6.2 Rayleigh waves

The radial action integrals (8.87) and (8.100) governing the propagation of Rayleigh waves may be written in the form

$$\mathcal{I} = \tfrac{1}{2}(\omega^2 I_1 - k^2 I_2 - k I_3 - I_4), \tag{11.68}$$

$$\mathcal{I}' = \tfrac{1}{2}(\omega^2 I_1 - k^2 I_2' - k I_3' - I_4'), \tag{11.69}$$

where

$$I_1 = \int_0^a \rho(U^2 + V^2)\, r^2 dr, \tag{11.70}$$

$$I_2 = \int_0^a [\mu U^2 + (\kappa + \tfrac{4}{3}\mu)V^2]\, dr, \tag{11.71}$$

$$I_3 = \int_0^a [\tfrac{4}{3}\mu V(\dot{U} - r^{-1}U) - 2\kappa V(\dot{U} + 2r^{-1}U) + 2\mu U(\dot{V} - r^{-1}V) + \rho(VP + 2gUV)]\, r dr, \tag{11.72}$$

$$I_4 = \int_0^a [(\kappa(\dot{U} + 2r^{-1}U)^2 + \tfrac{4}{3}\mu(\dot{U} - r^{-1}U)^2 + \mu(\dot{V} - r^{-1}V)^2 - 2\mu r^{-2}V^2 + \rho(4\pi G\rho U^2 + U\dot{P} - 4r^{-1}gU^2)]\, r^2 dr, \tag{11.73}$$

$$I_2' = I_2 + \frac{1}{4\pi G}\int_0^\infty P^2\, dr, \tag{11.74}$$

$$I_3' = I_3 + \int_0^a \rho V P\, r dr, \tag{11.75}$$

$$I_4' = I_4 + \int_0^a \rho U \dot{P}\, r^2 dr + \frac{1}{4\pi G}\int_0^\infty \dot{P}^2\, r^2 dr. \tag{11.76}$$

The Rayleigh-wave equipartition or dispersion relation $\mathcal{I} = \mathcal{I}' = 0$, analogous to equation (11.59), is

$$\omega^2 I_1 - k^2 I_2 - kI_3 - I_4 = \omega^2 I_1 - k^2 I_2' - kI_3' - I_4' = 0. \tag{11.77}$$

The Rayleigh-wave energy (8.125) is given by

$$\mathcal{E} = \tfrac{1}{2}(\omega^2 I_1 + k^2 I_2 + kI_3 + I_4) = \tfrac{1}{2}(\omega^2 I_1 + k^2 I_2' + kI_3' + I_4'). \tag{11.78}$$

The total energy of either a Love or Rayleigh wave is equal to two times its kinetic energy, $\mathcal{E} = \omega^2 I_1$, by virtue of the dispersion relations (11.59) and (11.77).

Upon considering two waves with angular frequencies ω and $\omega + \delta\omega$, wavenumbers k and $k + \delta k$, and associated radial eigenfunctions U, V, P and $U + \delta U$, $V + \delta V$, $P + \delta P$ and applying Rayleigh's principle as in Section 11.6.1, we obtain an analytical expression analogous to equation (11.67) for the Rayleigh-wave group speed:

$$C = \frac{I_2' + \frac{1}{2}k^{-1} I_3'}{cI_1}. \tag{11.79}$$

It is simplest to regard the potential perturbation P as one of the independent variables in obtaining the result (11.79); in varying the unprimed action $\mathcal{I}$ it is necessary to contend with the dependence of P upon the wavenumber k. It is conventional to use the two relations (11.67) and (11.79) to define the group speed of a free oscillation or standing wave, even in the case of modes ${}_n\mathrm{T}_l$ and ${}_n\mathrm{S}_l$ that are not equivalent to surface waves in the sense $n \ll l/4$. Mathematically, the quantity ${}_nC_l$ defined in this way is the slope of the smooth dispersion curve passing through the discrete eigenfrequencies $\ldots, {}_{n-1}\omega_l, {}_n\omega_l, {}_{n+1}\omega_l, \ldots$. We shall obtain a physical interpretation of the group speed of a body-wave equivalent mode in Chapter 12.

Figure 11.6 shows the variation of Rayleigh-wave group speed aC and phase speed ac as a function of period along the fundamental ($n = 0$) and first four overtone ($n = 1$–4) dispersion branches. Fundamental-mode Rayleigh waves on the PREM model exhibit a broad local minimum in

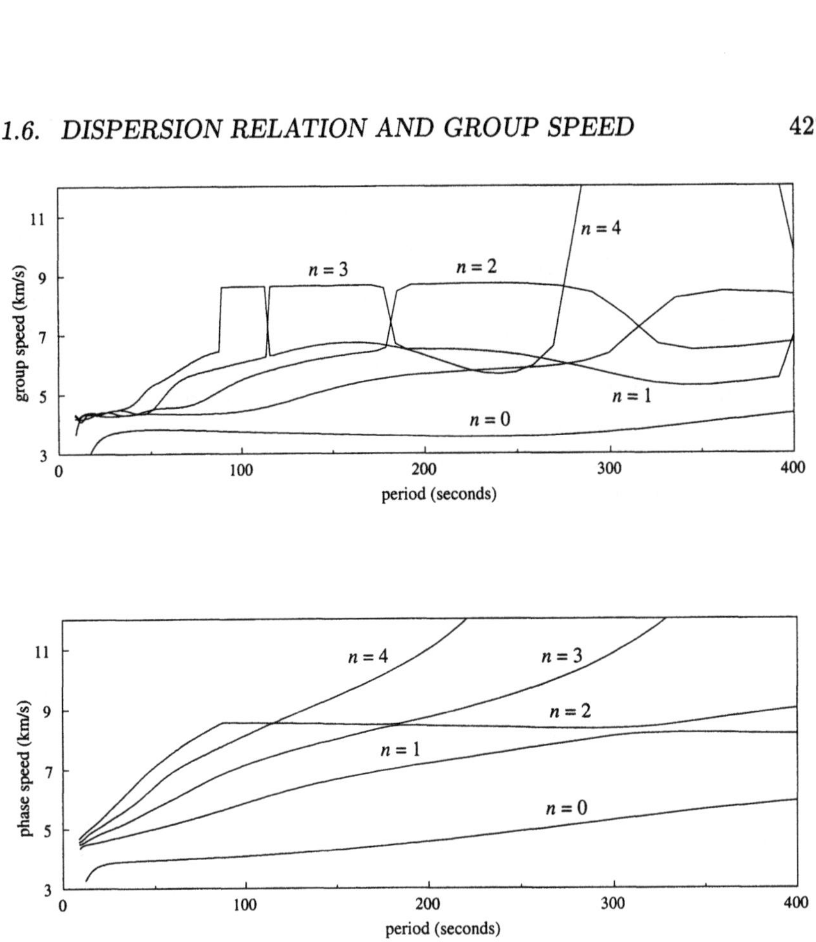

Figure 11.6. Group speed aC (*top*) and phase speed ac (*bottom*) of fundamental ($n = 0$) and first four overtone ($n =$ 1–4) Rayleigh waves. The piecewise linear character of the group-speed curves above $2\pi/\omega \approx 200$ s is an artifact due to insufficient sampling. At periods shorter than $2\pi/\omega \approx 40$ seconds, the dispersion illustrated here is unlike that anywhere on the Earth because of the unusual hybrid crustal structure of the spherically averaged PREM model—seventy percent oceanic and thirty percent continental.

group speed $aC_{\text{min}} \approx 3.6$ km/s at a period $2\pi/\omega \approx 240$ s and a broad local maximum $aC_{\text{max}} \approx 3.8$ km/s at $2\pi/\omega \approx 50$ s. The fastest waves are those with very long periods, which penetrate into the deeper high-speed regions of the upper mantle; the group speed aC exceeds 3.8 km/s at a period of approximately 300 seconds. Fundamental Rayleigh wavegroups with periods between 50 and 300 seconds propagate somewhat more slowly than 50–300 second fundamental Love wavegroups; it takes these long-period Rayleigh waves almost three hours to circumnavigate the globe, a half-hour longer than the quasi-monochromatic G wave. The clean separation between the $n = 1, 2, \ldots$ and $n = 0$ group speeds makes it easy to isolate a relatively pure fundamental-mode Rayleigh wavegroup from the earlier-arriving overtones by application of a window in the time domain. The group arrival times of $n = 1, 2, \ldots$ and $n = 0$ Love waves are more nearly coincident, particularly at periods shorter than 100–150 seconds (see Figure 11.5). For this reason, fundamental-mode Love-wave dispersion measurements are more likely to be contaminated by higher-mode interference than fundamental-mode Rayleigh-wave measurements are.

The nearly constant phase-speed curve crossing at $ac \approx 8.6$ km/s is the core-mantle-boundary Stoneley-mode branch, which is inextricably intertwined with the Rayleigh-wave overtone branches. The peculiar jumps in group speed and the group-speed plateau $aC \approx 8.6$ km/s are a manifestation of the same phenomenon; the coincidence of the Stoneley phase and group speeds is an expected consequence of their non-dispersive character. In most instances, such as the sharp $n = 3$–4 avoided crossing at about 120 seconds, it would be a simple matter to excise the Stoneley modes and re-assign the Rayleigh-wave branch numbers accordingly. The gradual character of the $n = 2$–3 and, particularly, the $n = 1$–2 osculations would, however, make this a rash scheme. In using either equation (10.63) or (11.33) to synthesize surface-wave accelerograms upon a spherically symmetric Earth, it is preferable simply to retain the Stoneley branch in the overtone sum; of course, these modes will contribute very little to the response, since both the source and receiver are far out on the exponential tail of the associated eigenfunctions.

★11.6.3 Tsunamis

In Section 8.8.11 we noted that an ocean-covered Earth model such as PREM supports tsunami modes with spheroidal displacement eigenfunctions U, V that are primarily confined to the surficial fluid layer. Figure 11.7 shows the phase and group speeds of these modes as a function of period for water depths $h = 2$ km, $h = 4$ km and $h = 6$ km. For each depth, the angular degrees $l = 10$, $l = 100$ and $l = 1000$ of the associated oscillations are indicated along the very top. The classical phase and group speeds of a surface-gravity wave in an ocean overlying a rigid seafloor are (Lamb 1932; Lighthill 1978)

$$ac = \sqrt{gk^{-1}\tanh kh}, \qquad aC = \tfrac{1}{2}ac\left(1 + \frac{2kh}{\sinh 2kh}\right). \tag{11.80}$$

In the shallow-water limit $kh \ll 1$, the propagation is non-dispersive,

$$ac = aC = \sqrt{gh}, \tag{11.81}$$

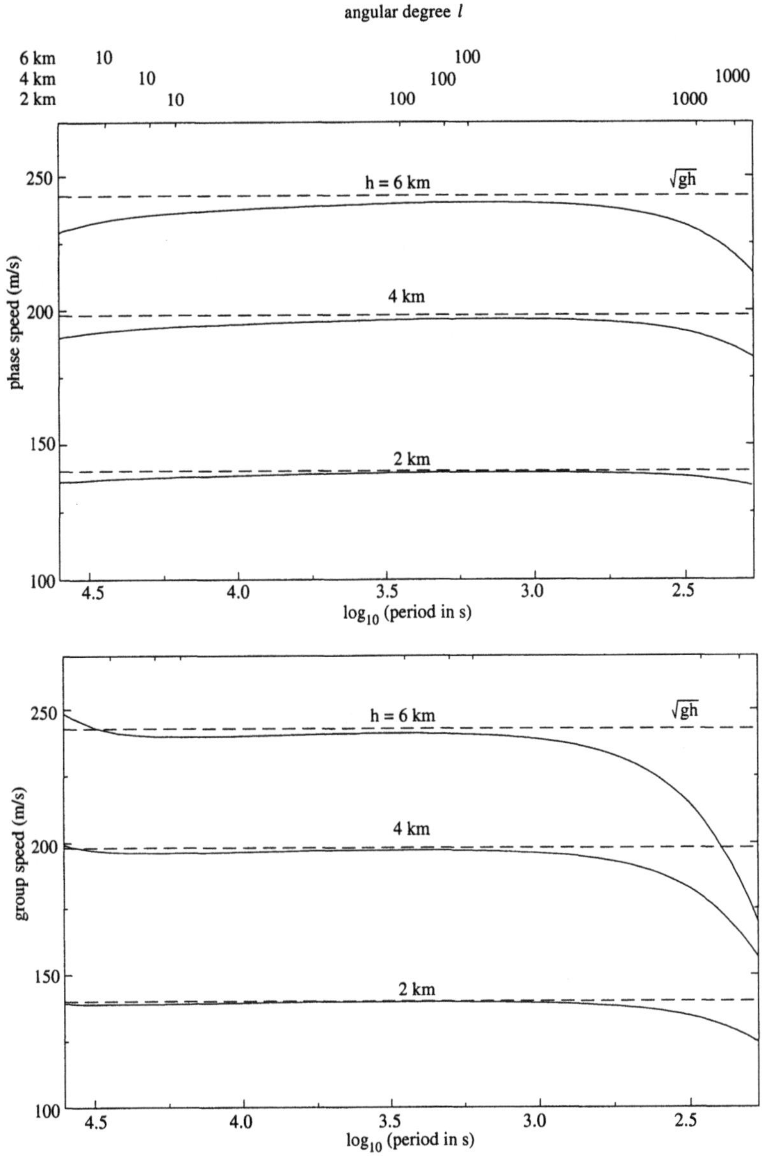

Figure 11.7. Phase speed ac (*top*) and group speed aC (*bottom*) of tsunami waves on a modified version of PREM, with a surficial ocean of depth $h = 2$ km, $h = 4$ km or $h = 6$ km. Dashed lines show the constant speed $ac = aC = \sqrt{gh}$ of a non-dispersive shallow-water surface-gravity wave for comparison.

whereas in the deep-water limit $kh \gg 1$, the phase speed is twice as large as the group speed,

$$ac = 2aC = \sqrt{gk^{-1}}. \tag{11.82}$$

At periods longer than 1000 seconds, tsunamis are essentially non-dispersive shallow-water waves with speeds in the range 150 m/s to 250 m/s, depending upon the water depth. The propagation time across the Pacific Ocean basin is approximately ten hours—enough time after an earthquake to evacuate exposed coastal areas. The dispersion at periods shorter than 1000 seconds is consistent with the classical relations (11.80); the departure from the rigid-bottom limit $ac = aC = \sqrt{gh}$ at very long periods is due to the elastic flexing of the seafloor.

★11.6.4 Transverse isotropy

The above results are valid in the case of a transversely isotropic Earth model provided that the Love-wave radial integrals (11.57)–(11.58) are replaced by

$$I_2 = \int_0^a NW^2\, dr, \tag{11.83}$$

$$I_3 = \int_0^a [L(\dot{W} - r^{-1}W)^2 - 2Nr^{-2}W^2]\, r^2 dr, \tag{11.84}$$

and the Rayleigh-wave integrals (11.71)–(11.73) are replaced by

$$I_2 = \int_0^a [AU^2 + LV^2]\, dr, \tag{11.85}$$

$$\begin{aligned} I_3 = \int_0^a & [2LU(\dot{V} - r^{-1}V) - 2F\dot{U}V \\ & - 4(A - N)r^{-1}UV + \rho(VP + 2gUV)]\, r dr, \end{aligned} \tag{11.86}$$

$$\begin{aligned} I_4 = \int_0^a & [C\dot{U}^2 + 4Fr^{-1}\dot{U}U + 4(A - N)r^{-2}U^2 \\ & + L(\dot{V} - r^{-1}V)^2 - 2Nr^{-2}V^2 \\ & \quad +\rho(4\pi G\rho U^2 + U\dot{P} - 4r^{-1}gU^2)]\, r^2 dr. \end{aligned} \tag{11.87}$$

The quantities C, A, L, N and F in (11.83)–(11.87) are the five elastic parameters introduced in Section 8.9.

11.7 Surface-Wave Seismograms

Synthetic surface-wave accelerograms can be computed by inverse fast Fourier transformation of the the double sum (11.33). Such a method is, however, strictly limited to epicentral distances in the range $0 \ll \Theta \ll \pi$. Because of this restriction, it is preferable simply to sum the associated normal modes ${}_n\mathrm{T}_l$ and ${}_n\mathrm{S}_l$ along each Love and Rayleigh dispersion branch $n = 0, 1, 2, \ldots$; we use this uniformly valid procedure to calculate the surface-wave accelerograms $\mathbf{a}(\mathbf{x}, t)$ and displacement seismograms $\mathbf{s}(\mathbf{x}, t)$ displayed here.

11.7.1 Mantle waves and X waves

Figure 11.8 depicts the radial, longitudinal and transverse components of the acceleration at station TAU in Tasmania, Australia following the October 4, 1994 Kuril Islands event. This large ($M_0 = 4 \times 10^{21}$ N m) shallow-focus earthquake excited spectacular multi-orbit wavetrains. Clear fundamental-mode Rayleigh-wave arrivals can be seen with the naked eye up to R6, and Love-wave arrivals can be discerned up to G8. These long-period surface waves—known as *mantle waves*—have circumnavigated the Earth four times! The minor-arc and major-arc waves G1, G2 or R1, R2 arrive simultaneously and interfere at the antipode of an earthquake, $\Theta = \pi$. To illustrate this, we plot a Rayleigh-wave record section in the vicinity of the antipode in Figure 11.9. Note the remarkable focusing of energy that occurs at this focal point.

The nature of long-period Love-wave propagation can be more clearly discerned on the synthetic transverse-component record section displayed in Figure 11.10. The source is a hypothetical strike-slip earthquake at a depth of 300 km; only the fundamental toroidal modes ${}_0\mathrm{T}_l$ have been summed to synthesize the transverse displacement seismograms $\hat{\boldsymbol{\Phi}} \cdot \mathbf{s}(\mathbf{x}, t)$. The source-receiver geometry is such that no spheroidal modes ${}_n\mathrm{S}_l$ are excited. The first five Love-wave arrivals G1–G5 may be readily identified. Two features are noteworthy: the impulsive character of the waveform, due to the near constancy of the group speed, $aC \approx 4.4$ km/s, for periods greater than $2\pi/\omega \approx 40$ seconds, and the non-alignment of the lines of constant phase—or individual wave crests and troughs—with the overall wavegroup arrival. The latter phenomenon may be best appreciated by viewing the figure at a grazing angle, with one's eye near the plane of the page; the wavefronts cut across the group with a gentler inclination, because the phase speed of mantle Love waves is higher than their group speed: $c > C$. Figure 11.11 shows the same record section $\hat{\boldsymbol{\Phi}} \cdot \mathbf{s}(\mathbf{x}, t)$ with the toroidal overtones ${}_1\mathrm{T}_l, {}_2\mathrm{T}_l, \ldots$ as well as the fundamental modes ${}_0\mathrm{T}_l$ included in the sum. The higher-mode arrivals precede the fundamental-mode G wave, because of their higher group speeds. Clear phases corresponding to the direct and multiply reflected SH, $\mathrm{SS_{SH}}, \mathrm{SSS_{SH}}, \ldots$ body waves may be identified. Note the exponential falloff in amplitude of the diffracted SH waves beyond the core shadow boundary at $\Theta \approx 100^\circ$.

Figures 11.12 and 11.13 show the synthetic radial-component displacement seismograms $\hat{\mathbf{r}} \cdot \mathbf{s}(\mathbf{x}, t)$ for the same 300 km deep strike-slip event. These have been obtained by summing the fundamental spheroidal modes ${}_0\mathrm{S}_l$ and the fundamental plus the higher modes ${}_1\mathrm{S}_l, {}_2\mathrm{S}_l, \ldots$, respectively. The dispersive character of fundamental-mode Rayleigh-wave propagation is evident: the periods of the earliest and latest waves in each arrival R1–R4 are $2\pi/\omega \approx 50$ s and $2\pi/\omega \approx 240$ s, respectively; these are the Airy phases

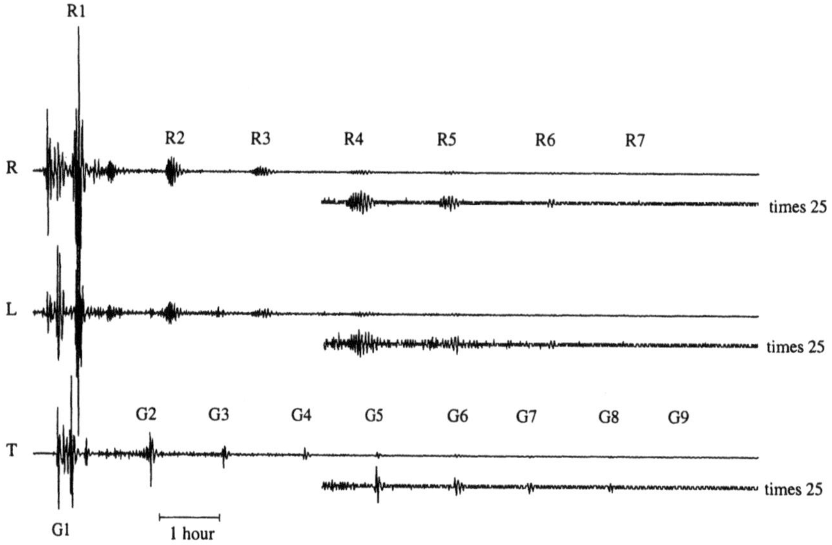

Figure 11.8. Synthetic multi-orbit surface-wave accelerograms. The source is the great shallow-focus Kuril Islands earthquake of October 4, 1994; the receiver is situated in Tasmania, Australia. Each record is a full mode sum, bandpass filtered between 50 and 250 seconds to accentuate the strongly excited fundamental Love and Rayleigh waves; the radial (R), longitudinal (L) and transverse (T) components and the wavegroup arrivals R1–R7 and G1–G9 are indicated. The faster Love-wave group speed is evident: $aC_{\mathrm{R}} = 3.6$–3.8 km/s, whereas $aC_{\mathrm{L}} = 4.4$ km/s. The higher-orbit waves R4–R6 and G5–G8 are much weaker than R1 and G1 as a result of anelastic attenuation; however, they are clearly visible when plotted on a 25× magnified scale.

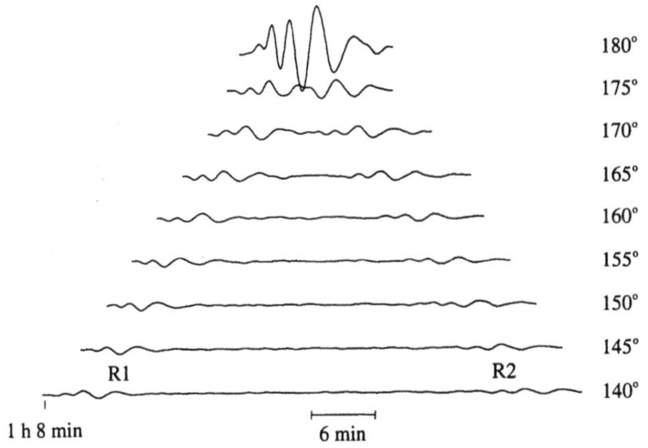

Figure 11.9. Radial-component Rayleigh-wave accelerograms in the vicinity of the antipode. The hypothetical equatorial source has the same moment tensor $\mathbf{M}$ as the October 4, 1994 Kuril Islands earthquake; the receivers are also situated on the equator, due east of the source, at epicentral distances ranging from $\Theta = 140^\circ$ to $\Theta = 180^\circ$. All records are plotted on the same scale, to illustrate the antipodal amplification of the minor-arc and major-arc arrivals R1 and R2.

associated with the two Rayleigh group-speed extrema $aC_{\rm max} \approx 3.8$ km/s and $aC_{\rm min} \approx 3.6$ km/s. Individual crests and troughs travel faster than the wavegroup itself, since the phase speed of mantle Rayleigh waves exceeds their group speed, $c > C$, as in the case of Love waves. Once again, this non-alignment of the phase and group arrivals can be most easily discerned by adopting a "flounder's-eye" view.

The higher-mode arrivals which precede R1–R4 are much more pronounced than in the case of Love waves; these long-period Rayleigh overtones have group speeds in the range $aC = 5-7$ km/s. Their ubiquitous presence upon both radial and longitudinal broad-band recordings following large earthquakes was first pointed out by Jobert, Gaulon, Dieulin & Roult (1977). These investigators correctly identified the high-speed arrivals as "une superposition d'harmoniques sphéroïdaux"; despite this certitude, they rather mysteriously christened them *X waves*. Several circumnavigations of the complex, multi-mode X wavegroup can be traced in the synthetic record section; the equivalent body-wave phases consist of SV surface multiples and SV-to-P conversions that are spawned upon each reflection.

The quantity plotted in each of the above four record sections is the product $\sqrt{t} \times \mathbf{s}(\mathbf{x}, t)$ rather than the raw particle displacement $\mathbf{s}(\mathbf{x}, t)$. This

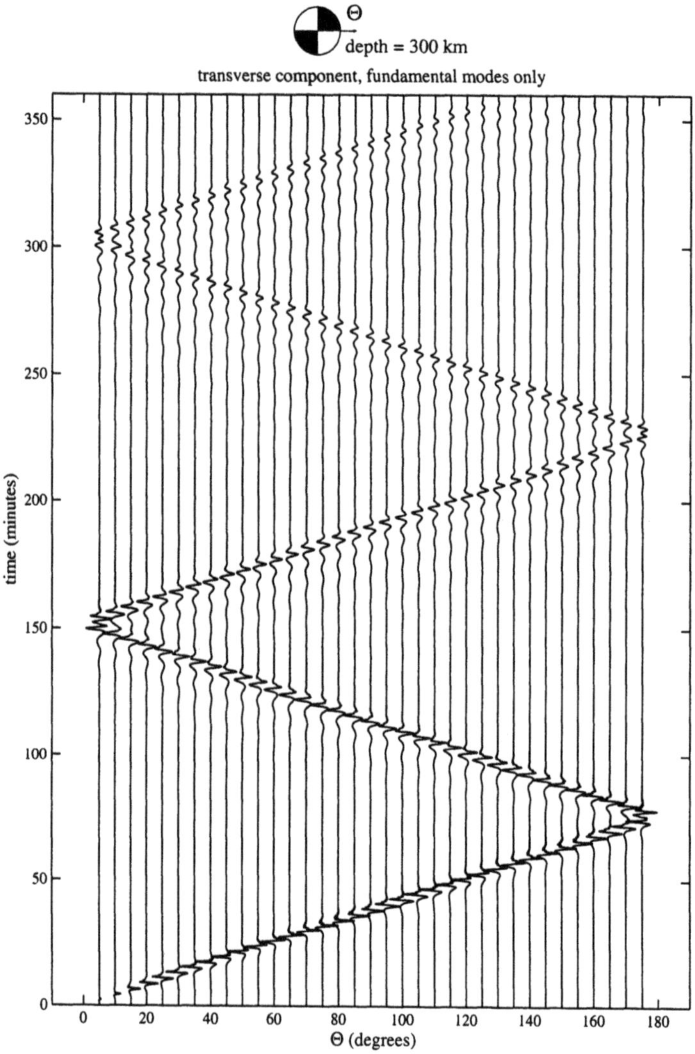

Figure 11.10. Record section of transverse-component fundamental-mode displacement seismograms $\hat{\mathbf{\Phi}} \cdot \mathbf{s}(\mathbf{x}, t)$ excited by a 300 km deep strike-slip fault, with an impulsive source time function $\dot{m}(t)$, in the PREM Earth model. The earthquake source mechanism is depicted schematically at the top; black and white areas on the beachball correspond to compressional and dilatational P-wave quadrants on the lower focal hemisphere, respectively. The stations are situated at 5° increments in epicentral distance Θ due east of the source, as indicated by the arrow attached to the beachball.

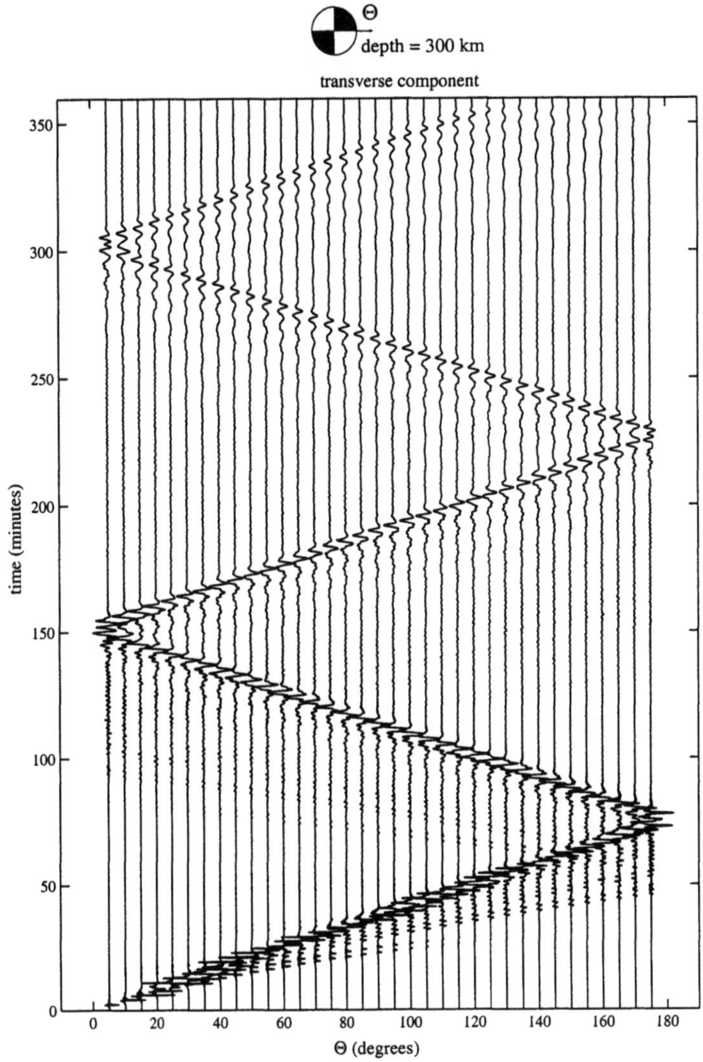

Figure 11.11. Same as Figure 11.10, except that the higher-mode Love waves have been included. All toroidal multiplets ${}_n\mathrm{T}_l$ with frequencies less than 50 mHz have been summed to synthesize these seismograms. Note the faster moving higher-mode arrivals.

square-root-of-time gain adjustment enhances the later arrivals, enabling them to be seen more clearly. Figure 11.14 illustrates that both the X

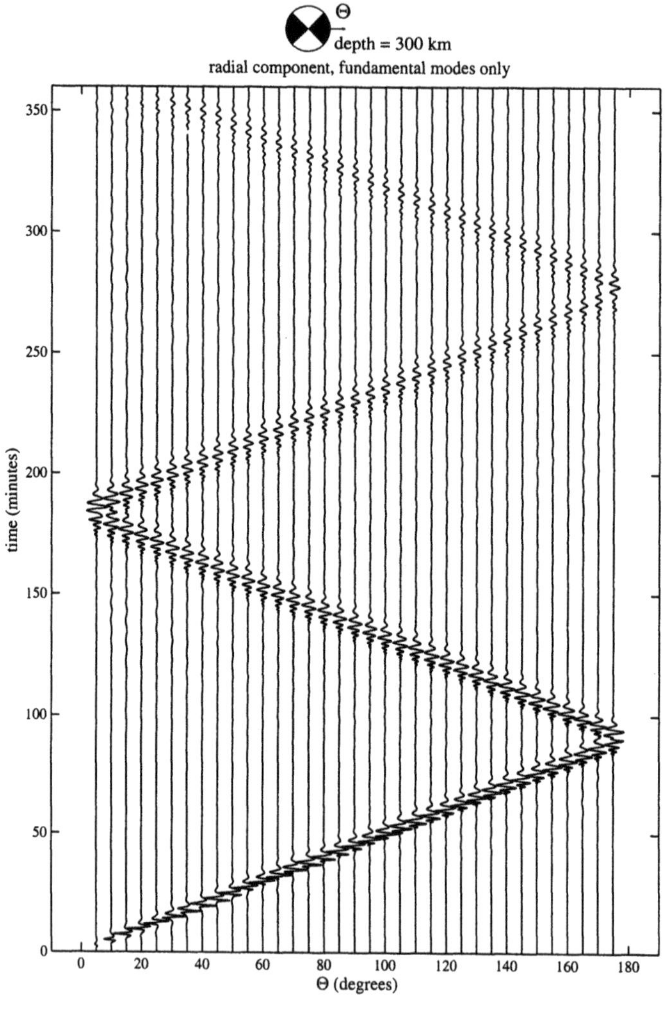

Figure 11.12. Record section of radial-component fundamental-mode displacement seismograms $\hat{\mathbf{r}} \cdot \mathbf{s}(\mathbf{x}, t)$ excited by a 300 km deep strike-slip fault, with an impulsive source time function $\dot{m}(t)$, in the PREM Earth model. The earthquake source mechanism is depicted schematically at the top; black and white areas on the beachball correspond to compressional and dilatational P-wave quadrants on the lower focal hemisphere, respectively. The stations are situated at 5° increments in epicentral distance Θ due east of the source, as indicated by the arrow attached to the beachball. The source-receiver geometry, with the fault striking N45°E as shown, insures that no toroidal modes ${}_n\mathrm{T}_l$ are excited.

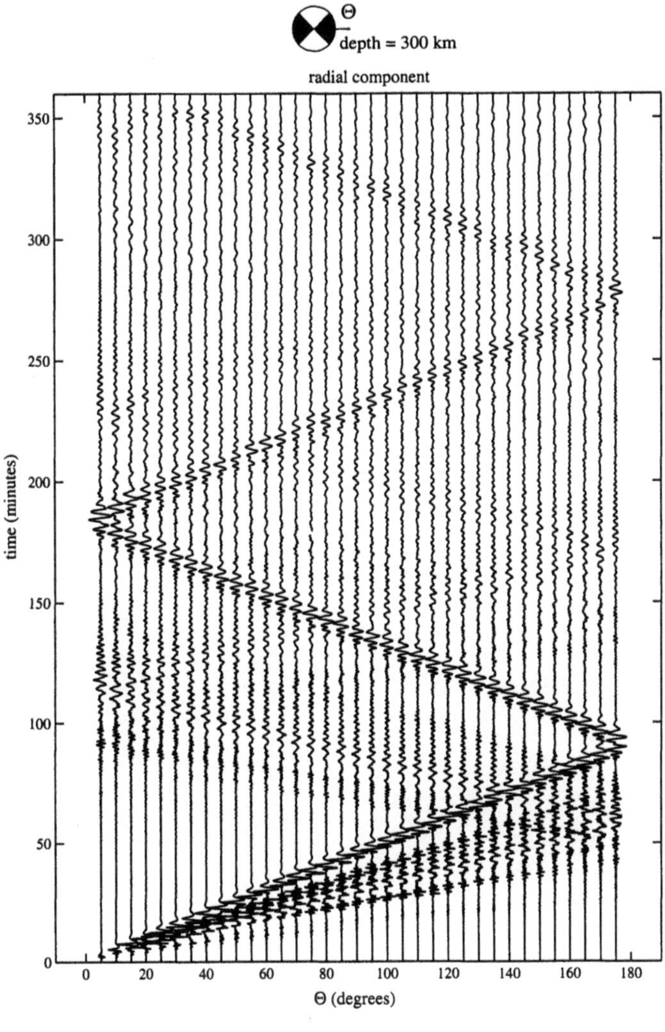

Figure 11.13. Same as Figure 11.12, except that the higher-mode Rayleigh waves have been included. The superposition of these $aC = 5-7$ km/s waves is known as the X phase. All spheroidal multiplets ${}_n\mathrm{S}_l$ with frequencies less than 50 mHz have been summed to synthesize the seismograms.

phase and the non-alignment of fundamental-mode Rayleigh-wave phase and group arrivals are evident in stacked, radial-component accelerograms; the black-or-white plotting format together with the $\sqrt{t}$ gain ranging results

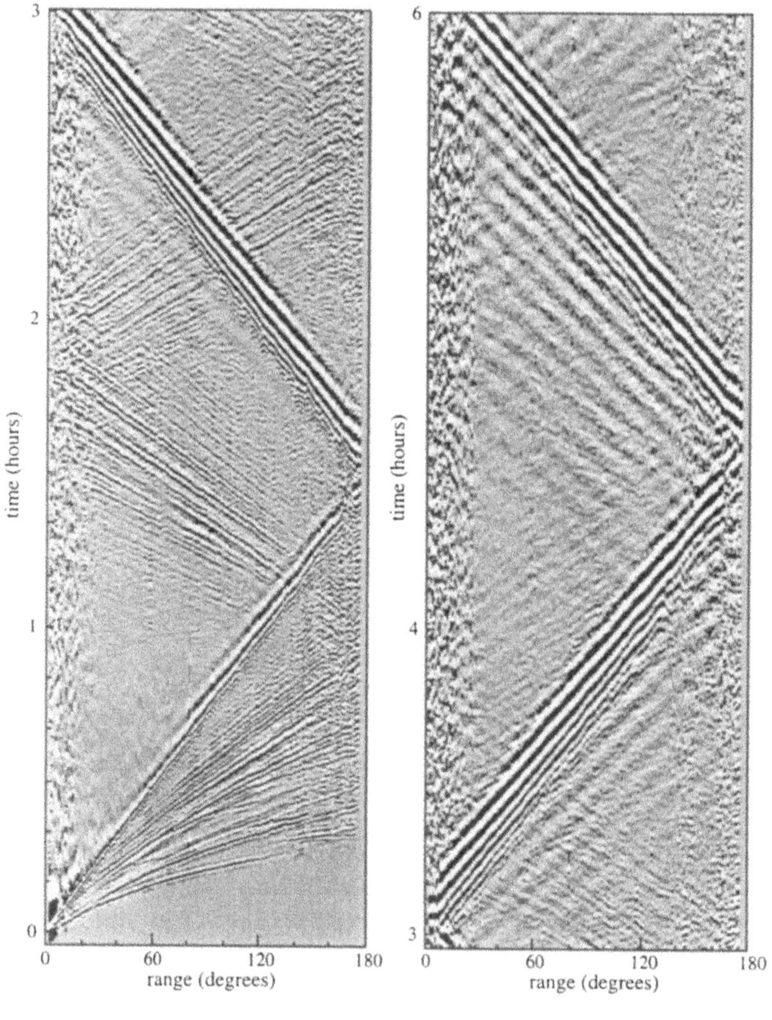

Figure 11.14. Stacked record section of radial-component accelerograms $\hat{\mathbf{r}} \cdot \mathbf{a}(\mathbf{x}, t)$ recorded by the IDA network during the eleven-year period 1981–1991. (*Left*) Time interval $t = 0 - 3$ hours after earthquake. (*Right*) Time interval $t = 3 - 6$ hours. The plotting format is that commonly employed in the seismic reflection industry: positive amplitudes are black and negative amplitudes are white. Each stacked accelerogram is scaled to have the same root-mean-square amplitude within the first 100 minutes. Multi-orbit arrivals have been enhanced by the application of a $\sqrt{t}$ gain-adjustment factor, as in Figures 11.10–11.13. (Courtesy of P. Shearer.)

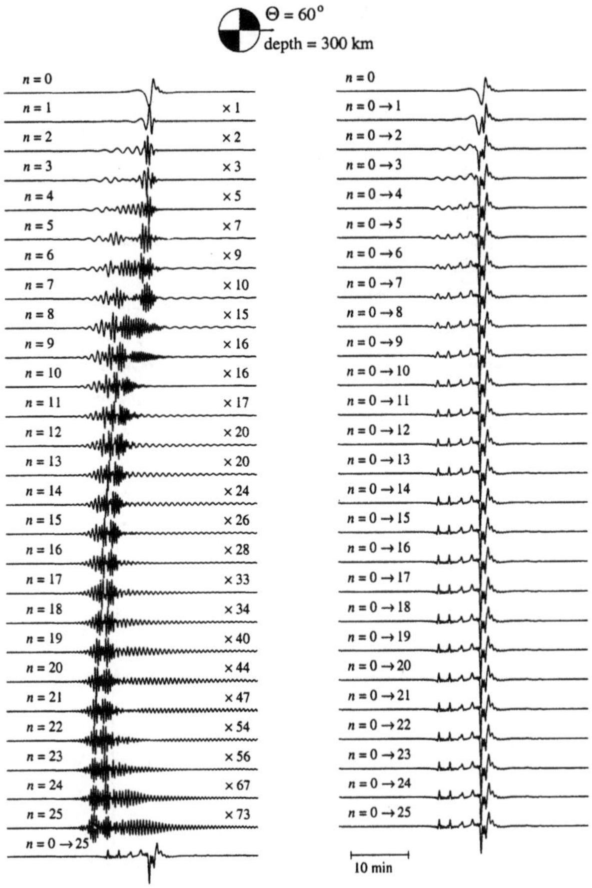

Figure 11.15. Transverse-component displacement response $\hat{\boldsymbol{\Phi}} \cdot \mathbf{s}(\mathbf{x}, t)$ to an intermediate-focus (h = 300 km) strike-slip fault. (*Top*) Schematic depiction of source mechanism and source-receiver geometry. (*Left*) Contribution of the fundamental and first 25 overtone toroidal-mode branches to the seismogram, starting with $n = 0$ at the top, and ending with $n = 25$ just above the bottom. The amplification factor of each overtone branch is indicated; for example, the $n = 10$ branch seismogram is plotted on a scale that is enhanced by a factor of 16 relative to the fundamental-mode seismogram. The complete $n = 0 \rightarrow 25$ synthetic seismogram is shown at the very bottom. (*Right*) Cumulative effect of adding one branch at a time. The topmost seismogram includes only the fundamental ($n = 0$) mode, the second shows the effect of adding the first overtone, the third includes branches $n = 0 \rightarrow 2$, and so on. The final trace is again the complete $n = 0 \rightarrow 25$ synthetic seismogram.

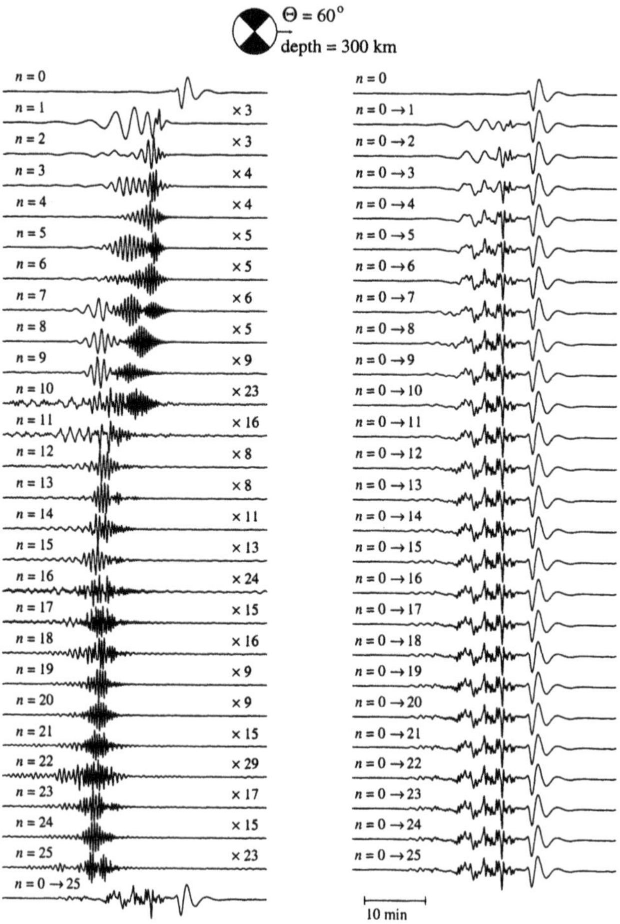

Figure 11.16. Radial-component displacement response $\hat{\mathbf{r}} \cdot \mathbf{s}(\mathbf{x}, t)$ to a 300 km deep strike-slip fault. (*Top*) Schematic depiction of source mechanism and source-receiver geometry. (*Left*) Contribution of the fundamental and first 25 overtone spheroidal-mode branches to the seismogram, starting with $n = 0$ at the top, and ending with $n = 25$ just above the bottom. The amplification factor of each overtone branch is indicated; for example, the $n = 10$ branch seismogram is plotted on a scale that is enhanced by a factor of 23 relative to the fundamental-mode seismogram. The complete $n = 0 \rightarrow 25$ synthetic seismogram is shown at the very bottom. (*Right*) Cumulative effect of adding one branch at a time. The topmost seismogram includes only the fundamental ($n = 0$) mode, the second shows the effect of adding the first overtone, the third includes branches $n = 0 \rightarrow 2$, and so on. The final trace is again the complete $n = 0 \rightarrow 25$ synthetic seismogram.

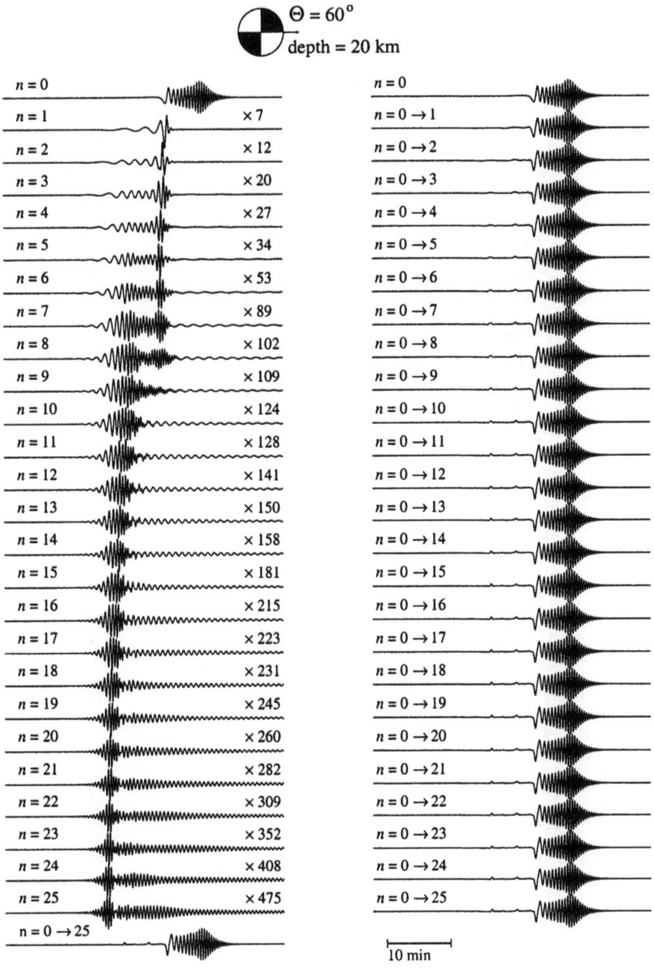

Figure 11.17. Same as Figure 11.15, for a shallow-focus ($h = 20$ km) strike-slip earthquake. Note that the overtone amplification factors are significantly larger, indicating the dominance of the fundamental ($n = 0$) mode.

in a visual image of the Earth's long-period surface-wave response that is reasonably uniform over the entire displayed range $0^\circ \leq \Theta \leq 180^\circ$ and $0 \leq t \leq 6$ hours (Shearer 1994a).

In Figures 11.15 and 11.16 we dissect the minor-arc Love and Rayleigh responses $\hat{\mathbf{\Phi}} \cdot \mathbf{s}(\mathbf{x}, t)$ and $\hat{\mathbf{r}} \cdot \mathbf{s}(\mathbf{x}, t)$ at an epicentral distance $\Theta = 60^\circ$, to il-

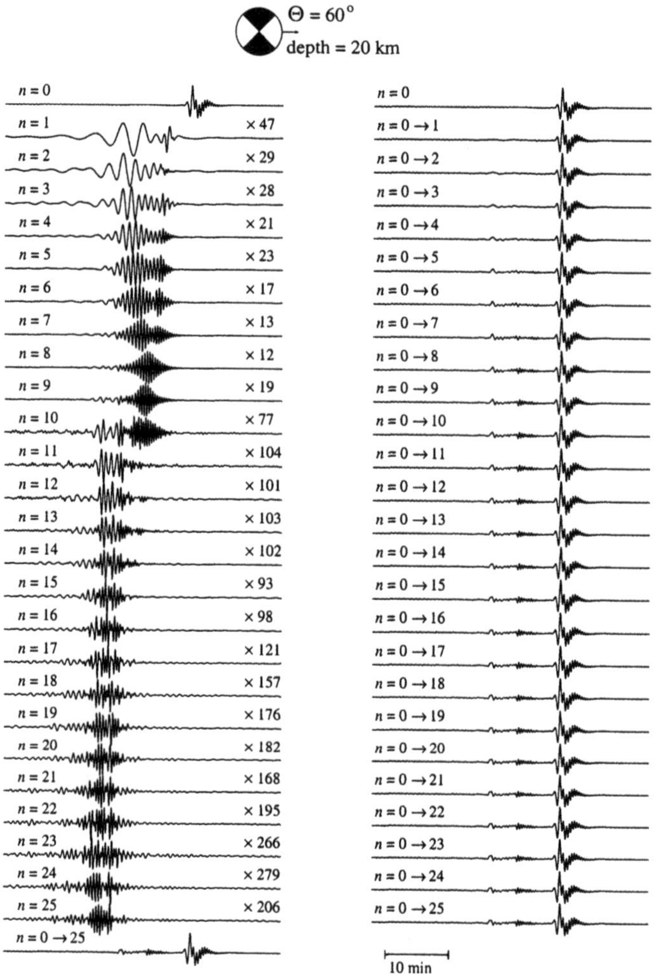

Figure 11.18. Same as Figure 11.16, for a shallow-focus (h = 20 km) strike-slip earthquake. Note that the overtone amplification factors are significantly larger, indicating the dominance of the fundamental ($n = 0$) mode.

lustrate the relative contributions of the individual toroidal and spheroidal mode branches ${}_0\mathrm{T}_l$–${}_{25}\mathrm{T}_l$ and ${}_0\mathrm{S}_l$–${}_{25}\mathrm{S}_l$. The hypothetical source is a 300 km deep strike-slip fault, as before. The left column in each figure displays the constituent single-branch seismograms whereas the right column displays the cumulative branch sums; this plotting scheme reveals in detail how the SH, SS_{SH}, SSS_{SH}, ... body waves and the SV and P-SV multiple reflections comprising the X phase are slowly built up by the superposition of the $n = 1, 2, \ldots$ overtone branches. Figures 11.17 and 11.18 show the analogous branch-by-branch dissection of the response to a shallow-focus ($h = 20$ km) strike-slip source. The fundamental modes ${}_0\mathrm{T}_l$ and ${}_0\mathrm{S}_l$ provide the dominant contribution to the seismograms in this case. The pulse-like arrival at the onset of the Love wavetrain is the $aC \approx 4.4$ km/s G wave; the highly dispersed tail is composed of slower, shorter-period ($2\pi/\omega < 40$ s) waves that are trapped in the 24.4 km thick PREM crust. The tail of the radial-component seismogram is likewise composed of short-period crustal Rayleigh waves. These short-period waves are not strongly excited by an intermediate-focus ($h = 300$ km) source.

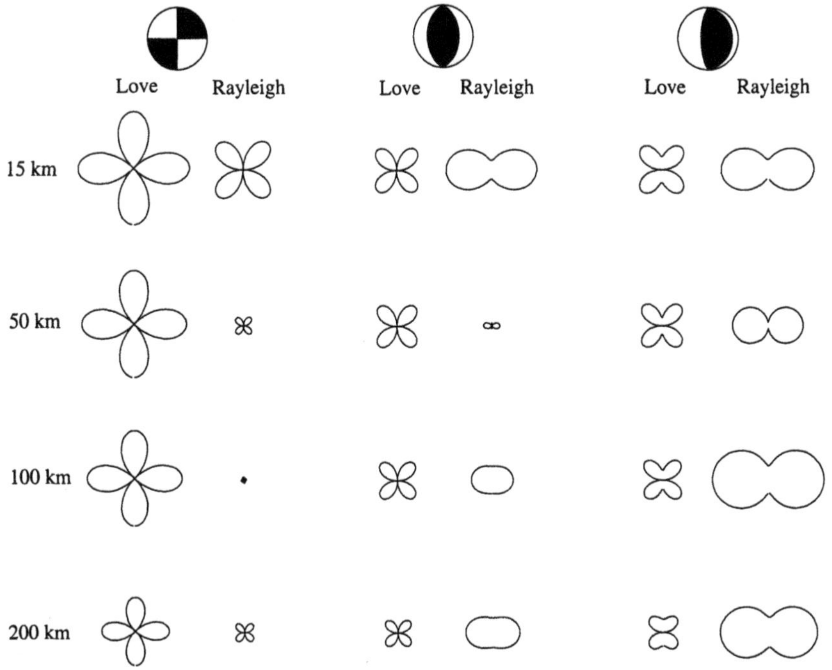

Figure 11.19. Azimuthal amplitude of the Love and Rayleigh waves radiated by a strike-slip fault (*left*), a 45° dipping thrust fault (*middle*), and a shallow-angle thrust or steeply dipping normal fault (*right*). Top row shows the associated mechanism; underlying rows show polar plots of the radiation pattern $|R(\Phi)|$ for sources at depths $h = 15$–200 km. The wave period is in all cases $2\pi/\omega = 100$ s; all patterns are plotted using the same scale.

11.7.2 Effect of source mechanism

In Figure 11.19 we display the azimuthal variation of surface-wave radiation for a variety of idealized earthquake source mechanisms and source depths $h = a - r_s$. The plots exhibit the modulus $|R(\Phi)|$ of the radiation pattern (11.34) for 100 second fundamental-mode Love and Rayleigh waves; this quantity—the *radiation amplitude*—is an even function of azimuth for any moment-tensor source:

$$|R(\Phi)| = |R(\Phi + \pi)|. \tag{11.88}$$

Note, furthermore, that since the absolute value $|R(\Phi)|$ is invariant under the substitution $\mathbf{M} \to -\mathbf{M}$, the patterns pertain either to the mechanism

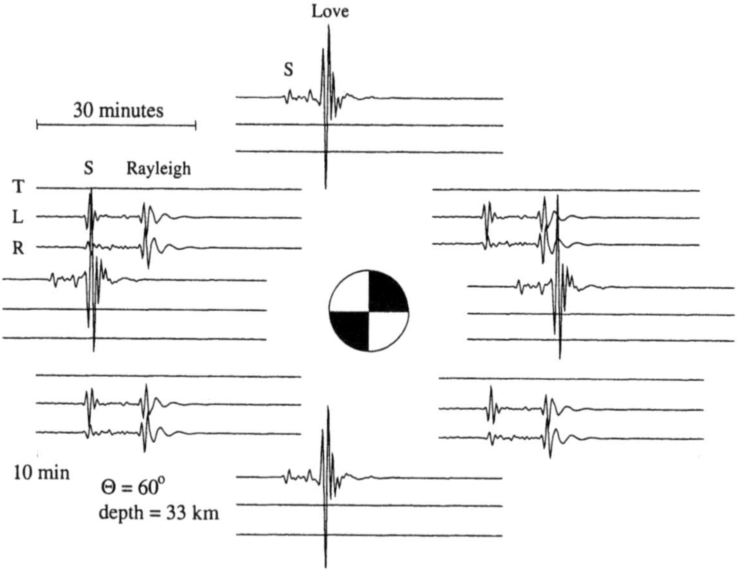

Figure 11.20. Illustration of the effect of the radiation pattern $R(\Phi)$ upon surface-wave accelerograms. The source is a vertical strike-slip fault situated at a focal depth $h = 33$ km; black and white areas on the central beachball correspond to compressional and dilatational P-wave quadrants on the lower focal hemisphere, respectively. The receivers are located (*clockwise from top*) to the north, northeast, east, southeast, south, southwest, west and northwest, all at an epicentral distance $\Theta = 60°$ from the event. The transverse (T), longitudinal (L) and radial (R) components and the S-wave, Love-wave and Rayleigh-wave arrivals are indicated. Identical to Figure 10.16, except that the mode sums have now been bandpass filtered between 50 and 250 seconds to accentuate the surface waves.

shown or to one with the black and white quadrants of the beachball interchanged; thus, for example, the middle column characterizes either a 45° dipping thrust fault or a 45° normal fault.

In the case of a pure vertical strike-slip event, the only non-zero elements of the moment tensor are $M_{\theta\theta}$, $M_{\phi\phi}$ and $M_{\theta\phi}$. As a result, both the Love and Rayleigh wave radiation patterns $|R(\Phi)|$ exhibit a pure quadrupolar $\sin 2\Phi$, $\cos 2\Phi$ dependence; Love-wave maxima in the along-strike and fault-perpendicular directions coincide with azimuthal nodes in the Rayleigh-wave pattern, and vice versa. The strength of the radiated Love waves decreases monotonically with increasing source depth; this simply reflects the quasi-exponential decay of the eigenfunction W_{s} (see Figure 8.7). The tangential Rayleigh-wave eigenfunction V_{s}, on the other hand, has a node at $h \approx 100$ km (see Figure 8.11); because of this, no 100 s Rayleigh waves whatsoever are excited by a strike-slip fault at this depth. Shorter-period and longer-period Rayleigh waves exhibit similar nodes for strike-slip sources at shallower and deeper depths, respectively. The Love waves excited by the idealized thrust and normal faults also display a predominantly quadrupolar radiation pattern; however, the Rayleigh waves exhibit a combined isotropic-plus-dipolar pattern. Note that it is possible for a double-couple

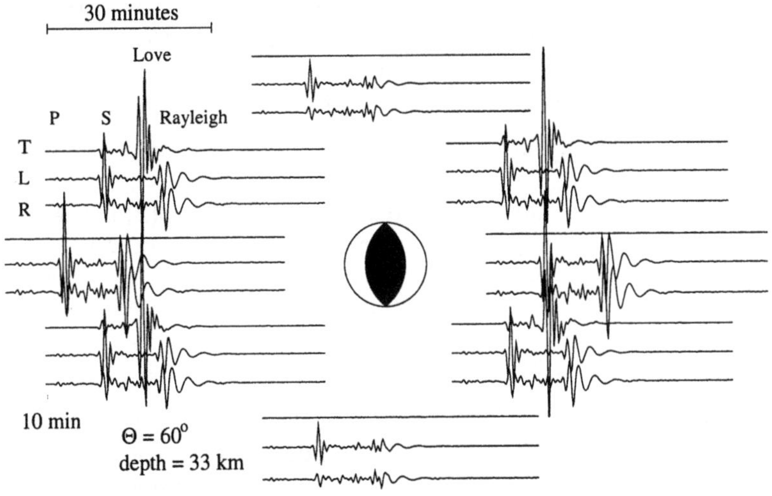

Figure 11.21. Same as Figure 11.20 except that the source is a shallow-focus (h = 33 km) thrust fault. The transverse (T), longitudinal (L) and radial (R) components and the P-wave, S-wave, Love-wave and Rayleigh-wave arrivals are indicated. Surface-wave version of Figure 10.17.

source to have no azimuthal nodes: the radiation of 100 s Rayleigh waves by a 100 km deep 45° dipping fault is very nearly isotropic. At shallow depths, both Love and Rayleigh waves have radiation minima coincident with the B axes of either a thrust or normal fault.

Figures 11.20 and 11.21 show the Love-wave and Rayleigh-wave accelerograms excited by a strike-slip fault and a 45° dipping thrust fault, respectively. The format is identical to that in Figures 10.16 and 10.17, except that the synthetic mode sums have been bandpass filtered between 50 and 250 seconds rather than between 20 and 80 seconds, in order to accentuate the fundamental surface waves. The north, east, south and west Love-wave nodes and the northeast, southeast, southwest and northwest Rayleigh-wave nodes are readily apparent in the case of the strike-slip fault. Note also the northeast, southeast, southwest and northwest Love-wave nodes and the non-nodal but predominantly east-west Rayleigh-wave radiation of the thrust fault.

11.8 Surface-Wave Perturbation Theory

The first-order effect of a spherical isotropic perturbation $\delta\kappa$, $\delta\mu$, $\delta\rho$, δd or $\delta\alpha$, $\delta\beta$, $\delta\rho$, δd upon the eigenfrequency ${}_n\omega_l$ of a free oscillation or standing wave is given in equations (9.12) and (9.20). The effect of a transversely isotropic perturbation δC, δA, δL, δN, δF or $\delta\alpha_{\mathrm{v}}$, $\delta\alpha_{\mathrm{h}}$, $\delta\beta_{\mathrm{v}}$, $\delta\beta_{\mathrm{h}}$, $\delta\eta$ is likewise given in (9.30) and (9.39). If we express the travelling-wave dispersion relations upon the perturbed Earth model in the form

$$\omega = \omega_n(k) + \delta\omega_n(k), \tag{11.89}$$

then these expressions can be used to find the change $\delta\omega_n(k)$ in the angular frequency of waves along the nth overtone branch *at a fixed wavenumber* k. Since quantitative surface-wave analyses are conducted in the frequency domain, it is preferable to rewrite (11.89) in the form

$$k = k_n(\omega) + \delta k_n(\omega), \tag{11.90}$$

and consider instead the change $\delta k_n(\omega)$ in the wavenumber *at a fixed angular frequency* ω. The first-order relation between the two perturbations $\delta k_n(\omega)$ and $\delta\omega_n(k)$ may be readily determined by expanding the perturbed dispersion relation (11.89) about the unperturbed wavenumber $k_n(\omega)$:

$$\omega = \omega_n(k_n) + C_n(k - k_n) + \delta\omega_n(k_n) + \cdots, \tag{11.91}$$

where $C_n = d\omega_n/dk$ is the group speed. The left side of equation (11.91) is cancelled by the first term on the right, because $\omega = \omega_n(k_n)$ is the

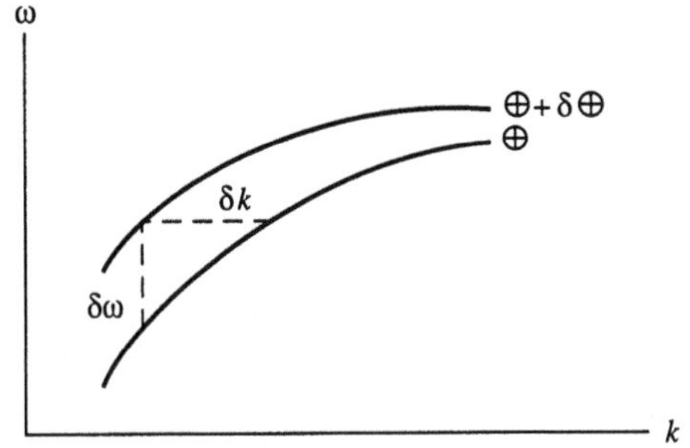

Figure 11.22. Schematic dispersion curves for two Earth models ⊕ and ⊕ + δ⊕. If the perturbation $\delta\omega$ in frequency at fixed wavenumber k is positive, then the perturbation δk in wavenumber at fixed frequency is negative. The two first-order perturbations are related by $\delta k = -C^{-1}\delta\omega$, where $C = d\omega/dk$ is the slope of the unperturbed curve.

dispersion relation upon the unperturbed Earth. The first-order change $\delta k_n = k - k_n$ in the wavenumber is therefore given by

$$\delta k = -C^{-1}\delta\omega, \tag{11.92}$$

where we have dropped the overtone index n for simplicity. The perturbation in the phase speed of a wave at a fixed angular frequency ω is simply $\delta c = -\omega k^{-2}\delta k$, so that the three fractional perturbations $\delta k/k$, $\delta c/c$ and $\delta\omega/\omega$ are related by

$$\left(\frac{\delta k}{k}\right)_{\omega} = -\left(\frac{\delta c}{c}\right)_{\omega} = -\frac{c}{C}\left(\frac{\delta\omega}{\omega}\right)_{k}, \tag{11.93}$$

where the subscripts indicate the variable that remains invariant in each instance. A schematic illustration of the frequency and wavenumber perturbations $\delta\omega$ and δk is shown in Figure 11.22; the proportionality relation (11.92) can alternatively be derived by a simple geometrical analysis based upon this diagram.

11.8.1 Fréchet derivatives of phase speed

Equation (11.93) enables us to calculate the Fréchet derivatives of the wavenumber k or phase speed c with respect to the parameters used to specify the spherically symmetric Earth model. On an isotropic model, for instance, the derivatives of the phase speed with respect to the compressional-

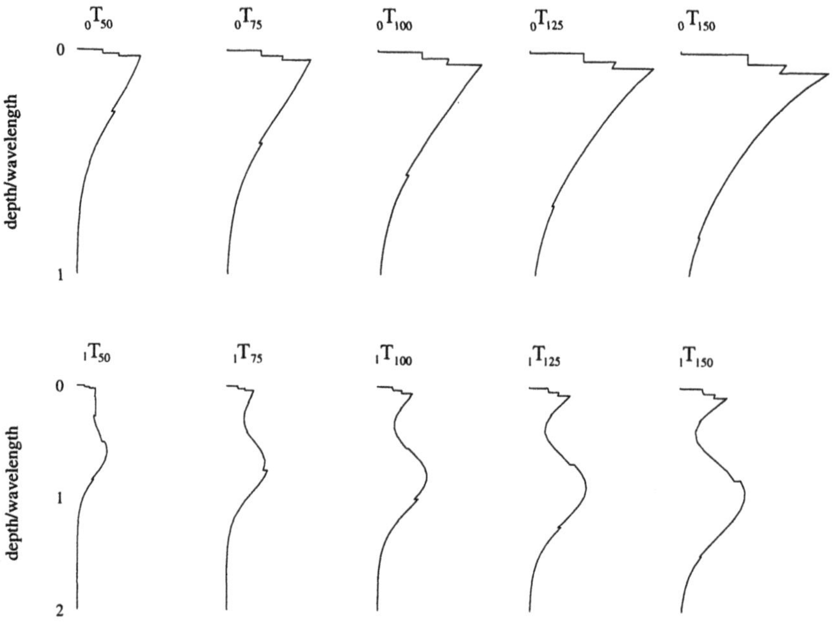

Figure 11.23. Dependence of the Fréchet derivative $(\partial c/\partial\beta)_{\alpha,\rho,d}$ upon the dimensionless depth z/λ for several fundamental (*top*) and first-overtone (*bottom*) Love-wave-equivalent modes.

wave speed α, the shear-wave speed β, the density ρ and the discontinuity radii d are given in terms of the kernels (9.13)–(9.16) and (9.21)–(9.23) by

$$\left(\frac{\partial c}{\partial\alpha}\right)_{\beta,\rho,d} = \left(\frac{c^2}{C\omega}\right) K_\alpha, \qquad \left(\frac{\partial c}{\partial\beta}\right)_{\alpha,\rho,d} = \left(\frac{c^2}{C\omega}\right) K_\beta, \tag{11.94}$$

$$\left(\frac{\partial c}{\partial\rho}\right)_{\alpha,\beta,d} = \left(\frac{c^2}{C\omega}\right) K_\rho', \qquad \left(\frac{\partial c}{\partial d}\right)_{\alpha,\beta,\rho} = \left(\frac{c^2}{C\omega}\right) K_d, \tag{11.95}$$

where the subscripts specify the variables—in addition to the angular frequency ω—that are held constant during the differentiation. The "shapes" of the derivatives of c, k and ω are identical, indicating that the extent to which a travelling or standing wave is sensitive to changes in the properties of the Earth at different depths is the same. Only the absolute magnitude of the three sensitivities is different; compare equations (11.94)–(11.95) with (9.24)–(9.25).

In Figures 11.23 and 11.24 we depict the Fréchet derivatives $(\partial c/\partial\alpha)_{\beta,\rho,d}$ and $(\partial c/\partial\beta)_{\alpha,\rho,d}$ for a number of fundamental and first-overtone Love- and

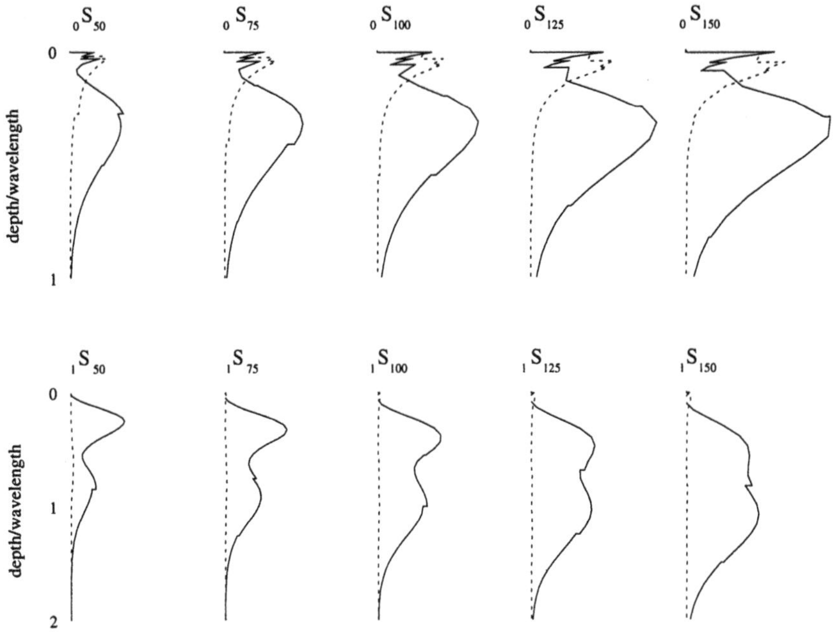

Figure 11.24. Dependence of the Fréchet derivatives $(\partial c/\partial\alpha)_{\beta,\rho,d}$ (*dashed line*) and $(\partial c/\partial\beta)_{\alpha,\rho,d}$ (*solid line*) upon the dimensionless depth z/λ for several fundamental (*top*) and first-overtone (*bottom*) Rayleigh-wave-equivalent modes.

Rayleigh-wave-equivalent modes; the independent variable in each case is the depth divided by the asymptotic surface wavelength,

$$\frac{z}{\lambda} = \frac{a - r}{2\pi k^{-1}}. \tag{11.96}$$

When re-scaled in this dimensionless manner, we see that all of the sensitivity kernels along a given surface-wave dispersion branch look essentially the same! A "rule of thumb" worth remembering is that fundamental Love and Rayleigh waves "feel" perturbations in the shear-wave speed β down to about one-quarter wavelength and one-half wavelength, respectively:

$$z_{\mathrm{L}} \approx \lambda/4, \qquad z_{\mathrm{R}} \approx \lambda/2. \tag{11.97}$$

Love waves are for this reason more strongly affected by the pronounced lateral variations in shear-wave speed in the lithosphere than Rayleigh waves are. First-overtone Love and Rayleigh waves provide significantly improved resolution at depth; they are sensitive to perturbations in β

down to about one wavelength and one and one-half wavelengths, respectively. This makes it highly advantageous to incorporate either higher-mode dispersion measurements or constraints obtained by fitting the equivalent SS + SSS + ⋯ waves in upper-mantle tomographic inversions. Fundamental Rayleigh waves have a limited sensitivity to shallow variations in the compressional-wave speed α—down to about one-eighth of a wavelength. Higher-overtone Rayleigh waves are essentially independent of α.

11.8.2 Fréchet derivatives of group speed

The effect of a perturbation upon the group speed may be determined by Fréchet differentiation of the relation (11.41). The Fréchet derivative of C with respect to a generic model parameter $m = \alpha, \beta, \rho, d, \ldots$ is

$$\frac{\partial c}{\partial m} = \frac{C}{c}\left(2 - \frac{C}{c}\right)\frac{\partial c}{\partial m} + \omega\left(\frac{C}{c}\right)^2 \frac{\partial}{\partial\omega}\left(\frac{\partial c}{\partial m}\right). \tag{11.98}$$

The final term in equation (11.98) may be computed by first-difference numerical differentiation of the phase-speed sensitivity kernel $\partial c/\partial m$ (Rodi, Glover, Li & Alexander 1975). Alternatively, it is possible to express $\partial_\omega(\partial c/\partial m)$ in terms of the quantities $\partial_\omega U$, $\partial_\omega V$, $\partial_\omega W$ and $\partial_\omega \phi$, and to calculate these frequency derivatives of the eigenfunctions by differentiation of the governing radial equations and boundary conditions (Gilbert 1976b). There are significant differences in detail between the shear-wave sensitivity kernels $(\partial c/\partial\beta)_{\alpha,\rho,d}$ and $(\partial C/\partial\beta)_{\alpha,\rho,d}$; however, the "rule of thumb" (11.97) for the fundamental-mode Love and Rayleigh "skin depth" pertains to the group-speed as well the phase-speed derivatives.

Chapter 12

Mode-Ray Duality

In the limit of high frequency, the far-field response to an earthquake can be conveniently expressed in terms of propagating SH and P-SV body waves. The body-wave portion of a seismogram can be modelled as a sequence of pulses or arrivals which travel between the source and the receiver along different rays. The presence of the free surface and discontinuities such as the core-mantle boundary gives rise to reflected phases such as pP, PP, PcP and sS, SS, ScS. Geometrical optics or ray theory provides a means of calculating the times of arrival and amplitudes of the various body-wave phases. In this chapter, we examine the correspondence between this classical ray-theoretical representation of the response and the representation as a sum of orthonormal modes. There are two complementary aspects to this *mode-ray duality*, both of which are considered here. We first take up the problem of going "from rays to modes", and demonstrate that each toroidal or spheroidal free oscillation can be regarded as a superposition of propagating body waves. Using simple physical arguments based upon the notion of constructive interference, we derive asymptotic formulae for the eigenfrequencies of a SNREI Earth model, which embody the principal qualitative features of the dispersion diagrams discussed in Chapter 8. To corroborate these formulae, we next conduct a more formal JWKB analysis of the governing radial differential equations and boundary conditions; this analysis yields asymptotic formulae for the associated toroidal and spheroidal eigenfunctions as well. Finally, we close the logical loop by considering the problem of going "from modes to rays" rather than vice versa. Using the JWKB eigenfrequencies and eigenfunctions, we show that the mode-sum representation of the response obtained in Chapter 10 is asymptotically equivalent to a superposition of SH and P-SV body waves. All possible reflections, refractions, conversions and reverberations are included.

12.1 Ray Theory Primer

We begin with a brief review of seismic ray theory in a SNREI Earth model. This discussion is not intended to replace either the classic introduction to this topic by Bullen (1963) or the elegant and authoritative treatments by Aki & Richards (1980) and Ben-Menahem & Singh (1981). Derivations that are given in these comprehensive references will not be repeated; our aim is simply to establish a consistent ray-theoretical notation and compile a compendium of formulae that will be required in the ensuing asymptotic analysis. Some very brief remarks regarding the extension to a transversely isotropic Earth model may be found in Section 12.4.5.

12.1.1 Nomenclature

We adhere to the standard seismological convention for labelling direct, reflected, refracted and converted seismic rays, as in Chapter 8. A compressional wave within the mantle, fluid outer core or solid inner core is denoted by P, K or I, respectively, whereas a shear wave within the mantle or inner core is denoted by S or J, respectively. Compound phases that propagate in more than one solid or fluid region of the Earth have concatenated names, such as SKS and PKIKP. A lower-case c and i are used to denote reflections off of the topside of the core-mantle boundary and inner-core boundary, respectively, as in ScS and PKiKP. Repeated capitals are used to denote reflections off of the underside of the two boundaries, such as SKKS or PKIIKP. Waves that leave the source going up and experience an immediate reflection off of the underside of the free surface are referred to as pP, sP, pS or sS, whereas waves that leave the source going down and are reflected off of the free surface halfway between the source and receiver are referred to as PP, SP, PS or SS. For simplicity, we shall not consider any of the complications introduced by the presence of an oceanic layer, the crust or any upper-mantle discontinuities; this eliminates seafloor multiples and solid-solid reflected or converted phases such as PmP, S660S or S410p. The only two internal discontinuities that are incorporated explicitly in our mode-ray duality analysis are the inner-core and core-mantle boundaries. These two fluid-solid boundaries are by far the most dramatic seismic discontinuities within the Earth. As we shall see, they are fundamentally responsible for the qualitative features of the asymptotic toroidal and spheroidal eigenfrequency spectra. The elastic properties and density are assumed to be smooth within the solid inner core $0 \leq r \leq c$, the fluid outer core $c \leq r \leq b$ and the mantle $b \leq r \leq a$. Subscripts a, b and c with distinguishing signs $+$ and $-$ will be used to signify evaluation of a variable on the top and bottom sides of the various boundaries.

12.1.2 Ray parameter

Seismic rays within a spherically symmetric Earth model are confined to the source-receiver great-circle plane. Let v denote either the compressional-wave speed α or the shear-wave speed β, and let i be the *angle of incidence* between a ray and the local upward vertical, as shown in Figure 12.1. The unit slowness vector in the direction of propagation of a wave along either the downgoing or upgoing leg of a ray is

$$\hat{\mathbf{p}} = \hat{\mathbf{r}} \cos i + \hat{\boldsymbol{\Theta}} \sin i, \tag{12.1}$$

where r, Θ, Φ are a system of epicentral spherical polar coordinates. The three quantities r, i and v all vary as we move from one point within the Earth to another; however the *ray parameter*

$$p = \frac{r \sin i}{v} \tag{12.2}$$

is constant along a ray. As a result, the various rays which leave a source can be identified and distinguished by their ray parameters. The path taken by a ray of ray parameter p can be determined by solving the three first-order equations

$$\frac{dr}{ds} = \cos i, \qquad \frac{d\Theta}{ds} = r^{-1} \sin i, \qquad \frac{di}{ds} = pr^{-1}(\dot{v} - v/r). \tag{12.3}$$

The independent variable, which is the angular arclength s along the ray, can be eliminated from (12.3), leading to a pair of coupled equations

$$\frac{dr}{d\Theta} = r \cot i, \qquad \frac{di}{d\Theta} = r(\dot{v}/v) - 1, \tag{12.4}$$

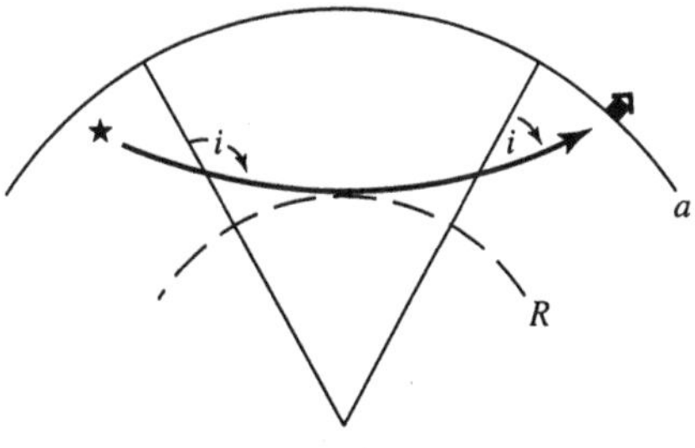

Figure 12.1. The angle of incidence is in the range $\pi \geq i \geq \pi/2$ on the downgoing leg and $\pi/2 \geq i \geq 0$ on the upgoing leg of a turning ray between a source (*star*) and a receiver (*doghouse*).

which can be integrated numerically to obtain $r(\Theta)$ and $i(\Theta)$, starting with the initial conditions $r(0) = r'$ and $i(0) = i'$, where a prime denotes evaluation at the location of the source $\mathbf{x}'$. Since the angle of incidence is $i = \pi/2$ at the *turning point* or deepest radius R, where $dr/d\Theta = 0$, the ray parameter there is simply $p = R/v(R)$. Whenever it is necessary to distinguish between the turning points of P, S, K, I and J waves, we shall do so by labelling them R_{P}, R_{S}, R_{K}, R_{I} and R_{J}.

The ray parameter p is related to the travel time T of a wave and the angular distance Θ that it travels by *Benndorf's relation*:

$$p = \frac{dT}{d\Theta}. \tag{12.5}$$

Equation (12.5) asserts that p is the *angular slowness* or slope of the travel-time curve. The total slowness in the direction of propagation is simply v^{-1}, so that the magnitude of the *radial slowness* of a seismic wave propagating along either a downgoing or upgoing leg of a ray is

$$q = \sqrt{v^{-2} - p^2 r^{-2}}. \tag{12.6}$$

In the evanescent region beneath the turning radius $r = R$, we choose the branch cut for the square root in equation (12.6) so that $\operatorname{Im} q \geq 0$, for a wave with positive angular frequency $\omega > 0$. Subscripts q_α and q_β will be used to distinguish the radial slownesses of compressional and shear waves, whenever it is necessary for clarity.

12.1.3 Travel time and distance

The quantities T and Θ can be determined by integrating the differential travel time $dT = v^{-1} ds$ and angular distance $d\Theta = r^{-1} \sin i \, ds$ along the associated ray path. In the simplest case of a turning ray with both endpoints situated on the Earth's surface, we find that

$$T = 2\int_R^a \frac{v^{-2}}{q}\, dr, \qquad \Theta = 2\int_R^a \frac{pr^{-2}}{q}\, dr, \tag{12.7}$$

where the factors of two account for both the downgoing and upgoing legs. The lower limit of integration R is replaced by either b or c in the case of a reflected wave such as ScS or PKiKP. More complicated cases involving multiple reflections or a buried source or receiver may be easily treated by combining integrals with the upper limit of integration a altered on the appropriate leg.

Equation (12.7) and its generalizations give the travel time $T(p)$ and angular distance $\Theta(p)$ as functions of the ray parameter p; these two relations define the associated travel-time curve $T(\Theta)$ parametrically. Benndorf's relation (12.5) can be verified by considering the ratio of the two derivatives dT/dp and $d\Theta/dp$. We shall assume for simplicity that there is not a strong decrease in the speed of either compressional waves or shear waves with depth in the asthenosphere, so that $\dot{v} < v/r$ everywhere within the Earth model. Equation (12.4) then guarantees that $di/d\Theta < 0$, so that there is a ray that turns at every radius in the mantle $b \leq r \leq a$; there are *no shadow zones*. Any sufficiently steep increase in wave speed with depth, of the sort that may be present in the upper-mantle transition zone, gives rise to a fold or *triplication* of the travel-time curve, as illustrated schematically in Figure 12.2. The cusps in the travel-time curve are associated with caustics; the third-arriving waves along the back branch have passed through the caustic, whereas the first and second arrivals have not (see Section 12.1.8).

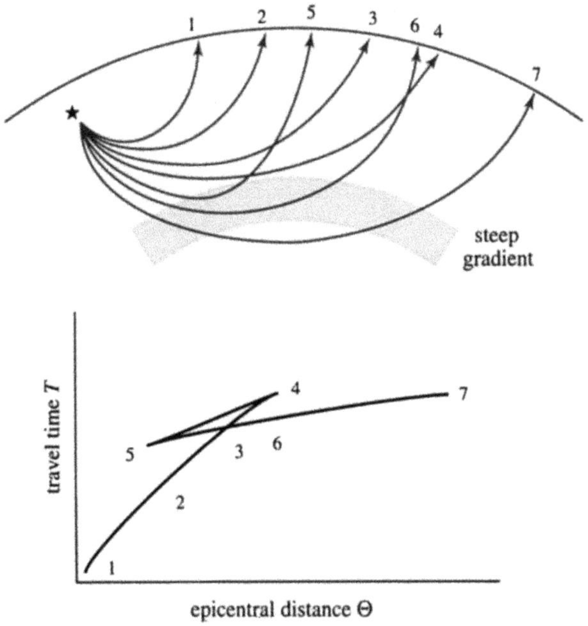

Figure 12.2. Schematic illustration of a travel-time triplication caused by a steep gradient $\dot{v} < 0$ in wave speed. (*Top*) Waves 1 through 7 have monotonically decreasing ray parameters $p_1 > \cdots > p_7$; the epicentral distance $\Theta(p)$ is smooth but multi-valued. (*Bottom*) Corresponding travel-time curve $T(\Theta)$; receivers situated between Θ_4 and Θ_5 have three ray-theoretical arrivals.

12.1.4 Intercept time

A tangent to the travel-time curve intercepts the vertical axis at a point

$$\tau = T - p\Theta, \tag{12.8}$$

as shown in Figure 12.3. Upon making use of equations (12.7) we find that this *intercept time* is the integrated radial slowness:

$$\tau = 2\int_R^a q\,dr. \tag{12.9}$$

It is noteworthy that the q^{-1} turning-point singularity which characterizes both the travel time T and epicentral distance Θ is eliminated in this integral for τ. For reflected waves or waves that do not begin and end on the Earth's surface, it is necessary to alter the limits of integration in equation (12.9); the intercept time of the inner-core J waves, for example, is two times the integral of the radial shear-wave slowness q_β from R_{J} to the inner-core boundary radius c.

An additional advantage of the $T-\Theta$ to $\tau-p$ transformation is that it "unfolds" upper-mantle triplications; the intercept time τ is always a *monotonically decreasing* function of the ray parameter p, with slope

$$\frac{d\tau}{dp} = -\Theta. \tag{12.10}$$

Figure 12.4 exhibits the $\tau-p$ curves for all of the principal seismic waves in model 1066A (Gilbert & Dziewonski 1975). The crustal layer has been stripped and the mantle properties α, β, ρ have been smoothly extrapolated

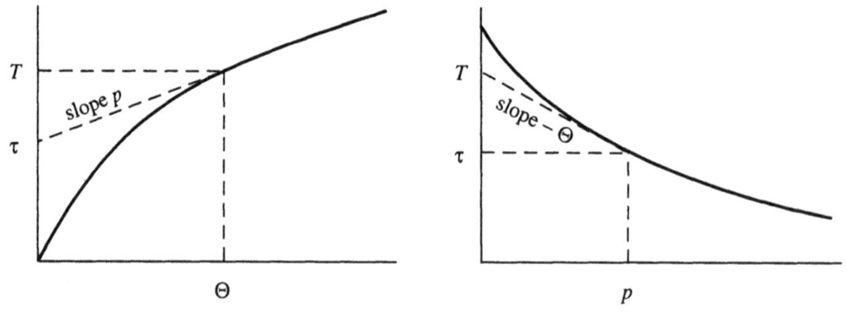

Figure 12.3. (*Left*) Schematic plot of T versus Θ showing the intercept time τ. (*Right*) Corresponding plot of τ versus the ray parameter p. The derivatives of the two curves are $dT/d\Theta = p$ and $d\tau/dp = -\Theta$.

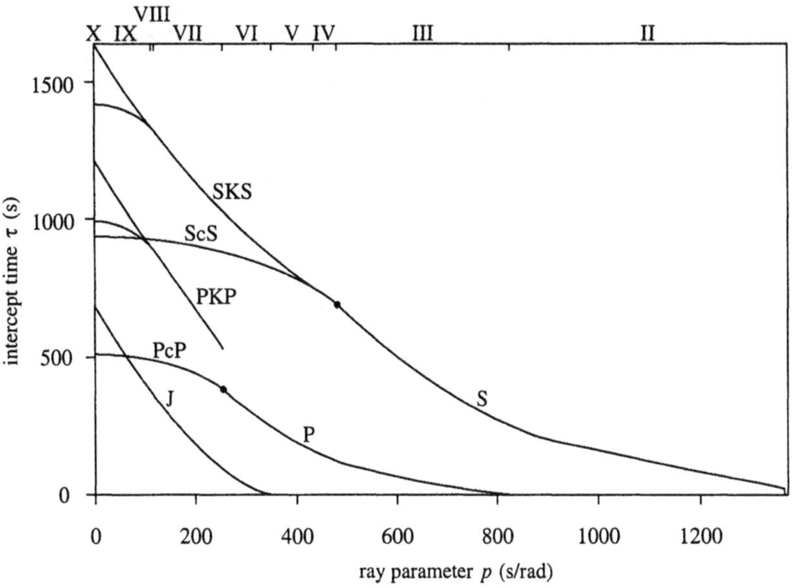

Figure 12.4. Intercept time τ versus ray parameter p for the major waves in a crustless version of model 1066A. The P-SV regimes II–X are indicated along the top. The Regime I boundary $p = a/\beta_a$ is situated at the right edge. Dots denote the P–PcP and S–ScS inflection points. The PKP curve bifurcates into PKiKP (*below*) and PKIKP (*above*) at low values of the ray parameter; likewise, the SKS curve bifurcates into SKiKS (*below*) and SKIKS (*above*).

to the surface $r = a$ to obtain a model with only two internal discontinuities. We shall employ subscripts such as $\tau_{\rm P}$ or $\tau_{\rm ScS}$ to distinguish geometrical arrivals or phases in the analysis that follows. Compound phases such as PP or SKKS are not shown; however, their properties can easily be obtained by addition and subtraction of the principal curves.

12.1.5 Polarization

The sign convention used to specify the direction of a *positive particle motion* accompanying a P, SV or SH wave is illustrated in Figure 12.5. Along either a downgoing or an upgoing leg of a ray traced in the direction of increasing Θ, the three polarization vectors are given explicitly by

$$\hat{\boldsymbol{\eta}}_{\rm P} = \hat{\mathbf{r}} \cos i + \hat{\boldsymbol{\Theta}} \sin i, \qquad \hat{\boldsymbol{\eta}}_{\rm SV} = \hat{\mathbf{r}} \sin i - \hat{\boldsymbol{\Theta}} \cos i,$$

$$\hat{\boldsymbol{\eta}}_{\rm SH} = -\hat{\boldsymbol{\Phi}}. \tag{12.11}$$

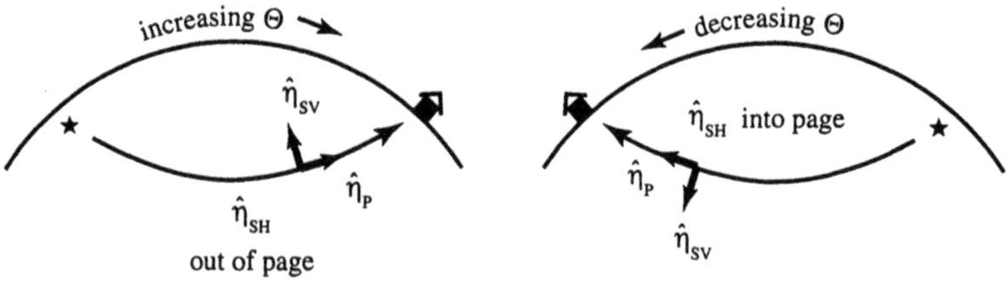

Figure 12.5. Proper ray bookkeeping requires that the positive polarization $\hat{\eta}$ of a P, SV or SH wave be defined in a consistent manner. This diagram illustrates our convention, for a wave going in the direction of increasing Θ (*left*) or decreasing Θ (*right*).

All of the polarizations are reversed for a wave travelling in the opposite direction: $\hat{\eta} \rightarrow -\hat{\eta}$ if $\Theta \rightarrow -\Theta$, where $\hat{\eta}$ denotes either $\hat{\eta}_{\mathrm{P}}$, $\hat{\eta}_{\mathrm{SV}}$, or $\hat{\eta}_{\mathrm{SH}}$. The general rule is that a positive compressional-wave polarization $\hat{\eta}_{\mathrm{P}}$ points *forward* whereas a positive SV-wave polarization $\hat{\eta}_{\mathrm{SV}}$ points to the *left* and a positive SH-wave polarization $\hat{\eta}_{\mathrm{SH}}$ points to the *right*, as one faces in the direction of propagation.

12.1.6 Reflection and transmission coefficients

We label the classical reflection and transmission coefficients for plane SH and P-SV waves incident upon the Earth's free surface and core-mantle and inner-core boundaries using a variation of the mnemonic notation employed by Aki & Richards (1980). Coefficients governing both the wave *amplitude* and the associated *energy flux* play a role in our considerations; italic letters P, S, K, I and J or script letters $\mathcal{P}$, $\mathcal{S}$, $\mathcal{K}$, $\mathcal{I}$ and $\mathcal{J}$ are used to denote the wave types in the two different cases, and grave and acute accents are used to denote the downgoing and upgoing wave directions. An SH wave that is incident upon either the free surface or a fluid-solid boundary is totally internally reflected. The two types of reflection coefficients are in that case identical: $\acute{S}\grave{S} = \grave{S}\acute{S} = \acute{J}\grave{J} = \acute{\mathcal{S}}\grave{\mathcal{S}} = \grave{\mathcal{S}}\acute{\mathcal{S}} = \acute{\mathcal{J}}\grave{\mathcal{J}} = 1$.

It is convenient in the P-SV constructive interference analysis to employ italic-letter coefficients that are ratios $(\hat{\mathbf{r}}\cdot\mathbf{s})_{\mathrm{out}}/(\hat{\mathbf{r}}\cdot\mathbf{s})_{\mathrm{inc}}$ of the *upward radial component* of the outgoing and incoming wave displacement rather than the ratios of the full vector displacement defined by Aki & Richards (1980). The radial-component reflection coefficients of a P-SV wave at the free surface of the Earth are given explicitly by

$$\acute{P}\grave{P} = \frac{-4p^2a^{-2}q_\alpha(a)q_\beta(a) + (\beta_a^{-2} - 2p^2a^{-2})^2}{4p^2a^{-2}q_\alpha(a)q_\beta(a) + (\beta_a^{-2} - 2p^2a^{-2})^2}, \tag{12.12}$$

$$\acute{S}\grave{P} = \frac{4(\beta_a^{-2} - 2p^2a^{-2})q_\alpha(a)q_\beta(a)}{4p^2a^{-2}q_\alpha(a)q_\beta(a) + (\beta_a^{-2} - 2p^2a^{-2})^2}, \tag{12.13}$$

$$\acute{P}\grave{S} = \frac{4(\beta_a^{-2} - 2p^2a^{-2})p^2a^{-2}}{4p^2a^{-2}q_\alpha(a)q_\beta(a) + (\beta_a^{-2} - 2p^2a^{-2})^2}, \tag{12.14}$$

$$\acute{S}\grave{S} = \frac{4p^2a^{-2}q_\alpha(a)q_\beta(a) - (\beta_a^{-2} - 2p^2a^{-2})^2}{4p^2a^{-2}q_\alpha(a)q_\beta(a) + (\beta_a^{-2} - 2p^2a^{-2})^2}, \tag{12.15}$$

where $p = a \sin i_{\mathrm{P}}/\alpha_a = a \sin i_{\mathrm{S}}/\beta_a$ is the ray parameter. The four coefficients (12.12)–(12.15) satisfy the symmetry relations

$$\acute{P}\grave{P} = -\acute{S}\grave{S}, \qquad \acute{P}\grave{P}\,\acute{S}\grave{S} - \acute{P}\grave{S}\,\acute{S}\grave{P} = -1. \tag{12.16}$$

Similar but more complicated formulae and symmetry relations for the core-mantle boundary and inner-core boundary P-SV coefficients are provided by Zhao & Dahlen (1993).

The P-SV body-wave Green tensor is more conveniently expressed in terms of script-letter reflection and transmission coefficients that are ratios $[|\rho v \cos i|^{1/2}(\hat{\boldsymbol{\eta}} \cdot \mathbf{s})]_{\mathrm{out}}/[|\rho v \cos i|^{1/2}(\hat{\boldsymbol{\eta}} \cdot \mathbf{s})]_{\mathrm{inc}}$ of the *signed square root* of the kinetic plus potential wave energy. The positive polarization vector $\hat{\boldsymbol{\eta}}$ of a P wave points in the direction of propagation, whereas that of an SV wave points to the left side of both the incoming and outgoing ray path, as illustrated in Figure 12.6. This choice of signs, which is consistent with that adopted for a turning wave, differs from the corresponding convention of Aki & Richards (1980). The (energy)$^{1/2}$ coefficients at the free surface $r = a$ are

$$\acute{\mathcal{P}}\grave{\mathcal{P}} = \acute{\mathcal{S}}\grave{\mathcal{S}} = \frac{4p^2a^{-2}q_\alpha(a)q_\beta(a) - (\beta_a^{-2} - 2p^2a^{-2})^2}{4p^2a^{-2}q_\alpha(a)q_\beta(a) + (\beta_a^{-2} - 2p^2a^{-2})^2}, \tag{12.17}$$

$$\acute{\mathcal{P}}\grave{\mathcal{S}} = -\acute{\mathcal{S}}\grave{\mathcal{P}} = \frac{4(\beta_a^{-2} - 2p^2a^{-2})^2pa^{-1}\sqrt{q_\alpha(a)q_\beta(a)}}{4p^2a^{-2}q_\alpha(a)q_\beta(a) + (\beta_a^{-2} - 2p^2a^{-2})^2}. \tag{12.18}$$

Equations (12.17)–(12.18) pertain to a wave travelling in the direction of increasing Θ; for a wave travelling in the opposite direction $\acute{\mathcal{P}}\grave{\mathcal{P}}$ and $\acute{\mathcal{S}}\grave{\mathcal{S}}$ are unchanged, but $\acute{\mathcal{P}}\grave{\mathcal{S}}$ and $\acute{\mathcal{S}}\grave{\mathcal{P}}$ undergo a change in sign. If we arrange the coefficients into forward and backward *scattering matrices*

$$\mathsf{S}_\pm = \begin{pmatrix} \acute{\mathcal{P}}\grave{\mathcal{P}} & \pm\acute{\mathcal{P}}\grave{\mathcal{S}} \\ \pm\acute{\mathcal{S}}\grave{\mathcal{P}} & \acute{\mathcal{S}}\grave{\mathcal{S}} \end{pmatrix}, \tag{12.19}$$

then this transformation can be written in the succinct form

$$\mathsf{S}_- = \mathsf{S}_+^{\mathrm{T}}, \qquad \mathsf{S}_+ = \mathsf{S}_-^{\mathrm{T}}, \tag{12.20}$$

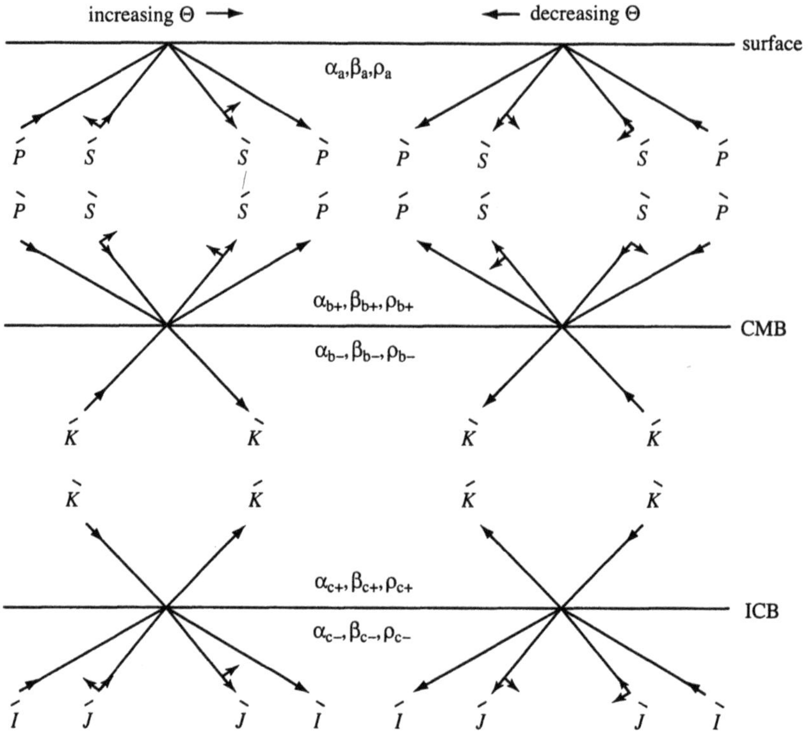

Figure 12.6. Schematic diagram illustrating the notation used in expressing the P-SV reflection and transmission coefficients at the free surface, the core-mantle boundary (CMB) and the inner-core boundary (ICB). Both the roman-letter and italic-letter coefficients adhere to this convention. The positive polarization $\hat{\eta}$ of an incoming or outgoing P, SV, K, I or J_{SV} wave is indicated.

where the subscripts plus and minus denote the direction of angular propagation. It is also readily verified that the forward and backward scattering matrices are inverses of each other in the sense

$$\mathsf{S}_+\mathsf{S}_- = \mathsf{S}_-\mathsf{S}_+ = \mathsf{I}. \tag{12.21}$$

Equation (12.21) expresses the *conservation of incoming and outgoing energy* at the interface:

$$\acute{\mathcal{P}}\grave{\mathcal{P}}^2 + \acute{\mathcal{P}}\grave{\mathcal{S}}^2 = \acute{\mathcal{S}}\grave{\mathcal{S}}^2 + \acute{\mathcal{S}}\grave{\mathcal{P}}^2 = 1. \tag{12.22}$$

The partition of P-SV energy at the core-mantle boundary is governed by a pair of 3×3 scattering matrices analogous to (12.19):

$$\mathsf{S}_{\pm} = \begin{pmatrix} \grave{\mathcal{P}}\acute{\mathcal{P}} & \pm\grave{\mathcal{P}}\acute{\mathcal{S}} & \grave{\mathcal{P}}\grave{\mathcal{K}} \\ \pm\grave{\mathcal{S}}\acute{\mathcal{P}} & \grave{\mathcal{S}}\acute{\mathcal{S}} & \pm\grave{\mathcal{S}}\grave{\mathcal{K}} \\ \acute{\mathcal{K}}\acute{\mathcal{P}} & \pm\acute{\mathcal{K}}\acute{\mathcal{S}} & \acute{\mathcal{K}}\grave{\mathcal{K}} \end{pmatrix}. \tag{12.23}$$

The corresponding matrices at the inner-core boundary are

$$\mathsf{S}_{\pm} = \begin{pmatrix} \grave{\mathcal{K}}\acute{\mathcal{K}} & \grave{\mathcal{K}}\grave{\mathcal{I}} & \pm\grave{\mathcal{K}}\grave{\mathcal{J}} \\ \acute{\mathcal{I}}\acute{\mathcal{K}} & \acute{\mathcal{I}}\grave{\mathcal{I}} & \pm\acute{\mathcal{I}}\grave{\mathcal{J}} \\ \pm\acute{\mathcal{J}}\acute{\mathcal{K}} & \pm\acute{\mathcal{J}}\grave{\mathcal{I}} & \acute{\mathcal{J}}\grave{\mathcal{J}} \end{pmatrix}. \tag{12.24}$$

At a solid-solid boundary, an incident SH wave gives rise to both a reflected and a transmitted wave, and an incident P or SV wave spawns four outgoing waves. Upon combining the P-SV and SH results, we may express the overall energy partition in terms of the 6×6 scattering matrices

$$\mathsf{S}_{\pm} = \begin{pmatrix} \mathsf{S}_{\pm}^{\text{P-SV}} & 0 \\ 0 & \mathsf{S}_{\pm}^{\text{SH}} \end{pmatrix}, \tag{12.25}$$

where $\mathsf{S}_{\pm}^{\text{P-SV}}$ and $\mathsf{S}_{\pm}^{\text{SH}}$ are 4×4 and 2×2, respectively. The free-surface and fluid-solid scattering matrices (12.19) and (12.23)–(12.24) can likewise be augmented to form the associated P-SV plus SH matrices by suitable infilling of the sub-matrices $\mathsf{S}_{\pm}^{\text{P-SV}}$ and $\mathsf{S}_{\pm}^{\text{SH}} = \mathsf{I}$ with zeroes and ones. Every such pair of forward and backward 6×6 scattering matrices $\mathsf{S}_{\pm}$ satisfies the two dynamical symmetry relations (12.20)–(12.21). A comprehensive account of the properties of both interfacial and more general elastic scattering matrices of the form (12.25) is given by Kennett (1983). His normalization convention, which is selected for convenience in the calculation of synthetic body-wave seismograms by means of the reflectivity method, differs from that employed here. Our geometrical convention, in which every wave "carries" its polarization with it whenever it turns or reflects, is the natural one to use in ray-theoretical calculations.

12.1.7 Geometrical spreading

In the smooth regions of the Earth between major discontinuities, the amplitude variations of high-frequency waves are governed by the focusing and defocusing of so-called *ray tubes* surrounding the associated rays. The flux of energy in a transient P, SV or SH wave is $\mathbf{K} = \omega^2 \rho v A^2 \hat{\mathbf{p}}$, where A is the displacement amplitude in the frequency domain. Consider a ray that departs from a source at $\mathbf{x}'$ and subsequently passes through two sequential points $\mathbf{x}_1$ and $\mathbf{x}_2$, as shown in Figure 12.7. In the absence of anelastic dissi-

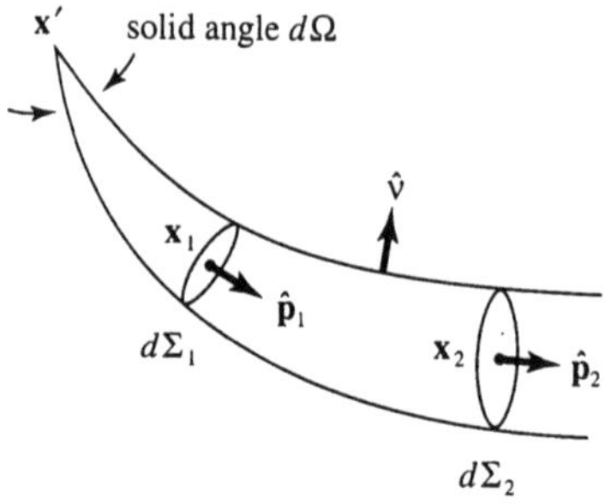

Figure 12.7. Schematic illustration of a ray tube subtending a differential solid angle $d\Omega$ at the source $\mathbf{x}'$. The unit slowness vectors at points $\mathbf{x}_1$ and $\mathbf{x}_2$ are $\hat{\mathbf{p}}_1$ and $\hat{\mathbf{p}}_2$, and the differential cross-sectional areas are $d\Sigma_1$ and $d\Sigma_2$, respectively.

pation, conservation of energy requires that $\|\mathbf{K}\|_1\, d\Sigma_1 = \|\mathbf{K}\|_2\, d\Sigma_2$, where $d\Sigma$ denotes the differential cross-sectional area of the surrounding ray tube; it follows that the amplitudes of the waves at $\mathbf{x}_1$ and $\mathbf{x}_2$ are related by

$$\frac{A_2}{A_1} = \left(\frac{\rho_2 v_2}{\rho_1 v_1}\right)^{-1/2} \left|\frac{d\Sigma_2}{d\Sigma_1}\right|^{-1/2}. \tag{12.26}$$

The absolute value is introduced in order to generalize the amplitude variation law (12.26) to the case in which the wave passes through one or more caustics (see Section 12.1.8).

It is convenient to define a point-source *geometrical spreading factor* $\mathcal{R}(\mathbf{x}, \mathbf{x}')$, which is analogous to the source-receiver distance $\|\mathbf{x} - \mathbf{x}'\|$ in a homogeneous medium, by

$$\mathcal{R} = \sqrt{|d\Sigma|/d\Omega}, \tag{12.27}$$

where $d\Omega$ is the differential solid angle subtended by the ray tube at the source $\mathbf{x}'$, and $\mathbf{x}$ denotes either $\mathbf{x}_1$ or $\mathbf{x}_2$. Upon rewriting the ratio (12.26) in terms of this coefficient, we obtain the geometrical amplitude variation law governing a high-frequency P, SV or SH wave in a perfectly elastic Earth:

$$A \sim (\rho v)^{-1/2} \mathcal{R}^{-1}. \tag{12.28}$$

The symbol $\sim$ in equation (12.28) stands for "varies along a ray as".

To calculate $\mathcal{R}$, we write the differential solid angle subtended at the source $\mathbf{x}'$ in the form $d\Omega = \sin i'\, |di'|\, d\Phi = r'^{-2} v' q'^{-1} p\, |dp|\, d\Phi$, and the differential cross-sectional area of the ray tube at the receiver $\mathbf{x}$ in the form

$d\Sigma = r^2|\cos i| \sin\Theta\,|d\Theta|\,d\Phi = r^2 vq \sin\Theta\,|d\Theta|\,d\Phi$, where $d\Phi$ is the differential azimuth, measured perpendicular to the ray plane. Upon making use of these results together with the identity (12.10) we find that

$$v'\mathcal{R} = rr'\sqrt{\frac{vv'qq'\sin\Theta}{p}\left|\frac{d^2\tau}{dp^2}\right|}. \tag{12.29}$$

It is evident from the form of (12.29) that the spreading coefficient of every simple or compound wave satisfies the dynamical *reciprocity relation*

$$v(\mathbf{x}')\mathcal{R}(\mathbf{x},\mathbf{x}') = v(\mathbf{x})\mathcal{R}(\mathbf{x}',\mathbf{x}). \tag{12.30}$$

The results (12.29)–(12.30) pertain not only to the direct waves P and S, but also to all concatenated or compound waves such as PcP and SKKS, regardless of the number of reflections, refractions and conversions; the absolute direction cosines $v'q' = |\cos i'|$ and $vq = |\cos i|$ are those of the waves leaving $\mathbf{x}'$ and arriving at $\mathbf{x}$ even if there has been a change in type along the intervening path, as in the case of PS or ScP. Figure 12.4 shows that the curvature $d^2\tau/dp^2$ in (12.29) is positive for all turning rays such as P, S, PKP and SKS but negative for the reflected rays PcP, ScS, PKiKP and SKiKS.

For a fixed source point $\mathbf{x}'$ the geometrical spreading factor $\mathcal{R}(\mathbf{x},\mathbf{x}')$ is a continuous function of the receiver location $\mathbf{x}$ along the ray, except at interfaces where it may suffer a jump discontinuity. The product $\mathcal{R}\,|\cos i|^{-1/2}$ is continuous; the offsetting jump in the quantity $|\cos i|^{-1/2} = (vq)^{-1/2}$ accounts for the change in ray-tube area upon reflection or refraction. The amplitude variation law (12.28) is generalized in the case of a compound wave to

$$A \sim (\rho v)^{-1/2}\Pi\mathcal{R}^{-1}, \tag{12.31}$$

where Π is the product of all of the (energy)$^{1/2}$ reflection and transmission coefficients $\grave{\mathcal{P}}\acute{\mathcal{S}}$, $\grave{\mathcal{P}}\acute{\mathcal{K}}$, etc. along the ray path. The relation (12.31) guarantees that the kinetic plus potential energy leaving the source in every infinitesimal solid angle $d\Omega$ is conserved; the reflection-transmission product Π accounts for the partition of the incoming energy among the various outgoing waves at each interface.

12.1.8 Caustic phase shift

The absolute values in equations (12.26) and (12.27) are necessary to account for the possible presence of *caustics*, where the differential cross-sectional area $d\Sigma$ of the ray tube goes to zero, as illustrated in Figure 12.8. Every time a wave passes through a caustic, it undergoes a non-geometrical

$\pi/2$ advance in phase; mnemonically, we may think of this as arising from the change in sign of the ray-tube area, as in equation (11.36):

$$\left(\frac{d\Sigma_2}{d\Sigma_1}\right)^{1/2} = \left|\frac{d\Sigma_2}{d\Sigma_1}\right|^{1/2} \exp(i\pi/2). \tag{12.32}$$

A more rigorous wave-theoretical derivation of the $\pi/2$ phase advance upon passage through a caustic is given by Landau & Lifshitz (1971). Upon incorporating caustic phase shifts into (12.31) we obtain the final form of the ray-theoretical body-wave amplitude variation law:

$$A \sim (\rho v)^{-1/2}\Pi\mathcal{R}^{-1}\exp(iM\pi/2). \tag{12.33}$$

The integer M is the so-called *Maslov index*, which keeps track of the number of caustic passages. Normally M increases by one every time that a wave passes through a caustic; in the rare case that a wave passes through a focal point, M increases by two. The number of caustic or focal-point encounters along every ray and reversed ray is the same:

$$M(\mathbf{x},\mathbf{x}') = M(\mathbf{x}',\mathbf{x}). \tag{12.34}$$

For example, both the SS ray and its reciprocal pass through a single caustic at the turning point of their second leg.

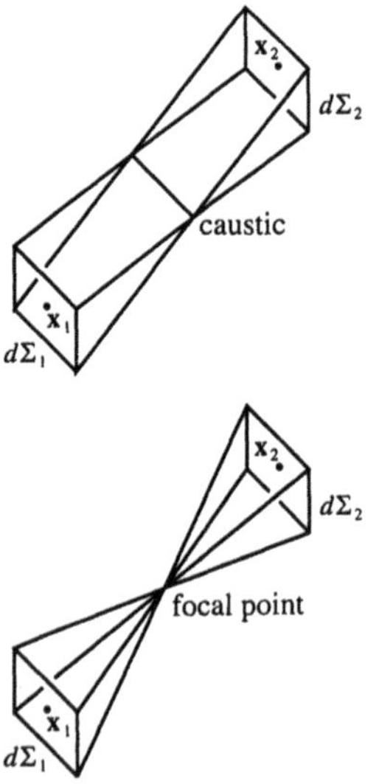

Figure 12.8. (*Top*) The cross-sectional area of a ray tube goes to zero in one but not both of its transverse dimensions at a caustic. If $d\Sigma_1$ at $\mathbf{x}_1$ is positive, then $d\Sigma_2$ at $\mathbf{x}_2$ may be regarded as negative. (*Bottom*) Focal points where the ray tube "collapses" in both transverse directions simultaneously (and where the phase advance of a wave is π rather than $\pi/2$) are exceedingly rare in a SNREI Earth model.

12.2 Constructive-Interference Principle

The free oscillations of a SNREI Earth are standing waves produced by the constructive interference of propagating SH and P-SV body waves *having the same ray parameter p*. In this section we derive asymptotic formulae for the normal-mode eigenfrequencies of a SNREI Earth model using elementary physical arguments based upon this notion. Increasingly complete accounts of the material presented here may be found in the original papers by Brune (1964; 1966), Odaka (1978), Levshin (1981) and Zhao & Dahlen (1993). The phases of all waves in the present summary are opposite in sign to those in the last analysis because of a difference in the Fourier transform convention.

12.2.1 Jeans relation

A family of turning waves with the same ray parameter p is depicted on the left in Figure 12.9. Every ray turns at the same radius; the spherical surface with this radius R is an envelope of the rays, or a *caustic*. If we suppose that all of the rays exhibit the same inclination to the $\hat{\mathbf{z}}$ axis, then there are also conical caustics associated with the northern-hemisphere and southern-hemisphere turning colatitudes θ_0 and $\pi - \theta_0$, as shown on the right. Fundamentally, the problem of finding the asymptotic normal modes

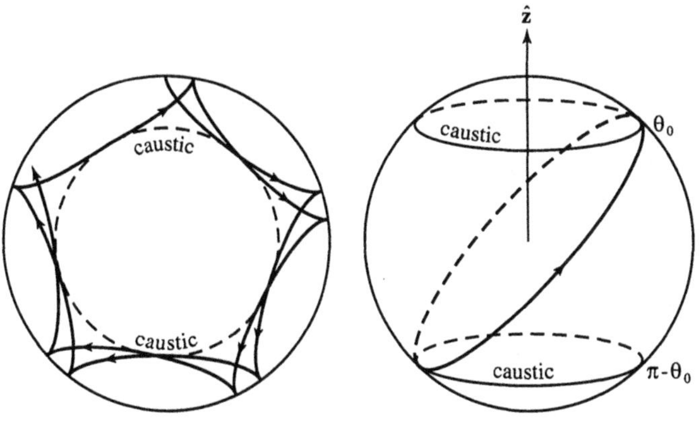

Figure 12.9. (*Left*) Schematic cross-section of the Earth showing the spherical caustic formed by a family of turning waves with the same ray parameter p. (*Right*) Conical caustics formed by a family of waves that all turn at the northern and southern colatitudes θ_0 and $\pi - \theta_0$.

of the Earth is a *quantization* problem—the requirement for constructive interference is that an integral number of oscillations must "fit" onto the surface of the sphere as well as between the two radii $r = R$ and $r = a$. The conditions which guarantee this are analogous to the *Bohr-Sommerfeld* quantization conditions in semi-classical quantum mechanics (Born 1927; Landau & Lifshitz 1965). Because of the three-dimensional geometry, there are three independent quantization conditions and three independent quantum numbers; we consider the simplest part of the problem—the angular quantization—in the remainder of this section.

The angular rate of propagation of the waves or phase fronts associated with the family of rays shown in Figure 12.9 is $d\Theta/dT = p^{-1}$; the corresponding spatial rate of change of phase experienced by a wave of angular frequency ω is $\omega(dT/d\Theta) = \omega p$. The total phase accumulated in a single complete passage of such a wave around the Earth, $\Theta \to \Theta + 2\pi$, is

$$\Delta\Psi = -2\pi\omega p + \pi, \tag{12.35}$$

where the minus sign is dictated by our Fourier convention, and the second term accounts for the two $\pi/2$ phase shifts produced by the passage through the northern-hemisphere and southern-hemisphere caustics. The Bohr-Sommerfeld condition for constructive interference of the propagating waves is

$$|\Delta\Psi| = 2\pi l, \tag{12.36}$$

where l is a non-negative integer. Comparison of (12.35) and (12.36) yields the well-known angular quantization relation

$$\omega p = l + \tfrac{1}{2}. \tag{12.37}$$

The quantum number l is—as we shall see next—the angular *degree* of the mode ${}_n\mathrm{T}_l$ or ${}_n\mathrm{S}_l$. Equation (12.37), which is valid for any value of l, including $l = 0$, was first obtained by Jeans (1927) in his classic investigation of the propagation of earthquake waves within a sphere.

In the short-wave limit, $l \gg 1$, the Jeans relation is consistent with the asymptotic representation of a surface spherical harmonic of order $m \geq 0$, obtained in Appendix B.7:

$$\begin{aligned} X_{lm}(\theta) \approx\ & \frac{1}{\pi}(\sin^2\theta - \sin^2\theta_0)^{-1/4} \\ & \times \cos\left[(l + \tfrac{1}{2})\arccos(\cos\theta/\cos\theta_0)\right. \\ & \qquad \left. - \, m\arccos(\cot\theta/\cot\theta_0) + m\pi - \tfrac{1}{4}\pi\right], \end{aligned} \tag{12.38}$$

where

$$\theta_0 = \arcsin\left(\frac{m}{l + \frac{1}{2}}\right). \tag{12.39}$$

Every standing-wave eigenfunction of the form $\sqrt{2}\, X_{lm}(\theta)\cos m\phi$, $X_{l0}(\theta)$ or $\sqrt{2}\, X_{lm}(\theta)\sin m\phi$ is composed of clockwise and counterclockwise travelling waves having a common inclination to the $\hat{\mathbf{z}}$ axis, as shown in Figure 12.9. The local wavenumber of each of these constructively interfering waves is a half-integer:

$$k = \sqrt{l(l+1)} \approx l + \tfrac{1}{2}. \tag{12.40}$$

Nevertheless, there are an integral number of oscillations—exactly l—about every great circle tangent to the turning colatitudes θ and $\pi - \theta_0$, because of the two $\pi/2$ caustic phase shifts. In the old quantum theory, it was noticed that better agreement between theory and observation could frequently be achieved by employing half-integer rather than integer quantum numbers, but no theoretical principle was available for deciding whether a half-integer or integer was more appropriate. This deficiency was not fully rectified until long after the theory was superseded by modern quantum mechanics, when Keller (1958) and Keller & Rubinow (1960) showed how to resolve the issue by counting the number of caustic passages of the classical trajectories.

Equation (12.39) can be regarded as the Bohr-Sommerfeld condition for fitting $l - m$ oscillations between the two turning colatitudes θ_0 and $\pi - \theta_0$. The second angular quantum number m is of course the number of longitudinal oscillations around any line of latitude, $0 \leq \phi \leq 2\pi$. This number, the *order* of the oscillation, does not play a role in the asymptotic eigenfrequency analysis, because of the spherical symmetry of the Earth model. The Jeans relation (12.37) provides one constraint upon the degenerate toroidal and spheroidal frequencies ${}_n\omega_l^{\mathrm{T}}$ and ${}_n\omega_l^{\mathrm{S}}$; the radial quantization condition—which is really the crux of the problem—provides the other.

12.2.2 Toroidal modes

We consider first the case of a toroidal mode produced by the constructive interference of SH waves that turn in the mantle; the ray parameter of these waves lies in the range $b/\beta_{b+} < p < a/\beta_a$. A segment of the associated SSS$\cdots$ ray path, connecting two points A and B situated on the Earth's surface $r = a$, is depicted on the left side of Figure 12.10. The phase accumulated by an SH wave during its propagation along this path is

$$\Delta\Psi = -\omega T + \pi/2, \tag{12.41}$$

where $T(p)$ is the travel time, and the second term represents the caustic phase shift which the wave experiences upon turning. In order that the wave arriving at B interfere constructively with the family of waves having ray parameter p, it is required that

$$|\Delta\Psi| = \omega p\Theta + 2n'\pi, \tag{12.42}$$

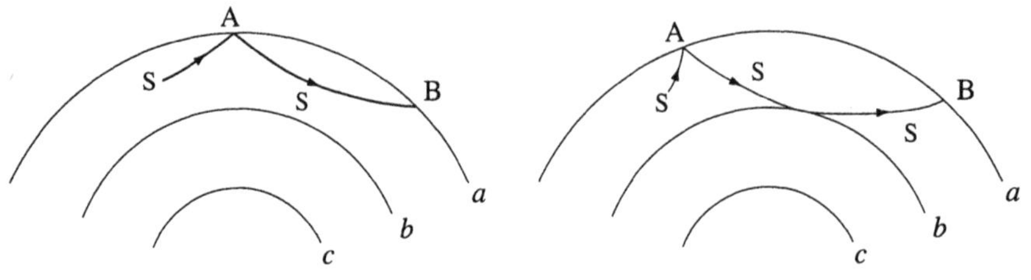

Figure 12.10. Schematic ray path of a turning SH wave in Regime II (*left*) and a reflected $\mathrm{ScS_{SH}}$ wave in Regime III (*right*). The phase difference between the waves at points B and A is quantized in the constructive-interference analysis.

where $\Theta(p)$ is the angular distance and n' is a positive integer. Upon comparing the two results (12.41) and (12.42) we find that

$$\omega(T - p\Theta) = \omega\tau = 2\pi(n' + \tfrac{1}{4}), \tag{12.43}$$

where $\tau(p)$ is the intercept time. To insure agreement with the conventional numbering scheme, which refers to the fundamental as the zeroth mode rather than the first, we set $n' = n + 1$. The final form of the Bohr-Sommerfeld radial quantization condition (12.43) is then

$$\omega\tau_{\mathrm{S}} = 2\pi(n + \tfrac{5}{4}), \tag{12.44}$$

where n is the usual overtone index, and we have appended a subscript S to serve as a reminder that this is for SH-equivalent modes.

The constructive-interference analysis of the $\mathrm{ScS_{SH}}$-equivalent modes, depicted on the right side of Figure 12.10, can be carried out in an analogous manner. The only difference is that the factor of $\pi/2$ is no longer present in equation (12.41), because a transversely polarized wave does not suffer any phase change upon reflection off of a fluid-solid interface. The caustic phase shift of a turning wave is responsible for the quarter-integer quantum number $n' + \frac{1}{4}$ in equation (12.43); the radial quantization condition for an $\mathrm{ScS_{SH}}$ mode, replacing (12.44), is therefore

$$\omega\tau_{\mathrm{ScS}} = 2\pi(n + 1). \tag{12.45}$$

To make certain there is no confusion, we reiterate that the $\pi/2$ phase shift in the SH-equivalent case arises because the associated waves *all have the same ray parameter p*. The turning SH waves excited by an earthquake or other point source all have different ray parameters; they do not exhibit a caustic phase shift. We consider the caustic phase shifts of multi-leg earthquake phases such as $\mathrm{SS_{SH}}$ in Section 12.5.1.

Altogether, there are four distinct SH ray-parameter regimes in the range $0 < p < \infty$; the oscillatory waves present in each of these regimes and the associated asymptotic eigenfrequency equations are summarized in Table 12.1. The interference analysis for the inner-core $\mathrm{J_{SH}}$ modes is identical to that for the mantle SH modes, because the waves are totally internally reflected with no change in phase at the inner-core boundary. There is also a limiting ray parameter $p = a/\beta_a$ beyond which no oscillatory waves of any type can propagate; there are no asymptotic toroidal modes with ray parameters exceeding this value.

The asymptotic and exact dispersion diagrams of the mantle toroidal modes of model 1066A are compared in Figure 12.11. The heavy diagonals labelled S and ScS are lines of constant ray parameter that separate the first three regimes in Table 12.1. The boundary between Regimes I and II corresponds to the limit of SH-wave criticality, whereas that between Regimes II and III coincides with the core-grazing ray; the $\mathrm{J_{SH}}$ inner-core modes in Regime IV are not shown. The agreement between the exact and asymptotic eigenfrequencies is obviously excellent, even along the fundamental and first few overtone branches. From the present perspective, these Love-wave-equivalent modes are produced by the constructive interference of SH waves turning in the upper mantle. The obvious inflection in each of the $\omega - l$ curves at $p = b/\beta_{b+}$ is caused by the corresponding inflection separating the contiguous $\tau - p$ curves of ScS and S waves (see Figure 12.4). The principal discrepancy between the two diagrams occurs in the vicinity of this inflection; every asymptotic $\omega - l$ curve exhibits a discontinuity across

Regime	Ray Parameter	Oscillatory Waves	Eigenfrequency
I	$a/\beta_a < p < \infty$	none	—
II	$b/\beta_{b+} < p < a/\beta_a$	turning SH	$\omega\tau_{\mathrm{S}} = 2\pi(n + \frac{5}{4})$
III	$0 < p < b/\beta_{b+}$	$\mathrm{ScS_{SH}}$	$\omega\tau_{\mathrm{ScS}} = 2\pi(n + 1)$
IV	$0 < p < c/\beta_{c-}$	turning $\mathrm{J_{SH}}$	$\omega\tau_{\mathrm{J}} = 2\pi(n + \frac{5}{4})$

Table 12.1. Ray-parameter regimes for SH waves in an Earth model with two internal discontinuities. There is a separate asymptotic toroidal eigenfrequency equation in each regime.

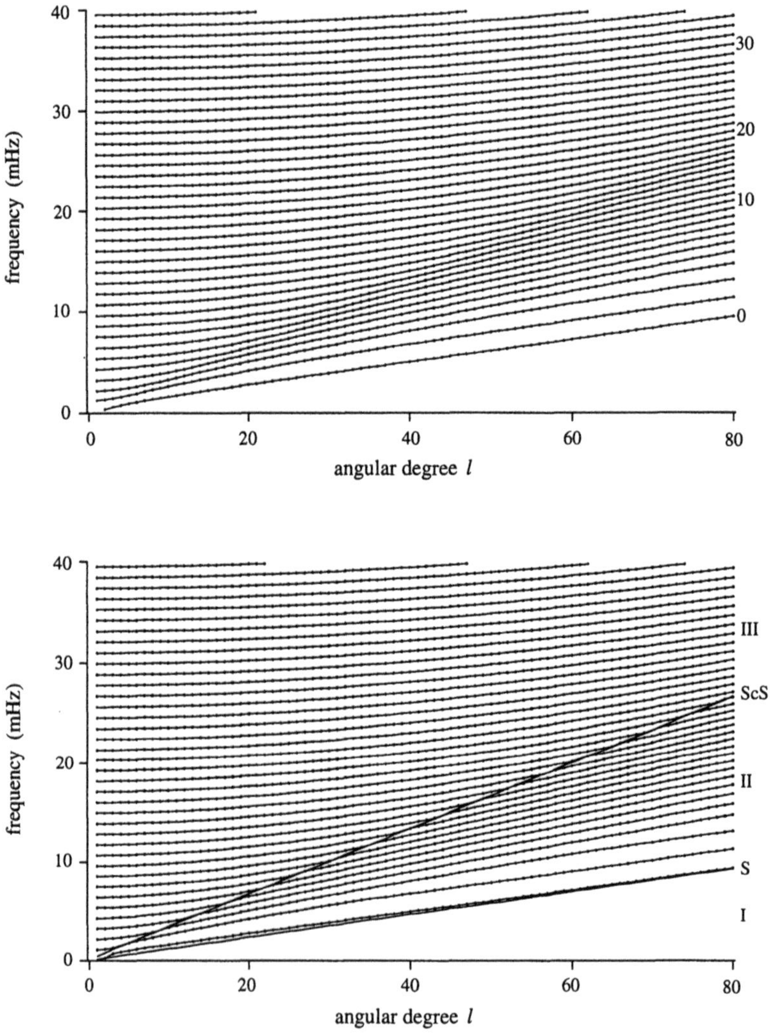

Figure 12.11. Toroidal-mode dispersion diagrams for the crustless version of model 1066A. (*Top*) Exact degenerate eigenfrequencies ${}_n\omega_l^{\mathrm{T}}$ calculated by numerical integration. (*Bottom*) Asymptotic eigenfrequencies calculated using equations (12.44)–(12.45). The diagonal lines labelled S and ScS have ray parameters $p = a/\beta_a$ and $p = b/\beta_{b+}$, respectively. (Courtesy of L. Zhao.)

the line $p = b/\beta_{b+}$ as a result of the jump $n+1 \rightarrow n+5/4$ in the quantum number. This is an artifact of the geometrical ray approximation, which is strictly valid only in the limit of infinite frequency, $\omega \to \infty$. Tunnelling of the core-grazing waves and other finite-frequency effects blur the distinction between finite-frequency SH and $\mathrm{ScS_{SH}}$ modes in this vicinity.

The influence of the Mohorovičić and upper-mantle discontinuities has been considered by a number of investigators; their principal effect is to give rise to a slight undulation of the toroidal ω–l curves that is analogous to the osculation and terracing of the spheroidal curves but much more subdued. This so-called *solotone effect* is not readily apparent at the level of resolution of the PREM dispersion diagram exhibited in Figure 8.3; it can only be discerned by close inspection of the differences between the exact numerical eigenfrequencies and those calculated using the asymptotic relations (12.44)–(12.45). Fundamentally, the solotone effect is a resonance phenomenon associated with the existence of reflected phases such as $\mathrm{S_{410}S_{SH}}$ and $\mathrm{S_{660}S_{SH}}$; a summary of the theory as well as references to the original literature may be found in Lapwood & Usami (1981). Any quantitative algorithm that seeks to calculate accurate asymptotic eigenfrequencies of a realistic SNREI Earth model such as PREM needs to account not only for reflections off of the various discontinuities, but also for tunnelling interactions of near-grazing rays that turn within each smooth layer. As noted earlier, our goal in this chapter is much more modest: we seek simply to illuminate the distinguishing physical characteristics of the various SH and P-SV mode types. The presence of more than the two principal fluid-solid discontinuities affects the quantitative details, but it does not alter the overall physical picture.

We conclude this section by presenting an alternative derivation of the toroidal-mode eigenfrequency relations, using an argument that is more readily extended to the spheroidal modes, which we consider next. We focus in this more general argument not simply upon the phases of the waves in Figure 12.10, but rather upon their complex amplitudes. Restricting attention initially to the SH-equivalent case, and denoting the amplitude of the transverse component at point A by a_{S}, we calculate the amplitude at point B in two different ways:

1. Since the rate of variation of phase along the spherical free surface is $p^{-1} = d\Theta/dT$, the amplitude at B must be $a_{\mathrm{S}} \exp(-i\omega p\Theta)$.

2. Considering instead the variation in phase along the turning ray, we deduce that the amplitude must also be $a_{\mathrm{S}} \exp(-i\omega T_{\mathrm{S}} + i\pi/2)$, where T_{S} is the SH-wave travel time and the factor of $\pi/2$ represents the caustic phase shift. The geometrical spreading factor is identical at points A and B, because the associated ray tube is composed of rays that *all have the same ray parameter p*.

Upon equating the above two results, we find the condition for constructive interference at point B to be

$$a_{\rm S}\exp(-i\omega T_{\rm S}+i\pi/2)=a_{\rm S}\exp(-i\omega p\Theta). \tag{12.46}$$

A similar relation pertains to the $\mathrm{ScS_{SH}}$-equivalent case, with the travel time $T_{\rm S}$ replaced by $T_{\rm ScS}$ and the phase shift $\pi/2$ replaced by zero:

$$a_{\rm S}\exp(-i\omega T_{\rm ScS})=a_{\rm S}\exp(-i\omega p\Theta). \tag{12.47}$$

Boiling the above argument down to its essentials, we divide both sides of (12.46)–(12.47) by $\exp(-i\omega p\Theta)$ to obtain the form of the SH and $\mathrm{ScS_{SH}}$ constructive-interference conditions that we shall subsequently generalize to the P-SV case:

$$a_{\rm S}\exp(-i\Psi_{\rm S})=a_{\rm S},\qquad a_{\rm S}\exp(-i\Psi_{\rm ScS})=a_{\rm S}, \tag{12.48}$$

where we have defined

$$\Psi_{\rm S}=\omega\tau_{\rm S}-\pi/2,\qquad \Psi_{\rm ScS}=\omega\tau_{\rm ScS}. \tag{12.49}$$

Both the left and right sides of (12.48) represent the complex amplitude of the associated SH and $\mathrm{ScS_{SH}}$ waves, measured relative to a "carrier" wave $\exp(-i\omega p\Theta)$ that propagates with uniform angular speed $p^{-1}=d\Theta/dT$. This relative amplitude is *independent of position on the sphere*, i.e., it is the same at both points A and B. Upon eliminating $a_{\rm S}$ from (12.48) and invoking standard trigonometric identities, we obtain a pair of real S and $\mathrm{ScS_{SH}}$ eigenfrequency relations,

$$\sin\tfrac{1}{2}\Psi_{\rm S}=0,\qquad \sin\tfrac{1}{2}\Psi_{\rm ScS}=0, \tag{12.50}$$

which are equivalent to the results (12.44)–(12.45).

12.2.3 Spheroidal modes

In a piecewise smooth Earth model with an inner-core and core-mantle boundary, there are ten independent P-SV ray-parameter regimes. As in the SH case, the constructive-interference principle must be applied one regime at a time. The bounding values of p, which correspond to critical and grazing rays, and the oscillatory waves that are present within each regime are summarized in Table 12.2. We illustrate the method by considering the mantle spheroidal modes in Regime III and the PKIKP, $\mathrm{ScS_{SV}}$ and $\mathrm{J_{SV}}$ modes in Regime IX. It is possible to conduct the analysis in terms of either the radial or longitudinal component of the particle motion; we choose the former alternative.

Regime	Ray Parameter	Oscillatory Seismic Body Waves
I	$a/\beta_a < p < \infty$	none
II	$a/\alpha_a < p < a/\beta_a$	turning SV
III	$b/\beta_{b+} < p < a/\alpha_a$	turning P and SV
IV	$b/\alpha_{b-} < p < b/\beta_{b+}$	turning P and $\mathrm{ScS_{SV}}$
V	$c/\beta_{c-} < p < b/\alpha_{b-}$	turning P, $\mathrm{ScS_{SV}}$ and SKS
VI	$b/\alpha_{b+} < p < c/\beta_{c-}$	turning P, $\mathrm{ScS_{SV}}$, SKS and turning $\mathrm{J_{SV}}$
VII	$c/\alpha_{c+} < p < b/\alpha_{b+}$	PKP, PcP, PKS, PcS, SKS, $\mathrm{ScS_{SV}}$, SKP, ScP and turning $\mathrm{J_{SV}}$
VIII	$c/\alpha_{c-} < p < c/\alpha_{c+}$	PKiKP, PKJKP, PcP, PKiKS, PKJKS, PcS, SKiKS, SKJKS, $\mathrm{ScS_{SV}}$, SKiKP, SKJKP and ScP
IX	$0 < p < c/\alpha_{c-}$	PKIKP, PKJKP, PKiKP, PcP, PKIKS, PKJKS, PKiKS, PcS, SKIKS, SKJKS, SKiKS, $\mathrm{ScS_{SV}}$, SKIKP, SKJKP, SKiKP and ScP
X	$p = 0$ (radial modes)	PKIKP, PKiKP and PcP

Table 12.2. Ray-parameter regimes for P-SV waves in an Earth model with two internal discontinuities. The regime boundaries are illustrated along the top of the τ–p diagram in Figure 12.4. There is a separate asymptotic spheroidal eigenfrequency equation corresponding to each regime. The core-mantle boundary and inner-core boundary Stoneley modes occur in Regimes III and V, respectively.

We denote the upward radial amplitudes of the incoming P and SV waves in Regime III by a_{P} and a_{S}, respectively. Referring to Figure 12.12, we may write the conditions for the constructive interference at the two

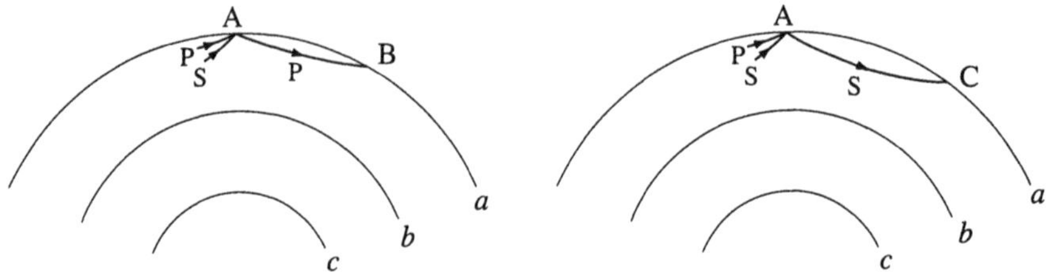

Figure 12.12. Schematic ray paths of the turning P and SV waves in Regime III. The P wave at point B and the SV wave at point C are composed of waves that combine and interact with the free surface at point A. The left and right "life-history" or "Feynman" diagrams give rise to equations (12.51) and (12.52), respectively.

points B and C in the form

$$(a_{\mathrm{P}}\acute{P}\grave{P} + a_{\mathrm{S}}\acute{S}\grave{P})\exp(-i\Psi_{\mathrm{P}}) = a_{\mathrm{P}}, \tag{12.51}$$

$$(a_{\mathrm{P}}\acute{P}\grave{S} + a_{\mathrm{S}}\acute{S}\grave{S})\exp(-i\Psi_{\mathrm{S}}) = a_{\mathrm{S}}, \tag{12.52}$$

where

$$\Psi_{\mathrm{P}} = \omega\tau_{\mathrm{P}} + \pi/2, \qquad \Psi_{\mathrm{S}} = \omega\tau_{\mathrm{S}} - \pi/2. \tag{12.53}$$

In each of equations (12.51)–(12.52), the terms on the left can be thought of as a chronological description of the "life history" of the waves that impinge upon the point A, combine and interact with the boundary, and eventually arrive at B and C. All amplitudes are measured relative to the angular "carrier" $\exp(-i\omega p\Theta)$ as in the SH–ScS$_{\mathrm{SH}}$ case (12.48); the relative amplitudes at points A, B and C are all the same. The four free-surface reflection coefficients $\acute{P}\grave{P}$, $\acute{P}\grave{S}$, $\acute{S}\grave{P}$ and $\acute{S}\grave{S}$ are the ratios of the upward radial components of the outgoing and incoming displacements, as discussed in Section 12.1.6. The second term in the definition (12.53) of Ψ_{P} represents the combined effect of the $\pi/2$ caustic phase shift and the sign reversal of the upward radial component of a P wave upon passage through its turning radius $r = R_{\mathrm{P}}$, as shown in Figure 12.13. The sign of the upward radial component of an SV wave is unchanged upon passing through $r = R_{\mathrm{S}}$, so that Ψ_{S} is the same as in the case of a turning SH wave, equation (12.49). The two constructive-interference conditions (12.51)–(12.52) constitute a homogeneous linear system of equations for the unknown relative amplitudes a_{P} and a_{S} of the form $\mathsf{B}\mathsf{a} = 0$, where $\mathsf{a} = (a_{\mathrm{P}}\;\, a_{\mathrm{S}})^{\mathrm{T}}$ and

$$\mathsf{B} = \begin{pmatrix} \acute{P}\grave{P}\exp(-i\Psi_{\mathrm{P}}) - 1 & \acute{S}\grave{P}\exp(-i\Psi_{\mathrm{P}}) \\ \acute{P}\grave{S}\exp(-i\Psi_{\mathrm{S}}) & \acute{S}\grave{S}\exp(-i\Psi_{\mathrm{S}}) - 1 \end{pmatrix}. \tag{12.54}$$

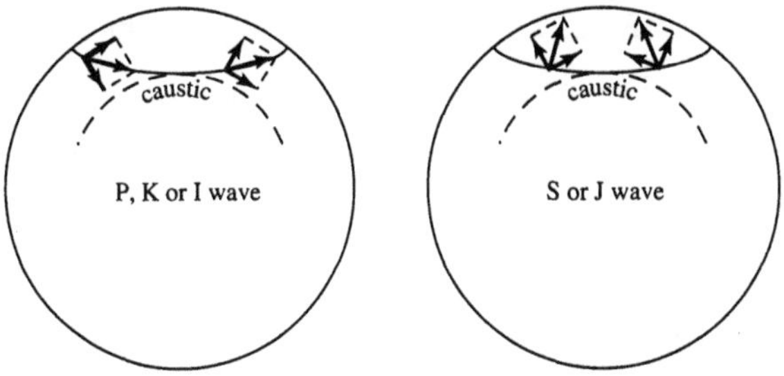

Figure 12.13. The radial component of a P, K or I wave (*left*) and the tangential component of an SV or J_{SV} wave (*right*) vanish at their turning points. The tangential component of a P, K or I wave and the radial component of an SV or J_{SV} wave exhibit a maximum. The diagonal vectors in each displacement rectangle are the polarizations $\hat{\eta}_P$ (*left*) and $\hat{\eta}_{SV}$ (*right*).

The condition that this system have a solution is $\det \mathsf{B} = 0$. Upon applying the free-surface symmetry relations (12.16), we can reduce this determinant equation to

$$\sin \tfrac{1}{2}(\Psi_S + \Psi_P) + \acute{S}\grave{S} \sin \tfrac{1}{2}(\Psi_S - \Psi_P) = 0. \tag{12.55}$$

This is the real asymptotic secular equation defining the Regime III eigenfrequencies. The free-surface shear-wave reflection coefficient $\acute{S}\grave{S}$ in this regime is real and negative.

The corresponding results in the other P-SV regimes can be derived in a similar manner. In each case it is advantageous to construct a series of diagrams that enumerate all of the possible ray-path combinations contributing to an upcoming P, S, K, I or J wave beneath the various boundaries. These "life-history" diagrams are analogous to Feynman diagrams in quantum mechanics or condensed-matter physics in that they can be translated directly into the governing dynamical equations. In Regime IX, which is the most general case, there are oscillatory I and J waves in the solid inner core and oscillatory K waves in the fluid outer core in addition to mantle P and S waves. The constructive-interference conditions obtained by inspection of Figure 12.14 can be written in the form $\mathsf{B}\mathsf{a} = 0$, where $\mathsf{a} = (a_P \; a_S \; a_K \; a_I \; a_J)^T$ is the column vector of unknown amplitudes and B is a 5×5 matrix too complicated to reproduce here. The secular equation $\det \mathsf{B} = 0$ can be reduced with the aid of reflection-transmission symmetries—and plenty of sharp pencils—to the final real relation

$$\cos \tfrac{1}{2}(\Psi_{PcP} + \Psi_{ScS} + \Psi_{KiK} + \Psi_I + \Psi_J)$$
$$- \acute{P}\grave{P}\,\grave{P}\acute{P} \cos \tfrac{1}{2}(\Psi_{PcP} - \Psi_{ScS} - \Psi_{KiK} - \Psi_I - \Psi_J)$$

$$
\begin{aligned}
&- \acute{P}\grave{P}\,\grave{P}\acute{P}\,\grave{K}\acute{K}\cos\tfrac{1}{2}(\Psi_{\mathrm{PcP}} - \Psi_{\mathrm{ScS}} - \Psi_{\mathrm{KiK}} + \Psi_{\mathrm{I}} + \Psi_{\mathrm{J}}) \\
&+ \acute{P}\grave{P}\,\grave{P}\acute{P}\,\acute{I}\grave{I}\cos\tfrac{1}{2}(\Psi_{\mathrm{PcP}} - \Psi_{\mathrm{ScS}} - \Psi_{\mathrm{KiK}} + \Psi_{\mathrm{I}} - \Psi_{\mathrm{J}}) \\
&+ \acute{P}\grave{P}\,\grave{P}\acute{P}\,\acute{J}\grave{J}\cos\tfrac{1}{2}(\Psi_{\mathrm{PcP}} - \Psi_{\mathrm{ScS}} - \Psi_{\mathrm{KiK}} - \Psi_{\mathrm{I}} + \Psi_{\mathrm{J}}) \\
&- \acute{S}\grave{S}\,\grave{S}\acute{S}\cos\tfrac{1}{2}(\Psi_{\mathrm{PcP}} - \Psi_{\mathrm{ScS}} + \Psi_{\mathrm{KiK}} + \Psi_{\mathrm{I}} + \Psi_{\mathrm{J}}) \\
&- \acute{S}\grave{S}\,\grave{S}\acute{S}\,\grave{K}\acute{K}\cos\tfrac{1}{2}(\Psi_{\mathrm{PcP}} - \Psi_{\mathrm{ScS}} + \Psi_{\mathrm{KiK}} - \Psi_{\mathrm{I}} - \Psi_{\mathrm{J}}) \\
&+ \acute{S}\grave{S}\,\grave{S}\acute{S}\,\acute{I}\grave{I}\cos\tfrac{1}{2}(\Psi_{\mathrm{PcP}} - \Psi_{\mathrm{ScS}} + \Psi_{\mathrm{KiK}} - \Psi_{\mathrm{I}} + \Psi_{\mathrm{J}}) \\
&+ \acute{S}\grave{S}\,\grave{S}\acute{S}\,\acute{J}\grave{J}\cos\tfrac{1}{2}(\Psi_{\mathrm{PcP}} - \Psi_{\mathrm{ScS}} + \Psi_{\mathrm{KiK}} + \Psi_{\mathrm{I}} - \Psi_{\mathrm{J}}) \\
&- (\gamma^{-1}\acute{P}\grave{S}\,\grave{S}\acute{P} + \gamma\acute{S}\grave{P}\,\grave{P}\acute{S})\cos\tfrac{1}{2}(\Psi_{\mathrm{KiK}} + \Psi_{\mathrm{I}} + \Psi_{\mathrm{J}}) \\
&- (\gamma^{-1}\acute{P}\grave{S}\,\grave{S}\acute{P} + \gamma\acute{S}\grave{P}\,\grave{P}\acute{S})\grave{K}\acute{K}\cos\tfrac{1}{2}(\Psi_{\mathrm{KiK}} - \Psi_{\mathrm{I}} - \Psi_{\mathrm{J}}) \\
&+ (\gamma^{-1}\acute{P}\grave{S}\,\grave{S}\acute{P} + \gamma\acute{S}\grave{P}\,\grave{P}\acute{S})\acute{I}\grave{I}\cos\tfrac{1}{2}(\Psi_{\mathrm{KiK}} - \Psi_{\mathrm{I}} + \Psi_{\mathrm{J}}) \\
&+ (\gamma^{-1}\acute{P}\grave{S}\,\grave{S}\acute{P} + \gamma\acute{S}\grave{P}\,\grave{P}\acute{S})\acute{J}\grave{J}\cos\tfrac{1}{2}(\Psi_{\mathrm{KiK}} + \Psi_{\mathrm{I}} - \Psi_{\mathrm{J}}) \\
&+ \grave{K}\acute{K}\cos\tfrac{1}{2}(\Psi_{\mathrm{PcP}} + \Psi_{\mathrm{ScS}} + \Psi_{\mathrm{KiK}} - \Psi_{\mathrm{I}} - \Psi_{\mathrm{J}}) \\
&- \acute{K}\grave{K}\cos\tfrac{1}{2}(\Psi_{\mathrm{PcP}} + \Psi_{\mathrm{ScS}} - \Psi_{\mathrm{KiK}} - \Psi_{\mathrm{I}} - \Psi_{\mathrm{J}}) \\
&- \grave{K}\acute{K}\,\acute{K}\grave{K}\cos\tfrac{1}{2}(\Psi_{\mathrm{PcP}} + \Psi_{\mathrm{ScS}} - \Psi_{\mathrm{KiK}} + \Psi_{\mathrm{I}} + \Psi_{\mathrm{J}}) \\
&+ \acute{K}\grave{K}\,\acute{I}\grave{I}\cos\tfrac{1}{2}(\Psi_{\mathrm{PcP}} + \Psi_{\mathrm{ScS}} - \Psi_{\mathrm{KiK}} + \Psi_{\mathrm{I}} - \Psi_{\mathrm{J}}) \\
&+ \acute{K}\grave{K}\,\acute{J}\grave{J}\cos\tfrac{1}{2}(\Psi_{\mathrm{PcP}} + \Psi_{\mathrm{ScS}} - \Psi_{\mathrm{KiK}} - \Psi_{\mathrm{I}} + \Psi_{\mathrm{J}}) \\
&- \acute{I}\grave{I}\cos\tfrac{1}{2}(\Psi_{\mathrm{PcP}} + \Psi_{\mathrm{ScS}} + \Psi_{\mathrm{KiK}} - \Psi_{\mathrm{I}} + \Psi_{\mathrm{J}}) \\
&- \acute{J}\grave{J}\cos\tfrac{1}{2}(\Psi_{\mathrm{PcP}} + \Psi_{\mathrm{ScS}} + \Psi_{\mathrm{KiK}} + \Psi_{\mathrm{I}} - \Psi_{\mathrm{J}}) = 0, \qquad (12.56)
\end{aligned}
$$

where

$$
\begin{aligned}
\Psi_{\mathrm{PcP}} = \omega\tau_{\mathrm{PcP}}, \qquad \Psi_{\mathrm{ScS}} = \omega\tau_{\mathrm{ScS}}, \qquad \Psi_{\mathrm{KiK}} = \omega\tau_{\mathrm{KiK}}, \\
\Psi_{\mathrm{I}} = \omega\tau_{\mathrm{I}} + \pi/2, \qquad \Psi_{\mathrm{J}} = \omega\tau_{\mathrm{J}} - \pi/2. \qquad (12.57)
\end{aligned}
$$

The dimensionless ratio $\gamma = (b/a)[q_\alpha(b+)q_\beta(b+)/q_\alpha(a)q_\beta(a)]^{1/2}$ and its reciprocal γ^{-1} account for the difference in geometrical spreading between the ScP and PcS waves that travel from point B to C and point A to D, respectively. As in the SH–$\mathrm{ScS_{SH}}$ case, the relevant spreading is that due to the focusing and defocusing of a tube of rays having the same ray parameter p; the corresponding ratio of spreading factors is unity for all P, S and K turning rays and all like-type reflections PcP, ScS and KiK.

The coupling between propagating P and SV waves becomes negligible in the limit of nearly normal incidence, $p \to 0$. The limiting values of the radial-component reflection coefficients in equation (12.56) are

$$\acute{P}\grave{P} \to 1,$$

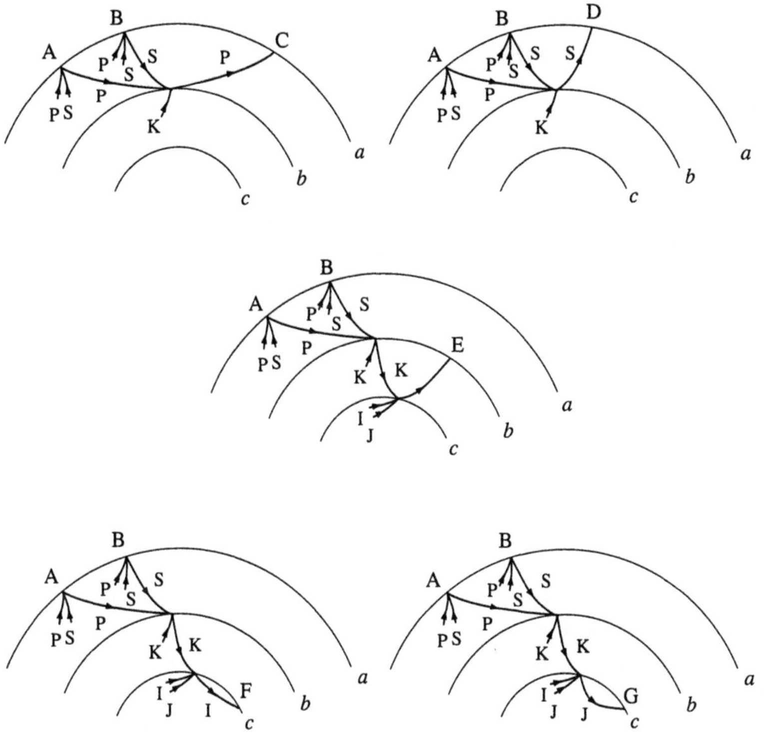

Figure 12.14. "Life-history" or "Feynman" diagrams that form the basis for the constructive-interference analysis in P-SV Regime IX. Each diagram gives rise to a separate linear equation governing the relative amplitudes a_{P}, a_{S}, a_{K}, a_{I} and a_{J} at points C through G.

$$\begin{aligned} &\acute{S}\grave{S} \to \grave{S}\acute{S} \to \acute{J}\grave{J} \to -1, \\ &\acute{P}\grave{S} \to \acute{S}\grave{P} \to \grave{P}\acute{S} \to \grave{S}\acute{P} \to 0, \end{aligned} \tag{12.58}$$

and

$$\begin{aligned} &\grave{P}\acute{P} \to -\acute{K}\grave{K} \to \frac{\rho_{b+}\alpha_{b+} - \rho_{b-}\alpha_{b-}}{\rho_{b+}\alpha_{b+} + \rho_{b-}\alpha_{b-}} \approx 0, \\ &\grave{K}\acute{K} \to -\acute{I}\grave{I} \to \frac{\rho_{c+}\alpha_{c+} - \rho_{c-}\alpha_{c-}}{\rho_{c+}\alpha_{c+} + \rho_{c-}\alpha_{c-}} \approx 0. \end{aligned} \tag{12.59}$$

The conditions (12.58)–(12.59) imply that a normally incident PKIKP wave is transmitted almost unimpeded through the Earth, whereas an ScS wave and a J wave suffer nearly total internal reflection at the core-mantle and inner-core boundaries. Upon imposing these $p \approx 0$ conditions and combining terms using trigonometric identities, we find that the daunting sum of twenty cosines in equation (12.56) can be reduced to a much simpler product of three terms:

$$(\sin\tfrac{1}{2}\Psi_{\rm ScS})(\cos\tfrac{1}{2}\Psi_{\rm J})(\sin\tfrac{1}{2}\Psi_{\rm PKIKP}) \approx 0, \tag{12.60}$$

where $\Psi_{\rm PKIKP} = \Psi_{\rm PcP} + \Psi_{\rm KiK} + \Psi_{\rm I} = \omega\tau_{\rm PKIKP} + \pi/2$. We noted in Section 8.8.10 that the spheroidal modes in Regime IX could be empirically subdivided into distinct $\mathrm{ScS_{SV}}$, $\mathrm{J_{SV}}$ and PKIKP families; equation (12.60) provides a strikingly clear confirmation and physical explanation of this phenomenon. The asymptotic eigenfrequencies of the three mode types are given, respectively, by

$$\omega\tau_{\rm ScS} \approx 2\pi n', \tag{12.61}$$

$$\omega\tau_{\rm J} \approx 2\pi(n'' - \tfrac{1}{4}), \tag{12.62}$$

$$\omega\tau_{\rm PKIKP} \approx 2\pi(n''' - \tfrac{1}{4}), \tag{12.63}$$

where n', n'' and n''' are positive integers. The overtone index n can be found by accounting for the graver modes in Regimes II through VIII and re-ordering the frequencies associated with these three primed quantum numbers.

Comparison of equations (12.61)–(12.62) with the results in Table 12.1 confirms that the $\mathrm{ScS_{SV}}$ and $\mathrm{ScS_{SH}}$ modes have asymptotically identical eigenfrequencies, whereas the eigenfrequencies of the $\mathrm{J_{SV}}$ and $\mathrm{J_{SH}}$ modes are offset or interleaved, as illustrated in Figure 8.13. This staggering of the two types of inner-core modes is caused in the present analysis by the difference between the $\mathrm{J_{SV}}$ and $\mathrm{J_{SH}}$ reflection coefficients. If we were instead to conduct the constructive-interference analysis in terms of the longitudinal component of the motion, these two inner-core-boundary reflection coefficients would be identical in the limit $p \to 0$; the staggering in that case would be due to the sign reversal of the $\mathrm{J_{SV}}$ wave upon turning. The predominantly $\mathrm{J_{SV}}$ modes in the present regime are the smooth continuation of a set of pure $\mathrm{J_{SV}}$ modes which are present in regimes VI and VII. The asymptotic eigenfrequencies of the latter are given by $\omega\tau_{\rm J} = 2\pi(n''+\frac{1}{4})-\psi_{\rm J}$, where $\acute{J}\grave{J} = \exp(-i\psi_{\rm J})$ is the total internal reflection coefficient of the postcritical $\mathrm{J_{SV}}$ wave.

The asymptotic radial modes, which are generated by the constructive interference of P, K and I waves with ray parameter $p = 0$, are purely compressional in character; there are no purely radial $\mathrm{ScS_{SV}}$ or $\mathrm{J_{SV}}$ modes. Since the center of the Earth $r = 0$ is a degenerate focal point at which the width of a ray tube shrinks to zero in two directions rather than one, the caustic phase shift is π rather than $\pi/2$. This non-geometrical phase shift is cancelled by the sign reversal of the upward radial component of the motion, so that the net change in phase upon "turning" of a radially convergent wave is zero. The asymptotic eigenfrequency of a radial mode is, as a result, given by

$$\omega T_{\mathrm{PKIKP}} \approx 2\pi(n+1), \tag{12.64}$$

where T_{PKIKP} is the one-way transit time of a compressional wave along a straight ray passing through the center of the Earth, and $n = n''' - 1$ is the conventional overtone index. The radial-mode eigenfrequencies (12.64) are the $p = 0$ continuation of the PKIKP eigenfrequencies (12.63); the quantity $\omega\tau_{\mathrm{PKIKP}}$ tends smoothly to $\omega T_{\mathrm{PKIKP}} - \pi/2$ in the limit $p \to 0$, since $\omega p\Theta \to \pi/2$ at the antipode $\Theta = \pi$ by virtue of the Jeans relation (12.37). The approximate nature of the equality in (12.64) stems from the reflection-coefficient approximations $\grave{P}\acute{P} \approx \grave{K}\acute{K} \approx 0$ made in equation (12.59). A more complete analysis yields an improved radial-mode eigenfrequency equation which accounts for the slight core solotone effect:

$$\begin{aligned}\sin\tfrac{1}{2}(\Psi_{\mathrm{PcP}} + \Psi_{\mathrm{KiK}} + \Psi_{\mathrm{I}}) \\ + \grave{P}\acute{P}\sin\tfrac{1}{2}(\Psi_{\mathrm{PcP}} - \Psi_{\mathrm{KiK}} - \Psi_{\mathrm{I}}) \\ + \grave{P}\acute{P}\,\grave{K}\acute{K}\sin\tfrac{1}{2}(\Psi_{\mathrm{PcP}} - \Psi_{\mathrm{KiK}} + \Psi_{\mathrm{I}}) \\ + \grave{K}\acute{K}\sin\tfrac{1}{2}(\Psi_{\mathrm{PcP}} + \Psi_{\mathrm{KiK}} - \Psi_{\mathrm{I}}) = 0.\end{aligned} \tag{12.65}$$

Fundamentally, the Earth exhibits "whole-Earth" PKIKP modes rather than quasi-independent PcP, KiK and I modes because of the weak contrast in compressional-wave impedance at the core-mantle and inner-core boundaries: $\rho_{b+}\alpha_{b+} \approx \rho_{b-}\alpha_{b-}$ and $\rho_{c+}\alpha_{c+} \approx \rho_{c-}\alpha_{c-}$.

The final constructive-interference relation in each of the spheroidal-mode ray-parameter regimes can be written in a form analogous to equations (12.55) and (12.56):

$$\sum_{\nu} A_\nu \sin\tfrac{1}{2}\Psi_\nu = 0 \qquad \text{or} \qquad \sum_{\nu} A_\nu \cos\tfrac{1}{2}\Psi_\nu = 0. \tag{12.66}$$

The real factors A_ν are multiplicative combinations of the absolute reflection coefficients and fixed-p spreading-coefficient ratios at the various boundaries, whereas the arguments Ψ_ν are linear combinations of the relative phases of the oscillatory body waves. To find the asymptotic eigenfrequencies, we substitute the Jeans relation $\omega = (l + \frac{1}{2})p^{-1}$ into the radial quantization relation (12.66) and solve the resulting equation for the ray parameter p in each regime. This results in a decreasing sequence of roots ${}_0p_l > {}_1p_l > \cdots > {}_np_l > \cdots$, which can be converted into an increasing

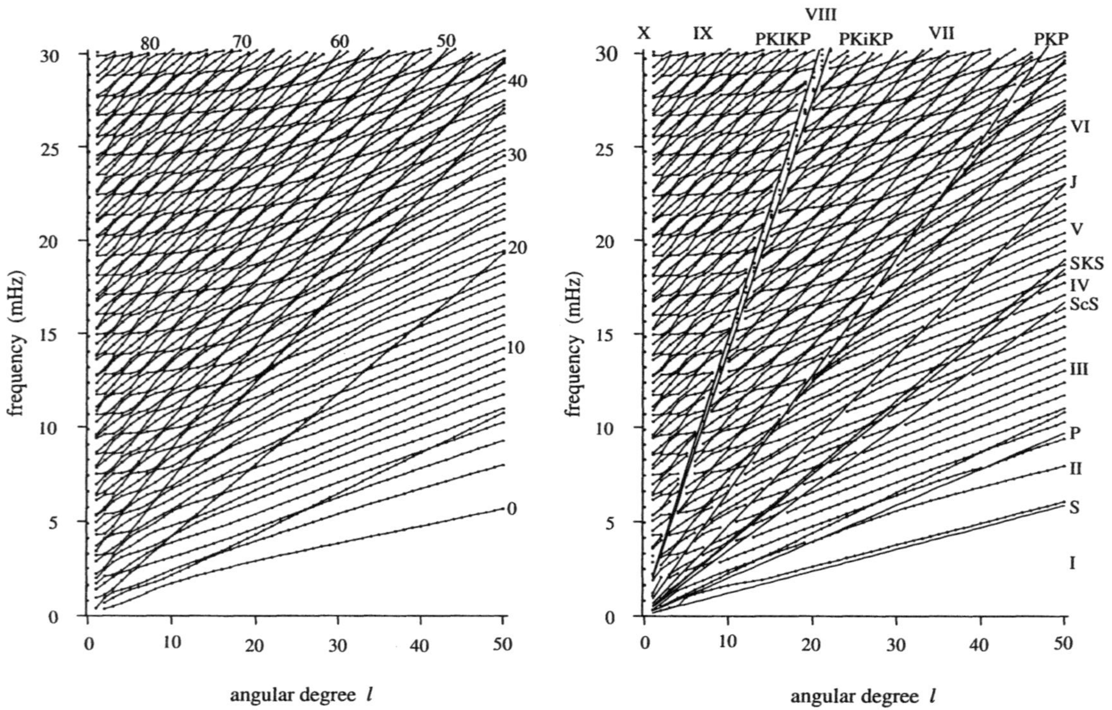

Figure 12.15. Spheroidal dispersion diagrams for the crustless version of model 1066A. (*Left*) Exact eigenfrequencies calculated by numerical integration. (*Right*) Asymptotic eigenfreqencies in P-SV Regimes II–X. See Figure 12.16 for a magnified view of the upper left corner. (Courtesy of L. Zhao.)

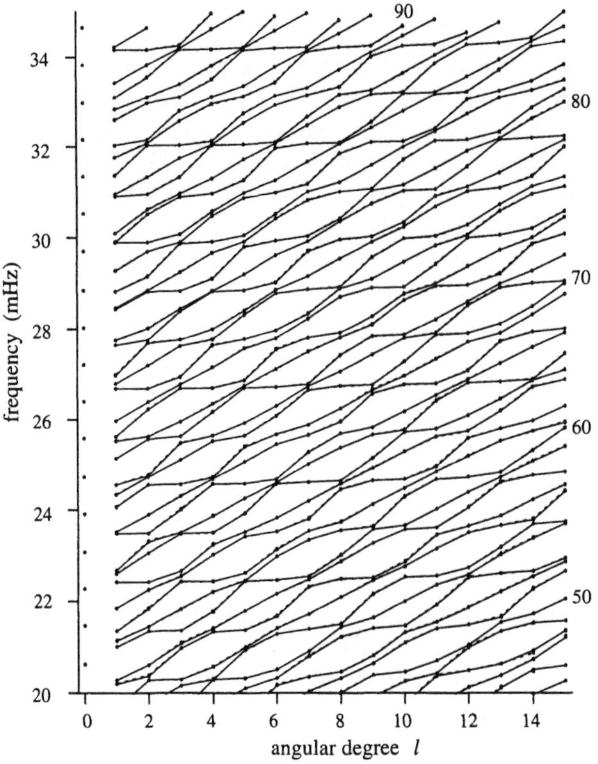

Figure 12.16. Blowup of the upper left corner of the spheroidal dispersion diagram in Figure 12.15. The exact eigenfrequencies calculated by numerical integration and the asymptotic Regime IX and X eigenfrequencies calculated using equations (12.56) and (12.65) are shown as filled circles connected by solid lines and open circles connected by dotted lines, respectively. Integers n = 50–90 are the overtone branch numbers. (Courtesy of L. Zhao.) Yes, there are open circles and dotted lines; the nearly perfect agreement makes them difficult to see!

sequence of eigenfrequencies ${}_0\omega_l < {}_1\omega_l < \cdots < {}_n\omega_l < \cdots$ for each integer degree $l > 0$. The inner-core and core-mantle boundary Stoneley modes, whose asymptotic ray parameters satisfy

$$\rho_{b+}(\beta_{b+}^{-2} - 2p^2b^{-2})^2 q_\alpha(b-) + \rho_{b-}\beta_{b+}^{-4} q_\alpha(b+) \\ + 4\rho_{b+}p^2b^{-2} q_\alpha(b+) q_\beta(b+) q_\alpha(b-) = 0 \tag{12.67}$$

and

$$\rho_{c-}(\beta_{c-}^{-2} - 2p^2c^{-2})^2 q_\alpha(c+) + \rho_{c+}\beta_{c-}^{-4} q_\alpha(c-)$$

$$+\,4\rho_{c-}p^2c^{-2}q_\alpha(c-)q_\beta(c-)q_\alpha(c+) = 0, \qquad (12.68)$$

respectively, must be inserted into the sequence in order to obtain the conventional overtone branch numbers. The quantities on the left of equations (12.67)–(12.68) are the common denominators of the reflection and transmission coefficients at the two boundaries. The radial-mode eigenfrequencies ${}_0\omega_0 < {}_1\omega_0 < \cdots < {}_n\omega_0 < \cdots$ may be computed directly, inasmuch as (12.65) is an equation for the single variable ω, in the case $p = 0$.

Figure 12.15 illustrates the excellent agreement between the asymptotic and exact spheroidal dispersion diagrams for the crustless version of model 1066A. The heavy diagonals labelled S, P, ScS, SKS, J, PKP, PKiKP and PKIKP are lines of constant ray parameter separating the various asymptotic regimes in Table 12.2. As in the case of the toroidal modes, the principal discrepancies occur near these boundaries, where the asymptotic eigenfrequencies exhibit discontinuities due to caustic and other phase-shift differences on either side. Tunnelling and related finite-frequency phenomena that affect the constituent near-grazing and near-critical waves smooth out these discontinuities in the exact dispersion diagram. The terracing of the PKIKP, $\mathrm{ScS_{SV}}$ and $\mathrm{J_{SV}}$ modes in Regime IX is captured by equation (12.56) in remarkable detail, as illustrated in Figure 12.16. The resulting "avoided crossings" of the fixed-n overtone branches $\ldots,\ {}_n\omega_{l-1},\ {}_n\omega_l,\ {}_n\omega_{l+1},\ldots$ are a characteristic feature of weakly coupled spectra in a wide variety of classical and quantum-mechanical systems (Arnold 1978).

12.3 Formal Asymptotic Analysis

The constructive-interference analysis summarized in the previous section provides a clear physical picture of the duality between the high-frequency toroidal and spheroidal modes and the propagating SH and P-SV body waves in an Earth model with two internal discontinuities. We next undertake a more rigorous mathematical analysis which confirms the above results and yields asymptotic formulae for the associated radial eigenfunctions as well. The procedure is to seek solutions to the governing ordinary differential equations and boundary conditions that are valid in the limit of high frequency, $\omega \to \infty$, for a fixed value of the SH or P-SV ray parameter p. The fundamental assumption which underlies the analysis is that the properties α, β, ρ of the Earth are smooth except for jump discontinuities at the inner-core boundary $r = c$, the core-mantle boundary $r = b$ and the free surface $r = a$. We follow the general approach elaborated by Woodhouse (1978); however, instead of looking for asymptotic eigenfunctions directly, we shall seek to transform the equations to a simpler system that is asymptotically equivalent. To facilitate the passage to the high-frequency limit, we first group like powers of ω^{-1} in a systematic manner. It is immaterial, for positive angular degrees $l > 0$, whether we define the ray parameter by the Jeans relation $\omega p = l + \frac{1}{2}$ or by

$$\omega p = k = \sqrt{l(l+1)}, \tag{12.69}$$

since the two relations are equivalent in the limit $\omega \to \infty$. We adopt the square-root definition (12.69) in the treatment that follows, because it is slightly more convenient. A dot is used to denote differentiation with respect to radius, as usual.

12.3.1 Toroidal modes

The linear system of first-order equations (8.114)–(8.115) governing the toroidal modes can be rewritten in matrix notation in the form

$$\dot{\mathsf{f}} = \omega(\mathsf{A}_0 + \omega^{-1}\mathsf{A}_1 + \omega^{-2}\mathsf{A}_2)\mathsf{f}. \tag{12.70}$$

The unknown displacement-traction two-vector f in equation (12.70) is defined by

$$\mathsf{f} = \begin{pmatrix} W \\ \omega^{-1}T \end{pmatrix}, \tag{12.71}$$

and the 2×2 matrices A_0, A_1 and A_2 are given by

$$\mathsf{A}_0 = \begin{pmatrix} 0 & \mu^{-1} \\ -\rho + p^2r^{-2}\mu & 0 \end{pmatrix},$$

$$\mathsf{A}_1 = \begin{pmatrix} r^{-1} & 0 \\ 0 & -3r^{-1} \end{pmatrix},$$

$$\mathsf{A}_2 = \begin{pmatrix} 0 & 0 \\ -2r^{-2}\mu & 0 \end{pmatrix}. \tag{12.72}$$

The scaling implicit in the definition (12.71) guarantees that the two elements of f will be of the same order in the limit $\omega \to \infty$.

Guided by the asymptotic potential representation of Richards (1974), we anticipate that the high-frequency eigenfunctions will be governed by a radial Schrödinger equation for an *SH wavefunction* H of the form

$$\ddot{H} + \omega^2 q_\beta^2 H = 0, \tag{12.73}$$

where $q_\beta^2 = \beta^{-2} - p^2r^{-2}$. Upon defining the quantities

$$\mathsf{g} = \begin{pmatrix} H \\ \omega^{-1}\dot{H} \end{pmatrix}, \qquad \mathsf{Q} = \begin{pmatrix} 0 & 1 \\ -q_\beta^2 & 0 \end{pmatrix}, \tag{12.74}$$

we can rewrite equation (12.73) in a matrix notation analogous to (12.70):

$$\dot{\mathsf{g}} = \omega \mathsf{Q} \mathsf{g}. \tag{12.75}$$

We seek a solution f to (12.70) of the form

$$\mathsf{f} = [\mathsf{Y}^{(0)} + \omega^{-1}\mathsf{Y}^{(1)} + \cdots]\mathsf{g}, \tag{12.76}$$

where the matrices $\mathsf{Y}^{(0)}$, $\mathsf{Y}^{(1)}, \ldots$ must be determined. Upon substituting the expansion (12.76) and equating like powers of ω^{-1} we obtain a series of equations, the first two of which are

$$\mathsf{A}_0\mathsf{Y}^{(0)} - \mathsf{Y}^{(0)}\mathsf{Q} = 0, \tag{12.77}$$

$$\mathsf{A}_0\mathsf{Y}^{(1)} - \mathsf{Y}^{(1)}\mathsf{Q} = \dot{\mathsf{Y}}^{(0)} - \mathsf{A}_1\mathsf{Y}^{(0)}. \tag{12.78}$$

Defining the ray parameter by $p = (l + \frac{1}{2})/\omega$ rather than by $p = k/\omega$ as we have done only alters the matrix A_2, which does not appear in the zeroth-order and first-order relations (12.77)–(12.78).

It is noteworthy that the matrices Q and A_0 are related to each other by a similarity transformation:

$$\mathsf{Q} = \mathsf{R}^{-1}\mathsf{A}_0\mathsf{R}, \tag{12.79}$$

where

$$\mathsf{R} = \begin{pmatrix} \mu^{-1/2} & 0 \\ 0 & \mu^{1/2} \end{pmatrix}, \qquad \mathsf{R}^{-1} = \begin{pmatrix} \mu^{1/2} & 0 \\ 0 & \mu^{-1/2} \end{pmatrix}. \tag{12.80}$$

This makes it convenient to rewrite the matrices $\mathsf{Y}^{(0)}$, $\mathsf{Y}^{(1)}, \ldots$ in terms of this transformation:

$$\mathsf{Y}^{(0)} = \mathsf{R}\Gamma^{(0)}, \qquad \mathsf{Y}^{(1)} = \mathsf{R}\Gamma^{(1)}, \qquad \ldots, \tag{12.81}$$

and to regard the quantities $\Gamma^{(0)}$, $\Gamma^{(1)}, \ldots$ as the new unknowns. Upon inserting the representation (12.81) into equations (12.77)–(12.78) we find that $\Gamma^{(0)}$, $\Gamma^{(1)}, \ldots$ must satisfy

$$[\mathsf{Q}, \Gamma^{(0)}] = 0, \tag{12.82}$$

$$[\mathsf{Q}, \Gamma^{(1)}] = \dot{\Gamma}^{(0)} + (\mathsf{R}^{-1}\dot{\mathsf{R}} - \mathsf{R}^{-1}\mathsf{A}_1\mathsf{R})\Gamma^{(0)}, \tag{12.83}$$

where the symbol $[\cdot, \cdot]$ denotes the commutator of the enclosed matrices. Equation (12.82) stipulates that the lowest-order unknown multiplier $\Gamma^{(0)}$ must commute with the Schrödinger matrix Q. It is readily verified that Q and the 2×2 identity I constitute a complete family of commuting matrices. It follows that we may write $\Gamma^{(0)}$ in the form

$$\Gamma^{(0)} = \gamma_1 \mathsf{I} + \gamma_2 \mathsf{Q}, \tag{12.84}$$

where the scalar functions of radius γ_1 and γ_2 remain to be determined.

To find these functions we must consider the first-order relation (12.83). The necessary and sufficient condition for the existence of a particular solution $\Gamma^{(1)}$ to this inhomogeneous equation is that the right side must be orthogonal to the null space of the commutation operator on the left:

$$\operatorname{tr}[\dot{\Gamma}^{(0)} + (\mathsf{R}^{-1}\dot{\mathsf{R}} - \mathsf{R}^{-1}\mathsf{A}_1\mathsf{R})\Gamma^{(0)}] = 0, \tag{12.85}$$

$$\operatorname{tr}[\mathsf{Q}\dot{\Gamma}^{(0)} + \mathsf{Q}(\mathsf{R}^{-1}\dot{\mathsf{R}} - \mathsf{R}^{-1}\mathsf{A}_1\mathsf{R})\Gamma^{(0)}] = 0. \tag{12.86}$$

These existence conditions reduce to a pair of first-order differential equations for the unknown scalars:

$$\dot{\gamma}_1 + r^{-1}\gamma_1 = 0, \qquad \dot{\gamma}_2 + (q_\beta^{-1}\dot{q}_\beta + r^{-1})\gamma_2 = 0. \tag{12.87}$$

The solutions to (12.87) are of the form

$$\gamma_1 = ar^{-1}, \qquad \gamma_2 = br^{-1}q_\beta^{-1}, \tag{12.88}$$

where a and b are arbitrary constants.

Combining these results, we find that a zeroth-order asymptotic solution to equation (12.70) can be written in the form

$$\mathsf{f} = r^{-1}\mathsf{R}(a\mathsf{I} + bq_\beta^{-1}\mathsf{Q})\mathsf{g}. \tag{12.89}$$

If the Schrödinger two-vector g is regarded as completely prescribed, then the constants a and b in the representation (12.89) provide the two degrees of freedom needed to satisfy the toroidal boundary conditions. Alternatively, we can prescribe a and b and encapsulate the degrees of freedom in a pair of linearly independent solutions to the Schrödinger equation (12.75). We adopt the latter alternative, setting $a = 1$ and $b = 0$, for simplicity; equation (12.89) reduces in that case to

$$\mathsf{f} = r^{-1}\mathsf{R}\mathsf{g}, \tag{12.90}$$

or, equivalently,

$$W = \mu^{-1/2}r^{-1}H, \qquad T = \mu^{1/2}r^{-1}\dot{H}. \tag{12.91}$$

Higher-order corrections to the representation (12.91) can be obtained in an analogous manner, as demonstrated by Woodhouse (1978). At every order of ω^{-1} it is necessary to consider the solubility condition for the next succeeding order to determine the asymptotic representation fully, as above. The first-order corrections depend upon the first derivatives of the shear-wave speed and density $\dot{\beta}$ and $\dot{\rho}$, the second-order corrections depend upon the second derivatives $\ddot{\beta}$ and $\ddot{\rho}$, etc. The zeroth-order representation (12.91) is valid for any Earth model whose properties are sufficiently smooth, as measured by the magnitude of these derivatives.

In summary, we have expressed the high-frequency toroidal displacement W and associated traction T in terms of an SH wavefunction H that satisfies the Schrödinger equation (12.73). We shall apply these results to an Earth model without a Mohorovičić or any upper-mantle discontinuities; the quantity H must then be regular at the center of the Earth $r = 0$, and it must satisfy the boundary condition $\dot{H} = 0$ on the inner-core boundary $r = c$, the core-mantle boundary $r = b$ and the free surface $r = a$. To extend the analysis to a more general model, it is simply necessary to require that the traction $\mu^{1/2}r^{-1}\dot{H}$ be continuous across any solid-solid boundaries $r = d_{\mathrm{SS}}$. If the model has an oceanic layer, then the upper boundary condition must be imposed upon the seafloor $r = s$ rather than upon $r = a$. In principle, we could solve for the zeroth-order toroidal eigenfrequencies and eigenfunctions by numerical integration of the Schrödinger equation subject to these boundary and continuity conditions. Such an approach would be perverse, however, since the computational labor required to solve the Schrödinger equation would be no less than that needed to solve the exact equations (8.114)–(8.116). The advantage of the asymptotic theory presented in this section is that it enables us to obtain an *analytical representation* of W and T by making use of a JWKB solution of the Schrödinger equation which is valid in the limit $\omega \to \infty$. We present a review of the classical one-dimensional JWKB method in Section 12.3.3 and apply the results to the toroidal modes in Section 12.3.4.

12.3.2 Spheroidal modes

The effect of self-gravitation upon the spheroidal normal modes can be safely neglected in the limit $\omega \to \infty$. The 4×4 system of coupled first-order ordinary differential equations (8.158)–(8.161) governing the solid regions of a non-gravitating Earth can be rewritten in matrix notation in a form analogous to equation (12.70):

$$\dot{\mathsf{f}} = \omega(\mathsf{A}_0 + \omega^{-1}\mathsf{A}_1 + \omega^{-2}\mathsf{A}_2)\mathsf{f}. \tag{12.92}$$

The appropriately scaled four-vector of displacements and tractions in this case is

$$\mathsf{f} = \begin{pmatrix} U \\ V \\ \omega^{-1}R \\ \omega^{-1}S \end{pmatrix}, \tag{12.93}$$

and the matrices A_0, A_1 and A_2 are given by

$$\mathsf{A}_0 = \begin{pmatrix} 0 & pr^{-1}\lambda\sigma^{-1} & \sigma^{-1} & 0 \\ -pr^{-1} & 0 & 0 & \mu^{-1} \\ -\rho & 0 & 0 & pr^{-1} \\ 0 & -\rho + 4p^2r^{-2}\mu\eta\sigma^{-1} & -pr^{-1}\lambda\sigma^{-1} & 0 \end{pmatrix},$$

$$\mathsf{A}_1 = \begin{pmatrix} -2r^{-1}\lambda\sigma^{-1} & 0 & 0 & 0 \\ 0 & r^{-1} & 0 & 0 \\ 0 & -6pr^{-2}\kappa\mu\sigma^{-1} & -4r^{-1}\mu\sigma^{-1} & 0 \\ -6pr^{-2}\kappa\mu\sigma^{-1} & 0 & 0 & -3r^{-1} \end{pmatrix},$$

$$\mathsf{A}_2 = \begin{pmatrix} 0 & 0 & 0 & 0 \\ 0 & 0 & 0 & 0 \\ 12r^{-2}\kappa\mu\sigma^{-1} & 0 & 0 & 0 \\ 0 & -2r^{-2}\mu & 0 & 0 \end{pmatrix}, \tag{12.94}$$

where $\lambda = \kappa - \frac{2}{3}\mu$, $\sigma = \kappa + \frac{4}{3}\mu$ and $\eta = \kappa + \frac{1}{3}\mu$.

We seek to express the eigenfunctions U, V, R and S in terms of the solutions to a *pair* of Schrödinger equations. Denoting the P wavefunction by P and the SV wavefunction by B, we specify that

$$\ddot{P} + \omega^2 q_\alpha^2 P = 0, \qquad \ddot{B} + \omega^2 q_\beta^2 B = 0, \tag{12.95}$$

where $q_\alpha^2 = \alpha^{-2} - p^2r^{-2}$ and $q_\beta^2 = \beta^{-2} - p^2r^{-2}$. The independence of the two equations (12.95) signifies that P and SV waves propagate independently through the smooth regions of the Earth model; the only P-SV coupling occurs at the external and internal boundaries in the limit $\omega \to \infty$. The wavefunctions P and B should not be confused with the incremental gravitational variables denoted by the same symbols in the full 6×6 system of spheroidal equations (8.135)–(8.140). Upon defining the matrices

$$\mathsf{g} = \begin{pmatrix} P \\ \omega^{-1}\dot{P} \\ B \\ \omega^{-1}\dot{B} \end{pmatrix}, \qquad \mathsf{Q} = \begin{pmatrix} 0 & 1 & 0 & 0 \\ -q_\alpha^2 & 0 & 0 & 0 \\ 0 & 0 & 0 & 1 \\ 0 & 0 & -q_\beta^2 & 0 \end{pmatrix}, \tag{12.96}$$

we can rewrite the pair of uncoupled Schrödinger equations in a manner identical to equation (12.75):

$$\dot{\mathsf{g}} = \omega\mathsf{Q}\mathsf{g}. \tag{12.97}$$

We now follow exactly the same approach as in the SH case, seeking an asymptotic solution f to equation (12.92) of the form (12.76). The transformation matrices R and R^{-1} relating Q and A_0 are given in this case by

$$\mathsf{R} = \rho^{-1/2}\begin{pmatrix} 0 & 1 & pr^{-1} & 0 \\ pr^{-1} & 0 & 0 & 1 \\ -\rho + 2p^2r^{-2}\mu & 0 & 0 & 2pr^{-1}\mu \\ 0 & 2pr^{-1}\mu & -\rho + 2p^2r^{-2}\mu & 0 \end{pmatrix},$$

$$\mathsf{R}^{-1} = \rho^{-1/2}\begin{pmatrix} 0 & 2pr^{-1}\mu & -1 & 0 \\ \rho - 2p^2r^{-2}\mu & 0 & 0 & pr^{-1} \\ 2pr^{-1}\mu & 0 & 0 & -1 \\ 0 & \rho - 2p^2r^{-2}\mu & pr^{-1} & 0 \end{pmatrix}. \tag{12.98}$$

The most general zeroth-order asymptotic relation between f and g depends upon four arbitrary constants a_α, b_α, a_β, b_β, one corresponding to each member of the following family of linearly independent matrices that commute with Q:

$$\mathsf{I}_\alpha = \begin{pmatrix} 1 & 0 & 0 & 0 \\ 0 & 1 & 0 & 0 \\ 0 & 0 & 0 & 0 \\ 0 & 0 & 0 & 0 \end{pmatrix}, \qquad \mathsf{Q}_\alpha = \begin{pmatrix} 0 & 1 & 0 & 0 \\ -q_\alpha^2 & 0 & 0 & 0 \\ 0 & 0 & 0 & 0 \\ 0 & 0 & 0 & 0 \end{pmatrix},$$

$$\mathsf{I}_\beta = \begin{pmatrix} 0 & 0 & 0 & 0 \\ 0 & 0 & 0 & 0 \\ 0 & 0 & 1 & 0 \\ 0 & 0 & 0 & 1 \end{pmatrix}, \qquad \mathsf{Q}_\beta = \begin{pmatrix} 0 & 0 & 0 & 0 \\ 0 & 0 & 0 & 0 \\ 0 & 0 & 0 & 1 \\ 0 & 0 & -q_\beta^2 & 0 \end{pmatrix}. \tag{12.99}$$

Upon setting $a_\alpha = a_\beta = 1$ and $b_\alpha = b_\beta = 0$ we obtain a representation analogous to equation (12.90):

$$\mathsf{f} = r^{-1}\mathsf{R}\mathsf{g}, \tag{12.100}$$

or, equivalently,

$$U = \rho^{-1/2}r^{-1}(\omega^{-1}\dot{P} + pr^{-1}B), \tag{12.101}$$

$$V = \rho^{-1/2}r^{-1}(\omega^{-1}\dot{B} + pr^{-1}P), \tag{12.102}$$

$$R = -\rho^{1/2}r^{-1}\omega[P - 2pr^{-1}\beta^2(\omega^{-1}\dot{B} + pr^{-1}P)], \tag{12.103}$$

$$S = -\rho^{1/2}r^{-1}\omega[B - 2pr^{-1}\beta^2(\omega^{-1}\dot{P} + pr^{-1}B)]. \tag{12.104}$$

The four linearly independent solutions of the P and SV Schrödinger equations (12.95) provide the degrees of freedom required to satisfy the associated boundary conditions.

In the fluid outer core the traction scalar S vanishes, and the tangential displacement is given by the algebraic equation $V = -(kr^{-1}R)/(\omega^2\rho)$. The 2×2 system of ordinary differential equations (8.162)–(8.163) can be rewritten in matrix notation as

$$\dot{\mathsf{f}} = \omega(\mathsf{A}_0 + \omega^{-1}\mathsf{A}_1)\mathsf{f}, \tag{12.105}$$

where

$$\mathsf{f} = \begin{pmatrix} U \\ \omega^{-1}R \end{pmatrix} \tag{12.106}$$

and

$$\mathsf{A}_0 = \begin{pmatrix} 0 & \kappa^{-1} - \rho^{-1}p^2r^{-2} \\ -\rho & 0 \end{pmatrix}, \tag{12.107}$$

$$\mathsf{A}_1 = \begin{pmatrix} -2r^{-1} & 0 \\ 0 & 0 \end{pmatrix}. \tag{12.108}$$

We seek to relate U and R to a P wavefunction P which satisfies the Schrödinger equation (12.95). Upon defining the quantities

$$\mathsf{g} = \begin{pmatrix} P \\ \omega^{-1}\dot{P} \end{pmatrix}, \qquad \mathsf{Q} = \begin{pmatrix} 0 & 1 \\ -q_\alpha^2 & 0 \end{pmatrix}, \tag{12.109}$$

we can rewrite this equation in the form

$$\dot{\mathsf{g}} = \omega\mathsf{Q}\mathsf{g}. \tag{12.110}$$

The matrices R and R^{-1} which relate Q and A_0 in this case are

$$\mathsf{R} = \begin{pmatrix} 0 & \rho^{-1/2} \\ -\rho^{1/2} & 0 \end{pmatrix}, \qquad \mathsf{R}^{-1} = \begin{pmatrix} 0 & -\rho^{1/2} \\ \rho^{-1/2} & 0 \end{pmatrix}. \tag{12.111}$$

A suitable asymptotic relation between f and g, which embeds the two degrees of freedom needed to satisfy the continuity conditions on the fluid-solid discontinuities in the Schrödinger equation, is

$$\mathsf{f} = r^{-1}\mathsf{R}\mathsf{g}. \tag{12.112}$$

The corresponding complete set of zeroth-order asymptotic relations between U, V, R, S and P in the fluid is

$$U = \rho^{-1/2}r^{-1}\omega^{-1}\dot{P}, \tag{12.113}$$

$$V = \rho^{-1/2} p r^{-2} P, \tag{12.114}$$

$$R = -\rho^{1/2} r^{-1} \omega P, \tag{12.115}$$

$$S = 0. \tag{12.116}$$

Obviously, we could also have obtained the results (12.113)–(12.116) immediately by setting the shear-wave speed β and the SV wavefunction B equal to zero in equations (12.101)–(12.104).

The radial modes are purely compressional; the displacement and traction scalars are related to the P wavefunction P by $U = \rho^{-1/2} r^{-1} \omega^{-1} \dot{P}$, $V = 0$, $R = -\rho^{1/2} r^{-1} \omega P$ and $S = 0$. These results follow from an asymptotic analysis of the governing equations (8.149)–(8.150); alternatively, they can be obtained by setting the ray parameter p and the SV wavefunction B equal to zero in equations (12.101)–(12.104) and (12.113)–(12.116).

In summary, we have expressed the displacements and tractions in both the solid and fluid parts of the Earth in terms of a pair of high-frequency wavefunctions P, B satisfying the Schrödinger equations (12.95). The kinematic and dynamic boundary conditions require that

$$P - 2pr^{-1}\beta^2(\omega^{-1}\dot{B} + pr^{-1}P) = 0, \tag{12.117}$$

$$B - 2pr^{-1}\beta^2(\omega^{-1}\dot{P} + pr^{-1}B) = 0 \tag{12.118}$$

on the free surface $r = a$, and

$$B - 2pr^{-1}\beta^2(\omega^{-1}\dot{P} + pr^{-1}B) = 0, \tag{12.119}$$

$$[\rho^{-1/2}(\omega^{-1}\dot{P} + pr^{-1}B)]_-^+ = 0, \tag{12.120}$$

$$[\rho^{1/2}\{P - 2pr^{-1}\beta^2(\omega^{-1}\dot{B} + pr^{-1}P)\}]_-^+ = 0 \tag{12.121}$$

on the core-mantle and inner-core boundaries $r = b$ and $r = c$. If the Earth model has an oceanic layer, then (12.119)–(12.121) must be satisfied on the seafloor $r = s$ as well; in addition, if there are any solid-solid discontinuities, then we must have (12.120)–(12.121) as well as $[\rho^{-1/2}(\omega^{-1}\dot{B} + pr^{-1}P)]_-^+ = 0$ on $r = d_{\mathrm{SS}}$. The appearance of both the wavefunctions P and B in these conditions reflects the conversion of P waves into SV waves and vice versa upon reflection and transmission at the free surface and the various internal interfaces.

More generally, we could have conducted the above analysis starting with a modified version of the 4×4 and 2×2 systems of equations (12.92) and (12.105), in which the perturbation to the Earth's gravitational potential is ignored, but the initial acceleration of gravity g is retained (see

Section 8.8.6). The lowest-order matrices A_0 are unaltered in this case; only A_1 and A_2 are changed. These higher-order matrices have no effect upon the zeroth-order representation (12.101)–(12.104), (12.113)–(12.116) and (12.117)–(12.121); hence, that representation is also valid in this *Cowling approximation*. We shall use these results to calculate the asymptotic eigenfrequencies and eigenfunctions of the mantle spheroidal modes in Section 12.3.5.

12.3.3 JWKB approximation

The next step in the analysis of both SH and P-SV waves is to seek asymptotic solutions to the radial Schrödinger equation

$$\ddot{X} + \omega^2 q^2 X = 0, \tag{12.122}$$

by means of classical JWKB theory. The unknown function of radius X in equation (12.122) represents any of the three scalar wavefunctions H, P or B in (12.73) or (12.95). The quantity $q^2 = v^{-2} - p^2 r^{-2}$ is the squared radial slowness of the associated SH, P or SV waves; the product ωq is the local radial wavenumber. Equation (12.122) is a *singular perturbation* problem, because the differentiated term is negligible compared to the undifferentiated term in the limit $\omega \to \infty$. We restrict attention to the case of a single turning point $r = R$, at which $q^2(R) = 0$; it is convenient to distinguish three overlapping regions, illustrated in Figure 12.17:

1. $r \ll R$ and $q^2 < 0$;
2. $r \approx R$ and $q^2 \approx \gamma(r - R)$, where $\gamma = 2v_R^{-3}(R^{-1}v_R - \dot{v}_R) > 0$;
3. $r \gg R$ and $q^2 > 0$.

Following Bender & Orszag (1978) we conduct an independent asymptotic analysis in each of these regions and connect the results using the *method of matched asymptotic expansions*; it is essential that the three regions be considered in the order listed above, starting beneath the turning point and concluding above it.

In region 1 we look for a solution in the form of a JWKB expansion:

$$X = [\mathcal{A}^{(0)} + \omega^{-1}\mathcal{A}^{(1)} + \cdots] \exp(\omega\Psi), \tag{12.123}$$

where we seek to determine the amplitudes $\mathcal{A}^{(0)}$, $\mathcal{A}^{(1)}, \ldots$ and the argument of the exponential Ψ. Upon substituting the ansatz (12.123) into the Schrödinger equation (12.122) and equating like powers of ω^{-1}, we obtain the classical scalar *eikonal* and *transport equations*

$$\dot{\Psi}^2 + q^2 = 0, \qquad \ddot{\Psi}\mathcal{A} + 2\dot{\Psi}\dot{\mathcal{A}} = 0. \tag{12.124}$$

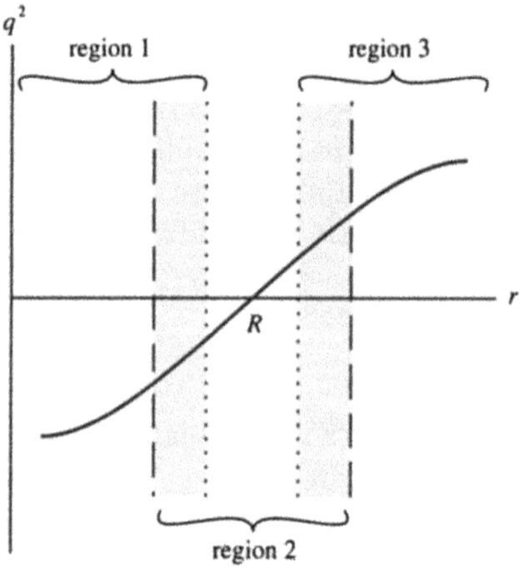

Figure 12.17. Schematic plot of the squared radial slowness $q^2(r)$ in equation (12.122). There is a single zero at $r = R$ where the associated waves turn. A separate JWKB analysis is conducted in each of the three regions; the solutions X_1, X_2 and X_3 are connected in the (*shaded*) overlap zones.

We have eliminated the superscript upon the amplitude $\mathcal{A}^{(0)} = \mathcal{A}$, since we shall only be concerned with the lowest-order approximation. The solutions to equations (12.124) are

$$\Psi = \pm \int_r^R |q|\, dr, \qquad \mathcal{A} = A|q|^{-1/2}, \tag{12.125}$$

where $|q|$ is the positive square root of $-q^2$ and A is an undetermined constant. The plus sign in the first of (12.125) corresponds to exponential growth away from the turning point, which is physically impermissible. Eliminating this possibility, we deduce that the lowest-order JWKB solution in region 1 is

$$X_1 = A|q|^{-1/2} \exp\left(-\omega \int_r^R |q|\, dr\right). \tag{12.126}$$

Equation (12.126) represents an exponentially decaying or *evanescent wave* beneath the turning point $r = R$.

In the vicinity of the turning point, the Schrödinger equation (12.122) reduces to the *Airy equation*

$$\ddot{X} + \omega^2 \gamma (r - R) X = 0. \tag{12.127}$$

The general solution to equation (12.127) can be written in terms of the two linearly independent Airy functions Ai and Bi in the form

$$X_2 = D\,\mathrm{Ai}\,[-\omega^{2/3}\gamma^{1/3}(r - R)] + E\,\mathrm{Bi}\,[-\omega^{2/3}\gamma^{1/3}(r - R)]. \tag{12.128}$$

The constants D and E can be related to the amplitude A of the evanescent wave by matching the two solutions (12.126) and (12.128). The divergence of the amplitude factor $|q|^{-1/2}$ renders equation (12.126) inapplicable in the immediate vicinity of $r = R$; likewise, equation (12.128) is invalid if we stray too far beneath $r = R$. There is, nevertheless, an overlap region in which both representations are valid. In this region, where $|q| \approx [\gamma(R-r)]^{1/2}$, we can approximate the evanescent solution by

$$X_1 \approx A[\gamma(R-r)]^{-1/4} \exp[-\tfrac{2}{3}\omega\gamma^{1/2}(R-r)^{3/2}], \tag{12.129}$$

and we can make use of the $x \to \infty$ approximations

$$\mathrm{Ai}(x) \approx \tfrac{1}{2}\pi^{-1/2}x^{-1/4}\exp(-\tfrac{2}{3}x^{3/2}), \tag{12.130}$$

$$\mathrm{Bi}(x) \approx \pi^{-1/2}x^{-1/4}\exp(\tfrac{2}{3}x^{3/2}) \tag{12.131}$$

to reduce the turning-point solution (12.128) to

$$\begin{aligned} X_2 \approx{}& \pi^{-1/2}\gamma^{-1/12}\omega^{-1/6}(R-r)^{-1/4} \\ &\times \left\{\tfrac{1}{2}D\exp[-\tfrac{2}{3}\omega\gamma^{1/2}(R-r)^{3/2}] \right. \\ &\qquad \left. +E\exp[\tfrac{2}{3}\omega\gamma^{1/2}(R-r)^{3/2}]\right\}. \end{aligned} \tag{12.132}$$

Upon comparing the two overlapping results (12.129) and (12.132) we find that $D = 2\pi^{1/2}\gamma^{-1/6}\omega^{1/6}A$ and $E = 0$. The solution in the vicinity of the turning point $r \approx R$ is therefore

$$X_2 = 2\pi^{1/2}A\gamma^{-1/6}\omega^{1/6}\mathrm{Ai}\,[-\omega^{2/3}\gamma^{1/3}(r-R)]. \tag{12.133}$$

The matching process guarantees that (12.133) is the smooth continuation into region 2 of the evanescent solution (12.126) in region 1.

In the region $r \gg R$, we again substitute the JWKB ansatz (12.123) into the Schrödinger equation (12.122). Because the squared radial slowness in this region is positive, the solution is oscillatory rather than evanescent:

$$\begin{aligned} X_3 ={}& Fq^{-1/2}\exp\left(i\omega\int_R^r q\,dr\right) \\ &+ Gq^{-1/2}\exp\left(-i\omega\int_R^r q\,dr\right), \end{aligned} \tag{12.134}$$

where q denotes the positive square root of q^2. The constants F and G are determined by matching the two solutions (12.133) and (12.134) in their overlap region. In the limit $x \to -\infty$, the Airy function in (12.133) can be approximated by

$$\mathrm{Ai}(x) \approx \pi^{-1/2}(-x)^{-1/4}\sin[\tfrac{2}{3}(-x)^{3/2} + \pi/4], \tag{12.135}$$

so that

$$X_2 \approx 2A[\gamma(r-R)]^{-1/4}\sin[\tfrac{2}{3}\omega\gamma^{1/2}(r-R)^{3/2}+\pi/4]. \tag{12.136}$$

On the other hand, equation (12.134) in the overlap region has the form

$$\begin{aligned} X_3 \approx\; & F[\gamma(r-R)]^{-1/4}\exp[\tfrac{2}{3}i\omega\gamma^{1/2}(r-R)^{3/2}] \\ & + G[\gamma(r-R)]^{-1/4}\exp[-\tfrac{2}{3}i\omega\gamma^{1/2}(r-R)^{3/2}]. \end{aligned} \tag{12.137}$$

Upon comparing (12.136) and (12.137) we deduce that $F = -iA\exp(i\pi/4)$ and $G = iA\exp(-i\pi/4)$. The asymptotic solution above the turning point is therefore

$$X_3 = 2Aq^{-1/2}\cos\left(\omega\int_R^r q\,dr - \frac{\pi}{4}\right). \tag{12.138}$$

The relation $A|q|^{-1/2}\exp(-\omega\int_r^R |q|\,dr) \Longrightarrow 2Aq^{-1/2}\cos(\omega\int_R^r q\,dr - \pi/4)$ between the two representations (12.126) and (12.138) is frequently referred to as the *JWKB connection formula*, since it determines the amplitude and phase of the oscillatory solution above the turning point in terms of the evanescent solution below. The final result of the asymptotic analysis in each of the three regions is summarized in Table 12.3.

It is noteworthy that the three solutions X_1, X_2 and X_3 can be combined into a single *uniformly valid* asymptotic representation known as the *Langer*

Region	Location	Asymptotic Solution
1	$r \ll R$	$X_1 = A\|q\|^{-1/2}\exp\left(-\omega\int_r^R \|q\|\,dr\right)$
2	$r \approx R$	$X_2 = 2\pi^{1/2}A\gamma^{-1/6}\omega^{1/6}\mathrm{Ai}\left[-\omega^{2/3}\gamma^{1/3}(r-R)\right]$
3	$r \gg R$	$X_3 = 2Aq^{-1/2}\cos\left(\omega\int_R^r q\,dr - \frac{\pi}{4}\right)$

Table 12.3. Asymptotic solutions to the Schrödinger equation (12.122) in the three regions shown in Figure 12.17. The JWKB wavefunction X is evanescent beneath the turning radius $r = R$ and oscillatory above it. The normalization constant A is arbitrary.

approximation:

$$X = 2\pi^{1/2}A\,\chi^{1/6}(-q^2)^{-1/4}\mathrm{Ai}(\chi^{2/3}), \tag{12.139}$$

where $\chi = -\frac{3}{2}\omega\int_R^r(-q^2)^{1/2}\,dr$. The validity of this approximation, which is by no means obvious, is most easily verified by demonstrating that it is asymptotically equivalent to the results in Table 12.3 in each of the three regions. Equation (12.139) is applicable only if the turning radius $r = R$ is far away from a discontinuity such as the core-mantle boundary $r = b$; more generally, it is necessary to include $\mathrm{Bi}(\chi^{2/3})$ as well as $\mathrm{Ai}(\chi^{2/3})$ to account for the tunnelling of nearly grazing waves (Woodhouse 1978).

12.3.4 Toroidal modes revisited

In this section we use the JWKB asymptotic analysis in Sections 12.3.1 and 12.3.3 to determine the toroidal eigenfrequencies and eigenfunctions of an Earth model whose only internal discontinuities are the inner-core and core-mantle boundaries. We consider first the modes in Regime II, which are equivalent to turning SH waves in the mantle. The squared radial slowness $q_\beta^2 = \beta^{-2} - p^2r^{-2}$ in this regime has a single zero at the turning radius $r = R_{\mathrm{S}}$; the SH wavefunction and its derivative in the oscillatory region $R_{\mathrm{S}} \ll r \leq a$ above the turning radius are of the form

$$H = 2Aq_\beta^{-1/2}\cos\left(\omega\int_{R_{\mathrm{S}}}^r q_\beta\,dr - \frac{\pi}{4}\right), \tag{12.140}$$

$$\dot{H} = -2\omega Aq_\beta^{1/2}\sin\left(\omega\int_{R_{\mathrm{S}}}^r q_\beta\,dr - \frac{\pi}{4}\right), \tag{12.141}$$

where we have ignored terms of higher order in ω^{-1} in performing the differentiation. The asymptotic eigenfrequencies are determined by the free-surface boundary condition $\dot{H} = 0$. Upon evaluating (12.141) at $r = a$, we obtain the asymptotic quantization condition

$$\omega\tau_{\mathrm{S}} = 2\pi(n + 5/4), \tag{12.142}$$

where τ_{S} is the intercept time and n is the overtone index. This is the same result we found in Section 12.2.2 by means of the more elementary and physically intuitive constructive-interference argument.

The oscillatory JWKB eigenfunctions associated with the eigenfrequencies (12.142) are given in the region $R_{\mathrm{S}} \ll r \leq a$ by

$$W = 2A\mu^{-1/2}r^{-1}q_\beta^{-1/2}\cos\left(\omega\int_{R_{\mathrm{S}}}^r q_\beta\,dr - \frac{\pi}{4}\right). \tag{12.143}$$

Below the turning radius, in the region $r \ll R_{\mathrm{S}}$, the behavior is evanescent:

$$W = A\mu^{-1/2}r^{-1}|q_\beta|^{-1/2}\exp\left(-\omega\int_r^{R_{\mathrm{S}}}|q_\beta|\,dr\right). \tag{12.144}$$

The constant A can be determined from the normalization relation for the toroidal modes:

$$\int_b^a \rho W^2\, r^2 dr = 1. \tag{12.145}$$

The evaluation of JWKB normalization integrals in the presence of a turning point is discussed by Bender & Orszag (1978); they show that the contribution from all three regions is asymptotically equivalent to integrating from $r = R_{\mathrm{S}}$, with the oscillatory term $\cos^2(\omega\int_{R_{\mathrm{S}}}^a q_\beta\, dr - \pi/4)$ replaced by its mean value 1/2, in accordance with the Riemann-Lebesgue lemma:

$$\int_b^a \rho W^2\, r^2 dr = 2A^2\int_{R_{\mathrm{S}}}^a \beta^{-2}q_\beta^{-1}\, dr. \tag{12.146}$$

It follows that $A = T_{\mathrm{S}}^{-1/2}$, where T_{S} is the S-wave travel time.

In Regime III the SH wavefunction is oscillatory throughout the entire mantle $b \le r \le a$:

$$H = Aq_\beta^{-1/2}\exp\left(i\omega\int_b^r q_\beta\, dr\right) + Bq_\beta^{-1/2}\exp\left(-i\omega\int_b^r q_\beta\, dr\right), \tag{12.147}$$

$$\dot{H} = i\omega Aq_\beta^{1/2}\exp\left(i\omega\int_b^r q_\beta\, dr\right) - i\omega Bq_\beta^{1/2}\exp\left(-i\omega\int_b^r q_\beta\, dr\right), \tag{12.148}$$

where A and B are arbitrary constants. The boundary condition $\dot{H} = 0$ on the core-mantle boundary $r = b$ implies that $B = A$; the corresponding condition on the free surface $r = a$ then yields the quantization condition

$$\omega\tau_{\mathrm{ScS}} = 2\pi(n+1), \tag{12.149}$$

which agrees with equation (12.45) as expected. The JWKB representation of an $\mathrm{ScS_{SH}}$-mode eigenfunction is

$$W = 2A\mu^{-1/2}r^{-1}q_\beta^{-1/2}\cos\left(\omega\int_b^r q_\beta\, dr\right). \tag{12.150}$$

The Riemann-Lebesgue lemma can again be used to evaluate the normalization integral (12.145), with the result $A = T_{\mathrm{ScS}}^{-1/2}$.

We summarize the oscillatory JWKB eigenfunctions in each of the four toroidal-mode regimes in Table 12.4. The results for the inner-core modes in Regime IV are identical to those in Regime II, with τ_{S} and T_{S} replaced by τ_{J} and T_{J}. The factor of $\pi/4$ in the standing-wave representation of the eigenfunctions in these two regimes is consistent with the $\pi/2$ caustic phase shift experienced by a turning SH wave. Figure 12.18 compares the exact and approximate eigenfunctions W for a number of mantle toroidal modes. The agreement is generally excellent even for the Love-wave equivalent modes ${}_n\mathrm{T}_l$ with $n \ll l/4$. The principal discrepancy occurs near the turning radius $r = R_{\mathrm{S}}$, where the asymptotic amplitude factor $q_\beta^{-1/2}$ diverges; this divergence is an inherent feature of the non-uniform validity of the JWKB approximation.

Kennett & Nolet (1979) have used the Langer approximation described in Section 12.3.3 to obtain improved results for the transitional modes near the boundary $p = b/\beta_{b+}$ between Regime II and Regime III. The discontin-

Regime	Type	Normalized Displacement Eigenfunction
I	—	evanescent SH waves only
II	SH	$W = 2T_{\mathrm{S}}^{-1/2}\mu^{-1/2}r^{-1}q_\beta^{-1/2}\cos\left(\omega\int_{R_{\mathrm{S}}}^{r} q_\beta\,dr - \frac{\pi}{4}\right)$
III	$\mathrm{ScS_{SH}}$	$W = 2T_{\mathrm{ScS}}^{-1/2}\mu^{-1/2}r^{-1}q_\beta^{-1/2}\cos\left(\omega\int_{b}^{r} q_\beta\,dr\right)$
IV	$\mathrm{J_{SH}}$	$W = 2T_{\mathrm{J}}^{-1/2}\mu^{-1/2}r^{-1}q_\beta^{-1/2}\cos\left(\omega\int_{R_{\mathrm{J}}}^{r} q_\beta\,dr - \frac{\pi}{4}\right)$

Table 12.4. Asymptotically normalized eigenfunction W in each of the SH ray-parameter regimes in an Earth model with two internal discontinuities. The expressions in Regime III are valid throughout the mantle $b \leq r \leq a$, whereas those in Regimes II and IV are only valid in the oscillatory region $R_{\mathrm{S}} \ll r \leq a$ or $R_{\mathrm{J}} \ll r \leq c$ above the mantle and inner-core turning points. There is a corresponding evanescent SH and $\mathrm{J_{SH}}$ eigenfunction beneath the turning points, in the regions $r \ll R_{\mathrm{S}}$ and $r \ll R_{\mathrm{J}}$.

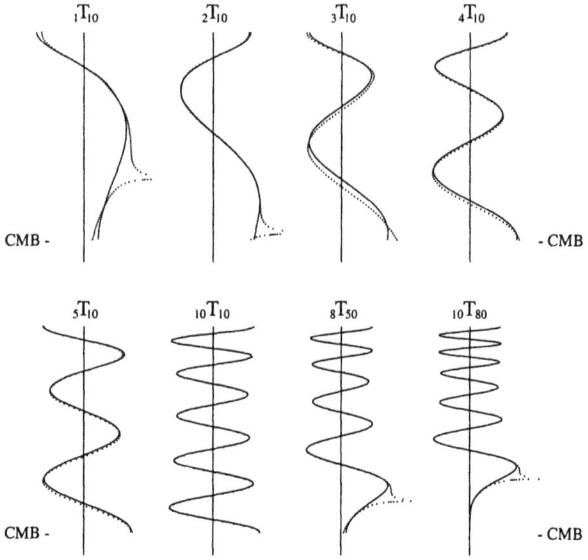

Figure 12.18. Exact (*solid line*) and asymptotic (*dotted line*) eigenfunctions W for a number of toroidal modes. Vertical axis is depth beneath the free surface of crustless model 1066A; the location of the core-mantle boundary (CMB) is shown. Modes ${}_1T_{10}$, ${}_2T_{10}$, ${}_8T_{50}$ and ${}_{10}T_{80}$ are SH-equivalent with turning points—where $W_{\rm JWKB}$ diverges—at $r = R_{\rm S}$ in the lower mantle. Modes ${}_3T_{10}$, ${}_4T_{10}$, ${}_5T_{10}$ and ${}_{10}T_{10}$ are ScS$_{\rm SH}$-equivalent, with no turning points. (Courtesy of L. Zhao.)

uous eigenfrequency formulae $\omega\tau_{\rm S} = 2\pi(n+5/4)$ and $\omega\tau_{\rm ScS} = 2\pi(n+1)$ are replaced by a single asymptotic formula $\omega\tau_{\rm S+ScS} = 2\pi(n+\delta)$, where δ varies smoothly between 5/4 and 1. The quantity $2\pi(\delta-1)$ is the phase shift of an SH+ScS$_{\rm SH}$ wave of frequency ω due to its interaction with the core-mantle boundary. The first-order effect of the transition-zone derivatives $\dot{\beta}$ and $\dot{\rho}$ is also accounted for in their analysis.

12.3.5 Spheroidal modes revisited

The asymptotic analysis of the spheroidal modes must be conducted in the same manner—one ray-parameter regime at a time. Since the eigenfunction expressions are lengthy, we only present the results for the modes in Regime III, which are equivalent to turning P and SV waves in the mantle. As before, we restrict attention to an oceanless Earth model with no solid-solid discontinuities. The oscillatory JWKB solutions to the Schrödinger equations (12.95) can be written in the form

$$P = 2Dq_\alpha^{-1/2} \sin\left(\omega \int_{R_P}^{r} q_\alpha \, dr + \frac{\pi}{4}\right), \tag{12.151}$$

$$\dot{P} = 2\omega D q_\alpha^{1/2} \cos\left(\omega \int_{R_P}^{r} q_\alpha \, dr + \frac{\pi}{4}\right) \tag{12.152}$$

and

$$B = 2Eq_\beta^{-1/2} \cos\left(\omega \int_{R_S}^{r} q_\beta \, dr - \frac{\pi}{4}\right), \tag{12.153}$$

$$\dot{B} = -2\omega E q_\beta^{1/2} \sin\left(\omega \int_{R_S}^{r} q_\beta \, dr - \frac{\pi}{4}\right), \tag{12.154}$$

where D and E are constants and $q_\alpha^2 = \alpha^{-2} - p^2 r^{-2}$ and $q_\beta^2 = \beta^{-2} - p^2 r^{-2}$. Upon substituting the representation (12.151)–(12.154) into the free-surface dynamical boundary conditions (12.117)–(12.118) we obtain two equations relating the constants D and E:

$$\begin{pmatrix} \zeta_\alpha \sin\frac{1}{2}\Psi_{\mathrm{P}} & \xi_\beta \sin\frac{1}{2}\Psi_{\mathrm{S}} \\ -\xi_\alpha \cos\frac{1}{2}\Psi_{\mathrm{P}} & \zeta_\beta \cos\frac{1}{2}\Psi_{\mathrm{S}} \end{pmatrix} \begin{pmatrix} D \\ E \end{pmatrix} = \begin{pmatrix} 0 \\ 0 \end{pmatrix}, \tag{12.155}$$

where $\Psi_{\mathrm{P}} = \omega\tau_{\mathrm{P}} + \pi/2$ and $\Psi_{\mathrm{S}} = \omega\tau_{\mathrm{S}} - \pi/2$. For both P and SV waves the quantities ζ and ξ are defined by

$$\zeta = (\beta_a^{-2} - 2p^2 a^{-2}) q^{-1/2}(a), \qquad \xi = 2pa^{-1} q^{1/2}(a). \tag{12.156}$$

Equation (12.155) has a solution only if the determinant of the matrix multiplying D and E vanishes; this solubility condition can be reduced to

$$\sin\tfrac{1}{2}(\Psi_{\mathrm{S}} + \Psi_{\mathrm{P}}) + \left[\frac{\xi_\alpha \xi_\beta - \zeta_\alpha \zeta_\beta}{\xi_\alpha \xi_\beta + \zeta_\alpha \zeta_\beta}\right] \sin\tfrac{1}{2}(\Psi_{\mathrm{S}} - \Psi_{\mathrm{P}}) = 0. \tag{12.157}$$

Equation (12.157) is equivalent to the asymptotic Regime III eigenfrequency relation (12.55); it is readily verified that the quantity in brackets is the free-surface SV-wave reflection coefficient $\acute{S}\grave{S}$. It is noteworthy that this coefficient (12.15), governing the reflection of a *plane* wave off a *plane* boundary, appears naturally in the present analysis. Fundamentally, of course, this is a consequence of the locally plane character of the high-frequency boundary interaction.

Once the eigenfrequencies have been calculated using equation (12.157), the constants D and E can be constrained by solving either of the two equations (12.155). We find $D = A\zeta_\beta \cos\frac{1}{2}\Psi_{\mathrm{S}}$ and $E = A\xi_\alpha \cos\frac{1}{2}\Psi_{\mathrm{P}}$, where

A is a normalization constant that remains undetermined. Making use of the representations (12.101) and (12.102), we then obtain the spheroidal radial eigenfunctions:

$$U = 2A\rho^{-1/2}r^{-1}\left[(\zeta_\beta \cos\tfrac{1}{2}\Psi_{\rm S})q_\alpha^{1/2}\cos\left(\omega\int_{R_{\rm P}}^{r} q_\alpha\,dr + \frac{\pi}{4}\right)\right.$$
$$\left. + (\xi_\alpha \cos\tfrac{1}{2}\Psi_{\rm P})pr^{-1}q_\beta^{-1/2}\cos\left(\omega\int_{R_{\rm S}}^{r} q_\beta\,dr - \frac{\pi}{4}\right)\right], \tag{12.158}$$

$$V = 2A\rho^{-1/2}r^{-1}\left[(\zeta_\beta \cos\tfrac{1}{2}\Psi_{\rm S})pr^{-1}q_\alpha^{-1/2}\sin\left(\omega\int_{R_{\rm P}}^{r} q_\alpha\,dr + \frac{\pi}{4}\right)\right.$$
$$\left. - (\xi_\alpha \cos\tfrac{1}{2}\Psi_{\rm P})q_\beta^{1/2}\sin\left(\omega\int_{R_{\rm S}}^{r} q_\beta\,dr - \frac{\pi}{4}\right)\right]. \tag{12.159}$$

The constant A can be determined from the normalization relation for the spheroidal modes:

$$\int_0^a \rho(U^2 + V^2)\,r^2 dr = 1. \tag{12.160}$$

Upon integrating from the two turning radii $r = R_{\rm P}$ and $r = R_{\rm S}$, making use of the Riemann-Lebesgue lemma, as in the toroidal case, we find that

$$A = [(\zeta_\beta \cos\tfrac{1}{2}\Psi_{\rm S})^2 T_{\rm P} + (\xi_\alpha \cos\tfrac{1}{2}\Psi_{\rm P})^2 T_{\rm S}]^{-1/2}, \tag{12.161}$$

where $T_{\rm P}$ and $T_{\rm S}$ are the P-wave and SV-wave travel times.

We can rewrite the normalization relation (12.160) in the form

$$f_{\rm P} + f_{\rm S} = 1, \tag{12.162}$$

where

$$f_{\rm P} = \frac{(\zeta_\beta \cos\frac{1}{2}\Psi_{\rm S})^2 T_{\rm P}}{(\zeta_\beta \cos\frac{1}{2}\Psi_{\rm S})^2 T_{\rm P} + (\xi_\alpha \cos\frac{1}{2}\Psi_{\rm P})^2 T_{\rm S}}, \tag{12.163}$$

$$f_{\rm S} = \frac{(\xi_\alpha \cos\frac{1}{2}\Psi_{\rm P})^2 T_{\rm S}}{(\zeta_\beta \cos\frac{1}{2}\Psi_{\rm S})^2 T_{\rm P} + (\xi_\alpha \cos\frac{1}{2}\Psi_{\rm P})^2 T_{\rm S}}. \tag{12.164}$$

The quantities $f_{\rm P}$ and $f_{\rm S}$ are the *fractional kinetic energies* of the mode present in the form of *P waves* and *SV waves*, respectively; these energies can only be defined asymptotically, in the context of mode-ray duality. The normalized oscillatory eigenfunctions can be written in terms of $f_{\rm P}$ and $f_{\rm S}$ in the form

$$U = 2\rho^{-1/2}r^{-1}\left[f_{\rm P}^{1/2}T_{\rm P}^{-1/2}q_\alpha^{1/2}\cos\left(\omega\int_{R_{\rm P}}^{r} q_\alpha\,dr + \frac{\pi}{4}\right)\right.$$
$$\left. + f_{\rm S}^{1/2}T_{\rm S}^{-1/2}pr^{-1}q_\beta^{-1/2}\cos\left(\omega\int_{R_{\rm S}}^{r} q_\beta\,dr - \frac{\pi}{4}\right)\right], \tag{12.165}$$

$$V = 2\rho^{-1/2}r^{-1}\left[f_{\mathrm{P}}^{1/2}T_{\mathrm{P}}^{-1/2}pr^{-1}q_\alpha^{-1/2}\sin\left(\omega\int_{R_{\mathrm{P}}}^{r} q_\alpha\,dr + \frac{\pi}{4}\right)\right.$$
$$\left. - f_{\mathrm{S}}^{1/2}T_{\mathrm{S}}^{-1/2}q_\beta^{1/2}\sin\left(\omega\int_{R_{\mathrm{S}}}^{r} q_\beta\,dr - \frac{\pi}{4}\right)\right]. \qquad (12.166)$$

It is noteworthy that the radial component of the displacement U diverges at $r = R_{\mathrm{S}}$ and vanishes at $r = R_{\mathrm{P}}$; in contrast, the tangential or horizontal displacement V diverges at $r = R_{\mathrm{P}}$ and vanishes at $r = R_{\mathrm{S}}$. This behavior is readily understood by considering the associated turning rays: the particle motion accompanying a P wave is purely longitudinal at $r = R_{\mathrm{P}}$, whereas that accompanying an SV wave is purely radial at $r = R_{\mathrm{S}}$. The P-wave and SV-wave terms in the JWKB asymptotic representation (12.165)–(12.166) are valid only in the oscillatory regions $R_{\mathrm{P}} \ll r \leq a$ and $R_{\mathrm{S}} \ll r \leq a$, respectively.

The asymptotic eigenfunctions U, V in the other spheroidal-mode ray-parameter regimes may be obtained in an analogous manner; in general, it is necessary to impose the boundary conditions (12.119)–(12.121) on the core-mantle and inner-core boundaries $r = b$ and $r = c$ in addition to the free-surface conditions (12.117)–(12.118). A complete catalogue of results for an Earth model with two fluid-solid discontinuities, including the effects of all quantitatively significant evanescent waves, is given by Zhao & Dahlen (1995a). In Regimes IX and X we are obliged to account for downgoing and upgoing P and SV waves in the mantle, downgoing and upgoing K waves in the fluid outer core, and turning and evanescent I and $\mathrm{J_{SV}}$ waves in the solid inner core. The resulting JWKB formulae for U and V are quite complicated; however, they may be reduced to a set of compact results for the decoupled $\mathrm{ScS_{SV}}$, $\mathrm{J_{SV}}$ and PKIKP modes upon making the steep-incidence-angle approximation $\grave{P}\acute{P} \approx \grave{K}\acute{K} \approx 0$. We summarize these simplified Regime IX and X eigenfunctions in Table 12.5; the radial modes are the $p \rightarrow 0$ limit of the PKIKP modes, as expected. It is noteworthy that the tangential displacements of the $\mathrm{J_{SV}}$ and $\mathrm{J_{SH}}$ inner-core modes are of the form $V \sim \sin(\omega\int_{R_J}^{r} q_\beta\,dr - \pi/4)$ and $W \sim \cos(\omega\int_{R_J}^{r} q_\beta\,dr - \pi/4)$, respectively. This quadrature relationship between the spheroidal and toroidal inner-core eigenfunctions reflects the different behavior of the associated waves upon turning: the tangential component of a $\mathrm{J_{SV}}$ wave exhibits a sign reversal (see Figure 12.13) whereas that of a $\mathrm{J_{SH}}$ wave does not.

Figure 12.19 compares the asymptotic and exact eigenfunctions for a number of spheroidal modes in Regimes III–IX. The divergence of the radial displacement U at $r = R_{\mathrm{S}}$ and the tangential displacement V at $r = R_{\mathrm{P}}$ are evident for the Regime III mode ${}_5\mathrm{S}_{25}$; note also the evanescent tail of this mode, which extends beneath the core-mantle boundary. Higher-overtone modes have more deeply penetrating eigenfunctions and a richer

Mode Type	Normalized Displacement Eigenfunctions
$\mathrm{ScS_{SV}}$	$U \approx 2T_{\mathrm{ScS}}^{-1/2}\rho^{-1/2}pr^{-2}q_\beta^{-1/2}\sin\left(\omega\int_b^r q_\beta\,dr\right)$ $V \approx 2T_{\mathrm{ScS}}^{-1/2}\rho^{-1/2}r^{-1}q_\beta^{1/2}\cos\left(\omega\int_b^r q_\beta\,dr\right)$
$\mathrm{J_{SV}}$	$U \approx -2T_{\mathrm{J}}^{-1/2}\rho^{-1/2}pr^{-2}q_\beta^{-1/2}\cos\left(\omega\int_{R_\mathrm{J}}^r q_\beta\,dr - \frac{\pi}{4}\right)$ $V \approx 2T_{\mathrm{J}}^{-1/2}\rho^{-1/2}r^{-1}q_\beta^{1/2}\sin\left(\omega\int_{R_\mathrm{J}}^r q_\beta\,dr - \frac{\pi}{4}\right)$
PKIKP	$U \approx 2T_{\mathrm{PKIKP}}^{-1/2}\rho^{-1/2}r^{-1}q_\alpha^{1/2}\cos\left(\omega\int_{R_\mathrm{I}}^r q_\alpha\,dr + \frac{\pi}{4}\right)$ $V \approx 2T_{\mathrm{PKIKP}}^{-1/2}\rho^{-1/2}pr^{-2}q_\alpha^{-1/2}\sin\left(\omega\int_{R_\mathrm{I}}^r q_\alpha\,dr + \frac{\pi}{4}\right)$
Radial	$U \approx 2T_{\mathrm{PKIKP}}^{-1}\rho^{-1/2}r^{-1}\alpha^{-1/2}\cos\left(\omega\int_0^r \alpha^{-1}\,dr\right)$ $V = 0$

Table 12.5. Asymptotically normalized eigenfunctions of the modes in Regimes IX and X. The $\mathrm{ScS_{SV}}$ expressions are valid throughout the mantle $b \leq r \leq a$; there is an associated evanescent form in the underlying fluid core $c \leq r \leq b$. The $\mathrm{J_{SV}}$ expressions are valid in the oscillatory region $R_\mathrm{J} \ll r \leq c$; in this case there is an evanescent contribution beneath the turning point $0 \leq r \ll R_\mathrm{J}$, as well as in the overlying fluid core $c \leq r \leq b$. The PKIKP expressions are valid in the oscillatory region $R_\mathrm{I} \ll r \leq a$; the radial-mode expressions are valid everywhere in $0 \ll r \leq a$.

assortment of constituent waves. The Regime V oscillation ${}_{15}\mathrm{S}_{25}$, for example, is composed of turning P, $\mathrm{ScS_{SV}}$ and turning SKS waves; its

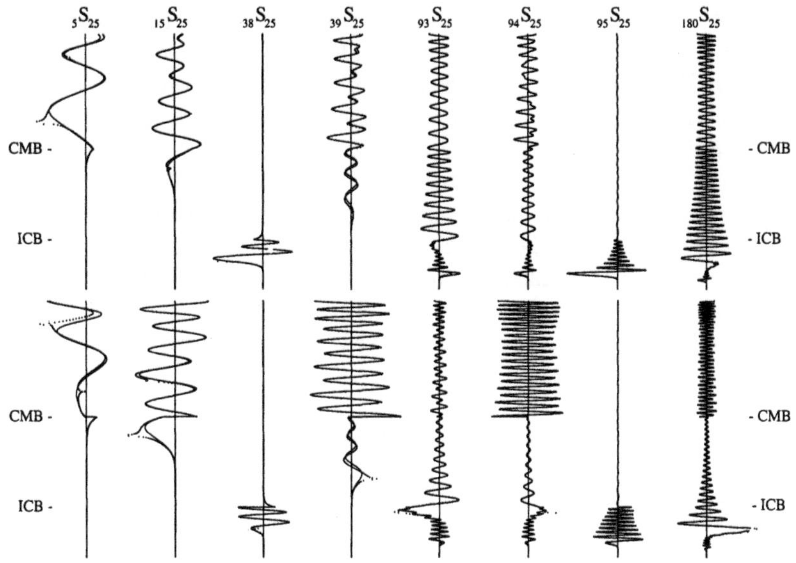

Figure 12.19. Exact (*solid line*) and asymptotic (*dotted line*) eigenfunctions for a number of ${}_n\mathrm{S}_{25}$ spheroidal modes. (*Top row*) Radial displacement U. (*Bottom row*) Tangential displacement V. Vertical axis extends from $r = 0$ to $r = a$; the locations of the core-mantle boundary (CMB) and inner-core boundary (ICB) are indicated. The asymptotic approximation is excellent away from the various turning points R_P, R_S, R_K, R_I and R_J. (Courtesy of L. Zhao.)

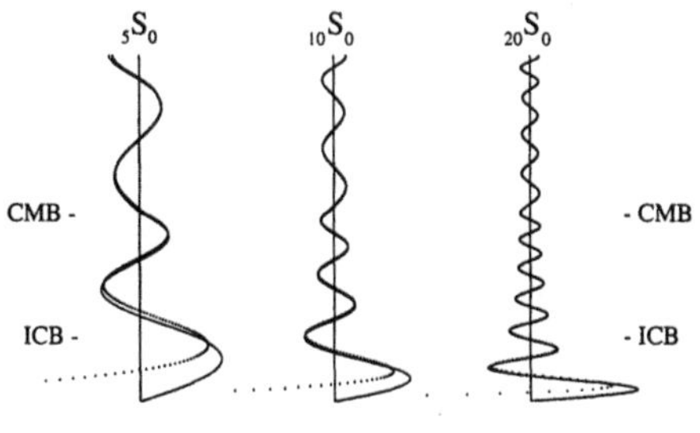

Figure 12.20. Exact (*solid line*) and asymptotic (*dotted line*) eigenfunctions U for the radial modes ${}_5\mathrm{S}_0$, ${}_{10}\mathrm{S}_0$ and ${}_{20}\mathrm{S}_0$. Vertical axis extends from $r = 0$ to $r = a$; the locations of the core-mantle boundary (CMB) and inner-core boundary (ICB) are shown. (Courtesy of L. Zhao.)

tangential displacement V diverges twice, at both $r = R_{\mathrm{P}}$ in the mantle and $r = R_{\mathrm{K}}$ in the fluid outer core. The two $\mathrm{ScS}_{\mathrm{SV}}$-equivalent modes ${}_{39}\mathrm{S}_{25}$ and ${}_{94}\mathrm{S}_{25}$ have tangential displacements that diverge at $r = R_{\mathrm{K}}$ and $r = R_{\mathrm{I}}$ in the outer and inner cores, respectively. Modes ${}_{38}\mathrm{S}_{25}$ and ${}_{95}\mathrm{S}_{25}$ are J_{SV}-equivalent, whereas modes ${}_{93}\mathrm{S}_{25}$ and ${}_{180}\mathrm{S}_{25}$ are PKIKP-equivalent. The extent to which the asymptotic representations capture the detailed oscillatory and evanescent character of such a wide variety of high-frequency modes is remarkable. Figure 12.20 compares the asymptotic and exact eigenfunctions for a number of Regime X radial modes. The agreement is excellent except near the center of the Earth, where U exhibits an r^{-1} divergence because the associated $p = 0$ ray "turns" there.

★12.4 Asymptotic Miscellany

We can make use of the above results to gain an improved understanding of the energy balance and group speed of a mode, and to find an asymptotic representation of the Fréchet kernels expressing the first-order dependence of the eigenfrequency upon the spherical-Earth model parameters. For convenience in what follows, we denote the quantities (12.66) whose zeroes determine the asymptotic eigenfrequencies by

$$f = \sum_{\nu} A_{\nu} \sin \tfrac{1}{2}\Psi_{\nu} \qquad \text{or} \qquad f = \sum_{\nu} A_{\nu} \cos \tfrac{1}{2}\Psi_{\nu}. \tag{12.167}$$

Most of the results in this section are due to Zhao & Dahlen (1995b).

★12.4.1 P-wave and S-wave energy

Equation (12.162) expressing the kinetic energy of a spheroidal mode in Regime III as a sum of the fractional P-wave and SV-wave energies can be readily generalized to the modes in other regimes. In each case we substitute the JWKB asymptotic representation of the eigenfunctions U, V into the normalization relation (12.160), and isolate the terms associated with the various oscillatory wave types. The result of this procedure can always be written in the form

$$\sum_{\nu} f_{\nu} = 1, \tag{12.168}$$

where

$$f_{\nu} = \frac{(\partial f/\partial \Psi_{\nu})T_{\nu}}{\sum_{\nu'}(\partial f/\partial \Psi_{\nu'})T_{\nu'}}. \tag{12.169}$$

The quantity f_ν is the *fractional kinetic energy* associated with the constituent wave of type ν. It is readily verified that equation (12.169) agrees with the previous result (12.163)–(12.164) in the case of a Regime III mode formed by the constructive interference of turning P and SV waves in the mantle. A mode in Regime IX, where there is the greatest proliferation of reflected and turning waves, has

$$f_{\mathrm{PcP}} + f_{\mathrm{ScS}} + f_{\mathrm{KiK}} + f_{\mathrm{I}} + f_{\mathrm{J}} = 1. \tag{12.170}$$

The expression f that must be differentiated to find the five fractional energies f_{PcP}, f_{ScS}, f_{KiK}, f_{I} and f_{J} is given in equation (12.56). The result (12.170) can be substantially simplified upon making the weak-contrast approximation $\grave{P}\acute{P} \approx \grave{K}\acute{K} \approx 0$. An $\mathrm{ScS_{SV}}$ mode in that case has $f_{\mathrm{ScS}} \approx 1$ whereas a $\mathrm{J_{SV}}$ mode has $f_{\mathrm{J}} \approx 1$. A PKIKP mode has

$$f_{\mathrm{PcP}} \approx \frac{T_{\mathrm{PcP}}}{T_{\mathrm{PKIKP}}}, \qquad f_{\mathrm{KiK}} \approx \frac{T_{\mathrm{KiK}}}{T_{\mathrm{PKIKP}}}, \qquad f_{\mathrm{I}} \approx \frac{T_{\mathrm{I}}}{T_{\mathrm{PKIKP}}}; \tag{12.171}$$

i.e., the fractional kinetic energy of the three compressional-wave types is simply equal to the fractional travel time spent in each region. This intuitively appealing approximation pertains to a radial mode in Regime X as well. The asymptotic energy-balance relation (12.168) reduces to a tautology in the case of the toroidal modes; an SH mode has $f_{\mathrm{S}} = 1$, an ScS mode has $f_{\mathrm{ScS}} = 1$ and a J mode has $f_{\mathrm{J}} = 1$.

*12.4.2 Group speed

To find the asymptotic group speed of a mode, we regard the constructive-interference relation (12.66) as an equation of the form

$$f(\omega, p) = f(\omega, k/\omega) = 0, \tag{12.172}$$

where $k = l + \frac{1}{2}$ is the asymptotic wavenumber. Upon differentiating this equation with respect to ω and k using the chain rule, we obtain

$$C = \frac{d\omega}{dk} = -\frac{1}{\omega}\left(\frac{\partial f}{\partial p}\right)\left(\frac{\partial f}{\partial \omega} - \frac{p}{\omega}\frac{\partial f}{\partial p}\right)^{-1}. \tag{12.173}$$

For a toroidal mode, $f = \sin\frac{1}{2}\Psi$ where Ψ denotes either Ψ_{S}, Ψ_{ScS} or Ψ_{J}, so that

$$\frac{\partial f}{\partial \omega} = \tfrac{1}{2}\tau\cos\tfrac{1}{2}\Psi, \qquad \frac{1}{\omega}\frac{\partial f}{\partial p} = -\tfrac{1}{2}\Theta\cos\tfrac{1}{2}\Psi, \tag{12.174}$$

where we have used the ray-theoretical relation $d\tau/dp = -\Theta$. Upon substituting (12.174) into equation (12.173) and using $T = \tau + p\Theta$, we find that the asymptotic group speed of a toroidal mode is simply the ratio of the epicentral distance and the travel time:

$$C = \Theta/T. \tag{12.175}$$

It is instructive to contrast this physically appealing result with the corresponding phase speed $c = p^{-1}$ of the same mode:

$$c = d\Theta/dT. \tag{12.176}$$

The phase speed c is a measure of the *apparent* angular rate of progression of the constructively interfering SH, $\mathrm{ScS_{SH}}$ or $\mathrm{J_{SH}}$ waves that comprise the mode, whereas the group speed C is a measure of their *absolute* angular rate of progression. It is noteworthy that C depends only upon the ray parameter p; this implies that all of the modes lying along a diagonal $\omega = ck$ in the toroidal dispersion diagram should have the same group speed C. A rudimentary visual inspection of Figure 8.3 reveals that this is indeed the case. Equation (12.175) can also be derived by substituting the JWKB eigenfunctions in Table 12.3 into the exact expression (11.67) and evaluating the resulting integrals with the aid of the Riemann-Lebesgue lemma, or, alternatively and even more simply, by differentiation of the explicit equation defining a toroidal eigenfrequency: $\omega\tau(k/\omega) = \text{constant}$.

For a spheroidal mode with $f = \sum_\nu A_\nu \sin\frac{1}{2}\Psi_\nu$ or $f = \sum_\nu A_\nu \cos\frac{1}{2}\Psi_\nu$, we have

$$\frac{\partial f}{\partial\omega} = \sum_\nu \left(\frac{\partial f}{\partial\Psi_\nu}\right)\tau_\nu, \qquad \frac{1}{\omega}\frac{\partial f}{\partial p} = -\sum_\nu \left(\frac{\partial f}{\partial\Psi_\nu}\right)\Theta_\nu, \tag{12.177}$$

correct to first order in ω^{-1}. Upon substituting these relations into equation (12.173) and making use of the definition (12.169), we find that

$$C = \sum_\nu f_\nu(\Theta_\nu/T_\nu), \tag{12.178}$$

which generalizes the elementary result (12.175). Equation (12.178) shows that, in general, the asymptotic group speed of a mode is a *weighted average* of the group speeds Θ_ν/T_ν of its constituent waves. It makes sense that the weighting factor f_ν is the *fractional kinetic energy*, inasmuch as the energy of a wave propagates with the group speed. The result (12.178) can also be derived by substituting the JWKB asymptotic eigenfunctions U, V in each spheroidal-mode regime into the exact integral expression (11.79) for the group speed. A mantle spheroidal mode in Regime III has a group speed $C = f_{\mathrm{P}}(\Theta_{\mathrm{P}}/T_{\mathrm{P}}) + f_{\mathrm{S}}(\Theta_{\mathrm{S}}/T_{\mathrm{S}})$, whereas a whole-Earth mode in Regime IX has

$$\begin{aligned} C = {} & f_{\mathrm{PcP}}(\Theta_{\mathrm{PcP}}/T_{\mathrm{PcP}}) + f_{\mathrm{ScS}}(\Theta_{\mathrm{ScS}}/T_{\mathrm{ScS}}) \\ & + f_{\mathrm{KiK}}(\Theta_{\mathrm{KIK}}/T_{\mathrm{KIK}}) + f_{\mathrm{I}}(\Theta_{\mathrm{I}}/T_{\mathrm{I}}) + f_{\mathrm{J}}(\Theta_{\mathrm{J}}/T_{\mathrm{J}}). \end{aligned} \tag{12.179}$$

The result (12.179) can be simplified in the case of seismological interest, $\grave{P}\acute{P} \approx \grave{K}\acute{K} \approx 0$; an $\mathrm{ScS}_{\mathrm{SV}}$ mode in that case has $C \approx \Theta_{\mathrm{ScS}}/T_{\mathrm{ScS}}$, a J_{SV} mode has $C \approx \Theta_{\mathrm{J}}/T_{\mathrm{J}}$ and a PKIKP mode has $C \approx \Theta_{\mathrm{PKIKP}}/T_{\mathrm{PKIKP}}$. For the radial modes, since the equivalent waves pass exactly through the center of the Earth, $\Theta_{\mathrm{PKIKP}} = \pi$. The group speed is not the same for all of the spheroidal modes of the same ray parameter p, because the fractional kinetic energies f_ν depend strongly upon the mode type.

★12.4.3 Fréchet kernels

The asymptotic Fréchet kernels of either a toroidal or spheroidal mode may be obtained in an analogous manner by rewriting the constructive-interference relation (12.172) in the form

$$f(\omega, k/\omega; \alpha, \beta, d) = 0, \tag{12.180}$$

where the arguments α, β and d express the dependence upon the compressional and shear wave speeds and the radii of the discontinuities c, b and a. Upon taking the first variation of this relation, holding the wavenumber k fixed, we obtain

$$\delta\omega = -\left[\left(\frac{\partial f}{\partial \alpha}\right)\delta\alpha + \left(\frac{\partial f}{\partial \beta}\right)\delta\beta + \left(\frac{\partial f}{\partial d}\right)\delta d\right] \times \left(\frac{\partial f}{\partial \omega} - \frac{p}{\omega}\frac{\partial f}{\partial p}\right)^{-1}. \tag{12.181}$$

Equation (12.181) gives the first-order perturbation $\delta\omega$ in the eigenfrequency of a mode ${}_n\mathrm{T}_l$ or ${}_n\mathrm{S}_l$ due to a slight perturbation $\delta\alpha$, $\delta\beta$, δd in the properties of the Earth model. Since the compressional and shear wave speeds appear in the integrands of the phase integrals Ψ_ν in equation (12.180), the variational expressions $(\partial f/\partial\alpha)\,\delta\alpha$ and $(\partial f/\partial\beta)\,\delta\beta$ are purely symbolic.

To illustrate the use of (12.181), we consider first an SH-equivalent toroidal mode, which has $f = \sin(\omega\int_{R_{\mathrm{S}}}^{a} q_\beta\, dr - \pi/4)$. Upon taking the indicated variations with respect to the shear-wave speed β and radius a, we find that

$$\delta\omega = 2\omega T_{\mathrm{S}}^{-1}\int_{R_{\mathrm{S}}}^{a} \beta^{-3} q_\beta^{-1}\delta\beta\, dr - 2\omega T_{\mathrm{S}}^{-1} q_\beta(a)\,\delta a. \tag{12.182}$$

We can rewrite this using the notation introduced in Chapter 9 in the form

$$\delta\omega = \int_0^a (\delta\alpha\, K_\alpha + \delta\beta\, K_\beta)\, dr + \sum_d \delta d\, [K_d]_-^+, \tag{12.183}$$

where $K_\alpha = 0$ and

$$K_\beta = \begin{cases} 2\omega T_{\mathrm{S}}^{-1}\beta^{-3}q_\beta^{-1}, & R_{\mathrm{S}} \ll r \le a \\ 0, & 0 \le r \ll R_{\mathrm{S}}, \end{cases} \tag{12.184}$$

$$[K_a]_-^+ = -2\omega T_{\mathrm{S}}^{-1}q_\beta(a), \qquad [K_b]_-^+ = [K_c]_-^+ = 0. \tag{12.185}$$

For an $\mathrm{ScS_{SH}}$-equivalent mode the non-zero Fréchet kernels are

$$K_\beta = \begin{cases} 2\omega T_{\mathrm{ScS}}^{-1}\beta^{-3}q_\beta^{-1}, & b{+} \le r \le a \\ 0, & 0 \le r \le b{-}, \end{cases} \tag{12.186}$$

$$[K_a]_-^+ = -2\omega T_{\mathrm{ScS}}^{-1}q_\beta(a), \qquad [K_b]_-^+ = 2\omega T_{\mathrm{ScS}}^{-1}q_\beta(b{+}). \tag{12.187}$$

The asymptotic shear-speed kernel K_β of an SH-equivalent oscillation diverges like q_β^{-1} above the turning radius $r = R_{\mathrm{S}}$, and it is identically zero beneath this radius. An $\mathrm{ScS_{SH}}$-equivalent mode is sensitive to the radius $r = b$ of the core-mantle boundary as well to the radius $r = a$ of the free surface.

The non-zero Fréchet kernels governing a mantle spheroidal mode in Regime III are

$$K_\alpha = \begin{cases} 2\omega f_{\mathrm{P}}T_{\mathrm{P}}^{-1}\alpha^{-3}q_\alpha^{-1}, & R_{\mathrm{P}} \ll r \le a \\ 0, & 0 \le r \ll R_{\mathrm{P}}, \end{cases} \tag{12.188}$$

$$K_\beta = \begin{cases} 2\omega f_{\mathrm{S}}T_{\mathrm{S}}^{-1}\beta^{-3}q_\beta^{-1}, & R_{\mathrm{S}} \ll r \le a \\ 0, & 0 \le r \ll R_{\mathrm{S}}, \end{cases} \tag{12.189}$$

$$[K_a]_-^+ = -2\omega[f_{\mathrm{P}}T_{\mathrm{P}}^{-1}q_\alpha(a) + f_{\mathrm{S}}T_{\mathrm{S}}^{-1}q_\beta(a)], \tag{12.190}$$

whereas those governing a whole-Earth mode in Regime IX are

$$K_\alpha = \begin{cases} 2\omega f_{\mathrm{PcP}}T_{\mathrm{PcP}}^{-1}\alpha^{-3}q_\alpha^{-1}, & b{+} \le r \le a \\ 2\omega f_{\mathrm{KiK}}T_{\mathrm{KiK}}^{-1}\alpha^{-3}q_\alpha^{-1}, & c{+} \le r \le b{-} \\ 2\omega f_{\mathrm{I}}T_{\mathrm{I}}^{-1}\alpha^{-3}q_\alpha^{-1}, & R_{\mathrm{I}} \ll r \le c{-} \\ 0, & 0 \le r \ll R_{\mathrm{I}}, \end{cases} \tag{12.191}$$

$$K_\beta = \begin{cases} 2\omega f_{\mathrm{ScS}}T_{\mathrm{ScS}}^{-1}\beta^{-3}q_\beta^{-1}, & b{+} \le r \le a \\ 0, & c{+} \le r \le b{-} \\ 2\omega f_{\mathrm{J}}T_{\mathrm{J}}^{-1}\beta^{-3}q_\beta^{-1}, & R_{\mathrm{J}} \ll r \le c{-} \\ 0, & 0 \le r \ll R_{\mathrm{J}}, \end{cases} \tag{12.192}$$

$$[K_a]_-^+ = -2\omega[f_{\mathrm{PcP}}T_{\mathrm{PcP}}^{-1}q_\alpha(a) + f_{\mathrm{ScS}}T_{\mathrm{ScS}}^{-1}q_\beta(a)], \tag{12.193}$$

$$[K_b]_-^+ = 2\omega[f_{\mathrm{PcP}}T_{\mathrm{PcP}}^{-1}q_\alpha(b+) + f_{\mathrm{ScS}}T_{\mathrm{ScS}}^{-1}q_\beta(b+) - f_{\mathrm{KiK}}T_{\mathrm{KiK}}^{-1}q_\alpha(b-)], \tag{12.194}$$

$$[K_c]_-^+ = 2\omega[f_{\mathrm{KiK}}T_{\mathrm{KiK}}^{-1}q_\alpha(c+) - f_{\mathrm{I}}T_{\mathrm{I}}^{-1}q_\beta(c-) - f_{\mathrm{J}}T_{\mathrm{J}}^{-1}q_\alpha(c-)]. \tag{12.195}$$

As usual the results (12.191)–(12.195) can be further simplified in the case $\grave{P}\acute{P} \approx \grave{K}\acute{K} \approx 0$. The asymptotic Fréchet kernels of a PKIKP mode reduce, for example, to $K_\beta \approx 0$ and

$$K_\alpha \approx \begin{cases} 2\omega T_{\mathrm{PKIKP}}^{-1}\alpha^{-3}q_\alpha^{-1}, & R_{\mathrm{I}} \ll r \leq a \\ 0, & 0 \leq r \ll R_{\mathrm{I}}, \end{cases} \tag{12.196}$$

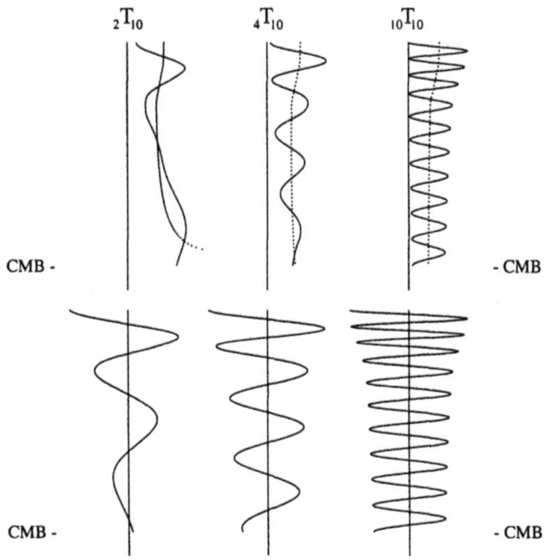

Figure 12.21. Exact (*solid line*) and asymptotic (*dotted line*) Fréchet kernels for selected $l = 10$ toroidal oscillations. (*Top row*) Shear-wave speed kernel K_β. (*Bottom row*) Density kernel K'_ρ. Vertical axis is depth beneath the free surface of crustless model 1066A; the location of the core-mantle boundary (CMB) is shown. Mode ${}_2T_{10}$ is SH-equivalent with a turning point at $r = R_{\mathrm{S}}$ in the lower mantle, whereas ${}_4T_{10}$ and ${}_{10}T_{10}$ are $\mathrm{ScS_{SH}}$-equivalent, with no turning point. The asymptotic shear-speed kernels are all strictly positive above $r = R_{\mathrm{S}}$, whereas the asymptotic density kernels are identically zero everywhere. (Courtesy of L. Zhao.)

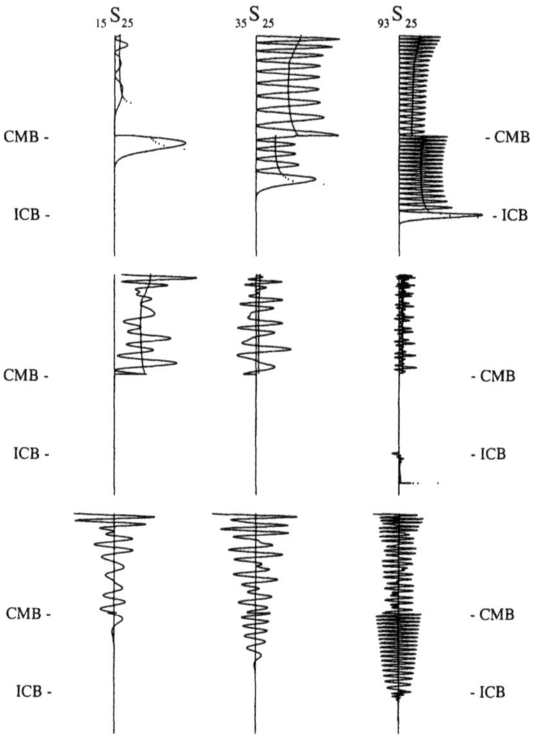

Figure 12.22. Exact (*solid line*) and asymptotic (*dotted line*) Fréchet kernels for selected $l = 25$ spheroidal oscillations. (*Top row*) Compressional-wave speed kernel K_α. (*Middle row*) Shear-wave speed kernel K_β. (*Bottom row*) Density kernel K'_ρ. Vertical axis extends from the center of crustless model 1066A to the free surface; the locations of the core-mantle boundary (CMB) and inner-core boundary (ICB) are shown. Mode ${}_{15}S_{25}$ is $\mathrm{ScS_{SV}}$-SKS-equivalent, mode ${}_{35}S_{25}$ is PKP-equivalent, and ${}_{93}S_{25}$ is PKIKP-equivalent. The asymptotic wave-speed kernels are all strictly positive above the turning point $r = R$; the asymptotic density kernels are identically zero everywhere. (Courtesy of L. Zhao.)

$$K_d \approx 2\omega T_{\mathrm{PKIKP}}^{-1} q_\alpha(d). \tag{12.197}$$

A radial mode has $K_\alpha \approx 2\omega T_{\mathrm{PKIKP}}^{-1}\alpha^{-2}$ and $K_d \approx 2\omega T_{\mathrm{PKIKP}}^{-1}\alpha_{\mathrm{d}}^{-1}$ in this weak-impedance approximation.

It is noteworthy that the asymptotic eigenfrequencies do not depend upon the density distribution ρ of the Earth model; the density kernel K'_ρ is identically zero for all modes. An asymptotic dependence upon the density does not arise until the next order of the expansion in the reciprocal frequency ω^{-1}. Interestingly, it is the *gradient* of the density $\dot{\rho}$ rather than the density itself that then appears; this is in accordance with our discussion regarding the density dependence in Section 9.3. In general, there is an excellent agreement between the asymptotic Fréchet kernels K_α, K_β and K'_ρ and the "running means" of the corresponding exact kernels, as illustrated for a number of toroidal modes in Figure 12.21 and a number of spheroidal modes in Figure 12.22. An alternative derivation, based upon substituting the JWKB asymptotic eigenfunctions into the exact expressions for the Fréchet kernels, makes this explicit, inasmuch as the squares of the oscillatory terms are replaced by $\frac{1}{2}$ when the integrals are evaluated by means of the Riemann-Lebesgue lemma. The asymptotic results are based upon the premise that the radial structure of the Earth is smooth enough between discontinuities to permit this averaging; this resolves the apparent contradiction between the degree of resolution provided by the two sets of kernels. To lowest order in ω^{-1}, the constraints imposed upon the Earth's internal structure by the high-frequency normal modes and body-wave travel times are essentially identical.

*12.4.4 Compressional and shear energy

The fractional compressional and shear energies of a spheroidal mode are defined by

$$f_\kappa = 2\omega^{-1}\int_0^a \kappa K_\kappa\, dr, \qquad f_\mu = 2\omega^{-1}\int_0^a \mu K_\mu\, dr, \tag{12.198}$$

where

$$K_\kappa = (2\rho\alpha)^{-1}K_\alpha, \qquad K_\mu = (2\rho\beta)^{-1}K_\beta + \tfrac{4}{3}(2\rho\alpha)^{-1}K_\alpha. \tag{12.199}$$

Upon inserting the asymptotic forms for K_α and K_β, we find that a mantle mode in Regime III has

$$f_\kappa \approx \tfrac{5}{9}f_{\mathrm{P}}, \qquad f_\mu \approx \tfrac{4}{9}f_{\mathrm{P}} + f_{\mathrm{S}}, \tag{12.200}$$

whereas a whole-Earth mode in Regime IX has

$$f_\kappa \approx \tfrac{5}{9}f_{\mathrm{PcP}} + f_{\mathrm{KiK}} + \tfrac{5}{9}f_{\mathrm{I}}, \tag{12.201}$$

$$f_\mu \approx \tfrac{4}{9}f_{\mathrm{PcP}} + f_{\mathrm{ScS}} + \tfrac{4}{9}f_{\mathrm{I}} + f_{\mathrm{J}}. \tag{12.202}$$

To simplify the results (12.200)–(12.202) we have made the Poisson approximation $\alpha^2 \approx 3\beta^2$ in the solid regions of the Earth model. It is noteworthy that the sum of the fractional compressional and shear energies is unity:

$$f_\kappa + f_\mu \approx 1, \tag{12.203}$$

as expected on a non-gravitating Earth model. An $\mathrm{ScS_{SV}}$ or $\mathrm{J_{SV}}$ mode has $f_\kappa \approx 0$ and $f_\mu \approx 1$, whereas a PKIKP or radial mode has

$$f_\kappa \approx T^{-1}_{\mathrm{PKIKP}}(\tfrac{5}{9}T_{\mathrm{PcP}} + T_{\mathrm{KiK}} + \tfrac{5}{9}T_{\mathrm{I}}), \tag{12.204}$$

$$f_\mu \approx T^{-1}_{\mathrm{PKIKP}}(\tfrac{4}{9}T_{\mathrm{PcP}} + \tfrac{4}{9}T_{\mathrm{I}}) \tag{12.205}$$

in the case $\grave{P}\acute{P} \approx \grave{K}\acute{K} \approx 0$. The ratio $f_\kappa : f_\mu = 5/9 : 4/9$ is the classical energy partition of a plane P wave in a homogeneous Poisson medium.

★12.4.5 Transversely isotropic Earth model

All of the arguments that we have used to obtain the asymptotic eigenfrequencies and eigenfunctions of a SNREI Earth can be extended to the case of a transversely isotropic Earth model in a relatively straightforward fashion. The first step is to calculate the travel time T, angular epicentral distance Θ and intercept time τ of a propagating P, SV or SH body wave; Woodhouse (1981b) shows that the kinematic prescriptions (12.7) and (12.9) must be replaced by generalized formulae of the form

$$T = 2\int_R^a q_T\, dr, \qquad \Theta = 2\int_R^a q_\Theta\, dr, \qquad \tau = 2\int_R^a q_\tau\, dr, \tag{12.206}$$

where the limits of integration need to be altered in the usual way to account for reflected rather than turning waves, multiple legs, and non-surficial endpoints. The integrands q_T, q_Θ and q_τ for an SH wave are

$$q_T = \frac{(\beta_{\mathrm{h}}\beta_{\mathrm{v}})^{-1}}{\sqrt{\beta_{\mathrm{h}}^{-2} - p^2r^{-2}}}, \qquad q_\Theta = \frac{(\beta_{\mathrm{h}}/\beta_{\mathrm{v}})pr^{-2}}{\sqrt{\beta_{\mathrm{h}}^{-2} - p^2r^{-2}}},$$

$$q_\tau = q_T - pq_\Theta = (\beta_{\mathrm{h}}/\beta_{\mathrm{v}})\sqrt{\beta_{\mathrm{h}}^{-2} - p^2r^{-2}}, \tag{12.207}$$

where β_{h} and β_{v} are the vertical and horizontal propagation speeds, respectively. The turning radius of an SH wave of ray parameter p is $R_{\mathrm{S}} = p\beta_{\mathrm{h}}$; the appearance of β_{h} rather than β_{v} in the square roots in (12.207) is natural, inasmuch as a wave is propagating horizontally upon turning. It is evident that the results (12.206)–(12.207) reduce to (12.7) and (12.9) in an isotropic Earth, which has $\beta_{\mathrm{h}} = \beta_{\mathrm{v}} = \beta$. More complicated formulae for q_T, q_Θ and q_τ pertain in the case of P and SV waves; the important relationships $p = dT/d\Theta$ and $d\tau/dp = -\Theta$ remain valid for all waves in a transversely isotropic Earth.

The formal asymptotic analysis of the toroidal equations and boundary conditions on a transversely isotropic Earth has been conducted by

Mochizuki (1992). He shows that the asymptotic eigenfrequencies in Table 12.1 remain valid provided that τ_{S}, τ_{ScS} and τ_{J} are all calculated using (12.206)–(12.207) rather than (12.9); the ray-parameter boundaries separating the four SH regimes are determined by the horizontal propagation speed β_{h}. To generalize the asymptotic eigenfunctions in Table 12.4, it is necessary to replace q_β by q_τ and the rigidity μ by the transversely isotropic parameter L in addition to recalculating the travel times T_{S}, T_{ScS} and T_{J}. The more complicated P-SV case has been considered by Mochizuki (1994); he ignores the presence of the inner and outer core, and obtains results only for the mantle spheroidal modes in Regime III. The asymptotic eigenfrequency relation for these modes is of the form (12.55), with Ψ_{P} and Ψ_{S} calculated using (12.206)–(12.207) and with $\acute{S}\grave{S}$ replaced by a generalized reflection coefficient depending upon the five transversely isotropic elastic parameters C, A, L, N and F. Analogous results for the modes in any other regime could be obtained using the methods outlined above, if desired.

12.5 Body-Wave Response

Thus far we have considered only the first aspect of the duality between modes and rays, namely, the problem of expressing the free oscillations of the Earth as a superposition of constructively interfering body waves. In this final section we take up the second aspect, namely, the inverse problem of going "from modes to rays". We shall show that the representation of the response of a SNREI Earth as a superposition of toroidal and spheroidal normal modes is asymptotically equivalent, in the limit $\omega \to \infty$, to a superposition of multiply reflected and refracted SH and P-SV body waves. Our procedure for effecting this mode-sum to ray-sum transformation is elementary; we treat the simpler SH case in full detail, and merely quote the final result for P-SV waves. In Chapter 15 we generalize the ray-theoretical response obtained here to a laterally heterogeneous Earth, by means of a systematic application of the JWKB approximation to the non-gravitating equation of motion $-\omega^2\rho\mathbf{s} - \boldsymbol{\nabla}\cdot\mathbf{T} = \mathbf{0}$, where $\mathbf{T} = \kappa(\boldsymbol{\nabla}\cdot\mathbf{s})\mathbf{I} + 2\mu\mathbf{d}$.

12.5.1 SH Green Tensor

The starting point of the present analysis is the travelling-wave representation of the frequency-domain Green tensor, equation (11.14). We ignore attenuation and set the decay rate $\gamma_n(k)$ along every dispersion branch equal to zero; in addition, we focus attention on the $s = 1$ waves that propagate less than halfway around the Earth, for the sake of brevity. With these provisos, we can rewrite the *exact* SH Green tensor $\mathbf{G}_{\mathrm{SH}}(\mathbf{x}, \mathbf{x}'; \omega)$ in the form

$$\mathbf{G} = \frac{1}{2\pi}(\hat{\mathbf{r}} \times \boldsymbol{\nabla}_1)(\hat{\mathbf{r}}' \times \boldsymbol{\nabla}_1') \sum_{n=0}^{\infty} \int_{-\infty}^{\infty} \left(\frac{W_n W_n'}{\omega_n^2 - \omega^2} \right) \times Q_{k-\frac{1}{2}}^{(1)}(\cos\Theta)\, k^{-1} dk, \qquad (12.208)$$

where $\Theta = \arccos(\hat{\mathbf{r}} \cdot \hat{\mathbf{r}}')$ is the angular epicentral distance, and a prime denotes evaluation at the source point $\mathbf{x}'$. To facilitate the conversion to a body-wave representation, we change the variable of integration from the wavenumber k to the ray parameter p; the relationship between these two quantities *at fixed overtone number* n is $k^{-1}dk = (1 - C/c)^{-1}p^{-1}dp$, where $C = d\omega/dk$ and $c = \omega/k$ are the group and phase speeds. After implementing this coordinate transformation, we interchange the order of summation and integration to obtain

$$\mathbf{G} = \frac{1}{2\pi}(\hat{\mathbf{r}} \times \boldsymbol{\nabla}_1)(\hat{\mathbf{r}}' \times \boldsymbol{\nabla}_1') \int_{-\infty}^{\infty} \sum_{n=0}^{\infty} \left(\frac{W_n W_n'}{\omega_n^2 - \omega^2} \right) \times Q_{\omega p-\frac{1}{2}}^{(1)}(\cos\Theta)\,(1 - C/c)^{-1}\, p^{-1} dp. \qquad (12.209)$$

The path of integration in (12.209) runs just beneath the real p axis. We assume that the angular frequency is positive, $\omega > 0$, for concreteness; positive real ray parameters, $\operatorname{Re} p > 0$, and negative real ray parameters, $\operatorname{Re} p < 0$, then correspond to waves travelling in the directions of increasing and decreasing Θ, respectively.

The next step is to insert the JWKB asymptotic expressions for the eigenfrequencies ω_n and associated receiver and source eigenfunctions W_n and W_n' into equation (12.209), and to evaluate the sum over the overtone index n. That sum is now at *fixed ray parameter* p, so that all of the modes that must be considered are in the same asymptotic regime. The $\mathrm{J_{SH}}$ inner-core modes are not excited by a point source situated in the mantle; hence, we only need to consider two sums, one for the turning SH modes in Regime II and another for the $\mathrm{ScS_{SH}}$ modes in Regime III. In the $\mathrm{ScS_{SH}}$ case it is convenient to define the three radial slowness integrals X, X' and $\overline{X}$ by

$$\omega X = \omega \int_b^r q\, dr + \pi/2, \qquad \omega X' = \omega \int_b^{r'} q\, dr + \pi/2,$$

$$\omega \overline{X} = \omega \int_b^a q\, dr = \tfrac{1}{2}\omega \tau_{\mathrm{ScS}}. \qquad (12.210)$$

Upon making use of the results in Tables 12.1 and 12.4, we can write the overtone sum in Regime III in the form

$$\sum_{n=0}^{\infty}\left(\frac{W_nW_n'}{\omega_n^2-\omega^2}\right)=4T_{\mathrm{ScS}}^{-1}(rr'\beta\beta')^{-1}(\rho\rho' q_\beta q_\beta')^{-1/2}$$
$$\times\sum_{n'=1}^{\infty}\frac{\sin n'\pi(X/\overline{X})\sin n'\pi(X'/\overline{X})}{(n'\pi/\overline{X})^2-\omega^2}, \tag{12.211}$$

where $n'=n+1$. The sum over n' in equation (12.211) can be evaluated analytically, with the result

$$\sum_{n'=1}^{\infty}\frac{\sin n'\pi(X/\overline{X})\sin n'\pi(X'/\overline{X})}{(n'\pi/\overline{X})^2-\omega^2}$$
$$=\frac{\overline{X}}{4\omega}\left[\frac{\cos\omega(\overline{X}-(X+X'))-\cos\omega(\overline{X}-|X-X'|)}{\sin\omega\overline{X}}\right]. \tag{12.212}$$

This identity can be readily verified by considering the Fourier cosine series of the expression on the right, using orthogonality to evaluate the coefficients in the customary manner. Upon decomposing the sines and cosines in equation (12.212) into complex exponentials and using the binomial expansion, we can express the sum (12.211) over the $\mathrm{ScS_{SH}}$ modes in Regime III in the form

$$\sum_{n=0}^{\infty}\left(\frac{W_nW_n'}{\omega_n^2-\omega^2}\right)=\tfrac{1}{2}(i\omega)^{-1}(\tau_{\mathrm{ScS}}/T_{\mathrm{ScS}})(rr'\beta\beta')^{-1}$$
$$\times(\rho\rho' q_\beta q_\beta')^{-1/2}\sum_{j=1}^{\infty}\exp(-i\omega\tau_j), \tag{12.213}$$

where

$$\tau_1=\left|\int_{r'}^{r}q\,dr\right|,\qquad \tau_2=\int_b^{r'}q\,dr+\int_b^{r}q\,dr,$$
$$\tau_3=\int_{r'}^{a}q\,dr+\int_{r}^{a}q\,dr,\qquad \tau_4=2\int_b^{a}q\,dr-\left|\int_{r'}^{r}q\,dr\right|,$$
$$\tau_j=\tau_{j-4}+2\int_b^{a}q\,dr\quad\text{for } j\geq 5. \tag{12.214}$$

The sum over the turning SH modes in Regime II can be evaluated in a similar manner by redefining the radial slowness integrals in equation (12.210):

$$\omega X=\omega\int_b^{r}q\,dr+\pi/4,\qquad \omega X'=\omega\int_b^{r'}q\,dr+\pi/4,$$

$$\omega\overline{X} = \omega \int_{R_\mathrm{S}}^{a} q\,dr - \pi/4 = \tfrac{1}{2}(\omega\tau_\mathrm{S} - \pi/2). \tag{12.215}$$

We obtain instead of equation (12.213) the result

$$\sum_{n=0}^{\infty} \frac{W_n W_n'}{\omega_n^2 - \omega^2} = \tfrac{1}{2}(i\omega)^{-1}(\tau_\mathrm{S}/T_\mathrm{S})(rr'\beta\beta')^{-1}$$
$$\times\, (\rho\rho' q_\beta q_\beta')^{-1/2} \sum_{j=1}^{\infty} \exp(-i\omega\tau_j + iN_j\pi/2), \tag{12.216}$$

where

$$\tau_1 = \left| \int_{r'}^{r} q\,dr \right|, \qquad \tau_2 = \int_{R_\mathrm{S}}^{r'} q\,dr + \int_{R_\mathrm{S}}^{r} q\,dr,$$

$$\tau_3 = \int_{r'}^{a} q\,dr + \int_{r}^{a} q\,dr, \qquad \tau_4 = 2\int_{R_\mathrm{S}}^{a} q\,dr - \left| \int_{r'}^{r} q\,dr \right|,$$

$$\tau_j = \tau_{j-4} + 2\int_{R_\mathrm{S}}^{a} q\,dr \quad \text{for } j \geq 5 \tag{12.217}$$

and

$$N_1 = 0, \qquad N_2 = 1, \qquad N_3 = 0, \qquad N_4 = 1,$$

$$N_j = N_{j-4} + 1 \quad \text{for } j \geq 5. \tag{12.218}$$

We have ignored a term of relative order ω^{-1} in setting $\overline{X}/T_\mathrm{S} = \frac{1}{2}(\tau_\mathrm{S}/T_\mathrm{S})$ in equation (12.216). This is permissible since the JWKB eigenfrequencies and eigenfunctions are high-frequency approximations in the first place.

Equations (12.213) and (12.216) are identical, except for the presence of the terms $\exp(iN_j\pi/2)$; we can combine the two results into a single formula that is valid in both Regimes II and III by dropping the subscripts ScS and S on τ and T, and specifying that the integers N_j are identically zero in the $\mathrm{ScS_{SH}}$ case. The quantities τ_j defined in equations (12.214) and (12.217) may be identified as the intercept times of the various waves that can propagate between the source and receiver radii r' and r, whereas the quantities N_j are the associated number of times that each wave turns. Figure 12.23 is a schematic illustration of the first six ray paths s, S, ss, Ss, sS and SS in Regime II; the source has been presumed to lie below the receiver, $r' < r$, since this is the usual situation in observational seismology. The corresponding "capitalized" paths ScS, ScSs, sScS and $\mathrm{ScS_2}$ in Regime III reflect off of the core-mantle boundary $r = b$ rather than turn,

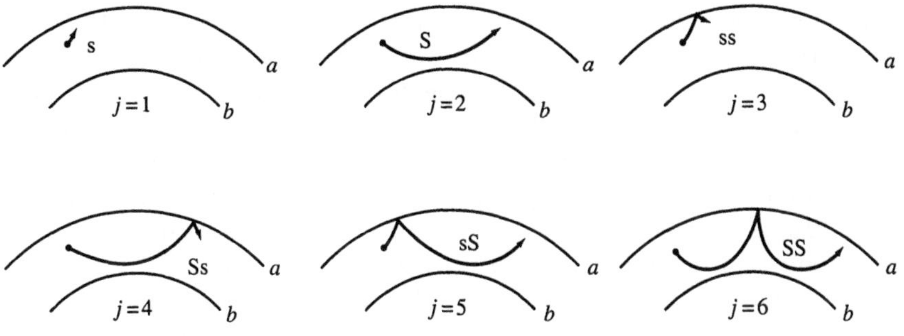

Figure 12.23. Direct, turning and surface-reflected ray paths s, S, ss, Ss, sS and SS, corresponding to $j = 1-6$, in the Regime II rainbow expansion (12.216). The source (*dot*) is assumed to lie below the receiver ($r' < r$). We use s and ss to denote the direct and surface-reflected waves that never turn, and Ss and sS to denote the surface-reflected waves that turn on their first and second legs, respectively; this represents a slight (but logical) extension to the classical ray nomenclature summarized in Section 12.1.1. Compare with Figure 12.24.

as illustrated in Figure 12.24. The $\pi/2$ discontinuity in the turning index N_j at the regime boundary $p = b/\beta_{b+}$ is a consequence of the $\pi/2$ caustic phase shift of a turning wave of fixed ray parameter p. The remaining quantities in equations (12.213) and (12.216), notably the ratio τ/T and the

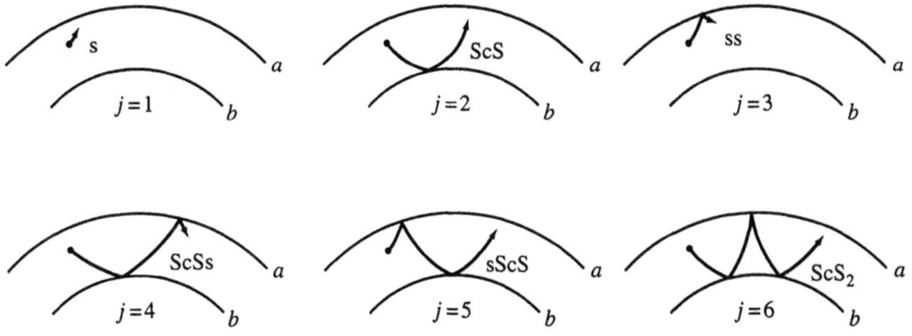

Figure 12.24. Direct, surface-reflected and CMB-reflected ray paths s, ScS, ss, ScSs, sScS and ScS_2, corresponding to $j = 1-6$, in the Regime III expansion (12.213). We use ScSs and sScS to denote the waves that reflect first off the CMB and then off the free surface, or vice versa; this is another slight extension to the classical ray nomenclature summarized in Section 12.1.1. Compare with Figure 12.23.

source-to-receiver intercept times τ_j, are all continuous across the SH–ScS$_{\rm SH}$ boundary. The infinite sum of exponentials $\sum_{j=0}^{\infty} \exp(-i\omega\tau_j + iN_j\pi/2)$ converges for $\omega > 0$ on the lower-half-plane integration contour $\operatorname{Im} p < 0$ since $\operatorname{Im}\tau_j \approx (\operatorname{Im} p)\Theta_j$, where Θ_j is the source-to-receiver epicentral distance associated with the intercept time τ_j, in the vicinity of the real ray-parameter axis. Any overtone-sum to ray-sum transformation of the form (12.213) or (12.216) is referred to as a *Debye* or *rainbow expansion*, after an analogous result that arises in the theory of the rainbow.

Upon substituting the ScS$_{\rm SH}$ and SH rainbow expansions in the Green tensor (12.209), we find that the factor $(1 - C/c)^{-1}$ is cancelled by the factor τ/T. The travelling-wave Legendre function can be approximated, in the limit $\omega \to \infty$, by the asymptotic expansion

$$Q^{(1)}_{\omega p - \frac{1}{2}}(\cos\Theta) \approx (2\pi\omega p \sin\Theta)^{-1/2} \exp(-i\omega p\Theta + i\pi/4). \tag{12.219}$$

To lowest order in ω^{-1}, the operators $\hat{\mathbf{r}} \times \boldsymbol{\nabla}_1$ and $\hat{\mathbf{r}}' \times \boldsymbol{\nabla}_1'$ act only upon the exponential $\exp(-i\omega p\Theta)$, giving the result $(\hat{\mathbf{r}} \times \boldsymbol{\nabla}_1)(\hat{\mathbf{r}}' \times \boldsymbol{\nabla}_1') \to \omega^2 p^2 \hat{\boldsymbol{\Phi}}\hat{\boldsymbol{\Phi}}'$, where $\hat{\boldsymbol{\Phi}} = \hat{\boldsymbol{\Phi}}'$ is the unit perpendicular to the ray plane. The upshot is that the high-frequency SH Green tensor can be written in the form

$$\begin{aligned} \mathbf{G} = {} & \frac{1}{4\pi}\hat{\boldsymbol{\Phi}}\hat{\boldsymbol{\Phi}}'(rr'\beta\beta')^{-1}(2\pi\rho\rho'\sin\Theta)^{-1/2} \\ & \times \int_{-\infty}^{\infty} (q_\beta q_\beta')^{-1/2} \sum_{j=1}^{\infty} \exp[-i\omega(\tau_j + p\Theta)] \\ & \qquad \times (\omega p)^{1/2} \exp[i(N_j\pi/2 - \pi/4)]\, dp. \end{aligned} \tag{12.220}$$

The lower limit of integration $-\infty$ in equation (12.220) is purely symbolic at this point, since the approximation (12.219) is only valid for $\operatorname{Re} p > 0$; we have employed this approximation in anticipation of the fact that the dominant contribution to $\mathbf{G}$ in the limit $\omega \to \infty$ comes from one or more *saddle points* situated on the real positive ray-parameter axis.

The locations p_k of these saddle points are determined by the *stationary phase condition*

$$\frac{d}{dp}(\tau_j + p\Theta)_{p=p_k} = 0. \tag{12.221}$$

Since $d\tau_j/dp = -\Theta_j$, equation (12.221) reduces to $\Theta_j(p_k) = \Theta$, so that there are saddle points at just the ray parameters corresponding to the SH body waves s, ScS, S, ss, ScSs, Ss, sScS, sS, ScS$_2$, SS,... between the source and receiver. The configuration of the first few saddle points is depicted schematically in Figure 12.25. In every case the orientation, which is determined by the sign of $(d^2\tau_j/dp^2)_{p=p_k}$, is favorable so that the original

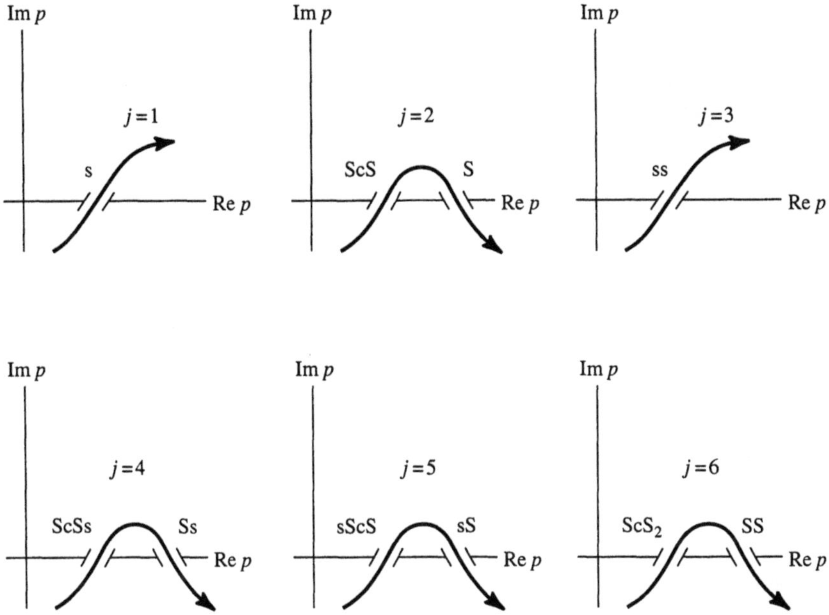

Figure 12.25. Schematic depiction of the complex ray-parameter plane, showing the location and orientation of the saddle points p_k for the first six terms in the sum (12.220). The order (*top left to bottom right*) of the diagrams $j = 1-6$ is the same as in Figures 12.23 and 12.24. In each case the contour of integration, which initially runs from $-\infty - i0$ to $\infty - i0$, can be deformed over the saddles as shown.

integration contour may be deformed to run over the saddles as shown. Upper-mantle triplications give rise to additional saddles, as illustrated in Figure 12.26. Upon evaluating each term in the representation (12.220) using the classical saddle-point approximation (Lighthill 1978), we obtain

$$\mathbf{G} = \frac{1}{4\pi}\hat{\mathbf{\Phi}}\hat{\mathbf{\Phi}}'(rr'\beta\beta')^{-1}(\rho\rho'\sin\Theta)^{-1/2}$$
$$\times \sum_j \sum_k \left[p^{1/2}(q_\beta q'_\beta)^{-1/2}|d^2\tau_j/dp^2|^{-1/2}\right]_{p=p_k}$$
$$\times \exp(-i\omega T_{jk} + iM_{jk}\pi/2), \qquad (12.222)$$

where $T_{jk} = \tau_j(p_k) + p_k\Theta$ and $M_{jk} = N_j - \frac{1}{2} - \frac{1}{2}\mathrm{sgn}(d^2\tau_j/dp^2)_{p=p_k}$. Physically, the double sum in equation (12.222) is over all of the geometrical SH ray paths between the source $\mathbf{x}'$ and receiver $\mathbf{x}$. Dropping the indices

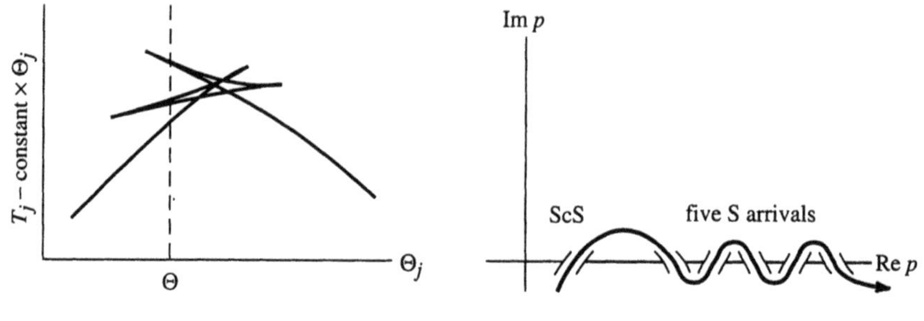

Figure 12.26. (*Left*) Reduced travel-time curve for turning SH waves with two upper-mantle triplications. A receiver at epicentral distance Θ has five geometrical SH arrivals. (*Right*) The resulting $j = 2$ integrand in equation (12.220) has five SH saddle points—in addition to the ScS saddle—oriented as shown.

j and k for simplicity and making use of the representation (12.29) of the geometrical spreading coefficient, we may rewrite the result (12.222) in the final compact form

$$\mathbf{G} = \frac{1}{4\pi}\hat{\mathbf{\Phi}}\hat{\mathbf{\Phi}}'(\rho\rho'\beta\beta'^3)^{-1/2}\sum_{\text{rays}}\mathcal{R}^{-1}\exp(-i\omega T + iM\pi/2). \qquad (12.223)$$

Equation (12.223) is precisely the classical JWKB representation of the SH body-wave Green tensor (Červený, Molotkov & Pšenčík 1977; Červený 1985). The quantity T is the travel time along a given ray, and the Maslov index M is the number of times that a wave propagating along this ray has passed through a caustic. Thus, $M = 0$ for an ScS_{SH} wave or an SH wave that turns in the lower mantle, whereas $M = 1$ for an SS_{SH} wave or a wave on the back branch of an upper-mantle SH triplication. It is noteworthy that the non-geometrical $\pi/2$ phase advance of an SH wave upon passage through a caustic arises naturally in the transformation from a mode-sum representation to a ray-sum representation of $\mathbf{G}_{\text{SH}}(\mathbf{x}, \mathbf{x}'; \omega)$.

Thus far we have restricted attention to the $s = 1$ waves whose angular propagation distance Δ is less than 180°; however, it may readily be verified that the result (12.223) pertains to the $s = 2, 3, 4, \ldots$ waves in equation (11.14) as well. The quantity $\sin\Theta$ in the geometrical spreading factor (12.29) can be replaced by $|\sin\Delta|$, if desired. The Maslov index M for these multi-orbit waves keeps track of "polar" phase shifts produced by any angular passages through the epicenter or its antipode as well. The phase shift of an SS_{SH} wave is associated with a radial "inversion" in the area of the ray tube, due to a change in sign $d^2\tau_j/dp^2 \rightarrow -d^2\tau_j/dp^2$, whereas an epicentral or antipodal phase shift is associated with a geographical "inversion", due to a change in sign $\sin\Delta \rightarrow -\sin\Delta$, just as for surface waves.

12.5.2 P-SV Green tensor

The P-SV Green tensor analogous to equation (12.209) is a sum of four terms $\mathbf{G}_1(\mathbf{x},\mathbf{x}';\omega) + \mathbf{G}_2(\mathbf{x},\mathbf{x}';\omega) + \mathbf{G}_3(\mathbf{x},\mathbf{x}';\omega) + \mathbf{G}_4(\mathbf{x},\mathbf{x}';\omega)$, given by

$$\mathbf{G}_1 = \frac{\omega^2}{2\pi}\hat{\mathbf{r}}\hat{\mathbf{r}}'\int_{-\infty}^{\infty}\sum_{n=0}^{\infty}\left(\frac{U_nU_n'}{\omega_n^2-\omega^2}\right)Q_{\omega p-\frac{1}{2}}^{(1)}(\cos\Theta)\,(1-C/c)^{-1}\,p\,dp,$$

$$\mathbf{G}_2 = \frac{\omega}{2\pi}\hat{\mathbf{r}}\boldsymbol{\nabla}_1'\int_{-\infty}^{\infty}\sum_{n=0}^{\infty}\left(\frac{U_nV_n'}{\omega_n^2-\omega^2}\right)Q_{\omega p-\frac{1}{2}}^{(1)}(\cos\Theta)\,(1-C/c)^{-1}\,dp,$$

$$\mathbf{G}_3 = \frac{\omega}{2\pi}\boldsymbol{\nabla}_1\hat{\mathbf{r}}'\int_{-\infty}^{\infty}\sum_{n=0}^{\infty}\left(\frac{V_nU_n'}{\omega_n^2-\omega^2}\right)Q_{\omega p-\frac{1}{2}}^{(1)}(\cos\Theta)\,(1-C/c)^{-1}\,dp,$$

$$\mathbf{G}_4 = \frac{1}{2\pi}\boldsymbol{\nabla}_1\boldsymbol{\nabla}_1'\int_{-\infty}^{\infty}\sum_{n=0}^{\infty}\left(\frac{V_nV_n'}{\omega_n^2-\omega^2}\right)Q_{\omega p-\frac{1}{2}}^{(1)}(\cos\Theta)\,(1-C/c)^{-1}p^{-1}dp.$$

(12.224)

To obtain a complete body-wave representation of $\mathbf{G}_{\text{P-SV}}(\mathbf{x},\mathbf{x}';\omega)$ we need to find the rainbow expansions of the overtone sums involving the source-receiver displacement products U_nU_n', U_nV_n', V_nU_n' and V_nV_n' in each of the asymptotic ray-parameter regimes. There is no elementary Fourier identity comparable to that used in the SH case that enables these overtone-sum to ray-sum transformations to be carried out in a straightforward manner; particularly in the higher regimes, the analytical expressions for the P-SV asymptotic eigenfrequencies and eigenfunctions are extremely complicated. Zhao & Dahlen (1996) show how the required sums over the compound mantle-turning waves may be obtained in Regime III; the corresponding results in other regimes may then be written down by analogy. The fundamental building block is the binomial expansion; repeated application yields all possible reverberations of the constituent P, SV, K, I and $\mathrm{J_{SV}}$ waves. Once all of the rainbow expansions have been found, the integrals over ray parameter p in $\mathbf{G}_1$ through $\mathbf{G}_4$ can be evaluated in the limit $\omega \rightarrow \infty$ by deforming the contour over the saddle points situated on the real positive axis as before. The final expression for the P-SV Green tensor obtained in this manner can be written in the form

$$\mathbf{G} = \frac{1}{4\pi}\sum_{\text{rays}}\hat{\boldsymbol{\eta}}\hat{\boldsymbol{\eta}}'(\rho\rho'vv'^3)^{-1/2}\Pi\mathcal{R}^{-1}\exp(-i\omega T + iM\pi/2), \qquad (12.225)$$

where the sum is over all the possible ray paths between the source $\mathbf{x}'$ and the receiver $\mathbf{x}$. The quantity T is the total travel time of the various waves along a ray, $\mathcal{R}$ is the associated geometrical spreading coefficient, and M is the Maslov index that keeps track of radial and geographic caustic phase shifts, as before. The speeds v' and v are those of the waves that depart from the source and arrive at the receiver along a given ray path, whereas $\hat{\boldsymbol{\eta}}'$ and $\hat{\boldsymbol{\eta}}$ are the corresponding polarizations, given by equation (12.11). The new factor Π, which does not appear in the SH expression (12.223), is the product of the $(\text{energy})^{1/2}$ reflection and transmission coefficients at the various interfaces encountered during propagation; for example, $\Pi = \grave{\mathcal{S}}\acute{\mathcal{P}}$ in the case of an ScP converted phase, whereas $\Pi = \grave{\mathcal{P}}\grave{\mathcal{K}}\,\grave{\mathcal{K}}\grave{\mathcal{I}}\,\acute{\mathcal{I}}\acute{\mathcal{K}}\,\acute{\mathcal{K}}\acute{\mathcal{P}}$ in the case of a throughgoing PKIKP wave.

We can regard (12.225) as a general formula for the JWKB body-wave Green tensor $\mathbf{G}_{\mathrm{SH}} + \mathbf{G}_{\text{P-SV}}$ by simply letting the summation index account for both SH and P-SV waves. The particle motion of an SH-polarized wave is always in the transverse direction, $\hat{\boldsymbol{\eta}}\hat{\boldsymbol{\eta}}' = \hat{\boldsymbol{\Phi}}\hat{\boldsymbol{\Phi}}'$, regardless of the direction of angular propagation, and the concatenated $(\text{energy})^{1/2}$ coefficient is always unity, $\Pi = 1$. The dependence of $\mathbf{G}(\mathbf{x},\mathbf{x}';\omega)$ upon the receiver location $\mathbf{x}$ along each ray path, for a fixed source location $\mathbf{x}'$, is in agreement with the geometrical amplitude-variation law (12.33), as expected. It is noteworthy that the JWKB Green tensor is consistent with the principle of dynamical source-receiver reciprocity,

$$\mathbf{G}(\mathbf{x},\mathbf{x}';\omega) = \mathbf{G}^{\mathrm{T}}(\mathbf{x}',\mathbf{x};\omega), \tag{12.226}$$

by virtue of the reflection-transmission, geometrical-spreading and Maslov-index symmetries (12.20), (12.30) and (12.34). Energy that propagates along a given ray path returns along the same path, with a reversal of the polarizations, $\hat{\boldsymbol{\eta}} \to -\hat{\boldsymbol{\eta}}'$ and $\hat{\boldsymbol{\eta}}' \to -\hat{\boldsymbol{\eta}}$, upon interchange of the source and receiver, $\mathbf{x} \to \mathbf{x}'$ and $\mathbf{x}' \to \mathbf{x}$. Any conversions upon reflection or transmission at interfaces occur in the opposite sense on the reversed ray; for example, a PcS wave is reciprocal to an ScP wave and vice versa.

*12.5.3 Hilbert transform glossary

To find the inverse Fourier transform of the JWKB Green tensor (12.225) and generalize this time-domain result so that it remains valid in the vicinity of caustics, we require some additional mathematical notation, which we develop in this section. All of the relations summarized here are either well known or easily demonstrated, so we do not present any proofs; for a more systematic account of these matters see Bracewell (1965).

The *Hilbert transform* of a real time-domain signal $f(t)$ is defined by

$$f_{\rm H}(t) = \mathcal{H}f(t) = \frac{1}{\pi} \fint_{-\infty}^{\infty} \frac{f(t')}{t'-t}\,dt'. \tag{12.227}$$

We may express $f(t)$ in terms of $f_{\rm H}(t)$ by means of the inverse Hilbert transformation

$$f(t) = \mathcal{H}^{-1}f_{\rm H}(t) = -\frac{1}{\pi} \fint_{-\infty}^{\infty} \frac{f_{\rm H}(t')}{t'-t}\,dt'. \tag{12.228}$$

The barred integrals in (12.227)–(12.228) denote the Cauchy principal values, defined by excision of the $(t'-t)^{-1}$ singularity, as in equation (6.52). The effect of M sequential Hilbert transforms will be denoted by

$$f_{\rm H}^{(M)}(t) = \underbrace{\mathcal{H}\cdots\mathcal{H}}_{M\ \text{times}}f(t). \tag{12.229}$$

It is noteworthy that $f_{\rm H}^{(0)}(t) = f(t)$, $f_{\rm H}^{(1)}(t) = f_{\rm H}(t)$ and $f_{\rm H}^{(2)}(t) = -f(t)$.

The complex *analytic signal* associated with $f(t)$ is defined by

$$F(t) = f(t) - if_{\rm H}(t). \tag{12.230}$$

The Hilbert transform of (12.230) is $F_{\rm H}(t) = f_{\rm H}(t) + if(t)$. Hilbert transformation advances the phase of every Fourier component of a signal by $\pi/2$. Thus, if $f(\omega)$ is the Fourier transform of $f(t)$ then $f(\omega)\exp(iM\pi/2)$ is the Fourier transform of $f_{\rm H}^{(M)}(t)$; that is,

$$f_{\rm H}^{(M)}(t) = \frac{1}{\pi}\,\mathrm{Re}\int_0^{\infty} f(\omega)\exp i(\omega t + M\pi/2)\,d\omega. \tag{12.231}$$

In particular, the multiply transformed Dirac delta function is

$$\delta_{\rm H}^{(M)}(t) = \frac{1}{\pi}\,\mathrm{Re}\int_0^{\infty} \exp i(\omega t + M\pi/2)\,d\omega. \tag{12.232}$$

The first few transforms in this case are

$$\delta_{\rm H}^{(0)}(t) = \delta(t), \qquad \delta_{\rm H}^{(1)}(t) = -(\pi t)^{-1}, \qquad \delta_{\rm H}^{(2)}(t) = -\delta(t). \tag{12.233}$$

The *analytic delta function* is $\Delta(t) = \delta(t) - i\delta_{\rm H}(t) = \delta(t) + i(\pi t)^{-1}$.

The *convolution* of two real time-domain signals $f(t)$ and $g(t)$ is, of course,

$$f(t) * g(t) = \int_{-\infty}^{\infty} f(t')g(t-t')\,dt'. \tag{12.234}$$

Convolution in the time domain is equivalent to multiplication in the frequency domain; that is, the Fourier transform of $f(t)*g(t)$ is $f(\omega)g(\omega)$. The

Hilbert transform (12.227) and its inverse (12.228) can be written formally as convolution products: $f_{\rm H}(t) = -f(t) * (\pi t)^{-1}$ and $f(t) = f_{\rm H}(t) * (\pi t)^{-1}$. The convolution product is commutative: $f(t) * g(t) = g(t) * f(t)$. Time differentiation (denoted by a dot) and Hilbert transformation can be shifted from one signal in a convolution to the other: $f(t) * \dot{g}(t) = \dot{f}(t) * g(t)$ and $f(t) * g_{\rm H}(t) = f_{\rm H}(t) * g(t)$. The convolution of two Hilbert transforms is $f_{\rm H}(t) * g_{\rm H}(t) = -f(t) * g(t)$. The Fourier transform of a time-lagged signal $f(t-T)$ is $f(\omega)\exp(-i\omega T)$. It is immaterial which signal in a convolution is lagged: $f(t) * g(t-T) = f(t-T) * g(t)$. If $f(t)$ and $g(t)$ have analytic signals $F(t)$ and $G(t)$, then $f(t) * \mathrm{Re}\,[G(t)] = \mathrm{Re}\,[F(t)] * g(t)$. The inverse Fourier transform of a complex frequency-independent constant $\mathcal{C} = \mathcal{A}+i\mathcal{B}$ is $\mathrm{Re}\,[\mathcal{C}\Delta(t)] = \mathcal{A}\delta(t) + \mathcal{B}\delta_{\rm H}(t)$. More generally, the inverse Fourier transform of a constant times $f(\omega)$ is $\mathrm{Re}\,[\mathcal{C}F(t)] = \mathcal{A}f(t) + \mathcal{B}f_{\rm H}(t)$.

Two Hilbert-transform pairs which prove to be useful in the present context are

$$v(t) = \frac{1}{\pi}\,\mathrm{Re}\int_0^\infty \omega^{-1/2}\exp i(\omega t - \pi/4)\,d\omega = \frac{H(t)}{\sqrt{\pi t}}, \tag{12.235}$$

$$v_{\rm H}(t) = \frac{1}{\pi}\,\mathrm{Re}\int_0^\infty \omega^{-1/2}\exp i(\omega t + \pi/4)\,d\omega = \frac{H(-t)}{\sqrt{-\pi t}} \tag{12.236}$$

and

$$\lambda(t) = \frac{1}{\pi}\,\mathrm{Re}\int_0^\infty \omega^{1/2}\exp i(\omega t - \pi/4)\,d\omega = -\frac{d}{dt}\frac{H(-t)}{\sqrt{-\pi t}}, \tag{12.237}$$

$$\lambda_{\rm H}(t) = \frac{1}{\pi}\,\mathrm{Re}\int_0^\infty \omega^{1/2}\exp i(\omega t + \pi/4)\,d\omega = \frac{d}{dt}\frac{H(t)}{\sqrt{\pi t}}, \tag{12.238}$$

where $H(t)$ is the Heaviside step function. It is evident that $\lambda(t) = -\dot{v}_{\rm H}(t)$ and $\lambda_{\rm H}(t) = \dot{v}(t)$. Convolution relations of interest include

$$v(t) * v(t-T) = -v_{\rm H}(t) * v_{\rm H}(t-T) = H(t-T), \tag{12.239}$$

$$\lambda(t) * v_{\rm H}(t-T) = -\lambda_{\rm H}(t) * v(t-T) = \delta(t-T), \tag{12.240}$$

$$\lambda(t) * v(t-T) = -\lambda_{\rm H}(t) * v_{\rm H}(t-T) = -\delta_{\rm H}(t-T), \tag{12.241}$$

$$\lambda(t) * \lambda(t-T) = -\lambda_{\rm H}(t) * \lambda_{\rm H}(t-T) = -\dot{\delta}(t-T). \tag{12.242}$$

The one-sided character of the signals (12.235)–(12.238) is noteworthy. The complex analytic signals $\Lambda(t) = \lambda(t) - i\lambda_{\rm H}(t)$ and $\Upsilon(t) = v(t) - iv_{\rm H}(t)$ are, in contrast, two-sided.

*12.5.4 Time-domain Green tensor

The ray-theoretical Green tensor $\mathbf{G}(\mathbf{x},\mathbf{x}';t)$ obtained by inverse Fourier transformation of equation (12.225) is

$$\mathbf{G} = \frac{1}{4\pi}\sum_{\text{rays}} \hat{\boldsymbol{\eta}}\hat{\boldsymbol{\eta}}'(\rho\rho' v v'^3)^{-1/2}\Pi\mathcal{R}^{-1}\delta_{\mathrm{H}}^{(M)}(t-T), \tag{12.243}$$

where $\delta_{\mathrm{H}}^{(M)}(t)$ is given by (12.232) and we have assumed that all of the concatenated reflection-and-transmission coefficients in the product Π are real. All $M=0$ body waves travel along least-time ray paths and exhibit a causal $\delta(t-T)$ time dependence, whereas $M=1$ waves travel along minimax ray paths and have a non-causal $\delta_{\mathrm{H}}(t-T) = -[\pi(t-T)]^{-1}$ time dependence. This Hilbert transformation accounts for the well-known emergent character of PP, SS and other multiply reflected and refracted phases that have passed through a caustic (Choy & Richards 1975).

Ray theory *per se* breaks down in the immediate vicinity of a caustic because of the divergence of the inverse geometrical spreading factor $\mathcal{R}^{-1}$. However, it is a simple matter to extend the result (12.243) so that it is regular everywhere, including near caustics. We restrict attention once again to the $s=1$ waves that propagate less than halfway around the Earth, for simplicity. Returning to the integral representation (12.220) of the SH Green tensor $\mathbf{G}_{\mathrm{SH}}(\mathbf{x},\mathbf{x}';\omega)$, we note that the only frequency dependence is that explicitly indicated: $\omega^{1/2}\exp[-i\omega(\tau_j + p\Theta)]$. Because of this it is possible to calculate the inverse Fourier transform $\mathbf{G}_{\mathrm{SH}}(\mathbf{x},\mathbf{x}';t)$ *exactly*. Designating the p-dependent argument in the exponential by

$$\Gamma_j(p) = \tau_j(p) + p\Theta = T_j(p) + p[\Theta - \Theta_j(p)], \tag{12.244}$$

we obtain an infinite sum of convolutions:

$$\mathbf{G} = \frac{1}{4\pi}\hat{\mathbf{\Phi}}\hat{\mathbf{\Phi}}'(rr'\beta\beta')^{-1}(2\rho\rho'\sin\Theta)^{-1/2} \times \sum_{j=1}^{\infty}\lambda_{\mathrm{H}}^{(N_j)}(t) * \sigma_j(t), \tag{12.245}$$

where

$$\lambda_{\mathrm{H}}^{(N_j)}(t) = \frac{1}{\pi}\,\mathrm{Re}\int_0^{\infty}\omega^{1/2}\exp i(\omega t + N_j\pi/2 - \pi/4)\,d\omega, \tag{12.246}$$

$$\sigma_j(t) = \int_{-\infty}^{\infty} p^{1/2}(\pi q_\beta q_\beta')^{-1/2}\,\delta(t-\Gamma_j)\,dp. \tag{12.247}$$

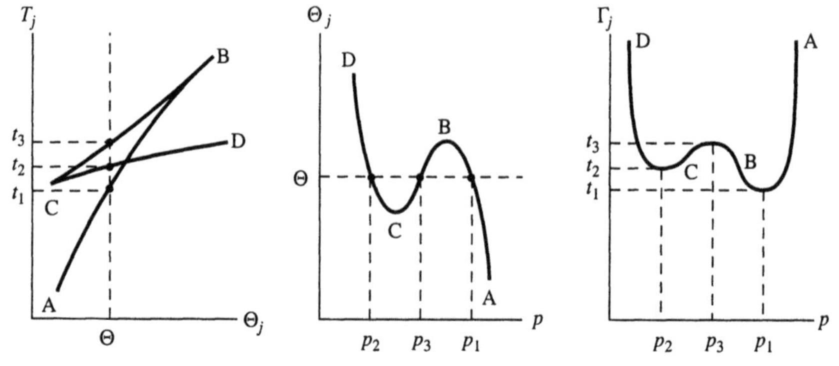

Figure 12.27. Schematic diagram depicting the evaluation of the function $\sigma_j(t)$. (*Left*) Triplicated travel-time curve $T_j(p)$ versus $\Theta_j(p)$, exhibiting caustics at points B and C. At a fixed epicentral distance Θ there are three geometrical arrivals at times t_1, t_2, t_3. (*Middle*) The associated ray parameters p_1, p_2, p_3 satisfy $\Theta_j(p_k) = \Theta$. (*Right*) For $t < t_1$ there are no intersections $t = \Gamma_j(p)$, for $t_1 < t < t_2$ there are two, and for $t_2 < t < t_3$ there are four. The caustics B and C are manifested as extrema in $\Theta_j(p)$ and as inflections in $\Gamma_j(p)$. The slope $d\Gamma_j/dp = \Theta - \Theta_j$ of the phase function vanishes at p_1, p_2, p_3 so that $\sigma_j(t)$ is singular at the ray-theoretical arrival times t_1, t_2, t_3. Note that if $t_1 < t_2 < t_3$ then $p_2 < p_3 < p_1$.

The first term in each convolution is the N_j-times Hilbert transform of the one-sided signal (12.237). The integral in equation (12.247) can be evaluated using the replication property of the Dirac delta function:

$$\begin{aligned}\sigma_j(t) &= \sum_{t=\Gamma_j} p^{1/2}(\pi q_\beta q'_\beta)^{-1/2}|d\Gamma_j/dp|^{-1} \\ &= \sum_{t=\Gamma_j} p^{1/2}(\pi q_\beta q'_\beta)^{-1/2}|\Theta - \Theta_j|^{-1},\end{aligned} \tag{12.248}$$

where the second equality follows from $d\Gamma_j/dp = -\Theta_j$. The summation in (12.248) is over all the values of ray parameter p that satisfy the equation $t = \Gamma_j(p)$, as illustrated in Figure 12.27. The $|\Theta - \Theta_j|^{-1}$ singularity at the ray-theoretical arrival times $T_{jk} = \tau_j(p_k) + p_k\Theta$ can be quelled by averaging $\sigma_j(t)$ over the digitization interval (Dey-Sarkar & Chapman 1978; Chapman, Chu & Lyness 1988).

We can formally recover ray theory by approximating the quantity $\Gamma_j(p)$ in the vicinity of its extrema p_k by a parabola:

$$\Gamma_j(p) = T_{jk} + \tfrac{1}{2}(d^2\tau_j/dp^2)_{p=p_k}(p - p_k)^2. \tag{12.249}$$

The intersections $t = \Gamma_j(p)$ occur in this approximation at the symmetrically spaced points

$$p = p_k \pm \left| \frac{2(t - T_{jk})}{(d^2\tau_j/dp^2)_{p=p_k}} \right|^{1/2}. \tag{12.250}$$

Upon inserting (12.250) into equation (12.248) we obtain

$$\sigma_j(t) = \sum_k \left[p^{1/2}(q_\beta q_\beta')^{-1/2} |d^2\tau_j/dp^2|^{-1/2} \right]_{p=p_k} \times \begin{cases} \sqrt{2}\, \upsilon(t - T_{jk}) & \text{if } \operatorname{sgn}(d^2\tau_j/dp^2)_{p=p_k} > 0 \\ \sqrt{2}\, \upsilon_{\mathrm{H}}(t - T_{jk}) & \text{if } \operatorname{sgn}(d^2\tau_j/dp^2)_{p=p_k} < 0. \end{cases} \tag{12.251}$$

Making use of the identities

$$\lambda_{\mathrm{H}}^{(N_j)}(t) * \upsilon(t - T_{jk}) = -\delta_{\mathrm{H}}^{(N_j+1)}(t - T_{jk}), \tag{12.252}$$

$$\lambda_{\mathrm{H}}^{(N_j)}(t) * \upsilon_{\mathrm{H}}(t - T_{jk}) = \delta_{\mathrm{H}}^{(N_j)}(t - T_{jk}), \tag{12.253}$$

we find that (12.245) reduces to the time-domain equivalent of the double sum (12.222):

$$\mathbf{G} = \frac{1}{4\pi} \hat{\mathbf{\Phi}} \hat{\mathbf{\Phi}}' (rr'\beta\beta')^{-1} (\rho\rho' \sin\Theta)^{-1/2} \times \sum_j \sum_k \left[p^{1/2}(q_\beta q_\beta')^{-1/2} |d^2\tau_j/dp^2|^{-1/2} \right]_{p=p_k} \times \delta_{\mathrm{H}}^{(M_{jk})}(t - T_{jk}). \tag{12.254}$$

The Hilbert transformation $\delta_{\mathrm{H}}(t - T_{jk}) = -[\pi(t - T_{jk})]^{-1}$ of $\mathrm{SS_{SH}}$ and other $M = 1$ minimax phases once again arises naturally. The quadratic approximation (12.250) becomes inaccurate whenever adjacent $\Gamma_j(p)$ extrema such as p_2, p_3 or p_3, p_1 in Figure 12.27 become close enough to "interfere" with each other and the intervening caustics C or B. The representation (12.245) is, on the other hand, uniformly valid, regardless of whether the receiver is near a caustic or not. Even overlapping triplications and multiple close caustics are handled automatically. Moreover, the numerical effort required to evaluate the uniformly valid expression (12.245) is only slightly greater than that required to evaluate the JWKB result (12.254).

The search for intersections $t = \Gamma_j(p)$ in equation (12.248) is over the full SH ray-parameter range $0 < p < a/\beta_a$, so that $\sigma_2(t)$, for example, contains arrivals corresponding to $\mathrm{ScS_{SH}}$ as well as turning—and possibly triplicated—SH waves. In practice, the method is seldom applied in this way; rather, seismologically "distinct" arrivals such as $\mathrm{ScS_{SH}}$ and triplicated

SH away from the fluid-core shadow are synthesized separately. Adopting this alternative point of view, we shall rewrite equation (12.245) using a more generic notation:

$$\mathbf{G} = \frac{1}{4\pi}\hat{\mathbf{\Phi}}\hat{\mathbf{\Phi}}'(rr'\beta\beta')^{-1}(2\rho\rho'\sin\Theta)^{-1/2} \times \sum_{\text{rays}} \lambda_{\mathrm{H}}^{(N)}(t) * \sigma(t), \tag{12.255}$$

where N is the number of distinct turning legs along a ray, and SH counts as a single "ray" even when it is triplicated. The representation (12.255) may be readily generalized to the P-SV case, by allowing for changes in polarization and wave speed at the source and receiver and accounting for interface interactions with a concatenated reflection-and-transmission coefficient Π. The full time-domain Green tensor $\mathbf{G}_{\mathrm{SH}} + \mathbf{G}_{\mathrm{P\text{-}SV}}$ comprising all $s = 1$ waves can be written in the form

$$\mathbf{G} = \frac{1}{4\pi}\sum_{\text{rays}} \hat{\boldsymbol{\eta}}\hat{\boldsymbol{\eta}}'(rr'vv')^{-1}(2\rho\rho'\sin\Theta)^{-1/2} \times \mathrm{Re}\,[\Lambda_{\mathrm{H}}^{(N)}(t) * \sigma(t)], \tag{12.256}$$

where

$$\Lambda_{\mathrm{H}}^{(N)}(t) = \underbrace{\mathcal{H}\cdots\mathcal{H}}_{N\text{ times}}\Lambda(t) = \lambda_{\mathrm{H}}^{(N)}(t) - i\lambda_{\mathrm{H}}^{(N+1)}(t), \tag{12.257}$$

$$\sigma(t) = \sum_{t=\Gamma} p^{1/2}(\pi q q')^{-1/2}\Pi\,|d\Gamma/dp|^{-1}. \tag{12.258}$$

The quantity $\mathrm{Re}\,[\Lambda(t)_{\mathrm{H}}^{(N)} * \sigma(t)]$ reduces to $\lambda(t)_{\mathrm{H}}^{(N)} * \sigma(t)$ as in (12.255) whenever $\mathrm{Im}\,\Pi = 0$. The more general form (12.256) involving the N-times Hilbert transform of the analytic signal $\Lambda(t) = \lambda(t) - i\lambda_{\mathrm{H}}(t)$ is applicable to post-critical (complex Π) as well as pre-critical (real Π) arrivals.

The first uniformly valid ray sum of the form (12.256) was obtained by Chapman (1976; 1978). He suggested calling such asymptotic representations "WKBJ seismograms" in recognition of the fact that they follow directly from the radial JWKB approximation. In fact (12.256) is an unusual implementation of a general technique for extending any JWKB result so that it is valid in the vicinity of caustics (Maslov 1972; Maslov & Fedoriuk 1981; Chapman & Drummond 1982; Liu & Tromp 1996). We prefer to use the unmodified acronym JWKB to refer to the strict ray-theoretical result (12.243), and shall refer to the uniformly valid representation (12.256) as the *Chapman-Maslov Green tensor*.

★12.5.5 JWKB and Chapman-Maslov seismograms

We consider, finally, the high-frequency body-wave response to an earthquake point source at a hypocentral location $\mathbf{x}_s$. We allow for the finite duration of rupture, modelling the source as synchronous, with a frequency-dependent moment tensor of the form

$$\mathbf{M}(\omega) = \sqrt{2}M_0\hat{\mathbf{M}}\, m(\omega). \tag{12.259}$$

The quantities M_0 and $\hat{\mathbf{M}}$ are the scalar moment and unit mechanism tensor, respectively, and $m(\omega)$ is the Fourier transform of the normalized source time function $\dot{m}(t)$, as described in Section 5.4.5. The exact frequency-domain displacement response is given in terms of the Green tensor by

$$\mathbf{s}(\mathbf{x},\omega) = (i\omega)^{-1}\mathbf{M}(\omega)\!:\!\boldsymbol{\nabla}_s\mathbf{G}^{\mathrm{T}}(\mathbf{x},\mathbf{x}_s;\omega). \tag{12.260}$$

Correct to first order in ω^{-1}, the gradient $\boldsymbol{\nabla}_s$ with respect to the hypocentral coordinates acts only upon the rapidly oscillating term $\exp(-i\omega T)$ in the JWKB representation (12.225), yielding a multiplier $\boldsymbol{\nabla}_s \rightarrow i\omega v_s^{-1}\hat{\mathbf{p}}_s$ where $\hat{\mathbf{p}}_s$ is the unit takeoff slowness vector. Upon collecting the receiver and source terms into factors

$$\Xi = (\rho v)^{-1/2}(\hat{\boldsymbol{\nu}}\cdot\hat{\boldsymbol{\eta}}), \tag{12.261}$$

$$\Sigma = \sqrt{2}M_0(\rho_s v_s^5)^{-1/2}[\hat{\mathbf{M}}\!:\!\tfrac{1}{2}(\hat{\mathbf{p}}_s\hat{\boldsymbol{\eta}}_s + \hat{\boldsymbol{\eta}}_s\hat{\mathbf{p}}_s)], \tag{12.262}$$

we can write the scalar displacement $s(\omega) = \hat{\boldsymbol{\nu}}\cdot\mathbf{s}(\mathbf{x},\omega)$ in the form

$$s(\omega) = \frac{1}{4\pi}\sum_{\mathrm{rays}}\Xi\Sigma\Pi\mathcal{R}^{-1}m(\omega)\exp(-i\omega T + iM\pi/2). \tag{12.263}$$

The time-domain response obtained by inverse Fourier transformation of equation (12.263) is

$$s(t) = \frac{1}{4\pi}\sum_{\mathrm{rays}}\Xi\Sigma\Pi\mathcal{R}^{-1}\dot{m}_{\mathrm{H}}^{(M)}(t-T). \tag{12.264}$$

The *pulse shape* of all least-time waves such as P and S is $\dot{m}(t)$; every passage through a caustic acts to Hilbert transform this far-field source-time function. The quantity $\hat{\mathbf{M}}:\frac{1}{2}(\hat{\mathbf{p}}_s\hat{\boldsymbol{\eta}}_s + \hat{\boldsymbol{\eta}}_s\hat{\mathbf{p}}_s)$ is the *radiation pattern* of the outgoing waves upon the focal sphere surrounding the source; note, in particular, that the P-wave amplitude is proportional to $\hat{\mathbf{p}}_s\cdot\hat{\mathbf{M}}\cdot\hat{\mathbf{p}}_s$, in accordance with our remarks regarding beachballs in Section 5.4.4.

Anelastic attenuation and the associated dispersion can be accounted for by introducing a complex wave speed:

$$v = v_0[1 + \tfrac{1}{2}iQ^{-1} + \tfrac{1}{\pi}Q^{-1}\ln(\omega/\omega_0)], \tag{12.265}$$

where v_0 is the reference speed—either α_0 or β_0—at the frequency ω_0, and Q denotes either Q_α or Q_β. We ignore the effect of anelasticity upon the ray geometry, but incorporate it in the calculation of the travel time by substituting in the frequency-domain ray sum (12.263):

$$T \to T - \tfrac{1}{2}iT^* - \tfrac{1}{\pi}T^* \ln(\omega/\omega_0). \tag{12.266}$$

The frequency-independent *attenuation time* T^* in equation (12.266) is given by

$$T^* = \int_{\mathbf{x}'}^{\mathbf{x}} \frac{ds}{v_0 Q} = \int_{\mathbf{x}'}^{\mathbf{x}} \frac{v_0^{-2}}{q_0 Q}\, dr, \tag{12.267}$$

where $q_0 = (v_0^{-2} - p^2 r^{-2})^{1/2}$ and the integral is taken along all of the legs of a compound ray. Typically, $T^* \approx 1-2$ seconds for teleseismic P waves whereas $T^* \approx 4-8$ seconds for teleseismic S waves. In the time domain, attenuation and dispersion can be incorporated with an additional convolution; the JWKB response (12.264) is replaced by

$$s(t) = \frac{1}{4\pi} \sum_{\text{rays}} \Xi \Sigma \Pi \mathcal{R}^{-1} \dot{m}_{\mathrm{H}}^{(M)}(t - T) * a(t), \tag{12.268}$$

where

$$a(t) = \frac{1}{\pi} \operatorname{Re} \int_0^\infty \exp i\omega\big[t + \tfrac{1}{2}iT^* + \tfrac{1}{\pi}T^* \ln(\omega/\omega_0)\big]\, d\omega. \tag{12.269}$$

High-frequency waves decay much more rapidly than low-frequency waves do, because of the factor $\exp(-\frac{1}{2}\omega T^*)$ in equation (12.269). This gives rise to a broadening as well as a damping of the $M = 0$ elastic pulse $\dot{m}(t - T)$. Anelastic dispersion acts to delay the pulse by an amount which depends not only upon T^* but also upon the location τ_{m} of the short-period absorption-band edge (Minster 1978; 1980).

It is a straightforward matter to extend the result (12.268) so that it is valid in the vicinity of caustics:

$$\begin{aligned} s(t) = \frac{1}{4\pi} \sum_{\text{rays}} & [r^{-1}(2v \sin\Theta)^{-1/2}\Xi][r_{\mathrm{s}}^{-1} v_{\mathrm{s}}^{1/2}\Sigma] \\ & \times \operatorname{Re}[\Lambda_{\mathrm{H}}^{(N)}(t) * \sigma(t)] * \dot{m}(t) * a(t). \end{aligned} \tag{12.270}$$

It is readily verified that the *Chapman-Maslov seismogram* (12.270) reduces to (12.268) wherever ray theory is applicable. Both of these expressions must be modified slightly if the receiver is situated upon the free surface of the Earth $r = a$, as is usual in observational seismology. Every upward-travelling SH wave is then accompanied by a downward-travelling $\mathrm{Ss}_{\mathrm{SH}}$

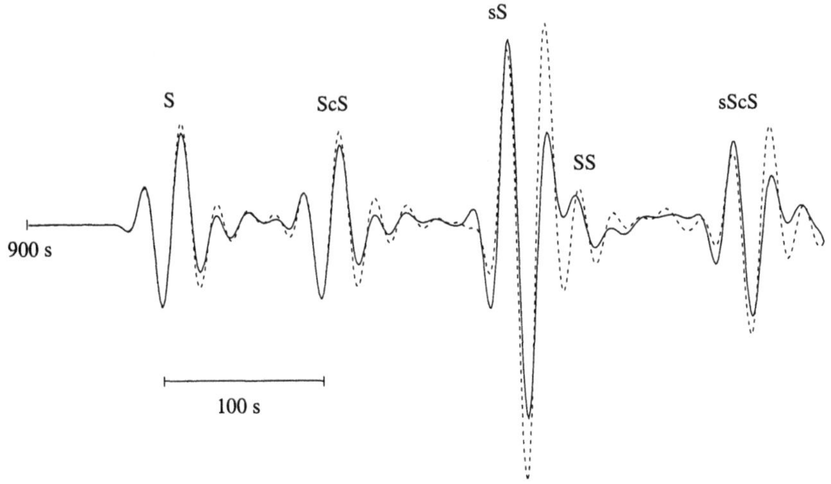

Figure 12.28. Comparison of the ray-theoretical result (12.268) with the exact PREM mode sum (10.51). The source is the June 9, 1994 deep-focus earthquake in Bolivia; the source time function is a Dirac delta function or instantaneous impulse: $\dot{m}(t) = \dot{\delta}(t)$. Both the ray-theoretical (*solid line*) and exact (*dotted line*) traces show the transverse component of ground acceleration $a(t) = \ddot{s}(t)$ at station CCM in Cathedral Cave, Missouri; an identical 20 − 80 second bandpass filter has been applied to the two accelerograms to accentuate the body-wave arrivals. Only the labelled phases SH, $\mathrm{ScS_{SH}}$, $\mathrm{sS_{SH}}$, $\mathrm{SS_{SH}}$ and $\mathrm{sScS_{SH}}$ are included in the ray sum.

reflection that arrives simultaneously and with the same phase; the net result is that the direct SH-wave sum must be multiplied by two to account for the effect of the free surface. A direct P or SV wave is accompanied by both $\mathrm{Pp{+}Ps_{SV}}$ or $\mathrm{Sp{+}Ss_{SV}}$ reflections; the free-surface correction factor in this case depends upon the type and ray parameter of the direct wave as well as the orientation of the receiver. Figure 12.28 shows a comparison between a partial SH ray-sum and an exact mode-sum seismogram for a source-receiver pair on the Preliminary Reference Earth Model; the epicentral distance $\Theta = 56°$ is beyond the upper-mantle triplication zone, so that the JWKB and Chapman-Maslov representations (12.268) and (12.270) yield results that are indistinguishable. Unmodelled finite-frequency effects (see Section 12.5.6) are unimportant for the six waves considered, all of which either turn far above the core-mantle boundary or reflect steeply off of it; the discrepancies are due to the neglect of crustal and other inter-layer reverberations.

It is instructive to compare the high-frequency body-wave response to an earthquake with the corresponding result for surface waves, given in equation (11.33). Both of these JWKB expressions are sums over all possible rays between the source $\mathbf{x}_s$ and the receiver $\mathbf{x}$. The propagation of body waves is three-dimensional whereas that of surface waves is two-dimensional; the two ray-theoretical responses have many common ingredients, including geometrical amplitude variations due to ray-tube focusing and defocusing, caustic phase shifts, an outgoing radiation pattern that depends upon the source geometry, and an incoming polarization that depends upon the type of wave. The major difference is the manner in which the radial structure α, β ρ of the Earth is taken into account. In the case of surface waves, the dependence upon the Earth model is encapsulated in the dispersion relation $k_n(\omega)$ and the associated radial eigenfunctions U_n, V_n and W_n along the various overtone branches; each of the branches $n = 0, 1, 2, \ldots$ must be considered separately and summed numerically as in (11.33) to find the full multi-mode surface-wave response. In contrast, the overtones have been summed analytically, as we have seen, to obtain the body-wave response. There are no "body-wave eigenfunctions" analogous to U_n, V_n and W_n that satisfy the boundary conditions upon the free surface $r = a$; nevertheless, the JWKB solution (12.263) is consistent with those conditions in an asymptotic sense. The surface-wave representation is applicable to the modes ${}_n\mathrm{T}_l$ and ${}_n\mathrm{S}_l$, with $n \ll l/4$, in the upper right portion of the $\omega - l$ diagram, whereas the body-wave representation is applicable to the modes ${}_n\mathrm{T}_l$ and ${}_n\mathrm{S}_l$, with $n \approx l$ or $n \gg l/4$, in the middle and upper left.

*12.5.6 Beyond the JWKB approximation

Both the JWKB ray sum (12.268) and its extension (12.270) are high-frequency approximations which are strictly valid only in the limit $\omega \to \infty$. There are many situations in global seismology where these results are insufficiently accurate for quantitative purposes; in particular, they do not account for a plethora of finite-frequency diffraction effects such as the tunnelling of near-grazing waves, interference head waves and whispering-gallery waves. A variety of computational techniques have been developed for dealing with these "full-wave" as opposed to ray-theoretical phenomena. The basic requirement is a more accurate means of evaluating the overtone sums such as $\sum_n (\omega_n^2 - \omega^2)^{-1} W_n W_n'$ in equation (12.209); this can be done either by straightforward numerical integration, by an iterative technique that accounts for higher-order "reflections" and "conversions", or by invoking the Langer approximation. Synthesis of a time-domain seismogram always involves the calculation of an inverse Fourier transform in addition to a contour integral over ray parameter:

$$s(t) = \frac{1}{\pi} \operatorname{Re} \int_0^{\infty} \int_{\infty}^{-\infty} s(p, \omega) \exp(i\omega t) \, dp \, d\omega. \tag{12.271}$$

To obtain the ray-theoretical response (12.268) we replaced the integrand $s(p, \omega)$ by its JWKB approximation $s_{\mathrm{JWKB}}(p, \omega)$ and evaluated the p integral first by means of the saddle-point method; to obtain the Chapman-Maslov response (12.270) we interchanged the order of integration and evaluated the ω integral exactly. Multiple strategies are also available for the evaluation of $s(t)$ in the more general case that $s(p, \omega) \neq s_{\mathrm{JWKB}}(p, \omega)$. Which of the two integrals is evaluated first and whether the ray-parameter contour is deformed away from the real axis during the evaluation are again the prime considerations. We make no attempt to survey the wide range of "full-wave" procedures, which constitute much of the corpus of quantitative body-wave seismology. Authoritative reviews of the various methods of calculating $s(p, \omega)$ and evaluating the double integral (12.271) are provided by Aki & Richards (1980), Kennett (1983) and Chapman & Orcutt (1985); the last authors, in particular, present a number of illustrative comparisons of the most commonly employed numerical algorithms.

Part III

The Aspherical Earth

Chapter 13

Perturbation Theory

In this chapter, we turn our attention to the effects of the Earth's rotation, hydrostatic ellipticity and lateral heterogeneity. In most global seismological applications, particularly at long periods, these departures from spherical symmetry can be regarded as *slight perturbations*. In this case, normal-mode perturbation theory can be used to calculate the singlet eigenfrequencies and associated eigenfunctions of the perturbed Earth, as well as the perturbed normal-mode response to a prescribed moment-tensor source. We consider the effect of an arbitrary perturbation to a spherically symmetric Earth model from a general and rather idealized point of view in this chapter. The practical application of normal-mode perturbation theory to long-period seismic data is discussed in greater detail in Chapter 14.

13.1 Isolated Mode

The fundamental problem, which we address first, is that of finding the perturbation to a non-degenerate eigenfrequency of a mode that is well isolated in the seismic spectrum. The solution to this classical *non-degenerate* perturbation problem serves as the basis for the degenerate and quasi-degenerate coupled-mode theories which we shall take up in the remainder of the chapter. We begin by ignoring the effect of anelastic attenuation, and treat the case of a purely elastic perturbation to an elastic non-hydrostatic Earth model. We then specialize to the case of a spherical hydrostatic initial model, and incorporate the Earth's slight anelasticity as an additional perturbation. As in Chapter 9, the principal difficulty is determining the effect of a slight change in the location of the boundaries. We follow the original treatment by Woodhouse & Dahlen (1978).

13.1.1 Recapitulation

We continue to dispense with the clear but cluttered notation used to develop the general results in Part I. The frequency-domain momentum equation and associated boundary conditions (4.3)–(4.6) governing the eigenfrequencies and eigenfunctions of a general non-rotating Earth can be written using this new notation in the form

$$-\omega^2\rho\mathbf{s} - \boldsymbol{\nabla}\cdot\widetilde{\mathbf{T}} + \rho\boldsymbol{\nabla}\phi + \rho\mathbf{s}\cdot\boldsymbol{\nabla}\boldsymbol{\nabla}\Phi = \mathbf{0} \quad \text{in } \oplus, \tag{13.1}$$

$$\hat{\mathbf{n}}\cdot\widetilde{\mathbf{T}} = \mathbf{0} \quad \text{on } \partial\oplus, \tag{13.2}$$

$$[\hat{\mathbf{n}}\cdot\widetilde{\mathbf{T}}]_-^+ = \mathbf{0} \quad \text{on } \Sigma_{\mathrm{SS}}, \tag{13.3}$$

$$[\widetilde{\mathbf{t}}]_-^+ = \hat{\mathbf{n}}[\hat{\mathbf{n}}\cdot\widetilde{\mathbf{t}}]_-^+ = \mathbf{0} \quad \text{on } \Sigma_{\mathrm{FS}}. \tag{13.4}$$

The incremental first Piola-Kirchhoff stress $\widetilde{\mathbf{T}}$ and the auxiliary vector $\widetilde{\mathbf{t}}$ are defined in terms of the displacement eigenfunction $\mathbf{s}$ by $\widetilde{\mathbf{T}} = \boldsymbol{\Lambda} : \boldsymbol{\nabla}\mathbf{s}$ and $\widetilde{\mathbf{t}} = \hat{\mathbf{n}}\cdot\widetilde{\mathbf{T}} + \boldsymbol{\nabla}^{\Sigma}\cdot(\varpi\mathbf{s}) - \varpi(\boldsymbol{\nabla}^{\Sigma}\mathbf{s})\cdot\hat{\mathbf{n}}$, where $\varpi = p - \hat{\mathbf{n}}\cdot\boldsymbol{\tau}\cdot\hat{\mathbf{n}}$. The corresponding equations and boundary conditions (4.149)–(4.152) in a non-rotating hydrostatic Earth are

$$\begin{aligned}-\omega^2\rho\mathbf{s} - \boldsymbol{\nabla}\cdot\mathbf{T} &+ \boldsymbol{\nabla}(\rho\mathbf{s}\cdot\boldsymbol{\nabla}\Phi) \\ &+ \rho\boldsymbol{\nabla}\phi - [\boldsymbol{\nabla}\cdot(\rho\mathbf{s})]\boldsymbol{\nabla}\Phi = \mathbf{0} \quad \text{in } \oplus,\end{aligned} \tag{13.5}$$

$$\hat{\mathbf{n}}\cdot\mathbf{T} = \mathbf{0} \quad \text{on } \partial\oplus, \tag{13.6}$$

$$[\hat{\mathbf{n}}\cdot\mathbf{T}]_-^+ = \mathbf{0} \quad \text{on } \Sigma_{\mathrm{SS}}, \tag{13.7}$$

$$[\hat{\mathbf{n}}\cdot\mathbf{T}]_-^+ = \hat{\mathbf{n}}[\hat{\mathbf{n}}\cdot\mathbf{T}\cdot\hat{\mathbf{n}}]_-^+ = \mathbf{0} \quad \text{on } \Sigma_{\mathrm{FS}}, \tag{13.8}$$

where $\mathbf{T} = \boldsymbol{\Gamma} : \boldsymbol{\varepsilon}$. The incremental Eulerian potential ϕ in either a non-hydrostatic or a hydrostatic Earth is found by solving the linearized Poisson boundary-value problem

$$\boldsymbol{\nabla}\cdot\boldsymbol{\xi} = 0 \quad \text{in } \bigcirc, \tag{13.9}$$

$$[\phi]_-^+ = 0 \quad \text{and} \quad [\hat{\mathbf{n}}\cdot\boldsymbol{\xi}]_-^+ = 0 \quad \text{on } \Sigma, \tag{13.10}$$

where $\boldsymbol{\xi} = (4\pi G)^{-1}\boldsymbol{\nabla}\phi + \rho\mathbf{s}$. The solution to this potential problem is

$$\phi = -G\int_{\oplus} \frac{\rho'\mathbf{s}'\cdot(\mathbf{x}-\mathbf{x}')}{\|\mathbf{x}-\mathbf{x}'\|^3}\, dV', \tag{13.11}$$

where the prime denotes evaluation at the dummy integration variable $\mathbf{x}'$. The rotation of the Earth is accounted for by adding a Coriolis-force term $2i\omega\boldsymbol{\Omega}\times\mathbf{s}$ and substituting the geopotential $\Phi + \psi$ for Φ in equations (13.1) and (13.5).

13.1.2 General elastic perturbation

An initial, unperturbed Earth model is completely characterized by its mass density ρ and associated gravitational potential Φ, its initial hydrostatic pressure p and deviatoric stress tensor $\boldsymbol{\tau}$, its isotropic or anisotropic elastic tensor $\mathbf{\Gamma}$, and the geometrical configuration of its internal and external boundaries $\Sigma = \partial\oplus \cup \Sigma_{\mathrm{SS}} \cup \Sigma_{\mathrm{FS}}$. The perturbations to these properties are then specified as follows:

(1) the density is perturbed, $\rho \rightarrow \rho + \delta\rho$;

(2) and so is the gravitational potential, $\Phi \rightarrow \Phi + \delta\Phi$;

(3) the initial hydrostatic pressure is perturbed, $p \rightarrow p + \delta p$;

(4) and so is the deviatoric stress, $\boldsymbol{\tau} \rightarrow \boldsymbol{\tau} + \boldsymbol{\delta\tau}$;

(5) the elastic tensor is perturbed, $\mathbf{\Gamma} \rightarrow \mathbf{\Gamma} + \boldsymbol{\delta}\mathbf{\Gamma}$;

(6) finally, the boundary Σ is displaced in a direction normal to itself by an amount δd, and the perturbed physical properties $\rho + \delta\rho$, $\Phi + \delta\Phi$, $p + \delta p$, $\boldsymbol{\tau} + \boldsymbol{\delta\tau}$, and $\mathbf{\Gamma} + \boldsymbol{\delta}\mathbf{\Gamma}$ are all redefined in the neighborhood of Σ to be smooth on either side of the unperturbed boundary.

As a result of these alterations in the Earth model, the eigenfrequency ω of a non-degenerate normal mode is perturbed by an amount $\delta\omega$, and the associated displacement and potential eigenfunctions $\mathbf{s}$ and ϕ are perturbed by amounts $\boldsymbol{\delta}\mathbf{s}$ and $\delta\phi$, respectively. As with the physical properties, the perturbed fields $\mathbf{s} + \boldsymbol{\delta}\mathbf{s}$ and $\phi + \delta\phi$ are redefined in the neighborhood of Σ to be smooth on either side of the unperturbed boundary.

We shall use Δq to denote the difference between any perturbed quantity $q + \delta q$ on the perturbed boundary and the corresponding unperturbed quantity q on the unperturbed boundary. Correct to first order in the perturbation δd, we have

$$\Delta q = \delta q + \delta d(\partial_n q), \tag{13.12}$$

where $\partial_n = \hat{\mathbf{n}} \cdot \boldsymbol{\nabla}$ denotes the normal derivative, and q can be a scalar, vector or tensor of any order. The normal $\hat{\mathbf{n}}$ to the boundary is perturbed as a result of the displacement, by an amount

$$\Delta\hat{\mathbf{n}} = -\boldsymbol{\nabla}^{\Sigma}(\delta d). \tag{13.13}$$

Combining equations (13.12) and (13.13), we find that for any vector or tensor quantity $\mathbf{q}$, we can write

$$\Delta q_n = \hat{\mathbf{n}} \cdot \boldsymbol{\delta}\mathbf{q} + \delta d(\partial_n q_n) - \boldsymbol{\nabla}^{\Sigma}(\delta d) \cdot \mathbf{q}, \tag{13.14}$$

where $q_n = \hat{\mathbf{n}} \cdot \mathbf{q}$. The corresponding perturbation in the surface gradient $\boldsymbol{\nabla}^\Sigma q = \boldsymbol{\nabla} q - \hat{\mathbf{n}}(\partial_n q)$ is given by

$$\Delta(\boldsymbol{\nabla}^\Sigma q) = \boldsymbol{\nabla}^\Sigma(\Delta q) + [\hat{\mathbf{n}}\boldsymbol{\nabla}^\Sigma(\delta d) - \delta d(\boldsymbol{\nabla}^\Sigma \hat{\mathbf{n}})] \cdot \boldsymbol{\nabla}^\Sigma q, \tag{13.15}$$

for any scalar, vector or tensor q on Σ.

The Earth model perturbations $\delta\rho$, $\delta\Phi$, δp, $\boldsymbol{\delta\tau}$ and δd cannot be specified independently. In the first place, the perturbation $\delta\Phi$ in the gravitational potential must be that due to the density perturbation $\delta\rho$ and the boundary displacement δd. We can find $\delta\Phi$ by solving the boundary-value problem

$$\nabla^2(\delta\Phi) = \begin{cases} 4\pi G\,\delta\rho & \text{in } \oplus_{\mathrm{S}} \\ 0 & \text{in } \bigcirc - \oplus, \end{cases} \tag{13.16}$$

$$[\delta\Phi]_-^+ = 0 \quad \text{and} \quad [\hat{\mathbf{n}} \cdot \boldsymbol{\nabla}(\delta\Phi) + 4\pi G\rho\,\delta d]_-^+ = 0 \quad \text{on } \Sigma. \tag{13.17}$$

The solution to equations (13.16)–(13.17) can be written in the form

$$\delta\Phi = -G \int_\oplus \frac{\delta\rho'}{\|\mathbf{x} - \mathbf{x}'\|}\, dV' + G \int_\Sigma \frac{\delta d'\,[\rho']_-^+}{\|\mathbf{x} - \mathbf{x}'\|}\, d\Sigma', \tag{13.18}$$

where the first term accounts for the volumetric density perturbation, and the second term accounts for the displacement of the boundaries.

In addition, the new initial stress must be in mechanical equilibrium with the new gravitational body forces; this will be so provided that δp and $\boldsymbol{\delta\tau}$ satisfy the perturbed equilibrium condition

$$-\boldsymbol{\nabla}(\delta p) + \boldsymbol{\nabla} \cdot \boldsymbol{\delta\tau} = \delta\rho\boldsymbol{\nabla}\Phi + \rho\boldsymbol{\nabla}(\delta\Phi) \quad \text{in } \oplus, \tag{13.19}$$

together with the perturbed continuity condition $[\Delta(-\hat{\mathbf{n}}p + \hat{\mathbf{n}} \cdot \boldsymbol{\tau})]_-^+ = \mathbf{0}$ or, equivalently,

$$\begin{aligned}[-\hat{\mathbf{n}}\,\delta p + \hat{\mathbf{n}} \cdot \boldsymbol{\delta\tau}]_-^+ &= +\delta d\,[\hat{\mathbf{n}}\,\partial_n p - \hat{\mathbf{n}} \cdot \partial_n \boldsymbol{\tau}]_-^+ \\ &\quad + \boldsymbol{\nabla}^\Sigma(\delta d) \cdot [-p\mathbf{I} + \boldsymbol{\tau}]_-^+ \quad \text{on } \Sigma. \end{aligned} \tag{13.20}$$

The need for the perturbed level surfaces of $\rho + \delta\rho$, $\Phi + \delta\Phi$ and $p + \delta p$ to coincide in the fluid regions of the Earth $\oplus_{\mathrm{F}}$ places a significant restriction upon the allowable fluid density and pressure perturbations $\delta\rho$ and δp, which we shall consider in Section 13.1.7.

In the solid regions of the Earth $\oplus_{\mathrm{S}}$ we may specify the perturbation in the elastic tensor $\boldsymbol{\delta\Gamma}$ arbitrarily, subject only to the elastic symmetry relations $\delta\Gamma_{ijkl} = \delta\Gamma_{jikl} = \delta\Gamma_{ijlk} = \delta\Gamma_{klij}$. The corresponding perturbation $\boldsymbol{\delta\Lambda}$ in the fourth-order tensor relating the incremental first Piola-Kirchhoff stress $\widetilde{\mathbf{T}}$ to $\boldsymbol{\nabla}\mathbf{s}$ is given in terms of $\boldsymbol{\delta\Gamma}$, δp and $\boldsymbol{\delta\tau}$ by

$$\begin{aligned}\delta\Lambda_{ijkl} = \delta\Gamma_{ijkl} - \delta p(\delta_{ij}\,\delta_{kl} - \delta_{il}\,\delta_{jk}) + \tfrac{1}{2}(\delta\tau_{ij}\,\delta_{kl} + \delta\tau_{kl}\,\delta_{ij} \\ + \delta\tau_{ik}\,\delta_{jl} - \delta\tau_{jk}\,\delta_{il} - \delta\tau_{il}\,\delta_{jk} - \delta\tau_{jl}\,\delta_{ik}).\end{aligned} \tag{13.21}$$

In the fluid regions $\oplus_{\mathrm{F}}$ the perturbations $\boldsymbol{\delta\Gamma}$ and $\boldsymbol{\delta\Lambda}$ must be of the form $\delta\Gamma_{ijkl} = \delta\kappa\,\delta_{ij}\,\delta_{kl}$ and $\delta\Lambda_{ijkl} = \delta\kappa\,\delta_{ij}\,\delta_{kl} - \delta p(\delta_{ij}\,\delta_{kl} - \delta_{il}\,\delta_{jk})$, where $\delta\kappa$ is the perturbation in the isentropic incompressibility.

We shall also require the two perturbations $\Delta\varpi$ and $\Delta(\boldsymbol{\nabla}^{\Sigma}\varpi)$, where $\varpi = p - \hat{\mathbf{n}}\cdot\boldsymbol{\tau}\cdot\hat{\mathbf{n}}$; the first of these is given by

$$\begin{aligned}\Delta\varpi = \delta p - \hat{\mathbf{n}}\cdot\boldsymbol{\delta\tau}\cdot\hat{\mathbf{n}} - \delta d\,(-\partial_n p + \hat{\mathbf{n}}\cdot\partial_n\boldsymbol{\tau}\cdot\hat{\mathbf{n}})\\ + 2\boldsymbol{\nabla}^{\Sigma}(\delta d)\cdot(\hat{\mathbf{n}}\cdot\boldsymbol{\tau}),\end{aligned} \tag{13.22}$$

whereas the second can be found using equation (13.15):

$$\Delta(\boldsymbol{\nabla}^{\Sigma}\varpi) = \boldsymbol{\nabla}^{\Sigma}(\Delta\varpi) + [\hat{\mathbf{n}}\boldsymbol{\nabla}^{\Sigma}(\delta d) - \delta d(\boldsymbol{\nabla}^{\Sigma}\hat{\mathbf{n}})]\cdot\boldsymbol{\nabla}^{\Sigma}\varpi. \tag{13.23}$$

The quantity $\Delta\varpi$ is continuous, $[\Delta\varpi]_-^+ = 0$ on Σ, by virtue of the continuity of the initial traction $-p\hat{\mathbf{n}} + \hat{\mathbf{n}}\cdot\boldsymbol{\tau}$ and the boundary condition (13.20).

13.1.3 Application of Rayleigh's principle

Rayleigh's variational principle can be used to determine the first-order perturbation $\delta\omega$ in a non-degenerate eigenfrequency without solving simultaneously for the perturbations in the associated eigenfunctions. The calculation is slightly less intricate if $\boldsymbol{\delta}\mathbf{s}$ and $\delta\phi$ are regarded as independent perturbations; we adopt that viewpoint and make use of the displacement-potential version of Rayleigh's principle in the derivation that follows. The basic strategy, as in the case of a spherical perturbation, is to regard the modified action $\mathcal{I}'$ as a functional not only of the eigenfunctions $\mathbf{s}$ and ϕ, but also of the associated eigenfrequency ω, and all of the parameters needed to describe the Earth model. We denote the ensemble of volumetric Earth model parameters ρ, $\boldsymbol{\Lambda}$ and $\boldsymbol{\nabla\nabla}\Phi$ symbolically by $\oplus$, and the ensemble of surficial Earth model parameters ϖ, $\boldsymbol{\nabla}^{\Sigma}\varpi$ and $\hat{\mathbf{n}}$ by $\oplus^{\Sigma}$. The full dependence of $\mathcal{I}'$ can then be exhibited explicitly by rewriting equation (4.36) in the form

$$\mathcal{I}' = \int_{\bigcirc} L'(\mathbf{s},\boldsymbol{\nabla}\mathbf{s},\boldsymbol{\nabla}\phi;\omega,\oplus)\,dV + \int_{\Sigma}[L^{\Sigma}(\mathbf{s},\boldsymbol{\nabla}^{\Sigma}\mathbf{s};\oplus^{\Sigma})]_-^+\,d\Sigma, \tag{13.24}$$

where

$$\begin{aligned}L' = \tfrac{1}{2}[\omega^2\rho\mathbf{s}\cdot\mathbf{s} - \boldsymbol{\nabla}\mathbf{s}:\boldsymbol{\Lambda}:\boldsymbol{\nabla}\mathbf{s} - 2\rho\mathbf{s}\cdot\boldsymbol{\nabla}\phi\\ - \rho\mathbf{s}\cdot\boldsymbol{\nabla\nabla}\Phi\cdot\mathbf{s} - (4\pi G)^{-1}\boldsymbol{\nabla}\phi\cdot\boldsymbol{\nabla}\phi],\end{aligned} \tag{13.25}$$

$$L^{\Sigma} = \tfrac{1}{2}[(\hat{\mathbf{n}}\cdot\mathbf{s})\boldsymbol{\nabla}^{\Sigma}\cdot(\varpi\mathbf{s}) - \varpi\mathbf{s}\cdot(\boldsymbol{\nabla}^{\Sigma}\mathbf{s})\cdot\hat{\mathbf{n}}]. \tag{13.26}$$

For brevity, we shall write the surface integrals that arise in the ensuing derivation over all of $\Sigma = \partial\oplus \cup \Sigma_{\mathrm{SS}} \cup \Sigma_{\mathrm{FS}}$; note, however, that the integral in (13.24) is actually only over the fluid-solid discontinuities $\Sigma_{\rm FS}$, since the surface Lagrangian density is continuous, $[L^\Sigma]_-^+ = 0$, on both $\partial\oplus$ and $\Sigma_{\rm SS}$.

The value of the modified action at every eigensolution ω, $\mathbf{s}$, ϕ of the unperturbed Earth model is

$$\mathcal{I}' = 0. \tag{13.27}$$

As in Chapter 9, we consider the *total variation* of equation (13.27) with respect to all of the variables, including ω, $\oplus$ and $\oplus^\Sigma$:

$$\begin{aligned}\delta\mathcal{I}'_{\rm total} &= \int_\oplus [\delta\mathbf{s}\cdot(\partial_{\mathbf{s}}L') + \boldsymbol{\nabla}(\delta\mathbf{s})\cdot(\partial_{\boldsymbol{\nabla}\mathbf{s}}L')]dV \\ &+ \int_\bigcirc \boldsymbol{\nabla}(\delta\phi)\cdot(\partial_{\boldsymbol{\nabla}\phi}L')\,dV \\ &+ \int_\oplus [\delta\omega(\partial_\omega L') + \delta\oplus(\partial_\oplus L')]\,dV \\ &+ \int_\Sigma [\Delta\mathbf{s}\cdot(\partial_{\mathbf{s}}L^\Sigma) + \Delta(\boldsymbol{\nabla}^\Sigma\mathbf{s}):(\partial_{\boldsymbol{\nabla}^\Sigma\mathbf{s}}L^\Sigma)]_-^+\,d\Sigma \\ &+ \int_\Sigma [\Delta\oplus^\Sigma(\partial_{\oplus^\Sigma}L^\Sigma) + \delta d(\boldsymbol{\nabla}\cdot\hat{\mathbf{n}})L^\Sigma - \delta d\,L']_-^+\,d\Sigma = 0. \end{aligned} \tag{13.28}$$

We have introduced the shorthand notation $\delta\oplus(\partial_\oplus L')$ and $\Delta\oplus^\Sigma(\partial_{\oplus^\Sigma}L^\Sigma)$ to denote the variations with respect to the Earth model parameters:

$$\begin{aligned}\delta\oplus(\partial_\oplus L') = \delta\rho(\partial_\rho L') &+ \boldsymbol{\delta\Lambda}\,\vdots\,(\partial_{\boldsymbol{\Lambda}}L') \\ &+ \boldsymbol{\nabla\nabla}(\delta\Phi):(\partial_{\boldsymbol{\nabla\nabla}\Phi}L'),\end{aligned} \tag{13.29}$$

$$\begin{aligned}\Delta\oplus^\Sigma(\partial_{\oplus^\Sigma}L^\Sigma) = \Delta\varpi(\partial_\varpi L^\Sigma) &+ \Delta(\boldsymbol{\nabla}^\Sigma\varpi)\cdot(\partial_{\boldsymbol{\nabla}^\Sigma\varpi}L^\Sigma) \\ &+ \Delta\hat{\mathbf{n}}\cdot(\partial_{\hat{\mathbf{n}}}L^\Sigma).\end{aligned} \tag{13.30}$$

The domain of integration in the first and third volume integrals in equation (13.28) is the volume of the Earth $\oplus$ rather than all of space $\bigcirc$, since only the term involving $\boldsymbol{\nabla}\phi\cdot\boldsymbol{\nabla}\phi$ is non-zero outside of the Earth, in $\bigcirc - \oplus$. The last term $-\delta d\,L'$ in the final surface integral arises from the perturbation in the region of volume integration due to the shift in the boundaries, and the next-to-last term arises from the corresponding perturbation in the differential area element: $\Delta(d\Sigma) = \delta d(\boldsymbol{\nabla}\cdot\hat{\mathbf{n}})\,d\Sigma$. Upon applying both the three-dimensional and two-dimensional versions of Gauss' theorem to equation (13.28), we obtain

$$\begin{aligned}\delta\mathcal{I}'_{\rm total} &= \int_\oplus \delta\mathbf{s}\cdot[\partial_{\mathbf{s}}L' - \boldsymbol{\nabla}\cdot(\partial_{\boldsymbol{\nabla}\mathbf{s}}L')]\,dV \\ &+ \int_\Sigma [\delta\mathbf{s}\cdot\{\partial_{\mathbf{s}}L^\Sigma - \boldsymbol{\nabla}^\Sigma\cdot(\partial_{\boldsymbol{\nabla}^\Sigma\mathbf{s}}L^\Sigma) - \hat{\mathbf{n}}\cdot(\partial_{\boldsymbol{\nabla}\mathbf{s}}L')\}]_-^+\,d\Sigma\end{aligned}$$

$$- \int_{\bigcirc} \delta\phi\,[\boldsymbol{\nabla}\cdot(\partial_{\boldsymbol{\nabla}\phi}L')]\,dV - \int_{\Sigma} [\delta\phi\,\{\hat{\mathbf{n}}\cdot(\partial_{\boldsymbol{\nabla}\phi}L')\}]_-^+\,d\Sigma$$
$$+ \int_{\Sigma} \delta d\,[\partial_n\mathbf{s}\cdot\{\partial_{\mathbf{s}}L^{\Sigma} - \boldsymbol{\nabla}^{\Sigma}\cdot(\partial_{\boldsymbol{\nabla}^{\Sigma}\mathbf{s}}L^{\Sigma})\}]_-^+\,d\Sigma$$
$$+ \int_{\Sigma} [\{\hat{\mathbf{n}}\boldsymbol{\nabla}^{\Sigma}(\delta d) - \delta d(\boldsymbol{\nabla}^{\Sigma}\hat{\mathbf{n}})\}\cdot(\boldsymbol{\nabla}^{\Sigma}\mathbf{s}):(\partial_{\boldsymbol{\nabla}^{\Sigma}\mathbf{s}}L^{\Sigma})]_-^+\,d\Sigma$$
$$+ \int_{\oplus} [\delta\omega(\partial_{\omega}L') + \delta\oplus(\partial_{\oplus}L')]\,dV$$
$$+ \int_{\Sigma} [\Delta\oplus^{\Sigma}(\partial_{\oplus^{\Sigma}}L^{\Sigma}) + \delta d\,(\boldsymbol{\nabla}\cdot\hat{\mathbf{n}})L^{\Sigma} - \delta d\,L']_-^+\,d\Sigma = 0. \tag{13.31}$$

The term involving the eigenfrequency perturbation $\delta\omega$ is simply

$$\delta\omega \int_{\oplus} \partial_{\omega}L'\,dV = \omega\,\delta\omega \int_{\oplus} \rho\,\mathbf{s}\cdot\mathbf{s}\,dV = \omega\,\delta\omega, \tag{13.32}$$

where we have invoked the normalization (4.21) to obtain the last equality. We retain the desired quantity (13.32) on the left, move everything else to the right, and combine terms. The volume integrals involving the perturbations $\boldsymbol{\delta}\mathbf{s}$ and $\delta\phi$ in (13.31) vanish, exactly as in the derivation of the displacement-potential version of Rayleigh's principle presented in Section 4.1.3, since the quantities $\partial_{\mathbf{s}}L' - \boldsymbol{\nabla}\cdot(\partial_{\boldsymbol{\nabla}\mathbf{s}}L') = \mathbf{0}$ and $\boldsymbol{\nabla}\cdot(\partial_{\boldsymbol{\nabla}\phi}L') = 0$ are the frequency-domain momentum equation (13.1) and Poisson's equation (13.9), respectively. The first two surface integrals do not, however, now vanish, because the displacement δd of the boundary renders the variations $\boldsymbol{\delta}\mathbf{s}$ and $\delta\phi$ *inadmissible.*

The quantity $\partial_{\mathbf{s}}L^{\Sigma} - \boldsymbol{\nabla}^{\Sigma}\cdot(\partial_{\boldsymbol{\nabla}^{\Sigma}\mathbf{s}}L^{\Sigma}) - \hat{\mathbf{n}}\cdot(\partial_{\boldsymbol{\nabla}\mathbf{s}}L')$ multiplying $\boldsymbol{\delta}\mathbf{s}$ in the first surface integral is still equal to $\tilde{\mathbf{t}}$, but because of the boundary perturbation δd it is no longer true that $[\boldsymbol{\delta}\mathbf{s}\cdot\tilde{\mathbf{t}}]_-^+ = 0$. On the solid-solid boundaries Σ_{SS}, we must have $[\Delta\mathbf{s}]_-^+ = \mathbf{0}$ or, equivalently,

$$[\boldsymbol{\delta}\mathbf{s}]_-^+ = -\delta d\,[\partial_n\mathbf{s}]_-^+, \tag{13.33}$$

whereas on the fluid-solid boundaries Σ_{FS}, we must have $[\Delta(\hat{\mathbf{n}}\cdot\mathbf{s})]_-^+ = 0$ or, equivalently,

$$[\hat{\mathbf{n}}\cdot\boldsymbol{\delta}\mathbf{s}]_-^+ = -\delta d\,[\partial_n s_n]_-^+ + \boldsymbol{\nabla}^{\Sigma}(\delta d)\cdot[\mathbf{s}]_-^+. \tag{13.34}$$

Since $\tilde{\mathbf{t}}$ is continuous on all of Σ and a normal vector on Σ_{FS}, we are able to write on all boundaries:

$$[\boldsymbol{\delta}\mathbf{s}\cdot\tilde{\mathbf{t}}]_-^+ = -\delta d\,[\tilde{\mathbf{t}}\cdot\partial_n\mathbf{s}]_-^+ + \boldsymbol{\nabla}^{\Sigma}(\delta d)\cdot[(\hat{\mathbf{n}}\cdot\tilde{\mathbf{t}})\mathbf{s}]_-^+. \tag{13.35}$$

We also still have $\partial_{\boldsymbol{\nabla}\phi}L = -\boldsymbol{\xi}$, but it is no longer true that $[\delta\phi(\hat{\mathbf{n}}\cdot\boldsymbol{\xi})]_-^+ = 0$. The perturbed potential boundary condition is $[\Delta\phi]_-^+ = 0$, or, equivalently,

$$[\delta\phi]_-^+ = -\delta d\,[\partial_n\phi]_-^+. \tag{13.36}$$

This together with the continuity of $\hat{\mathbf{n}}\cdot\boldsymbol{\xi}$ enables us to write

$$[\delta\phi(\hat{\mathbf{n}}\cdot\boldsymbol{\xi})]_-^+ = -\delta d\,[(\hat{\mathbf{n}}\cdot\boldsymbol{\xi})\partial_n\phi]_-^+. \tag{13.37}$$

The conditions (13.35) and (13.37) are precisely the relations needed to evaluate the first two surface integrals in equation (13.31). The remaining six integrals, which depend only upon the volumetric and surficial Earth model perturbations $\delta\oplus$ and $\Delta\oplus^\Sigma$ and not upon the eigenfunction perturbations $\delta\mathbf{s}$ and $\delta\phi$, can be evaluated directly using the forms (13.25) and (13.26) for L' and L^Σ.

With some effort, the final result for the first-order eigenfrequency perturbation $\delta\omega$ of a non-degenerate mode can be written in the form

$$\begin{aligned}
\delta\omega = {} & \frac{1}{2\omega}\int_\oplus [\delta\rho\,(-\omega^2\mathbf{s}\cdot\mathbf{s} + 2\mathbf{s}\cdot\boldsymbol{\nabla}\phi + \mathbf{s}\cdot\boldsymbol{\nabla}\boldsymbol{\nabla}\Phi\cdot\mathbf{s}) \\
& \qquad + \boldsymbol{\nabla}\mathbf{s}\!:\!\delta\boldsymbol{\Lambda}\!:\!\boldsymbol{\nabla}\mathbf{s} + \rho\mathbf{s}\cdot\boldsymbol{\nabla}\boldsymbol{\nabla}(\delta\Phi)\cdot\mathbf{s}]\,dV \\
& + \frac{1}{2\omega}\int_\Sigma \delta d\,[2L' + 2(\hat{\mathbf{n}}\cdot\widetilde{\mathbf{T}})\cdot\partial_n\mathbf{s} + 2(\hat{\mathbf{n}}\cdot\boldsymbol{\xi})\partial_n\phi \\
& \qquad + \varpi\{(\boldsymbol{\nabla}^\Sigma\mathbf{s})\!:\!(\boldsymbol{\nabla}^\Sigma\mathbf{s})^{\mathrm{T}} - (\boldsymbol{\nabla}^\Sigma\cdot\mathbf{s})^2\} \\
& \qquad + (\boldsymbol{\nabla}^\Sigma\varpi)\cdot\{s_n(\mathbf{s}\cdot\boldsymbol{\nabla}^\Sigma\hat{\mathbf{n}}) - \mathbf{s}(\boldsymbol{\nabla}^\Sigma\cdot\mathbf{s})\} \\
& \qquad - (\boldsymbol{\nabla}^\Sigma\boldsymbol{\nabla}^\Sigma\varpi)\!:\!\mathbf{s}\mathbf{s}]_-^+\,d\Sigma \\
& - \frac{1}{2\omega}\int_{\Sigma_{\mathrm{FS}}} \boldsymbol{\nabla}^\Sigma(\delta d)\cdot[2(\hat{\mathbf{n}}\cdot\widetilde{\mathbf{t}})\mathbf{s}]_-^+\,d\Sigma \\
& + \frac{1}{2\omega}\int_{\Sigma_{\mathrm{FS}}} \Delta\varpi\,[2(\mathbf{s}\cdot\boldsymbol{\nabla}^\Sigma s_n) - \mathbf{s}\cdot(\boldsymbol{\nabla}^\Sigma\hat{\mathbf{n}})\cdot\mathbf{s}]_-^+\,d\Sigma.
\end{aligned} \tag{13.38}$$

Equation (13.38) allows the explicit calculation of $\delta\omega$ in terms of the unperturbed eigenfunctions $\mathbf{s}$ and ϕ and the various perturbations in the Earth model. The first integral accounts for the effect of the volumetric perturbations in the density $\delta\rho$, the fourth-order tensor $\delta\boldsymbol{\Lambda}$ relating the incremental first Piola-Kirchhoff stress $\widetilde{\mathbf{T}}$ to $\boldsymbol{\nabla}\mathbf{s}$, and the gravitational potential $\delta\Phi$, whereas the next two integrals account for the effect of the perturbation δd in the location of the boundaries. The perturbation $\Delta\varpi$ in the final integral also depends upon δd as well as upon δp and $\delta\boldsymbol{\tau}$, as indicated in equation (13.22). All of the terms involving the normal traction ϖ in the integral over Σ are continuous across $\partial\oplus$ and Σ_{SS}; hence, like the integrals involving $\boldsymbol{\nabla}^\Sigma(\delta d)$ and $\Delta\varpi$, they only need to be evaluated on the fluid-solid boundary Σ_{FS}. In carrying out the reduction of the surface integrals, we have employed equations (13.13) and (13.23), the symmetry of the curvature tensor $\boldsymbol{\nabla}^\Sigma\hat{\mathbf{n}} = (\boldsymbol{\nabla}^\Sigma\hat{\mathbf{n}})^{\mathrm{T}}$, and the readily established surface-gradient identity $\boldsymbol{\nabla}^\Sigma\boldsymbol{\nabla}^\Sigma - \hat{\mathbf{n}}(\boldsymbol{\nabla}^\Sigma\hat{\mathbf{n}})\cdot\boldsymbol{\nabla}^\Sigma = [\boldsymbol{\nabla}^\Sigma\boldsymbol{\nabla}^\Sigma - \hat{\mathbf{n}}(\boldsymbol{\nabla}^\Sigma\hat{\mathbf{n}})\cdot\boldsymbol{\nabla}^\Sigma]^{\mathrm{T}}$.

13.1.4 Hydrostatic starting model

The above result can be simplified considerably if the starting model is hydrostatic, with an unperturbed initial stress $-p\mathbf{I}$. The perturbation in the initial stress is then of the form $-\delta p\,\mathbf{I}+\boldsymbol{\tau}$, where $\boldsymbol{\tau}$ is the total deviatoric stress. Since the deviatoric stress in the Earth is everywhere very small, treating it as a perturbation in this manner is a very good approximation. The boundary condition (13.20) reduces to $[-\hat{\mathbf{n}}\,\delta p+\hat{\mathbf{n}}\cdot\boldsymbol{\tau}]_-^+ = \delta d\,\hat{\mathbf{n}}[\partial_n p]_-^+$. The quantity ϖ is simply the initial pressure p, and the perturbation (13.22) becomes $\Delta\varpi=\delta p-\hat{\mathbf{n}}\cdot\boldsymbol{\tau}\cdot\hat{\mathbf{n}}+\delta d(\partial_n p)$. The terms involving the surface gradients $\boldsymbol{\nabla}^\Sigma\varpi$ and $\boldsymbol{\nabla}^\Sigma\boldsymbol{\nabla}^\Sigma\varpi$ in equation (13.38) vanish, since ϖ is a constant on Σ_{FS}. In addition, since a fluid-solid boundary cannot support any shear traction, we must have $\hat{\mathbf{n}}\cdot\boldsymbol{\tau}=\hat{\mathbf{n}}(\hat{\mathbf{n}}\cdot\boldsymbol{\tau}\cdot\hat{\mathbf{n}})$ on Σ_{FS}. Another lengthy calculation is required to reduce the remaining terms in (13.38), and obtain the most convenient form for the perturbation in the eigenfrequency; the result of this endeavor can be expressed in terms of a Fréchet kernel notation analogous to equations (9.12)–(9.16):

$$\begin{aligned}\delta\omega &= \int_\oplus \delta\oplus\,K_\oplus\,dV+\int_\Sigma \delta d\,[K_{\mathrm{d}}]_-^+\,d\Sigma \\ &\quad+\int_{\Sigma_{\mathrm{FS}}}\boldsymbol{\nabla}^\Sigma(\delta d)\cdot[\mathbf{K}_{\mathrm{d}}]_-^+\,d\Sigma+\int_{\oplus_{\mathrm{S}}}\boldsymbol{\tau}:\mathbf{K}_{\boldsymbol{\tau}}\,dV \\ &\qquad+\int_{\Sigma_{\mathrm{FS}}}(\hat{\mathbf{n}}\cdot\boldsymbol{\tau})\cdot[\mathbf{K}_{\boldsymbol{\tau}}^\Sigma]_-^+\,d\Sigma+\int_{\Sigma_{\mathrm{FS}}}\boldsymbol{\nabla}^\Sigma(\Delta\varpi)\cdot[\mathbf{K}_\varpi]_-^+\,d\Sigma,\end{aligned}\tag{13.39}$$

where

$$\begin{aligned}2\omega\,\delta\oplus K_\oplus &= \delta\rho[-\omega^2\mathbf{s}\cdot\mathbf{s}+2\mathbf{s}\cdot\boldsymbol{\nabla}\phi+\mathbf{s}\cdot\boldsymbol{\nabla}\boldsymbol{\nabla}\Phi\cdot\mathbf{s} \\ &\quad+\boldsymbol{\nabla}\Phi\cdot(\mathbf{s}\cdot\boldsymbol{\nabla}\mathbf{s}-\mathbf{s}\boldsymbol{\nabla}\cdot\mathbf{s})]+\boldsymbol{\varepsilon}:\boldsymbol{\delta\Gamma}:\boldsymbol{\varepsilon} \\ &\qquad+\rho\boldsymbol{\nabla}(\delta\Phi)\cdot(\mathbf{s}\cdot\boldsymbol{\nabla}\mathbf{s}-\mathbf{s}\boldsymbol{\nabla}\cdot\mathbf{s})+\rho\mathbf{s}\cdot\boldsymbol{\nabla}\boldsymbol{\nabla}(\delta\Phi)\cdot\mathbf{s},\end{aligned}\tag{13.40}$$

$$\begin{aligned}2\omega K_{\mathrm{d}} &= \rho[\omega^2\mathbf{s}\cdot\mathbf{s}-2\mathbf{s}\cdot\boldsymbol{\nabla}\phi-\mathbf{s}\cdot\boldsymbol{\nabla}\boldsymbol{\nabla}\Phi\cdot\mathbf{s} \\ &\quad-\boldsymbol{\nabla}\Phi\cdot(\mathbf{s}\cdot\boldsymbol{\nabla}\mathbf{s}-\mathbf{s}\boldsymbol{\nabla}\cdot\mathbf{s})]-\boldsymbol{\varepsilon}:\boldsymbol{\Gamma}:\boldsymbol{\varepsilon}-(4\pi G)^{-1}\boldsymbol{\nabla}\phi\cdot\boldsymbol{\nabla}\phi \\ &\qquad+2(\hat{\mathbf{n}}\cdot\mathbf{T})\cdot\partial_n\mathbf{s}+2(\hat{\mathbf{n}}\cdot\boldsymbol{\xi})\partial_n\phi,\end{aligned}\tag{13.41}$$

$$2\omega\mathbf{K}_{\mathrm{d}} = -2(\hat{\mathbf{n}}\cdot\mathbf{T}\cdot\hat{\mathbf{n}})\mathbf{s},\tag{13.42}$$

$$\begin{aligned}2\omega\mathbf{K}_{\boldsymbol{\tau}} &= \mathbf{s}\cdot\boldsymbol{\nabla}\boldsymbol{\nabla}\mathbf{s}-\mathbf{s}\boldsymbol{\nabla}(\boldsymbol{\nabla}\cdot\mathbf{s}) \\ &\quad+\tfrac{1}{2}\boldsymbol{\nabla}\mathbf{s}\cdot(\boldsymbol{\nabla}\mathbf{s})^{\mathrm{T}}-\tfrac{1}{2}(\boldsymbol{\nabla}\mathbf{s})^{\mathrm{T}}\cdot\boldsymbol{\nabla}\mathbf{s},\end{aligned}\tag{13.43}$$

$$2\omega\mathbf{K}_{\boldsymbol{\tau}}^\Sigma = s_n(\partial_n\mathbf{s})-\mathbf{s}(\partial_n s_n),\tag{13.44}$$

$$2\omega\mathbf{K}_\varpi = -s_n\mathbf{s}.\tag{13.45}$$

The first term in equation (13.39) represents the first-order effect of the perturbations in the density $\delta\rho$, the elastic tensor $\delta\boldsymbol{\Gamma}$ and the gravitational potential $\delta\Phi$, the second and third terms account for the displacement δd of the boundaries Σ, and the fourth and fifth terms account for the initial deviatoric stress $\boldsymbol{\tau}$ in $\oplus_{\mathrm{S}}$ and on Σ_{FS}. The sixth and final term is also associated with the deviatoric stress; however, it can be calculated entirely in terms of the pressure perturbation δp in $\oplus_{\mathrm{F}}$, since $\Delta\varpi$ is continuous across Σ_{FS}.

★13.1.5 Hydrostatic perturbation

If the final as well as the initial model is hydrostatic, so that $\boldsymbol{\tau} = \mathbf{0}$, the eigenfrequency perturbation $\delta\omega$ reduces to

$$\delta\omega = \int_{\oplus} \delta\oplus K_{\oplus}\, dV + \int_{\Sigma} \delta d\, [K_{\mathrm{d}}]_{-}^{+}\, d\Sigma + \int_{\Sigma_{\mathrm{FS}}} \boldsymbol{\nabla}^{\Sigma}(\delta d) \cdot [\mathbf{K}_{\mathrm{d}}]_{-}^{+}\, d\Sigma. \tag{13.46}$$

The final surface integral in equation (13.39) vanishes in addition to the terms involving $\boldsymbol{\tau}$, since $\boldsymbol{\nabla}^{\Sigma}(\Delta\varpi) = \mathbf{0}$ on Σ_{FS}. It is noteworthy that it is unnecessary to specify either the initial hydrostatic pressure p in the starting model or the perturbation in the pressure δp; the only perturbations that must be considered are $\delta\rho$, $\delta\boldsymbol{\Gamma}$, $\delta\Phi$, and δd. Strictly speaking, equation (13.46) is valid only if the initial and final models are both non-rotating, hydrostatic spheres; more generally, however, it can be used to determine the effect of a general, non-hydrostatic perturbation in the quasi-hydrostatic approximation.

We can also obtain the result (13.46) by applying Rayleigh's principle to the modified action $\mathcal{I}'$ governing a *hydrostatic* Earth model. We rewrite equation (4.162) in the form

$$\mathcal{I}' = \int_{\bigcirc} L'(\mathbf{s}, \boldsymbol{\nabla}\mathbf{s}, \boldsymbol{\nabla}\phi; \omega, \oplus)\, dV = 0, \tag{13.47}$$

where the symbol $\oplus$ now denotes the ensemble of hydrostatic Earth model parameters ρ, $\boldsymbol{\Gamma}$, $\boldsymbol{\nabla}\Phi$ and $\boldsymbol{\nabla}\boldsymbol{\nabla}\Phi$, and the modified Lagrangian density L' is given by

$$L' = \tfrac{1}{2}[\omega^2\rho\mathbf{s}\cdot\mathbf{s} - \boldsymbol{\varepsilon}{:}\boldsymbol{\Gamma}{:}\boldsymbol{\varepsilon} - 2\rho\mathbf{s}\cdot\boldsymbol{\nabla}\phi - \rho\mathbf{s}\cdot\boldsymbol{\nabla}\boldsymbol{\nabla}\Phi\cdot\mathbf{s} - \rho\boldsymbol{\nabla}\Phi\cdot(\mathbf{s}\cdot\boldsymbol{\nabla}\mathbf{s} - \mathbf{s}\boldsymbol{\nabla}\cdot\mathbf{s}) - (4\pi G)^{-1}\boldsymbol{\nabla}\phi\cdot\boldsymbol{\nabla}\phi]. \tag{13.48}$$

The total variation of (13.47) with respect to all of its variables is

$$\delta\mathcal{I}'_{\mathrm{total}} = \int_{\oplus} [\delta\mathbf{s}\cdot(\partial_{\mathbf{s}}L') + \boldsymbol{\nabla}(\delta\mathbf{s})\cdot(\partial_{\boldsymbol{\nabla}\mathbf{s}}L')]\, dV$$

$$+ \int_{\bigcirc} \boldsymbol{\nabla}(\delta\phi) \cdot (\partial_{\boldsymbol{\nabla}\phi} L')\, dV$$
$$+ \int_{\oplus} [\delta\omega(\partial_\omega L') + \delta\oplus(\partial_\oplus L')]\, dV$$
$$- \int_{\Sigma} \delta d\, [L']_-^+\, d\Sigma, \tag{13.49}$$

where

$$\delta\oplus(\partial_\oplus L') = \delta\rho(\partial_\rho L') + \boldsymbol{\delta\Gamma} \,\vdots\, (\partial_{\boldsymbol{\Gamma}} L')$$
$$+ \boldsymbol{\nabla}(\delta\Phi)\cdot(\partial_{\boldsymbol{\nabla}\Phi} L') + \boldsymbol{\nabla\nabla}(\delta\Phi) : (\partial_{\boldsymbol{\nabla\nabla}\Phi} L') \tag{13.50}$$

denotes the perturbation with respect to the hydrostatic Earth model parameters, and the surface integral accounts for the change in the region of integration. Upon applying Gauss' theorem to equation (13.49) we obtain

$$\delta\mathcal{I}'_{\mathrm{total}} = \int_{\oplus} \boldsymbol{\delta}\mathbf{s} \cdot [\partial_{\mathbf{s}} L' - \boldsymbol{\nabla} \cdot (\partial_{\boldsymbol{\nabla}\mathbf{s}} L')]\, dV$$
$$- \int_{\Sigma} [\boldsymbol{\delta}\mathbf{s} \cdot \{\hat{\mathbf{n}} \cdot (\partial_{\boldsymbol{\nabla}\mathbf{s}} L')\}]_-^+\, d\Sigma$$
$$- \int_{\bigcirc} \delta\phi\, [\boldsymbol{\nabla} \cdot (\partial_{\boldsymbol{\nabla}\phi} L')]\, dV - \int_{\Sigma} [\delta\phi\, \{\hat{\mathbf{n}} \cdot (\partial_{\boldsymbol{\nabla}\phi} L')\}]_-^+\, d\Sigma$$
$$+ \int_{\oplus} [\delta\omega(\partial_\omega L') + \delta\oplus(\partial_\oplus L')]\, dV$$
$$- \int_{\Sigma} \delta d\, [L']_-^+\, d\Sigma = 0. \tag{13.51}$$

The volume integrals involving the eigenfunction perturbations $\boldsymbol{\delta}\mathbf{s}$ and $\delta\phi$ vanish, since $\partial_{\mathbf{s}} L' - \boldsymbol{\nabla}\cdot(\partial_{\boldsymbol{\nabla}\mathbf{s}} L') = 0$ and $\boldsymbol{\nabla}\cdot(\partial_{\boldsymbol{\nabla}\phi} L') = 0$ are the momentum equation (13.5) and Poisson's equation (13.9), respectively. The surface integral involving the potential perturbation $\delta\phi$ can be evaluated using equation (13.37), as in a non-hydrostatic Earth, whereas that involving $\boldsymbol{\delta}\mathbf{s}$ can be evaluated with the aid of the hydrostatic boundary condition

$$[\boldsymbol{\delta}\mathbf{s} \cdot (\hat{\mathbf{n}} \cdot \mathbf{T})]_-^+ = -\delta d\, [(\hat{\mathbf{n}} \cdot \mathbf{T}) \cdot \partial_n \mathbf{s}]_-^+$$
$$+ \boldsymbol{\nabla}^\Sigma(\delta d) \cdot [(\hat{\mathbf{n}} \cdot \mathbf{T} \cdot \hat{\mathbf{n}})\mathbf{s}]_-^+. \tag{13.52}$$

Equation (13.52), which is valid on all of Σ, follows from the perturbed kinematic conditions (13.33)–(13.34) together with the continuity and normality conditions (13.6)–(13.8) governing the incremental traction $\hat{\mathbf{n}} \cdot \mathbf{T}$.

We can write the first-order eigenfrequency perturbation $\delta\omega$ entirely in terms of the hydrostatic Lagrangian density L' and its derivatives in a manner that is independent of the normalization imposed upon the unperturbed eigenfunction $\mathbf{s}$:

$$\delta\omega \int_{\oplus} \partial_\omega L' \, dV = - \int_{\oplus} \delta\oplus (\partial_\oplus L') \, dV + \int_{\Sigma} \delta d \, [L' - \hat{\mathbf{n}} \cdot (\partial_{\boldsymbol{\nabla}\mathbf{s}} L') \cdot \partial_n \mathbf{s} - \hat{\mathbf{n}} \cdot (\partial_{\boldsymbol{\nabla}\phi} L') \partial_n \phi]_-^+ \, d\Sigma + \int_{\Sigma_{\mathrm{FS}}} \boldsymbol{\nabla}^\Sigma (\delta d) \cdot [\{\hat{\mathbf{n}} \cdot (\partial_{\boldsymbol{\nabla}\mathbf{s}} L') \cdot \hat{\mathbf{n}}\}\mathbf{s}]_-^+ \, d\Sigma. \tag{13.53}$$

It is readily verified that

$$\omega \delta\oplus K_\oplus = -\delta\oplus (\partial_\oplus L'), \tag{13.54}$$

$$\omega K_{\mathrm{d}} = L' - \hat{\mathbf{n}} \cdot (\partial_{\boldsymbol{\nabla}\mathbf{s}} L') \cdot \partial_n \mathbf{s} - \hat{\mathbf{n}} \cdot (\partial_{\boldsymbol{\nabla}\phi} L') \, \partial_n \phi, \tag{13.55}$$

$$\omega \mathbf{K}_{\mathrm{d}} = [\hat{\mathbf{n}} \cdot (\partial_{\boldsymbol{\nabla}\mathbf{s}} L') \cdot \hat{\mathbf{n}}]\mathbf{s}, \tag{13.56}$$

so that equation (13.53) is identical to the previous result (13.46), if we adopt the normalization (13.32). The resemblance between the three-dimensional expression (13.53) and the analogous formula (9.7) for a purely radial perturbation is noteworthy.

★13.1.6 An alternative derivation

We can also obtain the perturbation in the eigenfrequency $\delta\omega$ of both a hydrostatic and a general, non-hydrostatic Earth model using a straightforward brute-force approach, which eschews Rayleigh's principle. We begin by perturbing the linearized equations of motion and boundary conditions to find the relations governing the perturbed quantities $\delta\omega$, $\boldsymbol{\delta}\mathbf{s}$ and $\delta\phi$. The object is to find $\delta\omega$ without having to solve for $\boldsymbol{\delta}\mathbf{s}$ and $\delta\phi$; we can do this by manipulating the perturbed equations in a very straightforward manner.

In the hydrostatic case, the perturbed momentum equation is

$$-2\omega \, \delta\omega \, \rho \mathbf{s} - \omega^2 \delta\rho \, \mathbf{s} - \omega^2 \rho \, \boldsymbol{\delta}\mathbf{s} - \boldsymbol{\nabla} \cdot \boldsymbol{\delta}\mathbf{T} + \boldsymbol{\nabla}[\delta\rho \, \mathbf{s} \cdot \boldsymbol{\nabla}\Phi + \rho \, \boldsymbol{\delta}\mathbf{s} \cdot \boldsymbol{\nabla}\Phi + \rho \mathbf{s} \cdot \boldsymbol{\nabla}(\delta\Phi)] + \delta\rho \boldsymbol{\nabla}\phi + \rho \boldsymbol{\nabla}(\delta\phi) - [\boldsymbol{\nabla} \cdot (\delta\rho \, \mathbf{s} + \rho \, \boldsymbol{\delta}\mathbf{s})]\boldsymbol{\nabla}\Phi - [\boldsymbol{\nabla} \cdot (\rho \mathbf{s})]\boldsymbol{\nabla}(\delta\Phi) = \mathbf{0}, \tag{13.57}$$

and the perturbed Poisson's equation is

$$\boldsymbol{\nabla} \cdot (\boldsymbol{\delta\xi}) = 0. \tag{13.58}$$

The perturbed incremental Lagrangian Cauchy stress $\boldsymbol{\delta}\mathbf{T}$ and auxiliary gravity vector $\boldsymbol{\delta\xi}$ are given, respectively, by

$$\boldsymbol{\delta}\mathbf{T} = \boldsymbol{\delta\Gamma} : \boldsymbol{\varepsilon} + \boldsymbol{\Gamma} : \boldsymbol{\delta\varepsilon}, \tag{13.59}$$

$$\delta\boldsymbol{\xi} = (4\pi G)^{-1}\boldsymbol{\nabla}(\delta\phi) + \delta\rho\,\mathbf{s} + \rho\,\delta\mathbf{s}. \tag{13.60}$$

These equations must be solved subject to the perturbed dynamical and gravitational boundary conditions

$$[\hat{\mathbf{n}}\cdot\delta\mathbf{T}]_-^+ = -\delta d\,[\hat{\mathbf{n}}\cdot\partial_n\mathbf{T}]_-^+ + \boldsymbol{\nabla}^\Sigma(\delta d)\cdot[\mathbf{T}]_-^+, \tag{13.61}$$

$$[\hat{\mathbf{n}}\cdot\delta\boldsymbol{\xi}]_-^+ = -\delta d\,[\hat{\mathbf{n}}\cdot\partial_n\boldsymbol{\xi}]_-^+ + \boldsymbol{\nabla}^\Sigma(\delta d)\cdot[\boldsymbol{\xi}]_-^+ \tag{13.62}$$

on all of Σ.

We proceed to take the dot product of $\delta\mathbf{s}$ with (13.5), minus the dot product of $\mathbf{s}$ with (13.57), minus the product of $\delta\phi$ with (13.9), plus the product of ϕ with (13.58), and integrate the result over all of space. Upon applying Gauss' theorem, and substituting (13.59) and (13.60), we find that all of the terms involving the eigenfunction perturbations $\delta\mathbf{s}$ and $\delta\phi$ cancel, leaving

$$\begin{aligned}\delta\omega = \frac{1}{2\omega}\int_\oplus [\delta\rho\{-\omega^2\mathbf{s}\cdot\mathbf{s} + 2\mathbf{s}\cdot\boldsymbol{\nabla}\phi + \mathbf{s}\cdot\boldsymbol{\nabla}\boldsymbol{\nabla}\Phi\cdot\mathbf{s}\\ + \boldsymbol{\nabla}\Phi\cdot(\mathbf{s}\cdot\boldsymbol{\nabla}\mathbf{s} - \mathbf{s}\boldsymbol{\nabla}\cdot\mathbf{s})\} + \boldsymbol{\varepsilon}{:}\delta\boldsymbol{\Gamma}{:}\boldsymbol{\varepsilon}\\ + \rho\boldsymbol{\nabla}(\delta\Phi)\cdot(\mathbf{s}\cdot\boldsymbol{\nabla}\mathbf{s} - \mathbf{s}\boldsymbol{\nabla}\cdot\mathbf{s})\\ + \rho\mathbf{s}\cdot\boldsymbol{\nabla}\boldsymbol{\nabla}(\delta\Phi)\cdot\mathbf{s}]\,dV\\ + \frac{1}{2\omega}\int_\Sigma [\hat{\mathbf{n}}\cdot\delta\mathbf{T}\cdot\mathbf{s} - \hat{\mathbf{n}}\cdot\mathbf{T}\cdot\delta\mathbf{s} + (\hat{\mathbf{n}}\cdot\delta\boldsymbol{\xi})\phi - (\hat{\mathbf{n}}\cdot\boldsymbol{\xi})\delta\phi]_-^+\,d\Sigma.\end{aligned} \tag{13.63}$$

Since the traction $\hat{\mathbf{n}}\cdot\mathbf{T}$ is continuous on Σ and a normal vector on Σ_{FS}, we must have $[\Delta(\hat{\mathbf{n}}\cdot\mathbf{T}\cdot\mathbf{s})]_-^+ = 0$, or, equivalently,

$$\begin{aligned}&[\hat{\mathbf{n}}\cdot\delta\mathbf{T}\cdot\mathbf{s} + \hat{\mathbf{n}}\cdot\mathbf{T}\cdot\delta\mathbf{s}]_-^+\\ &\quad = -\delta d\,[\hat{\mathbf{n}}\cdot\partial_n\mathbf{T}\cdot\mathbf{s} + \hat{\mathbf{n}}\cdot\mathbf{T}\cdot\partial_n\mathbf{s}]_-^+ + \boldsymbol{\nabla}^\Sigma(\delta d)\cdot[\mathbf{T}\cdot\mathbf{s}]_-^+\end{aligned} \tag{13.64}$$

on all boundaries. The continuity of the two quantities ϕ and $\hat{\mathbf{n}}\cdot\boldsymbol{\xi}$ likewise guarantees that $[\Delta\{(\hat{\mathbf{n}}\cdot\boldsymbol{\xi})\phi\}]_-^+ = 0$, or, equivalently,

$$\begin{aligned}&[(\hat{\mathbf{n}}\cdot\delta\boldsymbol{\xi})\phi + (\hat{\mathbf{n}}\cdot\boldsymbol{\xi})\delta\phi]_-^+\\ &\quad = -\delta d\,[(\hat{\mathbf{n}}\cdot\boldsymbol{\xi})\partial_n\phi + (\hat{\mathbf{n}}\cdot\partial_n\boldsymbol{\xi})\phi]_-^+ + \boldsymbol{\nabla}^\Sigma(\delta d)\cdot[\boldsymbol{\xi}\phi]_-^+.\end{aligned} \tag{13.65}$$

We can eliminate the terms involving the more highly differentiated perturbations $\hat{\mathbf{n}}\cdot\delta\mathbf{T}$ and $\hat{\mathbf{n}}\cdot\delta\boldsymbol{\xi}$ from the surface integral in (13.63) by making use of the results (13.64)–(13.65); the remaining terms are then either of the same type that have already been evaluated in the previous derivation, or they can then be evaluated using the two-dimensional version of Gauss' theorem. The details of this reduction procedure are given by Dahlen (1976);

the final result is

$$\int_{\Sigma}[\hat{\mathbf{n}}\cdot\boldsymbol{\delta}\mathbf{T}\cdot\mathbf{s}-\hat{\mathbf{n}}\cdot\mathbf{T}\cdot\delta\mathbf{s}+(\hat{\mathbf{n}}\cdot\delta\boldsymbol{\xi})\phi-(\hat{\mathbf{n}}\cdot\boldsymbol{\xi})\delta\phi]_-^+\,d\Sigma$$
$$=\int_{\Sigma}\delta d\,[\rho\{\omega^2\mathbf{s}\cdot\mathbf{s}-2\mathbf{s}\cdot\boldsymbol{\nabla}\phi-\mathbf{s}\cdot\boldsymbol{\nabla\nabla}\Phi\cdot\mathbf{s}$$
$$-\boldsymbol{\nabla}\Phi\cdot(\mathbf{s}\cdot\boldsymbol{\nabla}\mathbf{s}-\mathbf{s}\boldsymbol{\nabla}\cdot\mathbf{s})\}\tag{13.66}$$
$$-\boldsymbol{\varepsilon}:\boldsymbol{\Gamma}:\boldsymbol{\varepsilon}-(4\pi G)^{-1}\boldsymbol{\nabla}\phi\cdot\boldsymbol{\nabla}\phi$$
$$+2(\hat{\mathbf{n}}\cdot\mathbf{T})\cdot\partial_n\mathbf{s}+2(\hat{\mathbf{n}}\cdot\boldsymbol{\xi})\partial_n\phi]_-^+\,d\Sigma$$
$$-\int_{\Sigma_{\mathrm{FS}}}\boldsymbol{\nabla}^{\Sigma}(\delta d)\cdot[2(\hat{\mathbf{n}}\cdot\mathbf{T}\cdot\hat{\mathbf{n}})\mathbf{s}]_-^+\,d\Sigma,$$

which shows that equation (13.63) agrees with (13.46) and (13.40)–(13.42).

On a general, non-hydrostatic Earth, the perturbed momentum equation is

$$-2\omega\,\delta\omega\,\rho\mathbf{s}-\omega^2\delta\rho\,\mathbf{s}-\omega^2\rho\,\delta\mathbf{s}-\boldsymbol{\nabla}\cdot\boldsymbol{\delta}\widetilde{\mathbf{T}}$$
$$+\,\delta\rho\boldsymbol{\nabla}\phi+\rho\boldsymbol{\nabla}(\delta\phi)+(\delta\rho\,\mathbf{s}+\rho\,\delta\mathbf{s})\cdot\boldsymbol{\nabla\nabla}\Phi$$
$$+\,\rho\mathbf{s}\cdot\boldsymbol{\nabla\nabla}(\delta\Phi)=\mathbf{0},\tag{13.67}$$

the perturbed elastic constitutive relation is

$$\boldsymbol{\delta}\widetilde{\mathbf{T}}=\boldsymbol{\delta\Lambda}:\boldsymbol{\nabla}\mathbf{s}+\boldsymbol{\Lambda}:\boldsymbol{\nabla}(\delta\mathbf{s}),\tag{13.68}$$

and the perturbed dynamical boundary conditions are

$$\hat{\mathbf{n}}\cdot\boldsymbol{\delta}\widetilde{\mathbf{T}}=-\delta d\,(\hat{\mathbf{n}}\cdot\partial_n\widetilde{\mathbf{T}})+\boldsymbol{\nabla}^{\Sigma}(\delta d)\cdot\widetilde{\mathbf{T}}\quad\text{on }\partial\oplus,\tag{13.69}$$

$$[\hat{\mathbf{n}}\cdot\boldsymbol{\delta}\widetilde{\mathbf{T}}]_-^+=-\delta d\,[\hat{\mathbf{n}}\cdot\partial_n\widetilde{\mathbf{T}}]_-^++\boldsymbol{\nabla}^{\Sigma}(\delta d)\cdot[\widetilde{\mathbf{T}}]_-^+\quad\text{on }\Sigma_{\mathrm{SS}},\tag{13.70}$$

$$[\boldsymbol{\delta}\widetilde{\mathbf{t}}]_-^+=\hat{\mathbf{n}}[\hat{\mathbf{n}}\cdot\boldsymbol{\delta}\widetilde{\mathbf{t}}]_-^+=-\delta d\,[\partial_n\widetilde{\mathbf{t}}]_-^+\quad\text{on }\Sigma_{\mathrm{FS}}.\tag{13.71}$$

Applying the same multiplication-integration-cancellation procedure as in the hydrostatic case, we obtain

$$\delta\omega=\frac{1}{2\omega}\int_{\oplus}[\delta\rho\,(-\omega^2\mathbf{s}\cdot\mathbf{s}+2\mathbf{s}\cdot\boldsymbol{\nabla}\phi+\mathbf{s}\cdot\boldsymbol{\nabla\nabla}\Phi\cdot\mathbf{s})\tag{13.72}$$
$$+\,\boldsymbol{\nabla}\mathbf{s}:\boldsymbol{\delta\Lambda}:\boldsymbol{\nabla}\mathbf{s}+\rho\mathbf{s}\cdot\boldsymbol{\nabla\nabla}(\delta\Phi)\cdot\mathbf{s}]\,dV$$
$$+\,\frac{1}{2\omega}\int_{\Sigma}[\hat{\mathbf{n}}\cdot\boldsymbol{\delta}\widetilde{\mathbf{T}}\cdot\mathbf{s}-\hat{\mathbf{n}}\cdot\widetilde{\mathbf{T}}\cdot\delta\mathbf{s}+(\hat{\mathbf{n}}\cdot\delta\boldsymbol{\xi})\phi-(\hat{\mathbf{n}}\cdot\boldsymbol{\xi})\delta\phi]_-^+\,d\Sigma.$$

The gravitational terms in the surface integral are identical to those in equation (13.63), and they can be evaluated in the same manner. It is more difficult to eliminate the displacement and traction perturbations $\delta\mathbf{s}$

and $\hat{\mathbf{n}}\cdot\delta\widetilde{\mathbf{T}}$ from the remaining terms; however, this can be accomplished with the aid of the perturbed boundary conditions (13.70)–(13.71). We find, after a tedious calculation, that

$$\begin{aligned}\int_{\Sigma}&[\hat{\mathbf{n}}\cdot\delta\widetilde{\mathbf{T}}\cdot\mathbf{s}-\hat{\mathbf{n}}\cdot\widetilde{\mathbf{T}}\cdot\delta\mathbf{s}+(\hat{\mathbf{n}}\cdot\delta\boldsymbol{\xi})\phi-(\hat{\mathbf{n}}\cdot\boldsymbol{\xi})\delta\phi]_-^+\,d\Sigma\\&=\int_{\Sigma}\delta d\,[2L'+2(\hat{\mathbf{n}}\cdot\widetilde{\mathbf{T}})\cdot\partial_n\mathbf{s}+2(\hat{\mathbf{n}}\cdot\boldsymbol{\xi})\partial_n\phi\\&\qquad+\varpi\{(\boldsymbol{\nabla}^{\Sigma}\mathbf{s}):(\boldsymbol{\nabla}^{\Sigma}\mathbf{s})^{\mathrm{T}}-(\boldsymbol{\nabla}^{\Sigma}\cdot\mathbf{s})^2\}\\&\qquad\quad+(\boldsymbol{\nabla}^{\Sigma}\varpi)\cdot\{s_n(\mathbf{s}\cdot\boldsymbol{\nabla}^{\Sigma}\hat{\mathbf{n}})-\mathbf{s}(\boldsymbol{\nabla}^{\Sigma}\cdot\mathbf{s})\}\\&\qquad\qquad-(\boldsymbol{\nabla}^{\Sigma}\boldsymbol{\nabla}^{\Sigma}\varpi):\mathbf{ss}]_-^+\,d\Sigma\\&\quad-\int_{\Sigma_{\mathrm{FS}}}\boldsymbol{\nabla}^{\Sigma}(\delta d)\cdot[2(\hat{\mathbf{n}}\cdot\widetilde{\mathbf{t}})\mathbf{s}]_-^+\,d\Sigma\\&\quad+\int_{\Sigma_{\mathrm{FS}}}\Delta\varpi\,[2(\mathbf{s}\cdot\boldsymbol{\nabla}^{\Sigma}s_n)-\mathbf{s}\cdot(\boldsymbol{\nabla}^{\Sigma}\hat{\mathbf{n}})\cdot\mathbf{s}]_-^+\,d\Sigma,\end{aligned}\tag{13.73}$$

so that equation (13.72) agrees with (13.38), as expected.

In addition to lessening the algebraic labor required to find the perturbation $\delta\omega$, the use of Rayleigh's principle also clarifies the origin of the "extra" terms proportional to δd and $\boldsymbol{\nabla}^{\Sigma}(\delta d)$. As we have seen, these terms arise whenever the perturbation to the Earth model includes a slight displacement of the boundary Σ, precisely because the resulting perturbations $\delta\mathbf{s}$ and $\delta\phi$ are then *inadmissible*.

13.1.7 Spherical starting model

Inasmuch as the departures of the Earth away from sphericity are small, it is natural to take the unperturbed model to be *spherically symmetric* as well as hydrostatic. The static equilibrium conditions (13.19)–(13.20) in the perturbed Earth model reduce in that case to

$$-\boldsymbol{\nabla}(\delta p)+\boldsymbol{\nabla}\cdot\boldsymbol{\tau}=\delta\rho\, g\hat{\mathbf{r}}+\rho\boldsymbol{\nabla}(\delta\Phi)\quad\text{in }\oplus,\tag{13.74}$$

$$[-\hat{\mathbf{r}}\,\delta p+\hat{\mathbf{r}}\cdot\boldsymbol{\tau}]_-^+=-\delta d\,\hat{\mathbf{r}}\,g[\rho]_-^+\quad\text{on }\Sigma,\tag{13.75}$$

where g is the acceleration of gravity, and where we have used the initial conditions $\dot{p}+\rho g=0$ in $\oplus$ and $[p]_-^+=0$ on Σ. The restrictions imposed upon the perturbations $\delta\rho$, $\delta\Phi$, δp, $\boldsymbol{\tau}$ and δd by equations (13.74)–(13.75) have been considered by Backus (1967), Woodhouse & Dahlen (1978) and Wahr & de Vries (1989).

In examining these restrictions, it is convenient to decompose a general perturbation δq into its *spherical* and *aspherical parts*:

$$\delta q=\delta\bar{q}+\delta\hat{q},\tag{13.76}$$

where $\delta\bar{q}(r)$ is the average of δq over the spherical shell of radius r; the average of the aspherical perturbation $\delta\hat{q}$ over every spherical shell is zero:

$$\int_\Omega \delta\hat{q}\,d\Omega = 0. \tag{13.77}$$

Both the density perturbation $\delta\rho = \delta\bar{\rho} + \delta\hat{\rho}$ in the solid regions of the Earth $\oplus_{\rm S}$ and the radial displacements $\delta d = \delta\bar{d} + \delta\hat{d}$ of the boundaries Σ can be specified arbitrarily (except for the special case noted below). In the fluid core and oceans $\oplus_{\rm F}$ the perturbation in the incompressibility $\delta\kappa = \delta\bar{\kappa} + \delta\hat{\kappa}$ and the spherical density perturbation $\delta\bar{\rho}$ can be specified, but the perturbation in the pressure $\delta p = \delta\bar{p} + \delta\hat{p}$ and the aspherical density perturbation $\delta\hat{\rho}$ cannot. The spherical pressure perturbation $\delta\bar{p}$ in $\oplus_{\rm F}$ is determined, to within an additive constant, by the spherically averaged version of the mechanical equilibrium condition (13.74):

$$\frac{d}{dr}(\delta\bar{p}) = -(\delta\bar{\rho})\,g - \rho\frac{d}{dr}(\delta\bar{\Phi}), \tag{13.78}$$

where

$$\begin{aligned}\delta\bar{\Phi} = {}& -\frac{4\pi G}{r}\left\{\int_0^r \delta\bar{\rho}\,r^2\,dr - \sum_{d<r}\delta\bar{d}\,d^2[\rho]_-^+\right\}\\ & - 4\pi G\left\{\int_r^a \delta\bar{\rho}\,r\,dr - \sum_{d\geq r}\delta\bar{d}\,d\,[\rho]_-^+\right\}.\end{aligned} \tag{13.79}$$

There is also a restriction placed by the aspherical part of this condition upon the perturbations $\delta\hat{p}$ and $\delta\hat{\rho}$; they must be related to $\delta\hat{\Phi}$ in $\oplus_{\rm F}$ by

$$\delta\hat{p} = -\rho\,\delta\hat{\Phi}, \qquad \delta\hat{\rho} = \dot{\rho}g^{-1}\delta\hat{\Phi}, \tag{13.80}$$

where the dot denotes the derivative with respect to r. The aspherical potential perturbation $\delta\hat{\Phi}$ can be determined everywhere by solving the boundary-value problem

$$\nabla^2(\delta\hat{\Phi}) = \begin{cases} 4\pi G\,\delta\hat{\rho} & \text{in } \oplus_{\rm S}\\ 4\pi G\,\dot{\rho}g^{-1}\delta\hat{\Phi} & \text{in } \oplus_{\rm F}\\ 0 & \text{in } \bigcirc - \oplus,\end{cases} \tag{13.81}$$

$$[\delta\hat{\Phi}]_-^+ = 0 \quad\text{and}\quad [\partial_r(\delta\hat{\Phi}) + 4\pi G\rho\,\delta\hat{d}]_-^+ = 0 \quad \text{on } \Sigma. \tag{13.82}$$

The requirement that the fluid regions be in hydrostatic equilibrium is seen to impose a strong constraint—the aspherical density perturbation $\delta\hat{\rho}$ is completely determined, and the pressure perturbation δp is determined to within an irrelevant constant, by equations (13.78)–(13.80). In the solid regions $\oplus_{\rm S}$, the equilibrium conditions (13.74)–(13.75) place only three restrictions upon the six degrees of freedom inherent in the perturbations δp, $\delta\boldsymbol{\tau}$. Backus (1967) shows how to construct the complete catalogue of equilbrium stress fields consistent with a specified $\delta\rho$ and δd, embodying the three remaining degrees of freedom. We note, finally, that if the outer free surface $\partial\oplus$ is underlain by a fluid ocean, then only the spherically averaged displacement $\delta\bar{d}$ and not $\delta\hat{d}$ may be specified for that boundary. Equation (13.75) then determines the aspherical perturbation

$$\delta\hat{d} = -g^{-1}\delta\hat{\Phi}, \tag{13.83}$$

and the boundary condition (13.82) must be modified accordingly.

The initial gravitational potential Φ of a spherical hydrostatic model satisfies the identities $\boldsymbol{\nabla}\Phi = g\hat{\mathbf{r}}$ and $\boldsymbol{\nabla}\boldsymbol{\nabla}\Phi = r^{-1}g(\mathbf{I} - 3\hat{\mathbf{r}}\hat{\mathbf{r}}) + 4\pi G\rho\,\hat{\mathbf{r}}\hat{\mathbf{r}}$, where $\mathbf{I}$ is the identity tensor. Upon inserting these simplifications into equations (13.40)–(13.45), we find that the eigenfrequency perturbation $\delta\omega$ is of the general form (13.39), with

$$\begin{aligned}2\omega\,\delta\oplus K_{\oplus} = {} & \delta\rho(-\omega^2\mathbf{s}\cdot\mathbf{s} + 2\mathbf{s}\cdot\boldsymbol{\nabla}\phi + 4\pi G\rho s_r^2 + g\Upsilon)\\ & + \boldsymbol{\varepsilon}\!:\!\boldsymbol{\delta\Gamma}\!:\!\boldsymbol{\varepsilon} + \rho\boldsymbol{\nabla}(\delta\Phi)\cdot(\mathbf{s}\cdot\boldsymbol{\nabla}\mathbf{s} - \mathbf{s}\boldsymbol{\nabla}\cdot\mathbf{s})\\ & + \rho\mathbf{s}\cdot\boldsymbol{\nabla}\boldsymbol{\nabla}(\delta\Phi)\cdot\mathbf{s},\end{aligned} \tag{13.84}$$

$$\begin{aligned}2\omega K_{\rm d} = {} & \rho(\omega^2\mathbf{s}\cdot\mathbf{s} - 2\mathbf{s}\cdot\boldsymbol{\nabla}\phi - 8\pi G\rho s_r^2 - g\Upsilon)\\ & - \boldsymbol{\varepsilon}\!:\!\boldsymbol{\Gamma}\!:\!\boldsymbol{\varepsilon} + 2\hat{\mathbf{r}}\cdot\mathbf{T}\cdot\partial_r\mathbf{s},\end{aligned} \tag{13.85}$$

$$2\omega\mathbf{K}_{\rm d} = -2(\hat{\mathbf{r}}\cdot\mathbf{T}\cdot\hat{\mathbf{r}})\mathbf{s}, \tag{13.86}$$

$$2\omega\mathbf{K}_{\boldsymbol{\tau}}^{\Sigma} = s_r(\partial_r\mathbf{s}) - \mathbf{s}(\partial_r s_r), \tag{13.87}$$

$$2\omega\mathbf{K}_{\varpi} = -s_r\mathbf{s}, \tag{13.88}$$

where we have defined the auxiliary quantity

$$\Upsilon = \mathbf{s}\cdot\boldsymbol{\nabla}s_r - s_r(\boldsymbol{\nabla}\cdot\mathbf{s}) - 2r^{-1}s_r^2. \tag{13.89}$$

We have used the continuity of the radial vector $\xi_r = (4\pi G)^{-1}\partial_r\phi + \rho s_r$ in reducing the kernel $K_{\rm d}$ to the form (13.85).

The perturbation $\Delta\varpi$ is continuous across $\Sigma_{\rm FS}$, and on the fluid side it is given by

$$\Delta\varpi = \delta\bar{p} - \rho_{\rm F}g\,\delta\bar{d} - \rho_{\rm F}(\delta\hat{\Phi} + g\,\delta\hat{d}), \tag{13.90}$$

where ρ_F is the fluid density. It follows, since $\boldsymbol{\nabla}^{\Sigma}(\delta\bar{p}) = \mathbf{0}$ and $\boldsymbol{\nabla}^{\Sigma}(\delta\bar{d}) = \mathbf{0}$, that the final integral involving the kernel $\mathbf{K}_{\varpi}$ in equation (13.39) can be written in terms of the aspherical perturbations $\delta\hat{\Phi}$ and $\delta\hat{d}$ in the form

$$\int_{\Sigma_{FS}} \boldsymbol{\nabla}^{\Sigma}(\Delta\varpi)\cdot[\mathbf{K}_{\varpi}]_-^+\,d\Sigma$$
$$= \frac{1}{2\omega}\int_{\Sigma_{FS}} \rho_F s_r \boldsymbol{\nabla}^{\Sigma}(\delta\hat{\Phi} + g\,\delta\hat{d})\cdot[\mathbf{s}]_-^+\,d\Sigma. \tag{13.91}$$

It is clear from (13.91) that this term arises only in the case of an aspherical boundary perturbation $\delta\hat{d}$ that requires a deviatoric initial stress $\boldsymbol{\tau}$ in $\oplus_S$ for its support. All of the internal boundaries are equipotential surfaces, i.e., equation (13.83) prevails on all of Σ in the case of a purely hydrostatic perturbation. At the present time, the deviatoric stress within the Earth is not considered well enough known to be independently specified, so its effect is usually ignored. In a consistent application of the quasi-hydrostatic approximation, the term involving $\mathbf{K}_{\varpi}$ should be ignored in addition to the terms involving $\mathbf{K}_{\boldsymbol{\tau}}$ and $\mathbf{K}_{\boldsymbol{\tau}}^{\Sigma}$, even though it can be calculated using equation (13.91) without explicit knowledge of $\boldsymbol{\tau}$. We shall adopt this point of view in the future, accounting only for the volumetric density and elasticity perturbations $\delta\rho$ and $\boldsymbol{\delta\Gamma}$ and the boundary perturbations δd.

*13.1.8 Spherical perturbation

If the final as well as the initial model is spherically symmetric, we may write, by analogy with the results for the corresponding unperturbed quantities, $\boldsymbol{\nabla}(\delta\Phi) = \delta g\,\hat{\mathbf{r}}$ and $\boldsymbol{\nabla\nabla}(\delta\Phi) = r^{-1}\delta g\,(\mathbf{I} - 3\hat{\mathbf{r}}\hat{\mathbf{r}}) + 4\pi G\,\delta\rho\,\hat{\mathbf{r}}\hat{\mathbf{r}}$. The volumetric perturbation kernel (13.84) reduces in this case to

$$2\omega\delta\oplus K_{\oplus} = \delta\rho(-\omega^2\mathbf{s}\cdot\mathbf{s} + 2\mathbf{s}\cdot\boldsymbol{\nabla}\phi + 8\pi G\rho s_r^2 + g\Upsilon)$$
$$+\,\boldsymbol{\varepsilon}\!:\!\boldsymbol{\delta\Gamma}\!:\!\boldsymbol{\varepsilon} + \rho\,\delta g\Upsilon. \tag{13.92}$$

We can write the eigenfrequency perturbation $\delta\omega$ in the form

$$\delta\omega\int_{\oplus}\partial_{\omega}L'\,dV = -\int_{\oplus}[\delta\rho(\partial_{\rho}L') + \boldsymbol{\delta\Gamma}\,\vdots\,(\partial_{\boldsymbol{\Gamma}}L') + \delta g(\partial_g L')]\,dV$$
$$+\int_{\Sigma}\delta d\,[L' - \partial_r\mathbf{s}\cdot(\partial_{\partial_r\mathbf{s}}L') - \partial_r\phi(\partial_{\partial_r\phi}L')]\,d\Sigma, \tag{13.93}$$

where

$$L' = \tfrac{1}{2}[\rho\omega^2\mathbf{s}\cdot\mathbf{s} - \boldsymbol{\varepsilon}\!:\!\boldsymbol{\Gamma}\!:\!\boldsymbol{\varepsilon} - 2\rho\mathbf{s}\cdot\boldsymbol{\nabla}\phi$$
$$-\,4\pi G\rho^2 s_r^2 - \rho g\Upsilon - (4\pi G)^{-1}\boldsymbol{\nabla}\phi\cdot\boldsymbol{\nabla}\phi] \tag{13.94}$$

is the modified Lagrangian density governing a *spherical hydrostatic* Earth model. We used the corresponding radial Lagrangian density to obtain the Fréchet kernels for a spherical perturbation in Chapter 9. The results (9.12) on an isotropic Earth model and (9.30) on a transversely isotropic Earth model can alternatively be derived by substituting the zeroth-order eigenfunction representations $\mathbf{s} = U\mathbf{P}_{lm} + V\mathbf{B}_{lm} + W\mathbf{C}_{lm}$ and $\phi = P\mathcal{Y}_{lm}$ into equations (13.93)–(13.94) and using the orthonormality (8.85) of the scalar and vector spherical harmonics $\mathcal{Y}_{lm}$ and $\mathbf{P}_{lm}$, $\mathbf{B}_{lm}$, $\mathbf{C}_{lm}$ to evaluate the integrals over the angular variables θ, ϕ.

13.1.9 Anelasticity

The first-order effect of the Earth's anelasticity can be accounted for by replacing the perturbation in the elastic tensor $\boldsymbol{\delta\Gamma}$ in equation (13.84) by a complex, frequency-dependent perturbation. For simplicity, we assume at the outset that the unperturbed, spherically symmetric, perfectly elastic model is isotropic, with an incompressibility κ_0 and a rigidity μ_0, where the subscript 0 indicates that these are deemed to be appropriate at a reference or fiducial frequency $\omega_0 > 0$; the case of a transversely isotropic starting model is considered briefly in Section 13.1.10.

The complex three-dimensional perturbations in the incompressibility and rigidity at the unperturbed eigenfrequency $\omega > 0$ are of the form

$$\kappa_0 \rightarrow \kappa_0 + \delta\kappa(\omega) + i\kappa_0 q_\kappa, \qquad \mu_0 \rightarrow \mu_0 + \delta\mu(\omega) + i\mu_0 q_\mu, \tag{13.95}$$

where

$$q_\kappa = Q_\kappa^{-1}, \qquad q_\mu = Q_\mu^{-1} \tag{13.96}$$

are the reciprocal bulk and shear quality factors, which are assumed to be frequency independent. The eigenfrequency perturbation now has an imaginary part:

$$\omega \rightarrow \omega + \delta\omega + i\gamma, \tag{13.97}$$

where

$$\gamma = \tfrac{1}{2}\omega Q^{-1}. \tag{13.98}$$

The quantity γ is the decay rate, and Q^{-1} is the reciprocal normal-mode quality factor, which is given correct to first order in q_κ and q_μ by

$$Q^{-1} = \omega^{-2} \int_\oplus [\kappa_0 q_\kappa (\boldsymbol{\nabla} \cdot \mathbf{s})^2 + 2\mu_0 q_\mu (\mathbf{d}:\mathbf{d})]\, dV. \tag{13.99}$$

Equation (13.99) follows immediately upon employing the complex representations (13.95)–(13.96) and (13.97)–(13.98) in equation (13.39); alternatively, it can be obtained by substituting the unperturbed eigenfunction s into the exact result (6.170). Both the spherical and aspherical anelastic perturbations are accounted for:

$$q_\kappa = \bar{q}_\kappa + \hat{q}_\kappa, \qquad q_\mu = \bar{q}_\mu + \hat{q}_\mu. \tag{13.100}$$

In the degenerate case of a purely spherical perturbation, $\hat{q}_\kappa = \hat{q}_\mu = 0$, the three-dimensional result (13.99) reduces to equation (9.54), as expected. The real perturbation in the eigenfrequency of an anelastic Earth is

$$\delta\omega = \int_\oplus \delta\oplus\, K_\oplus\, dV + \int_\Sigma \delta d\,[K_{\mathrm{d}}]_-^+\, d\Sigma + \int_{\Sigma_{\mathrm{FS}}} \boldsymbol{\nabla}^\Sigma(\delta d)\cdot[\mathbf{K}_{\mathrm{d}}]_-^+\, d\Sigma, \tag{13.101}$$

where

$$\begin{aligned} 2\omega\,\delta\oplus K_\oplus &= \delta\rho(-\omega^2\mathbf{s}\cdot\mathbf{s} + 2\mathbf{s}\cdot\boldsymbol{\nabla}\phi + 4\pi G\rho s_r^2 + g\Upsilon) \\ &\quad + \delta\kappa(\boldsymbol{\nabla}\cdot\mathbf{s})^2 + 2\delta\mu(\mathbf{d}:\mathbf{d}) + \boldsymbol{\varepsilon}:\boldsymbol{\gamma}:\boldsymbol{\varepsilon} \\ &\quad + \rho\boldsymbol{\nabla}(\delta\Phi)\cdot(\mathbf{s}\cdot\boldsymbol{\nabla}\mathbf{s} - \mathbf{s}\boldsymbol{\nabla}\cdot\mathbf{s}) \\ &\quad + \rho\mathbf{s}\cdot\boldsymbol{\nabla}\boldsymbol{\nabla}(\delta\Phi)\cdot\mathbf{s}, \end{aligned} \tag{13.102}$$

$$\begin{aligned} 2\omega K_{\mathrm{d}} &= \rho(\omega^2\mathbf{s}\cdot\mathbf{s} - 2\mathbf{s}\cdot\boldsymbol{\nabla}\phi - 8\pi G\rho s_r^2 - g\Upsilon) \\ &\quad - \kappa_0(\boldsymbol{\nabla}\cdot\mathbf{s})(\boldsymbol{\nabla}\cdot\mathbf{s} - 2\partial_r s_r) - 2\mu_0\mathbf{d}:(\mathbf{d} - 2\hat{\mathbf{r}}\partial_r\mathbf{s}), \end{aligned} \tag{13.103}$$

$$2\omega\mathbf{K}_{\mathrm{d}} = -2[\kappa_0(\boldsymbol{\nabla}\cdot\mathbf{s}) + 2\mu_0 d_{rr}]\mathbf{s}. \tag{13.104}$$

The quantity $\boldsymbol{\gamma}$ is the anisotropic perturbation to the elastic tensor, which is assumed to be frequency independent.

The real incompressibility and rigidity perturbations "seen" by a mode whose unperturbed positive eigenfrequency is ω are

$$\delta\kappa = \delta\bar{\kappa} + \delta\hat{\kappa}, \qquad \delta\mu = \delta\bar{\mu} + \delta\hat{\mu}, \tag{13.105}$$

where

$$\delta\bar{\kappa} = \delta\bar{\kappa}_0 + (2/\pi)\kappa_0\bar{q}_\kappa \ln(\omega/\omega_0), \tag{13.106}$$

$$\delta\hat{\kappa} = \delta\hat{\kappa}_0 + (2/\pi)\kappa_0\hat{q}_\kappa \ln(\omega/\omega_0), \tag{13.107}$$

$$\delta\bar{\mu} = \delta\bar{\mu}_0 + (2/\pi)\mu_0\bar{q}_\mu \ln(\omega/\omega_0), \tag{13.108}$$

$$\delta\hat{\mu} = \delta\hat{\mu}_0 + (2/\pi)\mu_0\hat{q}_\mu \ln(\omega/\omega_0). \tag{13.109}$$

The first term in each of (13.106)–(13.109) is the perturbation at the reference frequency ω_0, whereas the second term is a dispersive contribution associated with the spherical and aspherical anelasticity $\bar{q}_\kappa$, $\bar{q}_\mu$ and $\hat{q}_\kappa$, $\hat{q}_\mu$. If the aspherical elastic and anelastic perturbations are anti-correlated, so that $\delta\hat{\mu}_0 > 0 \Longleftrightarrow \hat{q}_\mu < 0$ and $\delta\hat{\mu}_0 < 0 \Longleftrightarrow \hat{q}_\mu > 0$, as seems likely on physical grounds, then the lateral heterogeneity of the Earth will appear more pronounced at lower frequencies, as illustrated in Figure 13.1.

It is straightforward to express the volumetric perturbations in terms of the P-wave and S-wave speed rather than the incompressibility and rigidity, using the first-order relations

$$\delta\kappa = \delta\rho(\alpha_0^2 - \tfrac{4}{3}\beta_0^2) + 2\rho(\alpha_0\,\delta\alpha - \tfrac{4}{3}\beta_0\,\delta\beta), \tag{13.110}$$

$$\delta\mu = \delta\rho\,\beta_0^2 + 2\rho\beta_0\,\delta\beta, \tag{13.111}$$

where $\alpha_0 = [(\kappa_0 + \tfrac{4}{3}\mu_0)/\rho]^{1/2}$ and $\beta_0 = (\mu_0/\rho)^{1/2}$. If we define reciprocal P-wave and S-wave quality factors by

$$q_\alpha = [1 - \tfrac{4}{3}(\beta_0/\alpha_0)^2]q_\kappa + \tfrac{4}{3}(\beta_0/\alpha_0)^2 q_\mu, \qquad q_\beta = q_\mu, \tag{13.112}$$

then it is easily demonstrated that

$$\delta\alpha = \delta\bar{\alpha} + \delta\hat{\alpha}, \qquad \delta\beta = \delta\bar{\beta} + \delta\hat{\beta}, \tag{13.113}$$

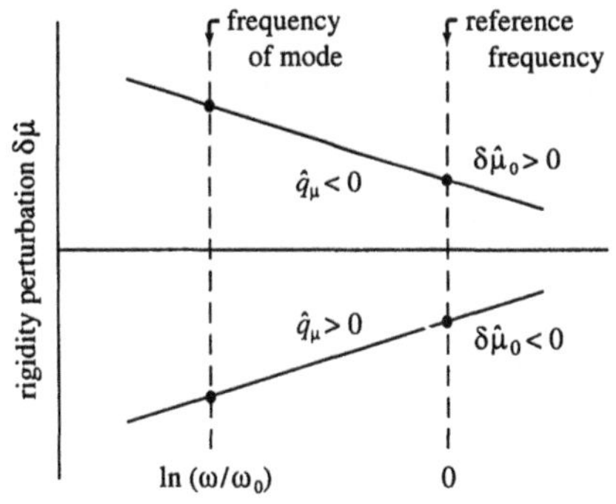

Figure 13.1. If the aspherical elastic and anelastic perturbations are due to lateral variations in temperature, then high-rigidity regions will exhibit less attenuation, and vice versa (Karato 1993). The lateral variations $\delta\hat{\mu}$ at the frequency ω of a mode (*left*) will in that case be more pronounced than the variations $\delta\hat{\mu}_0$ at a higher reference frequency ω_0 (*right*). The abscissa in this schematic plot is the logarithm of frequency; the perturbation in rigidity depends linearly upon this variable: $\delta\hat{\mu} = \delta\hat{\mu}_0 + (2/\pi)\mu_0\hat{q}_\mu \ln(\omega/\omega_0)$.

where

$$\delta\bar{\alpha} = \delta\bar{\alpha}_0 + (1/\pi)\alpha_0\bar{q}_\alpha \ln(\omega/\omega_0), \tag{13.114}$$

$$\delta\hat{\alpha} = \delta\hat{\alpha}_0 + (1/\pi)\alpha_0\hat{q}_\alpha \ln(\omega/\omega_0), \tag{13.115}$$

$$\delta\bar{\beta} = \delta\bar{\beta}_0 + (1/\pi)\beta_0\bar{q}_\beta \ln(\omega/\omega_0), \tag{13.116}$$

$$\delta\hat{\beta} = \delta\hat{\beta}_0 + (1/\pi)\beta_0\hat{q}_\beta \ln(\omega/\omega_0). \tag{13.117}$$

The tendency for the percent lateral heterogeneity to be accentuated at lower frequencies is more pronounced for $\delta\hat{\beta}/\beta_0$ than for $\delta\hat{\alpha}/\alpha_0$ by a factor of $\hat{q}_\beta/\hat{q}_\alpha \approx \frac{3}{4}(\alpha_0/\beta_0)^2 \approx \frac{9}{4}$, where the first approximation assumes that $\hat{q}_\kappa \ll \hat{q}_\mu$, and the final value is based upon the Poisson approximation.

*13.1.10 Transverse isotropy

Suppose now that the unperturbed spherical Earth is transversely isotropic with elastic parameters C_0, A_0, L_0, N_0 and F_0, where the subscript 0 denotes evaluation at the reference frequency ω_0. The most general anelastic perturbation $\boldsymbol{\delta\Gamma}(\omega)$ now consists of three parts:

1. The incompressibility $\kappa_0 = \frac{1}{9}(C_0 + 4A_0 - 4N_0 + 4F_0)$ and the rigidity $\mu_0 = \frac{1}{15}(C_0 + A_0 + 6L_0 + 5N_0 - 2F_0)$ exhibit three-dimensional perturbations of the form (13.95)–(13.96).

2. In addition, the parameters $C'_0 = C_0 - \kappa_0 - \frac{4}{3}\mu_0$, $A'_0 = A_0 - \kappa_0 - \frac{4}{3}\mu_0$, $L'_0 = L_0 - \mu_0$, $N'_0 = N_0 - \mu_0$ and $F'_0 = F_0 - \kappa_0 + \frac{2}{3}\mu_0$ which characterize the transverse isotropy may undergo spherically symmetric perturbations $\delta C'_0$, $\delta A'_0$, $\delta L'_0$, $\delta N'_0$ and $\delta F'_0$.

3. Finally, there may be a general three-dimensional anisotropic perturbation, which we shall continue to denote by $\boldsymbol{\gamma}$.

The perturbations $\delta\kappa(\omega) + i\kappa_0 q_\kappa$ and $\delta\mu(\omega) + i\mu_0 q_\mu$ to the "equivalent" isotropic parameters are complex and frequency dependent, whereas both of the anisotropic perturbations are assumed to be real and frequency independent. The first-order eigenfrequency perturbation is still of the form (13.101), with the discontinuity kernels replaced by

$$\begin{aligned}2\omega K_{\mathrm{d}} &= \rho(\omega^2\mathbf{s}\cdot\mathbf{s} - 2\mathbf{s}\cdot\boldsymbol{\nabla}\phi - 8\pi G\rho s_r^2 - g\Upsilon)\\ &\quad + C_0(\partial_r s_r)^2 - (A_0 - 2N_0)(\boldsymbol{\nabla}\cdot\mathbf{s} - \partial_r s_r)^2\\ &\quad\quad - 2N_0[\boldsymbol{\varepsilon}:\boldsymbol{\varepsilon} - 2|\hat{\mathbf{r}}\cdot\boldsymbol{\varepsilon}|^2 + (\partial_r s_r)^2]\\ &\quad\quad\quad - 4L_0(\hat{\mathbf{r}}\cdot\boldsymbol{\varepsilon})\cdot(\hat{\mathbf{r}}\cdot\boldsymbol{\varepsilon} - \partial_r\mathbf{s}),\end{aligned} \tag{13.118}$$

$$2\omega\mathbf{K}_{\rm d} = -2[C_0\partial_r s_r + F_0(\boldsymbol{\nabla}\cdot\mathbf{s} - \partial_r s_r)]\mathbf{s}. \tag{13.119}$$

The spherically symmetric anisotropic perturbation may be accounted for by adding a term $\int_0^a(\delta C_0' K_C + \delta A_0' K_A + \delta L_0' K_L + \delta N_0' K_N + \delta F_0' K_F)\,dr$ to the three-dimensional expression for $\delta\omega$. The one-dimensional Fréchet kernels K_C, K_A, K_L, K_N and K_F are given in equations (9.31)–(9.34). We shall eschew further consideration of transverse isotropy, and regard both the unperturbed spherical and perturbed aspherical Earth as isotropic throughout the remainder of this discussion.

★13.1.11 Rotation

The Coriolis force does not exert a first-order influence upon the eigenfrequency of a non-degenerate normal mode by virtue of the triple-product identity $\mathbf{s}\cdot(\boldsymbol{\Omega}\times\mathbf{s}) = 0$. The effect of the centrifugal force, which is of second order in the rotation rate $\Omega = \|\boldsymbol{\Omega}\|$, can be accounted for by replacing the gravitational potential perturbation $\delta\Phi$ by $\delta\Phi + \psi$ in the volumetric kernel (13.102). The centrifugal potential ψ has both a spherical part $\bar{\psi}$ and an aspherical part $\hat{\psi}$, and the argument in Section 13.1.7 regarding the determination of the hydrostatic pressure and density perturbations within the fluid regions of the Earth must be modified by replacing $\delta\bar{\Phi}$ by $\delta\bar{\Phi} + \bar{\psi}$ in equation (13.78) and $\delta\hat{\Phi}$ by $\delta\hat{\Phi} + \hat{\psi}$ in equations (13.80), (13.83) and the right side of (13.81) for this reason. The Coriolis force does act to split the degenerate eigenfrequencies of isolated spherical-Earth multiplets and to couple quasi-degenerate multiplets satisfying certain selection rules. These effects, which are much more important than the effects of the centrifugal force, inasmuch as they are of first order in Ω, are accounted for in Section 13.2. A more thorough discussion of the influence of the rotation and associated hydrostatic ellipticity of the Earth is given in Chapter 14.

13.1.12 Perturbed kinetic and potential energy

The final expression (13.101) for the eigenfrequency perturbation of a non-degenerate mode can be "derived" in a purely symbolic manner by taking the total variation of the energy equipartition relation

$$\tfrac{1}{2}(\omega^2\mathcal{T} - \mathcal{V}) = 0. \tag{13.120}$$

The variation with respect to the eigenfunction $\mathbf{s}$ vanishes by virtue of Rayleigh's principle, leaving the first-order result

$$\delta\omega = (2\omega)^{-1}(\delta\mathcal{V} - \omega^2\delta\mathcal{T}), \tag{13.121}$$

where we have made use of the normalization relation $\mathcal{T} = 1$. The perturbations in the kinetic and potential energies $\delta\mathcal{T}$, $\delta\mathcal{V}$ in equation (13.121) are due strictly to the perturbations $\delta\kappa$, $\delta\mu$, $\delta\rho$, $\delta\Phi$, δd and γ in the Earth model. The perturbation in the kinetic energy functional is given simply by

$$\delta\mathcal{T} = \int_{\oplus} \delta\rho(\mathbf{s}\cdot\mathbf{s})\,dV - \int_{\Sigma} \delta d[\rho\,\mathbf{s}\cdot\mathbf{s}]_-^+\,d\Sigma, \tag{13.122}$$

where the surface integral accounts for the change in the region of integration. Upon comparing the results (13.101) and (13.121) and extracting the expression (13.122) from equations (13.102)–(13.104), we infer that the corresponding perturbation in the elastic-gravitational potential-energy functional is

$$\begin{aligned}\delta\mathcal{V} = &\int_{\oplus} [\delta\kappa(\boldsymbol{\nabla}\cdot\mathbf{s})^2 + 2\delta\mu(\mathbf{d}\!:\!\mathbf{d}) + \delta\rho(2\mathbf{s}\cdot\boldsymbol{\nabla}\phi + 4\pi G\rho s_r^2 + g\Upsilon)\\ &\quad + \rho\boldsymbol{\nabla}(\delta\Phi)\cdot(\mathbf{s}\cdot\boldsymbol{\nabla}\mathbf{s} - \mathbf{s}\boldsymbol{\nabla}\cdot\mathbf{s}) + \rho\mathbf{s}\cdot\boldsymbol{\nabla}\boldsymbol{\nabla}(\delta\Phi)\cdot\mathbf{s} + \boldsymbol{\varepsilon}\!:\!\boldsymbol{\gamma}\!:\!\boldsymbol{\varepsilon}]\,dV\\ &- \int_{\Sigma} \delta d\,[\kappa_0(\boldsymbol{\nabla}\cdot\mathbf{s})(\boldsymbol{\nabla}\cdot\mathbf{s} - 2\partial_r s_r) + 2\mu_0\mathbf{d}\!:\!(\mathbf{d} - 2\hat{\mathbf{r}}\partial_r\mathbf{s})\\ &\quad + \rho(2\mathbf{s}\cdot\boldsymbol{\nabla}\phi + 8\pi G\rho s_r^2 + g\Upsilon)]_-^+\,d\Sigma\\ &- \int_{\Sigma_{\mathrm{FS}}} \boldsymbol{\nabla}^{\Sigma}(\delta d)\cdot[2\kappa_0(\boldsymbol{\nabla}\cdot\mathbf{s})\mathbf{s} + 4\mu_0 d_{rr}\mathbf{s}]_-^+\,d\Sigma.\end{aligned} \tag{13.123}$$

The perturbations $\delta\kappa$ and $\delta\mu$ in equation (13.123) are those "seen" by the mode of frequency ω. This decomposition of the eigenfrequency perturbation $\delta\omega$ into separate kinetic and potential energy perturbations $\delta\mathcal{T}$ and $\delta\mathcal{V}$ provides a convenient starting point for extending the above results to the case of actual seismological interest, as we shall see next.

13.2 Degeneracy and Quasi-Degeneracy

Thus far we have treated the unperturbed eigenfrequencies as if they were non-degenerate and well isolated in the Earth's normal-mode spectrum. In fact, the real eigenfrequencies of a non-rotating, spherically symmetric Earth model occur in $(2l+1)$-degenerate toroidal and spheroidal multiplets ${}_n\mathrm{T}_l$ and ${}_n\mathrm{S}_l$. The three-dimensional character of the structural perturbations $\delta\kappa$, $\delta\mu$, $\delta\rho$, $\delta\Phi$, δd and γ and the rotation of the Earth break the spherical symmetry and remove this eigenfrequency degeneracy. Multiplets whose unperturbed eigenfrequencies are in close juxtaposition may also be coupled by rotation and any three-dimensional perturbations. The simplest means of accounting for this *eigenfrequency splitting* and *quasi-degenerate mode coupling* is to employ a modified version of the Rayleigh-Ritz variational method discussed in Chapter 7. We use the real unperturbed *singlet* eigenfunctions $\mathbf{s}_k$ associated with the multiplets whose coupling we wish to investigate as the basis functions, and seek perturbed singlet eigenfunctions of the form

$$\mathbf{s} = \sum_k q_k \mathbf{s}_k, \tag{13.124}$$

where the expansion coefficients q_k must be determined. In principle, we should incorporate all of the unperturbed eigenfunctions in the basis set; in practice, of course, we must restrict consideration to a finite number of quasi-degenerate multiplets.

We treat the split singlet eigenfrequencies as small perturbations away from a positive reference or fiducial frequency ω_0; all significant coupling partners ${}_n\mathrm{T}_l$ and ${}_n\mathrm{S}_l$ with degenerate eigenfrequencies ω_k in the vicinity of the reference frequency ω_0 are presumed to be included in the spherical-Earth basis set. The orthonormality $\int_\oplus \rho\mathbf{s}_k \cdot \mathbf{s}_{k'}\, dV = \delta_{kk'}$ of the basis eigenfunctions guarantees that the unperturbed kinetic energy matrix is simply the identity I, whereas the unperturbed elastic-gravitational potential energy matrix is the diagonal matrix $\Omega^2 = \mathrm{diag}\,[\cdots\, \omega_k^2\, \cdots]$ of squared degenerate eigenfrequencies. Note that each squared eigenfrequency is repeated $2l+1$ times along the diagonal; the index k does "double duty" as a label of both basis eigenfunctions $\mathbf{s}_k$ and associated unperturbed eigenfrequencies ω_k.

Notable contributions to the theory of quasi-degenerate multiplet coupling have been made by Dahlen (1969), Luh (1973), Woodhouse (1980) and Park & Gilbert (1986); we summarize and extend these original treatments here. The most significant new feature in the following development is a renormalization procedure which yields singlet eigenfunctions (13.124) that are correctly orthonormalized or biorthonormalized in accordance with the fundamental principles enunciated in Chapters 4 and 6. We consider the simplest situation—a non-rotating, perfectly elastic perturbation—before dealing with the complications introduced by rotation and anelasticity, first in isolation and then in tandem.

13.2.1 Non-rotating elastic perturbation

The kinetic and potential energy matrices I and Ω^2 are altered as a result of the perturbations $\delta\kappa$, $\delta\mu$, $\delta\rho$, $\delta\Phi$, δd and γ in the Earth model. Adopting a slightly modified version of the notation employed in Chapter 7, we shall henceforth use T and V to denote the *first-order perturbation matrices*. The elements $T_{kk'}$ and $V_{kk'}$ of these matrices are symmetrized generalizations of the expressions (13.122) and (13.123) for the perturbed scalar kinetic and potential energy of a non-degenerate mode:

$$T_{kk'} = \int_{\oplus} \delta\rho(\mathbf{s}_k \cdot \mathbf{s}_{k'})\, dV - \int_{\Sigma} \delta d\, [\rho\, \mathbf{s}_k \cdot \mathbf{s}_{k'}]_-^+\, d\Sigma, \tag{13.125}$$

$$\begin{aligned} V_{kk'} = \int_{\oplus} & [\delta\kappa(\boldsymbol{\nabla} \cdot \mathbf{s}_k)(\boldsymbol{\nabla} \cdot \mathbf{s}_{k'}) + 2\delta\mu(\mathbf{d}_k : \mathbf{d}_{k'}) \\ & + \delta\rho\{\mathbf{s}_k \cdot \boldsymbol{\nabla}\phi_{k'} + \mathbf{s}_{k'} \cdot \boldsymbol{\nabla}\phi_k \\ & + 4\pi G\rho(\hat{\mathbf{r}} \cdot \mathbf{s}_k)(\hat{\mathbf{r}} \cdot \mathbf{s}_{k'}) + g\Upsilon_{kk'}\} \\ & + \tfrac{1}{2}\rho\boldsymbol{\nabla}(\delta\Phi) \cdot (\mathbf{s}_k \cdot \boldsymbol{\nabla}\mathbf{s}_{k'} + \mathbf{s}_{k'} \cdot \boldsymbol{\nabla}\mathbf{s}_k \\ & - \mathbf{s}_k \boldsymbol{\nabla} \cdot \mathbf{s}_{k'} - \mathbf{s}_{k'} \boldsymbol{\nabla} \cdot \mathbf{s}_k) + \rho\mathbf{s}_k \cdot \boldsymbol{\nabla}\boldsymbol{\nabla}(\delta\Phi) \cdot \mathbf{s}_{k'} \\ & + \boldsymbol{\varepsilon}_k : \boldsymbol{\gamma} : \boldsymbol{\varepsilon}_{k'}]\, dV \\ - \int_{\Sigma} & \delta d\, [\tfrac{1}{2}\kappa(\boldsymbol{\nabla} \cdot \mathbf{s}_k)(\boldsymbol{\nabla} \cdot \mathbf{s}_{k'} - 2\hat{\mathbf{r}} \cdot \partial_r \mathbf{s}_{k'}) \\ & + \tfrac{1}{2}\kappa(\boldsymbol{\nabla} \cdot \mathbf{s}_{k'})(\boldsymbol{\nabla} \cdot \mathbf{s}_k - 2\hat{\mathbf{r}} \cdot \partial_r \mathbf{s}_k) \\ & + \mu\mathbf{d}_k : (\mathbf{d}_{k'} - 2\hat{\mathbf{r}}\partial_r \mathbf{s}_{k'}) + \mu\mathbf{d}_{k'} : (\mathbf{d}_k - 2\hat{\mathbf{r}}\partial_r \mathbf{s}_k) \\ & + \rho\{\mathbf{s}_k \cdot \boldsymbol{\nabla}\phi_{k'} + \mathbf{s}_{k'} \cdot \boldsymbol{\nabla}\phi_k \\ & + 8\pi G\rho(\hat{\mathbf{r}} \cdot \mathbf{s}_k)(\hat{\mathbf{r}} \cdot \mathbf{s}_{k'}) + g\Upsilon_{kk'}\}]_-^+\, d\Sigma \\ - \int_{\Sigma_{\mathrm{FS}}} & \boldsymbol{\nabla}^{\Sigma}(\delta d) \cdot [\kappa(\boldsymbol{\nabla} \cdot \mathbf{s}_k)\mathbf{s}_{k'} + \kappa(\boldsymbol{\nabla} \cdot \mathbf{s}_{k'})\mathbf{s}_k \\ & + 2\mu(\hat{\mathbf{r}} \cdot \mathbf{d}_k \cdot \hat{\mathbf{r}})\mathbf{s}_{k'} + 2\mu(\hat{\mathbf{r}} \cdot \mathbf{d}_{k'} \cdot \hat{\mathbf{r}})\mathbf{s}_k]_-^+\, d\Sigma, \end{aligned} \tag{13.126}$$

where

$$\begin{aligned} \Upsilon_{kk'} = & \tfrac{1}{2}[\mathbf{s}_k \cdot \boldsymbol{\nabla}(\hat{\mathbf{r}} \cdot \mathbf{s}_{k'}) + \mathbf{s}_{k'} \cdot \boldsymbol{\nabla}(\hat{\mathbf{r}} \cdot \mathbf{s}_k)] \\ & - \tfrac{1}{2}[(\hat{\mathbf{r}} \cdot \mathbf{s}_k)(\boldsymbol{\nabla} \cdot \mathbf{s}_{k'}) + (\hat{\mathbf{r}} \cdot \mathbf{s}_{k'})(\boldsymbol{\nabla} \cdot \mathbf{s}_k)] \\ & - 2r^{-1}(\hat{\mathbf{r}} \cdot \mathbf{s}_k)(\hat{\mathbf{r}} \cdot \mathbf{s}_{k'}). \end{aligned} \tag{13.127}$$

The reality of the unperturbed eigenfunctions $\mathbf{s}_k$ guarantees that both of the perturbation matrices are real and symmetric: $\mathsf{T}^{\mathrm{T}} = \mathsf{T}$ and $\mathsf{V}^{\mathrm{T}} = \mathsf{V}$, where the superscript T denotes the transpose.

The perturbed eigenfrequencies of the singlets in the vicinity of the reference frequency ω_0 will be expressed in the form

$$\omega = \omega_0 + \delta\omega, \tag{13.128}$$

where it is assumed that $|\delta\omega| \ll \omega_0$. To determine the perturbations $\delta\omega$ and the associated column vectors q of eigenfunction expansion coefficients q_k, we must solve the generalized eigenvalue problem (7.13). Rewritten in the present notation, this equation takes the form

$$(\Omega^2 + \mathsf{V})\mathsf{q} = (\omega_0 + \delta\omega)^2(\mathsf{I} + \mathsf{T})\mathsf{q}. \tag{13.129}$$

We can reduce (13.129) to an ordinary eigenvalue problem for the singlet eigenfrequency perturbations $\delta\omega$ by defining the *renormalized eigenvectors*

$$\mathsf{z} = (\mathsf{I} + \tfrac{1}{2}\mathsf{T})\mathsf{q}. \tag{13.130}$$

Upon inserting the representation (13.130) into equation (13.129) and neglecting terms of second order in $\delta\omega$, $\omega_k - \omega_0$, T and V, we obtain

$$\mathsf{Hz} = \delta\omega\, \mathsf{z}, \tag{13.131}$$

where

$$\mathsf{H} = \Omega - \omega_0 \mathsf{I} + (2\omega_0)^{-1}(\mathsf{V} - \omega_0^2 \mathsf{T}). \tag{13.132}$$

Since H is real and symmetric, $\mathsf{H}^{\mathrm{T}} = \mathsf{H}$, its eigenvalues $\delta\omega$ are real and its eigenvectors are mutually orthogonal in the sense $\mathsf{z}^{\mathrm{T}}\mathsf{z}' = 0$ if $\delta\omega \neq \delta\omega'$. If we normalize the eigenvectors by requiring that $\mathsf{z}^{\mathrm{T}}\mathsf{z} = 1$, then they constitute an *orthonormal basis*:

$$\mathsf{Z}^{\mathrm{T}}\mathsf{Z} = \mathsf{I}, \tag{13.133}$$

where Z is the square matrix whose columns are the renormalized eigenvectors z. We can rewrite the eigenvalue problem (13.131) in the form

$$\mathsf{Z}^{\mathrm{T}}\mathsf{HZ} = \Delta, \tag{13.134}$$

where $\Delta = \mathrm{diag}\,[\cdots\ \delta\omega\ \cdots]$ is the diagonal matrix of eigenfrequency perturbations. Taken together, equations (13.133)–(13.134) show that Z is the orthogonal matrix that diagonalizes the *renormalized eigenfrequency perturbation matrix* H. Correct to first order in T, we may rewrite the orthonormality relation (13.133) in the form

$$\mathsf{Q}^{\mathrm{T}}(\mathsf{I} + \mathsf{T})\mathsf{Q} = \mathsf{I}, \tag{13.135}$$

where Q is the square matrix of original column vectors q.

In summary, we can find the first-order singlet eigenfrequency perturbations $\delta\omega$ and associated renormalized eigenvectors z of a non-rotating elastic Earth by solving the real symmetric eigenvalue problem $\mathsf{Hz} = \delta\omega\, \mathsf{z}$ subject to the requirement that $\mathsf{z}^{\mathrm{T}}\mathsf{z} = 1$. The original eigenvectors q are given in terms of the renormalized eigenvectors z by the inverse of equation (13.130):

$$\mathsf{q} = (\mathsf{I} - \tfrac{1}{2}\mathsf{T})\mathsf{z}. \tag{13.136}$$

This two-step procedure for determining the singlet eigenfunctions (13.124) renders them properly *orthonormal with respect to the perturbed Earth*, in the first-order sense

$$\int_{\oplus} (\rho + \delta\rho)\, \mathbf{s} \cdot \mathbf{s}'\, dV - \int_{\Sigma} \delta d\, [\rho\, \mathbf{s} \cdot \mathbf{s}']_{-}^{+}\, d\Sigma = 0 \quad \text{if } \delta\omega \neq \delta\omega', \tag{13.137}$$

$$\int_{\oplus}(\rho+\delta\rho)\,\mathbf{s}\cdot\mathbf{s}\,dV-\int_{\Sigma}\delta d\,[\rho\,\mathbf{s}\cdot\mathbf{s}]_{-}^{+}\,d\Sigma=1. \tag{13.138}$$

The three-dimensional equations (13.137)–(13.138) are equivalent to the matrix orthonormality relation (13.135). The surface-integral terms account for the change in the domain of integration due to the boundary perturbations δd.

★13.2.2 Rotating elastic perturbation

The effect of the centrifugal potential ψ can be accounted for by making the substitution $\delta\Phi\rightarrow\delta\Phi+\psi$ in equation (13.126), as discussed in Section 13.1.11. The perturbed kinetic energy matrix T is unchanged, whereas the elements of the perturbed potential energy matrix V are replaced by

$$\begin{aligned}V_{kk'}=&\int_{\oplus}[\delta\kappa(\boldsymbol{\nabla}\cdot\mathbf{s}_k)(\boldsymbol{\nabla}\cdot\mathbf{s}_{k'})+2\delta\mu(\mathbf{d}_k\!:\!\mathbf{d}_{k'})\\&+\delta\rho\{\mathbf{s}_k\cdot\boldsymbol{\nabla}\phi_{k'}+\mathbf{s}_{k'}\cdot\boldsymbol{\nabla}\phi_k\\&+4\pi G\rho(\hat{\mathbf{r}}\cdot\mathbf{s}_k)(\hat{\mathbf{r}}\cdot\mathbf{s}_{k'})+g\Upsilon_{kk'}\}\\&+\tfrac{1}{2}\rho\boldsymbol{\nabla}(\delta\Phi+\psi)\cdot(\mathbf{s}_k\cdot\boldsymbol{\nabla}\mathbf{s}_{k'}+\mathbf{s}_{k'}\cdot\boldsymbol{\nabla}\mathbf{s}_k\\&-\mathbf{s}_k\boldsymbol{\nabla}\cdot\mathbf{s}_{k'}-\mathbf{s}_{k'}\boldsymbol{\nabla}\cdot\mathbf{s}_k)+\rho\mathbf{s}_k\cdot\boldsymbol{\nabla}\boldsymbol{\nabla}(\delta\Phi+\psi)\cdot\mathbf{s}_{k'}\\&+\boldsymbol{\varepsilon}_k\!:\!\boldsymbol{\gamma}\!:\!\boldsymbol{\varepsilon}_{k'}]\,dV\\&-\int_{\Sigma}\delta d\,[\tfrac{1}{2}\kappa(\boldsymbol{\nabla}\cdot\mathbf{s}_k)(\boldsymbol{\nabla}\cdot\mathbf{s}_{k'}-2\hat{\mathbf{r}}\cdot\partial_r\mathbf{s}_{k'})\\&+\tfrac{1}{2}\kappa(\boldsymbol{\nabla}\cdot\mathbf{s}_{k'})(\boldsymbol{\nabla}\cdot\mathbf{s}_k-2\hat{\mathbf{r}}\cdot\partial_r\mathbf{s}_k)\\&+\mu\mathbf{d}_k\!:\!(\mathbf{d}_{k'}-2\hat{\mathbf{r}}\partial_r\mathbf{s}_{k'})+\mu\mathbf{d}_{k'}\!:\!(\mathbf{d}_k-2\hat{\mathbf{r}}\partial_r\mathbf{s}_k)\\&+\rho\{\mathbf{s}_k\cdot\boldsymbol{\nabla}\phi_{k'}+\mathbf{s}_{k'}\cdot\boldsymbol{\nabla}\phi_k\\&+8\pi G\rho(\hat{\mathbf{r}}\cdot\mathbf{s}_k)(\hat{\mathbf{r}}\cdot\mathbf{s}_{k'})+g\Upsilon_{kk'}\}]_{-}^{+}\,d\Sigma\\&-\int_{\Sigma_{\rm FS}}\boldsymbol{\nabla}^{\Sigma}(\delta d)\cdot[\kappa(\boldsymbol{\nabla}\cdot\mathbf{s}_k)\mathbf{s}_{k'}+\kappa(\boldsymbol{\nabla}\cdot\mathbf{s}_{k'})\mathbf{s}_k\\&+2\mu(\hat{\mathbf{r}}\cdot\mathbf{d}_k\cdot\hat{\mathbf{r}})\mathbf{s}_{k'}+2\mu(\hat{\mathbf{r}}\cdot\mathbf{d}_{k'}\cdot\hat{\mathbf{r}})\mathbf{s}_k]_{-}^{+}\,d\Sigma.\end{aligned} \tag{13.139}$$

As in Chapter 7, where the Rayleigh-Ritz method was applied to an arbitrarily heterogeneous Earth, we also introduce the *Coriolis matrix* W with elements

$$W_{kk'}=\int_{\oplus}\rho\mathbf{s}_k\cdot(i\boldsymbol{\Omega}\times\mathbf{s}_{k'})\,dV. \tag{13.140}$$

Both T and V are still real and symmetric, whereas the Coriolis matrix W is imaginary and anti-symmetric. All three perturbation matrices, if regarded as complex, are therefore Hermitian: $\mathsf{T}^{\rm H} = \mathsf{T}$, $\mathsf{V}^{\rm H} = \mathsf{V}$ and $\mathsf{W}^{\rm H} = \mathsf{W}$, where the superscript H denotes the complex conjugate transpose. The non-standard generalized eigenvalue problem (7.40) governing the normal modes of a rotating elastic Earth is

$$[\Omega^2 + \mathsf{V} + 2(\omega_0 + \delta\omega)\mathsf{W} - (\omega_0 + \delta\omega)^2(\mathsf{I} + \mathsf{T})]\mathsf{q} = 0. \tag{13.141}$$

Generalizing (13.130), we account for the effect of the Coriolis force by defining the renormalized eigenvectors

$$\mathsf{z} = (\mathsf{I} + \tfrac{1}{2}\mathsf{T} - \tfrac{1}{2}\omega_0^{-1}\mathsf{W})\mathsf{q}. \tag{13.142}$$

Upon inserting equation (13.142) into (13.141) and neglecting terms that are of second order in $\delta\omega$, $\omega_k - \omega_0$, T, V and W, we obtain an ordinary eigenvalue problem analogous to (13.131):

$$\mathsf{Hz} = \delta\omega\,\mathsf{z}, \tag{13.143}$$

where

$$\mathsf{H} = \Omega - \omega_0\mathsf{I} + \mathsf{W} + (2\omega_0)^{-1}(\mathsf{V} - \omega_0^2\mathsf{T}). \tag{13.144}$$

The Hermitian symmetry $\mathsf{H}^{\rm H} = \mathsf{H}$ of the complex renormalized eigenfrequency perturbation matrix (13.144) guarantees that all of the eigenfrequency perturbations $\delta\omega$ are real, and that the associated eigenvectors $\mathbf{z}$ may be orthonormalized in the sense

$$\mathsf{Z}^{\rm H}\mathsf{Z} = \mathsf{I}. \tag{13.145}$$

We can rewrite the eigenvalue problem (13.143) in terms of the matrix of complex eigenvectors Z and the diagonal matrix of real eigenfrequency perturbations $\Delta = \mathrm{diag}\,[\cdots\ \delta\omega\ \cdots]$ in a form analogous to equation (13.134):

$$\mathsf{Z}^{\rm H}\mathsf{HZ} = \Delta. \tag{13.146}$$

Equations (13.145) and (13.146) together assert that Z is the unitary transformation that diagonalizes H. The corresponding matrix Q of original eigenvectors q satisfies

$$\mathsf{Q}^{\rm H}(\mathsf{I} + \mathsf{T} - \omega_0^{-1}\mathsf{W})\mathsf{Q} = \mathsf{I}, \tag{13.147}$$

correct to first order in the perturbations T and W.

The above considerations show that it is possible to account for the effect of the Coriolis and centrifugal forces by means of a simple modification (13.144) to the eigenfrequency perturbation matrix H. The resulting ordinary eigenvalue problem $\mathsf{Hz} = \delta\omega\,\mathbf{z}$ for the eigenfrequency perturbations $\delta\omega$ and associated renormalized eigenvectors $\mathbf{z}$ in the vicinity of a reference frequency ω_0 is complex and Hermitian, rather than real and symmetric as in the non-rotating elastic case. The eigenvector back-transformation relation (13.136) is also modified as a consequence of the rotation:

$$\mathsf{q} = (\mathsf{I} - \tfrac{1}{2}\mathsf{T} + \tfrac{1}{2}\omega_0^{-1}\mathsf{W})\mathsf{z}. \tag{13.148}$$

If the eigenvectors z of H are constrained to satisfy $\mathsf{z}^{\mathrm{H}}\mathsf{z} = 1$, then equation (13.147) guarantees that the complex singlet eigenfunctions (13.124) are properly orthonormalized in the sense

$$\int_\oplus (\rho + \delta\rho)\,\mathbf{s}^* \cdot \mathbf{s}'\, dV - \int_\Sigma \delta d\,[\rho\,\mathbf{s}^* \cdot \mathbf{s}']_-^+\, d\Sigma$$
$$- \omega_0^{-1} \int_\oplus \rho\mathbf{s}^* \cdot (i\mathbf{\Omega} \times \mathbf{s}')\, dV = 0 \quad \text{if } \delta\omega \neq \delta\omega', \tag{13.149}$$

$$\int_\oplus (\rho + \delta\rho)\,\mathbf{s}^* \cdot \mathbf{s}\, dV - \int_\Sigma \delta d\,[\rho\,\mathbf{s}^* \cdot \mathbf{s}]_-^+\, d\Sigma$$
$$- \omega_0^{-1} \int_\oplus \rho\mathbf{s}^* \cdot (i\mathbf{\Omega} \times \mathbf{s})\, dV = 1. \tag{13.150}$$

The eigenfrequency perturbations $\delta\omega$ and associated eigenfunctions $\mathbf{s}$ obtained in this manner account for the effects of the Earth's rotation correct to first order in Ω. Such a perturbation-theoretical treatment should be a good approximation for all of the seismic modes, including the football mode ${}_0\mathrm{S}_2$, which has $\Omega/\omega_0 \approx 0.04$. It is noteworthy that $\delta\omega$, z^*, q^*, $\mathbf{s}^*$ is an eigensolution of the anti-Earth with a reversed sense of rotation, $\mathsf{W} \to -\mathsf{W}$, if and only if $\delta\omega$, z, q, $\mathbf{s}$ is an eigensolution of the actual Earth. Anelasticity renders the eigensolutions of the Earth and the anti-Earth independent, as we shall see in Section 13.2.4.

13.2.3 Non-rotating anelastic perturbation

It is convenient in the case of an anelastic perturbation to take the fiducial frequency ω_0 to be the frequency at which the incompressibility κ_0 and rigidity μ_0 of the unperturbed SNREI Earth model, as well as the spherical and aspherical perturbations $\delta\kappa_0 = \delta\bar{\kappa}_0 + \delta\hat{\kappa}_0$ and $\delta\mu_0 = \delta\bar{\mu}_0 + \delta\hat{\mu}_0$ to these quantities, are specified. The degenerate eigenfrequencies ω_k and associated basis eigenfunctions $\mathbf{s}_k$ employed in the Rayleigh-Ritz expansion (13.124) are in that case those of a model with *fixed* elastic properties κ_0 and μ_0. The gross anelastic dispersion *between* ω_0 and the "original" fiducial frequency at which the elastic parameters may have been specified (1 Hz in the case of PREM) is presumed to have been accounted for in the construction of the spherical-Earth basis-space catalogue. In principle, a free-oscillation program such as MINEOS or OBANI needs to be run anew to generate an updated mode catalogue each time that the reference frequency ω_0 is altered; in practice, a single catalogue of eigenfrequencies ω_k and eigenfunctions $\mathbf{s}_k$ is usually employed as a basis space for all coupling calculations in the interest of computational expediency. It is common to account for spherically symmetric dispersion in such a multi-purpose catalogue by assuming that each multiplet "sees" the Earth at its own degenerate eigenfrequency ω_k.

We write the complex angular eigenfrequency of a singlet in the vicinity of ω_0 in the form

$$\nu = \omega_0 + \delta\nu, \tag{13.151}$$

where

$$\delta\nu = \delta\omega + i\gamma = \delta\omega + \tfrac{1}{2}i\omega_0 Q^{-1}. \tag{13.152}$$

The quantity $\delta\omega$ is the real perturbation away from the fiducial frequency ω_0 as before, γ is the singlet decay rate, and Q is the associated quality factor. It is assumed that $|\delta\omega| \ll \omega_0$ and $|\gamma| \ll \omega_0$, or, equivalently, $Q \gg 1$. The complex eigenfrequency perturbations $\delta\nu$ and associated eigenvectors q are the solutions to the generalized eigenvalue problem (7.74):

$$[\Omega^2 + \mathsf{V}(\omega_0 + \delta\nu)]\mathsf{q} = (\omega_0 + \delta\nu)^2(\mathsf{I} + \mathsf{T})\mathsf{q}. \tag{13.153}$$

The dispersion *within* the narrow band of frequencies spanned by the coupled multiplets is sufficiently slight that it can be accounted for using the local approximation developed in Section 6.1.9. Written in the present notation, this approximation takes the form

$$\left(\frac{d\kappa}{d\nu}\right)_0 = \frac{2\kappa_0 q_\kappa}{\pi\omega_0}, \qquad \left(\frac{d\mu}{d\nu}\right)_0 = \frac{2\mu_0 q_\mu}{\pi\omega_0}, \tag{13.154}$$

where the subscript zero signifies evaluation at ω_0 as usual. Upon replacing the constant elastic parameters κ and μ in equation (13.126) by $\kappa(\nu)$ and $\mu(\nu)$ and invoking (13.154), we can write the complex frequency-dependent elastic-gravitational potential energy perturbation matrix in the vicinity of ω_0 as the sum of a zeroth-order and a first-order term:

$$\mathsf{V}(\omega_0 + \delta\nu) = \mathsf{V} + i\mathsf{A} + \tfrac{2}{\pi}(\delta\nu/\omega_0)\mathsf{A}, \tag{13.155}$$

where

$$\begin{aligned} V_{kk'} = \int_\oplus [&\delta\kappa_0(\boldsymbol{\nabla}\cdot\mathbf{s}_k)(\boldsymbol{\nabla}\cdot\mathbf{s}_{k'}) + 2\delta\mu_0(\mathbf{d}_k\!:\!\mathbf{d}_{k'}) \\ &+ \delta\rho\{\mathbf{s}_k\cdot\boldsymbol{\nabla}\phi_{k'} + \mathbf{s}_{k'}\cdot\boldsymbol{\nabla}\phi_k \\ &+ 4\pi G\rho(\hat{\mathbf{r}}\cdot\mathbf{s}_k)(\hat{\mathbf{r}}\cdot\mathbf{s}_{k'}) + g\Upsilon_{kk'}\} \end{aligned}$$

$$+\tfrac{1}{2}\rho\boldsymbol{\nabla}(\delta\Phi)\cdot(\mathbf{s}_k\cdot\boldsymbol{\nabla}\mathbf{s}_{k'}+\mathbf{s}_{k'}\cdot\boldsymbol{\nabla}\mathbf{s}_k$$
$$-\mathbf{s}_k\boldsymbol{\nabla}\cdot\mathbf{s}_{k'}-\mathbf{s}_{k'}\boldsymbol{\nabla}\cdot\mathbf{s}_k)+\rho\mathbf{s}_k\cdot\boldsymbol{\nabla}\boldsymbol{\nabla}(\delta\Phi)\cdot\mathbf{s}_{k'}$$
$$+\boldsymbol{\varepsilon}_k\!:\!\boldsymbol{\gamma}\!:\!\boldsymbol{\varepsilon}_{k'}]\,dV$$
$$-\int_\Sigma \delta d\,[\tfrac{1}{2}\kappa_0(\boldsymbol{\nabla}\cdot\mathbf{s}_k)(\boldsymbol{\nabla}\cdot\mathbf{s}_{k'}-2\hat{\mathbf{r}}\cdot\partial_r\mathbf{s}_{k'})$$
$$+\tfrac{1}{2}\kappa_0(\boldsymbol{\nabla}\cdot\mathbf{s}_{k'})(\boldsymbol{\nabla}\cdot\mathbf{s}_k-2\hat{\mathbf{r}}\cdot\partial_r\mathbf{s}_k)$$
$$+\mu_0\mathbf{d}_k\!:\!(\mathbf{d}_{k'}-2\hat{\mathbf{r}}\partial_r\mathbf{s}_{k'})+\mu_0\mathbf{d}_{k'}\!:\!(\mathbf{d}_k-2\hat{\mathbf{r}}\partial_r\mathbf{s}_k)$$
$$+\rho\{\mathbf{s}_k\cdot\boldsymbol{\nabla}\phi_{k'}+\mathbf{s}_{k'}\cdot\boldsymbol{\nabla}\phi_k$$
$$+8\pi G\rho(\hat{\mathbf{r}}\cdot\mathbf{s}_k)(\hat{\mathbf{r}}\cdot\mathbf{s}_{k'})+g\Upsilon_{kk'}\}]_-^+\,d\Sigma$$
$$-\int_{\Sigma_{\rm FS}}\boldsymbol{\nabla}^\Sigma(\delta d)\cdot[\kappa_0(\boldsymbol{\nabla}\cdot\mathbf{s}_k)\mathbf{s}_{k'}+\kappa_0(\boldsymbol{\nabla}\cdot\mathbf{s}_{k'})\mathbf{s}_k$$
$$+2\mu_0(\hat{\mathbf{r}}\cdot\mathbf{d}_k\cdot\hat{\mathbf{r}})\mathbf{s}_{k'}+2\mu_0(\hat{\mathbf{r}}\cdot\mathbf{d}_{k'}\cdot\hat{\mathbf{r}})\mathbf{s}_k]_-^+\,d\Sigma \qquad (13.156)$$

and

$$A_{kk'}=\int_\oplus[\kappa_0 q_\kappa(\boldsymbol{\nabla}\cdot\mathbf{s}_k)(\boldsymbol{\nabla}\cdot\mathbf{s}_{k'})+2\mu_0 q_\mu(\mathbf{d}_k\!:\!\mathbf{d}_{k'})]\,dV. \qquad (13.157)$$

Evidently, V and A are obtained by making the substitutions $\kappa\to\kappa_0$, $\mu\to\mu_0$ and $\delta\kappa\to\kappa_0 q_\kappa$, $\delta\mu\to\mu_0 q_\mu$, $\delta\rho\to 0$, $\delta d\to 0$, respectively, in equation (13.126). We have dropped the customary subscript zero on these two real matrices in order to avoid conflict with a hybrid-multiplet subscript notation which we shall introduce in Section 13.3.3.

A suitable choice for the renormalized eigenvectors upon a non-rotating anelastic Earth is

$$\mathsf{z}=(\mathsf{I}+\tfrac{1}{2}\mathsf{T}-\tfrac{1}{2\pi}\omega_0^{-2}\mathsf{A})\mathsf{q}. \qquad (13.158)$$

Inserting (13.155) and (13.158) into (13.153) and neglecting terms of second order in $\delta\omega$, γ, $\omega_k-\omega_0$, T, V and A, we obtain an ordinary eigenvalue problem governing the complex eigenfrequency perturbations:

$$\mathsf{H}\mathsf{z}=\delta\nu\,\mathsf{z}, \qquad (13.159)$$

where

$$\mathsf{H}=\Omega-\omega_0\mathsf{I}+(2\omega_0)^{-1}(\mathsf{V}+i\mathsf{A}-\omega_0^2\mathsf{T}). \qquad (13.160)$$

The renormalized eigenfrequency perturbation matrix (13.160) on a non-rotating anelastic Earth is neither real symmetric nor Hermitian; instead, it is *complex symmetric*: $\mathsf{H}^{\rm T}=\mathsf{H}$. Such a matrix is not necessarily diagonalizable; if there is any eigenvalue with a geometric multiplicity that is less than its algebraic multiplicity, then H is non-diagonalizable and is said to be *defective*. We shall eschew consideration of this case in accordance with the viewpoint enunciated in Section 6.2.3, and simply assume that H is non-defective. Since any non-defective complex symmetric matrix can be diagonalized by a *complex orthogonal transformation*, we can then write (Horn & Johnson 1985)

$$\mathsf{Z}^{\mathrm{T}}\mathsf{Z} = \mathsf{I}, \qquad \mathsf{Z}^{\mathrm{T}}\mathsf{H}\mathsf{Z} = \Delta, \tag{13.161}$$

where Z is the matrix of renormalized column vectors z as before, and where $\Delta = \mathrm{diag}\,[\cdots\,\delta\nu\,\cdots]$ is the diagonal matrix of singlet eigenfrequency perturbations. It is noteworthy that the two conditions (13.161) are identical to the corresponding elastic relations (13.133)–(13.134); the only difference is that the square matrices H, Z and Δ are complex in the case of an anelastic perturbation, whereas they are real in the absence of anelasticity. Correct to first order in T and A, we can rewrite the anelastic orthonormality relation in terms of the matrix Q of original column vectors q in the form

$$\mathsf{Q}^{\mathrm{T}}(\mathsf{I} + \mathsf{T} - \tfrac{1}{\pi}\omega_0^{-2}\mathsf{A})\mathsf{Q} = \mathsf{I}. \tag{13.162}$$

This result differs from the analogous relation (13.135) on a non-rotating elastic Earth by the presence of the dispersive term $\frac{1}{\pi}\omega_0^{-2}\mathsf{A}$.

To recapitulate, the eigenfrequency perturbations $\delta\nu = \delta\omega + i\gamma$ and associated renormalized eigenvectors z of a non-rotating anelastic Earth are found by solving the complex symmetric eigenvalue problem $\mathsf{H}\mathsf{z} = \delta\nu\,\mathsf{z}$ subject to the constraint $\mathsf{z}^{\mathrm{T}}\mathsf{z} = 1$. The original eigenvectors q are related to the renormalized eigenvectors z by the inverse of equation (13.158):

$$\mathsf{q} = (\mathsf{I} - \tfrac{1}{2}\mathsf{T} + \tfrac{1}{2\pi}\omega_0^{-2}\mathsf{A})\mathsf{z}. \tag{13.163}$$

Equation (13.162) stipulates that the singlet eigenfunctions $\mathbf{s}$ are orthonormal in the sense

$$\begin{aligned}\int_\oplus (\rho + \delta\rho)\, \mathbf{s}\cdot\mathbf{s}'\, dV &- \int_\Sigma \delta d\, [\rho\mathbf{s}\cdot\mathbf{s}']_-^+\, d\Sigma \\ &- \tfrac{1}{\pi}\omega_0^{-2}\int_\oplus [\kappa_0 q_\kappa(\boldsymbol{\nabla}\cdot\mathbf{s})(\boldsymbol{\nabla}\cdot\mathbf{s}') + 2\mu_0 q_\mu(\mathbf{d}:\mathbf{d}')]\, dV = 0 \\ &\qquad \text{if } \delta\nu \neq \delta\nu',\end{aligned} \tag{13.164}$$

$$\begin{aligned}\int_\oplus (\rho + \delta\rho)\, \mathbf{s}\cdot\mathbf{s}\, dV &- \int_\Sigma \delta d\, [\rho\mathbf{s}\cdot\mathbf{s}]_-^+\, d\Sigma \\ &- \tfrac{1}{\pi}\omega_0^{-2}\int_\oplus [\kappa_0 q_\kappa(\boldsymbol{\nabla}\cdot\mathbf{s})^2 + 2\mu_0 q_\mu(\mathbf{d}:\mathbf{d})]\, dV = 1.\end{aligned} \tag{13.165}$$

Equations (13.164)–(13.165), which pertain only to the singlet eigenfunctions (13.124) with associated eigenfrequencies in the vicinity of the fiducial frequency ω_0, can be regarded as a "narrow-band" version of the general anelastic orthonormality relations (6.143)–(6.144). It is noteworthy that there is no complex conjugation in (13.164)–(13.165) despite the fact that the eigenfunctions $\mathbf{s}$ are complex.

★13.2.4 Rotating anelastic perturbation

We can account for both rotation and anelasticity by judiciously combining the results in Sections 13.2.2 and 13.3.3. The real potential energy matrix V at the reference frequency ω_0 now incorporates the effect of the centrifugal potential ψ in addition to the structural perturbations $\delta\kappa_0$, $\delta\mu_0$, $\delta\rho$, $\delta\Phi$, δd and γ:

$$\begin{aligned}
V_{kk'} = \int_{\oplus} & [\delta\kappa_0(\boldsymbol{\nabla}\cdot\mathbf{s}_k)(\boldsymbol{\nabla}\cdot\mathbf{s}_{k'}) + 2\delta\mu_0(\mathbf{d}_k\!:\!\mathbf{d}_{k'}) \\
& + \delta\rho\{\mathbf{s}_k\cdot\boldsymbol{\nabla}\phi_{k'} + \mathbf{s}_{k'}\cdot\boldsymbol{\nabla}\phi_k \\
& + 4\pi G\rho(\hat{\mathbf{r}}\cdot\mathbf{s}_k)(\hat{\mathbf{r}}\cdot\mathbf{s}_{k'}) + g\Upsilon_{kk'}\} \\
& + \tfrac{1}{2}\rho\boldsymbol{\nabla}(\delta\Phi+\psi)\cdot(\mathbf{s}_k\cdot\boldsymbol{\nabla}\mathbf{s}_{k'} + \mathbf{s}_{k'}\cdot\boldsymbol{\nabla}\mathbf{s}_k \\
& - \mathbf{s}_k\boldsymbol{\nabla}\cdot\mathbf{s}_{k'} - \mathbf{s}_{k'}\boldsymbol{\nabla}\cdot\mathbf{s}_k) + \rho\mathbf{s}_k\cdot\boldsymbol{\nabla}\boldsymbol{\nabla}(\delta\Phi+\psi)\cdot\mathbf{s}_{k'} \\
& + \boldsymbol{\varepsilon}_k\!:\!\boldsymbol{\gamma}\!:\!\boldsymbol{\varepsilon}_{k'}]\,dV \\
- \int_{\Sigma} & \delta d\,[\tfrac{1}{2}\kappa_0(\boldsymbol{\nabla}\cdot\mathbf{s}_k)(\boldsymbol{\nabla}\cdot\mathbf{s}_{k'} - 2\hat{\mathbf{r}}\cdot\partial_r\mathbf{s}_{k'}) \\
& + \tfrac{1}{2}\kappa_0(\boldsymbol{\nabla}\cdot\mathbf{s}_{k'})(\boldsymbol{\nabla}\cdot\mathbf{s}_k - 2\hat{\mathbf{r}}\cdot\partial_r\mathbf{s}_k) \\
& + \mu_0\mathbf{d}_k\!:\!(\mathbf{d}_{k'} - 2\hat{\mathbf{r}}\partial_r\mathbf{s}_{k'}) + \mu_0\mathbf{d}_{k'}\!:\!(\mathbf{d}_k - 2\hat{\mathbf{r}}\partial_r\mathbf{s}_k) \\
& + \rho\{\mathbf{s}_k\cdot\boldsymbol{\nabla}\phi_{k'} + \mathbf{s}_{k'}\cdot\boldsymbol{\nabla}\phi_k \\
& + 8\pi G\rho(\hat{\mathbf{r}}\cdot\mathbf{s}_k)(\hat{\mathbf{r}}\cdot\mathbf{s}_{k'}) + g\Upsilon_{kk'}\}]_-^+\,d\Sigma \\
- \int_{\Sigma_{\mathrm{FS}}} & \boldsymbol{\nabla}^{\Sigma}(\delta d)\cdot[\kappa_0(\boldsymbol{\nabla}\cdot\mathbf{s}_k)\mathbf{s}_{k'} + \kappa_0(\boldsymbol{\nabla}\cdot\mathbf{s}_{k'})\mathbf{s}_k \\
& + 2\mu_0(\hat{\mathbf{r}}\cdot\mathbf{d}_k\cdot\hat{\mathbf{r}})\mathbf{s}_{k'} + 2\mu_0(\hat{\mathbf{r}}\cdot\mathbf{d}_{k'}\cdot\hat{\mathbf{r}})\mathbf{s}_k]_-^+\,d\Sigma.
\end{aligned} \tag{13.166}$$

More importantly, we must contend with a significant new complication: the *dual eigenfunctions* $\bar{\mathbf{s}}$ of the anti-Earth with $\mathsf{W} \to -\mathsf{W}$ are no longer simply the complex conjugates of the eigenfunctions $\mathbf{s}$ of the actual Earth; rather, $\mathbf{s}$ and $\bar{\mathbf{s}}$ must be represented by separate and independent expansions of the form

$$\mathbf{s} = \sum_k q_k\mathbf{s}_k, \qquad \bar{\mathbf{s}} = \sum_k \bar{q}_k\mathbf{s}_k. \tag{13.167}$$

To find the complex eigenfrequency perturbations $\delta\nu$ and the associated column vectors

$$\mathsf{q} = \begin{pmatrix} \vdots \\ q_k \\ \vdots \end{pmatrix}, \qquad \overline{\mathsf{q}} = \begin{pmatrix} \vdots \\ \overline{q}_k \\ \vdots \end{pmatrix} \tag{13.168}$$

in the vicinity of the reference frequency ω_0, we must solve the pair of generalized eigenvalue problems (7.86)–(7.87):

$$[\Omega^2 + \mathsf{V}(\omega_0 + \delta\nu) + 2(\omega_0 + \delta\nu)\mathsf{W} \\ - (\omega_0 + \delta\nu)^2(\mathsf{I} + \mathsf{T})]\mathsf{q} = 0, \tag{13.169}$$

$$[\Omega^2 + \mathsf{V}(\omega_0 + \delta\nu) - 2(\omega_0 + \delta\nu)\mathsf{W} \\ - (\omega_0 + \delta\nu)^2(\mathsf{I} + \mathsf{T})]\overline{\mathsf{q}} = 0. \tag{13.170}$$

Both the Coriolis force and narrow-band anelastic dispersion must now be accounted for in developing an appropriate renormalization procedure. Generalizing equations (13.142) and (13.158), we define z and $\overline{\mathsf{z}}$ by

$$\mathsf{z} = (\mathsf{I} + \tfrac{1}{2}\mathsf{T} - \tfrac{1}{2}\omega_0^{-1}\mathsf{W} - \tfrac{1}{2\pi}\omega_0^{-2}\mathsf{A})\mathsf{q}, \tag{13.171}$$

$$\overline{\mathsf{z}} = (\mathsf{I} + \tfrac{1}{2}\mathsf{T} + \tfrac{1}{2}\omega_0^{-1}\mathsf{W} - \tfrac{1}{2\pi}\omega_0^{-2}\mathsf{A})\overline{\mathsf{q}}. \tag{13.172}$$

Upon inserting (13.171)–(13.172) into equations (13.169)–(13.170) and ignoring terms that are of second order in $\delta\omega$, γ, $\omega_k - \omega_0$, T, V, A and W, we obtain the two ordinary eigenvalue problems

$$\mathsf{H}\mathsf{z} = \delta\nu\,\mathsf{z}, \qquad \overline{\mathsf{H}}\overline{\mathsf{z}} = \delta\nu\,\overline{\mathsf{z}}, \tag{13.173}$$

where

$$\mathsf{H} = \Omega - \omega_0\mathsf{I} + \mathsf{W} + (2\omega_0)^{-1}(\mathsf{V} + i\mathsf{A} - \omega_0^2\mathsf{T}), \tag{13.174}$$

$$\overline{\mathsf{H}} = \Omega - \omega_0\mathsf{I} - \mathsf{W} + (2\omega_0)^{-1}(\mathsf{V} + i\mathsf{A} - \omega_0^2\mathsf{T}). \tag{13.175}$$

The kinetic and potential energy perturbation matrices are both complex symmetric, $\mathsf{T}^{\mathrm{T}} = \mathsf{T}$ and $\mathsf{V}^{\mathrm{T}} = \mathsf{V}$, whereas the Coriolis matrix is complex anti-symmetric, $\mathsf{W}^{\mathrm{T}} = -\mathsf{W}$. The two renormalized eigenfrequency perturbation matrices (13.174)–(13.175) are therefore simply related by

$$\overline{\mathsf{H}} = \mathsf{H}^{\mathrm{T}}, \tag{13.176}$$

i.e., $\overline{\mathsf{H}}$ is simply the transpose of H. It is generally true that a complex matrix and its transpose have the same eigenvalues $\delta\nu$ by virtue of the elementary relation $\det(\mathsf{H}^{\mathrm{T}} - \delta\nu\,\mathsf{I}) = \det(\mathsf{H} - \delta\nu\,\mathsf{I})$. Upon taking the transpose of the relation $\mathsf{H}^{\mathrm{T}}\bar{\mathsf{z}} = \delta\nu\,\bar{\mathsf{z}}$, we see that we may alternatively consider the dual eigenvectors $\bar{\mathsf{z}}$ to be the *left eigenvectors* of H, rather than the *right eigenvectors* of $\overline{\mathsf{H}}$:

$$\bar{\mathsf{z}}^{\mathrm{T}}\mathsf{H} = \delta\nu\,\bar{\mathsf{z}}^{\mathrm{T}}. \tag{13.177}$$

By manipulating the product $\bar{\mathsf{z}}^{\mathrm{T}}\mathsf{H}\mathsf{z}'$ in two different ways, it is easy to verify that the dual (left) eigenvector $\bar{\mathsf{z}}$ and primal (right) eigenvector z' associated with distinct eigenvalues $\delta\nu \neq \delta\nu'$ are *biorthogonal* in the sense $\bar{\mathsf{z}}^{\mathrm{T}}\mathsf{z}' = 0$. If we normalize every z and its associated dual $\bar{\mathsf{z}}$ by stipulating that $\bar{\mathsf{z}}^{\mathrm{T}}\mathsf{z} = 1$, then we may write, by analogy with equations (13.161) on a non-rotating anelastic Earth,

$$\overline{\mathsf{Z}}^{\mathrm{T}}\mathsf{Z} = \mathsf{I}, \qquad \overline{\mathsf{Z}}^{\mathrm{T}}\mathsf{H}\mathsf{Z} = \Delta, \tag{13.178}$$

where Z and $\overline{\mathsf{Z}}$ are the matrices composed of right and left column vectors z and $\bar{\mathsf{z}}$, and Δ is the diagonal matrix of complex eigenfrequency perturbations $\delta\nu = \delta\omega + i\gamma$ in the vicinity of ω_0, as before. The biorthonormality relation $\overline{\mathsf{Z}}^{\mathrm{T}}\mathsf{Z} = \mathsf{I}$ may be written in terms of the matrices Q and $\overline{\mathsf{Q}}$ of original column vectors q and their duals $\bar{\mathsf{q}}$ in the form

$$\overline{\mathsf{Q}}^{\mathrm{T}}(\mathsf{I} + \mathsf{T} - \omega_0^{-1}\mathsf{W} - \tfrac{1}{\pi}\omega_0^{-2}\mathsf{A})\mathsf{Q} = \mathsf{I}. \tag{13.179}$$

It is noteworthy that the renormalized left and right eigenvector matrices are *each other's transposed inverse*:

$$\overline{\mathsf{Z}} = \mathsf{Z}^{-\mathrm{T}}. \tag{13.180}$$

This enables us to avoid overt consideration of the anti-Earth, and calculate $\overline{\mathsf{Z}}$ and $\overline{\mathsf{Q}}$ by numerical matrix inversion and transposition, if desired.

In summary, we are required to solve both the right and left eigenvalue problems $\mathsf{H}\mathsf{z} = \delta\nu\,\mathsf{z}$ and $\bar{\mathsf{z}}^{\mathrm{T}}\mathsf{H} = \delta\nu\,\bar{\mathsf{z}}$ subject to the requirement that $\bar{\mathsf{z}}^{\mathrm{T}}\mathsf{z} = 1$ to find the complex eigenfrequency perturbations $\delta\nu = \delta\omega + i\gamma$ and the associated renormalized eigenvectors z and their duals $\bar{\mathsf{z}}$ upon a rotating anelastic Earth. Alternatively, we may solve only the right eigenvalue problem $\mathsf{H}\mathsf{z} = \delta\nu\,\mathsf{z}$ and compute the duals $\bar{\mathsf{z}}$ using equation (13.180). The original right and left eigenvectors q and $\bar{\mathsf{q}}$ are related to the corresponding renormalized vectors z and $\bar{\mathsf{z}}$ by

$$\mathsf{q} = (\mathsf{I} - \tfrac{1}{2}\mathsf{T} + \tfrac{1}{2}\omega_0^{-1}\mathsf{W} + \tfrac{1}{2\pi}\omega_0^{-2}\mathsf{A})\mathsf{z}, \tag{13.181}$$

$$\bar{\mathsf{q}} = (\mathsf{I} - \tfrac{1}{2}\mathsf{T} - \tfrac{1}{2}\omega_0^{-1}\mathsf{W} + \tfrac{1}{2\pi}\omega_0^{-2}\mathsf{A})\bar{\mathsf{z}}. \tag{13.182}$$

The three-dimensional biorthonormality relations equivalent to the matrix equation (13.179) are

$$\int_{\oplus}(\rho+\delta\rho)\,\overline{\mathbf{s}}\cdot\mathbf{s}'\,dV-\int_{\Sigma}\delta d\,[\rho\,\overline{\mathbf{s}}\cdot\mathbf{s}']_{-}^{+}\,d\Sigma-\omega_0^{-1}\int_{\oplus}\rho\,\overline{\mathbf{s}}\cdot(i\boldsymbol{\Omega}\times\mathbf{s}')\,dV$$
$$-\tfrac{1}{\pi}\omega_0^{-2}\int_{\oplus}[\kappa_0 q_\kappa(\boldsymbol{\nabla}\cdot\overline{\mathbf{s}})(\boldsymbol{\nabla}\cdot\mathbf{s}')+2\mu_0 q_\mu(\overline{\mathbf{d}}:\mathbf{d}')]\,dV=0$$
$$\text{if }\delta\nu\neq\delta\nu',\tag{13.183}$$

$$\int_{\oplus}(\rho+\delta\rho)\,\overline{\mathbf{s}}\cdot\mathbf{s}\,dV-\int_{\Sigma}\delta d\,[\rho\,\overline{\mathbf{s}}\cdot\mathbf{s}]_{-}^{+}\,d\Sigma-\omega_0^{-1}\int_{\oplus}\rho\,\overline{\mathbf{s}}\cdot(i\boldsymbol{\Omega}\times\mathbf{s})\,dV$$
$$-\tfrac{1}{\pi}\omega_0^{-2}\int_{\oplus}[\kappa_0 q_\kappa(\boldsymbol{\nabla}\cdot\overline{\mathbf{s}})(\boldsymbol{\nabla}\cdot\mathbf{s})+2\mu_0 q_\mu(\overline{\mathbf{d}}:\mathbf{d})]\,dV=1.\tag{13.184}$$

The results of the overall calculation are correct to first order in the relative degenerate frequencies of the coupled multiplets $\omega_k-\omega_0$ and the full panoply of structural and rotational perturbations $\delta\kappa_0$, $\delta\mu_0$, $\delta\rho$, $\delta\Phi$, δd, γ, q_κ, q_μ and $\boldsymbol{\Omega}$. The errors in the real eigenfrequency perturbations $\delta\omega$, the decay rates γ, and the associated eigenfunctions $\mathbf{s}$ and their duals $\overline{\mathbf{s}}$ should be of second order in these small quantities, for all singlets in the vicinity of the reference frequency ω_0.

Of course, it is possible to conduct the calculation in terms of the perturbations in the P-wave and S-wave speeds $\delta\alpha_0$, $\delta\beta_0$ and the associated inverse quality factors q_α, q_β rather than in terms of $\delta\kappa_0$, $\delta\mu_0$ and q_κ, q_μ. The dependence of the real potential energy matrix V upon these wave-speed perturbations can be readily determined by substituting the first-order relations

$$\delta\kappa_0=\delta\rho(\alpha_0^2-\tfrac{4}{3}\beta_0^2)+2\rho(\alpha_0\,\delta\alpha_0-\tfrac{4}{3}\beta_0\,\delta\beta_0),\tag{13.185}$$

$$\delta\mu_0=\delta\rho\,\beta_0^2+2\rho\beta_0\,\delta\beta_0\tag{13.186}$$

and

$$\kappa_0 q_\kappa=\rho\alpha_0^2 q_\alpha-\tfrac{4}{3}\rho\beta_0^2 q_\beta,\qquad \mu_0 q_\mu=\rho\beta_0^2 q_\beta\tag{13.187}$$

into the defining relation (13.166). The effect of the initial deviatoric stress field $\boldsymbol{\tau}$ within the Earth can be accounted for by modifying V in accordance with equations (13.43) and (13.87)–(13.88) if desired. Transverse isotropy of the unperturbed Earth model can likewise be incorporated by means of a straightforward modification based upon equations (13.118)–(13.119).

13.2.5 Overview

Tables 13.1 and 13.2 summarize and compare the resulting ordinary eigenvalue problems that must be solved in the presence and absence of rotation and anelasticity. In each case the renormalized eigenfrequency perturbation matrix H is diagonalized by a *similarity transformation*:

$$\mathsf{Z}^{-1}\mathsf{H}\mathsf{Z} = \Delta, \tag{13.188}$$

where Z^{-1} denotes the inverse. The existence of such a diagonalizing transformation is guaranteed in the absence of anelasticity. On a non-rotating elastic Earth the matrix H is real symmetric, so that Z is orthogonal, $\mathsf{Z}^{-1} = \mathsf{Z}^{\mathrm{T}}$, whereas on a rotating elastic Earth H is complex Hermitian, so that Z is unitary, $\mathsf{Z}^{-1} = \mathsf{Z}^{\mathrm{H}}$. On an anelastic Earth, we cannot find a similarity transformation satisfying equation (13.188) unless the eigenfrequency perturbation matrix H is *non-defective*. However, any complex matrix H whose eigenvalues are *distinct* is non-defective and diagonalizable (Horn & Johnson 1985); inasmuch as a repeated eigenvalue $\delta\nu = \delta\omega + i\gamma$ can only be the result of a rare *accidental degeneracy*, it is reasonable to assume that a diagonalizing transformation Z can "almost always" be found. In the absence of rotation, the anelastic eigenfrequency perturbation matrix H is complex symmetric, $\mathsf{H}^{\mathrm{T}} = \mathsf{H}$, and the transformation Z is complex

Earth Model	Perturbation Matrices
Non-Rotating Elastic	$\mathsf{H} = \Omega - \omega_0\mathsf{I} + (2\omega_0)^{-1}(\mathsf{V} - \omega_0^2\mathsf{T})$
Rotating Elastic	$\mathsf{H} = \Omega - \omega_0\mathsf{I} + \mathsf{W} + (2\omega_0)^{-1}(\mathsf{V} - \omega_0^2\mathsf{T})$
Non-Rotating Anelastic	$\mathsf{H} = \Omega - \omega_0\mathsf{I} + (2\omega_0)^{-1}(\mathsf{V} + i\mathsf{A} - \omega_0^2\mathsf{T})$
Rotating Anelastic	$\mathsf{H} = \Omega - \omega_0\mathsf{I} + \mathsf{W} + (2\omega_0)^{-1}(\mathsf{V} + i\mathsf{A} - \omega_0^2\mathsf{T})$ $\overline{\mathsf{H}} = \Omega - \omega_0\mathsf{I} - \mathsf{W} + (2\omega_0)^{-1}(\mathsf{V} + i\mathsf{A} - \omega_0^2\mathsf{T})$

Table 13.1. Perturbation matrices H and $\overline{\mathsf{H}} = \mathsf{H}^{\mathrm{T}}$ which must be diagonalized in the presence and absence of rotation and anelasticity.

Earth Model	Matrix H	Diagonalization	Orthonormalization
Non-Rotating Elastic	Real Symmetric	$\mathsf{Z}^{\mathrm{T}}\mathsf{H}\mathsf{Z} = \Delta$	$\mathsf{Z}^{\mathrm{T}}\mathsf{Z} = \mathsf{I}$
Rotating Elastic	Complex Hermitian	$\mathsf{Z}^{\mathrm{H}}\mathsf{H}\mathsf{Z} = \Delta$	$\mathsf{Z}^{\mathrm{H}}\mathsf{Z} = \mathsf{I}$
Non-Rotating Anelastic	Complex Symmetric	$\mathsf{Z}^{\mathrm{T}}\mathsf{H}\mathsf{Z} = \Delta$	$\mathsf{Z}^{\mathrm{T}}\mathsf{Z} = \mathsf{I}$
Rotating Anelastic	General Complex	$\overline{\mathsf{Z}}^{\mathrm{T}}\mathsf{H}\mathsf{Z} = \Delta$	$\overline{\mathsf{Z}}^{\mathrm{T}}\mathsf{Z} = \mathsf{I}$

Table 13.2. Mathematical structure of the Rayleigh-Ritz eigenvalue problem that must be solved to find the diagonal matrix Δ of eigenfrequency perturbations $\delta\omega$ or $\delta\nu = \delta\omega + i\gamma$ in the vicinity of a fiducial frequency ω_0 and the associated matrices Z and $\overline{\mathsf{Z}}$ of renormalized eigenvectors z and dual eigenvectors $\overline{\mathsf{z}}$ in the presence and absence of rotation and anelasticity. On a non-rotating Earth the eigenvectors and their duals obviously coincide, $\overline{\mathsf{Z}} = \mathsf{Z}$, whereas on a rotating but perfectly elastic Earth they are each other's complex conjugates, $\overline{\mathsf{Z}} = \mathsf{Z}^*$.

orthogonal, $\mathsf{Z}^{-1} = \mathsf{Z}^{\mathrm{T}}$. In the presence of both rotation and anelasticity, H does not have any special symmetries or properties; in this, the most general case, we are required to solve for the Earth and anti-Earth eigenvectors separately, since $\mathsf{Z}^{-1} = \overline{\mathsf{Z}}^{\mathrm{T}}$.

The eigenvector back-transformation relations and the associated orthonormality or biorthonormality relations in each of the four cases are summarized in Tables 13.3 and 13.4. Renormalization guarantees that the singlet eigenfunctions $\mathbf{s}$ and their duals $\overline{\mathbf{s}}$ are properly orthonormalized or biorthonormalized with respect to the perturbed Earth, correct to first order in $\delta\rho$, δd, q_κ, q_μ and Ω. In each case, this first-order orthonormality or biorthonormality has been obtained by judicious neglect of some but not all second-order terms in the reduction to an ordinary eigenvalue problem (13.131), (13.143), (13.159) or (13.173). In the absence of any perturbations ($\mathsf{T} = 0$, $\mathsf{V} = \mathsf{A} = 0$, $\mathsf{W} = 0$) the eigenfrequency perturbation and eigenvector matrices reduce to $\Delta = \Omega - \omega_0\mathsf{I}$ and $\mathsf{Z} = \overline{\mathsf{Z}} = \mathsf{Q} = \overline{\mathsf{Q}} = \mathsf{I}$.

Efficient and stable numerical algorithms for solving ordinary eigenvalue problems such as those in Tables 13.1 and 13.2 are well documented and

Earth Model	Renormalization
Non-Rotating Elastic	$\mathsf{Q} = (\mathsf{I} - \frac{1}{2}\mathsf{T})\mathsf{Z}$
Rotating Elastic	$\mathsf{Q} = (\mathsf{I} - \frac{1}{2}\mathsf{T} + \frac{1}{2}\omega_0^{-1}\mathsf{W})\mathsf{Z}$
Non-Rotating Anelastic	$\mathsf{Q} = (\mathsf{I} - \frac{1}{2}\mathsf{T} + \frac{1}{2\pi}\omega_0^{-2}\mathsf{A})\mathsf{Z}$
Rotating Anelastic	$\mathsf{Q} = (\mathsf{I} - \frac{1}{2}\mathsf{T} + \frac{1}{2}\omega_0^{-1}\mathsf{W} + \frac{1}{2\pi}\omega_0^{-2}\mathsf{A})\mathsf{Z}$ $\overline{\mathsf{Q}} = (\mathsf{I} - \frac{1}{2}\mathsf{T} - \frac{1}{2}\omega_0^{-1}\mathsf{W} + \frac{1}{2\pi}\omega_0^{-2}\mathsf{A})\overline{\mathsf{Z}}$

Table 13.3. Renormalized-to-original eigenvector back-transformation relations in the presence and absence of rotation and anelasticity.

Earth Model	Orthonormality
Non-Rotating Elastic	$\mathsf{Q}^{\mathrm{T}}(\mathsf{I} + \mathsf{T})\mathsf{Q} = \mathsf{I}$
Rotating Elastic	$\mathsf{Q}^{\mathrm{T}}(\mathsf{I} + \mathsf{T} - \omega_0^{-1}\mathsf{W})\mathsf{Q} = \mathsf{I}$
Non-Rotating Anelastic	$\mathsf{Q}^{\mathrm{T}}(\mathsf{I} + \mathsf{T} - \frac{1}{\pi}\omega_0^{-2}\mathsf{A})\mathsf{Q} = \mathsf{I}$
Rotating Anelastic	$\overline{\mathsf{Q}}^{\mathrm{T}}(\mathsf{I} + \mathsf{T} - \omega_0^{-1}\mathsf{W} - \frac{1}{\pi}\omega_0^{-2}\mathsf{A})\mathsf{Q} = \mathsf{I}$

Table 13.4. Eigenvector orthonormality or biorthonormality relations in the presence and absence of rotation and anelasticity.

widely available (Smith & others 1976; Garbow & others 1977; Dongarra & Walker 1995). The optimal method to be used in any application depends upon the mathematical character of the matrix H that is to be diagonalized. Complex symmetric matrices and general complex matrices are more problematical than real symmetric or complex Hermitian matrices, since they may be technically non-defective but numerically defective or nearly so on a finite-precision computer. Despite this, the formalism developed here provides a practical means of determining the normal-mode eigenfrequencies and eigenfunctions of a laterally heterogeneous anelastic Earth model.

13.3 Singlet-Sum Synthetic Seismograms

Synthetic seismograms and spectra on a perturbed Earth can be calculated by singlet superposition; we devote the remainder of this chapter to a discussion of this topic. For maximum generality, we consider the case of a rotating anelastic Earth; however, only minor modifications are required to obtain the analogous results in the other three cases. In the absence of rotation, for example, it is simply necessary to set $\mathsf{H}^{\mathrm{T}} = \mathsf{H}$, and delete the "anti-Earth" overbars adorning the dual eigenvector matrices $\overline{\mathsf{Z}}$ and $\overline{\mathsf{Q}}$.

13.3.1 Narrow-band response

The seismic response to a step-function moment-tensor source depends upon the *receiver vector* r and *source vector* s, with real elements

$$r_k = \hat{\boldsymbol{\nu}} \cdot \mathbf{s}_k(\mathbf{x}), \qquad s_k = \mathbf{M} : \boldsymbol{\varepsilon}_k(\mathbf{x}_{\mathrm{s}}). \tag{13.189}$$

As usual, the quantities $\hat{\boldsymbol{\nu}}$ and $\mathbf{x}$ denote the polarization and geographical location of the receiver, whereas $\mathbf{M}$ and $\mathbf{x}_{\mathrm{s}}$ denote the moment tensor and hypocentral location of the source. The time-domain acceleration upon a rotating anelastic Earth is given by equation (7.114); rewritten in terms of the present notation, this result takes the form

$$a(t) = \mathrm{Re}\,[A(t)\exp(i\omega_0 t)], \tag{13.190}$$

where

$$A(t) = \mathsf{r}'^{\,\mathrm{T}} \exp(i\Delta t)\,\mathsf{s}'. \tag{13.191}$$

The primes denote *transformed* receiver and source vectors which are defined in terms of the primal and dual eigenvector matrices Q and $\overline{\mathsf{Q}}$ by

$$\mathsf{r}' = \mathsf{Q}^{\mathrm{T}}\mathsf{r}, \qquad \mathsf{s}' = \overline{\mathsf{Q}}^{\,\mathrm{T}}\mathsf{s}. \tag{13.192}$$

The fiducial origin time t_s of the earthquake source has been set equal to zero, so that equation (13.190) provides a description of the response for all times $t \geq 0$. We can alternatively express the primed receiver and source vectors in terms of the transformation matrices Z and $\overline{\mathsf{Z}}$ rather than Q and $\overline{\mathsf{Q}}$ in the form

$$\mathsf{r}' = \mathsf{Z}^{\mathrm{T}}\mathsf{u}, \qquad \mathsf{s}' = \overline{\mathsf{Z}}^{\mathrm{T}}\mathsf{v}, \tag{13.193}$$

where

$$\mathsf{u} = (\mathsf{I} - \tfrac{1}{2}\mathsf{T} - \tfrac{1}{2}\omega_0^{-1}\mathsf{W} + \tfrac{1}{2\pi}\omega_0^{-2}\mathsf{A})\mathsf{r}, \tag{13.194}$$

$$\mathsf{v} = (\mathsf{I} - \tfrac{1}{2}\mathsf{T} + \tfrac{1}{2}\omega_0^{-1}\mathsf{W} + \tfrac{1}{2\pi}\omega_0^{-2}\mathsf{A})\mathsf{s}. \tag{13.195}$$

The quantities u and v defined in equations (13.194)–(13.195) can be regarded as *renormalized receiver* and *source vectors* associated with the anti-Earth and the Earth, respectively. Because we have ignored terms of second order in $\delta\omega$ and $\omega_k - \omega_0$, the expression (13.190) is a representation of a *narrow-band accelerogram* which is valid only in the neighborhood of the reference frequency ω_0. The quantity $A(t)$ defined by the singlet sum (13.191) is a *slowly varying complex envelope* which modulates the reference "carrier" $\exp(i\omega_0 t)$.

13.3.2 Direct solution method

Upon making use of the eigenvector transformation relations (13.178), we can rewrite this narrow-band modulation function in terms of the original eigenfrequency perturbation matrix H, prior to diagonalization, in the form

$$A(t) = \mathsf{u}^{\mathrm{T}}\exp(i\mathsf{H}t)\,\mathsf{v}. \tag{13.196}$$

Equation (13.196) is an explicit or *direct* rather than a normal-mode representation of the response. We can regard the exponential

$$\mathsf{X}(t) = \exp(i\mathsf{H}t) \tag{13.197}$$

as a *slowly varying propagator matrix*, which can be found by solving the first-order system of linear ordinary differential equations

$$d\mathsf{X}/dt = i\mathsf{H}\mathsf{X}, \qquad \mathsf{X}(0) = \mathsf{I}. \tag{13.198}$$

This suggests the possibility of a *direct solution method* of calculating narrow-band modulation functions $A(t)$ which eliminates the need for large-scale matrix diagonalization. The slow variation of $\mathsf{X}(t)$ enables equation (13.198) to be integrated numerically using relatively coarse time steps.

A more efficient approach if one wishes to synthesize a suite of narrow-band accelerograms at a number of receivers $\hat{\boldsymbol{\nu}}$, $\mathbf{x}$ produced by a given seismic source $\mathbf{M}$, $\mathbf{x}_s$ is to solve the system of first-order equations

$$d\mathsf{d}/dt = i\mathsf{H}\mathsf{d}, \qquad \mathsf{d}(0) = \mathsf{v} \tag{13.199}$$

for the narrow-band *source response vector* $\mathsf{d}(t)$. The modulation function at each receiver is given in terms of $\mathsf{d}(t)$ by

$$A(t) = \mathsf{u}^{\mathrm{T}}\mathsf{d}(t). \tag{13.200}$$

Alternatively, if one seeks the response at a single receiver $\hat{\boldsymbol{\nu}}$, $\mathbf{x}$ due to a suite of earthquakes $\mathbf{M}$, $\mathbf{x}_s$ it is preferable to solve

$$d\mathsf{e}/dt = i\mathsf{H}^{\mathrm{T}}\mathsf{e}, \qquad \mathsf{e}(0) = \mathsf{u} \tag{13.201}$$

for the receiver response dual vector $\mathsf{e}(t)$. The response to each source is given in that case by

$$A(t) = \mathsf{e}^{\mathrm{T}}(t)\mathsf{v}. \tag{13.202}$$

Note that it is unnecessary to solve the two initial-value problems (13.199) and (13.201) on the Earth and the anti-Earth contemporaneously; a given narrow-band modulation function $A(t)$ can be calculated using either equation (13.200) or (13.202). In contrast, when using the diagonalization method, it is necessary to determine both the right and the left eigenvectors in order to evaluate the normal-mode sum (13.191). If one seeks the response to a number of different sources at a number of polarized receivers, it is possible to solve for either $\mathsf{d}(t)$ or $\mathsf{e}(t)$; if efficiency is the only criterion, the optimal choice obviously depends upon whether there are more receivers than sources, or vice versa.

It is straightforward to express the narrow-band acceleration response to a moment-tensor source in the frequency domain rather than the time domain. The normal-mode representation obtained either by rewriting equation (7.117) or by Fourier transforming equation (13.190) is

$$a(\omega) = \tfrac{1}{2}i\mathsf{r}'^{\mathrm{T}}[\Delta - (\omega - \omega_0)\mathsf{I}]^{-1}\mathsf{s}'. \tag{13.203}$$

Equation (13.203) is a weighted sum of Lorentzian resonance peaks centered upon the perturbed singlets in the vicinity of the positive reference frequency ω_0; the contribution from the corresponding negative-frequency peaks has been explicitly ignored. We can also write $a(\omega)$ directly in terms of the original perturbation matrix H and the renormalized receiver and source vectors u and v in the form

$$a(\omega) = \tfrac{1}{2}i\mathsf{u}^{\mathrm{T}}[\mathsf{H} - (\omega - \omega_0)\mathsf{I}]^{-1}\mathsf{v}. \tag{13.204}$$

The result (13.204) provides the theoretical basis for a *frequency-domain direct solution method* that is the narrow-band analogue of the method based upon equation (7.118). We can find the acceleration spectrum in the vicinity of the reference frequency ω_0 either by solving the system of linear algebraic equations

$$(\mathsf{H} - \omega\mathsf{I})\mathsf{d}(\omega) = i\mathsf{v} \tag{13.205}$$

for the source response vector $\mathsf{d}(\omega)$ and then forming the scalar product

$$a(\omega) = \tfrac{1}{2}\mathsf{u}^{\mathrm{T}}\mathsf{d}(\omega - \omega_0), \tag{13.206}$$

or by solving

$$(\mathsf{H}^{\mathrm{T}} - \omega\mathsf{I})\mathsf{e}(\omega) = i\mathsf{u} \tag{13.207}$$

for the dual receiver response vector $\mathsf{e}(\omega)$ and then forming the product

$$a(\omega) = \tfrac{1}{2}\mathsf{e}^{\mathrm{T}}(\omega - \omega_0)\mathsf{v}. \tag{13.208}$$

As the notation suggests, the two frequency-domain response vectors $\mathsf{d}(\omega)$ and $\mathsf{e}(\omega)$ are the Fourier transforms of the time-domain quantities $\mathsf{d}(t)$ and $\mathsf{e}(t)$. Note that it is necessary to form and solve a separate system of equations (13.205) or (13.207) at every frequency ω where the spectral response $a(\omega)$ is desired; the optimal spacing between samples is dictated by the degree of detail that must be resolved near the frequency ω_0. In general, it is preferable not to employ the numerical inverse matrices $(\mathsf{H} - \omega\mathsf{I})^{-1}$ or $(\mathsf{H}^{\mathrm{T}} - \omega\mathsf{I})^{-1}$, inasmuch as algorithms for calculating $\mathsf{d}(\omega)$ or $\mathsf{e}(\omega)$ based upon the QR factorization of $\mathsf{H} - \omega\mathsf{I}$ or $\mathsf{H}^{\mathrm{T}} - \omega\mathsf{I}$ are considerably more efficient. More extensive discussions of the direct solution method are provided by the originators of the technique (Hara, Tsuboi & Geller 1991; 1993).

Using the spherical-Earth eigenfunctions $\mathbf{s}_k$ as the basis vectors in the expansion (13.124) is but one of many available options, as we noted in Chapter 7. Geller & Ohminato (1994) and Cummins, Takeuchi & Geller (1997) have developed a promising variant of the frequency-domain direct solution method, in which the acceleration response $\mathbf{a}(\mathbf{x}, \omega)$ of the Earth is written in the form

$$\mathbf{a} = \sum_{klm} \left(p_{klm} X_k \mathbf{P}_{lm} + b_{klm} X_k \mathbf{B}_{lm} + c_{klm} X_k \mathbf{C}_{lm} \right), \tag{13.209}$$

where $\mathbf{P}_{lm}(\theta, \phi)$, $\mathbf{B}_{lm}(\theta, \phi)$ and $\mathbf{C}_{lm}(\theta, \phi)$ are the vector spherical harmonics, and $X_k(r)$ are the linear radial splines (10.86). The basis vectors $X_k\mathbf{P}_{lm}$, $X_k\mathbf{B}_{lm}$ and $X_k\mathbf{C}_{lm}$ in (13.209) are global in the geographical coordinates θ, ϕ but local in the radial coordinate r. The unknown expansion coefficients $p_{klm}(\omega)$, $b_{klm}(\omega)$ and $c_{klm}(\omega)$ are determined by solving a system of linear algebraic equations analogous to (13.205) or (13.207). In the special case of a SNREI Earth, the orthonormality of the harmonics $\mathbf{P}_{lm}$, $\mathbf{B}_{lm}$ and $\mathbf{C}_{lm}$ reduces this procedure to the one-dimensional spline-based formulation described in Section 10.7.

13.3.3 Response of a hybrid multiplet

In most circumstances, the singlet normal modes of a rotating anelastic Earth will be grouped into *hybrid multiplets* which can be identified as the perturbed counterparts of the spherical-Earth multiplets comprising the Rayleigh-Ritz basis. A complete broad-band accelerogram or spectrum must consist of exactly $2l+1$ decaying oscillations or Lorentzian resonance peaks per hybrid multiplet ${}_n\mathrm{T}_l$ or ${}_n\mathrm{S}_l$. Such a "seamless" singlet sum can be synthesized by means of a straightforward shifting and winnowing scheme which we now describe.

In order to isolate the portion of a narrow-band accelerogram (13.190) or spectrum (13.203) corresponding to a particular hybrid multiplet, it is convenient to employ a slightly modified notation, in which the index k is used as a generic label for *all* of the basis functions and/or singlets associated with a given ${}_n\mathrm{T}_l$ or ${}_n\mathrm{S}_l$. We now specify the reference frequency ω_0 to be the unperturbed degenerate eigenfrequency of the $k=0$ *target multiplet* which we wish to isolate, and use negative and positive values of k, respectively, to denote the neighboring basis-set multiplets with degenerate eigenfrequencies $\omega_k < \omega_0$ and $\omega_k > \omega_0$. The energy and anelasticity matrices T, V and A, the Coriolis matrix W, and the eigenfrequency perturbation matrix (13.174) can be expressed in terms of this revised notation in the form

$$\mathsf{H} = \begin{pmatrix} & \vdots & & \vdots & \\ \cdots & \mathsf{H}_{-1-1} & \mathsf{H}_{-10} & \mathsf{H}_{-11} & \cdots \\ & \mathsf{H}_{0-1} & \mathsf{H}_{00} & \mathsf{H}_{01} & \\ \cdots & \mathsf{H}_{1-1} & \mathsf{H}_{10} & \mathsf{H}_{11} & \cdots \\ & \vdots & & \vdots & \end{pmatrix}. \qquad (13.210)$$

If k denotes a basis multiplet of spherical-harmonic degree l, and k' denotes a basis multiplet of spherical-harmonic degree l', then the quantity $\mathsf{H}_{kk'}$ is a $(2l+1)\times(2l'+1)$ *submatrix* which governs the coupling between them. The diagonal matrix Δ of complex eigenfrequency perturbations consists of submatrices $\Delta_k = \mathrm{diag}\,[\cdots\ \delta\nu_k\ \cdots]$, one corresponding to each of the hybrid multiplets in the basis set:

$$\Delta = \begin{pmatrix} \ddots & & & & \\ & \Delta_{-1} & & & \\ & & \Delta_0 & & \\ & & & \Delta_1 & \\ & & & & \ddots \end{pmatrix}. \tag{13.211}$$

The associated eigenvector and dual-eigenvector matrices Z, Q and $\overline{\mathsf{Z}}$, $\overline{\mathsf{Q}}$ may likewise be decomposed into submatrices $\mathsf{Z}_{kk'}$, $\mathsf{Q}_{kk'}$ and $\overline{\mathsf{Z}}_{kk'}$, $\overline{\mathsf{Q}}_{kk'}$. The indices k and k' in this latter case are associated with the basis and hybrid multiplets, respectively. Transposed matrices such as H^{T} will be written in the form

$$\mathsf{H}^{\mathrm{T}} = \begin{pmatrix} & \vdots & & \vdots & \\ \cdots & \mathsf{H}^{\mathrm{T}}_{-1-1} & \mathsf{H}^{\mathrm{T}}_{-10} & \mathsf{H}^{\mathrm{T}}_{-11} & \cdots \\ & \mathsf{H}^{\mathrm{T}}_{0-1} & \mathsf{H}^{\mathrm{T}}_{00} & \mathsf{H}^{\mathrm{T}}_{01} & \cdots \\ \cdots & \mathsf{H}^{\mathrm{T}}_{1-1} & \mathsf{H}^{\mathrm{T}}_{10} & \mathsf{H}^{\mathrm{T}}_{11} & \\ & \vdots & & \vdots & \end{pmatrix}. \tag{13.212}$$

Note that $\mathsf{H}^{\mathrm{T}}_{kk'}$ denotes the kk' submatrix of the transposed matrix H^{T} rather than the transpose of the submatrix $\mathsf{H}_{kk'}$; in fact, $\mathsf{H}^{\mathrm{T}}_{kk'} = (\mathsf{H}_{k'k})^{\mathrm{T}}$. The same convention applies to all of the transposed eigenvector submatrices $\mathsf{Z}^{\mathrm{T}}_{kk'}$, $\mathsf{Q}^{\mathrm{T}}_{kk'}$ and $\overline{\mathsf{Z}}^{\mathrm{T}}_{kk'}$, $\overline{\mathsf{Q}}^{\mathrm{T}}_{kk'}$.

Determination of the ordered diagonal matrix (13.211) and the associated transformation matrices Z, Q and $\overline{\mathsf{Z}}$, $\overline{\mathsf{Q}}$ is a straightforward matter at sufficiently low values of the reference frequency ω_0, where the majority of the split hybrid multiplets are relatively well-isolated in the seismic spectrum. It is then straightforward to rank the real eigenfrequency perturbations $\delta\omega$ numerically, and assign the proper number of singlets to each $(2l+1)$-dimensional diagonal matrix Δ_k. Two or more strongly coupled multiplets $k, k', \ldots$ whose eigenfrequency perturbations are interlaced can be treated as a single *super-multiplet* of dimension $(2l+1)+(2l'+1)+\cdots$. The reference frequency ω_0 in that case may be taken to be the degenerate eigenfrequency of any one of them. We shall continue to speak of a target hybrid "multiplet" in what follows, with the understanding that the theory is applicable to such a strongly hybridized super-multiplet as well.

The time-domain acceleration response of the target multiplet can be written in terms of the above notation in the form

$$a_0(t) = \mathrm{Re}\,[A_0(t)\exp(i\omega_0 t)], \tag{13.213}$$

where

$$A_0(t) = \mathsf{r}_0'^{\mathrm{T}} \exp(i\Delta_0 t)\, \mathsf{s}_0'. \tag{13.214}$$

The quantities r_0' and s_0' are transformed $(2l+1)$-dimensional receiver and source vectors given by

$$\mathsf{r}_0' = \sum_k \mathsf{Q}_{0k}^{\mathrm{T}} \mathsf{r}_k = \sum_k \mathsf{Z}_{0k}^{\mathrm{T}} \mathsf{u}_k, \tag{13.215}$$

$$\mathsf{s}_0' = \sum_k \overline{\mathsf{Q}}_{0k}^{\mathrm{T}} \mathsf{s}_k = \sum_k \overline{\mathsf{Z}}_{0k}^{\mathrm{T}} \mathsf{v}_k. \tag{13.216}$$

We have appended a subscript zero to $a_0(t)$, $A_0(t)$ and r_0', s_0' to serve as a reminder that they represent the response of the $k = 0$ multiplet. The envelope $A_0(t)$ is a slowly varying *multiplet modulation function* which consists of $2l + 1$ complex exponentials. The corresponding frequency-domain response is a sum of $2l+1$ Lorentzians centered upon the split target eigenfrequencies:

$$a_0(\omega) = \tfrac{1}{2} i \mathsf{r}_0'^{\mathrm{T}} [\Delta_0 - (\omega - \omega_0)\mathsf{I}_{00}]^{-1} \mathsf{s}_0', \tag{13.217}$$

where I_{00} denotes the $(2l+1) \times (2l+1)$ identity. The amplitudes and phases of the target singlets in the sums (13.214) and (13.217) depend upon the source and receiver parameters $\mathbf{M}$, $\mathbf{x}_s$ and $\hat{\boldsymbol{\nu}}$, $\mathbf{x}$.

To synthesize a "seamless" broad-band accelerogram or spectrum, we select each hybrid multiplet ${}_n\mathrm{T}_l$ or ${}_n\mathrm{S}_l$ as the target in turn, compute and diagonalize a new matrix H in order to find the eigenfrequency perturbations Δ_0 and the transformed receiver and source vectors r_0' and s_0', and sum all of the target responses $a_0(t)$ or $a_0(\omega)$. To guarantee that the final result is accurate, the basis set must be continually updated to include all multiplets that are nearby in frequency and likely to be significantly coupled to the target. Hybrid-multiplet isolation cannot be implemented in either the time-domain or frequency-domain direct solution methods, because the normal modes are not calculated. Two narrow-band spectra $a(\omega)$ centered upon sufficiently nearby reference frequencies ω_0 should coincide in an overlap region between them; in principle, a "seamless" broad-band spectrum could be synthesized by continual trial-and-error verification of this reproductive property. We discuss a family of alternative normal-mode and direct solution methods, applicable to multiplets that are well-isolated or reasonably well-isolated in the seismic spectrum, in the next two sections.

13.3.4 Isolated multiplet approximation

The square matrix H_{00} at the center of equation (13.210) governs the coupling among the $2l+1$ spherical-Earth basis functions within the target multiplet ${}_n\mathrm{T}_l$ or ${}_n\mathrm{S}_l$. In the *isolated-multiplet approximation*, this so-called *self coupling* is accounted for, but the coupling between adjacent multiplets is ignored. The diagonal matrix of singlet eigenfrequency perturbations Δ_0 and the associated renormalized transformation matrices Z_{00} and $\overline{\mathsf{Z}}_{00}$ may then be found by solving a $(2l+1)\times(2l+1)$ eigenvalue problem:

$$\overline{\mathsf{Z}}_{00}^{\mathrm{T}}\mathsf{Z}_{00} = \mathsf{I}_{00}, \qquad \overline{\mathsf{Z}}_{00}^{\mathrm{T}}\mathsf{H}_{00}\mathsf{Z}_{00} = \Delta_0, \tag{13.218}$$

where

$$\mathsf{H}_{00} = \mathsf{W}_{00} + (2\omega_0)^{-1}(\mathsf{V}_{00} + i\mathsf{A}_{00} - \omega_0^2\mathsf{T}_{00}). \tag{13.219}$$

The transformed receiver and source vectors in this lowest-order approximation are given by

$$\mathsf{r}_0' = \mathsf{Q}_{00}^{\mathrm{T}}\mathsf{r}_0 = \mathsf{Z}_{00}^{\mathrm{T}}\mathsf{u}_0, \qquad \mathsf{s}_0' = \overline{\mathsf{Q}}_{00}^{\mathrm{T}}\mathsf{s}_0 = \overline{\mathsf{Z}}_{00}^{\mathrm{T}}\mathsf{v}_0, \tag{13.220}$$

where

$$\mathsf{u}_0 = (\mathsf{I}_{00} - \tfrac{1}{2}\mathsf{T}_{00} - \tfrac{1}{2}\omega_0^{-1}\mathsf{W}_{00} + \tfrac{1}{2\pi}\omega_0^{-2}\mathsf{A}_{00})\mathsf{r}_0, \tag{13.221}$$

$$\mathsf{v}_0 = (\mathsf{I}_{00} - \tfrac{1}{2}\mathsf{T}_{00} + \tfrac{1}{2}\omega_0^{-1}\mathsf{W}_{00} + \tfrac{1}{2\pi}\omega_0^{-2}\mathsf{A}_{00})\mathsf{s}_0. \tag{13.222}$$

The response of the target multiplet is given in terms of the quantities Δ_0 and r_0', s_0' by equations (13.213)–(13.214) and (13.217), as usual.

The target multiplet modulation function can be written in terms of the self-coupling eigenfrequency perturbation matrix (13.219), prior to diagonalization, in the explicit form $A_0(t) = \mathsf{u}_0^{\mathrm{T}}\exp(i\mathsf{H}_{00}t)\,\mathsf{v}_0$. The complex exponential $\mathsf{X}_{00} = \exp(i\mathsf{H}_{00}t)$ is a $(2l+1)\times(2l+1)$ propagator matrix satisfying $d\mathsf{X}_{00}/dt = i\mathsf{H}_{00}\mathsf{X}_{00}$ and $\mathsf{X}_{00}(0) = \mathsf{I}_{00}$. We can alternatively express $A_0(t)$ in terms of a target source response vector $\mathsf{d}_0(t)$ satisfying

$$d\mathsf{d}_0/dt = i\mathsf{H}_{00}\mathsf{d}_0, \qquad \mathsf{d}_0(0) = \mathsf{v}_0, \tag{13.223}$$

or in terms of a target receiver response vector $\mathsf{e}_0(t)$ satisfying

$$d\mathsf{e}_0/dt = i\mathsf{H}_0^{\mathrm{T}}\mathsf{e}_0, \qquad \mathsf{e}_0(0) = \mathsf{u}_0, \tag{13.224}$$

in the form

$$A_0(t) = \mathsf{u}_0^{\mathrm{T}}\mathsf{d}_0(t) = \mathsf{e}_0^{\mathrm{T}}(t)\mathsf{v}_0. \tag{13.225}$$

The explicit spectral result $a_0(\omega) = \frac{1}{2}i\mathsf{u}_0^{\mathrm{T}}[\mathsf{H}_{00} - (\omega - \omega_0)\mathsf{I}_{00}]^{-1}\mathsf{v}_0$ can likewise be rewritten in terms of a frequency-domain target source response vector $\mathsf{d}_0(\omega)$ satisfying

$$(\mathsf{H}_{00} - \omega\mathsf{I}_{00})\mathsf{d}_0(\omega) = i\mathsf{v}_0, \tag{13.226}$$

or in terms of a target receiver response vector $\mathsf{e}_0(\omega)$ satisfying

$$(\mathsf{H}_{00}^{\mathrm{T}} - \omega\mathsf{I}_{00})\mathsf{e}_0(\omega) = i\mathsf{u}_0, \tag{13.227}$$

in the form

$$a_0(\omega) = \tfrac{1}{2}\mathsf{u}_0^{\mathrm{T}}\mathsf{d}_0(\omega - \omega_0) = \tfrac{1}{2}\mathsf{e}_0^{\mathrm{T}}(\omega - \omega_0)\mathsf{v}_0. \tag{13.228}$$

The results (13.223)–(13.225) and (13.226)–(13.228) provide the basis for an efficient direct solution method, as first noted by Woodhouse & Girnius (1982). Regardless of whether we adopt a time-domain or a frequency-domain approach, we are required to solve a system of only $2l + 1$ simultaneous linear equations for each multiplet ${}_n\mathrm{T}_l$ or ${}_n\mathrm{S}_l$. Straightforward summation of the multiplet responses $a_0(t)$ or $a_0(\omega)$, no matter how they are calculated, yields a "seamless" broad-band accelerogram or spectrum, inasmuch as each self-coupled eigenvalue problem (13.218) gives rise to precisely $2l + 1$ perturbed singlets.

All of the original treatments of the splitting of the free oscillations of the Earth make use of lowest-order *degenerate Rayleigh-Schrödinger perturbation theory* (Backus & Gilbert 1961; Dahlen 1968; Zharkov & Lyubimov 1970a,b; Madariaga 1972). This classical technique, which is described in every good quantum mechanics textbook (Landau & Lifshitz 1965; Schiff 1968), is equivalent to the isolated multiplet approximation, with equations (13.221)–(13.222) replaced by the identities $\mathsf{u}_0 = \mathsf{r}_0$ and $\mathsf{v}_0 = \mathsf{s}_0$. The above results are superior to this traditional approximation, because the resulting singlet eigenfunctions within each target multiplet are properly biorthonormalized with respect to the perturbed Earth, at the expense of very little additional effort. We discuss a number of applications of the isolated multiplet approximation to the determination of the Earth's three-dimensional structure in Chapter 14.

★13.3.5 Quasi-isolated multiplet approximation

The advantage of the procedure outlined above is, of course, its computational efficiency; only matrices of modest dimension $(2l + 1) \times (2l + 1)$ need to be diagonalized or otherwise analyzed. The disadvantage is that multiplet coupling is completely ignored; this is only a reasonable approximation for multiplets ${}_n\mathrm{T}_l$ and ${}_n\mathrm{S}_l$ that are very well isolated in the seismic spectrum. The *quasi-isolated approximation*, which we describe next, provides a means of accounting for *weak coupling* of each target multiplet to its neighbors, without increasing the dimensions of the matrices that must be considered. The theory, which is referred to in the original literature as the *subspace-projection method*, is developed and described with increasing completeness by Park (1986), Dahlen (1987), Park (1990) and Um & Dahlen (1992). A closely related approach, which is applicable under the same conditions, is described by Lognonné & Romanowicz (1990) and Lognonné (1991). We present an independent analysis, which incorporates eigenfunction renormalization and is valid upon a rotating anelastic Earth, in what follows.

The original eigenvalue problem, which accounts for coupling among all of the multiplets in the basis set exactly, is $\overline{\mathsf{Z}}^{\mathrm{T}}\mathsf{Z} = \mathsf{I}$ and $\overline{\mathsf{Z}}^{\mathrm{T}}\mathsf{H}\mathsf{Z} = \Delta$. To implement the quasi-isolated multiplet approximation, we express the eigenvector and dual eigenvector matrices Z and $\overline{\mathsf{Z}}$ in the form

$$\mathsf{Z} = \mathsf{P}\mathsf{Z}^{\star}, \qquad \overline{\mathsf{Z}} = \overline{\mathsf{P}}\,\overline{\mathsf{Z}}^{\star}, \tag{13.229}$$

where

$$\overline{\mathsf{P}}^{\mathrm{T}}\mathsf{P} = \mathsf{I}, \qquad \overline{\mathsf{P}}^{\mathrm{T}}\mathsf{H}\mathsf{P} = \mathsf{H}^{\star}. \tag{13.230}$$

We then use analytical methods to find transformation matrices P and $\overline{\mathsf{P}}$ that render the various "starred" quantities $\mathsf{Z}^{\star}$, $\overline{\mathsf{Z}}^{\star}$ and $\mathsf{H}^{\star}$ *block diagonal*, of the form

$$\mathsf{H}^{\star} = \begin{pmatrix} \ddots & & & & \\ & \mathsf{H}^{\star}_{-1-1} & & & \\ & & \mathsf{H}^{\star}_{00} & & \\ & & & \mathsf{H}^{\star}_{11} & \\ & & & & \ddots \end{pmatrix}. \tag{13.231}$$

Such a transformation decomposes the fully coupled eigenvalue problem into a number of $(2l+1) \times (2l+1)$ eigenvalue problems, one corresponding to each of the hybrid multiplets of the perturbed Earth. To find the eigensolutions associated with the $k = 0$ multiplet, we only have to solve the *target problem*:

$$\overline{\mathsf{Z}}^{\star\mathrm{T}}_{00}\mathsf{Z}^{\star}_{00} = \mathsf{I}_{00}, \qquad \overline{\mathsf{Z}}^{\star\mathrm{T}}_{00}\mathsf{H}^{\star}_{00}\mathsf{Z}^{\star}_{00} = \Delta_0. \tag{13.232}$$

The biorthonormality relation $\overline{\mathsf{P}}^{\mathrm{T}}\mathsf{P} = \mathsf{I}$ and the requirement that the "starred" eigenfrequency perturbation matrix (13.231) be block diagonal can be written out in terms of submatrices in the form

$$\sum_{k''} \overline{\mathsf{P}}^{\mathrm{T}}_{kk''}\mathsf{P}_{k''k'} = \mathsf{I}_{kk'}, \tag{13.233}$$

$$\sum_{k''}\sum_{k'''}\overline{\mathsf{P}}^{\mathrm{T}}_{kk''}\mathsf{H}^{\star}_{k''k'''}\mathsf{P}_{k'''k'} = 0 \qquad \text{for } k \neq k'. \tag{13.234}$$

Here I_{kk} is the identity matrix associated with the kth hybrid multiplet, and we have defined $\mathsf{I}_{kk'} = 0$ whenever $k \neq k'$, for convenience. We seek perturbation-theoretical solutions

$$\mathsf{P}_{kk'} = \mathsf{I}_{kk'} + \mathsf{P}^{(1)}_{kk'} + \mathsf{P}^{(2)}_{kk'} + \tfrac{1}{2}\sum_{k''}\mathsf{P}^{(1)}_{kk''}\mathsf{P}^{(1)}_{k''k'} + \cdots, \tag{13.235}$$

$$\overline{\mathsf{P}}^{\mathrm{T}}_{kk'} = \mathsf{I}_{kk'} - \mathsf{P}^{(1)}_{kk'} - \mathsf{P}^{(2)}_{kk'} + \tfrac{1}{2}\sum_{k''}\mathsf{P}^{(1)}_{kk''}\mathsf{P}^{(1)}_{k''k'} + \cdots \tag{13.236}$$

to equations (13.233)–(13.234) in terms of a sequence of unknown matrices $\mathsf{P}^{(1)}_{kk'}$, $\mathsf{P}^{(2)}_{kk'}, \ldots$, where the superscripts denote terms of the indicated order in the multiplet coupling strength, which we shall define more precisely below. It is readily verified that the expansions (13.235)–(13.236) satisfy the relation (13.233), correct to second order in the superscripted quantities. Inserting (13.235)–(13.236) into equation (13.234) and equating terms of like order leads to a systematic determination of the successive off-diagonal terms $\mathsf{P}^{(1)}_{kk'}$, $\mathsf{P}^{(2)}_{kk'}, \ldots$, $k \neq k'$. The square matrices $\mathsf{P}^{(1)}_{kk}$, $\mathsf{P}^{(2)}_{kk}, \ldots$ are indeterminate, but it can be shown that the target response is independent of these matrices. The indeterminacy at every order reflects the non-uniqueness of the block-diagonalizing transformations P and $\overline{\mathsf{P}}$; a convenient choice, which determines the two transformations uniquely, is to require that they reduce to the identity in the limit of a vanishingly small perturbation:

$$\mathsf{P}^{(1)}_{kk} = \mathsf{P}^{(2)}_{kk} = \cdots = 0. \tag{13.237}$$

The final results may be conveniently expressed in terms of the auxiliary matrices

$$\mathsf{K}_{kk'} = W_{kk'} + (2\omega_0)^{-1}(\mathsf{V}_{kk'} + iA_{kk'} - \omega_0^2\mathsf{T}_{kk'}), \tag{13.238}$$

$$\mathsf{K}^{\mathrm{T}}_{kk'} = -W_{kk'} + (2\omega_0)^{-1}(\mathsf{V}_{kk'} + iA_{kk'} - \omega_0^2\mathsf{T}_{kk'}), \tag{13.239}$$

and

$$\mathsf{C}_{kk'} = -\tfrac{1}{2}\mathsf{T}_{kk'} + \tfrac{1}{2}\omega_0^{-1}\mathsf{W}_{kk'} + \tfrac{1}{2\pi}\omega_0^{-2}\mathsf{A}_{kk'}, \tag{13.240}$$

$$\mathsf{C}^{\mathrm{T}}_{kk'} = -\tfrac{1}{2}\mathsf{T}_{kk'} - \tfrac{1}{2}\omega_0^{-1}\mathsf{W}_{kk'} + \tfrac{1}{2\pi}\omega_0^{-2}\mathsf{A}_{kk'}. \tag{13.241}$$

Correct to first order in the coupling strength, we obtain

$$\mathsf{P}^{(1)}_{kk'} = (\omega_{k'} - \omega_k)^{-1}\mathsf{K}_{kk'}, \qquad k \neq k'. \tag{13.242}$$

Upon inserting equations (13.235)–(13.236) and (13.242) into (13.230), we see that the "starred" eigenfrequency perturbation matrix is given, correct to second order in the coupling, by

$$\mathsf{H}^{\star}_{00} = \mathsf{K}_{00} + \sum_{k\neq 0}(\omega_0 - \omega_k)^{-1}\mathsf{K}_{0k}\mathsf{K}_{k0}. \tag{13.243}$$

We can determine the eigenfrequency perturbations Δ_0 correct to the same order by solving the target eigenvalue problem (13.232). The transformed receiver and source vectors characterizing the target response are given in terms of the "starred" eigenvector and dual eigenvector matrices by

$$\mathsf{r}'_0 = \mathsf{Z}^{\star\mathrm{T}}_{00}\mathsf{u}^{\star}_0, \qquad \mathsf{s}'_0 = \overline{\mathsf{Z}}^{\star\mathrm{T}}_{00}\mathsf{v}^{\star}_0, \tag{13.244}$$

where

$$\mathsf{u}^{\star}_0 = (\mathsf{I}_{00} + \mathsf{C}^{\mathrm{T}}_{00})\mathsf{r}_0 + \sum_{k\neq 0}\left[\mathsf{C}^{\mathrm{T}}_{0k} + (\omega_0 - \omega_k)^{-1}\mathsf{K}^{\mathrm{T}}_{0k}\right]\mathsf{r}_k, \tag{13.245}$$

$$\mathsf{v}^{\star}_0 = (\mathsf{I}_{00} + \mathsf{C}_{00})\mathsf{s}_0 + \sum_{k\neq 0}\left[\mathsf{C}_{0k} + (\omega_0 - \omega_k)^{-1}\mathsf{K}_{0k}\right]\mathsf{s}_k. \tag{13.246}$$

The quantities $\mathsf{u}^{\star}_0$ and $\mathsf{v}^{\star}_0$ are "starred" receiver and source vectors which incorporate the effects of eigenfunction renormalization on a rotating anelastic Earth, correct to first order in the coupling. Upon ignoring the sums over $k \neq 0$ in equations (13.243) and (13.245)–(13.246), we recover the isolated multiplet approximation: $\mathsf{H}^{\star}_{00} \to \mathsf{H}_{00}$ and $\mathsf{u}^{\star}_0 \to \mathsf{u}_0$, $\mathsf{v}^{\star}_0 \to \mathsf{v}_0$, as expected.

It is straightforward but laborious to extend the above expansions to any desired order in the coupling strength. The "starred" eigenfrequency perturbation matrix is given, correct to third order in the coupling, by

$$\begin{aligned}\mathsf{H}^{\star}_{00} &= \mathsf{K}_{00} + \sum_{k\neq 0}(\omega_0 - \omega_k)^{-1}\mathsf{K}_{0k}\mathsf{K}_{k0} \\ &\quad - \tfrac{1}{2}\sum_{k\neq 0}(\omega_0 - \omega_k)^{-2}(\mathsf{K}_{00}\mathsf{K}_{0k}\mathsf{K}_{k0} + \mathsf{K}_{0k}\mathsf{K}_{k0}\mathsf{K}_{00}) \\ &\quad\quad + \sum_{k\neq 0}\sum_{k'\neq 0}(\omega_0 - \omega_k)^{-1}(\omega_0 - \omega_{k'})^{-1}\mathsf{K}_{0k}\mathsf{K}_{kk'}\mathsf{K}_{k'0},\end{aligned} \tag{13.247}$$

whereas the "starred" receiver and source vectors are given, correct to second order in the coupling, by

$$\mathsf{u}^{\star}_0 = \Big[\mathsf{I}_{00} + \mathsf{C}^{\mathrm{T}}_{00} + \sum_{k\neq 0}(\omega_0 - \omega_k)^{-1}\mathsf{K}^{\mathrm{T}}_{0k}\mathsf{C}^{\mathrm{T}}_{k0}$$

$$- \tfrac{1}{2} \sum_{k \neq 0} (\omega_0 - \omega_k)^{-2} \, \mathsf{K}_{0k}^{\mathrm{T}} \mathsf{K}_{k0}^{\mathrm{T}} \Big] \mathsf{r}_0 + \sum_{k \neq 0} \Big[\mathsf{C}_{0k}^{\mathrm{T}} + (\omega_0 - \omega_k)^{-1} \mathsf{K}_{0k}^{\mathrm{T}}$$

$$- (\omega_0 - \omega_k)^{-2} \mathsf{K}_{00}^{\mathrm{T}} \mathsf{K}_{0k}^{\mathrm{T}} + \sum_{k' \neq 0} (\omega_0 - \omega_{k'})^{-1} \mathsf{K}_{0k'}^{\mathrm{T}} \mathsf{C}_{k'k}^{\mathrm{T}}$$

$$+ (\omega_0 - \omega_k)^{-1} \sum_{k' \neq 0} (\omega_0 - \omega_{k'})^{-1} \mathsf{K}_{0k'}^{\mathrm{T}} \mathsf{K}_{k'k}^{\mathrm{T}} \Big] \mathsf{r}_k, \quad (13.248)$$

$$\mathsf{v}_0^\star = \Big[\mathsf{I}_{00} + \mathsf{C}_{00} + \sum_{k \neq 0} (\omega_0 - \omega_k)^{-1} \mathsf{K}_{0k} \mathsf{C}_{k0}$$

$$- \tfrac{1}{2} \sum_{k \neq 0} (\omega_0 - \omega_k)^{-2} \, \mathsf{K}_{0k} \mathsf{K}_{k0} \Big] \mathsf{s}_0 + \sum_{k \neq 0} \Big[\mathsf{C}_{0k} + (\omega_0 - \omega_k)^{-1} \mathsf{K}_{0k}$$

$$- (\omega_0 - \omega_k)^{-2} \mathsf{K}_{00} \mathsf{K}_{0k} + \sum_{k' \neq 0} (\omega_0 - \omega_{k'})^{-1} \mathsf{K}_{0k'} \mathsf{C}_{k'k}$$

$$+ (\omega_0 - \omega_k)^{-1} \sum_{k' \neq 0} (\omega_0 - \omega_{k'})^{-1} \mathsf{K}_{0k'} \mathsf{K}_{k'k} \Big] \mathsf{s}_k. \quad (13.249)$$

The only coupling which is taken into account in the lowest-order results (13.243) and (13.245)–(13.246) is that between the target multiplet and its neighbors, as specified by the matrices K_{0k} and K_{k0}. Both K_{kk} self coupling of neighboring multiplets and $\mathsf{K}_{kk'}$ coupling between neighbors are taken into account to the next order. The quasi-isolated multiplet approximation is a numerically superior but algebraically equivalent alternative to the application of higher-order degenerate Rayleigh-Schrödinger perturbation theory to the target multiplet, because it eliminates the problem of small divisors which arises in calculating the projections of the perturbed eigenfunctions onto the other zeroth-order eigenfunctions in the target subspace (Um, Dahlen & Park 1991). It is a general feature of both methods that the eigenfrequency perturbations are always calculated to one order higher than the associated eigenfunctions.

Once the perturbation expansions $\mathsf{H}_{00}^\star$ and $\mathsf{u}_0^\star$, $\mathsf{v}_0^\star$ have been determined to any desired order, we can either solve the eigenvalue problem (13.232) and calculate the target multiplet modulation function and the associated spectrum using the mode-sum representations $A_0(t) = \mathsf{r}_0'^{\mathrm{T}} \exp(i\Delta_0 t)\, \mathsf{s}_0'$ and $a_0(\omega) = \frac{1}{2} i \mathsf{r}_0'^{\mathrm{T}} [\Delta_0 - (\omega - \omega_0)\mathsf{I}_{00}]^{-1} \mathsf{s}_0'$, or we can employ a direct solution method in which equations (13.223)–(13.225) and (13.226)–(13.228) are replaced by their "starred" analogues. Broad-band accelerograms and spectra calculated using either approach are naturally "seamless" because each target contribution $a_0(t) = \mathrm{Re}\,[A_0(t) \exp(i\omega_0 t)]$ or $a_0(\omega)$ represents the response of a single hybrid multiplet, just as in the fully-isolated approximation.

The validity of the quasi-isolated multiplet approximation depends upon the strength of the coupling between the target or $k = 0$ multiplet and each of its $k = \pm 1,\ \pm 2, \ldots$ neighbors. A dimensionless measure of the strength of this coupling is provided by the parameter

$$\varepsilon_k = |\omega_0 - \omega_k|^{-1} \|\mathsf{K}_{0k}\mathsf{K}_{k0}\|^{1/2}, \tag{13.250}$$

where $\|\cdot\|$ denotes a suitable matrix norm. To a first approximation, we can consider the coupling strength to be the ratio of the real *splitting width* $\Delta\omega_0 = \max(\delta\omega_m) - \min(\delta\omega_m)$ of the target multiplet to the degenerate eigenfrequency spacing:

$$\varepsilon_k \approx |\omega_0 - \omega_k|^{-1} \Delta\omega_0. \tag{13.251}$$

A small value of this ratio, $\varepsilon_k \ll 1$, indicates that the approximation is applicable, because the target multiplet and its kth neighbor are well isolated in the perturbed spectrum. A large value, $\varepsilon_k \approx 1$, on the other hand, indicates that the two multiplets overlap appreciably; the "starred" perturbation expansions (13.247)–(13.249) will generally fail to converge in this case. We can account for this by using a modified perturbation theory in which the target multiplet and any of its strongly coupled neighbors are combined into a single *super-multiplet*, with the dimensions of the submatrices K_{00}, K_{0k}, K_{k0}, etc., enlarged accordingly. An adaptive scheme which decides whether to treat the coupling between any two prospective partners as weak or strong depending upon the magnitude of the parameters ε_k can easily be developed. In the conservative limit that the super-multiplet consists of the entire basis set, equation (13.232) reduces to the fully coupled eigenvalue problem $\overline{\mathsf{Z}}^{\mathrm{T}}\mathsf{Z} = \mathsf{I}$ and $\overline{\mathsf{Z}}^{\mathrm{T}}\mathsf{H}\mathsf{Z} = \Delta$. In the opposite extreme, if all of the coupling is simply ignored, it reduces to the isolated-multiplet self-coupling problem $\overline{\mathsf{Z}}_{00}^{\mathrm{T}}\mathsf{Z}_{00} = \mathsf{I}_{00}$ and $\overline{\mathsf{Z}}_{00}^{\mathrm{T}}\mathsf{H}_{00}\mathsf{Z}_{00} = \Delta_0$.

13.3.6 Born approximation

A central problem in global seismology is the determination of the three-dimensional elastic and anelastic structure of the Earth. In seeking to refine the values of the parameters $\delta\kappa_0$, $\delta\mu_0$, $\delta\rho$, $\delta\Phi$, δd, γ and q_κ, q_μ by iterative fitting of observed multiplet modulation functions or spectra, it is useful to know the *Fréchet derivatives* expressing the sensitivity of these observables to the Earth model. These partial derivatives can be determined by a straightforward application of the Born approximation, as we show next.

We consider the *differential spectrum* $\delta a_0(\omega)$ of a target multiplet ${}_n\mathrm{T}_l$ or ${}_n\mathrm{S}_l$ first, presuming that the isolated multiplet approximation is applicable for simplicity. An infinitesimal change in the three-dimensional Earth model changes the eigenfrequency perturbation matrix (13.219) and the renormalized receiver and source vectors (13.221)–(13.222) by amounts

$$\delta\mathsf{H}_{00} = (2\omega_0)^{-1}(\delta\mathsf{V}_{00} + i\delta\mathsf{A}_{00} - \omega_0^2\delta\mathsf{T}_{00}) = \delta\mathsf{H}_{00}^{\mathrm{T}}, \tag{13.252}$$

$$\delta\mathsf{u}_0 = (-\tfrac{1}{2}\delta\mathsf{T}_{00} + \tfrac{1}{2\pi}\omega_0^{-2}\delta\mathsf{A}_{00})\mathsf{r}_0, \tag{13.253}$$

$$\delta\mathsf{v}_0 = (-\tfrac{1}{2}\delta\mathsf{T}_{00} + \tfrac{1}{2\pi}\omega_0^{-2}\delta\mathsf{A}_{00})\mathsf{s}_0. \tag{13.254}$$

Note that we do not allow for any perturbation to the Coriolis matrix W_{00} inasmuch as the angular rate of rotation of the Earth is considered to be perfectly well determined. Upon perturbing equations (13.226)–(13.228), we find that the differential spectrum can be expressed either in the form

$$\delta a_0(\omega) = \tfrac{1}{2}[\delta\mathsf{u}_0^{\mathrm{T}}\,\mathsf{d}_0(\omega-\omega_0) + \mathsf{u}_0^{\mathrm{T}}\,\delta\mathsf{d}_0(\omega-\omega_0)], \tag{13.255}$$

where

$$(\mathsf{H}_{00} - \omega\mathsf{I}_{00})\,\delta\mathsf{d}_0(\omega) = i[\delta\mathsf{v}_0 - \delta\mathsf{H}_{00}\,\mathsf{d}_0(\omega)], \tag{13.256}$$

or in the form

$$\delta a_0(\omega) = \tfrac{1}{2}[\delta\mathsf{e}_0^{\mathrm{T}}(\omega-\omega_0)\,\mathsf{v}_0 + \mathsf{e}_0^{\mathrm{T}}(\omega-\omega_0)\,\delta\mathsf{v}_0], \tag{13.257}$$

where

$$(\mathsf{H}_{00}^{\mathrm{T}} - \omega\mathsf{I}_{00})\,\delta\mathsf{e}_0(\omega) = i[\delta\mathsf{u}_0 - \delta\mathsf{H}_{00}^{\mathrm{T}}\,\mathsf{e}_0(\omega)]. \tag{13.258}$$

Either of these two results may be used as the basis of a direct solution method of determining $\delta a_0(\omega)$; some of the computational considerations which arise in such a direct frequency-domain approach are elaborated by Geller & Hara (1993). The same QR factorization of $\mathsf{H}_{00} - \omega\mathsf{I}_{00}$ or $\mathsf{H}_{00}^{\mathrm{T}} - \omega\mathsf{I}_{00}$ needed to find $a_0(\omega)$ can be used to solve equation (13.256) or (13.258). The back-substitution step must be reiterated as many times as there are degrees of freedom in the three-dimensional Earth model, in order to find all of the partial derivatives. We can alternatively write the differential response explicitly in terms of the target source and receiver response vectors $\mathsf{d}_0(\omega)$ and $\mathsf{e}_0(\omega)$ in the form

$$\begin{aligned}\delta a_0(\omega) = \tfrac{1}{2}[&\delta\mathsf{u}_0^{\mathrm{T}}\,\mathsf{d}_0(\omega-\omega_0) + \mathsf{e}_0^{\mathrm{T}}(\omega-\omega_0)\,\delta\mathsf{v}_0 \\ &+ i\mathsf{e}_0^{\mathrm{T}}(\omega-\omega_0)\,\delta\mathsf{H}_{00}\,\mathsf{d}_0(\omega-\omega_0)].\end{aligned} \tag{13.259}$$

In this case, we are only required to solve equations (13.226) and (13.227) once for each source and receiver, respectively; the partial derivatives can then be determined by repeated evaluation of the vector and matrix multiplications in equation (13.259). Finally, we can rewrite the quantities

$\mathsf{d}_0(\omega)$ and $\mathsf{e}_0(\omega)$ in terms of the inverse matrix $[\mathsf{H}_{00} - (\omega - \omega_0)\mathsf{I}_{00}]^{-1}$, and use (13.218) to transform to a normal-mode representation, with the result

$$\begin{aligned}\delta a_0(\omega) = \tfrac{1}{2}i\{&\delta\mathsf{r}_0'^{\mathrm{T}}[\Delta_0 - (\omega - \omega_0)\mathsf{I}_{00}]^{-1}\mathsf{s}_0' \\ &+ \mathsf{r}_0'^{\mathrm{T}}[\Delta_0 - (\omega - \omega_0)\mathsf{I}_{00}]^{-1}\delta\mathsf{s}_0' \\ &- \mathsf{r}_0'^{\mathrm{T}}[\Delta_0 - (\omega - \omega_0)\mathsf{I}_{00}]^{-1}\delta\mathsf{H}_{00}'[\Delta_0 - (\omega - \omega_0)\mathsf{I}_{00}]^{-1}\mathsf{s}_0'\},\end{aligned} \tag{13.260}$$

where

$$\delta\mathsf{r}_0' = \mathsf{Z}_{00}^{\mathrm{T}}\,\delta\mathsf{u}_0, \qquad \delta\mathsf{s}_0' = \overline{\mathsf{Z}}_{00}^{\mathrm{T}}\,\delta\mathsf{v}_0, \qquad \delta\mathsf{H}_{00}' = \overline{\mathsf{Z}}_{00}^{\mathrm{T}}\,\delta\mathsf{H}_{00}\,\mathsf{Z}_{00}. \tag{13.261}$$

An expression analogous to equation (13.260), which does not properly account for the biorthonormality of the eigenfunctions, is given by Tsuboi & Geller (1987) and Geller, Hara & Tsuboi (1990). The final term involving the transformed matrix $\delta\mathsf{H}_{00}'$ is a double sum over all of the perturbed singlets in the target multiplet.

The *differential multiplet modulation function* $\delta A_0(t)$ can likewise be found by perturbing equations (13.223)–(13.225); the result can be written in the two equivalent forms

$$\delta A_0(t) = \delta\mathsf{u}_0^{\mathrm{T}}\,\mathsf{d}_0(t) + \mathsf{u}_0^{\mathrm{T}}\,\delta\mathsf{d}_0(t) = \delta\mathsf{e}_0^{\mathrm{T}}(t)\,\mathsf{v}_0 + \mathsf{e}_0^{\mathrm{T}}(t)\,\delta\mathsf{v}_0, \tag{13.262}$$

where

$$d(\delta\mathsf{d}_0)/dt = i(\mathsf{H}_{00}\,\delta\mathsf{d}_0 + \delta\mathsf{H}_{00}\,\mathsf{d}_0), \qquad \delta\mathsf{d}_0(0) = \delta\mathsf{v}_0, \tag{13.263}$$

$$d(\delta\mathsf{e}_0)/dt = i(\mathsf{H}_{00}^{\mathrm{T}}\,\delta\mathsf{e}_0 + \delta\mathsf{H}_{00}^{\mathrm{T}}\,\mathsf{d}_0), \qquad \delta\mathsf{e}_0(0) = \delta\mathsf{u}_0. \tag{13.264}$$

A direct solution approach based upon numerical time-stepping of either equation (13.263) or (13.264) requires as many integrations as there are degrees of freedom in the Earth model. Alternatively, we can integrate equations (13.223) and (13.224) once for each source and receiver, respectively, and calculate the partial derivatives by repeated evaluation of the explicit expression

$$\begin{aligned}\delta A_0(t) = \delta\mathsf{u}_0^{\mathrm{T}}\,\mathsf{d}_0(t) + \mathsf{e}_0^{\mathrm{T}}(t)\,\delta\mathsf{v}_0 \\ + i\int_0^t \mathsf{e}_0^{\mathrm{T}}(t - t')\,\delta\mathsf{H}_{00}\,\mathsf{d}_0(t')\,dt'.\end{aligned} \tag{13.265}$$

The perturbation to the propagator matrix $\mathsf{X}_{00} = \exp(i\mathsf{H}_{00}t)$ is given by a similar convolution integral:

$$\delta\mathsf{X}_{00}(t) = i\int_0^t \exp[i\mathsf{H}_{00}(t - t')]\,\delta\mathsf{H}_{00}\exp(i\mathsf{H}_{00}t')\,dt'. \tag{13.266}$$

Equation (13.265) is the time-domain analogue of the frequency-domain result (13.259); the normal-mode representation of the differential modulation function analogous to equation (13.260) can be written in the form

$$\delta A_0(t) = \delta \mathsf{r}_0'^{\mathrm{T}} \exp(i\Delta_0 t)\, \mathsf{s}_0' + \mathsf{r}_0'^{\mathrm{T}} \exp(i\Delta_0 t)\, \delta \mathsf{s}_0' + \mathsf{r}_0'^{\mathrm{T}}\, \delta \mathsf{X}_{00}'(t)\, \mathsf{s}_0', \tag{13.267}$$

where

$$\delta \mathsf{X}_{00}' = \overline{\mathsf{Z}}_{00}^{\mathrm{T}}\, \delta \mathsf{X}_{00}\, \mathsf{Z}_{00}. \tag{13.268}$$

The elements of the $(2l+1) \times (2l+1)$ transformed differential propagator (13.268) are given explicitly in terms of the complex frequency perturbations $\delta\nu_m$, $-l \leq m \leq l$, and the elements of $\delta \mathsf{H}_{00}' = \overline{\mathsf{Z}}_{00}^{\mathrm{T}}\, \delta \mathsf{H}_{00}\, \mathsf{Z}_{00}$ by

$$(\delta X_{00}')_{mm'} = \left[\frac{\exp(i\,\delta\nu_{m'}t) - \exp(i\,\delta\nu_m t)}{\delta\nu_{m'} - \delta\nu_m}\right] (\delta H_{00}')_{mm'}. \tag{13.269}$$

The diagonal elements give rise to a *secular* perturbation which grows linearly with time, since

$$\frac{\exp(i\,\delta\nu_{m'}t) - \exp(i\,\delta\nu_m t)}{\delta\nu_{m'} - \delta\nu_m} \to it \exp(i\,\delta\nu_m t) \tag{13.270}$$

in the limit $\delta\nu_{m'} \to \delta\nu_m$. The results (13.267)–(13.270) are due to Giardini, Li & Woodhouse (1988). They consider the singlet eigenfunctions to be biorthonormal with respect to the unperturbed rather than the perturbed Earth, so that the distinctions between the transformations Q_{00}, $\overline{\mathsf{Q}}_{00}$ and Z_{00}, $\overline{\mathsf{Z}}_{00}$ and the receiver and source vectors u_0, v_0 and r_0, s_0 are ignored; in that approximation, the differential multiplet modulation function is given by $\delta A_0(t) \approx \mathsf{r}_0'^{\mathrm{T}}\, \delta \mathsf{X}_{00}'(t)\, \mathsf{s}_0'$, where $\mathsf{r}_0' = \mathsf{Z}_{00}^{\mathrm{T}} \mathsf{r}_0$ and $\mathsf{s}_0' = \overline{\mathsf{Z}}_{00}^{\mathrm{T}} \mathsf{s}_0$.

Weak coupling can be accounted for in all of the procedures discussed in this section by using the quasi-isolated multiplet approximation instead. The "starred" perturbations $\delta \mathsf{H}_{00}^{\star}$ and $\delta \mathsf{u}_0^{\star}$, $\delta \mathsf{v}_0^{\star}$ may be readily obtained by varying equations (13.243) and (13.244)–(13.245) or their higher-order analogues. Clévédé & Lognonné (1996) use a variant of the quasi-isolated multiplet approximation to calculate the differential multiplet modulation function $\delta A_0(t)$ correct to third order in the eigenfrequencies and second order in the eigenfunctions.

⋆13.3.7 Complex basis formulation

We have emphasized the use of *real* Rayleigh-Ritz basis eigenfunctions proportional to the spherical harmonics $\mathcal{Y}_{lm}, -l \leq m \leq l$, throughout this book, because they highlight the fundamental distinctions among the four types of eigenvalue problems—with and without rotation and anelasticity. For historical reasons, however, most previous treatments of degenerate and quasi-degenerate perturbation theory have utilized *complex* basis eigenfunctions proportional to Y_{lm}, $-l \leq m \leq l$. To provide a link to these prior results, we briefly consider this case here.

We distinguish the real and complex basis eigenfunctions by the absence or presence of a tilde:

$$\mathbf{s}_k = U\hat{\mathbf{r}}\mathcal{Y}_{lm} + k^{-1}V\boldsymbol{\nabla}_1\mathcal{Y}_{lm} - k^{-1}W(\hat{\mathbf{r}} \times \boldsymbol{\nabla}_1\mathcal{Y}_{lm}), \tag{13.271}$$

$$\tilde{\mathbf{s}}_k = U\hat{\mathbf{r}}Y_{lm} + k^{-1}V\boldsymbol{\nabla}_1 Y_{lm} - k^{-1}W(\hat{\mathbf{r}} \times \boldsymbol{\nabla}_1 Y_{lm}), \tag{13.272}$$

where $k = \sqrt{l(l+1)}$ as usual. The elements of the complex receiver and source vectors $\tilde{\mathsf{r}}$ and $\tilde{\mathsf{s}}$ analogous to (13.189) are

$$\tilde{r}_k = \hat{\boldsymbol{\nu}} \cdot \tilde{\mathbf{s}}_k^*(\mathbf{x}), \qquad \tilde{s}_k = \mathbf{M}\!:\!\tilde{\boldsymbol{\varepsilon}}_k^*(\mathbf{x}_s). \tag{13.273}$$

Explicit formulae for the elements of the perturbation matrices $\tilde{\mathsf{T}}$, $\tilde{\mathsf{V}}$, $\tilde{\mathsf{A}}$ and $\tilde{\mathsf{W}}$ are given in Appendix D.2. The vectors $\tilde{\mathsf{r}}$, $\tilde{\mathsf{s}}$ and matrices $\tilde{\mathsf{T}}$, $\tilde{\mathsf{V}}$, $\tilde{\mathsf{A}}$, $\tilde{\mathsf{W}}$ are related to r, s and T, V, A, W by equations (D.178)–(D.179):

$$\tilde{\mathsf{r}} = \mathsf{U}\mathsf{r}, \qquad \tilde{\mathsf{s}} = \mathsf{U}\mathsf{s}, \tag{13.274}$$

$$\tilde{\mathsf{T}} = \mathsf{U}\mathsf{T}\mathsf{U}^{\mathrm{H}}, \qquad \tilde{\mathsf{V}} = \mathsf{U}\mathsf{V}\mathsf{U}^{\mathrm{H}}, \tag{13.275}$$

$$\tilde{\mathsf{A}} = \mathsf{U}\mathsf{A}\mathsf{U}^{\mathrm{H}}, \qquad \tilde{\mathsf{W}} = \mathsf{U}\mathsf{W}\mathsf{U}^{\mathrm{H}}. \tag{13.276}$$

Since both of the bases (13.271)–(13.272) are orthonormal, the transformation matrix U is unitary:

$$\mathsf{U}^{\mathrm{H}}\mathsf{U} = \mathsf{U}\mathsf{U}^{\mathrm{H}} = \mathsf{I}. \tag{13.277}$$

An explicit expression for U is given in equations (D.175)–(D.176).

Upon inserting (13.275)–(13.276) into equation (13.196) and invoking the result (13.277), we find that the narrow-band modulation function $A(t) = \mathsf{r}'^{\mathrm{T}}\exp(i\Delta t)\,\mathsf{s}' = \mathsf{u}^{\mathrm{T}}\exp(i\mathsf{H}t)\,\mathsf{v}$ can be rewritten in the form

$$A(t) = \tilde{\mathsf{r}}'^{\mathrm{H}}\exp(i\Delta t)\,\tilde{\mathsf{s}}' = \tilde{\mathsf{u}}^{\mathrm{H}}\exp(i\tilde{\mathsf{H}}t)\,\tilde{\mathsf{v}}, \tag{13.278}$$

where

$$\tilde{\mathsf{r}}' = \tilde{\mathsf{Z}}^{\mathrm{H}}\tilde{\mathsf{u}}, \qquad \tilde{\mathsf{s}}' = \tilde{\mathsf{Z}}^{-1}\tilde{\mathsf{v}}, \tag{13.279}$$

$$\tilde{\mathsf{u}} = (\mathsf{I} - \tfrac{1}{2}\tilde{\mathsf{T}} + \tfrac{1}{2}\omega_0^{-1}\tilde{\mathsf{W}} + \tfrac{1}{2\pi}\omega_0^{-2}\tilde{\mathsf{A}})\tilde{\mathsf{r}}, \tag{13.280}$$

$$\tilde{\mathsf{v}} = (\mathsf{I} - \tfrac{1}{2}\tilde{\mathsf{T}} + \tfrac{1}{2}\omega_0^{-1}\tilde{\mathsf{W}} + \tfrac{1}{2\pi}\omega_0^{-2}\tilde{\mathsf{A}})\tilde{\mathsf{s}}. \tag{13.281}$$

The quantity $\tilde{\mathsf{Z}}$ is the similarity transformation that diagonalizes the eigen-frequency perturbation matrix:

$$\tilde{\mathsf{Z}}^{-1}\tilde{\mathsf{Z}} = \mathsf{I}, \qquad \tilde{\mathsf{Z}}^{-1}\tilde{\mathsf{H}}\tilde{\mathsf{Z}} = \mathsf{I}. \tag{13.282}$$

It is noteworthy that the renormalized receiver and source vectors $\tilde{\mathsf{u}}$ and $\tilde{\mathsf{v}}$ are related to $\tilde{\mathsf{r}}$ and $\tilde{\mathsf{s}}$ by the *same* matrix $\mathsf{I} - \frac{1}{2}\tilde{\mathsf{T}} + \frac{1}{2}\omega_0^{-1}\tilde{\mathsf{W}} + \frac{1}{2\pi}\omega_0^{-2}\tilde{\mathsf{A}}$; compare equations (13.280)–(13.281) with the analogous real-basis relations (13.194)–(13.195).

Using (13.278)–(13.282), it is a straightforward matter to express all of the results in Sections 13.3.2–13.3.6 in terms of the complex basis eigenfunctions $\tilde{\mathbf{s}}_k$. The time-dependent amplitude of an isolated multiplet, for example, is given by

$$A_0(t) = \tilde{\mathsf{r}}_0'^{\mathrm{H}}\exp(i\Delta_0 t)\,\tilde{\mathsf{s}}_0' = \tilde{\mathsf{u}}_0^{\mathrm{H}}\exp(i\tilde{\mathsf{H}}_{00}t)\,\tilde{\mathsf{v}}_0, \tag{13.283}$$

where

$$\tilde{\mathsf{r}}_0' = \tilde{\mathsf{Z}}_{00}^{\mathrm{H}}\tilde{\mathsf{u}}_0, \qquad \tilde{\mathsf{s}}_0' = \tilde{\mathsf{Z}}_{00}^{-1}\tilde{\mathsf{v}}_0, \tag{13.284}$$

$$\tilde{\mathsf{u}}_0 = (\mathsf{I} - \tfrac{1}{2}\tilde{\mathsf{T}}_{00} + \tfrac{1}{2}\omega_0^{-1}\tilde{\mathsf{W}}_{00} + \tfrac{1}{2\pi}\omega_0^{-2}\tilde{\mathsf{A}}_{00})\tilde{\mathsf{r}}_0, \tag{13.285}$$

$$\tilde{\mathsf{v}}_0 = (\mathsf{I} - \tfrac{1}{2}\tilde{\mathsf{T}}_{00} + \tfrac{1}{2}\omega_0^{-1}\tilde{\mathsf{W}}_{00} + \tfrac{1}{2\pi}\omega_0^{-2}\tilde{\mathsf{A}}_{00})\tilde{\mathsf{s}}_0. \tag{13.286}$$

The quantity $\tilde{\mathsf{Z}}_{00}$ is the $(2l+1)\times(2l+1)$ diagonalizing transformation:

$$\tilde{\mathsf{Z}}_{00}^{-1}\tilde{\mathsf{Z}}_{00} = \mathsf{I}_{00}, \qquad \tilde{\mathsf{Z}}_{00}^{-1}\tilde{\mathsf{H}}_{00}\tilde{\mathsf{Z}}_{00} = \mathsf{I}_{00}. \tag{13.287}$$

The response of a quasi-isolated multiplet may likewise be expressed in terms of the "starred" vectors and matrices $\tilde{\mathsf{u}}_0^\star$, $\tilde{\mathsf{v}}_0^\star$ and $\tilde{\mathsf{H}}_{00}^\star$, $\tilde{\mathsf{Z}}_{00}^\star$.

Chapter 14

Mode Splitting and Coupling

The splitting of the free oscillations of the Earth was first observed following the great 1960 earthquake in Chile; the apparent doublet character of the ${}_0\mathrm{S}_2$ and ${}_0\mathrm{S}_3$ multiplets was immediately attributed to the rotation of the Earth, as we recounted in the historical introduction. Rotation and the associated hydrostatic ellipticity of figure contribute significantly to the splitting and coupling of the gravest free oscillations; however, splitting due to other perturbations is often much more significant. Observations of this non-hydrostatic splitting may be used to constrain the elastic lateral heterogeneity and anisotropy—and thus the internal dynamics—of the Earth. This chapter begins with a brief introduction to hydrostatic figure theory, and then considers the seismologically observable effects of rotation, hydrostatic ellipticity, large-scale mantle heterogeneity, and transverse isotropy of the inner core. We base our analysis of these effects upon the general formulation of degenerate and quasi-degenerate perturbation theory presented in Chapter 13, introducing a number of minor modifications and approximations for convenience. First, we take the unperturbed spherical Earth model to be the *anelastic terrestrial monopole*; spherically symmetric attenuation is incorporated into the monopole by modifying the degenerate reference eigenfrequencies: $\omega_k \rightarrow \omega_k + i\gamma_k$, where γ_k is the degenerate decay rate. The elastic and anelastic structural perturbations are thus considered to be *purely aspherical*. Second, the first-order renormalization terms introduced in the general analysis are small; in this chapter we drop them upon occasion for simplicity. Third, the contribution of aspherical anelasticity to the renormalization is particularly small; it will be ignored completely.

14.1 Hydrostatic Ellipticity

The shape of a slowly rotating planet in hydrostatic equilibrium is described by the classical theory of Clairaut (1743). Modern accounts of this venerable topic may be found in Jeffreys (1970) and Bullen (1975). We give our own brief summary of hydrostatic equilibrium theory here, following the treatment by Chandrasekhar & Roberts (1963). All of our considerations are valid to first order in the centrifugal-to-gravitational-force ratio Ω^2a^3/GM, where Ω is the sidereal rate of rotation, a and M are the mean radius and mass of the Earth, and G is the gravitational constant. Higher-order hydrostatic perturbation theories have been developed for application to rapidly rotating planets such as Jupiter and Saturn (Zharkov 1978); however, this increased level of sophistication is not required for terrestrial seismological applications.

14.1.1 Clairaut's equation

We demonstrated in Section 3.11.1 that the surfaces of constant density and geopotential must coincide in a rotating body in hydrostatic equilibrium. Classical hydrostatic figure theory addresses the question—how are the level surfaces of an initially spherical non-rotating model perturbed by a slow rotation $\boldsymbol{\Omega} = \Omega\hat{\mathbf{z}}$? The initial internal gravitational potential $\Phi(r)$ is related to the initial density $\rho(r)$ by

$$\Phi(r) = -4\pi G\left(\frac{1}{r}\int_0^r \rho'\, r'^2 dr' + \int_r^a \rho'\, r' dr'\right), \tag{14.1}$$

where the prime denotes evaluation at the dummy integration variable r'. We temporarily express the perturbations in the form

$$\rho(r) \to \rho(r) + \delta\rho(r)P_2(\cos\theta), \tag{14.2}$$

$$\Phi(r) \to \Phi(r) + \delta\Phi(r)P_2(\cos\theta), \tag{14.3}$$

where θ is the colatitude and $P_2(\cos\theta)$ is the Legendre polynomial of degree two. The geopotential $\Phi(r)+\delta\Phi(r)P_2(\cos\theta)+\psi(r,\theta)$ in the rotating Earth model is the sum of the gravitational potential and the centrifugal potential $\psi = -\frac{1}{3}\Omega^2r^2[1-P_2(\cos\theta)]$. A point r, θ on an initially spherical level surface will move radially inward or outward to a new position:

$$r \to r[1 - \tfrac{2}{3}\varepsilon(r)P_2(\cos\theta)]. \tag{14.4}$$

The constancy of the density and geopotential on the perturbed level surfaces (14.4) is guaranteed by the first-order conditions

$$\delta\rho = \tfrac{2}{3}r\varepsilon\dot{\rho}, \qquad \delta\Phi = \tfrac{2}{3}(r\varepsilon g - \tfrac{1}{2}\Omega^2r^2), \tag{14.5}$$

where a dot denotes differentiation with respect to radius r, and where $g(r) = \dot{\Phi}(r) = 4\pi G r^{-2} \int_0^r \rho' r'^2 dr'$ is the unperturbed, spherically symmetric acceleration of gravity. We seek to determine the radial dependence of the *hydrostatic ellipticity* or *flattening* $\varepsilon(r)$ of the perturbed Earth in the range $0 \leq r \leq a$. Geometrically, the quantity $\varepsilon(r)$ is the fractional difference $(r_{\mathrm{equator}} - r_{\mathrm{pole}})/r$ between the equatorial and polar radii of the level surface of mean radius r.

Equation (14.5) gives us one relation between the potential perturbation $\delta\Phi$ and the ellipticity; we may obtain another by solving the perturbed version of Poisson's equation

$$\delta\ddot{\Phi} + 2r^{-1}\delta\dot{\Phi} - 6r^{-2}\delta\Phi = \tfrac{8}{3}\pi G r\varepsilon\dot{\rho}, \tag{14.6}$$

subject to the boundary conditions

$$[\delta\Phi]_-^+ = 0 \quad \text{and} \quad [\delta\dot{\Phi} - \tfrac{8}{3}\pi G r\varepsilon\rho]_-^+ = 0 \tag{14.7}$$

at the internal and external discontinuity radii $r = d$. Equations (14.6) and (14.7) can be manipulated and combined into a single equation of the form

$$\left[\frac{d}{dr} - \frac{1}{r}\right]\left[\frac{1}{r^3}\frac{d}{dr}(r^3\delta\Phi)\right] = \tfrac{8}{3}\pi G\left\{r\varepsilon\dot{\rho} + \sum_d d\varepsilon[\rho]_-^+\,\delta(r-d)\right\}, \tag{14.8}$$

where the sum is over all of the discontinuities, and $\delta(r-d)$ is the Dirac delta distribution. Upon integrating (14.8) once we deduce that

$$\frac{1}{r^3}\frac{d}{dr}(r^3\delta\Phi) = -\tfrac{8}{3}\pi G r\left\{\int_r^a \varepsilon'\dot{\rho}'\,dr' + \sum_{d>r}\varepsilon[\rho]_-^+\right\}, \tag{14.9}$$

where the sum is over only those discontinuities d lying above the radius r. An additional integration by parts yields the explicit result

$$\delta\Phi(r) = \tfrac{8}{15}\pi G\left[\frac{1}{r^3}\int_0^r \rho'(r'\dot{\varepsilon}' + 5\varepsilon')\,r'^4 dr' + r^2\int_r^a \rho'\dot{\varepsilon}'\,dr'\right]. \tag{14.10}$$

The similarity between the relations (14.1) and (14.10) is noteworthy; both Φ and $\delta\Phi$ contain a separate integral contribution from the shells of matter below and above the radius r.

Straightforward differentiation of the second hydrostatic relation (14.5) yields the result

$$\frac{1}{r^3}\frac{d}{dr}(r^3\delta\Phi) = \tfrac{2}{3}\left[\frac{1}{r^3}\frac{d}{dr}(r^4\varepsilon g) - \tfrac{5}{2}\Omega^2 r\right]. \tag{14.11}$$

Upon equating (14.11) and the intermediate result (14.9) we obtain

$$4\pi G\left\{\int_r^a \varepsilon'\dot{\rho}'\,dr' + \sum_{d>r}\varepsilon[\rho]_-^+\right\} = -\frac{1}{r^4}\frac{d}{dr}(r^4\varepsilon g) + \tfrac{5}{2}\Omega^2. \tag{14.12}$$

Differentiation of this relation leads, after some algebra, to a second-order ordinary differential equation known as *Clairaut's equation*:

$$\ddot{\varepsilon} + 8\pi G\rho g^{-1}(\dot{\varepsilon} + r^{-1}\varepsilon) - 6r^{-2}\varepsilon = 0. \tag{14.13}$$

Clairaut's equation can be solved for the ellipticity ε, subject to the two restrictions at the endpoints

$$\dot{\epsilon}_0 = 0, \qquad \dot{\epsilon}_a = a^{-1}(\tfrac{5}{2}\Omega^2a^3/GM - 2\varepsilon_a), \tag{14.14}$$

where the subscripts denote evaluation at $r = 0$ and $r = a$, respectively. The first condition (14.14) is an obvious smoothness requirement at the center of the Earth, whereas the second is the limiting value of (14.12) at the Earth's surface.

14.1.2 Radau's approximation

It is possible to integrate Clairaut's equation numerically for a given density profile $\rho(r)$, using a shooting method. For seismological purposes, however, a remarkably accurate analytical approximation due to Radau (1885) is sufficient. Defining the auxiliary variable

$$\eta = (d\ln\varepsilon)/(d\ln r) = r\dot{\varepsilon}/\varepsilon, \tag{14.15}$$

we rewrite (14.13) in the alternative form

$$\frac{d}{dr}\left(r^4g\sqrt{1+\eta}\right) = 5gr^3f(\eta), \tag{14.16}$$

where

$$f(\eta) = \frac{1 + \frac{1}{2}\eta - \frac{1}{10}\eta^2}{\sqrt{1+\eta}}. \tag{14.17}$$

We focus our attention upon the variability of the function (14.17), noting first that it has a minimum value at the endpoint $f(0) = 1$ and a maximum value $f(1/3) = 1.00074$. The range of the dimensionless independent variable is $\eta_0 \leq \eta \leq \eta_a$ where $\eta_0 = 0$ and $\eta_a = \frac{5}{2}\varepsilon_a^{-1}(\Omega^2a^3/GM) - 2$. Making use of the still-to-be-determined fact that the Earth's surface ellipticity is $\varepsilon_a \approx 1/300$, we find that $\eta_a \approx 0.59$ and $f(\eta_a) \approx 0.99961$. We conclude from this rudimentary analysis that, within the Earth, $f(\eta)$ never differs from unity by more than a few parts in 10,000. As a result, we can make *Radau's approximation* and replace the second-order differential equation (14.16) by

$$\frac{d}{dr}\left(r^4 g\sqrt{1+\eta}\right) \approx 5gr^3. \tag{14.18}$$

This approximate relation can be readily integrated once with the result

$$\eta(r) \approx \tfrac{25}{4}\left(1 - \frac{\displaystyle\int_0^r \rho'\, r'^4 dr'}{r^2 \displaystyle\int_0^r \rho'\, r'^2 dr'}\right)^2 - 1. \tag{14.19}$$

The ellipticity is given in terms of the function (14.19) by

$$\varepsilon(r) \approx \varepsilon_a \exp\left(-\int_r^a \eta'\, r'^{-1} dr'\right). \tag{14.20}$$

From equation (14.14) we deduce that the hydrostatic surface ellipticity is $\varepsilon_a = \frac{5}{2}\Omega^2 a g_a^{-1}(\eta_a + 2)^{-1}$, or, equivalently,

$$\varepsilon_a \approx \frac{10\,\Omega^2 a^3/GM}{4 + 25(1 - \frac{3}{2}I/Ma^2)^2}, \tag{14.21}$$

where

$$I = \tfrac{8}{3}\pi \int_0^a \rho\, r^4 dr. \tag{14.22}$$

The quantity (14.22) is readily recognized to be the *mean moment of inertia* of the Earth.

Spherically symmetric Earth models are constrained to have the observed mean radius $a = 6371$ km, mass $M = 5.974 \times 10^{24}$ kg and moment of inertia $I = 0.3308\,Ma^2$ (Romanowicz & Lambeck 1977). All such models have the same hydrostatic surface ellipticity:

$$\varepsilon_a^{\mathrm{hyd}} = 1/299.8. \tag{14.23}$$

This is 0.5 percent smaller than the observed flattening of the best-fitting ellipsoid:

$$\varepsilon_a^{\mathrm{obs}} = 1/298.3. \tag{14.24}$$

This discrepancy—often referred to as an "excess equatorial bulge" of the Earth—was the first major discovery of artificial satellite geodesy (Henriksen 1960; Jeffreys 1963). Figure 14.1 shows the radial variation of the Radau ellipticity (14.20) and its logarithmic derivative (14.19). The corresponding results obtained by numerical integration of Clairaut's equation are virtually indistinguishable on this scale.

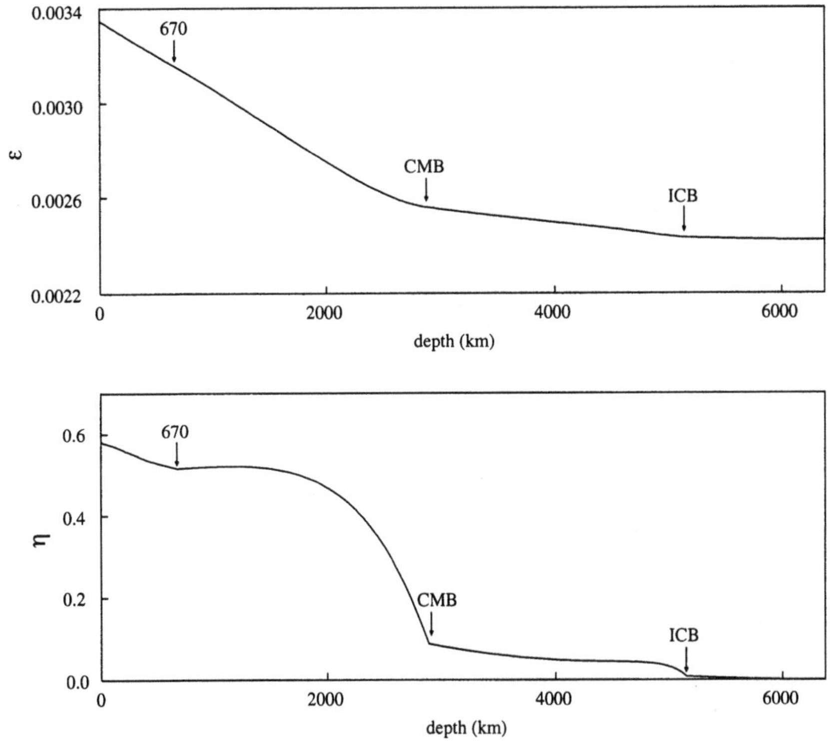

Figure 14.1. Depth dependence of the hydrostatic ellipticity ε (*top*) and its logarithmic derivative $\eta = r\dot{\varepsilon}/\varepsilon$ (*bottom*) within the PREM model. The locations of the 670 km discontinuity, the core-mantle boundary (CMB) and the inner-core boundary (ICB) are indicated.

14.1.3 Mass and moments of inertia

The mass of a hydrostatically flattened Earth model is the same as that of the undeformed sphere:

$$M = 4\pi \int_0^a \rho\, r^2 dr. \tag{14.25}$$

The polar and equatorial moments of inertia are given, respectively, by

$$C = \tfrac{8}{3}\pi \left[\int_0^a \rho\, r^4 dr + \tfrac{2}{15} \int_0^a \rho\varepsilon(\eta + 5)\, r^4 dr \right], \tag{14.26}$$

$$A = \tfrac{8}{3}\pi \left[\int_0^a \rho\, r^4 dr - \tfrac{1}{15} \int_0^a \rho\varepsilon(\eta + 5)\, r^4 dr \right]. \qquad (14.27)$$

The mean $I = \frac{1}{3}(C + 2A)$ of the polar and equatorial moments of inertia is equal to the unperturbed moment of inertia (14.22), as expected. The *precessional constant* or so-called dynamical ellipticity, is the ratio

$$H = \frac{C - A}{C} = \frac{\tfrac{1}{5} \displaystyle\int_0^a \rho\varepsilon(\eta + 5)\, r^4 dr}{\displaystyle\int_0^a \rho\, r^4 dr + \tfrac{2}{15} \int_0^a \rho\varepsilon(\eta + 5)\, r^4 dr}. \qquad (14.28)$$

The hydrostatic value of the precessional constant for any model having the observed mean radius, mass and moment of inertia of the Earth is

$$H^{\text{hyd}} = 1/308.8. \qquad (14.29)$$

The actual value determined from astronomical observations of the rate of precession of the equinoxes is one percent higher (Kinoshita 1977; Seidelmann 1982; Williams 1994):

$$H^{\text{obs}} = 1/305.4. \qquad (14.30)$$

Both of the geodetic discrepancies (14.23)–(14.24) and (14.29)–(14.30) are measures of the degree-two deviation of the density and shape of the Earth away from hydrostatic equilibrium (Defraigne 1997).

14.1.4 Elasticity variations

The results (14.20)–(14.21) describe the flattening of the surfaces of constant density and geopotential; strictly speaking, there is no fundamental reason why they should pertain to the surfaces of constant incompressibility and rigidity. It is conventional simply to *assume* that these elastic level surfaces *coincide* with those of the density and geopotential. Relinquishing our temporary notation, we may then write the complete catalogue of first-order ellipsoidal perturbations in the form

$$\delta\kappa = \tfrac{2}{3} r\varepsilon\dot{\kappa} P_2(\cos\theta), \qquad \delta\mu = \tfrac{2}{3} r\varepsilon\dot{\mu} P_2(\cos\theta),$$

$$\delta\rho = \tfrac{2}{3} r\varepsilon\dot{\rho} P_2(\cos\theta), \qquad \delta\Phi = \tfrac{2}{3}(r\varepsilon g - \tfrac{1}{2}\Omega^2 r^2) P_2(\cos\theta),$$

$$\delta d = -\tfrac{2}{3} d\varepsilon_d P_2(\cos\theta). \qquad (14.31)$$

Equations (14.20)–(14.21) and (14.31) will henceforth be regarded as the definition of a *rotating, hydrostatic, ellipsoidal Earth model.* Any additional aspherical perturbations $\delta\kappa$, $\delta\mu$, $\delta\rho$ and δd will be referred to as elastic *lateral heterogeneity*.

14.1.5 Geographic versus geocentric colatitude

The location of a seismic station situated upon the Earth's surface is conventionally specified by giving its elevation e above the geoid, its longitude ϕ with respect to Greenwich, and its *geographical colatitude* θ', which is the angle between the normal $\hat{\mathbf{n}}$ to the reference ellipsoid and the normal $\hat{\mathbf{z}}$ to the equatorial plane. Prior to calculating a synthetic accelerogram or spectrum by mode summation, it is necessary to convert the geographic colatitude θ' to the corresponding *geocentric colatitude* θ, which is the angle between the radius vector $\hat{\mathbf{r}}$ and $\hat{\mathbf{z}}$. Correct to first order in the ellipticity, the two colatitudes are related by

$$\tan\theta \approx (1 + 2\varepsilon_{\mathrm{a}})\tan\theta'. \tag{14.32}$$

Geometrically, the transformation $\theta' \to \theta$ projects a point on the reference ellipsoid to a point on the unperturbed sphere along a line through the origin. The original and projected points in Figure 14.2 are the feet of the unit normals $\hat{\mathbf{n}}$ and $\hat{\mathbf{r}}$, respectively. The hypocentral location of an earthquake source is likewise specified by giving its depth h beneath the geoid, its longitude ϕ_{s} and its geographical colatitude θ'_{s}. In calculating the real receiver and source vectors $r_k = \hat{\boldsymbol{\nu}} \cdot \mathbf{s}_k(\mathbf{x})$ and $s_k = \mathbf{M} : \boldsymbol{\varepsilon}_k(\mathbf{x}_{\mathrm{s}})$ we stipulate that

$$\mathbf{x} = (a, \theta, \phi), \qquad \mathbf{x}_{\mathrm{s}} = (a - h, \theta_{\mathrm{s}}, \phi_{\mathrm{s}}). \tag{14.33}$$

A similar geographic-to-geocentric coordinate transformation must precede the calculation of a spherical-Earth seismogram or spectrum, using equa-

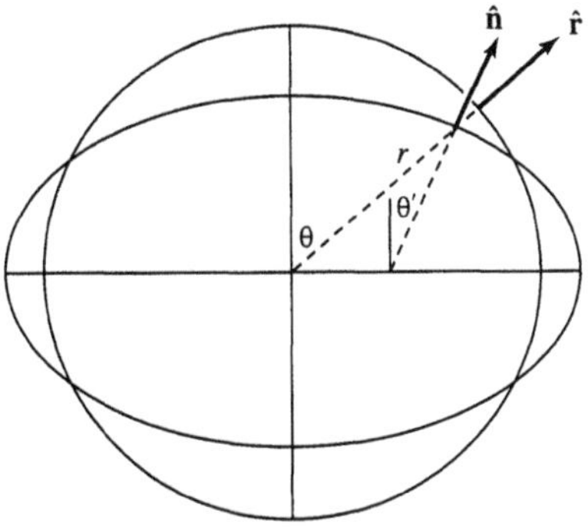

Figure 14.2. Cartoon illustrating the relation between the geographical colatitude θ and the geocentric colatitude θ'. The vectors $\hat{\mathbf{n}}$ and $\hat{\mathbf{r}}$ are the unit normals to the hydrostatic ellipsoid and the undeformed (equal-mass) sphere, respectively. Note that $\theta \geq \theta'$, with equality prevailing only at the poles and the equator.

tions (10.51)–(10.65). Station elevations e are generally ignored in normal-mode and long-period surface-wave seismology; however, they are routinely accounted for in short-period body-wave travel-time investigations.

14.2 Splitting of an Isolated Multiplet

At long periods several clearly isolated multiplets may be readily identified in normal-mode spectra, such as the one illustrated in Figure 14.3. Shorter-period free oscillations are not as well separated in frequency; however, a number of high-Q modes may be successfully isolated by attenuation filtering, as discussed in Section 10.5.2. In this section we discuss the splitting of an isolated multiplet, considering the effects of the Earth's rotation, hydrostatic ellipticity and lateral heterogeneity in that order. For simplicity, we shall drop the vector and matrix subscripts 0 and 00 used to denote the

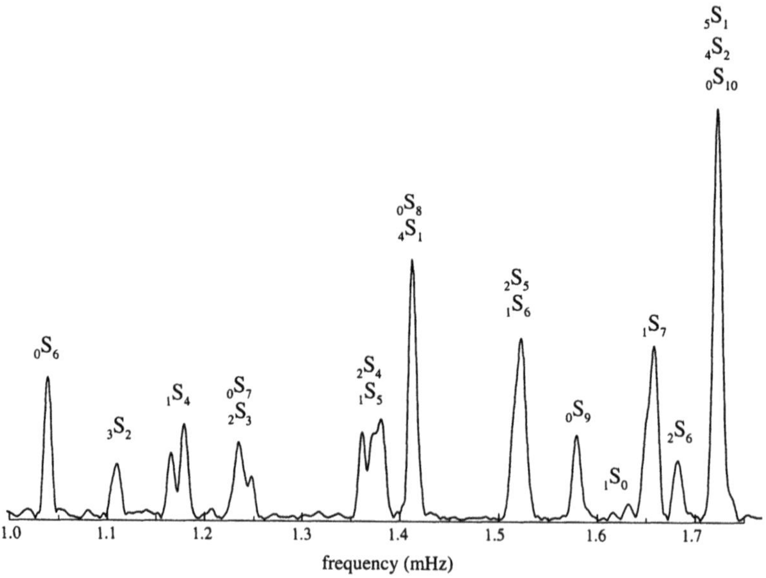

Figure 14.3. Radial-component, long-period amplitude spectrum $|\hat{\mathbf{r}} \cdot \mathbf{a}(\mathbf{x}, \omega)|$ of the ground acceleration recorded at station TUC in Tucson, Arizona following the June 9, 1994 deep-focus Bolivia earthquake. All of the peaks identified by a single (rather than double or triple) multiplet label are well isolated; mode $_1S_4$ is visibly split. A Hann taper has been applied to the 80-hour time series prior to Fourier transformation.

target multiplet in Section 13.3.4. The elements of the $(2l+1)\times(2l+1)$ self-coupling matrices are summarized in Appendix D.4.

14.2.1 First-order Coriolis splitting

To begin, we consider the first-order effect of the Earth's rotation, ignoring the centrifugal potential and associated ellipticity perturbation (14.31). The splitting is then governed by the $(2l+1)\times(2l+1)$ Coriolis matrix: $\mathsf{H}^{\rm rot}=\mathsf{W}$. This matrix is imaginary and anti-diagonal:

$$\mathsf{H}^{\rm rot}=i\chi\Omega\begin{pmatrix} & & & & \iddots \\ & & & -m & \\ & & 0 & & \\ & m & & & \\ \iddots & & & & \end{pmatrix}. \tag{14.34}$$

The quantity χ is the *Coriolis splitting parameter*, first given by Backus & Gilbert (1961):

$$\chi=k^{-2}\int_0^a\rho(V^2+2kUV+W^2)\,r^2dr, \tag{14.35}$$

where $k=\sqrt{l(l+1)}$. The unitary transformation that diagonalizes (14.34) is the complex-to-real basis transformation, i.e., the Hermitian transpose of the matrix (D.176):

$$\mathsf{Z}^{\rm H}\mathsf{Z}=\mathsf{I},\qquad \mathsf{Z}^{\rm H}\mathsf{H}^{\rm rot}\mathsf{Z}=\Delta, \tag{14.36}$$

where $\Delta=\chi\Omega\,{\rm diag}\,[\cdots\,-m\,\cdots\,0\,\cdots\,m\,\cdots]$ and

$$\mathsf{Z}=\begin{pmatrix} \ddots & & & & & & \iddots \\ & \frac{1}{\sqrt{2}} & & & & \frac{1}{\sqrt{2}} & \\ & & -\frac{1}{\sqrt{2}} & & \frac{1}{\sqrt{2}} & & \\ & & & 1 & & & \\ & & \frac{i}{\sqrt{2}} & & \frac{i}{\sqrt{2}} & & \\ & -\frac{i}{\sqrt{2}} & & & & \frac{i}{\sqrt{2}} & \\ \iddots & & & & & & \ddots \end{pmatrix}. \tag{14.37}$$

First-order Coriolis splitting is analogous to the *Zeeman splitting* of the quantum energy levels of a hydrogen atom in a magnetic field; the eigenfrequency perturbations are *uniformly spaced*:

$$\delta\omega_m=m\chi\Omega,\quad -l\leq m\leq l. \tag{14.38}$$

The normalization condition $\int_0^a \rho W^2 r^2 dr = 1$ implies that $\chi = [l(l+1)]^{-1}$ for every toroidal mode ${}_n\mathrm{T}_l$. Since the radial modes ${}_n\mathrm{S}_0$ are non-degenerate, they are unaffected by the Coriolis force, correct to first order in Ω/ω_0.

The transformed receiver and source vectors (13.220)–(13.222) in this approximation are

$$\mathsf{r}' = \mathsf{Z}^{\mathrm{T}}(\mathsf{I} - \tfrac{1}{2}\omega_0^{-1}\mathsf{W})\mathsf{r}, \qquad \mathsf{s}' = \mathsf{Z}^{\mathrm{H}}(\mathsf{I} + \tfrac{1}{2}\omega_0^{-1}\mathsf{W})\mathsf{s}, \tag{14.39}$$

where we have used the relation $\overline{\mathsf{Z}} = \mathsf{Z}^*$. Since Z is the real-to-complex basis transformation matrix, we find that r' and s' are simply related to the complex-basis receiver and source vectors defined in Appendix D.1:

$$\mathsf{r}' = (\mathsf{I} + \tfrac{1}{2}\omega_0^{-1}\Delta)\tilde{\mathsf{r}}^*, \qquad \mathsf{s}' = (\mathsf{I} + \tfrac{1}{2}\omega_0^{-1}\Delta)\tilde{\mathsf{s}}. \tag{14.40}$$

The acceleration response of a rotationally split isolated multiplet can be written in the form

$$a(t) = \mathrm{Re}\,[A_0(t)\exp(i\omega_0 t - \gamma_0 t)], \tag{14.41}$$

where we have incorporated the effect of spherically symmetric anelastic attenuation into the unperturbed eigenfrequency by making the substitution $\omega_0 \to \omega_0 + i\gamma_0$. The slowly varying multiplet modulation function (13.214) reduces to

$$A_0(t) = \sum_m [1 + m\chi(\Omega/\omega_0)]\tilde{r}_m^*\tilde{s}_m \exp(im\chi\Omega t). \tag{14.42}$$

The receiver-source vector-element product in (14.42) is given explicitly by

$$\begin{aligned}\tilde{r}_m^*\tilde{s}_m &= [\hat{\boldsymbol{\nu}} \cdot \tilde{\mathbf{s}}_m(\mathbf{x})][\mathbf{M} : \tilde{\boldsymbol{\varepsilon}}_m^*(\mathbf{x}_{\mathrm{s}})] \\ &= (\hat{\boldsymbol{\nu}} \cdot \mathbf{D})(\mathbf{M} : \mathbf{E}_{\mathrm{s}})\, Y_{lm}(\theta, \phi) Y_{lm}^*(\theta_{\mathrm{s}}, \phi_{\mathrm{s}}),\end{aligned} \tag{14.43}$$

where

$$\mathbf{D} = U\hat{\mathbf{r}} + k^{-1}V\boldsymbol{\nabla}_1 - k^{-1}W(\hat{\mathbf{r}} \times \boldsymbol{\nabla}_1), \tag{14.44}$$

$$\mathbf{E}_{\mathrm{s}} = \tfrac{1}{2}[\boldsymbol{\nabla}_{\mathrm{s}}\mathbf{D}_{\mathrm{s}} + (\boldsymbol{\nabla}_{\mathrm{s}}\mathbf{D}_{\mathrm{s}})^{\mathrm{T}}]. \tag{14.45}$$

The first-order factor $1 + m\chi(\Omega/\omega_0)$ insures that the eigenfunctions are properly orthonormalized with respect to the rotating Earth.

Of course, we could have obtained the above results much more economically by utilizing the complex basis eigenfunctions $\tilde{\mathbf{s}}_m$ rather than $\mathbf{s}_m$ at the outset. The splitting matrix in that case would have already been real and diagonal,

$$\tilde{\mathsf{H}}^{\mathrm{rot}} = \tilde{\mathsf{W}} = \Delta, \tag{14.46}$$

and the unitary transformation matrix $\tilde{\mathsf{Z}}$ would have been the identity. In quantum-mechanical parlance, the order $-l \leq m \leq l$ of the complex spherical harmonic Y_{lm} is said to be a *good quantum number* upon a slowly rotating Earth.

Following Backus & Gilbert (1961) we can obtain an elegant physical interpretation of the result (14.42), provided we ignore the renormalization factor $1 + m\chi(\Omega/\omega_0)$. Substituting (14.43) and applying the spherical-harmonic addition theorem (B.69), we then find that

$$A_0(t) = (\hat{\boldsymbol{\nu}} \cdot \mathbf{D})(\mathbf{M}:\mathbf{E}_{\mathrm{s}})\, P_l[\cos\Theta(t)], \tag{14.47}$$

where $\cos\Theta(t) = \cos\theta\cos\theta_{\mathrm{s}} + \sin\theta\sin\theta_{\mathrm{s}}\cos(\phi - \phi_{\mathrm{s}} + \chi\Omega t)$. This differs from the corresponding result upon a non-rotating, spherical Earth only by the substitution $\Theta \to \Theta(t)$ or

$$\phi - \phi_{\mathrm{s}} \to \phi - \phi_{\mathrm{s}} + \chi\Omega t. \tag{14.48}$$

Every multiplet in this approximation has an initial excitation amplitude $\mathbf{A}(r, \Theta, \Phi)$ identical to that upon a non-rotating, spherical Earth; subsequent to the excitation, however, this pattern *rotates westward relative to the Earth at a rate* $\chi\Omega$. The fastest observed rotation rate is that of the ${}_0\mathrm{S}_2$ football mode; its amplitude pattern $\mathbf{A}(r, \Theta, \Phi)$ circumnavigates the Earth in approximately 2.5 days.

14.2.2 Splitting due to rotation and ellipticity

Next, let us consider the combined effects of rotation and hydrostatic ellipticity. The $(2l+1) \times (2l+1)$ self-coupling matrix in this case is of the form

$$\mathsf{H}^{\mathrm{rot+ell}} = \mathsf{W} + (2\omega_0)^{-1}(\mathsf{V}^{\mathrm{ell+cen}} - \omega_0^2 \mathsf{T}^{\mathrm{ell}}). \tag{14.49}$$

Both the kinetic-energy and elliptical-plus-centrifugal potential-energy matrices are real and diagonal:

$$\mathsf{T}^{\mathrm{ell}} = \tau \begin{pmatrix} \ddots & & & & \\ & 1 - 3m^2/k^2 & & & \\ & & \ddots & & \\ & & & 1 & & \\ & & & & \ddots & \\ & & & & & 1 - 3m^2/k^2 \\ & & & & & & \ddots \end{pmatrix}, \tag{14.50}$$

$$\mathsf{V}^{\text{ell+cen}} = \tfrac{2}{3}\Omega^2(1 - k^2\chi)\mathsf{I} + \upsilon \begin{pmatrix} \ddots & & & & & \\ & 1 - 3m^2/k^2 & & & & \\ & & \ddots & & & \\ & & & 1 & & \\ & & & & \ddots & \\ & & & & & 1 - 3m^2/k^2 \\ & & & & & & \ddots \end{pmatrix}, \tag{14.51}$$

where

$$\tau = \frac{l(l+1)}{(2l+3)(2l-1)} \int_0^a \tfrac{2}{3}\varepsilon\rho\big[\bar{T}_\rho - (\eta+3)\check{T}_\rho\big] r^2 dr, \tag{14.52}$$

$$\upsilon = \frac{l(l+1)}{(2l+3)(2l-1)} \int_0^a \tfrac{2}{3}\varepsilon\Big\{\kappa\big[\bar{V}_\kappa - (\eta+1)\check{V}_\kappa\big] + \mu\big[\bar{V}_\mu - (\eta+1)\check{V}_\mu\big] + \rho\big[\bar{V}_\rho - (\eta+3)\check{V}_\rho\big]\Big\} r^2 dr. \tag{14.53}$$

The density, incompressibility and rigidity kernels $\bar{T}_\rho$, $\check{T}_\rho$ and $\bar{V}_\kappa$, $\check{V}_\kappa$, $\bar{V}_\mu$, $\check{V}_\mu$, $\bar{V}_\rho$, $\check{V}_\rho$ are defined in equations (D.183)–(D.189). The diagonalizing transformation Z is once again the real-to-complex basis transformation (14.37); if we had employed the complex basis functions $\tilde{\mathbf{s}}_m$ instead, the splitting matrix would have been diagonal: $\tilde{\mathsf{H}}^{\text{rot+ell}} = \Delta = \text{diag}[\cdots\ \delta\omega_m\ \cdots]$. The order m of the complex spherical harmonic Y_{lm} remains a good quantum number; however, the linear eigenfrequency perturbations (14.38) now exhibit a constant and quadratic term:

$$\delta\omega_m = \omega_0(a + bm + cm^2), \quad -l \le m \le l, \tag{14.54}$$

where

$$a = \tfrac{1}{3}(1 - k^2\chi)(\Omega/\omega_0)^2 + \tfrac{1}{2}\omega_0^{-2}(\upsilon - \omega_0^2\tau), \tag{14.55}$$

$$b = \chi(\Omega/\omega_0), \qquad c = -\tfrac{3}{2}\omega_0^{-2}k^{-2}(\upsilon - \omega_0^2\tau). \tag{14.56}$$

The first term in (14.55) represents the effect of the spherical part of the centrifugal potential $\bar{\psi} = \frac{1}{3}\Omega^2 r^2$, whereas the second term is due to the combined effects of the degree-two perturbations $\psi - \bar{\psi}$ and ε. Only $\bar{\psi}$ contributes to any shift in the mean frequency of the multiplet:

$$\frac{1}{2l+1}\sum_m \delta\omega_m = \tfrac{1}{3}(1 - k^2\chi)(\Omega^2/\omega_0), \tag{14.57}$$

by virtue of the identity $\sum_m(1 - 3m^2/k^2) = 0$. A toroidal multiplet ${}_n\mathrm{T}_l$ does not exhibit any net shift inasmuch as $\frac{1}{3}(1 - k^2\chi) = 0$. Every radial-mode eigenfrequency is increased by an amount $\omega_0 \to \omega_0[1 + \frac{1}{3}(\Omega/\omega_0)^2]$.

The acceleration response of an isolated multiplet upon a rotating, elliptical Earth is again of the form (14.41); the multiplet modulation function (14.42) is generalized to

$$A_0(t) = \sum_m [1 + m\chi(\Omega/\omega_0) - \tau(1 - 3m^2/k^2)] \times \tilde{r}_m^* \tilde{s}_m \exp[i\omega_0(a + bm + cm^2)t], \tag{14.58}$$

where the term $1 + m\chi(\Omega/\omega_0) - \tau(1 - 3m^2/k^2)$ arises from the renormalization. The spectral analogue of equation (14.58) is a superposition of $2l + 1$ Lorentzians centered upon the split singlet eigenfrequencies:

$$a(\omega) = \sum_m [1 + m\chi(\Omega/\omega_0) - \tau(1 - 3m^2/k^2)]\tilde{r}_m^* \tilde{s}_m \, \eta_m(\omega), \tag{14.59}$$

where $\eta_m = \frac{1}{2}[\gamma_0 + i(\omega - \omega_0 - \delta\omega_m)]^{-1}$. A radial-component sensor at the South Pole seismic station SPA has $\tilde{r}_m = 0$ for $m \neq 0$. This results in an *unsplit* spectrum: $a(\omega) = (1 - \tau)\tilde{r}_0^* \tilde{s}_0 \, \eta_0(\omega)$.

★14.2.3 Second-order Coriolis splitting

The Coriolis force exerts a perturbation upon the Earth's free oscillations of order Ω/ω_0, whereas the centrifugal force is a perturbation of order $(\Omega/\omega_0)^2$. A complete treatment of the Earth's rotation to order $(\Omega/\omega_0)^2$ must account for the effect of the Coriolis force to *second order*. We present a brief account of the requisite second-order analysis in this section. The matrix formulation developed in Chapter 13 is difficult to extend to higher order; for this reason, we resort to classical Rayleigh-Schrödinger perturbation theory, following the original treatment of the Coriolis splitting problem by Backus & Gilbert (1961).

Let $\mathcal{H}$ be the elastic-gravitational operator (4.154) governing the free oscillations of a spherically symmetric, non-rotating Earth model. We denote the real eigenfrequencies and associated *complex* eigenfunctions of this unperturbed Earth model by ω_k and $\tilde{\mathbf{s}}_k$. We seek to find perturbed real eigenfrequencies ω and associated complex eigenfunctions $\mathbf{u}$ which satisfy the equation

$$\mathcal{H}\mathbf{u} + 2i\omega\Omega\,\hat{\mathbf{z}} \times \mathbf{u} = \omega^2\mathbf{u}. \tag{14.60}$$

We look for ordinary perturbative solutions to (14.60) of the form

$$\omega/\omega_0 = 1 + \xi_1(\Omega/\omega_0) + \xi_2(\Omega/\omega_0)^2 + \cdots, \tag{14.61}$$

$$\mathbf{u} = \mathbf{u}_0 + \mathbf{u}_1(\Omega/\omega_0) + \mathbf{u}_2(\Omega/\omega_0)^2 + \cdots, \tag{14.62}$$

where the expansion parameter Ω/ω_0 is presumed to be small. Inserting (14.61)–(14.62) into (14.60) and equating powers of Ω/ω_0, we obtain a sequence of equations, of which we write only the first three:

$$(\mathcal{H} - \omega_0^2)\mathbf{u}_0 = \mathbf{0}, \tag{14.63}$$

$$(\mathcal{H} - \omega_0^2)\mathbf{u}_1 = 2\omega_0^2(\xi_1\mathbf{u}_0 - i\hat{\mathbf{z}} \times \mathbf{u}_0) \equiv \mathbf{e}_1, \tag{14.64}$$

$$\begin{aligned}(\mathcal{H} - \omega_0^2)\mathbf{u}_2 &= 2\omega_0^2(\xi_1\mathbf{u}_1 - i\hat{\mathbf{z}} \times \mathbf{u}_1) \\ &\quad + 2\omega_0^2\xi_1(\xi_1\mathbf{u}_0 - i\hat{\mathbf{z}} \times \mathbf{u}_0) + \omega_0^2(2\xi_2 - \xi_1^2)\mathbf{u}_0 \equiv \mathbf{e}_2.\end{aligned} \tag{14.65}$$

The zeroth-order equation (14.63) simply states that ω_0 and $\mathbf{u}_0$ must be an eigenfrequency and associated eigenfunction of the unperturbed, non-rotating Earth model. Knowing that the order m of Y_{lm} is a good quantum number, we take the zeroth-order eigenfunction to be a *single* complex basis eigenfunction:

$$\mathbf{u}_0 = \tilde{\mathbf{s}}_0. \tag{14.66}$$

The choice (14.66) is advantageous, since it allows us to proceed with an essentially *non-degenerate* version of perturbation theory. We employ the complex inner product $\langle \mathbf{s}, \mathbf{s}' \rangle = \int_\oplus \rho\mathbf{s}^* \cdot \mathbf{s}'\, dV$ introduced in equation (4.76). Orthogonality $\langle \mathbf{s}, \mathbf{s}' \rangle = 0$ of two complex functions will be indicated by the notation $\mathbf{s} \perp \mathbf{s}'$. Since the normalization of $\mathbf{u}$ is at our disposal, we may without loss of generality require that $\mathbf{u}_k \perp \tilde{\mathbf{s}}_0$ for $k \neq 0$.

The Hermitian character of the operator $\mathcal{H}$ assures us that (14.64) has a unique solution $\mathbf{u}_1 \perp \tilde{\mathbf{s}}_0$ if and only if $\mathbf{e}_1 \perp \tilde{\mathbf{s}}_0$, and that (14.65) has a unique solution $\mathbf{u}_2 \perp \tilde{\mathbf{s}}_0$ if and only if $\mathbf{e}_2 \perp \tilde{\mathbf{s}}_0$. The first of these solubility conditions determines the first-order Coriolis splitting parameter:

$$\xi_1 = \int_\oplus \rho\tilde{\mathbf{s}}_0^* \cdot (i\hat{\mathbf{z}} \times \tilde{\mathbf{s}}_0)\, dV. \tag{14.67}$$

Evaluation of the integral (14.67) yields the previous result $\xi_1 = m\chi$, as expected. The second condition reduces to

$$2\xi_2 - \xi_1^2 = 2\int_\oplus \rho\mathbf{u}_1^* \cdot (i\hat{\mathbf{z}} \times \tilde{\mathbf{s}}_0)\, dV. \tag{14.68}$$

As usual in Rayleigh-Schrödinger perturbation theory, we must solve for the first-order correction to the eigenfunction $\mathbf{u}_1$ before we can find the second-order correction to the eigenfrequency ξ_2. Since $\mathbf{u}_1 \perp \tilde{\mathbf{s}}_0$, we may expand it in terms of all of the other unperturbed basis eigenfunctions:

$$\mathbf{u}_1 = \sum_{k\neq 0} \langle \tilde{\mathbf{s}}_k, \mathbf{u}_1 \rangle \tilde{\mathbf{s}}_k. \tag{14.69}$$

To find the expansion coefficients $\langle \tilde{\mathbf{s}}_k, \mathbf{u}_1 \rangle$, we substitute (14.69) into equation (14.64) and make use of the orthonormality $\langle \tilde{\mathbf{s}}_k, \tilde{\mathbf{s}}_{k'} \rangle = \delta_{kk'}$. This leads

Mode	a	b	c	Mode	a	b	c
${}_0T_2$	−1.335	5.090	−0.231	${}_1S_1$	15.306	98.380	−0.554
${}_0T_3$	0.558	1.647	−0.279	${}_1S_2$	1.177	4.173	−0.428
${}_0T_4$	0.849	0.757	−0.162	${}_1S_3$	0.922	2.633	−0.215
${}_0T_5$	0.917	0.416	−0.102	${}_1S_4$	0.795	1.948	−0.122
${}_0T_6$	0.923	0.256	−0.069	${}_1S_5$	0.696	1.437	−0.075
${}_0T_7$	0.926	0.170	−0.050	${}_1S_6$	0.618	0.873	−0.049
${}_0T_8$	0.961	0.119	−0.038	${}_1S_7$	0.561	0.564	−0.033
				${}_1S_8$	0.500	0.427	−0.023
${}_1T_1$	−0.604	4.687	0.897	${}_1S_9$	0.446	0.349	−0.017
${}_1T_2$	0.250	1.463	−0.156				
${}_1T_3$	0.517	0.671	−0.128	${}_2S_3$	0.662	0.668	−0.153
${}_1T_4$	0.576	0.366	−0.084	${}_2S_4$	0.659	0.281	−0.093
${}_1T_5$	0.599	0.221	−0.058	${}_2S_5$	0.681	0.159	−0.064
${}_1T_6$	0.616	0.143	−0.042	${}_2S_6$	0.690	0.340	−0.047
${}_2T_2$	0.413	0.866	−0.207	${}_3S_1$	0.413	1.657	−0.353
${}_2T_3$	0.565	0.421	−0.141	${}_3S_2$	0.652	1.485	−0.292
${}_0S_0$	0.336	0.000	0.000	${}_6S_3$	0.548	−0.050	−0.136
${}_1S_0$	0.094	0.000	0.000	${}_8S_1$	0.893	0.081	−1.332
				${}_8S_5$	0.591	0.019	−0.059
${}_0S_2$	0.376	14.905	−0.267	${}_9S_3$	0.626	0.055	−0.156
${}_0S_3$	0.463	4.621	−0.118	${}_{11}S_4$	0.587	0.013	−0.088
${}_0S_4$	0.544	1.834	−0.075	${}_{11}S_5$	0.585	0.005	−0.058
${}_0S_5$	0.452	0.841	−0.047	${}_{13}S_1$	0.884	0.102	−1.323
${}_0S_6$	0.391	0.407	−0.033	${}_{13}S_2$	0.659	0.031	−0.329
${}_0S_7$	0.354	0.181	−0.025	${}_{18}S_3$	0.604	0.020	−0.151
${}_0S_8$	0.273	0.064	−0.020	${}_{18}S_4$	0.590	0.017	−0.088

Table 14.1. Rotational and elliptical splitting parameters for selected toroidal and spheroidal modes of Earth model 1066A (Gilbert & Dziewonski 1975). The parameters a and c incorporate the second-order Coriolis corrections of Dahlen & Sailor (1979). All of the tabulated values must be multiplied by 10^{-3}.

to the relation

$$\langle \tilde{\mathbf{s}}_k, \mathbf{u}_1 \rangle = \frac{2\omega_0^2}{\omega_0^2 - \omega_k^2} \int_\oplus \rho \tilde{\mathbf{s}}_k^* \cdot (i\boldsymbol{\Omega} \times \tilde{\mathbf{s}}_0)\, dV. \tag{14.70}$$

Upon substituting (14.69)–(14.70) into equation (14.68), we obtain an explicit expression for the second-order Coriolis splitting parameter:

$$2\xi_2 - \xi_1^2 = 4 \sum_{k \neq 0} \frac{\omega_0^2}{\omega_0^2 - \omega_k^2} \left| \int_\oplus \rho \tilde{\mathbf{s}}_k^* \cdot (i\boldsymbol{\Omega} \times \tilde{\mathbf{s}}_0)\, dV \right|^2. \tag{14.71}$$

Most of the terms in the sums (14.69) and (14.71) are zero, by virtue of the orthogonality of the basis eigenfunctions $\tilde{\mathbf{s}}_k$. In fact, correct to first order in Ω/ω_0, a toroidal multiplet ${}_n\mathrm{T}_l$ is only coupled to the adjacent spheroidal multiplets ${}_{n'}\mathrm{S}_{l\pm1}$, whereas a spheroidal multiplet ${}_n\mathrm{S}_l$ is coupled to the adjacent toroidal multiplets ${}_{n'}\mathrm{T}_{l\pm1}$ and the adjacent spheroidal multiplets ${}_{n'}\mathrm{S}_l$, $n' \neq n$. To insure completeness, it is necessary to include absolutely all of the possible coupling partners, including the rigid-body and geostrophic trivial modes discussed in Sections 8.7.2 and 8.8.2. A detailed recipe for evaluating the sum (14.71) is given by Dahlen & Sailor (1979). The final second-order splitting parameter consists of a constant plus a quadratic term:

$$\xi_2 = \lambda + m^2\zeta. \tag{14.72}$$

The upshot is that the eigenfrequency perturbations $\delta\omega_m$ are still of the form (14.54), with modified values of the coefficients:

$$a \to a + \lambda(\Omega/\omega_0)^2, \qquad c \to c + \zeta(\Omega/\omega_0)^2. \tag{14.73}$$

With the modification (14.73), the result (14.58) is correct to first order in the ellipticity and second order in the rotation. In Table 14.1 we list the splitting parameters a, b and c, for a number of moderately well-isolated toroidal and spheroidal multiplets of interest. The second-order Coriolis corrections are significant for several of the Earth's gravest observed modes, including the fundamental toroidal, spheroidal and radial modes ${}_0\mathrm{T}_2$, ${}_0\mathrm{S}_2$ and ${}_0\mathrm{S}_0$. Note that the as-yet-undetected Slichter triplet ${}_1\mathrm{S}_1$ is predicted to be severely split by the first-order Coriolis force.

14.2.4 Effect of lateral heterogeneity

The Earth's hydrostatic equatorial bulge is the most visible manifestation of its asphericity. Many low-frequency multiplets are, however, much more strongly split by non-hydrostatic *lateral heterogeneity*. These additional perturbations can be expanded in real surface spherical harmonics $\mathcal{Y}_{lm}$ in the form

$$\delta\kappa = \sum_{s=1}^{s_{\max}} \sum_{t=-s}^{s} \delta\kappa_{st}\mathcal{Y}_{st}, \qquad \delta\mu = \sum_{s=1}^{s_{\max}} \sum_{t=-s}^{s} \delta\mu_{st}\mathcal{Y}_{st},$$

$$\delta\rho = \sum_{s=1}^{s_{\max}} \sum_{t=-s}^{s} \delta\rho_{st}\mathcal{Y}_{st}, \qquad \delta\Phi = \sum_{s=1}^{s_{\max}} \sum_{t=-s}^{s} \delta\Phi_{st}\mathcal{Y}_{st},$$

$$\delta d = \sum_{s=1}^{s_{\max}} \sum_{t=-s}^{s} \delta d_{st}\mathcal{Y}_{st}. \tag{14.74}$$

The sums in (14.74) commence at $s = 1$ because of our presumption that the unperturbed Earth model is the terrestrial monopole. In forward-modelling applications the maximum degree $s_{\max}$ is limited only by computer storage considerations; in inversion studies, it is governed by the resolution of the available dataset. Ignoring the Earth's rotation and ellipticity for the moment, we shall denote the splitting matrix due to lateral heterogeneity alone by

$$\mathsf{H}^{\text{lat}} = (2\omega_0)^{-1}(\mathsf{V}^{\text{lat}} - \omega_0^2\mathsf{T}^{\text{lat}}). \tag{14.75}$$

The real elements of (14.75) may be written in the form

$$H_{mm'}^{\text{lat}} = \omega_0 \sum_{st} \sigma_{st} \int_\Omega \mathcal{Y}_{lm}\mathcal{Y}_{st}\mathcal{Y}_{lm'}\, d\Omega, \tag{14.76}$$

where

$$\sigma_{st} = \tfrac{1}{2}\omega_0^{-2}\left\{\int_0^a [\delta\kappa_{st}V_\kappa + \delta\mu_{st}V_\mu + \delta\rho_{st}(V_\rho - \omega_0^2 T_\rho)]\, r^2 dr \right.$$
$$\left. + \sum_d d^2\delta d_{st}[V_d - \omega_0^2 T_d]_-^+ \right\}. \tag{14.77}$$

Explicit expressions for the quantities V_κ, V_μ, $V_\rho - \omega_0^2 T_\rho$ and $V_d - \omega_0^2 T_d$ may be found in Appendix D.4.2. These kernels depend upon the degree s of the heterogeneity; however, they are independent of the order t. It is evident that the matrix H^{lat} is real and symmetric: $H_{mm'}^{\text{lat}} = H_{m'm}^{\text{lat}}$. The real *Gaunt integrals* in equation (14.76) satisfy the *selection rules*

$$\int_\Omega \mathcal{Y}_{lm}\mathcal{Y}_{st}\mathcal{Y}_{lm'}\, d\Omega = 0 \quad \text{unless} \quad \begin{cases} s \text{ is even} \\ 0 \le s \le 2l \\ t = m - m'. \end{cases} \tag{14.78}$$

The first rule stipulates that *the splitting of an isolated multiplet depends only upon the even-degree structure of the Earth.* This obviously has important implications for the lateral-heterogeneity inverse problem.

Lateral variations in anelasticity may be easily incorporated by allowing the incompressibility and rigidity to be complex: $\delta\kappa \to \delta\kappa + i\kappa_0 q_\kappa$ and $\delta\mu \to \delta\mu + i\kappa_0 q_\mu$, where

$$q_\kappa = \sum_{s=1}^{s_{\max}} \sum_{t=-s}^{s} q_{\kappa st} \mathcal{Y}_{st}, \qquad q_\mu = \sum_{s=1}^{s_{\max}} \sum_{t=-s}^{s} q_{\mu st} \mathcal{Y}_{st}. \tag{14.79}$$

The splitting matrix (14.75) is generalized to

$$\mathsf{H}^{\mathrm{lat}} = (2\omega_0)^{-1}(\mathsf{V}^{\mathrm{lat}} + i\mathsf{A} - \omega_0^2 \mathsf{T}^{\mathrm{lat}}). \tag{14.80}$$

The elements of the anelastic perturbation matrix A are given by

$$A_{mm'} = \omega_0 \sum_{st} \psi_{st} \int_\Omega \mathcal{Y}_{lm} \mathcal{Y}_{st} \mathcal{Y}_{lm'}\, d\Omega, \tag{14.81}$$

where

$$\psi_{st} = \tfrac{1}{2}\omega_0^{-2} \int_0^a (\kappa_0 q_{\kappa st} V_\kappa + \mu_0 q_{\mu st} V_\mu)\, r^2 dr. \tag{14.82}$$

In this case $\mathsf{H}^{\mathrm{lat}}$ is complex symmetric; the selection rules (14.78) are again applicable, so that there is no dependence of the splitting upon the odd-degree anelasticity.

14.2.5 Summary

The combined effects of rotation, ellipticity and lateral heterogeneity are governed by a juxtaposition of (14.49) and (14.80):

$$\mathsf{H} = \mathsf{W} + (2\omega_0)^{-1}[\mathsf{V}^{\mathrm{ell+cen}} + \mathsf{V}^{\mathrm{lat}} + i\mathsf{A} - \omega_0^2(\mathsf{T}^{\mathrm{ell}} + \mathsf{T}^{\mathrm{lat}})]. \tag{14.83}$$

The elements of this complete $(2l+1) \times (2l+1)$ self-coupling matrix are given by

$$\begin{aligned} H_{mm'} &= \omega_0[ibm\delta_{m\,-m'} + (a + cm^2)\delta_{mm'}] \\ &\quad + \omega_0 \sum_{st} (\sigma_{st} + i\psi_{st}) \int_\Omega \mathcal{Y}_{lm} \mathcal{Y}_{st} \mathcal{Y}_{lm'}\, d\Omega. \end{aligned} \tag{14.84}$$

Provided that H is non-defective, it may be diagonalized by a similarity transformation:

$$\mathsf{Z}^{-1}\mathsf{Z} = \mathsf{I}, \qquad \mathsf{Z}^{-1}\mathsf{HZ} = \Delta, \tag{14.85}$$

where $\Delta = \mathrm{diag}\,[\cdots\delta\nu_j\cdots]$ is the matrix of complex eigenfrequency perturbations $\delta\nu_j = \delta\omega_j + i\,\delta\gamma_j$, $j = 1, 2, \ldots, 2l+1$. The *columns* of the transformation matrix Z and the *rows* of its inverse Z^{-1} consist of the singlet eigenvectors z_j and $\bar{\mathsf{z}}_j$ of the Earth and the anti-Earth, respectively. In the absence of laterally heterogeneous anelasticity, $\mathsf{A} = 0$, the diagonalizing transformation is unitary: $\mathsf{Z}^{-1} = \mathsf{Z}^{\mathrm{H}}$.

The acceleration $a(t)$ of an isolated multiplet on a rotating, elliptical, laterally heterogeneous Earth model is a sum (14.41) of $2l+1$ slowly varying complex exponentials; the modulation function multiplying each complex "carrier" $\exp(i\omega_0 t - \gamma_0 t)$ is

$$A_0(t) = \mathsf{r}'^{\mathrm{T}} \exp(i\Delta t)\,\mathsf{s}' = \sum_j A_j \exp(i\,\delta\omega_j t - \delta\gamma_j t), \tag{14.86}$$

where $A_j = r'_j s'_j$. The renormalized receiver and source vectors r' and s' are related to their SNREI counterparts r and s by

$$\mathsf{r}' = \mathsf{Z}^{\mathrm{T}}(\mathsf{I} - \tfrac{1}{2}\mathsf{T}^{\mathrm{ell}} - \tfrac{1}{2}\mathsf{T}^{\mathrm{lat}} - \tfrac{1}{2}\omega_0^{-1}\mathsf{W})\mathsf{r}, \tag{14.87}$$

$$\mathsf{s}' = \mathsf{Z}^{-1}(\mathsf{I} - \tfrac{1}{2}\mathsf{T}^{\mathrm{ell}} - \tfrac{1}{2}\mathsf{T}^{\mathrm{lat}} + \tfrac{1}{2}\omega_0^{-1}\mathsf{W})\mathsf{s}, \tag{14.88}$$

where we have ignored the effect of laterally heterogeneous anelasticity. The frequency-domain response is a sum of $2l+1$ closely spaced Lorentzian resonance peaks:

$$a(\omega) = \tfrac{1}{2} i \mathsf{r}'^{\mathrm{T}}[\Delta - (\omega - \omega_0 - i\gamma_0)\mathsf{I}]^{-1}\mathsf{s}' = \sum_j A_j \eta_j(\omega), \tag{14.89}$$

where $\eta_j(\omega) = \frac{1}{2}[\gamma_0 + \delta\gamma_j + i(\omega - \omega_0 - \delta\omega_j)]^{-1}$. Each split singlet is centered upon its perturbed angular frequency $\omega_0 + \delta\omega_j$; lateral variations in anelasticity q_κ, q_μ give rise to distinct singlet decay rates $\gamma_0 + \delta\gamma_j$. The singlet index $j = 1, 2, \ldots, 2l+1$ simply serves as a counter; in the absence of axial symmetry, the order m of the complex spherical harmonic Y_{lm} is no longer a good quantum number.

14.2.6 Diagonal sum rule

The trace of the $(2l+1) \times (2l+1)$ splitting matrix is an invariant under the similarity transformation (14.85). This yields a formula for the sum of the complex eigenfrequency perturbations:

$$\sum_j \delta\omega_j + i\delta\gamma_j = \mathrm{tr}\,\mathsf{H}. \tag{14.90}$$

The traces of the constituent perturbation matrices are calculated in Appendices D.2.8 and D.3.2. The Earth's rotation, ellipticity and lateral heterogeneity are all purely aspherical perturbations, which satisfy the *diagonal sum rule*

$$\operatorname{tr}\mathsf{W}=0,\qquad \operatorname{tr}\mathsf{T}^{\mathrm{ell}}=\operatorname{tr}\mathsf{T}^{\mathrm{lat}}=0,$$
$$\operatorname{tr}\mathsf{V}^{\mathrm{ell}}=\operatorname{tr}\mathsf{V}^{\mathrm{lat}}=\operatorname{tr}\mathsf{A}=0. \tag{14.91}$$

However, the spherically averaged centrifugal potential $\bar{\psi}=-\frac{1}{3}\Omega^2r^2$ constitutes an $s=0$ perturbation, which renders the trace of $\mathsf{V}^{\mathrm{cen}}$ non-zero:

$$\operatorname{tr}\mathsf{V}^{\mathrm{cen}}=\tfrac{2}{3}(2l+1)\Omega^2(1-k^2\chi). \tag{14.92}$$

Allowing for the second-order effect of the Coriolis force, which also violates the diagonal sum rule, we obtain a generalization of equation (14.57):

$$\frac{1}{2l+1}\sum_j\delta\omega_j=\left[\tfrac{1}{3}(1-k^2\chi)+\lambda+\tfrac{1}{3}k^2\zeta\right](\Omega^2/\omega_0), \tag{14.93}$$

$$\frac{1}{2l+1}\sum_j\delta\gamma_j=0. \tag{14.94}$$

Equations (14.93)–(14.94) stipulate that the observed eigenfrequency ω_0^{obs} and decay rate γ_0^{obs} of an isolated split multiplet may be averaged to obtain the corresponding eigenfrequency ω_0^{mon} and decay rate γ_0^{mon} of the *terrestrial monopole*:

$$\omega_0^{\mathrm{mon}}=\omega_0^{\mathrm{obs}}\{1-\left[\tfrac{1}{3}(1-k^2\chi)+\lambda+\tfrac{1}{3}k^2\zeta\right](\Omega/\omega_0)^2\}, \tag{14.95}$$

$$\gamma_0^{\mathrm{mon}}=\gamma_0^{\mathrm{obs}}. \tag{14.96}$$

For a radial mode ${}_n\mathrm{S}_0$ the correction (14.95) reduces to $\omega_0^{\mathrm{mon}}=\omega_0^{\mathrm{obs}}(1-a)$. Referring to Table 14.1 we see that the measured eigenfrequencies of ${}_0\mathrm{S}_0$ and ${}_1\mathrm{S}_0$ must be reduced prior to spherical-Earth inversion by 336 and 94 parts per million, respectively. The two eigenfrequencies have been measured to ±5 and ±19 parts per million, as we have seen in Section 9.8, so that these corrections are extremely significant.

14.2.7 Singlet stripping

Let us suppose now that the Earth's elastic and anelastic lateral heterogeneity is *purely zonal*:

$$\sigma_{st}=\psi_{st}=0\quad\text{for } t\neq0. \tag{14.97}$$

The splitting matrix $\mathsf{H}^{\mathrm{zon}}$ in that case has elements of the form

$$H^{\mathrm{zon}}_{mm'} = \omega_0[ibm\delta_{m\,-m'} + (a + cm^2)\delta_{mm'}] + \omega_0\delta_{mm'}\sum_s(\sigma_{s0} + i\psi_{s0})\int_\Omega \mathcal{Y}_{lm}\mathcal{Y}_{s0}\mathcal{Y}_{lm}\,d\Omega. \tag{14.98}$$

The axial symmetry guarantees that the order m of the complex spherical harmonic Y_{lm} remains a good quantum number; hence (14.98) may be diagonalized by the real-to-complex similarity transformation (14.37), with the result

$$\delta\omega_m = \omega_0(a + bm + cm^2) + \omega_0\sum_s \sigma_{s0}\int_\Omega \mathcal{Y}_{lm}\mathcal{Y}_{s0}\mathcal{Y}_{lm}\,d\Omega, \tag{14.99}$$

$$\delta\gamma_m = \omega_0\sum_s \psi_{s0}\int_\Omega \mathcal{Y}_{lm}\mathcal{Y}_{s0}\mathcal{Y}_{lm}\,d\Omega. \tag{14.100}$$

The frequency domain response of an isolated multiplet may, under these circumstances, be written in the form

$$a(\omega) = \sum_m A_m\eta_m(\omega), \tag{14.101}$$

where $\eta_m(\omega) = \frac{1}{2}[\gamma_0 + \delta\gamma_m + i(\omega - \omega_0 - \delta\omega_m)]^{-1}$. The complex amplitude of each peak in the weighted sum (14.101) is

$$A_m = \Big[1 + m\chi(\Omega/\omega_0) - \tau(1 - 3m^2/k^2) - \sum_s \tau_s\int_\Omega \mathcal{Y}_{lm}\mathcal{Y}_{s0}\mathcal{Y}_{lm}\,d\Omega\Big]\tilde{r}^*_m\tilde{s}_m, \tag{14.102}$$

where we have defined the quantities

$$\tau_s = \int_0^a \delta\rho_{s0}T_\rho\, r^2 dr + \sum_d d^2\,\delta d_{s0}[T_d]^+_-. \tag{14.103}$$

The lengthy factor in brackets in (14.102) accounts for the renormalization.

Given a large number of spectra $a_p(\omega)$, $p = 1, 2, \ldots$, of the form (14.101) we may form the matrix equation

$$\mathsf{a}(\omega) = \mathsf{\Pi}\mathsf{\eta}(\omega), \tag{14.104}$$

where

$$\mathsf{a} = \begin{pmatrix} \vdots \\ a_p \\ \vdots \end{pmatrix}, \qquad \mathsf{\eta} = \begin{pmatrix} \vdots \\ \eta_m \\ \vdots \end{pmatrix}, \tag{14.105}$$

$$\Pi = \begin{pmatrix} & \vdots & \\ \cdots & A_{pm} & \cdots \\ & \vdots & \end{pmatrix}. \tag{14.106}$$

We denote the matrix (14.106) of excitation amplitudes by Π to avoid confusion with the anelasticity matrix A. The singlet Lorentzians $\eta_m(\omega)$ may be determined from the observed spectra $a_p(\omega)$ by inverting the relation (14.104) over a range of angular frequencies spanning the observed multiplet peak:

$$\eta(\omega) = \Pi^{-g}\mathsf{a}(\omega). \tag{14.107}$$

Here Π^{-g} is the *generalized inverse* of Π, which may be obtained by performing a singular-value decomposition. The split singlet eigenfrequencies $\omega_0 + \delta\omega_m$ and decay rates $\gamma_0 + \delta\gamma_m$ can be measured by least-squares fitting a Lorentzian to each $\eta_m(\omega)$; the perturbations $\delta\omega_m$ and $\delta\gamma_m$ can then be used in conjunction with equations (14.99)–(14.100) to constrain the zonal splitting coefficients σ_{s0} and ψ_{s0}. These coefficients are in turn linearly related to the zonal elastic and anelastic variations $\delta\kappa_{s0}$, $\delta\mu_{s0}$, $\delta\rho_{s0}$, δd_{s0} and $q_{\kappa s0}$, $q_{\mu s0}$ through equations (14.77) and (14.82).

This technique—which is known as *singlet stripping*—was developed by Buland, Berger & Gilbert (1979) and perfected and widely applied by Ritz-

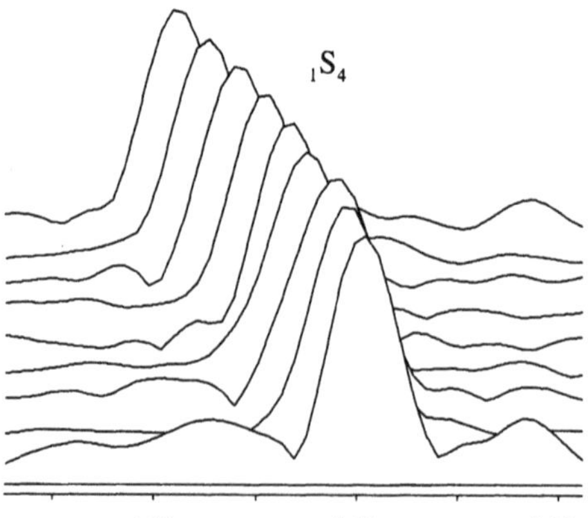

Figure 14.4. Singlet strips $|\eta_m(\omega)|$ of the spheroidal overtone $_1S_4$. The nine strips are displayed in order with $|\eta_{-4}(\omega)|$ in back and $|\eta_4(\omega)|$ in front. (Courtesy of G. Masters.)

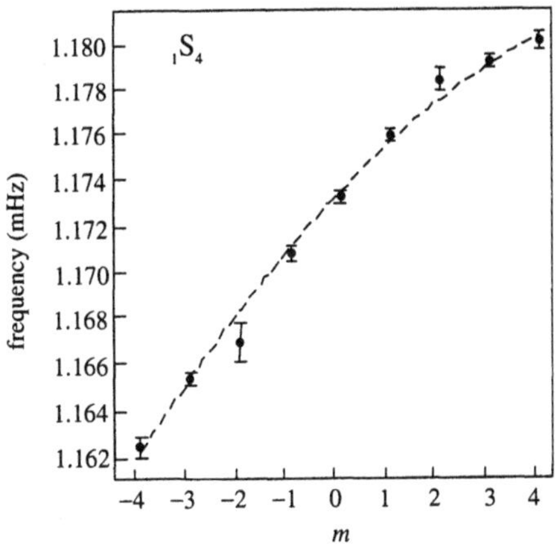

Figure 14.5. Observed singlet eigenfrequencies $\omega_0 + \delta\omega_m$ for mode $_1S_4$. These have been measured by fitting a separate Lorentzian to each of the nine strips in Figure 14.4. Each fit is performed in the *complex* spectral domain, i.e., to $\eta_m(\omega)$ rather than $|\eta_m(\omega)|$. Error bars denote the $\pm\sigma$ uncertainty. The predicted eigenfrequency splitting $\delta\omega_m = \omega_0(a + bm + cm^2)$ due to the Earth's rotation and hydrostatic ellipticity is indicated by the dashed line. (Courtesy of G. Masters.)

woller, Masters & Gilbert (1986) and Widmer (1991). The renormalization factor in (14.102) is commonly ignored in applications; that is, the amplitude of each singlet in (14.101) is approximated by $A_m \approx \tilde{r}_m^* \tilde{s}_m$. The method presumes that the lateral heterogeneity of the Earth is predominantly zonal; it happens there are a number of isolated multiplets for which this is an acceptable approximation. The spheroidal mode $_1S_4$, which is visibly split, as seen in Figure 14.3, is an excellent example. Figure 14.4 shows the resulting strips $|\eta_m(\omega)|$, arranged in order $-4 \leq m \leq 4$ from back to front. The nearly uniform peak spacing is an indication that the splitting of this multiplet is primarily a consequence of the first-order Coriolis force. In fact, the measured singlet eigenfrequencies $\omega_0 + \delta\omega_m$ agree extremely well with the theoretical peak distribution $\delta\omega_m = \omega_0(a + bm + cm^2)$ upon a rotating, elliptical Earth, as seen in Figure 14.5.

14.2.8 Anomalously split modes

Figure 14.6 shows the singlet strips of the PKIKP-equivalent multiplet $_{18}S_4$. The observed peak distribution of this mode *cannot* be attributed

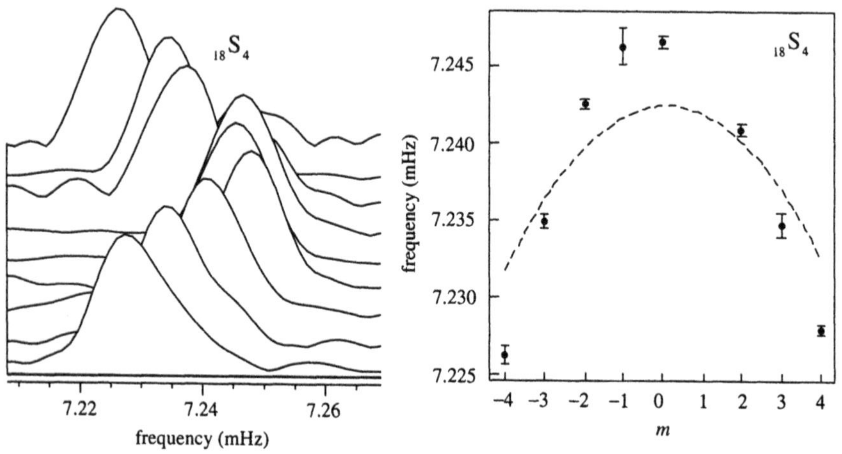

Figure 14.6. (*Left*) Singlet strips $|\eta_m(\omega)|$ of the anomalously split multiplet ${}_{18}S_4$. The nine strips are displayed in order with $|\eta_{-4}(\omega)|$ in back and $|\eta_4(\omega)|$ in front. (*Right*) The observed singlet eigenfrequencies $\omega_0 + \delta\omega_m$ exhibit a quasi-parabolic $\delta\omega_m \approx \omega_0(a' + c'm^2)$ dependence upon the order m; however, the strength of the splitting is much greater than that due to the Earth's rotation and hydrostatic ellipticity, indicated by the dashed line. (Courtesy of R. Widmer.)

to the Earth's rotation and hydrostatic ellipticity; the total splitting width $\Delta\omega \approx \delta\omega_0 - \delta\omega_{\pm 4}$ is almost twice as large as that predicted. The success of the stripping procedure is a clear indication that the causative perturbation is axially symmetric; however, very large coefficients $\sigma_{20}, \sigma_{40}, \sigma_{60}, \sigma_{80}$ are required if $\Delta\omega$ is attributed to non-hydrostatic zonal heterogeneity, as in equation (14.99). Similarly strong non-hydrostatic—and apparently zonal—splitting is a characteristic feature of approximately twenty other *core-sensitive* spheroidal multiplets, including ${}_3S_2$, ${}_9S_3$ and ${}_{13}S_2$. The mantle can be eliminated as a source, since its decidedly non-zonal structure is well constrained by other non-anomalously split modes, as we shall see in the next section.

The nature of this so-called *anomalous splitting* remained a controversial enigma for more than a decade, following its discovery by Masters & Gilbert (1981). A number of possible explanations were proposed and investigated during this period:

1. Some of the anomalously split modes have less than three percent of their compressional energy within the solid inner core. This led to speculation that the splitting might be produced by degree-two

heterogeneity of the compressional-wave speed $\delta\alpha_{20}$ within the *fluid outer core* (Ritzwoller, Masters & Gilbert 1988; Widmer, Masters & Gilbert 1992). Such a model can be tweaked to fit the observed anomalous splitting; however, the corresponding variations $\delta\rho_{20}$ in fluid-core density are ruled out by the hydrostatic considerations in Section 13.1.7 (Stevenson 1987).

2. Degree-two topographic variations δd_{st} on either the inner-core or core-mantle boundaries are also capable of explaining the observed anomalous splitting; however, such large non-hydrostatic boundary perturbations are inconsistent with body-wave travel-time data (Ritzwoller, Masters & Gilbert 1988; Widmer, Masters & Gilbert 1992).

3. Tanimoto (1989) showed conclusively that the Earth's magnetic field cannot be responsible for the observed anomalous splitting.

4. Gilbert (1994) demonstrated that the effect of large-scale convective flow in the fluid outer core is likewise insignificant.

5. Poupinet, Pillet & Souriau (1983) were the first to note that PKIKP waves traversing the inner core along a ray path parallel to the Earth's rotation axis arrive several seconds faster than PKIKP waves travelling in the equatorial plane. Motivated by this observation, Morelli, Dziewonski & Woodhouse (1986) and Woodhouse, Giardini & Li (1986) proposed *inner-core anisotropy* as an explanation for both the anomalous PKIKP travel times and the anomalous splitting of the core-sensitive spheroidal modes.

6. More comprehensive body-wave studies subsequently confirmed the presence of compressional-wave anisotropy of a few percent throughout the inner core (Creager 1992; Shearer 1994b; Song & Helmberger 1995). Nevertheless, the structure responsible for the anomalous multiplet splitting continued to be debated. After a thorough analysis of more than twenty PKIKP-equivalent modes, Widmer, Masters & Gilbert (1992) dismissed inner-core anisotropy as a viable explanation, because the model of Li, Giardini & Woodhouse (1991a) was inconsistent with their data.

Taking a new analytical approach, Tromp (1993) showed that all of the anomalous multiplet measurements are consistent with a simple transversely isotropic inner-core model having a symmetry axis aligned with the Earth's rotation. As demonstrated in Appendix D.4.4, this type of anisotropy gives rise to a diagonal splitting matrix $\mathsf{H}^{\mathrm{ani}} = (2\omega_0)^{-1}\mathsf{V}^{\mathrm{ani}}$, with elements

$$H^{\rm ani}_{mm'} = \delta_{mm'} \sum_{s=0,2,4} (-1)^m \left(\frac{2l+1}{2\omega_0}\right) \left(\frac{2s+1}{4\pi}\right)^{1/2}$$
$$\times \begin{pmatrix} l & s & l \\ -m & 0 & m \end{pmatrix} \sum_N \sum_I \int_0^c \Gamma_{NI}\, r^2 dr. \qquad (14.108)$$

The non-zero kernels Γ_{01}–Γ_{05}, Γ_{11}–Γ_{13}, Γ_{21}–Γ_{23}, Γ_{31} and Γ_{43} are defined in equations (D.209)–(D.221). The combined effects of rotation, ellipticity and co-axial inner-core anisotropy are governed by

$$\mathsf{H}^{\rm rot+ell+ani} = \mathsf{W} + (2\omega_0)^{-1}(\mathsf{V}^{\rm ell+cen} + \mathsf{V}^{\rm ani} - \omega_0^2 \mathsf{T}^{\rm ell}). \qquad (14.109)$$

The order m of the complex spherical harmonic Y_{lm} remains a good quantum number; the resulting eigenfrequency perturbations can be reduced with the aid of (C.229)–(C.231) to a compact form:

$$\delta\omega_m = \omega_0(a + bm + cm^2) + \omega_0(a' + c'm^2 + dm^4), \qquad (14.110)$$

where the first term accounts for the rotation and ellipticity and the second accounts for the anisotropy. The additional constant, quadratic and quartic splitting coefficients a', c', and d are given by

$$a' = \tfrac{1}{2}\omega_0^{-2}(-1)^l(2l+1)\left\{\left[\frac{(2l)!}{(2l+1)!}\right]^{1/2} K_0 \right.$$
$$- 2l(l+1)\left[\frac{(2l-2)!}{(2l+3)!}\right]^{1/2} K_2$$
$$\left. + 6(l+2)(l+1)l(l-1)\left[\frac{(2l-4)!}{(2l+5)!}\right]^{1/2} K_4 \right\}, \qquad (14.111)$$

$$c' = \tfrac{1}{2}\omega_0^{-2}(-1)^l(2l+1)\left\{6\left[\frac{(2l-2)!}{(2l+3)!}\right]^{1/2} K_2 \right.$$
$$\left. + [50 - 60l(l+1)]\left[\frac{(2l-4)!}{(2l+5)!}\right]^{1/2} K_4 \right\}, \qquad (14.112)$$

$$d = \tfrac{1}{2}\omega_0^{-2}(-1)^l(2l+1)\left\{70\left[\frac{(2l-4)!}{(2l+5)!}\right]^{1/2} K_4 \right\}, \qquad (14.113)$$

where

$$K_s = \sum_N \sum_I \left(\frac{2s+1}{4\pi}\right)^{1/2} \int_0^c \Gamma_{NI}\, r^2 dr, \qquad s = 0, 2, 4. \qquad (14.114)$$

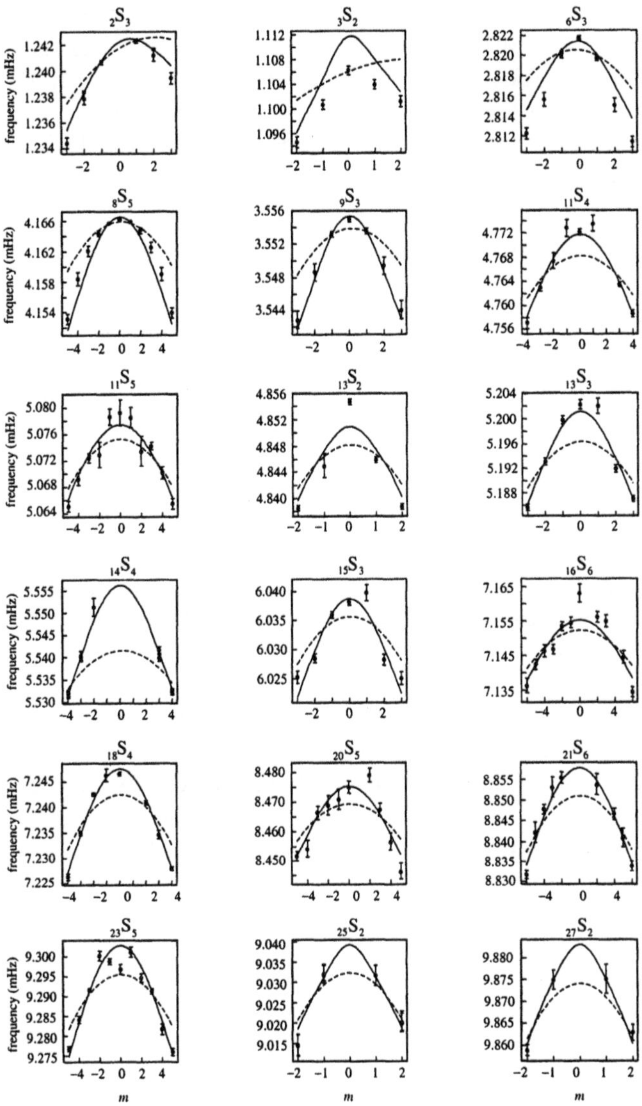

Figure 14.7. Observed singlet eigenfrequencies $\omega_0 + \delta\omega_m$ of eighteen anomalously split multiplets (courtesy of R. Widmer). The splitting of all of these modes is larger than that due to the Earth's rotation and hydrostatic ellipticity (*dashed line*). The combined effects $\delta\omega_m = \omega_0(a + bm + cm^2) + \omega_0(a' + c'm^2 + dm^4)$ of rotation, ellipticity and co-axial inner-core anisotropy (*solid line*) provide a much more satisfactory fit to the data.

The observed eigenfrequency perturbations $\delta\omega_m$, $-l \leq m \leq l$, of the anomalously split multiplets may be used to constrain the radial distribution of the five transversely isotropic perturbations δC, δA, δL, δN, δF within the inner core. Figure 14.7 shows the fit of such a transversely isotropic inner-core model to the splitting of eighteen well-characterized core-sensitive spheroidal modes (Tromp 1993).

14.2.9 Splitting function

Singlet stripping is based upon the presumption that the order m of a complex spherical harmonic Y_{lm} is a good quantum number; for this reason, it cannot be implemented whenever the predominant perturbation affecting an isolated multiplet is non-zonal. Instead, we must deal with the non-sparse $(2l+1) \times (2l+1)$ splitting matrix (14.83)–(14.84). Giardini, Li & Woodhouse (1987; 1988) and Ritzwoller, Masters & Gilbert (1986; 1988) developed an innovative two-step *spectral fitting* procedure which may be applied in this more general case. The method ignores lateral variations in anelasticity and eigenfunction renormalization; with those provisos, the spectral response (14.89) of an isolated multiplet can be written in terms of the original receiver and source vectors r and s in the form

$$a(\omega) = \tfrac{1}{2} i \mathsf{r}^{\mathrm{T}} [\mathsf{H} - (\omega - \omega_0 - i\gamma_0)\mathsf{I}]^{-1} \mathsf{s}, \tag{14.115}$$

where $\mathsf{H} = \mathsf{W} + (2\omega_0)^{-1}[\mathsf{V}^{\mathrm{ell+cen}} + \mathsf{V}^{\mathrm{lat}} - \omega_0^2(\mathsf{T}^{\mathrm{ell}} + \mathsf{T}^{\mathrm{lat}})]$. Each spectrum (14.115) depends upon the source $\mathbf{M}$, $\mathbf{x}_{\mathrm{s}}$, the receiver $\hat{\boldsymbol{\nu}}$, $\mathbf{x}$, the rotational and elliptical splitting coefficients a, b, c, and the elastic-structure coefficients σ_{st}, $s = 2, 4, \ldots, 2l$, $t = -s, \ldots, 0, \ldots, s$. Assuming that the source and receiver parameters and the rotation and ellipticity of the Earth are known, it is possible to solve for the $5 + 9 + \cdots + (4l+1) = l(2l+3)$ unknown coefficients σ_{st} by minimizing the misfit between a suite of observed and synthetic spectra $a_p(\omega)$, $p = 1, 2, \ldots$, in an iterative, non-linear least-squares inversion. Perturbation of (14.115) yields the derivatives $\partial a_p / \partial \sigma_{st}$ needed to converge upon the minimum:

$$\begin{aligned} \delta a(\omega) = -\tfrac{1}{2} i \mathsf{r}^{\mathrm{T}} [\mathsf{H} - (\omega - \omega_0 - i\gamma_0)\mathsf{I}]^{-1} \\ \times \delta\mathsf{H} [\mathsf{H} - (\omega - \omega_0 - i\gamma_0)\mathsf{I}]^{-1} \mathsf{s}, \end{aligned} \tag{14.116}$$

where we have ignored the renormalization terms in equation (13.259). Note that it is possible to combine spectra $a_p(\omega)$ from many different earthquakes $\mathbf{M}$, $\mathbf{x}_{\mathrm{s}}$ as well as receivers $\hat{\boldsymbol{\nu}}$, $\mathbf{x}$. Since the number of potentially available spectra is much larger than $l(2l+3)$, it is possible to obtain a strong constraint upon the coefficients σ_{st} characterizing a multiplet. This first step in the procedure is repeated for as many isolated modes as can be identified in the spectra. The quality of the spectral fits that can be obtained is illustrated in Figure 14.8. In the second step, the splitting coefficients of all of the analyzed modes are simultaneously inverted to find the even-degree lateral variations in incompressibility, rigidity, density and boundary topography. Equation (14.77) describes the strictly linear dependence of σ_{st} upon the perturbations $\delta\kappa_{st}$, $\delta\mu_{st}$, $\delta\rho_{st}$ and δd_{st}. The spectral-fitting method has been refined and applied by a number of investigators, including Li, Giardini & Woodhouse (1991a; 1991b), Widmer, Masters & Gilbert (1992), He & Tromp (1996) and Resovsky & Ritzwoller (1998); a comprehensive summary and comparison of results obtained prior to the occurrence of the June 9, 1994, Bolivia earthquake is given by Ritzwoller & Lavely (1995).

The splitting of an isolated multiplet may be conveniently visualized in terms of its geographical *splitting function*, defined by

$$\sigma = \sum_{\substack{s=2 \\ s \text{ even}}}^{2l} \sum_{t=-s}^{s} \sigma_{st} \mathcal{Y}_{st}. \tag{14.117}$$

At any location θ, ϕ on the Earth's surface, the function $\sigma(\theta, \phi)$ represents a local radial average of the underlying interior structure. The manner in which a given mode averages the parameters κ, μ, ρ and d is determined by the radial kernels V_κ, V_μ, $V_\rho - \omega_0^2 T_\rho$ and $V_d - \omega_0^2 T_d$ in equation (14.77). A plot of σ can be regarded as a representation of the geographical dependence of the Earth's three-dimensional elastic structure, as "seen" through these kernels. The selection rules (14.78) limit the sensitivity to even angular degrees $2 \leq s \leq 2l$. Mode ${}_8\mathrm{S}_1$, for example, is only sensitive to degree two, whereas mode ${}_0\mathrm{S}_6$ is sensitive to even-degree structure up to degree twelve. Figures 14.9 and 14.10 show the observed splitting functions σ^{obs} for a number of isolated fundamental-mode multiplets ${}_0\mathrm{S}_l$ and ${}_0\mathrm{T}_l$. The coherent, predominantly $\mathcal{Y}_{2\,\pm 2}$ pattern is one of the most characteristic features of the Earth's large-scale lateral heterogeneity—the shear-wave speed is faster than average ($\delta\beta > 0$) beneath the Americas and the Western Pacific and slower than average ($\delta\beta < 0$) beneath the Central Pacific and Africa. The four kernels V_κ, V_μ, $V_\rho - \omega_0^2 T_\rho$ and $V_d - \omega_0^2 T_d$ vary smoothly along a given mode branch, and so therefore do the splitting functions σ.

In principle, it is possible to allow for lateral variations in anelasticity in the spectral-fitting analysis by introducing a second splitting function

$$\psi = \sum_{\substack{s=2 \\ s \text{ even}}}^{2l} \sum_{t=-s}^{s} \psi_{st} \mathcal{Y}_{st}. \tag{14.118}$$

The measured coefficients ψ_{st} could be used to constrain $q_{\kappa st}$ and $q_{\mu st}$, utilizing equation (14.82). At the expense of adding two unknowns, it is

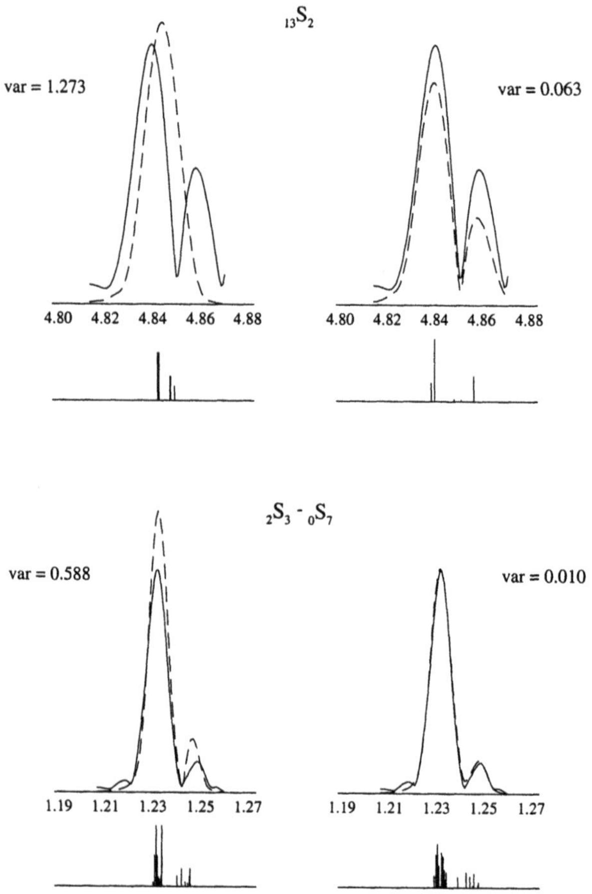

Figure 14.8. Two examples illustrating the first step of the spectral-fitting procedure. (*Top*) Amplitude spectra $|a(\omega)|$ of the multiplet ${}_{13}S_2$ recorded at station ATD in Djibouti following the June 9, 1994 deep-focus Bolivia earthquake. (*Bottom*) Spectra of the doublet ${}_2S_3 - {}_0S_7$ at station MAJO in Matsushiro, Japan following the October 4, 1994 Kuril Islands earthquake. Solid lines depict the observed radial-component spectra in both cases. Dashed lines on the left show the initial fit of a synthetic spectrum which incorporates only the effects of the Earth's rotation and hydrostatic ellipticity. Dashed lines on the right show the final synthetic spectrum after fitting for the expansion coefficients σ_{st}. The quality of each fit is indicated (var = squared misfit/squared data). Underlying line spectra show the split singlet eigenfrequencies $\omega_0 + \delta\omega_m$ or $\omega_0 + \delta\omega_j$ and excitation amplitudes $|A_m|$ or $|A_j|$.

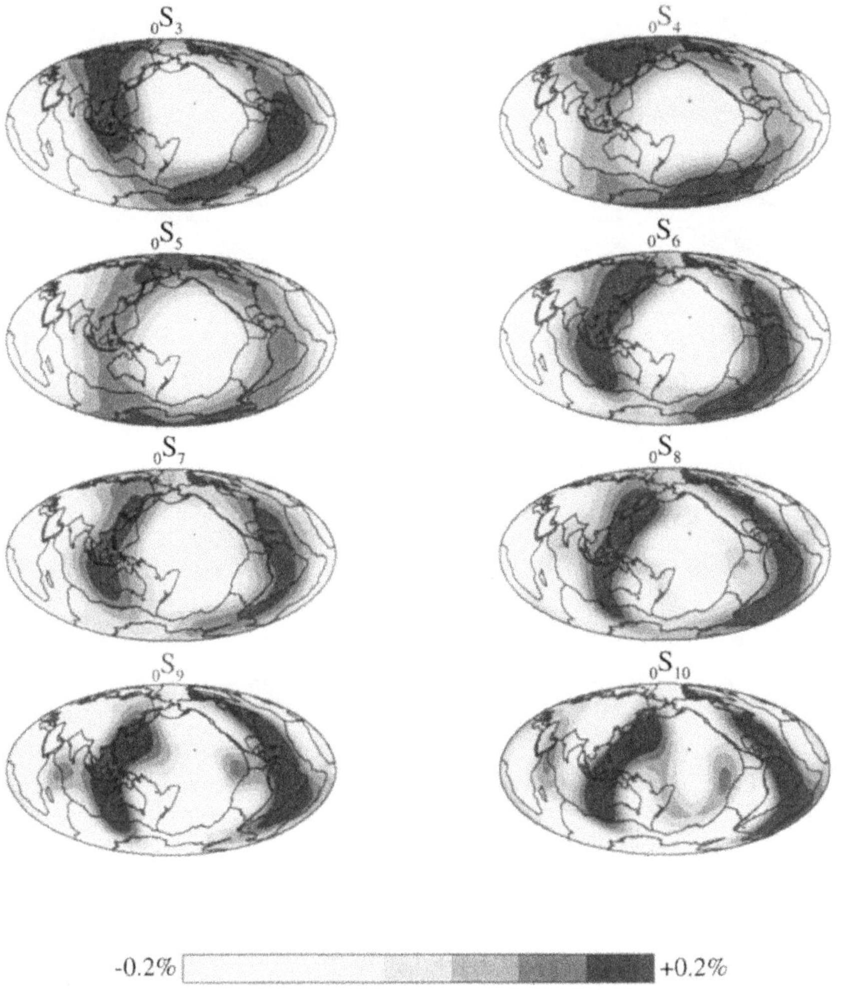

Figure 14.9. Observed splitting functions σ^{obs} of the fundamental spheroidal modes ${}_0\mathrm{S}_3 - {}_0\mathrm{S}_{10}$. Map projection is Aitoff equal-area; coastlines and tectonic plate boundaries have been superimposed for reference. The dark-shaded circum-Pacific ring, where $\sigma > 0$, is produced by faster-than-average shear-wave speeds, $\delta\beta > 0$, within the underlying mantle. The splitting functions have been truncated during the fitting, so that each is represented by only the three lowest even degrees $s = 2, 4, 6$.

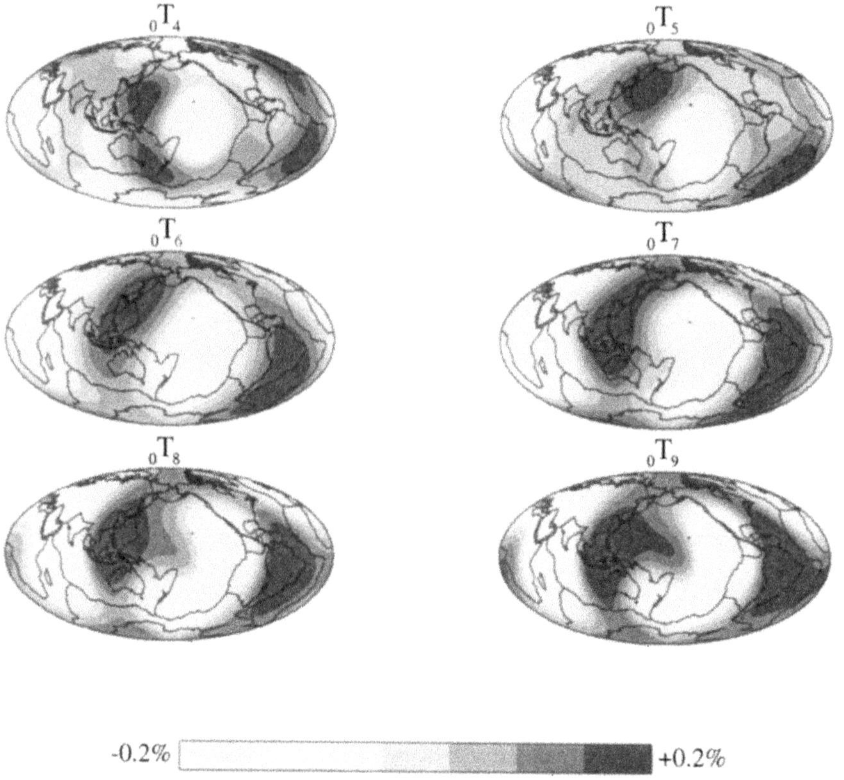

Figure 14.10. Observed splitting functions $\sigma^{\rm obs}$ of the fundamental toroidal modes ${}_0T_4 - {}_0T_9$. Map projection is Aitoff equal-area; coastlines and tectonic plate boundaries have been superimposed for reference. The characteristic circum-Pacific $\mathcal{Y}_{2\,\pm 2}$ pattern is very similar to that exhibited by the fundamental spheroidal modes (see Figure 14.9). The splitting functions have been truncated during the fitting, so that each is represented by only the three lowest even degrees $s = 2, 4, 6$.

also possible to allow for degree-zero perturbations $\delta\kappa_{00}$, $\delta\mu_{00}$, $\delta\rho_{00}$, δd_{00} and $q_{\kappa st}$ and $q_{\mu st}$. The measured coefficients σ_{00} and ψ_{00} can be used to refine the degenerate eigenfrequency $\omega_0^{\rm mon}$ and decay rate $\gamma_0^{\rm mon}$:

$$\omega^{\rm mon} = \omega_0\left(1 + \sqrt{4\pi}\,\sigma_{00}\right), \qquad \gamma^{\rm mon} = \gamma_0\left(1 + \sqrt{4\pi}\,\psi_{00}\right). \quad (14.119)$$

The resulting estimates can be used in conjunction with other global seismological data to invert for a new terrestrial monopole.

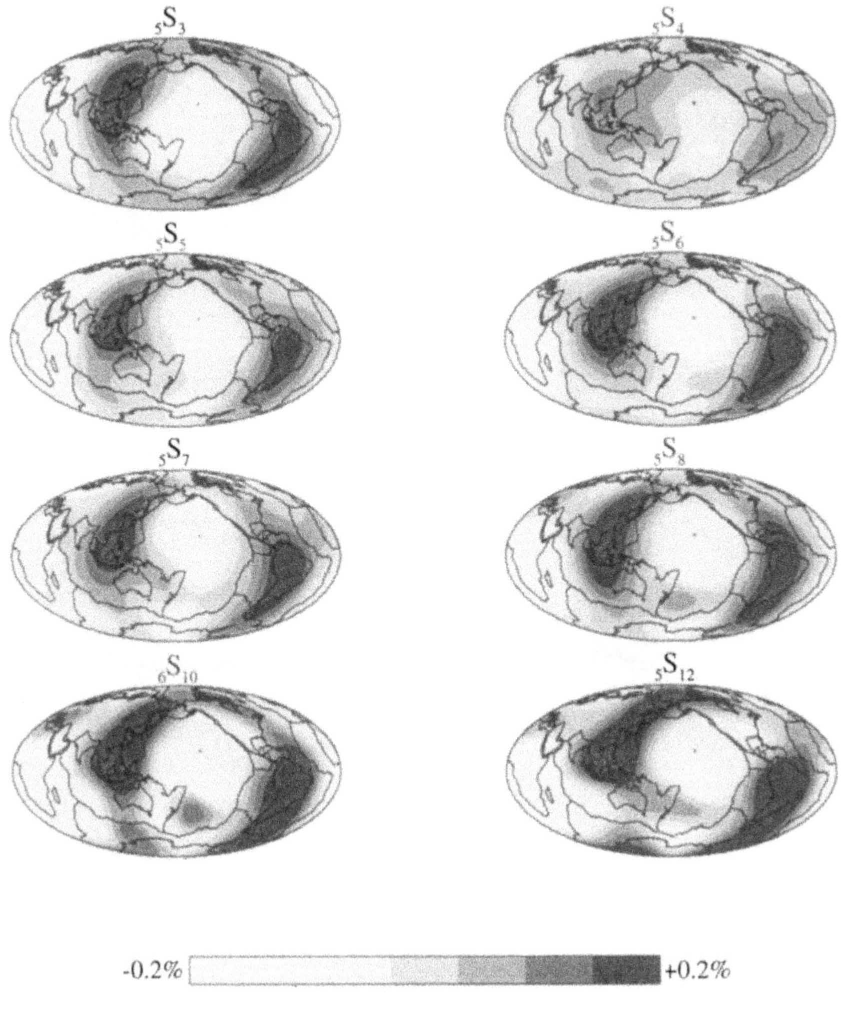

Figure 14.11. Observed splitting functions σ^{obs} of selected fifth-overtone spheroidal modes ${}_5\text{S}_3 - {}_5\text{S}_{12}$. Mode ${}_6\text{S}_{10}$ has been plotted instead of ${}_5\text{S}_{10}$ because the latter is an unobservable J_{SV} inner-core mode. Map projection is Aitoff equal-area; coastlines and tectonic plate boundaries have been superimposed for reference. Each of these mantle-sensitive overtones has a predominantly $\mathcal{Y}_{2\,\pm 2}$ splitting function, similar to those of the fundamental modes. Compare with the predicted splitting functions σ^{pre} of model SKS12WM13 in Figure 14.12.

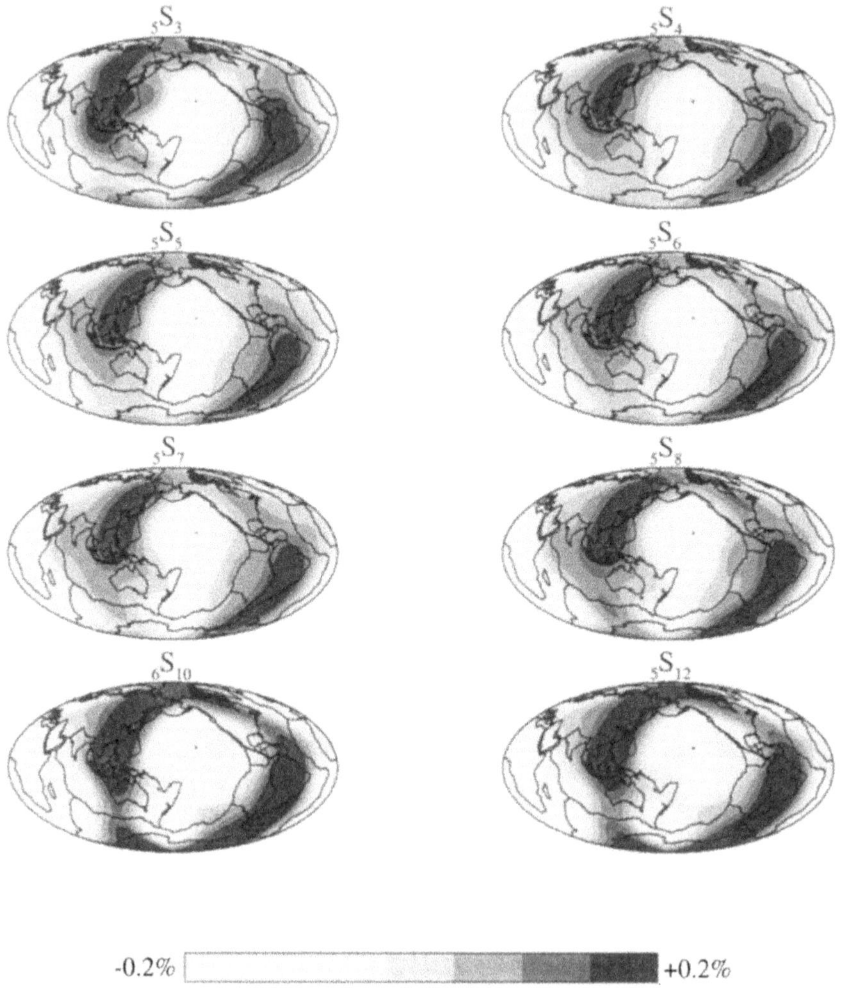

Figure 14.12. Predicted splitting functions $\sigma^{\rm pre}$ of the eight fifth-overtone modes in Figure 14.11. Model SKS12WM13 of Dziewonski, Liu & Su (1998) was employed in equations (14.77) and (14.117) to perform the calculations. Both the observed and predicted splitting functions have been been truncated, with only the three lowest even degrees $s = 2, 4, 6$ retained.

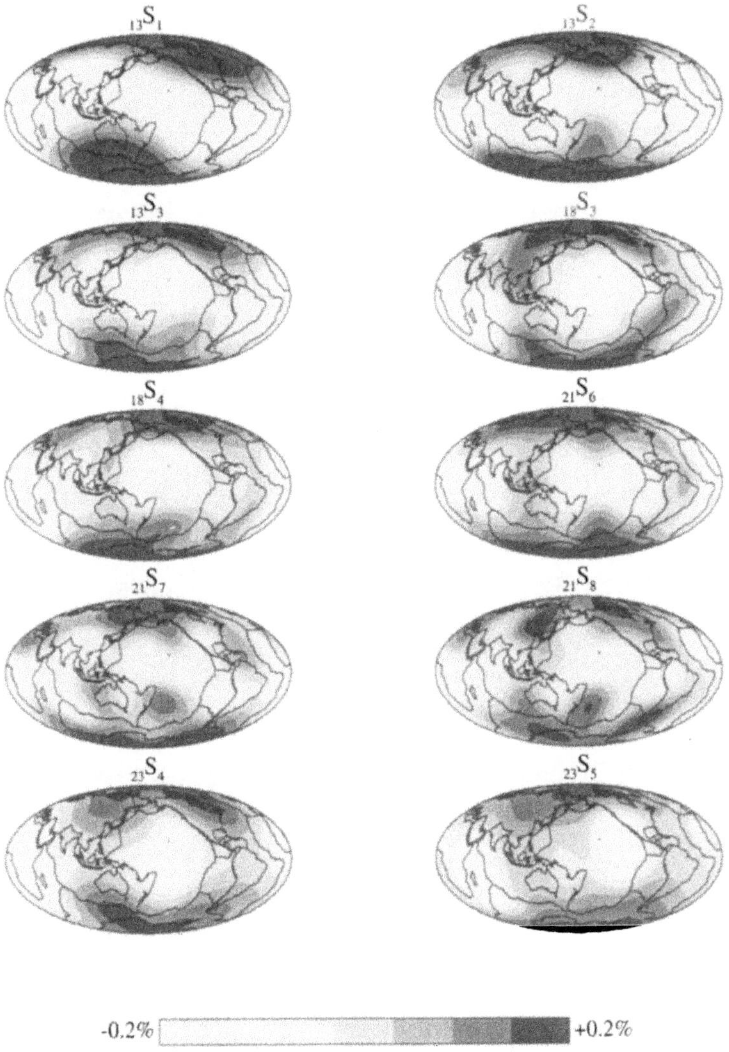

Figure 14.13. Core-sensitive spheroidal modes such as these exhibit an axially symmetric splitting function that is "fast" ($\sigma^{\mathrm{obs}} > 0$) in the vicinity of the poles and "slow" ($\sigma^{\mathrm{obs}} < 0$) in the vicinity of the equator. Map projection is Aitoff equal-area; coastlines and tectonic plate boundaries have been superimposed for reference.

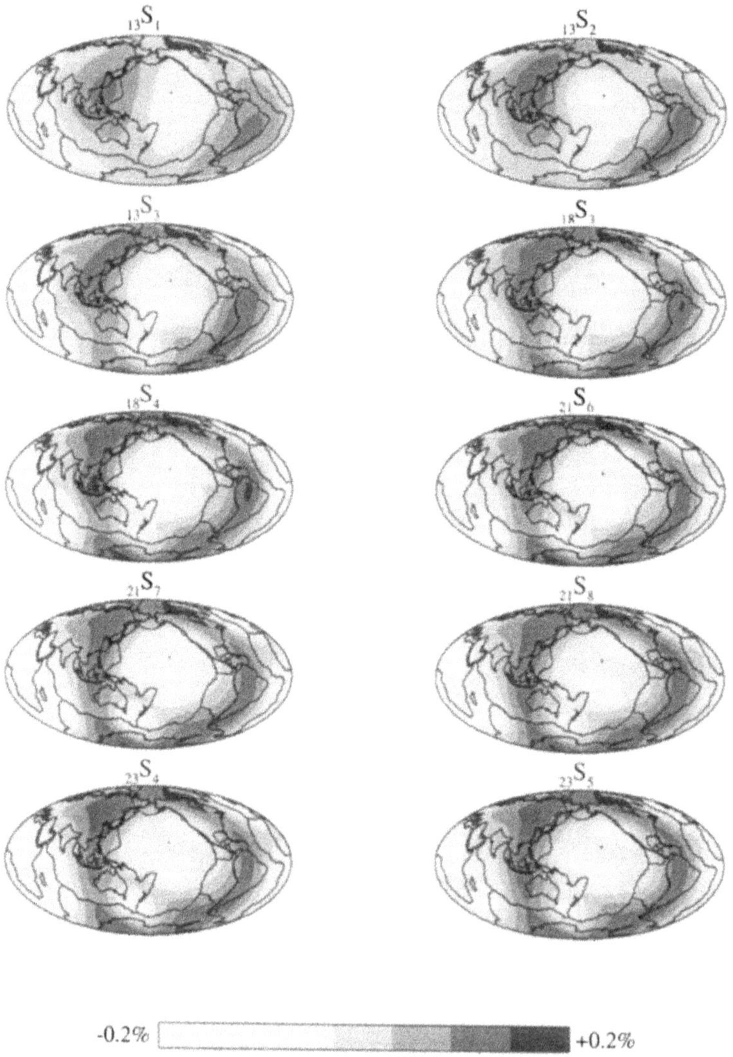

Figure 14.14. Predicted splitting functions $\sigma^{\rm pre}$ of the eight spheroidal multiplets in Figure 14.13. Model SKS12WM13 of Dziewonski, Liu & Su (1998) has been used to perform the calculations. The poor agreement between $\sigma^{\rm obs}$ and $\sigma^{\rm pre}$ is the reason for the appellation "anomalously split modes". Both the observed and predicted splitting functions have been been truncated, with only the three lowest even degrees $s = 2, 4, 6$ retained.

Current models of the lateral heterogeneity of the Earth's mantle agree remarkably well with the splitting functions of all of the mantle-sensitive spheroidal and toroidal modes. To illustrate this, we compare the observed and predicted variations σ^{obs} and σ^{pre} along the fifth spheroidal-overtone branch ${}_5\mathrm{S}_l$ in Figures 14.11 and Figure 14.12. The predicted splitting functions are those of the three-dimensional mantle model SKS12WM13 (Dziewonski, Liu & Su 1998). This model of the variations in shear-wave speed $\delta\beta$ was obtained by inverting a large collection of absolute and differential body-wave travel-time and phase-delay observations; no overt normal-mode data were used in its construction. The good agreement between σ^{obs} and σ^{pre} for these and many other mantle-sensitive modes is indicative of the robustness of our current understanding of large-scale mantle heterogeneity. The collection of available mantle-sensitive splitting functions may be used to constrain three-dimensional Earth structure.

The core-sensitive spheroidal oscillations displayed in Figure 14.13 are characterized by very different—predominantly $\mathcal{Y}_{20}$—splitting functions σ^{obs}. These observations are completely inconsistent with the predominantly $\mathcal{Y}_{2\,\pm2}$ predictions σ^{pre} of model SKS12WM13, shown in Figure 14.14. In fact, these modes are primarily split by the transverse isotropy of the solid inner core, as we have seen in Section 14.2.8. The splitting functions σ^{obs} of such anomalously split multiplets can be measured by fitting coefficients σ_{st} to a suite of spectra (14.115); however, these coefficients cannot be interpreted strictly in terms of isotropic heterogeneity $\delta\kappa_{st}$, $\delta\mu_{st}$, $\delta\rho_{st}$, δd_{st} using equation (14.77). It is a straightforward matter to incorporate *both* inner-core anisotropy and isotropic mantle heterogeneity into the spectral-fitting procedure; the splitting matrix in equation (14.115) must be replaced by $\mathsf{H} \rightarrow \mathsf{H} + (2\omega_0)^{-1}\mathsf{V}^{\mathrm{ani}}$. Every multiplet then has three more unknowns that are available to be adjusted in the fitting—the anisotropic coefficients a', c' and d in (14.111)–(14.113).

*14.2.10 Peak shifts

Many of the Earth's intermediate-frequency free oscillations exhibit unresolvably split spectra that may be well fit by a single Lorentzian; however, the peak locations are often systematically shifted with respect to the PREM model, as illustrated in Figure 14.15. The observed spectral peaks of mode ${}_0\mathrm{S}_6$ at stations ALE in Alert, Northwest Territories, Canada and KIP in Kipapa, Hawaii, following the great 1994 Bolivia earthquake, are obviously quite "singlet-like". The coincidence of the ALE peak with the degenerate PREM frequency is an indication that the average mantle shear-wave speed underlying the Bolivia-Alert great-circle path is similar

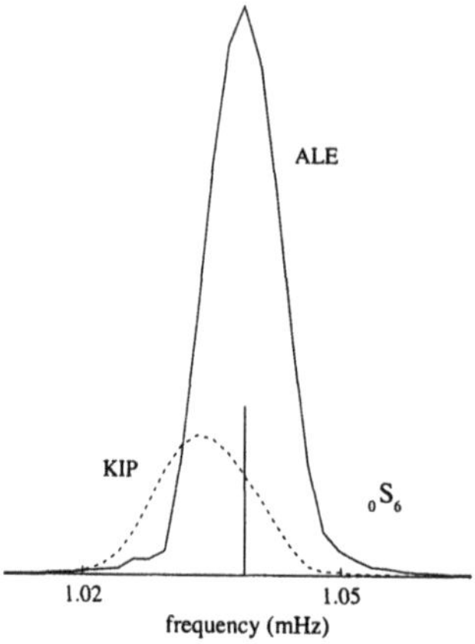

Figure 14.15. Radial-component amplitude spectra $|\hat{\mathbf{r}}\cdot\mathbf{a}(\mathbf{x},\omega)|$ of the ${}_0S_6$ multiplet at stations ALE in Alert, Canada (*solid curve*) and KIP in Kipapa, Hawaii (*dashed curve*), following the June 9, 1994 deep-focus earthquake in Bolivia. Vertical line denotes the degenerate PREM eigenfrequency.

to that of PREM. The low-frequency shift of the KIP peak indicates, on the other hand, that the average speed beneath the Bolivia-Kipapa path is slower than that of PREM. We discuss this *path-dependent peak-shift* phenomenon in this section. As a measure of the location of a spectral peak, we use the real *centroid* $\omega_\star$ of the spectrum $a(\omega)$, defined by

$$\mathrm{Re}\int_0^\infty (\omega-\omega_\star)a(\omega)\,d\omega = 0. \tag{14.120}$$

The *shift* in the peak location relative to the reference degenerate frequency ω_0 is the difference $\delta\omega_\star = \omega_\star - \omega_0$. We seek an asymptotic representation of $\delta\omega_\star$ valid in the limit of *smooth* lateral heterogeneity, $s_{\mathrm{max}} \ll l$.

To simplify matters, we ignore the rotation and aspherical anelasticity of the Earth; in addition, we ignore the renormalization of the singlet eigenfunctions, and treat the ellipticity as simply the hydrostatic part of the degree-two perturbation $\delta\kappa_{20}$, $\delta\mu_{20}$, $\delta\rho_{20}$, δd_{20}. The splitting matrix (14.83) may then be written, using an obvious notation, in the form

$$\mathsf{H} = \mathsf{V}^{\mathrm{ell+lat}} - \omega_0^2\mathsf{T}^{\mathrm{ell+lat}}. \tag{14.121}$$

Since (14.121) is real and symmetric it may be diagonalized by an orthogonal transformation: $\mathsf{Z}^{\mathrm{T}}\mathsf{Z} = \mathsf{I}$ and $\mathsf{Z}^{\mathrm{T}}\mathsf{H}\mathsf{Z} = \Delta$ where $\Delta = \mathrm{diag}\,[\cdots\delta\omega_j\cdots]$. The spectrum of an isolated multiplet may be written in terms of the transformed receiver and source vectors $\mathsf{r}' = \mathsf{Z}^{\mathrm{T}}\mathsf{r}$ and $\mathsf{s}' = \mathsf{Z}^{\mathrm{T}}\mathsf{s}$ in the form

$$a(\omega) = \tfrac{1}{2}i\mathsf{r}'^{\mathrm{T}}[\Delta - (\omega-\omega_0-i\gamma_0)\mathsf{I}]^{-1}\mathsf{s}' = \sum_j A_j\eta_j(\omega), \tag{14.122}$$

where

$$A_j = r_j' s_j' = \left(\sum_m Z_{mj} r_m\right)\left(\sum_m Z_{mj} s_m\right). \tag{14.123}$$

Upon substituting (14.122)–(14.123) into equation (14.120) we find that the centroid shift $\delta\omega_\star$ is given by

$$\delta\omega_\star = \frac{\sum_j \delta\omega_j A_j}{\sum_j A_j} = \frac{\mathsf{r}^{\mathrm{T}}\mathsf{H}\mathsf{s}}{\mathsf{r}^{\mathrm{T}}\mathsf{s}}. \tag{14.124}$$

The first equality in (14.124) is intuitively obvious—every unit Lorentzian in (14.122) has the same area $\int_0^\infty \eta_j(\omega)\,d\omega = \pi/2$; the centroid of $2l+1$ weighted peaks is therefore the centroid of their centers. The elements of the splitting matrix (14.121) are given by

$$H_{mm'} = \frac{1}{2\omega_0}\sum_{st}\left\{\int_0^a [\delta\kappa_{st}V_\kappa + \delta\mu_{st}V_\mu + \delta\rho_{st}(V_\rho - \omega_0^2 T_\rho)]\, r^2 dr \right.$$
$$\left. + \sum_d d^2 \delta d_{st}[V_d - \omega_0^2 T_d]_-^+\right\} \int_\Omega \mathcal{Y}_{lm}\mathcal{Y}_{st}\mathcal{Y}_{lm'}\, d\Omega. \tag{14.125}$$

Correct to order $(s/l)^2$ we may replace the radial kernels in (14.125) by those governing a spherically symmetric ($s = 0$) perturbation:

$$V_\kappa \approx V_\kappa^{s=0}, \qquad V_\mu \approx V_\mu^{s=0}, \qquad V_\rho \approx V_\rho^{s=0}, \tag{14.126}$$

$$T_\rho \approx T_\rho^{s=0}, \qquad V_d \approx V_d^{s=0}, \qquad T_d \approx T_d^{s=0}. \tag{14.127}$$

It is convenient to introduce a *local* eigenfrequency perturbation, defined at every point θ, ϕ on the Earth's surface by

$$\delta\omega(\theta,\phi) = \sum_{s,t} \delta\omega_{st}\mathcal{Y}_{st}(\theta,\phi), \tag{14.128}$$

where

$$\delta\omega_{st} = \frac{1}{2\omega_0}\left\{\int_0^a [\delta\kappa_{st}V_\kappa^{s=0} + \delta\mu_{st}V_\mu^{s=0} + \delta\rho_{st}(V_\rho^{s=0} - \omega_0^2 T_\rho^{s=0})]\, r^2 dr \right.$$
$$\left. + \sum_d d^2 \delta d_{st}[V_d^{s=0} - \omega_0^2 T_d^{s=0}]_-^+\right\}. \tag{14.129}$$

The quantity $\delta\omega(\theta,\phi)$ has a simple physical interpretation—it is the perturbation to ω_0 that would result if the entire Earth were to suffer a *spherically symmetric* perturbation identical to that beneath the point θ, ϕ. The approximations (14.126)–(14.127) allow us to write the elements (14.125) of H in terms of this local eigenfrequency perturbation in the form

$$H_{mm'} \approx \int_\Omega \mathcal{Y}_{lm}\, \delta\omega\, \mathcal{Y}_{lm'}\, d\Omega. \tag{14.130}$$

Upon substituting (14.130) into equation (14.124) and using the spherical-harmonic addition theorem (B.74) three times, we obtain the result

$$\begin{aligned} &\delta\omega_\star\, (\hat{\boldsymbol{\nu}}\cdot\mathbf{D})(\mathbf{M}\!:\!\mathbf{E}_{\mathrm{s}})\, X_{l0}(\Theta) \\ &\quad = \sqrt{\frac{2l+1}{4\pi}}\, (\hat{\boldsymbol{\nu}}\cdot\mathbf{D})(\mathbf{M}\!:\!\mathbf{E}_{\mathrm{s}}) \int_\Omega X_{l0}(\Theta_{\mathrm{r}})\, \delta\omega\, X_{l0}(\Theta_{\mathrm{s}})\, d\Omega. \end{aligned} \tag{14.131}$$

Here Θ denotes the epicentral distance between the source and the receiver, Θ_{r} denotes the angular distance between the integration point and the receiver, and Θ_{s} denotes the angular distance between the integration point and the source. The symbols $\mathbf{D}$ and $\mathbf{E}_{\mathrm{s}}$ denote the receiver displacement and source strain operators (14.44) and (14.45). For high-degree modes the zonal spherical harmonic X_{l0} may be approximated by the asymptotic representation (B.87). The method of stationary phase may then be used to evaluate the surface integral in equation (14.131). Retaining terms of order k^{-1}, we find

$$\begin{aligned} &\int_\Omega X_{l0}(\Theta_{\mathrm{r}})\, \delta\omega\, X_{l0}(\Theta_{\mathrm{s}})\, d\Omega \\ &\quad \approx \tfrac{1}{\pi}(\sin\Theta)^{-1/2}\big[\delta\bar{\omega}\cos(k\Theta - \pi/4) \\ &\qquad + \tfrac{1}{2}k^{-1}(\cot\Theta\, \partial_{\bar{\phi}}^2\delta\bar{\omega} - \partial_{\bar{\theta}}\partial_{\bar{\phi}}\delta\bar{\omega})\sin(k\Theta - \pi/4) \\ &\qquad + \tfrac{1}{8}k^{-1}\delta\bar{\omega}\cot\Theta\sin(k\Theta - \pi/4)\big]. \end{aligned} \tag{14.132}$$

The quantities $\bar{\theta}$ and $\bar{\phi}$ are the coordinates of the *positive pole of the source-receiver great circle*, and

$$\delta\bar{\omega}(\bar{\theta}, \bar{\phi}) = \frac{1}{2\pi}\oint_{\bar{\theta},\bar{\phi}} \delta\omega(\theta, \phi)\, d\Delta \tag{14.133}$$

is the *great-circular average* of the local eigenfrequency perturbation. All that remains is to let the operator product $(\hat{\boldsymbol{\nu}}\cdot\mathbf{D})(\mathbf{M}:\mathbf{E}_{\mathrm{s}})$ act upon the expression (14.132). Correct to order k^{-1}, for either a spheroidal mode observed on the radial component or a toroidal mode observed on the transverse component, we obtain

$$\begin{aligned} \delta\omega_\star \approx \delta\bar{\omega} + k^{-1}(\sin\Theta)^{-1}\Big\{ &x(\cos\Theta\, \partial_{\bar{\phi}}\delta\bar{\omega} - \sin\Theta\, \partial_{\bar{\theta}}\delta\bar{\omega}) \\ &+ \big[\tfrac{1}{2}(\cos\Theta\, \partial_{\bar{\phi}}^2\delta\bar{\omega} - \sin\Theta\, \partial_{\bar{\theta}}\partial_{\bar{\phi}}\delta\bar{\omega}) \end{aligned}$$

$$+ y(\cos\Theta\,\partial_{\bar{\phi}}\delta\bar{\omega} - \sin\Theta\,\partial_{\bar{\theta}}\delta\bar{\omega})]$$
$$\times \tan(k\Theta - \pi/4 + z)\Big\}, \tag{14.134}$$

where

$$x = \frac{(\Sigma_0 - \Sigma_2)\partial_{\bar{\phi}}\Sigma_1 + \Sigma_1\partial_{\bar{\phi}}\Sigma_2}{(\Sigma_0 - \Sigma_2)^2 + \Sigma_1^2}, \tag{14.135}$$

$$y = \frac{\Sigma_1\partial_{\bar{\phi}}\Sigma_1 - (\Sigma_0 - \Sigma_2)\partial_{\bar{\phi}}\Sigma_2}{(\Sigma_0 - \Sigma_2)^2 + \Sigma_1^2}, \tag{14.136}$$

$$z = \begin{cases} \arctan\left(\dfrac{\Sigma_1}{\Sigma_0 - \Sigma_2}\right) & \text{for a spheroidal mode} \\[2ex] \arctan\left(\dfrac{\partial_{\bar{\phi}}\Sigma_1}{\Sigma_1}\right) & \text{for a toroidal mode.} \end{cases} \tag{14.137}$$

The quantities $\Sigma_m(\bar{\phi})$, $m = 0, 1, 2$ are defined in terms of the moment-tensor excitation coefficients (10.54)–(10.59) by

$$\Sigma_m = (-1)^m \left[\frac{(l+m)!}{(l-m)!}\right]^{1/2} (A_m \cos m\bar{\phi} + B_m \sin m\bar{\phi}). \tag{14.138}$$

The above order k^{-1} analysis is due to Davis & Henson (1986) and Romanowicz & Roult (1986); the more suggestive lowest-order result

$$\delta\omega_\star \approx \frac{1}{2\pi}\oint_{\bar{\theta},\bar{\phi}} \delta\omega(\theta,\phi)\, d\Delta \tag{14.139}$$

was derived somewhat earlier by Jordan (1978) and Dahlen (1979a). The asymptotic influence of along-branch multiplet coupling has been considered by Park (1986) and Romanowicz (1987).

Equation (14.139) stipulates that the asymptotic location $\omega_0 + \delta\omega_\star$ of a peak depends only upon the average structure immediately underlying the source-receiver great-circle path. The mechanism responsible for the "singlet-like" appearance of the superposition (14.122) of $2l + 1$ closely spaced resonance peaks is mode-mode interference. The amplitudes (14.123) of the constituent Lorentzians are all real; however, the signs $\operatorname{sgn} A_j$ alternate between ± 1. Singlets with eigenfrequency perturbations $\delta\omega_j$ in the vicinity of $\delta\omega_\star$ tend to be strongly excited and of the same sign, so that they interfere constructively, whereas all other singlets are either weakly excited or interfere destructively. Dahlen (1979b) and Davis & Henson (1986) present a number of synthetic examples that illustrate this singlet interference "in action".

Upon making use of the Roberts-Ursell-Backus identity (B.111), we may write the spherical-harmonic representation of (14.139) in the form

$$\delta\omega_\star \approx \sum_{\substack{s=2 \\ s \text{ even}}}^{2l} \sum_{t=-s}^{s} \delta\omega_{st} P_s(0) \mathcal{Y}_{st}(\bar{\theta}, \bar{\phi}). \tag{14.140}$$

The expansion is limited to *even* degrees $s = 2, 4, \ldots, 2l$ by virtue of the fact that $P_s(0) = 0$ whenever s is odd. Equation (14.140) constitutes a linear inverse problem: peak-shift measurements $\delta\omega_\star$ may be used to determine the expansion coefficients $\delta\omega_{st}$, $s = 2, 4, \ldots, 2l$. The coefficients are in turn linearly related to the even-degree perturbations $\delta\rho_{st}$, $\delta\kappa_{st}$, $\delta\mu_{st}$, δd_{st} by equation (14.129). The selection rules (14.78) suggest that it should be possible to write the asymptotic splitting matrix in a manner that reflects its dependence only upon the even degrees s; in fact, the complex elements $\tilde{H}_{mm'} \approx \int_\Omega Y_{lm}^* \delta\omega Y_{lm'}\, d\Omega$ of the transformed matrix $\tilde{\mathsf{H}}$ are given by

$$\tilde{H}_{mm'} \approx \frac{1}{2\pi} \int_0^{2\pi} \delta\bar{\omega}(\bar{\theta}, \bar{\phi}) e^{-i(m-m')\bar{\phi}}\, d\bar{\phi}, \tag{14.141}$$

where

$$\cos\bar{\theta} = \frac{m+m'}{2l+1}. \tag{14.142}$$

The result (14.141)–(14.142) follows from (14.130) upon application of the Gaunt-integral asymptotic relation (C.237).

A more refined analysis allows for slow variations in the location and amplitude of a peak with time. The starting point in this case is the time-domain analogue of equation (14.122):

$$A_0(t) = \mathsf{r}^{\mathrm{T}} \exp(i\mathsf{H}t)\, \mathsf{s}. \tag{14.143}$$

We can rewrite this multiplet modulation function without further approximation as a single complex exponential:

$$A_0(t) = \mathsf{r}^{\mathrm{T}}\mathsf{s} \exp\left[i \int_0^t \delta\omega_\star(t')\, dt' \right]. \tag{14.144}$$

The integrand in equation (14.144) is a complex *instantaneous frequency shift* given by

$$\delta\omega_\star(t) = \frac{\mathsf{r}^{\mathrm{T}}\mathsf{H}\exp(i\mathsf{H}t)\,\mathsf{s}}{\mathsf{r}^{\mathrm{T}}\exp(i\mathsf{H}t)\,\mathsf{s}}. \tag{14.145}$$

At the outset, $t = 0$, this instantaneous shift coincides with (14.124); more generally, however, we may use (14.145) to express $\delta\omega_\star(t)$ as a slowly varying Taylor series:

$$\delta\omega_\star(t) = \delta\omega_\star + \delta\omega'_\star t + \tfrac{1}{2}\delta\omega''_\star t^2 + \cdots, \tag{14.146}$$

where

$$\delta\omega_\star = \frac{\mathsf{r}^{\mathrm{T}}\mathsf{H}\mathsf{s}}{\mathsf{r}^{\mathrm{T}}\mathsf{s}}, \qquad \delta\omega'_\star = i\left[\frac{\mathsf{r}^{\mathrm{T}}\mathsf{H}^2\mathsf{s}}{\mathsf{r}^{\mathrm{T}}\mathsf{s}} - \left(\frac{\mathsf{r}^{\mathrm{T}}\mathsf{H}\mathsf{s}}{\mathsf{r}^{\mathrm{T}}\mathsf{s}}\right)^2\right], \tag{14.147}$$

$$\delta\omega''_\star = -\left[\frac{\mathsf{r}^{\mathrm{T}}\mathsf{H}^3\mathsf{s}}{\mathsf{r}^{\mathrm{T}}\mathsf{s}} - \frac{3(\mathsf{r}^{\mathrm{T}}\mathsf{H}\mathsf{s})(\mathsf{r}^{\mathrm{T}}\mathsf{H}^2\mathsf{s})}{(\mathsf{r}^{\mathrm{T}}\mathsf{s})^2} + 2\left(\frac{\mathsf{r}^{\mathrm{T}}\mathsf{H}\mathsf{s}}{\mathsf{r}^{\mathrm{T}}\mathsf{s}}\right)^3\right]. \tag{14.148}$$

It is found empirically that most "singlet-like" peaks can be adequately fit by a single Lorentzian, without allowing explicitly for the temporal variations (14.146). The measured shift in that case is best interpreted as a weighted average $\langle\delta\omega_\star\rangle$ of the instantaneous shift $\delta\omega_\star(t)$ over the duration

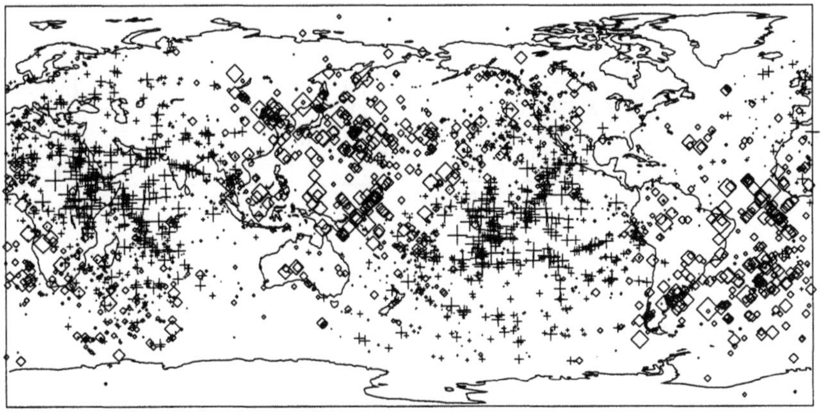

Figure 14.16. Observed peak shifts $\langle\delta\omega_\star\rangle$ of the fundamental spheroidal multiplet ${}_0\mathrm{S}_{23}$. Each of the 1400 measurements is plotted twice, at the two great-circle poles $\bar{\theta}, \bar{\phi}$ and $\pi - \bar{\theta}, \pi + \bar{\phi}$. Plus symbols + and diamonds ⋄ denote positive and negative shifts, respectively. The size of each symbol is proportional to the magnitude of the frequency shift, with maximum variations of ± 10 μHz. Paths with their poles in the Central Pacific are shifted to high frequency, because the shear-wave speed variations underlying the circum-Pacific are relatively fast. (Courtesy of R. Widmer.)

of the recording. The weighting function depends upon the decay rate γ_0 as well as the taper used in the analysis.

The above procedure was developed and utilized by Smith & Masters (1989a) to update the pioneering peak-shift measurements of Masters, Jordan, Silver & Gilbert (1982); the latter historic analysis provided the first unequivocal evidence for large-scale upper-mantle lateral heterogeneity, as we discussed in Section 1.6. Smith and Masters' last-gasp analysis fit nearly 3000 spectra, and extracted 1000 − 1500 peak shifts $\langle\delta\omega_\star\rangle$ for each of the fundamental spheroidal multiplets ${}_0S_{20} - {}_0S_{45}$. Figure 14.16 shows their raw measurements for mode ${}_0S_{23}$ plotted at the poles $\bar{\theta}, \bar{\phi}$ of the source-receiver great circles; note the remarkably coherent degree-two pattern that is vis-

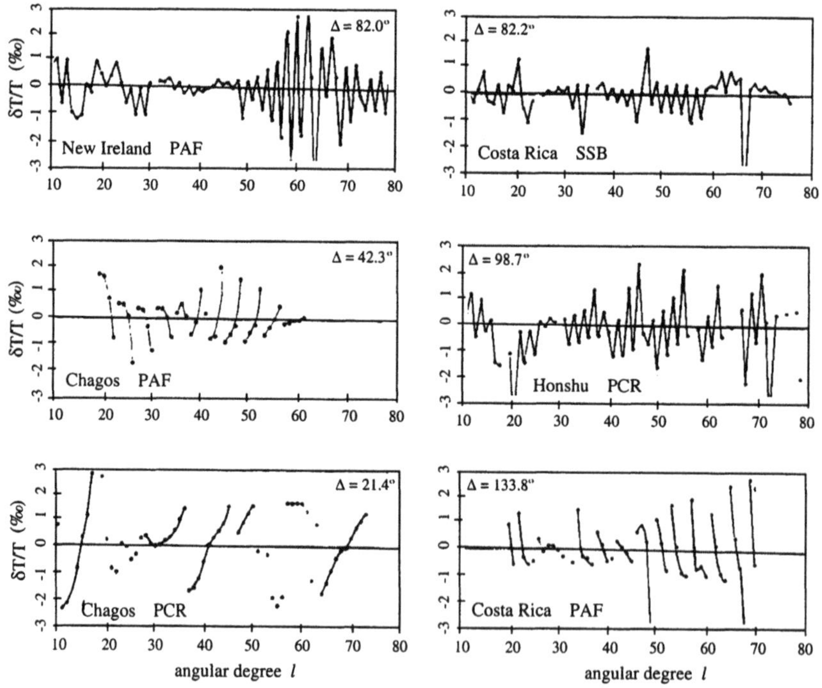

Figure 14.17. Fundamental spheroidal-mode peak-shift observations plotted versus angular degree l for a number of source-receiver combinations. The smooth $\delta\bar{\omega}$ variations have been removed to highlight the $\tan(k\Theta - \pi/4 + z)$ fluctuations. The event and station locations and epicentral distances are indicated in each panel. Note that the relative perturbation in period $\delta T/T$ is given in per mille. (Courtesy of B. Romanowicz & G. Roult.)

ible in the data! This aspect of the Earth's lateral heterogeneity has been confirmed—and resolved much more clearly—by numerous subsequent inversion studies using body and surface waves as well as normal modes (Masters 1989; Roult, Romanowicz & Montagner 1990; Woodward & Masters 1991; Masters, Johnson, Laske & Bolton 1996).

The order k^{-1} terms in (14.134) constitute a small correction, except at epicentral distances where $\tan(k\Theta - \pi/4 + z) \rightarrow \pm\infty$, corresponding to longitudinal *nodes* of the spherical-Earth excitation pattern (10.52). This divergence gives rise to a degree-dependent fluctuation or jitter of period $\Delta l \approx \pi/\Theta$ that is superimposed upon the expected smooth variation of $\delta\omega_\star$ along a fixed dispersion branch such as ${}_n\mathrm{S}_l$ or ${}_n\mathrm{T}_l$. Such a jitter is a characteristic feature of fundamental-mode peak-shift data, as illustrated in Figure 14.17. The instantaneous frequency shift $\delta\omega_\star(t)$ suffers from the same defect, since it is fundamentally due to the nodal divergence of the denominator $\mathsf{r}^{\mathrm{T}}\mathsf{s}$ in (14.124). This same spherical-Earth excitation amplitude appears in the denominators of (14.147)–(14.148); this renders the expansion (14.146) unsuitable near nodes. The ubiquity of these nodal fluctuations and—much more importantly—the insensitivity to odd-degree structure have led to the gradual abandonment of ${}_0\mathrm{S}_l$ and ${}_0\mathrm{T}_l$ peak-shift measurements. It is preferable to measure the phase speeds of the equivalent fundamental-mode surface waves, since these constrain the odd as well as the even part of the Earth's lateral heterogeneity.

⋆14.2.11 Spherical stacks

In Section 10.6 we discussed a spectral stacking procedure which may be used to isolate a target multiplet upon a spherically symmetric Earth and simultaneously reduce the noise. We return to this topic now and ask the question—how does the splitting due to the Earth's asphericity affect a spherical stack? An ideal dense-network stack consists of an integral (10.75) over the surface of the Earth:

$$\Sigma(\omega) = \int_\Omega \mathbf{A}(\mathbf{x}) \cdot \mathbf{a}(\mathbf{x},\omega)\, d\Omega, \tag{14.149}$$

where $\mathbf{A}(\mathbf{x})$ is the spherical-Earth excitation pattern. Upon evaluating this surface integral using the split-multiplet representation (14.122)–(14.123) of the stacked spectra $\mathbf{a}(\mathbf{x},\omega)$ we find that

$$\Sigma(\omega) = \sum_j \mathcal{A}_j \eta_j(\omega), \tag{14.150}$$

where

$$\mathcal{A}_j = (U_a^2 + V_a^2 + W_a^2)s_j'^2 = (U_a^2 + V_a^2 + W_a^2)\left(\sum_m Z_{mj}s_m\right)^2. \tag{14.151}$$

The subscript a denotes evaluation of the radial eigenfunctions U, V, W at the receiver radius $r = a$. An important distinction between the stacked spectrum (14.150)–(14.151) and a single-station spectrum (14.122)–(14.123) is apparent—the amplitude $\mathcal{A}_j$ of every singlet in $\Sigma(\omega)$ is *positive*. As a result, there is no destructive interference leading to a "singlet-like" peak, as in the case of $a(\omega)$; instead, the interference is purely constructive. This results in a peak that is significantly wider than any constituent Lorentzian $\eta_j(\omega) = \frac{1}{2}[\gamma_0 + i(\omega - \omega_0 - \delta\omega_j)]^{-1}$. Attenuation measurements made on a spherical stack $\Sigma(\omega)$ provide an *overestimate* $\gamma_\star$ of the degenerate decay rate γ_0 of a multiplet, for this reason. Comparisons with single-station measurements suggest that the resulting bias in the reciprocal quality factor of the fundamental spheroidal modes may be as high as forty percent, as

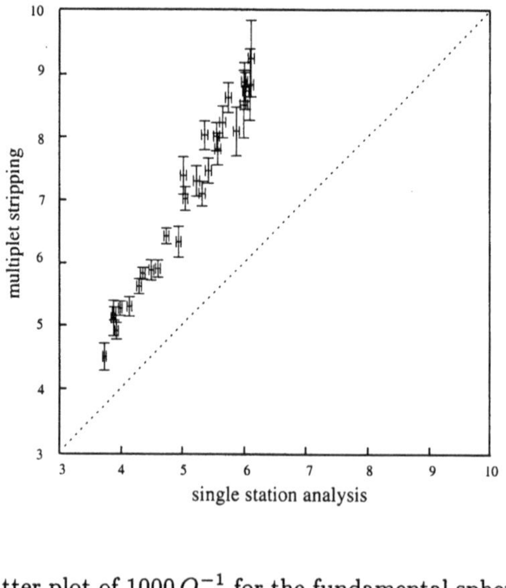

Figure 14.18. Scatter plot of $1000\,Q_0^{-1}$ for the fundamental spheroidal modes ${}_0S_l$. Ordinate shows measurements made on stripped spherical stacks $\Sigma(\omega)$; abscissa shows measurements made on single-station spectra $a(\omega)$. The latter are not expected to be severely biased; the stack measurements are systematically higher, since the spectra $\Sigma(\omega)$ are broadened by splitting as well as attenuation. Error bars denote the $\pm\sigma$ observational uncertainty. (Courtesy of R. Widmer.)

illustrated in Figure 14.18. The centroid $\omega_\star$ of a stack also provides an incorrect estimate of the degenerate frequency ω_0 as a consequence of the non-isotropic excitation: sectors of the Earth where the dipolar or quadrupolar excitation pattern $A(\mathbf{x})$ is nearly nodal will be underrepresented (Dahlen 1979a). In this case an obvious bias-reduction strategy is available—simply include as many events $\mathbf{M}$, $\mathbf{x}_s$ in the stack or stacks $\Sigma(\omega)$ as possible.

14.3 Multiplet Coupling

The isolated-multiplet approximation cannot be used to describe overlapping multiplets, such as the spheroidal-mode pairs ${}_0S_7 - {}_2S_3$, ${}_1S_5 - {}_2S_4$ and ${}_2S_5 - {}_1S_6$ in Figure 14.3. To account for the possibility of coupling between such pairs, it is necessary to treat them as a single *quasi-degenerate super-multiplet*. The resulting analysis is computationally more demanding; however, it has its advantages—as we shall see, super-multiplet splitting and coupling is sensitive to both even-degree and odd-degree lateral heterogeneity.

14.3.1 Formalities

The splitting and coupling of a super-multiplet upon a rotating, elliptical, laterally heterogeneous Earth model is governed by a super-version of the matrix (14.83):

$$\begin{aligned} \mathsf{H} &= \mathsf{N} - \nu_0 \mathsf{I} + \mathsf{W} \\ &\quad + (2\omega_0)^{-1}[\mathsf{V}^{\mathrm{ell+cen}} + \mathsf{V}^{\mathrm{lat}} + i\mathsf{A} - \omega_0^2(\mathsf{T}^{\mathrm{ell}} + \mathsf{T}^{\mathrm{lat}})]. \end{aligned} \tag{14.152}$$

The dimension of this matrix is $\sum_k(2l_k+1) \times \sum_k(2l_k+1)$, where l_k denotes the degree of the multiplet k. The quantity $\nu_0 = \omega_0 + i\gamma_0$ is a complex fiducial or reference frequency, and $\mathsf{N} = \mathrm{diag}\,[\cdots\ \nu_k\ \cdots]$ is the diagonal matrix of complex degenerate eigenfrequencies. The reference frequency is arbitrary, but it is typically chosen to be one of the degenerate frequencies $\nu_k = \omega_k + i\gamma_k$. The perturbations are considered to be superimposed upon the anelastic terrestrial monopole; note that H reduces to $\mathsf{N} - \nu_0\mathsf{I}$, so that the degenerate relative eigenfrequencies $\nu_k - \nu_0$ are recovered, when rotation, ellipticity and lateral heterogeneity are ignored.

More generally, the diagonal matrix $\Delta = \mathrm{diag}\,[\cdots\ \delta\nu_j\ \cdots]$ of split singlet eigenfrequencies $\delta\nu_j = \delta\omega_j + i\,\delta\gamma_j$ must be determined by means of a numerical similarity transformation:

$$\mathsf{Z}^{-1}\mathsf{Z} = \mathsf{I}, \qquad \mathsf{Z}^{-1}\mathsf{H}\mathsf{Z} = \Delta. \tag{14.153}$$

The complex modulation function multiplying the super-multiplet "carrier" $\exp(i\omega_0 t - \gamma_0 t)$ upon a rotating, elliptical, laterally heterogeneous Earth is

$$A_0(t) = \mathsf{r}'^{\,\mathrm{T}} \exp(i\Delta t)\,\mathsf{s}' = \sum_j A_j \exp(i\,\delta\omega_j t - \delta\gamma_j t), \tag{14.154}$$

where $A_j = r'_j s'_j$. The primed quantities are the transformed receiver and source vectors:

$$\mathsf{r}' = \mathsf{Z}^{\mathrm{T}}(\mathsf{I} - \tfrac{1}{2}\mathsf{T}^{\mathrm{ell}} - \tfrac{1}{2}\mathsf{T}^{\mathrm{lat}} - \tfrac{1}{2}\omega_0^{-1}\mathsf{W})\mathsf{r}, \tag{14.155}$$

$$\mathsf{s}' = \mathsf{Z}^{-1}(\mathsf{I} - \tfrac{1}{2}\mathsf{T}^{\mathrm{ell}} - \tfrac{1}{2}\mathsf{T}^{\mathrm{lat}} + \tfrac{1}{2}\omega_0^{-1}\mathsf{W})\mathsf{s}. \tag{14.156}$$

The spectral equivalent of equation (14.154) is

$$a(\omega) = \tfrac{1}{2} i\mathsf{r}'^{\,\mathrm{T}}[\Delta - (\omega - \omega_0 - i\gamma_0)\mathsf{I}]^{-1}\mathsf{s}' = \sum_j A_j \eta_j(\omega). \tag{14.157}$$

The results (14.153)–(14.157) are identical to (14.85)–(14.89); the only difference is the dimension of the matrices.

14.3.2 Rotation and ellipticity—selection rules

We focus attention now upon a rotating and elliptical, but laterally homogeneous, Earth model:

$$\mathsf{H}^{\mathrm{rot+ell}} = \mathsf{N} - \nu_0 \mathsf{I} + \mathsf{W} + (2\omega_0)^{-1}(\mathsf{V}^{\mathrm{ell+cen}} - \omega_0^2 \mathsf{T}^{\mathrm{ell}}). \tag{14.158}$$

Explicit expressions for the matrix elements of the Coriolis matrix W and the elliptical-plus-centrifugal matrices $\mathsf{V}^{\mathrm{ell+cen}}$ and $\mathsf{T}^{\mathrm{ell}}$ are given in Appendix D. Inspection of these matrices reveals two important facts. First, in the absence of elastic or anelastic lateral heterogeneity, the order m of the complex spherical harmonic Y_{lm} remains a good quantum number; that is, only complex basis singlets $\tilde{\mathbf{s}}_k$ having the same order m are coupled by the Earth's rotation and ellipticity. Second, the coupling is governed by the following angular-degree *selection rules*:

1. The Coriolis force gives rise to spheroidal-toroidal coupling between multiplets that differ by a *single* angular degree, i.e., pairs of the form ${}_n\mathrm{S}_l - {}_{n'}\mathrm{T}_{l\pm1}$ and ${}_n\mathrm{T}_l - {}_{n'}\mathrm{S}_{l\pm1}$.

2. The Earth's ellipticity gives rise to spheroidal-spheroidal and toroidal-toroidal coupling between multiplets that differ in angular degree by *two*, i.e., pairs of the form ${}_n\mathrm{S}_l - {}_{n'}\mathrm{S}_{l\pm2}$ and ${}_n\mathrm{T}_l - {}_{n'}\mathrm{T}_{l\pm2}$.

3. Ellipticity gives rise to coupling between toroidal multiplets of the *same* angular degree, i.e., pairs of the form ${}_n\mathrm{T}_l - {}_{n'}\mathrm{T}_l$. There is no rotational toroidal–toroidal coupling.

4. Rotation and ellipticity both give rise to coupling between spheroidal multiplets of the *same* angular degree, i.e., pairs of the form ${}_n\mathrm{S}_l - {}_{n'}\mathrm{S}_l$.

In general, the coupling between two multiplets k and k' is strong only if their degenerate eigenfrequencies $\nu_k = \omega_k + i\gamma_k$ and $\nu_{k'} = \omega_{k'} + i\gamma_{k'}$ are reasonably close.

As shown in Figure 14.19, the real eigenfrequencies of the fundamental-mode multiplet pairs ${}_0\mathrm{S}_l - {}_0\mathrm{T}_{l+1}$ or ${}_0\mathrm{T}_l - {}_0\mathrm{S}_{l+1}$ are roughly coincident within the *Coriolis coupling band*, situated between two and four mHz. We group the spheroidal and toroidal modes within this frequency band into triplex super-multiplets ${}_0\mathrm{T}_{l-1} - {}_0\mathrm{S}_l - {}_0\mathrm{T}_{l+1}$ and ${}_0\mathrm{S}_{l-1} - {}_0\mathrm{T}_l - {}_0\mathrm{S}_{l+1}$, and diagonalize the $3(2l+1) \times 3(2l+1)$ matrices (14.158) to find the split singlet eigenfre-

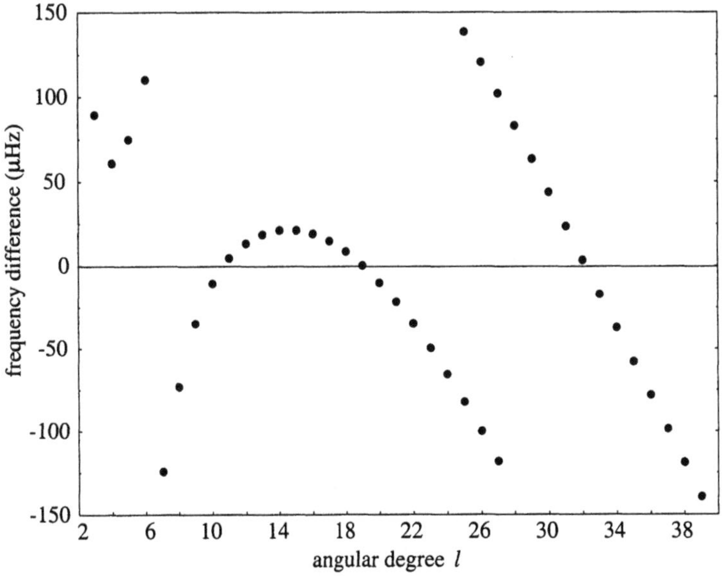

Figure 14.19. Degenerate eigenfrequency differences along the fundamental spheroidal and toroidal mode branches on the PREM model. The quantity plotted is ${}_0\omega_l^{\mathrm{S}} - {}_0\omega_{l+1}^{\mathrm{T}}$ in the range $l = 7$–28 and ${}_0\omega_l^{\mathrm{S}} - {}_0\omega_{l-1}^{\mathrm{T}}$ in the ranges $l = 3-6$ and $l = 25-39$. Coriolis coupling is most pronounced between the quasi-degenerate multiplet pairs ${}_0\mathrm{S}_{11} - {}_0\mathrm{T}_{12}$, ${}_0\mathrm{S}_{19} - {}_0\mathrm{T}_{20}$ and ${}_0\mathrm{T}_{31} - {}_0\mathrm{S}_{32}$.

quency perturbations $\delta\nu_j = \delta\omega_j + i\,\delta\gamma_j$ and associated eigenvectors $\mathbf{z}_j$ and dual eigenvectors $\overline{\mathbf{z}}_j$. The reference frequency ν_0 for each triplet is taken to be the degenerate eigenfrequency of the central or *target* multiplet. The spheroidal-toroidal coupling is particularly strong between the nearly degenerate "crossover" pairs ${}_0\mathrm{S}_{11} - {}_0\mathrm{T}_{12}$, ${}_0\mathrm{S}_{19} - {}_0\mathrm{T}_{20}$ and ${}_0\mathrm{T}_{31} - {}_0\mathrm{S}_{32}$; their singlet eigenfrequencies are intimately intermingled, and the associated eigenfunctions and dual eigenfunctions exhibit roughly equal spheroidal and toroidal characteristics. The other quasi-degenerate pairs within the band are coupled less strongly; their singlet eigenfrequencies are grouped into recognizable *hybrid multiplets* that are either predominantly spheroidal or predominantly toroidal.

In Figure 14.20 we plot the *mean* real eigenfrequency perturbations

$$\langle\delta\omega_0\rangle = \frac{1}{2l+1}\sum_j \delta\omega_j \tag{14.159}$$

of the hybrid target modes ${}_0\mathrm{S}_l$ and ${}_0\mathrm{T}_l$ versus degree l. The net effect of Coriolis coupling is always to shift the centroids $\omega_0 + \langle\delta\omega_0\rangle$ of the quasi-degenerate pairs ${}_0\mathrm{S}_l - {}_0\mathrm{T}_{l+1}$ or ${}_0\mathrm{T}_l - {}_0\mathrm{S}_{l+1}$ away from each other. This *repulsion* of the mean eigenfrequencies is apparent in the spherical-Earth multiplet strips displayed in Figures 10.21 and 10.22. The results in Figure 14.20 enable the measured strip frequencies to be corrected for Coriolis repulsion prior to inverting for the terrestrial monopole (Masters, Park & Gilbert 1983). The strongest repulsion occurs near the "crossovers" $l = 11-12$, $l = 19-20$ and $l = 31-32$, leading to the abrupt offsets or "tears" in the plots of $\langle\delta\omega_0\rangle$ versus degree.

The second-order Coriolis splitting result $\langle\delta\omega_0\rangle = \omega_0(\lambda + \frac{1}{3}k^2\zeta)(\Omega/\omega_0)^2$, obtained in Section 14.2.3, does a tolerable job of predicting the repulsive monopole correction for all of the fundamental-mode coupled multiplets except the quasi-degenerate pairs ${}_0\mathrm{S}_{11} - {}_0\mathrm{T}_{12}$, ${}_0\mathrm{S}_{19} - {}_0\mathrm{T}_{20}$ and ${}_0\mathrm{T}_{31} - {}_0\mathrm{S}_{32}$. The quadratic dependence of the second-order singlet eigenfrequency perturbations, $\delta\omega_m = \omega_0[m\chi(\Omega/\omega_0) + (\lambda + m^2\zeta)(\Omega/\omega_0)^2]$, $-l \le m \le l$, upon the order m of the complex spherical harmonic Y_{lm} is an indication that Coriolis coupling can also mimic the effect of *degree-two zonal* heterogeneity $\delta\kappa_{20}$, $\delta\mu_{20}$, $\delta\rho_{20}$, δd_{20}. The observed splitting coefficients σ_{20} of the fundamental spheroidal and toroidal modes ${}_0\mathrm{S}_l$ and ${}_0\mathrm{T}_l$ must be corrected for this effect, prior to employing them in any three-dimensional inversion. Smith & Masters (1989b) show that this correction significantly improves the along-branch continuity of the measured coefficients, particularly in the vicinity of the "crossover" pairs ${}_0\mathrm{S}_{11} - {}_0\mathrm{T}_{12}$, ${}_0\mathrm{S}_{19} - {}_0\mathrm{T}_{20}$ and ${}_0\mathrm{T}_{31} - {}_0\mathrm{S}_{32}$. The perturbation expansion (14.61) for these modes is divergent, indicating the need for the more general quasi-degenerate theory discussed here.

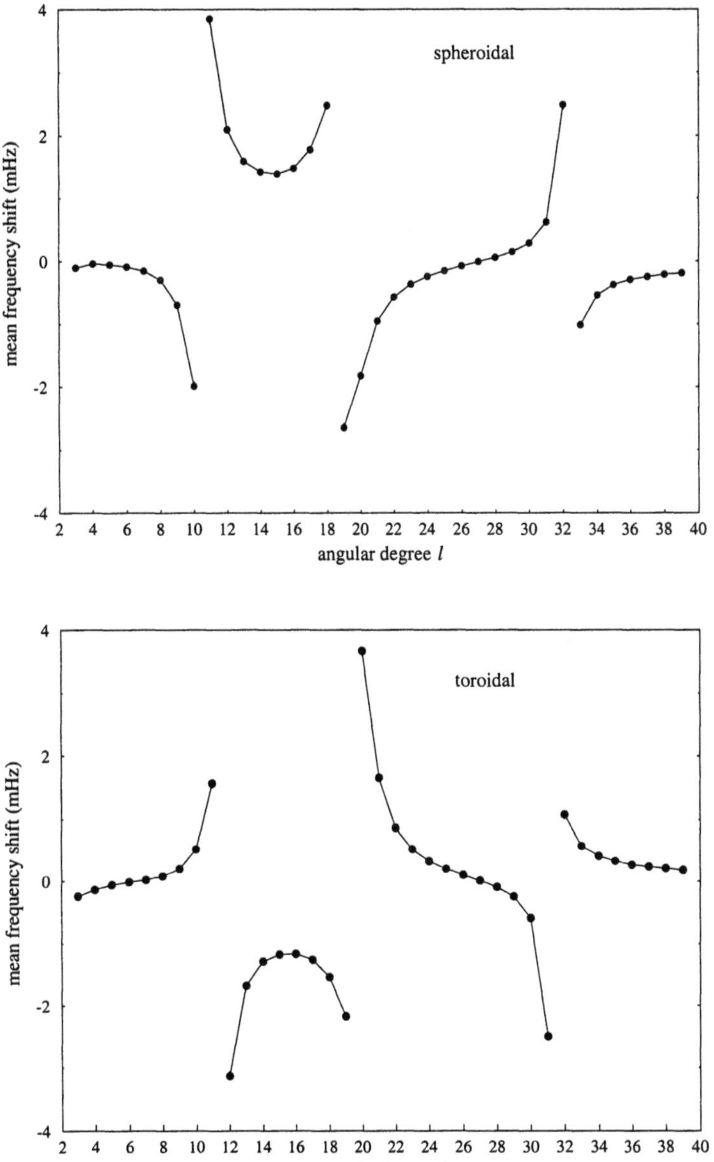

Figure 14.20. Mean real eigenfrequency perturbations $\langle\omega_0\rangle$ of the target hybrid multiplets ${}_0\mathrm{S}_l$ (*top*) and ${}_0\mathrm{T}_l$ (*bottom*).

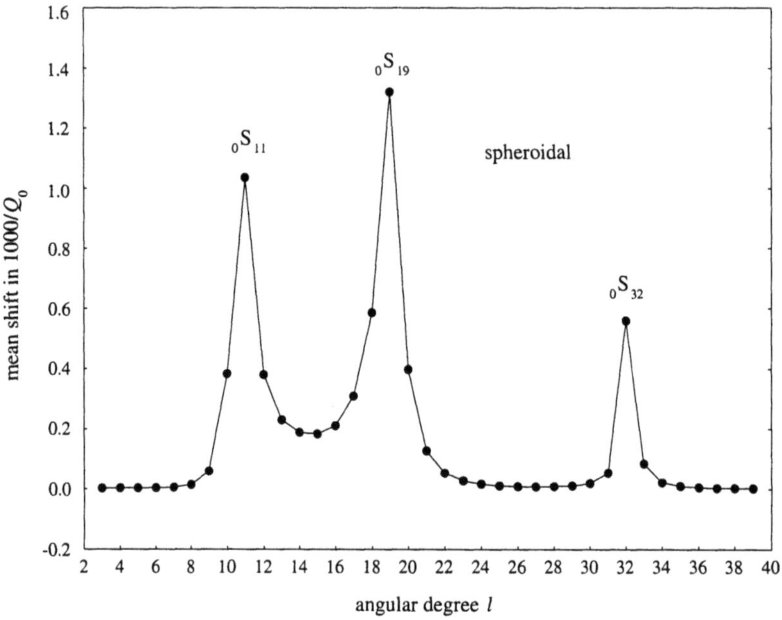

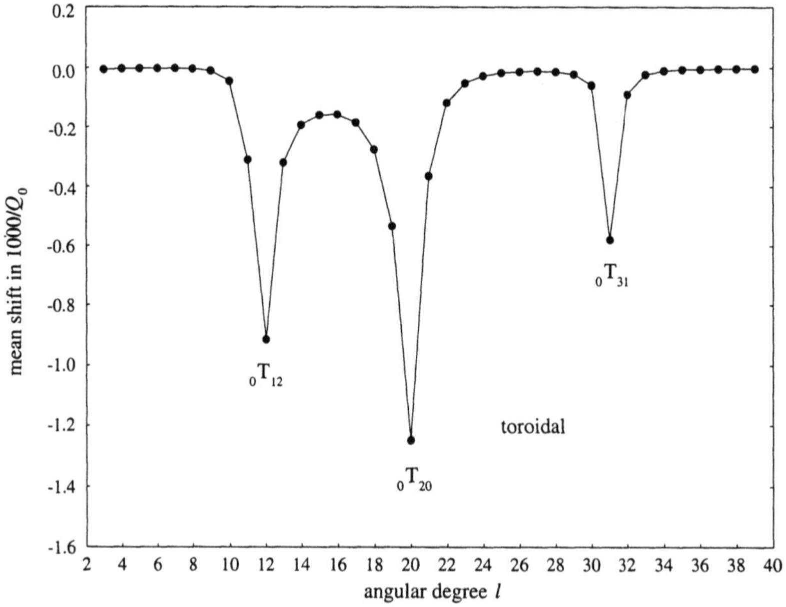

Figure 14.21. Perturbations $1000\,\langle\delta Q_0^{-1}\rangle = 2000\,\omega_0^{-1}\langle\delta\gamma_0\rangle$ in the mean reciprocal quality factors of the target hybrid multiplets $_0S_l$ (*top*) and $_0T_l$ (*bottom*).

Each hybrid spheroidal singlet on a rotating, elliptical Earth experiences a positive decay-rate perturbation, $\delta\gamma_j > 0$, because it has a lower-Q toroidal component, whereas each hybrid toroidal singlet experiences a negative perturbation, $\delta\gamma_j < 0$, because it it has a higher-Q spheroidal component. The *mean* decay rate of each target multiplet is therefore shifted by an amount

$$\langle\delta\gamma_0\rangle = \frac{1}{2l+1}\sum_j \delta\gamma_j. \tag{14.160}$$

This is illustrated in Figure 14.21, where we plot the dimensionless shifts $\langle\delta Q_0^{-1}\rangle = 2\omega_0^{-1}\langle\delta\gamma_0\rangle$ in the reciprocal quality factor Q_0^{-1}. Note that the decay-rate and reciprocal quality-factor centroids are *attracted*, whereas the eigenfrequency centroids are *repelled*.

Coriolis-coupling effects are most pronounced in the case of a source-receiver path great-circle that passes over or nearly over the North or South Pole. A simple physical argument reveals the reason for this phenomenon. A fundamental Love wave propagating along such a polar path has a transverse particle velocity $\partial_t\mathbf{s}$ that is everywhere perpendicular to the Earth's angular rotation $\boldsymbol{\Omega}$; as a result, it acquires a significant radial component $\boldsymbol{\Omega}\times\partial_t\mathbf{s}$ through the action of the Coriolis force. The radial component

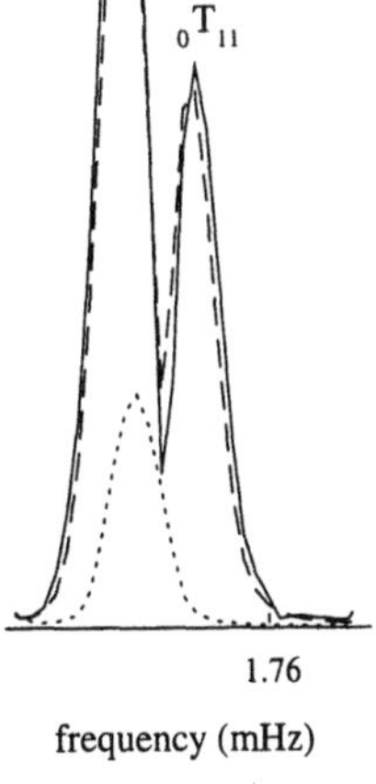

Figure 14.22. Radial-component amplitude spectra $|\hat{\mathbf{r}}\cdot\mathbf{a}(\mathbf{x},\omega)|$ showing the closely spaced hybrid multiplets $_0S_{10}$ and $_0T_{11}$ at station CAN in Canberra, Australia following the October 4, 1994 Kuril Islands earthquake. Solid line shows observed spectrum; dotted line shows synthetic spectrum for the spherical Earth model PREM; dashed line shows synthetic spectrum on the rotating, elliptical version of PREM, with the Coriolis coupling of the modes $_0S_{10}$ and $_0T_{11}$ accounted for.

of a north-south propagating Rayleigh wave likewise acquires a significant transverse component; the net result is that the two waves are strongly coupled by rotation. Figure 14.22 shows a radial-component amplitude spectrum $|a(\omega)|$ recorded in Canberra, Australia, nearly due south of the great 1994 Kuril Islands earthquake. A peak corresponding to the hybrid toroidal multiplet ${}_0\mathrm{T}_{11}$ is clearly visible on the right shoulder of the peak corresponding to the hybrid spheroidal mode ${}_0\mathrm{S}_{10}$. No such peak could occur on a non-rotating, spherical Earth, since the displacement associated with a toroidal mode is purely tangential. Coriolis coupling hybridizes the singlets within the split ${}_0\mathrm{T}_{11}$ multiplet, and endows them with an observable radial component; quasi-degenerate perturbation theory applied to the super-multiplet ${}_0\mathrm{S}_{10}-{}_0\mathrm{T}_{11}-{}_0\mathrm{S}_{12}$ does an excellent job of reproducing the observed spectrum.

14.3.3 Lateral heterogeneity—selection rules

Suppose now that the Earth is laterally heterogeneous, but neither rotating nor elliptical. The $\sum_k(2l_k+1) \times \sum_k(2l_k+1)$ matrix that must then be diagonalized is

$$\mathsf{H}^{\mathrm{lat}} = \mathsf{N} - \nu_0\mathsf{I} + (2\omega_0)^{-1}(\mathsf{V}^{\mathrm{lat}} - \omega_0^2\mathsf{T}^{\mathrm{lat}}). \tag{14.161}$$

In the case that all of the members of the super-multiplet are of the same type—either spheroidal or toroidal—the real elements of (14.161) may be written in a form analogous to (14.76)–(14.77):

$$H_{kk'}^{\mathrm{lat}\,mm'} = (\omega_k - \omega_0) + \omega_0 \sum_{st} \sigma_{st}^{kk'} \int_\Omega \mathcal{Y}_{lm}\mathcal{Y}_{st}\mathcal{Y}_{l'm'}\,d\Omega, \tag{14.162}$$

where

$$\sigma_{st}^{kk'} = \tfrac{1}{2}\omega_0^{-2}\begin{pmatrix} l & s & l' \\ 0 & 0 & 0 \end{pmatrix}^{-1} \left\{ \int_0^a [\delta\kappa_{st}V_\kappa + \delta\mu_{st}V_\mu \right.$$
$$\left. + \,\delta\rho_{st}(V_\rho - \omega_0^2 T_\rho)]\, r^2 dr + \sum_d d^2 \delta d_{st}[V_d - \omega_0^2 T_d]_-^+ \right\}. \tag{14.163}$$

Unlike-type (spheroidal-toroidal) coupling is somewhat more complicated; in that case, the elements of the complex-basis matrix $\tilde{\mathsf{H}}^{\mathrm{lat}}$ are given in terms of the expansion coefficients $\delta\tilde{\kappa}_{st}$, $\delta\tilde{\mu}_{st}$, $\delta\tilde{\rho}_{st}$ and $\delta\tilde{d}_{st}$ by

$$\tilde{H}_{kk'}^{\mathrm{lat}\,mm'} = (\omega_k - \omega_0) + \omega_0 \sum_{st} \tilde{\sigma}_{st}^{kk'} \Gamma_{st}^{kk'}, \tag{14.164}$$

where

$$\tilde{\sigma}_{st}^{kk'} = \tfrac{1}{2}\omega_0^{-2}\begin{pmatrix} l+1 & s+1 & l'+1 \\ 0 & 0 & 0 \end{pmatrix}^{-1} \left\{ \int_0^a [\delta\tilde{\kappa}_{st}V_\kappa + \delta\tilde{\mu}_{st}V_\mu \right.$$

$$\left. + \delta\tilde{\rho}_{st}(V_\rho - \omega_0^2 T_\rho)]\, r^2 dr + \sum_d d^2 \delta\tilde{d}_{st}[V_d - \omega_0^2 T_d]_-^+ \right\}, \qquad (14.165)$$

$$\Gamma_{st}^{kk'} = (-1)^m \left[\frac{(2l+1)(2s+1)(2l'+1)}{4\pi}\right]^{1/2}$$
$$\times \begin{pmatrix} l+1 & s+1 & l'+1 \\ 0 & 0 & 0 \end{pmatrix}\begin{pmatrix} l & s & l' \\ -m & t & m' \end{pmatrix}. \qquad (14.166)$$

The associated real elements $H_{kk'}^{\mathrm{lat}\,mm'}$ may be evaluated using the complex-to-real transformation relations (D.155)–(D.161). The order-independent *Woodhouse kernels* V_κ, V_μ, $V_\rho - \omega_0^2 T_\rho$, $V_d - \omega_0^2 T_d$ in (14.163) and (14.165) are defined in equations (D.46)–(D.53).

Laterally heterogeneous anelasticity may be incorporated by allowing the perturbed incompressibility and rigidity to be complex, as in (14.80):

$$\mathsf{H}^{\mathrm{lat}} = \mathsf{N} - \nu_0 \mathsf{I} + (2\omega_0)^{-1}(\mathsf{V}^{\mathrm{lat}} + i\mathsf{A} - \omega_0^2 \mathsf{T}^{\mathrm{lat}}). \qquad (14.167)$$

The spheroidal-spheroidal and toroidal-toroidal elements of the anelastic potential energy matrix A are given by

$$A_{kk'}^{\mathrm{lat}\,mm'} = \omega_0 \sum_{st} \psi_{st}^{kk'} \int_\Omega \mathcal{Y}_{lm}\mathcal{Y}_{st}\mathcal{Y}_{l'm'}\, d\Omega, \qquad (14.168)$$

where

$$\psi_{st}^{kk'} = \tfrac{1}{2}\omega_0^{-2}\begin{pmatrix} l & s & l' \\ 0 & 0 & 0 \end{pmatrix}^{-1}$$
$$\times \int_0^a (\kappa_0 q_{\kappa st} V_\kappa + \mu_0 q_{\mu st} V_\mu)\, r^2 dr. \qquad (14.169)$$

In the spheroidal-toroidal case, it is again simplest to specify the elements of the associated complex-basis matrix:

$$\tilde{A}_{kk'}^{\mathrm{lat}\,mm'} = \omega_0 \sum_{st} \tilde{\psi}_{st}^{kk'} \Gamma_{st}^{kk'}, \qquad (14.170)$$

where

$$\tilde{\psi}_{st}^{kk'} = \tfrac{1}{2}\omega_0^{-2}\begin{pmatrix} l+1 & s+1 & l'+1 \\ 0 & 0 & 0 \end{pmatrix}^{-1}$$
$$\times \int_0^a (\kappa_0 \tilde{q}_{\kappa st} V_\kappa + \mu_0 \tilde{q}_{\mu st} V_\mu)\, r^2 dr. \qquad (14.171)$$

The real elements $A_{kk'}^{\mathrm{lat},mm'}$ may be found, as before, using (D.155)–(D.161). If $\mathsf{A} = 0$ the splitting matrix (14.167) is real and symmetric, whereas if $\mathsf{A} \neq 0$ it is complex and symmetric.

Multiplet-multiplet coupling due to the Earth's elastic or anelastic lateral heterogeneity is governed by the following *selection rules*:

1. A multiplet ${}_n\mathrm{S}_l$ or ${}_n\mathrm{T}_l$ is coupled to a multiplet ${}_{n'}\mathrm{S}_{l'}$ or ${}_{n'}\mathrm{T}_{l'}$ by a lateral variation of degree s only if $|l - l'| \leq s \leq l + l'$.

2. Two spheroidal multiplets ${}_n\mathrm{S}_l$ and ${}_{n'}\mathrm{S}_{l'}$ are coupled by a lateral variation of degree s only if $l + l' + s$ is *even*.

3. Two toroidal multiplets ${}_n\mathrm{T}_l$ and ${}_{n'}\mathrm{T}_{l'}$ are coupled by a lateral variation of degree s only if $l + l' + s$ is *even*.

4. A spheroidal multiplet ${}_n\mathrm{S}_l$ is coupled to a toroidal multiplet ${}_{n'}\mathrm{T}_{l'}$ by a lateral variation of degree s only if $l + l' + s$ is *odd*.

The first rule, which governs both like-type and unlike-type coupling, is simply a restatement of the 3-j triangle condition (C.186). The next three rules are an elementary consequence of equation (C.219) and the "B-factor" identities (D.58)–(D.61).

14.3.4 Generalized diagonal sum rule

The traces of all but the centrifugal-potential matrix $\mathsf{V}^{\mathrm{cen}}$ satisfy the diagonal sum rule:

$$\mathrm{tr}\,\mathsf{W} = 0, \qquad \mathrm{tr}\,\mathsf{T}^{\mathrm{ell}} = \mathrm{tr}\,\mathsf{T}^{\mathrm{lat}} = 0,$$
$$\mathrm{tr}\,\mathsf{V}^{\mathrm{ell}} = \mathrm{tr}\,\mathsf{V}^{\mathrm{lat}} = \mathrm{tr}\,\mathsf{A} = 0. \tag{14.172}$$

From equation (D.147) we find that

$$\mathrm{tr}\,\mathsf{V}^{\mathrm{cen}} = \tfrac{2}{3}\Omega^2 \sum_k (2l_k + 1)[1 - l_k(l_k + 1)\chi_k], \tag{14.173}$$

where χ_k is the second-order Coriolis splitting parameter (14.35) of multiplet k. The toroidal multiplets ${}_n\mathrm{T}_l$ do not contribute to the trace (14.173) since $1 - l_k(l_k + 1)\chi_k = 0$. Noting that $\mathrm{tr}\,(\mathsf{N} - \nu_0\mathsf{I}) = \sum_k (2l_k + 1)(\nu_k - \nu_0)$, we obtain a super-multiplet generalization of equations (14.93)–(14.94):

$$\sum_j \delta\omega_j = \sum_k (2l_k + 1)\Big\{\omega_k - \omega_0 + \tfrac{2}{3}\Omega^2[1 - l_k(l_k + 1)\chi_k]\Big\}, \tag{14.174}$$

$$\sum_j \delta\gamma_j = \sum_k (2l_k + 1)(\gamma_k - \gamma_0). \tag{14.175}$$

Ignoring the relatively small effect of the spherically averaged centrifugal potential $\bar{\psi} = -\frac{1}{3}\Omega^2 r^2$, the results (14.174)–(14.175) stipulate that the averaged singlet eigenfrequencies and decay rates of the perturbed Earth model are equal to the corresponding averages of the terrestrial monopole:

$$\sum_j (\omega_0 + \delta\omega_j) = \sum_k (2l_k + 1)\omega_k, \tag{14.176}$$

$$\sum_j (\gamma_0 + \delta\gamma_j) = \sum_k (2l_k + 1)\gamma_k. \tag{14.177}$$

The contributions to the averages (14.176)–(14.177) from each multiplet are weighted by the number $2l_k + 1$ of singlets. This *generalized diagonal sum rule* is due to Woodhouse (1980).

14.3.5 Generalized splitting function

If we ignore lateral variations in anelasticity and eigenfunction renormalization, the spectral response (14.157) is given by the analogue of equation (14.115):

$$a(\omega) = \tfrac{1}{2} i \mathsf{r}^{\mathrm{T}} [\mathsf{H} - (\omega - \omega_0 - i\gamma_0)\mathsf{I}]^{-1} \mathsf{s}, \tag{14.178}$$

where $\mathsf{H} = \mathsf{W} + (2\omega_0)^{-1}[\mathsf{V}^{\mathrm{ell+cen}} + \mathsf{V}^{\mathrm{lat}} - \omega_0^2(\mathsf{T}^{\mathrm{ell}} + \mathsf{T}^{\mathrm{lat}})]$. In this approximation, the spectrum of a super-multiplet depends only upon the characteristics of the source $\mathbf{M}$, $\mathbf{x}_\mathrm{s}$ and receiver $\hat{\boldsymbol{\nu}}$, $\mathbf{x}$, the known rotation rate Ω and hydrostatic ellipticity $\varepsilon(r)$ of the Earth, and the unknown coefficients $\sigma_{st}^{kk'}$ and $\tilde{\sigma}_{st}^{kk'}$ defined in (14.163) and (14.165). By analogy with equation (14.117), it is customary to define the *generalized splitting function* $\sigma^{kk'}(\theta, \phi)$ of two coupled multiplets k and k' by

$$\sigma^{kk'} = \sum_{s=|l-l'|}^{l+l'} \sum_{t=-s}^{s} \sigma_{st}^{kk'} \mathcal{Y}_{st} = \sum_{s=|l-l'|}^{l+l'} \sum_{t=-s}^{s} \tilde{\sigma}_{st}^{kk'} Y_{st}. \tag{14.179}$$

This function of geographical position characterizes the coupling between two quasi-degenerate multiplets k, k' in the same manner that the splitting function σ characterizes the self coupling of a single isolated degenerate multiplet; in the case $k = k'$ the generalized and ordinary splitting functions coincide. Resovsky & Ritzwoller (1995; 1998) have developed and implemented a generalized spectral fitting procedure which first solves for the coefficients $\sigma_{st}^{kk'}$ by minimizing the misfit between a suite of observed and synthetic super-multiplet spectra $a_p(\omega)$, $p = 1, 2, \ldots$. The measured coefficients $\sigma_{st}^{kk'}$ of a number of super-multiplets may then be used as linear constraints in a subsequent inversion for the three-dimensional elastic

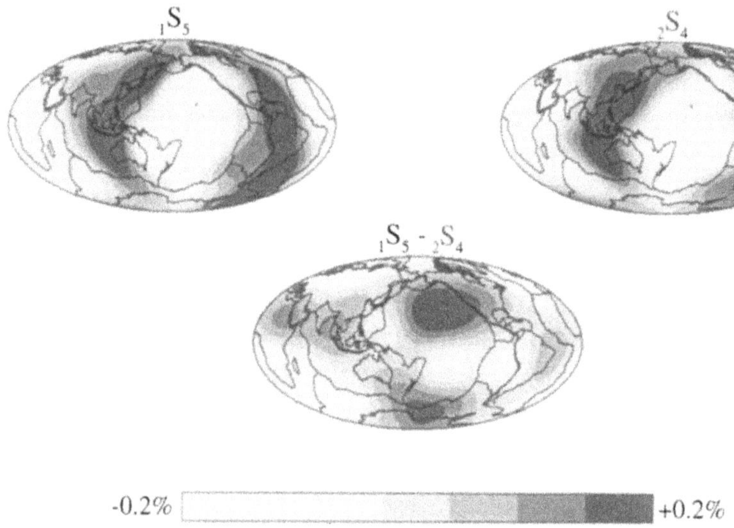

Figure 14.23. Observed generalized splitting functions for the quasi-degenerate multiplet pair ${}_1\mathrm{S}_5 - {}_2\mathrm{S}_4$. (*Top left*) Function σ^{kk} for the mode ${}_1\mathrm{S}_5$. (*Top right*) Function $\sigma^{k'k'}$ for the mode ${}_2\mathrm{S}_4$. (*Bottom*) Function $\sigma^{kk'}, k \neq k'$, governing the interaction between the two modes ${}_1\mathrm{S}_5$ and ${}_2\mathrm{S}_4$. All of the splitting functions have been truncated in the spectral-fitting process, so that σ^{kk}, $\sigma^{k'k'}$ incorporate only the three even degrees $s = 2, 4, 6$ and $\sigma^{kk'}, k \neq k'$, incorporates only the three odd degrees $s = 1, 3, 5$. (Courtesy of J. Resovsky & M. Ritzwoller.)

lateral heterogeneity $\delta\kappa_{st}$, $\delta\mu_{st}$, $\delta\rho_{st}$, δd_{st}. In order to describe the splitting and coupling of a cluster of K multiplets completely, it is necessary to solve for a total of $\frac{1}{2}K(K+1)$ generalized splitting functions. Figure 14.23 shows the three observed functions σ^{kk}, $\sigma^{k'k'}$ and $\sigma^{kk'}, k \neq k'$ for the $K = 2$ super-multiplet ${}_1\mathrm{S}_5 - {}_2\mathrm{S}_4$. The selection rules in this case of two *like-type multiplets* stipulate that both σ_{st}^{kk} and $\sigma_{st}^{k'k'}$ are zero whenever s is *even*, whereas $\sigma_{st}^{kk'}, k \neq k'$ is zero whenever s is *odd*. Spectra of ${}_1\mathrm{S}_5 - {}_2\mathrm{S}_4$ are sensitive to *both* the even-degree and odd-degree heterogeneity in the range $1 \leq s \leq 10$. A total of 164 generalized splitting coefficients must be measured in order to account for this sensitivity fully.

14.3.6 Whole-spectrum fitting

At higher frequencies, the spectrum becomes denser and grouping modes into identifiable super-multiplets becomes more difficult. In addition, the

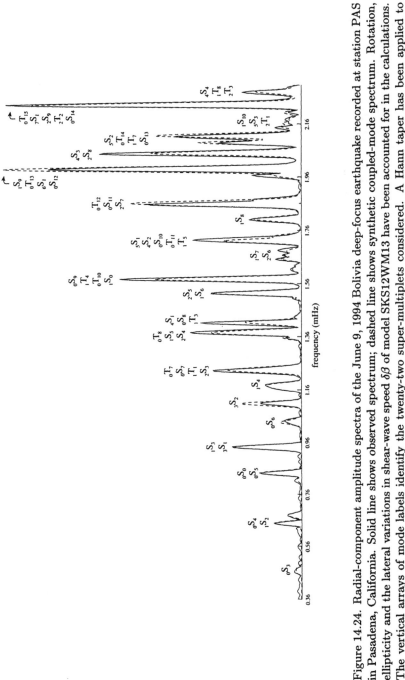

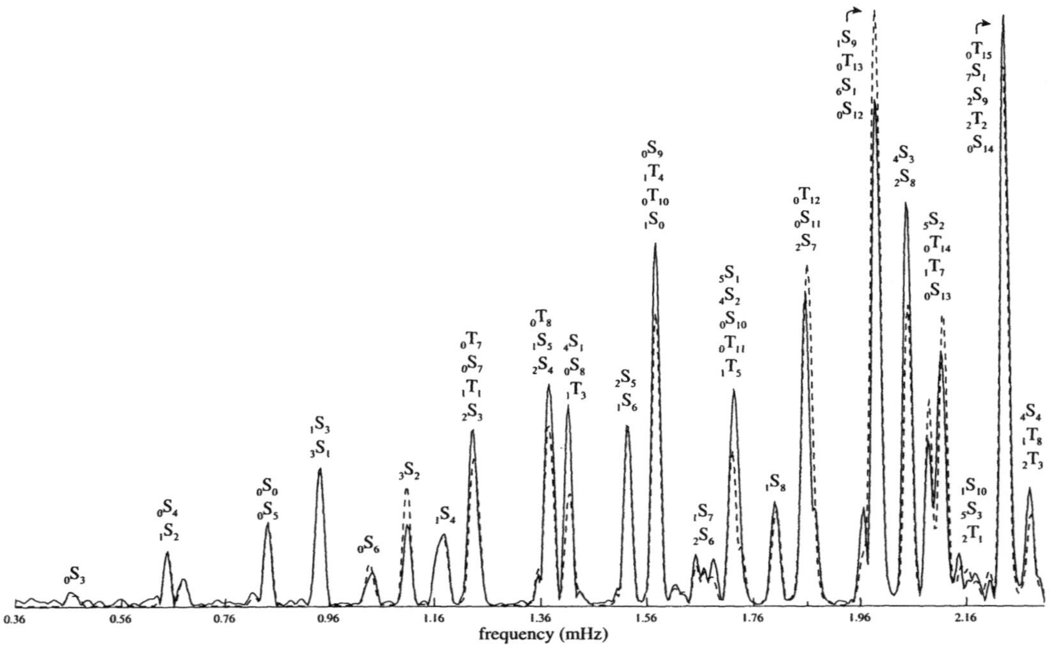

Figure 14.24. Radial-component amplitude spectra of the June 9, 1994 Bolivia deep-focus earthquake recorded at station PAS in Pasadena, California. Solid line shows observed spectrum; dashed line shows synthetic coupled-mode spectrum. Rotation, ellipticity and the lateral variations in shear-wave speed $\delta\beta$ of model SKS12WM13 have been accounted for in the calculations. The vertical arrays of mode labels identify the twenty-two super-multiplets considered. A Hann taper has been applied to both 35-hour time series prior to Fourier transformation.

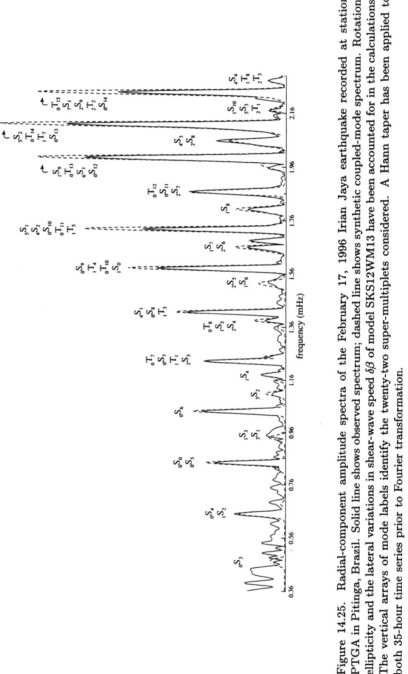

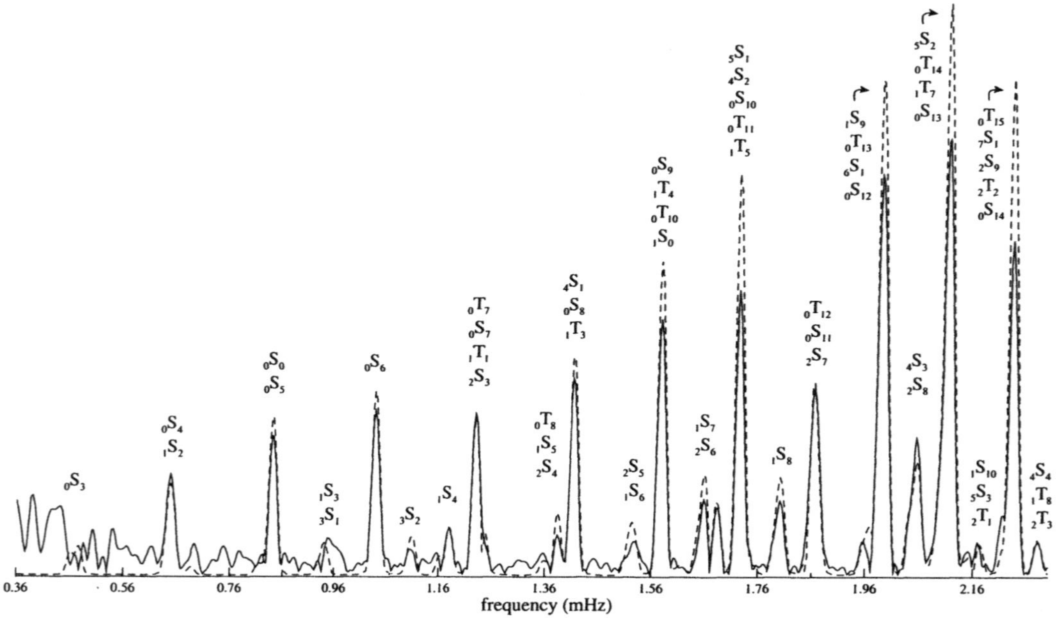

Figure 14.25. Radial-component amplitude spectra of the February 17, 1996 Irian Jaya earthquake recorded at station PTGA in Pitinga, Brazil. Solid line shows observed spectrum; dashed line shows synthetic coupled-mode spectrum. Rotation, ellipticity and the lateral variations in shear-wave speed $\delta\beta$ of model SKS12WM13 have been accounted for in the calculations. The vertical arrays of mode labels identify the twenty-two super-multiplets considered. A Hann taper has been applied to both 35-hour time series prior to Fourier transformation.

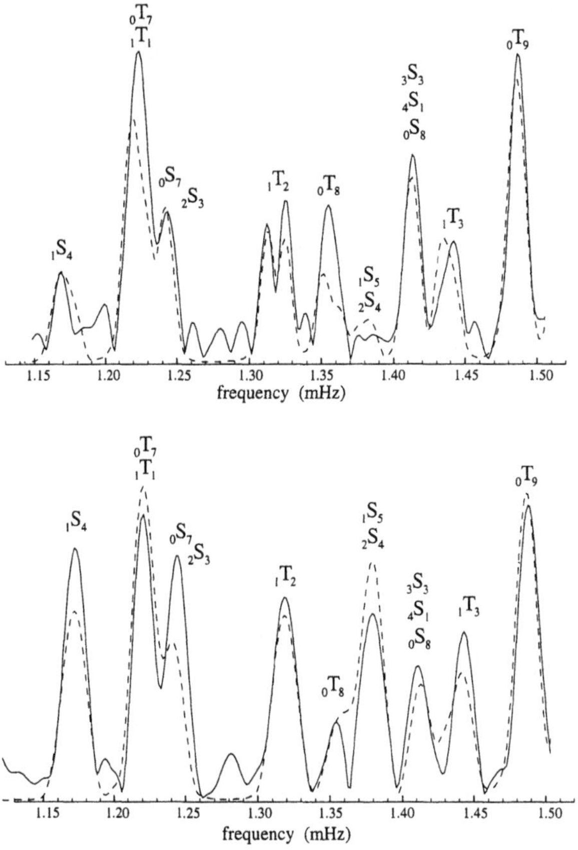

Figure 14.26. Amplitude spectra $|\hat{\mathbf{\Phi}} \cdot \mathbf{a}(\mathbf{x}, \omega)|$ of the June 9, 1994 Bolivia deep-focus earthquake recorded on the transverse components at stations ANMO in Albuquerque, New Mexico (*top*) and PAB in San Pablo, Spain (*bottom*). Solid lines show observed spectra; dashed lines show synthetic coupled-mode spectra. Rotation, ellipticity and the lateral variations in shear-wave speed $\delta\beta$ of model SKS12WM13 have been accounted for in the calculations. The vertical arrays of mode labels identify the super-multiplets considered. A Hann taper has been applied to all four 45-hour time series prior to Fourier transformation.

number of singlets within each "isolated" super-multiplet increases. For these reasons, the determination of generalized splitting functions rapidly becomes impractical. Rather than employing the two-step approach described above, it is preferable to fit whole portions of a suite of spectra

$a_p(\omega)$, $p = 1, 2, \ldots$, in a single-step inversion for the elastic lateral heterogeneity $\delta\kappa_{st}$, $\delta\mu_{st}$, $\delta\rho st$, δd_{st} of the mantle. Such *whole-spectrum fitting* is eminently feasible, given the power of contemporary computers and the high quality of modern digital seismic data. Current shear-wave speed models such as SKS12WM13 (Dziewonski, Liu & Su 1998) often provide quite acceptable initial fits, as we illustrate in Figure 14.24. The solid line shows the radial-component amplitude spectrum recorded in Pasadena, California, following the June 9, 1994 deep-focus earthquake in Bolivia; the dashed line shows the corresponding synthetic spectrum for a rotating, elliptical version of model SKS12WM13, calculated by grouping all of the identified modes into super-multiplets identified by the vertically stacked labels. It is remarkable how well this Earth model—which does not incorporate any normal-mode splitting constraints—fits the observed low-frequency spectrum. Needless to say, no model—not even the latest and greatest from Harvard—is a seismological panacea; the fit of SKS12WM13 to a more typical spectrum, recorded in Pitinga, Brazil following the February 17, 1996 Irian Jaya earthquake, is shown in Figure 14.25. The signal-to-noise ratio is significantly higher in this example, particularly beneath 1–1.5 mHz. A pair of transverse-component spectra recorded in Albuquerque, New Mexico and San Pablo, Spain after the 1994 Bolivia event are shown in Figure 14.26. A number of spheroidal modes such as the triplet ${}_3\mathrm{S}_3 - {}_4\mathrm{S}_1 - {}_0\mathrm{S}_8$ are a prominent feature of these spectra. With a few exceptions such as the doublet ${}_0\mathrm{S}_7 - {}_2\mathrm{S}_3$ at San Pablo, these multiplets are well fit by the SKS12WM13 coupled-mode synthetics. Note, finally, that the toroidal multiplet ${}_1\mathrm{T}_2$ is visibly split at Albuquerque.

★14.3.7 Along-branch coupling

Cross-branch coupling between surface-wave equivalent modes with distinct overtone numbers, $n \neq n'$, is weak or absent upon a smooth, laterally heterogeneous Earth model, by virtue of the triangle selection rule. In fact, two multiplets ${}_n\mathrm{S}_l$ or ${}_n\mathrm{T}_l$, $n \ll l$, and ${}_{n'}\mathrm{S}_{l'}$ or ${}_{n'}\mathrm{T}_{l'}$, $n' \ll l'$, are completely uncoupled by lateral heterogeneity of maximum angular degree $s_{\max} < |l - l'|$. This is a strong constraint, particularly at high frequency, where the surface-wave dispersion branches are well separated. At a period $2\pi/\omega \approx 100$ s, the spacing between the fundamental and first-overtone toroidal modes ${}_0\mathrm{T}_l$, ${}_1\mathrm{T}_{l'}$ is $|l - l'| \approx 20$, whereas that between the fundamental and first-overtone spheroidal modes ${}_0\mathrm{S}_l$, ${}_1\mathrm{S}_{l'}$ is $|l - l'| \approx 30$ (see Figures 8.3 and 8.9). Because of this circumstance, it is common to ignore cross-branch coupling when synthesizing surface-wave seismograms on a non-rotating Earth by means of normal-mode summation. Spheroidal-toroidal coupling is also negligible between surface-wave equivalent multiplets ${}_n\mathrm{S}_l$, $n \ll l$, and ${}_{n'}\mathrm{T}_{l'}$, $n' \ll l'$, as we shall see in Chapter 16. The only structural coupling which must be accounted for on a smooth isotropic Earth model is that along the individual dispersion branches $\cdots {}_n\mathrm{S}_{l-1} - {}_n\mathrm{S}_l - {}_n\mathrm{S}_{l+1} \cdots$ and $\cdots {}_n\mathrm{T}_{l-1} - {}_n\mathrm{T}_l - {}_n\mathrm{T}_{l+1} \cdots$. The shifting-and-winnowing procedure described in Section 13.3.3 can be used to compute a "seamless" single-branch accelerogram; each target multiplet ${}_n\mathrm{S}_l$ or ${}_n\mathrm{T}_l$ is coupled to its nearest $\pm L$ neighbors ${}_n\mathrm{S}_{l\pm1}, {}_n\mathrm{S}_{l\pm2}, \ldots, {}_n\mathrm{S}_{l\pm L}$ or ${}_n\mathrm{T}_{l\pm1}, {}_n\mathrm{T}_{l\pm2}, \ldots, {}_n\mathrm{T}_{l\pm L}$. This is referred to as $\pm L$ coupling. The method

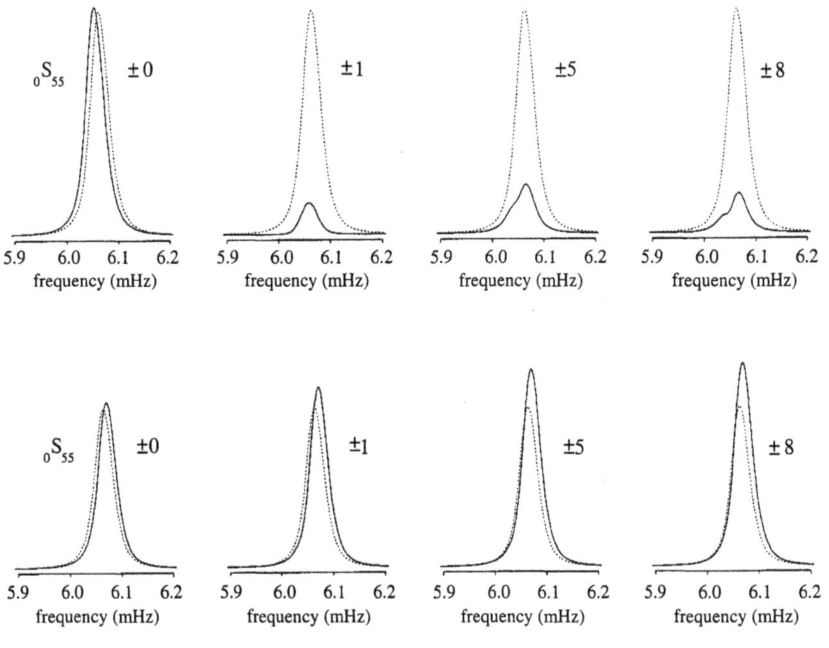

Figure 14.27. Synthetic radial-component acceleration spectra $|\hat{\mathbf{r}} \cdot \mathbf{a}(\mathbf{x}, \omega)|$ for the ${}_0\mathrm{S}_{55}$ multiplet at stations GAR (*top*) and KIP (*bottom*), following the 1978 Oaxaca, Mexico earthquake. The 111 singlet eigenfrequencies and eigenfunctions have been calculated by numerical diagonalization of the real symmetric splitting matrix (14.180). Rotation has been ignored, but the Earth's hydrostatic ellipticity has been accounted for. The coupling along the fundamental spheroidal branch has been truncated at various levels, ranging from ±0 (*left*) to ±8 (*right*). The spectra are calculated using a Hann taper on records of 20 hr duration. Dotted curves show the corresponding spectra on a spherically symmetric Earth for comparison.

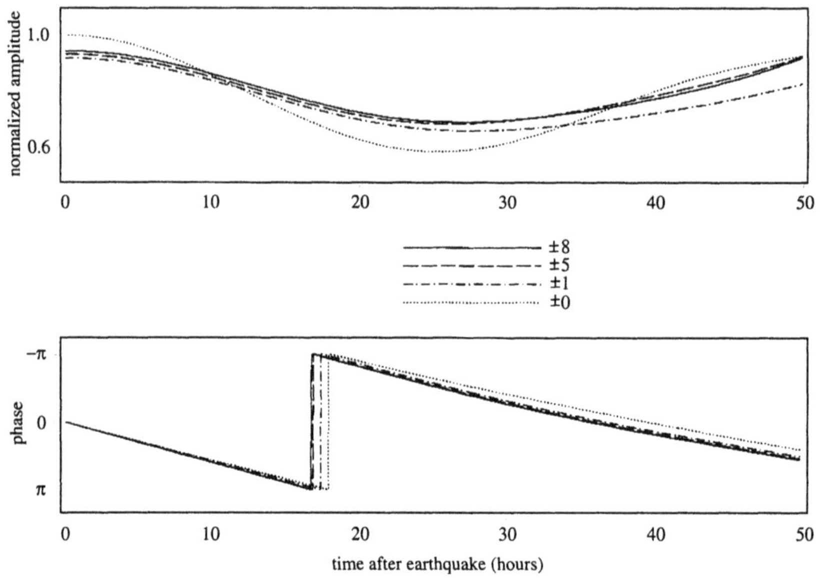

Figure 14.28. Instantaneous amplitude (*top*) and phase (*bottom*) of the normalized multiplet modulation function $A_0(t)/A_0$ of mode ${}_0S_{35}$ at station GAR following the 1978 Oaxaca earthquake. The radial acceleration is given in terms of $A_0(t)$ by $\hat{\mathbf{r}}\cdot\mathbf{a}(\mathbf{x},t) = \mathrm{Re}\,[A_0(t)\exp(i\omega_0 t - \gamma_0 t)]$. All of the modulation functions are calculated using an along-branch coupling scheme, with various levels of truncation: ± 0 (*dotted line*), ± 1 (*dot-dash line*), ± 5 (*dashed line*) and $\pm s_{\mathrm{max}} = \pm 8$ (*solid line*). In the ± 0 or self-coupling approximation, the initial value of the normalized modulation function is $A_0(0)/A_0 = 1$, as shown.

is profligate, inasmuch as $(2L+1)(2l+1)$ eigenvalues and eigenvectors must be computed for each target, and only $2l+1$ of these are retained. On an Earth model with spherically symmetric attenuation, $\mathsf{A} = 0$, it is possible to reduce the computational burden by ignoring the slight variation in the spherical-Earth decay rate along a dispersion branch. Upon making the approximation $\gamma_k = \gamma_0$ in equation (14.161) we obtain a *real* symmetric splitting matrix

$$\mathsf{H}^{\mathrm{lat}} = \Omega - \omega_0 \mathsf{I} + (2\omega_0)^{-1}(\mathsf{V}^{\mathrm{lat}} - \omega_0^2 \mathsf{T}^{\mathrm{lat}}), \tag{14.180}$$

where $\Omega = \mathrm{diag}\,[\cdots\omega_k\cdots]$ is the diagonal matrix of degenerate eigenfrequencies. At high frequencies and on rough Earth models, the split singlets of adjacent along-branch multiplets can overlap, and the ranking and winnowing of the eigenvalues $\delta\omega_j$ fails.

Figure 14.27 shows some representative synthetic acceleration spectra on the $s_{\max} = 8$ Earth model M84A (Woodhouse & Dziewonski 1984). The response of the ${}_0S_{55}$ fundamental spheroidal multiplet ($2\pi/\omega = 165$ s) to the November 29, 1978 Oaxaca, Mexico earthquake has been computed using a self-coupling or ± 0 as well as a ± 1, ± 5 and ± 8 coupling scheme; the dashed curve shows the corresponding spherical-Earth spectrum in each case. At station GAR in Garm, Tajikistan the interference of the 111 singlets gives rise to a somewhat misshapen spectral peak whose overall amplitude is much less than that on PREM; at station KIP in Kipapa, Hawaii the amplitude is slightly greater than that on PREM, and there is a significant shift in the centroid of the "singlet-like" peak toward higher frequency. In both cases, the results are seen to converge as the degree of along-branch coupling is increased. Self-coupling cannot account for the suppressed amplitude at GAR; however, truncation of the coupling at ± 5 yields virtually identical results to $\pm s_{\max} = \pm 8$. In Figures 14.28 and 14.29 we illustrate the ${}_0S_{35}$ and ${}_0S_{55}$ multiplet modulation functions; the quantities plotted are the instantaneous amplitude and phase of the dimensionless ratio $A_0(t)/A_0$, normalized by the real amplitude A_0 on a

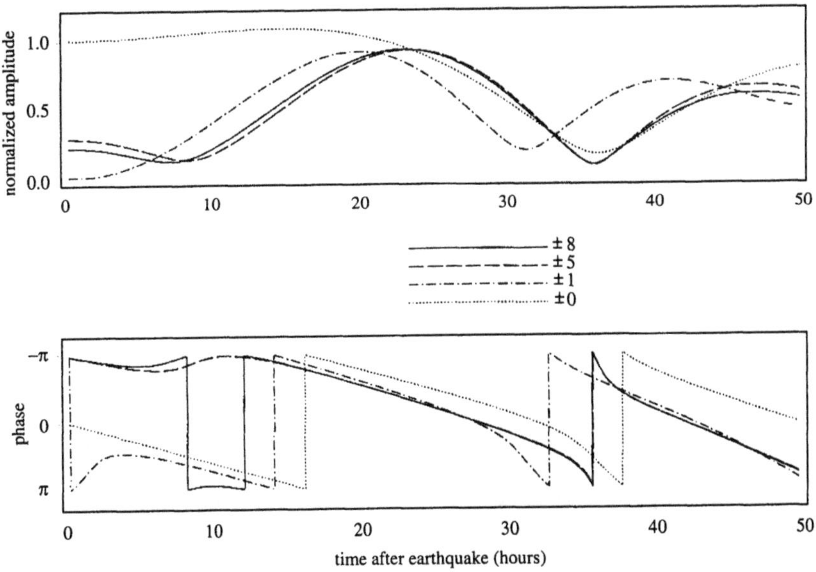

Figure 14.29. Same as Figure 14.28, except for the higher-frequency multiplet ${}_0S_{55}$.

spherical Earth. The time scale of the variations is more rapid for ${}_0S_{55}$ than for ${}_0S_{35}$, and the effect of along-branch truncation is more significant, as expected. As in the frequency domain, ± 5 along-branch coupling is seen to be a very good approximation, on a smooth Earth model such as M84A.

In Figure 14.30 we use the ± 8 coupled-mode results at KIP and GAR as "ground truth" to assess the accuracy of two alternative and much more efficient methods of synthesizing long-period surface-wave accelerograms: the lowest-order quasi-isolated multiplet approximation discussed in Section 13.3.5, and the path-average or *great-circle approximation* introduced by Woodhouse & Dziewonski (1984). We describe the theoretical rationale behind the Woodhouse-Dziewonski approximation in greater detail in Section 16.8.2; for now, we simply note that the multiplet modulation function

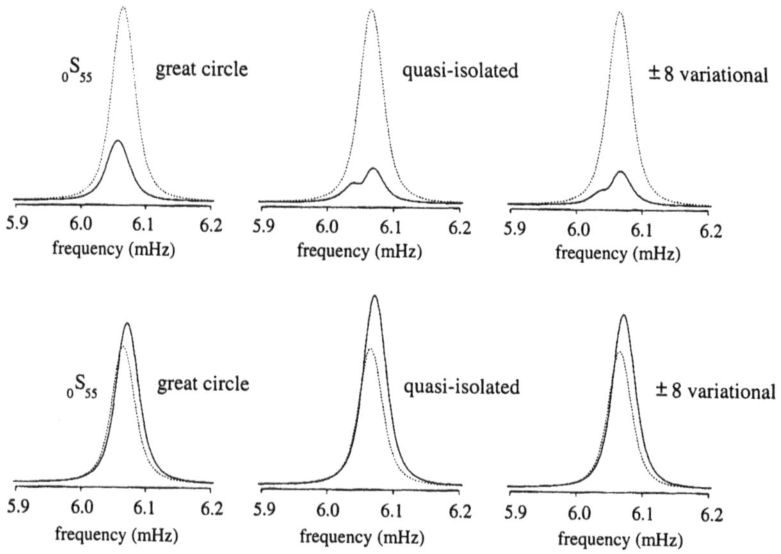

Figure 14.30. Synthetic radial-component acceleration spectra $|\hat{\mathbf{r}}\cdot\mathbf{a}(\mathbf{x},\omega)|$ for the ${}_0S_{55}$ multiplet at stations GAR (*top*) and KIP (*bottom*), following the 1978 Oaxaca earthquake. Spectra calculated using the Woodhouse-Dziewonski great-circle approximation (*left*), the quasi-isolated multiplet approximation (*middle*) and the full variational method truncated at ± 8 (*right*) are compared. The triangle selection rule insures that the lowest-order quasi-isolated multiplet approximation is naturally truncated at ± 8 on an $s_{\max} = 8$ Earth model such as M84A. All spectra are calculated using a Hann taper on records of 20 hr duration. Dashed curves show the corresponding spherical-Earth spectra for comparison.

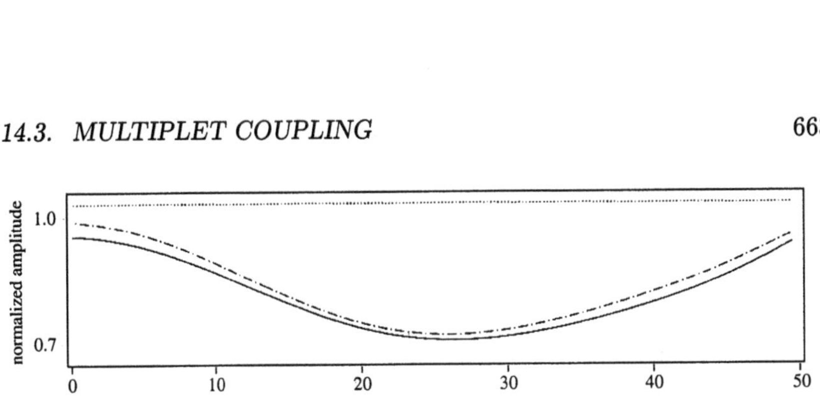

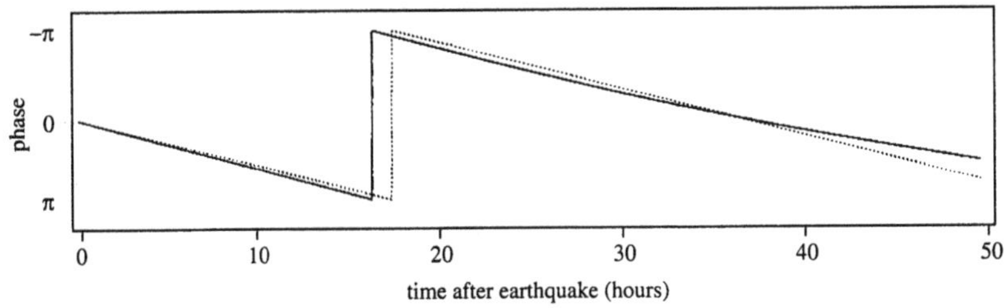

Figure 14.31. Instantaneous amplitude (*top*) and phase (*bottom*) of the normalized multiplet modulation function $A_0(t)/A_0$ of mode ${}_0\mathrm{S}_{35}$ at station GAR following the 1978 Oaxaca earthquake. Temporal variations calculated using the Woodhouse-Dziewonski great-circle approximation (*dotted line*), the quasi-isolated multiplet approximation (*dash-dot line*) and the ± 8 variational method (*solid line*) are compared.

is written as a *single complex exponential*:

$$A_0(t) = (A_0 + \delta A_0)\exp(i\,\delta\bar{\omega}\,t), \tag{14.181}$$

where $\delta\bar{\omega}$ is the great-circular average (14.133) of the local eigenfrequency perturbation (14.128)–(14.129) and δA_0 accounts for a fictitious shift in the location of the source. Both the spectral amplitudes and frequency shifts of the ${}_0\mathrm{S}_{55}$ multiplet are modelled reasonably well, though not perfectly, by the great-circle approximation. The quasi-isolated multiplet approximation is even more accurate, particularly at GAR, where the small "dimple" caused by singlet cancellation on the low-frequency shoulder of the peak is faithfully reproduced. Figures 14.31 and 14.32 compare the normalized multiplet modulation functions $A_0(t)/A_0$ for modes ${}_0\mathrm{S}_{35}$ and ${}_0\mathrm{S}_{55}$. In the case of the Woodhouse-Dziewonski single-exponential representation (14.181) the normalized amplitude $1 + \delta A_0/A_0$ is a constant, and the phase $\delta\bar{\omega}\,t$ varies linearly with time. Although the initial amplitude and

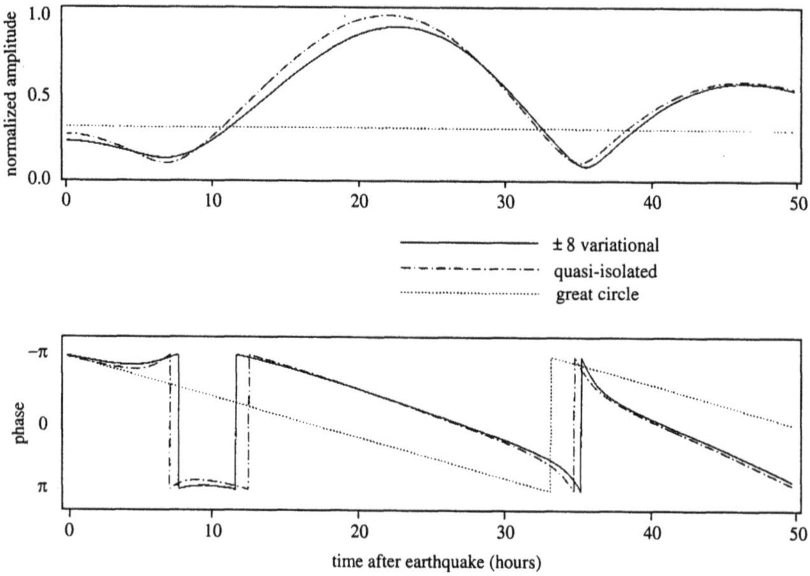

Figure 14.32. Same as Figure 14.31, but for the higher-frequency multiplet ${}_0S_{55}$.

phase of both ${}_0S_{35}$ and ${}_0S_{55}$ are fairly well modelled by this short-time approximation, the ensuing variations are not. The quasi-isolated multiplet approximation, on the other hand, reproduces $A_0(t)/A_0$ for both modes quite accurately over the entire 50 hr interval examined.

⋆14.3.8 A tale of two earthquakes

Prior to the occurrence of the great 1994 Bolivia earthquake, the Colombia earthquake of July 31, 1970 was the largest deep-focus event known. In a celebrated study, Dziewonski & Gilbert (1974) and Gilbert & Dziewonski (1975) used manually digitized recordings from the recently installed World-Wide Standard Seismographic Network to determine the moment tensor $\mathbf{M}(\omega)$ of this $M_0 = 1.8 \times 10^{21}$ N m event. They concluded on the basis of their spherical-Earth analysis that the moment tensor of the Colombia earthquake had a significant low-frequency *isotropic* component, which preceded the high-frequency deviatoric onset by approximately 100 seconds. These findings were interpreted as evidence for a precursory low-density to high-density phase transition within the source region. In general, the isotropic part of a deep-focus moment tensor is much less efficient at exciting the free oscillations of the Earth than the deviatoric part; for this reason, their provocative conclusions regarding the nature of deep-focus seismicity continued to incite controversy more than twenty years later (Kawakatsu 1996). In a recent re-analysis of the original data, Russakoff, Ekström & Tromp (1998) show that the isotropic component of the Colombia 1970 earthquake is an artifact of the Earth's rotation, ellipticity and lateral heterogeneity. Application of the centroid-moment tensor formalism upon a spherically symmetric Earth yields a statistically significant isotropic component; however, that component disappears when the effects of mode splitting and coupling in the band $2-3.5$ mHz are taken into account. The moment tensors on PREM and a rotating, elliptical version of model SKS12WM13 are compared in Figure 14.33. The incorporation of splitting and coupling into the inversion procedure greatly improves the fit to the data and obviates the need for either an isotropic or a large non-double-couple component. Similar or other artifacts may be present in other low-frequency source-mechanism determinations; a fully quantitative interpretation of $\mathbf{M}(\omega)$ requires a coupled-mode approach.

The $M_0 = 8.0 \times 10^{19}$ N m Landers, California earthquake of June 28, 1992 was characterized by right-lateral strike-slip motion on the Johnson Valley, Landers, Homestead Valley, Emerson Valley and Camp Rock strands of the southern San Andreas fault system (Wald & Heaton 1994). The moment tensor describing such a nearly vertical strike-slip source is of the form $M_{rr} \approx M_{r\theta} \approx M_{r\phi} \approx M_{\theta\theta} + M_{\phi\phi} \approx 0$. The long-period response of a non-rotating, spherical Earth should be nearly zero in the vicinity of the epicenter and its antipode, inasmuch as only the $m = 2$ coefficients multiplying $M_{\theta\theta} - M_{\phi\phi}$ and $M_{\theta\phi}$ are non-zero in equations (10.42) and (10.53), and the associated Legendre function of order two satisfies $P_{l2}(\cos\Theta) = 0$ at $\Theta = 0°$ and $\Theta = 180°$. From a travelling-wave perspective, this phenomenon is due to the destructive interference of the equivalent surface waves, which arrive simultaneously from every azimuth after circumnavigating the globe at the same phase speed along every great-circle path. Figure 14.34 shows that the observed radial-component spectra $|\hat{\mathbf{r}} \cdot \mathbf{s}(\mathbf{x}, \omega)|$ at stations MAJO in Matsushiro, Japan ($\Theta = 81°$) and BKS in Berkeley, California ($\Theta = 6°$) agree reasonably well with the corresponding spherical-Earth synthetic spectra; however, the spectrum at station PAS in Pasadena ($\Theta < 2°$) does not. Other nearby stations of the broad-band TERRAscope network operated by Caltech and the US Geological Survey exhibit a similar pattern: the amplitudes of the long-period R2-R3, R4-R5,... Rayleigh wave packets are, after propagation once or twice around the Earth, almost an order of magnitude larger than those calculated on PREM. Tsuboi & Um (1993) and Watada, Kanamori & Anderson (1993) have shown that this anomalous near-field amplification is a curious consequence of the lateral heterogeneity of the

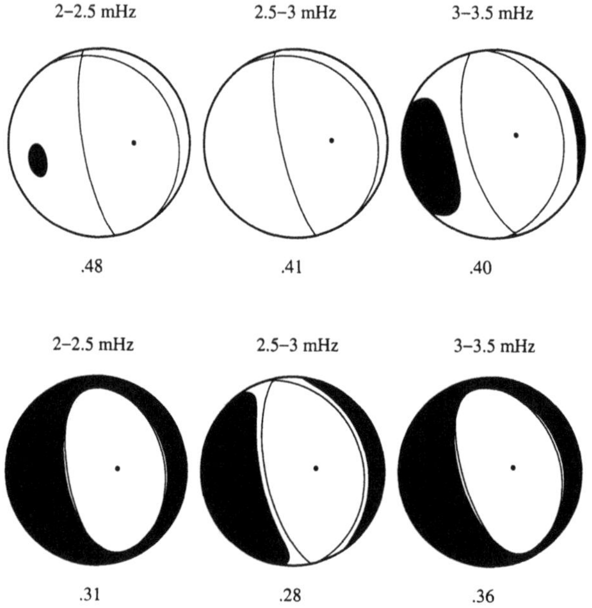

Figure 14.33. Retrieved focal mechanisms for the great 1970 Colombia earthquake in the three frequency ranges 2–2.5 mHz, 2.5–3mHz and 3–3.5mHz. (*Top row*) Results obtained upon the spherical Earth model PREM. (*Bottom row*) Results obtained upon the rotating, elliptical, laterally heterogeneous Earth model SKS12WM13 (Dziewonski, Liu & Su 1998). A total of nineteen super-multiplets, containing between two and six multiplets each, were incorporated in the coupled-mode analysis. Shading denotes compressional quadrants of the focal sphere; the P axes (*dots*) and the nodal planes of the best-fitting double couple are also shown. The residual variance is indicated beneath each focal mechanism.

Earth—because of the associated phase-speed perturbations, $c \to c + \delta c$, the surface waves travel at geographically variable rates along slightly perturbed great-circle paths, so that they no longer interfere destructively upon arriving back in the vicinity of the source. A complete quantitative theory of this phenomenon needs to account for spheroidal-toroidal Coriolis coupling as well as along-branch coupling due to the Earth's lateral heterogeneity. Figure 14.35 compares the observed Pasadena spectrum with synthetic coupled-mode spectra on a rotating, elliptical Earth and a rotating, elliptical and laterally heterogeneous Earth, respectively. Five adjacent fundamental spheroidal multiplets and five adjacent fundamental toroidal multiplets have been used as the basis set, to find the singlet eigenfrequen-

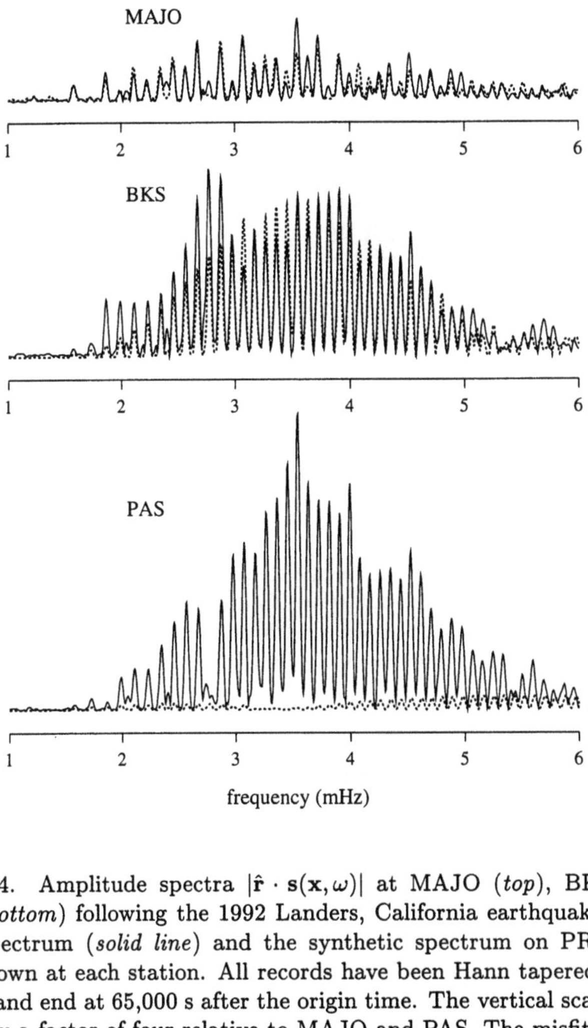

Figure 14.34. Amplitude spectra $|\hat{\mathbf{r}} \cdot \mathbf{s}(\mathbf{x}, \omega)|$ at MAJO (*top*), BKS (*middle*) and PAS (*bottom*) following the 1992 Landers, California earthquake. Both the observed spectrum (*solid line*) and the synthetic spectrum on PREM (*dotted line*) are shown at each station. All records have been Hann tapered, and begin at 15,000 s and end at 65,000 s after the origin time. The vertical scale of BKS is magnified by a factor of four relative to MAJO and PAS. The misfit of the BKS synthetic near 1.9 mHz and 2.8 mHz is due to the strong Coriolis coupling of the crossover modes ${}_0\mathrm{S}_{11}$ and ${}_0\mathrm{S}_{19}$. (Courtesy of S. Watada.)

cies and eigenfunctions associated with every Coriolis-coupled target mode pair ${}_0\mathrm{S}_l - {}_0\mathrm{T}_{l\pm1}$, $0 \leq l \leq 30$, below 4 mHz. Rotation and ellipticity alone are seen to have a significant effect upon the near-field response; however, they are unable to account for the observed anomalous amplification. The Rayleigh-wave refraction and attendant "non-cancellation" caused by the $s_{\max} = 8$ model SH8U4L8 of the Earth's lateral heterogeneity are, on the

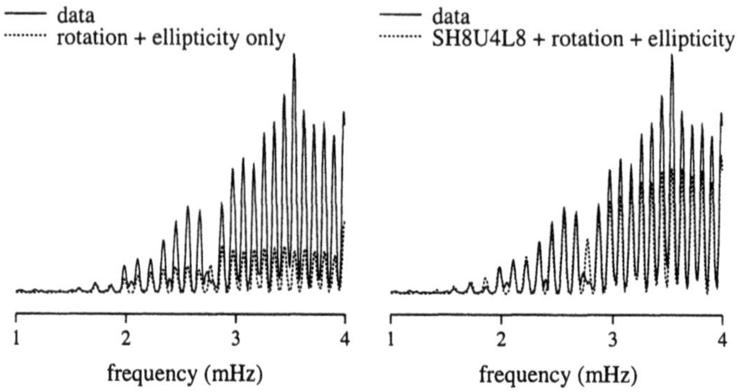

Figure 14.35. Observed (*solid line*) and synthetic (*dotted line*) spectra at the near-field station PAS following the 1992 Landers earthquake. (*Left*) Effect of the Earth's rotation and hydrostatic ellipticity. (*Right*) Combined effect of rotation, ellipticity and lateral heterogeneity model SH8U4L8 (Dziewonski & Woodward 1992). Comparison with Figure 14.34 shows that Coriolis coupling gives rise to a slight amplification of the response in the frequency range 2–4 mHz. Most of the additional effect of model SH8U4L8 is due to the self-splitting of the fundamental spheroidal modes; coupling along the ${}_0S_l$ dispersion branch is fairly weak on such a smooth Earth model. (Courtesy of S. Watada.)

other hand, able to explain almost the entire observed effect. More comprehensive studies by Watada, Kanamori & Anderson (1993) and Watada (1995) indicate that the theoretical amplification is extremely sensitive to the details of the large-scale three-dimensional structure of the crust and upper mantle; relatively slight adjustments to the shear-wave speed $\delta\beta$ are sufficient to bring the synthetic results into full agreement with the data. This strong sensitivity to the lateral heterogeneity is not unexpected, given the nature of the amplification mechanism. In closing, we note that the asymptotic, single great-circle approximation (14.134) or (14.139) is invalid in the vicinity of an earthquake epicenter or its antipode. A more general, uniformly valid asymptotic approximation, which accounts for the nearly simultaneous arrival of surface waves from all azimuths, has been developed by Dahlen (1980a).

Chapter 15

Body-Wave Ray Theory

Thus far we have regarded the departures of the Earth away from spherical symmetry as *slight*. In the remainder of this book, we shall consider a different class of approximations, which are applicable to an Earth model with arbitrarily large lateral variations, provided they are sufficiently *smooth*. By definition, a smooth variation is one satisfying $\lambda/\Lambda \ll 1$, where $\lambda = 2\pi k^{-1}$ is the wavelength of the wave of interest, and Λ is the radial or lateral distance over which the properties of the Earth change significantly. The approximation of interest in this situation goes by a variety of names, including *JWKB theory* and—more colloquially—*ray theory*. We devote this chapter to an analysis of high-frequency body-wave propagation in a non-rotating, elastic, isotropic Earth model. All of the results described here are well known; more detailed summaries are provided by Červený, Molotkov and Pšenčík (1977), Červený (1985) and Kravtsov & Orlov (1990). What we offer as a complement to these comprehensive accounts is a systematic treatment based upon the *slow variational principle* of Whitham (1965; 1974), Bretherton (1968) and Hayes (1973). An analogous variational analysis of surface-wave propagation upon a slowly varying, laterally heterogeneous Earth will be undertaken in Chapter 16.

15.1 Preliminaries

We continue to consider a general Earth model $\oplus = \oplus_S \cup \oplus_F$ composed of concentric solid and fluid regions separated by non-intersecting interfaces $\Sigma = \partial\oplus \cup \Sigma_{SS} \cup \Sigma_{FS}$ with unit outward normal $\hat{\mathbf{n}}$, as depicted in Figure 3.1. Long-range gravitational forces can be ignored in the limit $\lambda/\Lambda \ll 1$, as noted in Section 4.3.5; short-wavelength body-wave propagation is governed by the non-gravitating elastodynamic equation (4.174):

$$-\omega^2\rho\mathbf{s} = \boldsymbol{\nabla}\cdot\mathbf{T}, \tag{15.1}$$

where ρ is the density. The incremental stress is given by the isotropic version of Hooke's law:

$$\mathbf{T} = \kappa(\boldsymbol{\nabla}\cdot\mathbf{s})\mathbf{I} + 2\mu\mathbf{d}, \tag{15.2}$$

where κ and μ are the isentropic incompressibility and the rigidity, respectively, and $\mathbf{d} = \frac{1}{2}[\boldsymbol{\nabla}\mathbf{s} + (\boldsymbol{\nabla}\mathbf{s})^{\mathrm{T}}] - \frac{1}{3}(\boldsymbol{\nabla}\cdot\mathbf{s})\mathbf{I}$ is the strain deviator. The two equations (15.1) and (15.2) must be solved subject to the kinematic continuity conditions $[\mathbf{s}]_-^+ = \mathbf{0}$ on Σ_{SS}, $[\hat{\mathbf{n}}\cdot\mathbf{s}]_-^+ = 0$ on Σ_{FS} and the dynamic boundary conditions (4.150)–(4.152):

$$\hat{\mathbf{n}}\cdot\mathbf{T} = \mathbf{0} \quad \text{on } \partial\oplus, \tag{15.3}$$

$$[\hat{\mathbf{n}}\cdot\mathbf{T}]_-^+ = \mathbf{0} \quad \text{on } \Sigma_{\mathrm{SS}}, \tag{15.4}$$

$$[\hat{\mathbf{n}}\cdot\mathbf{T}]_-^+ = \hat{\mathbf{n}}[\hat{\mathbf{n}}\cdot\mathbf{T}\cdot\hat{\mathbf{n}}]_-^+ = \mathbf{0} \quad \text{on } \Sigma_{\mathrm{FS}}. \tag{15.5}$$

Our convention is, as always, that a negative exponential factor $\exp(-i\omega t)$ appears in the Fourier integral in transforming from time to frequency. Since we are now concerned with the propagation of transient waveforms, it is essential that we regard the Fourier transform of the displacement $\mathbf{s}(\mathbf{x},\omega)$ as *complex*, even upon a non-rotating Earth. Throughout this and the next chapter, we consider the angular frequency to be real and positive, $\omega > 0$; the corresponding results for negative frequencies can readily be obtained using the relation $q(\mathbf{x}, -\omega) = q^*(\mathbf{x},\omega)$ for any real scalar, vector or tensor time-domain field $q(\mathbf{x}, t)$.

The equation of motion (15.1) and boundary conditions (15.3)–(15.5) can be obtained from the variational principle $\delta\mathcal{I} = 0$, where

$$\mathcal{I} = \int_\oplus L(\mathbf{s}, \boldsymbol{\nabla}\mathbf{s}\,;\, \mathbf{s}^*, \boldsymbol{\nabla}\mathbf{s}^*)\, dV. \tag{15.6}$$

The Lagrangian density in equation (15.6) is given by

$$L = \tfrac{1}{2}[\omega^2\rho\mathbf{s}^*\cdot\mathbf{s} - \kappa(\boldsymbol{\nabla}\cdot\mathbf{s}^*)(\boldsymbol{\nabla}\cdot\mathbf{s}) - 2\mu\mathbf{d}^*\!:\!\mathbf{d}], \tag{15.7}$$

where we have introduced a complex conjugate into the isotropic versions of (4.161) and (4.175) in accordance with the above injunction. The variation of the frequency-domain action (15.6) can be written with the aid of Gauss' theorem in the form

$$\begin{aligned}\delta\mathcal{I} = {} & 2\,\mathrm{Re}\int_\oplus \delta\mathbf{s}^*\cdot[\partial_{\mathbf{s}^*}L - \boldsymbol{\nabla}\cdot(\partial_{\boldsymbol{\nabla}\mathbf{s}^*}L)]\, dV \\ & - 2\,\mathrm{Re}\int_\Sigma [\delta\mathbf{s}^*\cdot(\hat{\mathbf{n}}\cdot\partial_{\boldsymbol{\nabla}\mathbf{s}^*}L)]_-^+\, d\Sigma.\end{aligned} \tag{15.8}$$

This vanishes for arbitrary admissible variations $[\boldsymbol{\delta}\mathbf{s}]_-^+ = \mathbf{0}$ on $\Sigma_{\rm SS}$ and $[\hat{\mathbf{n}}\cdot\boldsymbol{\delta}\mathbf{s}]_-^+ = 0$ on $\Sigma_{\rm FS}$ if and only if s satisfies the Euler-Lagrange equation and associated boundary conditions

$$\partial_{\mathbf{s}^*}L - \boldsymbol{\nabla}\cdot(\partial_{\boldsymbol{\nabla}\mathbf{s}^*}L) = \mathbf{0} \quad \text{in } \oplus, \tag{15.9}$$

$$\hat{\mathbf{n}}\cdot(\partial_{\boldsymbol{\nabla}\mathbf{s}^*}L) = \mathbf{0} \quad \text{on } \partial\oplus, \tag{15.10}$$

$$[\hat{\mathbf{n}}\cdot(\partial_{\boldsymbol{\nabla}\mathbf{s}^*}L)]_-^+ = \mathbf{0} \quad \text{on } \Sigma_{\rm SS}, \tag{15.11}$$

$$[\hat{\mathbf{n}}\cdot(\partial_{\boldsymbol{\nabla}\mathbf{s}^*}L)]_-^+ = \hat{\mathbf{n}}[\hat{\mathbf{n}}\cdot(\partial_{\boldsymbol{\nabla}\mathbf{s}^*}L)\cdot\hat{\mathbf{n}}]_-^+ = \mathbf{0} \quad \text{on } \Sigma_{\rm FS}. \tag{15.12}$$

The partial derivative of the Lagrangian density with respect to the conjugate displacement gradient is $\partial_{\boldsymbol{\nabla}\mathbf{s}^*}L = -\mathbf{T}$, so that equations (15.9) and (15.10)–(15.12) are equivalent to (15.1) and (15.3)–(15.5). The value of the action at the stationary transient solution is $\mathcal{I} = 0$.

The kinetic-plus-potential energy density $E = \omega\partial_\omega L - L$ associated with the Lagrangian density (15.7) is

$$E = \tfrac{1}{2}[\omega^2\rho\mathbf{s}^*\cdot\mathbf{s} + \kappa(\boldsymbol{\nabla}\cdot\mathbf{s}^*)(\boldsymbol{\nabla}\cdot\mathbf{s}) + 2\mu\mathbf{d}^*:\mathbf{d}]. \tag{15.13}$$

The total energy density is twice the kinetic energy density: $E = \omega^2\rho\mathbf{s}^*\cdot\mathbf{s}$.

The final ingredient we need is a frequency-domain expression for the energy flux vector $\mathbf{K}$. To determine this, we note that the total energy radiated by a transient source in a smooth unbounded medium is

$$\int_0^\infty\int_\Sigma \hat{\mathbf{n}}\cdot(-\partial_t\mathbf{s}\cdot\mathbf{T})\,d\Sigma\,dt = \frac{1}{\pi}\,\mathrm{Re}\int_0^\infty\int_\Sigma \hat{\mathbf{n}}\cdot(i\omega\mathbf{s}^*\cdot\mathbf{T})\,d\Sigma\,d\omega, \tag{15.14}$$

where Σ is (temporarily) any surface completely surrounding the source, and we have used Parseval's identity to obtain the second representation. Generalizing this result, we shall consider

$$\mathbf{K} = \mathrm{Re}\,(i\omega\mathbf{s}^*\cdot\mathbf{T}) = \mathrm{Re}\,\{i\omega[\kappa\mathbf{s}^*(\boldsymbol{\nabla}\cdot\mathbf{s}) + 2\mu(\mathbf{s}^*\cdot\mathbf{d})]\} \tag{15.15}$$

to be the frequency-domain energy flux everywhere within a finite, piecewise continuous Earth model.

15.2 Whitham's Variational Principle

We seek asymptotic JWKB solutions to equations (15.1) and (15.3)–(15.5) of the form

$$\mathbf{s} = \mathbf{A}\exp(-i\omega T). \tag{15.16}$$

The quantities $\mathbf{A}$ and T may be interpreted as the real *amplitude* and *travel time* of the high-frequency waves, respectively. The associated *slowness vector* is defined in terms of the travel time by

$$\mathbf{p} = \boldsymbol{\nabla} T. \tag{15.17}$$

Upon inserting the representation (15.16) into equations (15.7), (15.13) and (15.15) we obtain the JWKB forms of the Lagrangian and energy densities and the energy flux vector:

$$\mathcal{L} = \tfrac{1}{2}\omega^2[(\rho - \mu\|\boldsymbol{\nabla} T\|^2)\|\mathbf{A}\|^2 - (\kappa + \tfrac{1}{3}\mu)(\boldsymbol{\nabla} T \cdot \mathbf{A})^2], \tag{15.18}$$

$$\mathcal{E} = \tfrac{1}{2}\omega^2[(\rho + \mu\|\boldsymbol{\nabla} T\|^2)\|\mathbf{A}\|^2 + (\kappa + \tfrac{1}{3}\mu)(\boldsymbol{\nabla} T \cdot \mathbf{A})^2], \tag{15.19}$$

$$\boldsymbol{\mathcal{K}} = \omega^2[\mu\|\mathbf{A}\|^2\boldsymbol{\nabla} T + (\kappa + \tfrac{1}{3}\mu)(\boldsymbol{\nabla} T \cdot \mathbf{A})\mathbf{A}], \tag{15.20}$$

where we have retained only the lowest-order terms in ω^{-1}. The three quantities (15.18)–(15.20) are said to be *slowly varying*, since the rapid (wavelength-scale) variations associated with the exponential $\exp(-i\omega T)$ have been eliminated. Each of $\mathcal{L}$, $\mathcal{E}$ and $\boldsymbol{\mathcal{K}}$ varies within the smooth sub-regions of the Earth $\oplus$ only on the slow scale Λ of the structure κ, μ, ρ. On the boundaries Σ between sub-regions, the parameters κ, μ, ρ and, therefore, the densities $\mathcal{L}$, $\mathcal{E}$ and the flux $\boldsymbol{\mathcal{K}}$ exhibit jump discontinuities.

In the JWKB approximation, the propagation is governed by the slowly varying action

$$\mathcal{I} = \int_{\oplus} \mathcal{L}(\mathbf{A}, \boldsymbol{\nabla} T)\, dV. \tag{15.21}$$

As indicated, the unknowns are now the amplitude $\mathbf{A}$ and travel time T. The variation of (15.21) with respect to these slowly varying fields is

$$\delta\mathcal{I} = \int_{\oplus} [\boldsymbol{\delta}\mathbf{A} \cdot \partial_{\mathbf{A}}\mathcal{L} - \delta T\, \boldsymbol{\nabla} \cdot (\partial_{\boldsymbol{\nabla} T}\mathcal{L})]\, dV$$
$$- \int_{\Sigma} [\delta T\, \hat{\mathbf{n}} \cdot (\partial_{\boldsymbol{\nabla} T}\mathcal{L})]_-^+\, d\Sigma. \tag{15.22}$$

We require that $\delta\mathcal{I}$ vanish for all kinematically *admissible* variations $\boldsymbol{\delta}\mathbf{A}$ and δT satisfying $[\boldsymbol{\delta}\mathbf{A}]_-^+ = \mathbf{0}$ on Σ_{SS}, $[\hat{\mathbf{n}} \cdot \boldsymbol{\delta}\mathbf{A}]_-^+ = 0$ on Σ_{FS} and $[\delta T]_-^+ = 0$ on all of Σ. This will be so if and only if $\mathbf{A}$ and T satisfy

$$\partial_{\mathbf{A}}\mathcal{L} = \mathbf{0} \quad \text{and} \quad \boldsymbol{\nabla} \cdot (\partial_{\boldsymbol{\nabla} T}\mathcal{L}) = 0 \quad \text{in } \oplus, \tag{15.23}$$

$$[\hat{\mathbf{n}} \cdot (\partial_{\boldsymbol{\nabla} T}\mathcal{L})]_-^+ = 0 \quad \text{on } \Sigma. \tag{15.24}$$

Written out explicitly, the slow Euler-Lagrange equations (15.23) are

$$[(\rho - \|\mathbf{p}\|^2\mu)\mathbf{I} - (\kappa + \tfrac{1}{3}\mu)\mathbf{pp}] \cdot \mathbf{A} = \mathbf{0}, \tag{15.25}$$

$$\boldsymbol{\nabla} \cdot [\mu\|\mathbf{A}\|^2\mathbf{p} + (\kappa + \tfrac{1}{3}\mu)(\mathbf{p}\cdot\mathbf{A})\mathbf{A}] = 0. \tag{15.26}$$

The boundary condition (15.24) is likewise

$$[\mu\|\mathbf{A}\|^2(\hat{\mathbf{n}}\cdot\mathbf{p}) + (\kappa + \tfrac{1}{3}\mu)(\mathbf{p}\cdot\mathbf{A})(\hat{\mathbf{n}}\cdot\mathbf{A})]_-^+ = 0. \tag{15.27}$$

Equation (15.25) exhibits non-zero solutions $\mathbf{A}$ if and only if

$$\begin{aligned}\det\,&[(\rho - \|\mathbf{p}\|^2\mu)\mathbf{I} - (\kappa + \tfrac{1}{3}\mu)\mathbf{pp}]\\ &= [\rho - \|\mathbf{p}\|^2(\kappa + \tfrac{4}{3}\mu)]\,[\rho - \|\mathbf{p}\|^2\mu]^2 = 0.\end{aligned} \tag{15.28}$$

The latter relation shows that there are two slowness roots, which we distinguish by subscripts:

$$\|\mathbf{p}_\mathrm{P}\|^2 = \alpha^{-2}, \qquad \|\mathbf{p}_\mathrm{S}\|^2 = \beta^{-2}, \tag{15.29}$$

where $\alpha = [(\kappa + \frac{4}{3}\mu)/\rho]^{1/2}$ and $\beta = (\mu/\rho)^{1/2}$ are the compressional-wave and shear-wave speeds, respectively. The decomposition of (15.28) into separate P-wave and S-wave *eikonal equations* (15.29) indicates that these two wave types propagate independently through every smooth sub-region of the Earth $\oplus$. Upon inserting (15.29) into (15.25) we deduce that the compressional-wave and shear-wave *polarizations* are parallel and perpendicular, respectively, to the direction of propagation:

$$\mathbf{A}_\mathrm{P} \parallel \mathbf{p}_\mathrm{P}, \qquad \mathbf{A}_\mathrm{S} \perp \mathbf{p}_\mathrm{S}. \tag{15.30}$$

The doublet character of the shear-wave root (15.28) is a reminder that the subspace of possible transverse particle motions $\mathbf{A}_\mathrm{S}$ is two-dimensional; we consider the polarization of shear waves in greater detail in Section 15.5.

The energy density (15.19) and flux (15.20) can be reduced with the aid of (15.29) and (15.30) to sums associated with compressional and shear waves, respectively. Denoting the scalar wave amplitudes by $A_\mathrm{P} = \|\mathbf{A}_\mathrm{P}\|$ and $A_\mathrm{S} = \|\mathbf{A}_\mathrm{S}\|$, we find that

$$\mathcal{E} = \mathcal{E}_\mathrm{P} + \mathcal{E}_\mathrm{S}, \qquad \boldsymbol{\mathcal{K}} = \boldsymbol{\mathcal{K}}_\mathrm{P} + \boldsymbol{\mathcal{K}}_\mathrm{S}, \tag{15.31}$$

where

$$\mathcal{E}_\mathrm{P} = \omega^2\rho A_\mathrm{P}^2, \qquad \mathcal{E}_\mathrm{S} = \omega^2\rho A_\mathrm{S}^2, \tag{15.32}$$

$$\boldsymbol{\mathcal{K}}_\mathrm{P} = \omega^2\rho\alpha^2 A_\mathrm{P}^2\mathbf{p}_\mathrm{P}, \qquad \boldsymbol{\mathcal{K}}_\mathrm{S} = \omega^2\rho\beta^2 A_\mathrm{S}^2\mathbf{p}_\mathrm{S}. \tag{15.33}$$

It is noteworthy that the magnitude of each flux vector is simply the associated energy density times the speed with which that energy is propagated:

$$\boldsymbol{\mathcal{K}}_{\mathrm{P}} = \alpha\mathcal{E}_{\mathrm{P}}\hat{\mathbf{p}}_{\mathrm{P}}, \qquad \boldsymbol{\mathcal{K}}_{\mathrm{S}} = \beta\mathcal{E}_{\mathrm{S}}\hat{\mathbf{p}}_{\mathrm{S}}. \tag{15.34}$$

Equations (15.26) and (15.27) can be rewritten in an easily interpreted form with the aid of (15.29) and (15.30):

$$\boldsymbol{\nabla}\cdot\boldsymbol{\mathcal{K}}_{\mathrm{P}} = 0 \quad \text{and} \quad \boldsymbol{\nabla}\cdot\boldsymbol{\mathcal{K}}_{\mathrm{S}} = 0 \quad \text{in } \oplus, \tag{15.35}$$

$$[\hat{\mathbf{n}}\cdot(\boldsymbol{\mathcal{K}}_{\mathrm{P}} + \boldsymbol{\mathcal{K}}_{\mathrm{S}})]_-^+ = 0 \quad \text{on } \Sigma. \tag{15.36}$$

We have written (15.35) as separate compressional and shear *transport equations* governing the independent propagation of these two wave types through the smooth sub-regions of the Earth. In contrast, the continuity condition (15.36) involves the full sum $\boldsymbol{\mathcal{K}}_{\mathrm{P}} + \boldsymbol{\mathcal{K}}_{\mathrm{S}}$, because conversions from one wave type to the other may occur at the boundaries. The slowly varying Lagrangian density (15.18) can also be decomposed in the form

$$\mathcal{L} = \mathcal{L}_{\mathrm{P}} + \mathcal{L}_{\mathrm{S}}, \tag{15.37}$$

where

$$\mathcal{L}_{\mathrm{P}} = \tfrac{1}{2}\omega^2\rho(1 - \alpha^2\|\boldsymbol{\nabla}T_{\mathrm{P}}\|^2)A_{\mathrm{P}}^2, \tag{15.38}$$

$$\mathcal{L}_{\mathrm{S}} = \tfrac{1}{2}\omega^2\rho(1 - \beta^2\|\boldsymbol{\nabla}T_{\mathrm{P}}\|^2)A_{\mathrm{S}}^2. \tag{15.39}$$

Equations (15.37)–(15.39) show that the value of the JWKB Lagrangian density at each of the stationary solutions (15.29) is $\mathcal{L} = 0$.

Note that it is not possible to vary the action (15.21) independently with respect to A_{P}, T_{P} and A_{S}, T_{S} because the compressional-wave and shear-wave amplitudes are *coupled at the boundaries*. In fact, an incident wave can give rise to as many as three transmitted and three reflected waves at every boundary, and the energy-flux sum $\boldsymbol{\mathcal{K}}_{\mathrm{P}} + \boldsymbol{\mathcal{K}}_{\mathrm{S}}$ in (15.36) should strictly include all of these. In the next two sections we show how to solve the decoupled eikonal equations (15.29) for the P-wave and S-wave travel times T_{P} and T_{S}, and the decoupled transport equations (15.35) for the associated amplitudes A_{P} and A_{S}. We restrict attention for the time being to the smooth sub-region of the Earth surrounding a source point $\mathbf{x}'$; the significant complications introduced by the presence of boundaries will be considered in Section 15.6.

15.3 Kinematic Ray Tracing

For a unified treatment of both compressional and shear waves, we drop the subscripts P and S, and replace α and β with a generic wave speed v. The generic eikonal equation is in that case

$$\|\mathbf{p}\|^2 = \|\boldsymbol{\nabla} T\|^2 = v^{-2}. \tag{15.40}$$

Equation (15.40) is a first-order, non-linear partial differential equation, which we may solve for the travel time T by the method of characteristics (Courant & Hilbert 1966). The characteristics are the *geometrical rays*, which are everywhere *perpendicular to the wavefronts* or surfaces of constant T. The variation of position $\mathbf{x}$ and slowness $\mathbf{p}$ along these rays is described by the *characteristic equations*

$$\frac{d\mathbf{x}}{d\sigma} = \mathbf{p}, \qquad \frac{d\mathbf{p}}{d\sigma} = \tfrac{1}{2}\boldsymbol{\nabla} v^{-2}. \tag{15.41}$$

The independent variable σ is the so-called *generating parameter*, which is related to the arclength s and travel time T along a ray by

$$d\sigma = v\,ds = v^2 dT. \tag{15.42}$$

Solving the first-order ordinary differential equations (15.41) subject to a given set of Cauchy initial conditions

$$\mathbf{x}(0) = \mathbf{x}', \qquad \mathbf{p}(0) = \mathbf{p}' \tag{15.43}$$

is referred to as *kinematic ray tracing*. The difference in travel time $T_2 - T_1$ between any two points $\mathbf{x}_1$ and $\mathbf{x}_2$ along a ray may be found by integration of equation (15.42):

$$T_2 - T_1 = \int_{T_1}^{T_2} dT = \int_{s_1}^{s_2} v^{-1} ds = \int_{\sigma_1}^{\sigma_2} v^{-2} d\sigma. \tag{15.44}$$

We can also write $T_2 - T_1$ in terms of the slowness $\mathbf{p}$ and differential position $d\mathbf{x}$ along the ray in the form

$$T_2 - T_1 = \int_{\mathbf{x}_1}^{\mathbf{x}_2} \mathbf{p} \cdot d\mathbf{x}. \tag{15.45}$$

To determine the travel time T between a source point $\mathbf{x}'$ and a *prescribed* receiver $\mathbf{x}$, it is necessary to find the initial slowness or slownesses $\mathbf{p}'$ such that $\mathbf{x}(\sigma_{\mathrm{gotcha}}) = \mathbf{x}$ for some σ_{gotcha}. We discuss a practical procedure for solving this two-point *ray-shooting* problem in Section 15.8.2

15.3.1 Hamiltonian formulation

Following Burridge (1976) we introduce the *Hamiltonian*

$$H = \tfrac{1}{2}[\mathbf{p}\cdot\mathbf{p} - v^{-2}(\mathbf{x})], \tag{15.46}$$

and rewrite the eikonal equation (15.40) in the form

$$H(\mathbf{x},\mathbf{p}) = 0. \tag{15.47}$$

The characteristic equations (15.41) are then readily identified as *Hamilton's equations*:

$$\frac{d\mathbf{x}}{d\sigma} = \frac{\partial H}{\partial \mathbf{p}}, \qquad \frac{d\mathbf{p}}{d\sigma} = -\frac{\partial H}{\partial \mathbf{x}}. \tag{15.48}$$

The Hamiltonian (15.46) is an integral of the first-order system (15.48) because

$$\frac{dH}{d\sigma} = \frac{\partial H}{\partial \mathbf{x}}\cdot\frac{d\mathbf{x}}{d\sigma} + \frac{\partial H}{\partial \mathbf{p}}\cdot\frac{d\mathbf{p}}{d\sigma} = 0. \tag{15.49}$$

In the language of classical mechanics, the slowness $\mathbf{p}$ is the generalized momentum *conjugate* to the position $\mathbf{x}$; the six-dimensional space $\mathbf{x}$, $\mathbf{p}$ is referred to as *phase space*.

It is noteworthy that the six equations (15.48) are not all independent. In the first place, the eikonal equation (15.47) imposes a constraint upon the length of the slowness vector; this reduces the number of independent equations to five. A further reduction is possible because one of the three spatial coordinates, namely, the position along the ray itself, can be shown to be *cyclic* or ignorable (Goldstein 1980). A systematic pursuit of these ideas leads ultimately to a *reduced* four-dimensional Hamiltonian, which depends only upon the so-called *ray-centered* coordinates; for a detailed account of this reduction procedure, see Červený (1985) and Farra & Madariaga (1987). We do not discuss ray-centered coordinates in this book; the advantage of having to solve four rather than six equations is offset by the somewhat cumbersome machinery needed to cope with the non-Euclidean nature of the reduced Hamiltonian. We present a simple and practical ray-tracing scheme, which also requires the solution of only four equations, in Section 15.8.1.

★15.3.2 Alternative forms

Many alternative versions of the ray-tracing equations may be enunciated. First, it is possible to use the arclength s or the travel time T as the independent variable instead of the parameter σ. This results in the forms

$$\frac{d\mathbf{x}}{ds} = \hat{\mathbf{p}}, \qquad \frac{d\mathbf{p}}{ds} = \boldsymbol{\nabla} v^{-1} \tag{15.50}$$

and

$$\frac{d\mathbf{x}}{dT} = v^2\mathbf{p}, \qquad \frac{d\mathbf{p}}{dT} = -\boldsymbol{\nabla}(\ln v). \tag{15.51}$$

Every such change of independent variable is associated with a transformation of the Hamiltonian (15.46). Thus (15.50) are Hamilton's equations $d\mathbf{x}/ds = \partial_{\mathbf{p}}H'$, $d\mathbf{p}/ds = -\partial_{\mathbf{x}}H'$ for the Hamiltonian

$$H' = \sqrt{\mathbf{p}\cdot\mathbf{p}} - v^{-1}(\mathbf{x}) = 0, \tag{15.52}$$

whereas (15.51) are Hamilton's equations $d\mathbf{x}/dT = \partial_{\mathbf{p}}H''$, $d\mathbf{p}/dT = -\partial_{\mathbf{x}}H''$ for the Hamiltonian

$$H'' = \tfrac{1}{2}[(\mathbf{p}\cdot\mathbf{p})v^2(\mathbf{x}) - 1] = 0. \tag{15.53}$$

Second, it is possible to combine the two first-order equations for $\mathbf{x}$ and $\mathbf{p}$ into a single second-order ordinary differential equation for the ray path:

$$\frac{d^2\mathbf{x}}{d\sigma^2} - \tfrac{1}{2}\boldsymbol{\nabla}v^{-2} = \mathbf{0}, \tag{15.54}$$

$$\frac{d}{ds}\left(v^{-1}\frac{d\mathbf{x}}{ds}\right) - \boldsymbol{\nabla}v^{-1} = \mathbf{0}, \tag{15.55}$$

$$\frac{d}{dT}\left(v^{-2}\frac{d\mathbf{x}}{dT}\right) + \boldsymbol{\nabla}(\ln v) = \mathbf{0}. \tag{15.56}$$

Finally, we can use (15.42) and the relation $d/ds = \hat{\mathbf{p}}\cdot\boldsymbol{\nabla}$ to eliminate the first derivatives $d\mathbf{x}/ds$ and $d\mathbf{x}/dT$ in equations (15.55) and (15.56); this leads to the forms

$$\frac{d^2\mathbf{x}}{ds^2} + \boldsymbol{\nabla}_{\perp}(\ln v) = \mathbf{0}, \qquad \frac{d^2\mathbf{x}}{dT^2} - \tfrac{1}{2}\boldsymbol{\nabla}_{\perp}v^2 = \mathbf{0}, \tag{15.57}$$

where $\boldsymbol{\nabla}_{\perp} = \boldsymbol{\nabla} - \hat{\mathbf{p}}\hat{\mathbf{p}}\cdot\boldsymbol{\nabla}$ is the gradient in the direction perpendicular to the ray path.

Each of the ray Hamiltonians H, H', H'' has an associated ray Lagrangian L, L', L''. Regarding σ as the independent variable and denoting differentiation $d/d\sigma$ by a dot, we define L in the customary manner (Goldstein 1980):

$$L(\mathbf{x},\dot{\mathbf{x}}) = \mathbf{p}\cdot\dot{\mathbf{x}} - H(\mathbf{x},\mathbf{p}). \tag{15.58}$$

The lack of any dependence of the Lagrangian (15.58) upon the slowness vector $\mathbf{p}$ can be confirmed by evaluating $\partial_{\mathbf{p}}L$ and invoking Hamilton's equations (15.48). Written out explicitly, we have

$$L = \tfrac{1}{2}[\dot{\mathbf{x}}\cdot\dot{\mathbf{x}} + v^{-2}(\mathbf{x})]. \tag{15.59}$$

The Hamiltonian H, Lagrangian L and equation of motion $\ddot{\mathbf{x}} - \frac{1}{2}\boldsymbol{\nabla} v^{-2} = \mathbf{0}$ are those of a particle of unit mass, moving in a static potential field $-\frac{1}{2}v^{-2}$. Note that no such simple mechanical interpretation is possible if either the arclength s or the travel time T is used as the independent variable instead of the parameter σ.

★15.3.3 Hamilton's and Fermat's principles

The ray-tracing equations can alternatively be obtained from *Hamilton's principle*:

$$\delta \int_{\sigma_1}^{\sigma_2} L(\mathbf{x}, \dot{\mathbf{x}})\, d\sigma = 0. \tag{15.60}$$

Equation (15.60) is a variational principle in $\mathbf{x}$ or *configuration space*; the corresponding Euler-Lagrange equation $d(\partial_{\dot{\mathbf{x}}} L)/d\sigma - \partial_{\mathbf{x}} L = \mathbf{0}$ is precisely the second-order ray-tracing equation (15.54). Alternatively, we can write Hamilton's principle in phase space in the form

$$\delta \int_{\sigma_1}^{\sigma_2} [\mathbf{p} \cdot \dot{\mathbf{x}} - H(\mathbf{x}, \mathbf{p})]\, d\sigma = 0. \tag{15.61}$$

In this case the Euler-Lagrange equations are the first-order Hamiltonian equations (15.48). Analogous statements of Hamilton's principle may be enunciated for the functionals L', H' and L'', H'' associated with the arclength s and travel time T.

Fermat's principle stipulates that the travel time along a geometrical ray between any two fixed points $\mathbf{x}_1$ and $\mathbf{x}_2$ is stationary:

$$\delta \int_{\mathbf{x}_1}^{\mathbf{x}_2} \mathbf{p} \cdot d\mathbf{x} = \delta \int_{s_1}^{s_2} v^{-1}\, ds = \delta \int_{T_1}^{T_2} dT = 0. \tag{15.62}$$

It is not permissible to regard the arclength s or the travel time T as independent variables in effecting the variation (15.62); rather, they must both be considered to be dependent variables, together with $\mathbf{x}$ and $d\mathbf{x}/ds$ or $d\mathbf{x}/dT$. Following Lanczos (1962), we can eliminate s and T by introducing a new independent variable ξ such that

$$\sqrt{\frac{d\mathbf{x}}{d\xi} \cdot \frac{d\mathbf{x}}{d\xi}} = \frac{ds}{d\xi} = v\frac{dT}{d\xi}. \tag{15.63}$$

Fermat's principle (15.62) then takes the form

$$\delta \int_{\xi_1}^{\xi_2} v^{-1}(\mathbf{x}) \sqrt{\frac{d\mathbf{x}}{d\xi} \cdot \frac{d\mathbf{x}}{d\xi}}\, d\xi = 0. \tag{15.64}$$

In classical mechanics, this is known as *Jacobi's form* of the principle of least action (Lanczos 1962; Goldstein 1980). The Euler-Lagrange equation equivalent to (15.64) is

$$\frac{d}{d\xi}\left[v^{-1}\left(\frac{d\mathbf{x}}{d\xi}\cdot\frac{d\mathbf{x}}{d\xi}\right)^{-1/2}\frac{d\mathbf{x}}{d\xi}\right]-\left(\frac{d\mathbf{x}}{d\xi}\cdot\frac{d\mathbf{x}}{d\xi}\right)^{1/2}\boldsymbol{\nabla}v^{-1}=\mathbf{0}. \qquad (15.65)$$

This equation determines the ray path in terms of the parameter ξ. Knowing $\mathbf{x}(\xi)$, we can find the arclength $s(\xi)$ and the travel time $T(\xi)$ by integrating equation (15.63). Alternatively, we can use (15.63) to convert (15.65) into equations involving s or T; in this way, we recover (15.55) and (15.56). The stationary integrals in (15.60)–(15.62) and (15.64) are all numerically equal to the travel-time difference $T_2 - T_1$ along a ray trajectory; the details of the variation process and the physical interpretation are different in each specific case.

*15.3.4 Serret-Frénet formulae

We present a brief review of the geometrical properties of three-dimensional curves in this section. It is conventional to use the arclength s as the independent variable in such discussions; for a more detailed treatment see any textbook on classical differential geometry, such as Willmore (1959). We note first that the unit slowness vector $\hat{\mathbf{p}}$ of a seismic ray is referred to in geometrical discussions as the *tangent vector*. The plane of the circle that coincides with a curve at any point $\mathbf{x}$ is known as the *osculating plane*. The line of intersection of the osculating plane and the plane normal to the tangent vector is called the principal normal to the curve. The *normal vector* $\hat{\boldsymbol{\nu}}$ is the unit vector along this line. The initial direction at a starting point $\mathbf{x}'$ may be chosen arbitrarily from the two possibilities; thereafter, $\hat{\boldsymbol{\nu}}$ is required to vary smoothly along the curve. The rate at which the tangent vector changes direction with arclength s is the *curvature* of a curve, denoted by κ. It may be readily demonstrated that $d\hat{\mathbf{p}}/ds$ lies in the osculating plane, and is perpendicular to $\hat{\mathbf{p}}$; hence, it is co-aligned with the normal $\hat{\boldsymbol{\nu}}$. The sign of the curvature, which is otherwise arbitrary, is fixed by stipulating that $d\hat{\mathbf{p}}/ds = \kappa\hat{\boldsymbol{\nu}}$. For concreteness, we take the curvature κ of an upward turning ray in a spherically symmetric Earth to be positive; the normal $\hat{\boldsymbol{\nu}}$ to such a ray then points upward as well—toward the center of curvature. The reciprocal $|\kappa|^{-1}$ of the absolute curvature of either an upward or downward turning ray is the radius of the tangent circle in the osculating plane. The *binormal* to a curve is the unit normal $\hat{\mathbf{b}} = \hat{\mathbf{p}} \times \hat{\boldsymbol{\nu}}$ to the osculating plane. Its arclength rate of change along the curve is likewise co-aligned with $\hat{\boldsymbol{\nu}}$; it is conventional to define the *torsion* or rate of twisting of the osculating plane by $d\hat{\mathbf{b}}/ds = -\tau\hat{\boldsymbol{\nu}}$. The minus sign renders the torsion τ positive whenever the plane twists in the direction of a right-handed screw pointed toward $\hat{\mathbf{p}}$. In summary, the tangent vector, normal and binormal $\hat{\mathbf{p}}$, $\hat{\boldsymbol{\nu}}$, $\hat{\mathbf{b}}$ comprise a right-handed set of orthonormal axes, whose variation along a three-dimensional curve is described by the *Serret-Frénet formulae*:

$$\frac{d\hat{\mathbf{p}}}{ds} = \kappa\hat{\boldsymbol{\nu}}, \qquad \frac{d\hat{\boldsymbol{\nu}}}{ds} = \tau\hat{\mathbf{b}} - \kappa\hat{\mathbf{p}}, \qquad \frac{d\hat{\mathbf{b}}}{ds} = -\tau\hat{\boldsymbol{\nu}}. \tag{15.66}$$

The curvature κ and torsion τ may be written explicitly in terms of the first three derivatives of the position vector $\mathbf{x}(s)$ in the form

$$\kappa^2 = \frac{d^2\mathbf{x}}{ds^2} \cdot \frac{d^2\mathbf{x}}{ds^2}, \tag{15.67}$$

$$\tau = \kappa^{-2} \left(\frac{d\mathbf{x}}{ds} \cdot \frac{d^2\mathbf{x}}{ds^2} \times \frac{d^3\mathbf{x}}{ds^3} \right). \tag{15.68}$$

Equations (15.66)–(15.68) pertain to any smooth curve in three-dimensional space; by combining these geometrical results with the ray-tracing equa-

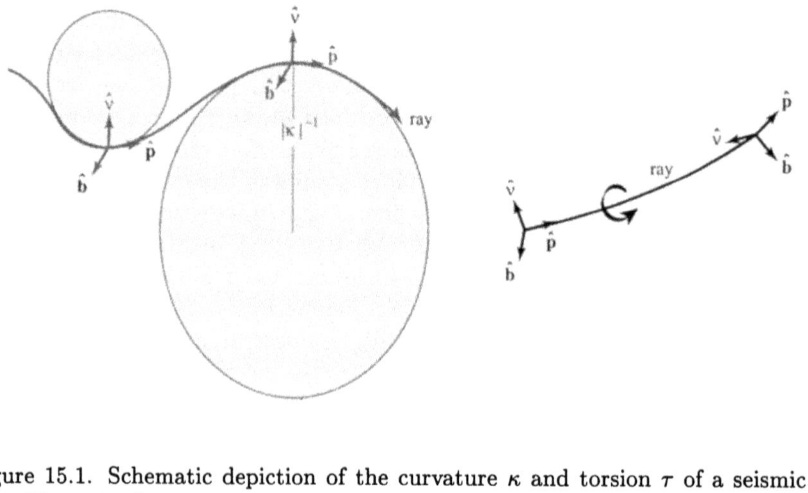

Figure 15.1. Schematic depiction of the curvature κ and torsion τ of a seismic ray. The unit slowness vector $\hat{\mathbf{p}}$ is everywhere tangent to the ray. (*Left*) The ray normal $\hat{\boldsymbol{\nu}}$ points either toward or away from the center of curvature of the shaded tangent circle, which lies within the osculating plane. The radius of the circle is the absolute reciprocal $|\kappa|^{-1}$ of the curvature. (*Right*) The torsion τ is the arclength rate at which the normal $\hat{\boldsymbol{\nu}}$ and binormal $\hat{\mathbf{b}} = \hat{\mathbf{p}} \times \hat{\boldsymbol{\nu}}$ twist in a right-hand sense about the ray.

tion (15.55), we can express the curvature and torsion of a seismic ray in terms of the logarithmic gradients of the wave speed:

$$\kappa = -\hat{\boldsymbol{\nu}} \cdot \boldsymbol{\nabla}(\ln v), \qquad \tau = -\kappa^{-1}\hat{\mathbf{p}} \cdot \boldsymbol{\nabla\nabla}(\ln v) \cdot \hat{\mathbf{b}}. \tag{15.69}$$

A ray twists in such a way that its normal $\hat{\boldsymbol{\nu}}$ is always parallel and its binormal $\hat{\mathbf{b}}$ is always perpendicular to the cross-path gradient:

$$\hat{\boldsymbol{\nu}} = -\kappa^{-1}\boldsymbol{\nabla}_{\perp}(\ln v), \qquad \hat{\mathbf{b}} \cdot \boldsymbol{\nabla}_{\perp}(\ln v) = 0. \tag{15.70}$$

With our sign convention, $\hat{\boldsymbol{\nu}}$ points in the direction of decreasing wave speed wherever the curvature κ is positive, and in the direction of increasing wave speed wherever it is negative. We illustrate the evolution of the Serret-Frénet triad $\hat{\mathbf{p}}$, $\hat{\boldsymbol{\nu}}$, $\hat{\mathbf{b}}$ along a seismic ray in Figure 15.1.

15.4 Amplitude Variation

The ray-tracing equations discussed above enable us to determine the ray trajectories and the variations in travel time T_{P} or T_{S} along them. The transport equations (15.35), which we consider next, determine the variations in amplitude A_{P} or A_{S} along the rays. Dropping the identifying subscripts P and S, as before, we rewrite (15.35) in the generic form

$$\boldsymbol{\nabla} \cdot \boldsymbol{\mathcal{K}} = 0, \tag{15.71}$$

where

$$\boldsymbol{\mathcal{K}} = v\mathcal{E}\hat{\mathbf{p}} = \omega^2 \rho v A^2 \hat{\mathbf{p}} = \omega^2 \rho v^2 A^2 \mathbf{p}. \tag{15.72}$$

Equation (15.71) can be solved for the amplitude variation A along a ray in a variety of ways; we describe several of these methods and indicate the relations between them in the following sections. Amplitude determination is an important consideration in a variety of applications, including wavefront extrapolation; we take a relatively limited view, restricting attention to the waves emitted by a point source.

15.4.1 Conservation of energy

We begin by noting that the differential relation $\boldsymbol{\nabla} \cdot \boldsymbol{\mathcal{K}} = 0$ has an obvious physical interpretation—it expresses the *conservation of body-wave energy*. This can be seen by considering an infinitesimally narrow tube of trajectories surrounding a ray path from $\mathbf{x}_1$ to $\mathbf{x}_2$, as shown in Figure 12.7. Let V be the volume of this ray-tube segment, and denote the outward unit normal on the boundary surface ∂V (temporarily) by $\hat{\boldsymbol{\nu}}$. Upon integrating (15.71) over V and invoking Gauss' theorem, we obtain

$$\int_V \boldsymbol{\nabla}\cdot\boldsymbol{\mathcal{K}}\,dV = \int_{\partial V} \hat{\boldsymbol{\nu}}\cdot\boldsymbol{\mathcal{K}}\,d\Sigma = \|\boldsymbol{\mathcal{K}}\|_2\,d\Sigma_2 - \|\boldsymbol{\mathcal{K}}\|_1\,d\Sigma_1 = 0, \qquad (15.73)$$

where $d\Sigma_1$ and $d\Sigma_2$ are the differential areas of the cross-sectional patches at $\mathbf{x}_1$ and $\mathbf{x}_2$, respectively, and we have used the fact that $\hat{\boldsymbol{\nu}}\cdot\boldsymbol{\mathcal{K}} = 0$ on the sidewalls of the ray tube. All of the wave energy that enters the ray tube at $\mathbf{x}_1$ leaves it at $\mathbf{x}_2$; there is no energy leakage through the sidewalls nor absorption within the perfectly elastic ray tube. Equation (15.73) yields the amplitude variation relation (12.26), which we repeat here for convenience:

$$\frac{A_2}{A_1} = \left(\frac{\rho_2 v_2}{\rho_1 v_1}\right)^{-1/2} \left|\frac{d\Sigma_2}{d\Sigma_1}\right|^{-1/2}. \qquad (15.74)$$

The absolute value is necessary to account for the change in sign of the differential area of a ray tube upon passage through a caustic or focal point, as we discuss in Sections 12.1.8 and 15.4.3.

★15.4.2 Ray-tube area

Following Kline & Kay (1979) we can obtain an explicit expression for the ray-tube area ratio $d\Sigma_2/d\Sigma_1$ in equation (15.74). Consider the volume integral

$$\int_V \boldsymbol{\nabla}\cdot\hat{\mathbf{p}}\,dV = \int_{\partial V} \hat{\boldsymbol{\nu}}\cdot\hat{\mathbf{p}}\,d\Sigma = d\Sigma_2 - d\Sigma_1, \qquad (15.75)$$

where the final equality is valid because V is an infinitesimally narrow ray tube. We can also express the left side of (15.75) as a line integral along the trajectory from $\mathbf{x}_1$ to $\mathbf{x}_2$:

$$\int_V \boldsymbol{\nabla}\cdot\hat{\mathbf{p}}\,dV = \int_{s_1}^{s_2} \boldsymbol{\nabla}\cdot\hat{\mathbf{p}}\,d\Sigma\,ds, \qquad (15.76)$$

where $d\Sigma(s)$ is the differential area at a running point along the ray. Upon comparing equations (15.75) and (15.76) we deduce that

$$d\Sigma_2 - d\Sigma_1 = \int_{s_1}^{s_2} \boldsymbol{\nabla}\cdot\hat{\mathbf{p}}\,d\Sigma\,ds. \qquad (15.77)$$

The solution to this linear integral equation is

$$\frac{d\Sigma_2}{d\Sigma_1} = \exp\left(\int_{s_1}^{s_2} \boldsymbol{\nabla}\cdot\hat{\mathbf{p}}\,ds\right). \qquad (15.78)$$

The integrand in equation (15.78) is the sum of the principal curvatures of the wavefront at a point s along the ray:

$$\boldsymbol{\nabla}\cdot\hat{\mathbf{p}} = \frac{1}{R_1} + \frac{1}{R_2}. \tag{15.79}$$

The quantities R_1 and R_2 are the corresponding *radii of curvature*.

Yet another expression for $d\Sigma_2/d\Sigma_1$ can be obtained by rewriting the transport equation $\boldsymbol{\nabla}\cdot(\rho v^2 A^2\mathbf{p}) = 0$ in the form

$$\frac{d}{d\sigma}\ln(\rho v^2 A^2) = -\boldsymbol{\nabla}\cdot\mathbf{p}, \tag{15.80}$$

where we have used the fact that $\mathbf{p}\cdot\boldsymbol{\nabla} = (d\mathbf{x}/d\sigma)\cdot\boldsymbol{\nabla} = d/d\sigma$ along a ray. The energy conservation law $\rho v A^2\, d\Sigma =$ constant enables us to convert this into a differential equation for the ray-tube area:

$$\frac{d}{d\sigma}\ln(v^{-1}d\Sigma) = \boldsymbol{\nabla}\cdot\mathbf{p}. \tag{15.81}$$

This equation can be integrated to yield the result

$$\frac{d\Sigma_2}{d\Sigma_1} = \frac{v_2}{v_1}\exp\left(\int_{\sigma_1}^{\sigma_2}\boldsymbol{\nabla}\cdot\mathbf{p}\, d\sigma\right). \tag{15.82}$$

The integrand $\boldsymbol{\nabla}\cdot\mathbf{p}$ in this case is the *Laplacian of the travel time* at a point σ along the ray:

$$\boldsymbol{\nabla}\cdot\mathbf{p} = \nabla^2 T. \tag{15.83}$$

The previous result (15.78) can be deduced from (15.82) by substituting $\boldsymbol{\nabla}\cdot\mathbf{p} = \boldsymbol{\nabla}\cdot(v^{-1}\hat{\mathbf{p}}) = v^{-1}\boldsymbol{\nabla}\cdot\hat{\mathbf{p}} - d(\ln v)/d\sigma$.

15.4.3 Point-source Jacobian

Only two parameters are needed to characterize the initial takeoff direction of a ray from a fixed point source $\mathbf{x}'$. For the time being, it is unnecessary to adopt a specific choice for these parameters; we shall refer to them as γ_1' and γ_2', where the primes serve as a reminder that they are measured at the point $\mathbf{x}'$. A given point in the ensemble of all rays shot from $\mathbf{x}'$ may be regarded as a function of the form $\mathbf{x} = \mathbf{x}(\sigma, \gamma_1', \gamma_2')$, where γ_1', γ_2' identify the ray and σ specifies the position along it. The partial derivatives with respect to the ray parameters $\partial_{\gamma_1'}\mathbf{x} = (\partial\mathbf{x}/\partial\gamma_1')_{\sigma,\gamma_2'}$ and $\partial_{\gamma_2'}\mathbf{x} = (\partial\mathbf{x}/\partial\gamma_2')_{\sigma,\gamma_1'}$ both lie within the instantaneous wavefront passing through the point $\mathbf{x}$; the condition that guarantees this is

$$\mathbf{p}\cdot\partial_{\gamma'}\mathbf{x} = 0, \tag{15.84}$$

where γ' denotes either γ_1' or γ_2'. The area of an infinitesimally narrow ray tube is given in terms of the differentials $d\gamma_1'$ and $d\gamma_2'$ by the classical geometrical relation (Willmore 1959):

$$d\Sigma = \hat{\mathbf{p}} \cdot (\partial_{\gamma_1'}\mathbf{x} \times \partial_{\gamma_2'}\mathbf{x})\, d\gamma_1'\, d\gamma_2'. \tag{15.85}$$

The cross product $\partial_{\gamma_1'}\mathbf{x} \times \partial_{\gamma_2'}\mathbf{x}$ in (15.85) is either parallel or anti-parallel to the direction of propagation $\hat{\mathbf{p}}$, depending upon the sign of $d\Sigma$.

We define the *point-source Jacobian* to be the determinant

$$J = \frac{\partial(x_1, x_2, x_3)}{\partial(\sigma, \gamma_1', \gamma_2')} = \det \begin{pmatrix} \dfrac{\partial x_1}{\partial \sigma} & \dfrac{\partial x_1}{\partial \gamma_1'} & \dfrac{\partial x_1}{\partial \gamma_2'} \\ \dfrac{\partial x_2}{\partial \sigma} & \dfrac{\partial x_2}{\partial \gamma_1'} & \dfrac{\partial x_2}{\partial \gamma_2'} \\ \dfrac{\partial x_3}{\partial \sigma} & \dfrac{\partial x_3}{\partial \gamma_1'} & \dfrac{\partial x_3}{\partial \gamma_2'} \end{pmatrix}. \tag{15.86}$$

This can be expanded with the aid of the kinematic relation $d\mathbf{x}/d\sigma = \mathbf{p}$:

$$J = \mathbf{p} \cdot (\partial_{\gamma_1'}\mathbf{x} \times \partial_{\gamma_2'}\mathbf{x}). \tag{15.87}$$

The quantities $d\Sigma$ and J change sign in unison, and are related at all points along a ray by

$$d\Sigma = vJ\, d\gamma_1'\, d\gamma_2'. \tag{15.88}$$

In other words, vJ is the Jacobian that relates the differential ray-tube area $d\Sigma$ to the ray parameters γ_1' and γ_2'. The absolute value of J can be written in the form

$$|J| = v^{-1}\|\partial_{\gamma_1'}\mathbf{x} \times \partial_{\gamma_2'}\mathbf{x}\| = v^{-1}\sqrt{EG - F^2}, \tag{15.89}$$

where we have let

$$E = \partial_{\gamma_1'}\mathbf{x} \cdot \partial_{\gamma_1'}\mathbf{x}, \qquad G = \partial_{\gamma_2'}\mathbf{x} \cdot \partial_{\gamma_2'}\mathbf{x}, \qquad F = \partial_{\gamma_1'}\mathbf{x} \cdot \partial_{\gamma_2'}\mathbf{x}. \tag{15.90}$$

Upon replacing the independent variable σ in (15.86) by s and T, we obtain the alternative point-source Jacobians $J' = vJ$ and $J'' = v^2 J$. The former is used as the basis of ray-amplitude calculations by many authors (e.g., Červený 1985) because of the convenient interpretation $d\Sigma = J'\, d\gamma_1'\, d\gamma_2'$.

*15.4.4 Smirnov's lemma

A general method of solving the transport equation (15.71) is based upon a mathematical result described by Smirnov (1964) and Thomson & Chapman (1985), which we reiterate here. Smirnov's lemma states that

$$\frac{d}{d\sigma}(\ln J) = \boldsymbol{\nabla} \cdot \mathbf{p} \quad \text{whenever} \quad \frac{d\mathbf{x}}{d\sigma} = \mathbf{p}. \tag{15.91}$$

To prove (15.91) we consider the quantity

$$\frac{dJ}{d\sigma} = \frac{\partial(p_1, x_2, x_3)}{\partial(\sigma, \gamma_1', \gamma_2')} + \frac{\partial(x_1, p_2, x_3)}{\partial(\sigma, \gamma_1', \gamma_2')} + \frac{\partial(x_1, x_2, p_3)}{\partial(\sigma, \gamma_1', \gamma_2')}, \tag{15.92}$$

where $p_i = dx_i/d\sigma$. The sum of three determinants (15.92) can be expanded into a sum of nine determinants by means of the substitutions

$$\frac{\partial p_i}{\partial \sigma} = \frac{\partial p_i}{\partial x_j}\frac{\partial x_j}{\partial \sigma}, \qquad \frac{\partial p_i}{\partial \gamma_1'} = \frac{\partial p_i}{\partial x_j}\frac{\partial x_j}{\partial \gamma_1'}, \qquad \frac{\partial p_i}{\partial \gamma_2'} = \frac{\partial p_i}{\partial x_j}\frac{\partial x_j}{\partial \gamma_2'}. \tag{15.93}$$

However, six of these nine determinants are identically zero, because their rows are linearly dependent. The remaining three determinants can be combined to give

$$\frac{dJ}{d\sigma} = \left(\frac{\partial p_j}{\partial x_j}\right)\frac{\partial(x_1, x_2, x_3)}{\partial(\sigma, \gamma_1', \gamma_2')} = (\boldsymbol{\nabla}\cdot\mathbf{p})\, J, \tag{15.94}$$

which is the desired result (15.91).

It is a straightforward matter to apply Smirnov's lemma; upon combining (15.91) and (15.80) we deduce that

$$\frac{d}{d\sigma}\ln(\rho v^2 A^2 J) = 0. \tag{15.95}$$

Equation (15.95) enables us to express the ratio of the amplitudes at two consecutive points $\mathbf{x}_1$ and $\mathbf{x}_2$ along a ray in terms of a ratio of Jacobians:

$$\frac{A_2}{A_1} = \left(\frac{\rho_2 v_2^2}{\rho_1 v_1^2}\right)^{-1/2} \left|\frac{J_2}{J_1}\right|^{-1/2}. \tag{15.96}$$

Of course, we could have obtained the result (15.96) from (15.74) by making use of the identification (15.88). However, Smirnov's lemma can be formulated and applied in a wide variety of other settings, as we shall see in Sections 15.8.8 and 16.4.4.

15.4.5 Geometrical spreading factor

As in the case of a spherically symmetric Earth, we define a positive *geometrical spreading factor* $\mathcal{R}(\mathbf{x}, \mathbf{x}')$, which is analogous to the source-receiver distance $\|\mathbf{x} - \mathbf{x}'\|$ in a homogeneous medium, by

$$\mathcal{R} = \sqrt{|d\Sigma|/d\Omega}, \tag{15.97}$$

where $d\Omega$ is the differential solid angle subtended by the ray tube at the source $\mathbf{x}'$. We can express this solid angle in terms of the partial derivatives of the unit slowness $\hat{\mathbf{p}}'$ at the source in the form:

$$d\Omega = \hat{\mathbf{p}}'\cdot(\partial_{\gamma_1'}\hat{\mathbf{p}}' \times \partial_{\gamma_2'}\hat{\mathbf{p}}')\, d\gamma_1'\, d\gamma_2' = \|\partial_{\gamma_1'}\hat{\mathbf{p}}' \times \partial_{\gamma_2'}\hat{\mathbf{p}}'\|\, d\gamma_1'\, d\gamma_2'. \tag{15.98}$$

The spreading factor is therefore related to the Jacobian (15.86) by

$$\mathcal{R} = \sqrt{\frac{v|J|}{\|\partial_{\gamma_1'}\hat{\mathbf{p}}' \times \partial_{\gamma_2'}\hat{\mathbf{p}}'\|}} = \sqrt{\frac{\|\partial_{\gamma_1'}\mathbf{x} \times \partial_{\gamma_2'}\mathbf{x}\|}{\|\partial_{\gamma_1'}\hat{\mathbf{p}}' \times \partial_{\gamma_2'}\hat{\mathbf{p}}'\|}}. \tag{15.99}$$

Equation (15.78) can be used to obtain another expression for the spreading factor, as a line integral along a ray:

$$\mathcal{R} = \lim_{s'\to 0} s' \left|\exp\left(\tfrac{1}{2}\int_{s'}^{s} \boldsymbol{\nabla}\cdot\hat{\mathbf{p}}\, ds\right)\right|, \tag{15.100}$$

where we have used the limiting relation $d\Sigma' \to s'^2 d\Omega$ in the vicinity of the source, $s' \to 0$. The outgoing wavefront in a homogeneous medium is a sphere with surface divergence $\boldsymbol{\nabla}\cdot\hat{\mathbf{p}} = 2s^{-1}$, so that (15.100) reduces to $\mathcal{R} = s = \|\mathbf{x} - \mathbf{x}'\|$, as expected.

15.4.6 Dynamical reciprocity

Let us rewrite the differential-area and solid-angle relationships (15.85) and (15.98) in terms of $\mathbf{p} = v^{-1}\hat{\mathbf{p}}$ using index notation:

$$d\Sigma = v\varepsilon_{imn}\, p_i(\partial_{\gamma_1'}x_m)(\partial_{\gamma_2'}x_n)\, d\gamma_1'\, d\gamma_2', \tag{15.101}$$

$$d\Omega = v'^3\varepsilon_{jkl}\, p_j'(\partial_{\gamma_1'}p_k')(\partial_{\gamma_2'}p_l')\, d\gamma_1'\, d\gamma_2'. \tag{15.102}$$

Upon regarding the travel time as a function $T(\mathbf{x},\mathbf{x}') = T(\mathbf{x}',\mathbf{x})$ of *both* endpoints, we can express the slowness derivatives $\partial_{\gamma'}\mathbf{p}'$ in terms of $\partial_{\gamma'}\mathbf{x}$ in the form $\partial_{\gamma'}\mathbf{p}' = -\partial_{\gamma'}\mathbf{x}\cdot\boldsymbol{\nabla}\boldsymbol{\nabla}'T$. Substituting this into (15.102) we obtain

$$d\Omega = v'^3\varepsilon_{jkl}\, p_j'(\partial_{\gamma_1'}x_m)(\partial_m\partial_k'T)(\partial_{\gamma_2'}x_n)(\partial_n\partial_l'T)\, d\gamma_1'\, d\gamma_2'. \tag{15.103}$$

It is convenient to introduce a tensor function of the two endpoints:

$$\mathbf{S}(\mathbf{x},\mathbf{x}') = (\det\boldsymbol{\nabla}\boldsymbol{\nabla}'T)(\boldsymbol{\nabla}\boldsymbol{\nabla}'T)^{-1} = \mathbf{S}^{\mathrm{T}}(\mathbf{x}',\mathbf{x}). \tag{15.104}$$

The components of (15.104) are given by Cramer's rule (A.36):

$$S_{ij}(\mathbf{x},\mathbf{x}') = S_{ji}(\mathbf{x}',\mathbf{x}) = \tfrac{1}{2}\varepsilon_{imn}\varepsilon_{jkl}(\partial_m\partial_k'T)(\partial_n\partial_l'T). \tag{15.105}$$

The easily verified relationship $\varepsilon_{imn}S_{ij} = \varepsilon_{jkl}(\partial_m\partial_k'T)(\partial_n\partial_l'T)$ enables us to write $d\Omega$ in terms of $d\Sigma$ in the form

$$d\Omega = v'^3\varepsilon_{imn}\, p_j'S_{ij}(\partial_{\gamma_1'}x_m)(\partial_{\gamma_2'}x_n) = vv'^3 p_i S_{ij} p_j'\, d\Sigma. \tag{15.106}$$

Reverting to invariant notation, we obtain another explicit expression for the geometrical spreading factor $\mathcal{R} = \sqrt{|d\Sigma|/d\Omega}$:

$$v(\mathbf{x}')\mathcal{R}(\mathbf{x},\mathbf{x}') = |\hat{\mathbf{p}}\cdot\mathbf{S}(\mathbf{x},\mathbf{x}')\cdot\hat{\mathbf{p}}'|^{-1/2}. \tag{15.107}$$

The symmetry (15.104) renders the right side of (15.107) invariant under an interchange of the source and receiver:

$$\mathbf{x} \to \mathbf{x}', \qquad \mathbf{x}' \to \mathbf{x}, \qquad \hat{\mathbf{p}} \to -\hat{\mathbf{p}}', \qquad \hat{\mathbf{p}}' \to -\hat{\mathbf{p}}. \tag{15.108}$$

It follows that $\mathcal{R}$ satisfies the dynamical symmetry or *reciprocity relation*

$$v(\mathbf{x}')\mathcal{R}(\mathbf{x},\mathbf{x}') = v(\mathbf{x})\mathcal{R}(\mathbf{x}',\mathbf{x}). \tag{15.109}$$

Equation (15.109) is identical to the relation governing geometrical spreading within a spherically symmetric Earth; see Section 12.1.7. The above geometrical argument, which generalizes this result to an arbitrary Earth, is given by Richards (1971); he attributes the proof to G. E. Backus.

A physical interpretation of the wave-speed factors $v(\mathbf{x})$ and $v(\mathbf{x}')$ in the reciprocity relation (15.109) is given by Snieder & Chapman (1998). Suppose the source $\mathbf{x}'$ lies directly below the receiver $\mathbf{x}$ in a medium whose speed depends only upon depth. If $v(\mathbf{x}') > v(\mathbf{x})$, the rays shot upward from $\mathbf{x}'$ diverge more slowly than the rays shot downward from $\mathbf{x}$, because rays are refracted away from high wave speeds. The spreading factor from the source to the receiver must therefore be less than that from the receiver to the source: $\mathcal{R}(\mathbf{x},\mathbf{x}') < \mathcal{R}(\mathbf{x}',\mathbf{x})$. This simple example shows why "pure" reciprocity $\mathcal{R}(\mathbf{x},\mathbf{x}') = \mathcal{R}(\mathbf{x}',\mathbf{x})$ of the geometrical spreading does not hold.

15.4.7 Caustics and focal points

Singular points along a ray where the differential area $d\Sigma$ vanishes are referred to as *caustics* or *focal points*. These two types of ray singularities are distinguished by the number of transverse dimensions in which the ray tube "collapses" simultaneously, as illustrated in Figure 12.8. Only one of the radii of curvature of the associated wavefront vanishes at a caustic, whereas both vanish at a focal point:

$$\begin{array}{ll} \text{caustic:} & R_1 R_2 = 0, \\ \text{focal point:} & R_1 = R_2 = 0. \end{array} \tag{15.110}$$

The ray-tube area $d\Sigma$ undergoes a change in sign upon passage through a caustic; upon passing through a focal point, it may be considered to change sign twice, so that its original sign is retained. Focal points are a design feature of telescopes and many other man-made imaging devices; however, they are extremely rare in seismology. Caustics, on the other hand, are a ubiquitous feature of seismic rays in strongly heterogeneous media. In three-dimensional space, the caustics lie on two-dimensional surfaces that are *envelopes* of the ray field. In addition to simple fold and cusp caustics, there are a number of other more complex types; the possible morphologies can be characterized using catastrophe theory (Poston & Stewart 1978). We do not discuss the morphological classification of these so-called *diffraction catastrophes* here; detailed treatments may be found in Berry & Upstill (1980) and Kravtsov & Orlov (1990).

The phase of a wave undergoes a non-geometrical $\pi/2$ phase advance upon every passage through a caustic, as discussed in Section 12.1.8. To keep track of these cumulative phase shifts, we introduce the *Maslov index*, a positive integer counter which starts at $M = 0$ and increases by one every time that a ray passes through a caustic. Rare passages through a focal point increase the Maslov index by two. The location and structure of the caustic surfaces or envelopes of the rays leaving a source $\mathbf{x}'$ and receiver $\mathbf{x}$ are obviously different, but the number of caustic passages along every ray and reversed ray is the same:

$$M(\mathbf{x}', \mathbf{x}) = M(\mathbf{x}, \mathbf{x}'). \tag{15.111}$$

A proof of this *Maslov reciprocity principle* can be based upon the observation that M is the number of sign changes of the differential area $d\Sigma$ along a ray; the considerations in Section 15.4.6 show that this number is independent of the direction in which the ray is traced.

15.4.8 Dynamical ray tracing

We can calculate the partial derivatives needed to evaluate (15.99) by differentiating Hamilton's equations (15.48) with respect to the ray parameters $\gamma' = \gamma_1', \gamma_2'$. This results in a system of linear equations governing the twelve derivatives $\partial_{\gamma'}\mathbf{x}$ and $\partial_{\gamma'}\mathbf{p}$:

$$\frac{d}{d\sigma}\begin{pmatrix} \partial_{\gamma'}\mathbf{x} \\ \partial_{\gamma'}\mathbf{p} \end{pmatrix} = \begin{pmatrix} \partial_{\mathbf{p}}\partial_{\mathbf{x}}H & \partial_{\mathbf{p}}\partial_{\mathbf{p}}H \\ -\partial_{\mathbf{x}}\partial_{\mathbf{x}}H & -\partial_{\mathbf{x}}\partial_{\mathbf{p}}H \end{pmatrix} \cdot \begin{pmatrix} \partial_{\gamma'}\mathbf{x} \\ \partial_{\gamma'}\mathbf{p} \end{pmatrix}. \tag{15.112}$$

For future reference, we denote the six-dimensional tensor on the right side of (15.112) by

$$\mathbf{A} = \begin{pmatrix} \partial_{\mathbf{p}}\partial_{\mathbf{x}}H & \partial_{\mathbf{p}}\partial_{\mathbf{p}}H \\ -\partial_{\mathbf{x}}\partial_{\mathbf{x}}H & -\partial_{\mathbf{x}}\partial_{\mathbf{p}}H \end{pmatrix} = \begin{pmatrix} \mathbf{0} & \mathbf{I} \\ \frac{1}{2}\boldsymbol{\nabla}\boldsymbol{\nabla}v^{-2} & \mathbf{0} \end{pmatrix}. \tag{15.113}$$

The rays leaving a point source are characterized by two requirements:

$$\partial_{\gamma'}\mathbf{x}' = \mathbf{0} \quad \text{and} \quad \mathbf{p}' \cdot \partial_{\gamma'}\mathbf{p}' = 0. \tag{15.114}$$

In accordance with (15.114), we must integrate (15.112) subject to the initial conditions

$$\partial_{\gamma'}\mathbf{x}(0) = \mathbf{0}, \qquad \partial_{\gamma'}\mathbf{p}(0) = (\mathbf{I} - \hat{\mathbf{p}}'\hat{\mathbf{p}}') \cdot \partial_{\gamma'}\mathbf{p}'. \tag{15.115}$$

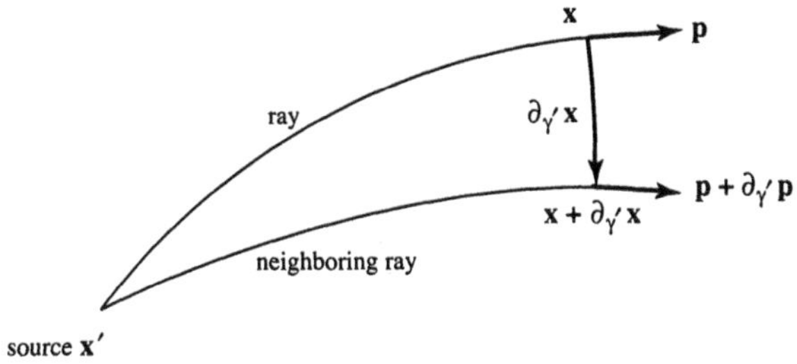

Figure 15.2. Schematic depiction of a central and neighboring or paraxial ray shot from a source $\mathbf{x}'$. The position vectors $\mathbf{x}$ and $\mathbf{x} + \partial_{\gamma'}\mathbf{x}$ are situated at the same value of σ along the two rays. The differential vector $\partial_{\gamma'}\mathbf{x}$ therefore lies within the wavefront. Wavefront curvature causes the central and paraxial slownesses $\mathbf{p}$ and $\mathbf{p} + \partial_{\gamma'}\mathbf{p}$ to be misaligned.

The two vectors $\mathbf{x} + \partial_{\gamma'}\mathbf{x}$ and $\mathbf{p} + \partial_{\gamma'}\mathbf{p}$ are the position and slowness along a pair of *neighboring or paraxial rays*, as illustrated in Figure 15.2. Equations (15.112) are referred to as the *dynamic ray-tracing equations*, to distinguish them from the kinematic equations (15.48). Alternative forms of these equations with the independent variable σ replaced by either s or T and the Hamiltonian H replaced by H' or H'' are easy to obtain.

Equation (15.84) imposes a constraint upon the solutions $\partial_{\gamma'}\mathbf{x}$, $\partial_{\gamma'}\mathbf{p}$ to the dynamic ray-tracing equations. Differentiation of the eikonal equation (15.47) yields another:

$$\begin{aligned}\partial_{\gamma'}H &= \partial_{\gamma'}\mathbf{x}\cdot\partial_{\mathbf{x}}H + \partial_{\gamma'}\mathbf{p}\cdot\partial_{\mathbf{p}}H \\ &= -\tfrac{1}{2}\partial_{\gamma'}\mathbf{x}\cdot\boldsymbol{\nabla}v^{-2} + \partial_{\gamma'}\mathbf{p}\cdot\mathbf{p} = 0.\end{aligned} \qquad (15.116)$$

Imposition of the two conditions (15.114) at the source $\sigma = 0$ guarantees that (15.84) and (15.116) are both satisfied for all $\sigma > 0$. Just as in the kinematic case, it is possible to use the constraints (15.84) and (15.116) to reduce (15.112) to a system of eight rather than twelve dynamic ray-tracing equations in ray-centered coordinates (Červený 1985; Farra & Madariaga 1987). We present an alternative system of eight dynamical equations in Section 15.8.5.

*15.4.9 Phase-space propagator

More general solutions to the dynamical ray-tracing equations (15.112), with arbitrary initial conditions, may be written in terms of the *propagator*

$\mathbf{P}(\sigma, 0)$ between the origin and a point σ along a ray, defined by (Gilbert & Backus 1966)

$$\frac{d\mathbf{P}}{d\sigma} = \mathbf{A} \cdot \mathbf{P}, \qquad \mathbf{P}(0, \sigma) \cdot \mathbf{P}(\sigma, 0) = \mathbf{P}(0, 0) = \mathbf{I}. \tag{15.117}$$

The elements of this six-dimensional propagator are the thirty-six partial derivatives of the final position and slowness coordinates with respect to the initial ones:

$$\mathbf{P}(\sigma, 0) = \begin{pmatrix} \dfrac{\partial x_1}{\partial x_1'} & \cdots & \dfrac{\partial x_1}{\partial p_3'} \\ \vdots & & \vdots \\ \dfrac{\partial p_3}{\partial x_1'} & \cdots & \dfrac{\partial p_3}{\partial p_3'} \end{pmatrix}. \tag{15.118}$$

As in any phase-space analysis, the initial position $\mathbf{x}'$ and slowness $\mathbf{p}'$ are regarded as *independent* variables. A given term such as the one in the upper right corner is thus of the form $(\partial x_1/\partial p_3')_{x_1, x_2, x_3, p_1', p_2', \sigma}$. It is convenient to write (15.118) in terms of four three-dimensional sub-propagators:

$$\mathbf{P} = \begin{pmatrix} \mathbf{X_x} & \mathbf{X_p} \\ \mathbf{P_x} & \mathbf{P_p} \end{pmatrix}. \tag{15.119}$$

The upper-right term $\mathbf{X_p}$ is the only one that is required to find the Jacobian (15.86) and geometrical spreading factor (15.99) of a point source:

$$\partial_{\gamma'}\mathbf{x} = \mathbf{X_p} \cdot (\mathbf{I} - \hat{\mathbf{p}}'\hat{\mathbf{p}}') \cdot \partial_{\gamma'}\mathbf{p}'. \tag{15.120}$$

Of course, it is not possible to solve for $\mathbf{X_p}$ alone, since it is coupled to the other three-dimensional propagators $\mathbf{X_x}$, $\mathbf{P_x}$ and $\mathbf{P_p}$.

It is easily demonstrated by means of an argument similar to that in Section 15.4.4 that the determinant of the phase-space propagator (15.118) satisfies

$$\frac{d}{d\sigma}(\det \mathbf{P}) = (\operatorname{tr} \mathbf{A})(\det \mathbf{P}). \tag{15.121}$$

Equation (15.121) can be integrated to yield the explicit result

$$\det \mathbf{P}(\sigma, 0) = \exp\left(\int_0^\sigma \operatorname{tr} \mathbf{A}\, d\sigma\right), \tag{15.122}$$

where we have made use of the initial condition $\det \mathbf{P}(0,0) = \det \mathbf{I} = 1$. The six-tensor (15.113) has zero trace: $\operatorname{tr} \mathbf{A} = 0$. It follows that we must have

$$\frac{\partial(x_1, x_2, x_3, p_1, p_2, p_3)}{\partial(x_1', x_2', x_3', p_1', p_2', p_3')} = \det \begin{pmatrix} \dfrac{\partial x_1}{\partial x_1'} & \cdots & \dfrac{\partial x_1}{\partial p_3'} \\ \vdots & & \vdots \\ \dfrac{\partial p_3}{\partial x_1'} & \cdots & \dfrac{\partial p_3}{\partial p_3'} \end{pmatrix} = 1 \tag{15.123}$$

at every point σ along a ray. Equation (15.123) stipulates that the size of a differential volume element in phase space is conserved; this is known as *Liouville's theorem* (Goldstein 1980).

★15.4.10 Symplectic structure

The kinematic and dynamic ray-tracing equations can be written in a succinct manner in terms of the six-vectors

$$\mathbf{y} = \begin{pmatrix} \mathbf{x} \\ \mathbf{p} \end{pmatrix}, \qquad \partial_{\gamma'}\mathbf{y} = \begin{pmatrix} \partial_{\gamma'}\mathbf{x} \\ \partial_{\gamma'}\mathbf{p} \end{pmatrix}. \tag{15.124}$$

We use a dot to denote differentiation with respect to σ, and introduce the anti-symmetric six-tensor

$$\mathbf{J} = \begin{pmatrix} \mathbf{0} & \mathbf{I} \\ -\mathbf{I} & \mathbf{0} \end{pmatrix} = -\mathbf{J}^{\mathrm{T}}. \tag{15.125}$$

Equations (15.48) and (15.112) are then equivalent to

$$\dot{\mathbf{y}} = \mathbf{J} \cdot \partial_{\mathbf{y}} H, \qquad \partial_{\gamma'}\dot{\mathbf{y}} = \mathbf{J} \cdot \partial_{\mathbf{yy}} H \cdot \partial_{\gamma'}\mathbf{y}. \tag{15.126}$$

This is known as the *symplectic* form of the phase-space equations (Goldstein 1980). The symmetric tensor

$$\partial_{\mathbf{yy}} H = \begin{pmatrix} \partial_{\mathbf{x}}\partial_{\mathbf{x}} H & \partial_{\mathbf{x}}\partial_{\mathbf{p}} H \\ \partial_{\mathbf{p}}\partial_{\mathbf{x}} H & \partial_{\mathbf{p}}\partial_{\mathbf{p}} H \end{pmatrix} = (\partial_{\mathbf{yy}} H)^{\mathrm{T}} \tag{15.127}$$

is the *Hessian* of the Hamiltonian. It is readily verified that $\mathbf{A} = \mathbf{J} \cdot \partial_{\mathbf{yy}} H$. The propagator (15.118) and its transpose satisfy

$$\dot{\mathbf{P}} = \mathbf{J} \cdot \partial_{\mathbf{yy}} H \cdot \mathbf{P}, \qquad \dot{\mathbf{P}}^{\mathrm{T}} = \mathbf{P}^{\mathrm{T}} \cdot \partial_{\mathbf{yy}} H \cdot \mathbf{J}^{\mathrm{T}}. \tag{15.128}$$

Making use of (15.128) and the identities $\mathbf{J}\cdot\mathbf{J} = \mathbf{J}^{\mathrm{T}}\cdot\mathbf{J}^{\mathrm{T}} = -\mathbf{I}$, we find that

$$\begin{aligned}\frac{d}{d\sigma}\Big(\mathbf{P}^{\mathrm{T}}\cdot\mathbf{J}\cdot\mathbf{P}\Big) &= \mathbf{P}^{\mathrm{T}}\cdot\mathbf{J}\cdot\dot{\mathbf{P}} - \dot{\mathbf{P}}^{\mathrm{T}}\cdot\mathbf{J}^{\mathrm{T}}\cdot\mathbf{P} \\ &= -\mathbf{P}^{\mathrm{T}}\cdot\partial_{\mathbf{yy}}H\cdot\mathbf{P} + \mathbf{P}^{\mathrm{T}}\cdot\partial_{\mathbf{yy}}H\cdot\mathbf{P} = \mathbf{0}.\end{aligned} \tag{15.129}$$

The initial value of the propagator is $\mathbf{P}(0,0) = \mathbf{I}$, so that the initial value of the product $\mathbf{P}^{\mathrm{T}}\cdot\mathbf{J}\cdot\mathbf{P}$ is $\mathbf{J}$. Equation (15.129) states that this value is preserved along a ray:

$$\mathbf{P}^{\mathrm{T}}\cdot\mathbf{J}\cdot\mathbf{P} = \mathbf{J}. \tag{15.130}$$

Any transformation $\mathbf{P}$ from $\mathbf{x}', \mathbf{p}'$ to $\mathbf{x}, \mathbf{p}$ having the property (15.130) is said to be symplectic. Upon taking the determinant of (15.130) and noting that $\det\mathbf{J} = 1$, we deduce that $(\det\mathbf{P})^2 = 1$. The initial condition dictates that we must choose the positive square root; this is an independent proof of Liouville's theorem.

Suppose now that we dot equation (15.130) on the left with $\mathbf{J}$ and on the right with the inverse propagator $\mathbf{P}^{-1}(\sigma,0) = \mathbf{P}(0,\sigma)$. This yields

$$\mathbf{P}^{-1} = -\mathbf{J}\cdot\mathbf{P}^{\mathrm{T}}\cdot\mathbf{J}. \tag{15.131}$$

Upon inserting the decomposition (15.119) and performing the indicated transpositions and multiplications, we find that

$$\mathbf{P}^{-1} = \begin{pmatrix} \mathbf{P}_{\mathbf{p}}^{\mathrm{T}} & -\mathbf{X}_{\mathbf{p}}^{\mathrm{T}} \\ -\mathbf{P}_{\mathbf{x}}^{\mathrm{T}} & \mathbf{X}_{\mathbf{x}}^{\mathrm{T}} \end{pmatrix}. \tag{15.132}$$

The point-source sub-propagator $\mathbf{X}_{\mathbf{p}}$, in particular, satisfies

$$\mathbf{X}_{\mathbf{p}}(0,\sigma) = -\mathbf{X}_{\mathbf{p}}^{\mathrm{T}}(\sigma,0). \tag{15.133}$$

Following Kendall, Guest & Thomson (1992) we can use the symplectic symmetry (15.133) to verify the dynamical reciprocity (15.109) of the geometrical spreading factor. We begin by substituting the representation (15.120) into (15.87); this yields an expression for the point-source Jacobian, which we write using index notation:

$$\begin{aligned} vv'^2 J &= \varepsilon_{jkl}\,\hat{p}_j X_{km} X_{ln}(\partial_{\gamma_1'}\hat{p}_m')(\partial_{\gamma_2'}\hat{p}_n') \\ &= \tfrac{1}{2}\varepsilon_{imn}\varepsilon_{jkl}\,\hat{p}_i'\hat{p}_j X_{km}X_{ln}\|\partial_{\gamma_1'}\hat{\mathbf{p}}' \times \partial_{\gamma_2'}\hat{\mathbf{p}}'\|, \end{aligned} \tag{15.134}$$

where we have dropped the subscript on $\mathbf{X}_{\mathbf{p}}$ for simplicity. The Levi-Cività product $\frac{1}{2}\varepsilon_{imn}\varepsilon_{jkl}\,\hat{p}_i'\hat{p}_j X_{km}X_{ln}$ can be expressed in terms of $\det\mathbf{X}_{\mathbf{p}}$ and the three-dimensional inverse $\mathbf{X}_{\mathbf{p}}^{-1}$ with the aid of Cramer's rule (A.36):

$$vv'^2 J = (\det\mathbf{X}_{\mathbf{p}})(\hat{\mathbf{p}}'\cdot\mathbf{X}_{\mathbf{p}}^{-1}\cdot\hat{\mathbf{p}})\|\partial_{\gamma_1'}\hat{\mathbf{p}}' \times \partial_{\gamma_2'}\hat{\mathbf{p}}'\|. \tag{15.135}$$

The initial cross-product magnitude $\|\partial_{\gamma_1'}\hat{\mathbf{p}}' \times \partial_{\gamma_2'}\hat{\mathbf{p}}'\|$ cancels upon substituting (15.135) into (15.99) to find the spreading factor:

$$v'\mathcal{R} = |(\det \mathbf{X}_{\mathbf{p}})(\hat{\mathbf{p}}' \cdot \mathbf{X}_{\mathbf{p}}^{-1} \cdot \hat{\mathbf{p}})|^{1/2}. \tag{15.136}$$

The right side of (15.136) is invariant under an interchange of source and receiver (15.108), so that the reciprocity relation (15.109) is confirmed. The two-point gradient of the travel time and the point-source sub-propagator are related by $(\det \mathbf{X}_{\mathbf{p}})(\det \boldsymbol{\nabla}\boldsymbol{\nabla}'T)[\hat{\mathbf{p}}' \cdot \mathbf{X}_{\mathbf{p}}^{-1} \cdot \hat{\mathbf{p}}][\hat{\mathbf{p}}' \cdot (\boldsymbol{\nabla}\boldsymbol{\nabla}'T)^{-1} \cdot \hat{\mathbf{p}}] = 1$. The geometrical argument given in Section 15.4.6 is more general than the above analytical proof, because it focuses only upon the endpoints $\mathbf{x}, \mathbf{x}'$ without regard for the intervening "life history" along a ray. The result (15.109) is valid even in a piecewise discontinuous Earth model, with an arbitrary number of boundary interactions between the source and receiver.

*15.5 Polarization

It is well known that an SV or SH wave retains its polarization as it propagates through a smooth sub-region of a spherically symmetric Earth. This raises the question—how does the polarization of a shear wave evolve along a ray in a smooth sub-region of a more general laterally heterogeneous Earth? The answer to this question cannot be ascertained from the slow variational analysis presented in Section 15.2. We must resort to an alternative method, which we turn to next.

*15.5.1 Classical JWKB analysis

We focus attention in this classical approach not upon the Lagrangian density (15.7), but rather upon the elastodynamic equation of motion (15.1). We substitute a more general JWKB ansatz,

$$\mathbf{s} = \left[\mathbf{A}^{(0)} + \omega^{-1}\mathbf{A}^{(1)} + \cdots\right] \exp(-i\omega T), \tag{15.137}$$

into this equation, and collect and equate like powers of ω^{-1}. The two leading equations obtained in this manner are

$$\left[(\rho - \|\mathbf{p}\|^2\mu)\mathbf{I} - (\kappa + \tfrac{1}{3}\mu)\mathbf{p}\mathbf{p}\right] \cdot \mathbf{A} = \mathbf{0}, \tag{15.138}$$

$$\begin{aligned}\boldsymbol{\nabla}(\kappa - \tfrac{2}{3}\mu)(\mathbf{p}\cdot\mathbf{A}) + \boldsymbol{\nabla}\mu \cdot (\mathbf{p}\mathbf{A} + \mathbf{A}\mathbf{p}) \\ + (\kappa + \tfrac{1}{3}\mu)[\boldsymbol{\nabla}\cdot(\mathbf{p}\cdot\mathbf{A}) + (\boldsymbol{\nabla}\cdot\mathbf{A})\mathbf{p}] \\ + \mu[(\boldsymbol{\nabla}\cdot\mathbf{p})\mathbf{A} + 2\mathbf{p}\cdot\boldsymbol{\nabla}\mathbf{A})] = \mathbf{0},\end{aligned} \tag{15.139}$$

where we have introduced the slowness $\mathbf{p} = \boldsymbol{\nabla} T$, and eliminated the superscript upon $\mathbf{A}^{(0)} = \mathbf{A}$, since we are only interested in the lowest-order approximation. The first of these results is the eikonal equation (15.25), whereas the second is a *vector* version of the *transport equation*. The scalar equation (15.26) can be recovered by dotting (15.139) with $\mathbf{A}$.

A compressional wave has $\|\mathbf{p}_\mathrm{P}\|^2 = \alpha^{-2}$ and an associated amplitude of the form

$$\mathbf{A}_\mathrm{P} = A_\mathrm{P}\hat{\boldsymbol{\eta}}_\mathrm{P} \quad \text{where} \quad \hat{\boldsymbol{\eta}}_\mathrm{P} = \hat{\mathbf{p}}_\mathrm{P}. \tag{15.140}$$

Upon substituting (15.140) into equation (15.139) we obtain the compressional energy conservation law $\boldsymbol{\nabla}\cdot(\rho\alpha A_\mathrm{P}^2\hat{\boldsymbol{\eta}}_\mathrm{P}) = 0$, as expected. A shear wave has $\|\mathbf{p}_\mathrm{S}\|^2 = \beta^{-2}$ and

$$\mathbf{A}_\mathrm{S} = A_\mathrm{S}\hat{\boldsymbol{\eta}}_\mathrm{S} \quad \text{where} \quad \hat{\boldsymbol{\eta}}_\mathrm{S}\cdot\hat{\mathbf{p}}_\mathrm{S} = 0. \tag{15.141}$$

Upon substituting (15.141) into (15.139) and taking the cross product with $A_\mathrm{S}\hat{\mathbf{p}}_\mathrm{S}$ we obtain, after some manipulation,

$$[\boldsymbol{\nabla}\cdot(\rho\beta A_\mathrm{S}^2\hat{\boldsymbol{\eta}}_\mathrm{S})](\hat{\mathbf{p}}_\mathrm{S}\times\hat{\boldsymbol{\eta}}_\mathrm{S}) + 2\rho\beta A_\mathrm{S}^2(\hat{\mathbf{p}}_\mathrm{S}\times d\hat{\boldsymbol{\eta}}_\mathrm{S}/ds) = \mathbf{0}, \tag{15.142}$$

where s is the arclength. The two terms in equation (15.142) are mutually perpendicular, so they must vanish individually. In addition to the energy conservation law $\boldsymbol{\nabla}\cdot(\rho\beta A_\mathrm{S}^2\hat{\boldsymbol{\eta}}_\mathrm{S}) = 0$, we obtain a constraint upon the shear-wave polarization:

$$\hat{\mathbf{p}}_\mathrm{S}\times\frac{d\hat{\boldsymbol{\eta}}_\mathrm{S}}{ds} = \mathbf{0}. \tag{15.143}$$

This equation stipulates that the polarization must evolve along a ray in such a way that its rate of change is always parallel to the direction of shear-wave propagation: $d\hat{\boldsymbol{\eta}}_\mathrm{S}/ds \parallel \hat{\mathbf{p}}_\mathrm{S}$.

★15.5.2 Shear-wave basis

Dropping the subscript S for simplicity, we define a mutually perpendicular pair of shear-wave polarization vectors by

$$\hat{\boldsymbol{\eta}}_1 = \hat{\boldsymbol{\nu}}\cos\psi + \hat{\mathbf{b}}\sin\psi, \qquad \hat{\boldsymbol{\eta}}_2 = -\hat{\boldsymbol{\nu}}\sin\psi + \hat{\mathbf{b}}\cos\psi. \tag{15.144}$$

It is noteworthy that the two basis vectors (15.144) differ from those first introduced in this context by Popov & Pšenčík (1976); our $\hat{\boldsymbol{\eta}}_1$ and $\hat{\boldsymbol{\eta}}_2$ are obtained by a *clockwise* rotation of the normal $\hat{\boldsymbol{\nu}}$ and binormal $\hat{\mathbf{b}}$ through an angle ψ, as illustrated in Figure 15.3. The three vectors $\hat{\mathbf{p}}$, $\hat{\boldsymbol{\eta}}_1$, $\hat{\boldsymbol{\eta}}_2$ form a right-handed orthonormal coordinate system: $\hat{\mathbf{p}}\cdot(\hat{\boldsymbol{\eta}}_1\times\hat{\boldsymbol{\eta}}_2) = 1$. Upon

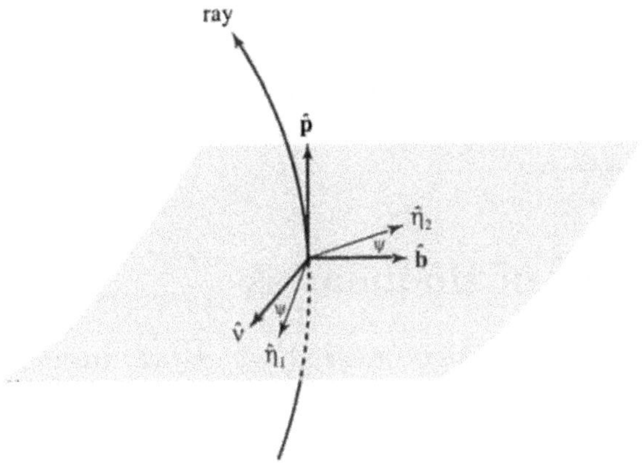

Figure 15.3. Relation of the shear-wave polarization vectors $\hat{\boldsymbol{\eta}}_1$ and $\hat{\boldsymbol{\eta}}_2$ to the ray normal $\hat{\boldsymbol{\nu}}$ and binormal $\hat{\mathbf{b}}$. The shaded plane is perpendicular to the ray tangent $\hat{\mathbf{p}}$; the four vectors $\hat{\boldsymbol{\nu}}$, $\hat{\mathbf{b}}$ and $\hat{\boldsymbol{\eta}}_1$, $\hat{\boldsymbol{\eta}}_2$ lie within this local wavefront plane.

differentiating equations (15.144) and invoking the Serret-Frénet formulae (15.66), we find that

$$\frac{d\hat{\boldsymbol{\eta}}_1}{ds} = -\kappa \cos\psi\,\hat{\mathbf{p}} + (\tau + d\psi/ds)\hat{\boldsymbol{\eta}}_2, \tag{15.145}$$

$$\frac{d\hat{\boldsymbol{\eta}}_2}{ds} = \kappa \sin\psi\,\hat{\mathbf{p}} - (\tau + d\psi/ds)\hat{\boldsymbol{\eta}}_1, \tag{15.146}$$

where κ and τ are the ray curvature and torsion, respectively. We see that we may render $d\hat{\boldsymbol{\eta}}_1/ds$ and $d\hat{\boldsymbol{\eta}}_2/ds$ parallel to $\mathbf{p}$, as required, by setting

$$\frac{d\psi}{ds} = -\tau. \tag{15.147}$$

At the source point $\mathbf{x}'$, we are free to choose the two initial shear-wave polarizations $\hat{\boldsymbol{\eta}}_1' = \hat{\boldsymbol{\nu}}'\cos\psi' + \hat{\mathbf{b}}'\sin\psi'$ and $\hat{\boldsymbol{\eta}}_2' = -\hat{\boldsymbol{\nu}}'\sin\psi' + \hat{\mathbf{b}}'\cos\psi'$ arbitrarily. The subsequent evolution of these independent vectors along a ray is then given by

$$\hat{\boldsymbol{\eta}}_1 = \hat{\boldsymbol{\eta}}_1' - \int_0^s \kappa\cos\psi\,\hat{\mathbf{p}}\,ds, \qquad \hat{\boldsymbol{\eta}}_2 = \hat{\boldsymbol{\eta}}_2' + \int_0^s \kappa\sin\psi\,\hat{\mathbf{p}}\,ds, \tag{15.148}$$

where

$$\psi = \psi' - \int_0^s \tau\,ds. \tag{15.149}$$

The basis vectors $\hat{\boldsymbol{\eta}}_1$ and $\hat{\boldsymbol{\eta}}_2$ twist around a ray relative to the normal $\hat{\boldsymbol{\nu}}$ and the binormal $\hat{\mathbf{b}}$ at a rate (15.147) that is *equal and opposite* to the rate at which those vectors twist themselves. In this sense, a shear wave in a laterally heterogeneous Earth may be said to "carry" its polarization with it as it propagates.

★15.6 Effect of Boundaries

Thus far our discussion of kinematic and dynamic ray tracing has been limited to the smooth sub-region of the Earth surrounding the source position. In this section we consider the effect of the external and internal boundaries $\Sigma = \partial\oplus \cup \Sigma_{\mathrm{SS}} \cup \Sigma_{\mathrm{FS}}$.

★15.6.1 Snell's law

The kinematic boundary conditions governing the displacement (15.16) require that the travel time along a ray must be continuous:

$$[T]_-^+ = 0. \tag{15.150}$$

It follows from (15.150) that $[\boldsymbol{\nabla}^\Sigma T]_-^+ = [\boldsymbol{\nabla} T - \hat{\mathbf{n}}\partial_n T]_-^+ = \mathbf{0}$. Written in terms of the slowness $\mathbf{p}$, this condition takes the form

$$[\mathbf{p}^\Sigma]_-^+ = [\mathbf{p} - \hat{\mathbf{n}}(\hat{\mathbf{n}}\cdot\mathbf{p})]_-^+ = \mathbf{0}. \tag{15.151}$$

This is *Snell's law*—the tangential component of the slowness must be continuous. The jump discontinuity in the normal component $[\hat{\mathbf{n}}\cdot\mathbf{p}]_-^+$ is determined by this law, together with the condition that the Hamiltonian must vanish on both sides of the boundary:

$$[H]_-^+ = \tfrac{1}{2}[\mathbf{p}\cdot\mathbf{p} - v^{-2}]_-^+ = 0. \tag{15.152}$$

The outgoing wave in equations (15.151)–(15.152) need not be of the same type as the incident wave; furthermore, both conditions pertain to reflected as well as transmitted waves, with an obvious re-interpretation of the "jump" symbol $[\,\cdot\,]_-^+$. In the case of a P-to-P or S-to-S reflection, the normal component of the slowness is reversed: $\hat{\mathbf{n}}\cdot\mathbf{p} \to -\hat{\mathbf{n}}\cdot\mathbf{p}$. Hamilton's equations (15.48) in the smooth sub-regions of the Earth, together with (15.151)–(15.152) on the boundaries Σ, enable us to find $\mathbf{x}$ and $\mathbf{p}$, and thus the travel time T, everywhere along a ray.

*15.6.2 Jump in geometrical spreading

The cross-sectional area of a ray tube $d\Sigma$, and therefore the geometrical spreading factor $\mathcal{R}$, suffer jump discontinuities at a boundary. To calculate $[\mathcal{R}]_-^+$ we must find the jumps in the paraxial vectors $[\partial_{\gamma'}\mathbf{x}]_-^+$ and $[\partial_{\gamma'}\mathbf{p}]_-^+$. We consider a specific geometry for concreteness—an incoming wave impinging upon the boundary Σ from below, giving rise to a transmitted wave in the overlying medium. We seek to determine the outgoing spreading factor given the incoming one: $\mathcal{R}_- \rightarrow \mathcal{R}_+$. The other transmitted case, $\mathcal{R}_+ \rightarrow \mathcal{R}_-$, and the two reflected cases, $\mathcal{R}_- \rightarrow \mathcal{R}_-$ and $\mathcal{R}_+ \rightarrow \mathcal{R}_+$, require only a trivial modification.

The situation under consideration is depicted in Figure 15.4; the central and paraxial rays strike the boundary at different points $\mathbf{x}$ and $\mathbf{x}+d\mathbf{x}_-$ and different "instants" σ and $\sigma+d\sigma_-$, as shown. We denote the slowness of the incident ray at $\mathbf{x}$ by $\mathbf{p}_-$, and the position and slowness of the paraxial ray at the "instant" σ by $\mathbf{x}+\partial_{\gamma'}\mathbf{x}_-$ and $\mathbf{p}_- + \partial_{\gamma'}\mathbf{p}_-$. The incident slowness at the paraxial intersection point $\mathbf{x} + d\mathbf{x}_-$ will be denoted by $\mathbf{p}_- + d\mathbf{p}_-$. We can relate the differential intersection vectors $d\mathbf{x}_-$, $d\mathbf{p}_-$ to the incoming vectors

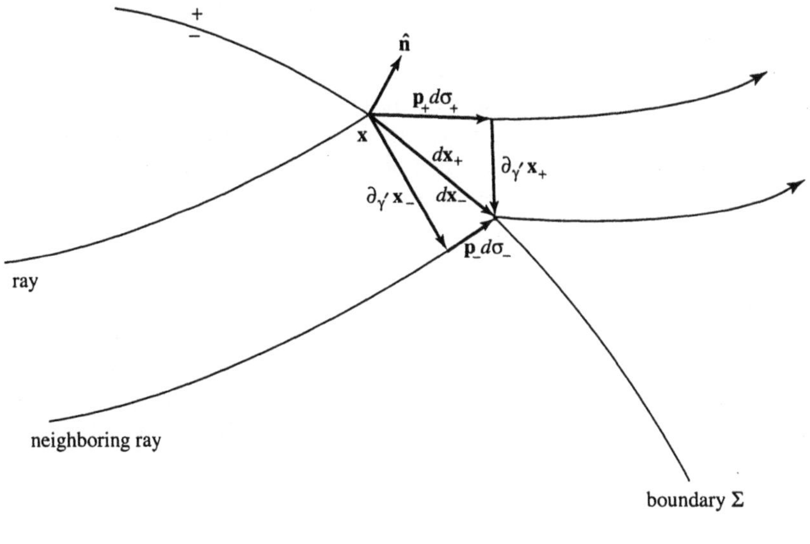

Figure 15.4. Schematic depiction of a ray tube impinging upon a boundary from below. The central ray and neighboring ray intersect Σ at the points $\mathbf{x}$ and $\mathbf{x}+d\mathbf{x}$, respectively. The unit normal to the boundary at the point $\mathbf{x}$ is $\hat{\mathbf{n}}$. The top-side and bottom-side differential vectors $\partial_{\gamma'}\mathbf{x}_\pm$ both lie within the wavefront.

$\partial_{\gamma'}\mathbf{x}_-$, $\partial_{\gamma'}\mathbf{p}_-$ by first-order integration of Hamilton's equations (15.48) in the underlying medium:

$$d\mathbf{x}_- = \partial_{\gamma'}\mathbf{x}_- + \partial_{\mathbf{p}}H_-\,d\sigma_- = \partial_{\gamma'}\mathbf{x}_- + \mathbf{p}_-\,d\sigma_-, \tag{15.153}$$

$$d\mathbf{p}_- = \partial_{\gamma'}\mathbf{p}_- - \partial_{\mathbf{x}}H_-\,d\sigma_- = \partial_{\gamma'}\mathbf{p}_- + \tfrac{1}{2}\boldsymbol{\nabla}v_-^{-2}\,d\sigma_-. \tag{15.154}$$

The propagation increment $d\sigma_-$ is determined by the first-order condition $\hat{\mathbf{n}}\cdot d\mathbf{x}_- = 0$ that both intersection points lie upon the boundary:

$$d\sigma_- = -\frac{\hat{\mathbf{n}}\cdot\partial_{\gamma'}\mathbf{x}_-}{\hat{\mathbf{n}}\cdot\mathbf{p}_-}. \tag{15.155}$$

Upon inserting equation (15.155) into (15.153)–(15.154) we obtain

$$\begin{pmatrix} d\mathbf{x}_- \\ d\mathbf{p}_- \end{pmatrix} = \begin{pmatrix} \boldsymbol{\Pi}_{1-} & \mathbf{0} \\ \boldsymbol{\Pi}_{2-} & \mathbf{I} \end{pmatrix}\cdot\begin{pmatrix} \partial_{\gamma'}\mathbf{x}_- \\ \partial_{\gamma'}\mathbf{p}_- \end{pmatrix}, \tag{15.156}$$

where

$$\boldsymbol{\Pi}_{1-} = \mathbf{I} - \frac{\mathbf{p}_-\hat{\mathbf{n}}}{\hat{\mathbf{n}}\cdot\mathbf{p}_-}, \qquad \boldsymbol{\Pi}_{2-} = -\frac{1}{2}\frac{\boldsymbol{\nabla}v_-^{-2}\hat{\mathbf{n}}}{\hat{\mathbf{n}}\cdot\mathbf{p}_-}. \tag{15.157}$$

For convenience in what follows, we also write (15.156)–(15.157) using an obvious six-vector notation: $d\mathbf{y}_- = \boldsymbol{\Pi}_-\cdot\partial_{\gamma'}\mathbf{y}_-$.

The central ray leaves the boundary from position $\mathbf{x}$ with slowness $\mathbf{p}_+$; we denote the departure position and outgoing slowness of the paraxial ray by $\mathbf{x}+d\mathbf{x}_+$ and $\mathbf{p}_+ + d\mathbf{p}_+$. Our next task is to relate the differential vectors $d\mathbf{x}_+$ and $d\mathbf{p}_+$ to $d\mathbf{x}_-$ and $d\mathbf{p}_-$. Since both the central and neighboring ray must be continuous, the constraint upon the paraxial position is simple:

$$[d\mathbf{x}]_-^+ = \mathbf{0}. \tag{15.158}$$

To find the jump in paraxial slowness $[d\mathbf{p}]_-^+$, we must account not only for the contrast in wave speed, but also for the curvature of the boundary. Perturbation of the continuity conditions (15.151)–(15.152) leads to two constraints. Resolving the perturbations $d\mathbf{p}_\pm = \hat{\mathbf{n}}(\hat{\mathbf{n}}\cdot d\mathbf{p}_\pm) + d\mathbf{p}_\pm^\Sigma$ and noting that the unit normal at $\mathbf{x}+d\mathbf{x}_\pm$ is $\hat{\mathbf{n}} + d\mathbf{x}_\pm\cdot\boldsymbol{\nabla}^\Sigma\hat{\mathbf{n}}$, we obtain

$$[d\mathbf{p}^\Sigma - (\hat{\mathbf{n}}\cdot\mathbf{p})(d\mathbf{x}\cdot\boldsymbol{\nabla}^\Sigma\hat{\mathbf{n}})]_-^+ = \mathbf{0}, \tag{15.159}$$

$$\begin{aligned}[dH]_-^+ &= [d\mathbf{x}\cdot\partial_{\mathbf{x}}H + d\mathbf{p}\cdot\partial_{\mathbf{p}}H]_-^+ \\ &= [-\tfrac{1}{2}\,d\mathbf{x}\cdot\boldsymbol{\nabla}v^{-2} + d\mathbf{p}\cdot\mathbf{p}]_-^+ = 0.\end{aligned} \tag{15.160}$$

We have made use of the tangent character of the surface curvature tensor, $\hat{\mathbf{n}}\cdot\boldsymbol{\nabla}^\Sigma\hat{\mathbf{n}} = \boldsymbol{\nabla}^\Sigma\hat{\mathbf{n}}\cdot\hat{\mathbf{n}} = \mathbf{0}$, in deriving the pariaxial version of Snell's law (15.159). Equation (15.160) can be regarded as the boundary analogue of the smooth-medium continuity condition (15.116). A moderate amount of straightforward algebra is required to manipulate these two constraints into the desired form:

$$\begin{pmatrix} d\mathbf{x}_+ \\ d\mathbf{p}_+ \end{pmatrix} = \begin{pmatrix} \mathbf{I} & \mathbf{0} \\ \mathbf{T}_1 & \mathbf{T}_2 \end{pmatrix} \cdot \begin{pmatrix} d\mathbf{x}_- \\ d\mathbf{p}_- \end{pmatrix}, \tag{15.161}$$

where

$$\mathbf{T}_1 = \hat{\mathbf{n}} \cdot (\mathbf{p}_+ - \mathbf{p}_-)\, \boldsymbol{\nabla}^\Sigma \hat{\mathbf{n}} - \left(1 - \frac{\hat{\mathbf{n}} \cdot \mathbf{p}_-}{\hat{\mathbf{n}} \cdot \mathbf{p}_+}\right) \hat{\mathbf{n}}\mathbf{p}_+ \cdot \boldsymbol{\nabla}^\Sigma \hat{\mathbf{n}}$$
$$+ \tfrac{1}{2} \left(\frac{1}{\hat{\mathbf{n}} \cdot \mathbf{p}_+}\right) \hat{\mathbf{n}} \boldsymbol{\nabla}(v_+^{-2} - v_-^{-2}), \tag{15.162}$$

$$\mathbf{T}_2 = \mathbf{I} - \hat{\mathbf{n}}\hat{\mathbf{n}} + \left(\frac{\hat{\mathbf{n}} \cdot \mathbf{p}_-}{\hat{\mathbf{n}} \cdot \mathbf{p}_+}\right) \hat{\mathbf{n}}\hat{\mathbf{n}}. \tag{15.163}$$

The six-vector abbreviated form of (15.162)–(15.163) is $d\mathbf{y}_+ = \mathbf{T} \cdot d\mathbf{y}_-$.

Thus far we have followed the treatment by Farra, Virieux & Madariaga (1989). We now take an additional step, and extrapolate the top-side vectors $d\mathbf{x}_+$ and $d\mathbf{p}_+$ in a manner analogous to (15.153)–(15.154) to find new paraxial vectors $\partial_{\gamma'}\mathbf{x}_+$ and $\partial_{\gamma'}\mathbf{p}_+$ in the overlying medium:

$$d\mathbf{x}_+ = \partial_{\gamma'}\mathbf{x}_+ + \partial_{\mathbf{p}} H_+\, d\sigma_+ = \partial_{\gamma'}\mathbf{x}_+ + \mathbf{p}_+\, d\sigma_+, \tag{15.164}$$

$$d\mathbf{p}_+ = \partial_{\gamma'}\mathbf{p}_+ - \partial_{\mathbf{x}} H_+\, d\sigma_+ = \partial_{\gamma'}\mathbf{p}_+ + \tfrac{1}{2}\boldsymbol{\nabla} v_+^{-2}\, d\sigma_+. \tag{15.165}$$

The outgoing increment $d\sigma_+$ can be determined by imposing the constraint $\mathbf{p}_+ \cdot \partial_{\gamma'}\mathbf{x}_+ = 0$ that both $\mathbf{x}$ and $\mathbf{x} + \partial_{\gamma'}\mathbf{x}_+$ lie upon the same wavefront:

$$d\sigma_+ = \frac{\mathbf{p}_+ \cdot d\mathbf{x}_+}{\mathbf{p}_+ \cdot \mathbf{p}_+}. \tag{15.166}$$

Upon inserting (15.166) into (15.164)–(15.165) we obtain

$$\begin{pmatrix} \partial_{\gamma'}\mathbf{x}_+ \\ \partial_{\gamma'}\mathbf{p}_+ \end{pmatrix} = \begin{pmatrix} \boldsymbol{\Pi}_{1+} & \mathbf{0} \\ \boldsymbol{\Pi}_{2+} & \mathbf{I} \end{pmatrix} \cdot \begin{pmatrix} d\mathbf{x}_+ \\ d\mathbf{p}_+ \end{pmatrix}, \tag{15.167}$$

where

$$\boldsymbol{\Pi}_{1+} = \mathbf{I} - \frac{\mathbf{p}_+\mathbf{p}_+}{\mathbf{p}_+ \cdot \mathbf{p}_+}, \qquad \boldsymbol{\Pi}_{2+} = -\tfrac{1}{2}\frac{\boldsymbol{\nabla} v_+^{-2}\mathbf{p}_+}{\mathbf{p}_+ \cdot \mathbf{p}_+}. \tag{15.168}$$

We shall abbreviate this as $d\mathbf{y}_+ = \boldsymbol{\Pi}_+ \cdot \partial_{\gamma'}\mathbf{y}_+$.

Upon concatenating the above results, we obtain the final relation

$$\partial_{\gamma'}\mathbf{y}_+ = \mathbf{B}\cdot\partial_{\gamma'}\mathbf{y}_- \quad \text{where} \quad \mathbf{B} = \boldsymbol{\Pi}_+\cdot\mathbf{T}\cdot\boldsymbol{\Pi}_-. \tag{15.169}$$

Written out explicitly, equation (15.169) is

$$\begin{pmatrix} \partial_{\gamma'}\mathbf{x}_+ \\ \partial_{\gamma'}\mathbf{p}_+ \end{pmatrix} = \begin{pmatrix} \mathbf{B}_1 & \mathbf{0} \\ \mathbf{B}_2 & \mathbf{B}_3 \end{pmatrix}\cdot\begin{pmatrix} \partial_{\gamma'}\mathbf{x}_- \\ \partial_{\gamma'}\mathbf{p}_- \end{pmatrix}, \tag{15.170}$$

where

$$\mathbf{B}_1 = \boldsymbol{\Pi}_{1+}\cdot\boldsymbol{\Pi}_{1-}, \tag{15.171}$$

$$\mathbf{B}_2 = \boldsymbol{\Pi}_{2+}\cdot\boldsymbol{\Pi}_{1-} + \mathbf{T}_1\cdot\boldsymbol{\Pi}_{1-} + \mathbf{T}_2\cdot\boldsymbol{\Pi}_{2-}, \tag{15.172}$$

$$\mathbf{B}_3 = \mathbf{T}_2. \tag{15.173}$$

Equation (15.170) is the relation needed to perform dynamical ray tracing in the presence of boundaries. We integrate the linear system (15.112) from the source point $\mathbf{x}'$ to the first boundary intersection, use (15.170) to step from the $-$ side to the $+$ side, continue integrating to the next boundary, and so on. Strictly speaking, the outgoing paraxial vectors $\partial_{\gamma'}\mathbf{x}_+$, $\partial_{\gamma'}\mathbf{p}_+$ are evaluated at $\sigma + d\sigma_- + d\sigma_+$ rather than at σ; however, such a "hiccup" is inconsequential in the limit of an infinitesimally narrow ray tube. It should be noted that $\mathbf{B}$ does *not* have the requisite six degrees of freedom needed to be a full-fledged boundary propagator, by virtue of the restriction $\mathbf{p}_+\cdot\partial_{\gamma'}\mathbf{x}_+ = 0$. For this reason, it is not symplectic: $\mathbf{B}^{\mathrm{T}}\cdot\mathbf{J}\cdot\mathbf{B} \neq \mathbf{J}$.

⋆15.6.3 Polarization and energy partition

In addition to (15.150) the JWKB ray sum at every ray-boundary intersection point is required to satisfy

$$\sum_{\text{rays}} [\mathbf{A}]_-^+ = \mathbf{0} \quad \text{on } \Sigma_{\mathrm{SS}}, \qquad \sum_{\text{rays}} [\hat{\mathbf{n}}\cdot\mathbf{A}]_-^+ = 0 \quad \text{on } \Sigma_{\mathrm{FS}}, \tag{15.174}$$

$$\sum_{\text{rays}} [\hat{\mathbf{n}}\cdot(\rho v A^2\hat{\mathbf{p}})]_-^+ = 0 \quad \text{on all of } \Sigma. \tag{15.175}$$

The first two relations (15.174) are obvious kinematic continuity conditions; the third (15.175) is the dynamical energy-flux conservation law (15.36). The sums are over the incident ray and all of the reflected and transmitted rays generated at the boundary. The coupling of incident and outgoing compressional and shear waves is completely specified at every boundary point $\mathbf{x}$ by equations (15.174)–(15.175) together with Snell's law (15.151).

Straightforward but tedious algebra is all that is required to find the outgoing wave amplitudes $\mathbf{A}^{\rm out}$ in terms of the incident amplitude $\mathbf{A}^{\rm inc}$. We do not give any details here, but simply outline the essential physical ideas; for a comprehensive treatment in ray-centered coordinates, see Červený, Molotkov and Pšenčík (1977) or Červený (1985).

We begin by noting that it is possible to define local "horizontal" and "vertical" shear-wave polarizations in terms of the unit slowness $\hat{\mathbf{p}}$ of either the incident or outgoing wave and the unit normal $\hat{\mathbf{n}}$ at every point on Σ:

$$\hat{\boldsymbol{\eta}}_{\rm SH} = \pm\frac{\hat{\mathbf{p}}\times\hat{\mathbf{n}}}{\|\hat{\mathbf{p}}\times\hat{\mathbf{n}}\|}, \qquad \hat{\boldsymbol{\eta}}_{\rm SV} = \pm(\hat{\boldsymbol{\eta}}_{\rm SH}\times\hat{\mathbf{p}}), \tag{15.176}$$

where the signs should be chosen to be consistent with the spherical-Earth conventions summarized in Figures 12.5 and 12.6. Of course, the terminology "horizontal" and "vertical" and the designations SV and SH are something of a misnomer. An incident P wave generates only outgoing P and SV waves, just as in a spherically symmetric Earth. The amplitudes $A_{\rm P}^{\rm out}$ and $A_{\rm SV}^{\rm out}$ of these are governed by the classical plane-wave (energy)$^{1/2}$ reflection and transmission coefficients, which we write using mnemonically accented script-letter combinations such as $\acute{\mathcal{P}}\grave{\mathcal{P}}$, $\acute{\mathcal{P}}\grave{\mathcal{S}}$, $\acute{\mathcal{K}}\grave{\mathcal{K}}$ and $\acute{\mathcal{K}}\acute{\mathcal{S}}$ (see Section 12.1.6). The amplitude of the outgoing SH waves is $A_{\rm SH}^{\rm out} = 0$. An incident shear wave is more problematical since it arrives with two independent polarizations $\hat{\boldsymbol{\eta}}_1^{\rm inc}$ and $\hat{\boldsymbol{\eta}}_2^{\rm inc}$ given by (15.144); normally, these incoming polarizations will not be aligned with $\hat{\boldsymbol{\eta}}_{\rm SV}$ and $\hat{\boldsymbol{\eta}}_{\rm SH}$. A general procedure which is suitable for any kind of incoming wave consists of the following steps:

1. Find the orthogonal transformation $\mathsf{Q}^{\rm inc}$ that rotates the incoming slowness-polarization triad $\hat{\mathbf{p}}^{\rm inc}$, $\hat{\boldsymbol{\eta}}_1^{\rm inc}$, $\hat{\boldsymbol{\eta}}_2^{\rm inc}$ to $\hat{\mathbf{p}}^{\rm inc}$, $\hat{\boldsymbol{\eta}}_{\rm SV}$, $\hat{\boldsymbol{\eta}}_{\rm SH}$.

2. Use the classical (energy)$^{1/2}$ reflection and transmission coefficients to calculate the amplitudes of all of the outgoing P, SV and SH waves.

3. Find the orthogonal transformation $\mathsf{Q}^{\rm out}$ that rotates every outgoing triad $\hat{\mathbf{p}}^{\rm out}$, $\hat{\boldsymbol{\eta}}_{\rm SV}$, $\hat{\boldsymbol{\eta}}_{\rm SH}$ to a new $\hat{\mathbf{p}}^{\rm out}$, $\hat{\boldsymbol{\eta}}_1^{\rm out}$, $\hat{\boldsymbol{\eta}}_2^{\rm out}$.

The first and third steps are unnecessary for an incoming P wave. The upshot of this procedure is a 3×3 forward scattering matrix

$$\mathsf{S}_+ = \mathsf{Q}^{\rm out}\cdot\mathsf{S}_+^{\rm sub}\cdot\mathsf{Q}^{\rm inc} \tag{15.177}$$

of *ray-specific* reflection and transmission coefficients that relate the outgoing and incoming amplitudes $A_{\rm P}^{\rm out}$, $A_{\rm S1}^{\rm out}$, $A_{\rm S2}^{\rm out}$ and $A_{\rm P}^{\rm inc}$, $A_{\rm S1}^{\rm inc}$, $A_{\rm S2}^{\rm inc}$. Every compound ray between a source $\mathbf{x}'$ and receiver $\mathbf{x}$ gives rise to a series of such matrices, one at every boundary intersection point along its path. Reversal of all of the rays likewise gives rise to a series of backward scattering matrices of the form

$$\mathsf{S}_- = (\mathsf{Q}^{\rm inc})^{\rm T} \cdot \mathsf{S}_-^{\rm sub} \cdot (\mathsf{Q}^{\rm out})^{\rm T}. \tag{15.178}$$

The interior sub-matrices $\mathsf{S}_\pm^{\rm sub}$ in (15.177)–(15.178) are composed of the appropriate elements of the full 6×6 scattering matrices (12.25); for example, in the case of the bottom-side reflections P660P, P660S, S660P, S660S off of the 660 km discontinuity,

$$\mathsf{S}_\pm^{\rm sub} = \begin{pmatrix} \acute{\mathcal{P}}\grave{\mathcal{P}} & \pm\acute{\mathcal{P}}\grave{\mathcal{S}} & 0 \\ \pm\acute{\mathcal{S}}\grave{\mathcal{P}} & \acute{\mathcal{S}}\grave{\mathcal{S}}_{\rm SV} & 0 \\ 0 & 0 & \acute{\mathcal{S}}\grave{\mathcal{S}}_{\rm SH} \end{pmatrix}. \tag{15.179}$$

The orthogonality $(\mathsf{Q}^{\rm inc})^{\rm T} \cdot \mathsf{Q}^{\rm inc} = (\mathsf{Q}^{\rm out})^{\rm T} \cdot \mathsf{Q}^{\rm out} = \mathsf{I}$ of the incident and outgoing coordinate transformations and the 6×6 ray-reversal symmetry relation (12.20) guarantee that every pair of forward and backward matrices (15.177)–(15.178) satisfies

$$\mathsf{S}_- = \mathsf{S}_+^{\rm T}, \qquad \mathsf{S}_+ = \mathsf{S}_-^{\rm T}. \tag{15.180}$$

The outgoing shear-wave polarizations may be selected arbitrarily at every boundary encounter; for example, it is possible simply to set $\mathsf{Q}^{\rm out} = \mathsf{I}$, in which case $\hat{\boldsymbol{\eta}}_1^{\rm out} = \hat{\boldsymbol{\eta}}_{\rm SV}$ and $\hat{\boldsymbol{\eta}}_2^{\rm out} = \hat{\boldsymbol{\eta}}_{\rm SH}$. The incoming polarizations $\hat{\boldsymbol{\eta}}_1^{\rm inc}$ and $\hat{\boldsymbol{\eta}}_2^{\rm inc}$ at the next boundary can be found by an obvious modification of equations (15.148)–(15.149). The freedom to choose $\hat{\boldsymbol{\eta}}_1^{\rm out}$ and $\hat{\boldsymbol{\eta}}_1^{\rm out}$ at each boundary is reminiscent of the freedom to choose the initial polarizations $\hat{\boldsymbol{\eta}}_1'$ and $\hat{\boldsymbol{\eta}}_2'$ at the source.

15.7 Ray-Theoretical Response

We have at last collected all of the ingredients needed to specify the ray-theoretical response. We first find the frequency-domain Green tensor $\mathbf{G}(\mathbf{x}, \mathbf{x}'; \omega)$, and then use it to determine the frequency-domain and time-domain response to a moment-tensor source in Section 15.7.2

15.7.1 Green tensor

Ignoring boundary interactions and multi-pathing for the moment, let us suppose that only a single compressional-wave ray and two shear-wave rays leave the source $\mathbf{x}'$ and arrive at the receiver $\mathbf{x}$. We may write the JWKB Green tensor or impulse response $\mathbf{G}(\mathbf{x}, \mathbf{x}'; \omega)$ as a sum over these three rays:

$$\mathbf{G} = \frac{1}{4\pi} \sum_{\text{rays}} \hat{\boldsymbol{\eta}}\hat{\boldsymbol{\eta}}' A'(\rho v)^{-1/2} \mathcal{R}^{-1} \exp(-i\omega T). \tag{15.181}$$

The quantities $\hat{\boldsymbol{\eta}}'$ and $\hat{\boldsymbol{\eta}}$ are the polarization vectors at the source and receiver, respectively. The compressional-wave polarization is everywhere in the direction of propagation,

$$\hat{\boldsymbol{\eta}}'_{\mathrm{P}} = \hat{\mathbf{p}}'_{\mathrm{P}}, \qquad \hat{\boldsymbol{\eta}}_{\mathrm{P}} = \hat{\mathbf{p}}_{\mathrm{P}}, \tag{15.182}$$

whereas the shear-wave polarizations at $\mathbf{x}'$ and $\mathbf{x}$ are connected by equations (15.148)–(15.149). The geometrical-spreading term $(\rho v)^{-1/2} \mathcal{R}^{-1}$ accounts for the variations in amplitude along the three rays, in accordance with the energy-conservation law (15.74) and the definition (15.97). The initial amplitudes A' of the outgoing waves remain to be determined.

We find these outgoing amplitudes by matching (15.181) to the far-field Green tensor of an infinite homogeneous medium in the vicinity of the source; this classical far-field impulse response $\mathbf{G}_{\mathrm{hom}}(\mathbf{x}, \mathbf{x}'; \omega)$ is given in terms of the source-receiver distance $R = \|\mathbf{x} - \mathbf{x}'\|$ by (Aki & Richards 1980; Ben-Menahem & Singh 1981)

$$\mathbf{G}_{\mathrm{hom}} = \frac{1}{4\pi} \sum_{\text{rays}} \hat{\boldsymbol{\eta}}'\hat{\boldsymbol{\eta}}' (\rho' v'^2 R)^{-1} \exp(-i\omega R/v'). \tag{15.183}$$

Noting that $\hat{\boldsymbol{\eta}} \rightarrow \hat{\boldsymbol{\eta}}'$, $\rho \rightarrow \rho'$, $v \rightarrow v'$, $\mathcal{R} \rightarrow R$ and $T \rightarrow R/v'$ in the near-source limit $\mathbf{x} \rightarrow \mathbf{x}'$, we find that $A' = (\rho' v'^3)^{-1/2}$. The Green tensor (15.181) of a slightly heterogeneous infinite medium is therefore

$$\mathbf{G} = \frac{1}{4\pi} \sum_{\text{rays}} \hat{\boldsymbol{\eta}}\hat{\boldsymbol{\eta}}' (\rho\rho' v v'^3)^{-1/2} \mathcal{R}^{-1} \exp(-i\omega T). \tag{15.184}$$

Three modifications are needed to generalize the result (15.184) so that it is valid for an Earth composed of several smooth heterogeneous sub-regions $\oplus = \oplus_{\mathrm{S}} \cup \oplus_{\mathrm{F}}$ separated by smooth boundaries $\Sigma = \partial\oplus \cup \Sigma_{\mathrm{SS}} \cup \Sigma_{\mathrm{FS}}$. First, we must sum over all of the rays between the source $\mathbf{x}'$ and the receiver $\mathbf{x}'$, including triplications and other multi-pathed arrivals as well as boundary reflections and conversions. Second, multi-pathing implies the presence of caustics, and we must incorporate the non-geometrical $\pi/2$ phase shifts that result from the changes in sign of the ray-tube area. Third, and finally, we must account for the partition of energy associated with the proliferation of rays at the boundaries. The complete JWKB Green tensor has the same form (12.225) as on a spherical Earth:

$$\mathbf{G} = \frac{1}{4\pi}\sum_{\text{rays}} \hat{\boldsymbol{\eta}}\hat{\boldsymbol{\eta}}'(\rho\rho' v v'^3)^{-1/2}\Pi\mathcal{R}^{-1}\exp(-i\omega T + iM\pi/2), \tag{15.185}$$

where M is the Maslov index, and Π is the appropriate product of ray-specific (energy)$^{1/2}$ reflection and transmission coefficients (15.177). The polarization vectors are interchanged and reversed, $\hat{\boldsymbol{\eta}} \to -\hat{\boldsymbol{\eta}}'$, $\hat{\boldsymbol{\eta}}' \to -\hat{\boldsymbol{\eta}}'$, upon an interchange of the source and receiver, $\mathbf{x} \to \mathbf{x}'$, $\mathbf{x}' \to \mathbf{x}$. The seismic reciprocity relation

$$\mathbf{G}(\mathbf{x},\mathbf{x}';\omega) = \mathbf{G}^{\mathrm{T}}(\mathbf{x}',\mathbf{x};\omega) \tag{15.186}$$

is guaranteed by the kinematic equivalence of the forward and reversed travel time, $T(\mathbf{x},\mathbf{x}') = T(\mathbf{x}',\mathbf{x})$, together with the dynamical spreading-factor, Maslov-index and scattering-matrix symmetries (15.109), (15.111) and (15.180). The initial shear-wave polarizations $\hat{\boldsymbol{\eta}}_1'$, $\hat{\boldsymbol{\eta}}_2'$ are arbitrary along every S ray leaving the source, as are the polarizations $\hat{\boldsymbol{\eta}}_1^{\text{out}}$, $\hat{\boldsymbol{\eta}}_2^{\text{out}}$ along every outgoing S ray at every boundary. The full ray-sum Green tensor (15.185) is, however, invariant with respect to these choices.

15.7.2 Moment-tensor response

The ray-theoretical displacement response $s(\omega) = \hat{\boldsymbol{\nu}} \cdot \mathbf{s}(\mathbf{x},\omega)$ to a synchronous moment-tensor source $\mathbf{M}(\omega) = \sqrt{2}M_0\hat{\mathbf{M}}\, m(\omega)$ at a hypocentral location $\mathbf{x}_{\mathrm{s}}$ is also of the same form (12.263) as on a spherical Earth:

$$s(\omega) = \frac{1}{4\pi}\sum_{\text{rays}} \Xi\Sigma\Pi\mathcal{R}^{-1}m(\omega)\exp(-i\omega T + iM\pi/2), \tag{15.187}$$

where

$$\Xi = (\rho v)^{-1/2}(\hat{\boldsymbol{\nu}}\cdot\hat{\boldsymbol{\eta}}), \tag{15.188}$$

$$\Sigma = \sqrt{2}M_0(\rho_{\mathrm{s}}v_{\mathrm{s}}^5)^{-1/2}[\hat{\mathbf{M}}:\tfrac{1}{2}(\hat{\mathbf{p}}_{\mathrm{s}}\hat{\boldsymbol{\eta}}_{\mathrm{s}} + \hat{\boldsymbol{\eta}}_{\mathrm{s}}\hat{\mathbf{p}}_{\mathrm{s}})]. \tag{15.189}$$

The quantities M_0 and $\hat{\mathbf{M}}$ are the scalar moment and unit mechanism tensor of the earthquake, and $m(\omega)$ is the Fourier transform of the normalized source time function $\dot{m}(t)$. The product $\hat{\mathbf{M}}:\frac{1}{2}(\hat{\mathbf{p}}_{\mathrm{s}}\hat{\boldsymbol{\eta}}_{\mathrm{s}} + \hat{\boldsymbol{\eta}}_{\mathrm{s}}\hat{\mathbf{p}}_{\mathrm{s}})$ is the radiation pattern of the waves upon leaving the source.

The time-domain moment-tensor response obtained by inverse Fourier transformation of equation (15.187) is

$$s(t) = \frac{1}{4\pi}\sum_{\text{rays}} \Xi\Sigma\Pi\mathcal{R}^{-1}\dot{m}_{\mathrm{H}}^{(M)}(t-T), \tag{15.190}$$

where

$$\dot{m}_{\rm H}^{(M)}(t) = \frac{1}{\pi}\,{\rm Re}\int_0^\infty m(\omega)\exp(i\omega t + iM\pi/2)\,d\omega. \tag{15.191}$$

All $M = 0$ waves travel along least-time ray paths, and exhibit a far-field pulse shape $\dot{m}(t)$; every passage through a caustic gives rise to a Hilbert transformation: $\dot{m}(t) \rightarrow \dot{m}_{\rm H}(t) \rightarrow \cdots \rightarrow \dot{m}_{\rm H}^{(\rm M)}(t)$. Anelastic attenuation and the associated physical dispersion can be accounted for by convolution with an additional causal function, as in (12.268):

$$s(t) = \frac{1}{4\pi}\sum_{\rm rays}\Xi\Sigma\Pi\mathcal{R}^{-1}\dot{m}_{\rm H}^{(M)}(t-T) * a(t), \tag{15.192}$$

where

$$a(t) = \frac{1}{\pi}\,{\rm Re}\int_0^\infty \exp i\omega\big[t + \tfrac{1}{2}iT^* + \tfrac{1}{\pi}T^*\ln(\omega/\omega_0)\big]\,d\omega. \tag{15.193}$$

The attenuation time T^* is given in terms of the wave speed v_0 at the reference frequency ω_0 and the anelastic quality factor Q by the ray integral

$$T^* = \int_{\mathbf{x}'}^{\mathbf{x}} \frac{ds}{v_0 Q}. \tag{15.194}$$

The quality factor Q is Q_α along every compressional-wave leg and Q_β along every shear-wave leg of a ray. The above results are valid provided that all of the ray-specific reflection-and-transmission coefficients in the product Π are real. Post-critical reflections can be synthesized using a straightforward generalization of the inverse Fourier transform (15.190):

$$s(t) = \frac{1}{4\pi}\,{\rm Re}\sum_{\rm rays}\Xi\Sigma\Pi\mathcal{R}^{-1}\dot{M}_{\rm H}^{(M)}(t-T) * a(t). \tag{15.195}$$

The quantity $\dot{M}_{\rm H}^{(M)}(t)$ in (15.195) is the M-times Hilbert transform of the *analytic* source time function $\dot{M}(t) = \dot{m}(t) - i\dot{m}_{\rm H}(t)$.

All of the above results break down in the vicinity of caustics, where the cross-sectional area of a ray tube vanishes. Uniformly valid extensions of (15.195) analogous to (12.270) can be developed; such *Chapman-Maslov* representations express the displacement $s(t)$ as either a one-dimensional or two-dimensional integral over a family of rays γ_1', γ_2' leaving the source (Chapman & Drummond 1982; Thomson & Chapman 1985; Liu & Tromp 1996). As a general rule, the amplitude of seismic signal $s(t)$ is particularly strong or intense in the vicinity of caustics, as might be expected from the divergence $\mathcal{R}^{-1} \rightarrow \infty$. Indeed, the term caustic comes from the Greek word $\kappa\alpha\upsilon\sigma\tau\acute{o}\sigma$ meaning "burnt".

15.8 Practical Numerical Implementation

We present a practical numerical scheme for calculating synthetic ray-theoretical seismograms $s(t)$ in this section. The seismic wave speed is presumed to be a specified function of the form $v(r,\theta,\phi)$, where r is the radius, θ is the colatitude and ϕ is the longitude. We restrict attention for simplicity to an Earth with spherically symmetric external and internal boundaries $\Sigma = \partial\oplus \cup \Sigma_{\mathrm{SS}} \cup \Sigma_{\mathrm{FS}}$; most contemporary global three-dimensional models have this character (Woodhouse & Dziewonski 1984; Su, Woodward & Dziewonski 1994; Masters, Johnson, Laske & Bolton 1996; Van der Hilst, Widiyantoro & Engdahl 1996; Dziewonski, Liu & Su 1998).

15.8.1 Kinematic ray tracing

We begin by expressing the slowness vector $\mathbf{p} = \boldsymbol{\nabla} T$ in terms of its *covariant* spherical polar components:

$$\mathbf{p} = p_r\hat{\mathbf{r}} + r^{-1}p_\theta\,\hat{\boldsymbol{\theta}} + (r\sin\theta)^{-1}p_\phi\,\hat{\boldsymbol{\phi}}, \tag{15.196}$$

where

$$p_r = \partial_r T, \qquad p_\theta = \partial_\theta T, \qquad p_\phi = \partial_\phi T. \tag{15.197}$$

The eikonal equation (15.47) can be written in terms of the position and conjugate momentum variables r,θ,ϕ and p_r,p_θ,p_ϕ in the form

$$H = \tfrac{1}{2}[p_r^2 + r^{-2}p_\theta^2 + (r\sin\theta)^{-2}p_\phi^2 - v^{-2}(r,\theta,\phi)] = 0. \tag{15.198}$$

Hamilton's equations for the Hamiltonian $H(r,\theta,\phi,p_r,p_\theta,p_\phi)$ are

$$\frac{dr}{d\sigma} = \frac{\partial H}{\partial p_r} = p_r, \tag{15.199}$$

$$\frac{d\theta}{d\sigma} = \frac{\partial H}{\partial p_\theta} = r^{-2}p_\theta, \tag{15.200}$$

$$\frac{d\phi}{d\sigma} = \frac{\partial H}{\partial p_\phi} = (r\sin\theta)^{-2}p_\phi, \tag{15.201}$$

$$\frac{dp_r}{d\sigma} = -\frac{\partial H}{\partial r} = \tfrac{1}{2}\partial_r v^{-2} + r^{-3}[p_\theta^2 + (\sin\theta)^{-2}p_\phi^2], \tag{15.202}$$

$$\frac{dp_\theta}{d\sigma} = -\frac{\partial H}{\partial \theta} = \tfrac{1}{2}\partial_\theta v^{-2} + r^{-2}\cot\theta\,(\sin\theta)^{-2}p_\phi^2, \tag{15.203}$$

$$\frac{dp_\phi}{d\sigma} = -\frac{\partial H}{\partial \phi} = \tfrac{1}{2}\partial_\phi v^{-2}. \tag{15.204}$$

We seek solutions to (15.199)–(15.204) subject to the six initial conditions $r(0) = r'$, $\theta(0) = \theta'$, $\phi(0) = \phi'$ and $p_r(0) = p'_r$, $p_\theta(0) = p'_\theta$, $p_\phi(0) = p'_\phi$.

We introduce the local *angle of incidence* i and *azimuth* ζ at every point r, θ, ϕ along a ray. Following Aki & Richards (1980), we measure i downward from vertical and ζ counterclockwise from due south, as illustrated in Figure 15.5. The slowness vector (15.196) can be written in terms of these two ray-direction angles as

$$\mathbf{p} = v^{-1}(\cos i\,\hat{\mathbf{r}} + \sin i \cos\zeta\,\hat{\boldsymbol{\theta}} + \sin i \sin\zeta\,\hat{\boldsymbol{\phi}}). \tag{15.205}$$

The covariant components (15.197) of the slowness

$$p_r = v^{-1}\cos i, \qquad p_\theta = rv^{-1}\sin i\cos\zeta,$$

$$p_\phi = rv^{-1}\sin\theta\sin i\sin\zeta \tag{15.206}$$

are readily shown to satisfy the eikonal equation (15.198). We can use the relations (15.206) to transform (15.199)–(15.204) into a system of first-order equations for r, θ, ϕ and i, ζ:

$$\frac{dr}{d\sigma} = v^{-1}\cos i, \tag{15.207}$$

$$\frac{d\theta}{d\sigma} = r^{-1}v^{-1}\sin i\cos\zeta, \tag{15.208}$$

$$\frac{d\phi}{d\sigma} = r^{-1}v^{-1}(\sin\theta)^{-1}\sin i\sin\zeta, \tag{15.209}$$

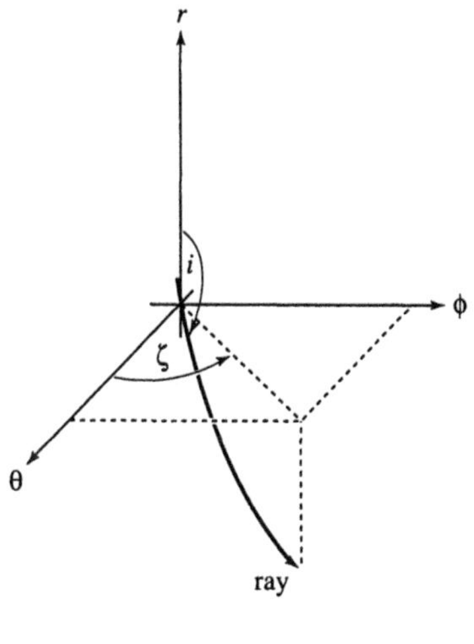

Figure 15.5. Schematic depiction of the local radial, colatitudinal and longitudinal directions r, θ, ϕ at an arbitrary point along a ray. The angle of incidence $0 \le i \le \pi$ is measured downward from vertical, whereas the azimuth $0 \le \zeta \le 2\pi$ is measured counterclockwise from due south. The instantaneous direction of propagation is completely specified by these two angles.

$$\frac{di}{d\sigma} = -\sin i\,(r^{-1}v^{-1} + \partial_r v^{-1}) + r^{-1}\cos i\,[\cos\zeta\,\partial_\theta v^{-1} + (\sin\theta)^{-1}\sin\zeta\,\partial_\phi v^{-1}], \qquad (15.210)$$

$$\frac{d\zeta}{d\sigma} = -r^{-1}(\sin i)^{-1}\sin\zeta\,\partial_\theta v^{-1} + r^{-1}(\sin\theta)^{-1}(\sin i)^{-1}\cos\zeta\,\partial_\phi v^{-1} - r^{-1}v^{-1}\cot\theta\sin i\sin\zeta. \qquad (15.211)$$

It is convenient to choose the two ray parameters to be the outgoing incidence angle and azimuth at the source:

$$\gamma_1' = i', \qquad \gamma_2' = \zeta'. \qquad (15.212)$$

The initial conditions associated with the five equations (15.207)–(15.211) are then $r(0) = r'$, $\theta(0) = \theta'$, $\phi(0) = \phi'$ and $i(0) = i'$, $\zeta(0) = \zeta'$.

We can reduce the above results even further by taking the longitude ϕ to be the independent variable. Using (15.209) to eliminate σ, we obtain a final system of *four kinematic ray-tracing equations*:

$$\frac{dr}{d\phi} = r\sin\theta\cot i\,(\sin\zeta)^{-1}, \qquad (15.213)$$

$$\frac{d\theta}{d\phi} = \sin\theta\cot\zeta, \qquad (15.214)$$

$$\frac{di}{d\phi} = \sin\theta\,(\sin\zeta)^{-1}(r\partial_r \ln v - 1) - \sin\theta\cot i\cot\zeta\,\partial_\theta \ln v - \cot i\,\partial_\phi \ln v, \qquad (15.215)$$

$$\frac{d\zeta}{d\phi} = -\cos\theta + \sin\theta\,(\sin i)^{-2}\partial_\theta \ln v - (\sin i)^{-2}\cot\zeta\,\partial_\phi \ln v. \qquad (15.216)$$

At every spherically symmetric discontinuity, these differential equations must be supplemented by the continuity conditions

$$[r]_-^+ = 0, \qquad [\theta]_-^+ = 0, \qquad [v^{-1}\sin i]_-^+ = 0, \qquad [\zeta]_-^+ = 0. \qquad (15.217)$$

Prior to integrating (15.213)–(15.217), it is advantageous to rotate the Earth model so that both the source and receiver are situated on the equator (see Appendix C.8.7):

$$\theta' = \pi/2,\ \phi' = 0 \quad \text{and} \quad \theta = \pi/2,\ \phi = \Theta, \qquad (15.218)$$

where $\cos\Theta = \cos\theta\cos\theta' + \sin\theta\sin\theta'\cos(\phi - \phi')$. The initial conditions in that case are

$$r(0) = r', \qquad \theta(0) = \pi/2, \qquad i(0) = i', \qquad \zeta(0) = \zeta'. \tag{15.219}$$

Since the longitude ϕ always increases along a ray in such an equatorial coordinate system, singularities associated with due north-south propagation $\zeta = 0, \pi$ do not arise.

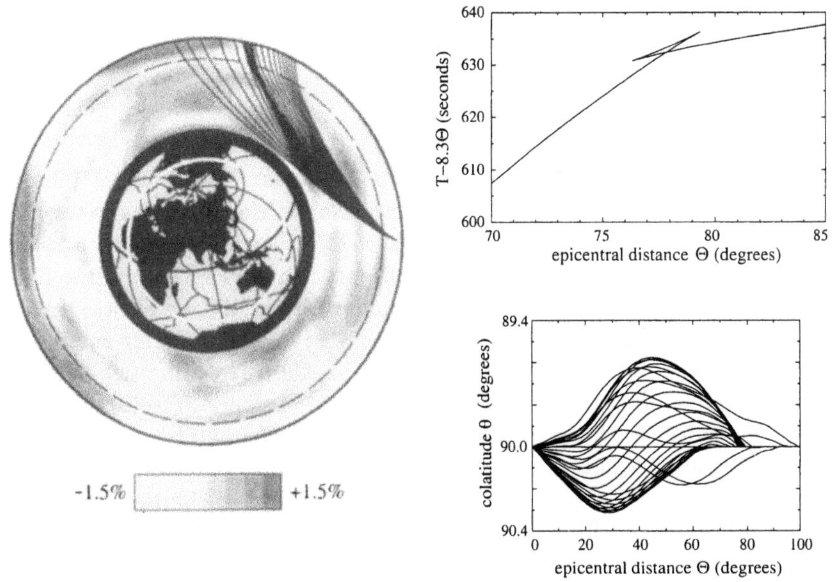

Figure 15.6. S-wave trajectories in the $s_{\max} = 12$ Earth model SKS12WM13. Rays are shot from a 200 km deep event in Fiji toward North America. (*Left*) White circle on interior map indicates the great circle through the source and the family of receivers; this circle is the equator in the rotated coordinate system used to trace rays. Outer cross-section shows mantle shear-wave speed perturbations in the source-receiver plane; long-dashed circle is the 670 km discontinuity. Scale bar shows the range $-1.5\% \leq \delta\beta/\beta \leq 1.5\%$ of relative perturbations; darker-shaded regions have higher speeds. All rays strike the surface of the Earth along the rotated equator; curves show the projection of these rays onto the cross-section. (*Top right*) The reduced travel-time curve for these perturbed equatorial rays exhibits a small-scale triplication; the associated caustics at $\Theta \approx 76.5°$ and $\Theta \approx 79°$ are apparent in the cross-section. (*Bottom right*) Rays projected onto the Earth's surface; the ordinate θ is the rotated colatitude. The equatorial location of the receivers is evident in this map view.

15.8.2 Shooting

The *two-point* ray-tracing problem requires us to find the initial takeoff angles i', ζ' that enable a ray to "hit" the receiver. For a receiver situated on the Earth's surface $r = a$, the endpoint conditions are

$$r(\Theta) = a, \qquad \theta(\Theta) = \pi/2. \tag{15.220}$$

A number of iterative schemes are available to solve this *ray-shooting* problem (Julian & Gubbins 1977). Whenever the geometrical spreading factor $\mathcal{R}$ must also be calculated, it is natural to use Newton's method; dropping the argument Θ for simplicity we update i' and ζ' using

$$\begin{pmatrix} \partial_{i'} r_n & \partial_{\zeta'} r_n \\ \partial_{i'} \theta_n & \partial_{\zeta'} \theta_n \end{pmatrix} \begin{pmatrix} i'_{n+1} - i'_n \\ \zeta'_{n+1} - \zeta'_n \end{pmatrix} = \begin{pmatrix} a - r_n \\ \pi/2 - \theta_n \end{pmatrix}, \tag{15.221}$$

where $n = 0, 1, 2, \ldots$ denotes the iteration number. The required partial derivatives $\partial_{i'} r_n$, $\partial_{\zeta'} r_n$ and $\partial_{i'} \theta_n$, $\partial_{\zeta'} \theta_n$ are available as solutions to the dynamical ray-tracing equations in Section 15.8.5. Convergence is expedited by a good first guess i'_0, ζ'_0; perturbation theory provides such an initial iterate, as we discuss in Section 15.9.3. Several examples of S-wave rays traced through the three-dimensional Earth model SKS12WM13 (Dziewonski, Liu & Su 1998) are shown in Figure 15.6. Small-scale travel-time triplications such as the one illustrated here are a ubiquitous feature at epicentral distances beyond $\Theta = 75°$, because of the strong lateral heterogeneity in the lowermost mantle. The third arrival of every such triplication has passed through the associated caustic, whereas the first and the second have not. The rays in model SKS12WM13 wander no more than 0.3°–0.4° out of the spherical-Earth ray plane. The trajectory of a typical SS ray is depicted in Figure 15.7. The minimax character of this phase is responsible for the pronounced ray-path deviations; the surface bounce points may be displaced as much as 2° away from the corresponding point in the spherical Earth.

15.8.3 Travel time and attenuation time

The variation of the travel time along a ray can be found from the differential relation $d\sigma = v^2 dT$ together with equation (15.209):

$$\frac{dT}{d\phi} = rv^{-1} \sin\theta \, (\sin i \, \sin\zeta)^{-1}. \tag{15.222}$$

This can be integrated to find the total travel time of any minor-arc body-wave phase in an elastic Earth:

$$T = \int_0^{\Theta} rv^{-1} \sin\theta \, (\sin i \, \sin\zeta)^{-1} \, d\phi. \tag{15.223}$$

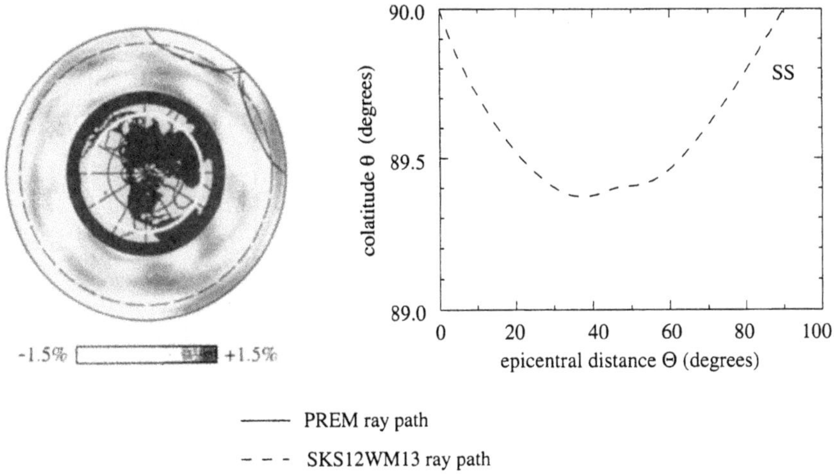

Figure 15.7. Comparison of an SS ray path in model SKS12WM13 and PREM. The source is situated on the equator and Greenwich Meridian, and the receiver is situated due east at an epicentral distance $\Theta = 90°$. The PREM surface bounce point is at $\Theta/2 = 45°$, in the vicinity of the East African Rift Valley. (*Left*) Unperturbed (*solid line*) and perturbed (*dashed line*) rays projected on a cross-section through model SKS12WM13; see caption to Figure 15.6 for explanation of interior map and other plotting conventions. (*Right*) Projection of the two ray paths onto the Earth's surface; the PREM ray is confined to the equatorial plane. The out-of-plane deviations are approximately twice as large as those of a typical teleseismic S wave.

The attenuation time (15.194) in an anelastic Earth is likewise given by

$$T^* = \int_0^\Theta r v_0^{-1} Q^{-1} \sin\theta\,(\sin i\,\sin\zeta)^{-1}\,d\phi, \tag{15.224}$$

where v_0 is the wave speed—either α_0 or β_0—at the reference frequency ω_0, and Q is the associated compressional-wave or shear-wave quality factor Q_α or Q_β. Equations (15.223) and (15.224) assume that the source and receiver are both situated on the equator, as in (15.218).

15.8.4 Geometrical spreading factor

The differential area of a ray tube (15.88) and the solid angle subtended by that tube at the source are given in terms of the initial takeoff angles i' and ζ' by

$$d\Sigma = vJ\,di'\,d\zeta', \qquad d\Omega = \sin i'\,di'\,d\zeta'. \tag{15.225}$$

The geometrical spreading factor $\mathcal{R} = \sqrt{|d\Sigma|/d\Omega}$ is therefore related to the Jacobian

$$J = \frac{\partial(x_1, x_2, x_3)}{\partial(\sigma, i', \zeta')} = \det \begin{pmatrix} \dfrac{\partial x_1}{\partial \sigma} & \dfrac{\partial x_1}{\partial i'} & \dfrac{\partial x_1}{\partial \zeta'} \\ \dfrac{\partial x_2}{\partial \sigma} & \dfrac{\partial x_2}{\partial i'} & \dfrac{\partial x_2}{\partial \zeta'} \\ \dfrac{\partial x_3}{\partial \sigma} & \dfrac{\partial x_3}{\partial i'} & \dfrac{\partial x_3}{\partial \zeta'} \end{pmatrix} \tag{15.226}$$

by $\mathcal{R} = \sqrt{v|J|/\sin i'}$. We can express J in terms of the spherical polar coordinates r, θ, ϕ using the composition rule for coordinate transformations:

$$\begin{aligned} J &= \frac{\partial(x_1, x_2, x_3)}{\partial(r, \theta, \phi)} \, \frac{\partial(r, \theta, \phi)}{\partial(i', \zeta', \phi)} \, \frac{\partial(i', \zeta', \phi)}{\partial(\sigma, i', \zeta')} \\ &= r v^{-1} \sin i \sin \zeta \, \frac{\partial(r, \theta)}{\partial(i', \zeta')}, \end{aligned} \tag{15.227}$$

where we have evaluated the three determinants and made use of (15.209) in deducing the second equality. The final spherical polar expression for the spreading factor obtained in this way is

$$\mathcal{R} = \sqrt{r \, (\sin i')^{-1} \sin i \sin \zeta \, |\Upsilon|}, \tag{15.228}$$

where we have introduced the two-dimensional spherical polar Jacobian

$$\Upsilon = \frac{\partial(r, \theta)}{\partial(i', \zeta')} = \frac{\partial r}{\partial i'} \frac{\partial \theta}{\partial \zeta'} - \frac{\partial r}{\partial \zeta'} \frac{\partial \theta}{\partial i'}. \tag{15.229}$$

Only four partial derivatives need to be evaluated: $\partial_{i'} r$, $\partial_{\zeta'} r$ and $\partial_{i'} \theta$, $\partial_{\zeta'} \theta$.

15.8.5 Dynamical ray tracing

To compute these, we differentiate (15.213)–(15.216) with respect to the initial takeoff angles i' and ζ'. This yields a system of *eight dynamical ray-tracing equations*:

$$\begin{aligned} \frac{d}{d\phi}(\partial_{\gamma'} r) = {} & \sin \theta \cot i \, (\sin \zeta)^{-1} \partial_{\gamma'} r \\ & + r \cos \theta \cot i \, (\sin \zeta)^{-1} \partial_{\gamma'} \theta \\ & - r \sin \theta \, (\sin i)^{-2} (\sin \zeta)^{-1} \partial_{\gamma'} i \\ & - r \sin \theta \cot i \cot \zeta \, (\sin \zeta)^{-1} \partial_{\gamma'} \zeta, \end{aligned} \tag{15.230}$$

$$\frac{d}{d\phi}(\partial_{\gamma'} \theta) = \cos \theta \cot \zeta \, \partial_{\gamma'} \theta - \sin \theta \, (\sin \zeta)^{-2} \partial_{\gamma'} \zeta, \tag{15.231}$$

$$\begin{aligned}\frac{d}{d\phi}(\partial_{\gamma'}i) &= [\sin\theta\,(\sin\zeta)^{-1}(r\partial_r^2\ln v + \partial_r\ln v)\\ &\qquad - \cot i\,(\sin\theta\cot\zeta\,\partial_r\partial_\theta\ln v + \partial_r\partial_\phi\ln v)]\,\partial_{\gamma'}r\\ &\quad + [\sin\theta\,(\sin\zeta)^{-1}r\partial_r\partial_\theta\ln v + \cos\theta\,(\sin\zeta)^{-1}(r\partial_r\ln v - 1)\\ &\qquad - \cot i\,(\sin\theta\cot\zeta\,\partial_\theta^2\ln v + \partial_\theta\partial_\phi\ln v\\ &\qquad + \cos\theta\cot\zeta\,\partial_\theta\ln v)]\,\partial_{\gamma'}\theta \qquad (15.232)\\ &\quad + (\sin i)^{-2}(\sin\theta\cot\zeta\,\partial_\theta\ln v + \partial_\phi\ln v)\partial_{\gamma'}i\\ &\quad - \sin\theta\,(\sin\zeta)^{-2}[\cos\zeta\,(r\partial_r\ln v - 1) - \cot i\,\partial_\theta\ln v]\,\partial_{\gamma'}\zeta,\end{aligned}$$

$$\begin{aligned}\frac{d}{d\phi}(\partial_{\gamma'}\zeta) &= (\sin i)^{-2}(\sin\theta\,\partial_r\partial_\theta\ln v - \cot\zeta\,\partial_r\partial_\phi\ln v)\,\partial_{\gamma'}r\\ &\quad + [(\sin i)^{-2}(\sin\theta\,\partial_\theta^2\ln v - \cot\zeta\,\partial_\theta\partial_\phi\ln v\\ &\qquad + \cos\theta\,\partial_\theta\ln v) + \sin\theta]\,\partial_{\gamma'}\theta \qquad (15.233)\\ &\quad - 2\cot i\,(\sin i)^{-2}(\sin\theta\,\partial_\theta\ln v - \cot\zeta\,\partial_\phi\ln v)\,\partial_{\gamma'}i\\ &\quad + (\sin i)^{-2}(\sin\zeta)^{-2}(\partial_\phi\ln v)\,\partial_{\gamma'}\zeta,\end{aligned}$$

where γ' denotes either i' or ζ'. These must be solved subject to the initial conditions

$$\partial_{i'}r(0) = \partial_{\zeta'}r(0) = 0, \qquad \partial_{i'}\theta(0) = \partial_{\zeta'}\theta(0) = 0,$$

$$\partial_{i'}\zeta(0) = \partial_{\zeta'}i(0) = 0, \qquad \partial_{i'}i(0) = \partial_{\zeta'}\zeta(0) = 1. \qquad (15.234)$$

The dependence upon the unknown partial derivatives in (15.230)–(15.233) is linear, and the equations governing $\partial_{i'}r$, $\partial_{i'}\theta$, $\partial_{i'}i$, $\partial_{i'}\zeta$ are decoupled from those governing $\partial_{\zeta'}r$, $\partial_{\zeta'}\theta$, $\partial_{\zeta'}i$, $\partial_{\zeta'}\zeta$, as in the Cartesian dynamical ray-tracing system (15.112).

At a discontinuity the differential equations (15.230)–(15.233) must be supplemented by conditions analogous to (15.217) that specify the jumps in the eight derivatives $\partial_{\gamma'}r$, $\partial_{\gamma'}\theta$, $\partial_{\gamma'}i$ and $\partial_{\gamma'}\zeta$. Let us denote the parameters of a central and neighboring ray at the intersection points $\mathbf{x}$ and $\mathbf{x}+d\mathbf{x}$ in Figure 15.4 by r, θ, ϕ, i, ζ and $r+dr$, $\theta+d\theta$, $\phi+d\phi$, $i+di$, $\zeta+d\zeta$. Since every point along a ray is uniquely determined by the initial takeoff angles i', ζ' and the longitude ϕ, we can express dr in terms of di', $d\zeta'$ and $d\phi$ in the form $dr = di'\,\partial_{i'}r + d\zeta'\,\partial_{\zeta'}r + d\phi\,\partial_\phi r$. At a spherical boundary we must have $dr = 0$; it follows that $d\phi = -(dr/d\phi)^{-1}(di'\,\partial_{i'}r + d\zeta'\,\partial_{\zeta'}r)$, where we have replaced $\partial_\phi r$ by $dr/d\phi$ in accordance with our usual convention. The continuity condition $[d\phi]_-^+ = 0$ and the independence of the perturbations di' and $d\zeta'$ imply that the quantities $(dr/d\phi)^{-1}\partial_{i'}r$ and $(dr/d\phi)^{-1}\partial_{\zeta'}r$ must both be continuous. The perturbed versions $[d\theta]_-^+ = 0$,

$[-v^{-2}dv\sin i + v^{-1}di\cos i]_-^+ = 0$ and $[d\zeta]_-^+ = 0$ of the remaining continuity conditions (15.217) likewise constrain the jumps in the other partial derivatives. The complete set of boundary conditions at every spherical discontinuity is

$$[(dr/d\phi)^{-1}\partial_{\gamma'}r]_-^+ = 0, \qquad [\partial_{\gamma'}\theta]_-^+ = 0,$$

$$[\cot i\,\partial_{\gamma'}i - (\partial_\theta \ln v)\,\partial_{\gamma'}\theta + (dr/d\phi)^{-1} \times \{(d\theta/d\phi)\,\partial_\theta \ln v + \partial_\phi \ln v - \cot i\,(di/d\phi)\}\,\partial_{\gamma'}r]_-^+ = 0,$$

$$[\partial_{\gamma'}\zeta - (dr/d\phi)^{-1}(d\zeta/d\phi)\partial_{\gamma'}r]_-^+ = 0. \tag{15.235}$$

In deriving (15.235) we have used Snell's law $[v^{-1}\sin i]_-^+ = 0$ in addition to the continuity condition $[d\theta/d\phi]_-^+ = 0$.

★15.8.6 Maslov index

The ray-tube singularity conditions (15.110) may be expressed in terms of spherical polar coordinates in the form

$$\begin{aligned} &\text{caustic:} && (\partial_{i'}r)(\partial_{\zeta'}\theta) = (\partial_{\zeta'}r)(\partial_{i'}\theta) \neq 0,\\ &\text{focal point:} && (\partial_{i'}r)(\partial_{\zeta'}\theta) = (\partial_{\zeta'}r)(\partial_{i'}\theta) = 0. \end{aligned} \tag{15.236}$$

These conditions allow the Maslov index M to be evaluated by monitoring the sign changes of the four partial derivatives $\partial_{i'}r$, $\partial_{\zeta'}r$ and $\partial_{i'}\theta$, $\partial_{\zeta'}\theta$.

★15.8.7 Shear-wave polarization

The normal $\hat{\boldsymbol{\nu}}$ and binormal $\hat{\mathbf{b}}$ of a body-wave ray can be expressed in terms of i, ζ and an additional twisting angle $0 \leq \xi \leq 2\pi$ in the form

$$\hat{\boldsymbol{\nu}} = \sin i\cos\xi\,\hat{\mathbf{r}} + (\sin\zeta\sin\xi - \cos i\cos\zeta\cos\xi)\,\hat{\boldsymbol{\theta}} - (\cos\zeta\sin\xi + \cos i\sin\zeta\cos\xi)\,\hat{\boldsymbol{\phi}}, \tag{15.237}$$

$$\hat{\mathbf{b}} = -\sin i\sin\xi\,\hat{\mathbf{r}} + (\sin\zeta\cos\xi + \cos i\cos\zeta\sin\xi)\,\hat{\boldsymbol{\theta}} - (cos\zeta\cos\xi - \cos i\sin\zeta\sin\xi)\,\hat{\boldsymbol{\phi}}. \tag{15.238}$$

The twist ξ is determined at any point along the ray by the geometrical relations (15.70), which together imply

$$\tan\xi = [\sin\theta\cos i\cos\zeta\,\partial_\theta \ln v + \cos i\sin\zeta\,\partial_\phi \ln v) - r\sin\theta\sin i\,\partial_r \ln v]^{-1}(\cos\zeta\,\partial_\phi \ln v - \sin\theta\sin\zeta\,\partial_\theta \ln v). \tag{15.239}$$

The curvature κ and torsion τ can be written in terms of r, θ, i, ζ and ξ using equations (15.69):

$$\begin{aligned}\kappa = -r^{-1}[&\sin i\cos\xi\, r\partial_r \ln v\\ &+ (\sin\zeta\sin\xi - \cos i\cos\zeta\cos\xi)\,\partial_\theta \ln v\\ &- (\sin\theta)^{-1}(\cos\zeta\sin\xi + \cos i\sin\zeta\cos\xi)\,\partial_\phi \ln v],\end{aligned} \tag{15.240}$$

$$\begin{aligned}\tau = -\kappa^{-1}r^{-2}\{&-\sin i\cos i\sin\xi\, r^2\partial_r^2 \ln v\\ &+ \sin i\cos\zeta(\cos i\cos\zeta\sin\xi + \sin\zeta\cos\xi)\\ &\qquad\times (r\partial_r \ln v + \partial_\theta^2 \ln v)\\ &+ \sin i\sin\zeta(\cos i\sin\zeta\sin\xi - \cos\zeta\cos\xi)\\ &\qquad\times [r\partial_r \ln v + \cot\theta\,\partial_\theta \ln v + (\sin\theta)^{-2}\partial_\phi^2 \ln v]\\ &+ [\sin^2 i\cos\zeta\sin\xi - \cos i(\cos i\cos\zeta\sin\xi + \sin\zeta\cos\xi)]\\ &\qquad\times (\partial_\theta \ln v - r\partial_r\partial_\theta \ln v)\\ &+ [\sin^2 i\sin\zeta\sin\xi - \cos i(\cos i\sin\zeta\sin\xi - \cos\zeta\cos\xi)]\\ &\qquad\times (\sin\theta)^{-1}(\partial_\phi \ln v - r\partial_r\partial_\phi \ln v)\\ &- [\sin i\cos\zeta(\cos i\sin\zeta\sin\xi - \cos\zeta\cos\xi)\\ &\qquad+ \sin i\sin\zeta(\cos i\cos\zeta\sin\xi + \sin\zeta\cos\xi)]\\ &\qquad\times (\sin\theta)^{-1}(\cot\theta\,\partial_\phi \ln v - \partial_\theta\partial_\phi \ln v)\}.\end{aligned} \tag{15.241}$$

To find the shear-wave polarization angle ψ, we integrate the equatorial version of equation (15.147):

$$\frac{d\psi}{d\phi} = -r\sin\theta(\sin i\sin\zeta)^{-1}\tau. \tag{15.242}$$

The relations (15.237)–(15.242) together with (15.144) determine the evolution of the shear-wave polarization vectors $\hat{\boldsymbol{\eta}}_1$ and $\hat{\boldsymbol{\eta}}_2$ along a ray.

*15.8.8 Son of Smirnov

In this section, we describe yet another method of determining the variation in body-wave amplitude along a ray; the starting point in this case is the spherical polar version of the transport equation (15.71):

$$\begin{aligned}&\partial_r(\rho v^2 A^2 r^2\sin\theta\, p_r) + \partial_\theta(\rho v^2 A^2\sin\theta\, p_\theta) + \partial_\phi[\rho v^2 A^2(\sin\theta)^{-1}p_\phi]\\ &\quad= \partial_r(\rho v^2 A^2 r^2\sin\theta\, dr/d\sigma) + \partial_\theta(\rho v^2 A^2 r^2\sin\theta\, d\theta/d\sigma)\\ &\qquad+ \partial_\phi(\rho v^2 A^2 r^2\sin\theta\, d\phi/d\sigma) = 0,\end{aligned} \tag{15.243}$$

where we have multiplied by $r^2 \sin\theta$ and used the scalar ray-tracing equations (15.199)–(15.204) to obtain the first and second forms, respectively. Using (15.209) to express (15.243) in terms of ϕ rather than σ, we find

$$\frac{d}{d\phi}\ln(\rho r v A^2 \sin i \sin\zeta) = -\partial_r(dr/d\sigma) - \partial_\theta(d\theta/d\sigma). \tag{15.244}$$

Equation (15.244) can be solved with the aid of Smirnov's lemma for the Jacobian (15.229):

$$\frac{d}{d\phi}(\ln\Upsilon) = \partial_r(dr/d\phi) + \partial_\theta(d\theta/d\phi), \tag{15.245}$$

whenever r, θ, ϕ satisfy the scalar ray-tracing equations (15.199)–(15.201). The proof of (15.245) is a straightforward calculation analogous to that in Section 15.4.4:

$$\begin{aligned}\frac{d\Upsilon}{d\phi} &= \frac{\partial(dr/d\phi,\theta)}{\partial(i',\zeta')} + \frac{\partial(r,d\theta/d\phi)}{\partial(i',\zeta')}\\ &= \frac{\partial(r^2\sin^2\theta\, p_r/p_\phi,\theta)}{\partial(i',\zeta')} + \frac{\partial(r,\sin^2\theta\, p_\theta/p_\phi)}{\partial(i',\zeta')}\\ &= \left[\partial_r(r^2\sin^2\theta\, p_r/p_\phi) + \partial_\theta(\sin^2\theta\, p_\theta/p_\phi)\right]\frac{\partial(r,\theta)}{\partial(i',\zeta')}\\ &= \left[\partial_r(dr/d\phi) + \partial_\theta(d\theta/d\phi)\right]\Upsilon. \end{aligned} \tag{15.246}$$

Using the lemma (15.245) we can rewrite equation (15.244) in the form

$$\frac{d}{d\phi}\ln(\rho r v A^2 \sin i \sin\zeta\,\Upsilon) = 0. \tag{15.247}$$

It follows from (15.247) that the amplitude ratio at two consecutive points r_1, θ_1, ϕ_1 and r_2, θ_2, ϕ_2 along a ray is

$$\frac{A_2}{A_1} = \left(\frac{\rho_2 r_2 v_2 \sin i_2 \sin\zeta_2}{\rho_1 r_1 v_1 \sin i_1 \sin\zeta_1}\right)^{-1/2}\left|\frac{\Upsilon_2}{\Upsilon_1}\right|^{-1/2}. \tag{15.248}$$

Equations (15.96) and (15.248) are consistent by virtue of the Jacobian relation (15.227).

To apply the amplitude variation law (15.248) we rewrite the Green tensor (15.181) in a smooth infinite medium in the form

$$\mathbf{G} = \frac{1}{4\pi}\sum_{\text{rays}}\hat{\boldsymbol{\eta}}\hat{\boldsymbol{\eta}}' B'(\rho r v \sin i \sin\zeta)^{-1/2}|\Upsilon|^{-1/2}\exp(-i\omega T). \tag{15.249}$$

The constant B' can be determined by matching equation (15.249) to the homogeneous-medium response (15.183) in the limit $\mathbf{x} \to \mathbf{x}'$, as before.

The partial derivatives obtained by integration of (15.230)–(15.231) in the vicinity of the source are $\partial_{i'}r \approx -r'\sin\theta'(\sin i')^{-2}(\sin\zeta')^{-1}\phi$, $\partial_{i'}\theta \approx 0$, $\partial_{\zeta'}r \approx -r'\sin\theta'\cot i'\cot\zeta'(\sin\zeta')^{-1}\phi$ and $\partial_{\zeta'}\theta \approx -\sin\theta'(\sin\zeta')^{-2}\phi$. The Jacobian (15.229) is therefore

$$\Upsilon \approx r'(\sin\theta')^2(\sin i')^{-2}(\sin\zeta')^{-3}\phi^2 \approx (r'\sin\zeta')^{-1}R^2, \tag{15.250}$$

where $R \approx r'\sin\theta'(\sin i'\sin\zeta')^{-1}\phi$ is the Cartesian distance from the source. Upon comparing equations (15.249)–(15.250) with (15.183) we find that $B' = (\rho' v'^3)^{-1/2}(\sin i')^{1/2}$. The resulting JWKB Green tensor is

$$\mathbf{G} = \frac{1}{4\pi}\sum_{\text{rays}} \hat{\boldsymbol{\eta}}\hat{\boldsymbol{\eta}}'(\rho\rho' v v'^3)^{-1/2}(\sin i')^{1/2}(r\sin i\sin\zeta)^{-1/2}$$
$$\times\, \Pi|\Upsilon|^{-1/2}\exp(-i\omega T + iM\pi/2). \tag{15.251}$$

The results (15.185) and (15.251) are identical, as of course they must be; the above argument is due to Liu & Tromp (1996).

15.8.9 Spherical Earth

In a spherically symmetric Earth, the above asymptotic ray theory reduces to the well-known results summarized in Sections 12.1 and 12.5, as we show next. To begin, we note that the ray-tracing equations (15.214) and (15.216) for θ and ζ can be integrated immediately, with the result

$$\theta(\phi) = \pi/2, \qquad \zeta(\phi) = \pi/2. \tag{15.252}$$

These relations simply state that every ray is confined to the plane containing the source, receiver and the center of the Earth. The remaining equations (15.213) and (15.215) governing r and i reduce to

$$\frac{dr}{d\phi} = r\cot i, \qquad \frac{di}{d\phi} = r(\dot{v}/v) - 1, \tag{15.253}$$

where the dot denotes differentiation with respect to r. The continuity conditions (15.217) require that $[r]_-^+ = 0$ and $[v^{-1}\sin i]_-^+ = 0$ at every discontinuity. It is readily demonstrated that the ray parameter $p = rv^{-1}\sin i$ is conserved along a ray:

$$\frac{dp}{d\phi} = 0 \quad \text{in } \oplus \quad \text{and} \quad [p]_-^+ = 0 \quad \text{on } \Sigma. \tag{15.254}$$

The travel time (15.223) reduces to

$$T = \int_0^\Theta \frac{r\,d\phi}{v\sin i}. \tag{15.255}$$

The results (15.253) and (15.255) agree with (12.4) and (12.7), as expected.

The dynamical ray-tracing equations (15.231) and (15.233) governing the partial derivatives $\partial_{\zeta'}\theta$ and $\partial_{\zeta'}\zeta$ can also be integrated, yielding

$$\partial_{\zeta'}\theta(\phi) = -\sin\phi, \qquad \partial_{\zeta'}\zeta(\phi) = \cos\phi. \tag{15.256}$$

The derivatives $\partial_{i'}r$ and $\partial_{i'}i$ are given by (15.230) and (15.232), which reduce to

$$\frac{d}{d\phi}(\partial_{i'}r) = \cot i\,\partial_{i'}r - r(\sin i)^{-2}\partial_{i'}i, \tag{15.257}$$

$$\frac{d}{d\phi}(\partial_{i'}i) = (rv^{-1}\ddot{v} - rv^{-2}\dot{v}^2 + v^{-1}\dot{v})\,\partial_{i'}r. \tag{15.258}$$

Equations (15.257)–(15.258) must be solved, subject to the initial conditions that $\partial_{i'}r(0) = 0$ and $\partial_{i'}i(0) = 1$. At every ray intersection with a boundary, the continuity conditions (15.235) require that

$$[\tan i\,\partial_{i'}r]_-^+ = 0, \qquad [\cot i\,\partial_{i'}i - (v^{-1}\dot{v} - r^{-1})\,\partial_{i'}r]_-^+ = 0. \tag{15.259}$$

In fact, it can be shown that the quantity

$$\cot i\,\partial_{i'}i - (v^{-1}\dot{v} - r^{-1})\,\partial_{i'}r = p^{-1}\partial_{i'}p = \cot i' \tag{15.260}$$

is conserved along a ray. The other four partial derivatives $\partial_{i'}\theta$, $\partial_{i'}\zeta$, $\partial_{\zeta'}r$ and $\partial_{\zeta'}i$ are all equal to zero on a spherically symmetric Earth. At the longitude $\phi = \Theta$ of the receiver the two-dimensional Jacobian (15.229) reduces to $\Upsilon = -\sin\Theta\,\partial_{i'}r$. The geometrical spreading factor (15.228) is therefore

$$\mathcal{R} = \sqrt{r(\sin i')^{-1}\sin i\sin\Theta\,|\partial_{i'}r|}. \tag{15.261}$$

The partial derivative $\partial_{i'}r$ is related to the reciprocal curvature $d\Theta/dp$ of the travel-time curve by

$$\partial_{i'}r = -rr'^2\cos i\cos i'\sin i\,(pv'^2\sin i)^{-1}(d\Theta/dp). \tag{15.262}$$

Upon substituting the geometrical identity (15.262) into (15.261), we obtain

$$v'\mathcal{R} = rr'\sqrt{\frac{|\cos i|\,|\cos i'|\sin\Theta}{p}\left|\frac{d\Theta}{dp}\right|}. \tag{15.263}$$

This expression agrees with equation (12.29), as expected.

The tangent, normal and binormal vectors of a spherical-Earth ray are given in rotated equatorial coordinates by

$$\hat{\mathbf{p}} = \cos i\,\hat{\mathbf{r}} + \sin i\,\hat{\boldsymbol{\phi}}, \qquad \hat{\boldsymbol{\nu}} = \sin i\,\hat{\mathbf{r}} - \cos i\,\hat{\boldsymbol{\phi}}, \qquad \hat{\mathbf{b}} = \hat{\boldsymbol{\theta}}. \tag{15.264}$$

Upon making use of (15.264) we find that the curvature and torsion (15.69) reduce to $\kappa = -pr^{-1}\dot{v}$, and $\tau = 0$. The curvature is positive or upward, $\kappa > 0$, in every region of downwardly increasing wave speed, $\dot{v} < 0$, and negative or downward, $\kappa < 0$, in every region of downwardly decreasing wave speed, $\dot{v} > 0$. Because a spherical-Earth ray is torsion-free, $\tau = 0$, the vertical and horizontal shear-wave polarization vectors are "conserved" in the sense $\hat{\boldsymbol{\eta}}_1 = \hat{\boldsymbol{\eta}}_{\rm SV} = \hat{\boldsymbol{\nu}}$, $\hat{\boldsymbol{\eta}}_2 = \hat{\boldsymbol{\eta}}_{\rm SH} = \hat{\mathbf{b}}$. Positive SV and SH polarizations point to the left and right of the ray, respectively, in accordance with the convention enunciated in Section 12.1.5.

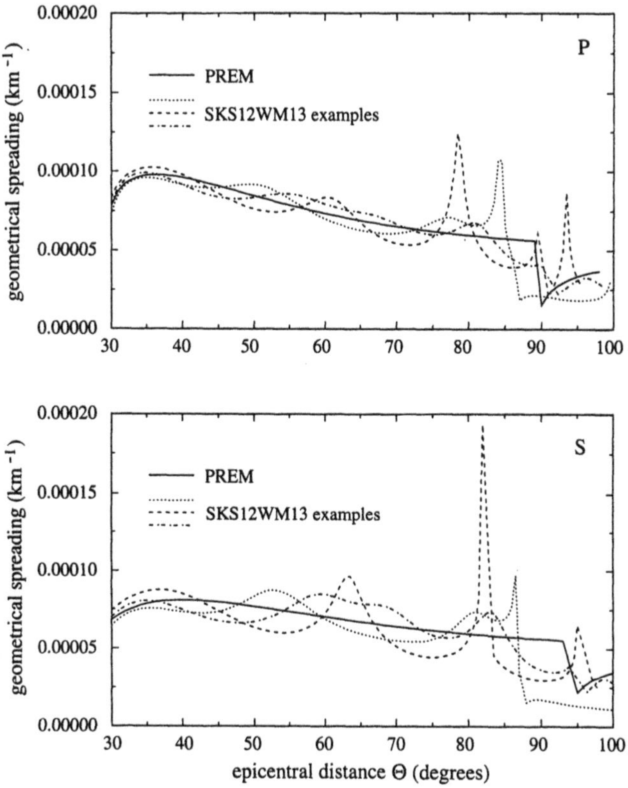

Figure 15.8. Variation of the reciprocal $\mathcal{R}^{-1}$ of the geometrical spreading factor (15.228) with epicentral distance Θ along a number of great-circular profiles in model SKS12WM13, compared with the corresponding variation (15.263) in PREM. (*Top*) P waves. (*Bottom*) S waves. The path differences in $\mathcal{R}^{-1}$ are due to focusing and defocusing of the ray tubes, caused by the lateral heterogeneity.

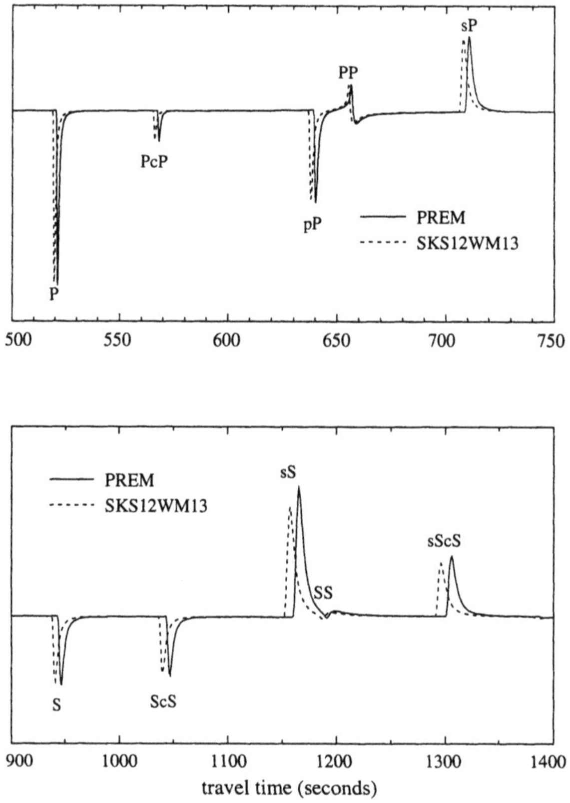

Figure 15.9. Ray-theoretical body-wave seismograms $s(t)$ in models SKS12WM13 and PREM. The location $\mathbf{x}_s$ and moment tensor $\mathbf{M}$ are those of the June 9, 1994 deep-focus earthquake in Bolivia; the receiver $\mathbf{x}$ is GSN station CCM in Cathedral Cave, Missouri, at an epicentral distance $\Theta = 56.3°$. (*Top*) Radial ($\hat{\boldsymbol{\nu}} = \hat{\mathbf{r}}$) component. (*Bottom*) Transverse ($\hat{\boldsymbol{\nu}} = \hat{\mathbf{r}} \times \hat{\boldsymbol{\Theta}}$) component. Travel-time anomalies such as those shown here are the basis of body-wave tomography. The twisting of the shear-wave polarization has been ignored.

15.8.10 Numerical examples

Figure 15.8 compares a number of great-circular profiles of $\mathcal{R}^{-1}$ versus epicentral distance Θ in the three-dimensional Earth model SKS12WM13 (Dziewonski, Liu & Su 1998) with the corresponding variation in the spherical Earth model PREM (Dziewonski & Anderson 1981). The singularities where $\mathcal{R}^{-1} \rightarrow \infty$ beyond $\Theta = 75°$ are due to the presence of small-scale triplications, produced by the strong lateral heterogeneity in the lowermost mantle (see Figure 15.6). Synthetic seismograms $s(t)$ on models SKS12WM13 and PREM, computed using (15.192), are compared in Figure 15.9. The source is presumed to be impulsive, so that the phases P, PcP, pP, sP on the radial component and S, ScS, sS, sScS on the transverse component are all anelastically broadened Dirac delta pulses of the form $\delta(t-T) * a(t)$. The surface reflections PP and SS are, on the other hand, Hilbert-transformed emergent arrivals of the form $\delta_{\mathrm{H}}(t) * a(t)$, due to having passed through a caustic. All of the paths between the great 1994 Bolivia earthquake and the receiver at station CCM in Cathedral Cave, Missouri are slightly faster in model SKS12WM13 than in PREM; there are also slight perturbations in the amplitudes of the pulses, due to variations in the focusing and defocusing of the associated ray tubes, produced by the lateral heterogeneity.

15.9 Ray Perturbation Theory

Ray theory in an Earth model $\oplus + \delta\oplus$ that is very nearly spherically symmetric constitutes a special case, which we investigate in this section. The deviations from spherical symmetry must be *slight as well as smooth* in order for the results of this *ray perturbation theory* to be applicable.

15.9.1 Travel time

As we have seen, Fermat's principle (15.62) stipulates that the travel time is a stationary functional of the path between a fixed source $\mathbf{x}'$ and receiver $\mathbf{x}$. This allows us to calculate the travel-time perturbation $\delta T = T_{\oplus+\delta\oplus} - T_{\oplus}$, correct to first order, by integration along the *unperturbed ray path* in the spherically symmetric Earth (Julian & Anderson 1968):

$$\delta T = -\int_{\mathbf{x}'}^{\mathbf{x}} v^{-2}\, \delta v\, ds = -p^{-1} \int_0^{\Theta} r^2 v^{-3}\, \delta v\, d\phi. \tag{15.265}$$

The final form assumes that the source and receiver are situated on the equator, as in (15.218). If there are perturbations δd in the locations of the boundaries in addition to the volumetric wave-speed perturbations δv, then (15.265) must be amended:

$$\delta T \to \delta T + \sum_d \delta T_d, \tag{15.266}$$

where the sum is over all of the boundaries encountered by the ray. The nature of the boundary delay δT_d depends upon the type of interaction:

$$\delta T_d = \begin{cases} -\delta d\,[(v_d^{-2} - p^2 d^{-2})^{1/2}]_-^+ & \text{transmitted ray} \\ -2\,\delta d\,(v_{d+}^{-2} - p^2 d^{-2})^{1/2} & \text{topside reflection} \\ +2\,\delta d\,(v_{d-}^{-2} - p^2 d^{-2})^{1/2} & \text{bottomside reflection.} \end{cases} \tag{15.267}$$

Equations (15.265)–(15.267) provide a *linear* relationship between measured travel-time anomalies δT and the lateral heterogeneity δv and δd of the Earth. These relations are the basis of linear travel-time tomography.

Figure 15.10 compares the linearized prediction (15.265) with the results obtained by exact ray tracing between a number of source-receiver pairs in Earth model SKS12WM13. First-order perturbation theory consistently overpredicts the travel-time anomaly δT of all refracted and core-reflected phases such as P, PcP, PKIKP and S, ScS, SKS. This phenomenon is an anticipated consequence of Fermat's principle—the geometrical ray path is the *least-time* path of any wave that has not passed through a caustic. Minimax phases such as PP and SS do not exhibit such a *Fermat bias*. The generally excellent agreement between the exact and first-order

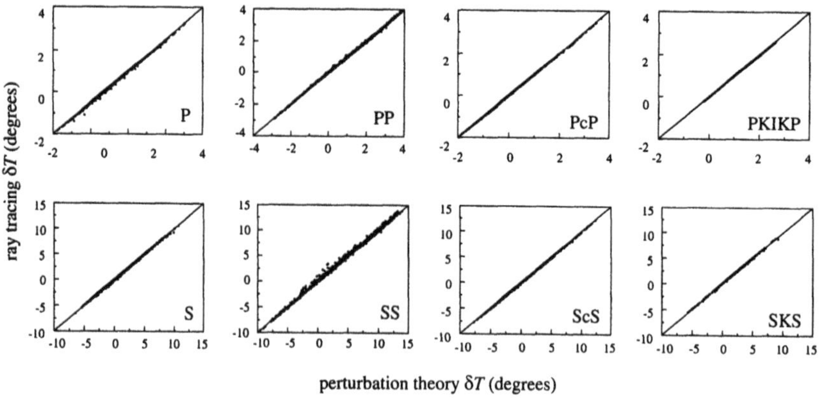

Figure 15.10. Scatter-plot comparison of the travel-time perturbation δT computed using first-order perturbation theory and exact ray tracing in model SKS12WM13. Each graph displays results for 1000 source-receiver paths, between randomly selected events in the Harvard moment-tensor catalogue and randomly selected stations in the Global Seismic Network. Epicentral distance ranges are as follows—P: $30° \le \Theta \le 95°$, S: $30° \le \Theta \le 80°$, PP and SS: $60° \le \Theta \le 179°$, PcP and ScS: $10° \le \Theta \le 75°$, PKIKP: $130° \le \Theta \le 170°$, SKS: $85° \le \Theta \le 130°$.

results in Figure 15.10 justifies the continued use of (15.265)–(15.267) in large-scale global tomographic investigations. Non-linear inversion schemes have begun to be developed, for use in higher-resolution regional-scale investigations characterized by strong lateral heterogeneity (Sambridge 1990; Papazachos & Nolet 1997a; 1997b).

★15.9.2 Ellipticity correction

Ellipticity corrections have routinely been applied to body-wave travel-time measurements for over sixty years (Bullen 1937; 1963). A modern treatment of the problem has been given by Dziewonski & Gilbert (1976); we summarize their findings here. Let ψ and Ψ be the colatitude and epicentral distance of the running point along the ray path from $\mathbf{x}'$ to $\mathbf{x}$, as illustrated in Figure 15.11. These two angular arclengths are related to the source colatitude θ' and the receiver azimuth ζ' by

$$\cos\psi = \cos\theta'\cos\Psi - \sin\theta'\sin\Psi\cos\zeta'. \tag{15.268}$$

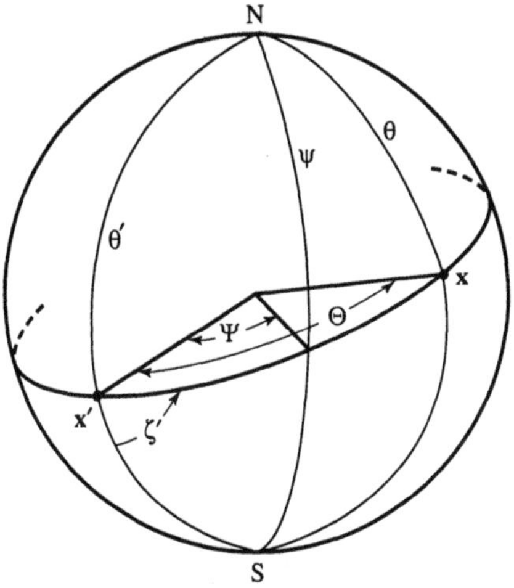

Figure 15.11. Geometrical notation used in the derivation of the ellipticity correction formula (15.271)–(15.272). The angles θ', θ and ψ are the colatitudes of the source, the receiver and the integration point along the ray, respectively. The epicentral coordinates of the integration point are Ψ and ζ'; the first of these is the angular distance from the source, and the second is the takeoff azimuth of the path to the receiver, measured counterclockwise from due south as usual.

The Earth's hydrostatic ellipticity is a degree-two zonal perturbation of the form (14.31):

$$\delta v = \tfrac{2}{3} r \varepsilon \dot{v} P_2(\cos\psi), \qquad \delta d = -\tfrac{2}{3} d\varepsilon_d P_2(\cos\psi). \tag{15.269}$$

The Legendre polynomial $P_2(\cos\psi)$ can be expressed in terms of the running epicentral distance Ψ and the fixed colatitude θ' and azimuth ζ' using the spherical-harmonic addition theorem (B.74):

$$P_2(\cos\psi) = \sum_{m=0}^{2} (-1)^m (2-\delta_{m0}) \left[\frac{(2-m)!}{(2+m)!}\right] \times P_{2m}(\cos\Psi) P_{2m}(\cos\theta') \cos m\zeta'. \tag{15.270}$$

Upon substituting (15.269)–(15.270) into (15.265)–(15.267) we can write the ellipticity travel-time perturbation as a sum of three associated Legendre functions:

$$\delta T^{\mathrm{ell}} = \sum_{m=0}^{2} \delta T_m(\Theta, h)\, P_{2m}(\cos\theta') \cos m\zeta'. \tag{15.271}$$

The three coefficients δT_0, δT_1, δT_2 are ray-specific functions of the epicentral distance Θ and source depth h, given by

$$\begin{aligned}\delta T_m = {} & -\tfrac{2}{3}(-1)^m (2-\delta_{m0}) \left[\frac{(2-m)!}{(2+m)!}\right] \\ & \times \Biggl\{ p^{-1} \int_0^{\Theta} r^3 v^{-3} \varepsilon \dot{v} P_2(\cos\Psi)\, d\Psi \\ & - \sum_d^{\mathrm{tran}} d\varepsilon_d \left[(v_d^{-2} - p^2 d^{-2})^{1/2}\right]_-^+ P_2(\cos\Psi_d) \\ & \mp \sum_d^{\mathrm{refl}} 2d\varepsilon_d (v_{d\pm}^{-2} - p^2 d^{-2})^{1/2} P_2(\cos\Psi_d) \Biggr\}.\end{aligned} \tag{15.272}$$

The first sum is over all of the boundaries through which the ray is transmitted, whereas the second is over all of the boundaries from which it is reflected; the top and bottom signs correspond to topside and bottomside reflections, respectively. Computer code to evaluate the coefficients (15.272) is documented in Doornbos (1988). The *IASPEI 1991 Seismological Tables* (Kennett 1991) include listings of δT_0, δT_1, δT_2 for selected phases, computed using this code. Kennett & Gudmundsson (1996) discuss the modifications needed to extend the results (15.271)–(15.272) to non-geometrical phases such as $\mathrm{P_{diff}}$ in the core shadow; these authors also show how to compute the ellipticity corrections for compound phases such as PP by combining results for the various legs. It should be noted that all of these references follow the original convention of Dziewonski & Gilbert (1976) rather than the one employed here; in particular, the azimuth is measured clockwise from north rather than counterclockwise from south, and the normalization of the Legendre functions in the equivalent of (15.271) is unusual.

⋆15.9.3 Ray geometry

Small changes in the geometry of a spherical-Earth ray can be determined by means of a perturbation analysis of Hamilton's equations (15.48). Alternatively, it is possible to perturb the equivalent system of four equations in ray-centered coordinates; the results of such a ray-centered perturbation analysis are presented by Farra & Madariaga (1987) and Coates & Chapman (1990). This section describes a third variant of ray perturbation theory, due to Liu & Tromp (1996). In this method, which is particularly well suited for global seismological applications, perturbation theory is applied to the four spherical polar ray-tracing equations (15.213)–(15.216). We shall restrict attention initially to the case of an Earth model with spherically symmetrical discontinuities; the effect of a slight boundary perturbation δd will be considered briefly in Section 15.9.4.

Supposing the unperturbed spherical-Earth ray to lie in the equatorial plane, with the local radius r and incidence angle i determined by (15.253), we consider perturbations of the form

$$r \to r + \delta r, \qquad \theta \to \pi/2 + \delta\theta, \qquad i \to i + \delta i, \qquad \zeta \to \pi/2 + \delta\zeta. \tag{15.273}$$

Upon substituting (15.273) into (15.213)–(15.216) and ignoring second-order terms, we obtain a system of linear equations governing the evolution of δr, $\delta\theta$, δi and $\delta\zeta$ along a perturbed ray:

$$\frac{d}{d\phi}\delta r = \cot i\,\delta r - r(\sin i)^{-2}\,\delta i, \tag{15.274}$$

$$\frac{d}{d\phi}\delta\theta = -\delta\zeta, \tag{15.275}$$

$$\begin{aligned}\frac{d}{d\phi}\delta i = {} & (rv^{-1}\ddot{v} - rv^{-2}\dot{v}^2 + v^{-1}\dot{v})\,\delta r + rv^{-1}\partial_r\delta v \\ & - v^{-1}\cot i\,\partial_\phi\delta v - rv^{-2}\dot{v}\,\delta v,\end{aligned} \tag{15.276}$$

$$\frac{d}{d\phi}\delta\zeta = \delta\theta + (\sin i)^{-2} v^{-1} \partial_\theta \delta v. \tag{15.277}$$

The associated perturbed boundary conditions require that

$$[\tan i\, \delta r]_-^+ = 0, \qquad [\delta\theta]_-^+ = 0, \qquad [\delta\zeta]_-^+ = 0,$$
$$[\cot i\, \delta i - (v^{-1}\dot{v} - r^{-1})\, \delta r - v^{-1}\delta v]_-^+ = 0 \tag{15.278}$$

at every spherical discontinuity. The term $-v^{-1}\delta v$ in the final condition (15.278) arises because the wave-speed perturbations experienced by the incident wave and the reflected or transmitted wave may differ. It is noteworthy that the equations and boundary conditions governing δr and δi are decoupled from those governing $\delta\theta$ and $\delta\zeta$.

Let us solve for the out-of-plane perturbations $\delta\theta$ and $\delta\zeta$ first. We begin by rewriting equations (15.275) and (15.277) in a convenient 2×2 matrix notation:

$$\frac{d\mathsf{y}}{d\phi} = \mathsf{A}\mathsf{y} + \mathsf{f}, \tag{15.279}$$

where

$$\mathsf{y} = \begin{pmatrix} \delta\theta \\ \delta\zeta \end{pmatrix}, \qquad \mathsf{f} = \begin{pmatrix} 0 \\ (\sin i)^{-2} v^{-1} \partial_\theta \delta v \end{pmatrix}, \tag{15.280}$$

$$\mathsf{A} = \begin{pmatrix} 0 & -1 \\ 1 & 0 \end{pmatrix}. \tag{15.281}$$

It is easily verified by direct substitution that the solution to the inhomogeneous equation (15.279) is

$$\mathsf{y}(\phi) = \mathsf{P}(\phi, 0) \left[\int_0^\phi \mathsf{P}^{-1}(\tilde{\phi}, 0)\, \mathsf{f}(\tilde{\phi})\, d\tilde{\phi} + \mathsf{y}(0) \right]$$
$$= \int_0^\phi \mathsf{P}(\phi, \tilde{\phi})\, \mathsf{f}(\tilde{\phi})\, d\tilde{\phi} + \mathsf{P}(\phi, 0)\, \mathsf{y}(0). \tag{15.282}$$

Here P is the 2×2 propagator matrix satisfying

$$\frac{d\mathsf{P}}{d\phi} = \mathsf{A}\mathsf{P}, \qquad \mathsf{P}(\phi, \phi) = \mathsf{I}. \tag{15.283}$$

The second equality in (15.282) follows from the inverse-propagator identity

$$\mathsf{P}(\phi, 0)\, \mathsf{P}^{-1}(\tilde{\phi}, 0) = \mathsf{P}(\phi, 0)\, \mathsf{P}(0, \tilde{\phi}) = \mathsf{P}(\phi, \tilde{\phi}). \tag{15.284}$$

The propagator from one arbitrary point $0 \leq \tilde{\phi} \leq \Theta$ to another $0 \leq \phi \leq \Theta$ is given explicitly by

$$\mathsf{P}(\phi, \tilde{\phi}) = \begin{pmatrix} \cos(\phi - \tilde{\phi}) & -\sin(\phi - \tilde{\phi}) \\ \sin(\phi - \tilde{\phi}) & \cos(\phi - \tilde{\phi}) \end{pmatrix}. \tag{15.285}$$

The perturbed ray is required to emanate from the same source and hit the same receiver as the unperturbed ray; that is,

$$\mathsf{y}(0) = \begin{pmatrix} 0 \\ \delta\zeta' \end{pmatrix}, \qquad \mathsf{y}(\Theta) = \begin{pmatrix} 0 \\ \delta\zeta \end{pmatrix}, \tag{15.286}$$

where $\delta\zeta'$ and $\delta\zeta$ denote the perturbed out-of-plane takeoff and arrival angles, respectively. Upon inserting (15.286) into (15.282), we obtain the closed-form representations

$$\delta\zeta' = -(\sin\Theta)^{-1} \int_0^{\Theta} \sin(\Theta - \phi)(\sin i)^{-2} v^{-1} \partial_\theta \delta v \, d\phi, \tag{15.287}$$

$$\delta\zeta = (\sin\Theta)^{-1} \int_0^{\Theta} \sin\phi \, (\sin i)^{-2} v^{-1} \partial_\theta \delta v \, d\phi. \tag{15.288}$$

Equations (15.287) and (15.288) determine the initial and final perturbations $\delta\zeta'$ and $\delta\zeta$ in terms of the wave-speed gradient $\partial_\theta \delta v$ perpendicular to the spherical-Earth ray plane. The integration is performed along the unperturbed ray $r(\phi)$, $i(\phi)$. The complete solution (15.282) at intermediate points $0 \leq \phi \leq \Theta$ is

$$\delta\theta(\phi) = -\int_0^{\phi} \sin(\phi - \tilde{\phi})(\sin\tilde{\imath})^{-2} \tilde{v}^{-1} \partial_\theta \delta\tilde{v} \, d\tilde{\phi} - \delta\zeta' \sin\phi, \tag{15.289}$$

$$\delta\zeta(\phi) = \int_0^{\phi} \cos(\phi - \tilde{\phi})(\sin\tilde{\imath})^{-2} \tilde{v}^{-1} \partial_\theta \delta\tilde{v} \, d\tilde{\phi} + \delta\zeta' \cos\phi, \tag{15.290}$$

where the tildes denote evaluation at the dummy variable $\tilde{\phi}$. Note that $\delta\theta(0) = \delta\theta(\Theta) = 0$ whereas $\delta\zeta(0) = \delta\zeta'$ and $\delta\zeta(\Theta) = \delta\zeta$, as expected.

Next, we determine the in-plane perturbations δr and δi. The governing equations (15.274) and (15.276) may be written in a matrix form analogous to (15.279):

$$\frac{d\mathsf{y}}{d\phi} = \mathsf{A}\mathsf{y} + \mathsf{f}. \tag{15.291}$$

To account for the jumps $[\delta r]_-^+$ and $[\delta i]_-^+$ at interfaces, we are required to solve equation (15.291) subject to an inhomogeneous continuity condition at every boundary interaction point along the unperturbed ray:

$$[\mathsf{B}\mathsf{y} + \mathsf{b}]_-^+ = 0. \tag{15.292}$$

The 2×1 column vectors y, b, f and the 2×2 matrices A, B in equations (15.291)–(15.292) are defined by

$$\mathsf{y} = \begin{pmatrix} \delta r \\ \delta i \end{pmatrix}, \qquad \mathsf{b} = \begin{pmatrix} 0 \\ -v^{-1}\delta v \end{pmatrix}, \tag{15.293}$$

$$\mathsf{f} = \begin{pmatrix} 0 \\ rv^{-1}\partial_r \delta v - \cot i\, v^{-1}\partial_\phi \delta v - rv^{-2}\dot{v}\,\delta v \end{pmatrix}, \tag{15.294}$$

$$\mathsf{A} = \begin{pmatrix} \cot i & -r(\sin i)^{-2} \\ rv^{-1}\ddot{v} - rv^{-2}\dot{v}^2 + v^{-1}\dot{v} & 0 \end{pmatrix}, \tag{15.295}$$

$$\mathsf{B} = \begin{pmatrix} \tan i & 0 \\ -v^{-1}\dot{v} + r^{-1} & \cot i \end{pmatrix}. \tag{15.296}$$

In this case the 2×2 propagator matrix P must satisfy an additional boundary constraint; the complete set of defining relations is

$$\frac{d\mathsf{P}}{d\phi} = \mathsf{AP}, \qquad [\mathsf{BP}]_-^+ = 0, \qquad \mathsf{P}(\phi, \phi) = \mathsf{I}. \tag{15.297}$$

The elements of the in-plane propagator are the four partial derivatives

$$\mathsf{P}(\phi, \tilde{\phi}) = \begin{pmatrix} \partial_{\tilde{r}} r(\phi, \tilde{\phi}) & \partial_{\tilde{\imath}} r(\phi, \tilde{\phi}) \\ \partial_{\tilde{r}} i(\phi, \tilde{\phi}) & \partial_{\tilde{\imath}} i(\phi, \tilde{\phi}) \end{pmatrix}, \tag{15.298}$$

where $\tilde{r} = r(\tilde{\phi})$ and $\tilde{\imath} = i(\tilde{\phi})$. The solution to equations (15.291)–(15.292) can be written in terms of the matrix (15.298) and its inverse,

$$\mathsf{P}^{-1}(\phi, \tilde{\phi}) = \frac{1}{\det \mathsf{P}(\phi, \tilde{\phi})} \begin{pmatrix} \partial_{\tilde{\imath}} i(\phi, \tilde{\phi}) & -\partial_{\tilde{\imath}} r(\phi, \tilde{\phi}) \\ -\partial_{\tilde{r}} i(\phi, \tilde{\phi}) & \partial_{\tilde{r}} r(\phi, \tilde{\phi}) \end{pmatrix}, \tag{15.299}$$

in the form

$$\mathsf{y}(\phi) = \mathsf{P}(\phi, 0) \left[\int_0^\phi \mathsf{P}^{-1}(\tilde{\phi}, 0)\, \mathsf{f}(\tilde{\phi})\, d\tilde{\phi} + \mathsf{y}(0) \right.$$

$$+\sum_d \mathsf{P}^{-1}(\phi_d,0)\left(\mathsf{B}_d^{\mathrm{out}}\right)^{-1}\left(\mathsf{b}_d^{\mathrm{inc}}-\mathsf{b}_d^{\mathrm{out}}\right)\Bigg]$$

$$= \int_0^{\phi} \mathsf{P}(\phi,\tilde{\phi})\,\mathsf{f}(\tilde{\phi})\,d\tilde{\phi} + \mathsf{P}(\phi,0)\,\mathsf{y}(0)$$

$$+\sum_d \mathsf{P}(\phi,\phi_d)\left(\mathsf{B}_d^{\mathrm{out}}\right)^{-1}\left(\mathsf{b}_d^{\mathrm{inc}}-\mathsf{b}_d^{\mathrm{out}}\right). \qquad (15.300)$$

The summation is over all of the boundaries encountered by the unperturbed ray; the superscripts inc and out denote evaluation upon the incident and outgoing sides of the discontinuity. Only when δv^{inc} and δv^{out} are different do the boundary terms contribute; in that case, their effect can be quite significant.

Once again we demand that the perturbed and unperturbed rays start from the same source and hit the same receiver:

$$\mathsf{y}(0) = \begin{pmatrix} 0 \\ \delta i' \end{pmatrix}, \qquad \mathsf{y}(\Theta) = \begin{pmatrix} 0 \\ \delta i \end{pmatrix}. \qquad (15.301)$$

Upon inserting the boundary conditions (15.301) into (15.300) and noting that $\tilde{r}(0) = r'$ and $\tilde{\imath}(0) = i'$, we find that the perturbed takeoff angle $\delta i'$ and arrival angle δi are given by

$$\begin{aligned}\delta i' = {} & \frac{1}{\partial_{i'} r(\Theta)} \int_0^{\Theta} D^{-1}(\phi)\,[\partial_{r'} r(\Theta)\partial_{i'} r(\phi) - \partial_{i'} r(\Theta)\partial_{r'} r(\phi)] \\ & \times [r v^{-1}\partial_r \delta v - \cot i\, v^{-1}\partial_\phi \delta v - r v^{-2}\dot{v}\,\delta v]\,d\phi \\ & + \frac{1}{\partial_{i'} r(\Theta)} \sum_d D^{-1}(\phi_d^{\mathrm{out}}) \\ & \times [\partial_{r'} r(\Theta)\partial_{i'} r(\phi_d^{\mathrm{out}}) - \partial_{i'} r(\Theta)\partial_{r'} r(\phi_d^{\mathrm{out}})] \\ & \times \tan i_d^{\mathrm{out}}[(v^{-1}\delta v)_d^{\mathrm{out}} - (v^{-1}\delta v)_d^{\mathrm{inc}}], \end{aligned} \qquad (15.302)$$

$$\begin{aligned}\delta i = {} & \frac{D(\Theta)}{\partial_{i'} r(\Theta)} \int_0^{\Theta} D^{-1}(\phi)\,\partial_{i'} r(\phi) \\ & \times [r v^{-1}\partial_r \delta v - \cot i\, v^{-1}\partial_\phi \delta v - r v^{-2}\dot{v}\,\delta v]\,d\phi \\ & + \frac{D(\Theta)}{\partial_{i'} r(\Theta)} \sum_d D^{-1}(\phi_d^{\mathrm{out}})\,\partial_{i'} r(\phi_d^{\mathrm{out}}) \\ & \tan i_d^{\mathrm{out}}[(v^{-1}\delta v)_d^{\mathrm{out}} - (v^{-1}\delta v)_d^{\mathrm{inc}}], \end{aligned} \qquad (15.303)$$

where

$$D(\phi) = \partial_{r'} r(\phi)\partial_{i'} i(\phi) - \partial_{i'} r(\phi)\partial_{r'} i(\phi). \qquad (15.304)$$

Equations (15.302)–(15.303) determine the perturbations $\delta i'$, δi in terms of the in-plane wave-speed gradients $\partial_r \delta v$, $\partial_\phi \delta v$ and the boundary contrasts $(v^{-1}\delta v)_d^{\mathrm{out}} - (v^{-1}\delta v)_d^{\mathrm{inc}}$. In shooting to find the geometrical rays between a given source and receiver, it is advantageous to use (15.287) and (15.302) as first estimates of the initial takeoff azimuth $\zeta_0' = \pi/2 + \delta\zeta'$ and incidence angle $i_0' = i' + \delta i'$; this can substantially decrease the number of iterations of (15.221) needed to hit the receiver.

The complete geometry of the perturbed ray is determined by equations (15.289)–(15.290) together with

$$\begin{aligned}\delta r(\phi) = & \int_0^\phi D^{-1}(\tilde{\phi})\,[-\partial_{r'} r(\phi)\partial_{i'} r(\tilde{\phi}) + \partial_{i'} r(\phi)\partial_{r'} r(\tilde{\phi})] \\ & \qquad \times [r v^{-1}\partial_r \delta v - \cot i\, v^{-1}\partial_\phi \delta v - r v^{-2}\dot{v}\,\delta v]\,d\phi \\ & + \sum_d D^{-1}(\phi_d^{\mathrm{out}}) \\ & \qquad \times [-\partial_{r'} r(\phi)\partial_{i'} r(\phi_d^{\mathrm{out}}) + \partial_{i'} r(\phi)\partial_{r'} r(\phi_d^{\mathrm{out}})] \\ & \qquad \times \tan i_d^{\mathrm{out}}[(v^{-1}\delta v)_d^{\mathrm{out}} - (v^{-1}\delta v)_d^{\mathrm{inc}}] + \delta i'\,\partial_{i'} r(\phi), \end{aligned} \qquad (15.305)$$

$$\begin{aligned}\delta i(\phi) = & \int_0^\phi D^{-1}(\tilde{\phi})\,[-\partial_{r'} i(\phi)\partial_{i'} r(\tilde{\phi}) + \partial_{i'} i(\phi)\partial_{r'} r(\tilde{\phi})] \\ & \qquad \times [r v^{-1}\partial_r \delta v - \cot i\, v^{-1}\partial_\phi \delta v - r v^{-2}\dot{v}\,\delta v]\,d\phi \\ & + \sum_d D^{-1}(\phi_d^{\mathrm{out}}) \\ & \qquad \times [-\partial_{r'} i(\phi)\partial_{i'} r(\phi_d^{\mathrm{out}}) + \partial_{i'} i(\phi)\partial_{r'} r(\phi_d^{\mathrm{out}})] \\ & \qquad \times \tan i_d^{\mathrm{out}}[(v^{-1}\delta v)_d^{\mathrm{out}} - (v^{-1}\delta v)_d^{\mathrm{inc}}] + \delta i'\,\partial_{i'} i(\phi), \end{aligned} \qquad (15.306)$$

where the summations are now carried out only over the discontinuities lying between the source and the instantaneous point ϕ. It is easily verified that $\delta r(0) = \delta r(\Theta) = 0$ and $\delta i(0) = \delta i'$, $\delta i(\Theta) = \delta i$. Figure 15.12 compares the results of exact and first-order ray tracing between a fixed source and receiver in model SKS12WM13. In general, perturbation theory predicts the geometry of rays in such a smooth ($s_{\max} = 12$) Earth model very well.

Figures 15.13 and 15.14 compare the exact and first-order incidence angle and azimuth anomalies $\delta i'$ and $\delta\zeta'$ at the source, for a number of ray paths in model SKS12WM13. Figures 15.15 and 15.16 show a similar comparison of the two arrival-angle anomalies δi and $\delta\zeta$ at the receiver. The large magnitude of the term $(\sin i)^{-2}$ makes the out-of-plane perturbations (15.287) and (15.288) of steeply propagating waves particularly sensitive to transverse gradients $\partial_\theta \delta v$. For this reason, the reflected phases PcP and ScS exhibit takeoff and arrival anomalies $\delta\zeta'$, $\delta\zeta$ that are approximately twice as large as those of the turning phases P and S.

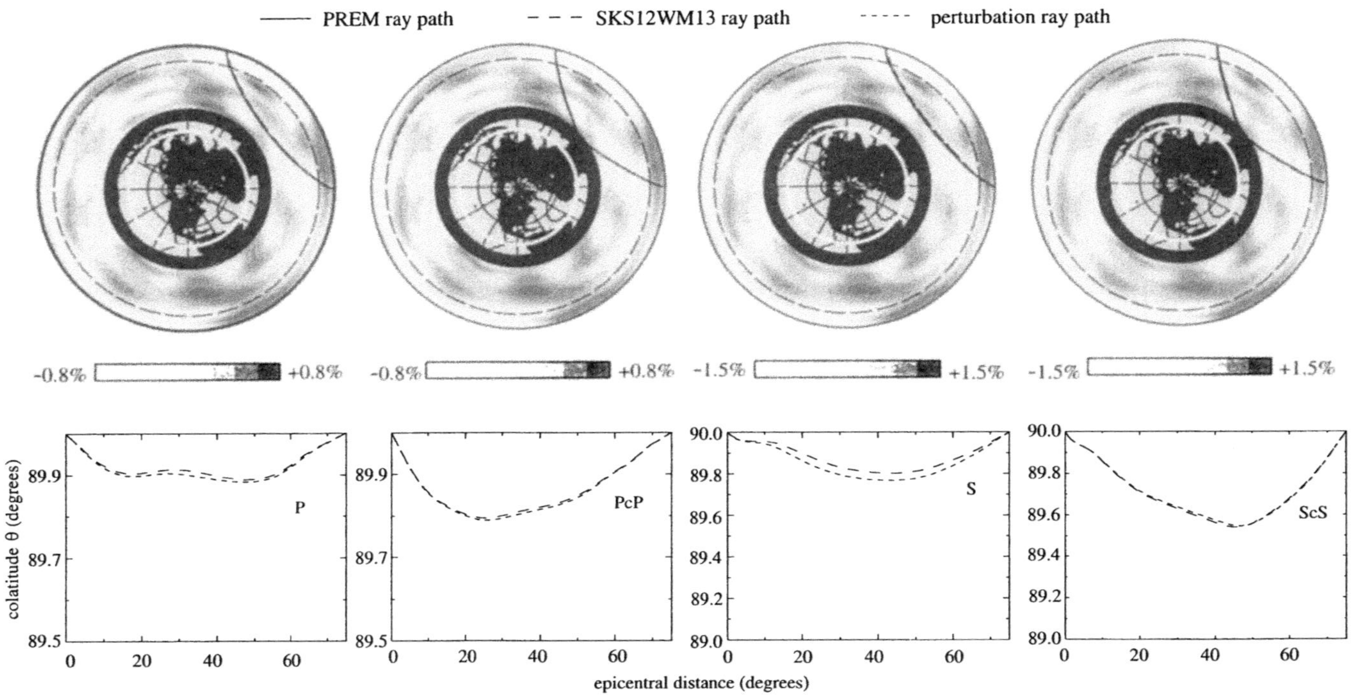

Figure 15.12. Comparison of the exact (*long dashed line*) and perturbation-theoretical (*short dashed line*) ray paths in model SKS12WM13. The unperturbed ray path in PREM is also shown (*solid line*). The source is situated on the equator and Greenwich Meridian, and the receiver is situated due east at an epicentral distance $\Theta = 75°$. (*Top*) P, PcP, S and ScS rays projected onto a cross-section of the source-receiver great-circle plane. Shading depicts the relative wave-speed perturbations $-0.8\% \leq \delta\alpha/\alpha \leq 0.8\%$ and $-1.5\% \leq \delta\beta/\beta \leq 1.5\%$. (*Bottom*) Projection of the same ray paths onto the surface of the Earth.

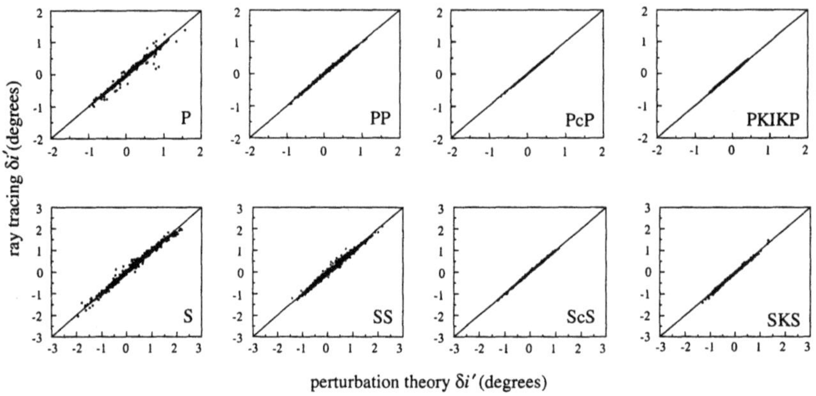

Figure 15.13. Scatter-plot comparison of the incidence-angle anomaly $\delta i'$ at the source, computed using first-order perturbation theory and exact ray tracing in model SKS12WM13. Each graph displays results for 1000 randomly selected source-receiver paths; the associated travel-time perturbations δT are compared in Figure 15.10. See Figure 15.15 for the corresponding anomaly δi at the receiver.

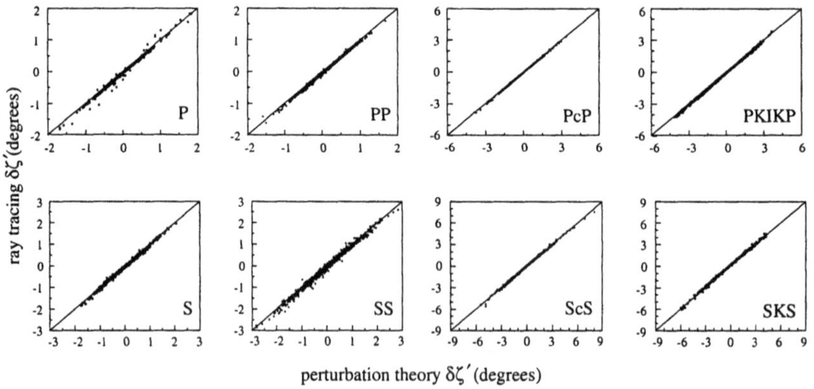

Figure 15.14. Scatter-plot comparison of the the azimuthal takeoff-angle anomaly $\delta\zeta'$, computed using first-order perturbation theory and exact ray tracing in model SKS12WM13. See Figure 15.16 for the corresponding anomaly $\delta\zeta$ at the receiver.

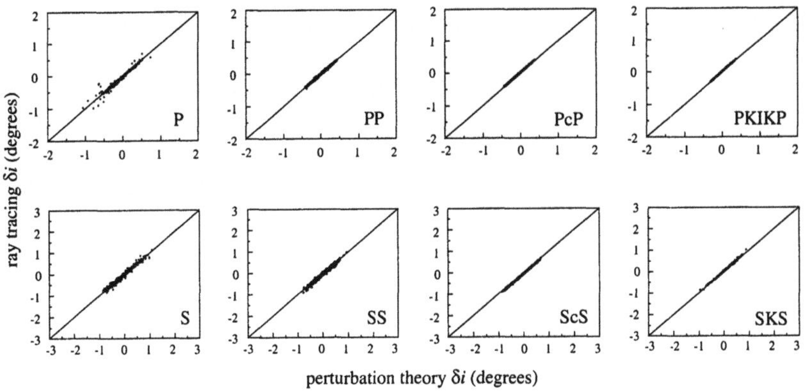

Figure 15.15. Same as Figure 15.13 for the observable incidence-angle anomaly δi at the receiver. Each scatter plot displays results for 1000 randomly selected source-receiver paths.

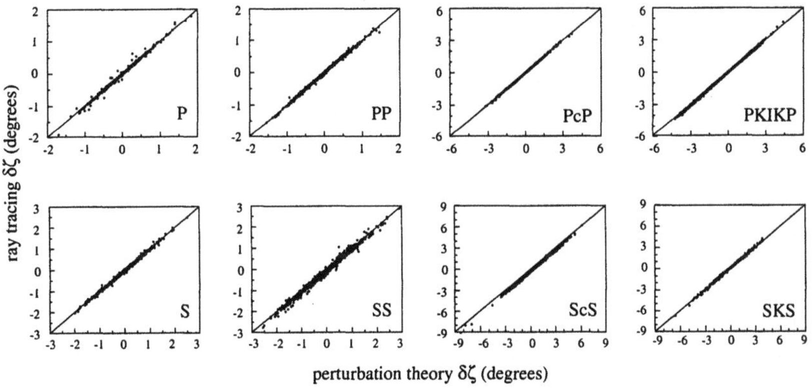

Figure 15.16. Same as Figure 15.14 for the azimuthal arrival-angle anomaly $\delta\zeta$. The epicentral distance ranges in all four Figures 15.13–15.16 are the same as in Figure 15.10—P: $30^\circ \leq \Theta \leq 95^\circ$, S: $30^\circ \leq \Theta \leq 80^\circ$, PP and SS: $60^\circ \leq \Theta \leq 179^\circ$, PcP and ScS: $10^\circ \leq \Theta \leq 75^\circ$, PKIKP: $130^\circ \leq \Theta \leq 170^\circ$, SKS: $85^\circ \leq \Theta \leq 130^\circ$.

*15.9.4 Boundary topography

The effects of a slight perturbation δd in the locations of the initially spherical discontinuities have been treated by Liu & Tromp (1996). We simply quote their results here. The takeoff-angle and arrival-angle perturbations (15.287)–(15.288) and (15.302)–(15.303) exhibit additional boundary contributions. The out-of-plane and in-plane deflections depend upon the transverse ∂_θ and longitudinal ∂_ϕ gradients in topography, respectively:

$$\delta\zeta' \to \delta\zeta' - (\sin\Theta)^{-1}\sum_d \sin(\Theta - \phi_d) \times (\cot i_d^{\rm out} - \cot i_d^{\rm inc})\,\partial_\theta \ln \delta d, \qquad (15.307)$$

$$\delta\zeta \to \delta\zeta + (\sin\Theta)^{-1}\sum_d \sin\phi_d(\cot i_d^{\rm out} - \cot i_d^{\rm inc})\,\partial_\theta \ln \delta d, \qquad (15.308)$$

$$\delta i' \to \delta i' - \frac{1}{\partial_{i'} r(\Theta)}\sum_d D^{-1}(\phi_d^{\rm out}) \times [\partial_{r'} r(\Theta)\partial_{i'} r(\phi_d^{\rm out}) - \partial_{i'} r(\Theta)\partial_{r'} r(\phi_d^{\rm out})] \times \tan i_d^{\rm out}(\cot i_d^{\rm out} - \cot i_d^{\rm inc})\,\partial_\phi \ln \delta d, \qquad (15.309)$$

$$\delta i \to \delta i - \frac{D(\Theta)}{\partial_{i'} r(\Theta)}\sum_d D^{-1}(\phi_d^{\rm out})\,\partial_{i'} r(\phi_d^{\rm out}) \times \tan i_d^{\rm out}(\cot i_d^{\rm out} - \cot i_d^{\rm inc})\,\partial_\phi \ln \delta d. \qquad (15.310)$$

The arrival-angle anomalies δi and $\delta\zeta$ at the receiver can be measured by means of a polarization analysis, applied to a single three-component recording or to a suite of recordings from a tightly spaced array. The above results enable such data to be used as supplemental constraints in global travel-time inversion studies; the dependence of $\delta\zeta$ and δi upon the volumetric and topographic gradients $\partial_r \delta v$, $\partial_\theta \delta v$, $\partial_\phi \delta v$ and $\partial_\theta \delta d$, $\partial_\phi \delta d$ is strictly linear. The effect of a possible source mislocation $\delta r'$, $\delta\theta'$ can be accounted for by adding terms $(\sin\Theta)^{-1}\delta\theta'$ and $-[D(\Theta)/\partial_{i'} r(\Theta)]\,\delta r'$ to the right sides of equations (15.308) and (15.310), respectively.

The ray-geometry equations (15.289)–(15.290) and (15.305)–(15.306) are likewise modified by the presence of boundary topography:

$$\delta\theta(\phi) \to \delta\theta(\phi) - \sum_d \sin(\phi - \phi_d) \times (\cot i_d^{\rm out} - \cot i_d^{\rm inc})\,\partial_\theta \ln \delta d, \qquad (15.311)$$

$$\delta\zeta(\phi) \to \delta\zeta(\phi) + \sum_d \cos(\phi - \phi_d) \times (\cot i_d^{\rm out} - \cot i_d^{\rm out})\, \partial_\theta \ln \delta d, \qquad (15.312)$$

$$\delta r(\phi) \to \delta r(\phi) - \sum_d D^{-1}(\phi_d^{\rm out}) \times [-\partial_{r'} r(\phi)\partial_{i'} r(\phi_d^{\rm out}) + \partial_{i'} r(\phi)\partial_{r'} r(\phi_d^{\rm out})] \times \tan i_d^{\rm out}(\cot i_d^{\rm out} - \cot i_d^{\rm inc})\, \partial_\phi \ln \delta d, \qquad (15.313)$$

$$\delta i(\phi) \to \delta i(\phi) - \sum_d D^{-1}(\phi_d^{\rm out}) \times [-\partial_{r'} i(\phi)\partial_{i'} r(\phi_d^{\rm out}) + \partial_{i'} i(\phi)\partial_{r'} r(\phi_d^{\rm out})] \times \tan i_d^{\rm out}(\cot i_d^{\rm out} - \cot i_d^{\rm inc})\, \partial_\phi \ln \delta d. \qquad (15.314)$$

The results (15.311)–(15.314) can be used, among other things, to correct for the effect of the Earth's ellipticity and variations in crustal thickness.

*15.9.5 Amplitude perturbation

The amplitude of every pulse in the elastic ray sum (15.190) is a product $A = \Xi\Sigma\Pi\mathcal{R}^{-1}$ of a receiver polarization factor Ξ, a point-source excitation factor Σ, a boundary interaction factor Π, and a geometrical spreading factor $\mathcal{R}^{-1}$. A three-dimensional perturbation $\delta\alpha$, $\delta\beta$, $\delta\rho$, δd of the Earth alters each of these four factors:

$$\delta A/A = \delta\Xi/\Xi + \delta\Sigma/\Sigma + \delta\Pi/\Pi - \delta\mathcal{R}/\mathcal{R}. \qquad (15.315)$$

The first two terms in (15.315) may be obtained by perturbing the defining equations (15.188) and (15.189):

$$\delta\Xi/\Xi = (\hat{\boldsymbol{\nu}}\cdot\delta\hat{\boldsymbol{\eta}})/(\hat{\boldsymbol{\nu}}\cdot\hat{\boldsymbol{\eta}}) - \tfrac{1}{2}(\delta\rho/\rho + \delta v/v), \qquad (15.316)$$

$$\delta\Sigma/\Sigma = [\mathbf{M}\!:\!\tfrac{1}{2}(\delta\hat{\mathbf{p}}_{\rm s}\hat{\boldsymbol{\eta}}_{\rm s} + \hat{\mathbf{p}}_{\rm s}\delta\hat{\boldsymbol{\eta}}_{\rm s} + \delta\hat{\boldsymbol{\eta}}_{\rm s}\hat{\mathbf{p}}_{\rm s} + \hat{\boldsymbol{\eta}}_{\rm s}\delta\hat{\mathbf{p}}_{\rm s})] \div [\mathbf{M}\!:\!\tfrac{1}{2}(\hat{\mathbf{p}}_{\rm s}\hat{\boldsymbol{\eta}}_{\rm s} + \hat{\boldsymbol{\eta}}_{\rm s}\hat{\mathbf{p}}_{\rm s})] - \tfrac{1}{2}(\delta\rho_{\rm s}/\rho_{\rm s} + 5\,\delta v_{\rm s}/v_{\rm s}). \qquad (15.317)$$

It is noteworthy that a local perturbation $\delta v_{\rm s}/v_{\rm s}$ in the wave speed at the source gives rise to an amplitude anomaly $\delta A/A$ that is five times larger than that produced by the corresponding perturbation $\delta v/v$ at the receiver. The perturbations $\delta\hat{\mathbf{p}}_{\rm s}$ and $\delta\hat{\boldsymbol{\eta}}_{\rm s}$ in the unit slowness and polarization at the source are easy to express in terms of the takeoff-angle perturbations $\delta i'$ and $\delta\zeta'$. It is also straightforward to relate the perturbation in the polarization $\delta\hat{\boldsymbol{\eta}}_{\rm P}$ of a compressional wave at the receiver to δi and $\delta\zeta$. Finding the perturbations $\delta\hat{\boldsymbol{\eta}}_{\mathrm{SV}}$ and $\delta\hat{\boldsymbol{\eta}}_{\mathrm{SH}}$ in the polarization of an incoming shear wave is more difficult; it is necessary to account for the twisting of the ray, by perturbing equations (15.239)–(15.242). The twisting also affects the perturbation in the concatenated reflection-and-transmission coefficient $\delta\Pi/\Pi$. The final term in (15.315) may be obtained by perturbing equation (15.261):

$$\delta\mathcal{R}/\mathcal{R} = \tfrac{1}{2}(\cot i\,\delta i - \cot i'\,\delta i') + \tfrac{1}{2}(\partial_{i'}r)^{-1}\delta(\partial_{i'}r) - \tfrac{1}{2}(\sin\Theta)^{-1}\delta(\partial_{\zeta'}\theta). \tag{15.318}$$

To find the two quantities $\delta(\partial_{i'}r)$ and $\delta(\partial_{\zeta'}\theta)$, it is necessary to perturb the dynamical ray-tracing equations (15.230)–(15.233). Liu & Tromp (1996) show how to compute $\delta\mathcal{R}/\mathcal{R}$ in terms of the first and second derivatives $\partial_r\delta v$, $\partial_\phi\delta v$, $\partial_r\delta d$, $\partial_\phi\delta d$ and $\partial_r^2\delta v$, $\partial_r\partial_\phi\delta v$, $\partial_\phi^2\delta v$, $\partial_r^2\delta d$, $\partial_r\partial_\phi\delta d$, $\partial_\phi^2\delta d$ of the lateral heterogeneity along the unperturbed ray path. The final results are extremely unwieldy, so we refrain from giving them here.

Laterally heterogeneous anelasticity gives rise to an additional term in equation (15.315):

$$\delta A/A = \delta\Xi/\Xi + \delta\Sigma/\Sigma + \delta\Pi/\Pi - \delta\mathcal{R}/\mathcal{R} + \exp(-\tfrac{1}{2}\omega\,\delta T^*) - 1, \tag{15.319}$$

where

$$\delta T^* = \int_{\mathbf{x}'}^{\mathbf{x}} \frac{\delta Q^{-1}}{Q^{-1}}\frac{ds}{v_0 Q}. \tag{15.320}$$

In principle, this result can be used to invert measured amplitude anomalies $\delta A/A$ for the three-dimensional variations in anelasticity $\delta Q^{-1}/Q^{-1}$, which need not be slight. However, it must be borne in mind that all of the other influences $\delta\Xi/\Xi$, $\delta\Sigma/\Sigma$, $\delta\Pi/\Pi$ and $\delta\mathcal{R}/\mathcal{R}$ upon the amplitude of a pulse may be present. In general, the amplitude of a body-wave pulse is one of its least robust characteristics; because of this, and because of the many phenomena that can affect $\delta A/A$, the general result (15.319) has received limited application.

Chapter 16

Surface-Wave JWKB Theory

In this final chapter we analyze the propagation of Rayleigh and Love waves on a non-rotating, hydrostatic, laterally heterogeneous Earth, using a surface-wave analogue of the body-wave JWKB theory developed in Chapter 15. The resulting surface-wave JWKB theory is valid for arbitrarily large three-dimensional variations in the elastic and anelastic properties of the Earth; however, the variations are required to be *laterally smooth* in the sense $\lambda/\Lambda_\Omega \ll 1$, where $\lambda = 2\pi k^{-1}$ is the wavelength of the surface wave on the unit sphere Ω, and Λ_Ω is the corresponding angular scale of the heterogeneity. Smooth topographic variations on the outer free surface and internal solid-solid and fluid-solid discontinuities are accounted for, as well as slowly varying lateral volumetric heterogeneity within the aspherical sub-regions of the Earth. As in our study of high-frequency body waves, we base our analysis upon a *slowly varying variational principle*, which yields local radial Love and Rayleigh eigenfunctions, local dispersion relations, and transport equations describing the conservation of surface-wave energy. The dispersion relations determine the geometry of surface-wave rays on Ω, whereas the transport equations determine the associated Love and Rayleigh amplitude variations.

The two-dimensional character of the propagation makes surface-wave ray theory simpler in many respects than body-wave ray theory. In addition, we presume there are no lateral boundaries, so that discontinuities in geometrical spreading and coupling between different dispersion branches need not be considered. On the other hand, there are some additional complications; in particular, the propagation is multi-modal and dispersive, and rays must be traced on the curved surface of the unit sphere rather than through three-dimensional Euclidean space. Important contributions to surface-wave JWKB theory have been made by Woodhouse (1974), Babich, Chikhachev & Yanovskaya (1976), Yomogida (1985), Yomogida & Aki (1985), Woodhouse & Wong (1986), Jobert & Jobert (1987) and Keilis-Borok, Levshin & others (1989). We follow the treatment by Tromp & Dahlen (1992a; 1992b) and Wang & Dahlen (1994; 1995), adhering to our usual convention that $\exp(-i\omega t)$ appears in the Fourier integral upon transforming from time t to angular frequency ω.

16.1 Preliminaries

As in the last chapter, we employ our customary set-theoretical notation to describe a piecewise continuous Earth model $\oplus = \oplus_{\mathrm{S}} \cup \oplus_{\mathrm{F}}$ composed of concentric solid and fluid regions separated by non-intersecting interfaces $\Sigma = \partial\oplus \cup \Sigma_{\mathrm{SS}} \cup \Sigma_{\mathrm{FS}}$ with unit outward normal $\hat{\mathbf{n}}$ (see Figure 3.1). The geographically variable radii of the outer free surface and interior solid-solid and fluid-solid discontinuities will be denoted by $d = a \cup d_{\mathrm{SS}} \cup d_{\mathrm{FS}}$ (see Section 8.2). Since we seek a theory that is applicable to mantle as well as crustal surface waves, we retain the effects of self-gravitation in the analysis. The displacement $\mathbf{s}$ and incremental Eulerian gravitational potential ϕ are governed by the linearized momentum equation (4.149) and Poisson's equation (3.96):

$$-\omega^2\rho\mathbf{s} + \boldsymbol{\nabla}(\rho\mathbf{s}\cdot\boldsymbol{\nabla}\phi) + \rho\boldsymbol{\nabla}\phi - [\boldsymbol{\nabla}\cdot(\rho\mathbf{s})]\boldsymbol{\nabla}\phi = \boldsymbol{\nabla}\cdot\mathbf{T}, \tag{16.1}$$

$$\nabla^2\phi = -\boldsymbol{\nabla}\cdot(\rho\mathbf{s}). \tag{16.2}$$

We restrict attention to an Earth having an isotropic elastic stress-strain relation, as in (15.2):

$$\mathbf{T} = \kappa(\boldsymbol{\nabla}\cdot\mathbf{s})\mathbf{I} + 2\mu\mathbf{d}, \tag{16.3}$$

where κ and μ are the isentropic incompressibility and the rigidity, respectively, and $\mathbf{d} = \frac{1}{2}[\boldsymbol{\nabla}\mathbf{s} + (\boldsymbol{\nabla}\mathbf{s})^{\mathrm{T}}] - \frac{1}{3}(\boldsymbol{\nabla}\cdot\mathbf{s})\mathbf{I}$ is the strain deviator. The system of elastic-gravitational equations (16.1)–(16.3) must be solved subject to the kinematic continuity conditions $[\mathbf{s}]_-^+ = \mathbf{0}$ on Σ_{SS}, $[\hat{\mathbf{n}}\cdot\mathbf{s}]_-^+ = 0$ on Σ_{FS} and $[\phi]_-^+ = 0$ on Σ, together with the dynamical and gravitational boundary conditions (4.150)–(4.152) and (3.97):

$$\hat{\mathbf{n}}\cdot\mathbf{T} = \mathbf{0} \quad \text{on } \partial\oplus, \tag{16.4}$$

$$[\hat{\mathbf{n}}\cdot\mathbf{T}]_-^+ = \mathbf{0} \quad \text{on } \Sigma_{\mathrm{SS}}, \tag{16.5}$$

$$[\hat{\mathbf{n}}\cdot\mathbf{T}]_-^+ = \hat{\mathbf{n}}[\hat{\mathbf{n}}\cdot\mathbf{T}\cdot\hat{\mathbf{n}}]_-^+ = \mathbf{0} \quad \text{on } \Sigma_{\mathrm{FS}}, \tag{16.6}$$

$$[\hat{\mathbf{n}}\cdot\boldsymbol{\nabla}\phi + 4\pi G\rho\hat{\mathbf{n}}\cdot\mathbf{s}]_-^+ = 0 \quad \text{on all of } \Sigma. \tag{16.7}$$

Since we are interested in the propagation of transient waves, it is essential to regard $\mathbf{s}(\mathbf{x},\omega)$, $\phi(\mathbf{x},\omega)$ and $\mathbf{T}(\mathbf{x},\omega)$ as *complex*. As in Chapter 15, we consider the real angular frequency to be positive: $\omega > 0$.

The equations of motion (16.1)–(16.2) and associated boundary conditions (16.4)–(16.7) can be obtained from the variational principle $\delta\mathcal{I}' = 0$, where

$$\mathcal{I}' = \int_{\bigcirc} L'(\mathbf{s},\boldsymbol{\nabla}\mathbf{s},\boldsymbol{\nabla}\phi\,;\,\mathbf{s}^*,\boldsymbol{\nabla}\mathbf{s}^*,\boldsymbol{\nabla}\phi^*)\,dV. \tag{16.8}$$

The integration in equation (16.8) is over all of space $\bigcirc$; the integrand is the Lagrangian density in an isotropic Earth, given by

$$\begin{aligned} L' = {} & \tfrac{1}{2}[\omega^2\rho\mathbf{s}^*\cdot\mathbf{s} - \kappa(\boldsymbol{\nabla}\cdot\mathbf{s}^*)(\boldsymbol{\nabla}\cdot\mathbf{s}) - 2\mu\mathbf{d}^*:\mathbf{d} \\ & - \rho(\mathbf{s}^*\cdot\boldsymbol{\nabla}\phi + \mathbf{s}\cdot\boldsymbol{\nabla}\phi^*) - \rho\mathbf{s}^*\cdot\boldsymbol{\nabla}\boldsymbol{\nabla}\Phi\cdot\mathbf{s} \\ & - \tfrac{1}{2}\rho\boldsymbol{\nabla}\Phi\cdot(\mathbf{s}^*\cdot\boldsymbol{\nabla}\mathbf{s} + \mathbf{s}\cdot\boldsymbol{\nabla}\mathbf{s}^* - \mathbf{s}^*\boldsymbol{\nabla}\cdot\mathbf{s} - \mathbf{s}\boldsymbol{\nabla}\cdot\mathbf{s}^*) \\ & - (4\pi G)^{-1}\boldsymbol{\nabla}\phi^*\cdot\boldsymbol{\nabla}\phi], \end{aligned} \tag{16.9}$$

where Φ is the initial gravitational potential and G is the gravitational constant. Equations (16.8) and (16.9) are the travelling-wave equivalents of the standing-wave action (4.162) and Lagrangian density (4.164). It is convenient to regard $\mathbf{s}$ and ϕ as independent variables in the present application; for that reason, we base our JWKB analysis upon the *modified* action $\mathcal{I}'$ and Lagrangian density L' rather than the travelling-wave versions of (4.161) and (4.163). The variation of the modified action (16.8) is

$$\begin{aligned} \delta\mathcal{I}' = {} & 2\,\mathrm{Re}\int_{\oplus} \delta\mathbf{s}^*\cdot[\partial_{\mathbf{s}^*}L' - \boldsymbol{\nabla}\cdot(\partial_{\boldsymbol{\nabla}\mathbf{s}^*}L')]\,dV \\ & - 2\,\mathrm{Re}\int_{\bigcirc} \delta\phi^*\,[\boldsymbol{\nabla}\cdot(\partial_{\boldsymbol{\nabla}\phi^*}L')]\,dV \\ & - 2\,\mathrm{Re}\int_{\Sigma} [\delta\mathbf{s}^*\cdot(\hat{\mathbf{n}}\cdot\partial_{\boldsymbol{\nabla}\mathbf{s}^*}L') + \delta\phi^*(\hat{\mathbf{n}}\cdot\partial_{\boldsymbol{\nabla}\phi^*}L')]_-^+\,d\Sigma. \end{aligned} \tag{16.10}$$

This vanishes for arbitrary independent variations $\boldsymbol{\delta}\mathbf{s}$ and $\delta\phi$ satisfying the admissibility constraints $[\boldsymbol{\delta}\mathbf{s}]_-^+ = \mathbf{0}$ on Σ_{SS}, $[\hat{\mathbf{n}}\cdot\boldsymbol{\delta}\mathbf{s}]_-^+ = 0$ on Σ_{FS} and $[\delta\phi]_-^+ = 0$ on Σ if and only if $\mathbf{s}$ and ϕ satisfy the Euler-Lagrange equations

$$\partial_{\mathbf{s}^*}L' - \boldsymbol{\nabla}\cdot(\partial_{\boldsymbol{\nabla}\mathbf{s}^*}L') \quad \text{in } \oplus, \tag{16.11}$$

$$\boldsymbol{\nabla}\cdot(\partial_{\boldsymbol{\nabla}\phi^*}L') \quad \text{in } \bigcirc \tag{16.12}$$

and associated boundary conditions

$$\hat{\mathbf{n}}\cdot(\partial_{\boldsymbol{\nabla}\mathbf{s}^*}L') = \mathbf{0} \quad \text{on } \partial\oplus, \tag{16.13}$$

$$[\hat{\mathbf{n}}\cdot(\partial_{\boldsymbol{\nabla}\mathbf{s}^*}L')]_-^+ = \mathbf{0} \quad \text{on } \Sigma_{\rm SS}, \tag{16.14}$$

$$[\hat{\mathbf{n}}\cdot(\partial_{\boldsymbol{\nabla}\mathbf{s}^*}L)]_-^+ = \hat{\mathbf{n}}[\hat{\mathbf{n}}\cdot(\partial_{\boldsymbol{\nabla}\mathbf{s}^*}L)\cdot\hat{\mathbf{n}}]_-^+ = \mathbf{0} \quad \text{on } \Sigma_{\rm FS}, \tag{16.15}$$

$$[\hat{\mathbf{n}}\cdot(\partial_{\boldsymbol{\nabla}\phi^*}L')]_-^+ = 0 \quad \text{on } \Sigma. \tag{16.16}$$

Equations (16.11)–(16.12) and (16.13)–(16.16) are identical to (16.1)–(16.2) and (16.4)–(16.7). The value of the modified action at the stationary transient solution is $\mathcal{I}' = 0$.

The volumetric energy density associated with the Lagrangian density (16.9) is $E = E' = \omega\partial_\omega L' - L'$. A more useful quantity in the present instance is the radially integrated *surface energy density*, defined by

$$E_\Omega = \int_0^\infty (\omega\partial_\omega L' - L')\, r^2 dr. \tag{16.17}$$

The transient, frequency-domain analogue of the elastic-gravitational energy flux (3.284) is

$$\mathbf{K}' = \mathrm{Re}\,\{i\omega[\kappa\mathbf{s}^*(\boldsymbol{\nabla}\cdot\mathbf{s}) + 2\mu(\mathbf{s}^*\cdot\mathbf{d}) \\ - \rho\mathbf{s}^*(\mathbf{s}\cdot\boldsymbol{\nabla}\Phi) + \rho\phi^*\mathbf{s} + (4\pi G)^{-1}\phi^*\boldsymbol{\nabla}\phi]\}. \tag{16.18}$$

In this case, the interesting quantity is the *total lateral flux of energy*:

$$\mathbf{K}'_\Omega = \int_0^\infty (\mathbf{I} - \hat{\mathbf{r}}\hat{\mathbf{r}})\cdot\mathbf{K}'\, r\,dr. \tag{16.19}$$

The surface through which the energy passes is a cone extending outward from the center of the Earth; the differential $r\,dr$ accounts for the radial dependence of the area of a patch on this cone. Both E_Ω and $\mathbf{K}'_\Omega$ may be regarded as functions of position $\hat{\mathbf{r}}$ on the unit sphere Ω. Written out explicitly, these surficial variables are

$$E_\Omega = \tfrac{1}{2}\int_0^\infty [\omega^2\rho\mathbf{s}^*\cdot\mathbf{s} + \kappa(\boldsymbol{\nabla}\cdot\mathbf{s}^*)(\boldsymbol{\nabla}\cdot\mathbf{s}) + 2\mu\mathbf{d}^*\!:\!\mathbf{d} \\ + \rho(\mathbf{s}^*\cdot\boldsymbol{\nabla}\phi + \mathbf{s}\cdot\boldsymbol{\nabla}\phi^*) + \rho\mathbf{s}^*\cdot\boldsymbol{\nabla}\boldsymbol{\nabla}\Phi\cdot\mathbf{s} \\ + \tfrac{1}{2}\rho\boldsymbol{\nabla}\Phi\cdot(\mathbf{s}^*\cdot\boldsymbol{\nabla}\mathbf{s} + \mathbf{s}\cdot\boldsymbol{\nabla}\mathbf{s}^* + \mathbf{s}^*\boldsymbol{\nabla}\cdot\mathbf{s} - \mathbf{s}\boldsymbol{\nabla}\cdot\mathbf{s}^*) \\ + (4\pi G)^{-1}\boldsymbol{\nabla}\phi^*\cdot\boldsymbol{\nabla}\phi]\, r^2 dr, \tag{16.20}$$

$$\mathbf{K}'_\Omega = \mathrm{Re}\int_0^\infty i\omega(\mathbf{I} - \hat{\mathbf{r}}\hat{\mathbf{r}})\cdot[\kappa\mathbf{s}^*(\boldsymbol{\nabla}\cdot\mathbf{s}) + 2\mu(\mathbf{s}^*\cdot\mathbf{d}) \\ - \rho\mathbf{s}^*(\mathbf{s}\cdot\boldsymbol{\nabla}\Phi) + \rho\phi^*\mathbf{s} + (4\pi G)^{-1}\phi^*\boldsymbol{\nabla}\phi]\, r\,dr. \tag{16.21}$$

The integration extends outward to infinity in both (16.20) and (16.21); however, only the gravitational-field self-energy $(8\pi G)^{-1}\boldsymbol{\nabla}\phi^* \cdot \boldsymbol{\nabla}\phi$ and the purely gravitational flux $\mathrm{Re}\,[i\omega(4\pi G)^{-1}\phi^*\boldsymbol{\nabla}\phi]$ are non-zero outside of the Earth, in $\bigcirc - \oplus$. The units of E_Ω and $\mathbf{K}'_\Omega$ are energy per steradian and energy per unit time per radian of wavefront arc on Ω, respectively.

16.2 Slow Variational Principle

Guided by the vector representation (B.157), we seek JWKB solutions to equations (16.1)–(16.2) and (16.4)–(16.7) of the form

$$\mathbf{s} = (U\hat{\mathbf{r}} + k_{\mathrm{R}}^{-1}V\boldsymbol{\nabla}_1)A_{\mathrm{R}}\exp(-i\psi_{\mathrm{R}}) \\ - k_{\mathrm{L}}^{-1}W(\hat{\mathbf{r}} \times \boldsymbol{\nabla}_1)A_{\mathrm{L}}\exp(-i\psi_{\mathrm{L}}), \tag{16.22}$$

$$\phi = PA_{\mathrm{R}}\exp(-i\psi_{\mathrm{R}}), \tag{16.23}$$

where the subscripts R and L stand for *Rayleigh* and *Love*, respectively. The amplitudes A_{R}, A_{L} and phases ψ_{R}, ψ_{L} are presumed to be real. On a spherically symmetric Earth the radial eigenfunctions U, V, W and P are functions of radius r only, and the complex quantities $A_{\mathrm{R}}\exp(-i\psi_{\mathrm{R}})$ and $A_{\mathrm{L}}\exp(-i\psi_{\mathrm{L}})$ are travelling-wave Legendre functions of the form

$$A\exp(-i\psi) \sim Q^{(1,2)}_{k-\frac{1}{2}\,m}(\cos\theta)\exp(im\phi), \tag{16.24}$$

as discussed in Chapter 11. On a laterally heterogeneous Earth, the real functions U, V, W and P are regarded as *local radial eigenfunctions*, which depend not only upon the radius r but also upon the geographical position $\hat{\mathbf{r}} = (\theta, \phi)$ on the unit sphere. The real *local Rayleigh and Love wavevectors* are defined in terms of the phases ψ_{R} and ψ_{L} by

$$\mathbf{k}_{\mathrm{R}} = \boldsymbol{\nabla}_1\psi_{\mathrm{R}}, \qquad \mathbf{k}_{\mathrm{L}} = \boldsymbol{\nabla}_1\psi_{\mathrm{L}}. \tag{16.25}$$

The associated scalar local wavenumbers are $k_{\mathrm{R}} = \|\mathbf{k}_{\mathrm{R}}\|$ and $k_{\mathrm{L}} = \|\mathbf{k}_{\mathrm{L}}\|$.

As in Section 15.2, we obtain a *slowly varying action* by substituting the JWKB representation (16.22)–(16.23) into (16.8) and retaining only the lowest-order terms in the small parameter λ/Λ_Ω. The resulting JWKB action separates naturally into distinct Rayleigh- and Love-wave parts:

$$\mathcal{I}' = \mathcal{I}'_{\mathrm{R}} + \mathcal{I}_{\mathrm{L}}, \tag{16.26}$$

where

$$\mathcal{I}'_{\mathrm{R}} = \int_{\bigcirc} \mathcal{L}'_{\mathrm{R}}(U, V, P, \partial_r U, \partial_r V, \partial_r P, A_{\mathrm{R}}, \boldsymbol{\nabla}_1\psi_{\mathrm{R}})\,dV, \tag{16.27}$$

$$\mathcal{I}_{\rm L} = \int_{\oplus} \mathcal{L}_{\rm L}(W, \partial_r W, A_{\rm L}, \boldsymbol{\nabla}_1\psi_{\rm L})\, dV. \tag{16.28}$$

We have retained the primes upon $\mathcal{I}'_{\rm R}$ and $\mathcal{L}'_{\rm R}$, but dropped them upon $\mathcal{I}_{\rm L}$ and $\mathcal{L}_{\rm L}$, since Love waves are unaccompanied by any gravitational perturbation. The integrands in (16.27) and (16.28) are *slowly varying Lagrangian densities* analogous to (15.18):

$$\begin{aligned}\mathcal{L}'_{\rm R} = \tfrac{1}{2}[&\omega^2\rho(U^2+V^2) - \kappa(\partial_r U + 2r^{-1}U - kr^{-1}V)^2 \\ &- \tfrac{1}{3}\mu(2\partial_r U - 2r^{-1}U + kr^{-1}V)^2 - \mu(\partial_r V - r^{-1}V - kr^{-1}U)^2 \\ &\quad - (k^2-2)\mu r^{-2}V^2 - 2\rho(U\partial_r P + kr^{-1}VP) \\ &\quad\quad - 4\pi G\rho^2U^2 + 2\rho g r^{-1}U(2U - kV) \\ &\quad\quad\quad - (4\pi G)^{-1}(\partial_r P^2 + k^2r^{-2}P^2)]\, A_{\rm R}^2,\end{aligned} \tag{16.29}$$

$$\mathcal{L}_{\rm L} = \tfrac{1}{2}[\omega^2\rho W^2 - \mu(\partial_r W - r^{-1}W)^2 - (k^2-2)\mu r^{-2}W^2]\,A_{\rm L}^2. \tag{16.30}$$

As in our discussion of body waves, we shall use script letters throughout this chapter to denote densities and other quantities that vary on the slow scale Λ_Ω of the lateral heterogeneity; the wavelength-scale variations associated with the exponentials $\exp(-i\psi_{\rm R})$ and $\exp(-i\psi_{\rm L})$ have been eliminated in (16.29) and (16.30). The Rayleigh-wave density $\mathcal{L}'_{\rm R}$ depends upon $U, V, P, \partial_r U, \partial_r V, \partial_r P, A_{\rm R}$ and the wavevector $\boldsymbol{\nabla}_1\psi_{\rm R}$, whereas the Love-wave density $\mathcal{L}_{\rm L}$ depends upon $W, \partial_r W, A_{\rm L}$ and $\boldsymbol{\nabla}_1\psi_{\rm L}$, as indicated by the arguments in equations (16.27)–(16.28). We have dropped the subscript R upon $k_{\rm R} = \|\boldsymbol{\nabla}_1\psi_{\rm R}\|$ in (16.29) and the subscript L upon $k_{\rm L} = \|\boldsymbol{\nabla}_1\psi_{\rm L}\|$ in (16.30) to facilitate the comparison of $\mathcal{L}'_{\rm R}$ and $\mathcal{L}_{\rm L}$ with the corresponding one-dimensional Lagrangian densities (8.101) and (8.90). Apart from the three-dimensionality and the extra amplitude factors $A_{\rm R}^2$ and $A_{\rm L}^2$, the only difference is the replacement of the ordinary radial derivatives $\dot{U}$, $\dot{V}$, $\dot{W}$ and $\dot{P}$ by the partial derivatives $\partial_r U$, $\partial_r V$, $\partial_r W$ and $\partial_r P$.

The decomposition (16.26) of the slowly varying action $\mathcal{I}'$ into separate terms $\mathcal{I}'_{\rm R}$ and $\mathcal{I}_{\rm L}$ is an indication that Rayleigh and Love waves propagate independently, to lowest order in the small parameter λ/Λ_Ω. Fundamental and overtone waves of the same type are likewise uncoupled; a surface wave of a given type and overtone number *clings to its identity* as it propagates. All of the information needed to trace surface-wave rays and connect the amplitudes $A_{\rm R}$, $A_{\rm L}$, the phases $\psi_{\rm R}$, $\psi_{\rm L}$ and the wavevectors $\mathbf{k}_{\rm R}$, $\mathbf{k}_{\rm L}$ at points along those rays is contained in the slowly varying Rayleigh and Love Lagrangian densities $\mathcal{L}'_{\rm R}$ and $\mathcal{L}_{\rm L}$. We consider the two types of surface waves separately in the next two sections, treating the easier case first.

16.2.1 Love waves

Dropping the subscripts upon $\mathcal{I}_{\mathrm{L}}$, $\mathcal{L}_{\mathrm{L}}$ and A_{L} for the sake of brevity, we consider the *slow Love-wave variational principle*

$$\begin{aligned}\delta\mathcal{I} &= \int_{\oplus} \delta W\left[\partial_W\mathcal{L} - r^{-2}\partial_r(r^2\partial_{\partial_r W}\mathcal{L})\right] dV \\ &\quad + \sum_d [\delta W\,(\partial_{\partial_r W}\mathcal{L})]_-^+ \\ &\quad + \int_\Omega \delta A\left[\int_0^a (\partial_A\mathcal{L})\, r^2 dr\right] d\Omega \\ &\quad + \int_\Omega \delta\psi\left\{\boldsymbol{\nabla}_1\cdot\left[\int_0^a (\partial_{\boldsymbol{\nabla}_1\psi}\mathcal{L})\, r^2 dr\right]\right\} d\Omega = 0. \end{aligned} \tag{16.31}$$

The second and third integrals involving the variations δA and $\delta\psi$ are over the unit sphere Ω, because the amplitude A and phase ψ are independent of the radius r. The variation (16.31) vanishes for arbitrary variations satisfying the admissibility constraint $[\delta W]_-^+ = 0$ on $r = d_{\mathrm{SS}}$ if and only if the fields W, A and ψ satisfy the three Euler-Lagrange equations

$$\partial_W\mathcal{L} - r^{-2}\partial_r(r^2\partial_{\partial_r W}\mathcal{L}) = 0 \quad \text{in } 0 \le r \le a, \tag{16.32}$$

$$\int_0^a (\partial_A\mathcal{L})\, r^2 dr = 0 \quad \text{on } \Omega, \tag{16.33}$$

$$\boldsymbol{\nabla}_1\cdot\left[\int_0^a (\partial_{\boldsymbol{\nabla}_1\psi}\mathcal{L})\, r^2 dr\right] = 0 \quad \text{on } \Omega, \tag{16.34}$$

together with the boundary conditions

$$\partial_{\partial_r W}\mathcal{L} = 0 \quad \text{on } r = a \text{ and } r = d_{\mathrm{FS}}, \tag{16.35}$$

$$[\partial_{\partial_r W}\mathcal{L}]_-^+ = 0 \quad \text{on } r = d_{\mathrm{SS}}. \tag{16.36}$$

Written out explicitly, equation (16.32) is

$$\begin{aligned} r^{-2}\partial_r[\mu r^2(\partial_r W - r^{-1}W)] + \mu r^{-1}(\partial_r W - r^{-1}W) \\ + [\omega^2\rho - (k^2-2)\mu r^{-2}]W = 0. \end{aligned} \tag{16.37}$$

The quantity $-\partial_{\partial_r W}\mathcal{L}$ is precisely the local traction $T = \mu(\partial_r W - r^{-1}W)$, so the local boundary conditions (16.35)–(16.36) are

$$T = 0 \quad \text{on } r = a \text{ and } r = d_{\mathrm{FS}}, \tag{16.38}$$

$$[T]_-^+ = 0 \quad \text{on } r = d_{\mathrm{SS}}. \tag{16.39}$$

Equations (16.37) and (16.38)–(16.39) are identical to the relations (8.45) and (8.50)–(8.52) that determine the Love-wave displacement W and traction T on a spherically symmetric Earth; the only difference is that differentiation d/dr is replaced by partial differentiation ∂_r. Recall that the free surface and interior discontinuity surfaces $d = a \cup d_{\mathrm{SS}} \cup d_{\mathrm{FS}}$ may exhibit slowly varying topographic variations.

The above results should be interpreted in the following manner. At every geographical position $\hat{\mathbf{r}} = (\theta, \phi)$ in a laterally heterogeneous Earth, we construct a hypothetical spherically symmetric model having the same structure κ, μ, ρ, d as that underlying the surface point $a\hat{\mathbf{r}}$. By repeatedly solving equations (16.37) and (16.38)–(16.39), we can determine the *local radial eigenfunction* W and the associated *local wavenumber* k at every point $\hat{\mathbf{r}}$ and every angular frequency ω. In normal-mode calculations we fix the angular degree l and therefore the wavenumber $k = \sqrt{l(l+1)}$, and we determine the fundamental and higher-overtone eigenfrequencies ω; in surface-wave calculations, it is preferable to fix ω and determine the wavenumbers k. We think of a monochromatic Love wave as propagating from point to point on the surface of the unit sphere Ω. As a wave propagates, it "carries" its local radial structure W and wavenumber k with it. In the language of classical mechanics, we say that W and k vary *adiabatically* along the surface-wave ray.

The second slow Euler-Lagrange equation (16.33) is the slowly varying analogue of the *Love-wave dispersion relation* (11.59):

$$\omega^2 I_1 - k^2 I_2 - I_3 = 0. \tag{16.40}$$

The local radial integrals I_1, I_2 and I_3 are identical to (11.56)–(11.58), with $\dot{W}$ replaced by $\partial_r W$:

$$I_1 = \int_0^a \rho W^2\, r^2 dr, \tag{16.41}$$

$$I_2 = \int_0^a \mu W^2\, dr, \tag{16.42}$$

$$I_3 = \int_0^a \mu[(\partial_r W - r^{-1}W)^2 - 2r^{-2}W^2]\, r^2 dr. \tag{16.43}$$

By subtracting the local dispersion relations for two waves with associated frequencies ω and $\omega + \delta\omega$, local wavenumbers k and $k + \delta k$, and local radial eigenfunctions W and $W + \delta W$, we find that the *local group speed* $C = d\omega/dk$ and *phase speed* $c = \omega/k$ of a Love wave are related by the slowly varying analogue of (11.67):

$$C = \frac{I_2}{cI_1}. \tag{16.44}$$

Both C and c are *angular* rates of propagation, measured in radians per second on the surface of the unit sphere Ω. We shall use the local dispersion relation (16.40) as the basis for kinematic Love-wave ray tracing in Section 16.3. In this sense, it is analogous to the eikonal equation (15.40) for body waves.

The slowly varying energy density and lateral energy flux obtained by substituting the JWKB representation $\mathbf{s} = -k^{-1}W(\hat{\mathbf{r}} \times \boldsymbol{\nabla}_1)A\exp(-i\psi)$ into equations (16.20) and (16.21) are

$$\mathcal{E}_\Omega = \tfrac{1}{2}(\omega^2 I_1 + k^2 I_2 + I_3)A^2 = \omega^2 I_1 A^2, \tag{16.45}$$

$$\boldsymbol{\mathcal{K}}_\Omega = \omega I_2 A^2 \mathbf{k} = C\mathcal{E}_\Omega \hat{\mathbf{k}}. \tag{16.46}$$

We have dropped the prime upon $\boldsymbol{\mathcal{K}}_\Omega = \boldsymbol{\mathcal{K}}_{\Omega\mathrm{L}}$, and used the dispersion relation (16.40) and the group-speed identity (16.44) to obtain the final results in (16.45) and (16.46), respectively. The energy density (16.45) is the sum of the kinetic energy density $\frac{1}{2}\omega^2 I_1 A^2$ and the elastic potential energy density $\frac{1}{2}(k^2 I_2 + I_3)A^2$. The dispersion relation stipulates that the energy of a propagating Love wave is equipartitioned; the total surface energy density is therefore twice the kinetic energy density. The magnitude of the Love-wave energy flux $\|\boldsymbol{\mathcal{K}}_\Omega\|$ is the product of the wave speed C and the energy density $\mathcal{E}_\Omega$, as in the case of body waves. It is noteworthy that it is C rather than the phase speed c that appears in equation (16.46); this is because the *energy of a dispersive wave propagates with the group speed.*

The final Euler-Lagrange equation (16.34) is the Love-wave analogue of the *transport equation* (15.35):

$$\boldsymbol{\nabla}_1 \cdot \boldsymbol{\mathcal{K}}_\Omega = 0. \tag{16.47}$$

This equation determines the variation in Love-wave amplitude along a ray, as we shall see in Section 16.4. The local dispersion relation (16.40) and the transport equation (16.47) can also be obtained from the *surface variational principle* $\delta\mathcal{I}_\Omega = 0$, where

$$\mathcal{I}_\Omega = \int_\Omega \mathcal{L}_\Omega(A, \boldsymbol{\nabla}_1\psi)\, d\Omega. \tag{16.48}$$

The *slowly varying surface Lagrangian density* in (16.48) is given by

$$\mathcal{L}_\Omega = \int_0^a \mathcal{L}\, r^2 dr = \tfrac{1}{2}(\omega^2 I_1 - k^2 I_2 - I_3)A^2. \tag{16.49}$$

The surface Lagrangian and energy densities are related to each other by a Legendre transformation, as usual: $\mathcal{E}_\Omega = \omega\partial_\omega\mathcal{L}_\Omega - \mathcal{L}_\Omega$. The flux (16.46) can be written in the form $\boldsymbol{\mathcal{K}}_\Omega = \partial_{\mathbf{k}}\mathcal{L}_\Omega$.

16.2.2 Rayleigh waves

The variation of the slowly varying Rayleigh-wave action (16.27) can be written in a manner analogous to (16.50):

$$\begin{aligned}
\delta\mathcal{I}' = & \int_{\oplus} \delta U \left[\partial_U \mathcal{L}' - r^{-2}\partial_r(r^2 \partial_{\partial_r U}\mathcal{L}')\right] dV \\
& + \int_{\oplus} \delta V \left[\partial_V \mathcal{L}' - r^{-2}\partial_r(r^2 \partial_{\partial_r V}\mathcal{L}')\right] dV \\
& + \int_{\bigcirc} \delta P \left[\partial_P \mathcal{L}' - r^{-2}\partial_r(r^2 \partial_{\partial_r P}\mathcal{L}')\right] dV \\
& + \sum_d \left[\delta U(\partial_{\partial_r U}\mathcal{L}') + \delta V(\partial_{\partial_r V}\mathcal{L}') + \delta P(\partial_{\partial_r P}\mathcal{L}')\right]_-^+ \\
& + \int_{\Omega} \delta A \left[\int_0^\infty (\partial_A \mathcal{L}')\, r^2 dr\right] d\Omega \\
& + \int_{\Omega} \delta\psi \left\{\boldsymbol{\nabla}_1 \cdot \left[\int_0^\infty (\partial_{\boldsymbol{\nabla}_1 \psi}\mathcal{L}')\, r^2 dr\right]\right\} d\Omega = 0,
\end{aligned} \tag{16.50}$$

where we have dropped the subscripts upon $\mathcal{I}'_{\mathrm{R}}$ and $\mathcal{L}'_{\mathrm{R}}$ for simplicity. This vanishes for arbitrary variations satisfying the admissibility constraints $[\delta U]_-^+ = 0$ on $r = d$, $[\delta V]_-^+ = 0$ on $r = d_{\mathrm{SS}}$ and $[\delta P]_-^+ = 0$ on $r = d$ if and only if U, V, P, A and ψ satisfy the Euler-Lagrange equations

$$\partial_U \mathcal{L}' - r^{-2}\partial_r(r^2 \partial_{\partial_r U}\mathcal{L}') = 0 \quad \text{in } 0 \le r \le a, \tag{16.51}$$

$$\partial_V \mathcal{L}' - r^{-2}\partial_r(r^2 \partial_{\partial_r V}\mathcal{L}') = 0 \quad \text{in } 0 \le r \le a, \tag{16.52}$$

$$\partial_P \mathcal{L}' - r^{-2}\partial_r(r^2 \partial_{\partial_r P}\mathcal{L}') = 0 \quad \text{in } 0 \le r \le \infty, \tag{16.53}$$

$$\int_0^a (\partial_A \mathcal{L}')\, r^2 dr = 0 \quad \text{on } \Omega, \tag{16.54}$$

$$\boldsymbol{\nabla}_1 \cdot \left[\int_0^a (\partial_{\boldsymbol{\nabla}_1 \psi}\mathcal{L}')\, r^2 dr\right] = 0 \quad \text{on } \Omega, \tag{16.55}$$

together with the boundary conditions

$$\partial_{\partial_r U}\mathcal{L}' = \partial_{\partial_r V}\mathcal{L}' = 0 \quad \text{on } r = a, \tag{16.56}$$

$$[\partial_{\partial_r U}\mathcal{L}']_-^+ = [\partial_{\partial_r V}\mathcal{L}']_-^+ = 0 \quad \text{on } r = d_{\mathrm{SS}}, \tag{16.57}$$

$$[\partial_{\partial_r U}\mathcal{L}']_-^+ = \partial_{\partial_r V}\mathcal{L}' = 0 \quad \text{on } r = d_{\mathrm{FS}}, \tag{16.58}$$

$$[\partial_{\partial_r P}\mathcal{L}']_-^+ = 0 \quad \text{on } r = d. \tag{16.59}$$

Written out explicitly, equations (16.51)–(16.53) are

$$\begin{aligned}
&r^{-2}\partial_r[r^2(\kappa+\tfrac{4}{3}\mu)\partial_r U+(\kappa-\tfrac{2}{3}\mu)r(2U-kV)]\\
&\quad+r^{-1}[(\kappa+\tfrac{4}{3}\mu)\partial_r U+(\kappa-\tfrac{2}{3}\mu)r^{-1}(2U-kV)]\\
&\quad-3\kappa r^{-1}(\partial_r U+2r^{-1}U-kr^{-1}V)\\
&\quad-k\mu r^{-1}(\partial_r V-r^{-1}V+kr^{-1}U)+\omega^2\rho U\\
&\quad-\rho[\partial_r P+(4\pi G\rho-4gr^{-1})U+kgr^{-1}V]=0,
\end{aligned} \tag{16.60}$$

$$\begin{aligned}
&r^{-2}\partial_r[\mu r^2(\partial_r V-r^{-1}V+kr^{-1}U)]\\
&\quad+\mu r^{-1}(\partial_r V-r^{-1}V+kr^{-1}U)\\
&\quad+k(\kappa-\tfrac{2}{3}\mu)r^{-1}\partial_r U+k(\kappa+\tfrac{1}{3}\mu)r^{-2}(2U-kV)\\
&\quad+[\omega^2\rho-(k^2-2)\mu r^{-2}]V-k\rho r^{-1}(P+gU)=0,
\end{aligned} \tag{16.61}$$

$$\begin{aligned}
&\partial_r^2 P+2r^{-1}\partial_r P-k^2r^{-2}P\\
&\quad=-4\pi G(\partial_r\rho)U-4\pi G\rho[\partial_r U+r^{-1}(2U-kV)].
\end{aligned} \tag{16.62}$$

The partial derivatives $-\partial_{\partial_r U}\mathcal{L}'$, $-\partial_{\partial_r V}\mathcal{L}'$ and $-\partial_{\partial_r P}\mathcal{L}$ are the tractions $R=(\kappa+\frac{4}{3}\mu)\partial_r U+(\kappa-\frac{2}{3}\mu)r^{-1}(2U-kV)$, $S=\mu(\partial_r V-r^{-1}V+kr^{-1}U)$ and the gravitational scalar $(4\pi G)^{-1}\partial_r P+\rho U$, respectively. The local boundary conditions (16.56)–(16.59) become

$$R=S=0 \quad \text{on } r=a, \tag{16.63}$$

$$[R]_-^+=[S]_-^+=0 \quad \text{on } r=d_{\mathrm{SS}}, \tag{16.64}$$

$$[R]_-^+=S=0 \quad \text{on } r=d_{\mathrm{FS}}, \tag{16.65}$$

$$[\partial_r P+4\pi G\rho U]_-^+=0 \quad \text{on } r=d. \tag{16.66}$$

As in the Love-wave case, these are identical to the corresponding relations (8.43)–(8.44) and (8.50)–(8.54) on a spherical Earth, with $\dot{U}$, $\dot{V}$ and $\dot{P}$ replaced by $\partial_r U$, $\partial_r V$ and $\partial_r P$. At every point $\hat{\mathbf{r}}=(\theta,\phi)$ on Ω, the elastic-gravitational boundary-value problem (16.60)–(16.66) determines the *local radial eigenfunctions* U, V and P, as well as the *local wavenumber* k. A monochromatic Rayleigh wave of angular frequency ω "carries" these adiabatically varying quantities with it as it propagates.

The first geographical Euler-Lagrange equation (16.54) is the slowly varying analogue of the Rayleigh-wave dispersion relation (11.77):

$$\omega^2 I_1-k^2I_2'-kI_3'-I_4'=0, \tag{16.67}$$

where

$$I_1 = \int_0^a \rho(U^2 + V^2)\, r^2 dr, \tag{16.68}$$

$$I_2' = \int_0^a [\mu U^2 + (\kappa + \tfrac{4}{3}\mu)V^2]\, dr + \frac{1}{4\pi G}\int_0^\infty P^2\, dr, \tag{16.69}$$

$$\begin{aligned} I_3' = \int_0^a & [\tfrac{4}{3}\mu V(\partial_r U - r^{-1}U) - 2\kappa V(\partial_r U + 2r^{-1}U) \\ & + 2\mu U(\partial_r V - r^{-1}V) + 2\rho(VP + gUV)]\, r dr, \end{aligned} \tag{16.70}$$

$$\begin{aligned} I_4' = \int_0^a & [(\kappa(\partial_r U + 2r^{-1}U)^2 + \tfrac{4}{3}\mu(\partial_r U - r^{-1}U)^2 \\ & + \mu(\partial_r V - r^{-1}V)^2 - 2\mu r^{-2}V^2 \\ & + \rho(4\pi G\rho U^2 + 2U\partial_r P - 4r^{-1}gU^2)]\, r^2 dr \\ & + \frac{1}{4\pi G}\int_0^\infty (\partial_r P)^2\, r^2 dr. \end{aligned} \tag{16.71}$$

The *local group speed* $C = d\omega/dk$ and *phase speed* $c = \omega/k$ of a Rayleigh wave are related by the slowly varying analogue of equation (11.79):

$$C = \frac{I_2' + \frac{1}{2}k^{-1}I_3'}{cI_1}. \tag{16.72}$$

The slowly varying energy density and lateral energy flux obtained by substituting the JWKB representations $\mathbf{s} = (U\hat{\mathbf{r}} + k^{-1}V\boldsymbol{\nabla}_1)A\exp(-i\psi)$ and $\phi = PA\exp(i\psi)$ into equations (16.20) and (16.21) are

$$\mathcal{E}_\Omega = \tfrac{1}{2}(\omega^2 I_1 + k^2 I_2' + kI_3' + I_4')A^2 = \omega^2 I_1 A^2, \tag{16.73}$$

$$\mathcal{K}_\Omega' = \omega(I_2' + \tfrac{1}{2}k^{-1}I_3')A^2\mathbf{k} = C\mathcal{E}_\Omega'\hat{\mathbf{k}}. \tag{16.74}$$

The total surface energy density is twice the kinetic energy density, and the magnitude of the energy flux $\|\mathcal{K}_\Omega'\|$ is the product of the group speed C and the energy density $\mathcal{E}_\Omega$, just as for Love waves. The final Euler-Lagrange equation (16.55) is the Rayleigh-wave *transport equation*:

$$\boldsymbol{\nabla}_1 \cdot \mathcal{K}_\Omega' = 0. \tag{16.75}$$

The local dispersion relation (16.67) and the transport equation (16.75) can alternatively be obtained from the surface variational principle $\delta\mathcal{I}_\Omega' = 0$, where

$$\mathcal{I}_\Omega' = \int_\Omega \mathcal{L}_\Omega'(A, \boldsymbol{\nabla}_1\psi)\, d\Omega. \tag{16.76}$$

The Rayleigh-wave analogue of the surface Lagrangian density (16.49) is

$$\mathcal{L}'_\Omega = \int_0^\infty \mathcal{L}'\, r^2 dr = \tfrac{1}{2}(\omega^2 I_1 - k^2 I'_2 - kI'_3 - I'_4)A^2. \tag{16.77}$$

The Rayleigh-wave surface energy density (16.73) and energy flux (16.74) are related to the surface Lagrangian density (16.77) by $\mathcal{E}'_\Omega = \omega\partial_\omega \mathcal{L}'_\Omega - \mathcal{L}'_\Omega$ and $\mathcal{K}'_\Omega = \partial_{\mathbf{k}} \mathcal{L}'_\Omega$.

16.3 Surface-Wave Ray Tracing

In this section we show how the local Love- and Rayleigh-wave dispersion relations (16.40) and (16.67) determine the geometry of surface-wave rays upon the unit sphere Ω. For maximum generality, we employ an arbitrary system of curvilinear coordinates x^1, x^2 upon Ω. Later, in Section 16.6, we shall introduce spherical polar coordinates and set $x^1 = \theta, x^2 = \phi$. For the moment, however, x^1 and x^2 need not even be orthogonal. We use Greek indices $\alpha, \beta, \ldots$ that take on the values 1 and 2 to denote the covariant and contravariant components of surface vectors and tensors upon Ω. Readers who are unfamiliar with the concepts of covariance and contravariance, and the use of the metric tensor to raise and lower indices, may wish to read Appendices A.6.3 and A.6.4 before proceeding.

16.3.1 Hamiltonian formulation

The wavevector $\mathbf{k} = \boldsymbol{\nabla}_1\psi$ of a Love or Rayleigh wave is a tangent vector upon Ω, with covariant components

$$k_1 = \frac{\partial\psi}{\partial x^1}, \qquad k_2 = \frac{\partial\psi}{\partial x^2}. \tag{16.78}$$

The local wavenumber $k = \sqrt{\mathbf{k}\cdot\mathbf{k}}$ may be written in terms of these components in the form

$$k = \sqrt{g^{\alpha\beta}k_\alpha k_\beta}. \tag{16.79}$$

The quantities $g^{\alpha\beta}$ are the contravariant components of the metric tensor $\mathbf{g} = \mathbf{I} - \hat{\mathbf{r}}\hat{\mathbf{r}}$ on the unit sphere: $g^{\alpha\beta} = (\boldsymbol{\nabla}_1 x^\alpha)\cdot(\boldsymbol{\nabla}_1 x^\beta)$. The local Love- and Rayleigh-wave dispersion relations (16.40) and (16.67) determine the wavenumber k as a function of position $\hat{\mathbf{r}} = (x^1, x^2)$ and angular frequency ω:

$$k = k(x^1, x^2, \omega). \tag{16.80}$$

By analogy with (15.46) we introduce the surface-wave *Hamiltonian*

$$H = \tfrac{1}{2}[g^{\alpha\beta}k_\alpha k_\beta - k^2(x^1, x^2)], \tag{16.81}$$

where the dependence upon frequency ω is henceforth understood. In the framework of classical mechanics, x^1 and x^2 are the *generalized coordinates* describing a surface-wave ray path, whereas k_1 and k_2 are the associated *conjugate momenta*. The Hamiltonian (16.81) vanishes along a ray path by virtue of the dispersion relation (16.80). In fact, we can regard the equation

$$H(x^1, x^2, k_1, k_2) = 0 \tag{16.82}$$

as an alternative statement of the Love or Rayleigh dispersion relation, just as we regard (15.47) as an alternative statement of the eikonal equation for body waves. The substitution (16.78) converts (16.82) into a first-order partial differential equation for the surface-wave phase ψ, which may be solved by the method of characteristics. The characteristic equations are *Hamilton's equations* for the Hamiltonian (16.81):

$$\frac{dx^\gamma}{d\sigma} = \frac{\partial H}{\partial k_\gamma}, \qquad \frac{dk_\gamma}{d\sigma} = -\frac{\partial H}{\partial x^\gamma}. \tag{16.83}$$

The only complication not present in the body-wave ray-tracing equations (15.48) is that we must account for the geographical variation of the metric tensor $\mathbf{g} = \mathbf{I} - \hat{\mathbf{r}}\hat{\mathbf{r}}$. The explicit form of the surface-wave ray-tracing equations (16.83) is

$$\frac{dx^\gamma}{d\sigma} = g^{\gamma\eta}k_\eta, \qquad \frac{dk_\gamma}{d\sigma} = -\tfrac{1}{2}\frac{\partial g^{\alpha\beta}}{\partial x^\gamma}k_\alpha k_\beta + \tfrac{1}{2}\frac{\partial k^2}{\partial x^\gamma}. \tag{16.84}$$

The dependent variable σ in (16.83) and (16.84) is the *generating parameter*, which is related to the phase ψ, the angular distance of propagation Δ and the travel time T along a surface-wave ray path by the analogue of equation (15.42):

$$d\psi = k^2\, d\sigma = k\, d\Delta = kC\, dT. \tag{16.85}$$

The ray-tracing equations (16.84) have a unique solution for every set of Cauchy initial conditions

$$x^\gamma(0) = x'^\gamma, \qquad k_\gamma(0) = k'_\gamma. \tag{16.86}$$

The Hamiltonian (16.81) is an integral of the system (16.84) because

$$\frac{dH}{d\sigma} = \frac{\partial H}{\partial x^\gamma}\frac{dx^\gamma}{d\sigma} + \frac{\partial H}{\partial k_\gamma}\frac{dk_\gamma}{d\sigma} = 0. \tag{16.87}$$

We can integrate (16.85) to find the phase difference between any two consecutive points $\hat{\mathbf{r}}_1$ and $\hat{\mathbf{r}}_2$ along a ray:

$$\psi_2 - \psi_1 = \int_{\sigma_1}^{\sigma_2} k^2\, d\sigma = \int_{\Delta_1}^{\Delta_2} k\, d\Delta = \int_{T_1}^{T_2} kC\, dt. \tag{16.88}$$

We can also write (16.88) in terms of the wavevector k_1, k_2 and differential position dx^1, dx^2 along the ray in a manner analogous to (15.45):

$$\psi_2 - \psi_1 = \int_{\hat{\mathbf{r}}_1}^{\hat{\mathbf{r}}_2} k_\gamma\, dx^\gamma = \int_{\hat{\mathbf{r}}_1}^{\hat{\mathbf{r}}_2} \mathbf{k}\cdot d\hat{\mathbf{r}}. \tag{16.89}$$

Mathematically, the surface-wave ray Hamiltonian (16.81) is identical to that of a *librating spherical pendulum* with kinetic energy $T = \frac{1}{2}g^{\alpha\beta}k_\alpha k_\beta$ and potential energy $V = -\frac{1}{2}k^2(x^1, x^2) < 0$. The pendulum librates rather than oscillates because its potential energy is negative everywhere on the surface of the unit sphere Ω. This reflects the fact that Love and Rayleigh waves must always propagate—they can never become trapped in a local minimum of the surface potential energy. The total energy is zero by virtue of the dispersion relation (16.82): $H = T + V = 0$.

⋆16.3.2 Alternative forms

As in the body-wave case, it is possible to rewrite the surface-wave ray-tracing equations in a myriad of alternative equivalent forms. We note first that there is a concise vector version of Hamilton's equations (16.84):

$$\frac{d\hat{\mathbf{r}}}{d\sigma} = \mathbf{k}, \qquad \frac{d\mathbf{k}}{d\sigma} = -k^2\hat{\mathbf{r}} + \tfrac{1}{2}\boldsymbol{\nabla}_1 k^2. \tag{16.90}$$

The vector form of the initial conditions (16.86) is $\hat{\mathbf{r}}(0) = \hat{\mathbf{r}}'$, $\mathbf{k}(0) = \mathbf{k}'$. The two first-order equations (16.90) may be combined into a single second-order vector equation,

$$\frac{d^2\hat{\mathbf{r}}}{d\sigma^2} + k^2\hat{\mathbf{r}} = \tfrac{1}{2}\boldsymbol{\nabla}_1 k^2. \tag{16.91}$$

This is the classical equation of motion of a forced spherical pendulum. The forcing term $\frac{1}{2}\boldsymbol{\nabla}_1 k^2$ vanishes in the absence of geographical wavenumber variations; the surface-wave trajectories on such a spherically symmetric Earth model are great circles: $\hat{\mathbf{r}}(\sigma) = \hat{\mathbf{r}}'\cos\sigma + \hat{\mathbf{k}}'\sin\sigma$. In index notation, equation (16.91) becomes

$$\frac{d}{d\sigma}\left(g_{\gamma\eta}\frac{dx^\eta}{d\sigma}\right) - \frac{1}{2}\frac{\partial g_{\alpha\beta}}{\partial x^\gamma}\frac{dx^\alpha}{d\sigma}\frac{dx^\beta}{d\sigma} = \frac{1}{2}\frac{\partial k^2}{\partial x^\gamma}. \tag{16.92}$$

The quantities $g_{\alpha\beta} = (\partial\hat{\mathbf{r}}/\partial x^\alpha)\cdot(\partial\hat{\mathbf{r}}/\partial x^\beta)$ are the covariant components of the metric tensor $\mathbf{g}$. In combining equations (16.84) to obtain the result (16.92), we have used the identity $\partial g^{\alpha\beta}/\partial x^\gamma = -g^{\alpha\mu}g^{\beta\nu}(\partial g_{\mu\nu}/\partial x^\gamma)$, which can be easily established by differentiation of the metric inverse relationship $g^{\alpha\eta}g_{\eta\beta} = \delta^\alpha{}_\beta$. If the local wavevector $\mathbf{k}$ is considered to be the conjugate momentum of a surface wave, then the corresponding *orbital angular momentum* is

$$\mathbf{L} = \hat{\mathbf{r}} \times \mathbf{k}. \tag{16.93}$$

The rate of change of the angular momentum along a ray is

$$\frac{d\mathbf{L}}{d\sigma} = \tfrac{1}{2}\hat{\mathbf{r}} \times \boldsymbol{\nabla}_1 k^2. \tag{16.94}$$

On a great-circular orbit about a spherically symmetric Earth, the angular momentum is conserved: $d\mathbf{L}/d\sigma = \mathbf{0}$.

We can also use the angular distance Δ rather than the generating parameter σ as the independent variable in the Greek-index version of Hamilton's equations (16.84):

$$\frac{dx^\gamma}{d\Delta} = \frac{g^{\gamma\eta}k_\eta}{k}, \qquad \frac{dk_\gamma}{d\Delta} = -\frac{1}{2k}\frac{\partial g^{\alpha\beta}}{\partial x^\gamma}k_\alpha k_\beta + \frac{\partial k}{\partial x^\gamma}. \tag{16.95}$$

The corresponding vector equations, analogous to (16.90), are

$$\frac{d\hat{\mathbf{r}}}{d\Delta} = \hat{\mathbf{k}}, \qquad \frac{d\mathbf{k}}{d\Delta} = -k\hat{\mathbf{r}} + \boldsymbol{\nabla}_1 k. \tag{16.96}$$

These first-order equations may be combined into the single second-order differential equation

$$\frac{d}{d\Delta}\left(k\frac{d\hat{\mathbf{r}}}{d\Delta}\right) + k\hat{\mathbf{r}} = \boldsymbol{\nabla}_1 k. \tag{16.97}$$

Using the relation $d/d\Delta = \hat{\mathbf{k}}\cdot\boldsymbol{\nabla}_1$, we can convert (16.97) into an alternative forced spherical pendulum analogous to (15.57):

$$\frac{d^2\hat{\mathbf{r}}}{d\Delta^2} + \hat{\mathbf{r}} = \boldsymbol{\nabla}_\perp \ln k, \tag{16.98}$$

where $\boldsymbol{\nabla}_\perp = \boldsymbol{\nabla}_1 - \hat{\mathbf{k}}\hat{\mathbf{k}}\cdot\boldsymbol{\nabla}_1$. It is noteworthy that only the cross-path wavenumber gradient acts to deflect a surface wave away from the great circle $\hat{\mathbf{r}}(\Delta) = \hat{\mathbf{r}}'\cos\Delta + \hat{\mathbf{k}}'\sin\Delta$. The arclength rate of change of the angular momentum is

$$\frac{d\mathbf{L}}{d\Delta} = \hat{\mathbf{r}} \times \boldsymbol{\nabla}_1 k. \tag{16.99}$$

Finally, we can use the travel time T as the independent variable; the governing equations in that case are

$$\frac{dx^\gamma}{dT} = C\left(\frac{g^{\gamma\eta}k_\eta}{k}\right), \tag{16.100}$$

$$\frac{dk_\gamma}{dT} = C\left(-\frac{1}{2k}\frac{\partial g^{\alpha\beta}}{\partial x^\gamma}k_\alpha k_\beta + \frac{\partial k}{\partial x^\gamma}\right) \tag{16.101}$$

or, equivalently,

$$\frac{d\hat{\mathbf{r}}}{dT} = C\hat{\mathbf{k}}, \qquad \frac{d\mathbf{k}}{dT} = C(-k\hat{\mathbf{r}} + \boldsymbol{\nabla}_1 k). \tag{16.102}$$

The analogues of equations (16.97)–(16.99) are

$$\frac{d}{dT}\left(\frac{k}{C}\frac{d\hat{\mathbf{r}}}{dT}\right) + kC\hat{\mathbf{r}} = C\boldsymbol{\nabla}_1 k, \tag{16.103}$$

$$\frac{d^2\hat{\mathbf{r}}}{dT^2} + C^2\hat{\mathbf{r}} = C^2\boldsymbol{\nabla}_\perp \ln k, \tag{16.104}$$

$$\frac{d\mathbf{L}}{dT} = C\hat{\mathbf{r}} \times \boldsymbol{\nabla}_1 k. \tag{16.105}$$

The derivative $d/dT = \mathbf{C}\cdot\boldsymbol{\nabla}_1$ is the time rate of change experienced by an observer moving along the ray with the group velocity $\mathbf{C} = C\hat{\mathbf{k}}$. When tracing rays in terms of T rather than σ or Δ, it is necessary to specify the group speed $C(x^1, x^2)$ as well as the wavenumber $k(x^1, x^2)$.

★16.3.3 Hamilton's and Fermat's principles

Using a dot to denote differentiation with respect to σ, we define the ray Lagrangian associated with the Hamiltonian (16.81) by

$$\begin{aligned} L(x^1, x^2, \dot{x}^1, \dot{x}^2) &= k_\gamma \dot{x}^\gamma - H(x^1, x^2, k_1, k_2) \\ &= \tfrac{1}{2}[g_{\alpha\beta}\dot{x}^\alpha\dot{x}^\beta + k^2(x^1, x^2)]. \end{aligned} \tag{16.106}$$

The ray trajectories $x^\gamma(\sigma)$ can be derived from *Hamilton's principle*

$$\delta\int_{\sigma_1}^{\sigma_2} L(x^1, x^2, \dot{x}^1, \dot{x}^2)\, d\sigma = 0. \tag{16.107}$$

The associated Euler-Lagrange equation $d(\partial_{\dot{x}^\gamma}L)/d\sigma - \partial_{x^\gamma}L = 0$ is precisely the configuration-space ray-tracing equation (16.92). Alternatively, we can write Hamilton's principle in phase space in the form

$$\delta\int_{\sigma_1}^{\sigma_2} [k_\gamma\dot{x}^\gamma - H(x^1, x^2, k_1, k_2)]\, d\sigma = 0. \tag{16.108}$$

In this case the Euler-Lagrange equations are Hamilton equations (16.84).

Fermat's principle stipulates that the phase difference $\psi_2 - \psi_1$ between any two fixed points on Ω is stationary:

$$\delta \int_{\hat{\mathbf{r}}_1}^{\hat{\mathbf{r}}_2} \mathbf{k} \cdot d\hat{\mathbf{r}} = \delta \int_{\Delta_1}^{\Delta_2} k\, d\Delta = \delta \int_{T_1}^{T_2} kC\, dT = 0. \tag{16.109}$$

As in the body-wave case, we are not permitted to regard the angular arclength Δ or the travel time T as the independent variables in taking the variation (16.109); instead, we must eliminate Δ and T by introducing a new independent variable ξ such that

$$\sqrt{g_{\alpha\beta} \frac{dx^\alpha}{d\xi} \frac{dx^\beta}{d\xi}} = \frac{d\Delta}{d\xi} = C \frac{dT}{d\xi}. \tag{16.110}$$

The substitution (16.110) transforms (16.109) into a variational relation analogous to (15.64):

$$\delta \int_{\xi_1}^{\xi_2} k(x^1, x^2) \sqrt{g_{\alpha\beta} \frac{dx^\alpha}{d\xi} \frac{dx^\beta}{d\xi}}\, d\xi = 0. \tag{16.111}$$

This is Jacobi's form of the principle of least action for surface waves. The associated Euler-Lagrange equation, analogous to (15.65), is

$$\begin{aligned} &\frac{d}{d\xi} \left[k \left(g_{\alpha\beta} \frac{dx^\alpha}{d\xi} \frac{dx^\beta}{d\xi} \right)^{-1/2} g_{\gamma\eta} \frac{dx^\eta}{d\xi} \right] \\ &\qquad - \tfrac{1}{2} k \left(g_{\alpha\beta} \frac{dx^\alpha}{d\xi} \frac{dx^\beta}{d\xi} \right)^{-1/2} \frac{\partial g_{\mu\nu}}{\partial x^\gamma} \frac{dx^\mu}{d\xi} \frac{dx^\nu}{d\xi} \\ &\qquad - \left(g_{\alpha\beta} \frac{dx^\alpha}{d\xi} \frac{dx^\beta}{d\xi} \right)^{1/2} \frac{\partial k}{\partial x^\gamma} = 0. \end{aligned} \tag{16.112}$$

This equation determines the ray path in terms of the parameter ξ. Knowing $x^\sigma(\xi)$ we can obtain the propagation distance $\Delta(\xi)$ and travel time $T(\xi)$ by integrating (16.110). Alternatively, we can use (16.110) to convert (16.112) directly into the Greek-index versions of (16.97) and (16.103). The above considerations show that Hamilton's and Fermat's principles are essentially equivalent, as in the body-wave case.

16.4 Amplitude Variation

Surface-wave amplitudes vary as a result of the geometrical spreading of the associated rays on the surface of the unit sphere Ω. For a general treatment that applies to both Love and Rayleigh waves, we drop the prime in equation (16.75) and consider a generic transport equation of the form

$$\boldsymbol{\nabla}_1 \cdot \boldsymbol{\mathcal{K}}_\Omega = 0, \tag{16.113}$$

where

$$\boldsymbol{\mathcal{K}}_\Omega = C\mathcal{E}_\Omega \hat{\mathbf{k}} = \omega^2 C I_1 A^2 \hat{\mathbf{k}} = \omega c C I_1 A^2 \mathbf{k}. \tag{16.114}$$

The correspondence between (16.113)–(16.114) and the body-wave transport relations (15.71)–(15.72) is obvious, and we shall conduct a parallel analysis. Attention will again be restricted to the waves excited by a point source.

16.4.1 Conservation of energy

The quantity $\boldsymbol{\mathcal{K}}_\Omega$ is the radially integrated, lateral energy flux of a Love or Rayleigh wave, and equation (16.113) expresses the *conservation of surface-wave energy*. To verify this, we consider an infinitesimally narrow tube of surface-wave trajectories enclosing the ray between two consecutive points $\hat{\mathbf{r}}_1$ and $\hat{\mathbf{r}}_2$, as illustrated in Figure 16.1. Let Σ be the surface area of this ray-tube segment, and let $\partial\Sigma$ be its boundary. The unit outward normal $\hat{\boldsymbol{\nu}}$ to Σ is a tangent vector: $\hat{\mathbf{r}} \cdot \hat{\boldsymbol{\nu}} = 0$. Note, furthermore, that $\hat{\boldsymbol{\nu}} \cdot \boldsymbol{\mathcal{K}}_\Omega = 0$ on the sidewalls of the ray tube. Upon integrating the energy conservation law (16.113) over Σ and making use of Gauss' theorem (A.77) on Ω, we obtain the surface-wave analogue of (15.73):

$$\begin{aligned}\int_\Sigma \boldsymbol{\nabla}_1 \cdot \boldsymbol{\mathcal{K}}_\Omega \, d\Sigma &= \int_{\partial\Sigma} \hat{\boldsymbol{\nu}} \cdot \boldsymbol{\mathcal{K}}_\Omega \, dl \\ &= \|\boldsymbol{\mathcal{K}}_\Omega\|_2 \, dw_2 - \|\boldsymbol{\mathcal{K}}_\Omega\|_1 \, dw_1 = 0,\end{aligned} \tag{16.115}$$

where dw_1 and dw_2 are the differential widths of the ray tube at $\hat{\mathbf{r}}_1$ and $\hat{\mathbf{r}}_2$, respectively. All of the elastic-gravitational energy that enters the patch Σ at $\hat{\mathbf{r}}_1$ leaves it at $\hat{\mathbf{r}}_2$; as a result, the energy of a Love or Rayleigh wave on a perfectly elastic Earth is conserved. The amplitude variation law obtained from equation (16.115) is

$$\frac{A_2}{A_1} = \left[\frac{(cCI_1)_2 k_2}{(cCI_1)_1 k_1}\right]^{-1/2} \left|\frac{dw_2}{dw_1}\right|^{-1/2}. \tag{16.116}$$

The absolute value is necessary to account for the change in sign of the width of a surface-wave ray tube upon passage through a caustic, as we discuss in Sections 11.4 and 16.4.7.

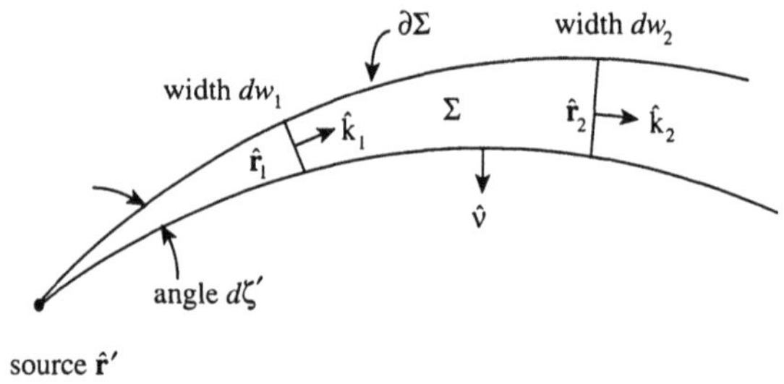

Figure 16.1. Schematic depiction of a Love or Rayleigh ray tube on the surface of the unit sphere Ω. The tube subtends a differential takeoff angle $d\zeta'$ at the source point $\hat{\mathbf{r}}'$. The unit wavevectors at the two points $\hat{\mathbf{r}}_1$ and $\hat{\mathbf{r}}_2$ are $\hat{\mathbf{k}}_1$ and $\hat{\mathbf{k}}_2$, respectively. The differential ray-tube widths dw_1 and dw_2 at these two points are measured in radians on Ω.

16.4.2 Surface-wave normalization

Thus far in this book, we have adhered strictly to the eigenfunction normalization convention $I_1 = 1$. Now, following Tromp & Dahlen (1992b), we shall depart from this, requiring instead that

$$cCI_1 = 1. \tag{16.117}$$

This new eigenfunction normalization relation is particularly convenient in the JWKB analysis of Love and Rayleigh surface waves on a laterally heterogeneous Earth. The amplitude variation law (16.116) is simplified to

$$\frac{A_2}{A_1} = \left(\frac{k_2}{k_1}\right)^{-1/2} \left|\frac{dw_2}{dw_1}\right|^{-1/2}. \tag{16.118}$$

Apart from a constant factor, the normalization (16.117) is identical to that introduced by Snieder & Nolet (1987). They show that it simplifies the expression for the surface-wave Green tensor $\mathbf{G}(\mathbf{x}, \mathbf{x}'; \omega)$ and moment-tensor response $\mathbf{a}(\mathbf{x}, \omega)$ on a spherically symmetric Earth; in particular, the reciprocal wave-speed product $(cC)^{-1}$ is replaced by unity in both (11.23) and (11.33).

★16.4.3 Ray-tube width

Explicit expressions for the ray-tube width ratio dw_2/dw_1 can be obtained using arguments analogous to those in Section 15.4.2. We shall not repeat the derivations, which are straightforward, but simply give the surface-wave analogues of equations (15.78) and (15.82) here:

$$\frac{dw_2}{dw_1} = \exp\left(\int_{\Delta_1}^{\Delta_2} \boldsymbol{\nabla}_1 \cdot \hat{\mathbf{k}}\, d\Delta\right), \tag{16.119}$$

$$\frac{k_2}{k_1}\frac{dw_2}{dw_1} = \exp\left(\int_{\sigma_1}^{\sigma_2} \boldsymbol{\nabla}_1 \cdot \mathbf{k}\, d\sigma\right). \tag{16.120}$$

The surface divergences in the integrands of (16.119) and (16.120) are related by $\boldsymbol{\nabla}_1 \cdot \mathbf{k} = \boldsymbol{\nabla}_1 \cdot (k\hat{\mathbf{k}}) = k\boldsymbol{\nabla}_1 \cdot \hat{\mathbf{k}} + d(\ln k)/d\sigma$.

16.4.4 Point-source Jacobian

The family of surface-wave rays shot from a source point $\hat{\mathbf{r}}'$ can be parameterized in terms of a single variable, which we shall denote by ζ'. We shall eventually take this ray parameter to be the takeoff azimuthal angle $0 \leq \zeta' \leq 2\pi$, measured counterclockwise from due south, in accordance with the convention established for body waves in Section 15.8. For the time being, however, this particular identification is not important; it is sufficient to regard ζ' as a label that distinguishes the various rays shot from $\hat{\mathbf{r}}'$. The partial derivative $\partial_{\zeta'}\hat{\mathbf{r}} = (\partial\hat{\mathbf{r}}/\partial\zeta')_\sigma$ is a tangent vector upon Ω that lies within the wavefront passing through the point $\hat{\mathbf{r}}$; the condition that guarantees this is

$$\hat{\mathbf{k}} \cdot \partial_{\zeta'}\hat{\mathbf{r}} = k_\gamma(\partial_{\zeta'}x^\gamma) = 0. \tag{16.121}$$

We define a two-dimensional *point-source Jacobian* by analogy with (15.86):

$$\begin{aligned} J = \frac{\partial(x^1, x^2)}{\partial(\sigma, \zeta')} &= \frac{\partial x^1}{\partial\sigma}\frac{\partial x^2}{\partial\zeta'} - \frac{\partial x^1}{\partial\zeta'}\frac{\partial x^2}{\partial\sigma} \\ &= (g^{11}k_1 + g^{12}k_2)(\partial_{\zeta'}x^2) - (\partial_{\zeta'}x^1)(g^{21}k_1 + g^{22}k_2). \end{aligned} \tag{16.122}$$

The amplitude variation along a ray can be expressed in terms of the Jacobian (16.122) by an application of Smirnov's lemma analogous to that in Section 15.4.4. Invoking the new eigenfunction normalization (16.117), we rewrite the surface-wave transport relation (16.113) in the form

$$\boldsymbol{\nabla}_1 \cdot (A^2\mathbf{k}) = 0, \tag{16.123}$$

or, equivalently,

$$\frac{d}{d\sigma}\ln A^2 = -\boldsymbol{\nabla}_1 \cdot \mathbf{k}. \tag{16.124}$$

Smirnov's lemma on a curved surface stipulates that

$$\frac{d}{d\sigma}\ln(gJ) = \boldsymbol{\nabla}_1\cdot\mathbf{k} \quad\text{whenever}\quad \frac{d\hat{\mathbf{r}}}{d\sigma} = \mathbf{k}. \tag{16.125}$$

The quantity g (not to be confused with the acceleration of gravity) is related to the covariant and contravariant components of the metric tensor $\mathbf{g} = \mathbf{I} - \hat{\mathbf{r}}\hat{\mathbf{r}}$ on the surface Ω by

$$g = \sqrt{\det g_{\alpha\beta}} = \frac{1}{\sqrt{\det g^{\alpha\beta}}}. \tag{16.126}$$

It may be readily verified by expansion of the respective determinants that

$$\frac{d}{d\sigma}\ln J = \frac{\partial k^\gamma}{\partial x^\gamma}, \qquad \frac{d}{d\sigma}\ln g = \tfrac{1}{2}g^{\alpha\beta}\left(\frac{dg_{\alpha\beta}}{d\sigma}\right). \tag{16.127}$$

To prove (16.125) we combine the results (16.127):

$$\begin{aligned}\frac{d}{d\sigma}\ln(gJ) &= \frac{\partial k^\gamma}{\partial x^\gamma} + \tfrac{1}{2}g^{\alpha\beta}\left(\frac{dg_{\alpha\beta}}{d\sigma}\right)\\ &= \frac{\partial k^\gamma}{\partial x^\gamma} + \tfrac{1}{2}g^{\alpha\beta}\left(\frac{dg_{\alpha\beta}}{dx^\gamma}\right)k^\gamma\\ &= \tfrac{1}{2}g^{\alpha\beta}\frac{d}{dx^\gamma}(g_{\alpha\beta}k^\gamma) = \boldsymbol{\nabla}_1\cdot\mathbf{k},\end{aligned} \tag{16.128}$$

where we have used $g^{\alpha\beta}g_{\alpha\beta} = 2$ in the penultimate step and (A.99) to obtain the final identification. Comparison of equations (16.124) and (16.125) now gives a result analogous to (15.95):

$$\frac{d}{d\sigma}\ln(gJA^2) = 0. \tag{16.129}$$

Equation (16.129) enables us to express the amplitude ratio (16.118) in terms of a ratio of Jacobians:

$$\frac{A_2}{A_1} = \left(\frac{g_2}{g_1}\right)^{-1/2}\left|\frac{J_2}{J_1}\right|^{-1/2}. \tag{16.130}$$

In fact, the differential ray-tube width dw and the Jacobian J are related everywhere along a ray by an equation analogous to (15.88):

$$k\,dw = gJ\,d\zeta'. \tag{16.131}$$

The amplitude variation laws (16.118) and (16.130) are equivalent by virtue of (16.131).

16.4.5 Geometrical spreading factor

It is convenient to define a *geometrical spreading factor* $S(\hat{\mathbf{r}},\hat{\mathbf{r}}')$ on the unit sphere Ω by

$$S = |dw|/d\zeta'. \tag{16.132}$$

The numerator $|dw|$ in (16.132) is the absolute differential width of a ray tube at the receiver $\hat{\mathbf{r}}$, whereas the denominator $d\zeta'$ is the differential solid angle subtended by this ray tube at the source $\hat{\mathbf{r}}'$, as illustrated in Figure 16.1. On a spherically symmetric Earth, this ratio reduces to $S = |\sin\Delta|$, where Δ is the total propagation distance around a great-circular orbit. The spreading factor S and the Jacobian J are related by

$$kS = g|J|. \tag{16.133}$$

We can express the differential width of a ray tube in a manner analogous to (15.101):

$$dw = \hat{\mathbf{k}}\cdot(\hat{\mathbf{r}}\times\partial_{\zeta'}\hat{\mathbf{r}})\,d\zeta'. \tag{16.134}$$

The geometrical spreading factor (16.132) is as a result given by

$$S = |\hat{\mathbf{k}}\cdot(\hat{\mathbf{r}}\times\partial_{\zeta'}\hat{\mathbf{r}})| = |(\hat{\mathbf{k}}\times\hat{\mathbf{r}})\cdot\partial_{\zeta'}\hat{\mathbf{r}}|. \tag{16.135}$$

It is readily demonstrated that $(\hat{\mathbf{k}}'\times\hat{\mathbf{r}}')\cdot\partial_{\zeta'}\hat{\mathbf{k}}' = -1$. The derivative of the takeoff wavevector $\partial_{\zeta'}\mathbf{k}' = k'\partial_{\zeta'}\hat{\mathbf{k}}'$ can be written in terms of the phase difference $\psi(\hat{\mathbf{r}},\hat{\mathbf{r}}') = \psi(\hat{\mathbf{r}}',\hat{\mathbf{r}})$ between the source and receiver in the form $\partial_{\zeta'}\mathbf{k}' = \partial_{\zeta'}\hat{\mathbf{r}}\cdot\boldsymbol{\nabla}_1\boldsymbol{\nabla}_1'\psi$. Upon combining these results and invoking (16.121) we find that

$$k'^{-1}[(\hat{\mathbf{k}}\times\hat{\mathbf{r}})\cdot\partial_{\zeta'}\hat{\mathbf{r}}]\,[(\hat{\mathbf{k}}\times\hat{\mathbf{r}})\cdot\boldsymbol{\nabla}_1\boldsymbol{\nabla}_1'\psi\cdot(\hat{\mathbf{k}}'\times\hat{\mathbf{r}}')] = -1. \tag{16.136}$$

Inserting the geometrical identity (16.136) into (16.135) we obtain an alternative representation of the spreading factor analogous to (15.107):

$$kS = kk'|(\hat{\mathbf{k}}\times\hat{\mathbf{r}})\cdot\boldsymbol{\nabla}_1\boldsymbol{\nabla}_1'\psi\cdot(\hat{\mathbf{k}}'\times\hat{\mathbf{r}}')|^{-1}. \tag{16.137}$$

The right side of equation (16.137) is invariant under an interchange of the source and receiver:

$$\hat{\mathbf{r}}\rightarrow\hat{\mathbf{r}}',\qquad \hat{\mathbf{r}}'\rightarrow\hat{\mathbf{r}},\qquad \hat{\mathbf{k}}\rightarrow-\hat{\mathbf{k}}',\qquad \hat{\mathbf{k}}'\rightarrow-\hat{\mathbf{k}}. \tag{16.138}$$

It follows that S satisfies the *dynamical reciprocity relation*

$$k(\hat{\mathbf{r}})S(\hat{\mathbf{r}},\hat{\mathbf{r}}') = k(\hat{\mathbf{r}}')S(\hat{\mathbf{r}}',\hat{\mathbf{r}}). \tag{16.139}$$

Equation (16.139) is the surface-wave analogue of (15.109); this result was first established by Woodhouse & Wong (1986).

16.4.6 Dynamical ray tracing

In order to calculate the Jacobian (16.122) we need to know the partial derivatives $\partial_{\zeta'}x^1$ and $\partial_{\zeta'}x^2$ along a ray. These derivatives satisfy a linear system of equations obtained by partial differentiation of (16.84):

$$\frac{d}{d\sigma}\left(\frac{\partial x^\gamma}{\partial \zeta'}\right) = \left(\frac{\partial g^{\gamma\eta}}{\partial x^\alpha}\right) k_\eta \left(\frac{\partial x^\alpha}{\partial \zeta'}\right) + g^{\gamma\eta}\left(\frac{\partial k_\eta}{\partial \zeta'}\right), \tag{16.140}$$

$$\begin{aligned}\frac{d}{d\sigma}\left(\frac{\partial k_\gamma}{\partial \zeta'}\right) = &-\tfrac{1}{2}\left(\frac{\partial^2 g^{\alpha\beta}}{\partial x^\gamma \partial x^\eta} k_\alpha k_\beta - \frac{\partial^2 k^2}{\partial x^\gamma \partial x^\eta}\right)\left(\frac{\partial x^\eta}{\partial \zeta'}\right) \\ &- \left(\frac{\partial g^{\alpha\beta}}{\partial x^\gamma}\right) k_\alpha \left(\frac{\partial k_\beta}{\partial \zeta'}\right).\end{aligned} \tag{16.141}$$

This system of four equations must be solved subject to the point-source initial conditions

$$\partial_{\zeta'}x^\gamma(0) = 0, \qquad \partial_{\zeta'}k_\gamma(0) = \partial_{\zeta'}k'_\gamma. \tag{16.142}$$

The two quantities $\hat{\mathbf{r}} + \partial_{\zeta'}\hat{\mathbf{r}}$ and $\mathbf{k} + \partial_{\zeta'}\mathbf{k}$ are the position and wavevector along a *neighboring or paraxial ray*.

16.4.7 Maslov index

Singular points on Ω where neighboring surface-wave rays cross are known as *caustics*, just as in the case of body waves. Caustic passages can be monitored by keeping track of sign changes in the Jacobian (16.122):

$$\text{caustic:} \quad J = 0. \tag{16.143}$$

We again introduce the *Maslov index* M, a positive integer counter that is incremented by one each time that the condition (16.143) is satisfied. The number of caustic passages is invariant under an interchange of the source and receiver:

$$M(\hat{\mathbf{r}}, \hat{\mathbf{r}}') = M(\hat{\mathbf{r}}', \hat{\mathbf{r}}). \tag{16.144}$$

On a spherical Earth the caustics are points situated at the source $\hat{\mathbf{r}}'$ and its antipode $-\hat{\mathbf{r}}'$. The rays shot from $\hat{\mathbf{r}}$ likewise exhibit caustics at $\pm\hat{\mathbf{r}}$. No matter how many full or partial orbits a ray makes, it is obvious that the reciprocity relation (16.144) is always satisfied. More generally, on a slightly heterogeneous Earth, the caustics are closed, multi-cusped curves in the vicinity of the source and its antipode, as we illustrate in Section 16.6.7. For a small enough perturbation, the relation (16.144) is still obvious; in fact, it pertains to any smooth Earth model, even one with with large lateral variations in surface-wave phase speed c. The phase of a Love or Rayleigh wave undergoes a non-geometrical $\pi/2$ phase advance upon every passage through a near-source or near-antipodal caustic. The existence of this so-called *polar phase shift* was first pointed out on a spherical Earth by Brune, Nafe & Alsop (1961).

16.4.8 Anelasticity

It is straightforward to incorporate the effects of slight, laterally variable anelasticity into surface-wave JWKB theory. As usual, we replace the perfectly elastic incompressibility and rigidity by complex, frequency-dependent moduli:

$$\kappa \rightarrow \kappa_0[1 + \tfrac{2}{\pi}Q_\kappa^{-1}\ln(\omega/\omega_0)] + i\kappa_0 Q_\kappa^{-1}, \tag{16.145}$$

$$\mu \rightarrow \mu_0[1 + \tfrac{2}{\pi}Q_\mu^{-1}\ln(\omega/\omega_0)] + i\mu_0 Q_\mu^{-1}, \tag{16.146}$$

where a subscript zero denotes evaluation at the reference frequency ω_0. The real parts of (16.145) and (16.146) are the real incompressibility and rigidity "seen" by a monochromatic Love or Rayleigh wave of angular frequency ω. We account for the effect of this anelastic dispersion upon the local real wavenumber k and the local radial eigenfunctions U, V, P and W by incorporating these frequency-dependent moduli directly into the defining equations (16.37) and (16.60)–(16.62) and associated boundary conditions (16.38)–(16.39) and (16.63)–(16.66). The effect of the imaginary perturbations $i\kappa_0 Q_\kappa^{-1}$ and $i\mu_0 Q_\mu^{-1}$ upon U, V, W and P will be ignored; their effect upon the local wavenumber can be found by perturbing the local dispersion relations (16.40) and (16.67). Rayleigh's principle is the critical cog in the analysis, as in Sections 9.2 and 9.7; the only difference is that we now perturb k at fixed ω rather than ω at fixed k. The distinction between these two types of perturbations is elaborated upon in Section 11.8. Writing the imaginary wavenumber perturbation in the form $k \rightarrow k - i\,\delta k$, we find, in terms of the present notation, that

$$\delta k_{\mathrm{L}} = \frac{k_{\mathrm{L}}^2\,\delta I_2 + \delta I_3}{2k_{\mathrm{L}} I_2}, \qquad \delta k_{\mathrm{R}} = \frac{k_{\mathrm{R}}^2\,\delta I_2' + k_{\mathrm{R}}\,\delta I_3' + \delta I_4'}{2k_{\mathrm{R}} I_2' + I_3'}, \tag{16.147}$$

where the subscripts L and R distinguish Love and Rayleigh waves, respectively. The perturbed radial integrals δI_2, δI_3 and $\delta I_2'$, $\delta I_3'$, $\delta I_4'$ in equations (16.147) are given by

$$\delta I_2 = \int_0^a \mu_0 Q_\mu^{-1} W^2\, dr, \tag{16.148}$$

$$\delta I_3 = \int_0^a \mu_0 Q_\mu^{-1}[(\partial_r W - r^{-1}W)^2 - 2r^{-2}W^2]\,r^2 dr, \tag{16.149}$$

$$\delta I_2' = \int_0^a [\mu_0 Q_\mu^{-1} U^2 + (\kappa_0 Q_\kappa^{-1} + \tfrac{4}{3}\mu_0 Q_\mu^{-1})V^2]\,dr, \tag{16.150}$$

$$\begin{aligned}\delta I_3' = \int_0^a & [\tfrac{4}{3}\mu_0 Q_\mu^{-1} V(\partial_r U - r^{-1}U) - 2\kappa_0 Q_\kappa^{-1} V(\partial_r U + 2r^{-1}U) \\ & + 2\mu_0 Q_\mu^{-1} U(\partial_r V - r^{-1}V)]\,r dr,\end{aligned} \tag{16.151}$$

$$\begin{aligned}\delta I_4' = \int_0^a & [(\kappa_0 Q_\kappa^{-1}(\partial_r U + 2r^{-1}U)^2 + \tfrac{4}{3}\mu_0 Q_\mu^{-1}(\partial_r U - r^{-1}U)^2 \\ & + \mu_0 Q_\mu^{-1}(\partial_r V - r^{-1}V)^2 - 2\mu_0 Q_\mu^{-1} r^{-2}V^2]\,r^2 dr.\end{aligned} \tag{16.152}$$

It is conventional to express the perturbations (16.147) in terms of the *temporal quality factors* of the associated standing waves. Making use of equations (16.44) and (16.72) we find that $\delta k = (ck)/(2CQ)$, where

$$Q_{\mathrm{L}}^{-1} = \frac{k_{\mathrm{L}}^2 \delta I_2 + \delta I_3}{\omega^2 I_{1\mathrm{L}}}, \qquad Q_{\mathrm{R}}^{-1} = \frac{k_{\mathrm{R}}^2 \delta I_2' + k_{\mathrm{R}} \delta I_3' + \delta I_4'}{\omega^2 I_{1\mathrm{R}}}. \tag{16.153}$$

The upshot is an exponential decay of the wave amplitude with angular distance from the source, of the form

$$\exp(-ik\Delta) \to \exp(-ik\Delta)\exp(-\omega\Delta/2CQ). \tag{16.154}$$

It is readily verified that the local Love and Rayleigh quality factors (16.153) are identical to (9.54). The factor C appears in the denominator of (16.154) because the wave energy—which is what is being dissipated—propagates with the group speed. Strictly speaking, C should be calculated in the reference-frequency Earth model, with laterally variable elastic parameters κ_0 and μ_0, for consistency with (16.148)–(16.152); we omit the customary subscript 0 used to denote this, for simplicity. In practice, it is possible to use the speed C at which a monochromatic wavegroup of frequency ω actually propagates, with negligible error.

16.5 JWKB Response

All of the ingredients needed to express the JWKB surface-wave response of a laterally heterogeneous Earth to a point source have now been assembled. We derive the JWKB Green tensor $\mathbf{G}(\mathbf{x}, \mathbf{x}'; \omega)$ and the acceleration response $\mathbf{a}(\mathbf{x}, \omega)$ to a moment-tensor source in the next two sections.

16.5.1 Green tensor

We begin by rewriting the far-field surface-wave Green tensor (11.23) on a spherically symmetric Earth, using the new eigenfunction normalization (16.117):

$$\begin{aligned} \mathbf{G}_{\mathrm{spher}}(\mathbf{x},\mathbf{x}';\omega) = \sum_{\mathrm{modes}}\sum_{\mathrm{rays}} (8\pi k|\sin\Delta|)^{-1/2} \\ \times [\hat{\mathbf{r}}U - i\hat{\mathbf{k}}V + i(\hat{\mathbf{r}}\times\hat{\mathbf{k}})W][\hat{\mathbf{r}}'U' + i\hat{\mathbf{k}}'V' - i(\hat{\mathbf{r}}'\times\hat{\mathbf{k}}')W'] \\ \times \exp i\,(-k\Delta + M\pi/2 - \pi/4)\exp(-\omega\Delta/2CQ). \end{aligned} \tag{16.155}$$

The first sum in (16.155) is over the various surface-wave *modes* or dispersion branches, whereas the second is over the multi-orbit *rays* or sequential arrivals, as illustrated in Figure 11.3. Both Love and Rayleigh waves are included in the double sum; $U = V = 0$ for the former, whereas $W = 0$ for the latter. The Maslov index M is simply $s - 1$, where $s = 1, 2, 3, \ldots$ is the numerical order of the arrival. We can write the JWKB Green tensor on a laterally heterogeneous Earth in a form analogous to (16.155):

$$\begin{aligned} \mathbf{G}(\mathbf{x},\mathbf{x}';\omega) = \sum_{\mathrm{modes}}\sum_{\mathrm{rays}} (kS)^{-1/2}[\hat{\mathbf{r}}U - i\hat{\mathbf{k}}V + i(\hat{\mathbf{r}}\times\hat{\mathbf{k}})W]\,\mathbf{A}' \\ \times \exp i\left(-\int_0^\Delta k\,d\Delta + M\frac{\pi}{2}\right)\exp\left(-\omega\int_0^\Delta \frac{d\Delta}{2CQ}\right), \end{aligned} \tag{16.156}$$

where we seek to determine the unknown vector $\mathbf{A}'$. The factor $(kS)^{-1/2}$ accounts for amplitude variations along the rays due to geometrical spreading, in accordance with the elastic energy-conservation law (16.118) and the definition (16.132). We find $\mathbf{A}'$ by matching the $s = 1$ ray in (16.156) to that in (16.155) in the vicinity of the source $\hat{\mathbf{r}}'$, where the Earth model may be considered locally homogeneous. The near-source geometrical spreading factor is simply $S \to \sin\Delta$, so that

$$\mathbf{A}' = (1/8\pi)^{1/2}[\hat{\mathbf{r}}'U' + i\hat{\mathbf{k}}'V' - i(\hat{\mathbf{r}}'\times\hat{\mathbf{k}}')W']\exp(-i\pi/4). \tag{16.157}$$

The JWKB Green tensor on a laterally heterogeneous Earth is therefore

$$\begin{aligned} \mathbf{G}(\mathbf{x},\mathbf{x}';\omega) = \sum_{\mathrm{modes}}\sum_{\mathrm{rays}} (8\pi kS)^{-1/2} \\ \times [\hat{\mathbf{r}}U - i\hat{\mathbf{k}}V + i(\hat{\mathbf{r}}\times\hat{\mathbf{k}})W][\hat{\mathbf{r}}'U' + i\hat{\mathbf{k}}'V' - i(\hat{\mathbf{r}}'\times\hat{\mathbf{k}}')W'] \\ \times \exp i\left(-\int_0^\Delta k\,d\Delta + M\frac{\pi}{2} - \frac{\pi}{4}\right)\exp\left(-\omega\int_0^\Delta \frac{d\Delta}{2CQ}\right). \end{aligned} \tag{16.158}$$

The JWKB response (16.158) is invariant under an interchange (16.138) of the source and receiver,

$$\mathbf{G}(\mathbf{x},\mathbf{x}';\omega) = \mathbf{G}^{\mathrm{T}}(\mathbf{x}',\mathbf{x};\omega), \tag{16.159}$$

by virtue of the geometrical reciprocity relations (16.139) and (16.144).

16.5.2 Moment-tensor response

The acceleration response to a frequency-dependent moment-tensor source $\mathbf{M}(\omega)$ at a hypocentral location $\mathbf{x}_{\mathrm{s}}$ can be expressed in terms of the Green tensor (16.158) in the form

$$\mathbf{a}(\mathbf{x},\omega) = i\omega\mathbf{M}:\boldsymbol{\nabla}_{\mathrm{s}}\mathbf{G}^{\mathrm{T}}(\mathbf{x},\mathbf{x}_{\mathrm{s}};\omega). \tag{16.160}$$

To lowest order, the gradient $\boldsymbol{\nabla}_{\mathrm{s}}$ with respect to the source coordinates $\mathbf{x}_{\mathrm{s}}$ acts only upon the oscillatory path integral $\exp(-i\int_0^\Delta k\,d\Delta)$ and the polarization vector $\hat{\mathbf{r}}_{\mathrm{s}}U_{\mathrm{s}} + i\hat{\mathbf{k}}_{\mathrm{s}}V_{\mathrm{s}} - i(\hat{\mathbf{r}}_{\mathrm{s}}\times\hat{\mathbf{k}}_{\mathrm{s}})W_{\mathrm{s}}$ of the surface wave leaving the source. We can write the surface-wave acceleration upon a laterally heterogeneous Earth in the form

$$\begin{aligned}\mathbf{a}(\mathbf{x},\omega) = i\omega \sum_{\text{modes}}\sum_{\text{rays}} &(8\pi kS)^{-1/2}[\hat{\mathbf{r}}U - i\hat{\mathbf{k}}V + i(\hat{\mathbf{r}}\times\hat{\mathbf{k}})W] \\ &\times (\mathbf{M}:\mathbf{E}_{\mathrm{s}}^*)\exp i\left(-\int_0^\Delta k\,d\Delta + M\frac{\pi}{2} - \frac{\pi}{4}\right) \\ &\times\exp\left(-\omega\int_0^\Delta \frac{d\Delta}{2CQ}\right),\end{aligned} \tag{16.161}$$

where the quantity

$$\begin{aligned}\mathbf{E}_{\mathrm{s}} = {}& \partial_r U_{\mathrm{s}}\hat{\mathbf{r}}_{\mathrm{s}}\hat{\mathbf{r}}_{\mathrm{s}} + r_{\mathrm{s}}^{-1}(U_{\mathrm{s}} - k_{\mathrm{s}}V_{\mathrm{s}})\hat{\mathbf{k}}_{\mathrm{s}}\hat{\mathbf{k}}_{\mathrm{s}} + r_{\mathrm{s}}^{-1}U_{\mathrm{s}}(\hat{\mathbf{r}}_{\mathrm{s}}\times\hat{\mathbf{k}}_{\mathrm{s}})(\hat{\mathbf{r}}_{\mathrm{s}}\times\hat{\mathbf{k}}_{\mathrm{s}}) \\ & -\tfrac{1}{2}i(\partial_r V_{\mathrm{s}} - r_{\mathrm{s}}^{-1}V_{\mathrm{s}} + k_{\mathrm{s}}r_{\mathrm{s}}^{-1}U_{\mathrm{s}})(\hat{\mathbf{r}}_{\mathrm{s}}\hat{\mathbf{k}}_{\mathrm{s}} + \hat{\mathbf{k}}_{\mathrm{s}}\hat{\mathbf{r}}_{\mathrm{s}}) \\ & + \tfrac{1}{2}i(\partial_r W_{\mathrm{s}} - r_{\mathrm{s}}^{-1}W_{\mathrm{s}})[\hat{\mathbf{r}}_{\mathrm{s}}(\hat{\mathbf{r}}_{\mathrm{s}}\times\hat{\mathbf{k}}_{\mathrm{s}}) + (\hat{\mathbf{r}}_{\mathrm{s}}\times\hat{\mathbf{k}}_{\mathrm{s}})\hat{\mathbf{r}}_{\mathrm{s}}] \\ & + \tfrac{1}{2}k_{\mathrm{s}}r_{\mathrm{s}}^{-1}W_{\mathrm{s}}[\hat{\mathbf{k}}_{\mathrm{s}}(\hat{\mathbf{r}}_{\mathrm{s}}\times\hat{\mathbf{k}}_{\mathrm{s}}) + (\hat{\mathbf{r}}_{\mathrm{s}}\times\hat{\mathbf{k}}_{\mathrm{s}})\hat{\mathbf{k}}_{\mathrm{s}}]\end{aligned} \tag{16.162}$$

is the JWKB strain tensor at the source. The results (16.158) and (16.161) are identical to the far-field surface-wave Green tensor (11.23) and moment-tensor response (11.33) on a spherically symmetric Earth, with the geometrical spreading factor $|\sin\Delta|^{-1/2}$ replaced by $S^{-1/2}$, and the source-to-receiver phase delay and attenuation $\exp(-ik\Delta)\exp(-\omega\Delta/2CQ)$ replaced by $\exp(-i\int_0^\Delta k\,d\Delta)\exp(-\omega\int_0^\Delta d\Delta/2CQ)$. The strain (16.162) is identical to (11.39) with the wavenumber k replaced by k_{s} and the ordinary radial derivatives $\dot{U}_{\mathrm{s}}$, $\dot{V}_{\mathrm{s}}$, $\dot{W}_{\mathrm{s}}$ replaced by $\partial_r U_{\mathrm{s}}$, $\partial_r V_{\mathrm{s}}$, $\partial_r W_{\mathrm{s}}$.

In preparation for the perturbation theory which we shall present in Section 16.8, it is convenient to write the $\hat{\boldsymbol{\nu}}$ component $a(\omega) = \hat{\boldsymbol{\nu}} \cdot \mathbf{a}(\mathbf{x}, \omega)$ of the acceleration (16.161) at the receiver $\mathbf{x}$ in the form

$$a = \sum_{\text{modes}} \sum_{\text{rays}} A \exp(-i\psi), \tag{16.163}$$

where

$$A = A_{\mathrm{r}} A_{\mathrm{p}} A_{\mathrm{s}}, \qquad \psi = \psi_{\mathrm{r}} + \phi_{\mathrm{p}} + \psi_{\mathrm{s}}. \tag{16.164}$$

The amplitudes A_{s}, A_{p}, A_{r} and phases ψ_{s}, ϕ_{p}, ψ_{r} are given explicitly by

$$A_{\mathrm{s}} \exp(-i\psi_{\mathrm{s}}) = i\omega(\mathbf{M}\!:\!\mathbf{E}_{\mathrm{s}}^*) \exp(-i\pi/4), \tag{16.165}$$

$$A_{\mathrm{p}} \exp(-i\psi_{\mathrm{p}}) = (8\pi k S)^{-1/2} \exp i \left(- \int_0^\Delta k\, d\Delta + M\frac{\pi}{2} \right)$$
$$\times \exp \left(- \int_0^\Delta \frac{\omega}{2CQ}\, d\Delta \right), \tag{16.166}$$

$$A_{\mathrm{r}} \exp(-i\psi_{\mathrm{r}}) = \hat{\boldsymbol{\nu}} \cdot [\hat{\mathbf{r}} U - i\hat{\mathbf{k}} V + i(\hat{\mathbf{r}} \times \hat{\mathbf{k}}) W]. \tag{16.167}$$

The subscripts s, p and r label contributions associated with the *source*, *path* and *receiver*, respectively. For Love waves, the contraction of the moment tensor $\mathbf{M}$ and the conjugate source strain $\mathbf{E}_{\mathrm{s}}^*$ is given by

$$\begin{aligned} \mathbf{M}\!:\!\mathbf{E}_{\mathrm{s}}^* &= i(\partial_r W_{\mathrm{s}} - r_{\mathrm{s}}^{-1} W_{\mathrm{s}})(M_{r\theta} \sin\zeta_{\mathrm{s}} - M_{r\phi} \cos\zeta_{\mathrm{s}}) \\ &\quad - k_{\mathrm{s}} r_{\mathrm{s}}^{-1} W_{\mathrm{s}} [\tfrac{1}{2}(M_{\theta\theta} - M_{\phi\phi}) \sin 2\zeta_{\mathrm{s}} - M_{\theta\phi} \cos 2\zeta_{\mathrm{s}}], \end{aligned} \tag{16.168}$$

whereas for Rayleigh waves it is

$$\begin{aligned} \mathbf{M}\!:\!\mathbf{E}_{\mathrm{s}}^* &= M_{rr} \partial_r U_{\mathrm{s}} + (M_{\theta\theta} + M_{\phi\phi}) r_{\mathrm{s}}^{-1} (U_{\mathrm{s}} - \tfrac{1}{2} k_{\mathrm{s}} V_{\mathrm{s}}) \\ &\quad + i(\partial_r V_{\mathrm{s}} - r_{\mathrm{s}}^{-1} V_{\mathrm{s}} + k_{\mathrm{s}} r_{\mathrm{s}}^{-1} U_{\mathrm{s}})(M_{r\phi} \sin\zeta_{\mathrm{s}} + M_{r\theta} \cos\zeta_{\mathrm{s}}) \\ &\quad - k_{\mathrm{s}} r_{\mathrm{s}}^{-1} V_{\mathrm{s}} [M_{\theta\phi} \sin 2\zeta_{\mathrm{s}} + \tfrac{1}{2}(M_{\theta\theta} - M_{\phi\phi}) \cos 2\zeta_{\mathrm{s}}]. \end{aligned} \tag{16.169}$$

The quantities M_{rr}, $M_{r\theta}$, $M_{r\phi}$, $M_{\theta\theta}$, $M_{\theta\phi}$, $M_{\phi\phi}$ are the six independent spherical polar components of $\mathbf{M}$, as displayed in equation (5.124), and the angle $0 \leq \zeta_{\mathrm{s}} \leq 2\pi$ is the takeoff azimuth of the surface-wave ray path on the unit sphere Ω, measured counterclockwise from due south at the earthquake epicenter $\hat{\mathbf{r}}_{\mathrm{s}}$. The source term

$$\begin{aligned} A_{\mathrm{s}} \exp(-i\psi_{\mathrm{s}}) &= \omega\big\{[M_{rr} \partial_r U_{\mathrm{s}} + (M_{\theta\theta} + M_{\phi\phi}) r_{\mathrm{s}}^{-1} (U_{\mathrm{s}} - \tfrac{1}{2} k_{\mathrm{s}} V_{\mathrm{s}})] e^{i\pi/4} \\ &\quad - (\partial_r V_{\mathrm{s}} - r_{\mathrm{s}}^{-1} V_{\mathrm{s}} + k_{\mathrm{s}} r_{\mathrm{s}}^{-1} U_{\mathrm{s}})(M_{r\phi} \sin\zeta_{\mathrm{s}} + M_{r\theta} \cos\zeta_{\mathrm{s}}) e^{-i\pi/4} \\ &\quad - k_{\mathrm{s}} r_{\mathrm{s}}^{-1} V_{\mathrm{s}} [M_{\theta\phi} \sin 2\zeta_{\mathrm{s}} + \tfrac{1}{2}(M_{\theta\theta} - M_{\phi\phi}) \cos 2\zeta_{\mathrm{s}}] e^{i\pi/4} \\ &\quad - (\partial_r W_{\mathrm{s}} - r_{\mathrm{s}}^{-1} W_{\mathrm{s}})(M_{r\theta} \sin\zeta_{\mathrm{s}} - M_{r\phi} \cos\zeta_{\mathrm{s}}) e^{-i\pi/4} \\ &\quad - k_{\mathrm{s}} r_{\mathrm{s}}^{-1} W_{\mathrm{s}} [\tfrac{1}{2}(M_{\theta\theta} - M_{\phi\phi}) \sin 2\zeta_{\mathrm{s}} - M_{\theta\phi} \cos 2\zeta_{\mathrm{s}}] e^{i\pi/4}\big\} \end{aligned} \tag{16.170}$$

is the *complex Love or Rayleigh radiation pattern*, denoted by the symbol $R(\zeta_{\mathrm{s}})$ in Sections 11.4 and 11.7.2. In fact, equation (16.170) is identical to equation (11.34), with the great-circular takeoff azimuths Φ (s odd) and $\Phi + \pi$ (s even) replaced by ζ_{s}, and with k and $\dot{U}_{\mathrm{s}}$, $\dot{V}_{\mathrm{s}}$, $\dot{W}_{\mathrm{s}}$ replaced by k_{s} and $\partial_r U_{\mathrm{s}}$, $\partial_r V_{\mathrm{s}}$, $\partial_r W_{\mathrm{s}}$.

16.6 Practical Numerical Implementation

In this section we outline a practical scheme for tracing surface-wave rays and computing phase and amplitude variations along them, analogous to the scheme for body waves in Section 15.8. We suppose the incompressibility, rigidity and density to be specified functions $\kappa(r,\theta,\phi)$, $\mu(r,\theta,\phi)$ and $\rho(r,\theta,\phi)$ of the radial distance from the center r, the colatitude θ and the longitude ϕ; likewise, the discontinuity radii are presumed to be functions of the form $d(\theta,\phi)$. The spherical polar coordinates of the receiver $\mathbf{x}$ are r,θ,ϕ whereas those of the source $\mathbf{x}'$ are r',θ',ϕ'.

16.6.1 Local modes

To compute the JWKB response, we must know the local displacement eigenfunctions U, V, W and their radial derivatives $\partial_r U$, $\partial_r V$, $\partial_r W$ at the source $\mathbf{x}'$ and receiver $\mathbf{x}$. To find these, it is necessary to run a modified version of a normal-mode code such as `MINEOS` or `OBANI` once at every geographical source and receiver position θ',ϕ' and θ,ϕ. Figure 16.2 shows the depth variation of U, V, W and $\partial_r U$, $\partial_r V$, $\partial_r W$ for 165-second fundamental-mode Love and Rayleigh waves at twenty randomly selected earthquake epicentral locations and fifty randomly selected GSN station locations on Earth model SH12WM13 (Su, Woodward & Dziewonski 1994). The variations in U, $\partial_r V$ and W in the uppermost 100 km are of order 15%; the variations in $\partial_r U$, V and $\partial_r W$ between 100 km and 300 km depth can be as large as 30%. Note that the quantities plotted are the JWKB-normalized eigenfunctions, which have $cCI_1 = 1$ at every point on the unit sphere Ω.

The local dispersion relation $k = k(\theta,\phi,\omega)$ and phase speed $c(\theta,\phi,\omega)$ must be known at every point θ,ϕ which may be visited by a surface wave. It is generally prohibitive to call `MINEOS` or `OBANI` at every step along a ray path; instead, we resort to the surface-wave perturbation theory described in Section 11.8. The local wavenumber and phase speed are regarded as small perturbations $k \to k + \delta k$ and $c \to c + \delta c$ away from a spherically averaged Earth. The perturbations in wavenumber, phase speed and eigen-

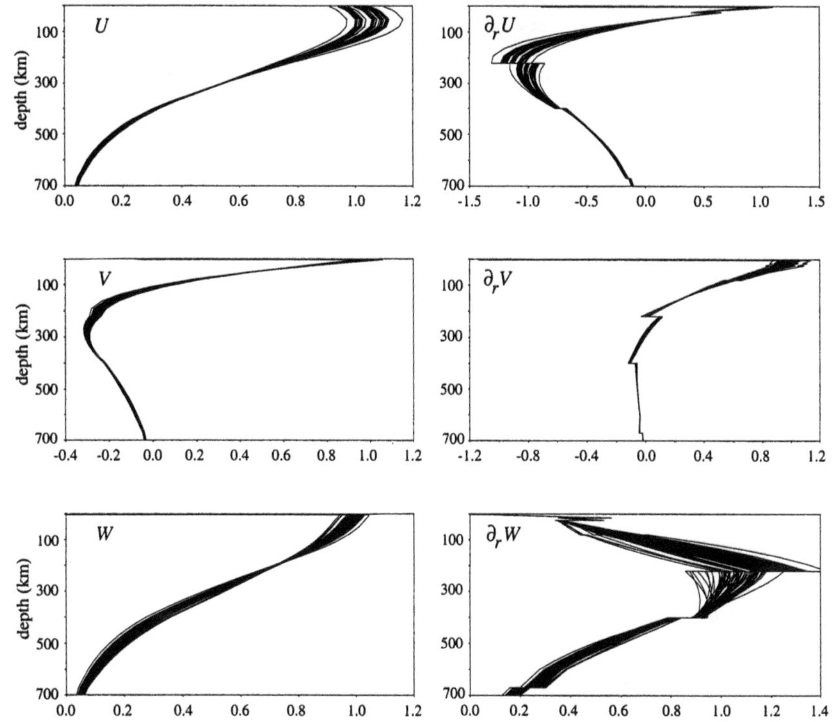

Figure 16.2. Depth variation of the eigenfunctions U, V, W and their radial derivatives $\partial_r U$, $\partial_r V$, $\partial_r W$ at the geographical locations θ, ϕ of twenty earthquakes and fifty GSN stations on model SH12WM13. The 165-second fundamental-mode eigenfunctions are JWKB normalized: $cCI_1 = 1$. The Rayleigh-wave quantities U, V and $\partial_r U$, $\partial_r V$ are scaled such that the average value on the free surface is unity. The average values of the Love-wave quantities W and $\partial_r W$ are scaled to unity on the seafloor and at a depth of 144 km, respectively.

frequency at every point on Ω are related by equation (11.93):

$$\frac{\delta k}{k} = -\frac{\delta c}{c} = -\frac{c}{C}\frac{\delta\omega}{\omega}, \tag{16.171}$$

where C is the unperturbed group speed. The resulting local dispersion relation and phase-speed variations are accurate to first order in the local structural perturbations $\delta\kappa$, $\delta\mu$, $\delta\rho$ and δd. This is an adequate approximation for long-period ($T > 100$ s) Love and Rayleigh waves that are not strongly affected by large lateral variations in the thickness of the crust.

16.6.2 Kinematic ray tracing

To trace surface-wave rays, it is convenient to use the colatitude and longitude as the generalized coordinates:

$$x^1 = \theta, \qquad x^2 = \phi. \tag{16.172}$$

The covariant and contravariant components of the metric tensor $\mathbf{g} = \mathbf{I} - \hat{\mathbf{r}}\hat{\mathbf{r}}$ are given in this case by

$$g_{\theta\theta} = 1, \qquad g_{\theta\phi} = g_{\phi\theta} = 0, \qquad g_{\phi\phi} = \sin^2\theta, \tag{16.173}$$

$$g^{\theta\theta} = 1, \qquad g^{\theta\phi} = g^{\phi\theta} = 0, \qquad g^{\phi\phi} = (\sin\theta)^{-2}. \tag{16.174}$$

The ray-tracing Hamiltonian (16.81) is therefore

$$H = \tfrac{1}{2}[k_\theta^2 + (\sin\theta)^{-2}k_\phi^2 - k^2(\theta,\phi)] = 0. \tag{16.175}$$

Hamilton's equations (16.83)–(16.84) for the spherical polar Hamiltonian $H(\theta, \phi, k_\theta, k_\phi)$ are

$$\frac{d\theta}{d\sigma} = k_\theta, \tag{16.176}$$

$$\frac{d\phi}{d\sigma} = (\sin\theta)^{-2}k_\phi, \tag{16.177}$$

$$\frac{dk_\theta}{d\sigma} = \tfrac{1}{2}\partial_\theta k^2 + \cot\theta(\sin\theta)^{-2}k_\phi^2, \tag{16.178}$$

$$\frac{dk_\phi}{d\sigma} = \tfrac{1}{2}\partial_\phi k^2. \tag{16.179}$$

The initial conditions for a ray shot from θ', ϕ' in a direction $0 \leq \zeta' \leq 2\pi$ are $\theta(0) = \theta'$, $\phi(0) = \phi'$ and $k_\theta(0) = k'\cos\zeta'$, $k_\phi(0) = k'\sin\zeta'$.

The order of the surface-wave ray-tracing system (16.176)–(16.179) can be reduced from four to two, in the same manner that the six body-wave ray-tracing equations (15.199)–(15.204) were reduced to four equations (15.213)–(15.216) in Section 15.8.1. To this end, we introduce the *local azimuth* or direction of propagation $0 \leq \zeta \leq 2\pi$ at an arbitrary point θ, ϕ along a ray (see Figure 16.3). The wavevector is given in terms of this azimuthal angle by

$$\mathbf{k} = k\cos\zeta\,\hat{\boldsymbol{\theta}} + k\sin\zeta\,\hat{\boldsymbol{\phi}}. \tag{16.180}$$

The covariant components of the wavevector,

$$k_\theta = k\cos\zeta, \qquad k_\phi = k\sin\theta\sin\zeta, \tag{16.181}$$

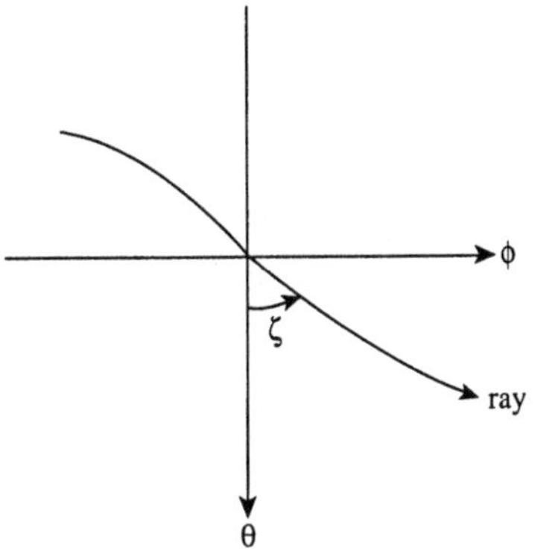

Figure 16.3. The propagation direction of a surface-wave ray at a point θ, ϕ is specified in terms of the *azimuth* $0 \leq \zeta \leq 2\pi$, measured counterclockwise from due south. The takeoff azimuth at the source θ', ϕ' is simply the primed value ζ' of this angle.

are readily shown to satisfy the dispersion relation (16.175). Using the relations (16.181), we can reduce the ray-tracing equations (16.176)–(16.179) to a system of three ordinary differential equations governing θ, ϕ and ζ:

$$\frac{d\theta}{d\sigma} = k \cos\zeta, \tag{16.182}$$

$$\frac{d\phi}{d\sigma} = k(\sin\theta)^{-1} \sin\zeta, \tag{16.183}$$

$$\frac{d\zeta}{d\sigma} = -\sin\zeta\, \partial_\theta k + (\sin\theta)^{-1} \cos\zeta\, \partial_\phi k - k \cot\theta \sin\zeta. \tag{16.184}$$

Equations (16.182)–(16.184) can be integrated, subject to the initial conditions $\theta(0) = \theta'$, $\phi(0) = \phi'$, $\zeta(0) = \zeta'$.

Finally, we can take the longitude ϕ to be the independent parameter, eliminating the generating parameter σ with the aid of (16.183). This results in a final system of two first-order differential equations describing the evolution of the colatitude θ and azimuth ζ:

$$\frac{d\theta}{d\phi} = \sin\theta \cot\zeta, \tag{16.185}$$

$$\frac{d\zeta}{d\phi} = -\cos\theta + \sin\theta\, \partial_\theta \ln c - \cot\zeta\, \partial_\phi \ln c, \tag{16.186}$$

where $c = \omega/k$ is the local surface-wave phase speed. As in the body-wave case, it is advantageous to rotate the Earth model so that both the source and receiver are situated upon the equator, in order to avoid the coordinate singularities at the poles:

$$\theta' = \pi/2,\ \phi' = 0 \quad \text{and} \quad \theta = \pi/2,\ \phi = \Theta, \tag{16.187}$$

where $\cos\Theta = \cos\theta\cos\theta' + \sin\theta\sin\theta'\cos(\phi - \phi')$. The Cauchy initial conditions associated with (16.185)–(16.186) are in that case

$$\theta(0) = \pi/2, \qquad \zeta(0) = \zeta'. \tag{16.188}$$

As expected, the body-wave ray-tracing equations (15.213)–(15.216) reduce to (16.185)–(16.186) upon setting $r = 1$, $i = \pi/2$ and substituting $v \to c$.

16.6.3 Shooting

To find the takeoff angle ζ' that enables a *minor-arc* G1 or R1 ray to "hit" a receiver situated at $\theta = \pi/2$, $\phi = \Theta$, we iteratively solve a one-dimensional analogue of equation (15.221):

$$\partial_{\zeta'}\theta_n(\Theta)\,(\zeta'_{n+1} - \zeta'_n) = \pi/2 - \theta_n(\Theta), \tag{16.189}$$

where $n = 0, 1, 2, \ldots$ is the iteration number. An optimal choice for the initial iterate ζ'_0 is derived using ray perturbation theory in Section 16.8. To trace a *major-arc* G2 or R2 ray in the direction of increasing ϕ, we perform a "backward" equatorial rotation,

$$\theta' = \pi/2,\ \phi' = 0 \quad \text{and} \quad \theta = \pi/2,\ \phi = 2\pi - \Theta, \tag{16.190}$$

and replace $\partial_{\zeta'}\theta_n(\Theta)$ and $\theta_n(\Theta)$ in (16.189) by $\partial_{\zeta'}\theta_n(2\pi-\Theta)$ and $\theta_n(2\pi-\Theta)$, respectively. Likewise, for G3, R3 and G4, R4 we trace "forward" and "backward" rays to longitudes $2\pi + \Theta$ and $4\pi - \Theta$, etc.

16.6.4 Phase and decay rate

The longitudinal rate of change in the phase ψ of a wave is

$$\frac{d\psi}{d\phi} = \frac{k\,d\Delta}{d\phi} = k\sin\theta(\cos\zeta)^{-1}. \tag{16.191}$$

This equation can be integrated to find the total accumulated phase and attenuation integrals along a minor-arc G1 or R1 ray:

$$\int_0^\Delta k\,d\Delta = \int_0^\Theta k\sin\theta(\cos\zeta)^{-1}\,d\phi, \tag{16.192}$$

$$\int_0^\Delta \frac{d\Delta}{2CQ} = \int_0^\Theta \frac{\sin\theta(\cos\zeta)^{-1}}{2CQ}\,d\phi. \tag{16.193}$$

These results may be readily generalized to find the phase and attenuation along a major-arc or multi-orbit G2, G3,... or R2, R3,... ray. Equations (16.192) and (16.193) are the surface-wave analogues of (15.223) and (15.224), respectively.

16.6.5 Geometrical spreading

With the choice $x^1 = \theta, x^2 = \phi$ and the source and receiver upon the equator, the point-source Jacobian (16.122) reduces to

$$\begin{aligned} J &= \frac{\partial(\theta,\phi)}{\partial(\sigma,\zeta')} = \frac{\partial(\theta,\phi)}{\partial(\phi,\zeta')}\,\frac{\partial(\phi,\zeta')}{\partial(\sigma,\zeta')} \\ &= -k\,(\sin\theta)^{-1}\sin\zeta\,(\partial_{\zeta'}\theta), \end{aligned} \tag{16.194}$$

where $\partial_{\zeta'}\theta$ now denotes the partial derivative at fixed longitude ϕ, and we have used (16.183) to obtain the final equality. The determinant of the surface metric tensor is $g = \sqrt{\det g_{\alpha\beta}} = \sin\theta$, so the geometrical spreading factor (16.133) is

$$S = k^{-1}g|J| = \sin\zeta\,|\partial_{\zeta'}\theta|. \tag{16.195}$$

To compute S we need to evaluate the partial derivative $\partial_{\zeta'}\theta = (\partial\theta/\partial\zeta')_\phi$ along the surface-wave trajectory. Upon differentiating the kinematic ray-tracing equations (16.185)–(16.186) with respect to the initial takeoff angle ζ', we obtain a linear system of *two dynamical ray-tracing equations*:

$$\frac{d}{d\phi}\,(\partial_{\zeta'}\theta) = \cos\theta\cot\zeta\,\partial_{\zeta'}\theta - \sin\theta\,(\sin\zeta)^{-2}\,\partial_{\zeta'}\zeta, \tag{16.196}$$

$$\begin{aligned} \frac{d}{d\phi}\,(\partial_{\zeta'}\zeta) &= [\sin\theta\,\partial_\theta^2\ln c - \cot\zeta\,\partial_\theta\partial_\phi\ln c + \cos\theta\,\partial_\theta\ln c \\ &\qquad + \sin\theta]\,\partial_{\zeta'}\theta + (\cos\zeta)^{-2}(\partial_\phi\ln c)\,\partial_{\zeta'}\zeta. \end{aligned} \tag{16.197}$$

These coupled equations can be integrated to find $\partial_{\zeta'}\theta$ and $\partial_{\zeta'}\zeta$, subject to the initial conditions

$$\partial_{\zeta'}\theta(0) = 0, \qquad \partial_{\zeta'}\zeta(0) = 1. \tag{16.198}$$

The singularity condition (16.143) for a surface-wave ray tube reduces to

$$\text{caustic:}\quad \partial_{\zeta'}\theta = 0. \tag{16.199}$$

This condition specifies the geographical location of surface-wave *caustics*, where neighboring rays cross. The Maslov index M can be evaluated by monitoring the sign changes of the derivative $\partial_{\zeta'}\theta$ along a ray.

16.6.6 Spherical Earth

On a spherically symmetric Earth, the solution to the two-point ray-tracing problem (16.185)–(16.187) is the minor arc of the equatorial great circle:

$$\theta(\phi) = \pi/2, \qquad \zeta(\phi) = \pi/2, \qquad 0 \leq \phi \leq \Theta. \tag{16.200}$$

A G1 or R1 wave traverses the geodesic or shortest possible path between the source and receiver, as expected. A G2 or R2 wave propagates in the opposite direction along the major arc, a G3 or R3 wave circumnavigates the equator once before arriving at the receiver, etc. The angular distances

$$\Delta = \begin{cases} \Theta + (s-1)\pi, & s \text{ odd} \\ s\pi - \Theta, & s \text{ even} \end{cases} \tag{16.201}$$

travelled by the later-arriving $s = 2, 3, \ldots$ wavegroups are *stationary*, in accordance with Fermat's principle (16.109). However, the associated great-circular arcs are situated at saddle points rather than local minima in the infinite-dimensional space of all possible ray paths. This *minimax* character of the higher-orbit surface-wave arrivals is geometrically obvious in the case of an equatorial source and receiver separated by an epicentral distance $\pi/2 < \Theta < \pi$. On the one hand, the G2 or R2 major-arc ray path, of length $2\pi - \Theta$, can obviously be lengthened by superimposing a small sinusoidal oscillation; on the other hand, it can be shortened by ironing out any oscillations, but increasing the inclination so that the midpoint is situated north or south of the equator.

The unique solution to the initial-value problem (16.196)–(16.198) on a spherically symmetric Earth is

$$\partial_{\zeta'}\theta(\phi) = -\sin\phi, \qquad \partial_{\zeta'}\zeta(\phi) = \cos\phi. \tag{16.202}$$

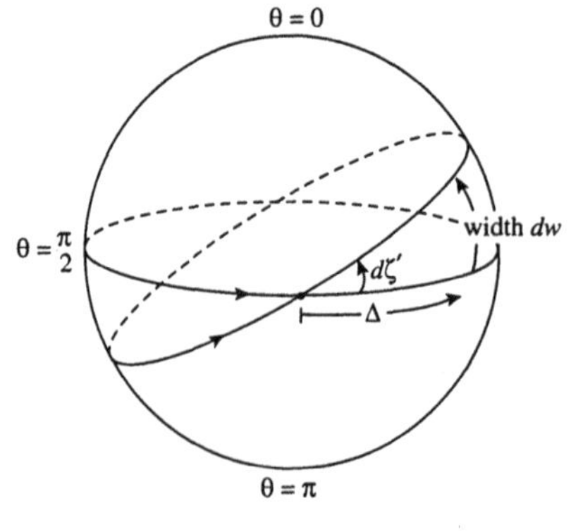

Figure 16.4. On a spherical Earth model the trajectories are great circles. Neighboring rays cross—first at the antipodal caustic, then again upon returning to the source, and so on. The geometrical spreading factor is $S = |dw|/d\zeta' = |\sin\Delta|$, where Δ is the angular distance of propagation.

The geometrical spreading factor (16.195) of a wave that has propagated a distance Δ is therefore $S = |\sin\Delta|$, as is obvious from elementary geometrical considerations; the caustics are degenerate points at the source and its antipode, as illustrated in Figure 16.4. There is an intimate connection between these caustics and the stationary character of the associated ray paths: the first-arriving G1 or R1 waves are those that have travelled along the geodesic ray path and *have not* passed through a caustic, whereas the later G2, G3,... or R2, R3,... arrivals travel along minimax ray paths and *have* passed through a caustic. Roughly speaking, the Maslov index M is the number of linearly independent directions in path space in which it is possible to decrease rather than increase the accumulated phase $\int_0^\Delta k\,d\Delta$ along a ray. This association between the number of caustic passages and the character of the stationarity in Fermat's principle is maintained in the presence of lateral heterogeneity; in fact, it is a universal feature of linear wave propagation in the JWKB approximation (Gutzwiller 1990).

*16.6.7 Caustic morphology

We illustrate the nature of the surface-wave caustics upon a laterally heterogeneous Earth by considering a specific example: 150-second fundamental-mode Rayleigh waves on model M84A (Woodhouse & Dziewonski 1984). The phase speed $c + \delta c$ of these waves is a spherical-harmonic sum of maximum angular degree $s_{\rm max} = 8$, with point-to-point variations of order $\delta c/c = \pm 2-3$ percent. To trace surface-wave rays, we rotate the model so that the unperturbed path lies along the equator, and integrate equations (16.185)–(16.186) numerically using a variable-order Runge-Kutta scheme. Figure 16.5 shows a family of rays shot out at 4° intervals from a hypothetical source situated on the equator and Greenwich meridian $\theta' = 90°$, $\phi' = 0°$. The first three orbits of the rays shot toward the west $(180° < \zeta' < 360°)$ are shown on the left and above, whereas the first three orbits of the rays shot toward the east $(0° < \zeta' < 180°)$ are shown on the right and below; the figure should be regarded as a single continuous strip that has been cut into three panels at $\phi = \pm 400°$ for convenience of display. The caustics in the vicinity of the antipode and the source are evident; half of each caustic appears on the left and above at each location, whereas the other half appears on the right and below.

Magnified views of the central portion of the first antipodal caustic, where R1 and R2 interfere, and the first source caustic, where R2 and R3 interfere, are shown in Figure 16.6. Each caustic is a single closed curve with sixteen cusps and a continuously evolving tangent (Wang, Dahlen & Tromp 1993). The gross features of this morphology can be understood on the basis of catastrophe theory; surface-wave caustics are closed curves

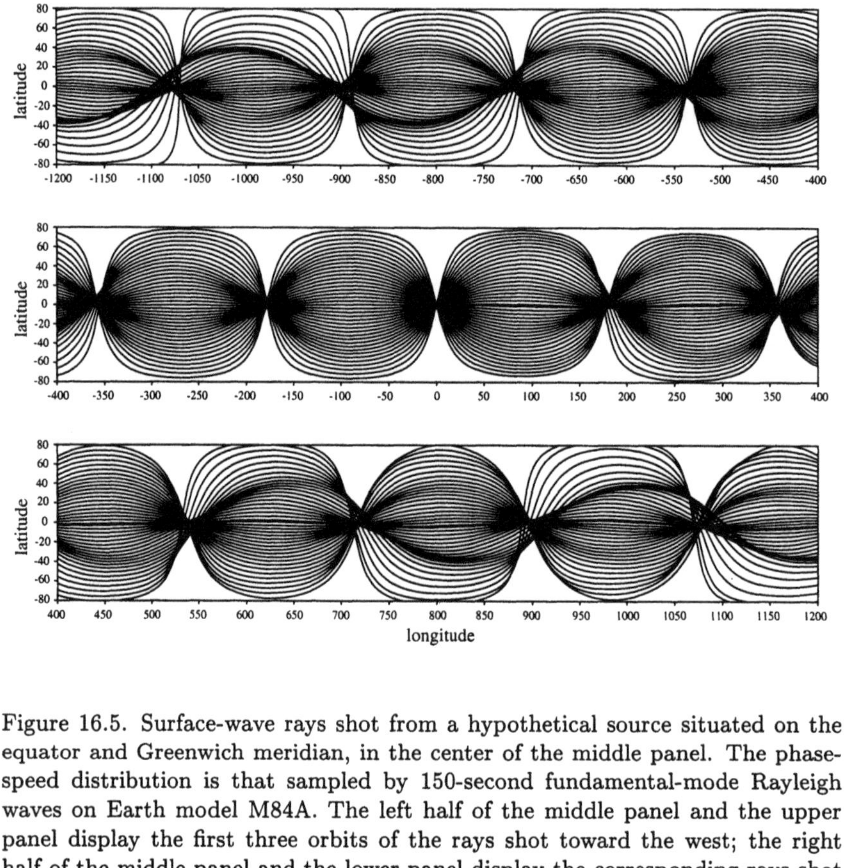

Figure 16.5. Surface-wave rays shot from a hypothetical source situated on the equator and Greenwich meridian, in the center of the middle panel. The phase-speed distribution is that sampled by 150-second fundamental-mode Rayleigh waves on Earth model M84A. The left half of the middle panel and the upper panel display the first three orbits of the rays shot toward the west; the right half of the middle panel and the lower panel display the corresponding rays shot toward the east. The map is a repeating linear cylindrical projection; the right edge of the top panel (longitude -400°) and the left edge of the bottom panel (longitude 400°) coincide with the left and right edges, respectively, of the middle panel. Rays that pass near the North and South Poles are not shown.

with multiple cusps because a fold and a cusp are the only two structurally stable catastrophes in two dimensions (Poston & Stewart 1978; Berry & Upstill 1980). Every sufficiently strong local minimum or maximum in the surface-wave phase speed $c + \delta c$ gives rise to a cusp; the total number is for this reason equal to twice the maximum angular degree of the Earth model. At any point that is sufficiently distant from the source or antipode, there are only two arrivals associated with the R1–R2 and R2–R3 caustics. Receivers in the vicinity of the source or antipode may have, however, four, six or more geometrical arrivals, because of the multiplicity of the folding.

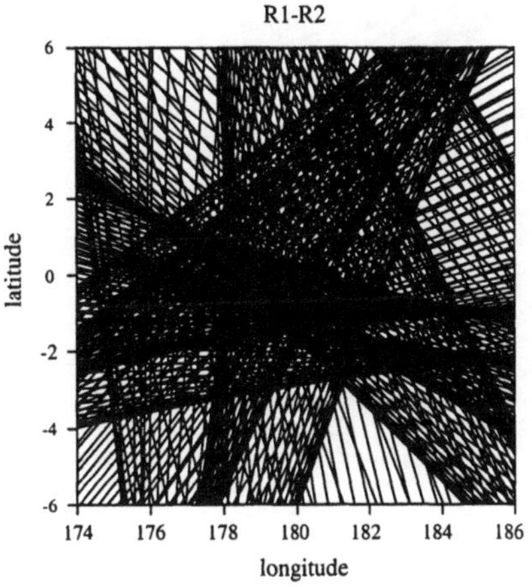

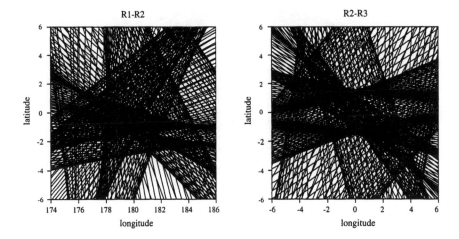

Figure 16.6. Blown-up bird's-eye view of the first antipodal (R1–R2) and source (R2–R3) caustics formed by 150-second fundamental-mode Rayleigh waves on model M84A. Each caustic is an envelope of the rays shot from the source at $\theta' = \pi/2$, $\phi' = 0$. Every ray passes through these and all succeeding caustics at a single point of tangency.

A surprisingly large fraction of the Earth's surface area is occupied by these complex, intricately folded caustics. Several of the cusps of the R1–R2 and R2–R3 caustics extend out to more than $20° - 30°$ from the antipode and source, respectively. The R3–R4, R4–R5 and succeeding caustics are even more extensive, because of the increasing lateral refraction of the waves. The divergence $S^{-1} \to \infty$ of the geometrical spreading factor renders the JWKB representation (16.161) of the surface-wave response invalid in the vicinity of the source and antipodal caustics. An alternative Maslov representation has been developed by Tromp & Dahlen (1993).

★16.6.8 Tsunamis

We noted in Section 8.8.11 that tsunamis, or earthquake-generated surface gravity waves, are the true fundamental modes of an Earth model with a neutrally stratified fluid outer core and surficial ocean. As we showed in Section 11.6.3 the surface phase speed of a tsunami is very nearly that of a shallow-water wave: $ac = \sqrt{gh}$, where g is the acceleration of gravity and $h(\theta, \phi)$ is the geographically variable water depth. We can use this relation together with our knowledge of seafloor bathymetry to determine a global tsunami phase-speed map $c(\theta, \phi)$. Such a map was first used to

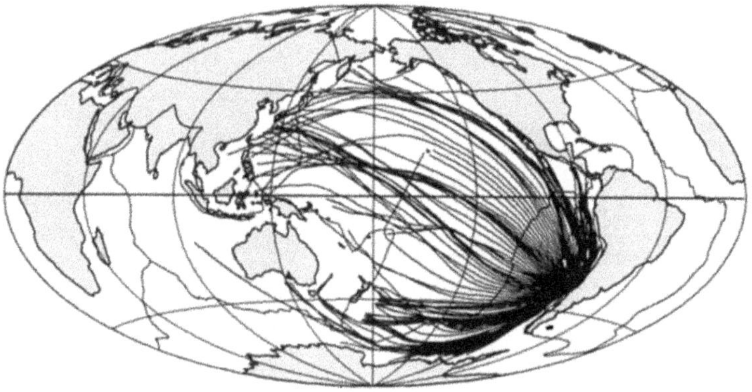

Figure 16.7. Tsunami trajectories shot at equally spaced takeoff-angle increments from the epicenter of the 1960 Chile earthquake.

trace tsunami rays within the world's ocean basins by Woods & Okal (1987) and Satake (1988). In Figure 16.7 we show the trajectories of the waves generated by the great 1960 earthquake in Chile. Note the strong focusing and defocusing of tsunami energy caused by the lateral variations in Pacific Ocean bathymetry. Regions of intense focusing, such as New Zealand, Japan, Nicaragua and Costa Rica are most likely to experience coastal devastation associated with runup effects.

For any given coastal site, we can determine the epicentral locations of potentially hazardous tsunamigenic earthquakes by exploiting the principle of source-receiver reciprocity. We consider a hypothetical tsunami source at the site itself, and trace rays to the periphery of the adjacent or surrounding ocean basin, as illustrated for the island of Tahiti in Figure 16.8. Those active seismic regions that exhibit strong "reciprocal focusing" are the most likely birthplaces of tsunamis that could affect the site in question. Tahiti, for example, is seen to be particularly susceptible to tsunamis generated by earthquakes in the Philippine and Kuril Islands and along selected regions of the South American coast. This *reciprocal map* technique was introduced by Woods & Okal (1987).

As a tsunami propagates, its local eigenfunctions U and V adiabatically adjust to the variations in water depth h, stretching vertically to fill the deep ocean where the wave travels rapidly, and compressing and increasing in amplitude over a shallower shelf, plateau or spreading ridge where the wave slows down. We illustrate these variations in the local radial structure of the tsunami mode along a cross-section of the South Pacific in Figure 16.9. As noted in Section 8.8.11, the displacement is predominantly horizon-

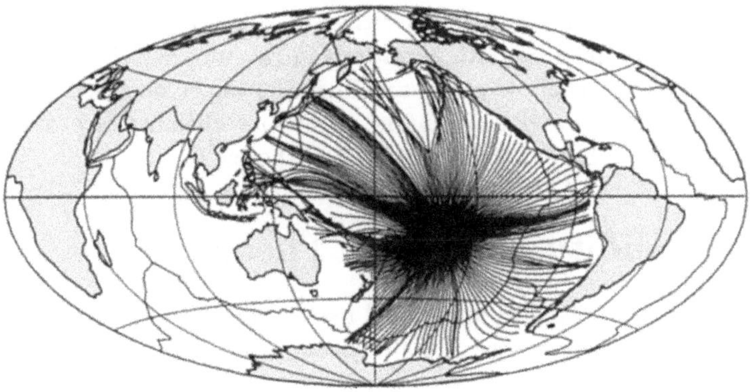

Figure 16.8. Trajectories of tsunamis generated by a fictitious earthquake on the island of Tahiti. Rays are shot at equally spaced takeoff-angle increments; the seismic reciprocity principle may be used to identify potentially dangerous tsunamigenic source regions along the Pacific rim.

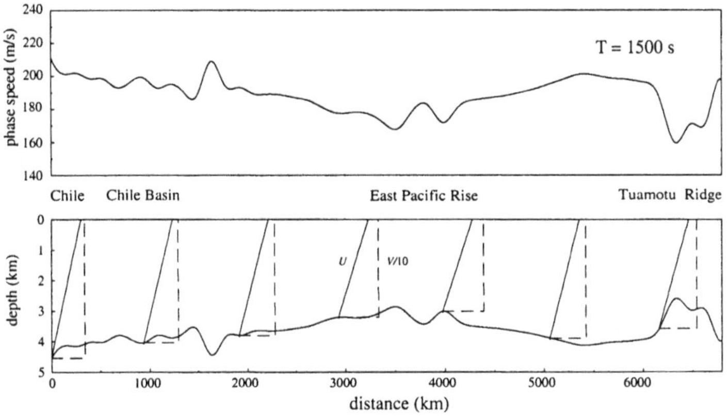

Figure 16.9. (*Top*) Shallow-water gravity-wave speed $ac = \sqrt{gh}$ along a 6500 km cross-section extending from Chile to the Tuamoto Ridge across the East Pacific Rise. (*Bottom*) Variation of the local tsunami eigenfunctions U (*solid line*) and $V/10$ (*dashed line*) along the same cross-section. The local modes adiabatically adapt themselves to the slowly varying bathymetry h; the tangential displacement is very nearly uniform throughout the water column. The quantities plotted are the JWKB-normalized ($cCI_1 = 1$) displacements for a $2\pi/\omega = 1500$ s tsunami.

tal; note that $V/10$ is plotted rather than V. The open-ocean height of a tsunami is determined by both variations in ray-tube width due to lateral refraction and the shallow-water amplification effect depicted here. Local harbor geometry and bathymetry govern its ultimate fate during the non-linear runup phase, once it reaches the shore.

*16.7 Validity of JWKB Theory

Surface-wave JWKB theory is a useful means of computing synthetic long-period seismograms, because of its physical simplicity and numerical efficiency. Like other ray-based methods, however, its applicability is subject to severe restrictions. First, as we have seen, it is invalid in the vicinity of the source and antipodal caustics, where neighboring rays cross. Second, even away from caustics, JWKB theory is valid only in the limit of a *smooth* lateral heterogeneity. The requirement that the angular wavelength $\lambda = 2\pi k^{-1}$ of the wave must be small compared to the characteristic angular scale Λ_Ω of the heterogeneity is necessary but not sufficient. A necessary and sufficient condition is that the perturbation in phase speed must not change appreciably across the width of the *first Fresnel zone* surrounding a surface-wave trajectory; this zone is the locus of single-scattering points $\hat{\mathbf{r}}''$ for which (Kravtsov & Orlov 1990)

$$|\psi(\hat{\mathbf{r}}'',\hat{\mathbf{r}}') + \psi(\hat{\mathbf{r}},\hat{\mathbf{r}}'') - \psi(\hat{\mathbf{r}},\hat{\mathbf{r}}')| \leq \pi, \tag{16.203}$$

where $\psi(\hat{\mathbf{r}}'',\hat{\mathbf{r}}') = \int_{\hat{\mathbf{r}}'}^{\hat{\mathbf{r}}''} \mathbf{k}\cdot d\hat{\mathbf{r}}$, $\psi(\hat{\mathbf{r}},\hat{\mathbf{r}}'') = \int_{\hat{\mathbf{r}}''}^{\hat{\mathbf{r}}} \mathbf{k}\cdot d\hat{\mathbf{r}}$ and $\psi(\hat{\mathbf{r}},\hat{\mathbf{r}}') = \int_{\hat{\mathbf{r}}'}^{\hat{\mathbf{r}}} \mathbf{k}\cdot d\hat{\mathbf{r}}$ are the phases accumulated along the source-scatterer, scatterer-receiver and source-receiver ray paths, respectively. The angular width of a surface-wave Fresnel zone at a typical scattering point midway between the source and receiver on a slightly heterogeneous Earth is $\delta \approx \sqrt{\lambda/2}$; a simple heuristic condition for the validity of surface-wave JWKB theory away from caustics is therefore (Wang & Dahlen 1995)

$$\sqrt{\lambda/2\Lambda_\Omega^2} \ll 1. \tag{16.204}$$

Because of the spherical geometry, the Fresnel-zone width $\delta \to 0$ at both the source-to-receiver and receiver-to-source caustics, rather than growing indefinitely as the number of orbits increases; for this reason, the restriction (16.204) is applicable to G2, G3,... and R2, R3,... waves as well as to the minor-arc arrivals G1 and R1.

The validity of surface-wave JWKB theory can also be assessed by comparing its predictions against more accurate coupled-mode calculations. Love-to-Rayleigh scattering and scattering between one Love or Rayleigh mode branch and another are neglected in the along-branch coupling scheme described in Section 14.3.7; however, diffraction and other finite-frequency wave-propagation effects that are ignored in JWKB theory are fully accounted for. In Figure 16.10, we present a comparison between the JWKB phase, arrival-angle and amplitude anomalies on model S12WM13 with the corresponding results obtained by processing synthetic single-branch accelerograms. All comparisons are carried out for 150-second fundamental-mode Rayleigh waves recorded on the radial component at fifty GSN station locations following twenty earthquakes that occurred between 1977 and 1984—a total of 1000 source-receiver paths. The accelerograms $a(t)$ are calculated using the real symmetric variational method, with each target multiplet along the ${}_0\mathrm{S}_l$ branch coupled to its nearest ± 5 neighbors. The phase and amplitude anomalies $\psi - \psi_{\mathrm{spher}\,\oplus}$ and $(A - A_{\mathrm{spher}\,\oplus})/A_{\mathrm{spher}\,\oplus}$ are measured by cross-correlation with the corresponding spherical-Earth accelerograms $a_{\mathrm{spher}\,\oplus}(t)$, using a one-hour time window centered upon the arrival of the R1 wavetrain at the period of interest. The arrival-angle anomalies $\zeta - \zeta_{\mathrm{spher}\,\oplus}$ are measured using a simple algorithm that minimizes the power on the transverse component; the north-south and east-west components are rotated to new components oriented at $\pm 45^\circ$ to the unperturbed ray-arrival direction prior to processing, to reduce numerical errors. To avoid the overlapping of neighboring wavetrains at the source and antipodal caustics, we only consider paths with source-receiver dis-

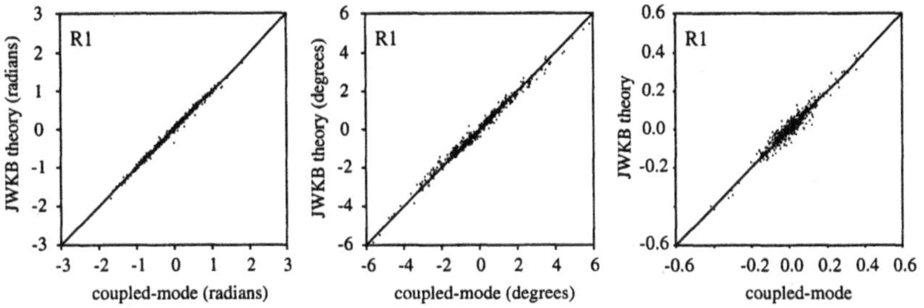

Figure 16.10. Comparison of JWKB observables with those measured by processing synthetic coupled-mode accelerograms on model S12WM13 of Su, Woodward & Dziewonski (1994). All results are for 150-second R1 Rayleigh waves in the epicentral distance range $60^\circ < \Theta < 120^\circ$; each dot represents one minor-arc path. (*Left*) Phase anomalies $\psi - \psi_{\mathrm{spher}\,\oplus}$, measured in radians. (*Middle*) Azimuthal arrival-angle anomalies $\zeta - \zeta_{\mathrm{spher}\,\oplus}$, measured in degrees, counterclockwise from due south at the receiver. (*Right*) Dimensionless (fractional) amplitude anomalies $(A - A_{\mathrm{spher}\,\oplus})/A_{\mathrm{spher}\,\oplus}$.

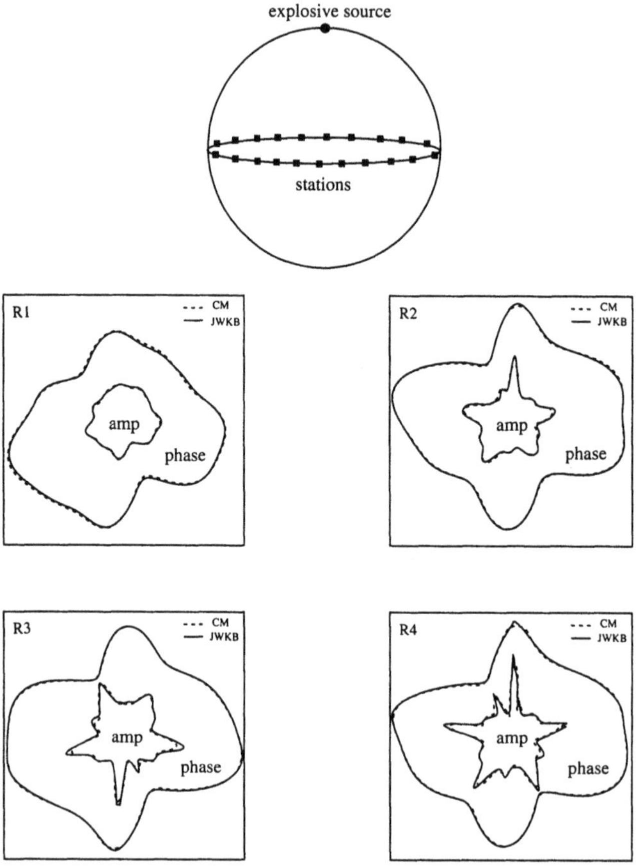

Figure 16.11. Polar plot of the phase anomaly $\psi - \psi_{\mathrm{spher}\,\oplus}$ (*outer curve*) and amplitude anomaly $A - A_{\mathrm{spher}\,\oplus}$ (*inner curve*) of 150-second R1–R4 Rayleigh waves at an epicentral distance of $\Theta = 90^\circ$ away from an explosive source located near the coast of Peru on model S12WM13. The outer curves are effectively a snapshot of a 150-second wavefront, whereas the inner curves show the variation of the amplitude along this wavefront. Cartoon (*top*) depicts the source-receiver geometry. The abbreviation CM stands for coupled-mode.

tances in the range $60^\circ < \Theta < 120^\circ$; we also discard a small number of paths having small unperturbed amplitudes compared with those at other stations for the same earthquake; these seismograms correspond to waves leaving the source near the nodes of the surface-wave radiation pattern. The final dataset consists of slightly more than 500 minor-arc paths. In general, the JWKB results are in good agreement with the coupled-mode measurements on model S12WM13, which has $s_{\rm max} = 12$ and $\Lambda_\Omega \approx 60^\circ$. The phase-anomaly predictions $\psi - \psi_{\rm spher\,\oplus}$ are more accurate than the arrival-angle predictions $\zeta - \zeta_{\rm spher\,\oplus}$, and these in turn are more accurate than the amplitude-anomaly predictions $(A - A_{\rm spher\,\oplus})/A_{\rm spher\,\oplus}$. This is an expected feature of the approximation on a slightly heterogeneous Earth, as we shall see in Section 16.8. The first-order phase, arrival-angle and amplitude perturbations $\delta\psi$, $\delta\zeta$ and $\delta A/A$ depend upon the zeroth, first and second cross-path derivatives of the phase-speed anomaly, respectively; thus, in a sense, each of these observables "sees" an increasingly "rough" image of the Earth.

There is a tendency for JWKB theory to overestimate absolute phase and amplitude anomalies $|\psi - \psi_{\rm spher\,\oplus}|$ and $|A - A_{\rm spher\,\oplus}|/A_{\rm spher\,\oplus}$. This bias is due to a finite-frequency *wavefront smoothing* phenomenon—diffraction acts to smooth out irregularities that would be present in the infinite-frequency wavefronts that are governed by ray theory. To illustrate this effect, we describe the results of a simple numerical experiment: a ring of receivers is deployed at an epicentral distance $\Theta = 90^\circ$ away from a hypothetical explosive source, with an isotropic moment tensor $\mathbf{M} = M_0\mathbf{I}$. The use of an isotropic source serves to eliminate radiation-pattern effects. Figure 16.11 compares the JWKB and coupled-mode phase and amplitude anomalies of 150-second fundamental-mode R1–R4 Rayleigh waves on model S12WM13, due to an explosion located near the coast of Peru; note that JWKB theory predicts the phase and amplitude of the radiated

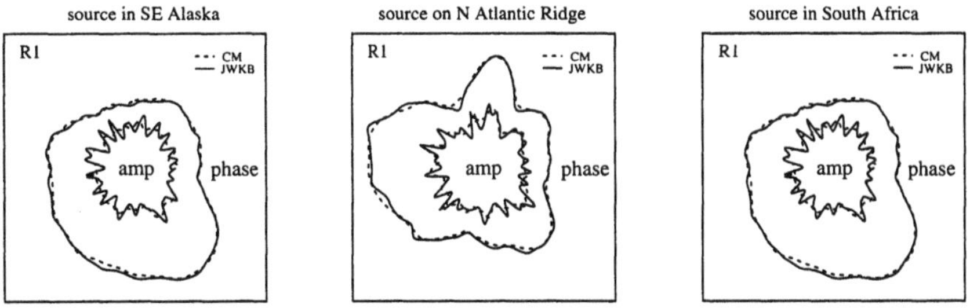

Figure 16.12. Same as Figure 16.11 for three explosive sources located in south-eastern Alaska, on the North Atlantic Ridge and in South Africa, on a moderately roughened version of model S12WM13. The 150-second R1 waves have propagated an unperturbed epicentral distance $\Theta = 90^\circ$. Finite-frequency diffraction effects have a tendency to smooth out ray-theoretical "crinkles" in the shape and—particularly—the amplitude of a wavefront.

wavefield quite accurately. To investigate the effect of rougher lateral variation on surface-wave propagation, we contrive a "model" with maximum angular degree $s_{\mathrm{max}} = 36$ and $\Lambda_\Omega \approx 45°$ by adding pseudo-random higher-degree structure to model S12WM13; the added power accounts for less than 10% of the total perturbation in the phase speed δc of a 150-second Rayleigh wave. Figure 16.12 shows a comparison of the Rayleigh waves generated by three explosions located in southeastern Alaska, on the North Atlantic Ridge and in South Africa, on this slightly rougher version of model S12WM13. The agreement between the JWKB and coupled-mode phase anomalies is not as good as for the R1 waves in Figure 16.11; both discrepancies are consistent with the heuristic validity criterion (16.204). The wavefront smoothing effect is particularly pronounced in this example for the amplitude anomaly $A - A_{\mathrm{spher}\,\oplus}$. On even rougher models, the JWKB phase anomaly $\psi - \psi_{\mathrm{spher}\,\oplus}$ would appear to be similarly "crinkled" compared to the true finite-frequency phase.

16.8 Ray Perturbation Theory

Suppose now that the lateral heterogeneity of the Earth is *slight as well as smooth*, so that the phase speed of a Love or Rayleigh wave can be treated as a first-order perturbation away from the uniform speed upon a spherically symmetric Earth:

$$c \to c + \delta c \quad \text{where} \quad |\delta c| \ll c. \tag{16.205}$$

The rays upon such an Earth model will deviate slightly from the great-circular rays upon a spherically symmetric Earth; as a result, the amplitude and phase of a surface wave will be perturbed:

$$A \to A + \delta A, \qquad \psi \to \psi + \delta\psi, \tag{16.206}$$

where

$$\frac{\delta A}{A} = \frac{\delta A_{\mathrm{r}}}{A_{\mathrm{r}}} + \frac{\delta A_{\mathrm{p}}}{A_{\mathrm{p}}} + \frac{\delta A_{\mathrm{s}}}{A_{\mathrm{s}}}, \tag{16.207}$$

$$\delta\psi = \delta\psi_{\mathrm{r}} + \delta\psi_{\mathrm{p}} + \delta\psi_{\mathrm{s}}. \tag{16.208}$$

The subscripts s, p and r identify the terms associated with the source, path and receiver, respectively. A surface-wave analogue of the *ray perturbation theory* developed in Section 15.9 can be used to find the geometry of the perturbed rays, as well as the first-order amplitude and phase perturbations $\delta A_{\mathrm{s}}/A_{\mathrm{s}}$, $\delta A_{\mathrm{p}}/A_{\mathrm{p}}$, $\delta A_{\mathrm{r}}/A_{\mathrm{r}}$ and $\delta\psi_{\mathrm{s}}$, $\delta\psi_{\mathrm{p}}$, $\delta\psi_{\mathrm{r}}$. We follow our usual custom, and employ a prime and a subscript s interchangeably to identify quantities such as $\zeta' = \zeta_{\mathrm{s}}$ evaluated at the source.

16.8.1 Fermat phase

As in the body-wave case, Fermat's principle enables us to determine the first-order perturbation in the accumulated phase between the source and receiver by integration along the *unperturbed great-circular ray path*:

$$\delta\psi_{\mathrm{p}} = \int_0^{\Delta} \delta k\, d\phi = -\omega c^{-2} \int_0^{\Delta} \delta c\, d\phi, \tag{16.209}$$

where the source and receiver are presumed to be situated on the equator, as before. The result (16.209) is valid for all G1, G2,... and R1, R2,... waves; the integration is over the full ray path of length $\Theta, 2\pi - \Theta, \ldots$ in the case of a multi-orbit wave. Figure 16.13 compares the first-order Fermat phase delays $\delta\psi_{\mathrm{p}}$ of a number of first-arriving and second-arriving Love

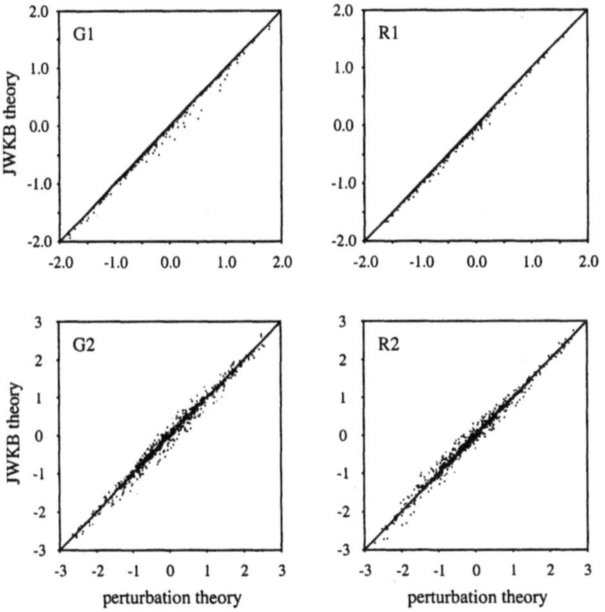

Figure 16.13. Scatter-plot comparison of the first-order path phase anomaly ψ_{p} with the exact JWKB phase anomaly $\psi_{\mathrm{p}} - \psi_{\mathrm{p}}^{\mathrm{spher}\,\oplus}$ for 165-second fundamental-mode surface waves on model S12WM13. Each of the 850 points represents a single G1, R1, G2 or R2 path. The phase anomalies are plotted in radians. There is an obvious Fermat bias for the first-arriving G1 and R1 waves (*top*), but none for the later G2 and R2 arrivals that have passed through an antipodal caustic (*bottom*).

and Rayleigh waves with the exact or true-path phase delays $\psi_{\mathrm{p}} - \psi_{\mathrm{p}}^{\mathrm{spher}\,\oplus}$ calculated by means of two-point ray tracing. The exact delays of all G1 and R1 waves are always less than the approximate Fermat delays,

$$\psi_{\mathrm{p}} - \psi_{\mathrm{p}}^{\mathrm{spher}\,\oplus} < \delta\psi_{\mathrm{p}}, \tag{16.210}$$

in accordance with the variational principle—the true ray path followed by a minor-arc wave is the minimum-phase path. The stationary ray paths of G2, R2 and other higher-orbit waves are not minimum-phase; there is thus no such *Fermat bias* for these minimax arrivals.

16.8.2 Fictitious frequency and source shift

In the so-called *path-average* or *great-circle approximation* to the moment-tensor response (16.163)–(16.164), the perturbation to the spherical-Earth amplitude $A = A_{\mathrm{r}}A_{\mathrm{p}}A_{\mathrm{s}}$ is ignored, and only the path contribution (16.209) to the first-order perturbation in the phase $\psi_{\mathrm{r}} + \psi_{\mathrm{p}} + \psi_{\mathrm{s}}$ is accounted for:

$$\delta A = 0, \qquad \delta\psi = \delta\psi_{\mathrm{p}}. \tag{16.211}$$

We can rewrite the first-order phase delay (16.209) for the odd-arriving and even-arriving waves in terms of the minor-arc and full great-circular averages of δk in the form

$$\delta\psi = \begin{cases} \delta\hat{k}\,\Theta + \delta\bar{k}\,(s-1)\pi, & s \text{ odd} \\ \delta\bar{k}\,s\pi - \delta\hat{k}\,\Theta, & s \text{ even,} \end{cases} \tag{16.212}$$

where

$$\delta\hat{k} = \frac{1}{\Theta}\int_0^\Theta \delta k\,d\phi, \qquad \delta\bar{k} = \frac{1}{2\pi}\oint \delta k\,d\phi. \tag{16.213}$$

This travelling-wave phase shift may alternatively be expressed in terms of a fictitious perturbation in the wavenumber $\delta\bar{k}$ and epicentral distance $\delta\Theta$ *upon a spherical Earth* in the form

$$\delta\psi = \begin{cases} \delta\bar{k}\,\Delta + k\,\delta\Theta, & s \text{ odd} \\ \delta\bar{k}\,\Delta - k\,\delta\Theta, & s \text{ even.} \end{cases} \tag{16.214}$$

Upon equating (16.212) and (16.214) we deduce that

$$\frac{\delta\Theta}{\Theta} = -\frac{\delta\hat{k} - \delta\bar{k}}{k} = \frac{\delta\hat{c} - \delta\bar{c}}{c} = \frac{c}{C}\frac{\delta\hat{\omega} - \delta\bar{\omega}}{\omega}. \tag{16.215}$$

The quantities $\delta\hat{\omega}$ and $\delta\bar{\omega}$ are the minor-arc and great-circular averages of the local perturbation $\delta\omega$ in the eigenfrequency of the associated normal mode or standing wave:

$$\delta\hat{\omega} = \frac{1}{\Theta}\int_0^\Theta \delta\omega\, d\phi, \qquad \delta\bar{\omega} = \frac{1}{2\pi}\oint \delta\omega\, d\phi. \tag{16.216}$$

These results enable us to calculate an approximate synthetic accelerogram upon a laterally heterogeneous Earth by means of a modest modification to the mode-summation procedure employed upon a spherically symmetric Earth (Woodhouse & Dziewonski 1984). It is simply necessary to replace the eigenfrequency ω of each mode and the epicentral distance Θ in equations (10.51)–(10.60) by $\omega + \delta\bar{\omega}$ and $\Theta + \delta\Theta$, respectively. Note that this must be done on both a mode-by-mode and a path-by-path basis; nevertheless, the method is extremely efficient, requiring very little more time than the corresponding summation on a spherical Earth. Even though the fictitious source shift (16.215) is determined by requiring that the first-order phase perturbations (16.212) and (16.214) must match for all possible orbits, the great-circle approximation is fundamentally a *short-time approximation*, because of the tendency of the higher-orbit ray paths to diverge farther and farther from the unperturbed great-circle (see Sections 14.3.7 and 16.8.4).

★16.8.3 Ellipticity correction

We demonstrated in Section 14.2.2 that the combined effect of the Earth's hydrostatic ellipticity ε and aspherical centrifugal potential $\psi - \bar{\psi}$ upon an isolated normal-mode multiplet is governed by a diagonal $(2l+1)\times(2l+1)$ splitting matrix of the form

$$H^{\mathrm{ell}}_{mm'} = \omega a^{\mathrm{ell}}(1 - 3m^2/k^2)\,\delta_{mm'}, \tag{16.217}$$

where ω is the unperturbed eigenfrequency and $k = \sqrt{l(l+1)}$. The quantity a^{ell} is the dimensionless ellipticity splitting parameter, given by

$$\begin{aligned} a^{\mathrm{ell}} &= \frac{l(l+1)}{2(2l+3)(2l-1)\omega^2} \\ &\quad\times \int_0^a \tfrac{2}{3}\varepsilon\Big\{\kappa\big[\bar{V}_\kappa - (\eta+1)\check{V}_\kappa\big] + \mu\big[\bar{V}_\mu - (\eta+1)\check{V}_\mu\big] \\ &\qquad + \rho\big[(\bar{V}_\rho - \omega^2\bar{T}_\rho) - (\eta+3)(\check{V}_\rho - \omega^2\check{T}_\rho\big]\Big\} r^2\,dr, \end{aligned} \tag{16.218}$$

where $\eta = r\dot{\varepsilon}/\varepsilon$, and $\bar{V}_\kappa$, $\check{V}_\kappa$, $\bar{V}_\mu$, $\check{V}_\mu$ and $\bar{V} - \omega^2\bar{T}_\rho$, $\check{V}_\rho - \omega^2\check{T}_\rho$ are the elliptical incompressibility, rigidity and density kernels defined in equations

(D.183)–(D.189). In the present context, it is more convenient to regard the ellipticity as a degree-two zonal perturbation to either the local eigenfrequency ω of an $n = 0, 1, 2, \ldots$ toroidal or spheroidal mode or the phase speed c of the equivalent Love or Rayleigh wave:

$$\delta\omega^{\rm ell} = \delta\omega_{20}^{\rm ell} X_{20}(\theta), \qquad \delta c^{\rm ell} = \delta c_{20}^{\rm ell} X_{20}(\theta), \tag{16.219}$$

where $X_{20}(\theta) = \frac{1}{4}\sqrt{5/\pi}\,(3\cos^2\theta - 1)$ and $\delta c_{20}^{\rm ell}/c = (c/C)(\delta\omega_{20}^{\rm ell}/\omega)$ as usual. We can relate the coefficients in (16.219) to the ellipticity splitting parameter (16.218) by making use of the asymptotic representation (14.141) of the splitting matrix (16.217):

$$H_{mm'}^{\rm ell} \approx \tfrac{1}{8}\sqrt{5/\pi}\,\delta\omega_{20}^{\rm ell}\,(1 - 3m^2/k^2)\,\delta_{mm'}. \tag{16.220}$$

This approximation is valid provided that $k \gg 1$. Upon comparing (16.217) and (16.220) we find that

$$\delta\omega_{20}^{\rm ell} = 4\sqrt{4\pi/5}\,\omega a^{\rm ell}, \qquad \delta c_{20}^{\rm ell} = 4\sqrt{4\pi/5}\,(c^2/C)a^{\rm ell}. \tag{16.221}$$

Prior to computing synthetic JWKB accelerograms (16.161) on a hydrostatic elliptical, laterally heterogeneous Earth, it is necessary to convert the event and station locations from geographic to geocentric coordinates as discussed in Section 14.1.5, in addition to accounting for the degree-two zonal contribution $\delta c^{\rm ell} = 4\sqrt{4\pi/5}\,(c^2/C)a^{\rm ell}X_{20}(\theta)$ to the surface-wave phase speed. Ellipticity can be accounted for in the Woodhouse-Dziewonski great-circle approximation by adding terms

$$\delta\bar{\omega}^{\rm ell} = \omega a^{\rm ell}(1 - 3\cos^2\bar{\theta}), \tag{16.222}$$

$$\delta\Theta^{\rm ell} = -3(c/C)a^{\rm ell}\sin\Theta\sin^2\bar{\theta}\cos 2\phi_{\rm mp} \tag{16.223}$$

to the fictitious frequency and source shift of each spherical-Earth mode. Equation (16.223) is the result of substituting the perturbation (16.219) into (16.215); the angle $\phi_{\rm mp}$ is the azimuth, measured counterclockwise from due south, of the midpoint of the minor arc from the pole $\bar{\theta}, \bar{\phi}$ of the source-receiver great circle.

⋆16.8.4 Ray geometry

To determine the perturbation in the geometry of a G1 or R1 surface-wave trajectory, we make the following substitutions in the kinematic ray-tracing equations (16.185) and (16.186):

$$\theta \to \pi/2 + \delta\theta, \qquad \zeta \to \pi/2 + \delta\zeta. \tag{16.224}$$

The source and receiver are assumed to be situated on the equator at longitudes $\phi = 0$ and $\phi = \Theta$, respectively, so that the unperturbed ray path is given by (16.200). Correct to first order in the small perturbations $\delta\theta$ and $\delta\zeta$, we obtain a system of linear equations governing the perturbed ray:

$$\frac{d}{d\phi}\delta\theta = -\delta\zeta, \qquad \frac{d}{d\phi}\delta\zeta = \delta\theta + c^{-1}\partial_\theta\delta c. \tag{16.225}$$

Equations (16.225) may be rewritten using a 2×2 matrix notation analogous to (15.279):

$$\frac{d\mathsf{y}}{d\phi} = \mathsf{A}\mathsf{y} + \mathsf{f}, \tag{16.226}$$

where

$$\mathsf{y} = \begin{pmatrix} \delta\theta \\ \delta\zeta \end{pmatrix}, \qquad \mathsf{f} = \begin{pmatrix} 0 \\ c^{-1}\delta c \end{pmatrix}, \qquad \mathsf{A} = \begin{pmatrix} 0 & -1 \\ 1 & 0 \end{pmatrix}. \tag{16.227}$$

The coefficient matrix A is identical, and so therefore is the propagator:

$$\mathsf{P}(\phi,\tilde{\phi}) = \begin{pmatrix} \cos(\phi-\tilde{\phi}) & -\sin(\phi-\tilde{\phi}) \\ \sin(\phi-\tilde{\phi}) & \cos(\phi-\tilde{\phi}) \end{pmatrix}. \tag{16.228}$$

The solution to (16.226) can be written in terms of the propagator (16.228) in the form (15.282):

$$\mathsf{y}(\phi) = \int_0^\phi \mathsf{P}(\phi,\tilde{\phi})\,\mathsf{f}(\tilde{\phi})\,d\tilde{\phi} + \mathsf{P}(\phi,0)\,\mathsf{y}(0). \tag{16.229}$$

The endpoint conditions stipulate that the perturbed ray must emanate from the same source and hit the same receiver as the unperturbed ray:

$$\mathsf{y}(0) = \begin{pmatrix} 0 \\ \delta\zeta' \end{pmatrix}, \qquad \mathsf{y}(\Theta) = \begin{pmatrix} 0 \\ \delta\zeta \end{pmatrix}, \tag{16.230}$$

where $\delta\zeta'$ and $\delta\zeta$ denote the perturbed takeoff and arrival angles, respectively. Upon inserting (16.230) into (16.229) we find that

$$\delta\zeta' = -(c\sin\Theta)^{-1}\int_0^\Theta \sin(\Theta-\phi)\,\partial_\theta\delta c\,d\phi, \tag{16.231}$$

$$\delta\zeta = (c\sin\Theta)^{-1}\int_0^\Theta \sin\phi\,\partial_\theta\delta c\,d\phi. \tag{16.232}$$

Equations (16.231) and (16.232) are the surface-wave analogues of (15.287) and (15.288). An interchange $\phi \longleftrightarrow \Theta - \phi$ of the source and receiver interchanges the angles $\delta\zeta \longleftrightarrow \delta\zeta'$, as expected. The complete solution at intermediate points $0 \leq \phi \leq \Theta$ analogous to (15.289) and (15.290) is

$$\delta\theta(\phi) = -c^{-1} \int_0^\phi \sin(\phi - \tilde{\phi})\, \partial_\theta \delta\tilde{c}\, d\tilde{\phi} - \delta\zeta' \sin\phi, \tag{16.233}$$

$$\delta\zeta(\phi) = c^{-1} \int_0^\phi \cos(\phi - \tilde{\phi})\, \partial_\theta \delta\tilde{c}\, d\tilde{\phi} + \delta\zeta' \cos\phi, \tag{16.234}$$

where the tildes denote evaluation at the dummy integration variable $\tilde{\phi}$. Note that $\delta\theta(0) = \delta\theta(\Theta) = 0$ whereas $\delta\zeta(0) = \delta\zeta'$ and $\delta\zeta(\Theta) = \delta\zeta$, as expected. The first-order takeoff-angle perturbation (16.231) provides a useful zeroth iterate $\zeta_0' = \pi/2 + \delta\zeta'$ in exact two-point ray tracing, based upon equation (16.189).

Figure 16.14 compares the first-order takeoff-angle and arrival-angle perturbations (16.231)–(16.232) to the exact JWKB azimuthal deflections

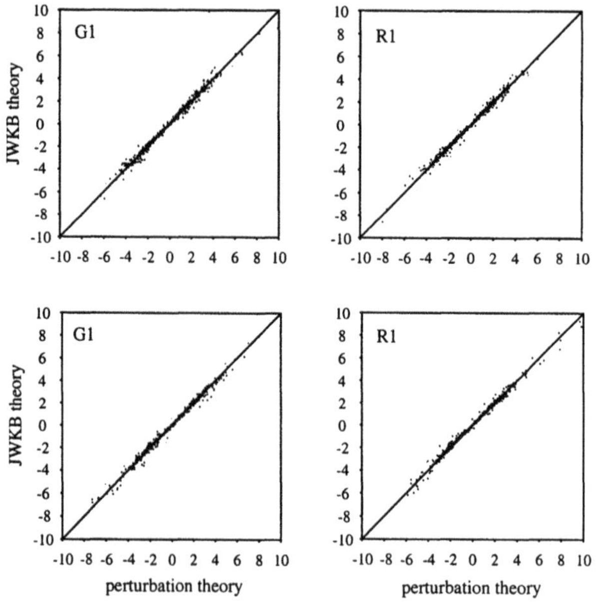

Figure 16.14. Scatter-plot comparison of the first-order azimuthal anomalies with the exact JWKB deflections suffered by 165-second fundamental-mode surface waves on model S12WM13. (*Top*) Takeoff-angle anomalies $\delta\zeta'$ versus $\zeta' - \zeta'_{\mathrm{spher}\,\oplus}$. (*Bottom*) Arrival-angle anomalies $\delta\zeta$ versus $\zeta - \zeta_{\mathrm{spher}\,\oplus}$. Each of the 850 points represents a single G1 or R1 path. The anomalies are measured in degrees.

$\zeta' - \zeta'_{\rm spher\,\oplus}$ and $\zeta - \zeta_{\rm spher\,\oplus}$ suffered by 165-second R1 and G1 waves, for a number of source-receiver combinations on model S12WM13. Perturbation theory provides a reasonably accurate prediction of the majority of the azimuthal deviations, which are in the range $|\zeta' - \zeta'_{\rm spher\,\oplus}| < 5°\!-\!6°$ and $|\zeta - \zeta_{\rm spher\,\oplus}| < 5°\!-\!6°$; the error is larger for a relatively small number of more significantly deflected rays. The largest deflections occur along paths having large transverse phase-speed gradients $\partial_\theta \delta c$, in accordance with equations (16.231)–(16.232).

The above results are easily generalized to later-arriving surface waves. The observable arrival-angle anomaly of a G1, G2,... or R1, R2,... wave at the receiver is given by (Woodhouse & Wong 1986)

$$\delta\zeta = \begin{cases} \delta\hat{\zeta}\,\Theta + \delta\bar{\zeta}\,(s-1)\pi, & s \text{ odd} \\ \delta\hat{\zeta}\,\Theta - \delta\bar{\zeta}\,s\pi, & s \text{ even,} \end{cases} \tag{16.235}$$

where $\delta\hat{\zeta}$ and $\delta\bar{\zeta}$ are the minor-arc and great-circular averages of the quantity $(c\sin\Theta)^{-1}\sin\phi\,\partial_\theta\delta c$. The arrival azimuth of an incoming wave is seen to increase linearly from one orbit to the next; a positive counterclockwise perturbation, $\delta\zeta > 0$, corresponds to an incoming $s = 1, 3, \ldots$ wave that is propagating in a direction slightly north of due east or to an incoming $s = 2, 4, \ldots$ wave that is propagating in a direction slightly south of due west, so that the cumulative sense of the azimuthal deviation is opposite for the odd and even orbits. The takeoff-angle anomaly at the source exhibits an analogous odd-left, even-right or odd-right, even-left dependence upon the orbit number:

$$\delta\zeta' = \begin{cases} \delta\hat{\zeta}'\,\Theta + \delta\bar{\zeta}'\,(s-1)\pi, & s \text{ odd} \\ \delta\hat{\zeta}'\,\Theta - \delta\bar{\zeta}'\,s\pi, & s \text{ even,} \end{cases} \tag{16.236}$$

where $\delta\hat{\zeta}'$ and $\delta\bar{\zeta}'$ are the minor-arc and great-circular averages of the quantity $(c\sin\Theta)^{-1}\sin(\Theta-\phi)\,\partial_\theta\delta c$. The results (16.235) and (16.236) are only correct to first order in the phase-speed perturbation δc; however, the tendency for the odd- and even-arriving waves to diverge and travel along systematically different ray paths is confirmed by exact ray tracing.

*16.8.5 Geometrical spreading

Recalling that the partial derivatives $\partial_{\zeta'}\theta$ and $\partial_{\zeta'}\zeta$ on a spherically symmetric Earth are given by equation (16.202), we stipulate that on a slightly heterogeneous Earth

$$\partial_{\zeta'}\theta \to -\sin\phi + \delta(\partial_{\zeta'}\theta), \qquad \partial_{\zeta'}\zeta \to \cos\phi + \delta(\partial_{\zeta'}\zeta). \tag{16.237}$$

The first-order perturbation in the geometrical amplitude of a G1 or R1 surface wave is found from (16.166) and (16.195) to be

$$\frac{\delta A_{\rm p}}{A_{\rm p}} = -\tfrac{1}{2}\left(\frac{\delta k}{k} + \frac{\delta S}{S}\right) = -\tfrac{1}{2}\left[\frac{\delta k}{k} - \frac{\delta(\partial_{\zeta'}\theta)}{\sin\Theta}\right]. \tag{16.238}$$

To compute the quantity $\delta(\partial_{\zeta'}\theta)$, we need to perturb the dynamical ray-tracing equations (16.196)–(16.197). Upon making use of (16.224) as well as (16.237), we obtain a linear system of equations governing the perturbed partial derivatives:

$$\frac{d}{d\phi}\delta(\partial_{\zeta'}\theta) = -\delta(\partial_{\zeta'}\zeta), \tag{16.239}$$

$$\frac{d}{d\phi}\delta(\partial_{\zeta'}\zeta) = \delta(\partial_{\zeta'}\theta) + c^{-1}(-\sin\phi\,\partial_\theta^2\delta c + \cos\phi\,\partial_\phi\delta c). \tag{16.240}$$

To solve (16.239)–(16.240) we note that it can be written in the form

$$\frac{d\mathsf{y}}{d\phi} = \mathsf{A}\mathsf{y} + \mathsf{f}, \tag{16.241}$$

where the 2×2 matrix A is again given in equation (16.227), and where

$$\mathsf{y} = \begin{pmatrix} \delta(\partial_{\zeta'}\theta) \\ \delta(\partial_{\zeta'}\zeta) \end{pmatrix}, \qquad \mathsf{f} = \begin{pmatrix} 0 \\ c^{-1}(-\sin\phi\,\partial_\theta^2\delta c + \cos\phi\,\partial_\phi\delta c) \end{pmatrix}. \tag{16.242}$$

The perturbed initial conditions (16.198) are $\delta(\partial_{\zeta'}\theta)(0) = 0$, $\delta(\partial_{\zeta'}\zeta)(0) = 0$ or, equivalently, $\mathsf{y}(0) = 0$. Making use of (16.229) we find that

$$\delta(\partial_{\zeta'}\theta)(\phi) = -c^{-1}\int_0^\phi \sin(\phi - \tilde{\phi})(-\sin\tilde{\phi}\,\partial_\theta^2\delta\tilde{c} + \cos\tilde{\phi}\,\partial_\phi\delta\tilde{c})\,d\tilde{\phi}, \tag{16.243}$$

$$\delta(\partial_{\zeta'}\zeta)(\phi) = c^{-1}\int_0^\phi \cos(\phi - \tilde{\phi})(-\sin\tilde{\phi}\,\partial_\theta^2\delta\tilde{c} + \cos\tilde{\phi}\,\partial_\phi\delta\tilde{c})\,d\tilde{\phi}, \tag{16.244}$$

where the tildes denote evaluation at the dummy integration variable $\tilde{\phi}$ as before. Upon evaluating the result (16.243) at the endpoint $\phi = \Theta$ and substituting into equation (16.238), we obtain a final explicit expression for the amplitude perturbation due to focusing and defocusing of the ray tube:

$$\frac{\delta A_{\mathrm{p}}}{A_{\mathrm{p}}} = \frac{\delta c' + \delta c}{2c} \qquad (16.245)$$
$$+ (2c\sin\Theta)^{-1} \int_0^{\Theta} [\sin(\Theta-\phi)\sin\phi\,\partial_\theta^2\delta c - \cos(\Theta-2\phi)\,\delta c]\,d\phi.$$

Both of the kernels $\sin(\Theta-\phi)\sin\phi$ and $\cos(\Theta-2\phi)$ are symmetric about the midpoint of the ray path, so that the amplitude perturbation (16.245) is invariant under an interchange $0 \longleftrightarrow \Theta$; we have integrated the term involving the along-path gradient $\partial_\phi\delta c$ by parts in order to expose this *source-receiver reciprocity*. Propagation in a low-speed channel, $\partial_\theta^2\delta c > 0$, leads to focusing and amplification, $\delta A_{\mathrm{p}} > 0$, whereas propagation in a high-speed channel, $\partial_\theta^2\delta c < 0$, leads to defocusing and deamplification, $\delta A_{\mathrm{p}} < 0$, as we might expect.

To extend (16.245) to higher-orbit G2, G3,... and R2, R3,... waves, it is simply necessary to replace the epicentral distance Θ by the distance $\Delta = 2\pi - \Theta, 2\pi + \Theta, \ldots$ which the wave has propagated. We can write this result in a manner analogous to (16.209) and (16.235), in order to highlight the dependence of the amplitude upon orbit number:

$$\frac{\delta A_{\mathrm{p}}}{A_{\mathrm{p}}} = \begin{cases} \frac{\delta c'+\delta c}{2c} + \frac{\delta\hat{A}_{\mathrm{p}}}{A_{\mathrm{p}}}\Theta + \frac{\delta\bar{A}_{\mathrm{p}}}{A_{\mathrm{p}}}(s-1)\pi, & s \text{ odd} \\ \frac{\delta c'+\delta c}{2c} + \frac{\delta\hat{A}_{\mathrm{p}}}{A_{\mathrm{p}}}\Theta - \frac{\delta\bar{A}_{\mathrm{p}}}{A_{\mathrm{p}}}\,s\pi, & s \text{ even,} \end{cases} \qquad (16.246)$$

where $\delta\hat{A}_{\mathrm{p}}/A_{\mathrm{p}}$ and $\delta\bar{A}_{\mathrm{p}}/A_{\mathrm{p}}$ are the minor-arc and great-circular averages of the integrand in equation (16.245). The result (16.246) provides a first-order explanation of the frequently observed alternation of high-amplitude and low-amplitude multi-orbit surface-wave arrivals on long-period seismograms, first noted by Lay & Kanamori (1985). The actual dependence of amplitude upon orbit number is more complicated, because the validity of ray perturbation theory diminishes with propagation distance, as we have seen; nevertheless, there is a general tendency for the $s = 2, 4, \ldots$ waves to be amplified whenever the $s = 1, 3, \ldots$ waves are deamplified, and vice versa.

*16.8.6 Initial amplitude and phase

The source and receiver contributions to the phase and amplitude perturbations $\delta\psi$ and δA can be determined from equations (16.165) and (16.167):

$$\delta\psi_{\mathrm{s}} = \mathrm{Im}\left[\frac{\mathbf{M}:\delta\mathbf{E}_{\mathrm{s}}^*}{\mathbf{M}:\mathbf{E}_{\mathrm{s}}^*}\right], \qquad \frac{\delta A_{\mathrm{s}}}{A_{\mathrm{s}}} = \mathrm{Re}\left[\frac{\mathbf{M}:\delta\mathbf{E}_{\mathrm{s}}^*}{\mathbf{M}:\mathbf{E}_{\mathrm{s}}^*}\right], \qquad (16.247)$$

$$\delta\psi_{\mathrm{r}} = \mathrm{Im}\left[\frac{\hat{\boldsymbol{\nu}}\cdot\delta\mathbf{s}}{\hat{\boldsymbol{\nu}}\cdot\mathbf{s}}\right], \qquad \frac{\delta A_{\mathrm{r}}}{A_{\mathrm{r}}} = \mathrm{Re}\left[\frac{\hat{\boldsymbol{\nu}}\cdot\delta\mathbf{s}}{\hat{\boldsymbol{\nu}}\cdot\mathbf{s}}\right], \tag{16.248}$$

where we have let $\mathbf{s} = \hat{\mathbf{r}}U - i\hat{\mathbf{k}}V + i(\hat{\mathbf{r}}\times\hat{\mathbf{k}})W$. In perturbing the conjugate source strain $\mathbf{E}_{\mathrm{s}}^*$ and receiver displacement $\mathbf{s}$, we shall ignore the radial eigenfunction perturbations δU, δV and δW, since they cannot be computed in closed form by a simple application of Rayleigh's principle. The geometrical perturbations in the tangential source and receiver polarization vectors $\hat{\mathbf{k}}_{\mathrm{s}}$, $\hat{\mathbf{r}}_{\mathrm{s}}\times\hat{\mathbf{k}}_{\mathrm{s}}$ and $\hat{\mathbf{k}}$, $\hat{\mathbf{r}}\times\hat{\mathbf{k}}$ are given in terms of the takeoff-angle and arrival-angle perturbations $\delta\zeta_{\mathrm{s}}$ and $\delta\zeta$ by

$$\delta\hat{\mathbf{k}}_{\mathrm{s}} = \delta\zeta_{\mathrm{s}}(\hat{\mathbf{r}}_{\mathrm{s}}\times\hat{\mathbf{k}}_{\mathrm{s}}), \qquad \delta(\hat{\mathbf{r}}_{\mathrm{s}}\times\hat{\mathbf{k}}_{\mathrm{s}}) = -\delta\zeta_{\mathrm{s}}\,\hat{\mathbf{k}}_{\mathrm{s}}, \tag{16.249}$$

$$\delta\hat{\mathbf{k}} = \delta\zeta(\hat{\mathbf{r}}\times\hat{\mathbf{k}}), \qquad \delta(\hat{\mathbf{r}}\times\hat{\mathbf{k}}) = -\delta\zeta\,\hat{\mathbf{k}}. \tag{16.250}$$

For Love waves we find that

$$\begin{aligned}\mathbf{M}:\boldsymbol{\delta}\mathbf{E}_{\mathrm{s}}^* &= i(\partial_r W_{\mathrm{s}} - r_{\mathrm{s}}^{-1}W_{\mathrm{s}})(M_{r\theta}\cos\zeta_{\mathrm{s}} + M_{r\phi}\sin\zeta_{\mathrm{s}})\,\delta\zeta_{\mathrm{s}}\\ &\quad - k_{\mathrm{s}}r_{\mathrm{s}}^{-1}W_{\mathrm{s}}[(M_{\theta\theta}-M_{\phi\phi})\cos 2\zeta_{\mathrm{s}} + 2M_{\theta\phi}\sin 2\zeta_{\mathrm{s}}]\,\delta\zeta_{\mathrm{s}}\\ &\quad - r_{\mathrm{s}}^{-1}W_{\mathrm{s}}[\tfrac{1}{2}(M_{\theta\theta}-M_{\phi\phi})\sin 2\zeta_{\mathrm{s}} - M_{\theta\phi}\cos 2\zeta_{\mathrm{s}}]\,\delta k_{\mathrm{s}},\end{aligned} \tag{16.251}$$

$$\delta\psi_{\mathrm{r}} = 0, \qquad \frac{\delta A_{\mathrm{r}}}{A_{\mathrm{r}}} = -\left[\frac{\hat{\boldsymbol{\nu}}\cdot\hat{\mathbf{k}}}{\hat{\boldsymbol{\nu}}\cdot(\hat{\mathbf{r}}\times\hat{\mathbf{k}})}\right]\delta\zeta, \tag{16.252}$$

whereas for Rayleigh waves

$$\begin{aligned}\mathbf{M}:\boldsymbol{\delta}\mathbf{E}_{\mathrm{s}}^* &= i(\partial_r V_{\mathrm{s}} - r_{\mathrm{s}}^{-1}V_{\mathrm{s}} + k_{\mathrm{s}}r_{\mathrm{s}}^{-1}U_{\mathrm{s}})(M_{r\phi}\cos\zeta_{\mathrm{s}} - M_{r\theta}\sin\zeta_{\mathrm{s}})\,\delta\zeta_{\mathrm{s}}\\ &\quad - k_{\mathrm{s}}r_{\mathrm{s}}^{-1}V_{\mathrm{s}}[2M_{\theta\phi}\cos 2\zeta_{\mathrm{s}} - (M_{\theta\theta}-M_{\phi\phi})\sin 2\zeta_{\mathrm{s}}]\,\delta\zeta_{\mathrm{s}}\\ &\quad - [\tfrac{1}{2}(M_{\theta\theta}+M_{\phi\phi})r_{\mathrm{s}}^{-1}V_{\mathrm{s}} - ir_{\mathrm{s}}^{-1}U_{\mathrm{s}}(M_{r\phi}\sin\zeta_{\mathrm{s}} + M_{r\theta}\cos\zeta_{\mathrm{s}})]\,\delta k_{\mathrm{s}}\\ &\quad - r_{\mathrm{s}}^{-1}V_{\mathrm{s}}[M_{\theta\phi}\sin 2\zeta_{\mathrm{s}} + \tfrac{1}{2}(M_{\theta\theta}-M_{\phi\phi})\cos 2\zeta_{\mathrm{s}}]\,\delta k_{\mathrm{s}},\end{aligned} \tag{16.253}$$

$$\delta\psi_{\mathrm{r}} = -\mathrm{Im}\left[\frac{\hat{\boldsymbol{\nu}}\cdot iV(\hat{\mathbf{r}}\times\hat{\mathbf{k}})}{\hat{\boldsymbol{\nu}}\cdot(U\hat{\mathbf{r}} - iV\hat{\mathbf{k}})}\right]\delta\zeta, \tag{16.254}$$

$$\frac{\delta A_{\mathrm{r}}}{A_{\mathrm{r}}} = -\mathrm{Re}\left[\frac{\hat{\boldsymbol{\nu}}\cdot iV(\hat{\mathbf{r}}\times\hat{\mathbf{k}})}{\hat{\boldsymbol{\nu}}\cdot(U\hat{\mathbf{r}} - iV\hat{\mathbf{k}})}\right]\delta\zeta. \tag{16.255}$$

If the polarization of the receiver is either radial ($\hat{\boldsymbol{\nu}} = \hat{\mathbf{r}}$) or horizontally coincident with the unperturbed polarization ($\hat{\boldsymbol{\nu}} = \hat{\mathbf{r}}\times\hat{\mathbf{k}}$ for a Love wave and $\hat{\boldsymbol{\nu}} = \hat{\mathbf{k}}$ for a Rayleigh wave) there is *no first-order receiver phase or amplitude perturbation*; this will of course normally be the case in any observational analysis. The dominant contributions to the source perturbations $\delta\psi_{\mathrm{s}}$ and $\delta A_{\mathrm{s}}/A_{\mathrm{s}}$ are due to the phenomenon illustrated schematically

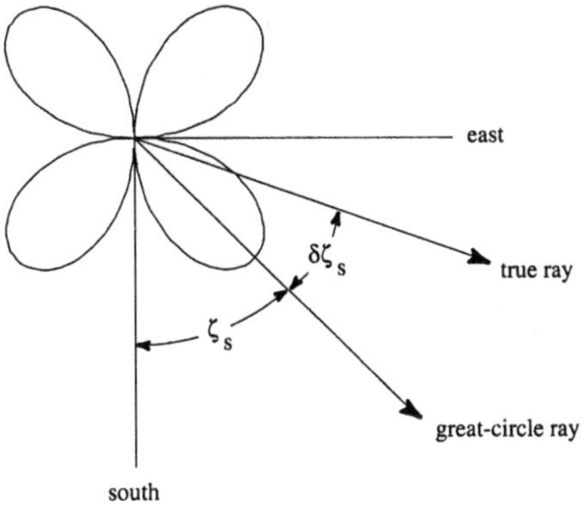

Figure 16.15. Schematic amplitude radiation pattern $|R(\zeta_s)|$ of a typical surface wave. A great-circle ray on a spherically symmetric Earth leaves the source at an azimuth ζ_s, whereas the corresponding perturbed ray on a laterally heterogeneous leaves at an azimuth $\zeta_s + \delta\zeta_s$.

in Figure 16.15—the perturbed ray samples the complex radiation pattern (16.170) at an azimuth $\zeta_s + \delta\zeta_s$ that is slightly different from the unperturbed great-circular azimuth ζ_s. The largest perturbations tend to occur in the vicinity of radiation nodes, for this reason.

Figure 16.16 shows a comparison of the first-order amplitude anomalies $\delta A/A = \delta A_p/A_p + \delta A_s/A_s$ obtained using ray perturbation theory with the exact ray-theoretical anomalies $(A - A_{\mathrm{spher}\,\oplus})/A_{\mathrm{spher}\,\oplus}$ for a number of G1 and R1 minor-arc paths on model S12WM13. It is evident that $\delta A/A$ is more discrepant than the first-order takeoff-angle and arrival-angle anoma-

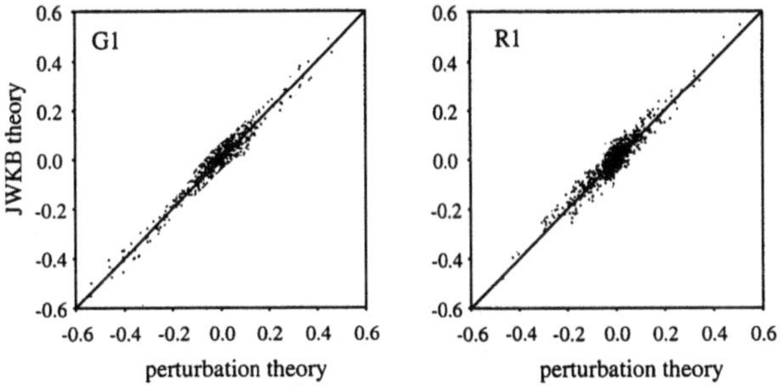

Figure 16.16. Scatter-plot comparison of the first-order amplitude anomaly $\delta A/A$ (due to both path and source effects) with the exact JWKB anomaly $(A - A_{\mathrm{spher}\,\oplus})/A_{\mathrm{spher}\,\oplus}$ for 165-second fundamental-mode surface waves on model S12WM13. Each of the 850 points represents a single G1 or R1 path. The exact anomalies incorporate the geographical variations in the eigenfunctions U_s, V_s, W_s and U, V, W at the source and receiver, which have been neglected in the perturbation analysis.

lies $\delta\zeta_{\rm s}$ and $\delta\zeta$ exhibited in Figure 16.14, and that these in turn are more discrepant than the first-order Fermat phase anomalies $\delta\psi_{\rm p}$ in Figure 16.13; this reflects the respective dependence of the dominant contributions to these quantities upon the second, first and zeroth cross-path derivatives $\partial_\theta^2\delta c$, $\partial_\theta\delta c$ and δc.

*16.8.7 Synthetic seismogram comparisons

Figure 16.17 shows two examples of surface-wave synthetic seismograms computed upon Earth model S12WM13 using the methods described in this chapter. The left column shows the ray paths and radial-component waveforms of R1 Rayleigh waves recorded at station CTAO in Charters Towers, Australia, following the February 16, 1979 earthquake located near the coast of Peru, whereas the right column shows the ray paths and transverse-component waveforms of G1 Love waves recorded at station BJT in Beijing, China, following the November 29, 1978 Oaxaca, Mexico event. The top panels show the exact ray deviations from the unperturbed great-circle paths, superimposed upon a map of the fractional phase-speed perturbation $\delta c/c$ for165-second G1 and R1 waves along a strip from source to receiver. The bottom three panels compare three methods of computation: (1) exact JWKB theory based upon Runge-Kutta integration of the kinematic and dynamical ray-tracing equations, with the geographical variations in the radial eigenfunctions $U_{\rm s}, V_{\rm s}, W_{\rm s}$ and U, V, W at the source and receiver fully accounted for; (2) first-order ray perturbation theory, with the lateral variations in $U_{\rm s}, V_{\rm s}, W_{\rm s}$ and U, V, W ignored; (3) the Woodhouse-Dziewonski great-circle approximation, which accounts for the lateral heterogeneity by means of a fictitious frequency and source shift $\omega \rightarrow \omega + \delta\bar{\omega}$ and $\Theta \rightarrow \Theta + \delta\Theta$ in a spherical-Earth normal-mode summation code. The synthetic displacement seismograms $\hat{\boldsymbol{\nu}} \cdot \mathbf{s}(\mathbf{x}, t)$ incorporate all fundamental-mode waves with periods $T = 2\pi/\omega$ in the range 50–500 seconds; the group-speed windows are 3.2–4.7 km/s for the Love waves and 3.5–5.0 km/s for the Rayleigh waves, respectively. The corresponding spherical-Earth seismogram is used as the standard of comparison in each panel. The R1 ray path from Peru to CTAO is deflected by only about 2° from the unperturbed great circle; the JWKB waveform in this case is quite well approximated by both first-order perturbation theory and the great-circle approximation. There is a slight discrepancy in amplitude; however, the phases agree with each other nicely throughout the waveform. This particular ray path traverses a low-speed region in the South Pacific, so there is a marked phase delay relative to the spherical Earth. The G1 ray path from Oaxaca to Beijing travels along the Pacific coastal region, and is deflected by about 5° from the unperturbed great circle. In this case, there is a significant discrepancy between

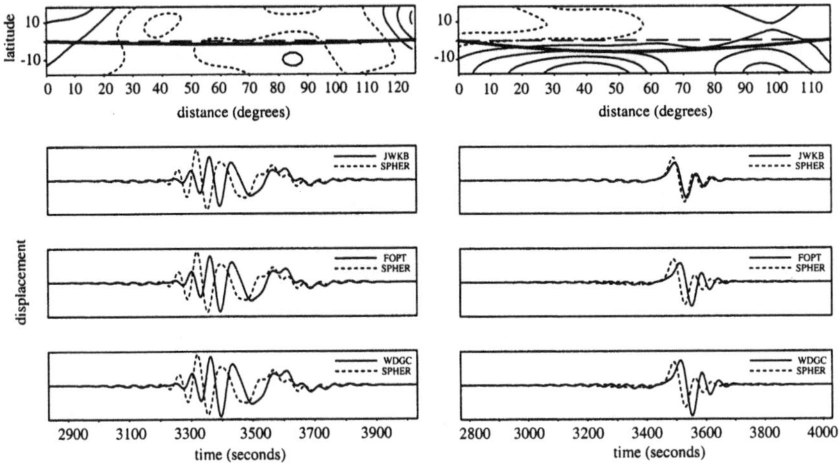

Figure 16.17. (*Top left*) JWKB ray path for R1 waves from the February 16, 1979 Peru event to station CTAO in Charters Towers, Australia, superimposed on a contour map of $\delta c/c$ for 165-second fundamental-mode Rayleigh waves on model S12WM13. (*Top right*) JWKB ray path for G1 waves from the November 11, 1978 Oaxaca event to station BJT in Beijing, China, superimposed on a contour map of $\delta c/c$ for 165-second fundamental-mode Love waves on model S12WM13. The maps have been rotated so that the source and receiver are situated on the equator at the left and right, respectively; this is the coordinate system used to trace surface-wave rays. The JWKB and great-circle ray paths are indicated by the heavy solid line and the long-dashed line, whereas positive and negative values of δc are indicated by the light solid contours and the short-dashed contours, respectively; the contour interval is 1%. The remaining panels show the JWKB synthetic displacement seismograms (*second from top*), the corresponding seismograms computed using first-order ray perturbation theory (*third from top*), and the seismograms computed using the Woodhouse-Dziewonski great-circle approximation (*bottom*). Full scale on the radial-component R1 seismograms (*left*) is 0.28 mm, whereas that on the transverse-component G1 seismograms (*right*) is 4.06 mm. The waveforms on model S12WM13 (*solid lines*) are in each case compared with the fundamental-mode spherical-Earth seismograms (*dashed lines*). The label SPHER designates the spherical-Earth synthetic, FOPT is an acronym for first-order ray-perturbation theory, and WDGC is an acronym for the Woodhouse-Dziewonski great-circle approximation. A 10% cosine taper has been applied to both ends of the spectrum in the frequency domain in order to suppress ringing in the time series.

the JWKB waveform and the waveforms predicted by first-order perturbation theory and the great-circle approximation. The Oaxaca to Beijing path is transitional in character, with a large cross-path gradient $\partial_\theta \delta c$. In general, the two approximations perform poorly for such ray paths; they perform best for locally fast or slow paths that tend to cross contours of δc at relatively steep angles, rather than run along them.

16.9 Surface-Wave Tomography

Our emphasis in this chapter has been upon the development of JWKB methods for solving the *forward problem* of synthesizing synthetic long-period surface-wave seismograms upon a smooth, laterally heterogeneous Earth. In this final section, we describe a number of approaches that have been used to address the surface-wave *inverse problem*:

> Given a suite of observed waveforms $\mathbf{a}_{\text{obs}}(\mathbf{x}, t)$ excited by earthquakes with known origin times t_{s}, hypocentral locations $\mathbf{x}_{\text{s}}$ and moment tensors $\mathbf{M}$, find the three-dimensional perturbations $\delta\kappa$, $\delta\mu$, $\delta\rho$, δd or $\delta\alpha$, $\delta\beta$, $\delta\rho$, δd to the known average spherically symmetrical structure of the Earth.

Our emphasis is upon the assumptions and approximations that underlie the inversion procedures; we refrain from presenting or comparing any results, inasmuch as these are more fruitfully discussed within the broader context of global seismic tomography and geodynamics—topics that are beyond the scope of this book.

16.9.1 Woodhouse-Dziewonski method

The modern era of three-dimensional mantle structural studies was inaugurated by Woodhouse & Dziewonski (1984), as we noted in Chapter 1. This pioneering investigation developed a method of *waveform fitting*, based upon the *great-circle approximation* discussed in Section 16.8.2, that is still in use today. A long-period accelerogram $a(t) = \hat{\boldsymbol{\nu}} \cdot \mathbf{a}(\mathbf{x}, t)$ is written as a sum over normal-mode multiplets ${}_n\mathrm{S}_l$ and ${}_n\mathrm{T}_l$ of the form (10.63):

$$a(t) = \sum_{\text{multiplets}} A \cos\omega t \exp(-\gamma t). \tag{16.256}$$

Elastic lateral heterogeneity is taken into account by means of a fictitious frequency shift $\delta\bar{\omega}$ and source shift $\delta\Theta$. The resulting perturbation to the accelerogram (16.256) is

$$\delta a(t) = \sum_{\text{multiplets}} [\delta\Theta\, \partial_\Theta A \cos\omega t - \delta\bar{\omega}\, A \sin\omega t] \exp(-\gamma t). \tag{16.257}$$

The partial derivative $\partial_\Theta A$ may be readily evaluated by making use of equations (10.52)–(10.60). Upon making use of (16.215) we can rewrite (16.257) in terms of the minor-arc plus the great-circular average of the local eigenfrequency perturbation:

$$\begin{aligned}\delta a(t) = & \sum_{\text{multiplets}} \delta\hat{\omega}\,(c/C\omega)\partial_\Theta A \cos\omega t \exp(-\gamma t) \\ & - \sum_{\text{multiplets}} \delta\bar{\omega}\,[A\sin\omega t + (c/C\omega)\partial_\Theta A\cos\omega t]\exp(-\gamma t). \qquad (16.258)\end{aligned}$$

The perturbation $\delta\omega(\theta,\phi)$ is identical to that of a spherical Earth having the same structure as that underlying the point a,θ,ϕ. For brevity in what follows, we abbreviate the dependence (9.12) upon $\delta\kappa$, $\delta\mu$, $\delta\rho$, δd or (9.20) upon $\delta\alpha$, $\delta\beta$, $\delta\rho$, δd using the symbolic notation

$$\delta\omega = \int_0^a \delta\oplus\, K_\oplus\, dr. \qquad (16.259)$$

Upon expanding the structural perturbations in terms of real surface spherical harmonics,

$$\delta\oplus = \sum_{st} \delta\oplus_{st}\mathcal{Y}_{st}, \qquad (16.260)$$

we may express the unknown quantities in (16.258) in the form

$$\delta\hat{\omega} = \sum_{st}\left[\int_0^a \delta\oplus_{st}\, K_\oplus\, dr\right]\hat{\mathcal{Y}}_{st}, \qquad (16.261)$$

$$\delta\bar{\omega} = \sum_{st}\left[\int_0^a \delta\oplus_{st}\, K_\oplus\, dr\right]\bar{\mathcal{Y}}_{st}. \qquad (16.262)$$

The minor-arc and great-circular averages $\hat{\mathcal{Y}}_{st}$ and $\bar{\mathcal{Y}}_{st}$ of $\mathcal{Y}_{st}$ are known functions, which may be readily calculated by rotation of the source-receiver path to the equator, as discussed in Appendices B.9 and C.8.7. Upon substituting (16.261)–(16.262) we obtain a linearized relation between the perturbation $\delta a(t)$ and the coefficients $\delta\oplus_{st}(r)$ describing a three-dimensional Earth model:

$$\delta a(t) = \sum_{st}\int_0^a \delta\oplus_{st}(r)\, K_{\oplus st}(r,t)\, dr, \qquad (16.263)$$

where

$$\begin{aligned}K_{\oplus st} = & \sum_{\text{multiplets}} (c/C\omega)\partial_\Theta A\cos\omega t\exp(-\gamma t)\, K_\oplus\,\hat{\mathcal{Y}}_{st} \qquad (16.264) \\ & - \sum_{\text{multiplets}} [A\sin\omega t + (c/C\omega)\partial_\Theta A\cos\omega t]\exp(-\gamma t)\, K_\oplus\,\bar{\mathcal{Y}}_{st}.\end{aligned}$$

Equations (16.263)–(16.264) provide the basis for an iterative least-squares inversion, which seeks to minimize the residual $\sum_{\rm paths}\int_{t_1}^{t_2}[a(t)-a_{\rm obs}(t)]^2\,dt$ between a suite of synthetic and observed seismograms; the fictitious frequency and source location are updated, $\omega\rightarrow\omega+\delta\bar{\omega}$, $\Theta\rightarrow\Theta+\delta\Theta$, and the kernels $K_{\oplus st}(r,t)$ are computed anew at each iteration. The principal advantage of the procedure is that it enables the entire long-period waveform $a_{\rm obs}(t)$ to be used as a constraint upon the unknown parameters $\delta\oplus_{st}$; in particular, all of the higher-overtone ($n=1,2,\ldots$) modes which propagate with very nearly the same group speed are included in addition to the multi-orbit, fundamental-mode ($n=0$) surface waves.

16.9.2 Partitioned-waveform method

In regions with many peripheral sources and stations, giving rise to good crossing-path coverage, it is possible to obtain higher-resolution models of crustal and upper-mantle structure by incorporating higher-frequency body and surface waves. The so-called *partitioned-waveform method* was developed by Nolet (1990), expressly for use in such applications. In this approach, attention is restricted to the minor-arc ($s=1$) arrivals in (16.161). Fermat's principle is invoked to calculate the path contribution to the phase, and amplitude variations due to ray-tube focusing and defocusing are ignored; the frequency-domain acceleration response $a(\omega)$ is written as a sum over surface-wave modes or dispersion branches of the form

$$a(\omega)=\sum_{\rm modes}A_{\rm r}\exp(-i\psi_{\rm r})A_{\rm s}\exp(-i\psi_{\rm s})(8\pi k\sin\Theta)^{-1/2}$$
$$\times\exp(-ik\Theta)\exp(-i\,\delta\hat{k}\,\Theta)\exp(-\omega\Theta/2CQ),\qquad(16.265)$$

where k, C and Q^{-1} are the unperturbed spherical-Earth wavenumber, group speed and reciprocal quality factor, and the known source and receiver contributions $A_{\rm s}\exp(-i\psi_{\rm s})$ and $A_{\rm r}\exp(-i\psi_{\rm r})$ are given by equations (16.165) and (16.167). The only unknown in the JWKB-Fermat representation (16.265) is the minor-arc average

$$\delta\hat{k}=\frac{1}{\Theta}\int_0^{\Theta}\delta k\,d\Delta\qquad(16.266)$$

of the local wavenumber perturbation δk. It can be expressed in terms of the corresponding average

$$\delta\hat{\oplus}=\frac{1}{\Theta}\int_0^{\Theta}\delta\oplus\,d\Delta\qquad(16.267)$$

of the three-dimensional structure along the unperturbed great-circular ray path in the symbolic form

$$\delta\hat{k} = -C^{-1} \int_0^a \delta\hat{\oplus}\, K_{\oplus}\, dr. \tag{16.268}$$

In the first step, the time-domain misfit $\int_{t_1}^{t_2} [a(t) - a_{\rm obs}(t)]^2\, dt$ of a suitably windowed portion of every seismogram is minimized using a nonlinear optimization procedure to find the best-fitting *path-average perturbation* $\delta\hat{\oplus}(r)$. Since the dimension of the radially parameterized model space is relatively small, it is possible to conduct a robust and efficient nonlinear search using the conjugate gradient algorithm. Local minima can be avoided by filtering $a_{\rm obs}(t)$, and fitting the long-period portion of the waveform first. In the second step, the averages along a collection of paths are combined to determine the three-dimensional structure $\delta\oplus(r, \theta, \phi)$ in the region. Every seismogram provides a separate *linear* constraint via the relation (16.267). Diagonalization of the Hessian matrix of second partial derivatives of the misfit function yields a system of *independent* linear equations with known variances on the right side, enabling a resolution analysis to be performed. A major advantage of the partitioned-waveform method is its great flexibility. Different elastic and anelastic background models $\oplus$, Q^{-1} may be used along different paths, and different eigenfunctions $U_{\rm s}$, $V_{\rm s}$, $W_{\rm s}$ and U, V, W may be used at every source and receiver, if it is appropriate. Higher overtones ($n = 1, 2, \ldots$) are included in the JWKB-Fermat sum (16.265), just as they are included in the Woodhouse-Dziewonski great-circle approximation (16.256), so that triplicated S waves in the epicentral distance range $0^\circ \leq \Theta \leq 30^\circ$ and SS, SSS,... multiples can be incorporated in the fitting procedure as well as fundamental-mode G1 and R1 waves. The method has been used by Nolet and his co-workers to obtain high-quality images of the upper-mantle shear-wave speed $\delta\beta$ beneath central Europe (Zielhuis & Nolet 1994), North America (Van der Lee 1996; Van der Lee & Nolet 1997) and the Philippine Sea plate (Lebedev, Nolet & Van der Hilst 1997).

16.9.3 Phase-speed tomography

Rather than fitting a single three-dimensional Earth model $\delta\oplus$ or a suite of separate path-average models $\delta\hat{\oplus}$ to a suite of observed frequency-domain seismograms

$$a_{\rm obs}(t) = \sum_{\rm modes} \sum_{\rm rays} A_{\rm obs} \exp(-i\psi_{\rm obs}), \tag{16.269}$$

many workers prefer to extract the amplitudes $A_{\rm obs}$ and phases $\psi_{\rm obs}$ of the constituent surface waves. Indeed, this is the classical approach to surface-wave tomography, as we noted in Chapter 1. The method is most easily applied to the long-period fundamental-mode waves, since they are relatively well isolated from the earlier-arriving overtones. The strong dispersion over long source-receiver paths gives rise to a rapid variation of phase $\psi_{\rm obs}$ with frequency ω; a straightforward Fourier or periodogram analysis is known to yield a biased spectral estimate in such circumstances. To reduce this bias, it is necessary to "de-disperse" the waveform; this is commonly done using a cross-spectral or matched-filter technique, which measures the residual amplitude $\delta A = A_{\rm obs} - A_{\rm spher\,\oplus}$ and phase $\delta\psi = \psi_{\rm obs} - \psi_{\rm spher\,\oplus}$ relative to those of a spherical-Earth synthetic seismogram

$$a_{\rm spher\,\oplus}(t) = \sum_{\rm modes}\sum_{\rm rays} A_{\rm spher\,\oplus}\exp(-i\psi_{\rm spher\,\oplus}). \tag{16.270}$$

The source contribution $\delta\psi_{\rm s}$ to the phase difference is generally ignored; in that case, $\delta\psi = \delta\psi_{\rm p}$ can be related to the surface-wave phase-speed perturbation δc either by the first-order Fermat approximation

$$\delta\psi = -\omega c^{-2}\underbrace{\int_{\hat{\mathbf{r}}_{\rm s}}^{\hat{\mathbf{r}}} \delta c\, d\Delta}_{\text{great circle}} \tag{16.271}$$

or the exact JWKB relation

$$\delta\psi = \omega\underbrace{\int_{\hat{\mathbf{r}}_{\rm s}}^{\hat{\mathbf{r}}} \frac{d\Delta}{c+\delta c}}_{\text{true ray path}} - \frac{\omega\Delta}{c}. \tag{16.272}$$

Equations (16.271) and (16.272) provide the foundation for phase-speed tomography; the perturbation δc may be either expanded in surface spherical harmonics $\mathcal{Y}_{st}$ (Wong 1989) or parameterized in terms of local basis functions using a block or triangular tesselation of the unit sphere (Zhang & Tanimoto 1993; Wang, Tromp & Ekström 1998). In practice, most tomographic analyses do not go beyond the Fermat approximation (16.271). If the geographical variations in δc are pronounced, an initial Fermat model may be iteratively refined by tracing rays through $c + \delta c$ and making use of (16.272). Independent phase-delay measurements $\delta\psi$ can be collected not only for $s = 1$ minor-arc paths, but also for $s = 2, 3, \ldots$ multi-orbit arrivals. Great-circle phase delays $\delta\psi_{s+2} - \delta\psi_s$ can also be measured by analysis of circumnavigational waves; these data have the added advantage that they are independent of the moment tensor $\mathbf{M}$. At periods $T = 2\pi/\omega$ less than 100–150 seconds, the lateral perturbations in phase speed δc are large enough to cause path variations (16.271)–(16.272) of several cycles in phase; an isolated measurement of the phase anomaly $\delta\psi$ at a single period is in this case indistinguishable from $\delta\psi \pm 2N\pi$. This so-called *cycle-skip* ambiguity can be eliminated by means of a "bootstrap" technique, in which the correct ($N = 0$) cycle count is anchored at longer periods, where it is almost always the case that $|\psi_{\rm obs} - \psi_{\rm PREM}| \leq \pi$; the measurements can subsequently be extended to shorter periods by requiring that the observed dispersion relation be continuous. Uncertainties can be assigned to measurements of $\delta\psi$ in a straightforward manner; the resulting geographical maps of δc can be regarded as intermediate "observables", which can be compared and utilized in conjunction with other data to constrain the three-dimensional elastic structure of the Earth $\delta\oplus$.

Modern surface-wave phase-speed studies are able to exploit the large and ever-growing data set of broad-band digital recordings from several global and regional seismographic networks. Table 16.1 compares the characteristics of some recent tomographic analyses; the total number of source-receiver paths for which the phase difference $\delta\psi$ has been measured is truly impressive! The study by Laske & Masters (1996) is unique in that it incorporates surface-wave arrival-angle measurements $\delta\zeta$ obtained by means of a multi-taper polarization analysis of three-component recordings, as well as phase measurements $\delta\psi$. The first-order relation (16.235) was used to fit the arrival-angle data; the dependence of $\delta\zeta$ upon the cross-path derivative of δc enhances the resolution of high-degree structure. As an unexpected by-product of their analysis, Laske & Masters detected $5^\circ - 15^\circ$ misalignments of the "north-south" and "east-west" components at several stations of the Global Seismographic Network!

Figures 16.18 and 16.19 illustrate the relative variations $\delta c/c$ in Love and Rayleigh phase speed obtained by Ekström, Tromp & Larson (1997).

Reference	Period Range	Number of Paths
Trampert & Woodhouse (1995)	40–150 s	24,000
Trampert & Woodhouse (1996)	40–150 s	62,000
Zhang & Lay (1996)	85–250 s	30,000
Ekström, Tromp & Larson (1997)	35–150 s	56,000
Laske & Masters (1996)	75–250 s	11,000

Table 16.1. Period range and number of source-receiver paths incorporated in some recent fundamental-mode Love and Rayleigh phase-speed analyses.

These $s_{\max} = 40$ models explain $70-96\%$ of the variance in the observed phase residuals $\psi_{\text{obs}} - \psi_{\text{PREM}}$, testifying to the high quality and internal consistency of the original dispersion measurements. Many of the most prominent features in these maps—such as the slow wave speeds in the vicinity of the Iceland hotspot and Red Sea rift zone—are seen in other studies as well, and have an obvious tectonic explanation or interpretation. At longer periods ($2\pi/\omega > 150$ s for Love waves and $2\pi/\omega > 75$ s for Rayleigh waves) the stable continental cratons exhibit fast speeds, whereas the mid-oceanic ridges are comparatively slow. The expected increase in wave speed in the older and deeper ocean basins, due to the cooling of the spreading lithosphere, is also apparent.

Prior to utilizing maps such as those in Figures 16.18 and 16.19 as constraints upon the three-dimensional structure $\delta\alpha$, $\delta\beta$, $\delta\rho$, δd of the upper mantle, it is necessary to correct for the effect of lateral variations in the thickness and structure of the crust. First-order perturbation theory may not provide an adequate correction at short and intermediate periods; to illustrate this, we account for crustal variations by stripping the spherically symmetric crust from PREM and replacing it with the laterally variable model CRUST 5.1 (Mooney, Laske & Masters 1998). Subsequently, at selected angular frequencies, we calculate the exact local modes on a grid of points θ, ϕ and fit surface spherical harmonics $\mathcal{Y}_{st}$ to the associated distribution of local surface-wave phase speed δc_{crust}. The resulting crustal phase-speed anomalies are depicted in Figure 16.20. In active compressional orogens, such as the Alpine-Himalayan belt and the Tibetan Plateau, where the crustal thickness exceeds 70 km, the phase-speed correction for 50 s Love waves is $-12\% \leq \delta c_{\text{crust}} \leq -10\%$. In fact, the phase speed of these short-period waves is dominated by the effects of crustal structure; the Fréchet kernel K_β for 50 s Love waves is negligible below a depth of 50–60 km, so it is no surprise that the observed and predicted maps in Figures 16.18 and 16.20 look almost the same. At periods greater than 150 s the crustal correction is much less significant, particularly for Rayleigh waves.

The phase residuals $\psi_{\text{obs}} - \psi_{\text{PREM}}$ of higher-mode Love and Rayleigh waves are much more difficult to measure reliably, because of the near coincidence of the $2\pi/\omega \leq 50-100$ s group arrivals in the time domain (see Sections 11.6.1–11.6.2 and 11.7.1). The pioneering analyses of Nolet (1977) and Cara (1978), which we discussed in the introduction, employed multi-station array methods to disentangle the $n = 1, 2, \ldots$ overtone branches. In a subsequent study, Stutzmann & Montagner (1993; 1994) made use of a single-station method which exploits recordings from a number of closely spaced earthquake sources. This technique is most effective when the sources have significantly different focal depths, so that the excitation amplitudes of the target branches are very different; unfortunately, the number

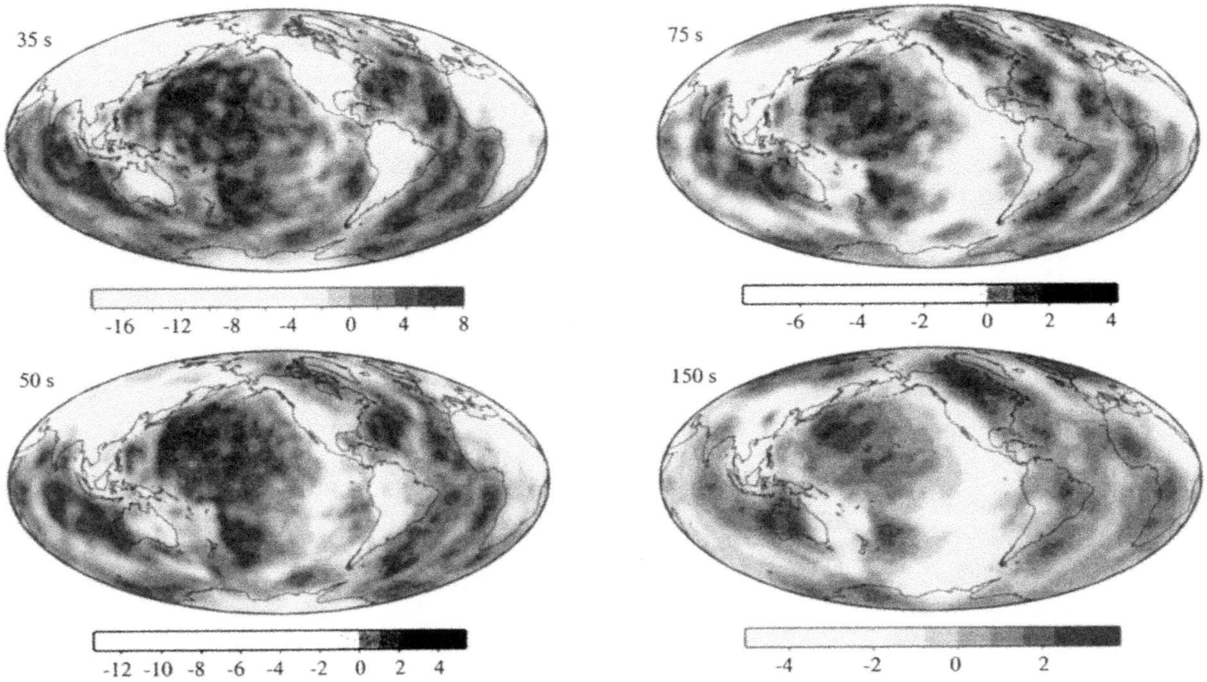

Figure 16.18. Global Love-wave phase-speed maps $\delta c/c$ (in percent) at periods of 35, 50, 75 and 150 seconds. Notice that the grey scales differ from map to map.

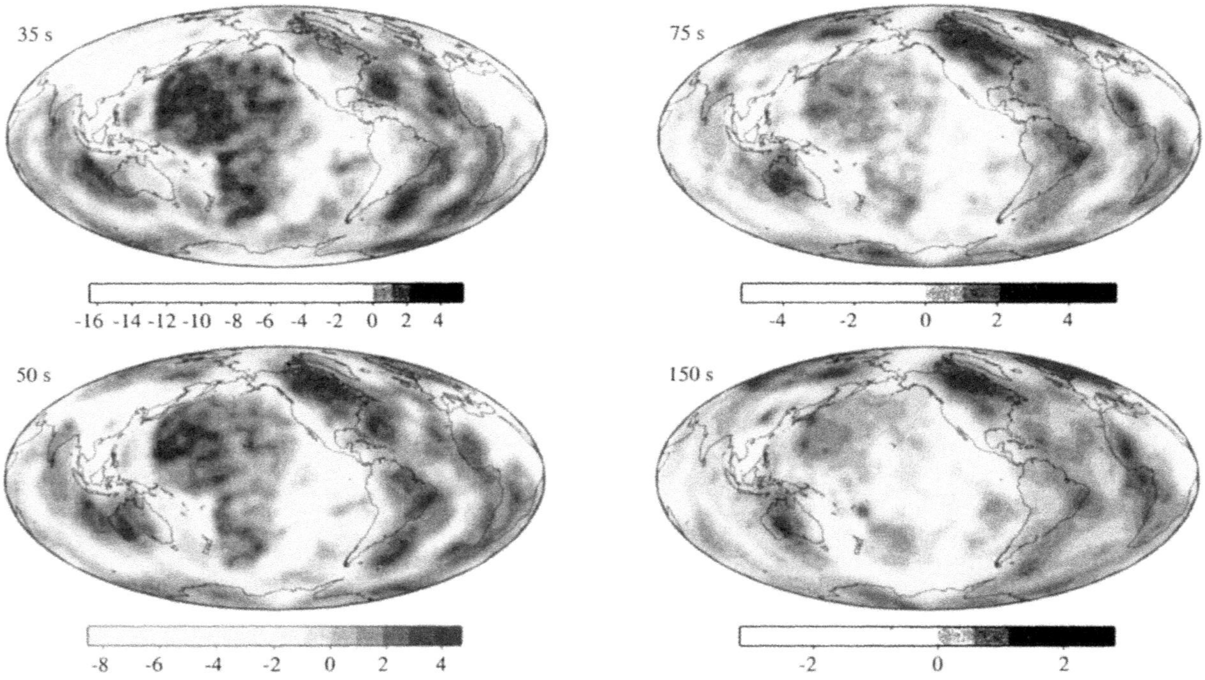

Figure 16.19. Global Rayleigh-wave phase-speed maps $\delta c/c$ (in percent) at periods of 35, 50, 75 and 150 s. Notice that the grey scales differ from map to map.

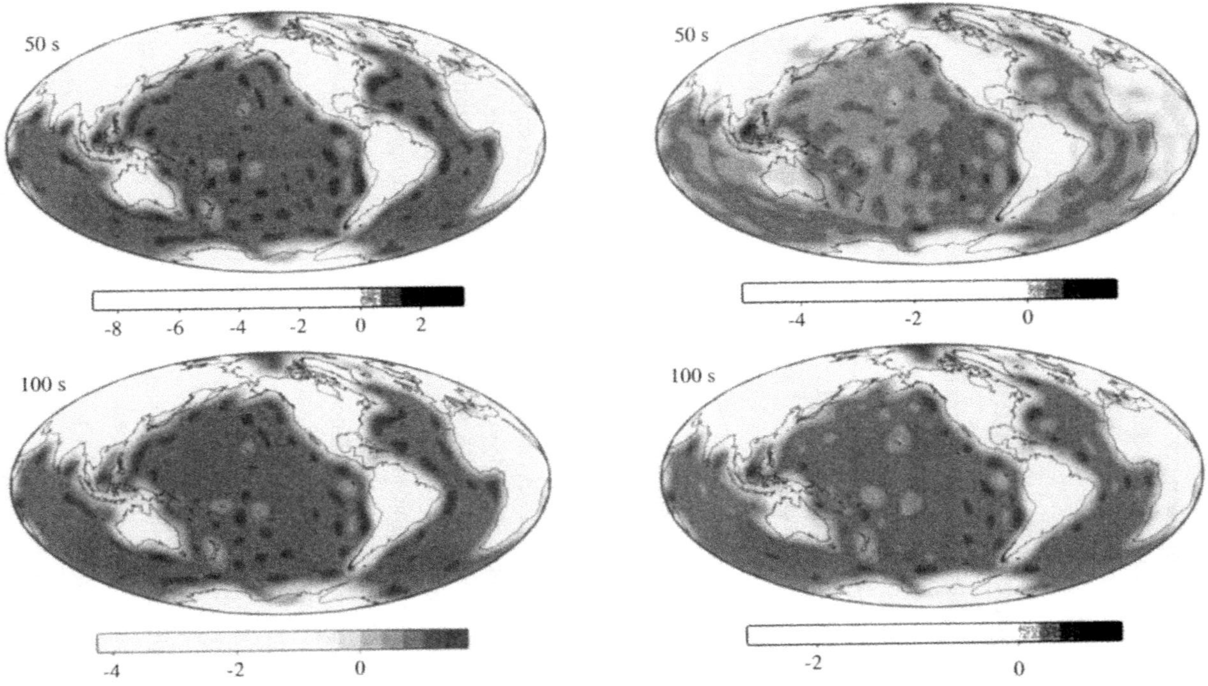

Figure 16.20. Relative perturbation $\delta c/c$ in the phase speed of fundamental-mode Love waves (*left*) and Rayleigh waves (*right*) due to lateral variations in the thickness and structure of the Earth's crust. The perturbations are shown at two periods: 35 s (*top*) and 75 s (*bottom*). Note that the grey scales differ from map to map.

of paths having the requisite multi-source geometry is limited, leading to poor lateral resolution of the phase-speed perturbations δc. The development of single-station, single-source methods based upon cross-correlation with single-branch spherical-Earth seismograms would obviously be desirable; Van Heijst & Woodhouse (1997) have made some recent progress toward this daunting goal.

16.9.4 Anelastic tomography

A number of investigators have attempted to use measured G1, G2,... and R1, R2,... amplitude anomalies $\delta A = A_{\mathrm{obs}} - A_{\mathrm{PREM}}$ to constrain global lateral variations in the anelasticity $\delta Q^{-1}/Q^{-1}$ of the Earth. This has proven to be an extremely intractable problem, because of the large number of additional effects which can give rise to comparable surface-wave amplitude variations, including uncertainties in the locations $\mathbf{x}_{\mathrm{s}}$ and mechanisms $\mathbf{M}$ of earthquake sources, as well as ray-tube focusing and defocusing, and other influences of the imperfectly known elastic lateral heterogeneity δc, such as the radiation-pattern effect illustrated in Figure 16.15. The total fractional amplitude perturbation is given by

$$\frac{\delta A}{A} = \frac{\delta A_{\mathrm{g}}}{A_{\mathrm{g}}} + \exp\left(-\frac{\omega}{2CQ}\int_0^\Delta \frac{\delta Q^{-1}}{Q^{-1}}\, d\Delta\right) - 1, \qquad (16.273)$$

where the term $\delta A_{\mathrm{g}}/A_{\mathrm{g}} = \delta A_{\mathrm{p}}/A_{\mathrm{p}} + \delta A_{\mathrm{s}}/A_{\mathrm{s}} + \cdots$ encapsulates the competing geometric path, source and other effects, and the integral is over the unperturbed great-circular ray path of a minor-arc or higher-orbit wave. Several approaches have been used to deal with the geometrical variations in global anelastic tomographic studies. It is not feasible simply to correct for the effects of lateral wave-speed variations δc using existing phase-speed maps, because of the strong dependence of $\delta A_{\mathrm{p}}/A_{\mathrm{p}}$ upon short-wavelength structure, which is very poorly constrained. Early efforts sought to exploit the observation that, in first-order ray perturbation theory,

$$\left(\frac{\delta A_{\mathrm{p}}}{A_{\mathrm{p}}}\right)_{s=4} - \left(\frac{\delta A_{\mathrm{p}}}{A_{\mathrm{p}}}\right)_{s=2} + \left(\frac{\delta A_{\mathrm{p}}}{A_{\mathrm{p}}}\right)_{s=3} - \left(\frac{\delta A_{\mathrm{p}}}{A_{\mathrm{p}}}\right)_{s=1} = 0, \qquad (16.274)$$

by virtue of the alternating odd and even dependence (16.246) upon orbit number s. The source term satisfies an analogous relation,

$$\left(\frac{\delta A_{\mathrm{s}}}{A_{\mathrm{s}}}\right)_{s=4} - \left(\frac{\delta A_{\mathrm{s}}}{A_{\mathrm{s}}}\right)_{s=2} + \left(\frac{\delta A_{\mathrm{s}}}{A_{\mathrm{s}}}\right)_{s=3} - \left(\frac{\delta A_{\mathrm{s}}}{A_{\mathrm{s}}}\right)_{s=1} = 0, \qquad (16.275)$$

because of the alternating dependence (16.236) of the takeoff-angle perturbation $\delta\zeta_{\mathrm{s}}$. The composite four-orbit datum

$$D = \frac{a_{s=4}}{a_{s=2}} \times \frac{a_{s=3}}{a_{s=1}} \qquad (16.276)$$

should in this approximation depend only upon the great-circular average of the anelastic perturbation:

$$\delta D = \exp\left(-\frac{\omega}{CQ}\oint \frac{\delta Q^{-1}}{Q^{-1}}\,d\Delta\right). \tag{16.277}$$

Romanowicz (1990) made use of equation (16.277) to obtain the first degree-two global model of upper-mantle anelasticity. Synthetic studies by Durek, Ritzwoller & Woodhouse (1993) subsequently demonstrated that departures from first-order ray perturbation theory are sufficient to vitiate the linearized relations (16.274)–(16.275) as the sole basis of quantitative anelastic tomography. They consider the quantity D to be "de-sensitized" to phase-speed variations. Prior to inverting for an $s = 2, 4, 6$ model of δQ^{-1} they applied an additional "de-biasing" correction, based upon an existing map of δc. Romanowicz (1995) used single-orbit R1 and R2 amplitude measurements to obtain the first global model of δQ^{-1} with odd as well as even degrees ($s = 1-6$). She sought to minimize the importance of $\delta A_{\rm s}/A_{\rm s}$ and $\delta A_{\rm p}/A_{\rm p}$ by heavy winnowing of the dataset, to eliminate low-amplitude nearly nodal recordings and measurements thought to have been strongly affected by focusing and defocusing, based upon a subjective criterion.

16.9.5 Beyond the path-average approximation

All waveform inversion schemes based upon the JWKB approximation share a common feature: the $s = 1$ portion of a seismogram $a(t)$ depends only upon the minor-arc path average $\hat{\oplus}$ of the laterally heterogeneous elastic structure $\oplus$. A change in the waveform is related to a change in the Earth model via a *one-dimensional* (radial) Fréchet kernel:

$$\delta a = \int_0^a \delta\hat{\oplus}\, K_{\oplus}^{\rm 1D}\, dr. \tag{16.278}$$

This is not an unreasonable approximation for the fundamental-mode Love and Rayleigh waves, but it is questionable for the earlier-arriving overtones. Intuitively, one expects the sensitivity of the early portion of a transverse-component waveform $a(t) = \hat{\boldsymbol{\Phi}}\cdot\mathbf{a}(\mathbf{x}, t)$ to be concentrated along the associated SS, $\mathrm{SS_{SH}}$, $\mathrm{SSS_{SH}}, \ldots$ body-wave rays. Li & Tanimoto (1993) demonstrated that the dependence upon $\hat{\oplus}$ is fundamentally a consequence of the absence of $n' \neq n$ cross-branch coupling in the JWKB approximation. By accounting for coupling between ${}_n\mathrm{T}_l$ and ${}_n\mathrm{S}_l$ multiplets using a normal-mode representation of $a(t)$, they obtained a *two-dimensional* Fréchet-kernel relationship of the form

$$\delta a = \int_\Sigma \delta\oplus\, K_{\oplus}^{\rm 2D}\, dA, \tag{16.279}$$

where Σ denotes the unperturbed ray plane. Equation (16.279) has been used as the basis of a global tomographic study by Li & Romanowicz (1995; 1996). Marquering & Snieder (1995; 1996) have developed a similar two-dimensional ray-plane sensitivity relationship using a more economical travelling-wave representation that accounts for coupling between $n' \neq n$ dispersion branches. Zhao & Jordan (1998) have used (16.279) to develop two-dimensional Fréchet kernels for functionals obtained by processing $a_{\mathrm{obs}}(t)$, such as travel-time anomalies measured by cross-correlation with a spherical-Earth synthetic seismogram $a_{\mathrm{spher}\,\oplus}(t)$. All of these investigations make use of a stationary-phase approximation which is valid only under the strong assumption that the lateral variations in the direction perpendicular to the ray plane are smooth. To overcome this objection, Marquering, Nolet & Dahlen (1998) have used the Born approximation together with a travelling-wave representation of $a(t)$ to obtain a fully three-dimensional waveform sensitivity-kernel relationship of the form

$$\delta a = \int_{\oplus} \delta\oplus \, K_{\oplus}^{\mathrm{3D}} \, dV. \tag{16.280}$$

These efforts to develop waveform and travel-time sensitivity kernels which go beyond the one-dimensional path-average approximation represent the current theoretical frontier in global seismic tomography. The subject is still in an active state of development, and it is premature to attempt to provide even a superficial synthesis at the present time.

Appendixes

Appendix A

Vectors and Tensors

The physical quantities that appear in the equations governing the free oscillations of the Earth are vectors and tensors such as displacement, velocity, strain and stress; our discussion of these equations presupposes a knowledge of elementary vector and tensor analysis. The notation and some of the basic results we employ are summarized in this appendix. No proofs are given, inasmuch as we do not intend this review to be a substitute for a more systematic mathematical treatment, such as those provided by Willmore (1959) and Marsden & Hughes (1983). Unlike these authors, we adopt a fundamentally Cartesian viewpoint, distinguishing between covariance and contravariance only when we introduce a system of curvilinear coordinates upon a two-dimensional surface in three-dimensional space.

A.1 Tensors as Multilinear Functionals

By definition, a tensor is a multilinear functional on the space of ordinary three-dimensional vectors. Alternatively, we may regard a tensor of order q as a linear operator which acts upon a tensor of order p to produce a tensor of order $q-p$. To illustrate the meaning of these two notions, it is convenient to utilize a more abstract notation than the one—due to Gibbs—that we eventually employ.

A.1.1 Vectors

A vector is a geometrical object—usually visualized as an arrow—which has both magnitude and direction. We write vectors using a bold font; the magnitude of the vector $\mathbf{u}$ is denoted by $\|\mathbf{u}\|$. A vector of unit length is distinguished by a hat or caret: $\|\hat{\mathbf{n}}\| = 1$. The scalar or *dot product* of two vectors $\mathbf{u}$ and $\mathbf{v}$ is $\mathbf{u}\cdot\mathbf{v} = \|\mathbf{u}\|\,\|\mathbf{v}\|\cos\theta$, where θ is the acute angle between the two vectors. Note that $\mathbf{u}\cdot\mathbf{v} = \mathbf{v}\cdot\mathbf{u}$ and $\|\mathbf{u}\| = (\mathbf{u}\cdot\mathbf{u})^{1/2}$. The vector or *cross product* is a vector of magnitude $\|\mathbf{u}\times\mathbf{v}\| = \|\mathbf{u}\|\,\|\mathbf{v}\|\sin\theta$, whose direction is given by the "right-hand rule". In general, $\mathbf{u}\times\mathbf{v} = -\mathbf{v}\times\mathbf{u}$ and $\mathbf{u}\times\mathbf{u} = \mathbf{0}$, where $\mathbf{0}$ is the zero vector. The *triple product* of three vectors satisfies $\mathbf{u}\cdot(\mathbf{v}\times\mathbf{w}) = \mathbf{w}\cdot(\mathbf{u}\times\mathbf{v}) = \mathbf{v}\cdot(\mathbf{w}\times\mathbf{u})$, whereas the double cross product is given by $\mathbf{u}\times(\mathbf{v}\times\mathbf{w}) = (\mathbf{u}\cdot\mathbf{w})\mathbf{v} - (\mathbf{u}\cdot\mathbf{v})\mathbf{w}$.

The *components* of a vector $\mathbf{u}$ with respect to a Cartesian axis system $\hat{\mathbf{x}}$, $\hat{\mathbf{y}}$, $\hat{\mathbf{z}}$ are given by $u_x = \hat{\mathbf{x}}\cdot\mathbf{u}$, $u_y = \hat{\mathbf{y}}\cdot\mathbf{u}$, $u_z = \hat{\mathbf{z}}\cdot\mathbf{u}$. We may write $\mathbf{u}$ in terms of its three components in the form $\mathbf{u} = u_x\hat{\mathbf{x}} + u_y\hat{\mathbf{y}} + u_z\hat{\mathbf{z}}$. A more succinct notation employs numbered axes $\hat{\mathbf{x}}_1$, $\hat{\mathbf{x}}_2$, $\hat{\mathbf{x}}_3$ rather than $\hat{\mathbf{x}}$, $\hat{\mathbf{y}}$, $\hat{\mathbf{z}}$:

$$u_i = \hat{\mathbf{x}}_i\cdot\mathbf{u}, \qquad \mathbf{u} = u_i\hat{\mathbf{x}}_i. \tag{A.1}$$

Equations (A.1) exemplify the use of the *summation convention*, in which a single index stands for three distinct equations, whereas any repeated index is summed over. The dot product of two vectors is given in terms of the components by $\mathbf{u}\cdot\mathbf{v} = u_iv_i$, whereas the cross product $\mathbf{w} = \mathbf{u}\times\mathbf{v}$ is given by $w_i = \varepsilon_{ijk}u_jv_k$. The quantity ε_{ijk} is the *Levi-Cività alternating symbol*, which takes on the value 1 if $\{i,j,k\}$ is an even permutation of $\{1,2,3\}$, the value -1 if $\{i,j,k\}$ is an odd permutation of $\{1,2,3\}$, and the value zero otherwise. This symbol satisfies the identities

$$\begin{aligned}\varepsilon_{ijk}\varepsilon_{lmn} &= \delta_{il}\delta_{jm}\delta_{kn} + \delta_{in}\delta_{jl}\delta_{km} + \delta_{im}\delta_{jn}\delta_{kl}\\ &\quad - \delta_{il}\delta_{jn}\delta_{km} - \delta_{in}\delta_{jm}\delta_{kl} - \delta_{im}\delta_{jl}\delta_{kn},\end{aligned} \tag{A.2}$$

$$\varepsilon_{ijk}\varepsilon_{imn} = \delta_{jm}\delta_{kn} - \delta_{jn}\delta_{km}, \tag{A.3}$$

$$\varepsilon_{ijk}\varepsilon_{ijn} = 2\delta_{kn}, \qquad \varepsilon_{ijk}\varepsilon_{ijk} = 6, \tag{A.4}$$

where δ_{ij} is the *Kronecker delta*, defined to be 1 if $i = j$ and zero if $i \neq j$. We adhere to index notation throughout most of this appendix; however, there are instances such as in Section A.7 and elsewhere when we find the "traditional" axes $\hat{\mathbf{x}}$, $\hat{\mathbf{y}}$, $\hat{\mathbf{z}}$ to be more convenient.

A.1.2 Linear functionals

A *linear functional* is simply a linear, scalar-valued function defined upon the space of three-dimensional vectors. Denoting the scalar assigned by the functional f to the vector $\mathbf{u}$ by $f(\mathbf{u})$, we require that

$$f(a\mathbf{u} + b\mathbf{v}) = af(\mathbf{u}) + bf(\mathbf{v}). \tag{A.5}$$

Like a vector, a linear functional is a *geometrical object*, in the sense that it exists independently of any Cartesian axis system. Using the fanciful but physically appealing analogy of Misner, Thorne & Wheeler (1973) we can visualize a linear functional f as a machine with one slot. Whenever a vector is inserted into the slot, the machine spews out a scalar:

$$\underset{\text{vector}}{f(\underset{\uparrow}{\cdot})} \to \text{scalar.} \tag{A.6}$$

If a linear combination of vectors $a\mathbf{u}+b\mathbf{v}$ is inserted, the machine generates a linear combination of scalars $af(\mathbf{u}) + bf(\mathbf{v})$, in accordance with (A.5). A linear combination of linear functionals is defined in the obvious way, namely, $(af + bg)(\mathbf{u}) = af(\mathbf{u}) + bg(\mathbf{u})$.

For any linear functional f, there is a unique vector $\mathbf{f}$ satisfying

$$f(\mathbf{u}) = \mathbf{f} \cdot \mathbf{u}, \tag{A.7}$$

for every vector $\mathbf{u}$. We say that $\mathbf{f}$ represents f. It is easy to show that if $\mathbf{f}$ represents f and $\mathbf{g}$ represents g, then $a\mathbf{f} + b\mathbf{g}$ represents $af + bg$. Any correspondence such as (A.7) which preserves the rules of vector arithmetic is known as an *isomorphism*. Because linear functionals and the vectors that represent them are algebraically indistinguishable, we can and will consider them to be the *same* geometrical objects, simply viewed from a different perspective, in the future. Visualizing a vector $\mathbf{f}$ as a machine (A.6) with one linear slot rather than as an arrow may seem at first like a curious thing to do. The advantage of this alternative viewpoint is that it can be readily generalized to tensors of higher order, as we discuss next.

A.1.3 Multilinear functionals

A multilinear functional of order q—also known as a *tensor of order* q—is a scalar-valued machine with q linear slots that accept vectors:

$$T(\underbrace{\cdot, \cdots\cdots, \cdot}_{q\ \text{vector slots}}) \to \text{scalar.} \tag{A.8}$$

If an ordered sequence of vectors $\mathbf{u}_1, \ldots, \mathbf{u}_q$ is inserted into these slots, the tensor T generates a scalar which we denote by $T(\mathbf{u}_1, \ldots, \mathbf{u}_q)$. If a linear combination of vectors is inserted into any of the slots, we obtain a linear combination of scalars:

$$T(\cdots, a\mathbf{u} + b\mathbf{v}, \cdots) = aT(\cdots, \mathbf{u}, \cdots) + bT(\cdots, \mathbf{v}, \cdots). \tag{A.9}$$

Tensors of the same order q can be multiplied by scalars and added to form linear combinations such as $aT + bP$. A first-order tensor is simply a vector, whereas a zeroth-order tensor is by default a scalar.

The outer or *tensor product* of a tensor T of order q and a tensor P of order p is a tensor TP of order $q+p$, defined by

$$TP(\underbrace{\cdot,\cdots\cdots,\cdot}_{q+p \text{ slots}}) = T(\underbrace{\cdot,\cdots,\cdot}_{q \text{ slots}})P(\underbrace{\cdot,\cdots,\cdot}_{p \text{ slots}}). \tag{A.10}$$

Since the order of the slots in a tensor is important, it is *not* generally true that $PT = TP$. The tensor product fg of two linear functionals is referred to as a *dyad*, whereas the tensor product $f_1 \cdots f_q$ of q linear functionals is referred to as a *polyad of order* q.

The *trace* or *contraction* of a second-order tensor is given by

$$\operatorname{tr} T = T(\hat{\mathbf{x}}_i, \hat{\mathbf{x}}_i), \tag{A.11}$$

where the repeated index implies summation, as usual. More generally, we define the contraction upon the rth and sth slots of a tensor of order q by

$$\operatorname{tr}_{rs} T(\cdot,\cdots\cdots,\cdot) = T(\cdots,\underset{r\text{th slot}}{\underset{\uparrow}{\hat{\mathbf{x}}_i}},\cdots,\underset{s\text{th slot}}{\underset{\uparrow}{\hat{\mathbf{x}}_i}},\cdots). \tag{A.12}$$

It is easy to show that both $\operatorname{tr} T$ and $\operatorname{tr}_{rs} T$ are independent of the Cartesian axis system $\hat{\mathbf{x}}_1$, $\hat{\mathbf{x}}_2$, $\hat{\mathbf{x}}_3$ used to calculate them. Contraction always reduces the order of a tensor by two, so that the machine $\operatorname{tr}_{rs} T$ has $q-2$ slots.

The *transpose* of a second-order tensor is obtained by interchanging the order of the two slots:

$$T^{\mathrm{T}}(\mathbf{u}, \mathbf{v}) = T(\mathbf{v}, \mathbf{u}). \tag{A.13}$$

More generally, we denote the transpose of the rth and sth slots of a tensor of order q by

$$\Pi_{rs} T(\cdots,\underset{r\text{th slot}}{\underset{\uparrow}{\mathbf{u}}},\cdots,\underset{s\text{th slot}}{\underset{\uparrow}{\mathbf{v}}},\cdots) = T(\cdots,\underset{r\text{th slot}}{\underset{\uparrow}{\mathbf{v}}},\cdots,\underset{s\text{th slot}}{\underset{\uparrow}{\mathbf{u}}},\cdots). \tag{A.14}$$

A *symmetric* second-order tensor is one satisfying $S^{\mathrm{T}} = S$ whereas an *anti-symmetric* tensor is one satisfying $A^{\mathrm{T}} = -A$. The second-order *identity tensor* and the third-order *alternating tensor* are defined by

$$I(\mathbf{u}, \mathbf{v}) = \mathbf{u} \cdot \mathbf{v}, \qquad \Lambda(\mathbf{u}, \mathbf{v}, \mathbf{w}) = \mathbf{u} \cdot (\mathbf{v} \times \mathbf{w}). \tag{A.15}$$

These two tensors are symmetric and anti-symmetric in every slot, respectively: $I^{\mathrm{T}} = I$, whereas $\Lambda = -\Pi_{12}\Lambda = -\Pi_{13}\Lambda = -\Pi_{23}\Lambda$.

A.1.4 Components

The three components $f_i = \hat{\mathbf{x}}_i \cdot \mathbf{f}$ of a vector $\mathbf{f}$ are given in terms of the corresponding linear functional f by $f_i = f(\hat{\mathbf{x}}_i)$. The 3^q components of a tensor of order q are defined in an analogous manner:

$$T_{i_1\cdots i_q} = T(\hat{\mathbf{x}}_{i_1}, \ldots, \hat{\mathbf{x}}_{i_q}). \tag{A.16}$$

Just as every vector $\mathbf{f} = f_i\hat{\mathbf{x}}_i$ is completely determined by its components relative to any Cartesian axis system, so is every tensor; in fact,

$$T = T_{i_1\cdots i_q}\hat{\mathbf{x}}_{i_1}\cdots\hat{\mathbf{x}}_{i_q}. \tag{A.17}$$

The machines on both sides of equation (A.17) assign the same scalar to any ordered sequence of vectors: $T(\mathbf{u}_1, \ldots, \mathbf{u}_q) = T_{i_1\cdots i_q}\hat{\mathbf{x}}_{i_1}\cdots\hat{\mathbf{x}}_{i_q}(\mathbf{u}_1, \ldots, \mathbf{u}_q)$. The 3^q polyads $\hat{\mathbf{x}}_{i_1}\cdots\hat{\mathbf{x}}_{i_q}$ constitute a basis for the space of all tensors of order q. A second-order tensor, for example, can be written in terms of the nine dyads $\hat{\mathbf{x}}_i\hat{\mathbf{x}}_j$ in the form $T = T_{ij}\hat{\mathbf{x}}_i\hat{\mathbf{x}}_j$.

Tensor arithmetic can be performed in terms of components; the outer product of a tensor of order q and a tensor of order p is given by

$$(TP)_{i_1\cdots i_q j_1\cdots j_p} = T_{i_1\cdots i_q}P_{j_1\cdots j_p}, \tag{A.18}$$

and the trace of a second-order tensor is given by

$$\operatorname{tr} T = T_{ii}. \tag{A.19}$$

More generally, for a tensor of order q,

$$(\operatorname{tr}_{rs}T)_{i_1\cdots i_{q-2}} = T_{i_1\cdots \underset{\substack{\uparrow\\ r\text{th index}}}{j} \cdots \underset{\substack{\uparrow\\ s\text{th index}}}{j} \cdots i_{q-2}}. \tag{A.20}$$

The transposed tensors T^{T} and $\Pi_{rs}T$ have transposed indices: $T^{\mathrm{T}}_{ij} = T_{ji}$ and $(\Pi_{rs}T)_{\ldots j\ldots k\ldots} = T_{\ldots k\ldots j\ldots}$, where the unaffected indices in the latter expression are represented by ellipses. The components of a symmetric second-order tensor S satisfy $S_{ij} = S_{ji}$, whereas those of an anti-symmetric tensor A satisfy $A_{ij} = -A_{ji}$.

The 3^q components $T'_{i_1\cdots i_q}$ of a tensor of order q with respect to a new primed Cartesian axis system $\hat{\mathbf{x}}'_1$, $\hat{\mathbf{x}}'_2$, $\hat{\mathbf{x}}'_3$ are related to the components $T_{j_1\cdots j_q}$ with respect to the original unprimed system $\hat{\mathbf{x}}_1$, $\hat{\mathbf{x}}_2$, $\hat{\mathbf{x}}_3$ by

$$\begin{aligned} T'_{i_1\cdots i_q} &= T(\hat{\mathbf{x}}'_{i_1}, \ldots, \hat{\mathbf{x}}'_{i_q}) \\ &= T((\hat{\mathbf{x}}'_{i_1}\cdot\hat{\mathbf{x}}_{j_1})\hat{\mathbf{x}}_{j_1}, \ldots, (\hat{\mathbf{x}}'_{i_q}\cdot\hat{\mathbf{x}}_{j_q})\hat{\mathbf{x}}_{j_q}) \\ &= (\hat{\mathbf{x}}'_{i_1}\cdot\hat{\mathbf{x}}_{j_1})\cdots(\hat{\mathbf{x}}'_{i_q}\cdot\hat{\mathbf{x}}_{j_q})T(\hat{\mathbf{x}}_{j_1}, \ldots, \hat{\mathbf{x}}_{j_q}) \\ &= (\hat{\mathbf{x}}'_{i_1}\cdot\hat{\mathbf{x}}_{j_1})\cdots(\hat{\mathbf{x}}'_{i_q}\cdot\hat{\mathbf{x}}_{j_q})T_{j_1\cdots j_q}. \end{aligned} \tag{A.21}$$

The transformation relation (A.21) between components under a rigid rotation of the axes serves as the starting point in many introductory discussions of tensors; the present treatment in terms of machines with linear slots is completely equivalent, but has a stronger geometrical flavor.

A.1.5 Isotropic tensors

The components of the second-order identity tensor and the third-order alternating tensor (A.15) are the Kronecker delta and the Levi-Cività alternating symbol, respectively:

$$I_{ij} = \delta_{ij}, \qquad \Lambda_{ijk} = \varepsilon_{ijk}. \tag{A.22}$$

Tensors such as I and Λ which have the same components in every right-handed coordinate system $\hat{\mathbf{x}}_1$, $\hat{\mathbf{x}}_2$, $\hat{\mathbf{x}}_3$ are said to be *isotropic*. There are no isotropic first-order tensors except the zero vector. Every isotropic second-order tensor is a multiple of the identity tensor I, whereas every isotropic third-order tensor is a multiple of the alternating tensor Λ. Isotropic tensors of higher order are linear combinations of products of I and Λ with permuted indices. For example, the most general isotropic fourth-order tensor is of the form $aII + b\Pi_{23}(II) + c\Pi_{24}(II)$, where a, b and c are scalars. We summarize the above results for tensors of order one through four using index notation in Table A.1.

Order	Form of Isotropic Tensor
0	all scalars
1	only the zero vector
2	$a\delta_{ij}$
3	$a\varepsilon_{ijk}$
4	$a\delta_{ij}\delta_{kl} + b\delta_{ik}\delta_{jl} + c\delta_{il}\delta_{jk}$

Table A.1. Most general isotropic tensor of order one through four. Technically, the third-order alternating tensor is chirally isotropic rather than completely isotropic, since its components change sign in transforming from a right-handed to a (shudder) left-handed Cartesian axis system. There are no completely isotropic tensors of any odd order q except the zero tensor.

A.1.6 Wedge operator

The wedge operator $\wedge$ acts upon a second-order tensor T to produce a vector $\wedge T$, given by

$$\wedge T = \mathrm{tr}_{23}\mathrm{tr}_{35}(\Lambda T). \tag{A.23}$$

This can be expressed in a more comprehensible form using index notation:

$$(\wedge T)_i = \varepsilon_{ijk}T_{jk}. \tag{A.24}$$

Application of the wedge operator to a dyad yields the cross product of the constituent vectors:

$$\wedge(\mathbf{fg}) = \mathbf{f} \times \mathbf{g}. \tag{A.25}$$

In fact, this provides the motivation for the terminology, since an alternative notation for the cross product $\mathbf{f} \times \mathbf{g}$ is $\mathbf{f} \wedge \mathbf{g}$. The vector returned by the wedge operator vanishes, $\wedge T = \mathbf{0}$, if and only if the tensor that is acted upon is symmetric, $T^{\mathrm{T}} = T$.

A.2 Tensors as Linear Operators

We may alternatively consider a tensor of order q to be a machine with a single linear slot that accepts tensors of order p and returns tensors of order $q - p$. This geometrical picture of a tensor as a linear operator is frequently the more natural one from a physical point of view. We begin by considering tensors of order $q = 2$, for which $p = q - p = 1$.

A.2.1 Second-order tensors

A vector-valued *linear operator* ψ is a single-slot machine of the form

$$\underset{\substack{\uparrow\\ \text{vector}}}{\psi(\,\cdot\,)} \rightarrow \text{vector}. \tag{A.26}$$

Insertion of a linear combination of vectors produces a linear combination of output vectors:

$$\psi(a\mathbf{u} + b\mathbf{v}) = a\psi(\mathbf{u}) + b\psi(\mathbf{v}). \tag{A.27}$$

A linear combination of linear operators is defined in the now familiar manner: $(a\psi + b\chi)(\mathbf{u}) = a\psi(\mathbf{u}) + b\chi(\mathbf{u})$. Every linear operator ψ generates a unique second-order tensor which we denote by T_ψ via the correspondence

$$T_\psi(\mathbf{u},\mathbf{v}) = \mathbf{u} \cdot \psi(\mathbf{v}). \tag{A.28}$$

This correspondence is an isomorphism, since $T_{a\psi+b\chi} = aT_\psi + bT_\chi$. The identity tensor I is generated by the *identity operator*, which simply gives the input vector back unchanged: $\psi(\mathbf{u}) = \mathbf{u}$. This is of course the reason for the terminology.

The *transpose* ψ^{T} of a linear operator is defined by

$$\psi^{\mathrm{T}}(\mathbf{u}) \cdot \mathbf{v} = \mathbf{u} \cdot \psi(\mathbf{v}). \tag{A.29}$$

It is obvious that $(\psi^{\mathrm{T}})^{\mathrm{T}} = \psi$. The transpose of an operator and the transpose of the tensor that it generates coincide since $T_{\psi^{\mathrm{T}}} = (T_\psi)^{\mathrm{T}}$. A *symmetric* linear operator satisfies $\psi^{\mathrm{T}} = \psi$ whereas an anti-symmetric operator satisfies $\psi^{\mathrm{T}} = -\psi$. The *product* of two operators is defined by

$$\psi\chi(\mathbf{u}) = \psi(\chi(\mathbf{u})). \tag{A.30}$$

The convention is that the operator on the right acts first, i.e., we take the output of χ and feed it to ψ. Operator products are not generally commutative: $\psi\chi \neq \chi\psi$. The order of the operators is interchanged upon taking the transpose: $(\psi\chi)^{\mathrm{T}} = \chi^{\mathrm{T}}\psi^{\mathrm{T}}$. What is the tensor $T_{\psi\chi}$ generated by $\psi\chi$? It is not $T_\psi T_\chi$, since that is a fourth-order tensor. In fact,

$$T_{\psi\chi} = \mathrm{tr}_{23}(T_\psi T_\chi). \tag{A.31}$$

Having established these results, we may proceed to confuse second-order tensors and vector linear operators just as we confuse vectors and linear functionals. We shall no longer distinguish between ψ and T_ψ in what follows; a common symbol will be used to denote both a linear operator and the tensor that it generates.

A.2.2 Components of a second-order tensor

The *components* T_{ij} of a linear operator T are simply the components of the corresponding tensor. The linear relation $\mathbf{u} = T(\mathbf{v})$ takes a familiar form in index notation, namely,

$$u_i = T_{ij}v_j. \tag{A.32}$$

If we arrange the components T_{ij} in a 3×3 array, we can write equation (A.32) out explicitly as

$$\begin{pmatrix} u_1 \\ u_2 \\ u_3 \end{pmatrix} = \begin{pmatrix} T_{11} & T_{12} & T_{13} \\ T_{21} & T_{22} & T_{23} \\ T_{31} & T_{32} & T_{33} \end{pmatrix} \begin{pmatrix} v_1 \\ v_2 \\ v_3 \end{pmatrix}. \tag{A.33}$$

The result (A.33) enables a linear operator $T = T_{ij}\hat{\mathbf{x}}_i\hat{\mathbf{x}}_j$ to be evaluated at a vector $\mathbf{v} = v_j\hat{\mathbf{x}}_j$ by means of matrix multiplication. The effect of a linear operator upon the Cartesian unit vectors $\hat{\mathbf{x}}_1, \hat{\mathbf{x}}_2, \hat{\mathbf{x}}_3$ can be expressed in terms of the components of the tensor in the form

$$T(\hat{\mathbf{x}}_i) = T_{ji}\hat{\mathbf{x}}_j. \tag{A.34}$$

The appearance of T_{ij} in (A.32) and T_{ji} in (A.34) is an unavoidable feature, which Halmos (1958) has dubbed the "perversity of the indices".

A.2.3 Determinant and inverse

The *determinant* of a linear operator T is the determinant of its matrix of components relative to any Cartesian axis, defined by

$$\det T = \tfrac{1}{6}\varepsilon_{ijk}\varepsilon_{lmn}T_{il}T_{jm}T_{kn}. \tag{A.35}$$

The same result is obtained in (A.35) no matter which axes $\hat{\mathbf{x}}_1$, $\hat{\mathbf{x}}_2$, $\hat{\mathbf{x}}_3$ are used; this ensures that $\det T$ is a geometric object. The determinant of an operator product is the product of determinants: $\det TP = (\det T)(\det P)$. A non-singular operator T is one whose determinant is non-zero: $\det T \neq 0$. Every such operator has a unique *inverse* T^{-1} having the property that $\mathbf{v} = T^{-1}(\mathbf{u})$ if and only if $\mathbf{u} = T(\mathbf{v})$; in other words, $T^{-1}T = TT^{-1} = I$. If two linear operators T and P are invertible then so is their product, and $(TP)^{-1} = P^{-1}T^{-1}$. The inverse of the transpose of an operator is the transpose of the inverse: $(T^{\mathrm{T}})^{-1} = (T^{-1})^{\mathrm{T}}$. In view of this we shall employ the condensed notation $T^{-\mathrm{T}}$. It is evident that $\det T^{\mathrm{T}} = \det T$ and $\det T^{-1} = (\det T)^{-1}$. The components T_{ij}^{-1} of the inverse operator T^{-1} are given in terms of the determinant $\det T$ and the components T_{ij} of T by

$$T_{ij}^{-1} = \tfrac{1}{2}(\det T)^{-1}\varepsilon_{imn}\varepsilon_{jkl}T_{km}T_{ln}. \tag{A.36}$$

Equation (A.36) is a succinct index version of the well-known algorithm for solving a system of three simultaneous linear equations, *Cramer's rule*.

A.2.4 Higher-order tensors

There is also an isomorphism between tensors or multilinear functionals Γ of order $q > 2$ and single-slot linear operators of the form

$$\underset{\substack{\uparrow \\ \text{tensor of order } p}}{\Gamma(\,\cdot\,)} \rightarrow \text{tensor of order } q - p. \tag{A.37}$$

As with the correspondence for tensors of order two, it is important to keep the order of the slots straight; by analogy with equation (A.32) the tensor $T = \Gamma(\varepsilon)$ has components

$$T_{i_1\cdots i_{q-p}} = \Gamma_{i_1\cdots i_{q-p}j_1\cdots j_p}\varepsilon_{j_i\cdots j_p}. \tag{A.38}$$

Any linear constitutive relation is of the form (A.38); for example, Hooke's "law" relating the stress T_{ij} and strain ε_{kl} in a classical elastic medium is $T_{ij} = \Gamma_{ijkl}\varepsilon_{kl}$. The symmetries $\Gamma_{ijkl} = \Gamma_{jikl} = \Gamma_{ijlk} = \Gamma_{klij}$ of the fourth-order elastic tensor allow us to shuffle the order of the indices in this particular example; more generally, however, we cannot.

A.3 Gibbs Notation

The invariant notation developed by Gibbs (1901) provides a particularly convenient means of dealing with vectors and second-order tensors. We summarize this notation, which is used throughout this book, here. Both vectors and tensors are written using a bold font. Generally, vectors are lower case whereas tensors are upper case. (We flout this convention whenever it conflicts with a well-established seismological tradition; for example, we use a lower-case $\boldsymbol{\varepsilon}$ to denote the infinitesimal strain tensor.) The transpose of a tensor $\mathbf{T}$ is written as $\mathbf{T}^{\mathrm{T}}$, the inverse is written as $\mathbf{T}^{-1}$, and the inverse transpose is written as $\mathbf{T}^{-\mathrm{T}}$. The outer product of two tensors $\mathbf{T}$ and $\mathbf{P}$ is denoted by $\mathbf{TP}$, without an intervening "product" symbol. A polyad of order q formed from the vectors $\mathbf{f}_1, \ldots, \mathbf{f}_q$ is thus written as $\mathbf{f}_1 \cdots \mathbf{f}_q$; we have already used this notation in stipulating that the 3^q polyads $\hat{\mathbf{x}}_{i_1} \cdots \hat{\mathbf{x}}_{i_q}$ constitute a basis for the space of tensors of order q. The scalar assigned by a bilinear functional $\mathbf{T}$ to the vectors $\mathbf{u}$ and $\mathbf{v}$ is denoted by $\mathbf{u} \cdot \mathbf{T} \cdot \mathbf{v}$, whereas the vector assigned by the linear operator $\mathbf{T}$ to $\mathbf{u}$ is denoted by $\mathbf{T} \cdot \mathbf{u}$. The vector assigned by the transposed operator $\mathbf{T}^{\mathrm{T}}$ to $\mathbf{u}$ is denoted by $\mathbf{T}^{\mathrm{T}} \cdot \mathbf{u} = \mathbf{u} \cdot \mathbf{T}$. Finally, the operator product of the two tensors $\mathbf{T}$ and $\mathbf{P}$ is denoted by $\mathbf{T} \cdot \mathbf{P} = \mathrm{tr}_{23}(\mathbf{TP})$.

The beauty of this notation is that the dot does "quadruple duty". Between two vectors $\mathbf{u}$ and $\mathbf{v}$ it is the ordinary dot product $\mathbf{u} \cdot \mathbf{v}$. Dotting a vector $\mathbf{u}$ into a tensor $\mathbf{T}$ on the right yields the resulting value of the associated linear operator $\mathbf{T} \cdot \mathbf{u}$, whereas dotting the same vector into $\mathbf{T}$ on the left yields the value of the transposed linear operator $\mathbf{u} \cdot \mathbf{T} = \mathbf{T}^{\mathrm{T}} \cdot \mathbf{u}$. Finally, a dot between two tensors $\mathbf{T}$ and $\mathbf{P}$ denotes the operator product $\mathbf{T} \cdot \mathbf{P}$. Such a multi-purpose dot product enables us to eliminate or add parentheses in expressions such as $\mathbf{u} \cdot \mathbf{T} \cdot \mathbf{v} = \mathbf{u} \cdot (\mathbf{T} \cdot \mathbf{v}) = (\mathbf{u} \cdot \mathbf{T}) \cdot \mathbf{v}$ at will. We can see immediately, for example, that the operator product $(\mathbf{uf}) \cdot (\mathbf{gv})$ is the weighted dyad $\mathbf{u}(\mathbf{f} \cdot \mathbf{g})\mathbf{v}$. The inverse of a non-singular tensor satisfies $\mathbf{T}^{-1} \cdot \mathbf{T} = \mathbf{T} \cdot \mathbf{T}^{-1} = \mathbf{I}$, where $\mathbf{I}$ is the identity. The transpose and the inverse of an operator dot product are given by $(\mathbf{T} \cdot \mathbf{P})^{\mathrm{T}} = \mathbf{P}^{\mathrm{T}} \cdot \mathbf{T}^{\mathrm{T}}$ and $(\mathbf{T} \cdot \mathbf{P})^{-1} = \mathbf{P}^{-1} \cdot \mathbf{T}^{-1}$.

Every dot signifies contraction over adjacent slots, so that it is trivial to translate an invariant expression written using Gibbs notation into index notation, and vice versa. The scalar $\mathbf{u}\cdot\mathbf{T}\cdot\mathbf{v}$ is $u_iT_{ij}v_j$, whereas the components of the vectors $\mathbf{T}\cdot\mathbf{u}$ and $\mathbf{u}\cdot\mathbf{T}=\mathbf{T}^{\mathrm{T}}\cdot\mathbf{u}$ are $T_{ij}u_j$ and $u_jT_{ji}=T^{\mathrm{T}}_{ij}u_j$, respectively. We also use a double dot product to signify contraction over two adjacent indices:

$$\mathbf{T}:\mathbf{P}=\mathrm{tr}(\mathbf{T}^{\mathrm{T}}\cdot\mathbf{P})=\mathrm{tr}(\mathbf{T}\cdot\mathbf{P}^{\mathrm{T}})=T_{ij}P_{ij}. \tag{A.39}$$

There are relatively few occasions in this book when we have to specify the norm of a second-order tensor; on these occasions, we shall always mean the Frobenius or ordinary Euclidean norm $\|\mathbf{T}\|=(\mathbf{T}:\mathbf{T})^{1/2}$.

The *cross product* of a vector and a dyad is defined in a way that also allows us to eliminate parentheses:

$$\mathbf{u}\times\mathbf{vw}=(\mathbf{u}\times\mathbf{v})\mathbf{w},\qquad \mathbf{vw}\times\mathbf{u}=\mathbf{v}(\mathbf{w}\times\mathbf{u}). \tag{A.40}$$

To find the cross product of a vector with a general second-order tensor, it is simply necessary to expand $\mathbf{u}=u_k\hat{\mathbf{x}}_k$ and $\mathbf{T}=T_{ij}\hat{\mathbf{x}}_i\hat{\mathbf{x}}_j$ and make use of the linearity. We find in this manner that the components of

$$\mathbf{P}=\mathbf{u}\times\mathbf{T},\qquad \mathbf{M}=\mathbf{T}\times\mathbf{u} \tag{A.41}$$

are given by

$$P_{ij}=\varepsilon_{ikl}u_kT_{lj},\qquad M_{ij}=\varepsilon_{jlk}T_{il}u_k. \tag{A.42}$$

The effect of transposing such vector-times-tensor or tensor-times-vector cross products is $(\mathbf{u}\times\mathbf{T})^{\mathrm{T}}=-\mathbf{T}^{\mathrm{T}}\times\mathbf{u}$ and $(\mathbf{T}\times\mathbf{u})^{\mathrm{T}}=-\mathbf{u}\times\mathbf{T}^{\mathrm{T}}$. The single dot or cross product of two higher-order tensors can be interpreted in an analogous manner; in any combination such as $\mathbf{T}\cdot\mathbf{P}$ or $\mathbf{T}\times\mathbf{P}$, it is the last slot of the left tensor and the first slot of the right tensor that are contracted or crossed.

In addition to the above strictly Gibbsian notation, we employ two minor modifications. First, we generalize the double dot product slightly, using $\mathbf{T}=\boldsymbol{\Gamma}:\boldsymbol{\varepsilon}$ to mean $T_{ij}=\Gamma_{ijkl}\varepsilon_{kl}$ and $\boldsymbol{\varepsilon}:\boldsymbol{\Gamma}:\boldsymbol{\varepsilon}$ to mean $\varepsilon_{ij}\Gamma_{ijkl}\varepsilon_{kl}$. Second, we use a triple dot to signify sequential contraction over *all* of the indices of two higher-order tensors:

$$\mathbf{T}\vdots\mathbf{P}=\mathrm{tr}_{12}\mathrm{tr}_{13}\cdots\mathrm{tr}_{1q}\mathrm{tr}_{1\,q+1}(\mathbf{TP})=T_{i_1\cdots i_q}P_{i_1\cdots i_q}. \tag{A.43}$$

The 3^q components of a tensor are given in terms of this invariant notation by $T_{i_1\cdots i_q}=(\hat{\mathbf{x}}_{i_1}\cdots\hat{\mathbf{x}}_{i_q})\vdots\mathbf{T}$. The Euclidean norm of $\mathbf{T}$ is $\|\mathbf{T}\|=(\mathbf{T}\vdots\mathbf{T})^{1/2}$.

Many more complicated constructions involving higher-order tensors cannot be written gracefully using Gibbs notation, even with the above extensions. The tensors with components u_jT_{ijk} and $P_{jk}M_{ijkl}$, for example, are $\mathbf{u}\cdot(\Pi_{12}\mathbf{T})$ and $\mathbf{P}:(\Pi_{23}\Pi_{12}\mathbf{M})$. Most people, upon confronting such unwieldy expressions, would simply translate them into index notation to figure out what they mean. We use Gibbs notation as frequently as possible in this book; however, we do not hesitate to resort to index notation whenever we feel that it aids comprehension.

A.4 Cartesian and Polar Decomposition

Every *symmetric* tensor $\mathbf{S} = \mathbf{S}^{\mathrm{T}}$ has three real *eigenvalues* λ_1, λ_2, λ_3 and three associated mutually perpendicular unit *eigenvectors* $\hat{\eta}_1$, $\hat{\eta}_2$, $\hat{\eta}_3$ satisfying

$$\mathbf{S}\cdot\hat{\eta}_1 = \lambda_1\hat{\eta}_1, \qquad \mathbf{S}\cdot\hat{\eta}_2 = \lambda_2\hat{\eta}_2, \qquad \mathbf{S}\cdot\hat{\eta}_3 = \lambda_3\hat{\eta}_3. \tag{A.44}$$

We can write such a tensor in *diagonal form* as a sum of three dyads:

$$\mathbf{S} = \lambda_1\hat{\eta}_1\hat{\eta}_1 + \lambda_2\hat{\eta}_2\hat{\eta}_2 + \lambda_3\hat{\eta}_3\hat{\eta}_3. \tag{A.45}$$

The matrix of components relative to the basis $\hat{\eta}_1$, $\hat{\eta}_2$, $\hat{\eta}_3$ is, of course,

$$\begin{pmatrix} S_{11} & S_{12} & S_{13} \\ S_{21} & S_{22} & S_{23} \\ S_{31} & S_{32} & S_{33} \end{pmatrix} = \begin{pmatrix} \lambda_1 & 0 & 0 \\ 0 & \lambda_2 & 0 \\ 0 & 0 & \lambda_3 \end{pmatrix}. \tag{A.46}$$

A *positive definite* symmetric tensor is one satisfying $\mathbf{u}\cdot\mathbf{S}\cdot\mathbf{u} > 0$ for every vector $\mathbf{u}$. A symmetric tensor is positive definite if and only if all of its eigenvalues are positive: $\lambda_1 > 0$, $\lambda_2 > 0$, $\lambda_3 > 0$. If $\mathbf{u}\cdot\mathbf{S}\cdot\mathbf{u} \geq 0$ instead, then $\mathbf{S}$ is said to be positive semi-definite; in this case, the eigenvalues are simply non-negative: $\lambda_1 \geq 0$, $\lambda_2 \geq 0$, $\lambda_3 \geq 0$. Every positive definite symmetric tensor has a unique positive definite *square root* which satisfies the relation $\mathbf{S}^{1/2}\cdot\mathbf{S}^{1/2} = \mathbf{S}$. To find $\mathbf{S}^{1/2}$ we diagonalize $\mathbf{S}$ and take the square root of the eigenvalues:

$$\mathbf{S}^{1/2} = \sqrt{\lambda_1}\hat{\eta}_1\hat{\eta}_1 + \sqrt{\lambda_2}\hat{\eta}_2\hat{\eta}_2 + \sqrt{\lambda_3}\hat{\eta}_3\hat{\eta}_3. \tag{A.47}$$

Two symmetric tensors $\mathbf{S}$ and $\mathbf{S}'$ have the same eigenvectors $\hat{\eta}_1$, $\hat{\eta}_2$, $\hat{\eta}_3$ if and only if they *commute*: $\mathbf{S}\cdot\mathbf{S}' = \mathbf{S}'\cdot\mathbf{S}$.

Every *anti-symmetric* tensor $\mathbf{A} = -\mathbf{A}^{\mathrm{T}}$ can be associated with a unique vector $\mathbf{a}$, and vice versa, via the isomomorphism

$$\mathbf{a} = -\tfrac{1}{2}\wedge\mathbf{A}, \qquad \mathbf{A} = -\mathbf{a}\cdot\mathbf{\Lambda}, \tag{A.48}$$

where $\wedge$ denotes the wedge operator and $\mathbf{\Lambda}$ is the third-order alternating tensor. In index notation, equations (A.48) take the form

$$a_i = -\tfrac{1}{2}\varepsilon_{ijk}A_{jk}, \qquad A_{jk} = -a_i\varepsilon_{ijk}. \tag{A.49}$$

The component matrix of $\mathbf{A}$ can be written in terms of the components of the vector $\mathbf{a}$ in the form

$$\begin{pmatrix} A_{11} & A_{12} & A_{13} \\ A_{21} & A_{22} & A_{23} \\ A_{31} & A_{32} & A_{33} \end{pmatrix} = \begin{pmatrix} 0 & -a_3 & a_2 \\ a_3 & 0 & -a_1 \\ -a_2 & a_1 & 0 \end{pmatrix}. \tag{A.50}$$

Taking the dot product of an anti-symmetric tensor with an arbitrary vector $\mathbf{u}$ is equivalent to taking the cross product with the associated vector:

$$\mathbf{A} \cdot \mathbf{u} = \mathbf{a} \times \mathbf{u}. \tag{A.51}$$

Dotting with the transpose corresponds to reversing the order of the cross product: $\mathbf{A}^{\mathrm{T}} \cdot \mathbf{u} = \mathbf{u} \cdot \mathbf{A} = \mathbf{u} \times \mathbf{a}$.

Every second-order tensor can be written as the sum of a symmetric tensor and an anti-symmetric tensor in the form

$$\mathbf{T} = \mathbf{S} + \mathbf{A}, \tag{A.52}$$

where

$$\mathbf{S} = \tfrac{1}{2}(\mathbf{T} + \mathbf{T}^{\mathrm{T}}), \qquad \mathbf{A} = \tfrac{1}{2}(\mathbf{T} - \mathbf{T}^{\mathrm{T}}). \tag{A.53}$$

Equation (A.52) is sometimes referred to as the *Cartesian decomposition* of a tensor $\mathbf{T}$, because of the analogy to the representation of a complex number as a sum of real and imaginary parts: $z = x + iy$ where $x = \frac{1}{2}(z + z^*)$ and $y = \frac{1}{2i}(z - z^*)$.

An *orthogonal tensor* is one satisfying $\mathbf{Q} \cdot \mathbf{Q}^{\mathrm{T}} = \mathbf{Q}^{\mathrm{T}} \cdot \mathbf{Q} = \mathbf{I}$ or, equivalently, $\mathbf{Q}^{-1} = \mathbf{Q}^{\mathrm{T}}$. Since $\det \mathbf{Q}^{\mathrm{T}} = \det \mathbf{Q}$ and $\det \mathbf{I} = 1$, it follows that $(\det \mathbf{Q})^2 = 1$. An orthogonal tensor having $\det \mathbf{Q} = 1$ is said to be *proper* whereas one having $\det \mathbf{Q} = -1$ is said to be *improper*. Every proper orthogonal tensor corresponds to a *rigid rotation*, with a component matrix of the form

$$\begin{pmatrix} Q_{11} & Q_{12} & Q_{13} \\ Q_{21} & Q_{22} & Q_{23} \\ Q_{31} & Q_{32} & Q_{33} \end{pmatrix} = \begin{pmatrix} \cos\gamma & -\sin\gamma & 0 \\ \sin\gamma & \cos\gamma & 0 \\ 0 & 0 & 1 \end{pmatrix} \tag{A.54}$$

in some Cartesian axis system $\hat{\boldsymbol{\sigma}}_1$, $\hat{\boldsymbol{\sigma}}_2$, $\hat{\boldsymbol{\sigma}}_3$. Applying $\mathbf{Q}$ to an arbitrary vector $\mathbf{u}$ yields a vector $\mathbf{Q} \cdot \mathbf{u}$ which has been rotated through an angle γ about the $\hat{\boldsymbol{\sigma}}_3$ axis; the sense of rotation for $\gamma > 0$ is given by the right-hand rule. The converse statement that every rigid rotation can be represented by a proper orthogonal tensor of the form (A.54) in some Cartesian axis system $\hat{\boldsymbol{\sigma}}_1$, $\hat{\boldsymbol{\sigma}}_2$, $\hat{\boldsymbol{\sigma}}_3$ is known as *Euler's theorem*. The finite rotation angle γ may be found, given the component matrix of $\mathbf{Q}$ in an arbitrary Cartesian axis system, using the invariant relation $\operatorname{tr} \mathbf{Q} = Q_{11} + Q_{22} + Q_{33} = 1 + 2\cos\gamma$.

Every improper orthogonal tensor corresponds to a rigid rotation either followed or preceded by a mirror reflection.

Every non-singular tensor can be written as the product of a positive-definite symmetric tensor and an orthogonal tensor in two ways:

$$\mathbf{T} = \mathbf{Q} \cdot \mathbf{R} = \mathbf{L} \cdot \mathbf{Q}. \tag{A.55}$$

These two decompositions are unique; the right and left symmetric tensors are given explicitly by

$$\mathbf{R} = (\mathbf{T}^{\mathrm{T}} \cdot \mathbf{T})^{1/2} = \mathbf{R}^{\mathrm{T}}, \qquad \mathbf{L} = (\mathbf{T} \cdot \mathbf{T}^{\mathrm{T}})^{1/2} = \mathbf{L}^{\mathrm{T}}, \tag{A.56}$$

whereas the orthogonal tensor is

$$\mathbf{Q} = \mathbf{T} \cdot \mathbf{R}^{-1} = \mathbf{L}^{-1} \cdot \mathbf{T}. \tag{A.57}$$

It is readily verified that $\mathbf{Q}$ satisfies $\mathbf{Q} \cdot \mathbf{Q}^{\mathrm{T}} = \mathbf{Q}^{\mathrm{T}} \cdot \mathbf{Q} = \mathbf{I}$; whether it is proper or improper depends upon whether $\det \mathbf{T} > 0$ or $\det \mathbf{T} < 0$. Equation (A.55) is analogous to the polar representation of a complex number: $z = re^{i\gamma}$ where $r = (x^2 + y^2)^{1/2}$ and $\gamma = \arctan(y/x)$. For this reason, it is referred to as the *polar decomposition*.

A.5 Grad, Div and All That

A *tensor field* is a rule that assigns a tensor $\mathbf{T}$ to every point $\mathbf{r}$ in all or a portion of three-dimensional space. We say that $\mathbf{T}$ is differentiable at a point $\mathbf{r}$ if there exists a tensor field $\boldsymbol{\nabla}\mathbf{T}$ having the property that

$$\mathbf{T}(\mathbf{r} + d\mathbf{r}) = \mathbf{T}(\mathbf{r}) + d\mathbf{r} \cdot \boldsymbol{\nabla}\mathbf{T}(\mathbf{r}) + \cdots, \tag{A.58}$$

where the ellipsis denotes terms that are of second order or higher in the differential position $\|d\mathbf{r}\|$. If $\mathbf{T}$ is a tensor field of order q with components $T_{j_1 \cdots j_q}$, then $\boldsymbol{\nabla}\mathbf{T}$ is a tensor field of order $q+1$ with components $\partial_i T_{j_1 \cdots j_q}$. We refer to $\boldsymbol{\nabla}\mathbf{T}$ as the *gradient* of the tensor field $\mathbf{T}$; the del or grad symbol can be regarded as a vector operator of the form $\boldsymbol{\nabla} = \hat{\mathbf{x}}_i \partial_i$. The gradient of the product of two tensors satisfies the *chain rule*:

$$\boldsymbol{\nabla}(\mathbf{T}\mathbf{P}) = (\boldsymbol{\nabla}\mathbf{T})\mathbf{P} + \mathbf{T}(\boldsymbol{\nabla}\mathbf{P}) \tag{A.59}$$

or, equivalently, $\partial_i (T_{j_1 \cdots j_q} P_{k_1 \cdots k_p}) = (\partial_i T_{j_1 \cdots j_q}) P_{k_1 \cdots k_p} + T_{j_1 \cdots j_q} (\partial_i P_{k_1 \cdots k_p})$.

The gradient of a scalar field is a vector $\boldsymbol{\nabla}\psi$ which is normal to the surfaces of constant ψ. The double gradient of ψ is a symmetric second-order tensor $\boldsymbol{\nabla}\boldsymbol{\nabla}\psi = (\boldsymbol{\nabla}\boldsymbol{\nabla}\psi)^{\mathrm{T}}$ with components $\partial_i \partial_j \psi = \partial_j \partial_i \psi$. We adhere scrupulously to the convention that a differential operator such as $\boldsymbol{\nabla}$ acts only upon the field *to its immediate right*, writing $\boldsymbol{\nabla}\psi \cdot \boldsymbol{\nabla}\chi$ rather than the parenthesis-laden $(\boldsymbol{\nabla}\psi)\cdot(\boldsymbol{\nabla}\chi)$. Parentheses are required to express the gradient of a product such as $\boldsymbol{\nabla}(\mathbf{u}\cdot\boldsymbol{\nabla}\psi)$. We never allow an operator to act upon a field to its left; thus, we write the strain accompanying a displacement $\mathbf{u}$ in the form $\frac{1}{2}[\boldsymbol{\nabla}\mathbf{u}+(\boldsymbol{\nabla}\mathbf{u})^{\mathrm{T}}]$ rather than $\frac{1}{2}(\boldsymbol{\nabla}\mathbf{u}+\mathbf{u}\boldsymbol{\nabla})$.

The *divergence* of a tensor field of order q is a tensor field of order $q-1$, defined by

$$\boldsymbol{\nabla}\cdot\mathbf{T}=\mathrm{tr}_{12}(\boldsymbol{\nabla}\mathbf{T}). \tag{A.60}$$

The *Laplacian* is a tensor field of the same order, given by

$$\nabla^2\mathbf{T}=\boldsymbol{\nabla}\cdot(\boldsymbol{\nabla}\mathbf{T})=\mathrm{tr}_{12}(\boldsymbol{\nabla}\boldsymbol{\nabla}\mathbf{T}). \tag{A.61}$$

The components of $\boldsymbol{\nabla}\cdot\mathbf{T}$ and $\nabla^2\mathbf{T}$ are $\partial_i T_{ij_1\cdots j_{q-1}}$ and $\partial_i^2 T_{j_1\cdots j_q}$, respectively. Symbolically, we may write $\nabla^2=\boldsymbol{\nabla}\cdot\boldsymbol{\nabla}=\partial_i^2$. The divergence of a vector field $\mathbf{u}$ is a scalar field $\boldsymbol{\nabla}\cdot\mathbf{u}=\partial_i u_i$, and its Laplacian is a vector field $\nabla^2\mathbf{u}$ with components $\partial_i^2 u_j$. The divergence of a second-order tensor field $\mathbf{T}$ is a vector field $\boldsymbol{\nabla}\cdot\mathbf{T}$ with components $\partial_i T_{ij}$.

The *curl* $\boldsymbol{\nabla}\times\mathbf{u}$ of a vector field is a vector field with components $\varepsilon_{ijk}\partial_j u_k$. The curl of a gradient vanishes, $\boldsymbol{\nabla}\times\boldsymbol{\nabla}\psi=\mathbf{0}$, as does the divergence of a curl, $\boldsymbol{\nabla}\cdot(\boldsymbol{\nabla}\times\mathbf{u})=0$. The divergence of a vector cross product is $\boldsymbol{\nabla}\cdot(\mathbf{u}\times\mathbf{v})=\mathbf{v}\cdot\boldsymbol{\nabla}\times\mathbf{u}-\mathbf{u}\cdot\boldsymbol{\nabla}\times\mathbf{v}$, whereas the curl of a cross product is $\boldsymbol{\nabla}\times(\mathbf{u}\times\mathbf{v})=\mathbf{v}\cdot\boldsymbol{\nabla}\mathbf{u}+(\boldsymbol{\nabla}\cdot\mathbf{v})\mathbf{u}-\mathbf{u}\cdot\boldsymbol{\nabla}\mathbf{v}-(\boldsymbol{\nabla}\cdot\mathbf{u})\mathbf{v}$. The curl of a curl is given by $\boldsymbol{\nabla}\times(\boldsymbol{\nabla}\times\mathbf{u})=\boldsymbol{\nabla}(\boldsymbol{\nabla}\cdot\mathbf{u})-\nabla^2\mathbf{u}$. The curl $\boldsymbol{\nabla}\times\mathbf{T}$ of a second-order tensor is a tensor with components $\varepsilon_{ijk}\partial_j T_{kl}$. The curl of even higher-order tensors may be defined accordingly; the general rule is that $\boldsymbol{\nabla}$ is crossed with the first slot of $\mathbf{T}$.

If $\mathbf{T}$ is a continuously differentiable tensor field in some connected region V with surface ∂V and unit outward normal $\hat{\mathbf{n}}$, then

$$\int_V \boldsymbol{\nabla}\cdot\mathbf{T}\,dV=\int_{\partial V}\hat{\mathbf{n}}\cdot\mathbf{T}\,d\Sigma. \tag{A.62}$$

This is, of course, *Gauss' theorem*—by far the most frequently exploited mathematical result in this book. It is noteworthy that both $\boldsymbol{\nabla}$ and $\hat{\mathbf{n}}$ are contracted with the first index of the tensor $\mathbf{T}$; in component notation, equation (A.62) is

$$\int_V \partial_i T_{ij_1\cdots j_{q-1}}\,dV=\int_{\partial V}\hat{n}_i T_{ij_1\cdots j_{q-1}}\,d\Sigma. \tag{A.63}$$

The "if" clause of the theorem is important: whenever $\mathbf{T}$ is piecewise continuously differentiable within a composite region $V_1\cup V_2$, it is necessary to integrate over the two sub-volumes separately. If Σ is the boundary separating V_1 and V_2, then Gauss' theorem (A.62) is generalized to

$$\int_V \boldsymbol{\nabla}\cdot\mathbf{T}\,dV=\int_{\partial V}\hat{\mathbf{n}}\cdot\mathbf{T}\,d\Sigma-\int_\Sigma[\hat{\mathbf{n}}\cdot\mathbf{T}]_-^+\,d\Sigma, \tag{A.64}$$

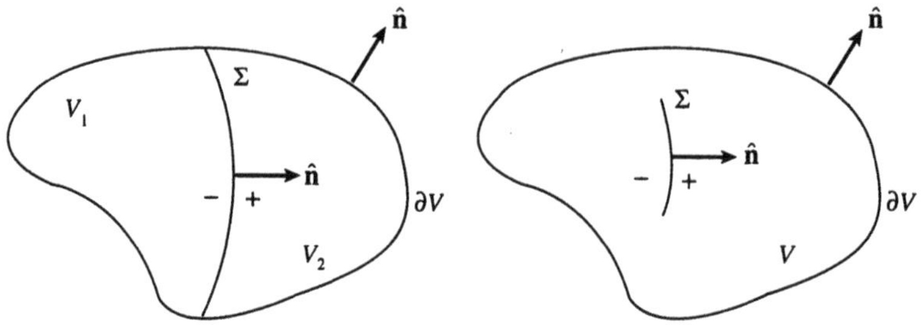

Figure A.1. (*Left*) Composite domain of integration consisting of two contiguous regions V_1 and V_2 separated by a surface Σ across which the quantity $\hat{\mathbf{n}}\cdot\mathbf{T}$ suffers a jump discontinuity. (*Right*) Integration region V with an embedded oriented surface Σ across which $\hat{\mathbf{n}}\cdot\mathbf{T}$ suffers a jump discontinuity. The generalized version of Gauss' theorem (A.64) pertains to both situations.

where the unit normal $\hat{\mathbf{n}}$ points toward the + side of Σ, as shown in Figure A.1. This leads to jump terms $[\hat{\mathbf{n}}\cdot\mathbf{T}]_-^+ = [\hat{n}_i T_{ij_1\cdots j_{q-1}}]_-^+$ over the various internal interfaces within the Earth in many of our applications. The result (A.64) is also valid if Σ is an oriented surface (e.g., a fault) embedded within V; this may be easily shown by means of a limiting argument, integrating over a punctured volume $V - V_\varepsilon$ with a "hole" V_ε that collapses onto both faces of Σ.

A.6 Surfaces

Thus far we have restricted attention to vectors and tensors and the gradient and related operators that act upon them in three-dimensional Euclidean space. Algebraic and differential calculations involving vectors and tensors defined upon curved two-dimensional surfaces are also encountered frequently in quantitative global seismology. In this section, we discuss the rudimentary notions from classical differential geometry which we use to perform such calculations.

A.6.1 Tangent vectors and tensors

An arbitrary vector defined upon an oriented surface Σ with unit normal $\hat{\mathbf{n}}$ can be written in the form

$$\mathbf{u} = \hat{\mathbf{n}}u_n + \mathbf{u}^\Sigma, \tag{A.65}$$

where $u_n = \hat{\mathbf{n}} \cdot \mathbf{u}$ and $\hat{\mathbf{n}} \cdot \mathbf{u}^\Sigma = 0$. The quantity u_n is of course the normal component of $\mathbf{u}$. Any vector $\mathbf{u}^\Sigma$ having the property that $\hat{\mathbf{n}} \cdot \mathbf{u}^\Sigma = 0$ is referred to as a *tangent vector*, for obvious reasons. The representation (A.65) decomposes $\mathbf{u}$ into its normal and tangential parts. An arbitrary second-order tensor can be decomposed in an analogous, but slightly more complicated, manner:

$$\mathbf{T} = \hat{\mathbf{n}}\hat{\mathbf{n}}T_{nn} + \hat{\mathbf{n}}\mathbf{T}^{n\Sigma} + \mathbf{T}^{\Sigma n}\hat{\mathbf{n}} + \mathbf{T}^{\Sigma\Sigma}, \tag{A.66}$$

where $T_{nn} = \hat{\mathbf{n}} \cdot \mathbf{T} \cdot \hat{\mathbf{n}}$. The quantities $\mathbf{T}^{n\Sigma}$ and $\mathbf{T}^{\Sigma n}$ are tangent vectors satisfying $\hat{\mathbf{n}} \cdot \mathbf{T}^{n\Sigma} = \hat{\mathbf{n}} \cdot \mathbf{T}^{\Sigma n} = 0$, whereas $\mathbf{T}^{\Sigma\Sigma}$ is a so-called *tangent tensor*, defined by $\hat{\mathbf{n}} \cdot \mathbf{T}^{\Sigma\Sigma} = \mathbf{T}^{\Sigma\Sigma} \cdot \hat{\mathbf{n}} = \mathbf{0}$. The transpose of $\mathbf{T}$ is given by

$$\mathbf{T}^{\mathrm{T}} = \hat{\mathbf{n}}\hat{\mathbf{n}}T_{nn} + \mathbf{T}^{n\Sigma}\hat{\mathbf{n}} + \hat{\mathbf{n}}\mathbf{T}^{\Sigma n} + (\mathbf{T}^{\Sigma\Sigma})^{\mathrm{T}}. \tag{A.67}$$

If $\mathbf{T}$ is a symmetric tensor, then $\mathbf{T}^{\Sigma n} = \mathbf{T}^{n\Sigma}$ and $(\mathbf{T}^{\Sigma\Sigma})^{\mathrm{T}} = \mathbf{T}^{\Sigma\Sigma}$.

The dot product of two vectors is

$$\mathbf{u} \cdot \mathbf{v} = u_n v_n + \mathbf{u}^\Sigma \cdot \mathbf{v}^\Sigma, \tag{A.68}$$

and the left or right dot product of a tensor and a vector is

$$\mathbf{T} \cdot \mathbf{u} = \hat{\mathbf{n}}(T_{nn}u_n + \mathbf{T}^{n\Sigma} \cdot \mathbf{u}^\Sigma) + \mathbf{T}^{\Sigma n}u_n + \mathbf{T}^{\Sigma\Sigma} \cdot \mathbf{u}^\Sigma, \tag{A.69}$$

$$\mathbf{u} \cdot \mathbf{T} = \hat{\mathbf{n}}(u_n T_{nn} + \mathbf{u}^\Sigma \cdot \mathbf{T}^{\Sigma n}) + u_n\mathbf{T}^{n\Sigma} + \mathbf{u}^\Sigma \cdot \mathbf{T}^{\Sigma\Sigma}. \tag{A.70}$$

Obviously, $\mathbf{T} \cdot \hat{\mathbf{n}} = \hat{\mathbf{n}}T_{nn} + \mathbf{T}^{\Sigma n}$ whereas $\hat{\mathbf{n}} \cdot \mathbf{T} = \hat{\mathbf{n}}T_{nn} + \mathbf{T}^{n\Sigma}$. Finally, the double dot product of two tensors is

$$\mathbf{T}:\mathbf{P} = T_{nn}P_{nn} + \mathbf{T}^{\Sigma n} \cdot \mathbf{P}^{\Sigma n} + \mathbf{T}^{n\Sigma} \cdot \mathbf{P}^{n\Sigma} + \mathbf{T}^{\Sigma\Sigma}:\mathbf{P}^{\Sigma\Sigma}. \tag{A.71}$$

The simplest example of a tangent tensor is the *surface identity tensor* $\mathbf{I}^{\Sigma\Sigma}$, defined by

$$\mathbf{I} = \hat{\mathbf{n}}\hat{\mathbf{n}} + \mathbf{I}^{\Sigma\Sigma}. \tag{A.72}$$

This tensor deserves its name inasmuch as $\mathbf{I}^{\Sigma\Sigma} \cdot \mathbf{u}^\Sigma = \mathbf{u}^\Sigma \cdot \mathbf{I}^{\Sigma\Sigma} = \mathbf{u}^\Sigma$ for any tangent vector $\mathbf{u}^\Sigma$. It is obvious that the surface identity tensor is symmetric: $(\mathbf{I}^{\Sigma\Sigma})^{\mathrm{T}} = \mathbf{I}^{\Sigma\Sigma}$.

A.6.2 Surface gradient

The three-dimensional gradient operator can also be decomposed into a normal and a tangential part:

$$\boldsymbol{\nabla} = \hat{\mathbf{n}}\partial_n + \boldsymbol{\nabla}^\Sigma, \tag{A.73}$$

where $\partial_n = \hat{\mathbf{n}} \cdot \boldsymbol{\nabla}$ and $\hat{\mathbf{n}} \cdot \boldsymbol{\nabla}^\Sigma = 0$. The tangential part $\boldsymbol{\nabla}^\Sigma$ is called the *surface gradient* operator. Since $\boldsymbol{\nabla}^\Sigma$ involves only differentiations in directions tangent to the surface Σ, it can be applied to any scalar, vector or tensor field defined upon Σ, whether that field is defined elsewhere or not. The surface gradient of a scalar field $\boldsymbol{\nabla}^\Sigma\psi$ is a measure of the rate of change of ψ upon Σ in the tangential direction in which it is changing most rapidly, just as $\boldsymbol{\nabla}\psi$ is a measure of the rate of change of ψ in the three-dimensional direction in which it is changing most rapidly. In applying the operator (A.73) to decomposed vectors and second-order tensors of the form (A.65)–(A.66), it must be remembered that the surface gradient acts upon the variable unit normal $\hat{\mathbf{n}}$; the divergence of a vector field is given, for example, by $\boldsymbol{\nabla}\cdot\mathbf{u} = (\partial_n + \boldsymbol{\nabla}^\Sigma\cdot\hat{\mathbf{n}})u_n + \boldsymbol{\nabla}^\Sigma\cdot\mathbf{u}^\Sigma$, where $\boldsymbol{\nabla}^\Sigma\cdot\hat{\mathbf{n}} = \mathrm{tr}\,(\boldsymbol{\nabla}^\Sigma\hat{\mathbf{n}})$. It is also noteworthy that the concatenated operator $\boldsymbol{\nabla}^\Sigma\boldsymbol{\nabla}^\Sigma$ is not symmetric; in fact, it is easily demonstrated that

$$[\boldsymbol{\nabla}^\Sigma\boldsymbol{\nabla}^\Sigma - \hat{\mathbf{n}}\,(\boldsymbol{\nabla}^\Sigma\hat{\mathbf{n}})\cdot\boldsymbol{\nabla}^\Sigma]^{\mathrm{T}} = \boldsymbol{\nabla}^\Sigma\boldsymbol{\nabla}^\Sigma - \hat{\mathbf{n}}\,(\boldsymbol{\nabla}^\Sigma\hat{\mathbf{n}})\cdot\boldsymbol{\nabla}^\Sigma. \tag{A.74}$$

The three-dimensional Laplacian operator $\nabla^2 = \boldsymbol{\nabla}\cdot\boldsymbol{\nabla}$ can be written in terms of the normal derivative ∂_n and the surface gradient $\boldsymbol{\nabla}^\Sigma$ in the form

$$\nabla^2 = \partial_n^2 + (\boldsymbol{\nabla}^\Sigma\cdot\hat{\mathbf{n}})\partial_n + (\nabla^\Sigma)^2. \tag{A.75}$$

Using a non-standard but obvious terminology, we shall refer to the two-dimensional, scalar differential operator

$$(\nabla^\Sigma)^2 = \boldsymbol{\nabla}^\Sigma\cdot\boldsymbol{\nabla}^\Sigma = \mathrm{tr}\,(\boldsymbol{\nabla}^\Sigma\boldsymbol{\nabla}^\Sigma) \tag{A.76}$$

as the *surface Laplacian*; in more technical mathematical accounts (A.76) is known as the *Beltrami operator*.

If $\mathbf{v}^\Sigma$ is a continuously differentiable tangent vector field on an oriented curved surface Σ, the *two-dimensional generalization of Gauss' theorem* stipulates that

$$\int_\Sigma \boldsymbol{\nabla}^\Sigma\cdot\mathbf{v}^\Sigma\,d\Sigma = \int_{\partial\Sigma} \hat{\mathbf{b}}\cdot\mathbf{v}^\Sigma\,dL. \tag{A.77}$$

The line integral is taken over the boundary $\partial\Sigma$ of the surface Σ; the unit normal $\hat{\mathbf{b}}$ is tangent to Σ and pointing out of Σ on $\partial\Sigma$. There is an analogous result for a second-order tangent tensor field $\mathbf{T}^{\Sigma\Sigma}$, namely,

$$\int_\Sigma \boldsymbol{\nabla}^\Sigma\cdot\mathbf{T}^{\Sigma\Sigma}\,d\Sigma = \int_{\partial\Sigma} \hat{\mathbf{b}}\cdot\mathbf{T}^{\Sigma\Sigma}\,dL. \tag{A.78}$$

If the surface Σ is closed equations (A.77) and (A.78) reduce to

$$\int_\Sigma \boldsymbol{\nabla}^\Sigma\cdot\mathbf{v}^\Sigma\,d\Sigma = 0, \qquad \int_\Sigma \boldsymbol{\nabla}^\Sigma\cdot\mathbf{T}^{\Sigma\Sigma}\,d\Sigma = \mathbf{0}, \tag{A.79}$$

since a closed surface does not have a boundary.

A.6.3 Covariance and contravariance

Let x^1, x^2 be a two-dimensional system of (not necessarily orthogonal) curvilinear coordinates upon Σ, and let ∂_α denote the ordinary partial derivative (not the covariant derivative) with respect to x^α. The differential tangent vector connecting a surface point $\mathbf{r}$ with coordinates x^1, x^2 to a nearby point $\mathbf{r} + d\mathbf{r}$ with coordinates $x^1 + dx^1$, $x^2 + dx^2$ is

$$d\mathbf{r} = dx^\alpha(\partial_\alpha \mathbf{r}). \tag{A.80}$$

The surface gradient operator $\boldsymbol{\nabla}^\Sigma$ and the partial derivatives ∂_α are related to each other by

$$\boldsymbol{\nabla}^\Sigma = (\boldsymbol{\nabla}^\Sigma x^\alpha)\partial_\alpha, \qquad \partial_\alpha = (\partial_\alpha \mathbf{r}) \cdot \boldsymbol{\nabla}^\Sigma. \tag{A.81}$$

Greek indices in equations (A.80)–(A.81) and the ensuing discussion take on the values 1 and 2; any repeated index—which can only appear once in the upper level and and once in the lower level—is summed over.

Both $\boldsymbol{\nabla}^\Sigma x^1$, $\boldsymbol{\nabla}^\Sigma x^2$ and $\partial_1 \mathbf{r}$, $\partial_2 \mathbf{r}$ constitute two-dimensional bases which can be used to represent arbitrary tangent vectors and tangent tensors on the surface Σ. Neither of these bases is orthonormal; however, each is the *dual* of the other, in the sense

$$(\boldsymbol{\nabla}^\Sigma x^\alpha) \cdot (\partial_\beta \mathbf{r}) = \delta^\alpha{}_\beta, \qquad (\partial_\alpha \mathbf{r}) \cdot (\boldsymbol{\nabla}^\Sigma x^\beta) = \delta_\alpha{}^\beta. \tag{A.82}$$

Both $\delta^\alpha{}_\beta$ and $\delta_\alpha{}^\beta$ denote the two-dimensional Kronecker delta, which is equal to one whenever $\alpha = \beta$ and zero otherwise; two distinct but identical symbols are employed in accordance with the convention that both the level—up or down—and the order of the indices must now be distinguished. We restrict attention to right-handed curvilinear coordinate systems satisfying $\hat{\mathbf{n}} \cdot (\partial_1 \mathbf{r} \times \partial_2 \mathbf{r}) > 0$ and $\hat{\mathbf{n}} \cdot (\boldsymbol{\nabla}^\Sigma x^1 \times \boldsymbol{\nabla}^\Sigma x^2) > 0$.

The *covariant components* u_α and the *contravariant components* u^α of a tangent vector $\mathbf{u}^\Sigma$ are defined by

$$\mathbf{u}^\Sigma = u_\alpha(\boldsymbol{\nabla}^\Sigma x^\alpha) = u^\alpha(\partial_\alpha \mathbf{r}). \tag{A.83}$$

The covariant components $T_{\alpha\beta}$, the contravariant components $T^{\alpha\beta}$, and the two *mixed components* $T_\alpha{}^\beta$ and $T^\alpha{}_\beta$ of a tangent tensor $\mathbf{T}^{\Sigma\Sigma}$ are likewise defined by

$$\begin{aligned} \mathbf{T}^{\Sigma\Sigma} &= T^{\alpha\beta}(\partial_\alpha \mathbf{r})(\partial_\beta \mathbf{r}) = T_{\alpha\beta}(\boldsymbol{\nabla}^\Sigma x^\alpha)(\boldsymbol{\nabla}^\Sigma x^\beta) \\ &= T^\alpha{}_\beta(\partial_\alpha \mathbf{r})(\boldsymbol{\nabla}^\Sigma x^\beta) = T_\alpha{}^\beta(\boldsymbol{\nabla}^\Sigma x^\alpha)(\partial_\beta \mathbf{r}). \end{aligned} \tag{A.84}$$

To find the components of a given vector $\mathbf{u}^\Sigma$ or tensor $\mathbf{T}^{\Sigma\Sigma}$, we dot with the dual or duals and invoke equations (A.82):

$$u_\alpha = (\partial_\alpha \mathbf{r}) \cdot \mathbf{u}^\Sigma, \qquad u^\alpha = (\boldsymbol{\nabla}^\Sigma x^\alpha) \cdot \mathbf{u}^\Sigma \tag{A.85}$$

and

$$T_{\alpha\beta} = (\partial_\alpha \mathbf{r}) \cdot \mathbf{T}^{\Sigma\Sigma} \cdot (\partial_\beta \mathbf{r}), \tag{A.86}$$

$$T^{\alpha\beta} = (\boldsymbol{\nabla}^\Sigma x^\alpha) \cdot \mathbf{T}^{\Sigma\Sigma} \cdot (\boldsymbol{\nabla}^\Sigma x^\beta), \tag{A.87}$$

$$T_\alpha{}^\beta = (\partial_\alpha \mathbf{r}) \cdot \mathbf{T}^{\Sigma\Sigma} \cdot (\boldsymbol{\nabla}^\Sigma x^\beta), \tag{A.88}$$

$$T^\alpha{}_\beta = (\boldsymbol{\nabla}^\Sigma x^\alpha) \cdot \mathbf{T}^{\Sigma\Sigma} \cdot (\partial_\beta \mathbf{r}). \tag{A.89}$$

The covariant, contravariant and mixed components of higher-order tensors that are tangent in every slot can be defined in an analogous manner.

A.6.4 Metric tensor

For historical reasons the components of the surface identity tensor $\mathbf{I}^{\Sigma\Sigma}$ are usually written as

$$g_{\alpha\beta} = (\partial_\alpha \mathbf{r}) \cdot (\partial_\beta \mathbf{r}), \qquad g^{\alpha\beta} = (\boldsymbol{\nabla}^\Sigma x^\alpha) \cdot (\boldsymbol{\nabla}^\Sigma x^\beta), \tag{A.90}$$

$$g_\alpha{}^\beta = (\partial_\alpha \mathbf{r}) \cdot (\boldsymbol{\nabla}^\Sigma x^\beta), \qquad g^\alpha{}_\beta = (\boldsymbol{\nabla}^\Sigma x^\alpha) \cdot (\partial_\beta \mathbf{r}). \tag{A.91}$$

The duality relation (A.82) implies that $g_\alpha{}^\beta = \delta_\alpha{}^\beta$ and $g^\alpha{}_\beta = \delta^\alpha{}_\beta$; therefore

$$\mathbf{I}^{\Sigma\Sigma} = (\partial_\alpha \mathbf{r})(\boldsymbol{\nabla}^\Sigma x^\beta) = (\boldsymbol{\nabla}^\Sigma x^\alpha)(\partial_\beta \mathbf{r}) = \boldsymbol{\nabla}^\Sigma \mathbf{r}. \tag{A.92}$$

Contraction of equation (A.92) yields

$$\operatorname{tr} \mathbf{I}^{\Sigma\Sigma} = \boldsymbol{\nabla}^\Sigma \cdot \mathbf{r} = 2. \tag{A.93}$$

The results (A.92) and (A.93) generalize the familiar three-dimensional identities $\mathbf{I} = \boldsymbol{\nabla}\mathbf{r}$ and $\operatorname{tr}\mathbf{I} = \boldsymbol{\nabla} \cdot \mathbf{r} = 3$.

Upon applying equation (A.85) to the two basis vector fields $\partial_\alpha \mathbf{r}$ and $\boldsymbol{\nabla}^\Sigma x^\alpha$ we find that

$$\partial_\alpha \mathbf{r} = g_{\alpha\beta}(\boldsymbol{\nabla}^\Sigma x^\beta), \qquad \boldsymbol{\nabla}^\Sigma x^\alpha = g^{\alpha\beta}(\partial_\beta \mathbf{r}). \tag{A.94}$$

Equations (A.94) show that the symmetric covariant and contravariant component matrices $g_{\alpha\beta}$ and $g^{\alpha\beta}$ are the inverse of each other:

$$g_{\alpha\gamma} g^{\gamma\beta} = g_\alpha{}^\beta, \qquad g^{\alpha\gamma} g_{\gamma\beta} = g^\alpha{}_\beta. \tag{A.95}$$

This is a special case of the index lowering-and-raising rule which can be used to convert contravariant into covariant components, and vice versa:

$$u_\alpha = g_{\alpha\beta} u^\beta, \qquad u^\alpha = g^{\alpha\beta} u_\beta \tag{A.96}$$

for a tangent vector $\mathbf{u}^\Sigma$, and

$$T_{\alpha\beta} = g_{\alpha\gamma}T^{\gamma}{}_{\beta} = g_{\alpha\gamma}g_{\beta\eta}T^{\gamma\eta}, \qquad T^{\alpha\beta} = g^{\alpha\gamma}T_{\gamma}{}^{\beta} = g^{\alpha\gamma}g^{\beta\eta}T_{\gamma\eta} \quad \text{(A.97)}$$

for a tangent tensor $\mathbf{T}^{\Sigma\Sigma}$. The dot product of two tangent vectors can be written in any of the equivalent forms

$$\mathbf{u}^\Sigma \cdot \mathbf{v}^\Sigma = u_\alpha v^\alpha = u^\alpha v_\alpha = g_{\alpha\beta}u^\alpha v^\beta = g^{\alpha\beta}u_\alpha v_\beta. \quad \text{(A.98)}$$

Dot products involving higher-order tangent tensors can likewise be written in terms of either the covariant, contravariant or mixed components of those tensors using the index raising-and-lowering rule. The *surface divergence* of a tangent vector is given by

$$\boldsymbol{\nabla}^\Sigma \cdot \mathbf{u}^\Sigma = \tfrac{1}{2}g^{\alpha\beta}\partial_\gamma(g_{\alpha\beta}u^\gamma). \quad \text{(A.99)}$$

The Greek-index version of equation (A.93) is $g^{\alpha\beta}g_{\alpha\beta} = g^{\alpha}{}_{\alpha} = 2$.

The squared distance ds^2 between a point $\mathbf{r}$ with coordinates x^1, x^2 and a nearby point $\mathbf{r} + d\mathbf{r}$ with coordinates $x^1 + dx^1$, $x^2 + dx^2$ on the surface Σ is given by

$$ds^2 = d\mathbf{r} \cdot d\mathbf{r} = g_{\alpha\beta}\, dx^\alpha dx^\beta. \quad \text{(A.100)}$$

Equation (A.100) shows that the surface identity tensor $\mathbf{I}^{\Sigma\Sigma}$ is the *metric tensor* on Σ. The differential-geometric expression $g_{\alpha\beta}\, dx^\alpha dx^\beta$ is known as the *first fundamental form*.

A.6.5 Curvature tensor

The classical diagnostic of the curvature of the surface Σ is the quantity

$$\mathbf{F}^{\Sigma\Sigma} = (\boldsymbol{\nabla}^\Sigma\mathbf{I}^{\Sigma\Sigma}) \cdot \hat{\mathbf{n}} = (\boldsymbol{\nabla}^\Sigma\boldsymbol{\nabla}^\Sigma\mathbf{r}) \cdot \hat{\mathbf{n}}. \quad \text{(A.101)}$$

It is obvious from the definition (A.101) that $\mathbf{F}^{\Sigma\Sigma}$ is a symmetric tangent tensor; i.e., $\hat{\mathbf{n}} \cdot \mathbf{F}^{\Sigma\Sigma} = \mathbf{F}^{\Sigma\Sigma} \cdot \hat{\mathbf{n}} = \mathbf{0}$ and $(\mathbf{F}^{\Sigma\Sigma})^{\mathrm{T}} = \mathbf{F}^{\Sigma\Sigma}$. The perpendicular distance dn of the point $\mathbf{r} + d\mathbf{r}$ with coordinates $x^1 + dx^1$, $x^2 + dx^2$ from the plane tangent to Σ at the point $\mathbf{r}$ with coordinates x^1, x^2 is given by

$$dn = \tfrac{1}{2}(d\mathbf{r} \cdot \mathbf{F}^{\Sigma\Sigma} \cdot d\mathbf{r}) = \tfrac{1}{2}f_{\alpha\beta}\, dx^\alpha dx^\beta, \quad \text{(A.102)}$$

where $f_{\alpha\beta} = (\partial_\alpha\mathbf{r}) \cdot \mathbf{F}^{\Sigma\Sigma} \cdot (\partial_\beta\mathbf{r})$ are the covariant components of $\mathbf{F}^{\Sigma\Sigma}$. A positive differential distance dn is measured in the direction of the unit normal $\hat{\mathbf{n}}$. Because the geometrical construction of the tangent plane explicitly recognizes that the surface Σ is embedded in three-dimensional Euclidean space, $\mathbf{F}^{\Sigma\Sigma}$ is referred to as the "extrinsic" *curvature tensor*. The quadratic expression $\frac{1}{2}f_{\alpha\beta}dx^\alpha dx^\beta$ is known as the *second fundamental form*.

Upon taking the surface gradient of $\mathbf{I}^{\Sigma\Sigma}\cdot\hat{\mathbf{n}} = (\boldsymbol{\nabla}^{\Sigma}\mathbf{r})\cdot\hat{\mathbf{n}} = \mathbf{0}$, we obtain an alternative representation of the curvature tensor:

$$\mathbf{F}^{\Sigma\Sigma} = -\boldsymbol{\nabla}^{\Sigma}\hat{\mathbf{n}}. \tag{A.103}$$

The symmetry of $\mathbf{F}^{\Sigma\Sigma}$ implies that

$$(\boldsymbol{\nabla}^{\Sigma}\hat{\mathbf{n}})^{\mathrm{T}} = \boldsymbol{\nabla}^{\Sigma}\hat{\mathbf{n}}. \tag{A.104}$$

The divergence of the unit normal is given by

$$\boldsymbol{\nabla}^{\Sigma}\cdot\hat{\mathbf{n}} = \frac{1}{R_1} + \frac{1}{R_2}, \tag{A.105}$$

where $1/R_1$ and $1/R_2$ are the two *principal curvatures* of Σ at the point $\mathbf{r}$. A final relation of interest in this context is $\boldsymbol{\nabla}^{\Sigma}\mathbf{I}^{\Sigma\Sigma} = -\boldsymbol{\nabla}^{\Sigma}(\hat{\mathbf{n}}\hat{\mathbf{n}})$.

A.7 Spherical Polar Coordinates

Let $r = \|\mathbf{r}\|$ be the radial distance of a point $\mathbf{r} = x\hat{\mathbf{x}} + y\hat{\mathbf{y}} + z\hat{\mathbf{z}}$ from the origin, let $0 \leq \theta \leq \pi$ be the angle between the polar or $\hat{\mathbf{z}}$ axis and $\mathbf{r}$, and let $0 \leq \phi \leq 2\pi$ be the angle between the $\hat{\mathbf{x}}$ axis and the equatorial projection $x\hat{\mathbf{x}} + y\hat{\mathbf{y}}$ of $\mathbf{r}$. Every point $\mathbf{r}$ can be uniquely specified in terms of these *spherical polar coordinates*; the angles θ and ϕ are the geocentric *colatitude* and *longitude*, respectively. The Cartesian coordinates x, y, z and the spherical polar coordinates r, θ, ϕ of a point $\mathbf{r}$ are related by

$$x = r\sin\theta\cos\phi, \qquad y = r\sin\theta\sin\phi, \qquad z = r\cos\theta. \tag{A.106}$$

The converse relations expressing r, θ, ϕ in terms of x, y, z are

$$r = \sqrt{x^2+y^2+z^2}, \qquad \theta = \arctan(\sqrt{x^2+y^2}/z),$$
$$\phi = \arctan(y/x). \tag{A.107}$$

The unit vectors $\hat{\mathbf{r}}$, $\hat{\boldsymbol{\theta}}$, $\hat{\boldsymbol{\phi}}$ in the direction of increasing r, θ, ϕ form a local right-handed orthonormal basis related to the Cartesian basis $\hat{\mathbf{x}}$, $\hat{\mathbf{y}}$, $\hat{\mathbf{z}}$ by

$$\hat{\mathbf{r}} = \hat{\mathbf{x}}\sin\theta\cos\phi + \hat{\mathbf{y}}\sin\theta\sin\phi + \hat{\mathbf{z}}\cos\theta, \tag{A.108}$$

$$\hat{\boldsymbol{\theta}} = \hat{\mathbf{x}}\cos\theta\cos\phi + \hat{\mathbf{y}}\cos\theta\sin\phi - \hat{\mathbf{z}}\sin\theta, \tag{A.109}$$

$$\hat{\boldsymbol{\phi}} = -\hat{\mathbf{x}}\sin\phi + \hat{\mathbf{y}}\cos\phi. \tag{A.110}$$

Conversely,

$$\hat{\mathbf{x}} = \hat{\mathbf{r}}\sin\theta\cos\phi + \hat{\boldsymbol{\theta}}\cos\theta\cos\phi - \hat{\boldsymbol{\phi}}\sin\phi, \tag{A.111}$$

$$\hat{\mathbf{y}} = \hat{\mathbf{r}} \sin\theta \sin\phi + \hat{\boldsymbol{\theta}} \cos\theta \sin\phi + \hat{\boldsymbol{\phi}} \cos\phi, \tag{A.112}$$

$$\hat{\mathbf{z}} = \hat{\mathbf{r}} \cos\theta - \hat{\boldsymbol{\theta}} \sin\theta. \tag{A.113}$$

The partial derivatives of the vectors $\hat{\mathbf{r}}$, $\hat{\boldsymbol{\theta}}$, $\hat{\boldsymbol{\phi}}$ are given by

$$\partial_r \hat{\mathbf{r}} = \mathbf{0}, \qquad \partial_\theta \hat{\mathbf{r}} = \hat{\boldsymbol{\theta}}, \qquad \partial_\phi \hat{\mathbf{r}} = \hat{\boldsymbol{\phi}} \sin\theta, \tag{A.114}$$

$$\partial_r \hat{\boldsymbol{\theta}} = \mathbf{0}, \qquad \partial_\theta \hat{\boldsymbol{\theta}} = -\hat{\mathbf{r}}, \qquad \partial_\phi \hat{\boldsymbol{\theta}} = \hat{\boldsymbol{\phi}} \cos\theta, \tag{A.115}$$

$$\partial_r \hat{\boldsymbol{\phi}} = \mathbf{0}, \qquad \partial_\theta \hat{\boldsymbol{\phi}} = \mathbf{0}, \qquad \partial_\phi \hat{\boldsymbol{\phi}} = -\hat{\mathbf{r}} \sin\theta - \hat{\boldsymbol{\theta}} \cos\theta. \tag{A.116}$$

The three-dimensional identity tensor can be written in either of the two equivalent forms $\mathbf{I} = \hat{\mathbf{x}}\hat{\mathbf{x}} + \hat{\mathbf{y}}\hat{\mathbf{y}} + \hat{\mathbf{z}}\hat{\mathbf{z}} = \hat{\mathbf{r}}\hat{\mathbf{r}} + \hat{\boldsymbol{\theta}}\hat{\boldsymbol{\theta}} + \hat{\boldsymbol{\phi}}\hat{\boldsymbol{\phi}}$.

A.7.1 Unit sphere

It is convenient to regard θ, ϕ as curvilinear coordinates on the *unit sphere*; we denote this sphere, consisting of the points $\|\hat{\mathbf{r}}\| = 1$, by Ω. The three-dimensional gradient operator $\boldsymbol{\nabla} = \hat{\mathbf{x}}\partial_x + \hat{\mathbf{y}}\partial_y + \hat{\mathbf{z}}\partial_z$ can be written in terms of the partial derivatives ∂_r, ∂_θ, ∂_ϕ in the form

$$\boldsymbol{\nabla} = \hat{\mathbf{r}}\partial_r + r^{-1}\boldsymbol{\nabla}_1, \tag{A.117}$$

where

$$\boldsymbol{\nabla}_1 = \hat{\boldsymbol{\theta}}\partial_\theta + \hat{\boldsymbol{\phi}}(\sin\theta)^{-1}\partial_\phi. \tag{A.118}$$

The operator $\boldsymbol{\nabla}_1$ is the dimensionless surface gradient on the unit sphere. The surface gradients of the three unit vectors $\hat{\mathbf{r}}$, $\hat{\boldsymbol{\theta}}$, $\hat{\boldsymbol{\phi}}$ on Ω are

$$\boldsymbol{\nabla}_1 \hat{\mathbf{r}} = \hat{\boldsymbol{\theta}}\hat{\boldsymbol{\theta}} + \hat{\boldsymbol{\phi}}\hat{\boldsymbol{\phi}}, \tag{A.119}$$

$$\boldsymbol{\nabla}_1 \hat{\boldsymbol{\theta}} = -\hat{\boldsymbol{\theta}}\hat{\mathbf{r}} + \hat{\boldsymbol{\phi}}\hat{\boldsymbol{\phi}} \cot\theta, \tag{A.120}$$

$$\boldsymbol{\nabla}_1 \hat{\boldsymbol{\phi}} = -\hat{\boldsymbol{\phi}}\hat{\mathbf{r}} - \hat{\boldsymbol{\phi}}\hat{\boldsymbol{\theta}} \cot\theta. \tag{A.121}$$

Upon taking the trace of equations (A.119)–(A.121) we obtain

$$\boldsymbol{\nabla}_1 \cdot \hat{\mathbf{r}} = 2, \qquad \boldsymbol{\nabla}_1 \cdot \hat{\boldsymbol{\theta}} = \cot\theta, \qquad \boldsymbol{\nabla}_1 \cdot \hat{\boldsymbol{\phi}} = 0. \tag{A.122}$$

Application of the wedge operator $\wedge$ yields the analogous relations

$$\boldsymbol{\nabla}_1 \times \hat{\mathbf{r}} = \mathbf{0}, \qquad \boldsymbol{\nabla}_1 \times \hat{\boldsymbol{\theta}} = \hat{\boldsymbol{\phi}}, \qquad \boldsymbol{\nabla}_1 \times \hat{\boldsymbol{\phi}} = -\hat{\boldsymbol{\theta}} + \hat{\mathbf{r}} \cot\theta. \tag{A.123}$$

The first equation in (A.122) is in accordance with the general results (A.93) and (A.105), as expected.

The operator $\mathbf{r}\times\boldsymbol{\nabla} = \hat{\mathbf{x}}(y\partial_z - z\partial_y) + \hat{\mathbf{y}}(z\partial_x - x\partial_z) + \hat{\mathbf{z}}(x\partial_z - z\partial_x)$ can also be considered to act on the unit sphere, since $\mathbf{r}\times\boldsymbol{\nabla} = \hat{\mathbf{r}}\times\boldsymbol{\nabla}_1$. The spherical polar representation of the dimensionless cross product $\hat{\mathbf{r}}\times\boldsymbol{\nabla}_1$ is

$$\hat{\mathbf{r}}\times\boldsymbol{\nabla}_1 = -\hat{\boldsymbol{\theta}}(\sin\theta)^{-1}\partial_\phi + \hat{\boldsymbol{\phi}}\,\partial_\theta. \tag{A.124}$$

The scalar triple product $d\boldsymbol{\omega}\cdot\hat{\mathbf{r}}\times\boldsymbol{\nabla}_1\psi$ can be regarded as a measure of the change in a scalar field ψ at an infinitesimally rotated point $\hat{\mathbf{r}}+d\boldsymbol{\omega}\times\hat{\mathbf{r}}$ just as the dot product $d\mathbf{r}\cdot\boldsymbol{\nabla}_1\psi$ is a measure of the change at an infinitesimally displaced point $\hat{\mathbf{r}}+d\hat{\mathbf{r}}$:

$$\psi(\hat{\mathbf{r}}+d\hat{\mathbf{r}}) = \psi(\hat{\mathbf{r}}) + d\hat{\mathbf{r}}\cdot\boldsymbol{\nabla}_1\psi(\hat{\mathbf{r}}) + \cdots, \tag{A.125}$$

$$\psi(\hat{\mathbf{r}}+d\boldsymbol{\omega}\times\hat{\mathbf{r}}) = \psi(\hat{\mathbf{r}}) + d\boldsymbol{\omega}\cdot\hat{\mathbf{r}}\times\boldsymbol{\nabla}_1\psi(\hat{\mathbf{r}}) + \cdots. \tag{A.126}$$

We refer to the dimensionless operator $\hat{\mathbf{r}}\times\boldsymbol{\nabla}_1$ as the *surface curl*. The quantity dotting the rotation vector $d\boldsymbol{\omega}$ in (A.126) can be written in any of the equivalent forms $\hat{\mathbf{r}}\times\boldsymbol{\nabla}_1\psi = \mathbf{r}\times\boldsymbol{\nabla}\psi = -\boldsymbol{\nabla}\times(\mathbf{r}\psi) = -\boldsymbol{\nabla}_1\times(\hat{\mathbf{r}}\psi)$. The two operators $\boldsymbol{\nabla}_1$ and $\hat{\mathbf{r}}\times\boldsymbol{\nabla}_1$ play a parallel role in the scalar representation of tangent vector fields upon Ω, as we shall see in Appendix B.12.

Repeated applications of the surface gradient $\boldsymbol{\nabla}_1$ and curl $\hat{\mathbf{r}}\times\boldsymbol{\nabla}_1$ give rise to higher-order tensor operators, the first four of which are

$$\begin{aligned}\boldsymbol{\nabla}_1\boldsymbol{\nabla}_1 = {} & -\hat{\boldsymbol{\theta}}\hat{\mathbf{r}}\,\partial_\theta - \hat{\boldsymbol{\phi}}\hat{\mathbf{r}}\,(\sin\theta)^{-1}\partial_\phi + \hat{\boldsymbol{\theta}}\hat{\boldsymbol{\theta}}\,\partial_\theta^2 \\ & + (\hat{\boldsymbol{\theta}}\hat{\boldsymbol{\phi}} + \hat{\boldsymbol{\phi}}\hat{\boldsymbol{\theta}})(\sin\theta)^{-1}(\partial_\theta\partial_\phi - \cot\theta\,\partial_\phi) \\ & + \hat{\boldsymbol{\phi}}\hat{\boldsymbol{\phi}}\,[(\sin\theta)^{-2}\partial_\phi^2 + \cot\theta\,\partial_\theta],\end{aligned} \tag{A.127}$$

$$\begin{aligned}\boldsymbol{\nabla}_1(\hat{\mathbf{r}}\times\boldsymbol{\nabla}_1) = {} & \hat{\boldsymbol{\theta}}\hat{\mathbf{r}}\,(\sin\theta)^{-1}\partial_\phi - \hat{\boldsymbol{\phi}}\hat{\mathbf{r}}\,\partial_\theta \\ & - (\hat{\boldsymbol{\theta}}\hat{\boldsymbol{\theta}} - \hat{\boldsymbol{\phi}}\hat{\boldsymbol{\phi}})(\sin\theta)^{-1}(\partial_\theta\partial_\phi - \cot\theta\,\partial_\phi) \\ & + \hat{\boldsymbol{\theta}}\hat{\boldsymbol{\phi}}\,\partial_\theta^2 - \hat{\boldsymbol{\phi}}\hat{\boldsymbol{\theta}}\,[(\sin\theta)^{-2}\partial_\phi^2 + \cot\theta\,\partial_\theta],\end{aligned} \tag{A.128}$$

$$\begin{aligned}(\hat{\mathbf{r}}\times\boldsymbol{\nabla}_1)\boldsymbol{\nabla}_1 = {} & \hat{\boldsymbol{\theta}}\hat{\mathbf{r}}\,(\sin\theta)^{-1}\partial_\phi - \hat{\boldsymbol{\phi}}\hat{\mathbf{r}}\,\partial_\theta \\ & - (\hat{\boldsymbol{\theta}}\hat{\boldsymbol{\theta}} - \hat{\boldsymbol{\phi}}\hat{\boldsymbol{\phi}})(\sin\theta)^{-1}(\partial_\theta\partial_\phi - \cot\theta\,\partial_\phi) \\ & - \hat{\boldsymbol{\theta}}\hat{\boldsymbol{\phi}}\,[(\sin\theta)^{-2}\partial_\phi^2 + \cot\theta\,\partial_\theta] + \hat{\boldsymbol{\phi}}\hat{\boldsymbol{\theta}}\,\partial_\theta^2,\end{aligned} \tag{A.129}$$

$$\begin{aligned}(\hat{\mathbf{r}}\times\boldsymbol{\nabla}_1)(\hat{\mathbf{r}}\times\boldsymbol{\nabla}_1) = {} & \hat{\boldsymbol{\theta}}\hat{\mathbf{r}}\,\partial_\theta + \hat{\boldsymbol{\phi}}\hat{\mathbf{r}}\,(\sin\theta)^{-1}\partial_\phi \\ & + \hat{\boldsymbol{\theta}}\hat{\boldsymbol{\theta}}\,[(\sin\theta)^{-2}\partial_\phi^2 + \cot\theta\,\partial_\theta] + \hat{\boldsymbol{\phi}}\hat{\boldsymbol{\phi}}\,\partial_\theta^2 \\ & - (\hat{\boldsymbol{\theta}}\hat{\boldsymbol{\phi}} + \hat{\boldsymbol{\phi}}\hat{\boldsymbol{\theta}})(\sin\theta)^{-1}(\partial_\theta\partial_\phi - \cot\theta\,\partial_\phi).\end{aligned} \tag{A.130}$$

Neither the dimensionless double gradient $\boldsymbol{\nabla}_1\boldsymbol{\nabla}_1$ nor the dimensionless double curl $(\hat{\mathbf{r}}\times\boldsymbol{\nabla}_1)(\hat{\mathbf{r}}\times\boldsymbol{\nabla}_1)$ is symmetric; in fact,

$$(\boldsymbol{\nabla}_1\boldsymbol{\nabla}_1-\hat{\mathbf{r}}\boldsymbol{\nabla}_1)^{\mathrm{T}}=\boldsymbol{\nabla}_1\boldsymbol{\nabla}_1-\hat{\mathbf{r}}\boldsymbol{\nabla}_1, \tag{A.131}$$

$$\begin{aligned}&[(\hat{\mathbf{r}}\times\boldsymbol{\nabla}_1)(\hat{\mathbf{r}}\times\boldsymbol{\nabla}_1)-(\hat{\mathbf{r}}\boldsymbol{\nabla}_1)^{\mathrm{T}}]^{\mathrm{T}}\\&\quad=(\hat{\mathbf{r}}\times\boldsymbol{\nabla}_1)(\hat{\mathbf{r}}\times\boldsymbol{\nabla}_1)-(\hat{\mathbf{r}}\boldsymbol{\nabla}_1)^{\mathrm{T}},\end{aligned} \tag{A.132}$$

$$[\boldsymbol{\nabla}_1(\hat{\mathbf{r}}\times\boldsymbol{\nabla}_1)-\hat{\mathbf{r}}(\hat{\mathbf{r}}\times\boldsymbol{\nabla}_1)]^{\mathrm{T}}=(\hat{\mathbf{r}}\times\boldsymbol{\nabla}_1)\boldsymbol{\nabla}_1-\hat{\mathbf{r}}(\hat{\mathbf{r}}\times\boldsymbol{\nabla}_1). \tag{A.133}$$

Equation (A.131) is a special case of the symmetry relation (A.74) on a general curved surface Σ.

The spherical polar representation of the three-dimensional Laplacian operator $\nabla^2=\boldsymbol{\nabla}\cdot\boldsymbol{\nabla}$ is

$$\nabla^2=\partial_r^2+2r^{-1}\partial_r+r^{-2}\nabla_1^2. \tag{A.134}$$

The *surface Laplacian* $\nabla_1^2=\boldsymbol{\nabla}_1\cdot\boldsymbol{\nabla}_1=(\hat{\mathbf{r}}\times\boldsymbol{\nabla}_1)\cdot(\hat{\mathbf{r}}\times\boldsymbol{\nabla}_1)$ obtained by contracting equations (A.127) and (A.130) is

$$\nabla_1^2=\partial_\theta^2+\cot\theta\,\partial_\theta+(\sin\theta)^{-2}\partial_\phi^2. \tag{A.135}$$

Contraction of (A.128) and (A.129) confirms that the dimensionless gradient and curl are everywhere orthogonal: $\boldsymbol{\nabla}_1\cdot(\hat{\mathbf{r}}\times\boldsymbol{\nabla}_1)=(\hat{\mathbf{r}}\times\boldsymbol{\nabla}_1)\cdot\boldsymbol{\nabla}_1=0$. Finally, we note that $\boldsymbol{\nabla}_1\times\boldsymbol{\nabla}_1=-(\hat{\mathbf{r}}\times\boldsymbol{\nabla}_1)\times(\hat{\mathbf{r}}\times\boldsymbol{\nabla}_1)=\hat{\mathbf{r}}\times\boldsymbol{\nabla}_1$ whereas $\boldsymbol{\nabla}_1\times(\hat{\mathbf{r}}\times\boldsymbol{\nabla}_1)=-(\hat{\mathbf{r}}\times\boldsymbol{\nabla}_1)\times\boldsymbol{\nabla}_1=\hat{\mathbf{r}}\,\nabla_1^2-\boldsymbol{\nabla}_1$.

The covariant and contravariant components of the surface identity tensor $\mathbf{I}^{\Omega\Omega}=\mathbf{I}-\hat{\mathbf{r}}\hat{\mathbf{r}}=\hat{\boldsymbol{\theta}}\hat{\boldsymbol{\theta}}+\hat{\boldsymbol{\phi}}\hat{\boldsymbol{\phi}}$ upon Ω are $g_{11}=1$, $g_{12}=g_{21}=0$, $g_{22}=\sin^2\theta$ and $g^{11}=1$, $g^{12}=g^{21}=0$, $g^{22}=(\sin\theta)^{-2}$ where we have set $x^1=\theta$ and $x^2=\phi$. We could use these results to define and manipulate the covariant and contravariant components u_α and u^α of tangent vectors $\mathbf{u}^\Omega$ and the covariant, contravariant and mixed components $T_{\alpha\beta}$, $T^{\alpha\beta}$, $T_\alpha{}^\beta$ and $T^\alpha{}_\beta$ of tangent tensors $\mathbf{T}^{\Omega\Omega}$; however, it is physically more appealing and mathematically no more laborious to conduct calculations instead using the "ordinary" or *physical spherical polar components*, as we discuss next.

A.7.2 Physical components

A three-dimensional vector field can be written in the form

$$\mathbf{u}=u_r\hat{\mathbf{r}}+u_\theta\hat{\boldsymbol{\theta}}+u_\phi\hat{\boldsymbol{\phi}}, \tag{A.136}$$

where $u_r = \hat{\mathbf{r}}\cdot\mathbf{u}$, $u_\theta = \hat{\boldsymbol{\theta}}\cdot\mathbf{u}$ and $u_\phi = \hat{\boldsymbol{\phi}}\cdot\mathbf{u}$. The corresponding representation of a second-order tensor field is

$$\begin{aligned}\mathbf{T} = T_{rr}\hat{\mathbf{r}}\hat{\mathbf{r}} + T_{r\theta}\hat{\mathbf{r}}\hat{\boldsymbol{\theta}} + T_{r\phi}\hat{\mathbf{r}}\hat{\boldsymbol{\phi}} \\ + T_{\theta r}\hat{\boldsymbol{\theta}}\hat{\mathbf{r}} + T_{\theta\theta}\hat{\boldsymbol{\theta}}\hat{\boldsymbol{\theta}} + T_{\theta\phi}\hat{\boldsymbol{\theta}}\hat{\boldsymbol{\phi}} \\ + T_{\phi r}\hat{\boldsymbol{\phi}}\hat{\mathbf{r}} + T_{\phi\theta}\hat{\boldsymbol{\phi}}\hat{\boldsymbol{\theta}} + T_{\phi\phi}\hat{\boldsymbol{\phi}}\hat{\boldsymbol{\phi}},\end{aligned} \tag{A.137}$$

where $T_{rr} = \hat{\mathbf{r}}\cdot\mathbf{T}\cdot\hat{\mathbf{r}}$, $T_{r\theta} = \hat{\mathbf{r}}\cdot\mathbf{T}\cdot\hat{\boldsymbol{\theta}}, \ldots, T_{\phi\phi} = \hat{\boldsymbol{\phi}}\cdot\mathbf{T}\cdot\hat{\boldsymbol{\phi}}$. Generally, when we write an expression such as u_ϕ or $T_{\phi\phi}$ using spherical polar coordinates in this book, we shall mean the physical components of $\mathbf{u}$ and $\mathbf{T}$ as in equations (A.136) and (A.137), rather than the corresponding covariant components $\sin\theta\, u_\phi$ and $\sin^2\theta\, T_{\phi\phi}$. Vector and tensor products can be readily evaluated in terms of the physical components by making use of the orthonormality of the local basis vectors $\hat{\mathbf{r}}$, $\hat{\boldsymbol{\theta}}$, $\hat{\boldsymbol{\phi}}$. The dot product of two vectors $\mathbf{u}$ and $\mathbf{v}$, for example, is $\mathbf{u}\cdot\mathbf{v} = u_r v_r + u_\theta v_\theta + u_\phi v_\phi$ whereas the cross product is $\mathbf{u}\times\mathbf{v} = \hat{\mathbf{r}}(u_\theta v_\phi - u_\phi v_\theta) + \hat{\boldsymbol{\theta}}(u_\phi v_r - u_r v_\phi) + \hat{\boldsymbol{\phi}}(u_r v_\theta - u_\theta v_r)$. The double dot product of two tensors is $\mathbf{T}:\mathbf{P} = T_{rr}P_{rr} + T_{r\theta}P_{\theta r} + \cdots + T_{\phi\phi}P_{\phi\phi}$.

Three-dimensional derivatives can be calculated in a straightforward manner using the decomposition (A.117)–(A.118) together with the differential basis-vector relations (A.119)–(A.121). The gradient of a vector field $\boldsymbol{\nabla}\mathbf{u}$ and the associated symmetric tensor $\boldsymbol{\varepsilon} = \frac{1}{2}[\boldsymbol{\nabla}\mathbf{u} + (\boldsymbol{\nabla}\mathbf{u})^{\mathrm{T}}]$ are given by

$$\begin{aligned}\boldsymbol{\nabla}\mathbf{u} = (\partial_r u_r)\hat{\mathbf{r}}\hat{\mathbf{r}} + r^{-1}(\partial_\theta u_\theta + u_r)\hat{\boldsymbol{\theta}}\hat{\boldsymbol{\theta}} \\ + r^{-1}[(\sin\theta)^{-1}\partial_\phi u_\phi + u_r + u_\theta\cot\theta]\hat{\boldsymbol{\phi}}\hat{\boldsymbol{\phi}} \\ + (\partial_r u_\theta)\hat{\mathbf{r}}\hat{\boldsymbol{\theta}} + r^{-1}(\partial_\theta u_r - u_\theta)\hat{\boldsymbol{\theta}}\hat{\mathbf{r}} \\ + (\partial_r u_\phi)\hat{\mathbf{r}}\hat{\boldsymbol{\phi}} + r^{-1}[(\sin\theta)^{-1}\partial_\phi u_r - u_\phi]\hat{\boldsymbol{\phi}}\hat{\mathbf{r}} \\ + r^{-1}(\partial_\theta u_\phi)\,\hat{\boldsymbol{\theta}}\hat{\boldsymbol{\phi}} + r^{-1}[(\sin\theta)^{-1}\partial_\phi u_\theta - u_\phi\cot\theta]\hat{\boldsymbol{\phi}}\hat{\boldsymbol{\theta}}\end{aligned} \tag{A.138}$$

and

$$\begin{aligned}\boldsymbol{\varepsilon} = (\partial_r u_r)\hat{\mathbf{r}}\hat{\mathbf{r}} + r^{-1}(\partial_\theta u_\theta + u_r)\hat{\boldsymbol{\theta}}\hat{\boldsymbol{\theta}} \\ + r^{-1}[(\sin\theta)^{-1}\partial_\phi u_\phi + u_r + u_\theta\cot\theta]\hat{\boldsymbol{\phi}}\hat{\boldsymbol{\phi}} \\ + \tfrac{1}{2}[\partial_r u_\theta + r^{-1}(\partial_\theta u_r - u_\theta)](\hat{\mathbf{r}}\hat{\boldsymbol{\theta}} + \hat{\boldsymbol{\theta}}\hat{\mathbf{r}}) \\ + \tfrac{1}{2}\{\partial_r u_\phi + r^{-1}[(\sin\theta)^{-1}\partial_\phi u_r - u_\phi]\}(\hat{\mathbf{r}}\hat{\boldsymbol{\phi}} + \hat{\boldsymbol{\phi}}\hat{\mathbf{r}}) \\ + \tfrac{1}{2}r^{-1}[\partial_\theta u_\phi + (\sin\theta)^{-1}\partial_\phi u_\theta - u_\phi\cot\theta](\hat{\boldsymbol{\theta}}\hat{\boldsymbol{\phi}} + \hat{\boldsymbol{\theta}}\hat{\boldsymbol{\phi}}).\end{aligned} \tag{A.139}$$

The divergence $\boldsymbol{\nabla}\cdot\mathbf{u} = \operatorname{tr}\boldsymbol{\varepsilon}$ obtained by contracting (A.138) or (A.139) is

$$\begin{aligned}\boldsymbol{\nabla}\cdot\mathbf{u} = \partial_r u_r + 2r^{-1}u_r \\ + r^{-1}[\partial_\theta u_\theta + u_\theta\cot\theta + (\sin\theta)^{-1}\partial_\phi u_\phi].\end{aligned} \tag{A.140}$$

The bracketed expression in (A.140) is the surface divergence:

$$\boldsymbol{\nabla}_1\cdot\mathbf{u} = \partial_\theta u_\theta + u_\theta\cot\theta + (\sin\theta)^{-1}\partial_\phi u_\phi. \tag{A.141}$$

The curl $\boldsymbol{\nabla}\times\mathbf{u}$ of a vector field is

$$\begin{aligned}\boldsymbol{\nabla}\times\mathbf{u} &= r^{-1}[\partial_\theta u_\phi + u_\phi\cot\theta - (\sin\theta)^{-1}\partial_\phi u_\theta]\hat{\mathbf{r}}\\ &+ [r^{-1}(\sin\theta)^{-1}\partial_\phi u_r - \partial_r u_\phi - r^{-1}u_\phi]\hat{\boldsymbol{\theta}}\\ &+ (\partial_r u_\theta + r^{-1}u_\theta - r^{-1}\partial_\theta u_r)\hat{\boldsymbol{\phi}}\end{aligned} \tag{A.142}$$

and the Laplacian $\nabla^2\mathbf{u} = \boldsymbol{\nabla}(\boldsymbol{\nabla}\cdot\mathbf{u}) - \boldsymbol{\nabla}\times(\boldsymbol{\nabla}\times\mathbf{u})$ is

$$\begin{aligned}\nabla^2\mathbf{u} &= \{[\partial_r^2 + 2r^{-1}\partial_r + r^{-2}(\partial_\theta^2 + \cot\theta\,\partial_\theta + (\sin\theta)^{-2}\partial_\phi^2)]u_r\\ &- 2r^{-2}[u_r + \partial_\theta u_\theta + u_\theta\cot\theta + (\sin\theta)^{-1}\partial_\phi u_\phi]\}\,\hat{\mathbf{r}}\\ &+ \{[\partial_r^2 + 2r^{-1}\partial_r + r^{-2}(\partial_\theta^2 + \cot\theta\,\partial_\theta + (\sin\theta)^{-2}\partial_\phi^2)]u_\theta\\ &+ r^{-2}[2\partial_\theta u_r - (\sin\theta)^{-2}u_\theta - 2(\sin\theta)^{-1}\cot\theta\,\partial_\phi u_\phi]\}\,\hat{\boldsymbol{\theta}}\\ &+ \{[\partial_r^2 + 2r^{-1}\partial_r + r^{-2}(\partial_\theta^2 + \cot\theta\,\partial_\theta + (\sin\theta)^{-2}\partial_\phi^2)]u_\phi\\ &+ r^{-2}(\sin\theta)^{-1}[2\partial_\phi u_r + 2\cot\theta\,\partial_\phi u_\theta - (\sin\theta)^{-1}u_\phi]\}\,\hat{\boldsymbol{\phi}}.\end{aligned} \tag{A.143}$$

The first, third and fifth lines of equation (A.143) are the scalar Laplacians $\nabla^2 u_r$, $\nabla^2 u_\theta$ and $\nabla^2 u_\phi$, respectively. Finally, the divergence $\boldsymbol{\nabla}\cdot\mathbf{T}$ of a second-order tensor field is

$$\begin{aligned}\boldsymbol{\nabla}\cdot\mathbf{T} &= \{\partial_r T_{rr} + r^{-1}[\partial_\theta T_{\theta r} + (\sin\theta)^{-1}\partial_\phi T_{\phi r}\\ &+ 2T_{rr} - T_{\theta\theta} - T_{\phi\phi} + \cot\theta\,T_{\theta r}]\}\,\hat{\mathbf{r}}\\ &+ \{\partial_r T_{r\theta} + r^{-1}[\partial_\theta T_{\theta\theta} + (\sin\theta)^{-1}\partial_\phi T_{\phi\theta}\\ &+ 2T_{r\theta} + T_{\theta r} + \cot\theta(T_{\theta\theta} - T_{\phi\phi})]\}\,\hat{\boldsymbol{\theta}}\\ &+ \{\partial_r T_{r\phi} + r^{-1}[\partial_\theta T_{\theta\phi} + (\sin\theta)^{-1}\partial_\phi T_{\phi\phi}\\ &+ 2T_{r\phi} + T_{\phi r} + \cot\theta(T_{\theta\phi} + T_{\phi\theta})]\}\,\hat{\boldsymbol{\phi}}.\end{aligned} \tag{A.144}$$

The lengthiness of the above expressions—both $\boldsymbol{\nabla}(\boldsymbol{\nabla}\cdot\mathbf{u})$ and $\boldsymbol{\nabla}\times(\boldsymbol{\nabla}\times\mathbf{u})$ are even more unpalatable—provides much of the rationale for the introduction of the alternative representations of $\mathbf{u}$ and $\mathbf{T}$ which we discuss in Appendices B and C.

Appendix B

Spherical Harmonics

Spherical harmonics are orthonormal basis functions on the surface of the unit sphere. Not surprisingly, these functions play an important role in many branches of geophysics, because of the near sphericity of the Earth. Both the results of large-scale data reduction procedures, such as the external geopotential and main geomagnetic fields, and models or theoretical predictions, such as the three-dimensional compressional or shear wave speed within the Earth or the flow velocity of the molten iron-rich material just beneath the core-mantle boundary, are commonly expanded in spherical harmonics. In addition, spherical harmonics arise naturally in the analysis of the elastic-gravitational free oscillations of a spherically symmetric, non-rotating Earth model; they enable the separation of the equations of motion into toroidal and spheroidal systems of ordinary radial scalar equations which can be integrated numerically.

We have emphasized the use of real spherical harmonics $\mathcal{Y}_{lm}$ throughout this book, since these are the most natural and convenient basis to use in mode splitting and coupling calculations. We begin this appendix, however, by defining the *complex* spherical harmonics Y_{lm}. The operational method we use to construct and analyze these functions has its foundation in the quantum-mechanical theory of angular momentum. The slim classic monograph by Edmonds (1960) provides an excellent, accessible introduction to this topic; a more exhaustive discussion is given by Varshalovich, Moskalev & Khersonskii (1988). Our summary in this appendix is similar to another recent account which is also intended for geophysicists (Backus, Parker & Constable 1996). Their treatment has a more pedagogic air than ours, since they disdain the phrase "it can be shown that". We discuss a number of topics which they omit, but which are of interest in global seismology, including the real vector spherical harmonics $\mathbf{P}_{lm}, \mathbf{B}_{lm}, \mathbf{C}_{lm}$ and the complex-degree travelling-wave harmonics $Q^{(1)}_{\lambda m}$, $Q^{(2)}_{\lambda m}$.

One further notational ingredient is useful: a complex scalar function ψ on the unit sphere Ω is said to be *square integrable* if

$$\int_\Omega \psi^*\psi\, d\Omega < \infty, \tag{B.1}$$

where the asterisk denotes complex conjugation. The *inner product* of two square-integrable functions ψ and χ is defined by

$$\langle\psi,\chi\rangle = \int_\Omega \psi^*\chi\, d\Omega. \tag{B.2}$$

Two functions ψ and χ are said to be *orthogonal* if $\langle\psi,\chi\rangle = 0$. The *norm* $\|\psi\|$ of a single function ψ is $\|\psi\| = \langle\psi,\psi\rangle^{1/2}$. If the functions ψ and χ are real, then the asterisks in equations (B.1) and (B.2) may be omitted. A review of the geometrical properties of the unit sphere which we employ here may be found in Section A.7.

B.1 Harmonic Homogeneous Polynomials

A *homogeneous polynomial* of degree l is a real or complex function of position $\mathbf{r} = x\hat{\mathbf{x}} + y\hat{\mathbf{y}} + z\hat{\mathbf{z}}$ in three-dimensional space of the form

$$H_l(\mathbf{r}) = \sum_{\alpha,\beta,\gamma} C_{\alpha\beta\gamma}x^\alpha y^\beta z^\gamma, \tag{B.3}$$

where α, β and γ are non-negative integers satisfying

$$\alpha+\beta+\gamma = l. \tag{B.4}$$

The restriction (B.4) stipulates that every monomial in the sum (B.3) has the same total degree l; hence the designation "homogeneous". A complex polynomial has real coefficients $C_{\alpha\beta\gamma}$ whereas a real polynomial has complex coefficients $C_{\alpha\beta\gamma}$. The value of every homogeneous polynomial upon the sphere of radius r can be determined from its value upon the unit sphere Ω by means of the relation

$$H_l(\mathbf{r}) = r^l \sum_{\alpha,\beta,\gamma} C_{\alpha\beta\gamma} \left(\frac{x}{r}\right)^\alpha \left(\frac{y}{r}\right)^\beta \left(\frac{z}{r}\right)^\gamma = r^l H_l(\hat{\mathbf{r}}). \tag{B.5}$$

The dimension of the space of real or complex homogeneous polynomials of degree l is $\frac{1}{2}(l+1)(l+2)$, since that is the number of independent non-negative integer combinations α, β, γ satisfying the constraint (B.4).

A *harmonic homogeneous polynomial* of degree l is a homogeneous polynomial $Y_l(\mathbf{r})$ of degree l which satisfies Laplace's equation:

$$\nabla^2 Y_l = 0. \tag{B.6}$$

Following Kellogg (1967) we can determine the dimension of the space of harmonic homogeneous polynomials by considering the representation

$$Y_l = a_l + a_{l-1}z + \cdots + a_0 z^l, \tag{B.7}$$

where each a_s is a homogeneous polynomial of degree s in x and y. The requirement (B.6) that (B.7) be harmonic yields a sequence of recursion relations governing the polynomials a_s. As a result we can write

$$Y_l = a_l + a_{l-1}z - \frac{1}{2!}z^2\nabla^2 a_l - \frac{1}{3!}z^3\nabla^2 a_{l-1} + \frac{1}{4!}z^4\nabla^2\nabla^2 a_l + \frac{1}{5!}z^5\nabla^2\nabla^2 a_{l-1} - \cdots. \tag{B.8}$$

We see from (B.8) that Y_l is completely determined by the two leading polynomials $a_l(x,y)$ and $a_{l-1}(x,y)$. There are $l+1$ linearly independent choices for a_l and l linearly independent choices for a_{l-1}; hence the dimension of the space of real or complex harmonic homogeneous polynomials is $2l+1$. An alternative name for a harmonic homogeneous polynomial Y_l of degree l is a *solid spherical harmonic.*

It can be demonstrated by induction that every homogeneous polynomial of degree l can be written as a unique linear combination of solid spherical harmonics $Y_l, Y_{l-2}, \ldots$ in the form

$$H_l = \begin{cases} Y_l + r^2 Y_{l-2} + \cdots + r^l Y_0 & \text{if } l \text{ is even} \\ Y_l + r^2 Y_{l-2} + \cdots + r^{l-1} Y_1 & \text{if } l \text{ is odd.} \end{cases} \tag{B.9}$$

By combining terms, we can also write H_l in terms of a unique solid spherical harmonic of degree l and a unique non-harmonic homogeneous polynomial of degree $l-2$:

$$H_l = Y_l + r^2 H_{l-2}. \tag{B.10}$$

The spaces containing the functions H_l and $r^2 H_{l-2}$ in the direct-sum decomposition (B.10) have dimension $\frac{1}{2}(l+1)(l+2)$ and $\frac{1}{2}(l-1)l$, respectively; this provides an alternative demonstration that the dimension of the space of solid or surface spherical harmonics Y_l is $2l+1$.

The restriction of a solid spherical harmonic to points $\hat{\mathbf{r}} = (\theta, \phi)$ upon the surface of the unit sphere Ω via the relation $Y_l(\mathbf{r}) = r^l Y_l(\hat{\mathbf{r}}) = r^l Y_l(\theta, \phi)$ is known as a *surface spherical harmonic.* We use the same symbol to represent a solid and surface spherical harmonic, distinguishing the functions

$Y_l(\mathbf{r})$ and $Y_l(\hat{\mathbf{r}}) = Y_l(\theta,\phi)$ only by the argument, or by the context in the case that no argument is given. Upon expanding Laplace's equation (B.6) in the form $[\partial_r^2 + 2r^{-1}\partial_r + r^{-2}\nabla_1^2][r^l Y_l(\theta,\phi)] = 0$, we find that every surface harmonic of degree l satisfies the *spherical Helmholtz equation*

$$\begin{aligned}\nabla_1^2 Y_l &+ l(l+1)Y_l \\ &= [\partial_\theta^2 + \cot\theta\,\partial_\theta + (\sin\theta)^{-2}\partial_\phi^2 + l(l+1)]Y_l = 0. \qquad \text{(B.11)}\end{aligned}$$

We could, if we wished, construct a basis consisting of $2l+1$ linearly independent surface spherical harmonics of degree l by seeking separable solutions to the partial differential equation (B.11). We describe instead a *purely algebraic* construction procedure, which exploits the properties of the quantum-mechanical angular-momentum operator. In Appendix C we show how this algebraic method can be generalized to vector and higher-order tensor spherical harmonics by accounting for spin as well as orbital angular momentum.

B.2 Angular-Momentum Operator

Observable quantities in quantum mechanics are represented by Hermitian linear operators; the momentum operator is $-i\hbar\boldsymbol{\nabla}$ and the corresponding angular-momentum operator is $-i\hbar(\mathbf{r}\times\boldsymbol{\nabla}) = -i\hbar(\hat{\mathbf{r}}\times\boldsymbol{\nabla}_1)$. The latter operator is the essential ingredient in the analysis of the complex surface spherical harmonics; since we are not interested in quantum-mechanical applications, we set Planck's constant $\hbar$ equal to unity, and refer to the dimensionless vector

$$\mathbf{L} = -i(\hat{\mathbf{r}}\times\boldsymbol{\nabla}_1) = i[\hat{\boldsymbol{\theta}}(\sin\theta)^{-1}\partial_\phi - \hat{\boldsymbol{\phi}}\partial_\theta] \qquad \text{(B.12)}$$

as the *angular-momentum operator*. We may regard $\mathbf{L}$ either as a three-dimensional operator acting upon scalar functions of r,θ,ϕ or as a two-dimensional operator restricted to the surface of the unit sphere Ω; in the latter case it maps scalar fields dependent only upon θ,ϕ into tangent vector fields: $\hat{\mathbf{r}}\cdot\mathbf{L} = 0$. The square $L^2 = \mathbf{L}\cdot\mathbf{L}$ of the angular-momentum operator is the negative surface Laplacian:

$$L^2 = -\nabla_1^2 = -[\partial_\theta^2 + \cot\theta\,\partial_\theta + (\sin\theta)^{-2}\partial_\phi^2]. \qquad \text{(B.13)}$$

In Cartesian coordinates x,y,z we can write the operators $\mathbf{L}$ and L^2 as

$$\mathbf{L} = \hat{\mathbf{x}}L_x + \hat{\mathbf{y}}L_y + \hat{\mathbf{z}}L_z, \qquad L^2 = L_x^2 + L_y^2 + L_z^2, \qquad \text{(B.14)}$$

where

$$\begin{aligned}L_x = -i(y\partial_z - z\partial_y), &\qquad L_y = -i(z\partial_x - x\partial_z), \\ L_z = -i(x\partial_y &- y\partial_x). \qquad \text{(B.15)}\end{aligned}$$

It is straightforward to demonstrate using either the spherical polar representation (B.12) or the Cartesian representation (B.14) that

$$\mathbf{L} \times \mathbf{L} = i\mathbf{L}. \tag{B.16}$$

The *commutator* of two arbitrary scalar or vector linear operators $\mathcal{A}$ and $\mathcal{B}$ is defined by

$$[\mathcal{A}, \mathcal{B}] = \mathcal{A}\mathcal{B} - \mathcal{B}\mathcal{A}. \tag{B.17}$$

It is obvious that $[\mathcal{A}, \mathcal{B}] = -[\mathcal{B}, \mathcal{A}]$. Whenever $[\mathcal{A}, \mathcal{B}]$ vanishes we say that the two operators $\mathcal{A}$ and $\mathcal{B}$ *commute*. The notion of commutativity plays a fundamental role in quantum mechanics: two observables have a common basis of eigenfunctions and can thus be simultaneously specified without violating the Heisenberg uncertainty principle if and only if their operators commute. The angular-momentum cross-product relationship (B.16) can be written using the commutator notation (B.17) in the form

$$[L_x, L_y] = iL_z, \qquad [L_y, L_z] = iL_x, \qquad [L_z, L_x] = iL_y. \tag{B.18}$$

Equations (B.18) show that the Cartesian components L_x, L_y, L_z do not commute with each other. On the other hand, it is easily verified that L^2 commutes with $\mathbf{L}$, and thus with any of its components:

$$[L^2, L_x] = [L^2, L_y] = [L^2, L_z] = 0. \tag{B.19}$$

Two orthogonal components of the angular momentum $\mathbf{L}$ cannot therefore be simultaneously specified; however, its square L^2 and any single component can. It is evident from the definition (B.12) that $\mathbf{L}$ commutes with ∂_r and with any scalar or vector function $f(r)$ of radius alone: $[\partial_r, \mathbf{L}] = \mathbf{0}$ and $[f(r), \mathbf{L}] = \mathbf{0}$. It follows from this and the spherical polar representation of the Laplacian $\nabla^2 = \partial_r^2 + 2r^{-1}\partial_r + r^{-2}\nabla_1^2$ that $[\nabla^2, \mathbf{L}] = \mathbf{0}$. Equations (B.19) stipulate that the surface Laplacian and the surface curl operators commute: $[\nabla_1^2, \hat{\mathbf{r}} \times \boldsymbol{\nabla}_1] = \mathbf{0}$. On the other hand, the surface Laplacian and the surface gradient do not; in fact $[\nabla_1^2, \boldsymbol{\nabla}_1] = -2\hat{\mathbf{r}}\,\nabla_1^2$.

In the next section we shall construct an orthonormal set of basis functions upon the unit sphere, with the aid of the two *ladder operators*

$$L_+ = L_x + iL_y, \qquad L_- = L_x - iL_y. \tag{B.20}$$

The commutation relations governing the operators $L_\pm$ are

$$[L^2, L_\pm] = 0, \qquad [L_z, L_\pm] = \pm L_\pm, \qquad [L_+, L_-] = 2L_z. \tag{B.21}$$

We can express L^2 in terms of $L_\pm$ and L_z in either of the two forms

$$L^2 = L_+L_- + L_z^2 - L_z = L_-L_+ + L_z^2 + L_z. \tag{B.22}$$

Explicit representations of L_x, L_y and L_z in spherical polar coordinates can be obtained using (B.12) and the geometrical relations (A.109)–(A.110):

$$L_x = i(\sin\phi\,\partial_\theta + \cot\theta\cos\phi\,\partial_\phi), \tag{B.23}$$

$$L_y = i(-\cos\phi\,\partial_\theta + \cot\theta\sin\phi\,\partial_\phi), \tag{B.24}$$

$$L_z = -i\partial_\phi. \tag{B.25}$$

The corresponding representation of the ladder operators (B.20) is

$$L_\pm = e^{\pm i\phi}(\pm\partial_\theta + i\cot\theta\,\partial_\phi). \tag{B.26}$$

Upon taking the complex conjugate of (B.26) we see that $(L_\pm)^* = -L_\mp$.

The *adjoint* $\mathcal{A}^\dagger$ of a scalar or vector operator $\mathcal{A}$ that maps the unit sphere Ω into itself is defined by $\langle\psi,\mathcal{A}\chi\rangle = \langle\mathcal{A}^\dagger\psi,\chi\rangle$. It is easy to see, upon taking the complex conjugate of this relation, that $(\mathcal{A}^\dagger)^\dagger = \mathcal{A}$. The angular-momentum operator (B.12) is *self-adjoint* or *Hermitian*:

$$\mathbf{L}^\dagger = \mathbf{L}. \tag{B.27}$$

The proof of (B.27) is most straightforward in the case of the L_z component:

$$\begin{aligned}\langle\psi,L_z\chi\rangle &= \int_\Omega \psi^*(L_z\chi)\,d\Omega = \int_\Omega \psi^*(-i\partial_\phi\chi)\,d\Omega\\ &= \int_\Omega (-i\partial_\phi\psi)^*\chi\,d\Omega = \int_\Omega (L_z\psi)^*\chi\,d\Omega = \langle L_z\psi,\chi\rangle.\end{aligned} \tag{B.28}$$

We have integrated by parts with respect to the longitude ϕ to obtain the third equality. Because the axes $\hat{\mathbf{x}}$, $\hat{\mathbf{y}}$, $\hat{\mathbf{z}}$ may be chosen arbitrarily, the operators L_x and L_y are Hermitian as well; this may also be verified directly by using the representations (B.23)–(B.24) and integrating by parts with respect to both θ and ϕ. The squared angular-momentum operator may likewise be shown to be Hermitian: $(L^2)^\dagger = (\mathbf{L}\cdot\mathbf{L})^\dagger = \mathbf{L}^\dagger\cdot\mathbf{L}^\dagger = \mathbf{L}\cdot\mathbf{L} = L^2$. The two ladder operators are each other's adjoints: $(L_\pm)^\dagger = L_\mp$.

B.3 Construction of a Basis

The eigenvalues of a Hermitian operator are real, and the associated eigenfunctions are orthogonal. For this reason we will use the Hermitian operators L^2 and L_z to construct a basis of complex surface spherical harmonics. The commutation relation $[L^2, L_z] = 0$ guarantees that we can find simultaneous eigenfunctions of these two operators. We distinguish these eigenfunctions by a second index m, known as the *order* of the spherical harmonic, writing

$$L^2 Y_{lm} = l(l+1)Y_{lm}, \qquad L_z Y_{lm} = mY_{lm}. \tag{B.29}$$

Upon substituting the spherical polar representation (B.25) of L_z into the second of the eigenvalue equations (B.29), we find that every basis element must be of the form

$$Y_{lm}(\theta,\phi) = X_{lm}(\theta)\exp(im\phi). \tag{B.30}$$

We see from (B.30) that the order m must be an integer in order to guarantee the single-valuedness of Y_{lm}.

Using the commutation relations (B.21) it is easily demonstrated that the functions $L_\pm Y_{lm}$ are simultaneous eigenfunctions of L^2 and L_z with associated eigenvalues $l(l+1)$ and $m \pm 1$:

$$L^2(L_\pm Y_{lm}) = L_\pm(L^2 Y_{lm}) = l(l+1)(L_\pm Y_{lm}), \tag{B.31}$$

$$L_z(L_\pm Y_{lm}) = (L_\pm L_z \pm L_\pm)Y_{lm} = (m\pm 1)(L_\pm Y_{lm}). \tag{B.32}$$

We now see the reason for the nomenclature—the *ascending* ladder operator L_+ transforms Y_{lm} into a constant times $Y_{l\,m+1}$, whereas the *descending* ladder operator L_- transforms Y_{lm} into a constant times $Y_{l\,m-1}$:

$$L_\pm Y_{lm} = c_\pm Y_{l\,m\pm 1}. \tag{B.33}$$

The squared norms of the transformed functions $L_\pm Y_{lm}$ are related to the squared norm of Y_{lm} by

$$\begin{aligned}\|L_\pm Y_{lm}\|^2 &= \langle L_\pm Y_{lm}, L_\pm Y_{lm}\rangle = \langle Y_{lm}, L_\mp L_\pm Y_{lm}\rangle \\ &= \langle Y_{lm}, (L^2 - L_z^2 \mp L_z)Y_{lm}\rangle \\ &= (l\mp m)(l\pm m+1)\,\|Y_{lm}\|^2.\end{aligned} \tag{B.34}$$

Since the norm of a function vanishes if and only if the function is identically zero, we must have $L_\pm Y_{l\,\pm l} = 0$. Whether we ladder up or down there is a natural end to the process; the eigenvalues of L_z are confined to the range $-l \leq m \leq l$. There are thus $2l+1$ basis surface spherical harmonics Y_{lm} for every integer degree $0 \leq l \leq \infty$.

The Hermitian character of L^2 and L_z guarantees that harmonics Y_{lm} and $Y_{l'm'}$ of differing degree $l \neq l'$ and order $m \neq m'$ are orthogonal. We stipulate in addition that every element must be of unit length; this renders the basis *orthonormal*:

$$\langle Y_{lm}, Y_{l'm'}\rangle = \int_\Omega Y_{lm}^* Y_{l'm'}\, d\Omega = \delta_{ll'}\delta_{mm'}. \tag{B.35}$$

Upon setting $\|Y_{lm}\|^2 = 1$ for $-l \leq m \leq l$ in (B.34) and making use of (B.33) we find that $|c_\pm|^2 = (l \mp m)(l \pm m + 1)$. The signs of the square roots $c_\pm$ may be chosen arbitrarily.

The conditions $L_\pm Y_{l\,\pm l} = e^{\pm i\phi}(\pm\partial_\theta + i\cot\theta\,\partial_\phi)[X_{l\,\pm l}(\theta)e^{\pm il\phi}] = 0$ determine the colatitudinal dependence of the highest-order and lowest-order eigenfunctions via the ordinary differential equations

$$\left(\frac{d}{d\theta} - l\cot\theta\right)X_{l\,\pm l} = 0. \tag{B.36}$$

The solutions to (B.36) are $X_{l\,\pm l} = A_\pm(\sin\theta)^l$. The integration constants $A_{\pm l}$ can be determined, again up to a sign, by the normalization condition $\|Y_{l\,\pm l}\|^2 = 1$. Taking the lowest-order eigenfunction $Y_{l\,-l}$ to be the first element of our orthonormal basis, and using the ascending ladder operator to construct the remaining members $Y_{l\,-l-1},\ldots,Y_{l0},\ldots,Y_{ll}$, we define

$$Y_{l-l} = \left(\frac{2l+1}{4\pi}\right)^{1/2} \frac{\sqrt{(2l)!}}{2^l l!}(\sin\theta)^l\, e^{-il\phi} \tag{B.37}$$

and

$$Y_{lm} = \left[\frac{(l-m)!}{(l+m)!}\right]^{1/2} \left[\frac{1}{(2l)!}\right]^{1/2} (L_+)^{l+m}\, Y_{l\,-l}, \tag{B.38}$$

where we have stipulated that $\operatorname{sgn} A_- = 1$ and $\operatorname{sgn} c_+ = 1$. We could instead have made the highest-order eigenfunction Y_{ll} the first element of the basis, and used the descending ladder operator to construct the remaining members $Y_{l\,l-1},\ldots,Y_{l0},\ldots,Y_{l\,-l}$; the sign choices $\operatorname{sgn} A_+ = (-1)^l$ and $\operatorname{sgn} c_- = 1$ lead to the alternative defining relations

$$Y_{ll} = (-1)^l \left(\frac{2l+1}{4\pi}\right)^{1/2} \frac{\sqrt{(2l)!}}{2^l l!}(\sin\theta)^l\, e^{il\phi} \tag{B.39}$$

and

$$Y_{lm} = \left[\frac{(l+m)!}{(l-m)!}\right]^{1/2} \left[\frac{1}{(2l)!}\right]^{1/2} (L_-)^{l-m}\, Y_{ll}, \tag{B.40}$$

which are equivalent to (B.37)–(B.38). The resulting basis elements satisfy the eigenvalue and orthonormality equations (B.29) and (B.35) and the ladder relations

$$L_\pm Y_{lm} = \sqrt{(l \mp m)(l \pm m + 1)}\, Y_{l\,m\pm 1}. \tag{B.41}$$

Equation (B.41) can be made valid for all m by defining $Y_{lm} = 0$ for $|m| > l$. The sign choices we have made above are those of Condon & Shortley (1935) and Edmonds (1960).

To obtain explicit expressions for the surface spherical harmonics Y_{lm} we need to determine the repeated effect of the operators L_- and L_+ on Y_{ll} and $Y_{l\,-l}$, respectively. It is straightforward to prove by induction that

$$(L_\pm)^{l\pm m}\left[(\sin\theta)^l e^{\mp il\phi}\right] = (\sin\theta)^{\pm m} \times \left(\pm\frac{1}{\sin\theta}\frac{d}{d\theta}\right)^{l\pm m}(\sin\theta)^{2l}e^{im\phi}. \tag{B.42}$$

l m	$Y_{lm}(\hat{\mathbf{r}})$	$Y_{lm}(\mathbf{r})$
0 0	$\frac{1}{2\sqrt{\pi}}$	$\frac{1}{2\sqrt{\pi}}$
1 0	$\frac{1}{2}\sqrt{\frac{3}{\pi}}\cos\theta$	$\frac{1}{2}\sqrt{\frac{3}{\pi}}\,z$
1 ±1	$\mp\frac{1}{2}\sqrt{\frac{3}{2\pi}}\sin\theta\, e^{\pm i\phi}$	$\mp\frac{1}{2}\sqrt{\frac{3}{2\pi}}\,(x \pm iy)$
2 0	$\frac{1}{4}\sqrt{\frac{5}{\pi}}\,(3\cos^2\theta - 1)$	$\frac{1}{4}\sqrt{\frac{5}{\pi}}\,(2z^2 - x^2 - y^2)$
2 ±1	$\mp\frac{1}{2}\sqrt{\frac{15}{2\pi}}\sin\theta\cos\theta\, e^{\pm i\phi}$	$\mp\frac{1}{2}\sqrt{\frac{15}{2\pi}}\,z(x \pm iy)$
2 ±2	$\frac{1}{4}\sqrt{\frac{15}{2\pi}}\sin^2\theta\, e^{\pm 2i\phi}$	$\frac{1}{4}\sqrt{\frac{15}{2\pi}}\,(x \pm iy)^2$
3 0	$\frac{1}{4}\sqrt{\frac{7}{\pi}}\cos\theta\,(5\cos^2\theta - 3)$	$\frac{1}{4}\sqrt{\frac{7}{\pi}}\,z(2z^2 - 3x^2 - 3y^2)$
3 ±1	$\mp\frac{1}{8}\sqrt{\frac{21}{\pi}}\sin\theta\,(5\cos^2\theta - 1)\,e^{\pm i\phi}$	$\mp\frac{1}{8}\sqrt{\frac{21}{\pi}}\,(x \pm iy)(4z^2 - x^2 - y^2)$
3 ±2	$\frac{1}{4}\sqrt{\frac{105}{2\pi}}\cos\theta\sin^2\theta\, e^{\pm 2i\phi}$	$\frac{1}{4}\sqrt{\frac{105}{2\pi}}\,z(x \pm iy)^2$
3 ±3	$\mp\frac{1}{8}\sqrt{\frac{35}{\pi}}\sin^3\theta\, e^{\pm 3i\phi}$	$\mp\frac{1}{8}\sqrt{\frac{35}{\pi}}\,(x \pm iy)^3$

Table B.1. Complex surface spherical harmonics $Y_{lm}(\hat{\mathbf{r}})$ and associated solid spherical harmonics $Y_{lm}(\mathbf{r}) = r^l Y_{lm}(\hat{\mathbf{r}})$ of degrees zero through three.

As a result, equations (B.37)–(B.38) and (B.39)–(B.40) are equivalent to

$$\begin{aligned} Y_{lm} &= \left(\frac{2l+1}{4\pi}\right)^{1/2} \frac{1}{2^l l!} \left[\frac{(l-m)!}{(l+m)!}\right]^{1/2} \\ &\quad \times (\sin\theta)^m \left(\frac{1}{\sin\theta}\frac{d}{d\theta}\right)^{l+m} (\sin\theta)^{2l} e^{im\phi} \\ &= (-1)^l \left(\frac{2l+1}{4\pi}\right)^{1/2} \frac{1}{2^l l!} \left[\frac{(l+m)!}{(l-m)!}\right]^{1/2} \\ &\quad \times (\sin\theta)^{-m} \left(-\frac{1}{\sin\theta}\frac{d}{d\theta}\right)^{l-m} (\sin\theta)^{2l} e^{im\phi}. \end{aligned} \tag{B.43}$$

Upon comparing these two expressions we obtain the important relation

$$Y_{l\,-m} = (-1)^m Y_{lm}^{*}. \tag{B.44}$$

Since the associated solid spherical harmonics $Y_{lm}(\mathbf{r}) = r^l Y_{lm}(\hat{\mathbf{r}})$ are homogeneous polynomials of degree l, we must also have $Y_{lm}(-\hat{\mathbf{r}}) = (-1)^l Y_{lm}(\hat{\mathbf{r}})$, or, equivalently,

$$Y_{lm}(\pi-\theta, \phi+\pi) = (-1)^l Y_{lm}(\theta, \phi). \tag{B.45}$$

The result (B.45) shows that every Y_{lm} is symmetric or anti-symmetric with respect to inversions (reflections through the origin), depending upon whether the degree l is even or odd. In quantum-mechanical parlance we say that the *parity* of Y_{lm} is $(-1)^l$. Explicit formulae for the first few surface and solid spherical harmonics are given in Table B.1.

B.4 Associated Legendre Functions

Upon substituting the representation $Y_{lm}(\theta,\phi) = X_{lm}(\theta)e^{im\phi}$ into the spherical Helmholtz equation $\nabla_1^2 Y_{lm} + l(l+1)Y_{lm} = 0$, we find that the colatitudinal functions $X_{lm}(\theta)$ satisfy the ordinary differential equation

$$\frac{d^2X}{d\theta^2} + \cot\theta\,\frac{dX}{d\theta} + \left[l(l+1) - \frac{m^2}{\sin^2\theta}\right]X = 0. \tag{B.46}$$

The substitution $\mu = \cos\theta$ transforms this into *Legendre's equation*:

$$(1-\mu^2)\frac{d^2X}{d\mu^2} - 2\mu\frac{dX}{d\mu} + \left[l(l+1) - \frac{m^2}{1-\mu^2}\right]X = 0. \tag{B.47}$$

The regular solutions to equation (B.47) in the interval $-1 \le \mu \le 1$ are the *associated Legendre functions* of degree l and order $-l \le m \le l$, defined by

$$P_{lm}(\mu) = \frac{1}{2^l l!}\,(1-\mu^2)^{m/2} \left(\frac{d}{d\mu}\right)^{l+m} (\mu^2-1)^l. \tag{B.48}$$

These functions satisfy the fixed-order orthonormality relation

$$\int_{-1}^{1} P_{lm}(\mu)P_{l'm}(\mu)\,d\mu = \frac{2}{2l+1}\frac{(l+m)!}{(l-m)!}\delta_{ll'} \tag{B.49}$$

and a welter of three-term recursion relations, including

$$(l-m+1)P_{l+1\,m} - (2l+1)\mu P_{lm} + (l+m)P_{l-1\,m} = 0, \tag{B.50}$$

$$(1-\mu^2)^{1/2}P_{l\,m+1} - 2m\mu P_{lm} + (l+m)(l-m+1)(1-\mu^2)^{1/2}P_{l\,m-1} = 0, \tag{B.51}$$

$$P_{l+1\,m} - \mu P_{lm} - (l+m)(1-\mu^2)^{1/2}P_{l\,m-1} = 0, \tag{B.52}$$

$$\mu P_{lm} - P_{l-1\,m} - (l-m+1)(1-\mu^2)^{1/2}P_{l\,m-1} = 0, \tag{B.53}$$

$$(l-m+1)P_{l+1\,m} + (1-\mu^2)^{1/2}P_{l\,m+1} - (l+m+1)\mu P_{lm} = 0, \tag{B.54}$$

$$(l-m)\mu P_{lm} - (l+m)P_{l-1\,m} + (1-\mu^2)^{1/2}P_{l\,m+1} = 0.$$

The derivative of P_{lm} is given by any of the equivalent formulae

$$\begin{aligned}(1-\mu^2)\frac{dP_{lm}}{d\mu} &= (1-\mu^2)^{1/2}P_{l\,m+1} - m\mu P_{lm} \\ &= m\mu P_{lm} - (l+m)(l-m+1)(1-\mu^2)^{1/2}P_{l\,m-1} \\ &= (l+1)\mu P_{lm} - (l-m+1)P_{l+1m} \\ &= (l+m)P_{l-1\,m} - l\mu P_{lm}.\end{aligned} \tag{B.55}$$

The effect of a change in the sign of the order or the argument is

$$P_{l\,-m}(\mu) = (-1)^m\frac{(l-m)!}{(l+m)!}P_{lm}(\mu), \tag{B.56}$$

$$P_{lm}(-\mu) = (-1)^{l+m}P_{lm}(\mu). \tag{B.57}$$

We can express the surface spherical harmonics $Y_{lm}(\theta,\phi) = X_{lm}(\theta)e^{im\phi}$ in terms of the associated Legendre functions (B.48) in the form

$$X_{lm}(\theta) = (-1)^m\left(\frac{2l+1}{4\pi}\right)^{1/2}\left[\frac{(l-m)!}{(l+m)!}\right]^{1/2} P_{lm}(\cos\theta). \tag{B.58}$$

The orthonormality and symmetry relations analogous to the results (B.49) and (B.56)–(B.57) are

$$\int_0^\pi X_{lm}(\theta)X_{l'm}(\theta)\,\sin\theta\,d\theta = \tfrac{1}{2\pi}\,\delta_{ll'} \tag{B.59}$$

and

$$X_{l\,-m}(\theta) = (-1)^m X_{lm}(\theta), \tag{B.60}$$

$$X_{lm}(\pi-\theta) = (-1)^{l+m} X_{lm}(\theta). \tag{B.61}$$

Equation (B.61) shows that X_{lm} is symmetric or anti-symmetric across the equator $\theta = \pi/2$, depending upon whether $l+m$ is even or odd.

The limiting behavior of X_{lm} near the poles $\theta \approx 0$ and $\theta \approx \pi$ is

$$X_{lm}(\theta) \approx \begin{cases} (-1)^m b_{lm}\theta^{|m|} & \text{if } -l \le m < 0 \\ b_{l0}[1 - \frac{1}{4}l(l+1)\theta^2] & \text{if } m = 0 \\ b_{lm}\theta^m & \text{if } 0 < m \le l, \end{cases} \tag{B.62}$$

$$X_{lm}(\pi-\theta) \approx \begin{cases} (-1)^l b_{lm}\theta^{|m|} & \text{if } -l \le m < 0 \\ (-1)^l b_{l0}[1 - \frac{1}{4}l(l+1)\theta^2] & \text{if } m = 0 \\ (-1)^{l+m} b_{lm}\theta^m & \text{if } 0 < m \le l, \end{cases} \tag{B.63}$$

where

$$b_{lm} = \frac{(-1)^m}{2^{|m|}|m|!}\left(\frac{2l+1}{4\pi}\right)^{1/2}\left[\frac{(l+|m|)!}{(l-|m|)!}\right]^{1/2}. \tag{B.64}$$

Only the harmonics of order zero are non-zero right at the poles:

$$X_{lm}(0) = X_{lm}(\pi) = 0 \quad \text{for } 0 < |m| < l, \tag{B.65}$$

whereas

$$X_{l0}(0) = \sqrt{(2l+1)/4\pi}, \qquad X_{l0}(\pi) = (-1)^l\sqrt{(2l+1)/4\pi}. \tag{B.66}$$

The results (B.62)–(B.64) were used by Gilbert & Dziewonski (1975) to find the normal-mode response of a spherical Earth to a moment-tensor source in epicentral coordinates; we obtain the same results in Chapter 10 by means of an alternative operational method.

B.5 Legendre Polynomials

The associated Legendre function of order $m = 0$ is the classical *Legendre polynomial of degree* l:

$$P_l(\mu) = \frac{1}{2^l l!}\left(\frac{d}{d\mu}\right)^l (\mu^2 - 1)^l. \tag{B.67}$$

The defining relation (B.67) is known as *Rodrigues' formula*. The orthogonality of the Legendre polynomials on the unit interval makes them useful in interpolation. They also arise in the representation of the gravitational potential of a point mass via the expansion

$$\frac{1}{\|\mathbf{r} - \mathbf{r}'\|} = \frac{1}{r}\sum_{l=0}^{\infty}\left(\frac{r'}{r}\right)^l P_l(\cos\Theta), \tag{B.68}$$

where $\cos\Theta = \hat{\mathbf{r}}\cdot\hat{\mathbf{r}}' = \cos\theta\cos\theta' + \sin\theta\sin\theta'\cos(\phi - \phi')$.

The *spherical-harmonic addition theorem*

$$\sum_{m=-l}^{l} Y_{lm}^*(\hat{\mathbf{r}}')Y_{lm}(\hat{\mathbf{r}}) = \left(\frac{2l+1}{4\pi}\right) P_l(\cos\Theta) \tag{B.69}$$

enables us to write the result (B.68) as a full-fledged spherical-harmonic expansion:

$$\frac{1}{\|\mathbf{r} - \mathbf{r}'\|} = \frac{1}{r}\sum_{l=0}^{\infty}\left(\frac{4\pi}{2l+1}\right)\left(\frac{r'}{r}\right)^l \sum_{m=-l}^{l} Y_{lm}^*(\hat{\mathbf{r}}')Y_{lm}(\hat{\mathbf{r}}). \tag{B.70}$$

Equations (B.68) and (B.70) are valid for $r' < r$; the corresponding results for $r' > r$ can be obtained by a simple interchange, $r \longleftrightarrow r'$. We verify the result (B.69) in Appendix C.8, by showing it to be a special case of a more general addition theorem governing the matrix elements describing two successive finite rotations. Backus, Parker & Constable (1996) give a more elementary proof, first establishing the existence of a *self-reproducing kernel* $K(\hat{\mathbf{r}}, \hat{\mathbf{r}}')$ with the property $Y_l(\hat{\mathbf{r}}) = \frac{1}{4\pi}\int_\Omega K(\hat{\mathbf{r}}, \hat{\mathbf{r}}')Y_l(\hat{\mathbf{r}}')\, d\Omega'$, and then showing that $K(\hat{\mathbf{r}}, \hat{\mathbf{r}}') = (2l+1)P_l(\hat{\mathbf{r}}\cdot\hat{\mathbf{r}}')$.

Upon comparing equations (B.48) and (B.67) we see that the associated Legendre functions of *positive* order, $m > 0$, can be written in terms of $P_l(\mu)$ in the form

$$P_{lm}(\mu) = (1 - \mu^2)^{m/2}\left(\frac{d}{d\mu}\right)^m P_l(\mu). \tag{B.71}$$

Equation (B.71) shows that $P_{lm}(\mu)$ is a polynomial of degree l for even orders $m = 2, 4, \ldots$, but it is not a polynomial for odd orders $m = 1, 3, \ldots$.

B.6 Real Spherical Harmonics

The $2l+1$ complex surface spherical harmonics $Y_{lm}(\theta,\phi) = X_{lm}(\theta)e^{im\phi}$, $-l \leq m \leq l$, provide a natural basis for the inherently complex wavefunctions that arise in the quantum theory of atomic and molecular angular momentum. In many geophysical applications, on the other hand, we seek to expand fields that are inherently real; a convenient basis in this case consists of the $2l+1$ real functions $\sqrt{2}X_{ll}(\theta)\cos l\phi, \ldots, X_{l0}(\theta), \ldots, \sqrt{2}X_{ll}(\theta)\sin l\phi$. We retain the range $-l \leq m \leq l$ of the order index and use a script font to denote these real surface spherical harmonics:

$$\mathcal{Y}_{lm}(\theta,\phi) = \begin{cases} \sqrt{2}X_{l|m|}(\theta)\cos m\phi & \text{if } -l \leq m < 0 \\ X_{l0}(\theta) & \text{if } m = 0 \\ \sqrt{2}X_{lm}(\theta)\sin m\phi & \text{if } 0 < m \leq l. \end{cases} \tag{B.72}$$

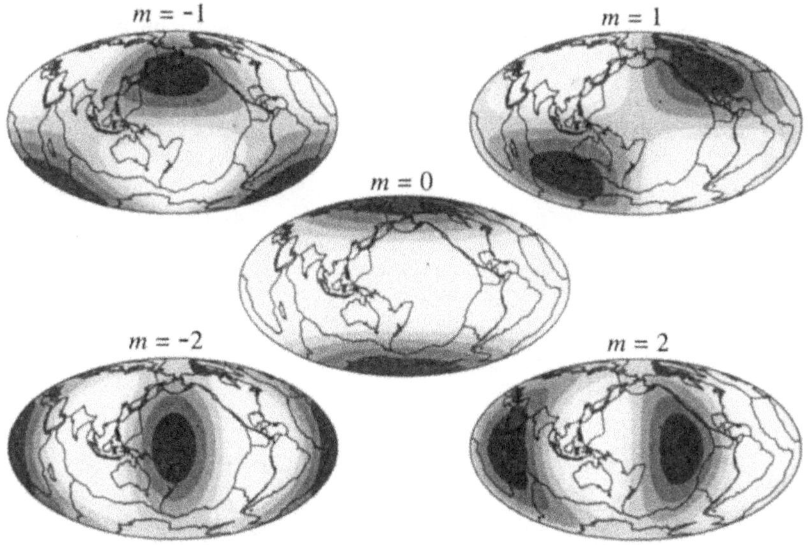

Figure B.1. The five real surface spherical harmonics $\mathcal{Y}_{2\,-2}, \mathcal{Y}_{2\,-1}, \mathcal{Y}_{20}, \mathcal{Y}_{21}, \mathcal{Y}_{22}$. Shaded and unshaded regions are negative and positive, respectively. The entire surface of the unit sphere is shown in an Aitoff equal-area projection centered on $\phi = \pi$; the Greenwich Meridian coincides with the boundary ellipse. The Earth's coastlines and plate boundaries have been superimposed for reference.

The script harmonics $\mathcal{Y}_{lm}$, $-l \leq m \leq l$, defined by (B.72) are orthonormal in the sense

$$\int_\Omega \mathcal{Y}_{lm}\mathcal{Y}_{l'm'}\,d\Omega = \delta_{ll'}\delta_{mm'}. \tag{B.73}$$

Furthermore, they satisfy the real spherical-harmonic addition theorem

$$\sum_{m=-l}^{l} \mathcal{Y}_{lm}(\theta',\phi')\mathcal{Y}_{lm}(\theta,\phi) = \left(\frac{2l+1}{4\pi}\right) P_l(\cos\Theta), \tag{B.74}$$

where $\cos\Theta = \cos\theta\cos\theta' + \sin\theta\sin\theta'\cos(\phi-\phi')$. We use the script harmonics (B.72) as the real basis of choice throughout this book; the eigenfunctions of a spherical elastic and anelastic Earth are expressed in terms of $\mathcal{Y}_{lm}$, $-l \leq m \leq l$, in Chapter 8 and Section 9.9, respectively.

A real spherical harmonic $\mathcal{Y}_{l0}$ of order zero has l colatitudinal nodes and no longitudinal nodes, and is said to be *zonal*, whereas a real spherical harmonic $\mathcal{Y}_{l\,\pm l}$ of order $\pm l$ has no colatitudinal nodes and $2l$ longitudinal nodes, and is said to be *sectoral*. A real harmonic $\mathcal{Y}_{l\,\pm m}$ of order $-l < m < l$ has $l - |m|$ colatitudinal nodes and $2|m|$ longitudinal nodes, and is said to be *tesseral*. This terminology is easy to remember: the positive and

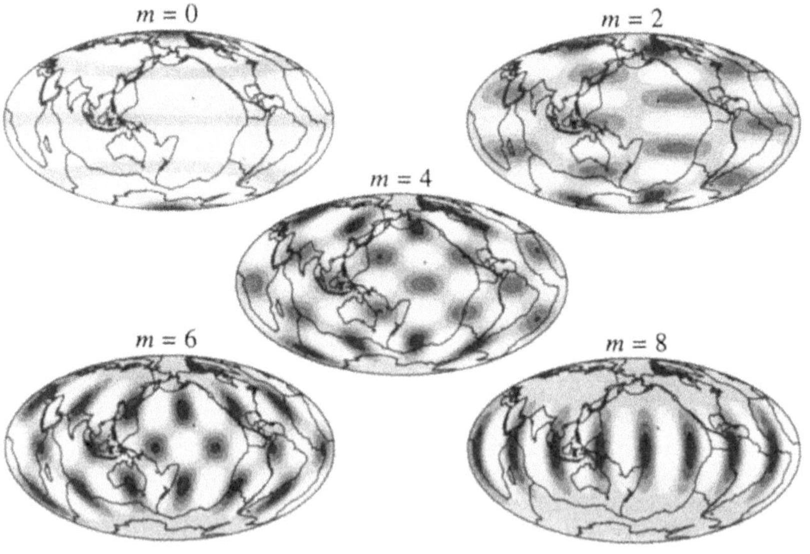

Figure B.2. The five real even-order surface spherical harmonics $\mathcal{Y}_{80}$, $\mathcal{Y}_{82}$, $\mathcal{Y}_{84}$, $\mathcal{Y}_{86}$, $\mathcal{Y}_{88}$. Map projection and other details are the same as in Figure B.1.

negative regions of a sectoral harmonic are shaped like the sectors of an orange, whereas those of a tesseral harmonic tesselate the surface of the unit sphere. In Figure B.1 we show the five real spherical harmonics $\mathcal{Y}_{2\,-2}$, $\mathcal{Y}_{2\,-1}$, $\mathcal{Y}_{20}$, $\mathcal{Y}_{21}$, $\mathcal{Y}_{22}$ of degree two, and in Figure B.2 we show the five even-order harmonics $\mathcal{Y}_{80}$, $\mathcal{Y}_{82}$, $\mathcal{Y}_{84}$, $\mathcal{Y}_{86}$, $\mathcal{Y}_{88}$ of degree eight.

B.7 Asymptotic Representation

An asymptotic representation of $X_{lm}(\theta)$ that is valid in the limit $l \gg 1$ may be obtained by an application of the JWKB approximation to the governing differential equation (B.47). The substitution

$$X = (1-\mu^2)^{-1/2}Z \tag{B.75}$$

converts this Legendre equation into a one-dimensional Schrödinger equation:

$$\varepsilon^2\,\frac{d^2Z}{d\mu^2} + \left[\frac{a^2-\mu^2+\varepsilon^2}{(1-\mu^2)^2}\right]Z = 0, \tag{B.76}$$

where

$$\varepsilon = \frac{1}{\sqrt{l(l+1)}}, \qquad a = \left[1 - \frac{m^2}{l(l+1)}\right]^{1/2}. \tag{B.77}$$

Following Brussaard & Tolhoek (1957) we seek solutions to equation (B.76) of the form

$$Z = A\exp(i\varepsilon^{-1}\psi), \tag{B.78}$$

in the limit $\varepsilon \ll 1$. Upon substituting (B.78) and equating powers of ε we find, to lowest order, that the JWKB phase ψ and amplitude A satisfy the *eikonal* and *transport equations*

$$\left(\frac{d\psi}{d\mu}\right)^2 - \frac{a^2-\mu^2}{(1-\mu^2)^2} = 0, \qquad \frac{d}{d\mu}\left(A^2\frac{d\psi}{d\mu}\right) = 0, \tag{B.79}$$

respectively. There are two turning points $\mu = \pm a$ in the interval of interest $-1 \leq \mu \leq 1$. The corresponding *turning colatitudes* are situated in the northern and southern hemispheres at θ_0 and $\pi - \theta_0$, where

$$\theta_0 = \arcsin\left(\frac{|m|}{\sqrt{l(l+1)}}\right) \approx \arcsin\left(\frac{|m|}{l+\frac{1}{2}}\right). \tag{B.80}$$

The asymptotic solution is oscillatory in the interior region $\theta_0 \ll \theta \ll \pi - \theta_0$ and evanescent in the two polar regions $0 \leq \theta \ll \theta_0$ and $\pi - \theta_0 \ll \theta \leq \pi$. The oscillatory solution is of the form

$$X \approx C_+(a^2 - \mu^2)^{-1/4} \exp\left(i\varepsilon^{-1} \int_0^\mu \frac{\sqrt{a^2 - \nu^2}}{1 - \nu^2} d\nu\right)$$
$$+ C_-(a^2 - \mu^2)^{-1/4} \exp\left(-i\varepsilon^{-1} \int_0^\mu \frac{\sqrt{a^2 - \nu^2}}{1 - \nu^2} d\nu\right), \qquad \text{(B.81)}$$

where the complex constants C_+ and C_- remain to be determined. Upon demanding that (B.81) be smoothly connected to the evanescent solutions at the turning points $\mu = \pm a$ and invoking the JWKB normalization condition $\int_{-a}^{a} X^2 \, d\mu \approx 1/2\pi$, we obtain the real result

$$X \approx \frac{1}{\pi}(a^2 - \mu^2)^{-1/4} \cos\left[\varepsilon^{-1} \int_0^\mu \frac{\sqrt{a^2 - \nu^2}}{1 - \nu^2} d\nu - (l + m)\frac{\pi}{2}\right]. \qquad \text{(B.82)}$$

The phase integral in (B.82) can be evaluated analytically by rewriting the integrand in the form

$$\frac{\sqrt{a^2 - \nu^2}}{1 - \nu^2} = \frac{1}{\sqrt{a^2 - \nu^2}} - \frac{1 - a^2}{(1 - \nu^2)\sqrt{a^2 - \nu^2}}. \qquad \text{(B.83)}$$

The first term in (B.83) may then be integrated immediately, whereas the second may be integrated by making the substitution $\xi = \mu(a^2 - \nu^2)^{1/2}$. With some rearrangement, we may then write the asymptotic representation of the normalized associated Legendre function of degree l and non-negative order $m \geq 0$ in the final form

$$X_{lm}(\theta) \approx \frac{1}{\pi}(\sin^2\theta - \sin^2\theta_0)^{-1/4}$$
$$\times \cos\left[(l + \tfrac{1}{2}) \arccos(\cos\theta/\cos\theta_0)\right.$$
$$\left. - m \arccos(\cot\theta/\cot\theta_0) + m\pi - \tfrac{1}{4}\pi\right]. \qquad \text{(B.84)}$$

The corresponding results for negative order $m < 0$ follow from the symmetry $X_{l\,-m} = (-1)^m X_{lm}$. Figure B.3 shows that equation (B.84) provides an excellent approximation to the exact harmonic X_{lm} in the oscillatory regime $\theta_0 \ll \theta \ll \pi - \theta_0$, even for degrees as low as $l = 8$. The divergence in the vicinity of the turning colatitudes θ_0 and $\pi - \theta_0$ is a characteristic feature of the JWKB approximation; a more complete treatment yields not only the matched asymptotic representations in the evanescent regions

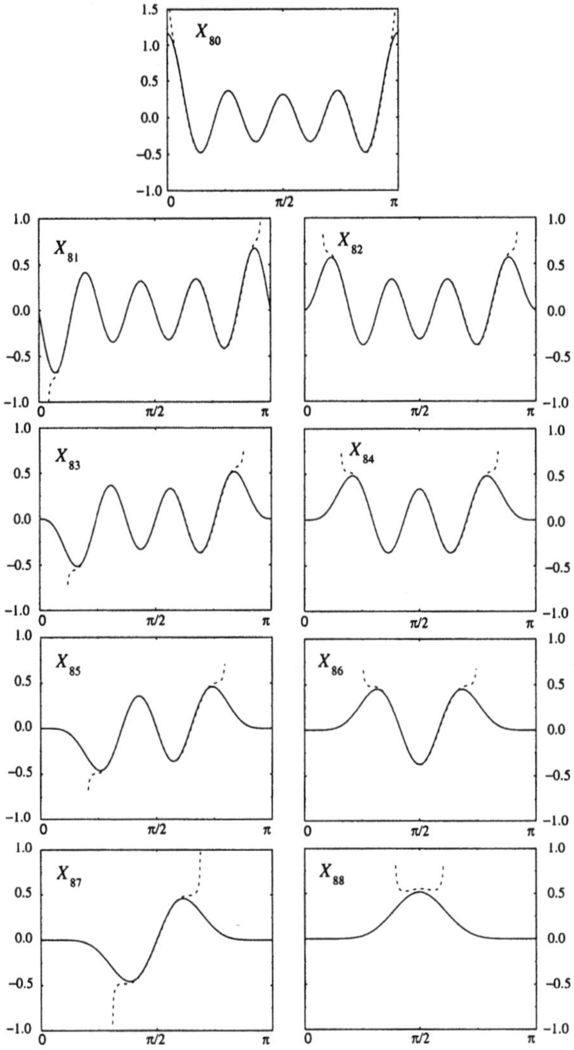

Figure B.3. Exact (*solid line*) and asymptotic (*dashed line*) colatitudinal harmonics X_{80},X_{81},X_{82},X_{83},X_{84}, X_{85},X_{86}, X_{87},X_{88}. The $l-m$ nodes of $X_{lm}, m \geq 0$, are evident (see also Figure B.2). Except in the non-oscillatory case $m = l$, the asymptotic approximation (B.84) provides a faithful representation of $X_{lm}, l \gg 1$, between the turning colatitudes $\theta_0 \ll \theta \ll \pi - \theta_0$.

$0 \leq \theta \ll \theta_0$ and $\pi - \theta_0 \ll \theta \leq \pi$, but also a characterization of the transitional behavior near the turning points in terms of Airy functions (Backus, Parker & Constable 1996).

In quantum mechanics we can regard the asymptotic representation

$$\begin{aligned} Y_{lm}(\theta, \phi) \approx \frac{1}{\pi}(\sin^2\theta - \sin^2\theta_0)^{-1/4} \\ \times \cos\left[(l + \tfrac{1}{2}) \arccos(\cos\theta/\cos\theta_0)\right. \\ \left. - \, m \arccos(\cot\theta/\cot\theta_0) + m\pi - \tfrac{1}{4}\pi\right] e^{im\phi} \end{aligned} \tag{B.85}$$

as the semi-classical wavefunction of an orbiting particle with angular momentum $L = l + \frac{1}{2}$ and $L_z = m$, as illustrated in Figure B.4. The two equatorial components L_x and L_y remain unspecified; hence, all great-circular orbits that turn at θ_0 and $\pi - \theta_0$ contribute equally to Y_{lm}. The squared modulus $|Y_{lm}|^2 \approx (1/2\pi^2)(\sin^2\theta - \sin^2\theta_0)^{-1/2}$ is the classical probability of finding the particle at a colatitude between θ and $\theta + d\theta$, in accordance with the quantum-mechanical correspondence principle.

In the limit $m \ll l$ equation (B.84) can be reduced even further:

$$X_{lm}(\theta) \approx \frac{1}{\pi}(\sin\theta)^{-1/2} \cos\left[(l + \tfrac{1}{2})\theta + m\pi/2 - \pi/4\right], \tag{B.86}$$

in the oscillatory regime $0 \ll \theta \ll \pi$. This low-order asymptotic approximation is valid for both positive and negative m, just like the original

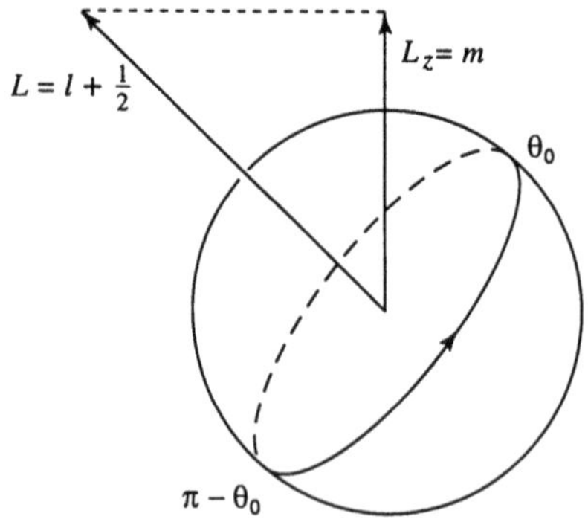

Figure B.4. Semi-classical characterization of a quantum-mechanical particle orbiting on the surface of the unit sphere Ω between the northern-hemisphere and southern-hemisphere turning colatitudes θ_0 and $\pi - \theta_0$. The magnitude $L = l + \frac{1}{2}$ and the length of the z component $L_z = m$ of the angular momentum $\mathbf{L}$ may be specified simultaneously. Since L_x and L_y are indeterminate, the orbit must be thought of as "precessing" in an indefinite manner about the $\hat{\mathbf{z}}$ axis.

phase-integral representation (B.82). The corresponding complex surface spherical harmonic $Y_{lm}(\theta,\phi) = X_{lm}(\theta)e^{im\phi}$ can be considered to be the semi-classical wavefunction of a particle in a highly inclined, nearly polar orbit: $\theta_0 \approx 0$ and $\pi - \theta_0 \approx \pi$. The neglected terms in equation (B.86) are of order $(l+\frac{1}{2})^{-1}$; for the special case of zonal harmonic, $m = 0$, the next term in the expansion is (Robin 1958)

$$X_{l0}(\theta) \approx \frac{1}{\pi}(\sin\theta)^{-1/2}\Big\{\cos\left[(l+\tfrac{1}{2})\theta - \pi/4\right] + \tfrac{1}{8}(l+\tfrac{1}{2})^{-1}\cot\theta\sin\left[(l+\tfrac{1}{2})\theta - \pi/4\right]\Big\}. \tag{B.87}$$

The extended relation (B.87) provides an improved approximation in the vicinity of the colatitudinal nodes $(l+\frac{1}{2})\theta \approx \frac{3}{4}\pi, \frac{7}{4}\pi, \ldots, (l-\frac{1}{4})\pi$.

B.8 Spherical-Harmonic Expansions

Let $\psi(\theta,\phi)$ be a complex square-integrable function on the surface of the unit sphere, with *Laplace coefficients*

$$\psi_{lm} = \langle Y_{lm}, \psi\rangle = \int_\Omega Y_{lm}^*\psi\, d\Omega. \tag{B.88}$$

Consider the problem of approximating ψ by a *finite* linear combination of surface spherical harmonics. Adopting a least-squares criterion, we seek to find coefficients $\hat{\psi}_{lm}$ satisfying

$$\|\psi - \sum_{l=0}^{L}\sum_{m=-l}^{l}\hat{\psi}_{lm}Y_{lm}\|^2 = \text{minimum}. \tag{B.89}$$

Upon making use of the orthonormality relation (B.35) we can rewrite the left side of (B.89) in the form

$$\begin{aligned}\|\psi - \sum_{l=0}^{L}\sum_{m=-l}^{l}\hat{\psi}_{lm}Y_{lm}\|^2 &= \int_\Omega |\psi - \sum_{l=0}^{L}\sum_{m=-l}^{l}\hat{\psi}_{lm}Y_{lm}|^2\, d\Omega \\ &= \|\psi\|^2 + \sum_{l=0}^{L}\sum_{m=-l}^{l}|\hat{\psi}_{lm} - \psi_{lm}|^2 - \sum_{l=0}^{L}\sum_{m=-l}^{l}|\psi_{lm}|^2. \end{aligned} \tag{B.90}$$

It is evident from this decomposition that the minimum is attained when $\hat{\psi}_{lm} = \psi_{lm}$. The Lth partial Laplace sum

$$\psi_L = \sum_{l=0}^{L}\sum_{m=-l}^{l}\psi_{lm}Y_{lm}, \tag{B.91}$$

consisting of $1+3+\cdots+(2L+1)=(L+1)^2$ terms, is therefore the best finite spherical-harmonic approximation to ψ in a least-squares sense. The squared error associated with this approximation is

$$\|\psi-\psi_L\|^2=\int_\Omega|\psi-\psi_L|^2\,d\Omega=\|\psi\|^2-\sum_{l=0}^{L}\sum_{m=-l}^{l}|\psi_{lm}|^2. \tag{B.92}$$

It can be demonstrated that this truncation error can be made as small as one likes: $\lim_{L\to\infty}\|\psi-\psi_L\|^2=0$ for any square-integrable function ψ. The infinite Laplace sum,

$$\psi=\sum_{l=0}^{\infty}\sum_{m=-l}^{l}\psi_{lm}Y_{lm}\quad\text{where}\quad\psi_{lm}=\int_\Omega Y_{lm}^*\psi\,d\Omega, \tag{B.93}$$

is said to *converge in the mean*. The squared norm of ψ is the infinite sum of squared expansion coefficients:

$$\|\psi\|^2=\int_\Omega|\psi|^2\,d\Omega=\sum_{l=0}^{\infty}\sum_{m=-l}^{l}|\psi_{lm}|^2. \tag{B.94}$$

The result (B.94) is the spherical-harmonic version of *Parseval's theorem*. The normalized interior sum

$$\sigma_l^2=\frac{1}{2l+1}\sum_{m=-l}^{l}|\psi_{lm}|^2 \tag{B.95}$$

is the variance or *power* per degree l and per unit area of the function ψ. A Dirac delta function on the unit sphere,

$$(\sin\theta)^{-1}\delta(\theta-\theta')\delta(\phi-\phi')=\sum_{l=0}^{\infty}\sum_{m=-l}^{l}Y_{lm}^*(\theta',\phi')Y_{lm}(\theta,\phi), \tag{B.96}$$

has a *flat* power spectrum: $\sigma_l^2=1/4\pi$ for all $l\geq 0$.

A *real* function ψ can of course be expanded in the form (B.93); the complex Laplace coefficients in that case satisfy $\psi_{l\,-m}=(-1)^m\psi_{lm}^*$. Alternatively, and more economically, we may expand ψ in terms of the real surface spherical harmonics (B.72):

$$\psi=\sum_{l=0}^{\infty}\sum_{m=-l}^{l}\Psi_{lm}\mathcal{Y}_{lm}\quad\text{where}\quad\Psi_{lm}=\int_\Omega\mathcal{Y}_{lm}\psi\,d\Omega. \tag{B.97}$$

The real and complex expansion coefficients Ψ_{lm} and ψ_{lm} are related by

$$\Psi_{lm} = \begin{cases} \sqrt{2}\,\mathrm{Re}\,\psi_{l|m|} & \text{if } -l \le m < 0 \\ \psi_{l0} & \text{if } m = 0 \\ -\sqrt{2}\,\mathrm{Im}\,\psi_{lm} & \text{if } 0 < m \le l. \end{cases} \tag{B.98}$$

In many applications it is convenient to eschew negative orders and rewrite the expansion (B.97) in the longer but more transparent form

$$\psi = \sum_{l=0}^{\infty} \left[a_{l0}X_{l0} + \sqrt{2} \sum_{m=1}^{l} X_{lm}(a_{lm}\cos m\phi + b_{lm}\sin m\phi) \right], \tag{B.99}$$

where

$$a_{l0} = \int_\Omega X_{l0}\,\psi\,d\Omega, \tag{B.100}$$

$$a_{lm} = \sqrt{2} \int_\Omega (X_{lm}\cos m\phi)\,\psi\,d\Omega \quad \text{if } 1 \le m \le l, \tag{B.101}$$

$$b_{lm} = \sqrt{2} \int_\Omega (X_{lm}\sin m\phi)\,\psi\,d\Omega \quad \text{if } 1 \le m \le l. \tag{B.102}$$

Parseval's theorem (B.94) for a real field (B.97) or (B.99)–(B.102) is

$$\|\psi\|^2 = \sum_{l=0}^{\infty} \sum_{m=-l}^{l} \Psi_{lm}^2 = \sum_{l=0}^{\infty} \left[a_{l0}^2 + \sum_{m=1}^{l} (a_{lm}^2 + b_{lm}^2) \right]. \tag{B.103}$$

The power per degree l and per unit area of such an expansion is

$$\sigma_l^2 = \frac{1}{2l+1} \sum_{m=-l}^{l} \Psi_{lm}^2 = \frac{1}{2l+1} \left[a_{l0}^2 + \sum_{m=1}^{l} (a_{lm}^2 + b_{lm}^2) \right]. \tag{B.104}$$

All of the above considerations pertain to a field $\psi(\theta, \phi)$ upon the surface of the unit sphere Ω. A three-dimensional field $\psi(\mathbf{r})$ can be expanded in spherical harmonics Y_{lm}, $\mathcal{Y}_{lm}$ or $\sqrt{2}\,X_{ll}\cos l\phi, \ldots, X_{l0}, \ldots, X_{ll}\sin l\phi$ by simply allowing the coefficients ψ_{lm}, Ψ_{lm} or a_{l0}, a_{lm}, b_{lm} to be functions of radius r.

The standard convention used in geodesy is similar to (B.99)–(B.102) except that the real harmonics differ in sign for odd m and have a squared norm equal to 4π rather than unity (Lambeck 1988; Stacey 1992). A real field in that case is expanded in the form

$$\psi = \sum_{l=0}^{\infty} \sum_{m=0}^{l} p_{lm}(c_{lm}\cos m\phi + s_{lm}\sin m\phi), \tag{B.105}$$

where $p_{lm} = \sqrt{4\pi}\,(2 - \delta_{m0})(-1)^m X_{lm}$. The geodetic expansion coefficients are related to the coefficients (B.100)–(B.102) by

$$c_{lm} = \sqrt{4\pi}\,(-1)^m a_{lm}, \qquad s_{lm} = \sqrt{4\pi}\,(-1)^m b_{lm}. \tag{B.106}$$

The so-called quasi-normalized Schmidt harmonics which are commonly employed in geomagnetism differ from both X_{lm} and p_{lm} (Chapman & Bartels 1940; Backus, Parker & Constable 1996). It is a trifling but inescapable annoyance of geophysical life that spherical-harmonic sign and normalization conventions vary from one discipline—or even one study—to the next, so that care must be taken in comparing or utilizing results. Quantum mechanics is not as balkanized in this regard; the complex harmonics Y_{lm}, $-l \leq m \leq l$, of Condon & Shortley (1935) and Edmonds (1960) have become well-nigh universal.

B.9 Integration Around a Great Circle

We designate a *directed great-circular path* on the surface of the unit sphere Ω by its *positive pole* $\bar{\mathbf{r}}$, which is defined using the right-hand rule. If the path is the extension of the geodesic minor arc connecting two sequential points $\hat{\mathbf{r}}_1$ and $\hat{\mathbf{r}}_2$, as illustrated in Figure B.5, then $\bar{\mathbf{r}} = (\hat{\mathbf{r}}_1 \times \hat{\mathbf{r}}_2)/\|\hat{\mathbf{r}}_1 \times \hat{\mathbf{r}}_2\|$. The spherical polar coordinates $\bar{\theta}, \bar{\phi}$ of the great-circular pole are given in terms of the "source" and "receiver" coordinates θ_1, ϕ_1 and θ_2, ϕ_2 by

$$\cos\bar{\theta} = \frac{\sin\theta_2 \sin\theta_1 \sin(\phi_2 - \phi_1)}{\sin\Theta}, \tag{B.107}$$

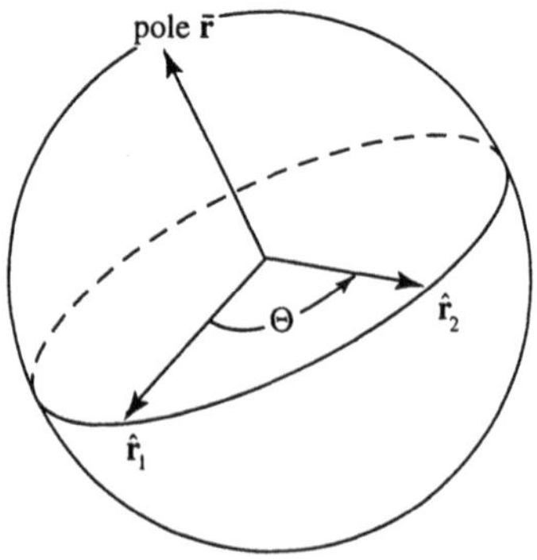

Figure B.5. Pole $\bar{\mathbf{r}} = (\hat{\mathbf{r}}_1 \times \hat{\mathbf{r}}_2)/\|\hat{\mathbf{r}}_1 \times \hat{\mathbf{r}}_2\|$ of a directed great circle passing sequentially through the points $\hat{\mathbf{r}}_1$ and $\hat{\mathbf{r}}_2$. The pole of the oppositely directed great circle is the antipode $-\bar{\mathbf{r}}$.

$$\tan\bar{\phi} = \frac{\sin\theta_2\cos\theta_1\cos\phi_2 - \cos\theta_2\sin\theta_1\cos\phi_1}{\cos\theta_2\sin\theta_1\sin\phi_1 - \sin\theta_2\cos\theta_1\sin\phi_2}, \tag{B.108}$$

where $\cos\Theta = \hat{\mathbf{r}}_1\cdot\hat{\mathbf{r}}_2 = \cos\theta_2\cos\theta_1 + \sin\theta_2\sin\theta_1\cos(\phi_2-\phi_1)$.

The *great-circular average* of a function $\psi(\theta,\phi)$ is defined by

$$\bar{\psi}(\bar{\theta},\bar{\phi}) = \frac{1}{2\pi}\oint_{\bar{\theta},\bar{\phi}} \psi(\theta,\phi)\,d\Delta, \tag{B.109}$$

where the integral is taken over the full circumference and $d\Delta$ is the differential angular arclength. We can regard $\bar{\psi}$ as a continuous function of position on the unit sphere, obtained by integrating over all possible directed great circles and assigning the average to the pole. Since the direction of integration about a great circle is immaterial, we must have

$$\bar{\psi}(\pi-\bar{\theta},\bar{\phi}+\pi) = \bar{\psi}(\bar{\theta},\bar{\phi}), \tag{B.110}$$

i.e., the great-circular average $\bar{\psi}$ of any field ψ must be an *even function* of position $\bar{\mathbf{r}}$ on Ω.

The great-circular average of a surface spherical harmonic of degree l is

$$\frac{1}{2\pi}\oint_{\bar{\theta},\bar{\phi}} Y_l(\theta,\phi)\,d\Delta = P_l(0)Y_l(\bar{\theta},\bar{\phi}). \tag{B.111}$$

Following Roberts & Ursell (1960) we prove this remarkably simple result by considering a spherical polar coordinate system whose $\hat{\mathbf{z}}$ axis coincides with the pole $\bar{\theta},\bar{\phi}$. The function $Y_l(\bar{\theta},\bar{\phi})$ can be written in this coordinate system as a linear combination of basis harmonics Y_{lm}, $-l\le m\le l$. The zonal harmonic Y_{l0} satisfies equation (B.111), each side having the value $\sqrt{(2l+1)/4\pi}\,P_l(0)$, and every non-zonal harmonic satisfies it, each side vanishing. It follows that (B.111) is valid for a general real or complex spherical harmonic Y_l. We present an alternative proof of this great-circular identity in Appendix C.8, using a more general argument first given by Backus (1964). This latter technique provides a means of evaluating the minor-arc integral between two points θ_1,ϕ_1 and θ_2,ϕ_2 as well.

The coefficients $\bar{\psi}_{lm}$ in the spherical-harmonic expansion of $\bar{\psi}(\bar{\theta},\bar{\phi})$ are related to the coefficients ψ_{lm} of the averaged function $\psi(\theta,\phi)$ by

$$\bar{\psi}_{lm} = P_l(0)\,\psi_{lm}. \tag{B.112}$$

The Legendre polynomial $P_l(0)$ is given explicitly by

$$P_l(0) = \begin{cases} 0 & \text{if } l \text{ is odd} \\ (-1)^{l/2}\,l!\,2^{-l}[(l/2)!]^{-2} & \text{if } l \text{ is even.} \end{cases} \tag{B.113}$$

Degree l	Coefficient $P_l(0)$
0	1
2	$-1/2$
4	$3/8$
6	$-5/16$
$\vdots$	
∞	$(-1)^{l/2}\sqrt{2/\pi l}$

Table B.2. Numerical value of the great-circular-average filter coefficients $P_l(0)$ for even degrees l. The $l \to \infty$ limit is an asymptotic representation obtained using Stirling's formula to approximate the factorials in (B.113).

The absence of any odd-degree coefficients $\bar{\psi}_{lm}$ is consistent with the elementary observation (B.110). Fundamentally, the integral of an odd spherical harmonic vanishes because of the cancellation of antipodal points $\pm\hat{\mathbf{r}}$ on the great-circular path. The asymptotic limit $P_l(0) \to (-1)^{l/2}\sqrt{2/\pi l}$ for high even degrees shows that great-circular integration is a smoothing process: if the power spectrum σ_l^2 of ψ decays like $l^{-\alpha}$ then that of $\bar{\psi}$ decays like $l^{-\alpha-1}$. Equation (B.111) is valid for the real spherical harmonics (B.72), so that (B.112) is valid for the associated coefficients $\bar{\Psi}_{lm}$, Ψ_{lm}. The values of $P_l(0)$ for $l = 0, 2, 4, 6, \ldots$ are shown in Table B.2.

B.10 Practical Considerations

The optimum procedure for calculating surface spherical harmonics depends upon the application. To synthesize normal-mode seismograms upon a spherically symmetric Earth, we only require the associated Legendre functions P_{l0}, P_{l1}, P_{l2} and their colatitudinal derivatives. These may be calculated using the fixed-order recurrence relation (B.50), which is stable when iterated in the direction of increasing degree l. In order to minimize the error in the vicinity of the endpoints $\mu = \pm 1$, and obtain a result that is uniformly valid in the interval $-1 \leq \mu \leq 1$, it is advantageous to define an auxiliary function R_{lm} by

$$P_{lm}(\mu) = (1 - \mu^2)^{m/2} R_{lm}(\mu). \tag{B.114}$$

The starting values for the pre-conditioned recursion relation

$$(l-m+1)R_{l+1\,m}-(2l+1)\mu R_{lm}+(l+m)R_{l-1\,m}=0 \tag{B.115}$$

are then $R_{00}=1$, $R_{10}=\mu$, $R_{11}=1$, $R_{21}=3\mu$ and $R_{22}=3$, $R_{32}=15\mu$. The derivatives $dP_{l0}(\cos\theta)/d\theta$, $dP_{l1}(\cos\theta)/d\theta$ and $dP_{l2}(\cos\theta)/d\theta$ can be found using either the second or fourth of equations (B.55):

$$\begin{aligned}\frac{dP_{lm}}{d\theta}&=-m\cot\theta\,P_{lm}+(l+m)(l-m+1)P_{l\,m-1}\\&=l\cot\theta\,P_{lm}-(l+m)\csc\theta\,P_{l-1\,m}.\end{aligned} \tag{B.116}$$

The roundoff error is multiplied by a factor of order unity upon each iteration of (B.115) so that the required quantities can be calculated accurately up to very high degree without significant loss of precision.

To calculate the real or complex source and receiver vectors upon an aspherical Earth we need X_{lm} and $dX_{lm}/d\theta$ for all orders $-l\le m\le l$. The simplest procedure in this case makes use of the fixed-degree recurrence relation (B.51), which is stable when iterated in the direction of decreasing order m. Defining an auxiliary function W_{lm} by

$$X_{lm}=(\sin\theta)^m W_{lm}, \tag{B.117}$$

we rewrite this relation in the pre-conditioned form

$$\begin{aligned}&(1-\mu^2)\sqrt{(l-m)(l+m+1)}\,W_{l\,m+1}+2m\mu W_{lm}\\&\quad+\sqrt{(l+m)(l-m+1)}\,W_{l\,m-1}=0,\end{aligned} \tag{B.118}$$

where $\mu=\cos\theta$. The starting values for (B.118) are

$$W_{ll}=(-1)^l\left(\frac{2l+1}{4\pi}\right)^{1/2}\frac{\sqrt{(2l)!}}{2^l l!},\qquad W_{l\,l-1}=-\sqrt{2l}\,\mu W_{ll}. \tag{B.119}$$

Note that it is unnecessary to run the recursion beyond $m=0$; the negative orders can be determined from the symmetry $X_{l\,-m}=(-1)^{l+m}X_{lm}$. The roundoff error near $m=l$ is multiplied by a factor of order $\sqrt{l}$ upon each iteration; because of this, a double-precision implementation of (B.118) is only able to provide satisfactory results up to degree $l\approx 200$. Libbrecht (1985) describes a slightly more complicated procedure based upon the variable order-and-degree recursion relation (B.52) which can be used to generate higher-degree harmonics X_{lm} if necessary. The colatitudinal derivative $dX_{lm}/d\theta$ can be calculated using

$$\begin{aligned}\frac{dX_{lm}}{d\theta}&=\tfrac{1}{2}\sqrt{(l-m)(l+m+1)}\,X_{l\,m+1}\\&\quad-\tfrac{1}{2}\sqrt{(l+m)(l-m+1)}\,X_{l\,m-1},\end{aligned} \tag{B.120}$$

which is a modified combination of the first two of equations (B.55).

An alternative scheme for computing X_{lm} and $dX_{lm}/d\theta$ simultaneously is described by Masters & Richards-Dinger (1998). They rewrite the second and first of (B.55) as a pair of coupled recurrence relations:

$$X_{l\,m-1} = -\frac{dX_{lm}/d\theta + m\cot\theta X_{lm}}{\sqrt{(l+m)(l-m+1)}}, \tag{B.121}$$

$$\frac{dX_{l\,m-1}}{d\theta} = (m-1)\cot\theta X_{l\,m-1} + \sqrt{(l+m)(l-m+1)}\,X_{lm}. \tag{B.122}$$

The stable iteration direction is again downward from $m = l$ to $m = 0$; the starting values in this case are

$$X_{ll} = (-1)^l \left(\frac{2l+1}{4\pi}\right)^{1/2} \frac{\sqrt{(2l)!}}{2^l l!} (\sin\theta)^l, \qquad \frac{dX_{ll}}{d\theta} = l\cot\theta X_{ll}. \tag{B.123}$$

The absence of any pre-conditioning causes (B.123) to underflow for large degrees $l \gg 1$ at polar colatitudes, $\theta \approx 0$ or $\theta \approx \pi$. In such instances, they fix the sign by starting the iteration with $X_{ll} = (-1)^l$, and rescale X_{lm} and its derivative $dX_{lm}/d\theta$ during recursion downward to $m = 0$ in order to avoid overflow. This rescaling is undone, allowing underflow to occur, at the end of the calculation; the final results are normalized to be consistent with the spherical-harmonic addition theorem:

$$\sum_{m=-l}^{l} X_{lm}^2 = (2l+1)/4\pi. \tag{B.124}$$

B.11 Complex Legendre Functions

It is possible to extend the definitions of the Legendre polynomials $P_l(\mu)$ and the associated Legendre functions $P_{lm}(\mu)$ considerably, allowing all three parameters l, m and μ to be complex. We have no need for such generality in this book; a comprehensive summary may be found in Erdélyi, Magnus, Oberhettinger & Tricomi (1953) and Robin (1958). Legendre functions of complex degree λ, non-negative integer order $m = 0, 1, 2, \ldots$ and real argument $-1 \leq \mu \leq 1$ arise in the analysis of travelling waves upon the surface of the unit sphere; we summarize the salient properties of these seismologically useful functions here.

B.11.1 Legendre functions of the first and second kind

Our analysis of the surface-wave and body-wave Green tensors in Chapters 11 and 12 makes use of an operator formalism which enables us to avoid explicit consideration of orders $m \neq 0$. For this reason we consider the azimuthally symmetric $m = 0$ functions first. Since Legendre's equation

$$(1-\mu^2)\frac{d^2X}{d\mu^2} - 2\mu\frac{dX}{d\mu} + \lambda(\lambda+1)X = 0 \tag{B.125}$$

is second order, it has two linearly independent solutions, which are conventionally denoted by P_λ and Q_λ, and referred to as the *Legendre functions of the first and second kind*, respectively. For our purposes, these are most conveniently defined by the integral representations

$$P_\lambda(\mu) = \frac{2}{\pi}\,\mathrm{Im}\int_0^\infty \left(\mu - i\sqrt{1-\mu^2}\cosh t\right)^{-\lambda-1} dt, \tag{B.126}$$

$$Q_\lambda(\mu) = \mathrm{Re}\int_0^\infty \left(\mu - i\sqrt{1-\mu^2}\cosh t\right)^{-\lambda-1} dt, \tag{B.127}$$

which are valid for $\mathrm{Re}\,\lambda > -1$. Both P_λ and Q_λ are made single-valued by choosing the principal value of the power in the integrand.

The Legendre function of the first kind coincides with the Legendre polynomial (B.67) for non-negative integer orders $\lambda = 0, 1, \ldots$. The corresponding Legendre functions of the second kind $Q_0, Q_1, \ldots$ are given by

$$Q_l(\mu) = \tfrac{1}{2}P_l(\mu)\ln\left(\frac{1+\mu}{1-\mu}\right) - M_{l-1}, \tag{B.128}$$

where $M_{-1} = 0$ and

$$M_{l-1} = \frac{2l-1}{1(l-0)}\,P_{l-1} + \frac{2l-5}{3(l-1)}\,P_{l-3} + \frac{2l-9}{5(l-2)}\,P_{l-5} + \cdots \tag{B.129}$$

for $l > 0$. The logarithmic singularity at the endpoints $\mu = \pm 1$ is a characteristic feature of Q_λ for all complex degrees, not just $\lambda = 0, 1, \ldots$; P_λ has a single logarithmic singularity at $\mu = -1$ for $\lambda \neq 0, \pm 1, \ldots$. The analytical continuations of equations (B.126)–(B.127) into the region $\mathrm{Re}\,\lambda \leq -1$ satisfy the symmetry relations

$$P_{-\lambda-1} = P_\lambda, \qquad Q_{-\lambda-1} = Q_\lambda - \pi\cot\lambda\pi\, P_\lambda. \tag{B.130}$$

The second of equations (B.130) shows that Q_λ has a simple pole at each of the negative integers $\lambda = -1, -2, \ldots$. The values of the Legendre functions at negative and positive arguments are related by

$$P_\lambda(-\mu) = \cos\lambda\pi\, P_\lambda(\mu) - \frac{2}{\pi}\sin\lambda\pi\, Q_\lambda(\mu), \tag{B.131}$$

$$Q_\lambda(-\mu) = -\cos\lambda\pi\, Q_\lambda(\mu) - \frac{\pi}{2}\sin\lambda\pi\, P_\lambda(\mu). \tag{B.132}$$

B.11.2 Travelling-wave Legendre functions

The *travelling-wave Legendre functions* used in this book are those described by Nussenzveig (1965):

$$Q_\lambda^{(1,2)} = \tfrac{1}{2}\left(P_\lambda \pm \frac{2i}{\pi}Q_\lambda\right), \tag{B.133}$$

where the concatenated superscripts on the left and right are associated with the concatenated signs on the top and bottom, respectively. For $\mathrm{Re}\,\lambda > -1$ we have the explicit integral representation

$$Q_\lambda^{(1,2)}(\mu) = \pm\frac{i}{\pi}\int_0^\infty \left(\mu \pm i\sqrt{1-\mu^2}\cosh t\right)^{-\lambda-1} dt. \tag{B.134}$$

From equations (B.130)–(B.132) together with the definitions (B.133) we find that

$$Q_{-\lambda-1}^{(1,2)} = Q_\lambda^{(1,2)} \mp i\cot\lambda\pi\, P_\lambda, \tag{B.135}$$

$$Q_\lambda^{(1,2)}(-\mu) = e^{\mp i\lambda\pi} Q_\lambda^{(2,1)}(\mu). \tag{B.136}$$

Both of the travelling-wave Legendre functions $Q_\lambda^{(1)}$ and $Q_\lambda^{(2)}$ have simple poles at the negative integers $\lambda = -1, -2, \ldots$, which cancel upon addition to form the Legendre function of the first kind P_λ. By combining the above results governing the interchanges $\lambda \to -\lambda - 1$ and $\mu \to -\mu$, we can obtain a number of other interrelations among the various Legendre functions, for example,

$$P_\lambda(-\mu) = e^{\mp i\lambda\pi} P_\lambda(\mu) \pm 2i\sin\lambda\pi\, Q_\lambda^{(1,2)}(\mu) \tag{B.137}$$

and

$$\begin{aligned} Q_{-\lambda-1}^{(1,2)}(\mu) &= -e^{\pm 2i\lambda\pi} Q_\lambda^{(1,2)}(\mu) \\ &\quad \pm ie^{\pm i\lambda\pi}\tan(\lambda + \tfrac{1}{2})\pi\, P_\lambda(-\mu). \end{aligned} \tag{B.138}$$

We use equation (B.138) to express the travelling-wave Green tensor of a spherically symmetric Earth in its most easily interpreted form (11.14).

The asymptotic representations of the four Legendre functions in the limit of large complex degree $|\lambda| \gg 1$ are

$$P_{\lambda-\frac{1}{2}}(\cos\theta) \approx \left(\frac{2}{\pi\lambda\sin\theta}\right)^{1/2} \cos(\lambda\theta - \pi/4), \tag{B.139}$$

$$Q_{\lambda-\frac{1}{2}}(\cos\theta) \approx -\left(\frac{\pi}{2\lambda\sin\theta}\right)^{1/2} \sin(\lambda\theta - \pi/4), \tag{B.140}$$

$$Q^{(1,2)}_{\lambda-\frac{1}{2}}(\cos\theta) \approx \left(\frac{1}{2\pi\lambda\sin\theta}\right)^{1/2} \exp\left[\mp i(\lambda\theta - \pi/4)\right]. \tag{B.141}$$

Equations (B.139)–(B.141) are valid everywhere within the oscillatory interval $0 \ll \theta \ll \pi$ and the sector $|\arg\lambda| \ll \pi$. It is evident that P_λ and Q_λ represent colatitudinal standing waves, whereas $Q^{(1)}_\lambda$ and $Q^{(2)}_\lambda$ represent waves travelling in the direction of increasing and decreasing θ, respectively.

B.11.3 Associated Legendre functions

The corresponding *associated Legendre functions* of positive integer order $m = 1, 2, \ldots$ are defined by analogy with equation (B.71):

$$P_{\lambda m}(\mu) = (1-\mu^2)^{m/2}\left(\frac{d}{d\mu}\right)^m P_\lambda(\mu), \tag{B.142}$$

$$Q_{\lambda m}(\mu) = (1-\mu^2)^{m/2}\left(\frac{d}{d\mu}\right)^m Q_\lambda(\mu). \tag{B.143}$$

The associated travelling-wave Legendre functions are given by the analogue of equation (B.133):

$$Q^{(1,2)}_{\lambda m} = \tfrac{1}{2}\left(P_{\lambda m} \pm \frac{2i}{\pi} Q_{\lambda m}\right). \tag{B.144}$$

Obviously, we may also write $Q^{(1,2)}_{\lambda m}$ in terms of $Q^{(1,2)}_\lambda$ in the form

$$Q^{(1,2)}_{\lambda m}(\mu) = (1-\mu^2)^{m/2}\left(\frac{d}{d\mu}\right)^m Q^{(1,2)}_\lambda(\mu). \tag{B.145}$$

The left-half-plane and right-half-plane functions (B.144) are related by a generalization of equation (B.138):

$$\begin{aligned} Q^{(1,2)}_{-\lambda-1\,m}(\mu) &= -e^{\pm 2i\lambda\pi} Q^{(1,2)}_{\lambda m}(\mu) \\ &\quad \pm i(-1)^m e^{\pm i\lambda\pi} \tan(\lambda + \tfrac{1}{2})\pi\, P_{\lambda m}(-\mu). \end{aligned} \tag{B.146}$$

Equation (B.146) is used in the derivation of the travelling-wave moment-tensor response (11.29). The asymptotic representations of the four associated Legendre functions are

$$\begin{aligned} P_{\lambda-\frac{1}{2}\,m}(\cos\theta) &\approx (-\lambda)^m \left(\frac{2}{\pi\lambda\sin\theta}\right)^{1/2} \\ &\quad \times \cos(\lambda\theta + m\pi/2 - \pi/4), \end{aligned} \tag{B.147}$$

$$Q_{\lambda-\frac{1}{2}\,m}(\cos\theta) \approx -(-\lambda)^m \left(\frac{\pi}{2\lambda\sin\theta}\right)^{1/2} \times \sin(\lambda\theta + m\pi/2 - \pi/4), \tag{B.148}$$

$$Q^{(1,2)}_{\lambda-\frac{1}{2}\,m}(\cos\theta) \approx (-\lambda)^m \left(\frac{1}{2\pi\lambda\sin\theta}\right)^{1/2} \times \exp\left[\mp i(\lambda\theta + m\pi/2 - \pi/4)\right]. \tag{B.149}$$

These results are valid for *small* positive integer orders $m = 1, 2, \ldots$ in the limit $|\lambda| \gg 1$. Alternative definitions of the two travelling-wave Legendre functions which retain the essential $\exp[\mp i(\lambda\theta + m\pi/2 - \pi/4)]$ asymptotic character are discussed by Clemmow (1961), Burridge (1966) and Ansell (1973).

B.12 Vector Spherical Harmonics

Thus far we have considered only the expansion of a scalar field ψ in surface spherical harmonics. In this section we discuss the problem of expanding a vector field $\mathbf{u}$ in *vector spherical harmonics*. For concreteness, we shall suppose that $\mathbf{u}$ is real, and define the surface spherical harmonics in terms of the real scalar harmonics $\mathcal{Y}_{lm}$, $-l \leq m \leq l$; however, it is clear that complex vector spherical harmonics could be defined in terms of Y_{lm}, $-l \leq m \leq l$, in an entirely analogous fashion. In Appendix C we consider the expansion of vector and higher-order tensor fields in terms of complex generalized spherical harmonics.

B.12.1 Helmholtz representation of a tangent vector

The basic idea is to represent $\mathbf{u}$ in terms of three scalar fields, each of which is expanded in terms of $\mathcal{Y}_{lm}$, $-l \leq m \leq l$. One obvious starting point is the three-component representation $\mathbf{u} = u_r\hat{\mathbf{r}} + u_\theta\hat{\boldsymbol{\theta}} + u_\phi\hat{\boldsymbol{\phi}}$; however, this has the undesirable feature that it is singular at the poles $\theta = 0$ and $\theta = \pi$. A more useful approach is based upon the *Helmholtz representation* of the tangent-vector field $\mathbf{u}^\Omega = \mathbf{u} - u_r\hat{\mathbf{r}}$ as the sum of a surface gradient and a surface curl:

$$\mathbf{u}^\Omega = \boldsymbol{\nabla}_1 V - \hat{\mathbf{r}} \times \boldsymbol{\nabla}_1 W, \tag{B.150}$$

where

$$\int_\Omega V\,d\Omega = \int_\Omega W\,d\Omega = 0. \tag{B.151}$$

For any tangent vector $\mathbf{u}^{\Omega}$ upon the surface of the unit sphere Ω, there exist unique scalar fields V and W which both satisfy equation (B.150) and average to zero in the sense (B.151). The minus sign preceding the surface curl term in (B.150) is introduced to facilitate agreement with an older vector spherical-harmonic convention (Morse & Feshbach 1953).

Upon dotting both sides of (B.150) with the operators $\boldsymbol{\nabla}_1$ and $\hat{\mathbf{r}} \times \boldsymbol{\nabla}_1$, and making use of the relations $\boldsymbol{\nabla}_1 \cdot \boldsymbol{\nabla}_1 = (\hat{\mathbf{r}} \times \boldsymbol{\nabla}_1) \cdot (\hat{\mathbf{r}} \times \boldsymbol{\nabla}_1) = \nabla_1^2$ and $\boldsymbol{\nabla}_1 \cdot (\hat{\mathbf{r}} \times \boldsymbol{\nabla}_1) = (\hat{\mathbf{r}} \times \boldsymbol{\nabla}_1) \cdot \boldsymbol{\nabla}_1 = 0$, we obtain the decoupled equations

$$\nabla_1^2 V = \boldsymbol{\nabla}_1 \cdot \mathbf{u}^{\Omega}, \qquad \nabla_1^2 W = -(\hat{\mathbf{r}} \times \boldsymbol{\nabla}_1) \cdot \mathbf{u}^{\Omega}. \tag{B.152}$$

In order to use these equations to find the scalar fields V and W it is necessary to invert the surface Laplacian, that is, we must be able to solve the inhomogeneous equation $\nabla_1^2 \psi = \chi$ for the quantity $\psi = \nabla_1^{-2} \chi$. Upon integrating $\nabla_1^2 \psi = \chi$ over Ω and applying Gauss' theorem, we find that a solution exists if and only if $\int_{\Omega} \chi \, d\Omega = 0$. The spherical-harmonic expansion of the right side χ therefore cannot have a term of degree zero:

$$\chi = \sum_{l=1}^{\infty} \sum_{m=-l}^{l} \chi_{lm} \mathcal{Y}_{lm}. \tag{B.153}$$

Imposing the constraint (B.153) and recalling that $\nabla_1^2 \mathcal{Y}_{lm} = -l(l+1)\mathcal{Y}_{lm}$, we can write the inverse $\psi = \nabla_1^{-2} \chi$ in the unique form

$$\psi = -\sum_{l=1}^{\infty} \sum_{m=-l}^{l} \frac{\chi_{lm}}{l(l+1)} \mathcal{Y}_{lm}. \tag{B.154}$$

The corresponding result in the spatial domain is (Backus 1958)

$$\psi(\hat{\mathbf{r}}) = \frac{1}{4\pi} \int_{\Omega} \chi(\hat{\mathbf{r}}') \ln(1 - \hat{\mathbf{r}} \cdot \hat{\mathbf{r}}') \, d\Omega'. \tag{B.155}$$

In summary, the operator ∇_1^{-2} is well defined by either (B.154) or (B.155) upon the space of all square-integrable functions χ that average to zero on the unit sphere Ω. This inverse operator is linear and, as long as χ is sufficiently smooth, $\nabla_1^2 \nabla_1^{-2} \chi = \nabla_1^{-2} \nabla_1^2 \chi = \chi$. Applying these results to equations (B.152) we find that

$$V = \nabla_1^{-2}(\boldsymbol{\nabla}_1 \cdot \mathbf{u}^{\Omega}), \qquad W = -\nabla_1^{-2}[(\hat{\mathbf{r}} \times \boldsymbol{\nabla}_1) \cdot \mathbf{u}^{\Omega}]. \tag{B.156}$$

There remains the question of whether $\mathbf{u}^{\Omega}$ is given in terms of V and W by equation (B.150). To answer this in the affirmative, it suffices to consider the residual tangent vector $\boldsymbol{\delta}^{\Omega} = \mathbf{u}^{\Omega} - \boldsymbol{\nabla}_1 V + \hat{\mathbf{r}} \times \boldsymbol{\nabla}_1 W$ and make use of the fact that $\boldsymbol{\nabla}_1 \cdot \boldsymbol{\delta}^{\Omega} = (\hat{\mathbf{r}} \times \boldsymbol{\nabla}_1) \cdot \boldsymbol{\delta}^{\Omega} = 0$ implies that $\boldsymbol{\delta}^{\Omega} = \mathbf{0}$.

B.12.2 Spheroidal and toroidal fields

Denoting the radial component u_r by U for the sake of uniformity, we can write an arbitrary vector field $\mathbf{u}$ on the surface of the unit sphere Ω in terms of three scalar fields in the form

$$\mathbf{u} = \hat{\mathbf{r}}\, U + \boldsymbol{\nabla}_1 V - \hat{\mathbf{r}} \times \boldsymbol{\nabla}_1 W, \tag{B.157}$$

where $\int_\Omega V\, d\Omega = \int_\Omega W\, d\Omega = 0$. A vector field of the form $\hat{\mathbf{r}}\, U + \boldsymbol{\nabla}_1 V$ is said to be *spheroidal* whereas one of the form $-\hat{\mathbf{r}} \times \boldsymbol{\nabla}_1 W$ is said to be *toroidal*. Thus, equation (B.157) represents the decomposition of an arbitrary vector field $\mathbf{u}$ into its spheroidal and toroidal parts. Note that a spheroidal field has both radial and tangential components, whereas a toroidal field is purely tangential. There is no commonly accepted name for the purely tangential part $\boldsymbol{\nabla}_1 V$ of a spheroidal field; Backus (1986) suggests calling it *consoidal*.

Expansion of the three fields U, V and W in terms of the real scalar harmonics $\mathcal{Y}_{lm}$, $-l \leq m \leq l$, would lead naturally to radial, consoidal and toroidal vector harmonics $\hat{\mathbf{r}}\, \mathcal{Y}_{lm}$, $\boldsymbol{\nabla}_1 \mathcal{Y}_{lm}$ and $-\hat{\mathbf{r}} \times \boldsymbol{\nabla}_1 \mathcal{Y}_{lm}$. It is preferable, however, to renormalize; we define the three real *vector spherical harmonics* $\mathbf{P}_{lm}$, $\mathbf{B}_{lm}$ and $\mathbf{C}_{lm}$ of degree $0 < l \leq \infty$ and order $-l \leq m \leq l$ by

$$\mathbf{P}_{lm} = \hat{\mathbf{r}}\, \mathcal{Y}_{lm}, \tag{B.158}$$

$$\mathbf{B}_{lm} = \frac{\boldsymbol{\nabla}_1 \mathcal{Y}_{lm}}{\sqrt{l(l+1)}} = \frac{[\hat{\boldsymbol{\theta}} \partial_\theta + \hat{\boldsymbol{\phi}} (\sin\theta)^{-1} \partial_\phi] \mathcal{Y}_{lm}}{\sqrt{l(l+1)}}, \tag{B.159}$$

$$\mathbf{C}_{lm} = \frac{-\hat{\mathbf{r}} \times \boldsymbol{\nabla}_1 \mathcal{Y}_{lm}}{\sqrt{l(l+1)}} = \frac{[\hat{\boldsymbol{\theta}} (\sin\theta)^{-1} \partial_\phi - \hat{\boldsymbol{\phi}} \partial_\theta] \mathcal{Y}_{lm}}{\sqrt{l(l+1)}}. \tag{B.160}$$

The degree- and order-zero harmonics are assumed to be purely radial: $\mathbf{P}_{00} = (4\pi)^{-1/2} \hat{\mathbf{r}}$ whereas $\mathbf{B}_{00} = \mathbf{C}_{00} = \mathbf{0}$. The extra factor of $1/\sqrt{l(l+1)}$ in the definitions (B.159)–(B.160) renders the vector spherical harmonics *orthonormal*:

$$\int_\Omega \mathbf{P}_{lm} \cdot \mathbf{P}_{l'm'}\, d\Omega = \delta_{ll'} \delta_{mm'}, \tag{B.161}$$

$$\int_\Omega \mathbf{B}_{lm} \cdot \mathbf{B}_{l'm'}\, d\Omega = \delta_{ll'} \delta_{mm'}, \tag{B.162}$$

$$\int_\Omega \mathbf{C}_{lm} \cdot \mathbf{C}_{l'm'}\, d\Omega = \delta_{ll'} \delta_{mm'}, \tag{B.163}$$

$$\int_\Omega \mathbf{P}_{lm} \cdot \mathbf{B}_{l'm'}\, d\Omega = 0, \tag{B.164}$$

$$\int_\Omega \mathbf{P}_{lm} \cdot \mathbf{C}_{l'm'}\, d\Omega = 0, \tag{B.165}$$

$$\int_\Omega \mathbf{B}_{lm} \cdot \mathbf{C}_{l'm'}\, d\Omega = 0. \tag{B.166}$$

Equation (B.161) is an immediate consequence of the orthonormality (B.73) of the scalar harmonics $\mathcal{Y}_{lm}$, whereas the results (B.164)–(B.166) are obvious pointwise identities. To verify (B.162)–(B.163) we note that

$$\int_\Omega \boldsymbol{\nabla}_1\mathcal{Y}_{lm} \cdot \boldsymbol{\nabla}_1\mathcal{Y}_{l'm'}\, d\Omega = -\int_\Omega (\nabla_1^2\mathcal{Y}_{lm})\mathcal{Y}_{l'm'}\, d\Omega$$
$$= l(l+1)\int_\Omega \mathcal{Y}_{lm}\mathcal{Y}_{l'm'}\, d\Omega = l(l+1)\delta_{ll'}\delta_{mm'}, \tag{B.167}$$

by virtue of Gauss' theorem upon a closed surface, equation (A.79). In the special case $l' = l$ and $m' = m$ the normalization relation (B.167) reduces to $\|\boldsymbol{\nabla}_1\mathcal{Y}_{lm}\| = \sqrt{l(l+1)}\,\|\mathcal{Y}_{lm}\| \approx (l+\frac{1}{2})\,\|\mathcal{Y}_{lm}\|$. This is consistent with our finding in Section B.7 that the asymptotic wavenumber of a high-degree scalar spherical harmonic is $l+\frac{1}{2}$.

The vector spherical harmonics $\mathbf{P}_{lm}$, $\mathbf{B}_{lm}$, and $\mathbf{C}_{lm}$ may be used to expand a vector field upon the unit sphere Ω in the same way that the scalar spherical harmonics $\mathcal{Y}_{lm}$ are used to expand a scalar field:

$$\mathbf{u} = \sum_{l=0}^{\infty}\sum_{m=-l}^{l} U_{lm}\mathbf{P}_{lm} + V_{lm}\mathbf{B}_{lm} + W_{lm}\mathbf{C}_{lm}. \tag{B.168}$$

The expansion coefficients U_{lm}, V_{lm}, W_{lm} obtained by systematic application of the orthonormality relations (B.161)–(B.166) are

$$U_{lm} = \int_\Omega (\mathbf{P}_{lm} \cdot \mathbf{u})\, d\Omega, \tag{B.169}$$

$$V_{lm} = \int_\Omega (\mathbf{B}_{lm} \cdot \mathbf{u})\, d\Omega, \tag{B.170}$$

$$W_{lm} = \int_\Omega (\mathbf{C}_{lm} \cdot \mathbf{u})\, d\Omega. \tag{B.171}$$

Note that $V_{00} = W_{00} = 0$, in agreement with the constraint (B.151). The equality in equation (B.168) is to be interpreted in the sense of convergence in the mean, for square-integrable vector fields satisfying $\int_\Omega \|\mathbf{u}\|^2\, d\Omega < \infty$, just as in the case of the real scalar spherical-harmonic expansion (B.97). The integral of $\|\mathbf{u}\|^2$ is the sum of the squared expansion coefficients:

$$\int_\Omega \|\mathbf{u}\|^2\, d\Omega = \sum_{l=0}^{\infty}\sum_{m=-l}^{l} (U_{lm}^2 + V_{lm}^2 + W_{lm}^2). \tag{B.172}$$

All of the above results, which pertain to a field $\mathbf{u}(\theta, \phi)$ upon the surface of the unit sphere Ω, may be immediately generalized to a three-dimensional vector field $\mathbf{u}(\mathbf{r})$. The only difference is that the coefficients U_{lm}, V_{lm}, W_{lm} are then functions of radius r.

A useful feature of the representation (B.168) is that it leads to a convenient decomposition of the divergence $\boldsymbol{\nabla} \cdot \mathbf{u}$, curl $\boldsymbol{\nabla} \times \mathbf{u}$ and related second derivatives $\boldsymbol{\nabla}(\boldsymbol{\nabla} \cdot \mathbf{u})$, $\boldsymbol{\nabla} \times (\boldsymbol{\nabla} \times \mathbf{u})$ and $\nabla^2 \mathbf{u} = \boldsymbol{\nabla}(\boldsymbol{\nabla} \cdot \mathbf{u}) - \boldsymbol{\nabla} \times (\boldsymbol{\nabla} \times \mathbf{u})$ of a three-dimensional vector field. The following ancillary results are easily demonstrated:

$$\boldsymbol{\nabla}_1 \cdot \mathbf{P}_{lm} = 2\mathcal{Y}_{lm}, \qquad \boldsymbol{\nabla}_1 \times \mathbf{P}_{lm} = \sqrt{l(l+1)}\,\mathbf{C}_{lm}, \tag{B.173}$$

$$\boldsymbol{\nabla}_1 \cdot \mathbf{B}_{lm} = -\sqrt{l(l+1)}\,\mathcal{Y}_{lm}, \qquad \boldsymbol{\nabla}_1 \times \mathbf{B}_{lm} = -\mathbf{C}_{lm}, \tag{B.174}$$

$$\boldsymbol{\nabla}_1 \cdot \mathbf{C}_{lm} = 0, \qquad \boldsymbol{\nabla}_1 \times \mathbf{C}_{lm} = \sqrt{l(l+1)}\,\mathbf{P}_{lm} + \mathbf{B}_{lm}, \tag{B.175}$$

$$\nabla_1^2 \mathbf{P}_{lm} = -[l(l+1)+2]\,\mathbf{P}_{lm} + 2\sqrt{l(l+1)}\,\mathbf{B}_{lm}, \tag{B.176}$$

$$\nabla_1^2 \mathbf{B}_{lm} = 2\sqrt{l(l+1)}\,\mathbf{P}_{lm} - l(l+1)\,\mathbf{B}_{lm}, \tag{B.177}$$

$$\nabla_1^2 \mathbf{C}_{lm} = -l(l+1)\,\mathbf{C}_{lm}. \tag{B.178}$$

Upon making use of equations (B.173)–(B.178) we obtain

$$\boldsymbol{\nabla} \cdot \mathbf{u} = \sum_{l=0}^{\infty} \sum_{m=-l}^{l} \left\{ \frac{dU_{lm}}{dr} + \frac{1}{r}\left[2U_{lm} - \sqrt{l(l+1)}\,V_{lm}\right] \right\} \mathcal{Y}_{lm}, \tag{B.179}$$

$$\begin{aligned} \boldsymbol{\nabla} \times \mathbf{u} = \sum_{l=0}^{\infty} \sum_{m=-l}^{l} \Bigg[& \left[\frac{\sqrt{l(l+1)}}{r}\,W_{lm}\right] \mathbf{P}_{lm} + \left(\frac{dW_{lm}}{dr} + \frac{W_{lm}}{r}\right) \mathbf{B}_{lm} \\ & - \left\{ \frac{dV_{lm}}{dr} + \frac{1}{r}\left[V_{lm} - \sqrt{l(l+1)}\,U_{lm}\right] \right\} \mathbf{C}_{lm}, \end{aligned} \tag{B.180}$$

$$\begin{aligned} \boldsymbol{\nabla}(\boldsymbol{\nabla} \cdot \mathbf{u}) = \sum_{l=0}^{\infty} \sum_{m=-l}^{l} & \left\{ \frac{d^2U_{lm}}{dr^2} + \frac{1}{r}\left[2\,\frac{dU_{lm}}{dr} - \sqrt{l(l+1)}\,\frac{dV_{lm}}{dr}\right] \right. \\ & \left. - \frac{1}{r^2}\left[2U_{lm} - \sqrt{l(l+1)}\,V_{lm}\right] \right\} \mathbf{P}_{lm} \\ & + \frac{\sqrt{l(l+1)}}{r} \left\{ \frac{dU_{lm}}{dr} + \frac{1}{r}\left[2U_{lm} - \sqrt{l(l+1)}\,V_{lm}\right] \right\} \mathbf{B}_{lm}, \end{aligned} \tag{B.181}$$

$$\begin{aligned}\boldsymbol{\nabla}\times(\boldsymbol{\nabla}\times\mathbf{u}) = \sum_{l=0}^{\infty}\sum_{m=-l}^{l}&\left\{-\frac{\sqrt{l(l+1)}}{r}\left[\frac{dV_{lm}}{dr}\right.\right.\\ &\left.\left.+\frac{1}{r}\left(V_{lm}-\sqrt{l(l+1)}\,U_{lm}\right)\right]\right\}\mathbf{P}_{lm}\\ &-\left[\frac{d^2V_{lm}}{dr^2}+\frac{2}{r}\frac{dV_{lm}}{dr}-\frac{\sqrt{l(l+1)}}{r}\frac{dU_{lm}}{dr}\right]\mathbf{B}_{lm}\\ &-\left[\frac{d^2W_{lm}}{dr^2}+\frac{2}{r}\frac{dW_{lm}}{dr}-\frac{l(l+1)}{r^2}W_{lm}\right]\mathbf{C}_{lm},\end{aligned}\tag{B.182}$$

$$\begin{aligned}\nabla^2\mathbf{u} = \sum_{l=0}^{\infty}\sum_{m=-l}^{l}&\left\{\frac{d^2U_{lm}}{dr^2}+\frac{2}{r}\frac{dU_{lm}}{dr}-\frac{2}{r^2}U_{lm}\right.\\ &\left.+\frac{\sqrt{l(l+1)}}{r^2}\left[2V_{lm}-\sqrt{l(l+1)}\,U_{lm}\right]\right\}\mathbf{P}_{lm}\\ &+\left\{\frac{d^2V_{lm}}{dr^2}+\frac{2}{r}\frac{dV_{lm}}{dr}+\frac{\sqrt{l(l+1)}}{r^2}\left[2U_{lm}-\sqrt{l(l+1)}V_{lm}\right]\right\}\mathbf{B}_{lm}\\ &+\left[\frac{d^2W_{lm}}{dr^2}+\frac{2}{r}\frac{dW_{lm}}{dr}-\frac{l(l+1)}{r^2}W_{lm}\right]\mathbf{C}_{lm}.\end{aligned}\tag{B.183}$$

We see from (B.179)–(B.180) that the divergence of a toroidal field is zero, and that a spheroidal field has a toroidal curl, and vice versa. As a result, none of the second derivatives (B.181)–(B.183) change the character of a spheroidal or a toroidal vector field. These properties of $\boldsymbol{\nabla}\cdot\mathbf{u}$, $\boldsymbol{\nabla}\times\mathbf{u}$, $\boldsymbol{\nabla}(\boldsymbol{\nabla}\cdot\mathbf{u})$, $\boldsymbol{\nabla}\times(\boldsymbol{\nabla}\times\mathbf{u})$ and $\nabla^2\mathbf{u}$ are responsible for the decoupling of the free oscillations of a spherically symmetric Earth into distinct spheroidal and toroidal oscillations, as we discuss in Sections 8.6.1 and 8.6.2.

Table B.3 gives a complete list of integrals involving double-dotted combinations of $\hat{\mathbf{r}}\mathbf{P}_{lm}$, $\hat{\mathbf{r}}\mathbf{B}_{lm}$, $\hat{\mathbf{r}}\mathbf{C}_{lm}$ and $\boldsymbol{\nabla}_1\mathbf{P}_{lm}$, $\boldsymbol{\nabla}_1\mathbf{B}_{lm}$, $\boldsymbol{\nabla}_1\mathbf{C}_{lm}$ which arise in the calculation of radial Lagrangian and energy densities upon a spherical Earth. These results are obtained using Gauss' theorem on the unit sphere Ω together with the surface operator identities (A.127)–(A.130). It is noteworthy that all integrals involving spheroidal-toroidal combinations such as $\boldsymbol{\nabla}_1\mathbf{B}_{lm}:\boldsymbol{\nabla}_1\mathbf{C}_{l'm'}$ are zero. This accounts for the decomposition of the Lagrangian governing a spherically symmetric Earth into separate spheroidal and toroidal terms, as we discuss in Section 8.6.4.

B.12.3 Poloidal field

A *solenoidal* vector field $\boldsymbol{\omega}$ is one whose divergence vanishes everywhere:

$$\boldsymbol{\nabla}\cdot\boldsymbol{\omega} = 0.\tag{B.184}$$

$$\int_\Omega (\hat{\mathbf{r}}\mathbf{B}_{lm})^{\mathrm{T}} : (\boldsymbol{\nabla}_1 \mathbf{B}_{l'm'})\, d\Omega = -\delta_{ll'}\delta_{mm'}$$

$$\int_\Omega (\hat{\mathbf{r}}\mathbf{C}_{lm})^{\mathrm{T}} : (\boldsymbol{\nabla}_1 \mathbf{C}_{l'm'})\, d\Omega = -\delta_{ll'}\delta_{mm'}$$

$$\int_\Omega (\hat{\mathbf{r}}\mathbf{B}_{lm})^{\mathrm{T}} : (\boldsymbol{\nabla}_1 \mathbf{P}_{l'm'})\, d\Omega = \sqrt{l(l+1)}\,\delta_{ll'}\delta_{mm'}$$

$$\int_\Omega (\boldsymbol{\nabla}_1 \mathbf{P}_{lm}) : (\boldsymbol{\nabla}_1 \mathbf{P}_{l'm'})\, d\Omega = [l(l+1)+2]\,\delta_{ll'}\delta_{mm'}$$

$$\int_\Omega (\boldsymbol{\nabla}_1 \mathbf{P}_{lm})^{\mathrm{T}} : (\boldsymbol{\nabla}_1 \mathbf{P}_{l'm'})\, d\Omega = 2\,\delta_{ll'}\delta_{mm'}$$

$$\int_\Omega (\boldsymbol{\nabla}_1 \mathbf{B}_{lm}) : (\boldsymbol{\nabla}_1 \mathbf{B}_{l'm'})\, d\Omega = l(l+1)\,\delta_{ll'}\delta_{mm'}$$

$$\int_\Omega (\boldsymbol{\nabla}_1 \mathbf{B}_{lm})^{\mathrm{T}} : (\boldsymbol{\nabla}_1 \mathbf{B}_{l'm'})\, d\Omega = [l(l+1)-1]\,\delta_{ll'}\delta_{mm'}$$

$$\int_\Omega (\boldsymbol{\nabla}_1 \mathbf{C}_{lm}) : (\boldsymbol{\nabla}_1 \mathbf{C}_{l'm'})\, d\Omega = l(l+1)\,\delta_{ll'}\delta_{mm'}$$

$$\int_\Omega (\boldsymbol{\nabla}_1 \mathbf{C}_{lm})^{\mathrm{T}} : (\boldsymbol{\nabla}_1 \mathbf{C}_{l'm'})\, d\Omega = -\delta_{ll'}\delta_{mm'}$$

$$\int_\Omega (\boldsymbol{\nabla}_1 \mathbf{P}_{lm}) : (\boldsymbol{\nabla}_1 \mathbf{B}_{l'm'})\, d\Omega = -2\sqrt{l(l+1)}\,\delta_{ll'}\delta_{mm'}$$

$$\int_\Omega (\boldsymbol{\nabla}_1 \mathbf{P}_{lm})^{\mathrm{T}} : (\boldsymbol{\nabla}_1 \mathbf{B}_{l'm'})\, d\Omega = -\sqrt{l(l+1)}\,\delta_{ll'}\delta_{mm'}$$

Table B.3. Surface integrals over the unit sphere Ω involving the surface gradients of the three vector spherical harmonics. A multitude of closely related integrals can be evaluated using the trivial identities $\mathbf{F}^{\mathrm{T}}:\mathbf{G} = \mathbf{F}:\mathbf{G}^{\mathrm{T}}$ and $\mathbf{F}:\mathbf{G} = \mathbf{F}^{\mathrm{T}}:\mathbf{G}^{\mathrm{T}}$, which are valid for arbitrary second-order tensors $\mathbf{F}$ and $\mathbf{G}$. All other integrals involving double-dotted combinations of $\hat{\mathbf{r}}\mathbf{P}_{lm}$, $\hat{\mathbf{r}}\mathbf{B}_{lm}$, $\hat{\mathbf{r}}\mathbf{C}_{lm}$, $\boldsymbol{\nabla}_1\mathbf{P}_{lm}$, $\boldsymbol{\nabla}_1\mathbf{B}_{lm}$, $\boldsymbol{\nabla}_1\mathbf{C}_{lm}$ and their transposes are identically zero.

Such a field can be written in the form $\boldsymbol{\omega} = \hat{\mathbf{r}}\,U + \boldsymbol{\nabla}_1 V + \hat{\mathbf{r}} \times \boldsymbol{\nabla}_1 Q$, where we have replaced the toroidal scalar W by $-Q$ for convenience in what follows. The scalar representation of the constraint (B.184) is

$$(r\partial_r + 2)U + \nabla_1^2 V = 0. \tag{B.185}$$

For any regular field V, there is a unique regular field P which vanishes at $r = 0$ and which satisfies

$$V = -\partial_r(rP). \tag{B.186}$$

If U and V satisfy equation (B.185) then $\partial_r[r^2(U - r^{-1}\nabla_1^2 P)] = 0$. This in turn implies that

$$U = r^{-1}\nabla_1^2 P, \tag{B.187}$$

where we have used the regularity of U at $r = 0$ to eliminate the constant of integration. If V averages to zero on every spherical surface then so do P and U. The above considerations show that a solenoidal vector field $\boldsymbol{\omega}$ is completely determined by the two scalars P and Q. Insertion of equations (B.186)–(B.187) into $\boldsymbol{\omega} = \hat{\mathbf{r}}\,U + \boldsymbol{\nabla}_1 V + \hat{\mathbf{r}} \times \boldsymbol{\nabla}_1 Q$ leads to the so-called *Mie representation* (Backus 1958; 1986; Backus, Parker & Constable 1996):

$$\begin{aligned}\boldsymbol{\omega} &= \boldsymbol{\nabla} \times (\mathbf{r} \times \boldsymbol{\nabla} P) + \mathbf{r} \times \boldsymbol{\nabla} Q \\ &= -\boldsymbol{\nabla} \times (\boldsymbol{\nabla} \times \mathbf{r}P) - \boldsymbol{\nabla} \times \mathbf{r}Q.\end{aligned} \tag{B.188}$$

For every solenoidal vector field $\boldsymbol{\omega}$ there are unique scalar fields P and Q satisfying equation (B.188) and $\int_\Omega P\,d\Omega = \int_\Omega Q\,d\Omega = 0$. Upon dotting the Mie representation with both $\mathbf{r}$ and $\mathbf{r} \times \boldsymbol{\nabla}$ we find that

$$P = \nabla_1^{-2}(\mathbf{r} \cdot \boldsymbol{\omega}), \qquad Q = \nabla_1^{-2}[\mathbf{r} \cdot (\boldsymbol{\nabla} \times \boldsymbol{\omega})]. \tag{B.189}$$

The radial component of any solenoidal field averages to zero on every spherical surface; this justifies the application of the inverse Laplacian operator ∇_1^{-2} in equation (B.189).

Any vector field of the form $\boldsymbol{\nabla} \times (\mathbf{r} \times \boldsymbol{\nabla} P)$ is said to be *poloidal*. The Mie representation (B.188) therefore decomposes an arbitrary solenoidal field $\boldsymbol{\omega}$ into its poloidal and toroidal parts. The curl of a toroidal field is, by definition, poloidal; conversely, the curl of a poloidal field is toroidal. In fact, for any $\boldsymbol{\omega}$ satisfying (B.188),

$$\boldsymbol{\nabla} \times \boldsymbol{\omega} = \boldsymbol{\nabla} \times (\mathbf{r} \times \boldsymbol{\nabla} Q) - \mathbf{r} \times \boldsymbol{\nabla}(\nabla^2 P). \tag{B.190}$$

The Laplacian $\nabla^2\boldsymbol{\omega} = -\boldsymbol{\nabla} \times (\boldsymbol{\nabla} \times \boldsymbol{\omega})$ of a solenoidal field is given by

$$\nabla^2\boldsymbol{\omega} = \boldsymbol{\nabla} \times [\mathbf{r} \times \boldsymbol{\nabla}(\nabla^2 P)] + \mathbf{r} \times \boldsymbol{\nabla}(\nabla^2 Q). \tag{B.191}$$

The property (B.191) renders the representation (B.188) particularly convenient for solenoidal fields that satisfy the vector Helmholtz equation

$$\nabla^2\boldsymbol{\omega} + k^2\boldsymbol{\omega} = \mathbf{0}, \tag{B.192}$$

in addition to the constraint $\boldsymbol{\nabla}\cdot\boldsymbol{\omega} = 0$. Both the electric field $\mathbf{E}$ and the magnetic field $\mathbf{B}$ satisfy (B.184) and (B.192) everywhere in free space; a toroidal electric field is associated with a spheroidal magnetic field, and vice versa (Mie 1908; Stratton 1941).

B.12.4 Harmonic potential field

If $\boldsymbol{\omega}$ is *irrotational*, $\boldsymbol{\nabla}\times\boldsymbol{\omega} = \mathbf{0}$, as well as solenoidal, $\boldsymbol{\nabla}\cdot\boldsymbol{\omega} = 0$, then it can be written in the form

$$\boldsymbol{\omega} = -\boldsymbol{\nabla}\psi \quad \text{where} \quad \nabla^2\psi = 0. \tag{B.193}$$

The harmonic potential ψ and the poloidal and toroidal scalars P and Q are related by (Backus 1986; Backus, Parker & Constable 1996)

$$\psi = \partial_r(rP), \qquad P = -\nabla_1^{-2}(r\partial_r\psi), \qquad Q = 0. \tag{B.194}$$

The general solution to Laplace's equation $\nabla^2\psi = 0$ in spherical polar coordinates is

$$\psi = \sum_{l=0}^{\infty}\sum_{m=-l}^{l}\left[g_{lm}(a/r)^{l+1} + h_{lm}(r/a)^{l}\right]\mathcal{Y}_{lm}(\theta,\phi), \tag{B.195}$$

where a is a characteristic radius such as that of the Earth or the core-mantle boundary. The real constants g_{lm} and h_{lm} are the so-called *Gauss coefficients* of the fields generated by internal sources in the region beneath r and external sources in the region above r, respectively. Equation (B.195) is valid for any ψ that is harmonic in a shell between the two source regions. If $\boldsymbol{\omega} = -\boldsymbol{\nabla}\psi$ is solenoidal *everywhere*, then the sum over degree begins at $l = 1$ rather than $l = 0$ by virtue of the constraint $\int_\Omega \hat{\mathbf{r}}\cdot\boldsymbol{\omega}\,d\Omega = 0$. The conventional representations of the external geopotential and main geomagnetic fields are both based upon the result (B.195); however, as we have noted in Section B.8, different spherical-harmonic normalizations are customarily employed in these two applications.

Appendix C

Generalized Spherical Harmonics

A means of extending the scalar and vector spherical-harmonic expansions discussed in Appendix B to tensor fields of higher order is obviously desirable. Backus (1967) developed a potential representation of a second-order tangent tensor field $\mathbf{T}^{\Omega\Omega}$ in terms of four scalar fields analogous to the two-potential representation $\mathbf{u}^{\Omega} = \boldsymbol{\nabla}_1 V - \hat{\mathbf{r}} \times \boldsymbol{\nabla}_1 W$ of a tangent vector field, and used it to give an economical derivation of the radial differential equations governing the toroidal and spheroidal free oscillations of a spherically symmetrical Earth model, as well as to obtain an exhaustive catalogue of possible initial static stress fields in the mantle. His representation can be used in conjunction with the tangent vector-plus-tensor decomposition $\mathbf{T} = \hat{\mathbf{r}}\hat{\mathbf{r}}T_{rr} + \hat{\mathbf{r}}\mathbf{T}^{r\Omega} + \mathbf{T}^{\Omega r}\hat{\mathbf{r}} + \mathbf{T}^{\Omega\Omega}$ to define nine second-order tensor spherical harmonics analogous to the three vector harmonics $\mathbf{P}_{lm}$, $\mathbf{B}_{lm}$, $\mathbf{C}_{lm}$. However, it is difficult to extend this approach, which is rooted in classical differential geometry, to higher-order tensors $\mathbf{T}$.

We describe in this appendix a more powerful, and at the same time more routine, procedure for expanding tensor fields of any order in terms of *generalized spherical harmonics*. The first systematic application of generalized spherical-harmonic expansions to the solution of classical tensor differential equations was made by Gelfand & Shapiro (1956). Their results were distilled and placed upon a more practical basis by Burridge (1969) and subsequently applied to the study of the free oscillations of the Earth by Phinney & Burridge (1973). There are intimate connections between the generalized spherical harmonics and the representations of the rotation group $O(3)$; however, we shall not dwell upon these more sophisticated aspects of the theory here. We have a specific and more limited goal—to develop a mathematical formalism for the manipulation of algebraic and differential relations involving vectors and higher-order tensors in spherical polar coordinates. Once a few basic concepts have been grasped, it is possible to use this formalism to conduct generalized spherical-harmonic calculations on a "turn-crank" basis, without any regard for the group-theoretical underpinnings of the analysis.

Before proceeding, it may be prudent to point out that the generalized vector spherical harmonics described in this appendix are *not the same* as the vector harmonics that are commonly used in quantum-theoretical investigations of electric and magnetic field interactions (Edmonds 1960). The latter are expanded in a basis consisting of three complex Cartesian vectors $-\frac{1}{\sqrt{2}}(\hat{\mathbf{x}}+i\hat{\mathbf{y}})$, $\hat{\mathbf{z}}$, $\frac{1}{\sqrt{2}}(\hat{\mathbf{x}}-i\hat{\mathbf{y}})$, whereas the Gelfand-Shapiro-Burridge harmonics are exressed in terms of the canonical spherical polar basis vectors $\hat{\mathbf{e}}_-$, $\hat{\mathbf{e}}_0$, $\hat{\mathbf{e}}_+$ and $\hat{\mathbf{e}}^-$, $\hat{\mathbf{e}}^0$, $\hat{\mathbf{e}}^+$, which we define in Appendix C.2.2. The quantum-electromagnetic vector-harmonic representation has been placed in a strictly classical framework, and extended to tensors of arbitrary order by James (1976); his eminently readable paper provides a comprehensive introduction to the topic. We do not discuss the properties of the Edmonds-James tensor spherical harmonics in this book since, for historical reasons, they have never been employed in theoretical global seismology.

By analogy with equation (B.2) we define the *inner product* of two (not necessarily tangent) tensor fields $\mathbf{T}$ and $\mathbf{P}$ on the unit sphere Ω by

$$\langle \mathbf{T}, \mathbf{P}\rangle = \int_\Omega \mathbf{T}^* \,\vdots\, \mathbf{P}\, d\Omega. \tag{C.1}$$

The triple dot product denotes the contraction over all of the spherical polar components: $\mathbf{T}^* \vdots \mathbf{P} = T^*_{rr\cdots r}P_{rr\cdots r} + T^*_{\theta r\cdots r}P_{\theta r\cdots r} + \cdots T^*_{\phi\phi\cdots\phi}P_{\phi\phi\cdots\phi}$. Two tensor fields $\mathbf{T}$ and $\mathbf{P}$ on the surface of the unit sphere are said to be *orthogonal* if $\langle \mathbf{T}, \mathbf{P}\rangle = 0$. The tensor *norm* generated by the inner product (C.1) will be denoted by $|||\mathbf{T}||| = \langle \mathbf{T}, \mathbf{T}\rangle^{1/2}$. The triple bar serves to distinguish this from the complex Euclidean norm $\|\mathbf{T}\| = (\mathbf{T}^* \vdots \mathbf{T})^{1/2}$.

C.1 Angular Momentum—Reprise

We begin by considering a heretofore unmentioned property of the angular-momentum operator $\mathbf{L}$, which was first introduced in Appendix B.2. As we shall see, $\mathbf{L}$ arises naturally in the analysis of the transformation properties of scalar fields under a rigid-body rotation; however, it must be generalized if we seek to describe the transformation properties of vector and higher-order tensor fields as well. A rigid rotation from an unprimed Cartesian axis system $\hat{\mathbf{x}}$, $\hat{\mathbf{y}}$, $\hat{\mathbf{z}}$ to a primed axis system $\hat{\mathbf{x}}'$, $\hat{\mathbf{y}}'$, $\hat{\mathbf{z}}'$ can be characterized by the matrix

$$\mathsf{R} = \begin{pmatrix} \hat{\mathbf{x}}' \cdot \hat{\mathbf{x}} & \hat{\mathbf{y}}' \cdot \hat{\mathbf{x}} & \hat{\mathbf{z}}' \cdot \hat{\mathbf{x}} \\ \hat{\mathbf{x}}' \cdot \hat{\mathbf{y}} & \hat{\mathbf{y}}' \cdot \hat{\mathbf{y}} & \hat{\mathbf{z}}' \cdot \hat{\mathbf{y}} \\ \hat{\mathbf{x}}' \cdot \hat{\mathbf{z}} & \hat{\mathbf{y}}' \cdot \hat{\mathbf{z}} & \hat{\mathbf{z}}' \cdot \hat{\mathbf{z}} \end{pmatrix}. \tag{C.2}$$

Every such matrix, composed of direction cosines, is proper orthogonal: $\mathsf{R}^{-1} = \mathsf{R}^{\mathrm{T}}$ and $\det \mathsf{R} = 1$. The primed and unprimed coordinates of a point $\hat{\mathbf{r}} = x\hat{\mathbf{x}} + y\hat{\mathbf{y}} + z\hat{\mathbf{z}} = x'\hat{\mathbf{x}}' + y'\hat{\mathbf{y}}' + z'\hat{\mathbf{z}}'$ upon the unit sphere Ω are related by

$$\begin{pmatrix} x' \\ y' \\ z' \end{pmatrix} = \mathsf{R} \begin{pmatrix} x \\ y \\ z \end{pmatrix}. \tag{C.3}$$

Euler's theorem asserts that every proper orthogonal matrix R is isomorphic to a vector $\boldsymbol{\omega}$ whose direction $\hat{\boldsymbol{\omega}}$ is the axis about which the rotation occurs, and whose length $\|\boldsymbol{\omega}\|$ is the right-handed angle of rotation measured in radians. The inverse or transpose matrix $\mathsf{R}^{-1} = \mathsf{R}^{\mathrm{T}}$ is associated with the vector $-\boldsymbol{\omega}$, whereas the identity I is associated with the zero vector.

It is important to emphasize that x', y', z' and x, y, z in equation (C.3) are the coordinates of the *same point* $\hat{\mathbf{r}}$ in the two Cartesian systems. Such a *passive* point of view is most convenient in geophysical applications, and will be utilized here exclusively; an *active* viewpoint, in which the orthonormal axes $\hat{\mathbf{x}}$, $\hat{\mathbf{y}}$, $\hat{\mathbf{z}}$ remain fixed but the observation point $\hat{\mathbf{r}} = x\hat{\mathbf{x}} + y\hat{\mathbf{y}} + z\hat{\mathbf{z}}$ moves to $\hat{\mathbf{r}}' = x'\hat{\mathbf{x}} + y'\hat{\mathbf{y}} + z'\hat{\mathbf{z}}$, is often employed in quantum mechanics to describe the effect of a rotation of a physical system upon the wavefunctions (Schiff 1968). A passive rotation $\boldsymbol{\omega}$ about a single fixed axis has the same effect as an equal and opposite active rotation $-\boldsymbol{\omega}$, as illustrated in Figure C.1.

The net effect of two successive rotations R_1 from $\hat{\mathbf{x}}$, $\hat{\mathbf{y}}$, $\hat{\mathbf{z}}$ to $\hat{\mathbf{x}}'$, $\hat{\mathbf{y}}'$, $\hat{\mathbf{z}}'$ followed by R_2 from $\hat{\mathbf{x}}'$, $\hat{\mathbf{y}}'$, $\hat{\mathbf{z}}'$ to $\hat{\mathbf{x}}''$, $\hat{\mathbf{y}}''$, $\hat{\mathbf{z}}''$ is described by

$$\begin{pmatrix} x'' \\ y'' \\ z'' \end{pmatrix} = \mathsf{R}_2\mathsf{R}_1 \begin{pmatrix} x \\ y \\ z \end{pmatrix}, \tag{C.4}$$

where

$$\mathsf{R}_2\mathsf{R}_1 = \begin{pmatrix} \hat{\mathbf{x}}'' \cdot \hat{\mathbf{x}}' & \hat{\mathbf{y}}'' \cdot \hat{\mathbf{x}}' & \hat{\mathbf{z}}'' \cdot \hat{\mathbf{x}}' \\ \hat{\mathbf{x}}'' \cdot \hat{\mathbf{y}}' & \hat{\mathbf{y}}'' \cdot \hat{\mathbf{y}}' & \hat{\mathbf{z}}'' \cdot \hat{\mathbf{y}}' \\ \hat{\mathbf{x}}'' \cdot \hat{\mathbf{z}}' & \hat{\mathbf{y}}'' \cdot \hat{\mathbf{z}}' & \hat{\mathbf{z}}'' \cdot \hat{\mathbf{z}}' \end{pmatrix} \begin{pmatrix} \hat{\mathbf{x}}' \cdot \hat{\mathbf{x}} & \hat{\mathbf{y}}' \cdot \hat{\mathbf{x}} & \hat{\mathbf{z}}' \cdot \hat{\mathbf{x}} \\ \hat{\mathbf{x}}' \cdot \hat{\mathbf{y}} & \hat{\mathbf{y}}' \cdot \hat{\mathbf{y}} & \hat{\mathbf{z}}' \cdot \hat{\mathbf{y}} \\ \hat{\mathbf{x}}' \cdot \hat{\mathbf{z}} & \hat{\mathbf{y}}' \cdot \hat{\mathbf{z}} & \hat{\mathbf{z}}' \cdot \hat{\mathbf{z}} \end{pmatrix}. \tag{C.5}$$

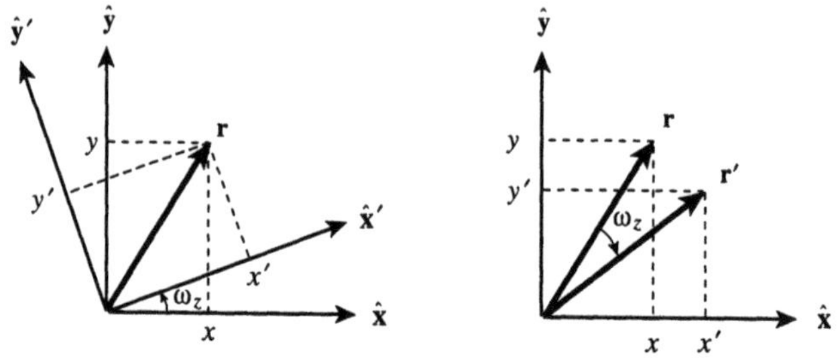

Figure C.1. (*Left*) Schematic depiction of a passive rotation through an angle ω_z about the $\hat{\mathbf{z}}$ axis. The position vector $\mathbf{r} = x\hat{\mathbf{x}} + y\hat{\mathbf{y}} = x'\hat{\mathbf{x}}' + y'\hat{\mathbf{y}}'$ of the observation point is unchanged. If $\omega_z > 0$, the coordinate axes $\hat{\mathbf{x}}$, $\hat{\mathbf{y}}$ are rotated counterclockwise to $\hat{\mathbf{x}}'$, $\hat{\mathbf{y}}'$, as shown. (*Right*) In the equivalent active viewpoint, the coordinate axes are invariant; every point $\mathbf{r} = x\hat{\mathbf{x}} + y\hat{\mathbf{y}}$ is rotated clockwise to a new position $\mathbf{r}' = x'\hat{\mathbf{x}} + y'\hat{\mathbf{y}}$.

Note that R_2 involves direction cosines such as $\hat{\mathbf{x}}'' \cdot \hat{\mathbf{y}}'$ rather than $\hat{\mathbf{x}}'' \cdot \hat{\mathbf{y}}$. In this sense, we should regard each passive rotation in an ordered sequence $\mathsf{R}_1, \mathsf{R}_2, \ldots$ as being about a sequence of axes $\boldsymbol{\omega}_1, \boldsymbol{\omega}_2, \ldots$ that have been previously *rotated*. The equivalent sequence of equal and opposite active rotations $-\boldsymbol{\omega}_1, -\boldsymbol{\omega}_2, \ldots$ is applied in the same order, but about axes that remain *fixed* (Wolf 1969). Since the order of two successive finite rotations is important, the associated matrices are not commutative: $\mathsf{R}_2\mathsf{R}_1 \neq \mathsf{R}_1\mathsf{R}_2$. The "undoing" of a previous rotation by application of the inverse is of course an exception: $\mathsf{R}^{-1}\mathsf{R} = \mathsf{R}\mathsf{R}^{-1} = \mathsf{I}$. Two or more sequential rotations must, as usual, be "undone" in reverse order: $(\mathsf{R}_2\mathsf{R}_1)^{-1} = \mathsf{R}_1^{-1}\mathsf{R}_2^{-1}$.

We seek to determine the effect of a rotation R or $\boldsymbol{\omega}$ upon a scalar, vector or higher-order tensor field upon the surface of the unit sphere Ω. Since the Euclidean length $x'^2 + y'^2 + z'^2 = x^2 + y^2 + z^2$ is invariant under a rigid rotation (C.3), the results that we obtain are equally applicable to fields in three-dimensional space. We shall restrict attention to two-dimensional fields $\psi(\hat{\mathbf{r}})$, $\mathbf{u}(\hat{\mathbf{r}})$ and $\mathbf{T}(\hat{\mathbf{r}})$ upon Ω in what follows, with the understanding that $\hat{\mathbf{r}}$ can be replaced by $\mathbf{r}$ by simply relaxing the constraint that $x'^2 + y'^2 + z'^2 = x^2 + y^2 + z^2 = 1$.

A real or complex scalar field $\psi(\hat{\mathbf{r}})$ has a definite numerical value at every point $\hat{\mathbf{r}} = x\hat{\mathbf{x}} + y\hat{\mathbf{y}} + z\hat{\mathbf{z}} = x'\hat{\mathbf{x}}' + y'\hat{\mathbf{y}}' + z'\hat{\mathbf{z}}'$. In the original unprimed coordinate system we denote this value by $\psi(x, y, z)$. In the rotated frame it will be given by a *different function* $\psi'(x', y', z')$. The condition that the values of the two functions be the same at every point $\hat{\mathbf{r}}$ is

$$\psi'(x', y', z') = \psi(x, y, z). \tag{C.6}$$

We regard the functions on the left and right of (C.6) as rules by which the values of the field are determined by the coordinates. The two rules differ, since they must yield the same values when evaluated at x', y', z' in the first instance and at x, y, z in the second. We seek to find the operator $\mathcal{D}(\mathsf{R})$ or $\mathcal{D}(\boldsymbol{\omega})$ that relates the primed and unprimed functions:

$$\psi'(x,y,z) = \mathcal{D}(\boldsymbol{\omega})\,\psi(x,y,z). \tag{C.7}$$

Note the fundamental difference between equations (C.6) and (C.7): the arguments x, y, z—or equivalently x', y', z'—are dummy variables which are the same on both sides of the latter. The operator $\mathcal{D}(\boldsymbol{\omega})$ can be thought of as *changing the rule* by which the values of the field are determined by the coordinates. If we wish to find $\psi'(x',y',z')$, then we must evaluate the original function ψ in equation (C.7) at the primed coordinates x', y', z' before applying the operator $\mathcal{D}(\boldsymbol{\omega})$. Upon carrying out this replacement of the dummy variables and comparing with (C.6), we find that

$$\mathcal{D}(\boldsymbol{\omega})\,\psi(x',y',z') = \psi(x,y,z). \tag{C.8}$$

Equation (C.8) can alternatively be written in terms of the *inverse rotation operator* $\mathcal{D}^{-1}(\boldsymbol{\omega}) = \mathcal{D}(-\boldsymbol{\omega})$ associated with the equal but opposite rotation:

$$\psi(x',y',z') = \mathcal{D}^{-1}(\boldsymbol{\omega})\,\psi(x,y,z). \tag{C.9}$$

The result (C.9) provides a second, equally valid interpretation of the rotation operator: the application of the inverse operator $\mathcal{D}^{-1}(\boldsymbol{\omega})$ to ψ can be thought of as leaving the rule unchanged, but *changing the coordinates* from x, y, z to x', y', z'. We shall adhere to the first interpretation, which is encapsulated in equation (C.7), throughout the remainder of this appendix.

It is easy to find the relation between ψ' and ψ in the case of a rotation through an *infinitesimal* angle $d\boldsymbol{\omega}$; the primed and unprimed unit vectors are related in that case by

$$\hat{\mathbf{x}}' \approx \hat{\mathbf{x}} + d\boldsymbol{\omega}\times\hat{\mathbf{x}}, \qquad \hat{\mathbf{y}}' \approx \hat{\mathbf{y}} + d\boldsymbol{\omega}\times\hat{\mathbf{y}}, \qquad \hat{\mathbf{z}}' \approx \hat{\mathbf{z}} + d\boldsymbol{\omega}\times\hat{\mathbf{z}}, \tag{C.10}$$

so that the rotation matrix (C.2) is of the form

$$\mathsf{R} \approx \begin{pmatrix} 1 & d\omega_z & -d\omega_y \\ -d\omega_z & 1 & d\omega_x \\ d\omega_y & -d\omega_x & 1 \end{pmatrix}. \tag{C.11}$$

Upon inserting (C.3) and (C.11) into (C.6) and ignoring terms of second order in $\|d\boldsymbol{\omega}\|$, we obtain $\psi' = \mathcal{D}(d\boldsymbol{\omega})\psi \approx \psi + (d\boldsymbol{\omega}\times\hat{\mathbf{r}})\cdot\boldsymbol{\nabla}_1\psi$ or, equivalently,

$$\mathcal{D}(d\boldsymbol{\omega})\psi \approx (1 + i\,d\boldsymbol{\omega}\cdot\mathbf{L})\psi, \tag{C.12}$$

where $\mathbf{L} = -i(\hat{\mathbf{r}} \times \boldsymbol{\nabla}_1) = -i[(y\partial_z - z\partial_y)\hat{\mathbf{x}} + (z\partial_x - x\partial_z)\hat{\mathbf{y}} + (x\partial_y - y\partial_x)\hat{\mathbf{z}}]$ is the angular-momentum operator.

Given the first-order result (C.12) it is a straightforward matter to find the operator (C.7) governing a *finite* rotation. Suppose that the axes have been subjected to an initial rotation $\boldsymbol{\omega}$, so that $\psi' = \mathcal{D}(\boldsymbol{\omega})\psi$. The effect of an additional infinitesimal rotation $d\boldsymbol{\omega}$ is then $\psi' = \mathcal{D}(\boldsymbol{\omega} + d\boldsymbol{\omega})\psi$, where

$$\mathcal{D}(\boldsymbol{\omega} + d\boldsymbol{\omega}) \approx (1 + i\, d\boldsymbol{\omega} \cdot \mathbf{L})\, \mathcal{D}(\boldsymbol{\omega}). \tag{C.13}$$

Upon taking the limit as $d\boldsymbol{\omega}$ tends to zero we see that $\mathcal{D}(\boldsymbol{\omega})$ satisfies the first-order ordinary differential equation

$$d\mathcal{D}/d\boldsymbol{\omega} = i\mathbf{L}\mathcal{D}. \tag{C.14}$$

The solution to (C.14) subject to the boundary condition that $\mathcal{D}(\mathbf{0}) = 1$ in the case of no rotation is

$$\mathcal{D}(\boldsymbol{\omega}) = \exp(i\boldsymbol{\omega} \cdot \mathbf{L}). \tag{C.15}$$

The angular-momentum operator $\mathbf{L}$ is said to be the *generator* of the finite-rotation operator (C.15) governing a scalar field.

A real or complex vector field $\mathbf{u}(\mathbf{r})$ is likewise represented by different functions $\mathbf{u}(x, y, z)$ and $\mathbf{u}'(x', y', z')$ in the two coordinate systems; by analogy with equation (C.6) we must have

$$\mathbf{u}'(x', y', z') = \mathbf{u}(x, y, z). \tag{C.16}$$

We again introduce a rotation operator $\mathcal{D}(\mathsf{R})$ or $\mathcal{D}(\boldsymbol{\omega})$ that transforms the unprimed field into the primed field:

$$\mathbf{u}'(x, y, z) = \mathcal{D}(\boldsymbol{\omega})\, \mathbf{u}(x, y, z). \tag{C.17}$$

To find $\mathcal{D}(\boldsymbol{\omega})$ we write the unprimed field in the form $\mathbf{u} = u_x\hat{\mathbf{x}} + u_y\hat{\mathbf{y}} + u_z\hat{\mathbf{z}}$. The components u_x, u_y, u_z are scalar fields whose transformations in the case of an infinitesimal rotation $d\boldsymbol{\omega}$ are described by the operator (C.12):

$$\mathcal{D}(d\boldsymbol{\omega})u_x \approx (1 + i\, d\boldsymbol{\omega} \cdot \mathbf{L})u_x, \qquad \mathcal{D}(d\boldsymbol{\omega})u_y \approx (1 + i\, d\boldsymbol{\omega} \cdot \mathbf{L})u_y,$$
$$\mathcal{D}(d\boldsymbol{\omega})u_z \approx (1 + i\, d\boldsymbol{\omega} \cdot \mathbf{L})u_z. \tag{C.18}$$

The unit vectors $\hat{\mathbf{x}}$, $\hat{\mathbf{y}}$, $\hat{\mathbf{z}}$ are transformed by the same rotation into

$$\mathcal{D}(d\boldsymbol{\omega})\hat{\mathbf{x}} \approx \hat{\mathbf{x}} - d\boldsymbol{\omega} \times \hat{\mathbf{x}}, \qquad \mathcal{D}(d\boldsymbol{\omega})\hat{\mathbf{y}} \approx \hat{\mathbf{y}} - d\boldsymbol{\omega} \times \hat{\mathbf{y}},$$
$$\mathcal{D}(d\boldsymbol{\omega})\hat{\mathbf{z}} \approx \hat{\mathbf{z}} - d\boldsymbol{\omega} \times \hat{\mathbf{z}}. \tag{C.19}$$

The sign difference between equations (C.10) and (C.19) reflects the passive nature of the transformation. If we replace the dummy Cartesian axis vectors $\hat{\mathbf{x}}$, $\hat{\mathbf{y}}$, $\hat{\mathbf{z}}$ in (C.19) by $\hat{\mathbf{x}}'$, $\hat{\mathbf{y}}'$, $\hat{\mathbf{z}}'$ then we may regard the transformed vectors $\mathcal{D}\hat{\mathbf{x}}'$, $\mathcal{D}\hat{\mathbf{y}}'$, $\mathcal{D}\hat{\mathbf{z}}'$ as the "vestigial remains" of the original vectors that are "left behind" by the infinitesimal transformation $\mathcal{D}(d\boldsymbol{\omega})$.

We can rewrite the results (C.19) in a manner analogous to (C.18):

$$\mathcal{D}(d\boldsymbol{\omega})\hat{\mathbf{x}} \approx (1 + i\, d\boldsymbol{\omega} \cdot \mathbf{S})\hat{\mathbf{x}}, \qquad \mathcal{D}(d\boldsymbol{\omega})\hat{\mathbf{y}} \approx (1 + i\, d\boldsymbol{\omega} \cdot \mathbf{S})\hat{\mathbf{y}},$$
$$\mathcal{D}(d\boldsymbol{\omega})\hat{\mathbf{z}} \approx (1 + i\, d\boldsymbol{\omega} \cdot \mathbf{S})\hat{\mathbf{z}}. \tag{C.20}$$

The quantity $\mathbf{S} = \hat{\mathbf{x}}S_x + \hat{\mathbf{y}}S_y + \hat{\mathbf{z}}S_z$ in equation (C.20) is a vector operator whose effect upon the unit vectors $\hat{\mathbf{x}}$, $\hat{\mathbf{y}}$, $\hat{\mathbf{z}}$ is given explicitly by

$$\mathbf{S}\hat{\mathbf{x}} = -i(\hat{\mathbf{y}}\hat{\mathbf{z}} - \hat{\mathbf{z}}\hat{\mathbf{y}}), \qquad \mathbf{S}\hat{\mathbf{y}} = -i(\hat{\mathbf{z}}\hat{\mathbf{x}} - \hat{\mathbf{x}}\hat{\mathbf{z}}),$$
$$\mathbf{S}\hat{\mathbf{z}} = -i(\hat{\mathbf{x}}\hat{\mathbf{y}} - \hat{\mathbf{y}}\hat{\mathbf{x}}). \tag{C.21}$$

It is easily demonstrated that the square of the operator (C.21) is

$$S^2 = \mathbf{S} \cdot \mathbf{S} = S_x^2 + S_y^2 + S_z^2 = 2, \tag{C.22}$$

whereas its self-cross product is

$$\mathbf{S} \times \mathbf{S} = i\mathbf{S}. \tag{C.23}$$

Upon combining the results (C.18) and (C.20) we obtain the vector transformation relation analogous to the scalar relation (C.12):

$$\begin{aligned}\mathcal{D}(d\boldsymbol{\omega})[u_x\hat{\mathbf{x}} + u_y\hat{\mathbf{y}} + u_z\hat{\mathbf{z}}]&\\ \approx \mathbf{u} + i\, d\boldsymbol{\omega} \cdot [(\mathbf{L}u_x)\hat{\mathbf{x}} &+ (\mathbf{L}u_y)\hat{\mathbf{y}} + (\mathbf{L}u_z)\hat{\mathbf{z}}\\ &+ u_x(\mathbf{S}\hat{\mathbf{x}}) + u_y(\mathbf{S}\hat{\mathbf{y}}) + u_z(\mathbf{S}\hat{\mathbf{z}})].\end{aligned} \tag{C.24}$$

We shall henceforth rewrite equation (C.24) and other similar relations using the convenient notation

$$\mathcal{D}(d\boldsymbol{\omega})\mathbf{u} \approx (1 + i\, d\boldsymbol{\omega} \cdot \mathbf{J})\mathbf{u} \quad \text{where} \quad \mathbf{J} = \mathbf{L} + \mathbf{S}. \tag{C.25}$$

In this more general context, the vector $\mathbf{L}$ is referred to as the *orbital* angular-momentum operator; it accounts for the change in the x, y, z dependence of either ψ or u_x, u_y, u_z due to an infinitesimal rotation $d\boldsymbol{\omega}$. The vector $\mathbf{S}$, which is known as the *spin* angular-momentum operator, acts to rearrange the Cartesian unit vectors $\hat{\mathbf{x}}$, $\hat{\mathbf{y}}$, $\hat{\mathbf{z}}$ in accordance with the definition (C.21) without affecting the components u_x, u_y, u_z. Combining $\mathbf{L}$ and $\mathbf{S}$ to form the *total* angular-momentum operator $\mathbf{J}$ may seem like a dubious thing to do, since the two vectors act upon independent entities; we form the sum $\mathbf{J} = \mathbf{L} + \mathbf{S}$ only with the express understanding that equation (C.25) is simply a convenient abbreviation of the longer result (C.24).

In the future we shall not bother to distinguish between orbital and spin angular momentum, but simply employ the total angular-momentum operator, writing

$$\mathbf{J}\mathbf{u} = (\mathbf{J}u_x)\hat{\mathbf{x}} + (\mathbf{J}u_y)\hat{\mathbf{y}} + (\mathbf{J}u_z)\hat{\mathbf{z}} \\ + u_x(\mathbf{J}\hat{\mathbf{x}}) + u_y(\mathbf{J}\hat{\mathbf{y}}) + u_z(\mathbf{J}\hat{\mathbf{z}}), \tag{C.26}$$

where it is taken for granted that $\mathbf{J}u_x = \mathbf{L}u_x$, $\mathbf{J}u_y = \mathbf{L}u_y$, $\mathbf{J}u_z = \mathbf{L}u_z$ and $\mathbf{J}\hat{\mathbf{x}} = \mathbf{S}\hat{\mathbf{x}}$, $\mathbf{J}\hat{\mathbf{y}} = \mathbf{S}\hat{\mathbf{y}}$, $\mathbf{J}\hat{\mathbf{z}} = \mathbf{S}\hat{\mathbf{z}}$. The explicit effect of $\mathbf{J} = \hat{\mathbf{x}}J_x + \hat{\mathbf{y}}J_y + \hat{\mathbf{z}}J_z$ upon the Cartesian unit vectors $\hat{\mathbf{x}}$, $\hat{\mathbf{y}}$, $\hat{\mathbf{z}}$ is

$$J_x\hat{\mathbf{x}} = \mathbf{0}, \qquad J_x\hat{\mathbf{y}} = i\hat{\mathbf{z}}, \qquad J_x\hat{\mathbf{z}} = -i\hat{\mathbf{y}}, \tag{C.27}$$

$$J_y\hat{\mathbf{x}} = -i\hat{\mathbf{z}}, \qquad J_y\hat{\mathbf{y}} = \mathbf{0}, \qquad J_y\hat{\mathbf{z}} = i\hat{\mathbf{x}}, \tag{C.28}$$

$$J_z\hat{\mathbf{x}} = i\hat{\mathbf{y}}, \qquad J_z\hat{\mathbf{y}} = -i\hat{\mathbf{x}}, \qquad J_z\hat{\mathbf{z}} = \mathbf{0}. \tag{C.29}$$

Equations (C.27)–(C.29) are simply a rewritten version of (C.21).

We can integrate the differential result $\mathcal{D}(d\boldsymbol{\omega}) \approx 1 + i\, d\boldsymbol{\omega}\cdot\mathbf{J}$ to find the transformation relation $\mathbf{u}' = \mathcal{D}(\boldsymbol{\omega})\mathbf{u}$ associated with a finite rotation using an argument identical to that leading from (C.12) to (C.15):

$$\mathcal{D}(\boldsymbol{\omega}) = \exp(i\boldsymbol{\omega}\cdot\mathbf{J}). \tag{C.30}$$

The total angular-momentum operator $\mathbf{J}$ is the generator of the finite-rotation operator (C.30) governing a vector field. A finite rotation through an angle ω_x, ω_y, ω_z about any of the coordinate axes leaves that axis unchanged so that, for example, $\exp(i\omega_z J_z)\hat{\mathbf{z}} = \hat{\mathbf{z}}$. Upon utilizing the Taylor series expansion of the exponential operator and rearranging terms we find, on the other hand, that $\exp(i\omega_z J_z)\hat{\mathbf{x}} = \hat{\mathbf{x}}\cos\omega_z - \hat{\mathbf{y}}\sin\omega_z$ and $\exp(i\omega_z J_z)\hat{\mathbf{y}} = \hat{\mathbf{x}}\sin\omega_z + \hat{\mathbf{y}}\cos\omega_z$. This corresponds to a passive *clockwise* rotation of $\hat{\mathbf{x}}$ and $\hat{\mathbf{y}}$ through an angle $\omega_z > 0$, as illustrated in Figure C.2.

The effect of two successive finite rotations, $\boldsymbol{\omega}_1$ followed by $\boldsymbol{\omega}_2$, is

$$\mathcal{D}(\boldsymbol{\omega}_2)\mathcal{D}(\boldsymbol{\omega}_1) = \exp(i\boldsymbol{\omega}_2\cdot\mathbf{J})\exp(i\boldsymbol{\omega}_1\cdot\mathbf{J}), \tag{C.31}$$

where the rightmost operator acts first followed by the leftmost, as usual. The order of two infinitesimal rotations $d\boldsymbol{\omega}_1$ and $d\boldsymbol{\omega}_2$ is immaterial, so that $\mathcal{D}(d\boldsymbol{\omega}_2)\mathcal{D}(d\boldsymbol{\omega}_1) \approx 1 + i(d\boldsymbol{\omega}_2 + d\boldsymbol{\omega}_1)\cdot\mathbf{J} \approx \mathcal{D}(d\boldsymbol{\omega}_1)\mathcal{D}(d\boldsymbol{\omega}_2)$. Since two finite rotations are not commutative, however, $\mathcal{D}(\boldsymbol{\omega}_2)\mathcal{D}(\boldsymbol{\omega}_1) \neq \mathcal{D}(\boldsymbol{\omega}_1)\mathcal{D}(\boldsymbol{\omega}_2)$. The inverse of the rotation operator is

$$\mathcal{D}^{-1}(\boldsymbol{\omega}) = \mathcal{D}(-\boldsymbol{\omega}) = \exp(-i\boldsymbol{\omega}\cdot\mathbf{J}). \tag{C.32}$$

Operator multiplication $\mathcal{D}(\boldsymbol{\omega}_2)\mathcal{D}(\boldsymbol{\omega}_1)$ and matrix multiplication $\mathsf{R}_2\mathsf{R}_1$ are synonymous, and $\mathcal{D}^{-1}(\boldsymbol{\omega})$ is the operator associated with R^{-1}.

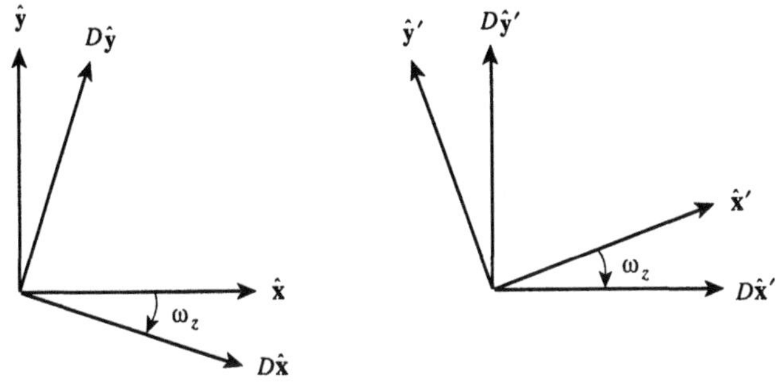

Figure C.2. (*Left*) Passive transformation of the unprimed unit vectors $\hat{\mathbf{x}}$, $\hat{\mathbf{y}}$ by an operator $\mathcal{D} = \exp(i\omega_z J_z)$. The transformed vectors $\mathcal{D}\hat{\mathbf{x}}$, $\mathcal{D}\hat{\mathbf{y}}$ are rotated clockwise through an angle $\omega_z > 0$. (*Right*) If $\mathcal{D} = \exp(i\omega_z J_z)$ is applied to the primed unit vectors $\hat{\mathbf{x}}'$, $\hat{\mathbf{y}}'$ instead, the results $\mathcal{D}\hat{\mathbf{x}}'$, $\mathcal{D}\hat{\mathbf{y}}'$ can be regarded as the "vestigial remains" of the vectors $\hat{\mathbf{x}}$, $\hat{\mathbf{y}}$ "left behind" by the rotation to $\hat{\mathbf{x}}'$, $\hat{\mathbf{y}}'$.

The finite-rotation operator (C.30) can be used to describe the transformation properties of a higher-order tensor as well; that is, if $\mathbf{T}(\mathbf{r})$ is a tensor field of any order satisfying $\mathbf{T}'(x', y', z') = \mathbf{T}(x, y, z)$, then

$$\mathbf{T}'(x, y, z) = \mathcal{D}(\boldsymbol{\omega})\,\mathbf{T}(x, y, z). \tag{C.33}$$

In the case of a second-order tensor $\mathbf{T} = T_{xx}\hat{\mathbf{x}}\hat{\mathbf{x}} + T_{xy}\hat{\mathbf{x}}\hat{\mathbf{y}} + \cdots + T_{zz}\hat{\mathbf{z}}\hat{\mathbf{z}}$ equation (C.26) is generalized to

$$\begin{aligned}\mathbf{J}\mathbf{T} = {} & (\mathbf{J}T_{xx})\hat{\mathbf{x}}\hat{\mathbf{x}} + (\mathbf{J}T_{xy})\hat{\mathbf{x}}\hat{\mathbf{y}} + \cdots + (\mathbf{J}T_{zz})\hat{\mathbf{z}}\hat{\mathbf{z}} \\ & + T_{xx}(\mathbf{J}\hat{\mathbf{x}})\hat{\mathbf{x}} + T_{xy}(\mathbf{J}\hat{\mathbf{x}})\hat{\mathbf{y}} + \cdots + T_{zz}(\mathbf{J}\hat{\mathbf{z}})\hat{\mathbf{z}} \\ & + T_{xx}\hat{\mathbf{x}}(\mathbf{J}\hat{\mathbf{x}}) + T_{xy}\hat{\mathbf{x}}(\mathbf{J}\hat{\mathbf{y}}) + \cdots + T_{zz}\hat{\mathbf{z}}(\mathbf{J}\hat{\mathbf{z}}).\end{aligned} \tag{C.34}$$

The extension of equation (C.34) to an arbitrary-order tensor is immediate; in general $\mathbf{J}$ acts sequentially upon each component as well as each unit vector in the polyadic representation. The effect upon the product of two tensors $\mathbf{T}$ and $\mathbf{P}$ of any order is

$$\mathbf{J}(\mathbf{T}\mathbf{P}) = (\mathbf{J}\mathbf{T})\mathbf{P} + \mathbf{T}(\mathbf{J}\mathbf{P}). \tag{C.35}$$

It is evident that the continued use of traditional symbols such as u_x, u_y, u_z and $T_{xx}, T_{xy}, \ldots, T_{zz}$ would be quite cumbersome in dealing with higher-order tensors $\mathbf{T}$. We recast the above results in spherical polar coordinates using a more streamlined index notation in Section C.2.

Since the orbital and spin angular-momentum operators act upon different entities, they satisfy $\mathbf{L} \times \mathbf{S} = -\mathbf{S} \times \mathbf{L}$. Upon combining this result with the identities $\mathbf{L} \times \mathbf{L} = i\mathbf{L}$ and $\mathbf{S} \times \mathbf{S} = i\mathbf{S}$ we find that the cross product of the total angular-momentum operator is

$$\mathbf{J} \times \mathbf{J} = i\mathbf{J}. \tag{C.36}$$

We can rewrite the result (C.36) as a set of *commutation relations* for the components J_x, J_y, J_z:

$$[J_x, J_y] = iJ_z, \qquad [J_y, J_z] = iJ_x, \qquad [J_z, J_x] = iJ_y. \tag{C.37}$$

The squared total angular-momentum operator $J^2 = \mathbf{J} \cdot \mathbf{J} = J_x^2 + J_y^2 + J_z^2$ commutes with $\mathbf{J}$ and thus with any of its components:

$$[J^2, J_x] = [J^2, J_y] = [J^2, J_z] = 0. \tag{C.38}$$

In addition, $\mathbf{J}$ commutes with the partial derivative operator ∂_r as well as with any scalar, vector or tensor function $f(r)$ of radius r alone: $[\partial_r, \mathbf{J}] = \mathbf{0}$ and $[f(r), \mathbf{J}] = \mathbf{0}$. By analogy with equation (B.20) we define the total angular-momentum *ladder operators* $J_\pm$ by

$$J_+ = J_x + iJ_y, \qquad J_- = J_x - iJ_y. \tag{C.39}$$

The commutation relations satisfied by these operators are

$$[J^2, J_\pm] = 0, \qquad [J_z, J_\pm] = \pm J_\pm, \qquad [J_+, J_-] = 2J_z. \tag{C.40}$$

We can express J^2 in terms of $J_\pm$ and J_z in either of the two forms

$$J^2 = J_+J_- + J_z^2 - J_z = J_-J_+ + J_z^2 + J_z. \tag{C.41}$$

Each of the total angular-momentum relations (C.37)–(C.41) is identical to its orbital angular-momentum counterpart (B.18)–(B.22) in Section B.2.

C.2 Spherical Polar Coordinates

Colatitudinal and longitudinal coordinates θ, ϕ and θ', ϕ' can be associated with the original and rotated Cartesian axis systems $\hat{\mathbf{x}}$, $\hat{\mathbf{y}}$, $\hat{\mathbf{z}}$ and $\hat{\mathbf{x}}'$, $\hat{\mathbf{y}}'$, $\hat{\mathbf{z}}'$ in the usual manner. The relation (C.33) between the primed and unprimed descriptions of a tensor field $\mathbf{T}(\hat{\mathbf{r}})$ upon the surface of the unit sphere $\boldsymbol{\Omega}$ can be rewritten in the form

$$\mathbf{T}'(\theta, \phi) = \mathcal{D}(\boldsymbol{\omega})\, \mathbf{T}(\theta, \phi). \tag{C.42}$$

A three-dimensional field $\mathbf{T}(\mathbf{r})$ transforms in an identical fashion, with the coordinates θ, ϕ in (C.42) replaced by r, θ, ϕ. The radius r is invariant under a rigid rotation $\boldsymbol{\omega}$ about the origin $\mathbf{0}$.

C.2.1 Transformation of the unit vectors

The spherical polar unit vectors $\hat{\mathbf{r}}$, $\hat{\boldsymbol{\theta}}$, $\hat{\boldsymbol{\phi}}$ are related to the Cartesian unit vectors by equations (A.108)–(A.110). Using these relations together with the spherical polar representation (B.23)–(B.25) of the orbital angular-momentum operators L_x, L_y, L_z we find that

$$J_x\hat{\mathbf{r}} = 0, \qquad J_x\hat{\boldsymbol{\theta}} = i(\sin\theta)^{-1}\cos\phi\,\hat{\boldsymbol{\phi}}, \qquad J_x\hat{\boldsymbol{\phi}} = -i(\sin\theta)^{-1}\cos\phi\,\hat{\boldsymbol{\theta}}, \tag{C.43}$$

$$J_y\hat{\mathbf{r}} = 0, \qquad J_y\hat{\boldsymbol{\theta}} = i(\sin\theta)^{-1}\sin\phi\,\hat{\boldsymbol{\phi}}, \qquad J_y\hat{\boldsymbol{\phi}} = -i(\sin\theta)^{-1}\sin\phi\,\hat{\boldsymbol{\theta}}, \tag{C.44}$$

$$J_z\hat{\mathbf{r}} = J_z\hat{\boldsymbol{\theta}} = J_z\hat{\boldsymbol{\phi}} = 0. \tag{C.45}$$

The corresponding relations in terms of the ladder operators $J_\pm$ are

$$J_\pm\hat{\mathbf{r}} = 0, \qquad J_\pm\hat{\boldsymbol{\theta}} = i(\sin\theta)^{-1}e^{\pm i\phi}\,\hat{\boldsymbol{\phi}}, \qquad J_\pm\hat{\boldsymbol{\phi}} = -i(\sin\theta)^{-1}e^{\pm i\phi}\,\hat{\boldsymbol{\theta}}. \tag{C.46}$$

Application of the squared angular-momentum operator $J^2 = J_x^2 + J_y^2 + J_z^2$ yields a scalar times the input vector itself:

$$J^2\hat{\mathbf{r}} = 0, \qquad J^2\hat{\boldsymbol{\theta}} = (\sin\theta)^{-2}\hat{\boldsymbol{\theta}}, \qquad J^2\hat{\boldsymbol{\phi}} = (\sin\theta)^{-2}\hat{\boldsymbol{\phi}}. \tag{C.47}$$

Loosely speaking, we may regard $\hat{\mathbf{r}}$, $\hat{\boldsymbol{\theta}}$, $\hat{\boldsymbol{\phi}}$ as generalized eigenvectors of J^2 with associated "eigenvalues" zero and $(\sin\theta)^{-2}$. We introduce two triads of complex unit vectors that are simultaneous eigenvectors of the operators J_z, $J_\pm$ and J^2 in the next section.

C.2.2 Dual canonical bases

To expedite matters, it is convenient to employ an index notation in which the spherical polar unit vectors are denoted by

$$\hat{\mathbf{e}}_1 = \hat{\mathbf{r}}, \qquad \hat{\mathbf{e}}_2 = \hat{\boldsymbol{\theta}}, \qquad \hat{\mathbf{e}}_3 = \hat{\boldsymbol{\phi}}. \tag{C.48}$$

Italic indices $i, j, k, \ldots$ which take on the values $\{1, 2, 3\}$ are used throughout this appendix to label the (ordinary physical) components of vectors and tensors relative to this basis. We define the lower and upper *canonical* basis vectors $\hat{\mathbf{e}}_-$, $\hat{\mathbf{e}}_0$, $\hat{\mathbf{e}}_+$ and $\hat{\mathbf{e}}^-$, $\hat{\mathbf{e}}^0$, $\hat{\mathbf{e}}^+$ by

$$\hat{\mathbf{e}}_- = \tfrac{1}{\sqrt{2}}(\hat{\boldsymbol{\theta}} - i\hat{\boldsymbol{\phi}}), \qquad \hat{\mathbf{e}}_0 = \hat{\mathbf{r}}, \qquad \hat{\mathbf{e}}_+ = -\tfrac{1}{\sqrt{2}}(\hat{\boldsymbol{\theta}} + i\hat{\boldsymbol{\phi}}) \tag{C.49}$$

and

$$\hat{\mathbf{e}}^- = \tfrac{1}{\sqrt{2}}(\hat{\boldsymbol{\theta}} + i\hat{\boldsymbol{\phi}}), \qquad \hat{\mathbf{e}}^0 = \hat{\mathbf{r}}, \qquad \hat{\mathbf{e}}^+ = -\tfrac{1}{\sqrt{2}}(\hat{\boldsymbol{\theta}} - i\hat{\boldsymbol{\phi}}). \tag{C.50}$$

Greek indices $\alpha, \beta, \gamma, \ldots$ which take on the values $\{-, 0, +\}$ or, equivalently, $\{-1, 0, 1\}$ will be used to refer to these two bases. As an example of this notation, we note that (C.49) and (C.50) are simply complex conjugates: $\hat{\mathbf{e}}^\alpha = \hat{\mathbf{e}}_\alpha^*$. In addition, each of the canonical bases is orthonormal, in the sense

$$\hat{\mathbf{e}}_\alpha^* \cdot \hat{\mathbf{e}}_\beta = \delta_{\alpha\beta}, \qquad \hat{\mathbf{e}}^{\alpha *} \cdot \hat{\mathbf{e}}^\beta = \delta^{\alpha\beta}. \tag{C.51}$$

Finally, the two bases are *each other's duals*; that is,

$$\hat{\mathbf{e}}_\alpha \cdot \hat{\mathbf{e}}^\beta = \delta_\alpha{}^\beta, \qquad \hat{\mathbf{e}}^\alpha \cdot \hat{\mathbf{e}}_\beta = \delta^\alpha{}_\beta. \tag{C.52}$$

The necessity to distinguish between lower and upper Greek indices obliges us to use all four Kronecker delta symbols $\delta_{\alpha\beta}$, $\delta^{\alpha\beta}$, $\delta_\alpha{}^\beta$ and $\delta^\alpha{}_\beta$ in equations (C.51)–(C.52).

We can rewrite the definitions (C.49)–(C.50) using index notation in the succinct form

$$\hat{\mathbf{e}}_\alpha = (\hat{\mathbf{e}}_i \cdot \hat{\mathbf{e}}_\alpha)\,\hat{\mathbf{e}}_i, \qquad \hat{\mathbf{e}}^\alpha = (\hat{\mathbf{e}}_i \cdot \hat{\mathbf{e}}_\alpha)\,\hat{\mathbf{e}}_i. \tag{C.53}$$

The inverse relations giving the spherical polar basis in terms of the canonical bases are

$$\hat{\mathbf{e}}_i = (\hat{\mathbf{e}}^\alpha \cdot \hat{\mathbf{e}}_i)\,\hat{\mathbf{e}}_\alpha = (\hat{\mathbf{e}}_\alpha \cdot \hat{\mathbf{e}}_i)\,\hat{\mathbf{e}}^\alpha. \tag{C.54}$$

The complex matrix elements $\hat{\mathbf{e}}^\alpha \cdot \hat{\mathbf{e}}_i$ and $\hat{\mathbf{e}}_\alpha \cdot \hat{\mathbf{e}}_i$ in (C.54) are given explicitly by

$$\begin{pmatrix} \hat{\mathbf{e}}^- \cdot \hat{\mathbf{e}}_1 & \hat{\mathbf{e}}^0 \cdot \hat{\mathbf{e}}_1 & \hat{\mathbf{e}}^+ \cdot \hat{\mathbf{e}}_1 \\ \hat{\mathbf{e}}^- \cdot \hat{\mathbf{e}}_2 & \hat{\mathbf{e}}^0 \cdot \hat{\mathbf{e}}_2 & \hat{\mathbf{e}}^+ \cdot \hat{\mathbf{e}}_2 \\ \hat{\mathbf{e}}^- \cdot \hat{\mathbf{e}}_3 & \hat{\mathbf{e}}^0 \cdot \hat{\mathbf{e}}_3 & \hat{\mathbf{e}}^+ \cdot \hat{\mathbf{e}}_3 \end{pmatrix} = \begin{pmatrix} 0 & 1 & 0 \\ \frac{1}{\sqrt{2}} & 0 & -\frac{1}{\sqrt{2}} \\ \frac{i}{\sqrt{2}} & 0 & \frac{i}{\sqrt{2}} \end{pmatrix} \tag{C.55}$$

and

$$\begin{pmatrix} \hat{\mathbf{e}}_- \cdot \hat{\mathbf{e}}_1 & \hat{\mathbf{e}}_0 \cdot \hat{\mathbf{e}}_1 & \hat{\mathbf{e}}_+ \cdot \hat{\mathbf{e}}_1 \\ \hat{\mathbf{e}}_- \cdot \hat{\mathbf{e}}_2 & \hat{\mathbf{e}}_0 \cdot \hat{\mathbf{e}}_2 & \hat{\mathbf{e}}_+ \cdot \hat{\mathbf{e}}_2 \\ \hat{\mathbf{e}}_- \cdot \hat{\mathbf{e}}_3 & \hat{\mathbf{e}}_0 \cdot \hat{\mathbf{e}}_3 & \hat{\mathbf{e}}_+ \cdot \hat{\mathbf{e}}_3 \end{pmatrix} = \begin{pmatrix} 0 & 1 & 0 \\ \frac{1}{\sqrt{2}} & 0 & -\frac{1}{\sqrt{2}} \\ -\frac{i}{\sqrt{2}} & 0 & -\frac{i}{\sqrt{2}} \end{pmatrix}. \tag{C.56}$$

The transformation to the complex canonical bases is unitary inasmuch as

$$(\hat{\mathbf{e}}_i \cdot \hat{\mathbf{e}}_\alpha)(\hat{\mathbf{e}}^\alpha \cdot \hat{\mathbf{e}}_j) = (\hat{\mathbf{e}}_i \cdot \hat{\mathbf{e}}^\alpha)(\hat{\mathbf{e}}_\alpha \cdot \hat{\mathbf{e}}_j) = \delta_{ij}. \tag{C.57}$$

Note the use of the summation convention in equations (C.54) and (C.57). Italic indices follow the usual elementary rules; a repeated Greek index must be present in both the upper and lower registers.

The two triads of canonical basis vectors $\hat{\mathbf{e}}_-$, $\hat{\mathbf{e}}_0$, $\hat{\mathbf{e}}_+$ and $\hat{\mathbf{e}}^-$, $\hat{\mathbf{e}}^0$, $\hat{\mathbf{e}}^+$ are, in fact, the sought-after simultaneous eigenvectors of the total angular-momentum operators J_z, $J_\pm$ and J^2. From equations (C.45)–(C.47) we deduce that

$$J_z\hat{\mathbf{e}}_\alpha = J_z\hat{\mathbf{e}}^\alpha = \mathbf{0}, \tag{C.58}$$

$$J_\pm\hat{\mathbf{e}}_\alpha = \alpha(\sin\theta)^{-1}e^{\pm i\phi}\,\hat{\mathbf{e}}_\alpha, \qquad J_\pm\hat{\mathbf{e}}^\alpha = -\alpha(\sin\theta)^{-1}e^{\pm i\phi}\,\hat{\mathbf{e}}^\alpha, \tag{C.59}$$

$$J^2\hat{\mathbf{e}}_\alpha = \alpha^2(\sin\theta)^{-2}\,\hat{\mathbf{e}}_\alpha, \qquad J^2\hat{\mathbf{e}}^\alpha = \alpha^2(\sin\theta)^{-2}\,\hat{\mathbf{e}}^\alpha. \tag{C.60}$$

The results (C.58)–(C.60) play a crucial role in the construction of the generalized spherical harmonics; indeed, they are the motivation for introducing the canonical bases.

C.2.3 Covariant and contravariant components

A tensor $\mathbf{T}$ of order q can be expressed in terms of its ordinary spherical polar components in the form

$$\mathbf{T} = T_{i_1\cdots i_q}\hat{\mathbf{e}}_{i_1}\cdots\hat{\mathbf{e}}_{i_q}, \tag{C.61}$$

where

$$T_{i_1\cdots i_q} = (\hat{\mathbf{e}}_{i_1}\cdots\hat{\mathbf{e}}_{i_q}) \,\vdots\, \mathbf{T}. \tag{C.62}$$

The same tensor can be written in terms of its *covariant* and *contravariant* components with respect to the canonical bases:

$$\mathbf{T} = T_{\alpha_1\cdots\alpha_q}\hat{\mathbf{e}}^{\alpha_1}\cdots\hat{\mathbf{e}}^{\alpha_q} = T^{\alpha_1\cdots\alpha_q}\hat{\mathbf{e}}_{\alpha_1}\cdots\hat{\mathbf{e}}_{\alpha_q}, \tag{C.63}$$

where

$$T_{\alpha_1\cdots\alpha_q} = (\hat{\mathbf{e}}_{\alpha_1}\cdots\hat{\mathbf{e}}_{\alpha_q}) \,\vdots\, \mathbf{T}, \tag{C.64}$$

$$T^{\alpha_1\cdots\alpha_q} = (\hat{\mathbf{e}}^{\alpha_1}\cdots\hat{\mathbf{e}}^{\alpha_q}) \,\vdots\, \mathbf{T}. \tag{C.65}$$

The components (C.62) and (C.64)–(C.65) are related to each other by the unitary transformation matrices (C.55)–(C.56):

$$T_{\alpha_1\cdots\alpha_q} = (\hat{\mathbf{e}}_{\alpha_1}\cdot\hat{\mathbf{e}}_{i_1})\cdots(\hat{\mathbf{e}}_{\alpha_q}\cdot\hat{\mathbf{e}}_{i_q})\,T_{i_1\cdots i_q}, \tag{C.66}$$

$$T^{\alpha_1\cdots\alpha_q} = (\hat{\mathbf{e}}^{\alpha_1}\cdot\hat{\mathbf{e}}_{i_1})\cdots(\hat{\mathbf{e}}^{\alpha_q}\cdot\hat{\mathbf{e}}_{i_q})\,T_{i_1\cdots i_q}. \tag{C.67}$$

A plethora of mixed covariant-contravariant canonical components such as $T_{\alpha_1}{}^{\alpha_2\cdots\alpha_q} = (\hat{\mathbf{e}}_{\alpha_1}\hat{\mathbf{e}}^{\alpha_2}\cdots\hat{\mathbf{e}}^{\alpha_q}){:}\mathbf{T} = (\hat{\mathbf{e}}_{\alpha_1}\cdot\hat{\mathbf{e}}_{i_1})(\hat{\mathbf{e}}^{\alpha_2}\cdot\hat{\mathbf{e}}_{i_2})\cdots(\hat{\mathbf{e}}^{\alpha_q}\cdot\hat{\mathbf{e}}_{i_q})\,T_{i_1\cdots i_q}$ can also be defined.

A vector can be written as $\mathbf{u} = u_\alpha\hat{\mathbf{e}}^\alpha = u^\alpha\hat{\mathbf{e}}_\alpha$, where $u_\alpha = \hat{\mathbf{e}}_\alpha\cdot\mathbf{u}$ and $u^\alpha = \hat{\mathbf{e}}^\alpha\cdot\mathbf{u}$. A second-order tensor has four canonical forms:

$$\mathbf{T} = T_{\alpha\beta}\hat{\mathbf{e}}^\alpha\hat{\mathbf{e}}^\beta = T^{\alpha\beta}\hat{\mathbf{e}}_\alpha\hat{\mathbf{e}}_\beta = T_\alpha{}^\beta\hat{\mathbf{e}}^\alpha\hat{\mathbf{e}}_\beta = T^\alpha{}_\beta\hat{\mathbf{e}}_\alpha\hat{\mathbf{e}}^\beta, \tag{C.68}$$

where

$$T_{\alpha\beta} = \hat{\mathbf{e}}_\alpha\cdot\mathbf{T}\cdot\hat{\mathbf{e}}_\beta, \qquad T^{\alpha\beta} = \hat{\mathbf{e}}^\alpha\cdot\mathbf{T}\cdot\hat{\mathbf{e}}^\beta, \tag{C.69}$$

$$T_\alpha{}^\beta = \hat{\mathbf{e}}_\alpha\cdot\mathbf{T}\cdot\hat{\mathbf{e}}^\beta, \qquad T^\alpha{}_\beta = \hat{\mathbf{e}}^\alpha\cdot\mathbf{T}\cdot\hat{\mathbf{e}}_\beta. \tag{C.70}$$

The components (C.69)–(C.70) of a symmetric tensor $\mathbf{T} = \mathbf{T}^{\mathrm{T}}$ satisfy the symmetry relations $T_{\alpha\beta} = T_{\beta\alpha}$, $T^{\alpha\beta} = T^{\beta\alpha}$ and $T_\alpha{}^\beta = T^\beta{}_\alpha$, $T^\alpha{}_\beta = T_\beta{}^\alpha$.

The covariant and contravariant components of the identity tensor $\mathbf{I}$ are given by

$$g_{\alpha\beta} = \hat{\mathbf{e}}_\alpha\cdot\hat{\mathbf{e}}_\beta, \qquad g^{\alpha\beta} = \hat{\mathbf{e}}^\alpha\cdot\hat{\mathbf{e}}^\beta, \tag{C.71}$$

or, equivalently,

$$\begin{pmatrix} g_{--} & g_{-0} & g_{-+} \\ g_{0-} & g_{00} & g_{0+} \\ g_{+-} & g_{+0} & g_{++} \end{pmatrix} = \begin{pmatrix} g^{--} & g^{-0} & g^{-+} \\ g^{0-} & g^{00} & g^{0+} \\ g^{+-} & g^{+0} & g^{++} \end{pmatrix} = \begin{pmatrix} 0 & 0 & -1 \\ 0 & 1 & 0 \\ -1 & 0 & 0 \end{pmatrix}. \tag{C.72}$$

The mixed components $g_\alpha{}^\beta = \hat{\mathbf{e}}_\alpha\cdot\hat{\mathbf{e}}^\beta$ and $g^\alpha{}_\beta = \hat{\mathbf{e}}^\alpha\cdot\hat{\mathbf{e}}_\beta$ are simply the Kronecker deltas $\delta_\alpha{}^\beta$ and $\delta^\alpha{}_\beta$, by virtue of the duality relation (C.52). We can regard the quantities $g_{\alpha\beta}$ and $g^{\beta\alpha}$ as the covariant and contravariant components of a *metric tensor*, which can be used to lower and raise Greek indices in the customary manner. Thus,

$$u_\alpha = g_{\alpha\beta}u^\beta, \qquad u^\alpha = g^{\alpha\beta}u_\beta \tag{C.73}$$

for a vector $\mathbf{u}$ whereas

$$T_{\alpha\beta} = g_{\alpha\gamma}T^\gamma{}_\beta = g_{\alpha\gamma}g_{\beta\eta}T^{\gamma\eta}, \qquad T^{\alpha\beta} = g^{\alpha\gamma}T_\gamma{}^\beta = g^{\alpha\gamma}g^{\beta\eta}T_{\gamma\eta} \tag{C.74}$$

for a second-order tensor $\mathbf{T}$. We can express $\mathbf{I}$ in terms of its components in any of the equivalent forms $\mathbf{I} = g_{\alpha\beta}\hat{\mathbf{e}}^\alpha\hat{\mathbf{e}}^\beta = g^{\alpha\beta}\hat{\mathbf{e}}_\alpha\hat{\mathbf{e}}_\beta = \hat{\mathbf{e}}^\alpha\hat{\mathbf{e}}_\alpha = \hat{\mathbf{e}}_\alpha\hat{\mathbf{e}}^\alpha$.

The covariant and contravariant components of the third-order alternating tensor $\mathbf{\Lambda}$ are

$$\varepsilon_{\alpha\beta\gamma} = \hat{\mathbf{e}}_\alpha \cdot (\hat{\mathbf{e}}_\beta \times \hat{\mathbf{e}}_\gamma), \qquad \varepsilon^{\alpha\beta\gamma} = \hat{\mathbf{e}}^\alpha \cdot (\hat{\mathbf{e}}^\beta \times \hat{\mathbf{e}}^\gamma). \tag{C.75}$$

Both $\varepsilon_{\alpha\beta\gamma}$ and $\varepsilon^{\alpha\beta\gamma}$ share the permutation properties of the conventional Levi-Cività alternating symbol ε_{ijk}; in fact,

$$\varepsilon_{\alpha\beta\gamma} = \begin{cases} i & \text{if } \alpha,\beta,\gamma \text{ is an even permutation of } -,0,+ \\ -i & \text{if } \alpha,\beta,\gamma \text{ is an odd permutation of } -,0,+ \\ 0 & \text{otherwise} \end{cases} \tag{C.76}$$

and

$$\varepsilon^{\alpha\beta\gamma} = \begin{cases} -i & \text{if } \alpha,\beta,\gamma \text{ is an even permutation of } -,0,+ \\ i & \text{if } \alpha,\beta,\gamma \text{ is an odd permutation of } -,0,+ \\ 0 & \text{otherwise.} \end{cases} \tag{C.77}$$

The alternating tensor can be written as $\mathbf{\Lambda} = \varepsilon_{\alpha\beta\gamma}\hat{\mathbf{e}}^\alpha\hat{\mathbf{e}}^\beta\hat{\mathbf{e}}^\gamma = \varepsilon^{\alpha\beta\gamma}\hat{\mathbf{e}}_\alpha\hat{\mathbf{e}}_\beta\hat{\mathbf{e}}_\gamma$.

C.2.4 Dot and cross products

The canonical representation of the dot product $\mathbf{u}\cdot\mathbf{v} = u_i v_i$ of two vectors can be written in any of the four equivalent forms

$$u_\alpha v^\alpha = u^\alpha v_\alpha = g_{\alpha\beta} u^\alpha v^\beta = g^{\alpha\beta} u_\alpha v_\beta. \tag{C.78}$$

Likewise, the double dot product $\mathbf{T}:\mathbf{P} = T_{ij}P_{ij}$ of two second-order tensors is given by

$$T_{\alpha\beta}P^{\alpha\beta} = T^{\alpha\beta}P_{\alpha\beta} = g_{\alpha\gamma}g_{\beta\eta}T^{\alpha\beta}P^{\gamma\eta} = g^{\alpha\gamma}g^{\beta\eta}T_{\alpha\beta}P_{\gamma\eta}, \tag{C.79}$$

as well as by a number of intermediate expressions involving the mixed components $T_\alpha{}^\beta$, $T^\alpha{}_\beta$ and $P_\alpha{}^\beta$, $P^\alpha{}_\beta$. The cross product $\mathbf{w} = \mathbf{u}\times\mathbf{v}$ of two vectors can be expressed in terms of the purely covariant and contravariant alternating symbols (C.76)–(C.77) in the form

$$w_\alpha = \varepsilon_{\alpha\beta\gamma} u^\beta v^\gamma, \qquad w^\alpha = \varepsilon^{\alpha\beta\gamma} u_\beta v_\gamma. \tag{C.80}$$

In this case, there is an even greater welter of equivalent intermediate relations involving the covariant and contravariant components of the metric tensor and the mixed components of the alternating tensor.

For simplicity we shall henceforth conduct all calculations involving physical vector and tensor variables $\mathbf{u}$ and $\mathbf{T}$ in terms of their *contravariant* components u^α and $T^{\alpha_1\cdots\alpha_q}$. The only covariant components that will be employed will be those of the identity and alternating tensors, $g_{\alpha\beta}$ and

$\varepsilon_{\alpha\beta\gamma}$; mixed components will be eschewed entirely. The explicit contravariant representations of the dot product $\mathbf{u}\cdot\mathbf{v} = g_{\alpha\beta}u^{\alpha}v^{\beta}$ of two vectors and the double dot product of two tensors $\mathbf{T}:\mathbf{P} = g_{\alpha\gamma}g_{\beta\eta}T^{\alpha\beta}P^{\gamma\eta}$ are

$$\mathbf{u}\cdot\mathbf{v} = -u^{-}v^{+} + u^{0}v^{0} - u^{+}v^{-}, \tag{C.81}$$

$$\begin{aligned}\mathbf{T}:\mathbf{P} = T^{--}P^{++} - T^{-0}P^{+0} + T^{-+}P^{+-} \\ - T^{0-}P^{0+} + T^{00}P^{00} - T^{0+}P^{0-} \\ + T^{+-}P^{-+} - T^{+0}P^{-0} + T^{++}P^{--}.\end{aligned} \tag{C.82}$$

The contravariant components $w^{\alpha} = g^{\alpha\eta}\varepsilon_{\eta\beta\gamma}u^{\beta}v^{\gamma}$ of a vector cross product are $w^{-} = i(u^{0}v^{-} - u^{-}v^{0})$, $w^{0} = i(u^{+}v^{-} - u^{-}v^{+})$ and $w^{+} = i(u^{+}v^{0} - u^{0}v^{+})$.

The contraction of two higher-order tensors can be written in terms of their contravariant components in the form

$$\mathbf{T}\vdots\mathbf{P} = g_{\alpha_1\beta_1}\cdots g_{\alpha_q\beta_q}T^{\alpha_1\cdots\alpha_q}P^{\beta_1\cdots\beta_q}. \tag{C.83}$$

Contractions involving a complex conjugate are simpler than (C.83) by virtue of the orthonormality relation (C.51):

$$\mathbf{T}^{*}\vdots\mathbf{P} = \delta_{\alpha_1\beta_1}\cdots\delta_{\alpha_q\beta_q}T^{\alpha_1\cdots\alpha_q}P^{\beta_1\cdots\beta_q}. \tag{C.84}$$

It is tempting to write the right side of equation (C.84) in the more succinct form $T^{\alpha_1\cdots\alpha_q}P^{\alpha_1\cdots\alpha_q}$; however, this would violate our convention that a repeated Greek index must be present in both registers.

C.2.5 Hermiticity of J

The component-wise representation of the inner product (C.1) of two complex tensor fields $\mathbf{T}$ and $\mathbf{P}$ upon the unit sphere Ω is

$$\langle\mathbf{T},\mathbf{P}\rangle = \int_{\Omega}\delta_{\alpha_1\beta_1}\cdots\delta_{\alpha_q\beta_q}T^{\alpha_1\cdots\alpha_q *}P^{\beta_1\cdots\beta_q}\,d\Omega. \tag{C.85}$$

Armed with (C.85) we are finally ready to establish the important result that the total angular-momentum operator is self-adjoint or Hermitian:

$$\mathbf{J}^{\dagger} = \mathbf{J}. \tag{C.86}$$

As in the case of the orbital angular momentum in Appendix B, the argument is simplest in the case of the polar component J_z. Since this operator has no effect upon either the canonical basis vectors or the metric tensor, an integration by parts with respect to ϕ as in equation (B.28) suffices to show that $\langle\mathbf{T}, J_z\mathbf{P}\rangle = \langle J_z\mathbf{T},\mathbf{P}\rangle$. A direct demonstration that $\langle\mathbf{T}, J_x\mathbf{P}\rangle = \langle J_x\mathbf{T},\mathbf{P}\rangle$ and $\langle\mathbf{T}, J_y\mathbf{P}\rangle = \langle J_y\mathbf{T},\mathbf{P}\rangle$ is more involved; however, we may again fall back upon the argument that $J_z^\dagger = J_z$ implies that $J_x^\dagger = J_x$ and $J_y^\dagger = J_y$, inasmuch as the Cartesian axes $\hat{\mathbf{x}}$, $\hat{\mathbf{y}}$, $\hat{\mathbf{z}}$ may be chosen arbitrarily. The squared angular-momentum operator is also Hermitian whereas the two ladder operators are each other's adjoints: $(J^2)^\dagger = J^2$ and $(J_\pm)^\dagger = J_\mp$. The adjoint of $\mathcal{D}(\boldsymbol{\omega}) = \exp(i\boldsymbol{\omega}\cdot\mathbf{J})$ is $\mathcal{D}^\dagger(\boldsymbol{\omega}) = \exp(-i\boldsymbol{\omega}\cdot\mathbf{J})$. Since the adjoint and the inverse coincide, that is,

$$\mathcal{D}^\dagger(\boldsymbol{\omega}) = \mathcal{D}^{-1}(\boldsymbol{\omega}), \tag{C.87}$$

the finite rotation operator is *unitary*.

C.3 Construction of a Basis

Following a procedure identical to that employed in Section B.3 we construct an orthonormal basis of *generalized surface spherical harmonics* that are simultaneous eigenfunctions of the commuting Hermitian operators J^2 and J_z. We denote these harmonics, which are confined to the surface of the unit sphere, by $\mathbf{Y}_{lm}^N(\theta,\phi)$ and stipulate by analogy with (B.29) that

$$J^2\mathbf{Y}_{lm}^N = l(l+1)\mathbf{Y}_{lm}^N, \qquad J_z\mathbf{Y}_{lm}^N = m\mathbf{Y}_{lm}^N. \tag{C.88}$$

Supposing $\mathbf{Y}_{lm}^N$ to be a tensor of order q, we expand it in terms of the canonical basis vectors in the form

$$\mathbf{Y}_{lm}^N = Y_{lm}^N\hat{\mathbf{e}}_1\cdots\hat{\mathbf{e}}_q. \tag{C.89}$$

We refer to $\mathbf{Y}_{lm}^N$ as the *tensor* harmonic and to Y_{lm}^N as the associated *scalar* harmonic. The lower indices on both the tensor and scalar are, as before, the *degree* l and *order* m of the harmonic; the upper index N depends upon the order q of the tensor $\mathbf{Y}_{lm}^N$, as we shall see. The second of the eigenvalue equations (C.88) implies that every scalar harmonic must be of the form

$$Y_{lm}^N(\theta,\phi) = X_{lm}^N(\theta)\exp(im\phi). \tag{C.90}$$

This shows that the order m must be an integer, to ensure that Y_{lm}^N is single valued upon the unit sphere. Every generalized surface spherical harmonic generates an associated solid spherical harmonic defined by the relationship $\mathbf{Y}_{lm}^N(\mathbf{r}) = r^l\mathbf{Y}_{lm}^N(\theta,\phi)$; however, these three-dimensional functions do not play any essential role in the theory.

By analogy with equations (B.31)–(B.32) we can use the commutation relations (C.40) to show that the tensors $J_\pm\mathbf{Y}_{lm}^N$ are eigenfunctions of J^2 and J_z with associated eigenvalues $l(l+1)$ and $m\pm 1$:

$$J^2(J_\pm\mathbf{Y}_{lm}^N) = J_\pm(J^2\mathbf{Y}_{lm}^N) = l(l+1)(J_\pm\mathbf{Y}_{lm}^N), \tag{C.91}$$

$$J_z(J_\pm \mathbf{Y}_{lm}^N) = (J_\pm J_z \pm J_\pm)\mathbf{Y}_{lm}^N = (m \pm 1)(J_\pm \mathbf{Y}_{lm}^N). \tag{C.92}$$

The *ascending* ladder operator J_+ transforms $\mathbf{Y}_{lm}^N$ into a constant times $\mathbf{Y}_{l\,m+1}^N$, whereas the *descending* ladder operator transforms it into a constant times $\mathbf{Y}_{l\,m-1}^N$:

$$J_\pm \mathbf{Y}_{lm}^N = c_\pm \mathbf{Y}_{l\,m\pm 1}^N. \tag{C.93}$$

The squared norms of the tensors $J_\pm \mathbf{Y}_{lm}^N$ and $\mathbf{Y}_{lm}^N$ are related using an argument identical to that given in (B.34):

$$|||J_\pm \mathbf{Y}_{lm}^N|||^2 = (l \mp m)(l \pm m + 1)\, |||\mathbf{Y}_{lm}^N|||^2. \tag{C.94}$$

Repeated application of the ladder operators $J_\pm$ to $\mathbf{Y}_{l\mp l}^N$ leads eventually to the conclusion that $J_\pm \mathbf{Y}_{l\pm l}^N = \mathbf{0}$. This cessation of the laddering after $2l+1$ steps implies that the degree l must be either a positive half-integer or an integer. Half-integer quantum numbers are needed to describe the angular momentum of particles and systems of particles with spin in quantum mechanics; in our case, however, l is required to be an integer because m is. There are therefore $2l+1$ integer orders, $-l \le m \le l$, for every non-negative integer degree l.

Generalized spherical harmonics $\mathbf{Y}_{lm}^N$ and $\mathbf{Y}_{l'm'}^{N'}$ having different degrees $l \ne l'$, orders $m \ne m'$ and upper indices $N \ne N'$ are orthogonal by virtue of the Hermiticity of the two operators J^2 and J_z. We render the basis *orthonormal* by demanding that

$$\langle \mathbf{Y}_{lm}^N, \mathbf{Y}_{l'm'}^{N'} \rangle = \int_\Omega \mathbf{Y}_{lm}^{N*} : \mathbf{Y}_{l'm'}^{N'}\, d\Omega = \delta_{ll'}\delta_{mm'}\delta_{NN'}. \tag{C.95}$$

Upon combining equations (C.93) and (C.95) we find that the constants $c_\pm$ satisfy $|c_\pm|^2 = (l \mp m)(l \pm m + 1)$.

The limiting harmonics $\mathbf{Y}_{l\pm l}^N$ can be found by explicitly expanding the ladder-termination relation $J_\pm \mathbf{Y}_{l\pm l}^N = \mathbf{0}$ in the form

$$\begin{aligned} J_\pm[X_{l\pm l}^N(\theta)\exp(im\phi)\hat{\mathbf{e}}_{\alpha_1}\cdots\hat{\mathbf{e}}_{\alpha_q}] &= [J_\pm X_{l\pm l}^N(\theta)\exp(im\phi)]\hat{\mathbf{e}}_{\alpha_1}\cdots\hat{\mathbf{e}}_{\alpha_q} \\ &\quad + X_{l\pm l}^N(\theta)\exp(im\phi)(J_\pm\hat{\mathbf{e}}_{\alpha_1})\cdots\hat{\mathbf{e}}_{\alpha_q} \\ &\quad\ \vdots \\ &\quad + X_{l\pm l}^N(\theta)\exp(im\phi)\hat{\mathbf{e}}_{\alpha_1}\cdots(J_\pm\hat{\mathbf{e}}_{\alpha_q}) = \mathbf{0}. \end{aligned} \tag{C.96}$$

Upon making use of the results (B.26) and (C.59) we obtain an ordinary differential equation analogous to (B.36) governing the scalar colatitudinal dependence $X_{l\pm l}^N(\theta)$:

$$\left[\frac{d}{d\theta} - l\cot\theta \pm N(\sin\theta)^{-1}\right] X_{l\pm l}^N = 0, \tag{C.97}$$

where—for the first time we obtain an identification of the upper index—

$$N = \alpha_1 + \cdots + \alpha_q. \tag{C.98}$$

The solution to equation (C.97) is $X_{l\pm l}^{N} = A_{\pm}(\sin\frac{1}{2}\theta)^{l-N}(\cos\frac{1}{2}\theta)^{l+N}$. The integer N must lie within the range $-l \leq N \leq l$ in order to ensure regularity. The moduli $|A_{\pm}|$ are determined by the requirement that $|||\mathbf{Y}_{l\pm l}^{N}||| = 1$. As in the scalar case the signs of the constants $c_{\pm}$ and $A_{\pm}$ are still at our disposal.

Choosing $\operatorname{sgn} A_{-} = 1$ and $\operatorname{sgn} c_{+} = 1$ and laddering up from the lowest-order basis element we obtain the generalized spherical harmonics

$$\mathbf{Y}_{l\,-l}^{N} = \left(\frac{2l+1}{4\pi}\right)^{1/2} \left[\frac{(2l)!}{(l+N)!(l-N)!}\right]^{1/2} \times (\sin\tfrac{1}{2}\theta)^{l+N}(\cos\tfrac{1}{2}\theta)^{l-N}e^{-il\phi}\,\hat{\mathbf{e}}_{\alpha_1}\cdots\hat{\mathbf{e}}_{\alpha_q} \tag{C.99}$$

and

$$\mathbf{Y}_{lm}^{N} = \left[\frac{(l-m)!}{(l+m)!}\right]^{1/2} \left[\frac{1}{(2l)!}\right]^{1/2} (J_{+})^{l+m}\,\mathbf{Y}_{l\,-l}^{N}. \tag{C.100}$$

Choosing $\operatorname{sgn} A_{+} = (-1)^{l+N}$ and $\operatorname{sgn} c_{-} = 1$ and laddering down from the highest-order element gives the alternative and equivalent representation

$$\mathbf{Y}_{ll}^{N} = (-1)^{l+N}\left(\frac{2l+1}{4\pi}\right)^{1/2} \left[\frac{(2l)!}{(l+N)!(l-N)!}\right]^{1/2} \times (\sin\tfrac{1}{2}\theta)^{l-N}(\cos\tfrac{1}{2}\theta)^{l+N}e^{il\phi}\,\hat{\mathbf{e}}_{\alpha_1}\cdots\hat{\mathbf{e}}_{\alpha_q} \tag{C.101}$$

and

$$\mathbf{Y}_{lm}^{N} = \left[\frac{(l+m)!}{(l-m)!}\right]^{1/2} \left[\frac{1}{(2l)!}\right]^{1/2} (J_{-})^{l-m}\,\mathbf{Y}_{ll}^{N}. \tag{C.102}$$

The definitions (C.99)–(C.100) and (C.101)–(C.102) are valid for all integer degrees l, orders m and upper indices $N = \alpha_1 + \cdots + \alpha_q$ in the range

$$0 \leq l \leq \infty, \qquad -l \leq m \leq l, \qquad -l \leq N \leq l. \tag{C.103}$$

The generalized spherical harmonics $\mathbf{Y}_{lm}^{N}$ defined in this manner satisfy the eigenvalue relations (C.88), the orthonormality relations (C.95) and the ladder relations

$$J_{\pm}\mathbf{Y}_{lm}^{N} = \sqrt{(l \mp m)(l \pm m + 1)}\,\mathbf{Y}_{l\,m\pm 1}^{N}. \tag{C.104}$$

Outside of the range (C.103) it is convenient to define $\mathbf{Y}_{lm}^{N} = \mathbf{0}$ for $|m| > l$ and $|N| > l$. This renders equation (C.104) valid for all values of m, just as its scalar analogue (B.41) is. Upon comparing equations (C.99)–(C.102) with the corresponding definitions (B.37)–(B.40) of the ordinary spherical harmonics, we see that $Y_{lm}^{0} = Y_{lm}$; this is the motivation for the sign choices we have made above.

The generalization of equation (B.42) can be shown by induction to be

$$\begin{aligned}(J_\pm)^{l\pm m}&\left[(\sin\tfrac{1}{2}\theta)^{l\pm N}(\cos\tfrac{1}{2}\theta)^{l\mp N}e^{\mp il\phi}\,\hat{\mathbf{e}}_{\alpha_1}\cdots\hat{\mathbf{e}}_{\alpha_q}\right]\\ &=\Bigg\{2^{l\pm m}(\sin\tfrac{1}{2}\theta)^{\pm m\mp N}(\cos\tfrac{1}{2}\theta)^{\pm m\pm N}\\ &\times\left(\pm\frac{1}{\sin\theta}\frac{d}{d\theta}\right)^{l\pm m}\left[(\sin\tfrac{1}{2}\theta)^{2l\pm 2N}(\cos\tfrac{1}{2}\theta)^{2l\mp 2N}\right]\Bigg\}\\ &\times e^{im\phi}\,\hat{\mathbf{e}}_{\alpha_1}\cdots\hat{\mathbf{e}}_{\alpha_q}.\end{aligned}\tag{C.105}$$

Upon utilizing the result (C.105) in (C.99)–(C.102) we obtain the explicit representations

$$\begin{aligned}\mathbf{Y}_{lm}^{N}&=\left(\frac{2l+1}{4\pi}\right)^{1/2}\left[\frac{1}{(l+N)!(l-N)!}\right]^{1/2}\left[\frac{(l-m)!}{(l+m)!}\right]^{1/2}\\ &\times\Bigg\{2^{l+m}(\sin\tfrac{1}{2}\theta)^{m-N}(\cos\tfrac{1}{2}\theta)^{m+N}\\ &\times\left(\frac{1}{\sin\theta}\frac{d}{d\theta}\right)^{l+m}\left[(\sin\tfrac{1}{2}\theta)^{2l+2N}(\cos\tfrac{1}{2}\theta)^{2l-2N}\right]\Bigg\}\\ &\times e^{im\phi}\,\hat{\mathbf{e}}_{\alpha_1}\cdots\hat{\mathbf{e}}_{\alpha_q}\\ &=(-1)^{l-N}\left(\frac{2l+1}{4\pi}\right)^{1/2}\left[\frac{1}{(l+N)!(l-N)!}\right]^{1/2}\left[\frac{(l+m)!}{(l-m)!}\right]^{1/2}\\ &\times\Bigg\{2^{l-m}(\sin\tfrac{1}{2}\theta)^{-m+N}(\cos\tfrac{1}{2}\theta)^{-m-N}\\ &\times\left(-\frac{1}{\sin\theta}\frac{d}{d\theta}\right)^{l-m}\left[(\sin\tfrac{1}{2}\theta)^{2l-2N}(\cos\tfrac{1}{2}\theta)^{2l+2N}\right]\Bigg\}\\ &\times e^{im\phi}\,\hat{\mathbf{e}}_{\alpha_1}\cdots\hat{\mathbf{e}}_{\alpha_q}.\end{aligned}\tag{C.106}$$

A comparison of these two expressions enables us to deduce that

$$\mathbf{Y}_{l\,-m}^{-N}=(-1)^{m}\mathbf{Y}_{lm}^{N*},\tag{C.107}$$

which generalizes the symmetry relation (B.44). In obtaining (C.107) we have made use of the fact that $\hat{\mathbf{e}}_{-\alpha_1}\cdots\hat{\mathbf{e}}_{-\alpha_q}=(-1)^{N}(\hat{\mathbf{e}}_{\alpha_1}\cdots\hat{\mathbf{e}}_{\alpha_q})^{*}$.

We conclude this section by calling attention to the ambiguity of the above notation. The symbol Y_{lm}^N denotes a unique scalar generalized spherical harmonic; however, $\mathbf{Y}_{lm}^N$ is really shorthand for an infinite horde of tensors of order $q > |N|$:

$$\begin{aligned}
\mathbf{Y}_{lm}^0 &= \{Y_{lm}^0,\, Y_{lm}^0\hat{\mathbf{e}}_0,\, Y_{lm}^0\hat{\mathbf{e}}_0\hat{\mathbf{e}}_0,\, Y_{lm}^0\hat{\mathbf{e}}_-\hat{\mathbf{e}}_+,\, Y_{lm}^0\hat{\mathbf{e}}_+\hat{\mathbf{e}}_-,\ldots\},\\
\mathbf{Y}_{lm}^{\pm1} &= \{Y_{lm}^{\pm1}\hat{\mathbf{e}}_\pm,\, Y_{lm}^{\pm1}\hat{\mathbf{e}}_0\hat{\mathbf{e}}_\pm,\, Y_{lm}^{\pm1}\hat{\mathbf{e}}_\pm\hat{\mathbf{e}}_0,\ldots\},\\
\mathbf{Y}_{lm}^{\pm2} &= \{Y_{lm}^{\pm2}\hat{\mathbf{e}}_\pm\hat{\mathbf{e}}_\pm,\ldots\},\\
&\;\;\vdots
\end{aligned} \tag{C.108}$$

Every ordered polyad $\hat{\mathbf{e}}_{\alpha_1}\cdots\hat{\mathbf{e}}_{\alpha_q}$ having $\alpha_1+\cdots+\alpha_q = N$ is associated with the same generalized scalar harmonic Y_{lm}^N. The scalar harmonics satisfy

$$Y_{l\,-m}^{-N} = (-1)^{m+N}Y_{lm}^{N*}. \tag{C.109}$$

Two generalized spherical harmonics $\mathbf{Y}_{lm}^N$ and $\mathbf{Y}_{l'm'}^{N'}$ with different upper indices $N \neq N'$ are orthogonal by virtue of the orthogonality of the associated polyads. For $N = N'$, on the other hand, equation (C.95) implies that

$$\int_\Omega Y_{lm}^{N*}\,Y_{l'm'}^N\,d\Omega = \delta_{ll'}\delta_{mm'}. \tag{C.110}$$

Surface integrals involving the product $Y_{lm}^{N*}\,Y_{l'm'}^{N'}$ of two scalar harmonics with $N \neq N'$ never arise in practice.

C.4 Generalized Legendre Functions

Upon substituting the representation $\mathbf{Y}_{lm}^N(\theta,\phi) = X_{lm}^N(\theta)e^{im\phi}\hat{\mathbf{e}}_{\alpha_1}\cdots\hat{\mathbf{e}}_{\alpha_q}$ into the eigenvalue problem $J^2\mathbf{Y}_{lm}^N = l(l+1)\mathbf{Y}_{lm}^N$, we find that the colatitudinal scalar X_{lm}^N satisfies the ordinary differential equation

$$\begin{aligned}
&(1-\mu^2)\frac{d^2X}{d\mu^2} - 2\mu\frac{dX}{d\mu}\\
&\qquad + \left[l(l+1) - \frac{m^2 - 2mN\mu + N^2}{1-\mu^2}\right]X = 0,
\end{aligned} \tag{C.111}$$

where $\mu = \cos\theta$. The so-called *generalized Legendre equation* (C.111) reduces to the ordinary Legendre equation (B.47) in the case $N = 0$, as expected. Following Phinney & Burridge (1973) we define the *generalized Legendre functions* of degree l, order $-l \leq m \leq l$ and upper index $-l \leq N \leq l$ by

$$P_{lm}^{N}(\mu) = \frac{1}{2^l}\left[\frac{1}{(l+N)!(l-N)!}\right]^{1/2}\left[\frac{(l+m)!}{(l-m)!}\right]^{1/2} \times (1-\mu)^{-\frac{1}{2}(m-N)}(1+\mu)^{-\frac{1}{2}(m+N)} \times \left(\frac{d}{d\mu}\right)^{l-m}\left[(\mu-1)^{l-N}(\mu+1)^{l+N}\right]. \tag{C.112}$$

These regular solutions to equation (C.111) in the interval $-1 \leq \mu \leq 1$ do *not reduce* to the associated Legendre functions $P_{lm}(\mu)$ in the case $N = 0$; in fact,

$$P_{lm}^{0} = (-1)^m\left[\frac{(l-m)!}{(l+m)!}\right]^{1/2} P_{lm}. \tag{C.113}$$

The normalization of the generalized Legendre functions (C.112) is designed instead to render

$$P_{lm}^{N}(1) = \delta_{Nm}, \tag{C.114}$$

where δ_{Nm} is the $(2l+1)\times(2l+1)$ Kronecker delta. This will prove to be convenient when we seek to determine the matrix elements of tensor products in Section C.7 and of finite rotations in Section C.8.

The generalized Legendre functions of degrees zero and one are $P_{00}^{0} = 1$ and

$$\begin{array}{ccccc} & N\!=\!-1 & N\!=\!0 & N\!=\!1 & \\ P_{1m}^{N} = & \left(\begin{array}{c} \frac{1}{2}(1+\cos\theta) \\ -\frac{1}{\sqrt{2}}\sin\theta \\ \frac{1}{2}(1-\cos\theta) \end{array}\right. & \begin{array}{c} \frac{1}{\sqrt{2}}\sin\theta \\ \cos\theta \\ -\frac{1}{\sqrt{2}}\sin\theta \end{array} & \left.\begin{array}{c} \frac{1}{2}(1-\cos\theta) \\ \frac{1}{\sqrt{2}}\sin\theta \\ \frac{1}{2}(1+\cos\theta) \end{array}\right) & \begin{array}{l} m\!=\!-1 \\ m\!=\!0 \\ m\!=\!1. \end{array} \end{array} \tag{C.115}$$

The orthonormality relation governing two functions of the same order m and upper index N is

$$\int_{-1}^{1} P_{lm}^{N}(\mu)P_{l'm}^{N}(\mu)\,d\mu = \left(\frac{2}{2l+1}\right)\delta_{ll'}. \tag{C.116}$$

Equation (C.116) generalizes the ordinary Legendre orthonormality relation (B.49). The scalar harmonics X_{lm}^N are given in terms of the generalized Legendre functions by

$$X_{lm}^N(\theta) = \left(\frac{2l+1}{4\pi}\right)^{1/2} P_{lm}^N(\cos\theta). \tag{C.117}$$

As we have noted previously, X_{lm}^0 *does* coincide with X_{lm}.

The generalized Legendre functions P_{lm}^N satisfy a number of symmetry relations, which are valid for X_{lm}^N as well:

$$P_{l\,-m}^N(\mu) = (-1)^{l+N} P_{lm}^N(-\mu), \qquad P_{lm}^{-N}(\mu) = (-1)^{l+m} P_{lm}^N(-\mu),$$

$$P_{l\,-m}^{-N}(\mu) = (-1)^{m+N} P_{lm}^N(\mu), \qquad P_{lN}^m(\mu) = (-1)^{m+N} P_{lm}^N(\mu),$$

$$P_{l\,-N}^{-m}(\mu) = P_{lm}^N(\mu). \tag{C.118}$$

From the tensor equation (C.104) we obtain the scalar differential relations

$$\left[\pm d/d\theta + N(\sin\theta)^{-1} - m\cot\theta\right] P_{lm}^N$$
$$= \sqrt{(l\mp m)(l\pm m+1)}\, P_{l\,m\pm 1}^N. \tag{C.119}$$

Upon combining (C.118) and (C.119) we may also deduce that

$$\left[\pm d/d\theta + N\cot\theta - m(\sin\theta)^{-1}\right] P_{lm}^N$$
$$= \sqrt{(l\pm N)(l\mp N+1)}\, P_{lm}^{N\mp 1}. \tag{C.120}$$

Upon adding the plus-and-minus versions of equations (C.119) and (C.120) we obtain the recursion relations

$$\left[N(\sin\theta)^{-1} - m\cot\theta\right] P_{lm}^N = \tfrac{1}{2}\sqrt{(l+m)(l-m+1)}\, P_{l\,m-1}^N$$
$$+ \tfrac{1}{2}\sqrt{(l-m)(l+m+1)}\, P_{l\,m+1}^N, \tag{C.121}$$

$$\left[N\cot\theta - m(\sin\theta)^{-1}\right] P_{lm}^N = \tfrac{1}{2}\sqrt{(l+N)(l-N+1)}\, P_{lm}^{N-1}$$
$$+ \tfrac{1}{2}\sqrt{(l-N)(l+N+1)}\, P_{lm}^{N+1}. \tag{C.122}$$

The argument of the generalized Legendre functions in all of these relations is of course $\mu = \cos\theta$.

Either equation (C.121) or (C.122) can be made the basis of a practical scheme for computing the $(2l+1)\times(2l+1)$ functions P_{lm}^N, $-l \le m \le l$, $-l \le N \le l$. Numerical stability considerations make it necessary to iterate the fixed-N recursion relation (C.121) upward from $m = -l$ as well as downward from $m = l$. The starting values for the downward recursion are

$$P_{ll}^{N} = (-1)^{l+N}\left[\frac{(2l)!}{(l+N)!(l-N)!}\right]^{1/2}(\sin\tfrac{1}{2}\theta)^{l-N}(\cos\tfrac{1}{2}\theta)^{l+N}, \tag{C.123}$$

$$P_{l\,l-1}^{N} = \sqrt{2l}\,(\sin\theta)^{-1}(N/l - \cos\theta)\,P_{ll}^{N}, \tag{C.124}$$

whereas those for the upward recursion are

$$P_{l\,-l}^{N} = \left[\frac{(2l)!}{(l+N)!(l-N)!}\right]^{1/2}(\sin\tfrac{1}{2}\theta)^{l+N}(\cos\tfrac{1}{2}\theta)^{l-N}, \tag{C.125}$$

$$P_{l\,-l+1}^{N} = \sqrt{2l}\,(\sin\theta)^{-1}(N/l + \cos\theta)\,P_{l\,-l}^{N}. \tag{C.126}$$

The optimal meeting point of the two recursions is $m = N\mu = N\cos\theta$. Other options are available; for example it is possible to iterate the fixed-m relation (C.122) downward from $N = l$ and upward from $N = -l$, meeting at $N = m\mu^{-1}$, or it is possible to iterate both (C.121) and (C.122) either downwards from $m = N = l$ or upwards from $m = N = -l$, and use the symmetry relations (C.118) to fill in the remaining values. In Section C.8.2 we derive the generalized Legendre addition theorem

$$\sum_{m=-l}^{l} P_{lm}^{N}(\mu)P_{lm}^{N'}(\mu) = \delta_{NN'}. \tag{C.127}$$

This result can be used to renormalize or verify the accuracy of the functions P_{lm}^{N}, $-l \leq m \leq l$, $-l \leq N \leq l$, just as equation (B.124) is used to renormalize or verify the accuracy of the ordinary spherical harmonics X_{lm}, $-l \leq m \leq l$. The colatitudinal derivative of P_{lm}^{N} is given in terms of $P_{l\,m\pm1}^{N}$ or $P_{lm}^{N\pm1}$ by

$$\begin{aligned}\frac{dP_{lm}^{N}}{d\theta} &= \tfrac{1}{2}\sqrt{(l-m)(l+m+1)}\,P_{l\,m+1}^{N} \\ &\qquad - \tfrac{1}{2}\sqrt{(l+m)(l-m+1)}\,P_{l\,m-1}^{N} \\ &= \tfrac{1}{2}\sqrt{(l+N)(l-N+1)}\,P_{lm}^{N-1} \\ &\qquad - \tfrac{1}{2}\sqrt{(l-N)(l+N+1)}\,P_{lm}^{N+1}.\end{aligned} \tag{C.128}$$

Masters & Richards-Dinger (1998) describe a scheme in which P_{lm}^{N} and $dP_{lm}^{N}/d\theta$ are calculated simultaneously by downward and upward iteration of a pair of coupled recursion relations analogous to (B.121)–(B.122).

C.5 Generalized Expansions

A complex square-integrable tensor field $\mathbf{T}(\theta,\phi)$ upon the surface of the unit sphere Ω can be expanded in terms of generalized surface spherical harmonics $\mathbf{Y}_{lm}^{N}(\theta,\phi)$ in a manner analogous to the expansion of a complex scalar field $\psi(\theta,\phi)$ in terms of ordinary spherical harmonics $Y_{lm}(\theta,\phi)$. We write $\mathbf{T}$ in terms of its contravariant components with respect to the canonical basis $\hat{\mathbf{e}}_-$, $\hat{\mathbf{e}}_0$, $\hat{\mathbf{e}}_+$, and expand each of the components in the form

$$T^{\alpha_1\cdots\alpha_n} = \sum_{l=0}^{\infty}\sum_{m=-l}^{l} T_{lm}^{\alpha_1\cdots\alpha_n} Y_{lm}^{N}, \tag{C.129}$$

where $N = \alpha_1 + \cdots + \alpha_q$. The resulting generalized spherical-harmonic representation of a tensor field $\mathbf{T} = T^{\alpha_1\cdots\alpha_q}\hat{\mathbf{e}}_{\alpha_1}\cdots\hat{\mathbf{e}}_{\alpha_q}$ of order q is

$$\begin{aligned}\mathbf{T} &= \sum_{l=0}^{\infty}\sum_{m=-l}^{l} T_{lm}^{\alpha_1\cdots\alpha_q} Y_{lm}^{N}\hat{\mathbf{e}}_{\alpha_1}\cdots\hat{\mathbf{e}}_{\alpha_q} \\ &= \sum_{l=0}^{\infty}\sum_{m=-l}^{l} T_{lm}^{\alpha_1\cdots\alpha_q}\mathbf{Y}_{lm}^{N},\end{aligned} \tag{C.130}$$

where the summation over $\alpha_1,\ldots,\alpha_q$ is understood in the final shorthand formula. The complex expansion coefficients $T_{lm}^{\alpha_1\cdots\alpha_q}$ are given by

$$T_{lm}^{\alpha_1\cdots\alpha_q} = \langle\mathbf{Y}_{lm}^{N},\mathbf{T}\rangle = \int_{\Omega} Y_{lm}^{N*} T^{\alpha_1\cdots\alpha_q}\,d\Omega, \tag{C.131}$$

by virtue of the orthonormality relation (C.95). It is noteworthy that $T_{lm}^{\alpha_1\cdots\alpha_q} = 0$ for $l < |N|$. A three-dimensional tensor field $\mathbf{T}(\mathbf{r})$ can be expressed in the form (C.129)–(C.130) by simply allowing the expansion coefficients $T_{lm}^{\alpha_1\cdots\alpha_q}$ to be functions of radius r. The coefficients of a real tensor field satisfy $T_{l\,-m}^{-\alpha_1\cdots-\alpha_q} = (-1)^m T_{lm}^{\alpha_1\cdots\alpha_q *}$. The tensor version of Parseval's theorem is

$$|||\mathbf{T}|||^2 = \int_{\Omega}\mathbf{T}^*:\mathbf{T}\,d\Omega = \sum_{l=0}^{\infty}\sum_{m=-l}^{l}|T_{lm}^{\alpha_1\cdots\alpha_q}|^2. \tag{C.132}$$

The above relations generalize the familiar scalar results (B.93)–(B.94).

The generalized spherical-harmonic representation of a vector field $\mathbf{u}$ is

$$\mathbf{u} = \sum_{l=0}^{\infty}\sum_{m=-l}^{l}\left(u_{lm}^{-}Y_{lm}^{-}\,\hat{\mathbf{e}}_- + u_{lm}^{0}Y_{lm}^{0}\,\hat{\mathbf{e}}_0 + u_{lm}^{+}Y_{lm}^{+}\,\hat{\mathbf{e}}_+\right), \tag{C.133}$$

where

$$u_{lm}^{\alpha} = \int_{\Omega} Y_{lm}^{\alpha *} u^{\alpha}\, d\Omega = \int_{\Omega} Y_{lm}^{\alpha *} \hat{\mathbf{e}}_{\alpha} \cdot \mathbf{u}\, d\Omega. \tag{C.134}$$

We can alternatively express $\mathbf{u}$ using the *complex version* of the vector spherical-harmonic representation (B.168)–(B.171):

$$\mathbf{u} = \sum_{l=0}^{\infty} \sum_{m=-l}^{l} \left\{ u_{lm} \hat{\mathbf{r}} Y_{lm} + \frac{1}{\sqrt{l(l+1)}} \times \left[v_{lm} \boldsymbol{\nabla}_1 Y_{lm} - w_{lm} (\hat{\mathbf{r}} \times \boldsymbol{\nabla}_1 Y_{lm}) \right] \right\}, \tag{C.135}$$

where

$$u_{lm} = \int_{\Omega} \hat{\mathbf{r}} Y_{lm}^{*} \cdot \mathbf{u}\, d\Omega, \tag{C.136}$$

$$v_{lm} = \frac{1}{\sqrt{l(l+1)}} \int_{\Omega} \boldsymbol{\nabla}_1 Y_{lm}^{*} \cdot \mathbf{u}\, d\Omega, \tag{C.137}$$

$$w_{lm} = -\frac{1}{\sqrt{l(l+1)}} \int_{\Omega} (\hat{\mathbf{r}} \times \boldsymbol{\nabla}_1 Y_{lm}^{*}) \cdot \mathbf{u}\, d\Omega. \tag{C.138}$$

How are the canonical expansion coefficients u_{lm}^{-}, u_{lm}^{0}, u_{lm}^{+} and the complex vector spherical-harmonic coefficients u_{lm}, v_{lm}, w_{lm} related? The effect of the surface gradient (A.118) and curl (A.124) upon a complex surface spherical harmonic Y_{lm} can be determined with the aid of the identity (C.120):

$$\boldsymbol{\nabla}_1 Y_{lm} = \sqrt{l(l+1)/2}\left(Y_{lm}^{-} \hat{\mathbf{e}}_{-} + Y_{lm}^{+} \hat{\mathbf{e}}_{+} \right), \tag{C.139}$$

$$\hat{\mathbf{r}} \times \boldsymbol{\nabla}_1 Y_{lm} = i\sqrt{l(l+1)/2}\left(Y_{lm}^{-} \hat{\mathbf{e}}_{-} - Y_{lm}^{+} \hat{\mathbf{e}}_{+} \right). \tag{C.140}$$

Upon making use of (C.139)–(C.140) we find that $u_{lm} = u_{lm}^{0}$ and

$$v_{lm} = \tfrac{1}{\sqrt{2}}\left(u_{lm}^{-} + u_{lm}^{+} \right), \qquad w_{lm} = \tfrac{i}{\sqrt{2}}\left(u_{lm}^{-} - u_{lm}^{+} \right), \tag{C.141}$$

$$u_{lm}^{-} = \tfrac{1}{\sqrt{2}}(v_{lm} - i w_{lm}), \qquad u_{lm}^{+} = \tfrac{1}{\sqrt{2}}(v_{lm} + i w_{lm}). \tag{C.142}$$

Equations (C.141)–(C.142) show that the spheroidal and toroidal parts of a tensor field $\mathbf{u}$ get mixed in the generalized spherical-harmonic representation (C.133). It is noteworthy—and reassuring—that $v_{00} = w_{00} = 0$ implies that $u_{00}^{-} = u_{00}^{+} = 0$, and vice versa. The real vector spherical-harmonic coefficients U_{lm}, V_{lm}, W_{lm} of a real vector field (B.168)–(B.171) are related to the complex coefficients u_{lm}, v_{lm}, w_{lm} by lower-to-upper-case equations analogous to (B.98). These equations together with (C.141)–(C.142) express the relation between U_{lm}, V_{lm}, W_{lm} and the generalized spherical-harmonic coefficients u_{lm}^{-}, u_{lm}^{0}, u_{lm}^{+}.

C.6 Gradient of a Tensor Field

In many continuum-mechanical applications in spherical polar coordinates, we are required to calculate the gradient of a three-dimensional tensor field,

$$\boldsymbol{\nabla}\mathbf{T}(\mathbf{r}) = \left[\hat{\mathbf{e}}_0\partial_r + r^{-1}\boldsymbol{\nabla}_1\right]\sum_{l=0}^{\infty}\sum_{m=-l}^{l} T_{lm}^{\alpha_1\cdots\alpha_n}(r)\mathbf{Y}_{lm}^N(\theta,\phi). \qquad \text{(C.143)}$$

The fundamental ingredient needed to evaluate (C.143) is the surface gradient of a generalized spherical harmonic,

$$\begin{aligned}\boldsymbol{\nabla}_1\mathbf{Y}_{lm}^N(\theta,\phi) &= \tfrac{1}{\sqrt{2}}\{\hat{\mathbf{e}}_-[\partial_\theta + i(\sin\theta)^{-1}\partial_\phi] \\ &\quad + \hat{\mathbf{e}}_+[-\partial_\theta + i(\sin\theta)^{-1}\partial_\phi]\}\left[X_{lm}^N(\theta)e^{im\phi}\hat{\mathbf{e}}_{\alpha_1}\cdots\hat{\mathbf{e}}_{\alpha_q}\right]. \qquad \text{(C.144)}\end{aligned}$$

It is easily demonstrated that $[\pm\partial_\theta + i(\sin\theta)^{-1}\partial_\phi]\hat{\mathbf{e}}_\alpha = \alpha\cot\theta\,\hat{\mathbf{e}}_\alpha - \sqrt{2}\,\hat{\mathbf{e}}_{\alpha\pm1}$, where $\hat{\mathbf{e}}_\alpha$ is defined to be zero for $|\alpha| > 1$. Using this result together with the recursion relation (C.120), it is a matter of some algebra to show that

$$\begin{aligned}\boldsymbol{\nabla}_1\mathbf{Y}_{lm}^N = \Omega_l^N Y_{lm}^{-1+N}\,&\hat{\mathbf{e}}_-\hat{\mathbf{e}}_{\alpha_1}\hat{\mathbf{e}}_{\alpha_2}\cdots\hat{\mathbf{e}}_{\alpha_q} \\ &- Y_{lm}^N\,\hat{\mathbf{e}}_-\hat{\mathbf{e}}_{(\alpha_1+1)}\hat{\mathbf{e}}_{\alpha_2}\cdots\hat{\mathbf{e}}_{\alpha_q} \\ &- Y_{lm}^N\,\hat{\mathbf{e}}_-\hat{\mathbf{e}}_{\alpha_1}\hat{\mathbf{e}}_{(\alpha_2+1)}\cdots\hat{\mathbf{e}}_{\alpha_q} \\ &\quad\vdots \\ &- Y_{lm}^N\,\hat{\mathbf{e}}_-\hat{\mathbf{e}}_{\alpha_1}\hat{\mathbf{e}}_{\alpha_2}\cdots\hat{\mathbf{e}}_{(\alpha_q+1)} \\ + \Omega_l^{-N}Y_{lm}^{1+N}\,&\hat{\mathbf{e}}_+\hat{\mathbf{e}}_{\alpha_1}\hat{\mathbf{e}}_{\alpha_2}\cdots\hat{\mathbf{e}}_{\alpha_q} \\ &- Y_{lm}^N\,\hat{\mathbf{e}}_+\hat{\mathbf{e}}_{(\alpha_1-1)}\hat{\mathbf{e}}_{\alpha_2}\cdots\hat{\mathbf{e}}_{\alpha_q} \\ &- Y_{lm}^N\,\hat{\mathbf{e}}_+\hat{\mathbf{e}}_{\alpha_1}\hat{\mathbf{e}}_{(\alpha_2-1)}\cdots\hat{\mathbf{e}}_{\alpha_q} \\ &\quad\vdots \\ &- Y_{lm}^N\,\hat{\mathbf{e}}_+\hat{\mathbf{e}}_{\alpha_1}\hat{\mathbf{e}}_{\alpha_2}\cdots\hat{\mathbf{e}}_{(\alpha_q-1)}, \qquad \text{(C.145)}\end{aligned}$$

where we have defined the coefficients

$$\Omega_l^{\pm N} = \sqrt{\tfrac{1}{2}(l\pm N)(l\mp N+1)}. \qquad \text{(C.146)}$$

The gradient (C.143) is accordingly

$$\begin{aligned}\boldsymbol{\nabla}\mathbf{T} = \sum_{l=0}^{\infty}\sum_{m=-l}^{l}\Bigg\{&\left[\frac{dT_{lm}^{\alpha_1\cdots\alpha_q}}{dr}\right]Y_{lm}^N\,\hat{\mathbf{e}}_0\hat{\mathbf{e}}_{\alpha_1}\hat{\mathbf{e}}_{\alpha_2}\cdots\hat{\mathbf{e}}_{\alpha_q} \\ &+ r^{-1}T_{lm}^{\alpha_1\cdots\alpha_q}\Big[\Omega_l^N Y_{lm}^{-1+N}\,\hat{\mathbf{e}}_-\hat{\mathbf{e}}_{\alpha_1}\hat{\mathbf{e}}_{\alpha_2}\cdots\hat{\mathbf{e}}_{\alpha_q}\end{aligned}$$

$$
\begin{aligned}
&- Y_{lm}^{N}\,\hat{\mathbf{e}}_{-}\hat{\mathbf{e}}_{(\alpha_1+1)}\hat{\mathbf{e}}_{\alpha_2}\cdots\hat{\mathbf{e}}_{\alpha_q}\\
&- Y_{lm}^{N}\,\hat{\mathbf{e}}_{-}\hat{\mathbf{e}}_{\alpha_1}\hat{\mathbf{e}}_{(\alpha_2+1)}\cdots\hat{\mathbf{e}}_{\alpha_q}\\
&\qquad\vdots\\
&- Y_{lm}^{N}\,\hat{\mathbf{e}}_{-}\hat{\mathbf{e}}_{\alpha_1}\hat{\mathbf{e}}_{\alpha_2}\cdots\hat{\mathbf{e}}_{(\alpha_q+1)}\\
+\,&\Omega_l^{-N} Y_{lm}^{1+N}\hat{\mathbf{e}}_{+}\hat{\mathbf{e}}_{\alpha_1}\hat{\mathbf{e}}_{\alpha_2}\cdots\hat{\mathbf{e}}_{\alpha_q}\\
&- Y_{lm}^{N}\,\hat{\mathbf{e}}_{+}\hat{\mathbf{e}}_{(\alpha_1-1)}\hat{\mathbf{e}}_{\alpha_2}\cdots\hat{\mathbf{e}}_{\alpha_q}\\
&- Y_{lm}^{N}\,\hat{\mathbf{e}}_{+}\hat{\mathbf{e}}_{\alpha_1}\hat{\mathbf{e}}_{(\alpha_2-1)}\cdots\hat{\mathbf{e}}_{\alpha_q}\\
&\qquad\vdots\\
&- Y_{lm}^{N}\,\hat{\mathbf{e}}_{+}\hat{\mathbf{e}}_{\alpha_1}\hat{\mathbf{e}}_{\alpha_2}\cdots\hat{\mathbf{e}}_{(\alpha_q-1)}\Big]\Big\}.
\end{aligned}
\tag{C.147}
$$

C.6.1 Contravariant derivative

Equation (C.147) can be rewritten in the less unwieldy form

$$
\begin{aligned}
\boldsymbol{\nabla}\mathbf{T} = \sum_{l=0}^{\infty}\sum_{m=-l}^{l}\Bigg\{&\left[\frac{dT_{lm}^{\alpha_1\cdots\alpha_q}}{dr}\right] Y_{lm}^{N}\,\hat{\mathbf{e}}_0\hat{\mathbf{e}}_{\alpha_1}\hat{\mathbf{e}}_{\alpha_2}\cdots\hat{\mathbf{e}}_{\alpha_q}\\
&+ r^{-1}\Big[\Omega_l^{N} T_{lm}^{\alpha_1\alpha_2\cdots\alpha_q} - T_{lm}^{(\alpha_1-1)\alpha_2\cdots\alpha_q} - T_{lm}^{\alpha_1(\alpha_2-1)\cdots\alpha_n}\\
&- \cdots - T_{lm}^{\alpha_1\alpha_2\cdots(\alpha_q-1)}\Big] Y_{lm}^{-1+N}\,\hat{\mathbf{e}}_{-}\hat{\mathbf{e}}_{\alpha_1}\hat{\mathbf{e}}_{\alpha_2}\cdots\hat{\mathbf{e}}_{\alpha_q}\\
&+ r^{-1}\Big[\Omega_l^{-N} T_{lm}^{\alpha_1\alpha_2\cdots\alpha_q} - T_{lm}^{(\alpha_1+1)\alpha_2\cdots\alpha_q} - T_{lm}^{\alpha_1(\alpha_2+1)\cdots\alpha_q}\\
&- \cdots - T_{lm}^{\alpha_1\alpha_2\cdots(\alpha_q+1)}\Big] Y_{lm}^{1+N}\,\hat{\mathbf{e}}_{+}\hat{\mathbf{e}}_{\alpha_1}\hat{\mathbf{e}}_{\alpha_2}\cdots\hat{\mathbf{e}}_{\alpha_q}\Bigg\},
\end{aligned}
\tag{C.148}
$$

where we stipulate that any coefficient $T_{lm}^{(\alpha_1\pm1)\alpha_2\cdots\alpha_q},\ldots,T_{lm}^{\alpha_1\alpha_2\cdots(\alpha_q\pm1)}$ for which $|\alpha_1\pm1|>1,\ldots,|\alpha_q\pm1|>1$ is zero. Suppose that we denote the contravariant components of the $(q+1)$th-order tensor $\boldsymbol{\nabla}\mathbf{T}$ by

$$
\begin{aligned}
\partial^{\sigma}T^{\alpha_1\cdots\alpha_q} &= (\hat{\mathbf{e}}^{\sigma}\hat{\mathbf{e}}^{\alpha_1}\cdots\hat{\mathbf{e}}^{\alpha_q})\,\vdots\,\boldsymbol{\nabla}\mathbf{T}\\
&= (\hat{\mathbf{e}}^{\sigma}\cdot\hat{\mathbf{e}}_i)(\hat{\mathbf{e}}^{\alpha_1}\cdot\hat{\mathbf{e}}_{j_1})\cdots(\hat{\mathbf{e}}^{\alpha_q}\cdot\hat{\mathbf{e}}_{j_q})\partial_i T_{j_1\cdots j_q}.
\end{aligned}
\tag{C.149}
$$

The generalized spherical-harmonic expansion of $\boldsymbol{\nabla}\mathbf{T}$ is then

$$
\begin{aligned}
\boldsymbol{\nabla}\mathbf{T} &= (\partial^{\sigma}T^{\alpha_1\cdots\alpha_q})\,\hat{\mathbf{e}}_{\sigma}\hat{\mathbf{e}}_{\alpha_1}\cdots\hat{\mathbf{e}}_{\alpha_e}\\
&= \sum_{l=0}^{\infty}\sum_{m=-l}^{l}(\partial^{\sigma}T_{lm}^{\alpha_1\cdots\alpha_q})\,Y_{lm}^{\sigma+N}\,\hat{\mathbf{e}}_{\sigma}\hat{\mathbf{e}}_{\alpha_1}\cdots\hat{\mathbf{e}}_{\alpha_e}
\end{aligned}
$$

$$= \sum_{l=0}^{\infty} \sum_{m=-l}^{l} (\partial^{\sigma} T_{lm}^{\alpha_1 \cdots \alpha_q}) \mathbf{Y}_{lm}^{\sigma+N}. \tag{C.150}$$

Equation (C.148) shows that the expansion coefficients of $\boldsymbol{\nabla}\mathbf{T}$ are related to those of $\mathbf{T}$ by

$$\partial^{-} T_{lm}^{\alpha_1 \cdots \alpha_q} = r^{-1} \Big[\Omega_l^N T_{lm}^{\alpha_1 \alpha_2 \cdots \alpha_q} - T_{lm}^{(\alpha_1 - 1)\alpha_2 \cdots \alpha_q} \\ - T_{lm}^{\alpha_1 (\alpha_2 - 1) \cdots \alpha_q} - \cdots - T_{lm}^{\alpha_1 \alpha_2 \cdots (\alpha_q - 1)} \Big], \tag{C.151}$$

$$\partial^{0} T_{lm}^{\alpha_1 \cdots \alpha_q} = \frac{d T_{lm}^{\alpha_1 \cdots \alpha_q}}{dr}, \tag{C.152}$$

$$\partial^{+} T_{lm}^{\alpha_1 \cdots \alpha_q} = r^{-1} \Big[\Omega_l^{-N} T_{lm}^{\alpha_1 \alpha_2 \cdots \alpha_q} - T_{lm}^{(\alpha_1 + 1)\alpha_2 \cdots \alpha_q} \\ - T_{lm}^{\alpha_1 (\alpha_2 + 1) \cdots \alpha_q} - \cdots - T_{lm}^{\alpha_1 \alpha_2 \cdots (\alpha_q + 1)} \Big]. \tag{C.153}$$

Formally, we may regard ∂^{σ} as a contravariant derivative operator defined by equations (C.151)–(C.153). This operator has the important property that it preserves the chain rule; that is, the invariant relation

$$\boldsymbol{\nabla}(\mathbf{TP}) = (\boldsymbol{\nabla}\mathbf{T})\mathbf{P} + \mathbf{T}(\boldsymbol{\nabla}\mathbf{P}) \tag{C.154}$$

implies that

$$\partial^{\sigma}(T^{\alpha_1 \cdots \alpha_q} P^{\beta_1 \cdots \beta_p}) \\ = (\partial^{\sigma} T^{\alpha_1 \cdots \alpha_q}) P^{\beta_1 \cdots \beta_p} + T^{\alpha_1 \cdots \alpha_q} (\partial^{\sigma} P^{\beta_1 \cdots \beta_p}). \tag{C.155}$$

It is noteworthy, however, that this relation does *not* in turn imply that $\partial^{\sigma}(T^{\alpha_1 \cdots \alpha_q} P^{\beta_1 \cdots \beta_p})_{lm} = (\partial^{\sigma} T_{lm}^{\alpha_1 \cdots \alpha_q}) P_{lm}^{\beta_1 \cdots \beta_p} + T_{lm}^{\alpha_1 \cdots \alpha_q} (\partial^{\sigma} P_{lm}^{\beta_1 \cdots \beta_p})$.

C.6.2 Special cases

Other derivative operators, notably the divergence and curl, and higher-order combinations of these derivatives, such as the Laplacian, can easily be written in terms of ∂^{σ} using the contraction and permutation properties of the metric and alternating tensors (C.72) and (C.76)–(C.77). In fact, invariant differential expressions involving single fields can be "translated" into generalized spherical-harmonic notation almost as easily as into Cartesian index notation. We illustrate this by determining the expansion coefficients for a number of common scalar, vector and second-order tensor derivatives, listed in Table C.1. The *gradient of a scalar field*

$$\psi = \sum_{l=0}^{\infty} \sum_{m=-l}^{l} \psi_{lm} Y_{lm}^{0} \qquad \text{(C.156)}$$

is

$$\boldsymbol{\nabla}\psi = \sum_{l=0}^{\infty} \sum_{m=-l}^{l} (\partial^{\alpha}\psi_{lm})\, Y_{lm}^{\alpha}\, \hat{\mathbf{e}}_{\alpha} \qquad \text{(C.157)}$$

Invariant Expression	Cartesian Index Notation	Generalized Spherical-Harmonic Expansion Coefficients
$\boldsymbol{\nabla}\psi$	$\partial_i\psi$	$\partial^{\alpha}\psi_{lm}$
$\boldsymbol{\nabla}\mathbf{u}$	$\partial_i u_j$	$\partial^{\alpha}u_{lm}^{\beta}$
$\frac{1}{2}[\boldsymbol{\nabla}\mathbf{u} + (\boldsymbol{\nabla}\mathbf{u})^{\mathrm{T}}]$	$\frac{1}{2}(\partial_i u_j + \partial_j u_i)$	$\frac{1}{2}(\partial^{\alpha}u_{lm}^{\beta} + \partial^{\beta}u_{lm}^{\alpha})$
$\boldsymbol{\nabla}\cdot\mathbf{u}$	$\partial_i u_i$	$g_{\alpha\beta}\,\partial^{\alpha}u_{lm}^{\beta}$
$\boldsymbol{\nabla}\times\mathbf{u}$	$\varepsilon_{ijk}\partial_j u_k$	$g^{\alpha\eta}\varepsilon_{\eta\beta\gamma}\,\partial^{\beta}u_{lm}^{\gamma}$
$\boldsymbol{\nabla}\cdot\mathbf{T}$	$\partial_i T_{ij}$	$g_{\sigma\alpha}\,\partial^{\sigma}T_{lm}^{\alpha\beta}$
$\boldsymbol{\nabla}^2\psi$	$\partial_i^2\psi$	$g_{\alpha\beta}\,\partial^{\alpha}\partial^{\beta}\psi_{lm}$
$\boldsymbol{\nabla}^2\mathbf{u}$	$\partial_i^2 u_j$	$g_{\sigma\alpha}\partial^{\sigma}\partial^{\alpha}u_{lm}^{\beta}$
$\boldsymbol{\nabla}(\boldsymbol{\nabla}\cdot\mathbf{u})$	$\partial_i\partial_j u_j$	$g_{\alpha\beta}\,\partial^{\sigma}\partial^{\alpha}u_{lm}^{\beta}$
$\boldsymbol{\nabla}\times(\boldsymbol{\nabla}\times\mathbf{u})$	$\varepsilon_{ijk}\varepsilon_{kpq}\partial_j\partial_p u_q$	$g^{\alpha\eta}g^{\gamma\sigma}\varepsilon_{\eta\beta\gamma}\,\varepsilon_{\sigma\mu\nu}\,\partial^{\beta}\partial^{\mu}u_{lm}^{\nu}$

Table C.1. Generalized spherical-harmonic representation of the gradient and related derivatives of a scalar field ψ, a vector field $\mathbf{u}$ and a second-order tensor field $\mathbf{T}$. The expansion coefficients of all except the last two entries are specified in greater detail in the text. More complicated combinations of derivatives can be "translated" from invariant or Cartesian index notation to generalized spherical-harmonic notation in a similar manner. Derivatives such as $\partial^{\sigma}g_{\alpha\beta}$ and $\partial^{\sigma}\varepsilon_{\alpha\beta\gamma}$ are zero since both the metric and alternating tensors are constant.

where

$$\partial^0\psi_{lm} = \frac{d\psi_{lm}}{dr}, \qquad \partial^\pm\psi_{lm} = \Omega_l^0 r^{-1}\psi_{lm}, \tag{C.158}$$

and the *gradient of a vector field*

$$\mathbf{u} = \sum_{l=0}^{\infty}\sum_{m=-l}^{l} u_{lm}^\beta Y_{lm}^\beta\, \hat{\mathbf{e}}_\beta \tag{C.159}$$

is

$$\boldsymbol{\nabla}\mathbf{u} = \sum_{l=0}^{\infty}\sum_{m=-l}^{l} (\partial^\alpha u_{lm}^\beta)\, Y_{lm}^{\alpha+\beta}\, \hat{\mathbf{e}}_\alpha\hat{\mathbf{e}}_\beta \tag{C.160}$$

where

$$\partial^0 u_{lm}^0 = \frac{du_{lm}^0}{dr}, \qquad \partial^0 u_{lm}^\pm = \frac{du_{lm}^\pm}{dr}, \tag{C.161}$$

$$\partial^\pm u_{lm}^0 = r^{-1}(\Omega_l^0 u_{lm}^0 - u_{lm}^\pm), \qquad \partial^\pm u_{lm}^\pm = \Omega_l^2 r^{-1} u_{lm}^\pm, \tag{C.162}$$

$$\partial^\pm u_{lm}^\mp = r^{-1}(\Omega_l^0 u_{lm}^\mp - u_{lm}^0). \tag{C.163}$$

In reducing (C.161)–(C.163) we have made use of the identity $\Omega_l^N = \Omega_l^{-N+1}$. The *symmetric gradient* $\boldsymbol{\varepsilon} = \frac{1}{2}[\boldsymbol{\nabla}\mathbf{u} + (\boldsymbol{\nabla}\mathbf{u})^{\mathrm{T}}]$ is given by

$$\boldsymbol{\varepsilon} = \sum_{l=0}^{\infty}\sum_{m=-l}^{l} \varepsilon_{lm}^{\alpha\beta}\, Y_{lm}^{\alpha+\beta}\, \hat{\mathbf{e}}_\alpha\hat{\mathbf{e}}_\beta, \tag{C.164}$$

where $\varepsilon_{lm}^{\alpha\beta} = \frac{1}{2}(\partial^\alpha u_{lm}^\beta + \partial^\beta u_{lm}^\alpha) = \varepsilon_{lm}^{\beta\alpha}$ or, equivalently,

$$\varepsilon_{lm}^{00} = \frac{du_{lm}^0}{dr}, \qquad \varepsilon_{lm}^{\pm\pm} = \Omega_l^2 r^{-1} u_{lm}^\pm, \tag{C.165}$$

$$\varepsilon_{lm}^{0\pm} = \varepsilon_{lm}^{\pm 0} = \tfrac{1}{2}\left(\frac{du_{lm}^\pm}{dr} - \frac{u_{lm}^\pm}{r} + \frac{\Omega_l^0 u_{lm}^0}{r}\right), \tag{C.166}$$

$$\varepsilon_{lm}^{\pm\mp} = \tfrac{1}{2}\Omega_l^0 r^{-1}(u_{lm}^- + u_{lm}^+) - r^{-1}u_{lm}^0. \tag{C.167}$$

The *divergence* and *curl of a vector field* are given by

$$\begin{aligned}\boldsymbol{\nabla}\cdot\mathbf{u} &= \sum_{l=0}^{\infty}\sum_{m=-l}^{l} (g_{\alpha\beta}\partial^\alpha u^\beta)\, Y_{lm}^0 \\ &= \sum_{l=0}^{\infty}\sum_{m=-l}^{l} \left[\frac{du_{lm}^0}{dr} + \frac{2u_{lm}^0}{r} - \frac{\Omega_l^0(u_{lm}^- + u_{lm}^+)}{r}\right] Y_{lm}^0\end{aligned} \tag{C.168}$$

and

$$\mathbf{w} = \boldsymbol{\nabla} \times \mathbf{u} = \sum_{l=0}^{\infty} \sum_{m=-l}^{l} w_{lm}^{\alpha}\, Y_{lm}^{\alpha}\, \hat{\mathbf{e}}_{\alpha} \tag{C.169}$$

where

$$w_{lm}^{0} = i\Omega_l^0 r^{-1}(u_{lm}^{-} - u_{lm}^{+}), \tag{C.170}$$

$$w_{lm}^{\pm} = \mp i \left(\frac{du_{lm}^{\pm}}{dr} + \frac{u_{lm}^{\pm}}{r} - \frac{\Omega_l^0 u_{lm}^{0}}{r} \right). \tag{C.171}$$

The *divergence of a second-order tensor field*

$$\mathbf{T} = \sum_{l=0}^{\infty} \sum_{m=-l}^{l} T_{lm}^{\alpha\beta}\, Y_{lm}^{\alpha+\beta}\, \hat{\mathbf{e}}_{\alpha}\hat{\mathbf{e}}_{\beta} \tag{C.172}$$

is a vector field

$$\mathbf{w} = \boldsymbol{\nabla} \cdot \mathbf{T} = \sum_{l=0}^{\infty} \sum_{m=-l}^{l} w_{lm}^{\beta}\, Y_{lm}^{\beta}\, \hat{\mathbf{e}}^{\beta} \tag{C.173}$$

with coefficients

$$w_{lm}^{0} = \frac{dT_{lm}^{00}}{dr} + \frac{T_{lm}^{-+} + 2T_{lm}^{00} + T_{lm}^{+-}}{r} - \frac{\Omega_l^0 (T_{lm}^{-0} + T_{lm}^{+0})}{r}, \tag{C.174}$$

$$w_{lm}^{\pm} = \frac{dT_{lm}^{0\pm}}{dr} + \frac{2T_{lm}^{0\pm} + T_{lm}^{\pm 0}}{r} - \frac{\Omega_l^0 T_{lm}^{\mp\pm} + \Omega_l^2 T_{lm}^{\pm\pm}}{r}. \tag{C.175}$$

If $\mathbf{T}$ is *symmetric*, so that $T_{lm}^{\alpha\beta} = T_{lm}^{\beta\alpha}$, then equations (C.174)–(C.175) reduce to

$$w_{lm}^{0} = \frac{dT_{lm}^{00}}{dr} + \frac{2(T_{lm}^{00} + T_{lm}^{-+})}{r} - \frac{\Omega_l^0 (T_{lm}^{0-} + T_{lm}^{0+})}{r}, \tag{C.176}$$

$$w_{lm}^{\pm} = \frac{dT_{lm}^{0\pm}}{dr} + \frac{3T_{lm}^{0\pm}}{r} - \frac{\Omega_l^0 T_{lm}^{-+} + \Omega_l^2 T_{lm}^{\pm\pm}}{r}. \tag{C.177}$$

Upon setting $\mathbf{u} = \boldsymbol{\nabla}\psi$ in equation (C.168) we obtain the *Laplacian of a scalar field*:

$$\nabla^2 \psi = \sum_{l=0}^{\infty} \sum_{m=-l}^{l} \left[\frac{d^2\psi_{lm}}{dr^2} + \frac{2}{r}\frac{d\psi_{lm}}{dr} - \frac{l(l+1)\psi_{lm}}{r^2} \right] Y_{lm}^{0}. \tag{C.178}$$

Finally, setting $T_{lm}^{\alpha\beta} = \partial^{\alpha} u_{lm}^{\beta}$ in equations (C.174)–(C.175) yields the *Laplacian of a vector field*:

$$\mathbf{w} = \nabla^2 \mathbf{u} = \sum_{l=0}^{\infty} \sum_{m=-l}^{l} w_{lm}^{\beta}\, Y_{lm}^{\beta}\, \hat{\mathbf{e}}_{\beta} \tag{C.179}$$

where

$$\begin{aligned} w_{lm}^{0} &= \frac{d^2 u_{lm}^{0}}{dr^2} + \frac{2}{r}\frac{du_{lm}^{0}}{dr} - \frac{[l(l+1)+2]u_{lm}^{0}}{r^2} \\ &\quad + \frac{2\Omega_l^0(u_{lm}^{-} + u_{lm}^{+})}{r^2}, \end{aligned} \tag{C.180}$$

$$w_{lm}^{\pm} = \frac{d^2 u_{lm}^{\pm}}{dr^2} + \frac{2}{r}\frac{du_{lm}^{\pm}}{dr} - \frac{l(l+1)u_{lm}^{\pm}}{r^2} + \frac{2\Omega_l^0 u_{lm}^{0}}{r^2}. \tag{C.181}$$

In reducing the final two expressions we have used the elementary identities $(\Omega_l^0)^2 + 1 = \frac{1}{2}[l(l+1)+2]$ and $(\Omega_l^0)^2 + (\Omega_l^2)^2 = l(l+1)$. The divergence, curl and Laplacian of a vector field can be rewritten in terms of the conventional vector spherical-harmonic expansion coefficients U_{lm}, V_{lm}, W_{lm} rather than u_{lm}^{-}, u_{lm}^{0}, u_{lm}^{+} using the relations (B.98) and (C.141)–(C.142). Upon effecting this transformation we find that equations (C.168)–(C.171) and (C.179)–(C.181) are equivalent to (B.179)–(B.180) and (B.183), as expected.

C.7 Tensor Products

We consider next the generalized spherical-harmonic representation of the product of a tensor field of order q with one of order p. This topic is closely related to the analysis of spin-spin, spin-orbital and orbital-orbital coupling in quantum mechanics; much of what follows is a transcription of the theory governing the addition of two angular-momentum operators $\mathbf{J}_1$ and $\mathbf{J}_2$ (Edmonds 1960) into language that is expunged of excessive quantum-mechanical jargon.

C.7.1 Product of two generalized spherical harmonics

We begin by considering a pair of generalized surface spherical harmonics $\mathbf{Y}_{l_1m_1}^{N_1} = Y_{l_1m_1}^{N_1}\hat{\mathbf{e}}_{\alpha_1}\cdots\hat{\mathbf{e}}_{\alpha_q}$ and $\mathbf{Y}_{l_2m_2}^{N_2} = Y_{l_2m_2}^{N_2}\hat{\mathbf{e}}_{\beta_1}\cdots\hat{\mathbf{e}}_{\beta_p}$, of orders q and p, respectively, with associated upper-register indices $N_1 = \alpha_1 + \cdots + \alpha_q$ and $N_2 = \beta_1 + \cdots + \beta_p$. The product of these two quantities,

$$\mathbf{Y}_{l_1m_1}^{N_1}\mathbf{Y}_{l_2m_2}^{N_2} = Y_{l_1m_1}^{N_1}Y_{l_2m_2}^{N_2}\hat{\mathbf{e}}_{\alpha_1}\cdots\hat{\mathbf{e}}_{\alpha_q}\hat{\mathbf{e}}_{\beta_1}\cdots\hat{\mathbf{e}}_{\beta_p}, \tag{C.182}$$

is a tensor field of order $q+p$ upon the unit sphere Ω. Like any such field, it can be written as a linear combination of generalized spherical harmonics $\mathbf{Y}_{lm}^{N} = Y_{lm}^{N}\hat{\mathbf{e}}_{\alpha_1}\cdots\hat{\mathbf{e}}_{\alpha_q}\hat{\mathbf{e}}_{\beta_1}\cdots\hat{\mathbf{e}}_{\beta_p}$, where $N = N_1 + N_2$, in the form

$$\mathbf{Y}_{l_1m_1}^{N_1}\mathbf{Y}_{l_2m_2}^{N_2} = \sum_{l=0}^{\infty}\sum_{m=-l}^{l}\langle\mathbf{Y}_{lm}^{N}, \mathbf{Y}_{l_1m_1}^{N_1}\mathbf{Y}_{l_2m_2}^{N_2}\rangle\,\mathbf{Y}_{lm}^{N}. \tag{C.183}$$

The orthonormality relation (C.95) has been used to determine the scalar expansion coefficients

$$\langle\mathbf{Y}_{lm}^{N}, \mathbf{Y}_{l_1m_1}^{N_1}\mathbf{Y}_{l_2m_2}^{N_2}\rangle = \int_{\Omega}\mathbf{Y}_{lm}^{N*}\,\vdots\,\mathbf{Y}_{l_1m_1}^{N_1}\mathbf{Y}_{l_2m_2}^{N_2}\,d\Omega. \tag{C.184}$$

We have employed the usual infinite summation limits in equation (C.183); however, it is evident that the interior sum consists of a single term with order $m = m_1 + m_2$. In fact, the exterior sum over the degrees l is finite as well. To see this and determine the actual limits, we note that for a fixed l_1, N_1 and l_2, N_2 there are $(2l_1+1)(2l_2+1)$ possible products $\mathbf{Y}_{l_1m_1}^{N_1}\mathbf{Y}_{l_2m_2}^{N_2}$, each of which can be expanded in the form (C.183). The only harmonics $\mathbf{Y}_{lm}^{N}$ that can appear in these expansions must have $-l \leq m \leq l$, where $m = m_1 + m_2$ for some $-l_1 \leq m_1 \leq l_1$ and $-l_2 \leq m_2 \leq l_2$. This restricts the degree l to lie between the positive difference and the sum of l_1 and l_2, i.e., $|l_1 - l_2| \leq l \leq l_1 + l_2$. The total number of terms needed to represent all of the variable-order products $\mathbf{Y}_{l_1m_1}^{N_1}\mathbf{Y}_{l_2m_2}^{N_2}$ is the same as the number of products, since

$$\sum_{l=|l_1-l_2|}^{l_1+l_2}\sum_{m=-l}^{l}(2l+1) = (2l_1+1)(2l_2+1). \tag{C.185}$$

Because the sum over m collapses, the number of terms in any particular expansion of the form (C.183) is only $2\min\{l_1, l_2\} + 1$.

In summary, the expansion coefficients in equation (C.183) satisfy the following *selection rules*:

$$\langle\mathbf{Y}_{lm}^{N}, \mathbf{Y}_{l_1m_1}^{N_1}\mathbf{Y}_{l_2m_2}^{N_2}\rangle = 0 \quad \text{unless} \quad \begin{cases} N = N_1 + N_2 \\ m = m_1 + m_2 \\ |l_1 - l_2| \leq l \leq l_1 + l_2. \end{cases} \tag{C.186}$$

The final restriction in (C.186) is referred to as the *triangle condition*, since it expresses the geometrical requirement that the integers l_1, l_2 and l must form the sides of a closed triangle. In certain cases coefficients that are consistent with (C.186) may also vanish, as we shall see. However, we can *always* replace equation (C.183) by

$$\mathbf{Y}_{l_1m_1}^{N_1}\mathbf{Y}_{l_2m_2}^{N_2} = \sum_{l=|l_1-l_2|}^{l_1+l_2} \sum_{m=-l}^{l} \langle \mathbf{Y}_{lm}^{N}, \mathbf{Y}_{l_1m_1}^{N_1}\mathbf{Y}_{l_2m_2}^{N_2} \rangle \, \mathbf{Y}_{lm}^{N}. \tag{C.187}$$

We shall henceforth suppress the summation limits in this section to avoid clutter; in all cases the limits are dictated by the selection rules (C.186). In many instances the sums over m or N are degenerate, as in equation (C.187).

We can rewrite (C.184) as a surface integral involving the three generalized scalar spherical harmonics:

$$\langle \mathbf{Y}_{lm}^{N}, \mathbf{Y}_{l_1m_1}^{N_1}\mathbf{Y}_{l_2m_2}^{N_2} \rangle = \int_{\Omega} Y_{lm}^{N*} Y_{l_1m_1}^{N_1} Y_{l_2m_2}^{N_2} \, d\Omega. \tag{C.188}$$

Upon substituting $Y_{lm}^{N}(\theta,\phi) = \sqrt{(2l+1)/4\pi}\, P_{lm}^{N}(\cos\theta)e^{im\phi}$ we can reduce this to an integral over the three associated generalized Legendre functions:

$$\langle \mathbf{Y}_{lm}^{N}, \mathbf{Y}_{l_1m_1}^{N_1}\mathbf{Y}_{l_2m_2}^{N_2} \rangle = \frac{1}{2}\left[\frac{(2l+1)(2l_1+1)(2l_2+1)}{4\pi}\right]^{1/2}$$
$$\times (-1)^{m+N} \int_{-1}^{1} P_{l\,-m}^{-N} P_{l_1m_1}^{N_1} P_{l_2m_2}^{N_2} \, d\mu. \tag{C.189}$$

In obtaining these results we have taken it for granted that $N = N_1 + N_2$ and $m = m_1 + m_2$. Equation (C.189) shows that all of the expansion coefficients are real:

$$\langle \mathbf{Y}_{lm}^{N}, \mathbf{Y}_{l_1m_1}^{N_1}\mathbf{Y}_{l_2m_2}^{N_2} \rangle^* = \langle \mathbf{Y}_{lm}^{N}, \mathbf{Y}_{l_1m_1}^{N_1}\mathbf{Y}_{l_2m_2}^{N_2} \rangle. \tag{C.190}$$

In addition, the symmetry $P_{lN}^{m} = (-1)^{m+N} P_{lm}^{N}$ enables us to deduce that

$$\langle \mathbf{Y}_{lN}^{m}, \mathbf{Y}_{l_1N_1}^{m_1}\mathbf{Y}_{l_2N_2}^{m_2} \rangle = \langle \mathbf{Y}_{lm}^{N}, \mathbf{Y}_{l_1m_1}^{N_1}\mathbf{Y}_{l_2m_2}^{N_2} \rangle. \tag{C.191}$$

Since the ladder operators $J_\pm$ are each other's adjoints we may write

$$\langle J_\pm \mathbf{Y}_{lm}^{N}, \mathbf{Y}_{l_1m_1}^{N_1}\mathbf{Y}_{l_2m_2}^{N_2} \rangle = \langle \mathbf{Y}_{lm}^{N}, J_\mp(\mathbf{Y}_{l_1m_1}^{N_1})\mathbf{Y}_{l_2m_2}^{N_2} \rangle$$
$$+ \langle \mathbf{Y}_{lm}^{N}, \mathbf{Y}_{l_1m_1}^{N_1} J_\mp(\mathbf{Y}_{l_2m_2}^{N_2}) \rangle. \tag{C.192}$$

Upon making use of the identity (C.104) we obtain the *recursion relation*

$$\sqrt{(l\mp m)(l\pm m+1)}\, \langle \mathbf{Y}_{l\,m\pm1}^{N}, \mathbf{Y}_{l_1m_1}^{N_1}\mathbf{Y}_{l_2m_2}^{N_2} \rangle$$
$$= \sqrt{(l_1\pm m_1)(l_1\mp m_1+1)}\, \langle \mathbf{Y}_{lm}^{N}, \mathbf{Y}_{l_1\,m_1\mp1}^{N_1}\mathbf{Y}_{l_2m_2}^{N_2} \rangle$$
$$+ \sqrt{(l_2\pm m_2)(l_2\mp m_2+1)}\, \langle \mathbf{Y}_{lm}^{N}, \mathbf{Y}_{l_1m_1}^{N_1}\mathbf{Y}_{l_2\,m_2\mp1}^{N_2} \rangle. \tag{C.193}$$

A similar relation with the indices N_1, N_2, N and m_1, m_2, m interchanged follows from the symmetry (C.191).

C.7.2 Product of two arbitrary tensors

Suppose now that $\mathbf{T}$ is an arbitrary tensor field of order q and that $\mathbf{P}$ is an arbitrary tensor field of order p:

$$\mathbf{T} = \sum_{l_1 m_1} T_{l_1 m_1}^{\alpha_1 \cdots \alpha_q} \mathbf{Y}_{l_1 m_1}^{N_1}, \qquad \mathbf{P} = \sum_{l_2 m_2} P_{l_2 m_2}^{\beta_1 \cdots \beta_p} \mathbf{Y}_{l_2 m_2}^{N_2}. \tag{C.194}$$

Their product $\mathbf{B} = \mathbf{TP}$ is a tensor field of order $q + p$, given by

$$\begin{aligned} \mathbf{B} &= \sum_{l_1 m_1} \sum_{l_2 m_2} T_{l_1 m_1}^{\alpha_1 \cdots \alpha_q} P_{l_2 m_2}^{\beta_1 \cdots \beta_p} \mathbf{Y}_{l_1 m_1}^{N_1} \mathbf{Y}_{l_2 m_2}^{N_2} \\ &= \sum_{lm} B_{lm}^{\alpha_1 \cdots \alpha_q \beta_1 \cdots \beta_p} \mathbf{Y}_{lm}^{N}, \end{aligned} \tag{C.195}$$

where

$$B_{lm}^{\alpha_1 \cdots \alpha_q \beta_1 \cdots \beta_p} = \sum_{l_1 m_1} \sum_{l_2 m_2} \langle \mathbf{Y}_{lm}^{N}, \mathbf{Y}_{l_1 m_1}^{N_1} \mathbf{Y}_{l_2 m_2}^{N_2} \rangle \, T_{l_1 m_1}^{\alpha_1 \cdots \alpha_q} P_{l_2 m_2}^{\beta_1 \cdots \beta_p}. \tag{C.196}$$

We see from (C.195)–(C.196) that we may regard $\langle \mathbf{Y}_{lm}^{N}, \mathbf{Y}_{l_1 m_1}^{N_1} \mathbf{Y}_{l_2 m_2}^{N_2} \rangle$ as the elements of a transformation "matrix" which relates the expansion coefficients $(TP)_{lm}^{\alpha_1 \cdots \alpha_q \beta_1 \cdots \beta_p}$ of the product $\mathbf{TP}$ to the product $T_{l_1 m_1}^{\alpha_1 \cdots \alpha_q} P_{l_2 m_2}^{\beta_1 \cdots \beta_p}$ of the coefficients of $\mathbf{T}$ and $\mathbf{P}$.

C.7.3 Wigner 3-j symbols

Motivated by the symmetry relation (C.191), we express the dependence of the transformation-matrix element $\langle \mathbf{Y}_{lm}^{N}, \mathbf{Y}_{l_1 m_1}^{N_1} \mathbf{Y}_{l_2 m_2}^{N_2} \rangle$ upon the three upper indices N_1, N_2, N and the three orders m_1, m_2, m as follows:

$$\begin{aligned} \langle \mathbf{Y}_{lm}^{N}, \mathbf{Y}_{l_1 m_1}^{N_1} \mathbf{Y}_{l_2 m_2}^{N_2} \rangle &= (-1)^{N+m} \left[\frac{(2l+1)(2l_1+1)(2l_2+1)}{4\pi} \right]^{1/2} \\ &\quad \times \begin{pmatrix} l & l_1 & l_2 \\ -N & N_1 & N_2 \end{pmatrix} \begin{pmatrix} l & l_1 & l_2 \\ -m & m_1 & m_2 \end{pmatrix}. \end{aligned} \tag{C.197}$$

The notation in (C.197) is that conventionally employed in quantum mechanics; the arrays, which are known as *Wigner 3-j symbols*, are real numbers specified by the enclosed indices. The top row of every 3-j symbol must be consistent with the triangle condition, whereas the bottom row must sum to zero, by virtue of the selection rules (C.186). Equation (C.188) allows us to rewrite (C.197) in terms of the associated scalar harmonics:

$$\int_{\Omega} Y_{lm}^{N*} Y_{l_1 m_1}^{N_1} Y_{l_2 m_2}^{N_2} \, d\Omega$$

$$= (-1)^{N+m}\left[\frac{(2l+1)(2l_1+1)(2l_2+1)}{4\pi}\right]^{1/2}$$
$$\times\begin{pmatrix} l & l_1 & l_2 \\ -N & N_1 & N_2 \end{pmatrix}\begin{pmatrix} l & l_1 & l_2 \\ -m & m_1 & m_2 \end{pmatrix}. \qquad \text{(C.198)}$$

Upon invoking the orthonormality (C.95) we obtain the equivalent relation

$$Y_{l_1m_1}^{N_1} Y_{l_2m_2}^{N_2} = \sum_l \sum_N \sum_m (-1)^{N+m}\left[\frac{(2l+1)(2l_1+1)(2l_2+1)}{4\pi}\right]^{1/2}$$
$$\times\begin{pmatrix} l & l_1 & l_2 \\ -N & N_1 & N_2 \end{pmatrix}\begin{pmatrix} l & l_1 & l_2 \\ -m & m_1 & m_2 \end{pmatrix} Y_{lm}^N. \qquad \text{(C.199)}$$

The analogous results for the generalized Legendre functions (C.112) are

$$\int_{-1}^{1} P_{lm}^N P_{l_1m_1}^{N_1} P_{l_2m_2}^{N_2}\, d\mu = 2\begin{pmatrix} l & l_1 & l_2 \\ N & N_1 & N_2 \end{pmatrix}$$
$$\times\begin{pmatrix} l & l_1 & l_2 \\ m & m_1 & m_2 \end{pmatrix} \qquad \text{(C.200)}$$

and

$$P_{l_1m_1}^{N_1} P_{l_2m_2}^{N_2} = \sum_l \sum_N \sum_m (2l+1)\begin{pmatrix} l & l_1 & l_2 \\ N & N_1 & N_2 \end{pmatrix}$$
$$\times\begin{pmatrix} l & l_1 & l_2 \\ m & m_1 & m_2 \end{pmatrix} P_{lm}^N. \qquad \text{(C.201)}$$

The inverse relationship expressing P_{lm}^N in terms of the product $P_{l_1m_1}^{N_1} P_{l_2m_2}^{N_2}$ is readily shown to be

$$(2l+1)^{-1}\delta_{ll'} P_{lm}^N = \sum_{N_1N_2} \sum_{m_1m_2} \begin{pmatrix} l & l_1 & l_2 \\ N & N_1 & N_2 \end{pmatrix}$$
$$\times\begin{pmatrix} l' & l_1 & l_2 \\ m & m_1 & m_2 \end{pmatrix} P_{l_1m_1}^{N_1} P_{l_2m_2}^{N_2}. \qquad \text{(C.202)}$$

Upon setting the argument $\mu = 1$ in equations (C.201)–(C.202) and making use of (C.114) we obtain the 3-j *orthonormality relations*

$$\sum_l \sum_m (2l+1)\begin{pmatrix} l & l_1 & l_2 \\ m & N_1 & N_2 \end{pmatrix}\begin{pmatrix} l & l_1 & l_2 \\ m & m_1 & m_2 \end{pmatrix}$$
$$= \delta_{N_1m_1}\delta_{N_2m_2}, \qquad \text{(C.203)}$$

$$\sum_{m_1m_2} \begin{pmatrix} l & l_1 & l_2 \\ N & m_1 & m_2 \end{pmatrix}\begin{pmatrix} l' & l_1 & l_2 \\ m & m_1 & m_2 \end{pmatrix}$$
$$= (2l+1)^{-1}\delta_{ll'}\delta_{Nm}. \qquad \text{(C.204)}$$

We have changed the sign of N and m in equations (C.200)–(C.202) in order to achieve a more symmetrical notation.

A scheme for calculating the 3-j symbols can be based upon the bottom-row recursion relation

$$
\begin{aligned}
-\sqrt{(l\mp m)(l\pm m+1)}&\begin{pmatrix} l & l_1 & l_2 \\ -m\mp 1 & m_1 & m_2 \end{pmatrix}\\
&=\sqrt{(l_1\pm m_1)(l_1\mp m_1+1)}\begin{pmatrix} l & l_1 & l_2 \\ -m & m_1\mp 1 & m_2 \end{pmatrix}\\
&+\sqrt{(l_2\pm m_2)(l_2\mp m_2+1)}\begin{pmatrix} l & l_1 & l_2 \\ -m & m_1 & m_2\mp 1 \end{pmatrix}, \qquad \text{(C.205)}
\end{aligned}
$$

which follows from (C.193). Setting $m=-l$ and $m_1=-l_1$ we obtain

$$
\begin{aligned}
&\begin{pmatrix} l & l_1 & l_2 \\ l & -l_1+1 & -l+l_1-1 \end{pmatrix} \qquad \text{(C.206)}\\
&\quad=\left[\frac{(-l+l_1+l_2)(l-l_1+l_2+1)}{2l_1}\right]^{1/2}\begin{pmatrix} l & l_1 & l_2 \\ l & -l_1 & -l+l_1 \end{pmatrix}.
\end{aligned}
$$

Successive application of this result l_1+m_1 times yields

$$
\begin{aligned}
\begin{pmatrix} l & l_1 & l_2 \\ l & m_1 & -l-m_1 \end{pmatrix}&=(-1)^{l_1+m_1}\\
&\times\left[\frac{(-l+l_1+l_2+1)!(l+l_2+m_1)!(l_1-m_1)!}{(2l_1)!(l-l_1+l_2)!(-l+l_2-m_1)!(l_1+m_1)!}\right]^{1/2}\\
&\times\begin{pmatrix} l & l_1 & l_2 \\ l & -l_1 & -l+l_1 \end{pmatrix}. \qquad \text{(C.207)}
\end{aligned}
$$

The orthonormality relation (C.204) stipulates that

$$
\sum_{m_1}\begin{pmatrix} l & l_1 & l_2 \\ l & -m_1 & -l+m_1 \end{pmatrix}^2=\frac{1}{2l+1}. \qquad \text{(C.208)}
$$

Upon combining equations (C.207) and (C.208) with the factorial identity

$$
\begin{aligned}
&\sum_{m_1}\frac{(l+l_2+m_1)!(l_1-m_1)!}{(-l+l_2-m_1)!(l_1+m_1)!}\\
&\quad=\frac{(l+l_1+l_2+1)!(l-l_1+l_2)!(l+l_1-l_2)!}{(2l+1)!(-l+l_1+l_2)!} \qquad \text{(C.209)}
\end{aligned}
$$

we find that

$$
\begin{pmatrix} l & l_1 & l_2 \\ l & -l_1 & -l+l_1 \end{pmatrix}^2=\frac{(2l)!(2l_1)!}{(l+l_1+l_2+1)!(l+l_1-l_2)!}. \qquad \text{(C.210)}
$$

The sign of the 3-j symbol (C.210) may be chosen arbitrarily. The conventional quantum-mechanical choice is that made by Condon & Shortley (1935) and Edmonds (1960):

$$\begin{pmatrix} l & l_1 & l_2 \\ l & -l_1 & -l+l_1 \end{pmatrix} = \sqrt{\frac{(2l)!(2l_1)!}{(l+l_1+l_2+1)!(l+l_1-l_2)!}}. \tag{C.211}$$

We could instead have set $m = l$ and $m_1 = l_1$ in equation (C.205); a different choice of sign would have been required in that case:

$$\begin{pmatrix} l & l_1 & l_2 \\ -l & l_1 & l-l_1 \end{pmatrix} = (-1)^{l+l_1+l_2}\sqrt{\frac{(2l)!(2l_1)!}{(l+l_1+l_2+1)!(l+l_1-l_2)!}}. \tag{C.212}$$

The remaining 3-j symbols can be calculated by using either of the two equivalent results (C.211)–(C.212) to seed the recursion relation (C.205). Since the starting symbols depend only upon l, l_1 and l_2 the sign convention has no effect upon the expansion coefficients (C.197).

The final result can be written in a number of forms, the most symmetrical of which is (Racah 1942; Edmonds 1960)

$$\begin{aligned} \begin{pmatrix} l & l_1 & l_2 \\ m & m_1 & m_2 \end{pmatrix} &= (-1)^{l_1+l_2+m} \\ &\times \frac{\sqrt{(l+l_1-l_2)!(l-l_1+l_2)!(-l+l_1+l_2)!}}{\sqrt{(l+l_1+l_2+1)!}} \\ &\times \sqrt{(l+m)!(l-m)!(l_1+m_1)!(l_1-m_1)!(l_2+m_2)!(l_2-m_2)!} \\ &\times \sum_k \frac{(-1)^k}{D_k}, \end{aligned} \tag{C.213}$$

where

$$\begin{aligned} D_k = k!(-l+l_1+l_2-k)!(l_1-m_1-k)!(l_2+m_2-k)! \\ (l-l_1-m_2+k)!(l-l_2+m_1+k)!. \end{aligned} \tag{C.214}$$

Wigner (1959) gives an alternative, less symmetrical, expression derived using group-theoretical methods:

$$\begin{aligned} \begin{pmatrix} l & l_1 & l_2 \\ m & m_1 & m_2 \end{pmatrix} &= (-1)^{l_1+l_2+m} \\ &\times \frac{\sqrt{(l+l_1-l_2)!(l-l_1+l_2)!(-l+l_1+l_2)!(l+m)!(l-m)!}}{\sqrt{(l+l_1+l_2+1)!(l_1+m_1)!(l_1-m_1)!(l_2+m_2)!(l_2-m_2)!}} \\ &\times \sum_k \frac{(-1)^k(l+l_2+m_1-k)!(l_1-m_1+k)!}{k!(l-m-k)!(l-l_1+l_2-k)!(l_1-l_2+m+k)!}. \end{aligned} \tag{C.215}$$

Either equation (C.213)–(C.214) or (C.215) can be used to evaluate the Wigner 3-j symbols numerically; judicious grouping can be used to ameliorate the problems of handling the large factorials. The square of every 3-j symbol is a rational fraction; this makes it possible to use exact arithmetic if desired. Rotenberg, Bivins, Metropolis & Wooten (1959) give a table of exact 3-j symbols up to degree eight. An alternative computational procedure which makes use of both upper-level and lower-level recursion relations is described by Schulten & Gordon (1975a; 1976).

An even permutation of the columns of a 3-j symbol (C.213)–(C.215) leaves its numerical value unchanged:

$$\begin{aligned}\begin{pmatrix} l_1 & l_2 & l \\ m_1 & m_2 & m \end{pmatrix} &= \begin{pmatrix} l_2 & l & l_1 \\ m_2 & m & m_1 \end{pmatrix} \\ &= \begin{pmatrix} l & l_1 & l_2 \\ m & m_1 & m_2 \end{pmatrix},\end{aligned} \tag{C.216}$$

whereas an odd permutation is equivalent to multiplication by $(-1)^{l+l_1+l_2}$:

$$\begin{aligned}\begin{pmatrix} l_1 & l & l_2 \\ m_1 & m & m_2 \end{pmatrix} = \begin{pmatrix} l & l_2 & l_1 \\ m & m_2 & m_1 \end{pmatrix} = \begin{pmatrix} l_2 & l_1 & l \\ m_2 & m_1 & m \end{pmatrix} \\ = (-1)^{l+l_1+l_2}\begin{pmatrix} l & l_1 & l_2 \\ m & m_1 & m_2 \end{pmatrix}.\end{aligned} \tag{C.217}$$

The effect of a change in the sign of the bottom indices is

$$\begin{pmatrix} l & l_1 & l_2 \\ -m & -m_1 & -m_2 \end{pmatrix} = (-1)^{l+l_1+l_2}\begin{pmatrix} l & l_1 & l_2 \\ m & m_1 & m_2 \end{pmatrix}. \tag{C.218}$$

In addition to equations (C.216)–(C.218) there are a number of subtler symmetries—a total of 72 in all (Regge 1958). The number of 3-j symbols (C.213)–(C.215) having fixed values of the indices l_1, m_1 and l_2, m_2 is $\min\{l_1, l_2\} + 1$. This provides an a posteriori justification for the representation of the inner product $\langle \mathbf{Y}_{lm}^{N}, \mathbf{Y}_{l_1m_1}^{N_1}\mathbf{Y}_{l_2m_2}^{N_2}\rangle$ by a *single* separable expression (C.197).

C.7.4 Special cases

The 3-j symbols with the bottom indices $m = m_1 = m_2 = 0$ are particularly simple. Equation (C.218) shows that

$$\begin{pmatrix} l & l_1 & l_2 \\ 0 & 0 & 0 \end{pmatrix} = 0 \quad \text{if } l + l_1 + l_2 \text{ is odd.} \tag{C.219}$$

On the other hand, if $\Sigma = l + l_1 + l_2$ is even, then

$$\begin{pmatrix} l & l_1 & l_2 \\ 0 & 0 & 0 \end{pmatrix} = (-1)^{\Sigma/2} \sqrt{\frac{(\Sigma - 2l)!(\Sigma - 2l_1)!(\Sigma - 2l_2)!}{(\Sigma + 1)!}}$$
$$\times \frac{(\Sigma/2)!}{(\Sigma/2 - l)!(\Sigma/2 - l_1)!(\Sigma/2 - l_2)!}. \tag{C.220}$$

Other special 3-j symbols of interest include

$$\begin{pmatrix} l & l_1 & l_2 \\ -1 & 0 & 1 \end{pmatrix} = -\frac{L + L_2 - L_1}{2\sqrt{LL_2}} \begin{pmatrix} l & l_1 & l_2 \\ 0 & 0 & 0 \end{pmatrix}, \tag{C.221}$$

$$\begin{pmatrix} l & l_1 & l_2 \\ -2 & 0 & 2 \end{pmatrix} = \frac{(L + L_2 - L_1)(L + L_2 - L_1 - 2) - 2LL_2}{2\sqrt{L(L-2)L_2(L_2 - 2)}}$$
$$\times \begin{pmatrix} l & l_1 & l_2 \\ 0 & 0 & 0 \end{pmatrix}, \tag{C.222}$$

if $l + l_1 + l_2$ is even, and

$$\begin{pmatrix} l & l_1 & l_2 \\ -1 & 0 & 1 \end{pmatrix} = -\frac{1}{2\sqrt{LL_2}}$$
$$\times \sqrt{(\Sigma + 1 - 2l)(\Sigma + 1 - 2l_1)(\Sigma + 1 - 2l_2)}$$
$$\times \sqrt{(\Sigma + 2)(\Sigma + 4)/(\Sigma + 3)}$$
$$\times \begin{pmatrix} l+1 & l_1+1 & l_2+1 \\ 0 & 0 & 0 \end{pmatrix}, \tag{C.223}$$

$$\begin{pmatrix} l & l_1 & l_2 \\ -2 & 0 & 2 \end{pmatrix} = \frac{L + L_2 - L_1 - 2}{2\sqrt{L(L-2)L_2(L_2 - 2)}}$$
$$\times \sqrt{(\Sigma + 1 - 2l)(\Sigma + 1 - 2l_1)(\Sigma + 1 - 2l_2)}$$
$$\times \sqrt{(\Sigma + 2)(\Sigma + 4)/(\Sigma + 3)}$$
$$\times \begin{pmatrix} l+1 & l_1+1 & l_2+1 \\ 0 & 0 & 0 \end{pmatrix}, \tag{C.224}$$

if $\Sigma = l + l_1 + l_2$ is odd. In writing (C.221)–(C.224) we have used the shorthand notation $L = l(l+1)$, $L_1 = l_1(l_1+1)$ and $L_2 = l_2(l_2+1)$.

C.7.5 Gaunt and Adams integrals

The integral over the unit sphere of three ordinary spherical harmonics is given in terms of the $m = m_1 = m_2 = 0$ symbols (C.219)–(C.220) by

$$\int_\Omega Y_{lm}^* Y_{l_1 m_1} Y_{l_2 m_2}\, d\Omega = (-1)^m \left[\frac{(2l+1)(2l_1+1)(2l_2+1)}{4\pi}\right]^{1/2}$$

$$\times \begin{pmatrix} l & l_1 & l_2 \\ 0 & 0 & 0 \end{pmatrix} \begin{pmatrix} l & l_1 & l_2 \\ -m & m_1 & m_2 \end{pmatrix}. \tag{C.225}$$

Further specialization gives the integral of three Legendre polynomials:

$$\int_{-1}^{1} P_l P_{l_1} P_{l_2}\, d\mu = 2 \begin{pmatrix} l & l_1 & l_2 \\ 0 & 0 & 0 \end{pmatrix}^2. \tag{C.226}$$

Triple-product integrals of the form (C.225) and (C.226) were first evaluated by Gaunt (1929) and Adams (1878), respectively.

The splitting of an isolated multiplet of degree l by lateral heterogeneity of degree s is governed by a special case of the result (C.225):

$$\int_\Omega Y_{lm}^* Y_{st} Y_{lm'}\, d\Omega = (-1)^m (2l+1) \left(\frac{2s+1}{4\pi}\right)^{1/2}$$
$$\times \begin{pmatrix} l & s & l \\ 0 & 0 & 0 \end{pmatrix} \begin{pmatrix} l & s & l \\ -m & t & m' \end{pmatrix}. \tag{C.227}$$

The degree and order *selection rules* governing the integral (C.227) are

$$\int_\Omega Y_{lm}^* Y_{st} Y_{lm'}\, d\Omega = 0 \quad \text{unless} \quad \begin{cases} s \text{ is even} \\ 0 \le s \le 2l \\ t = m - m'. \end{cases} \tag{C.228}$$

The effect of zonal or axially symmetric heterogeneity Y_{00}, Y_{20}, Y_{40} depends upon the symbols

$$\begin{pmatrix} l & 0 & l \\ -m & 0 & m \end{pmatrix} = (-1)^{l+m} \left[\frac{(2l)!}{(2l+1)!}\right]^{1/2}, \tag{C.229}$$

$$\begin{pmatrix} l & 2 & l \\ -m & 0 & m \end{pmatrix} = (-1)^{l+m} \left[\frac{(2l-2)!}{(2l+3)!}\right]^{1/2}$$
$$\times 2\left[3m^2 - l(l+1)\right], \tag{C.230}$$

$$\begin{pmatrix} l & 4 & l \\ -m & 0 & m \end{pmatrix} = (-1)^{l+m} \left[\frac{(2l-4)!}{(2l+5)!}\right]^{1/2} \tag{C.231}$$
$$\times \{6(l+2)(l+1)l(l-1)$$
$$+[50 - 60l(l+1)]m^2 + 70m^4\}.$$

In particular, equations (C.227) and (C.230) imply that

$$\int_\Omega Y_{lm}^* P_2 Y_{lm}\, d\Omega = \frac{l(l+1) - 3m^2}{(2l-1)(2l+3)}. \tag{C.232}$$

C.7.6 3-j asymptotics

Splitting and coupling on a *smooth Earth model* are governed by Wigner 3-j symbols having one degree—that of the heterogeneity—much smaller than the other two:

$$\begin{pmatrix} l & s & l' \\ -m & t & m' \end{pmatrix} \quad \text{where} \quad s \ll l,\ s \ll l'. \tag{C.233}$$

Symbols of the form (C.233) may be asymptotically approximated by

$$\begin{pmatrix} l & s & l' \\ -m & t & m' \end{pmatrix} \approx \frac{(-1)^{l+m}}{\sqrt{l+l'+1}}\, P_{st}^{l-l'}\!\left(\frac{m+m'}{l+l'+1}\right). \tag{C.234}$$

Brussaard & Tolhoek (1957) provide an elementary proof of a result equivalent to (C.234), based upon an application of Stirling's formula to the factorials in equation (C.213). We have modified the argument of the square root and the generalized Legendre function $P_{st}^{l-l'}$ to guarantee the invariance of the right as well as the left side under the interchanges $l \longleftrightarrow l'$, $m \longleftrightarrow m'$ and $t \longleftrightarrow -t$. In the special case $l = l'$ equation (C.234) reduces to

$$\begin{pmatrix} l & s & l \\ 0 & 0 & 0 \end{pmatrix} \approx \frac{(-1)^{l}}{\sqrt{2l+1}}\, P_{s}(0), \tag{C.235}$$

$$\begin{pmatrix} l & s & l \\ -m & t & m' \end{pmatrix} \approx \frac{(-1)^{l+m}}{\sqrt{2l+1}}\, P_{st}\!\left(\frac{m+m'}{2l+1}\right). \tag{C.236}$$

Upon combining (C.235) and (C.236) we obtain a useful asymptotic formula for the Gaunt integral (C.227), valid for $s \ll l$:

$$\int_\Omega Y_{lm}^{*}\, Y_{st}\, Y_{lm'}\, d\Omega \approx P_s(0) Y_{st}(\eta, 0) \quad \text{where} \quad \cos\eta = \frac{m+m'}{2l+1}. \tag{C.237}$$

Ponzano & Regge (1968) used heuristic reasoning to obtain a generalization of the result (C.234), valid when all three degrees—s as well as l and l'—are large. Their conjecture, which is too lengthy to repeat here, was subsequently proved by Miller (1974), by an analysis of the correspondence relations between quantum and classical mechanics, and by Schulten & Gordon (1975b), by means of a discrete JWKB analysis of the recursion relation (C.205).

C.8 Rotation of a Tensor Field

We began this appendix by considering the effect of a passive finite rotation upon a scalar, vector or tensor field. We return to this topic once again in this final section, and show explicitly how to rotate a field expanded in generalized spherical harmonics.

C.8.1 Euler angles

It is well known that every finite rotation can be described by its three *Euler angles*. There are a number of different ways in which these angles may be defined; see Goldstein (1980) for a full discussion. We follow the convention of Edmonds (1960) and specify a finite rotation from an unprimed coordinate system x, y, z to a primed system x', y', z' by the following ordered sequence (see Figure C.3):

1. A rotation about the z axis through an angle $0 \leq \alpha \leq 2\pi$ leading to an intermediate coordinate system ξ, η, $\zeta = z$. This rotation is represented by the matrix

$$\mathsf{A} = \begin{pmatrix} \cos\alpha & \sin\alpha & 0 \\ -\sin\alpha & \cos\alpha & 0 \\ 0 & 0 & 1 \end{pmatrix}. \tag{C.238}$$

2. A rotation about the η axis through an angle $0 \leq \beta \leq \pi$, leading to a second intermediate coordinate system ξ', $\eta' = \eta$, ζ'. This rotation about the so-called *line of nodes* is represented by the matrix

$$\mathsf{B} = \begin{pmatrix} \cos\beta & 0 & -\sin\beta \\ 0 & 1 & 0 \\ \sin\beta & 0 & \cos\beta \end{pmatrix}. \tag{C.239}$$

3. A rotation about the ζ' axis through an angle $0 \leq \gamma \leq 2\pi$, leading to the final coordinate system x', y', $z' = \zeta'$. The matrix describing this third rotation is

$$\mathsf{C} = \begin{pmatrix} \cos\gamma & \sin\gamma & 0 \\ -\sin\gamma & \cos\gamma & 0 \\ 0 & 0 & 1 \end{pmatrix}. \tag{C.240}$$

The net effect of this sequence of axial rotations is described by the product

$$\mathsf{R} = \mathsf{CBA} \tag{C.241}$$

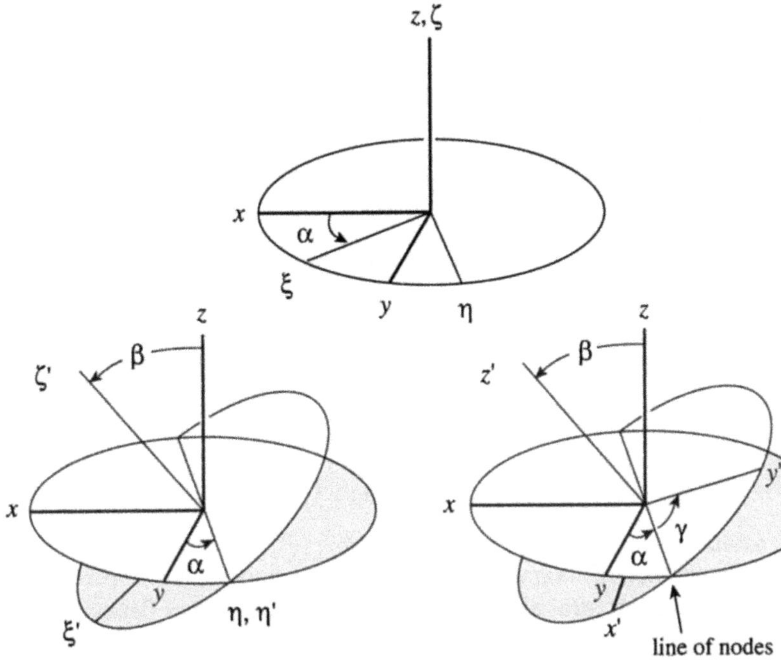

Figure C.3. Ordered sequence of axial rotations defining the three Euler angles: (*top*) rotation A through an angle α about the z axis; (*bottom left*) rotation B through an angle β about the line of nodes η; (*bottom right*) rotation C through an angle γ about the ζ' axis. The coordinates of a point $\mathbf{r}$ are transformed from x, y, z to x', y', z'.

of the three matrices (C.238)–(C.240). Upon performing the indicated multiplications, we find that

$$\mathsf{R} = \begin{pmatrix} \cos\alpha\cos\beta\cos\gamma - \sin\alpha\sin\gamma & \sin\alpha\cos\beta\cos\gamma + \cos\alpha\sin\gamma & -\sin\beta\cos\gamma \\ -\cos\alpha\cos\beta\sin\gamma - \sin\alpha\cos\gamma & -\sin\alpha\cos\beta\sin\gamma + \cos\alpha\cos\gamma & \sin\beta\sin\gamma \\ \cos\alpha\sin\beta & \sin\alpha\sin\beta & \cos\beta \end{pmatrix}. \tag{C.242}$$

The associated finite rotation operator (C.30) is

$$\mathcal{D} = \mathcal{D}_C\mathcal{D}_B\mathcal{D}_A = \exp(i\gamma J_z)\exp(i\beta J_y)\exp(i\alpha J_z). \tag{C.243}$$

It is straightforward to show that the final configuration in Figure C.3 can be obtained by means of an alternative ordered sequence of rotations:

1. A rotation through an angle $0 \le \gamma \le 2\pi$ about the z axis.

2. A rotation through an angle $0 \le \beta \le \pi$ about the *original* y axis.

3. A rotation through an angle $0 \le \alpha \le 2\pi$ about the *original* z axis.

We shall adhere in what follows to the illustrated situation, in which each successive rotation $1 \to 2 \to 3$ occurs about an axis that has been previously rotated. The order of the operations in equation (C.243) must, of course, be respected. The sign difference between (A.54) and (C.238) reflects the passive character of the latter.

C.8.2 Rotation of a generalized spherical harmonic

Let $\mathbf{Y}_{lm}^{N}(\theta,\phi)$ be a generalized spherical harmonic on the surface of the unit sphere Ω, and let

$$\mathbf{Y}'^{N}_{lm}(\theta,\phi) = \mathcal{D}\mathbf{Y}_{lm}^{N}(\theta,\phi) = \mathcal{D}_C\mathcal{D}_B\mathcal{D}_A\mathbf{Y}_{lm}^{N}(\theta,\phi) \tag{C.244}$$

be its rotated equivalent. By definition, $\mathbf{Y}'^{N}_{lm}(\theta',\phi') = \mathbf{Y}_{lm}^{N}(\theta,\phi)$, where θ',ϕ' and θ,ϕ denote the same point in the primed and unprimed coordinate systems. To find the rotated tensor field $\mathbf{Y}'^{N}_{lm}(\theta,\phi)$ we must determine the effect of the operator triple product (C.243) upon $\mathbf{Y}_{lm}^{N}(\theta,\phi)$. Because every generalized spherical harmonic is an eigenfunction of the z component of the total angular momentum, $J_z\mathbf{Y}_{lm}^{N} = m\mathbf{Y}_{lm}^{N}$, the rightmost operator $\mathcal{D}_A = \exp(i\alpha J_z)$ yields

$$\begin{aligned}\exp(i\alpha J_z)\mathbf{Y}_{lm}^{N} &= (1 + i\alpha J_z - \tfrac{1}{2}\alpha^2 J_z^2 + \cdots)\,\mathbf{Y}_{lm}^{N}\\ &= (1 + im\alpha - \tfrac{1}{2}m^2\alpha^2 + \cdots)\mathbf{Y}_{lm}^{N}\\ &= e^{im\alpha}\,\mathbf{Y}_{lm}^{N}.\end{aligned} \tag{C.245}$$

To determine the effect of the nodal rotation $\mathcal{D}_B = \exp(i\beta J_y)$ we write the associated generator in the form $J_y = \frac{1}{2}i(J_- - J_+)$ and make use of the ladder relations (C.104):

$$\begin{aligned}J_y\mathbf{Y}_{lm}^{N} &= \tfrac{1}{2}i\sqrt{(l+m)(l-m+1)}\,\mathbf{Y}_{l\,m-1}^{N}\\ &\quad - \tfrac{1}{2}i\sqrt{(l-m)(l+m+1)}\,\mathbf{Y}_{l\,m+1}^{N}.\end{aligned} \tag{C.246}$$

Repeated application of J_y to $\mathbf{Y}_{lm}^{N}$ always gives rise to a linear combination of harmonics $\{\mathbf{Y}_{l\,-l}^{N},\ldots,\mathbf{Y}_{l0}^{N},\ldots,\mathbf{Y}_{ll}^{N}\}$ with the same degree l and upper index N, as a consequence of the cessation of the laddering. It follows that $\exp(i\beta J_y)\,\mathbf{Y}_{lm}^{N} = (1 + i\beta J_y - \frac{1}{2}\beta^2 J_y^2 + \cdots)\,\mathbf{Y}_{lm}^{N}$ can be written in the form

$$\exp(i\beta J_y)\,\mathbf{Y}_{lm}^{N} = \sum_{m'=-l}^{l} d_{m'm}^{(l)}(\beta)\mathbf{Y}_{lm'}^{N}, \tag{C.247}$$

where the coefficients $d^{(l)}_{m'm}(\beta)$ remain to be determined. The effect of the third and final rotation operator $\mathcal{D}_C = \exp(i\gamma J_z)$ upon each of the harmonics $\mathbf{Y}^N_{lm'}$ in equation (C.247) can be found by an argument analogous to that in (C.245):

$$\exp(i\gamma J_z)\,\mathbf{Y}^N_{lm'} = e^{im'\gamma}\,\mathbf{Y}^N_{lm'}. \tag{C.248}$$

Upon combining the above results we obtain the fully rotated harmonic:

$$\mathbf{Y}'^N_{lm} = \sum_{m'=-l}^{l} \mathcal{D}^{(l)}_{m'm}(\alpha,\beta,\gamma)\mathbf{Y}^N_{lm'}, \tag{C.249}$$

where

$$\mathcal{D}^{(l)}_{m'm}(\alpha,\beta,\gamma) = e^{im'\gamma}\,d^{(l)}_{m'm}(\beta)\,e^{im\alpha}. \tag{C.250}$$

We may regard the quantities (C.250) as the elements of a $(2l+1)\times(2l+1)$ transformation matrix:

$$\mathcal{D}^{(l)}_{m'm}(\alpha,\beta,\gamma) = \langle\mathbf{Y}^N_{lm'},\mathbf{Y}'^N_{lm}\rangle = \langle\mathbf{Y}^N_{lm'},\mathcal{D}\mathbf{Y}^N_{lm}\rangle. \tag{C.251}$$

The inner-product representation (C.251) of $\mathcal{D}^{(l)}_{m'm}(\alpha,\beta,\gamma)$ follows from equation (C.249) and the tensor orthonormality relation (C.95).

We turn next to the determination of the nodal matrix elements

$$d^{(l)}_{m'm}(\beta) = \langle\mathbf{Y}^N_{lm'},\mathcal{D}_B\mathbf{Y}^N_{lm}\rangle. \tag{C.252}$$

The primed and unprimed coordinates are related in the case of a rotation $\{0,\beta,0\}$ about the $\hat{\mathbf{y}}$ axis by $\theta' = \theta-\beta$, $\phi' = \phi$. The value of a generalized spherical harmonic $\mathbf{Y}^N_{lm}$ at a point $\theta = \beta+\beta'$, $\phi = 0$ *upon the prime meridian* can be written in terms of the values of $\mathbf{Y}^N_{l\,-l},\ldots,\mathbf{Y}^N_{l0},\ldots,\mathbf{Y}^N_{ll}$ at the (same) point $\theta' = \beta'$, $\phi' = 0$ in the form

$$\begin{aligned}\mathbf{Y}^N_{lm}(\theta=\beta+\beta',\phi=0) &= \mathbf{Y}'^N_{lm}(\theta'=\beta',\phi'=0)\\ &= \sum_{m'=-l}^{l} d^{(l)}_{m'm}(\beta)\mathbf{Y}^N_{lm'}(\theta'=\beta',\phi'=0).\end{aligned} \tag{C.253}$$

It is intuitively obvious that a rotation about the $\hat{\mathbf{y}}$ axis does not affect the spherical polar unit vectors $\hat{\mathbf{r}}$, $\hat{\boldsymbol{\theta}}$, $\hat{\boldsymbol{\phi}}$ upon the meridian; this can be confirmed by substituting $\phi = 0$ in equations (C.44). The meridional basis vectors $\hat{\mathbf{e}}_-$, $\hat{\mathbf{e}}_0$, $\hat{\mathbf{e}}_+$ are likewise unaffected; this allows us to rewrite (C.253) in terms of the generalized Legendre functions (C.112) in the form

$$P^N_{lm}[\cos(\beta+\beta')] = \sum_{m'=-l}^{l} d^{(l)}_{m'm}(\beta)P^N_{lm'}(\cos\beta'). \tag{C.254}$$

Upon taking the limit $\beta' \to 0$ and making use of the normalization relation (C.114) we obtain the simple result

$$d^{(l)}_{m'm}(\beta) = P^{m'}_{lm}(\cos\beta). \tag{C.255}$$

Equation (C.255) is valid for a positive rotation $0 \le \beta \le \pi$. Upon substituting back we can rewrite equation (C.254) as a matrix-element relation:

$$d^{(l)}_{m'm}(\beta + \beta') = \sum_{m''=-l}^{l} d^{(l)}_{m'm''}(\beta')\, d^{(l)}_{m''m}(\beta). \tag{C.256}$$

This result expresses the effect of a rotation β followed by β' in terms of the individual effects of the two constituent rotations.

C.8.3 Properties of the matrix elements

The *inverse* of a finite rotation $\{\alpha, \beta, \gamma\}$ is $\{-\gamma, -\beta, -\alpha\}$; each of the axial rotations in Section C.8.1 must be "undone" in reverse order. The matrix describing this inverse rotation is the transpose $\mathsf{R}^{-1} = \mathsf{R}^{\mathrm{T}} = \mathsf{A}^{\mathrm{T}}\mathsf{B}^{\mathrm{T}}\mathsf{C}^{\mathrm{T}}$, and the associated rotation operator is the adjoint

$$\begin{aligned}\mathcal{D}^{-1} = \mathcal{D}^{\dagger} &= \mathcal{D}^{\dagger}_{A}\mathcal{D}^{\dagger}_{B}\mathcal{D}^{\dagger}_{C} \\ &= \exp(-i\alpha J_z)\exp(-i\beta J_y)\exp(-i\gamma J_z).\end{aligned} \tag{C.257}$$

The matrix elements analogous to (C.251) are

$$\begin{aligned}\mathcal{D}^{(l)}_{m'm}(-\gamma, -\beta, -\alpha) &= \langle \mathbf{Y}^N_{lm'}, \mathcal{D}^{\dagger}\mathbf{Y}^N_{lm}\rangle = \langle \mathcal{D}\mathbf{Y}^N_{lm'}, \mathbf{Y}^N_{lm}\rangle \\ &= \langle \mathbf{Y}^N_{lm}, \mathcal{D}\mathbf{Y}^N_{lm'}\rangle^* = \mathcal{D}^{(l)*}_{mm'}(\alpha, \beta, \gamma).\end{aligned} \tag{C.258}$$

Setting $\alpha = \gamma = 0$ and noting that $d^{(l)}_{mm'}(\beta)$ is real we find, in particular, that

$$d^{(l)}_{m'm}(-\beta) = d^{(l)}_{mm'}(\beta). \tag{C.259}$$

Equations (C.258) and (C.259) define $\mathcal{D}^{(l)}_{m'm}(\alpha, \beta, \gamma)$ and $d^{(l)}_{m'm}(\beta)$ for *negative* values of the Euler angles $-2\pi \le \alpha < 0$, $-\pi \le \beta < 0$, $-2\pi \le \gamma < 0$.

The orthonormality of the rotated harmonics guarantees that

$$\langle \mathcal{D}\mathbf{Y}^N_{lm}, \mathcal{D}\mathbf{Y}^N_{lm'}\rangle = \langle \mathbf{Y}^N_{lm}, \mathcal{D}^{\dagger}\mathcal{D}\mathbf{Y}^N_{lm'}\rangle = \delta_{mm'} \tag{C.260}$$

or, equivalently,

$$\sum_{m''=-l}^{l} \mathcal{D}^{(l)*}_{m''m}(\alpha, \beta, \gamma)\, \mathcal{D}^{(l)}_{m''m'}(\alpha, \beta, \gamma) = \delta_{mm'}. \tag{C.261}$$

Upon either setting $\alpha = \gamma = 0$ in this result, or setting $\beta' = -\beta$ in (C.256) and using (C.259), we obtain

$$\sum_{m''=-l}^{l} d^{(l)}_{m''m}(\beta)\, d^{(l)}_{m''m'}(\beta) = \delta_{mm'}. \tag{C.262}$$

Equations (C.261)–(C.262) stipulate that the $(2l+1) \times (2l+1)$ matrices (C.251)–(C.252) are *unitary*. The result (C.127) follows from (C.262) upon making the identification (C.255).

The symmetry relations satisfied by $d^{(l)}_{m'm}(\beta)$ follow immediately from those governing the generalized Legendre functions $P^{m'}_{lm}(\cos\beta)$, given in equations (C.118). For example, for a fixed rotation β, we have

$$d^{(l)}_{-m'\,-m}(\beta) = (-1)^{m+m'} d^{(l)}_{m'm}(\beta), \tag{C.263}$$

$$d^{(l)}_{mm'}(\beta) = (-1)^{m+m'} d^{(l)}_{m'm}(\beta), \tag{C.264}$$

$$d^{(l)}_{-m\,-m'}(\beta) = d^{(l)}_{m'm}(\beta). \tag{C.265}$$

It is a simple matter to extend (C.263)–(C.265) to the full matrix elements; in particular, we find that the complex conjugate of (C.250) is given by

$$\mathcal{D}^{(l)*}_{m'm}(\alpha,\beta,\gamma) = (-1)^{m'+m}\, \mathcal{D}^{(l)}_{-m'\,-m}(\alpha,\beta,\gamma). \tag{C.266}$$

Equations (C.201)–(C.202) can likewise be extended to yield the equivalent matrix-element relations

$$\mathcal{D}^{(l_1)}_{m_1'm_1}\mathcal{D}^{(l_2)}_{m_2'm_2} = \sum_l\sum_{m'}\sum_m (2l+1) \begin{pmatrix} l & l_1 & l_2 \\ m' & m_1' & m_2' \end{pmatrix} \times \begin{pmatrix} l & l_1 & l_2 \\ m & m_1 & m_2 \end{pmatrix} \mathcal{D}^{(l)*}_{m'm}, \tag{C.267}$$

$$(2l+1)^{-1}\,\delta_{ll'}\,\mathcal{D}^{(l)*}_{m'm} = \sum_{m_1'm_2'}\sum_{m_1m_2} \begin{pmatrix} l & l_1 & l_2 \\ m' & m_1' & m_2' \end{pmatrix} \times \begin{pmatrix} l' & l_1 & l_2 \\ m & m_1 & m_2 \end{pmatrix} \mathcal{D}^{(l_1)}_{m_1'm_1}\mathcal{D}^{(l_2)}_{m_2'm_2}, \tag{C.268}$$

where the common arguments α, β, γ are understood. Upon making use of the unitary nature of the transformation matrices (C.261) we obtain the fully symmetric relation

$$\sum_{m'}\sum_{m_1'}\sum_{m_2'} \mathcal{D}^{(l)}_{m'm}\mathcal{D}^{(l_1)}_{m_1'm_1}\mathcal{D}^{(l_2)}_{m_2'm_2} \begin{pmatrix} l & l_1 & l_2 \\ m' & m_1' & m_2' \end{pmatrix} = \begin{pmatrix} l & l_1 & l_2 \\ m & m_1 & m_2 \end{pmatrix}. \tag{C.269}$$

This in turn can be reduced to an analogous symmetric relation involving the generalized Legendre functions, if desired.

The full matrix elements are orthonormal with respect to integration over the three Euler angles $\{\alpha, \beta, \gamma\}$ in the sense

$$\frac{1}{8\pi^2}\int_0^{2\pi}\int_0^{\pi}\int_0^{2\pi} \mathcal{D}^{(l_1)*}_{m_1' m_1} \mathcal{D}^{(l_2)}_{m_2' m_2}\, d\alpha \sin\beta\, d\beta\, d\gamma$$
$$= (2l_1 + 1)^{-1}\delta_{l_1 l_2}\delta_{m_1' m_1}\delta_{m_2' m_2}. \tag{C.270}$$

The integral over three matrix elements obtained using equation (C.200) is

$$\frac{1}{8\pi^2}\int_0^{2\pi}\int_0^{\pi}\int_0^{2\pi} \mathcal{D}^{(l)}_{m' m}\mathcal{D}^{(l_1)}_{m_1' m_1} \mathcal{D}^{(l_2)}_{m_2' m_2}\, d\alpha \sin\beta\, d\beta\, d\gamma$$
$$= \begin{pmatrix} l & l_1 & l_2 \\ m' & m_1' & m_2' \end{pmatrix}\begin{pmatrix} l & l_1 & l_2 \\ m & m_1 & m_2 \end{pmatrix}. \tag{C.271}$$

Obviously, equation (C.271) can be rewritten in a form analogous to (C.198) with the first element conjugated by employing the symmetry (C.266). The factor $1/8\pi^2$ can be thought of as the total "volume" of angular integration.

C.8.4 Addition theorem

The net effect of two successive finite rotations, $\mathsf{R}_1 = \{\alpha_1, \beta_1, \gamma_1\}$ followed by $\mathsf{R}_2 = \{\alpha_2, \beta_2, \gamma_2\}$, is a *single* rotation $\mathsf{R} = \mathsf{R}_2\mathsf{R}_1 = \{\alpha, \beta, \gamma\}$, whose Euler angles are given by

$$\cos(\alpha - \alpha_1) = \frac{\cos\beta_2 - \cos\beta\cos\beta_1}{\sin\beta\sin\beta_1}, \tag{C.272}$$

$$\cos\beta = \cos\beta_1\cos\beta_2 - \sin\beta_1\sin\beta_2\cos(\alpha_2 + \gamma_1), \tag{C.273}$$

$$\cos(\gamma - \gamma_2) = \frac{\cos\beta_1 - \cos\beta\cos\beta_2}{\sin\beta\sin\beta_2}. \tag{C.274}$$

The operators associated with these rotations satisfy $\mathcal{D}(\mathsf{R}) = \mathcal{D}(\mathsf{R}_2)\mathcal{D}(\mathsf{R}_1)$ or, equivalently,

$$\mathcal{D}^{(l)}_{m' m}(\alpha, \beta, \gamma) = \sum_{m''=-l}^{l} \mathcal{D}^{(l)}_{m' m''}(\alpha_2, \beta_2, \gamma_2)\, \mathcal{D}^{(l)}_{m'' m}(\alpha_1, \beta_1, \gamma_1). \tag{C.275}$$

The matrix-element relation (C.275) is the most general form of the so-called *addition theorem*. It reduces to the uniaxial result (C.256) in the case $\{\alpha_1, \beta_1, \gamma_1\} = \{0, \beta, 0\}$ and $\{\alpha_2, \beta_2, \gamma_2\} = \{0, \beta', 0\}$, and to the unitary relation (C.261) in the case that the second rotation is the inverse of the first: $\{\alpha_2, \beta_2, \gamma_2\} = \{-\gamma_1, -\beta_1, -\alpha_1\}$.

Other choices of the Euler angles in (C.275) lead to a variety of more specialized addition theorems; for example, upon setting $m' = m = 0$ and $\{\alpha_1, \beta_1, \gamma_1\} = \{0, \theta_1, \phi_1\}$ and $\{\alpha_2, \beta_2, \gamma_2\} = \{\pi - \phi_2, \theta_2, 0\}$, we obtain

$$\left(\frac{2l+1}{4\pi}\right) P_l(\cos\Theta) = \sum_{m=-l}^{l} Y_{lm}^*(\theta_2, \phi_2) Y_{lm}(\theta_1, \phi_1), \tag{C.276}$$

where $\cos\Theta = \cos\theta_2 \cos\theta_1 + \sin\theta_2 \sin\theta_1 \cos(\phi_2 - \phi_1)$. This result—the classical spherical-harmonic addition theorem (B.69)—can be derived in a simpler manner by evaluating the left side of the rotationally invariant expression

$$\sum_{m=-l}^{l} {Y'}_{lm}^*(\theta_2', \phi_2') {Y'}_{lm}(\theta_1', \phi_1') = \sum_{m=-l}^{l} Y_{lm}^*(\theta_2, \phi_2) Y_{lm}(\theta_1, \phi_1) \tag{C.277}$$

in the limit that either of the two rotated points θ_1', ϕ_1' or θ_2', ϕ_2' approaches the North or South Pole.

C.8.5 Recursion relations

The matrix elements $d_{m'm}^{(l)}(\beta)$ can be calculated using the recursion relations (C.121)–(C.122) governing the generalized Legendre functions. We may either iterate the fixed-m' relation

$$\begin{aligned}\left[m'(\sin\beta)^{-1} - m\cot\beta\right] d_{m'm}^{(l)} &= \tfrac{1}{2}\sqrt{(l+m)(l-m+1)}\, d_{m'\,m-1}^{(l)} \\ &+ \tfrac{1}{2}\sqrt{(l-m)(l+m+1)}\, d_{m'\,m+1}^{(l)}\end{aligned} \tag{C.278}$$

downward and upward from $m = \pm l$, meeting at $m = m'\cos\beta$, or we may iterate the fixed-m relation

$$\begin{aligned}\left[m'\cot\beta - m(\sin\beta)^{-1}\right] d_{m'm}^{(l)} &= \tfrac{1}{2}\sqrt{(l+m')(l-m'+1)}\, d_{m'-1\,m}^{(l)} \\ &+ \tfrac{1}{2}\sqrt{(l-m')(l+m'+1)}\, d_{m'+1\,m}^{(l)}\end{aligned} \tag{C.279}$$

downward and upward from $m' = \pm l$, meeting at $m' = m(\cos\beta)^{-1}$. Other computational schemes are also available; for example, Edmonds (1960) shows how to express the elements $d_{m'm}^{(l)}(\beta)$ in terms of the Jacobi polynomials, which can be calculated using an alternative recursion relation.

C.8.6 Rotation of an arbitrary tensor

We consider finally the effect of a finite rotation $\{\alpha, \beta, \gamma\}$ upon an arbitrary tensor field having the generalized spherical-harmonic representation

$$\mathbf{T} = \sum_{l=0}^{\infty} \sum_{m=-l}^{l} T_{lm}^{\alpha_1 \cdots \alpha_n} \mathbf{Y}_{lm}^N. \tag{C.280}$$

The rotated field obtained using equation (C.249) is

$$\mathbf{T}' = \mathcal{D}\mathbf{T} = \sum_{l=0}^{\infty}\sum_{m=-l}^{l} T_{lm}^{\alpha_1\cdots\alpha_n}\left(\mathcal{D}\mathbf{Y}_{lm}^{N}\right)$$
$$= \sum_{l=0}^{\infty}\sum_{m=-l}^{l}\sum_{m'=-l}^{l} \mathcal{D}_{m'm}^{(l)}(\alpha,\beta,\gamma)\, T_{lm}^{\alpha_1\cdots\alpha_n}\mathbf{Y}_{lm'}^{N}. \qquad \text{(C.281)}$$

We can rewrite the result (C.281) in a form analogous to (C.280):

$$\mathbf{T}' = \sum_{l=0}^{\infty}\sum_{m=-l}^{l} T_{lm}^{\prime\alpha_1\cdots\alpha_q}\mathbf{Y}_{lm}^{N}. \qquad \text{(C.282)}$$

The *transformed expansion coefficients* $T_{lm}^{\prime\alpha_1\cdots\alpha_q}$ are related to the original coefficients $T_{lm}^{\alpha_1\cdots\alpha_q}$ by

$$T_{lm}^{\prime\alpha_1\cdots\alpha_q} = \sum_{m'=-l}^{l} \mathcal{D}_{mm'}^{(l)}(\alpha,\beta,\gamma)\, T_{lm'}^{\alpha_1\cdots\alpha_q}. \qquad \text{(C.283)}$$

Note the "perversity of the indices" in going from the basis transformation relation (C.249) to equation (C.283); the transformed coefficients $T_{lm}^{\prime\alpha_1\cdots\alpha_q}$ are calculated by *ordinary matrix multiplication*, in a manner analogous to equation (A.33). The inverse of (C.283) obtained using the unitary property (C.261) is

$$T_{lm}^{\alpha_1\cdots\alpha_q} = \sum_{m'=-l}^{l} \mathcal{D}_{m'm}^{(l)*}(\alpha,\beta,\gamma)\, T_{lm'}^{\prime\alpha_1\cdots\alpha_q}. \qquad \text{(C.284)}$$

The rotated and original partial derivatives $\partial/\partial T_{lm}^{\prime\alpha_1\cdots\alpha_n}$ and $\partial/\partial T_{lm}^{\alpha_1\cdots\alpha_q}$ are related by

$$\partial/\partial T_{lm}^{\prime\alpha_1\cdots\alpha_q} = \sum_{m'=-l}^{l} \mathcal{D}_{mm'}^{(l)*}(\alpha,\beta,\gamma)\,(\partial/\partial T_{lm'}^{\alpha_1\cdots\alpha_q}) \qquad \text{(C.285)}$$

and

$$\partial/\partial T_{lm}^{\alpha_1\cdots\alpha_q} = \sum_{m'=-l}^{l} \mathcal{D}_{m'm}^{(l)}(\alpha,\beta,\gamma)\,(\partial/\partial T_{lm'}^{\prime\alpha_1\cdots\alpha_q}). \qquad \text{(C.286)}$$

The transformation relations (C.285) and (C.286) are useful in inverse problems involving the tensor fields $\mathbf{T}$ and $\mathbf{T}' = \mathcal{D}\mathbf{T}$.

C.8.7 Rotation to the equator

To conclude this appendix, we consider the problem of rotating a "source" point θ_1, ϕ_1 and a "receiver" point θ_2, ϕ_2 to the equator. We demand that the rotated source point be situated upon the prime meridian, so that

$$\theta_1' = \pi/2,\ \phi_1' = 0 \quad \text{and} \quad \theta_2' = \pi/2,\ \phi_2' = \Theta, \tag{C.287}$$

where Θ is the geodesic angular distance between the two points, given by $\cos\Theta = \cos\theta_2\cos\theta_1 + \sin\theta_2\sin\theta_1\cos(\phi_2 - \phi_1)$. The Euler angles $\{\alpha, \beta, \gamma\}$ which effect the transformation (C.287) are given by

$$\tan\alpha = \frac{\sin\theta_2\cos\theta_1\cos\phi_2 - \cos\theta_2\sin\theta_1\cos\phi_1}{\cos\theta_2\sin\theta_1\sin\phi_1 - \sin\theta_2\cos\theta_1\sin\phi_2}, \tag{C.288}$$

$$\cos\beta = \frac{\sin\theta_2\sin\theta_1\sin(\phi_2 - \phi_1)}{\sin\Theta}, \tag{C.289}$$

$$\tan\gamma = \frac{\cos\theta_1\cos\Theta - \cos\theta_2}{\cos\theta_1\sin\Theta}. \tag{C.290}$$

Comparison of equations (C.288)–(C.289) and (B.107)–(B.108) enables us to make the identification

$$\alpha = \bar{\phi}, \qquad \beta = \bar{\theta}. \tag{C.291}$$

The first two rotations (C.291) carry the $\hat{\mathbf{z}}$ axis to the positive pole $\bar{\theta}, \bar{\phi}$ of the source-receiver great circle, whereas the third carries the meridian around the new equator to the source point. Figure C.4 illustrates the transformation (C.287); the endpoints θ_1, ϕ_1 and θ_2, ϕ_2 of the geodesic minor arc in this example are situated in Rio de Janeiro and Cairo, respectively. Note the passive nature of the rotation: the positions $\hat{\mathbf{r}}$ of the coastlines and other features on the globe remain unchanged, but the primed and unprimed coordinate lines are different.

Rotation of θ_1, ϕ_1 and θ_2, ϕ_2 to the equator can be used, among other things, to calculate the integral of a scalar field ψ along the geodesic path between the source and receiver:

$$\int_{\theta_1,\phi_1}^{\theta_2,\phi_2} \psi(\theta,\phi)\, d\Delta = \sum_{l=0}^{\infty}\sum_{m=-l}^{l} \psi_{lm} \int_{\theta_1,\phi_1}^{\theta_2,\phi_2} Y_{lm}(\theta,\phi)\, d\Delta. \tag{C.292}$$

Path integrals of the form (C.292) arise in the application of Fermat's principle to surface-wave tomography, as we discuss in Section 16.9; the field ψ

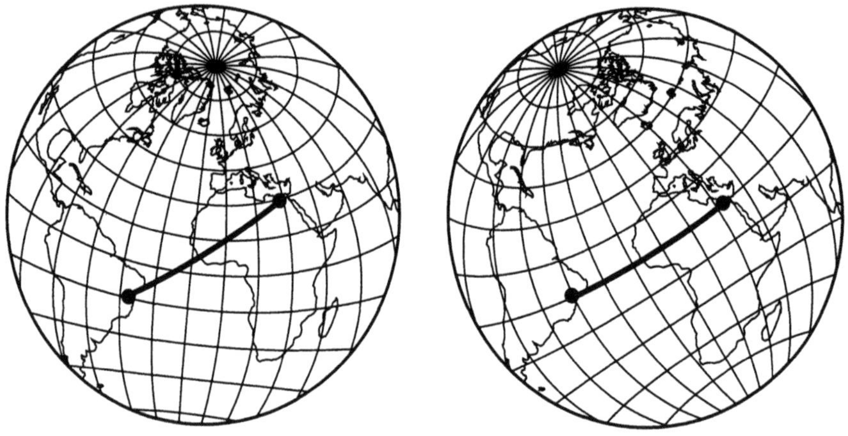

Figure C.4. Passive rotation of the Rio de Janeiro to Cairo minor-arc path. (*Left*) Heavy great-circular arc denotes the original path. (*Right*) In the new primed coordinate system, Rio de Janeiro is situated at the intersection of the equator and the prime meridian: $\theta_1' = \pi/2$, $\phi_1' = 0$. Cairo is also on the equator, a distance $\Theta \approx 90°$ to the east: $\theta_2' = \pi/2$, $\phi_2' = \Theta$.

in that case is the inverse phase speed of a monochromatic Love or Rayleigh wave. The integral of a complex surface spherical harmonic Y_{lm} is given by

$$\int_{\theta_1,\phi_1}^{\theta_2,\phi_2} Y_{lm}(\theta,\phi)\,d\Delta = \int_0^{\Theta} Y_{lm}'(\pi/2,\phi')\,d\phi'$$

$$= \sum_{m'=-l}^{l} (i/m')\,X_{lm'}(\pi/2)\,(1-e^{im'\Theta})\,\mathcal{D}_{m'm}^{(l)}(\alpha,\beta,\gamma), \qquad \text{(C.293)}$$

where we have invoked (C.249) and evaluated the integrals $\int_0^{\Theta} e^{im\phi'}\,d\phi'$ analytically. Upon substituting the result (C.293) into equation (C.292) we find that

$$\int_{\theta_1,\phi_1}^{\theta_2,\phi_2} \psi(\theta,\phi)\,d\Delta = \sum_{l=0}^{\infty}\sum_{m=-l}^{l} (i/m)\,X_{lm}(\pi/2)\,(1-e^{im\Theta})$$

$$\times \sum_{m'=-l}^{l} \mathcal{D}_{mm'}^{(l)}(\alpha,\beta,\gamma)\,\psi_{lm}. \qquad \text{(C.294)}$$

We have interchanged the indices m and m' to make it clear that the final sum in (C.294) is simply the transformed expansion coefficient ψ_{lm}'.

It is straightforward to express the integral of a real field ψ in terms of the real expansion coefficients Ψ_{lm} or a_{l0}, a_{lm}, b_{lm} using the elementary relations (B.98).

The quantity $(i/m')(1 - e^{im'\Theta})$ in equation (C.293) reduces to $2\pi\,\delta_{m'0}$ in the limit $\Theta \to 2\pi$. The average of a spherical harmonic Y_{lm} around a full great-circle path $\bar{\theta}, \bar{\phi}$ is therefore

$$\frac{1}{2\pi}\oint_{\bar{\theta},\bar{\phi}} Y_{lm}(\theta,\phi)\,d\Delta = P_l(0)Y_{lm}(\bar{\theta},\bar{\phi}). \tag{C.295}$$

The expansion coefficients $\bar{\psi}_{lm}$ of the great-circular average of an arbitrary field are related to the original coefficients ψ_{lm} by $\bar{\psi}_{lm} = P_l(0)\,\psi_{lm}$. These results agree with equations (B.111)–(B.112), as expected.

Appendix D

Whole Earth Catalogue

In this final appendix we present explicit expressions for the elements of the receiver and source vectors r and s and the kinetic-energy, potential-energy, anelasticity and Coriolis matrices T, V, A and W. These indigestible but important formulae provide a basis for the synthesis of coupled-mode seismograms, as we discuss in Section 13.2. The history of this subject is a checkered one, marred by a number of incorrect early treatments of the effect of a perturbation in the location of the free surface and other boundaries (Backus & Gilbert 1967; Dahlen 1968). Woodhouse (1976) pointed out these errors and gave a correct derivation of the Fréchet kernel governing a spherical boundary perturbation. The effect of the Earth's hydrostatic ellipticity upon an isolated multiplet was then calculated correctly by Dahlen (1976). The first correct analysis of a laterally heterogeneous perturbation was conducted by Woodhouse & Dahlen (1978); they gave a recipe for calculating perturbed eigenfrequencies and singlet eigenfunctions in the self coupling approximation, and derived a numerically superior formula for the ellipticity splitting parameter. A complete catalogue of matrix elements governing spheroidal-spheroidal, toroidal-toroidal and spheroidal-toroidal coupling due to the Earth's rotation, ellipticity and isotropic lateral heterogeneity was subsequently presented by Woodhouse (1980). The matrix elements needed to account for a volumetric anisotropic perturbation were first reduced to a numerically tractable form by Tanimoto (1986) and Mochizuki (1986), who considered self coupling and full coupling, respectively. Finally, the ellipticity and other boundary-perturbation formulae were extended to a transversely isotropic starting model by Henson (1989) and Shibata, Suda & Fukao (1990). Upon reducing their results to the case of a SNREI starting model, these last two investigations discovered a number of errors in the isotropic ellipticity kernels of Woodhouse (1980).

The elements of the kinetic-energy, potential-energy and anelasticity matrices are most readily expressed in terms of the generalized spherical harmonics discussed in Appendix C. Since these are complex, it is advantageous to use complex rather than real basis eigenfunctions in the initial algebraic reductions. For brevity and ease of comparison with previous treatments, it is also convenient to introduce new radial eigenfunctions which are related to the eigenfunctions used throughout the rest of this book by

$$u = U, \qquad v = k^{-1}V, \qquad w = k^{-1}W, \qquad p = P, \tag{D.1}$$

where $k = \sqrt{l(l+1)}$ as usual. The real and complex basis eigenfunctions may be expressed in terms of these lower-case scalars in the form

$$\mathbf{s}_k = u\hat{\mathbf{r}}\mathcal{Y}_{lm} + v\boldsymbol{\nabla}_1\mathcal{Y}_{lm} - w(\hat{\mathbf{r}} \times \boldsymbol{\nabla}_1\mathcal{Y}_{lm}), \qquad \phi_k = p\mathcal{Y}_{lm}, \tag{D.2}$$

$$\tilde{\mathbf{s}}_k = u\hat{\mathbf{r}}Y_{lm} + v\boldsymbol{\nabla}_1 Y_{lm} - w(\hat{\mathbf{r}} \times \boldsymbol{\nabla}_1 Y_{lm}), \qquad \tilde{\phi}_k = pY_{lm}, \tag{D.3}$$

where $\mathcal{Y}_{lm}$ and Y_{lm} are the real and complex spherical harmonics of degree l and order m defined by equations (B.72) and (B.30), respectively. A tilde will be used throughout this appendix to adorn quantities calculated using the complex eigenfunctions (D.3). We begin by expressing the elements of the complex receiver and source vectors $\tilde{\mathsf{r}}$, $\tilde{\mathsf{s}}$ and associated energy, anelasticity and Coriolis matrices $\tilde{\mathsf{T}}$, $\tilde{\mathsf{V}}$, $\tilde{\mathsf{A}}$ and $\tilde{\mathsf{W}}$; the desired vectors r, s and matrices T, V, A and W are then obtained by means of a complex-to-real transformation described in Section D.3.

D.1 Receiver and Source Vector

The spherical polar components $\tilde{s}_r = \hat{\mathbf{r}} \cdot \tilde{\mathbf{s}}_k$, $\tilde{s}_\theta = \hat{\boldsymbol{\theta}} \cdot \tilde{\mathbf{s}}_k$ and $\tilde{s}_\phi = \hat{\boldsymbol{\phi}} \cdot \tilde{\mathbf{s}}_k$ are

$$\tilde{s}_r = uY_{lm}, \tag{D.4}$$

$$\tilde{s}_\theta = v\partial_\theta Y_{lm} + imw(\sin\theta)^{-1}Y_{lm}, \tag{D.5}$$

$$\tilde{s}_\phi = imv(\sin\theta)^{-1}Y_{lm} - w\partial_\theta Y_{lm}. \tag{D.6}$$

The elements $\tilde{r}_k = \hat{\boldsymbol{\nu}} \cdot \tilde{\mathbf{s}}_k^*(\mathbf{x})$ of the complex receiver vector $\tilde{\mathsf{r}}$ are given in terms of the components (D.4)–(D.6) by

$$\tilde{r}_k = \nu_r\tilde{s}_r^* + \nu_\theta\tilde{s}_\theta^* + \nu_\phi\tilde{s}_\phi^*. \tag{D.7}$$

Practical procedures for calculating the complex spherical harmonic Y_{lm} and its derivative $\partial_\theta Y_{lm}$ are discussed in Appendix B.10.

A receiver at either of the two poles $\theta = 0$ or $\theta = \pi$ constitutes a special case; installation of an ocean-bottom seismometer at the North Pole would be a logistically difficult undertaking, but station SPA at South Pole, Antarctica has been the site of a three-component instrument since the inception of the World-Wide Standard Seismographic Network. Upon making use of the limiting Legendre relations (B.62)–(B.63) we find that equations (D.4)–(D.6) reduce to

$$\tilde{s}_r = \left(\frac{2l+1}{4\pi}\right)^{1/2} u\delta_{m0}, \tag{D.8}$$

$$\tilde{s}_\theta = \tfrac{1}{2}\left(\frac{2l+1}{4\pi}\right)^{1/2} k(v+imw)(\delta_{m\,-1}-\delta_{m1}), \tag{D.9}$$

$$\tilde{s}_\phi = \tfrac{1}{2}\left(\frac{2l+1}{4\pi}\right)^{1/2} k(imv-w)(\delta_{m\,-1}-\delta_{m1}) \tag{D.10}$$

and

$$\tilde{s}_r = (-1)^l\left(\frac{2l+1}{4\pi}\right)^{1/2} u\delta_{m0}, \tag{D.11}$$

$$\tilde{s}_\theta = \tfrac{1}{2}(-1)^l\left(\frac{2l+1}{4\pi}\right)^{1/2} k(v-imw)(\delta_{m\,-1}-\delta_{m1}), \tag{D.12}$$

$$\tilde{s}_\phi = \tfrac{1}{2}(-1)^l\left(\frac{2l+1}{4\pi}\right)^{1/2} k(-imv-w)(\delta_{m\,-1}-\delta_{m1}) \tag{D.13}$$

at $\theta = 0$ and $\theta = \pi$, respectively. The only non-zero terms at the poles are associated with the orders $-1 \leq m \leq 1$. In particular, note that a radial-component ($\hat{\boldsymbol{\nu}} = \hat{\mathbf{r}}$) sensor at station SPA senses only the azimuthally symmetric $m = 0$ basis singlet.

Upon making use of equation (A.139) we find that the spherical polar components of the strain tensor $\tilde{\boldsymbol{\varepsilon}}_k = \frac{1}{2}[\boldsymbol{\nabla}\tilde{\mathbf{s}}_k + (\boldsymbol{\nabla}\tilde{\mathbf{s}}_k)^{\mathrm{T}}]$ are

$$\tilde{\varepsilon}_{rr} = \dot{u}Y_{lm}, \tag{D.14}$$

$$\begin{aligned}\tilde{\varepsilon}_{\theta\theta} &= r^{-1}uY_{lm} - r^{-1}v[\cot\theta\,\partial_\theta Y_{lm} - m^2(\sin\theta)^{-2}Y + k^2Y_{lm}]\\ &\quad + imr^{-1}w(\sin\theta)^{-1}(\partial_\theta Y_{lm} - \cot\theta\, Y_{lm}),\end{aligned} \tag{D.15}$$

$$\begin{aligned}\tilde{\varepsilon}_{\phi\phi} &= r^{-1}uY_{lm} + r^{-1}v[\cot\theta\,\partial_\theta Y_{lm} - m^2(\sin\theta)^{-2}Y_{lm}]\\ &\quad - imr^{-1}w(\sin\theta)^{-1}(\partial_\theta Y_{lm} - \cot\theta\, Y_{lm}),\end{aligned} \tag{D.16}$$

$$\tilde{\varepsilon}_{r\theta} = \tfrac{1}{2}[x\partial_\theta Y_{lm} + imz(\sin\theta)^{-1}Y_{lm}], \tag{D.17}$$

$$\tilde{\varepsilon}_{r\phi} = \tfrac{1}{2}[imx(\sin\theta)^{-1}Y_{lm} - z\partial_\theta Y_{lm}], \tag{D.18}$$

$$\begin{aligned}\tilde{\varepsilon}_{\theta\phi} &= imr^{-1}v(\sin\theta)^{-1}(\partial_\theta Y_{lm} - \cot\theta\, Y_{lm}) \\ &\quad + r^{-1}w[\cot\theta\,\partial_\theta Y_{lm} - m^2(\sin\theta)^{-2}Y_{lm} + \tfrac{1}{2}k^2 Y_{lm}],\end{aligned} \tag{D.19}$$

where we have defined the auxiliary variables

$$x = \dot{v} - r^{-1}v + r^{-1}u, \qquad z = \dot{w} - r^{-1}w. \tag{D.20}$$

Legendre's equation (B.46) has been used to eliminate the dependence upon the second derivative $\partial_\theta^2 Y_{lm}$. The elements $\tilde{s}_k = \mathbf{M}:\tilde{\boldsymbol{\varepsilon}}_k^*(\mathbf{x}_s)$ of the complex source vector $\tilde{\mathbf{s}}$ are given by

$$\begin{aligned}\tilde{s}_k &= M_{rr}\tilde{\varepsilon}^*_{rrs} + M_{\theta\theta}\tilde{\varepsilon}^*_{\theta\theta s} + M_{\phi\phi}\tilde{\varepsilon}^*_{\phi\phi s} \\ &\quad + 2M_{r\theta}\tilde{\varepsilon}^*_{r\theta s} + 2M_{r\phi}\tilde{\varepsilon}^*_{r\phi s} + 2M_{\theta\phi}\tilde{\varepsilon}^*_{\theta\phi s},\end{aligned} \tag{D.21}$$

where the subscript s denotes evaluation at the hypocenter $\mathbf{x}_s$ as usual.

Using (B.62) once again it can be shown that at the North Pole ($\theta = 0$) equations (D.14)–(D.19) reduce to

$$\tilde{\varepsilon}_{rr} = \left(\frac{2l+1}{4\pi}\right)^{1/2} \dot{u}\,\delta_{m0}, \tag{D.22}$$

$$\begin{aligned}\tilde{\varepsilon}_{\theta\theta} = \tfrac{1}{2}\left(\frac{2l+1}{4\pi}\right)^{1/2} &\Big[f\,\delta_{m0} + \tfrac{1}{4}k\sqrt{k^2-2} \\ &\times\; r^{-1}(2v + imw)(\delta_{m\,-2} + \delta_{m2})\Big],\end{aligned} \tag{D.23}$$

$$\begin{aligned}\tilde{\varepsilon}_{\phi\phi} = \tfrac{1}{2}\left(\frac{2l+1}{4\pi}\right)^{1/2} &\Big[f\,\delta_{m0} - \tfrac{1}{4}k\sqrt{k^2-2} \\ &\times\; r^{-1}(2v + imw)(\delta_{m\,-2} + \delta_{m2})\Big],\end{aligned} \tag{D.24}$$

$$\tilde{\varepsilon}_{r\theta} = \tfrac{1}{4}\left(\frac{2l+1}{4\pi}\right)^{1/2} k(x + imz)(\delta_{m\,-1} - \delta_{m1}), \tag{D.25}$$

$$\tilde{\varepsilon}_{r\phi} = \tfrac{1}{4}\left(\frac{2l+1}{4\pi}\right)^{1/2} k(imx - z)(\delta_{m\,-1} - \delta_{m1}), \tag{D.26}$$

$$\tilde{\varepsilon}_{\theta\phi} = \tfrac{1}{8}\left(\frac{2l+1}{4\pi}\right)^{1/2} k\sqrt{k^2-2}$$
$$\times\, r^{-1}(imv - 2w)(\delta_{m\,-2} + \delta_{m2}), \tag{D.27}$$

where

$$f = r^{-1}(2u - k^2 v). \tag{D.28}$$

Gilbert & Dziewonski (1975) used equations (D.22)–(D.27) to calculate the response of a spherically symmetric Earth to a moment-tensor source in an epicentral system of coordinates ($\theta_s = 0$). It is noteworthy that only the singlets $-2 \leq m \leq 2$ are excited. We deduce this result using an alternative argument, which avoids passage to the limit, in Section 10.3.

D.2 Perturbation Matrices

The elements of the kinetic-energy matrix $\tilde{\mathsf{T}}$ are given in terms of the complex basis eigenfunctions (D.3) by

$$\tilde{T}_{kk'} = \int_{\oplus} \delta\rho(\tilde{\mathbf{s}}_k^* \cdot \tilde{\mathbf{s}}_{k'})\, dV - \int_{\Sigma} \delta d\,[\rho\, \tilde{\mathbf{s}}_k^* \cdot \tilde{\mathbf{s}}_{k'}]_-^+\, d\Sigma. \tag{D.29}$$

The quantity $\delta\rho$ is the volumetric perturbation in the density, and δd is the perturbation in the location of the boundaries. It is convenient to separate the elastic-gravitational potential-energy matrix into three terms:

$$\tilde{\mathsf{V}} = \tilde{\mathsf{V}}^{\text{iso}} + \tilde{\mathsf{V}}^{\text{ani}} + \tilde{\mathsf{V}}^{\text{cen}}. \tag{D.30}$$

The first term encapsulates all of the *isotropic* aspherical perturbations, the second allows for the possibility of an additional *anisotropic* perturbation, and the third accounts for the effect of the *centrifugal* potential. The components of these three matrices are given by

$$\tilde{V}_{kk'}^{\text{iso}} = \int_{\oplus} [\delta\kappa(\boldsymbol{\nabla}\cdot\tilde{\mathbf{s}}_k^*)(\boldsymbol{\nabla}\cdot\tilde{\mathbf{s}}_{k'}) + 2\delta\mu(\tilde{\mathbf{d}}_k^* : \tilde{\mathbf{d}}_{k'})$$
$$+ \delta\rho\{\tilde{\mathbf{s}}_k^* \cdot \boldsymbol{\nabla}\tilde{\phi}_{k'} + \tilde{\mathbf{s}}_{k'} \cdot \boldsymbol{\nabla}\tilde{\phi}_k^*$$
$$+ 4\pi G\rho(\hat{\mathbf{r}}\cdot\tilde{\mathbf{s}}_k^*)(\hat{\mathbf{r}}\cdot\tilde{\mathbf{s}}_{k'}) + g\tilde{\Upsilon}_{kk'}\}$$
$$+ \tfrac{1}{2}\rho\boldsymbol{\nabla}(\delta\Phi)\cdot(\tilde{\mathbf{s}}_k^* \cdot \boldsymbol{\nabla}\tilde{\mathbf{s}}_{k'} + \tilde{\mathbf{s}}_{k'} \cdot \boldsymbol{\nabla}\tilde{\mathbf{s}}_k^*$$
$$- \tilde{\mathbf{s}}_k^* \boldsymbol{\nabla}\cdot\tilde{\mathbf{s}}_{k'} - \tilde{\mathbf{s}}_{k'}\boldsymbol{\nabla}\cdot\tilde{\mathbf{s}}_k^*) + \rho\tilde{\mathbf{s}}_k^* \cdot \boldsymbol{\nabla}\boldsymbol{\nabla}(\delta\Phi)\cdot\tilde{\mathbf{s}}_{k'}]\, dV$$
$$- \int_{\Sigma} \delta d\,[\tfrac{1}{2}\kappa_0(\boldsymbol{\nabla}\cdot\tilde{\mathbf{s}}_k^*)(\boldsymbol{\nabla}\cdot\tilde{\mathbf{s}}_{k'} - 2\hat{\mathbf{r}}\cdot\partial_r\tilde{\mathbf{s}}_k)$$
$$+ \tfrac{1}{2}\kappa_0(\boldsymbol{\nabla}\cdot\tilde{\mathbf{s}}_{k'})(\boldsymbol{\nabla}\cdot\tilde{\mathbf{s}}_k^* - 2\hat{\mathbf{r}}\cdot\partial_r\tilde{\mathbf{s}}_k^*)$$

$$+ \mu_0\tilde{\mathbf{d}}_k^*:(\tilde{\mathbf{d}}_{k'} - 2\hat{\mathbf{r}}\partial_r\tilde{\mathbf{s}}_{k'}) + \mu_0\tilde{\mathbf{d}}_{k'}:(\tilde{\mathbf{d}}_k^* - 2\hat{\mathbf{r}}\partial_r\tilde{\mathbf{s}}_k^*)$$
$$+ \rho\{\tilde{\mathbf{s}}_k^* \cdot \boldsymbol{\nabla}\tilde{\phi}_{k'} + \tilde{\mathbf{s}}_{k'} \cdot \boldsymbol{\nabla}\tilde{\phi}_k^*$$
$$+ 8\pi G\rho(\hat{\mathbf{r}} \cdot \tilde{\mathbf{s}}_k^*)(\hat{\mathbf{r}} \cdot \tilde{\mathbf{s}}_{k'}) + g\tilde{\Upsilon}_{kk'}\}]_-^+ \, d\Sigma$$
$$- \int_{\Sigma_{\mathrm{FS}}} \boldsymbol{\nabla}^\Sigma(\delta d) \cdot [\kappa_0(\boldsymbol{\nabla} \cdot \tilde{\mathbf{s}}_k^*)\tilde{\mathbf{s}}_{k'} + \kappa_0(\boldsymbol{\nabla} \cdot \tilde{\mathbf{s}}_{k'})\tilde{\mathbf{s}}_k^*$$
$$+ 2\mu_0(\hat{\mathbf{r}} \cdot \tilde{\mathbf{d}}_k^* \cdot \hat{\mathbf{r}})\tilde{\mathbf{s}}_{k'} + 2\mu_0(\hat{\mathbf{r}} \cdot \tilde{\mathbf{d}}_{k'} \cdot \hat{\mathbf{r}})\tilde{\mathbf{s}}_k^*]_-^+ \, d\Sigma, \tag{D.31}$$

$$\tilde{V}_{kk'}^{\mathrm{ani}} = \int_\oplus (\tilde{\boldsymbol{\varepsilon}}_k^* : \boldsymbol{\gamma} : \tilde{\boldsymbol{\varepsilon}}_{k'}) \, dV, \tag{D.32}$$

$$\tilde{V}_{kk'}^{\mathrm{cen}} = \int_\oplus [\tfrac{1}{2}\rho\boldsymbol{\nabla}\psi \cdot (\tilde{\mathbf{s}}_k^* \cdot \boldsymbol{\nabla}\tilde{\mathbf{s}}_{k'} + \tilde{\mathbf{s}}_{k'} \cdot \boldsymbol{\nabla}\tilde{\mathbf{s}}_k^*$$
$$- \tilde{\mathbf{s}}_k^* \boldsymbol{\nabla} \cdot \tilde{\mathbf{s}}_{k'} - \tilde{\mathbf{s}}_{k'} \boldsymbol{\nabla} \cdot \tilde{\mathbf{s}}_k^*) + \rho\tilde{\mathbf{s}}_k^* \cdot \boldsymbol{\nabla\nabla}\psi \cdot \tilde{\mathbf{s}}_{k'}] \, dV. \tag{D.33}$$

The quantities $\tilde{\boldsymbol{\varepsilon}}_k = \frac{1}{2}[\boldsymbol{\nabla}\tilde{\mathbf{s}}_k + (\boldsymbol{\nabla}\tilde{\mathbf{s}}_k)^{\mathrm{T}}]$ and $\tilde{\mathbf{d}}_k = \tilde{\boldsymbol{\varepsilon}}_k - \frac{1}{3}(\boldsymbol{\nabla} \cdot \tilde{\mathbf{s}}_k)\mathbf{I}$ are the strain and deviatoric strain eigenfunctions, respectively, and

$$\tilde{\Upsilon}_{kk'} = \tfrac{1}{2}[\tilde{\mathbf{s}}_k^* \cdot \boldsymbol{\nabla}(\hat{\mathbf{r}} \cdot \tilde{\mathbf{s}}_{k'}) + \tilde{\mathbf{s}}_{k'} \cdot \boldsymbol{\nabla}(\hat{\mathbf{r}} \cdot \tilde{\mathbf{s}}_k^*)]$$
$$- \tfrac{1}{2}[(\hat{\mathbf{r}} \cdot \tilde{\mathbf{s}}_k^*)(\boldsymbol{\nabla} \cdot \tilde{\mathbf{s}}_{k'}) + (\hat{\mathbf{r}} \cdot \tilde{\mathbf{s}}_{k'})(\boldsymbol{\nabla} \cdot \tilde{\mathbf{s}}_k^*)]$$
$$- 2r^{-1}(\hat{\mathbf{r}} \cdot \tilde{\mathbf{s}}_k^*)(\hat{\mathbf{r}} \cdot \tilde{\mathbf{s}}_{k'}). \tag{D.34}$$

The parameters κ_0 and μ_0 are the isentropic incompressibility and rigidity at the reference frequency ω_0, and the scalars $\delta\kappa$ and $\delta\mu$ and the fourth-order tensor $\boldsymbol{\gamma}$ are the associated isotropic and anisotropic perturbations; the quantity $\delta\Phi$ is the perturbation in the gravitational potential, and ψ is the centrifugal potential. Finally, the elements of the anelasticity and Coriolis matrices are given by

$$\tilde{A}_{kk'} = \int_\oplus [\kappa_0 q_\kappa(\boldsymbol{\nabla} \cdot \tilde{\mathbf{s}}_k^*)(\boldsymbol{\nabla} \cdot \tilde{\mathbf{s}}_{k'}) \, dV + 2\mu_0 q_\mu(\tilde{\mathbf{d}}_k^* : \tilde{\mathbf{d}}_{k'})] \, dV, \tag{D.35}$$

$$\tilde{W}_{kk'} = \int_\oplus \rho\,\tilde{\mathbf{s}}_k^* \cdot (i\boldsymbol{\Omega} \times \tilde{\mathbf{s}}_{k'}) \, dV. \tag{D.36}$$

The quantities $q_\kappa = Q_\kappa^{-1}$ and $q_\mu = Q_\mu^{-1}$ are the reciprocal bulk and shear quality factors, and $\boldsymbol{\Omega}$ is the angular velocity of rotation.

D.2.1 Isotropic asphericity and anelasticity

It is convenient to expand the isotropic elastic and anelastic perturbations in complex spherical harmonics:

$$\delta\kappa = \sum_{s=1}^{s_{\max}} \sum_{t=-s}^{s} \delta\tilde{\kappa}_{st} Y_{st}, \qquad \delta\mu = \sum_{s=1}^{s_{\max}} \sum_{t=-s}^{s} \delta\tilde{\mu}_{st} Y_{st},$$

$$\delta\rho = \sum_{s=1}^{s_{\max}} \sum_{t=-s}^{s} \delta\tilde{\rho}_{st} Y_{st}, \qquad \delta\Phi = \sum_{s=1}^{s_{\max}} \sum_{t=-s}^{s} \delta\tilde{\Phi}_{st} Y_{st},$$

$$q_\kappa = \sum_{s=1}^{s_{\max}} \sum_{t=-s}^{s} \tilde{q}_{\kappa st} Y_{st}, \qquad q_\mu = \sum_{s=1}^{s_{\max}} \sum_{t=-s}^{s} \tilde{q}_{\mu st} Y_{st},$$

$$\delta d = \sum_{s=1}^{s_{\max}} \sum_{t=-s}^{s} \delta\tilde{d}_{st} Y_{st}. \tag{D.37}$$

We affix a tilde to the expansion coefficients $\delta\tilde{\kappa}_{st}$, $\delta\tilde{\mu}_{st}$, $\delta\tilde{\rho}_{st}$, $\delta\tilde{\Phi}_{st}$, $\tilde{q}_{\kappa st}$, $\tilde{q}_{\mu st}$ and $\delta\tilde{d}_{st}$ to distinguish them from the corresponding real expansion coefficients which we shall introduce in Section D.4.2. The commencement of the sums at $s = 1$ rather than $s = 0$ is noteworthy; this guarantees that the perturbation is *strictly aspherical*. The unperturbed spherically symmetric starting model κ_0, μ_0, ρ is presumed to be the *terrestrial monopole*. Formally, the maximum degree $s_{\max}$ may be taken to be infinity; truncation at a finite $s_{\max}$ is of course required in any numerical implementation. The matrix elements $\tilde{T}_{kk'}$ $\tilde{V}^{\rm iso}_{kk'}$ and $\tilde{A}_{kk'}$ are comprised of three-dimensional integrals over the volume of the Earth $\oplus$ and two-dimensional integrals over the external and internal boundaries Σ. The expansions (D.37) allow the integrations over the unit sphere Ω to be performed analytically using the Wigner 3-j formalism developed in Appendix C.7. We give an illustrative example to show how this reduction to a sum of radial integrals can be accomplished before presenting the complete results in Section D.2.3.

D.2.2 Example

The most cumbersome isotropic term to evaluate is the contribution to the isotropic potential-energy matrix element (D.31) from an aspherical *rigidity* perturbation:

$$\tilde{V}^{\rm rig}_{kk'} = \int_\oplus 2\delta\mu(\tilde{\mathbf{d}}^*_k : \tilde{\mathbf{d}}_{k'})\, dV. \tag{D.38}$$

The generalized spherical-harmonic representation of the deviatoric strain eigenfunction $\tilde{\mathbf{d}}_k$ is

$$\tilde{\mathbf{d}}_k = \tilde{d}^{\alpha\beta}\, Y^{\alpha+\beta}_{lm}\, \hat{\mathbf{e}}_\alpha \hat{\mathbf{e}}_\beta, \tag{D.39}$$

where

$$\tilde{d}^{00} = \tfrac{1}{3}(2\dot{u} - f), \qquad \tilde{d}^{\pm\pm} = \tfrac{1}{2}k\sqrt{k^2 - 2}\, r^{-1}(v \pm iw),$$
$$\tilde{d}^{0\pm} = d^{\pm 0} = \tfrac{\sqrt{2}}{4}k(x \pm iz), \qquad \tilde{d}^{\pm\mp} = \tfrac{1}{6}(2\dot{u} - f). \tag{D.40}$$

The contravariant components $\tilde{d}^{\alpha\beta}$ are functions of radius r, degree l and overtone number n, but they are independent of colatitude θ, longitude ϕ and order m, just like the radial eigenfunctions u, v and w. Dropping the explicit summation limits over the indices s and t of the asphericity for simplicity, and making use of equation (C.198), we manipulate the three-dimensional integral (D.38) as follows:

$$\begin{aligned}
\tilde{V}_{kk'}^{\mathrm{rig}} &= \sum_{st} \int_{\oplus} 2\delta\tilde{\mu}_{st} Y_{st} (\tilde{d}^{\alpha\beta} Y_{lm}^{\alpha+\beta} \hat{\mathbf{e}}_{\alpha}\hat{\mathbf{e}}_{\beta})^{*} : (\tilde{d}'^{\eta\sigma} Y_{l'm'}^{\eta+\sigma} \hat{\mathbf{e}}_{\eta}\hat{\mathbf{e}}_{\sigma})\, dV \\
&= \sum_{st} \int_{\oplus} 2\delta\tilde{\mu}_{st}\, \tilde{d}^{\alpha\beta *} \tilde{d}'^{\alpha\beta} Y_{lm}^{\alpha+\beta *} Y_{st} Y_{l'm'}^{\alpha+\beta}\, dV \\
&= \sum_{st} \int_{0}^{a} 2\delta\tilde{\mu}_{st}\, \tilde{d}^{\alpha\beta *} \tilde{d}'^{\alpha\beta}\, r^2 dr \int_{\Omega} Y_{lm}^{\alpha+\beta *} Y_{st} Y_{l'm'}^{\alpha+\beta}\, d\Omega \\
&= \sum_{st} (-1)^{\alpha+\beta+m} \left[\frac{(2l+1)(2s+1)(2l'+1)}{4\pi} \right]^{1/2} \\
&\quad \times \begin{pmatrix} l & s & l' \\ -\alpha-\beta & 0 & \alpha+\beta \end{pmatrix} \begin{pmatrix} l & s & l' \\ -m & t & m' \end{pmatrix} \\
&\quad \times \int_{0}^{a} 2\delta\tilde{\mu}_{st}\, \tilde{d}^{\alpha\beta *} \tilde{d}'^{\alpha\beta}\, r^2 dr \\
&= \sum_{st} (-1)^{m} \left[\frac{(2l+1)(2s+1)(2l'+1)}{4\pi} \right]^{1/2} \begin{pmatrix} l & s & l' \\ -m & t & m' \end{pmatrix} \\
&\quad \times \int_{0}^{a} \delta\tilde{\mu}_{st} \Big[\tfrac{1}{3}(2\dot{u} - f)(2\dot{u}' - f') B_{lsl'}^{(0)+} + (xx' + zz') B_{lsl'}^{(1)+} \\
&\qquad - i(xz' - zx') B_{lsl'}^{(1)-} + r^{-2}(vv' + ww') B_{lsl'}^{(2)+} \\
&\qquad - i r^{-2}(vw' - wv') B_{lsl'}^{(2)-} \Big] r^2 dr,
\end{aligned} \tag{D.41}$$

where the primes on $d'^{\alpha\beta}$ and u', v', w', x', z', f' denote association with the primed eigenfunction $\tilde{\mathbf{s}}_{k'}$, and we have defined the quantities

$$\begin{aligned}
B_{lsl'}^{(N)\pm} &= \tfrac{1}{2}(-1)^N \left[1 \pm (-1)^{l+s+l'} \right] \left[\frac{(l+N)!(l'+N)!}{(l-N)!(l'-N)!} \right]^{1/2} \\
&\quad \times \begin{pmatrix} l & s & l' \\ -N & 0 & N \end{pmatrix}.
\end{aligned} \tag{D.42}$$

In combining terms to arrive at the final result (D.41) we have used the symmetry (C.218) of the Wigner 3-j symbols.

D.2.3 Woodhouse kernels

The elements (D.29), (D.31) and (D.35) of the kinetic-energy matrix $\tilde{\mathsf{T}}$, the isotropic elastic-gravitational potential-energy matrix $\tilde{\mathsf{V}}^{\mathrm{iso}}$ and the anelasticity matrix $\tilde{\mathsf{A}}$ may be written in the final form

$$\tilde{T}_{kk'} = \sum_{st}(-1)^m \left[\frac{(2l+1)(2s+1)(2l'+1)}{4\pi}\right]^{1/2} \begin{pmatrix} l & s & l' \\ -m & t & m' \end{pmatrix} \times \left\{ \int_0^a \delta\tilde{\rho}_{st} T_\rho \, r^2 dr + \sum_d d^2 \delta\tilde{d}_{st} \, [T_d]_-^+ \right\}, \tag{D.43}$$

$$\tilde{V}^{\mathrm{iso}}_{kk'} = \sum_{st}(-1)^m \left[\frac{(2l+1)(2s+1)(2l'+1)}{4\pi}\right]^{1/2} \begin{pmatrix} l & s & l' \\ -m & t & m' \end{pmatrix} \times \left\{ \int_0^a \left(\delta\tilde{\kappa}_{st} V_\kappa + \delta\tilde{\mu}_{st} V_\mu + \delta\tilde{\rho}_{st} V_\rho + \delta\tilde{\Phi}_{st} V_\Phi + \delta\dot{\tilde{\Phi}}_{st} V_{\dot{\Phi}}\right) r^2 dr + \sum_d d^2 \delta\tilde{d}_{st} \, [V_d]_-^+ \right\}, \tag{D.44}$$

$$\tilde{A}_{kk'} = \sum_{st}(-1)^m \left[\frac{(2l+1)(2s+1)(2l'+1)}{4\pi}\right]^{1/2} \begin{pmatrix} l & s & l' \\ -m & t & m' \end{pmatrix} \times \int_0^a \left(\kappa_0 \tilde{q}_{\kappa st} V_\kappa + \mu_0 \tilde{q}_{\mu st} V_\mu\right) r^2 dr. \tag{D.45}$$

The *Woodhouse kernels* T_ρ, T_d, V_κ, V_μ, V_ρ, V_Φ, $V_{\dot{\Phi}}$ and V_d are given by

$$T_\rho = uu' B^{(0)+}_{lsl'} + (vv' + ww') B^{(1)+}_{lsl'} - i(vw' - wv') B^{(1)-}_{lsl'}, \tag{D.46}$$

$$T_d = -\rho T_\rho, \tag{D.47}$$

$$V_\kappa = (\dot{u} + f)(\dot{u}' + f') B^{(0)+}_{lsl'}, \tag{D.48}$$

$$V_\mu = \tfrac{1}{3}(2\dot{u} - f)(2\dot{u}' - f') B^{(0)+}_{lsl'} + (xx' + zz') B^{(1)+}_{lsl'} - i(xz' - zx') B^{(1)-}_{lsl'} + r^{-2}(vv' + ww') B^{(2)+}_{lsl'} - ir^{-2}(vw' - wv') B^{(2)-}_{lsl'}, \tag{D.49}$$

$$V_\rho = \left[u\dot{p}' + \dot{p}u' - \tfrac{1}{2}g(4r^{-1}uu' + fu' + uf') + 8\pi G\rho uu'\right] B^{(0)+}_{lsl'} + r^{-1}\left[(pv' + vp') + \tfrac{1}{2}g(uv' + vu')\right] B^{(1)+}_{lsl'} - ir^{-1}\left[(pw' - wp') + \tfrac{1}{2}g(uw' - wu')\right] B^{(1)-}_{lsl'}, \tag{D.50}$$

$$\begin{aligned} V_\Phi = {} & s(s+1)\rho r^{-2}uu'B^{(0)+}_{lsl'} \\ & + \tfrac{1}{2}\rho r^{-1}(u\dot{v}' - \dot{u}v' + r^{-1}uv' - 2fv')B^{(1)+}_{l'ls} \\ & + \tfrac{1}{2}i\rho r^{-1}(u\dot{w}' - \dot{u}w' + r^{-1}uw' - 2fw')B^{(1)-}_{l'ls} \\ & + \tfrac{1}{2}\rho r^{-1}(\dot{v}u' - v\dot{u}' + r^{-1}vu' - 2vf')B^{(1)+}_{ll's} \\ & - \tfrac{1}{2}i\rho r^{-1}(\dot{w}u' - w\dot{u}' + r^{-1}wu' - 2wf')B^{(1)-}_{ll's}, \end{aligned} \tag{D.51}$$

$$\begin{aligned} V_{\dot\Phi} = {} & -\rho(fu' + uf')B^{(0)+}_{lsl'} \\ & + \tfrac{1}{2}\rho r^{-1}uv'B^{(1)+}_{l'ls} + \tfrac{1}{2}i\rho r^{-1}uw'B^{(1)-}_{l'ls} \\ & + \tfrac{1}{2}\rho r^{-1}vu'B^{(1)+}_{ll's} - \tfrac{1}{2}i\rho r^{-1}wu'B^{(1)-}_{ll's}, \end{aligned} \tag{D.52}$$

$$\begin{aligned} V_d = {} & -\kappa_0 V_\kappa - \mu_0 V_\mu - \rho V_\rho \\ & + \kappa_0 \Big[(2\dot{u}\dot{u}' + \dot{u}f' + f\dot{u}')B^{(0)+}_{lsl'} \\ & \qquad - r^{-1}(\dot{u} + f)v'B^{(1)+}_{l'ls} - ir^{-1}(\dot{u} + f)w'B^{(1)-}_{l'ls} \\ & \qquad - r^{-1}v(\dot{u}' + f')B^{(1)+}_{ll's} + ir^{-1}w(\dot{u}' + f')B^{(1)-}_{ll's} \Big] \\ & + \mu_0 \Big[\tfrac{2}{3}(4\dot{u}\dot{u}' - \dot{u}f' - f\dot{u}')B^{(0)+}_{lsl'} \\ & \qquad + (\dot{v}x' + x\dot{v}' + \dot{w}z' + z\dot{w}')B^{(1)+}_{lsl'} \\ & \qquad - i(\dot{v}z' - z\dot{v}' + x\dot{w}' - \dot{w}x')B^{(1)-}_{lsl'} \\ & \qquad - \tfrac{2}{3}r^{-1}(2\dot{u} - f)v'B^{(1)+}_{l'ls} - \tfrac{2}{3}ir^{-1}(2\dot{u} - f)w'B^{(1)-}_{l'ls} \\ & \qquad - \tfrac{2}{3}r^{-1}v(2\dot{u}' - f')B^{(1)+}_{ll's} + \tfrac{2}{3}ir^{-1}w(2\dot{u}' - f')B^{(1)-}_{ll's} \Big]. \end{aligned} \tag{D.53}$$

These kernels depend upon the mode indices k, k' and the degree s, but they are independent of the order t. The terms in (D.44) involving the potential perturbations $\delta\Phi$ and $\delta\dot{\Phi}$ may be eliminated, if desired, by making use of the explicit representation

$$\begin{aligned} \delta\tilde{\Phi}_{st} = -\frac{4\pi G}{2s+1} \Bigg\{ & r^{-s-1}\left(\int_0^r r^{s+2}\delta\tilde{\rho}_{st}\,dr - \sum_{d<r} d^{s+2}\delta\tilde{d}_{st}[\rho]^+_- \right) \\ & + r^s \left(\int_r^a r^{-s+1}\delta\tilde{\rho}_{st}\,dr - \sum_{d>r} d^{-s+1}\delta\tilde{d}_{st}[\rho]^+_- \right) \Bigg\}. \end{aligned} \tag{D.54}$$

Upon inserting (D.54) and integrating by parts, we obtain an alternative expression for the potential-energy matrix element $\tilde{V}^{\rm iso}_{kk'}$. To avoid rewriting, we simply note that the combined effect of the substitutions

$$V_\rho \to V_\rho + \frac{4\pi G}{2s+1}\left\{ r^s \int_r^a r^{-s}\left[(s+1)V_{\dot{\Phi}} - rV_\Phi\right] dr \right.$$
$$\left. - r^{-s-1}\int_0^r r^{s+1}\left(sV_{\dot{\Phi}} + rV_\Phi\right) dr \right\}, \tag{D.55}$$

$$V_\Phi \to 0, \qquad V_{\dot{\Phi}} \to 0 \tag{D.56}$$

is to leave the final result unchanged. We have stipulated in (D.37) that the complex spherical-harmonic expansions of $\delta\kappa$, $\delta\mu$, $\delta\rho$, $\delta\Phi$, q_κ, q_μ and δd all begin at $s = 1$; however, the results (D.43)–(D.56) are applicable to an $s = 0$ perturbation of the terrestrial monopole as well. Inspection of these Woodhouse-kernel expressions reveals that the positive-superscript factors $B^{(1)+}$, $B^{(2)+}$ are associated with spheroidal-spheroidal and toroidal-toroidal coupling, whereas the negative-superscript factors $B^{(1)-}$, $B^{(2)-}$ are associated with spheroidal-toroidal coupling.

It is possible to write the kernels T_ρ, T_d, V_κ, V_μ, V_ρ, V_Φ, $V_{\dot{\Phi}}$ and V_d solely in terms of 3-j symbols of the form

$$\begin{pmatrix} l & s & l' \\ 0 & 0 & 0 \end{pmatrix} \quad \text{and} \quad \begin{pmatrix} l+1 & s+1 & l'+1 \\ 0 & 0 & 0 \end{pmatrix} \tag{D.57}$$

by making use of the identities

$$B_{lsl'}^{(1)+} = \tfrac{1}{2}(L + L' - S)B_{lsl'}^{(0)+}, \tag{D.58}$$

$$B_{lsl'}^{(2)+} = \tfrac{1}{2}[(L + L' - S)(L + L' - S - 2) - 2LL']B_{lsl'}^{(0)+}, \tag{D.59}$$

$$B_{lsl'}^{(1)-} = \tfrac{1}{2}\sqrt{(\Sigma + 1 - 2l)(\Sigma + 1 - 2l')(\Sigma + 1 - 2s)}$$
$$\times \sqrt{(\Sigma + 2)(\Sigma + 4)/(\Sigma + 3)}\, B_{l+1\,s+1\,l'+1}^{(0)+}, \tag{D.60}$$

$$B_{lsl'}^{(2)-} = \tfrac{1}{2}(L + L' - S - 2)$$
$$\times \sqrt{(\Sigma + 1 - 2l)(\Sigma + 1 - 2l')(\Sigma + 1 - 2s)}$$
$$\times \sqrt{(\Sigma + 2)(\Sigma + 4)/(\Sigma + 3)}\, B_{l+1\,s+1\,l'+1}^{(0)+}, \tag{D.61}$$

where we have let $\Sigma = l + l' + s$ and $L = l(l+1)$, $L' = l'(l'+1)$ and $S = s(s+1)$. These results follow from the 3-j identities (C.221)–(C.224).

All of the formulae tabulated in this section were first derived, correctly, by Woodhouse (1980). Our kernels (D.46)–(D.53) agree with his equations (A36)–(A42) after rectification of an obvious index transposition in the final line of the latter ($B_{ll''l'}^{(0)+} \to B_{l'l''l}^{(0)+}$).

D.2.4 Direct numerical integration

The spherical-harmonic representation of sharp lateral gradients such as those associated with continental margins, active or extinct subduction zones, or mid-oceanic ridges requires a large maximum angular degree $s_{\max}$ in the truncated expansions (D.37). The computational burden of calculating the matrix elements $\tilde{T}_{kk'}$, $\tilde{V}^{\rm iso}_{kk'}$ and $\tilde{A}_{kk'}$ in terms of Wigner 3-j symbols may then be substantial. To reduce this burden, Lognonné & Romanowicz (1990) have developed an alternative *spectral* method of evaluating the integrals in equations (D.29) and (D.31)–(D.35). We describe this numerical integration procedure briefly, considering as before the aspherical rigidity-perturbation element $\tilde{V}^{\rm rig}_{kk'}$. Forsaking the spherical-harmonic expansion of $\delta\mu$, let us rewrite the third line of equation (D.41) in the form

$$\tilde{V}^{\rm rig}_{kk'} = \int_0^{2\pi}\left[\int_0^{\pi} X^{\alpha+\beta}_{lm}(\theta)\, M(\theta,\phi)\sin\theta\, d\theta\right]e^{-i(m-m')\phi}\,d\phi, \qquad \text{(D.62)}$$

where

$$M(\theta,\phi) = \left[\int_0^{a} 2\delta\mu(r,\theta,\phi)\,d^{\alpha\beta*}(r)d'^{\alpha\beta}(r)\,r^2dr\right]X^{\alpha+\beta}_{l'm'}(\theta). \qquad \text{(D.63)}$$

We may regard (D.62) as a generalized *Legendre-Fourier transform* of the function (D.63). Every term in $\tilde{T}_{kk'}$, $\tilde{V}^{\rm iso}_{kk'}$ and $\tilde{A}_{kk'}$ may be given a similar interpretation; the multiplier in every transform is $X^{N}_{lm}(\theta)\,e^{-i(m-m')\phi}$ where $-2\leq N\leq 2$.

The interior integral over colatitude $0\leq\theta\leq\pi$ may be evaluated using *Gauss-Legendre quadrature*; we recall the salient features of this method (Press, Flannery, Teukolsky & Vetterling 1992). For every finite integer I, there is one and only one set of points or knots $\mu_i=\cos\theta_i$, $i=1,2,\ldots,I$, and associated weights w_i, $i=1,2,\ldots,I$, such that

$$\int_0^{\pi} Q(\cos\theta)\sin\theta\, d\phi = \int_{-1}^{1} Q(\mu)\,d\mu = \sum_{i=1}^{I} w_iQ(\mu_i) \qquad \text{(D.64)}$$

is an identity for every polynomial $Q(\mu)$ of degree less than $2I-1$. The knots μ_i are the roots of the Legendre polynomial $P_I(\mu)$ and the weights w_i are given by

$$w_i = \frac{2}{(1-\mu_i^2)[dP_l/d\mu]^2_{\mu=\mu_i}}. \qquad \text{(D.65)}$$

The integral of an arbitrary function of colatitude $f(\theta)$ may be approximated by a result analogous to (D.64):

$$\int_0^{\pi} f(\theta)\sin\theta\, d\theta \approx \sum_{i=1}^{I} w_if(\theta_i). \qquad \text{(D.66)}$$

To evaluate the colatitudinal term in (D.62) we set $f(\theta) = X_{lm}^{\alpha+\beta}(\theta)M(\theta,\phi)$.

The integral over longitude $0 \leq \phi \leq 2\pi$ must be evaluated for all orders $-l-l' \leq m-m' \leq l+l'$. This can be accomplished using the fast Fourier transform in an extremely efficient manner. The matrix element (D.62) is given in this case by

$$\tilde{V}_{kk'}^{\rm rig} \approx \frac{2\pi}{J} \sum_{j=0}^{J-1} \sum_{i=1}^{I} w_i X_{lm}^{\alpha+\beta}(\theta_i)\, M(\theta_i,\phi_j)\, e^{-i(m-m')\phi_j}, \tag{D.67}$$

where $\phi_j = 2\pi j/J$, $j = 0,\ldots,J-1$. The accuracy of the approximation (D.67) depends upon the number of Gauss-Legendre and Fourier knots. To avoid errors due to aliasing it is necessary to take $I \approx \frac{1}{2}(l + l' + s_{\rm max})$ and $J \approx 2(l+l'+s_{\rm max})$. A direct integration scheme based upon (D.67) and its generalizations is a more versatile method of calculating the matrix elements $\tilde{T}_{kk'}$, $\tilde{V}_{kk'}^{\rm iso}$ and $\tilde{A}_{kk'}$ because it is not restricted to spherical-harmonic models of the form (D.37). Any type of representation that allows the perturbations $\delta\kappa$, $\delta\mu$ $\delta\rho$, $\delta\Phi$, q_κ, q_μ, δd to be calculated at the geographical knots θ_i, ϕ_j may be used. In most applications it is advantageous to calculate all of the radial integrals such as $\int_0^a 2\delta\mu(r,\theta_i,\phi_j) d^{\alpha\beta *}(r) d'^{\alpha\beta}(r)\, r^2 dr$ prior to numerical Gauss-Legendre-Fourier transformation.

D.2.5 Rotation

The elements (D.36) of the complex Coriolis matrix may be reduced to

$$\begin{aligned}\tilde{W}_{kk'} &= m\Omega\delta_{ll'}\delta_{mm'} \int_0^a \rho W^{\rm S}\, r^2 dr \\ &\quad - i\Omega(S_{lm}\delta_{l\,l'+1} + S_{l'm}\delta_{l\,l'-1})\delta_{mm'} \int_0^a \rho W^{\rm A}\, r^2 dr,\end{aligned} \tag{D.68}$$

where

$$S_{lm} = \left[\frac{(l+m)(l-m)}{(2l+1)(2l-1)}\right]^{1/2}. \tag{D.69}$$

The symmetric kernel $W^{\rm S}$ governing self and spheroidal-spheroidal coupling and the anti-symmetric kernel $W^{\rm A}$ governing spheroidal-toroidal coupling are defined by

$$W^{\rm S} = vv' + uv' + vu' + ww', \tag{D.70}$$

$$\begin{aligned}W^{\rm A} &= \tfrac{1}{2}(k^2 - k'^2 - 2)uw' + \tfrac{1}{2}(k^2 - k'^2 + 2)wu' \\ &\quad + \tfrac{1}{2}(k^2 + k'^2 - 2)(vw' - wv').\end{aligned} \tag{D.71}$$

The elements (D.33) of the centrifugal-potential matrix may be written in the form

$$\begin{aligned} \tilde{V}^{\rm cen}_{kk'} &= \tfrac{2}{3}\Omega^2\delta_{\sigma\sigma'}\delta_{nn'}\delta_{ll'}\delta_{mm'} - \tfrac{2}{3}k^2\Omega^2\delta_{ll'}\delta_{mm'}\int_0^a \rho W^{\rm S}\, r^2 dr \\ &\quad + (-1)^m\sqrt{(2l+1)(2l'+1)}\begin{pmatrix} l & 2 & l' \\ -m & 0 & m \end{pmatrix}\delta_{mm'} \\ &\quad\quad \times \int_0^a \left(\tfrac{1}{3}\Omega^2 r^2 V_\Phi^{s=2} + \tfrac{2}{3}\Omega^2 r V_{\dot{\Phi}}^{s=2}\right) r^2 dr, \end{aligned} \tag{D.72}$$

where σ denotes either S or T, so that $\delta_{\sigma\sigma'}\delta_{nn'}\delta_{ll'}\delta_{mm'} = \delta_{kk'}$. The first two terms in equation (D.72) represent the contribution from the spherically averaged centrifugal potential $\bar{\psi} = -\frac{1}{3}\Omega^2 r^2$, and the third and final term is the contribution from the aspherical potential $\psi - \bar{\psi} = \frac{1}{3}\Omega^2 r^2 P_2(\cos\theta)$. The quantities $V_\Phi^{s=2}$ and $V_{\dot{\Phi}}^{s=2}$ are the degree-two Woodhouse kernels (D.50) and (D.51). It is noteworthy that there is no rotational coupling between two toroidal multiplets ${}_n\mathrm{T}_l$ and ${}_{n'}\mathrm{T}_{l'}$.

D.2.6 Ellipticity

The Earth's hydrostatic ellipticity is a degree-two aspherical perturbation of the form (14.31):

$$\delta\kappa = \tfrac{2}{3}r\varepsilon\dot{\kappa}P_2(\cos\theta), \qquad \delta\mu = \tfrac{2}{3}r\varepsilon\dot{\mu}P_2(\cos\theta),$$

$$\delta\rho = \tfrac{2}{3}r\varepsilon\dot{\rho}P_2(\cos\theta), \qquad \delta\Phi = \tfrac{2}{3}(r\varepsilon g - \tfrac{1}{2}\Omega^2 r^2)P_2(\cos\theta),$$

$$\delta d = -\tfrac{2}{3}d\varepsilon_d P_2(\cos\theta). \tag{D.73}$$

We decompose the kinetic and isotropic elastic-gravitational potential energy matrices into a part due to the *ellipticity* and a part due to any additional *lateral heterogeneity*:

$$\tilde{\mathsf{T}} = \tilde{\mathsf{T}}^{\rm ell} + \tilde{\mathsf{T}}^{\rm lat}, \qquad \tilde{\mathsf{V}}^{\rm iso} = \tilde{\mathsf{V}}^{\rm ell} + \tilde{\mathsf{V}}^{\rm lat}. \tag{D.74}$$

The ellipticity matrix elements obtained by substituting equations (D.73) into (D.43)–(D.44) are

$$\begin{aligned} \tilde{T}^{\rm ell}_{kk'} &= (-1)^m\sqrt{(2l+1)(2l'+1)}\begin{pmatrix} l & 2 & l' \\ -m & 0 & m \end{pmatrix}\delta_{mm'} \\ &\quad \times\left\{\int_0^a \tfrac{2}{3}r\varepsilon\dot{\rho}T_\rho^{s=2}\, r^2 dr - \sum_d \tfrac{2}{3}d^3\varepsilon_d [T_d^{s=2}]_-^+\right\}, \end{aligned} \tag{D.75}$$

$$\tilde{V}^{\text{ell}}_{kk'} = (-1)^m \sqrt{(2l+1)(2l'+1)} \begin{pmatrix} l & 2 & l' \\ -m & 0 & m \end{pmatrix} \delta_{mm'}$$
$$\times \Big\{ \int_0^a \Big(\tfrac{2}{3} r\varepsilon\dot{\kappa} V_\kappa^{s=2} + \tfrac{2}{3} r\varepsilon\dot{\mu} V_\mu^{s=2} + \tfrac{2}{3} r\varepsilon\dot{\rho} V_\rho^{s=2}$$
$$+ \tfrac{2}{3}(r\varepsilon g - \tfrac{1}{2}\Omega^2 r^2) V_\Phi^{s=2} + \tfrac{2}{3}[\varepsilon(\eta+1)g + r\varepsilon\dot{g} - \Omega^2 r] V_{\dot{\Phi}}^{s=2} \Big) r^2 dr$$
$$- \sum_d \tfrac{2}{3} d^3 \varepsilon_d [V_d^{s=2}]_-^+ \Big\}, \tag{D.76}$$

where we have let

$$\eta = r\dot{\varepsilon}/\varepsilon. \tag{D.77}$$

The sums over the discontinuities d may be incorporated into the radial integrals by noting that

$$\sum_d \tfrac{2}{3} d^3 \varepsilon_d [T_d^{s=2}]_-^+ = -\int_0^a \tfrac{2}{3}\big[\varepsilon(\eta+3) T_\rho^{s=2} + r\varepsilon \dot{T}_\rho^{s=2}\big] r^2 dr, \tag{D.78}$$

$$\sum_d \tfrac{2}{3} d^3 \varepsilon_d [V_d^{s=2}]_-^+ = -\int_0^a \tfrac{2}{3}\big[\varepsilon(\eta+3) V_\rho^{s=2} + r\varepsilon \dot{V}_\rho^{s=2}\big] r^2 dr. \tag{D.79}$$

The derivatives $\dot{T}_\rho^{s=2}$ and $\dot{V}_\rho^{s=2}$ in (D.78)–(D.79) can be eliminated by making use of the radial equations governing u, v, w and p. A significant amount of algebra is required to carry this reduction process to completion. The reward for this drudgery is, however, substantial: the model derivatives $\dot{\kappa}$, $\dot{\mu}$ and $\dot{\rho}$ are also eliminated! This enables the ellipticity matrix elements to be written in a form that does not require numerical differentiation:

$$\tilde{T}^{\text{ell}}_{kk'} = (R_{lm}\delta_{ll'} + \tfrac{3}{2} S_{lm} S_{l'+1\,m} \delta_{l\,l'+2} + \tfrac{3}{2} S_{l+1\,m} S_{l'm} \delta_{l\,l'-2}) \delta_{mm'}$$
$$\times \int_0^a \tfrac{2}{3}\varepsilon\rho\big[\bar{T}_\rho^{\text{S}} - (\eta+3)\check{T}_\rho^{\text{S}}\big] r^2 dr$$
$$- 3im(S_{lm}\delta_{l\,l'+1} + S_{l'm}\delta_{l\,l'-1})\delta_{mm'}$$
$$\times \int_0^a \tfrac{2}{3}\varepsilon\rho\big[\bar{T}_\rho^{\text{A}} - (\eta+3)\check{T}_\rho^{\text{A}}\big] r^2 dr, \tag{D.80}$$

$$\tilde{V}^{\text{ell}}_{kk'} = (R_{lm}\delta_{ll'} + \tfrac{3}{2} S_{lm} S_{l'+1\,m} \delta_{l\,l'+2} + \tfrac{3}{2} S_{l+1\,m} S_{l'm} \delta_{l\,l'-2}) \delta_{mm'}$$
$$\times \int_0^a \tfrac{2}{3}\varepsilon\Big\{\kappa\big[\bar{V}_\kappa^{\text{S}} - (\eta+1)\check{V}_\kappa^{\text{S}}\big] + \mu\big[\bar{V}_\mu^{\text{S}} - (\eta+1)\check{V}_\mu^{\text{S}}\big]$$
$$+ \rho\big[\bar{V}_\rho^{\text{S}} - (\eta+3)\check{V}_\rho^{\text{S}}\big]\Big\} r^2 dr$$
$$- 3im(S_{lm}\delta_{l\,l'+1} + S_{l'm}\delta_{l\,l'-1})\delta_{mm'}$$

$$\times \int_0^a \tfrac{2}{3}\varepsilon\Big\{-\kappa(\eta+2)\bar{V}_\kappa^{\rm A} + \mu\big[\bar{V}_\mu^{\rm A} - (\eta+1)\check{V}_\mu^{\rm A}\big]$$
$$+\rho\big[\bar{V}_\rho^{\rm A} - (\eta+3)\check{V}_\rho^{\rm A}\big]\Big\}r^2 dr$$
$$-(-1)^m\sqrt{(2l+1)(2l'+1)}\begin{pmatrix} l & 2 & l' \\ -m & 0 & m \end{pmatrix}\delta_{mm'}$$
$$\times \int_0^a \left(\tfrac{1}{3}\Omega^2 r^2 V_\Phi^{s=2} + \tfrac{2}{3}\Omega^2 r V_{\dot\Phi}^{s=2}\right) r^2 dr, \qquad \text{(D.81)}$$

where

$$R_{lm} = \frac{l(l+1)-3m^2}{(2l+3)(2l-1)} \qquad \text{(D.82)}$$

and

$$\bar{T}_\rho^{\rm S} = \tfrac{1}{2}(k^2 - k'^2 - 6)uv' - \tfrac{1}{2}(k^2 - k'^2 + 6)vu', \qquad \text{(D.83)}$$

$$\check{T}_\rho^{\rm S} = uu' + \tfrac{1}{2}(k^2 + k'^2 - 6)(vv' + ww'), \qquad \text{(D.84)}$$

$$\bar{T}_\rho^{\rm A} = uw' - wu', \qquad \text{(D.85)}$$

$$\check{T}_\rho^{\rm A} = vw' - wv', \qquad \text{(D.86)}$$

$$\bar{V}_\kappa^{\rm S} = -[\dot{u} + \tfrac{1}{2}(k^2 - k'^2 + 6)r^{-1}v](\dot{u}' + f')$$
$$-(\dot{u} + f)[\dot{u}' - \tfrac{1}{2}(k^2 - k'^2 - 6)r^{-1}v'], \qquad \text{(D.87)}$$

$$\bar{V}_\mu^{\rm S} = -\tfrac{1}{3}[2\dot{u} + \tfrac{1}{2}(k^2 - k'^2 + 6)(3\dot{v} - 4r^{-1}v)](2\dot{u}' - f')$$
$$-[\tfrac{1}{2}(k^2 - k'^2 - 6)\dot{u} + \tfrac{1}{2}(k^2 + k'^2 + 6)\dot{v}$$
$$+\tfrac{1}{2}(k^2 - k'^2 + 6)k'^2 r^{-1}v]x' + [\tfrac{1}{2}(k^2 - k'^2 + 6)k'^2$$
$$+3(k^2 + k'^2 - 6)]r^{-1}(v\dot{v}' + w\dot{w}') - \tfrac{1}{2}(k^2 + k'^2 - 6)z\dot{w}'$$
$$-\tfrac{1}{3}(2\dot{u} - f)[2\dot{u}' - \tfrac{1}{2}(k^2 - k'^2 - 6)(3\dot{v}' - 4r^{-1}v')]$$
$$+x[\tfrac{1}{2}(k^2 - k'^2 + 6)\dot{u}' - \tfrac{1}{2}(k^2 + k'^2 - 6)\dot{v}'$$
$$+\tfrac{1}{2}k^2(k^2 - k'^2 - 6)r^{-1}v'] - [\tfrac{1}{2}k^2(k^2 - k'^2 - 6)$$
$$-3(k^2 + k'^2 - 6)]r^{-1}(\dot{v}v' + \dot{w}w') - \tfrac{1}{2}(k^2 + k'^2 - 6)\dot{w}z', \qquad \text{(D.88)}$$

$$\bar{V}_\rho^{\rm S} = (r\dot{p} + 4\pi G\rho r u + gu)f' - \tfrac{1}{2}(k^2 - k'^2 + 6)r^{-1}gvu'$$
$$+3r^{-1}guu' + r^{-1}p[\tfrac{1}{2}(k^2 + k'^2 - 6)v' - k^2 u']$$
$$+f(r\dot{p}' + 4\pi G\rho r u' + gu') + \tfrac{1}{2}(k^2 - k'^2 - 6)r^{-1}guv'$$
$$+3r^{-1}guu' + r^{-1}[\tfrac{1}{2}(k^2 + k'^2 - 6)v - k'^2 u]p', \qquad \text{(D.89)}$$

$$\begin{aligned}\check{V}^{\rm S}_{\kappa} &= \tfrac{1}{2}[-\dot{u} + f + (k^2 - k'^2 + 6)r^{-1}v](\dot{u}' + f')\\ &\quad + \tfrac{1}{2}(\dot{u} + f)[-\dot{u}' + f' - (k^2 - k'^2 - 6)r^{-1}v'],\end{aligned} \tag{D.90}$$

$$\begin{aligned}\check{V}^{\rm S}_{\mu} &= \tfrac{1}{2}[(k^2 + k'^2 - 8)(k^2 + k'^2 - 6) - 2k^2k'^2]r^{-2}(vv' + ww')\\ &\quad + \tfrac{1}{2}(k^2 + k'^2 - 6)(xx' + zz' - \dot{v}x' - \dot{w}z' - x\dot{v}' - z\dot{w}')\\ &\quad - \tfrac{1}{3}[\dot{u} + \tfrac{1}{2}f - (k^2 - k'^2 + 6)r^{-1}v](2\dot{u}' - f')\\ &\quad - \tfrac{1}{3}(2\dot{u} - f)[\dot{u}' + \tfrac{1}{2}f + (k^2 - k'^2 - 6)r^{-1}v'],\end{aligned} \tag{D.91}$$

$$\begin{aligned}\check{V}^{\rm S}_{\rho} &= \tfrac{1}{2}u[2\dot{p}' + 8\pi G\rho u' + (k^2 - k'^2 - 6)gr^{-1}v']\\ &\quad + \tfrac{1}{2}[2\dot{p} + 8\pi G\rho u - (k^2 - k'^2 + 6)gr^{-1}v]u'\\ &\quad + \tfrac{1}{2}(k^2 + k'^2 - 6)r^{-1}(vp' + pv'),\end{aligned} \tag{D.92}$$

$$\begin{aligned}\bar{V}^{\rm A}_{\mu} &= \dot{w}[2\dot{v}' - \dot{u}' + 3r^{-1}u' + (k^2 - k'^2 - 7)r^{-1}v']\\ &\quad + r^{-1}w[\tfrac{5}{3}\dot{u}' - 7\dot{v}' + \tfrac{7}{3}k'^2r^{-1}v' - (k'^2 + \tfrac{8}{3})r^{-1}u']\\ &\quad - [2\dot{v} - \dot{u} + 3r^{-1}u - (k^2 - k'^2 + 7)r^{-1}v]\dot{w}'\\ &\quad - r^{-1}[\tfrac{5}{3}\dot{u} - 7\dot{v} + \tfrac{7}{3}k^2r^{-1}v - (k^2 + \tfrac{8}{3})r^{-1}u]w',\end{aligned} \tag{D.93}$$

$$\bar{V}^{\rm A}_{\rho} = r^{-1}(p + gu)w' - r^{-1}w(p' + gu'), \tag{D.94}$$

$$\check{V}^{\rm A}_{\kappa} = r^{-1}w(\dot{u}' + f') - r^{-1}(\dot{u} + f)w', \tag{D.95}$$

$$\begin{aligned}\check{V}^{\rm A}_{\mu} &= r^{-2}w(u' - v') + \tfrac{2}{3}r^{-1}w(2\dot{u}' - f') + \dot{w}\dot{v}'\\ &\quad - (k^2 + k'^2 - 8)r^{-2}wv' - r^{-2}(u - v)w'\\ &\quad - \tfrac{2}{3}r^{-1}(2\dot{u} - f)w - \dot{v}\dot{w}' + (k^2 + k'^2 - 8)r^{-2}vw',\end{aligned} \tag{D.96}$$

$$\check{V}^{\rm A}_{\rho} = \bar{V}^{\rm A}_{\rho}. \tag{D.97}$$

The misprints noted by Henson (1989) and Shibata, Suda & Fukao (1990) in equations (A23)–(A24), (A30)–(A31) and (A34) of Woodhouse (1980) have been confirmed and corrected here. It is noteworthy that a portion of the $s = 2$ gravitational potential perturbation (D.73) is equal and opposite to the aspherical centrifugal potential $\psi - \bar{\psi}$. This accounts for the cancellation of the final terms in equations (D.81) and (D.72) upon combining to form the elements of the composite matrix $\tilde{\mathsf{V}}^{\rm ell+cen}$.

D.2.7 Anisotropy

The generalized spherical-harmonic representation of the strain eigenfunction $\tilde{\boldsymbol{\varepsilon}}_k$ is

$$\tilde{\boldsymbol{\varepsilon}}_k = \tilde{\varepsilon}^{\alpha\beta}\mathbf{Y}_{lm}^{\alpha+\beta} = \tilde{\varepsilon}^{\alpha\beta}\,Y_{lm}^{\alpha+\beta}\,\hat{\mathbf{e}}_\alpha\hat{\mathbf{e}}_\beta, \tag{D.98}$$

where

$$\tilde{\varepsilon}^{00} = \dot{u}, \qquad \tilde{\varepsilon}^{\pm\pm} = \tfrac{1}{2}k\sqrt{k^2-2}\,r^{-1}(v \pm iw),$$
$$\tilde{\varepsilon}^{0\pm} = d^{\pm 0} = \tfrac{\sqrt{2}}{4}k(x \pm iz), \qquad \tilde{\varepsilon}^{\pm\mp} = -\tfrac{1}{2}f. \tag{D.99}$$

The fourth-order anisotropic elastic tensor γ can likewise be expanded in the form

$$\boldsymbol{\gamma} = \sum_{st} \gamma_{st}^{\alpha\beta\zeta\eta} Y_{st}^{\alpha+\beta+\zeta+\eta}\hat{\mathbf{e}}_\alpha\hat{\mathbf{e}}_\beta\hat{\mathbf{e}}_\zeta\hat{\mathbf{e}}_\eta. \tag{D.100}$$

The canonical contravariant components $\gamma^{\alpha\beta\zeta\eta}$ are related to the ordinary spherical polar components $\gamma_{rrrr}, \ldots, \gamma_{\phi\phi\phi\phi}$ by

$$\gamma^{0000} = \gamma_{rrrr}, \tag{D.101}$$

$$\gamma^{++--} = \tfrac{1}{4}\gamma_{\theta\theta\theta\theta} + \tfrac{1}{4}\gamma_{\phi\phi\phi\phi} - \tfrac{1}{2}\gamma_{\theta\theta\phi\phi} + \gamma_{\theta\phi\theta\phi}, \tag{D.102}$$

$$\gamma^{+-+-} = \tfrac{1}{4}\gamma_{\theta\theta\theta\theta} + \tfrac{1}{4}\gamma_{\phi\phi\phi\phi} + \tfrac{1}{2}\gamma_{\theta\theta\phi\phi}, \tag{D.103}$$

$$\gamma^{+-00} = -\tfrac{1}{2}(\gamma_{\theta\theta rr} + \gamma_{\phi\phi rr}), \tag{D.104}$$

$$\gamma^{+0-0} = -\tfrac{1}{2}(\gamma_{\theta r\theta r} + \gamma_{\phi r\phi r}), \tag{D.105}$$

$$\gamma^{\pm 000} = \mp\tfrac{1}{\sqrt{2}}\gamma_{\theta rrr} + \tfrac{i}{\sqrt{2}}\gamma_{\phi rrr}, \tag{D.106}$$

$$\gamma^{\pm\pm\mp 0} = \pm\tfrac{1}{2\sqrt{2}}(\gamma_{\theta\theta\theta r} + 2\gamma_{\theta\phi\phi r} - \gamma_{\phi\phi\theta r})$$
$$+ \tfrac{i}{2\sqrt{2}}(\gamma_{\theta\theta\phi r} - 2\gamma_{\theta\phi\theta r} - \gamma_{\phi\phi\phi r}), \tag{D.107}$$

$$\gamma^{+-\pm 0} = \pm\tfrac{1}{2\sqrt{2}}(\gamma_{\theta\theta\theta r} + \gamma_{\phi\phi\theta r}) - \tfrac{i}{2\sqrt{2}}(\gamma_{\theta\theta\phi r} + \gamma_{\phi\phi\phi r}), \tag{D.108}$$

$$\gamma^{\pm\pm 00} = \tfrac{1}{2}(\gamma_{\theta\theta rr} - \gamma_{\phi\phi rr}) \mp i\gamma_{\theta\phi rr}, \tag{D.109}$$

$$\gamma^{\pm 0\pm 0} = \tfrac{1}{2}(\gamma_{\theta r\theta r} - \gamma_{\phi r\phi r}) \mp i\gamma_{\theta r\phi r}, \tag{D.110}$$

$$\gamma^{\pm\pm +-} = -\tfrac{1}{4}(\gamma_{\theta\theta\theta\theta} - \gamma_{\phi\phi\phi\phi}) \pm \tfrac{i}{2}(\gamma_{\theta\theta\theta\phi} + \gamma_{\theta\phi\phi\phi}), \tag{D.111}$$

$$\begin{aligned}\gamma^{\pm\pm\pm 0} &= \mp\tfrac{1}{2\sqrt{2}}(\gamma_{\theta\theta\theta r} - 2\gamma_{\theta\phi\phi r} - \gamma_{\phi\phi\theta r}) \\ &\quad + \tfrac{i}{2\sqrt{2}}(\gamma_{\theta\theta\phi r} + 2\gamma_{\theta\phi\theta r} - \gamma_{\phi\phi\phi r}),\end{aligned} \tag{D.112}$$

$$\begin{aligned}\gamma^{\pm\pm\pm\pm} &= \tfrac{1}{4}\gamma_{\theta\theta\theta\theta} + \tfrac{1}{4}\gamma_{\phi\phi\phi\phi} - \tfrac{1}{2}\gamma_{\theta\theta\phi\phi} - \gamma_{\theta\dot{\varphi}\theta\phi} \\ &\quad \mp i(\gamma_{\theta\theta\theta\phi} - \gamma_{\theta\phi\phi\phi}).\end{aligned} \tag{D.113}$$

The mechanical and thermodynamical symmetries of the elastic tensor guarantee that

$$\gamma^{\alpha\beta\zeta\eta} = \gamma^{\beta\alpha\zeta\eta} = \gamma^{\alpha\beta\eta\zeta} = \gamma^{\zeta\eta\alpha\beta}. \tag{D.114}$$

The elements of the anisotropic perturbation matrix (D.32) are given by

$$\begin{aligned}\tilde{V}^{\rm ani}_{kk'} &= \sum_{st}\int_0^a \tilde{\varepsilon}^{\alpha\beta *}_{lm}\,\gamma^{\alpha\beta\zeta\eta}_{st}\,\tilde{\varepsilon}^{\zeta'\eta'}_{l'm'}\,g_{\zeta\zeta'}g_{\eta\eta'}\,r^2dr \\ &\qquad \times \int_\Omega Y^{\alpha+\beta *}_{lm}\,Y^{\alpha+\beta+\zeta+\eta}_{st}\,Y^{\zeta'+\eta'}_{l'm'}\,d\Omega \\ &= \sum_{st}\int_0^a \tilde{\varepsilon}^{\alpha\beta *}_{lm}\,\gamma^{\alpha\beta\zeta\eta}_{st}\,\tilde{\varepsilon}^{\zeta'\eta'}_{l'm'}\,g_{\zeta\zeta'}g_{\eta\eta'}\,r^2dr \\ &\qquad \times (-1)^{m+N}\left[\frac{(2l+1)(2s+1)(2l'+1)}{4\pi}\right]^{1/2} \\ &\qquad \times \begin{pmatrix} l & s & l' \\ -N & N-N' & N' \end{pmatrix}\begin{pmatrix} l & s & l' \\ -m & t & m' \end{pmatrix},\end{aligned} \tag{D.115}$$

where $N = \alpha + \beta$, $N' = \zeta' + \eta'$ and $N - N' = \alpha + \beta + \zeta + \eta$. It is straightforward to manipulate this expression into a form suitable for numerical computation:

$$\begin{aligned}\tilde{V}^{\rm ani}_{kk'} &= \sum_{st}(-1)^m\left[\frac{(2l+1)(2s+1)(2l'+1)}{4\pi}\right]^{1/2} \\ &\qquad \times \begin{pmatrix} l & s & l' \\ -m & t & m' \end{pmatrix}\sum_N\sum_I\int_0^a \Gamma_{NI}\,r^2dr.\end{aligned} \tag{D.116}$$

The sum over the generalized spherical-harmonic index N goes from -4 to 4, and the sum over I goes from 1 to I_N, where $I_0 = 5$, $I_{\pm 1} = 3$, $I_{\pm 2} = 3$, $I_{\pm 3} = 1$ and $I_{\pm 4} = 1$. There are thus 21 radial integrands Γ_{NI}, one for each of the 21 independent expansion coefficients $\gamma^{\alpha\beta\zeta\eta}_{st}$:

$$\Gamma_{01} = \dot{u}\dot{u}' B^{(0)+}_{lsl'}\,\gamma^{0000}_{st}, \tag{D.117}$$

$$\Gamma_{02} = \tfrac{1}{2}r^{-2}[(vv' + ww')B^{(2)+}_{lsl'} - i(vw' - wv')B^{(2)-}_{lsl'}]\gamma^{++--}_{st}, \tag{D.118}$$

$$\Gamma_{03} = ff'B^{(0)+}_{lsl'}\gamma^{+-+-}_{st}, \tag{D.119}$$

$$\Gamma_{04} = -(f\dot{u}' + \dot{u}f')B^{(0)+}_{lsl'}\gamma^{+-00}_{st}, \tag{D.120}$$

$$\Gamma_{05} = -[(xx' + zz')B^{(1)+}_{lsl'} - i(xz' - zx')B^{(1)-}_{lsl'}]\gamma^{+0-0}_{st}, \tag{D.121}$$

$$\Gamma_{11} = -\left[\Omega^0_l(x + iz)\dot{u}'\begin{pmatrix} l & s & l' \\ -1 & 1 & 0 \end{pmatrix} + \Omega^0_{l'}\dot{u}(x' + iz')\begin{pmatrix} l & s & l' \\ 0 & 1 & -1 \end{pmatrix}\right]\gamma^{+000}_{st}, \tag{D.122}$$

$$\Gamma_{-11} = -\left[\Omega^0_l(x - iz)\dot{u}'\begin{pmatrix} l & s & l' \\ 1 & -1 & 0 \end{pmatrix} + \Omega^0_{l'}\dot{u}(x' - iz')\begin{pmatrix} l & s & l' \\ 0 & -1 & 1 \end{pmatrix}\right]\gamma^{-000}_{st}, \tag{D.123}$$

$$\Gamma_{12} = -\Omega^0_l\Omega^0_{l'}\left[\Omega^2_l r^{-1}(v + iw)(x' - iz')\begin{pmatrix} l & s & l' \\ -2 & 1 & 1 \end{pmatrix} + \Omega^2_{l'} r^{-1}(x - iz)(v' + iw')\begin{pmatrix} l & s & l' \\ 1 & 1 & -2 \end{pmatrix}\right]\gamma^{++-0}_{st}, \tag{D.124}$$

$$\Gamma_{-12} = -\Omega^0_l\Omega^{l'}_0\left[\Omega^2_l r^{-1}(v - iw)(x' + iz')\begin{pmatrix} l & s & l' \\ 2 & -1 & -1 \end{pmatrix} + \Omega^2_{l'} r^{-1}(x + iz)(v' - iw')\begin{pmatrix} l & s & l' \\ -1 & -1 & 2 \end{pmatrix}\right]\gamma^{--+0}_{st}, \tag{D.125}$$

$$\Gamma_{13} = \left[\Omega^l_0(x + iz)f'\begin{pmatrix} l & s & l' \\ -1 & 1 & 0 \end{pmatrix} + \Omega^0_{l'}f(x' + iz')\begin{pmatrix} l & s & l' \\ 0 & 1 & -1 \end{pmatrix}\right]\gamma^{+-+0}_{st}, \tag{D.126}$$

$$\Gamma_{-13} = \left[\Omega^0_l(x - iz)f'\begin{pmatrix} l & s & l' \\ 1 & -1 & 0 \end{pmatrix} + \Omega^0_{l'}f(x' - iz')\begin{pmatrix} l & s & l' \\ 0 & -1 & 1 \end{pmatrix}\right]\gamma^{-+-0}_{st}, \tag{D.127}$$

$$\Gamma_{21} = \left[\Omega_l^0\Omega_l^2 r^{-1}(v+iw)\dot{u}' \begin{pmatrix} l & s & l' \\ -2 & 2 & 0 \end{pmatrix}\right.$$
$$\left. + \Omega_{l'}^0\Omega_{l'}^2\, r^{-1}\dot{u}(v'+iw') \begin{pmatrix} l & s & l' \\ 0 & 2 & -2 \end{pmatrix}\right]\gamma_{st}^{++00}, \tag{D.128}$$

$$\Gamma_{-21} = \left[\Omega_l^0\Omega_l^2 r^{-1}(v-iw)\dot{u}' \begin{pmatrix} l & s & l' \\ 2 & -2 & 0 \end{pmatrix}\right.$$
$$\left. + \Omega_{l'}^0\Omega_{l'}^2\, r^{-1}\dot{u}(v'-iw') \begin{pmatrix} l & s & l' \\ 0 & -2 & 2 \end{pmatrix}\right]\gamma_{st}^{--00}, \tag{D.129}$$

$$\Gamma_{22} = \Omega_l^0\Omega_{l'}^0[(xx'-zz')+i(xz'+zx')]$$
$$\times \begin{pmatrix} l & s & l' \\ -1 & 2 & -1 \end{pmatrix}\gamma_{st}^{+0+0}, \tag{D.130}$$

$$\Gamma_{-22} = \Omega_l^0\Omega_{l'}^0[(xx'-zz')-i(xz'+zx')]$$
$$\times \begin{pmatrix} l & s & l' \\ 1 & -2 & 1 \end{pmatrix}\gamma_{st}^{-0-0}, \tag{D.131}$$

$$\Gamma_{23} = -\left[\Omega_0^l\Omega_2^l r^{-1}(v+iw)f' \begin{pmatrix} l & s & l' \\ -2 & 2 & 0 \end{pmatrix}\right.$$
$$\left. + \Omega_{l'}^0\Omega_{l'}^2\, r^{-1}f(v'+iw') \begin{pmatrix} l & s & l' \\ 0 & 2 & -2 \end{pmatrix}\right]\gamma_{st}^{+++-}, \tag{D.132}$$

$$\Gamma_{-23} = -\left[\Omega_l^0\Omega_l^2 r^{-1}(v-iw)f' \begin{pmatrix} l & s & l' \\ 2 & -2 & 0 \end{pmatrix}\right.$$
$$\left. + \Omega_{l'}^0\Omega_{l'}^2\, r^{-1}f(v'-iw') \begin{pmatrix} l & s & l' \\ 0 & -2 & 2 \end{pmatrix}\right]\gamma_{st}^{---+}, \tag{D.133}$$

$$\Gamma_{31} = -\Omega_l^0\Omega_{l'}^0\left[\Omega_l^2 r^{-1}(v+iw)(x'+iz') \begin{pmatrix} l & s & l' \\ -2 & 3 & -1 \end{pmatrix}\right.$$
$$\left. + \Omega_{l'}^2\, r^{-1}(x+iz)(v'+iw') \begin{pmatrix} l & s & l' \\ -1 & 3 & -2 \end{pmatrix}\right]\gamma_{st}^{+++0}, \tag{D.134}$$

$$\Gamma_{-31} = -\Omega_l^0\Omega_{l'}^0\left[\Omega_l^2 r^{-1}(v-iw)(x'-iz') \begin{pmatrix} l & s & l' \\ 2 & -3 & 1 \end{pmatrix}\right.$$
$$\left. + \Omega_{l'}^2\, r^{-1}(x-iz)(v'-iw') \begin{pmatrix} l & s & l' \\ 1 & -3 & 2 \end{pmatrix}\right]\gamma_{st}^{---0}, \tag{D.135}$$

$$\Gamma_{41} = \Omega_l^0\Omega_l^2\Omega_{l'}^0\Omega_{l'}^2\, r^{-2}[(vv' - ww') + i(vw' + wv')] \begin{pmatrix} l & s & l' \\ -2 & 4 & -2 \end{pmatrix} \gamma_{st}^{++++}, \tag{D.136}$$

$$\Gamma_{-41} = \Omega_l^0\Omega_l^2\Omega_{l'}^0\Omega_{l'}^2\, r^{-2}[(vv' - ww') - i(vw' + wv')] \begin{pmatrix} l & s & l' \\ 2 & -4 & 2 \end{pmatrix} \gamma_{st}^{----}. \tag{D.137}$$

The coefficients of degree $s = 0$ describe a transversely isotropic perturbation of a SNREI starting model:

$$\gamma_{00}^{0000} = \delta C, \qquad \gamma_{00}^{++--} = 2\,\delta N, \qquad \gamma_{00}^{+-+-} = \delta A - \delta N,$$
$$\gamma_{00}^{+-00} = -\delta F, \qquad \gamma_{00}^{+0-0} = -\delta L. \tag{D.138}$$

Such a spherically symmetric perturbation acts to shift the degenerate eigenfrequency of every multiplet ${}_n\mathrm{S}_l$ or ${}_n\mathrm{T}_l$, but it does not give rise to any splitting or coupling between multiplets. The generalized spherical-harmonic expansion (D.100) of an *aspherical* anisotropic perturbation γ^{asp} begins at $s = 1$ rather than at $s = 0$. The results (D.116)–(D.137) were first given, correctly, by Mochizuki (1986).

D.2.8 Diagonal sum rule

Each of the matrix elements (D.43)–(D.45) and (D.116) is of the general form

$$\tilde{M}_{kk'} = \sum_{st} (-1)^{l+m} \begin{pmatrix} l & s & l' \\ -m & t & m' \end{pmatrix} \|\tilde{M}_{\sigma nl;\sigma'n'l'}\|_{st}, \tag{D.139}$$

where $\tilde{\mathsf{M}}$ denotes any of the matrices $\tilde{\mathsf{T}}$, $\tilde{\mathsf{V}}^{\mathrm{iso}}$, $\tilde{\mathsf{V}}^{\mathrm{ani+asp}}$ or $\tilde{\mathsf{A}}$ and σ is again a pseudonym for either S or T. The quantities $\|\tilde{M}_{\sigma nl;\sigma'n'l'}\|_{st}$ depend upon the multiplets ${}_n\mathrm{S}_l$ or ${}_n\mathrm{T}_l$ and ${}_{n'}\mathrm{S}_{l'}$ or ${}_{n'}\mathrm{T}_{l'}$, as well as upon the indices s and t, but they are independent of the orders m and m'. These so-called *reduced* or *double-bar* matrix elements are defined by equation (D.139); thus, for example,

$$\|\tilde{T}_{\sigma nl;\sigma'n'l'}\|_{st} = (-1)^l \left[\frac{(2l+1)(2s+1)(2l'+1)}{4\pi}\right]^{1/2} \times \left\{\int_0^a \delta\tilde{\rho}_{st} T_\rho\, r^2 dr + \sum_d d^2 \delta\tilde{d}_{st}\, [T_d]_-^+\right\}. \tag{D.140}$$

The decomposition of a general matrix element $\tilde{M}_{kk'}$ into a sum of order-independent terms times the explicit 3-j factors in (D.139) is a celebrated result, known in quantum mechanics as the *Wigner-Eckart theorem*; the factor of $(-1)^l$ in the definition of $\|\tilde{M}_{\sigma nl;\sigma'n'l'}\|_{st}$ is conventional (Edmonds 1960). Note that the double-bar symbol does not denote any kind of norm or modulus of the matrix $\tilde{\mathsf{M}}$; in fact the reduced matrix element $\|\tilde{M}_{\sigma nl;\sigma'n'l'}\|_{st}$ is in general complex.

The *trace* or sum of the diagonal elements of $\tilde{\mathsf{M}}$ is

$$\operatorname{tr}\tilde{\mathsf{M}} = \sum_{\sigma nl}\sum_{st}\sum_{m}(-1)^{l+m}\begin{pmatrix} l & s & l \\ -m & t & m \end{pmatrix}\|\tilde{M}_{\sigma nl;\sigma nl}\|_{st}. \tag{D.141}$$

The sum over m can be evaluated by invoking equation (C.227) together with the spherical-harmonic addition theorem (C.276):

$$\begin{aligned}\sum_m(-1)^m\begin{pmatrix} l & s & l \\ -m & t & m \end{pmatrix} &= \sum_m(2l+1)^{-1}\left(\frac{2s+1}{4\pi}\right)^{-1/2} \\ &\quad\times\begin{pmatrix} l & s & l \\ 0 & 0 & 0 \end{pmatrix}^{-1}\int_\Omega Y_{lm}^*Y_{st}Y_{lm}\,d\Omega \\ &= \frac{1}{4\pi}\left(\frac{2s+1}{4\pi}\right)^{-1/2}\begin{pmatrix} l & s & l \\ 0 & 0 & 0 \end{pmatrix}^{-1}\int_\Omega Y_{st}\,d\Omega \\ &= \begin{pmatrix} l & 0 & l \\ 0 & 0 & 0 \end{pmatrix}^{-1}\delta_{s0} = (-1)^l\sqrt{2l+1}\,\delta_{s0}.\end{aligned} \tag{D.142}$$

The important conclusion is that (D.142) vanishes for all degrees $s > 0$. As long as the unperturbed model of the Earth κ_0, μ_0, ρ is the terrestrial monopole, we must have

$$\operatorname{tr}\tilde{\mathsf{M}} = 0. \tag{D.143}$$

The result (D.143), which was first enunciated in this seismological context by Gilbert (1971a), is known as the *diagonal sum rule*. The $s = 2$ ellipticity matrices satisfy this relation, as well as the $1 \leq s \leq s_{\mathrm{max}}$ lateral-heterogeneity, anisotropy and anelasticity matrices:

$$\operatorname{tr}\tilde{\mathsf{T}}^{\mathrm{ell}} = \operatorname{tr}\tilde{\mathsf{T}}^{\mathrm{lat}} = 0, \tag{D.144}$$

$$\operatorname{tr}\tilde{\mathsf{V}}^{\mathrm{ell}} = \operatorname{tr}\tilde{\mathsf{V}}^{\mathrm{lat}} = \operatorname{tr}\tilde{\mathsf{V}}^{\mathrm{ani+asp}} = \operatorname{tr}\tilde{\mathsf{A}} = 0. \tag{D.145}$$

The Coriolis matrix (D.68) is readily shown to obey the diagonal sum rule:

$$\operatorname{tr}\tilde{\mathsf{W}} = 0. \tag{D.146}$$

However, the centrifugal-potential matrix (D.72) does *not*, because of the presence of the spherically symmetric potential perturbation $\bar{\psi} = -\frac{1}{3}\Omega^2r^2$.

In fact,

$$\operatorname{tr}\tilde{\mathsf{V}}^{\text{cen}} = \tfrac{2}{3}\Omega^2 \sum_{\sigma nl} (2l+1)\left[1 - k^2 \int_0^a \rho(v^2 + 2uv + w^2)\, r^2 dr\right]. \tag{D.147}$$

The toroidal multiplets ${}_n\mathrm{T}_l$ do not contribute to the trace (D.147), since $k^2 \int_0^a \rho w^2\, r^2 dr = 1$.

D.3 Complex-to-Real Basis Transformation

As promised in the introduction, we shall now demonstrate how to transform the elements of the vectors $\tilde{\mathsf{r}}$, $\tilde{\mathsf{s}}$ and matrices $\tilde{\mathsf{T}}$, $\tilde{\mathsf{V}}$, $\tilde{\mathsf{W}}$ into the corresponding elements of r, s and T, V, W. Changing notation slightly, we shall—in this section only—use k as a shorthand *multiplet* label, designating either ${}_n\mathrm{S}_l$ or ${}_n\mathrm{T}_l$. To avoid subscript clutter, it is also convenient to elevate the order index to the level of a superscript, writing the relation between the real and complex eigenfunctions in the form

$$\mathbf{s}_k^{-m} = \frac{1}{\sqrt{2}}(\tilde{\mathbf{s}}_k^{m*} + \tilde{\mathbf{s}}_k^{m}), \tag{D.148}$$

$$\mathbf{s}_k^{0} = \tilde{\mathbf{s}}_k^{0}, \tag{D.149}$$

$$\mathbf{s}_k^{m} = \frac{i}{\sqrt{2}}(\tilde{\mathbf{s}}_k^{m*} - \tilde{\mathbf{s}}_k^{m}), \tag{D.150}$$

where it is presumed that $m > 0$.

D.3.1 Receiver and source vector

Upon using (D.148)–(D.150) to express the elements of the real receiver vector r in terms of the elements (D.7) of the complex vector $\tilde{\mathsf{r}}$, we find

$$r_k^{-m} = \sqrt{2}\,\mathrm{Re}\,\tilde{r}_k^{m}, \qquad r_k^{0} = \tilde{r}_k^{0}, \qquad r_k^{m} = -\sqrt{2}\,\mathrm{Im}\,\tilde{r}_k^{m}. \tag{D.151}$$

The elements of the real and complex source vectors s and $\tilde{\mathsf{s}}$ are related in a similar fashion:

$$s_k^{-m} = \sqrt{2}\,\mathrm{Re}\,\tilde{s}_k^{m}, \qquad s_k^{0} = \tilde{s}_k^{0}, \qquad s_k^{m} = -\sqrt{2}\,\mathrm{Im}\,\tilde{s}_k^{m}. \tag{D.152}$$

Equations (D.151) and (D.152) are of course simply examples of the general scalar transformation relation (B.98).

D.3.2 Perturbation matrices

We shall use $\tilde{\mathsf{M}}$ in this section to denote any of the aspherical kinetic or potential energy matrices $\tilde{\mathsf{T}}^{\mathrm{ell}}$, $\tilde{\mathsf{T}}^{\mathrm{lat}}$ or $\tilde{\mathsf{V}}^{\mathrm{ell}}$, $\tilde{\mathsf{V}}^{\mathrm{cen}}$, $\tilde{\mathsf{V}}^{\mathrm{lat}}$, $\tilde{\mathsf{V}}^{\mathrm{ani+asp}}$, $\tilde{\mathsf{A}}$—but *not* the Coriolis matrix $\tilde{\mathsf{W}}$. These complex matrices satisfy the symmetries

$$\tilde{M}_{kk'}^{mm'} = \tilde{M}_{k'k}^{m'm*} = (-1)^{m+m'}\tilde{M}_{kk'}^{-m-m'*}, \tag{D.153}$$

$$\tilde{W}_{kk'}^{mm'} = \tilde{W}_{k'k}^{m'm*} = -(-1)^{m+m'}\tilde{W}_{kk'}^{-m-m'*}. \tag{D.154}$$

The first and third equalities simply assert that both $\tilde{\mathsf{M}}$ and $\tilde{\mathsf{W}}$ are Hermitian: $\tilde{\mathsf{M}} = \tilde{\mathsf{M}}^{\mathrm{H}}$ and $\tilde{\mathsf{W}} = \tilde{\mathsf{W}}^{\mathrm{H}}$. The second equality follows from the reality of the perturbations $\delta\kappa$, $\delta\mu$, $\delta\rho$, $\delta\Phi$, q_κ, q_μ, δd and γ^{asp}, together with the spherical-harmonic identity $Y_{l\,-m} = (-1)^m Y_{lm}^*$. The fourth and last is an easily verified property of the Coriolis matrix representation (D.68). The complex-to-real matrix transformation relations obtained with the aid of equations (D.148)–(D.150) and (D.153)–(D.154) are

$$M_{kk'}^{-m-m'} = \mathrm{Re}\,[\tilde{M}_{kk'}^{mm'} + (-1)^{m'}\tilde{M}_{kk'}^{m-m'}], \tag{D.155}$$

$$M_{kk'}^{mm'} = \mathrm{Re}\,[\tilde{M}_{kk'}^{mm'} - (-1)^{m'}\tilde{M}_{kk'}^{m-m'}], \tag{D.156}$$

$$M_{kk'}^{-mm'} = \mathrm{Im}\,[\tilde{M}_{kk'}^{mm'} - (-1)^{m'}\tilde{M}_{kk'}^{m-m'}], \tag{D.157}$$

$$M_{kk'}^{m-m'} = -\mathrm{Im}\,[\tilde{M}_{kk'}^{mm'} + (-1)^{m'}\tilde{M}_{kk'}^{m-m'}], \tag{D.158}$$

$$M_{kk'}^{-m0} = \sqrt{2}\,\mathrm{Re}\,\tilde{M}_{kk'}^{m0}, \qquad M_{kk'}^{0-m'} = \sqrt{2}\,\mathrm{Re}\,\tilde{M}_{kk'}^{0m'}, \tag{D.159}$$

$$M_{kk'}^{m0} = \sqrt{2}\,\mathrm{Im}\,\tilde{M}_{kk'}^{m0}, \qquad M_{kk'}^{0m'} = \sqrt{2}\,\mathrm{Im}\,\tilde{M}_{kk'}^{0m'}, \tag{D.160}$$

$$M_{kk'}^{00} = \tilde{M}_{kk'}^{00}, \tag{D.161}$$

$$W_{kk'}^{-m-m'} = W_{kk'}^{mm'} = i\,\mathrm{Im}\,[\tilde{W}_{kk'}^{mm'}], \tag{D.162}$$

$$W_{kk'}^{-mm'} = -W_{kk'}^{m-m'} = -i\,\mathrm{Re}\,[\tilde{W}_{kk'}^{mm'}], \tag{D.163}$$

$$W_{kk'}^{-m0} = W_{kk'}^{0-m'} = W_{kk'}^{m0} = W_{kk'}^{0m'} = W_{kk'}^{00} = 0, \tag{D.164}$$

where it is presumed that $m > 0$ and $m' > 0$. The resulting energy matrices are real and symmetric, whereas the Coriolis matrix is purely imaginary and anti-symmetric:

$$\mathrm{Im}\,M_{kk'}^{mm'} = 0, \qquad \mathrm{Re}\,W_{kk'}^{mm'} = 0, \tag{D.165}$$

$$M_{kk'}^{mm'} = M_{k'k}^{m'm}, \qquad W_{kk'}^{mm'} = -W_{k'k}^{m'm}. \tag{D.166}$$

The new elements of the Coriolis and elliptical-plus-centrifugal matrices are given explicitly by

$$\begin{aligned} W_{kk'}^{mm'} &= im\Omega\delta_{ll'}\delta_{m\,-m'} \int_0^a \rho W^{\mathrm{S}}\, r^2 dr \\ &\quad - i\Omega(S_{lm}\delta_{l\,l'+1} + S_{l'm}\delta_{l\,l'-1})\delta_{mm'} \int_0^a \rho W^{\mathrm{A}}\, r^2 dr, \end{aligned} \tag{D.167}$$

$$\begin{aligned} T_{kk'}^{\mathrm{ell}} &= (R_{lm}\delta_{ll'} + \tfrac{3}{2}S_{lm}S_{l'+1\,m}\delta_{l\,l'+2} + \tfrac{3}{2}S_{l+1\,m}S_{l'm}\delta_{l\,l'-2})\delta_{mm'} \\ &\qquad \times \int_0^a \tfrac{2}{3}\varepsilon\rho\big[\bar{T}_\rho^{\mathrm{S}} - (\eta+3)\check{T}_\rho^{\mathrm{S}}\big]\, r^2 dr \\ &\quad + 3m(S_{lm}\delta_{l\,l'+1} + S_{l'm}\delta_{l\,l'-1})\delta_{m\,-m'} \\ &\qquad \times \int_0^a \tfrac{2}{3}\varepsilon\rho\big[\bar{T}_\rho^{\mathrm{A}} - (\eta+3)\check{T}_\rho^{\mathrm{A}}\big]\, r^2 dr, \end{aligned} \tag{D.168}$$

$$\begin{aligned} V_{kk'}^{\mathrm{ell+cen}} &= \tfrac{2}{3}\Omega^2\delta_{\sigma\sigma'}\delta_{nn'}\delta_{ll'}\delta_{mm'} - \tfrac{2}{3}k^2\Omega^2\delta_{ll'}\delta_{mm'} \int_0^a \rho W^{\mathrm{S}}\, r^2 dr \\ &\quad + (R_{lm}\delta_{ll'} + \tfrac{3}{2}S_{lm}S_{l'+1\,m}\delta_{l\,l'+2} + \tfrac{3}{2}S_{l+1\,m}S_{l'm}\delta_{l\,l'-2})\delta_{mm'} \\ &\quad \times \int_0^a \tfrac{2}{3}\varepsilon\Big\{\kappa\big[\bar{V}_\kappa^{\mathrm{S}} - (\eta+1)\check{V}_\kappa^{\mathrm{S}}\big] + \mu\big[\bar{V}_\mu^{\mathrm{S}} - (\eta+1)\check{V}_\mu^{\mathrm{S}}\big] \\ &\qquad\quad + \rho\big[\bar{V}_\rho^{\mathrm{S}} - (\eta+3)\check{V}_\rho^{\mathrm{S}}\big]\Big\} r^2 dr \\ &\quad + 3m(S_{lm}\delta_{l\,l'+1} + S_{l'm}\delta_{l\,l'-1})\delta_{m\,-m'} \\ &\quad \times \int_0^a \tfrac{2}{3}\varepsilon\Big\{-\kappa(\eta+2)\bar{V}_\kappa^{\mathrm{A}} + \mu\big[\bar{V}_\mu^{\mathrm{A}} - (\eta+1)\check{V}_\mu^{\mathrm{A}}\big] \\ &\qquad\quad + \rho\big[\bar{V}_\rho^{\mathrm{A}} - (\eta+3)\check{V}_\rho^{\mathrm{A}}\big]\Big\} r^2 dr, \end{aligned} \tag{D.169}$$

where we have reinstated the convention that the subscript k is a singlet rather than a multiplet label. Equations (D.167)–(D.169) satisfy (D.166) by virtue of the symmetry and anti-symmetry, respectively, of the kernels distinguished by the superscripts S and A. We remark, finally, that since the basis transformation (D.148)–(D.150) is unitary, the diagonal-sum relations (D.144)–(D.147) are preserved:

$$\mathrm{tr}\,\mathsf{T}^{\mathrm{ell}} = \mathrm{tr}\,\mathsf{T}^{\mathrm{lat}} = 0, \tag{D.170}$$

$$\mathrm{tr}\,\mathsf{V}^{\mathrm{ell}} = \mathrm{tr}\,\mathsf{V}^{\mathrm{lat}} = \mathrm{tr}\,\mathsf{V}^{\mathrm{ani+asp}} = \mathrm{tr}\,\mathsf{A} = 0, \tag{D.171}$$

$$\mathrm{tr}\,\mathsf{W} = 0, \qquad \mathrm{tr}\,\mathsf{V}^{\mathrm{cen}} = \mathrm{tr}\,\tilde{\mathsf{V}}^{\mathrm{cen}}. \tag{D.172}$$

The values of $\mathrm{tr}\,\mathsf{W}$, $\mathrm{tr}\,\mathsf{T}^{\mathrm{ell}}$ and $\mathrm{tr}\,\mathsf{V}^{\mathrm{ell+cen}}$ may be verified by direct summation, using the explicit representations (D.167)–(D.169).

D.3.3 Transformation matrix

It is possible to write the complex-to-real basis transformation relations (D.151)–(D.152) and (D.155)–(D.164) using a compact matrix notation:

$$\mathsf{r} = \mathsf{U}^{\mathrm{H}}\tilde{\mathsf{r}}, \qquad \mathsf{s} = \mathsf{U}^{\mathrm{H}}\tilde{\mathsf{s}}, \tag{D.173}$$

$$\mathsf{M} = \mathsf{U}^{\mathrm{H}}\tilde{\mathsf{M}}\mathsf{U}, \qquad \mathsf{W} = \mathsf{U}^{\mathrm{H}}\tilde{\mathsf{W}}\mathsf{U}, \tag{D.174}$$

where the superscript H denotes the Hermitian transpose. The transformation matrix U is block diagonal:

$$\mathsf{U} = \begin{pmatrix} \ddots & & & & & & \\ & \mathsf{U}_{-2-2} & & & & & \\ & & \mathsf{U}_{-1-1} & & & & \\ & & & \mathsf{U}_{00} & & & \\ & & & & \mathsf{U}_{11} & & \\ & & & & & \mathsf{U}_{22} & \\ & & & & & & \ddots \end{pmatrix}, \tag{D.175}$$

where we have elevated $k = \ldots, -2, -1, 0, 1, 2, \ldots$ to the status of a multiplet label again. The central or *target* multiplet is denoted by $k = 0$. Each $(2l+1) \times (2l+1)$ submatrix in equation (D.175) is of the same form:

$$\mathsf{U}_{kk} = \begin{pmatrix} \ddots & & & & & & \iddots \\ & \frac{1}{\sqrt{2}} & & & & \frac{i}{\sqrt{2}} & \\ & & -\frac{1}{\sqrt{2}} & & -\frac{i}{\sqrt{2}} & & \\ & & & 1 & & & \\ & & \frac{1}{\sqrt{2}} & & -\frac{i}{\sqrt{2}} & & \\ & \frac{1}{\sqrt{2}} & & & & -\frac{i}{\sqrt{2}} & \\ \iddots & & & & & & \ddots \end{pmatrix}, \tag{D.176}$$

where the signs continue to alternate along the upper diagonal and anti-diagonal in the manner indicated. Since the transformation is unitary,

$$\mathsf{U}^{\mathrm{H}}\mathsf{U} = \mathsf{U}\mathsf{U}^{\mathrm{H}} = \mathsf{I}, \tag{D.177}$$

we can invert (D.173)–(D.174) to find the complex receiver and source vectors and perturbation matrices in terms of the real ones:

$$\tilde{\mathsf{r}} = \mathsf{U}\mathsf{r}, \qquad \tilde{\mathsf{s}} = \mathsf{U}\mathsf{s}, \tag{D.178}$$

$$\tilde{\mathsf{M}} = \mathsf{U}\mathsf{M}\mathsf{U}^{\mathrm{H}}, \qquad \tilde{\mathsf{W}} = \mathsf{U}\mathsf{W}\mathsf{U}^{\mathrm{H}}. \tag{D.179}$$

The energy and Coriolis matrices transform identically in (D.173)–(D.174) and (D.178)–(D.179) but differently in (D.155)–(D.164) because we have used the additional non-Hermitian symmetries (D.153)–(D.154) to reduce the latter.

D.4 Self Coupling

The splitting of an isolated spheroidal or toroidal multiplet ${}_n\mathrm{S}_l$ or ${}_n\mathrm{T}_l$ is governed by "self-coupling" matrices of dimension $(2l+1)\times(2l+1)$. We tabulate the real symmetric matrix elements $T^{\rm ell}_{mm'}$, $T^{\rm lat}_{mm'}$, $V^{\rm ell+cen}_{mm'}$, $V^{\rm lat}_{mm'}$, $V^{\rm ani}_{mm'}$, $A_{mm'}$ and the imaginary anti-symmetric elements $W_{mm'}$ for convenience in this section. Of course these results can be obtained by simply setting $\sigma'=\sigma$, $n'=n$ and $l'=l$.

D.4.1 Rotation and ellipticity

The matrix elements governing the effects of the Earth's rotation and hydrostatic ellipticity are given in the isolated-multiplet approximation by

$$W_{mm'} = im\Omega\delta_{m\,-m'}\int_0^a \rho(v^2+2uv+w^2)\,r^2dr, \tag{D.180}$$

$$T^{\rm ell}_{mm'} = R_{lm}\delta_{mm'}\int_0^a \tfrac{2}{3}\varepsilon\rho\big[\bar{T}_\rho-(\eta+3)\check{T}_\rho\big]r^2dr, \tag{D.181}$$

$$\begin{aligned} V^{\rm ell+cen}_{mm'} &= \tfrac{2}{3}\Omega^2\delta_{mm'}\left[1-k^2\int_0^a \rho(v^2+2uv+w^2)\,r^2dr\right] \\ &\quad + R_{lm}\delta_{mm'}\int_0^a \tfrac{2}{3}\varepsilon\Big\{\kappa\big[\bar{V}_\kappa-(\eta+1)\check{V}_\kappa\big] \\ &\quad + \mu\big[\bar{V}_\mu-(\eta+1)\check{V}_\mu\big]+\rho\big[\bar{V}_\rho-(\eta+3)\check{V}_\rho\big]\Big\}r^2dr, \end{aligned} \tag{D.182}$$

where R_{lm} is given by equation (D.82) and

$$\bar{T}_\rho = -6uv, \qquad \check{T}_\rho = u^2+(k^2-3)(v^2+w^2), \tag{D.183}$$

$$\bar{V}_\kappa = -2(\dot{u}+f)(\dot{u}+3r^{-1}v), \tag{D.184}$$

$$\begin{aligned} \bar{V}_\mu &= -\tfrac{2}{3}(2\dot{u}-f)(2\dot{u}+9\dot{v}-12r^{-1}v) \\ &\quad + 2x\big[3\dot{u}-(k^2-3)\dot{v}-3r^{-1}k^2v)\big] \\ &\quad + 18(k^2-2)r^{-1}(v\dot{v}+w\dot{w})-2(k^2-3)z\dot{w}, \end{aligned} \tag{D.185}$$

$$\bar{V}_\rho = 2f(r\dot{p} + 4\pi G\rho r u + gu) - 6r^{-1}guv + 6r^{-1}gu^2 + 2r^{-1}[(k^2-3)v - k^2u]p, \tag{D.186}$$

$$\check{V}_\kappa = -(\dot{u}+f)(\dot{u} - f - 6r^{-1}v), \tag{D.187}$$

$$\check{V}_\mu = (k^2-12)(k^2-2)r^{-2}(v^2+w^2) - (k^2-3)(x^2+z^2-2x\dot{v}-2z\dot{w}) - \tfrac{2}{3}(2\dot{u}-f)(\dot{u}+\tfrac{1}{2}f-6r^{-1}v), \tag{D.188}$$

$$\check{V}_\rho = 2(k^2-3)r^{-1}pv + u(2\dot{p} + 8\pi G\rho u - 6gr^{-1}v). \tag{D.189}$$

It is noteworthy that $\mathsf{T}^{\mathrm{ell}}$ and $\mathsf{V}^{\mathrm{ell+cen}}$ are diagonal, whereas the Coriolis matrix W is *anti-diagonal*.

D.4.2 Lateral heterogeneity and anelasticity

The isotropic lateral heterogeneity and anelasticity of the Earth can be expanded in real spherical harmonics $\mathcal{Y}_{st}$ in the form

$$\delta\kappa = \sum_{s=1}^{s_{\max}}\sum_{t=-s}^{s}\delta\kappa_{st}\mathcal{Y}_{st}, \qquad \delta\mu = \sum_{s=1}^{s_{\max}}\sum_{t=-s}^{s}\delta\mu_{st}\mathcal{Y}_{st},$$

$$\delta\rho = \sum_{s=1}^{s_{\max}}\sum_{t=-s}^{s}\delta\rho_{st}\mathcal{Y}_{st}, \qquad \delta\Phi = \sum_{s=1}^{s_{\max}}\sum_{t=-s}^{s}\delta\Phi_{st}\mathcal{Y}_{st},$$

$$q_\kappa = \sum_{s=1}^{s_{\max}}\sum_{t=-s}^{s}q_{\kappa st}\mathcal{Y}_{st}, \qquad q_\mu = \sum_{s=1}^{s_{\max}}\sum_{t=-s}^{s}q_{\mu st}\mathcal{Y}_{st},$$

$$\delta d = \sum_{s=1}^{s_{\max}}\sum_{t=-s}^{s}\delta d_{st}\mathcal{Y}_{st}. \tag{D.190}$$

The real expansion coefficients in (D.190) are related to the complex coefficients in (D.37) by

$$m_{st} = \begin{cases} \sqrt{2}\,\mathrm{Re}\,\tilde{m}_{s|t|} & \text{if } -s \le t < 0 \\ \tilde{m}_{s0} & \text{if } t = 0 \\ -\sqrt{2}\,\mathrm{Im}\,\tilde{m}_{st} & \text{if } 0 < t \le s, \end{cases} \tag{D.191}$$

where m_{st} denotes any of $\delta\kappa_{st}$, $\delta\mu_{st}$, $\delta\rho_{st}$, $\delta\Phi_{st}$, $q_{\kappa st}$, $q_{\mu st}$ or δd_{st}. The elements of the self-coupling matrices $\mathsf{T}^{\mathrm{lat}}$, $\mathsf{V}^{\mathrm{lat}}$ and A may be written in terms of an integral over three real spherical harmonics:

$$T^{\mathrm{lat}}_{mm'} = \sum_{st} \int_\Omega \mathcal{Y}_{lm}\mathcal{Y}_{st}\mathcal{Y}_{lm'}\, d\Omega \times \left\{ \int_0^a \delta\rho_{st} T_\rho\, r^2 dr + \sum_d d^2 \delta d_{st} \left[T_d\right]_-^+ \right\}, \tag{D.192}$$

$$V^{\mathrm{lat}}_{mm'} = \sum_{st} \int_\Omega \mathcal{Y}_{lm}\mathcal{Y}_{st}\mathcal{Y}_{lm'}\, d\Omega \times \left\{ \int_0^a \left(\delta\kappa_{st} V_\kappa + \delta\mu_{st} V_\mu + \delta\rho_{st} V_\rho + \delta\Phi_{st} V_\Phi + \delta\dot{\Phi}_{st} V_{\dot{\Phi}}\right) r^2 dr + \sum_d d^2 \delta d_{st} \left[V_d\right]_-^+ \right\}, \tag{D.193}$$

$$A_{mm'} = \sum_{st} \int_\Omega \mathcal{Y}_{lm}\mathcal{Y}_{st}\mathcal{Y}_{lm'}\, d\Omega \times \int_0^a \left(\kappa_0 q_{\kappa st} V_\kappa + \mu_0 q_{\mu st} V_\mu\right) r^2 dr, \tag{D.194}$$

where

$$T_\rho = u^2 + [k^2 - \tfrac{1}{2}s(s+1)](v^2 + w^2), \tag{D.195}$$

$$T_d = -\rho T_\rho, \tag{D.196}$$

$$V_\kappa = (\dot{u} + f)^2, \tag{D.197}$$

$$V_\mu = \tfrac{1}{3}(2\dot{u} - f)^2 + [k^2 - \tfrac{1}{2}s(s+1)](x^2 + z^2) + \{k^2(k^2 - 2) - \tfrac{1}{2}s(s+1)[4k^2 - s(s+1) - 2]\} \times r^{-2}(v^2 + w^2), \tag{D.198}$$

$$V_\rho = [k^2 - \tfrac{1}{2}s(s+1)]r^{-1}(2vp + guv) + 8\pi G\rho u^2 + 2u\dot{p} - gu(2r^{-1}u + f), \tag{D.199}$$

$$V_\Phi = \rho[\tfrac{1}{2}s(s+1)r^{-1}(u\dot{v} - v\dot{u} - 2vf + r^{-1}uv) + s(s+1)r^{-2}u^2], \tag{D.200}$$

$$V_{\dot{\Phi}} = \rho[\tfrac{1}{2}s(s+1)r^{-1}uv - 2uf], \tag{D.201}$$

$$\begin{aligned} V_d = {} & -\kappa_0 V_\kappa - \mu_0 V_\mu - \rho V_\rho \\ & + \kappa_0(\dot{u} + f)[2\dot{u} - s(s+1)r^{-1}v] \\ & + \mu_0\{2[k^2 - \tfrac{1}{2}s(s+1)](\dot{v}x + \dot{w}z) \\ & \quad + \tfrac{2}{3}(2\dot{u} - f)[2\dot{u} - s(s+1)r^{-1}v]\}. \end{aligned} \tag{D.202}$$

As in the general case, the perturbations in the gravitational potential $\delta\Phi$ and its derivative $\delta\dot{\Phi}$ may be eliminated from equation (D.193) by making the substitutions (D.55)–(D.56). In practice, the elements (D.192)–(D.194) are best calculated by starting with the complex expansions (D.37), using equation (C.227) to evaluate the complex integral $\int_\Omega Y_{lm}^* Y_{st} Y_{lm'}\, d\Omega$, and then invoking the transformation (D.155)–(D.161).

D.4.3 Spherically symmetric perturbation

The results (D.192)–(D.194) are applicable to an $s = 0$ perturbation of the terrestrial monopole as well. The energy and anelasticity matrices are in that case diagonal:

$$\mathsf{V}^{\mathrm{sph}} - \omega^2\mathsf{T}^{\mathrm{sph}} = \delta\omega\,\mathsf{I}, \qquad \mathsf{A}^{\mathrm{sph}} = \gamma\mathsf{I}, \tag{D.203}$$

where $\delta\omega$ and γ are the degenerate eigenfrequency perturbation and decay rate of the multiplet. Comparison of (D.195)–(D.202) and (9.13)–(9.16) permits the kernel identifications

$$V_\kappa = 2\omega r^{-2}K_\kappa, \qquad V_\mu = 2\omega r^{-2}K_\mu, \tag{D.204}$$

$$V_\rho - \omega^2 T_\rho = 2\omega r^{-2}K_\rho, \qquad V_d - \omega^2 T_d = 2\omega r^{-2}K_d, \tag{D.205}$$

where we have invoked the substitutions (D.55)–(D.56).

D.4.4 Inner-core anisotropy

The matrix elements $V_{mm'}^{\mathrm{ani}}$ governing a general anisotropic perturbation γ can be readily determined by specializing the results (D.116)–(D.137). We restrict attention in this final section to the physical situation of greatest practical interest: transverse isotropy of the solid inner core with an axis of symmetry $\hat{\mathbf{z}}$ parallel to the axis of rotation (Tromp 1995). The following nine Cartesian components of the fourth-order tensor γ are in that case non-zero:

$$\begin{aligned} &\gamma_{xxxx} = \gamma_{yyyy} = \delta A, \qquad \gamma_{zzzz} = \delta C, \\ &\gamma_{xyxy} = \delta N, \qquad \gamma_{xxyy} = \delta A - 2\,\delta N, \\ &\gamma_{xzxz} = \gamma_{yzyz} = \delta L, \qquad \gamma_{xxzz} = \gamma_{yyzz} = \delta F. \end{aligned} \tag{D.206}$$

To dispel any possible confusion, we note that the five coefficients δC, δA, δL, δN and δF describing a transversely isotropic perturbation with a *co-rotational* symmetry axis in equation (D.206) are different from those describing a transversely isotropic perturbation with a *radial* symmetry axis in equation (D.138). To determine the components of γ with respect to the complex basis $\hat{\mathbf{e}}_-, \hat{\mathbf{e}}_0, \hat{\mathbf{e}}_+$ we can either effect a direct transformation from Cartesian to canonical coordinates, or convert from Cartesian to spherical coordinates, and make use of the results (D.101)–(D.113) to obtain the canonical representation. The generalized spherical-harmonic expansion coefficients $\gamma_{st}^{\alpha\beta\zeta\eta}$ obtained using either procedure are listed in Table D.1. The parameters $\lambda_1, \lambda_2, \lambda_3, \lambda_4$ and λ_5 are defined by

$$\begin{aligned}
\lambda_1 &= \delta C + 6\,\delta A - 4\,\delta L - 10\,\delta N + 8\,\delta F,\\
\lambda_2 &= \delta C + \delta A + 6\,\delta L + 5\,\delta N - 2\,\delta F,\\
\lambda_3 &= \delta C - 6\,\delta A - 4\,\delta L + 14\,\delta N + 5\,\delta F,\\
\lambda_4 &= \delta C + \delta A + 3\,\delta L - 7\,\delta N - 2\,\delta F,\\
\lambda_5 &= \delta C + \delta A - 4\,\delta L - 2\,\delta F.
\end{aligned} \tag{D.207}$$

Every coefficient is real and of order $t = 0$; this is an expected consequence of the zonal symmetry of the perturbation (D.206). There are five coefficients of angular degree $s = 0$ determined by λ_1 and λ_2, eleven coefficients of degree $s = 2$ determined by λ_3 and λ_4, and thirteen coefficients of angular degree $s = 4$ determined by λ_5. The resulting $(2l+1) \times (2l+1)$ self-coupling matrix $\tilde{\mathsf{V}}^{\mathrm{ani}} = \mathsf{V}^{\mathrm{ani}}$ is *real and diagonal*:

$$\begin{aligned}
V_{mm'}^{\mathrm{ani}} = \delta_{mm'} &\sum_{s=0,2,4} (-1)^m (2l+1) \left(\frac{2s+1}{4\pi}\right)^{1/2}\\
&\times \begin{pmatrix} l & s & l \\ -m & 0 & m \end{pmatrix} \sum_N \sum_I \int_0^c \Gamma_{NI}\, r^2 dr,
\end{aligned} \tag{D.208}$$

where

$$\Gamma_{01} = \dot{u}^2 \begin{pmatrix} l & s & l \\ 0 & 0 & 0 \end{pmatrix} \gamma_{s0}^{0000}, \tag{D.209}$$

$$\Gamma_{02} = 2\Omega_l^0\Omega_l^2\Omega_l^0\Omega_l^2 r^{-2}(v^2+w^2) \begin{pmatrix} l & s & l \\ -2 & 0 & 2 \end{pmatrix} \gamma_{s0}^{++--}, \tag{D.210}$$

$$\Gamma_{03} = f^2 \begin{pmatrix} l & s & l \\ 0 & 0 & 0 \end{pmatrix} \gamma_{s0}^{+-+-}, \tag{D.211}$$

$$\Gamma_{04} = -2f\dot{u} \begin{pmatrix} l & s & l \\ 0 & 0 & 0 \end{pmatrix} \gamma_{s0}^{+-00}, \tag{D.212}$$

$$\Gamma_{05} = 2\Omega_l^0\Omega_l^0(x^2+z^2) \begin{pmatrix} l & s & l \\ -1 & 0 & 1 \end{pmatrix} \gamma_{s0}^{+0-0}, \tag{D.213}$$

Coefficient	$s=0$	$s=2$	$s=4$
$\sqrt{\frac{2s+1}{4\pi}}\,\gamma_{s0}^{0000}$	$\frac{1}{15}(\lambda_1+2\lambda_2)$	$\frac{4}{21}(\lambda_3+2\lambda_4)$	$\frac{8}{35}\lambda_5$
$\sqrt{\frac{2s+1}{4\pi}}\,\gamma_{s0}^{\pm\pm\mp\mp}$	$\frac{2}{15}\lambda_2$	$-\frac{4}{21}\lambda_4$	$\frac{2}{35}\lambda_5$
$\sqrt{\frac{2s+1}{4\pi}}\,\gamma_{s0}^{\pm\mp\pm\mp}$	$\frac{1}{15}(\lambda_1+\lambda_2)$	$-\frac{2}{21}(\lambda_3+\lambda_4)$	$\frac{2}{35}\lambda_5$
$\sqrt{\frac{2s+1}{4\pi}}\,\gamma_{s0}^{\pm\mp00}$	$-\frac{1}{15}\lambda_1$	$-\frac{1}{21}\lambda_3$	$\frac{4}{35}\lambda_5$
$\sqrt{\frac{2s+1}{4\pi}}\,\gamma_{s0}^{\pm0\mp0}$	$-\frac{1}{15}\lambda_2$	$-\frac{1}{21}\lambda_4$	$\frac{4}{35}\lambda_5$
$\sqrt{\frac{2s+1}{4\pi}}\,\gamma_{s0}^{\pm000}$	0	$\frac{1}{7\sqrt{3}}(\lambda_3+2\lambda_4)$	$\frac{4}{7\sqrt{10}}\lambda_5$
$\sqrt{\frac{2s+1}{4\pi}}\,\gamma_{s0}^{\pm\pm\mp0}$	0	$-\frac{2}{7\sqrt{3}}\lambda_4$	$\frac{2}{7\sqrt{10}}\lambda_5$
$\sqrt{\frac{2s+1}{4\pi}}\,\gamma_{s0}^{\pm\mp\pm0}$	0	$-\frac{1}{7\sqrt{3}}(\lambda_3+\lambda_4)$	$\frac{2}{7\sqrt{10}}\lambda_5$
$\sqrt{\frac{2s+1}{4\pi}}\,\gamma_{s0}^{\pm\pm00}$	0	$\frac{2}{7\sqrt{6}}\lambda_3$	$\frac{4}{7\sqrt{10}}\lambda_5$
$\sqrt{\frac{2s+1}{4\pi}}\,\gamma_{s0}^{\pm0\pm0}$	0	$\frac{2}{7\sqrt{6}}\lambda_4$	$\frac{4}{7\sqrt{10}}\lambda_5$
$\sqrt{\frac{2s+1}{4\pi}}\,\gamma_{s0}^{\pm\pm\pm\mp}$	0	$-\frac{2}{7\sqrt{6}}(\lambda_3+2\lambda_4)$	$\frac{2}{7\sqrt{10}}\lambda_5$
$\sqrt{\frac{2s+1}{4\pi}}\,\gamma_{s0}^{\pm\pm\pm0}$	0	0	$\frac{2}{\sqrt{70}}\lambda_5$
$\sqrt{\frac{2s+1}{4\pi}}\,\gamma_{20}^{\pm\pm\pm\pm}$	0	0	$\frac{4}{\sqrt{70}}\lambda_5$

Table D.1. Non-zero coefficients in the generalized spherical-harmonic expansion of the fourth-order elastic perturbation tensor $\boldsymbol{\gamma} = \sum_{st} \gamma_{st}^{\alpha\beta\zeta\eta} Y_{st}^{\alpha+\beta+\zeta+\eta} \hat{\mathbf{e}}_\alpha \hat{\mathbf{e}}_\beta \hat{\mathbf{e}}_\zeta \hat{\mathbf{e}}_\eta$ describing the co-rotational transverse isotropy of the solid inner core.

$$\Gamma_{11} = -4\Omega_l^0 x\dot{u}\begin{pmatrix} l & s & l \\ -1 & 1 & 0 \end{pmatrix}\gamma_{s0}^{+000}, \tag{D.214}$$

$$\Gamma_{12} = -4\Omega_l^0\Omega_l^2\Omega_l^0 r^{-1}(vx+wz)\begin{pmatrix} l & s & l \\ -2 & 1 & 1 \end{pmatrix}\gamma_{s0}^{++-0}, \tag{D.215}$$

$$\Gamma_{13} = 4\Omega_l^0 xf\begin{pmatrix} l & s & l \\ -1 & 0 & 1 \end{pmatrix}\gamma_{s0}^{+-+0}, \tag{D.216}$$

$$\Gamma_{21} = 4\Omega_l^0\Omega_l^2 r^{-1}v\dot{u}\begin{pmatrix} l & s & l \\ -2 & 2 & 0 \end{pmatrix}\gamma_{s0}^{++00}, \tag{D.217}$$

$$\Gamma_{22} = 2\Omega_l^0\Omega_l^0(x^2-z^2)\begin{pmatrix} l & s & l \\ -1 & 2 & -1 \end{pmatrix}\gamma_{s0}^{+0+0}, \tag{D.218}$$

$$\Gamma_{23} = -4\Omega_l^0\Omega_l^2 r^{-1}vf\begin{pmatrix} l & s & l \\ -2 & 2 & 0 \end{pmatrix}\gamma_{s0}^{+++-}, \tag{D.219}$$

$$\Gamma_{31} = -4\Omega_l^0\Omega_l^2\Omega_l^0 r^{-1}(vx-wz)\begin{pmatrix} l & s & l \\ -2 & 3 & -1 \end{pmatrix}\gamma_{s0}^{+++0}, \tag{D.220}$$

$$\Gamma_{41} = 2\Omega_l^0\Omega_l^2\Omega_l^0\Omega_l^2 r^{-2}(v^2-w^2)\begin{pmatrix} l & s & l \\ -2 & 4 & -2 \end{pmatrix}\gamma_{s0}^{++++}. \tag{D.221}$$

The plus-and-minus integrands $\Gamma_{N\pm I}$ in equation (D.116) have been combined in (D.208). The 3-j symbols

$$\begin{pmatrix} l & s & l \\ -m & 0 & m \end{pmatrix}, \qquad s = 0, 2, 4, \tag{D.222}$$

are given explicitly in terms of the degree l and order $-l \leq m \leq l$ in equations (C.229)–(C.231).

Bibliography

Adams, J. C., 1878. On the expression of the product of any two Legendre's coefficients by means of a series of Legendre's coefficients, *Proc. Roy. Soc. Lond.*, **27**, 63–71.

Agnew, D., Berger, J., Buland, R., Farrell, W. & Gilbert, F., 1976. International Deployment of Accelerometers: A network for very long period seismology, *EOS, Trans. Am. Geophys. Un.*, **57**, 180–188.

Aki, K., 1966. Generation and propagation of G waves from the Niigata earthquake of June 16, 1964—2. Estimation of earthquake moment, released energy, and strain-drop from the G wave spectrum, *Bull. Earthquake Res. Inst. Tokyo*, **44**, 23–88.

Aki, K. & Richards, P. G., 1980. *Quantitative Seismology*, Freeman, New York.

Akopyan, S. Ts., Zharkov, V. N. & Lyubimov, V. M., 1975. The dynamic shear modulus in the interior of the Earth, *Dokl. Akad. Nauk USSR, Earth Sci. Sect.*, **223**, 1–3.

Akopyan, S. Ts., Zharkov, V. N. & Lyubimov, V. M., 1976. Corrections to the eigenfrequencies of the Earth due to the dynamic shear modulus, *Izv. Bull. Akad. Sci. USSR, Phys. Solid Earth*, **12**, 625–630.

Alsop, L. E., Sutton, G. H. & Ewing, M., 1961. Free oscillations of the Earth observed on strain and pendulum seismographs, *J. Geophys. Res.*, **66**, 631–641.

Alterman, Z., Jarosch, H. & Pekeris, C. L., 1959. Oscillations of the Earth, *Proc. Roy. Soc. Lond., Ser. A*, **252**, 80–95.

Aly, J.-J. & Pérez, J., 1992. On the stability of a gaseous sphere against non-radial perturbations, *Mon. Not. Roy. Astron. Soc.*, **259**, 95–103.

Anderson, D. L., 1991. *Theory of the Earth*, Blackwell Scientific Publications, Boston, Massachusetts.

Anderson, D. L. & Minster, J. B., 1979. The frequency dependence of Q in the Earth and implications for mantle rheology and the Chandler wobble, *Geophys. J. Roy. Astron. Soc.*, **58**, 431–440.

Angenheister, G., 1906. Bestimmung der Fortpflanzungschwindigkeit und Absorption von Erdbebenwellen, die durch den Gegenpunkt des Herdes gegangen sind, *Nachr. Königlichen Gesell. Wissen. Göttingen, Math. Phys. Klasse*, pages 110–123.

Angenheister, G., 1921. Beobachtungen an pazifischen Beben, *Nachr. Königlichen Gesell. Wissen. Göttingen, Math. Phys. Klasse*, pages 113–146.

Ansell, J. H., 1973. Legendre functions, the Hilbert transform and surface waves on a sphere, *Geophys. J. Roy. Astron. Soc.*, **32**, 95–117.

Arnold, V. I., 1978. *Mathematical Methods of Classical Mechanics*, Springer-Verlag, New York.

Babich, V. M., Chikhachev, B. A. & Yanovskaya, T. B., 1976. Surface waves in a vertically inhomogeneous elastic half space with weak horizontal inhomogeneity, *Izv. Bull. Akad. Sci. USSR, Phys. Solid Earth*, **4**, 24–31.

Backus, G. E., 1958. A class of self-sustaining spherical dynamos, *Ann. Phys. New York*, **4**, 372–447.

Backus, G. E., 1964. Geographical interpretation of measurements of average phase velocities of surface waves over great circular and great semi-circular paths, *Bull. Seismol. Soc. Am.*, **54**, 571–610.

Backus, G. E., 1967. Converting vector and tensor equations to scalar equations in spherical coordinates, *Geophys. J. Roy. Astron. Soc.*, **13**, 71–101.

Backus, G. E., 1977a. Interpreting the seismic glut moments of total degree two or less, *Geophys. J. Roy. Astron. Soc.*, **51**, 1–25.

Backus, G. E., 1977b. Seismic sources with observable glut moments of spatial degree two, *Geophys. J. Roy. Astron. Soc.*, **51**, 27–45.

Backus, G. E., 1986. Poloidal and toroidal fields in geomagnetic field modeling, *Rev. Geophys.*, **24**, 75–109.

Backus, G. E. & Gilbert, F., 1961. The rotational splitting of the free oscillations of the Earth, *Proc. Nat. Acad. Sci. USA*, **47**, 362–371.

Backus, G. E. & Gilbert, F., 1967. Numerical applications of a formalism for geophysical inverse problems, *Geophys. J. Roy. Astron. Soc.*, **13**, 247–276.

Backus, G. E. & Gilbert, F., 1968. The resolving power of gross Earth data, *Geophys. J. Roy. Astron. Soc.*, **16**, 169–205.

Backus, G. E. & Gilbert, F., 1970. Uniqueness in the inversion of gross Earth data, *Phil. Trans. Roy. Soc. Lond., Ser. A*, **266**, 169–205.

Backus, G. E. & Mulcahy, M., 1976a. Moment tensors and other phenomenological descriptions of seismic sources—I. Continuous displacements, *Geophys. J. Roy. Astron. Soc.*, **46**, 341–361.

Backus, G. E. & Mulcahy, M., 1976b. Moment tensors and other phenomenological descriptions of seismic sources—II. Discontinuous displacements, *Geophys. J. Roy. Astron. Soc.*, **47**, 301–329.

Backus, G. E., Parker, R. L. & Constable, C., 1996. *Foundations of Geomagnetism*, Cambridge University Press, Cambridge.

Baker, T. F., Curtis, D. J. & Dodson, A. H., 1996. A new test of Earth tide models in central Europe, *Geophys. Res. Lett.*, **23**, 3559–3562.

Ben-Menahem, A. & Harkrider, D. G., 1964. Radiation patterns of seismic surface waves from buried dipolar point sources in a stratified Earth, *J. Geophys. Res.*, **69**, 2605–2620.

Ben-Menahem, A. & Singh, S. J., 1981. *Seismic Waves and Sources*, Springer-Verlag, New York.

Bender, C. M. & Orszag, S. A., 1978. *Advanced Mathematical Methods for Scientists and Engineers*, McGraw-Hill, New York.

Benioff, H., 1958. Long waves observed in the Kamchatka earthquake of November 4, 1952, *J. Geophys. Res.*, **63**, 589–593.

Benioff, H., Press, F. & Smith, S., 1961. Excitation of the free oscillations of the Earth by earthquakes, *J. Geophys. Res.*, **66**, 605–619.

Berry, M. V. & Upstill, C., 1980. Catastrophe optics: Morphologies of caustics and their diffraction patterns, *Prog. Opt.*, **18**, 257–346.

Binney, J. & Tremaine, S., 1987. *Galactic Dynamics*, Princeton University Press, Princeton, New Jersey.

Blackman, R. B. & Tukey, J. W., 1958. *The Measurement of Power Spectra*, Dover Publications, New York.

Bland, D. R., 1960. *The Theory of Linear Viscoelasticity*, Pergamon Press, Oxford.

Bolt, B. A. & Marussi, A., 1962. Eigenvibrations of the Earth observed at Trieste, *Geophys. J. Roy. Astron. Soc.*, **6**, 299–311.

Born, M., 1927. *The Mechanics of the Atom*, G. Bell, London.

Bracewell, R., 1965. *The Fourier Transform and Its Applications*, McGraw-Hill, New York.

Bretherton, F. P., 1968. Propagation in slowly varying waveguides, *Proc. Roy. Soc. Lond., Ser. A*, **302**, 555–576.

Bromwich, T., 1898. On the influence of gravity on elastic waves, and, in particular, on the vibrations of an elastic globe, *Proc. Lond. Math. Soc.*, **30**, 98–120.

Brune, J. N., 1964. Travel times, body waves and normal modes of the Earth, *Bull. Seismol. Soc. Am.*, **54**, 1315–1321.

Brune, J. N., 1966. P and S travel times and spheroidal normal modes of a homogeneous sphere, *J. Geophys. Res.*, **71**, 2959–2965.

Brune, J. N. & Dorman, J., 1963. Seismic waves and Earth structure in the Canadian shield, *Bull. Seismol. Soc. Am.*, **53**, 167–210.

Brune, J. N., Nafe, J. E. & Alsop, L. E., 1961. The polar phase shift of surface waves on a sphere, *Bull. Seismol. Soc. Am.*, **51**, 247–257.

Brune, J. N., Nafe, J. E. & Oliver, J., 1960. A simplified method for the analysis and synthesis of dispersed wavetrains, *J. Geophys. Res.*, **65**, 287–304.

Brussaard, P. J. & Tolhoek, H. A., 1957. Classical limits of Clebsch-Gordan coefficients, Racah coefficients and $\mathcal{D}^{l,m}_{mn}(\phi, \theta, \psi)$-functions, *Physica*, **23**, 955–971.

Bukchin, B. G., 1995. Determination of stress glut moments of total degree two from teleseismic surface wave amplitude spectra, *Tectonophysics*, **248**, 185–191.

Buland, R., 1981. Free oscillations of the Earth, *Ann. Rev. Earth Planet. Sci.*, **9**, 385–413.

Buland, R., Berger, J. & Gilbert, F., 1979. Observations from the IDA network of attenuation and splitting during a recent earthquake, *Nature*, **277**, 358–362.

Buland, R. & Gilbert, F., 1976. Matched filtering for the seismic moment tensor, *Geophys. Res. Lett.*, **3**, 205–206.

Buland, R., Yuen, D. A., Konstanty, K. & Widmer, R., 1985. Source phase shift: A new phenomenon in wave propagation due to anelasticity, *Geophys. Res. Lett.*, **12**, 569–572.

Bullen, K. E., 1937. The ellipticity correction to travel times of P and S earthquake waves, *Mon. Not. Roy. Astron. Soc., Geophys. Suppl.*, **4**, 143–157.

Bullen, K. E., 1963. *An Introduction to the Theory of Seismology*, Cambridge University Press, Cambridge.

Bullen, K. E., 1975. *The Earth's Density*, Chapman and Hall, London.

Burridge, R., 1966. The Legendre functions of the second kind with complex argument in the theory of wave propagation, *J. Math. Phys.*, **45**, 322–330.

Burridge, R., 1969. Spherically symmetric differential equations, the rotation group, and tensor spherical functions, *Proc. Camb. Phil. Soc.*, **65**, 157–175.

Burridge, R., 1976. *Some Mathematical Topics in Seismology*, Courant Institute of Mathematical Sciences, New York University, New York.

Burridge, R. & Knopoff, L., 1964. Body force equivalents for seismic dislocations, *Bull. Seismol. Soc. Am.*, **54**, 1875–1888.

Cara, M., 1978. Regional variations of higher Rayleigh-mode phase velocities: A spatial-filtering method, *Geophys. J. Roy. Astron. Soc.*, **54**, 439–460.

Carpenter, E. W. & Davies, D., 1966. Frequency dependent seismic phase velocities: An attempted reconciliation between the Jeffreys/Bullen and Gutenberg models of the upper mantle, *Nature*, **212**, 134–135.

Cavendish, H., 1798. Experiments to determine the density of the Earth, *Phil. Trans. Roy. Soc. Lond.*, **58**, 469–526.

Červený, V., 1985. The application of ray tracing to the numerical modeling of seismic wavefields in complex structures. In Dohr, G. P., editor, *Seismic Shear Waves, Part A: Theory, Handbook of Geophysical Exploration*, volume 15A, pages 1–124, Geophysical Press, London.

Červený, V., Molotkov, I. A. & Pšenčík, I., 1977. *Ray Method in Seismology*, Univerzita Karlova, Praha.

Chandrasekhar, S. & Roberts, P. H., 1963. The ellipticity of a slowly rotating configuration, *Astrophys. J.*, **138**, 801–808.

Chao, B. F. & Gross, R. S., 1995. Changes in the Earth's rotational energy induced by earthquakes, *Geophys. J. Int.*, **122**, 776–783.

Chao, B. F., Gross, R. S. & Dong, D.-N., 1995. Changes in global gravitational energy induced by earthquakes, *Geophys. J. Int.*, **122**, 784–789.

Chapman, C. H., 1976. A first-motion alternative to geometrical ray theory, *Geophys. Res. Lett.*, **3**, 153–156.

Chapman, C. H., 1978. A new method for computing synthetic seismograms, *Geophys. J. Roy. Astron. Soc.*, **54**, 481–518.

Chapman, C. H., Chu, J.-Y. & Lyness, D. G., 1988. The WKBJ seismogram algorithm. In Doornbos, D. J., editor, *Seismological Algorithms: Computational Methods and Computer Programs*, pages 47–74, Academic Press, New York.

Chapman, C. H. & Drummond, R., 1982. Body-wave seismograms in inhomogeneous media using Maslov asymptotic theory, *Bull. Seismol. Soc. Am.*, **72**, S277–S317.

Chapman, C. H. & Orcutt, J. A., 1985. The computation of body wave seismograms in laterally homogeneous media, *Rev. Geophys.*, **23**, 105–163.

Chapman, S. & Bartels, J., 1940. *Geomagnetism*, Oxford University Press, Oxford.

Choy, G. L. & Richards, P. G., 1975. Pulse distortion and Hilbert transformation in multiply reflected and refracted body waves, *Bull. Seismol. Soc. Am.*, **65**, 55–70.

Chree, C., 1889. The equations of an isotropic elastic solid in polar and cylindrical coordinates, their solution and applications, *Trans. Camb. Phil. Soc.*, **14**, 250–369.

Clairaut, A. C., 1743. *Théorie de la Figure de la Terre, Tirée des Principes de l'Hydrostatique*, David Fils, Paris.

Clemmow, P. C., 1961. An infinite Legendre integral transform and its inverse, *Proc. Camb. Phil. Soc.*, **57**, 547–560.

Clévédé, E. & Lognonné, P., 1996. Fréchet derivatives of coupled seismograms with respect to an anelastic rotating Earth, *Geophys. J. Int.*, **124**, 456–482.

Coates, R. T. & Chapman, C. H., 1990. Ray perturbation theory and the Born approximation, *Geophys. J. Int.*, **100**, 379–392.

Condon, E. U. & Shortley, G. H., 1935. *The Theory of Atomic Spectra*, Cambridge University Press, Cambridge.

Connes, J., Blum, P. A., Jobert, N. & Jobert, G., 1962. Observations des oscillations propres de la Terre, *Ann. Geophys.*, **18**, 260–268.

Courant, R. & Hilbert, D., 1966. *Methods of Mathematical Physics*, Wiley, London.

Cowling, T. G., 1941. The non-radial oscillations of polytropic stars, *Mon. Not. Roy. Astron. Soc.*, **101**, 369–373.

Cowling, T. G. & Newing, R. A., 1949. The oscillations of a rotating star, *Astrophys. J.*, **109**, 149–158.

Cox, J. P., 1980. *Theory of Stellar Pulsation*, Princeton University Press, Princeton, New Jersey.

Creager, K. C., 1992. Anisotropy of the inner core from differential travel times of the phases PKP and PKIKP, *Nature*, **356**, 309–314.

Cummins, P. R., 1997. Earthquake near field and W phase observations at teleseismic distances, *Geophys. Res. Lett.*, **24**, 2857–2860.

Cummins, P. R., Geller, R. J., Hatori, T. & Takeuchi, N., 1994. DSM complete synthetic seismograms: SH, spherically symmetric, case, *Geophys. Res. Lett.*, **21**, 533–536.

Cummins, P. R., Geller, R. J. & Takeuchi, N., 1994. DSM complete synthetic seismograms: P-SV, spherically symmetric, case, *Geophys. Res. Lett.*, **21**, 1633–1666.

Cummins, P. R., Takeuchi, N. & Geller, R. J., 1997. Computation of complete synthetic seismograms for laterally heterogeneous models using the direct solution method, *Geophys. J. Int.*, **130**, 1–16.

Dahlen, F. A., 1968. The normal modes of a rotating, elliptical Earth, *Geophys. J. Roy. Astron. Soc.*, **16**, 329–367.

Dahlen, F. A., 1969. The normal modes of a rotating, elliptical Earth—II. Near-resonance multiplet coupling, *Geophys. J. Roy. Astron. Soc.*, **18**, 397–436.

Dahlen, F. A., 1972. Elastic dislocation theory for a self-gravitating elastic configuration with an initial static stress field, *Geophys. J. Roy. Astron. Soc.*, **28**, 357–383.

Dahlen, F. A., 1973. Elastic dislocation theory for a self-gravitating elastic configuration with an initial static stress field—II. Energy release, *Geophys. J. Roy. Astron. Soc.*, **31**, 469–484.

Dahlen, F. A., 1976. Reply to comments by A. M. Dziewonski and R. V. Sailor on "The correction of great circular phase velocity measurements for the rotation and ellipticity of the Earth", *J. Geophys. Res.*, **81**, 4951–4956.

Dahlen, F. A., 1977. The balance of energy in earthquake faulting, *Geophys. J. Roy. Astron. Soc.*, **48**, 239–261.

Dahlen, F. A., 1978. Excitation of the normal modes of a rotating Earth by an earthquake fault, *Geophys. J. Roy. Astron. Soc.*, **54**, 1–9.

Dahlen, F. A., 1979a. The spectra of unresolved split normal mode multiplets, *Geophys. J. Roy. Astron. Soc.*, **58**, 1–33.

Dahlen, F. A., 1979b. Exact and asymptotic synthetic multiplet spectra on an ellipsoidal Earth, *Geophys. J. Roy. Astron. Soc.*, **59**, 19–42.

Dahlen, F. A., 1980a. A uniformly valid asymptotic representation of normal mode multiplet spectra on a laterally heterogeneous Earth, *Geophys. J. Roy. Astron. Soc.*, **62**, 225–247.

Dahlen, F. A., 1980b. Addendum to "Excitation of the normal modes of a rotating Earth by an earthquake fault", *Geophys. J. Roy. Astron. Soc.*, **62**, 719–721.

Dahlen, F. A., 1982. The effect of data windows on the estimation of free oscillation parameters, *Geophys. J. Roy. Astron. Soc.*, **69**, 537–549.

Dahlen, F. A., 1987. Multiplet coupling and the calculation of synthetic long-period seismograms, *Geophys. J. Roy. Astron. Soc.*, **91**, 241–254.

Dahlen, F. A., 1993. Single-force representation of shallow landslide sources, *Bull. Seismol. Soc. Am.*, **83**, 130–143.

Dahlen, F. A. & Sailor, R. V., 1979. Rotational and elliptical splitting of the free oscillations of the Earth, *Geophys. J. Roy. Astron. Soc.*, **58**, 609–623.

Dahlen, F. A. & Smith, M. L., 1975. The influence of rotation on the free oscillations of the Earth, *Phil. Trans. Roy. Soc. Lond., Ser. A*, **279**, 583–627.

Davis, J. P. & Henson, I. H., 1986. Validity of the great circle average approximation for inversion of normal mode measurements, *Geophys. J. Roy. Astron. Soc.*, **85**, 69–92.

Defraigne, P., 1997. Geophysical model of the dynamical flattening of the Earth in agreement with the precession constant, *Geophys. J. Int.*, **130**, 47–56.

DeMets, C., Gordon, R. G., Argus, D. F. & Stein, S., 1990. Current plate motions, *Geophys. J. Int.*, **101**, 525–478.

Derr, J. S., 1969. Free oscillation observations through 1978, *Bull. Seismol. Soc. Am.*, **59**, 2079–2099.

Dey-Sarkar, S. K. & Chapman, C. H., 1978. A simple method for the computation of body-wave seismograms, *Bull. Seismol. Soc. Am.*, **68**, 1577–1593.

Dickman, S. R., 1988. The self-consistent dynamic pole tide in non-global oceans, *Geophys. J.*, **94**, 519–543.

Dickman, S. R., 1993. Dynamic ocean-tide effects on Earth's rotation, *Geophys. J. Int.*, **112**, 448–470.

Dongarra, J. J. & Walker, D. W., 1995. Software libraries for linear algebra computations on high performance computers, *SIAM Review*, **37**, 151–180.

Doornbos, D. J., 1988. Asphericity and ellipticity corrections. In Doornbos, D. J., editor, *Seismological Algorithms: Computational Methods and Computer Programs*, pages 75–85, Academic Press, New York.

Dratler, J., Farrell, W. E., Block, B. & Gilbert, F., 1971. High-Q overtone modes of the Earth, *Geophys. J. Roy. Astron. Soc.*, **23**, 399–410.

Durek, J. J. & Ekström, G., 1995. Evidence of bulk attenuation in the asthenosphere from recordings of the Bolivian earthquake, *Geophys. Res. Lett.*, **22**, 2309–2312.

Durek, J. J. & Ekström, G., 1996. A radial model of anelasticity consistent with long-period surface-wave attenuation data, *Bull. Seismol. Soc. Am.*, **86**, 144–158.

Durek, J. J. & Ekström, G., 1997. Investigating discrepancies among measurements of travelling and standing wave attenuation, *J. Geophys. Res.*, **102**, 24529–24544.

Durek, J. J., Ritzwoller, M. H. & Woodhouse, J. H., 1993. Constraining upper mantle anelasticity using surface wave amplitude anomalies, *Geophys. J. Int.*, **114**, 249–272.

Dyson, J. & Schutz, B. F., 1979. Perturbations and stability of rotating stars—I. Completeness of normal modes, *Proc. Roy. Soc. Lond., Ser. A*, **368**, 389–410.

Dziewonski, A. M. & Anderson, D. L., 1981. Preliminary reference Earth model, *Phys. Earth Planet. Int.*, **25**, 297–356.

Dziewonski, A. M., Chou, T.-A. & Woodhouse, J. H., 1981. Determination of earthquake source parameters from waveform data for studies of global and regional seismicity, *J. Geophys. Res.*, **86**, 2825–2852.

Dziewonski, A. M. & Gilbert, F., 1971. Solidity of the inner core of the Earth inferred from normal mode observations, *Nature*, **234**, 465–466.

Dziewonski, A. M. & Gilbert, F., 1972. Observations of normal modes from 84 recordings of the Alaskan earthquake of 1964 March 28, *Geophys. J. Roy. Astron. Soc.*, **27**, 393–446.

Dziewonski, A. M. & Gilbert, F., 1973. Observations of normal modes from 84 recordings of the Alaskan earthquake of 1964 March 28—II. Further remarks based upon new spheroidal overtone data, *Geophys. J. Roy. Astron. Soc.*, **35**, 401–437.

Dziewonski, A. M. & Gilbert, F., 1974. Temporal variation of the seismic moment tensor and the evidence of precursive compression for two deep earthquakes, *Nature*, **257**, 185–188.

Dziewonski, A. M. & Gilbert, F., 1976. The effect of small, aspherical perturbations on travel times and a re-examination of the corrections for ellipticity, *Geophys. J. Roy. Astron. Soc.*, **44**, 7–17.

Dziewonski, A. M., Liu, X.-F. & Su, W.-J., 1997. Lateral heterogeneity in the lowermost mantle. In Crossley, D. J., editor, *Earth's Deep Interior, The Doornbos Memorial Volume*, pages 11–50, Gordon and Breach, Amsterdam.

Dziewonski, A. M. & Woodhouse, J. H., 1983. An experiment in systematic study of global seismicity: Centroid-moment tensor solutions for 201 moderate and large earthquakes of 1981, *J. Geophys. Res.*, **88**, 3247–3271.

Dziewonski, A. M. & Woodward, R. L., 1992. Acoustic imaging at the planetary scale, *Acoustical Imaging*, **19**, 785–797.

Eckart, C., 1960. *Hydrodynamics of Oceans and Atmospheres*, Pergamon Press, Oxford.

Edmonds, A. R., 1960. *Angular Momentum in Quantum Mechanics*, Princeton University Press, Princeton, New Jersey.

Ekström, G., 1989. A very broad band inversion method for the recovery of earthquake source parameters, *Tectonophysics*, **166**, 73–100.

Ekström, G., 1994. Anomalous earthquakes on volcano ring-fault structures, *Earth Planet. Sci. Lett.*, **128**, 707–712.

Ekström, G., 1995. Calculation of static deformation following the Bolivia earthquake by summation of the Earth's normal modes, *Geophys. Res. Lett.*, **22**, 2289–2292.

Ekström, G. & Dziewonski, A. M., 1985. Centroid-moment tensor solutions for 35 earthquakes in western North America, *Bull. Seismol. Soc. Am.*, **75**, 23–39.

Ekström, G., Tromp, J. & Larson, E. W. F., 1997. Measurements and global models of surface wave propagation, *J. Geophys. Res.*, **102**, 8137–8157.

Erdélyi, A., Magnus, W., Oberhettinger, F. & Tricomi, F. G., 1953. *Higher Transcendental Functions*, McGraw-Hill, New York.

Ewing, M., Jardetzky, W. S. & Press, F., 1957. *Elastic Waves in Layered Media*, McGraw-Hill, New York.

Ewing, M. & Press, F., 1954. An investigation of mantle Rayleigh waves, *Bull. Seismol. Soc. Am.*, **44**, 127–148.

Farra, V. & Madariaga, R., 1987. Seismic waveform modeling in heterogeneous media by ray perturbation theory, *J. Geophys. Res.*, **92**, 2697–2712.

Farra, V., Virieux, J. & Madariaga, R., 1989. Ray perturbation theory for interfaces, *Geophys. J. Int.*, **99**, 377–390.

Forsyth, D. W., 1975. The early structural evolution and anisotropy of the oceanic upper mantle, *Geophys. J. Roy. Astron. Soc.*, **43**, 103–162.

Freund, L. B., 1990. *Dynamic Fracture Mechanics*, Cambridge University Press, Cambridge.

Friederich, W. & Dalkolmo, J., 1995. Complete synthetic seismograms for a spherically symmetric Earth by a numerical computation of the Green's function in the frequency domain, *Geophys. J. Int.*, **122**, 537–550.

Frohlich, C., 1990. Note concerning non-double-couple source components from slip along surfaces of revolution, *J. Geophys. Res.*, **95**, 6861–6866.

Garbow, B., Boyle, J., Dongarra, J. & Moler, C., 1977. *Matrix Eigensystem Routines—EISPACK Guide Extension*, Lecture Notes in Computer Science, Springer-Verlag, New York.

Gaunt, J. A., 1929. The triplets of helium, *Phil. Trans. Roy. Soc. Lond.*, **228**, 151–196.

Gelfand, I. M. & Shapiro, Z. Y., 1956. Representations of the group of rotations in three-dimensional space and their applications, *Am. Math. Soc. Transl.*, **2**, 207–316.

Geller, R. J. & Hara, T., 1993. Two efficient algorithms for iterative linearized inversion of seismic waveform data, *Geophys. J. Int.*, **115**, 695–710.

Geller, R. J., Hara, T. & Tsuboi, S., 1990. On the equivalence of two methods for computing partial derivatives of seismic waveforms, *Geophys. J. Int.*, **100**, 153–156.

Geller, R. J. & Ohminato, T., 1994. Computation of synthetic seismograms and their partial derivatives for heterogeneous media with arbitrary natural boundary conditions using the direct solution method, *Geophys. J. Int.*, **116**, 421–446.

Geller, R. J. & Stein, S., 1979. Time domain attenuation measurements for fundamental spheroidal modes (${}_0\mathrm{S}_6$–${}_0\mathrm{S}_{28}$) for the 1977 Indonesian earthquake, *Bull. Seismol. Soc. Am.*, **169**, 1671–1691.

Giardini, D., Li, X.-D. & Woodhouse, J. H., 1987. Three-dimensional structure of the Earth from splitting in free oscillation spectra, *Nature*, **325**, 405–411.

Giardini, D., Li, X.-D. & Woodhouse, J. H., 1988. Splitting functions of long-period normal modes of the Earth, *J. Geophys. Res.*, **93**, 13716–13742.

Gibbs, J. W., 1901. *Vector Analysis*, Charles Scribner's Sons, New York. Transcribed by E. B. Wilson. Reprinted in 1960 by Dover Publications, New York.

Gilbert, F., 1967. Gravitationally perturbed elastic waves, *Bull. Seismol. Soc. Am.*, **57**, 783–794.

Gilbert, F., 1970. Excitation of the normal modes of the Earth by earthquake sources, *Geophys. J. Roy. Astron. Soc.*, **22**, 223–226.

Gilbert, F., 1971a. The diagonal sum rule and averaged eigenfrequencies, *Geophys. J. Roy. Astron. Soc.*, **23**, 119–123.

Gilbert, F., 1971b. Ranking and winnowing gross Earth data for inversion and resolution, *Geophys. J. Roy. Astron. Soc.*, **23**, 125–128.

Gilbert, F., 1973. Derivation of source parameters from low-frequency spectra, *Phil. Trans. Roy. Soc. Lond., Ser. A*, **274**, 369–371.

Gilbert, F., 1976a. The representation of seismic displacements in terms of travelling waves, *Geophys. J. Roy. Astron. Soc.*, **44**, 275–280.

Gilbert, F., 1976b. Differential kernels for group velocity, *Geophys. J. Roy. Astron. Soc.*, **44**, 649–660.

Gilbert, F., 1980. An introduction to low-frequency seismology. In Dziewonski, A. M. & Boschi, E., editors, *Fisica dell'interno della Terra, Rendiconti della Scuola Internazionale di Fisica "Enrico Fermi"*, Varenna, Italy, course LXXVII, pages 41–81, North-Holland, Amsterdam.

Gilbert, F., 1994. Splitting of the free-oscillation multiplets by steady flow, *Geophys. J. Int.*, **116**, 227–229.

Gilbert, F. & Backus, G. E., 1966. Propagator matrices in elastic wave and vibration problems, *Geophysics*, **31**, 326–333.

Gilbert, F. & Backus, G. E., 1969. A computational problem encountered in a study of the Earth's normal modes. In *Proc. Fall Joint Comp. Conf., Am. Fed. Inf. Proc. Soc.*, pages 1273–1277.

Gilbert, F. & Buland, R., 1976. An enhanced deconvolution procedure for retrieving the seismic moment tensor from a sparse network, *Geophys. J. Roy. Astron. Soc.*, **50**, 251–255.

Gilbert, F. & Dziewonski, A. M., 1975. An application of normal mode theory to the retrieval of structural parameters and source mechanisms from seismic spectra, *Phil. Trans. Roy. Soc. Lond., Ser. A*, **278**, 187–269.

Gilbert, F. & MacDonald, G. J. F., 1960. Free oscillations of the Earth—I. Toroidal oscillations, *J. Geophys. Res.*, **65**, 675–693.

Goldstein, H., 1980. *Classical Mechanics*, Addison-Wesley, Reading, Massachusetts.

Gross, B., 1953. *Mathematical Structure of the Theories of Viscoelasticity*, Hermann, Paris.

Gutenberg, B., 1924. Dispersion und Extinktion von seismischen Oberflächenwellen und der Aufbau der obersten Erdschichten, *Physik. Zeitschrift.*, **25**, 377–381.

Gutenberg, B. & Richter, C. F., 1934. On seismic waves (Part I), *Gerlands Beitr. Geophys.*, **43**, 56–133.

Gutzwiller, M. C., 1990. *Chaos in Classical and Quantum Mechanics*, Springer-Verlag, New York.

Halmos, P. R., 1958. *Finite-Dimensional Vector Spaces*, Van Nostrand, New York.

Hara, T., Kuge, K. & Kawakatsu, H., 1995. Determination of the isotropic component of the 1994 Bolivia deep earthquake, *Geophys. Res. Lett.*, **22**, 2265–2268.

Hara, T., Kuge, K. & Kawakatsu, H., 1996. Determination of the isotropic component of deep-focus earthquakes by inversion of normal-mode data, *Geophys. J. Int.*, **127**, 515–528.

Hara, T., Tsuboi, S. & Geller, R. J., 1991. Inversion for laterally heterogeneous Earth structure using a laterally heterogeneous starting model: Preliminary results, *Geophys. J. Int.*, **104**, 523–540.

Hara, T., Tsuboi, S. & Geller, R. J., 1993. Inversion for laterally heterogeneous Earth structure using iterative linearized waveform inversion, *Geophys. J. Int.*, **115**, 667–698.

Harris, F. J., 1978. On the use of windows for harmonic analysis with the discrete Fourier transform, *Proc. IEEE*, **66**, 51–83.

Haskell, N. A., 1964. Radiation pattern of surface waves from point sources in a layered medium, *Bull. Seismol. Soc. Am.*, **54**, 377–393.

Hayes, W. D., 1973. Group velocity and nonlinear dispersive wave propagation, *Proc. Roy. Soc. Lond., Ser. A*, **332**, 199–221.

He, X. & Tromp, J., 1996. Normal-mode constraints on the structure of the Earth, *J. Geophys. Res.*, **87**, 7772–7778.

Heinz, D., Jeanloz, R. & O'Connell, R. J., 1982. Bulk attenuation in a polycrystalline Earth, *J. Geophys. Res.*, **87**, 7772–7778.

Henriksen, S. W., 1960. The hydrostatic flattening of the Earth, *Ann. Int. Geophys. Year*, **12**, 197–198.

Henson, I. H., 1989. Multiplet coupling of the normal modes of an elliptical, transversely isotropic Earth, *Geophys. J.*, **98**, 457–459.

Horn, R. A. & Johnson, C. A., 1985. *Matrix Analysis*, Cambridge University Press, Cambridge.

Hoskins, L. M., 1920. The strain of a gravitating sphere of variable density and elasticity, *Trans. Am. Math. Soc.*, **21**, 1–43.

Ihmlé, P. F. & Jordan, T. H., 1994. Teleseismic search for slow precursors to large earthquakes, *Science*, **266**, 1547–1551.

Ihmlé, P. F. & Jordan, T. H., 1995. Source time function of the great 1994 Bolivia deep earthquake by waveform and spectral inversions, *Geophys. Res. Lett.*, **22**, 2253–2256.

Jackson, D. D., 1972. Interpretation of inaccurate, insufficient and inconsistent data, *Geophys. J. Roy. Astron. Soc.*, **28**, 97–110.

Jackson, I., 1993. Progress in the experimental study of seismic wave attenuation, *Ann. Rev. Earth Planet. Sci.*, **21**, 375–406.

Jackson, J. D., 1962. *Classical Electrodynamics*, Wiley, New York.

Jaerisch, P., 1880. Über der elastischen Schwingungen einer isotropen Kugel, *Crelle, J. Reine Angew. Math.*, **88**, 131–145.

James, R. W., 1976. New tensor spherical harmonics, for application to the partial differential equations of mathematical physics, *Phil. Trans. Roy. Soc. Lond., Ser. A*, **281**, 195–221.

Jeans, J. H., 1903. On the vibrations and stability of a gravitating planet, *Phil. Trans. Roy. Soc. Lond., Ser. A*, **27**, 157–184.

Jeans, J. H., 1927. The propagation of earthquake waves, *Proc. Roy. Soc. Lond., Ser. A*, **102**, 554–574.

Jeffreys, H., 1924. *The Earth*, first edition, Cambridge University Press, Cambridge.

Jeffreys, H., 1935. The surface waves of earthquakes, *Mon. Not. Roy. Astron. Soc., Geophys. Suppl.*, **3**, 253–261.

Jeffreys, H., 1958a. A modification of Lomnitz's law of creep in rocks, *Geophys. J. Roy. Astron. Soc.*, **1**, 92–95.

Jeffreys, H., 1958b. Rock creep, tidal friction and the moon's ellipticities, *Mon. Not. Roy. Astron. Soc.*, **118**, 14–17.

Jeffreys, H., 1961. Small corrections in the theory of surface waves, *Geophys. J. Roy. Astron. Soc.*, **6**, 115–117.

Jeffreys, H., 1963. On the hydrostatic theory of the figure of the Earth, *Geophys. J. Roy. Astron. Soc.*, **8**, 196–202.

Jeffreys, H., 1970. *The Earth*, fifth edition, Cambridge University Press, Cambridge.

Jiao, W., Wallace, T. C., Beck, S. L., Silver, P. G. & Zandt, G., 1995. Evidence for static displacements from the June 9, 1994 deep Bolivian earthquake, *Geophys. Res. Lett.*, **16**, 2285–2288.

Jobert, N., 1956. Évaluation de la période d'oscillation d'une sphère hétérogène, par application du principe de Rayleigh, *Comptes Rendus Acad. Sci. Paris*, **243**, 1230–1232.

Jobert, N., 1957. Sur la période propre des oscillations sphéroïdales de la Terre, *Comptes Rendus Acad. Sci. Paris*, **244**, 921–922.

Jobert, N., 1961. Calcul approché de la période des oscillations sphéroïdales de la Terre, *Geophys. J. Roy. Astron. Soc.*, **4**, 242–258.

Jobert, N., Gaulon, R., Dieulin, A. & Roult, G., 1977. Sur les ondes de trés longue période, caractéristiques du manteau supérieur, *Comptes Rendus Acad. Sci. Paris, Sér. B*, **285**, 49–51.

Jobert, N. & Jobert, G., 1987. Ray tracing for surface waves. In Nolet, G., editor, *Seismic Tomography, with Applications in Global Seismology and Exploration Geophysics*, pages 275–300, Reidel, Dordrecht.

Jordan, T. H., 1978. A procedure for estimating lateral variations from low-frequency eigenspectra data, *Geophys. J. Roy. Astron. Soc.*, **52**, 441–455.

Julian, B. R. & Anderson, D. L., 1968. Travel times, apparent velocities and amplitudes of body waves, *Bull. Seismol. Soc. Am.*, **58**, 339–366.

Julian, B. R. & Gubbins, D., 1977. Three-dimensional seismic ray tracing, *J. Geophys.*, **43**, 95–113.

Kanamori, H., 1970. Velocity and Q of mantle waves, *Phys. Earth Planet. Int.*, **2**, 259–275.

Kanamori, H., 1976. Re-examination of the Earth's free oscillations excited by the Kamchatka earthquake of November 4, 1952, *Phys. Earth Planet. Int.*, **11**, 216–226.

Kanamori, H., 1977. The energy release in great earthquakes, *J. Geophys. Res.*, **82**, 2981–2987.

Kanamori, H., 1993. W phase, *Geophys. Res. Lett.*, **20**, 1691–1694.

Kanamori, H. & Anderson, D. L., 1977. Importance of physical dispersion in surface wave and free oscillation problems: Review, *Rev. Geophys. Space Phys.*, **15**, 105–112.

Kanamori, H. & Given, J. W., 1982. Analysis of long-period seismic waves excited by the May 18, 1980 eruption of Mt. St. Helens: A terrestrial monopole?, *J. Geophys. Res.*, **87**, 5422–5432.

Kanamori, H. & Kikuchi, M., 1993. The 1992 Nicaragua Earthquake: A slow tsunami earthquake associated with subducted sediments, *Nature*, **361**, 714–716.

Kanamori, H. & Mori, J., 1992. Harmonic excitation of mantle Rayleigh waves by the 1992 eruption of Mount Pinatubo, Philippines, *Geophys. Res. Lett.*, **19**, 721–724.

Karato, S., 1993. Importance of anelasticity in the interpretation of seismic tomography, *Geophys. Res. Lett.*, **20**, 1623–1626.

Karato, S. & Spetzler, H. A., 1990. Defect microdynamics in crystals and solid-state mechanisms of seismic wave attenuation and velocity dispersion in the mantle, *Rev. Geophys.*, **29**, 399–421.

Kawakatsu, H., 1991. Insignificant isotropic component in the moment tensor of deep earthquakes, *Nature*, **351**, 50–53.

Kawakatsu, H., 1996. Observability of the isotropic component of a moment tensor, *Geophys. J. Int.*, **126**, 525–544.

Keilis-Borok, V. L., Levshin, A. L., Yanovskaya, T. B., Lander, A. V., Bukchin, B. G., Barmin, M. P., Ratnikova, L. I. & Its, E. N., 1989. *Seismic Surface Waves in a Laterally Inhomogeneous Earth*, Kluwer, Dordrecht.

Keller, J. B., 1958. Corrected Bohr-Sommerfeld quantum conditions for nonseparable systems, *Ann. Phys. New York*, **4**, 180–188.

Keller, J. B. & Rubinow, S. I., 1960. Asymptotic solution of eigenvalue problems, *Ann. Phys. New York*, **9**, 24–75.

Kellogg, O. D., 1967. *Foundations of Potential Theory*, Springer-Verlag, New York.

Kendall, J.-M., Guest, W. S. & Thomson, C. J., 1992. Ray-theory Green's function reciprocity and ray-centered coordinates in anisotropic media, *Geophys. J. Int.*, **108**, 364–371.

Kennett, B. L. N., 1983. *Seismic Wave Propagation in Stratified Media*, Cambridge University Press, Cambridge.

Kennett, B. L. N., 1991. *IASPEI 1991 Seismological Tables*, Research School of Earth Sciences, Australian National University, Canberra.

Kennett, B. L. N. & Gudmundsson, O., 1996. Ellipticity corrections for seismic phases, *Geophys. J. Int.*, **127**, 40–48.

Kennett, B. L. N. & Nolet, G., 1979. The influence of upper mantle discontinuities upon the free oscillations of the Earth, *Geophys. J. Roy. Astron. Soc.*, **56**, 283–308.

Kikuchi, M. & Kanamori, H., 1982. Inversion of complex body waves, *Bull. Seismol. Soc. Am.*, **72**, 491–506.

Kikuchi, M. & Kanamori, H., 1994. The mechanism of the deep Bolivia earthquake of June 9, 1994, *Geophys. Res. Lett.*, **21**, 2341–2344.

Kinoshita, H., 1977. Theory of the rotation of the rigid Earth, *Celest. Mech.*, **15**, 277–326.

Kjartansson, E., 1979. Constant Q wave propagation and attenuation, *J. Geophys. Res.*, **84**, 4737–4748.

Kline, M. & Kay, I. W., 1979. *Electromagnetic Theory and Geometrical Optics*, Krieger, New York.

Knopoff, L. & Randall, M. J., 1970. The compensated linear vector dipole: A possible mechanism for deep-focus earthquakes, *J. Geophys. Res.*, **75**, 4957–4963.

Kostrov, B. V., 1970. The theory of the focus for tectonic earthquakes, *Izv. Bull. Akad. Sci. USSR, Phys. Solid Earth*, **4**, 84–101.

Kostrov, B. V., 1974. Seismic moment and energy of earthquakes, and seismic flow of rock, *Izv. Bull. Akad. Sci. USSR, Phys. Solid Earth*, **1**, 23–40.

Kostrov, B. V. & Das, S., 1988. *Principles of Earthquake Source Mechanics*, Cambridge University Press, Cambridge.

Kravtsov, Yu. A. & Orlov, Y. I., 1990. *Geometrical Optics of Inhomogeneous Media*, Springer-Verlag, New York.

Kuge, K. & Lay, T., 1994. Systematic non-double-couple components of earthquake mechanisms: The role of fault zone irregularity, *J. Geophys. Res.*, **99**, 15457–15467.

Lamb, H., 1882. On the vibrations of an elastic sphere, *Proc. Lond. Math. Soc.*, **13**, 189–212.

Lamb, H., 1904. On the propagation of tremors over the surface of an elastic solid, *Phil. Trans. Roy. Soc. Lond.*, **203**, 1–42.

Lamb, H., 1932. *Hydrodynamics*, Cambridge University Press, Cambridge. Reprinted in 1945 by Dover Publications, New York.

Lambeck, K., 1988. *Geophysical Geodesy*, Oxford University Press, Oxford.

Lanczos, C., 1962. *The Variational Principles of Mechanics*, University of Toronto Press, Toronto.

Landau, L. D. & Lifshitz, E. M., 1965. *Quantum Mechanics: Non-Relativistic Theory*, Pergamon Press, Oxford.

Landau, L. D. & Lifshitz, E. M., 1971. *The Classical Theory of Fields*, Pergamon Press, Oxford.

Langston, C. A., 1981. Source inversion of seismic waveforms: The Koyna, India, earthquake of 13 September 1967, *Bull. Seismol. Soc. Am.*, **71**, 1–24.

Lapwood, E. R. & Usami, T., 1981. *Free Oscillations of the Earth*, Cambridge University Press, Cambridge.

Laske, G. & Masters, G., 1996. Constraints on global phase velocity maps from long-period polarization data, *J. Geophys. Res.*, **101**, 16059–16075.

Lay, T. & Kanamori, H., 1985. Geometrical effects of global lateral heterogeneity on long-period surface wave propagation, *J. Geophys. Res.*, **90**, 605–621.

Lebedev, S., Nolet, G. & Van der Hilst, R. D., 1997. The upper mantle beneath the Philippine Sea region from waveform inversions, *Geophys. Res. Lett.*, **24**, 1851–1854.

Ledoux, P., 1951. The non-radial oscillations of gaseous stars and the problem of Beta Canis Majoris, *Astrophys. J.*, **114**, 373–384.

Levshin, A. L., 1981. The relation between P and S travel times, phase velocities of higher Rayleigh waves, and frequencies of spheroidal oscillations in a radially inhomogeneous Earth, *Computational Seismology*, **13**, 103–109.

Li, X.-D., Giardini, D. & Woodhouse, J. H., 1991a. Large-scale three-dimensional even-degree structure of the Earth from splitting of long-period normal modes, *J. Geophys. Res.*, **96**, 551–557.

Li, X.-D., Giardini, D. & Woodhouse, J. H., 1991b. The relative amplitudes of mantle heterogeneity in P velocity, S velocity, and density from free oscillation data, *Geophys. J. Int.*, **105**, 649–657.

Li, X.-D. & Romanowicz, B. A., 1995. Comparison of global waveform inversions with and without considering cross-branch modal coupling, *Geophys. J. Int.*, **121**, 695–709.

Li, X.-D. & Romanowicz, B. A., 1996. Global mantle shear-velocity model developed using nonlinear asymptotic coupling theory, *J. Geophys. Res.*, **101**, 22245–22272.

Li, X.-D. & Tanimoto, T., 1993. Waveforms of long-period body waves in a slightly aspherical Earth model, *Geophys. J. Int.*, **112**, 92–102.

Libbrecht, K. G., 1985. Practical considerations for the generation of large order spherical harmonics, *Solar Physics*, **99**, 371–373.

Lighthill, J., 1978. *Waves in Fluids*, Cambridge University Press, Cambridge.

Liu, H.-P., Anderson, D. L. & Kanamori, H., 1976. Velocity dispersion due to anelasticity; implications for seismology and mantle composition, *Geophys. J. Roy. Astron. Soc.*, **47**, 41–58.

Liu, X.-F. & Tromp, J., 1996. Uniformly valid body-wave theory, *Geophys J. Int.*, **127**, 461–491.

Lognonné, P., 1991. Normal modes and seismograms in an anelastic rotating Earth, *J. Geophys. Res.*, **96**, 365–395.

Lognonné, P., Clévédé, E. & Kanamori, H., 1998. Normal mode summation of seismograms and barograms in a spherical Earth with a realistic atmosphere, *Geophys. J. Int.*, in press.

Lognonné, P. & Romanowicz, B. A., 1990. Fully coupled Earth's vibrations: The spectral method, *Geophys. J. Int.*, **102**, 20309–20319.

Lomnitz, C., 1956. Creep measurements in igneous rocks, *J. Geol.*, **64**, 473–479.

Lomnitz, C., 1957. Linear dissipation in solids, *J. Appl. Phys.*, **28**, 201–205.

Love, A. E. H., 1907. The gravitational stability of the Earth, *Phil. Trans. Roy. Soc. Lond., Ser. A*, **207**, 171–241.

Love, A. E. H., 1909. The yielding of the Earth to disturbing forces, *Proc. Roy. Soc. Lond., Ser. A*, **82**, 73–88.

Love, A. E. H., 1911. *Some Problems of Geodynamics*, Cambridge University Press, Cambridge. Reprinted in 1967 by Dover Publications, New York.

Love, A. E. H., 1927. *A Treatise on the Mathematical Theory of Elasticity*, Cambridge University Press, Cambridge. Reprinted in 1944 by Dover Publications, New York.

Luh, P. C., 1973. Free oscillations of the laterally inhomogeneous Earth: Quasi-degenerate multiplet coupling, *Geophys. J. Roy. Astron. Soc.*, **32**, 187–202.

Lyttleton, R. A., 1953. *The Stability of Rotating Liquid Masses*, Cambridge University Press, Cambridge.

Madariaga, R. I., 1972. Toroidal free oscillations of the laterally heterogeneous Earth, *Geophys. J. Roy. Astron. Soc.*, **27**, 81–100.

Malvern, L. E., 1969. *Introduction to the Mechanics of a Continuous Medium*, Prentice-Hall, Englewood Cliffs, New Jersey.

Marquering, H. & Snieder, R., 1995. Surface-wave mode coupling for efficient forward modelling and inversion of body-wave phases, *Geophys. J. Int.*, **120**, 186–208.

Marquering, H. & Snieder, R., 1996. Shear-wave velocity structure beneath Europe, the northeastern Atlantic and western Asia from waveform inversions including surface-wave mode coupling, *Geophys. J. Int.*, **127**, 283–304.

Marquering, H., Nolet, G. & Dahlen, F. A., 1998. Three-dimensional waveform sensitivity kernels, *Geophys. J. Int.*, **132**, 521–534.

Marsden, J. E. & Hughes, T. J. R., 1983. *Mathematical Foundations of Elasticity*, Prentice-Hall, Englewood Cliffs, New Jersey.

Maslov, V. P., 1972. *Théorie des Perturbations et Méthodes Asympotiques*, Dunod, Paris.

Maslov, V. P. & Fedoriuk, M. V., 1981. *Semi-Classical Approximation in Quantum Mechanics*, Reidel, Dordrecht.

Masters, G., 1979. Observational constraints on the chemical and thermal structure of the Earth's interior, *Geophys. J. Roy. Astron. Soc.*, **57**, 507–534.

Masters, G., 1989. Seismic modelling of the Earth's large-scale three-dimensional structure, *Phil. Trans. Roy. Soc. Lond., Ser. A*, **328**, 329–349.

Masters, G. & Gilbert, F., 1981. Structure of the inner core inferred from observations of its spheroidal shear modes, *Geophys. Res. Lett.*, **8**, 569–571.

Masters, G. & Gilbert, F., 1983. Attenuation in the Earth at low frequencies, *Phil. Trans. Roy. Soc. Lond., Ser. A*, **308**, 479–522.

Masters, G., Johnson, S., Laske, G. & Bolton, H., 1996. A shear-velocity model of the mantle, *Phil. Trans. Roy. Soc. Lond., Ser. A*, **354**, 1385–1411.

Masters, G., Jordan, T. H., Silver, P. G. & Gilbert, F., 1982. Aspherical Earth structure from fundamental spheroidal mode data, *Nature*, **298**, 609–613.

Masters, G., Park, J. & Gilbert, F., 1983. Observations of coupled spheroidal and toroidal modes, *J. Geophys. Res.*, **88**, 10285–10298.

Masters, G. & Richards-Dinger, K., 1998. On the efficient calculation of ordinary and generalized spherical harmonics, *Geophys. J. Int.*, in press.

Masters, G. & Widmer, R., 1995. Free oscillations: Frequencies and attenuations. In Ahrens, T. J., editor, *Global Earth Physics: A Handbook of Physical Constants*, pages 104–125, American Geophysical Union, Washington, D.C.

McCowan, D. C. & Dziewonski, A. M., 1977. An application of the energy-moment tensor relation to estimation of seismic energy by point and line sources, *Geophys. J. Roy. Astron. Soc.*, **51**, 531–544.

Meissner, E., 1926. Elastische Oberflächen-Querwellen, *Proc. 2d Int. Congr. Appl. Mech.,* Zürich, pages 3–11.

Mendiguren, J., 1973. Identification of free oscillation spectral peaks for 1970 July 31, Colombian deep shock using the excitation criterion, *Geophys. J. Roy. Astron. Soc.*, **33**, 281–321.

Menke, W., 1984. *Geophysical Data Analysis: Discrete Inverse Theory*, Academic Press, New York.

Michelson, A. A. & Gale, H. G., 1919. The rigidity of the Earth, *Astrophys. J.*, **30**, 330–345.

Mie, G., 1908. Beiträge zur Optic trüber Medien, speziell kolloidaler Metallösungen, *Ann. Phys.*, **25**, 377–445.

Mikumo, T., 1968. Atmospheric pressure waves and tectonic deformation associated with the Alaskan earthquake of March 28, 1964, *J. Geophys. Res.*, **73**, 2009–2025.

Miller, W. H., 1974. Classical-limit quantum mechanics and the theory of molecular collisions. In Prigogine, I. & Rice, S. R., editors, *Advances in Chemical Physics*, volume XXV, pages 69–177, Wiley, New York.

Minster, J. B., 1978. Transient and impulse responses of a one-dimensional linearly attenuating medium—I. Analytical results, II. A parametric study, *Geophys. J. Roy. Astron. Soc.*, **52**, 497–501, 503–524.

Minster, J. B., 1980. Anelasticity and attenuation. In Dziewonski, A. M. & Boschi, E., editors, *Fisica dell'interno della Terra, Rendiconti della Scuola Internazionale di Fisica "Enrico Fermi"*, Varenna, Italy, course LXXVII, pages 152–212, North-Holland, Amsterdam,

Minster, J. B. & Jordan, T. H., 1978. Present-day plate motions, *J. Geophys. Res.*, **83**, 5331–5354.

Misner, C. W., Thorne, K. S. & Wheeler, J. A., 1973. *Gravitation*, W. H. Freeman, San Francisco.

Mochizuki, E., 1986. The free oscillations of an anisotropic and heterogeneous Earth, *Geophys. J. Roy. Astron. Soc.*, **86**, 167–176.

Mochizuki, E., 1992. Toroidal oscillations and SH body waves of an aspherical Earth, *Geophys. J. Int.*, **111**, 497–504.

Mochizuki, E., 1994. Asymptotic spheroidal oscillations of a transversely isotropic Earth, *J. Phys. Earth*, **42**, 261–267.

Mooney, W. D., Laske, G. & Masters, G., 1998. CRUST 5.1: A global crustal model at 5 × 5 degrees, *J. Geophys. Res.*, **103**, 727–747.

Morelli, A., Dziewonski, A. M. & Woodhouse, J. H., 1986. Anisotropy of the inner core inferred from PKIKP travel times, *Geophys. Res. Lett.*, **13**, 1545–1548.

Morse, P. M. & Feshbach, H., 1953. *Methods of Theoretical Physics*, McGraw-Hill, New York.

Nábělek, J. L., 1985. Geometry and mechanism of faulting of the 1980 El Asnam, Algeria, earthquake from inversion of teleseismic body waves and comparison with field observations, *J. Geophys. Res.*, **90**, 12713–12728.

Nafe, J. & Brune, J. N., 1960. Observations of phase velocity for Rayleigh waves in the period range 100 to 400 seconds, *Bull. Seismol. Soc. Am.*, **50**, 427–439.

Ness, N. F., Harrison, J. C. & Slichter, L. B., 1961. Observations of the free oscillations of the Earth, *J. Geophys. Res.*, **66**, 621–629.

Nolet, G., 1976. Higher modes and the determination of upper mantle structure, Ph.D. thesis, University of Utrecht.

Nolet, G., 1977. The upper mantle under western Europe inferred from the dispersion of Rayleigh modes, *J. Geophys.*, **43**, 265–285.

Nolet, G., 1990. Partitioned waveform inversion and two-dimensional structure under the Network of Autonomously Recording Seismographs, *J. Geophys. Res.*, **95**, 8499–8512.

Nowick, A. S. & Berry, B. S., 1972. *Anelastic Relaxation in Crystalline Solids*, Academic Press, New York.

Nussenzveig, H. M., 1965. High-frequency scattering by an impenetrable sphere, *Ann. Phys. New York*, **23**, 23–95.

O'Connell, R. J. & Budiansky, B., 1978. Measures of dissipation in viscoelastic media, *Geophys. Res. Lett.*, **5**, 5–8.

Odaka, T., 1978. Derivation of asymptotic frequency equations in terms of ray and normal mode theory and some related problems—Radial and spheroidal oscillations of an elastic sphere, *J. Phys. Earth*, **26**, 105–121.

Okal, E. A., 1982. Mode-wave equivalence and other asymptotic problems in tsunami theory, *Phys. Earth Planet. Int.*, **30**, 1–11.

Okal, E. A., 1990. Single forces and double couples: A theoretical review of their relative efficiency for the excitation of seismic and tsunami waves, *J. Phys. Earth*, **38**, 445–474.

Okal, E. A., 1996. Radial modes from the great 1994 Bolivian earthquake: No evidence for an isotropic component to the source, *Geophys. Res. Lett.*, **23**, 431–434.

Papazachos, C. & Nolet, G., 1997a. Non-linear arrival time tomography, *Annali di Geofisica*, **XL**, 85–97.

Papazachos, C. & Nolet, G., 1997b. P and S deep velocity structure of the Hellenic area obtained by robust nonlinear inversion of travel times, *J. Geophys. Res.*, **102**, 8349–8367.

Park, J., 1986. Asymptotic coupled-mode expressions for multiplet amplitude anomalies and frequency shifts on a laterally heterogeneous Earth, *Geophys. J. Roy. Astron. Soc.*, **90**, 129–169.

Park, J., 1990. The subspace projection method for constructing coupled-mode synthetic seismograms, *Geophys. J. Int.*, **101**, 111–123.

Park, J. & Gilbert, F., 1986. Coupled free oscillations of an aspherical, dissipative, rotating Earth: Galerkin theory, *J. Geophys. Res.*, **91**, 7241–7260.

Parker, R. L., 1994. *Geophysical Inverse Theory*, Princeton University Press, Princeton, New Jersey.

Pekeris, C. L., 1948. Theory of propagation of explosive sound in shallow water, *Geol. Soc. Am. Mem.*, **27**, 117 pages.

Pekeris, C. L., Alterman, Z. & Jarosch, H., 1961a. Comparison of theoretical with observed values of the periods of the free oscillations of the Earth, *Proc. Nat. Acad. Sci. USA*, **47**, 91–98.

Pekeris, C. L., Alterman, Z. & Jarosch, H., 1961b. Rotational multiplets in the spectrum of the Earth, *Phys. Rev.*, **122**, 1692–1700.

Pekeris, C. L. & Jarosch, H., 1958. The free oscillations of the Earth. In Benioff, H., Ewing, M., Howell, Jr., B. F. & Press, F., editors, *Contributions in Geophysics in Honor of Beno Gutenberg*, pages 171–192, Pergamon, New York.

Phinney, R. A. & Burridge, R., 1973. Representation of the elastic-gravitational excitation of a spherical Earth model by generalized spherical harmonics, *Geophys. J. Roy. Astron. Soc.*, **34**, 451–487.

Poisson, S. D., 1829. Mémoire sur l'équilibre et le mouvement des corps élastiques, *Mém. Acad. Roy. Sci. Inst. France*, **8**, 357–570.

Ponzano, G. & Regge, T., 1968. Semiclassical limit of Racah coefficients. In Bloch, F., Cohen, S. G., De-Shalit, A., Tambursky, S. & Talmi, I., editors, *Spectroscopic and Group Theoretical Methods in Physics*, pages 1–58, North-Holland, Amsterdam.

Popov, M. M. & Pšenčík, I., 1976. Ray amplitudes in inhomogeneous media with curved interfaces, *Geofysikální Sborník*, **24**, 111–129.

Poston, T. & Stewart, I., 1978. *Catastrophe Theory and its Applications*, Pitman, Boston.

Poupinet, G. R., Pillet, R. & Souriau, A., 1983. Possible heterogeneity of the Earth's core deduced from PKIKP travel times, *Nature*, **305**, 204–206.

Press, F., 1956. Determination of crustal structure from phase velocity of Rayleigh waves—Part I. Southern California, *Bull. Geol. Soc. Am.*, **67**, 1647–1658.

Press, F. & Ewing, M., 1952. Two slow surface waves across North America, *Bull. Seismol. Soc. Am.*, **42**, 219–228.

Press, W. H., Flannery, B. P., Teukolsky, S. A. & Vetterling, W. T., 1992. *Numerical Recipes: The Art of Scientific Computing*, Cambridge University Press, Cambridge.

Racah, G., 1942. Theory of complex spectra, II, *Phys. Rev.*, **62**, 438–462.

Radau, R., 1885. Sur la loi des densités à l'intérieur de la Terre, *Comptes Rendus Acad. Sci. Paris*, **100**, 972–974.

Randall, M. J., 1976. Attenuative dispersion and frequency shifts of the Earth's free oscillations, *Phys. Earth Planet. Int.*, **12**, P1–P3.

Ray, R. D., Eanes, R. J. & Chao, B. F., 1996. Detection of tidal dissipation in the solid Earth by satellite tracking and altimetry, *Nature*, **381**, 595–597.

Rayleigh, J. W. S., 1877. *The Theory of Sound*, Macmillan, London. Reprinted in 1967 by Dover Publications, New York.

Rayleigh, J. W. S., 1885. On waves propagated along the plane surface of an elastic solid, *Proc. Lond. Math. Soc.*, **17**, 4–11.

Rayleigh, J. W. S., 1906. On the dilational stability of the Earth, *Proc. Roy. Soc. Lond., Ser. A*, **77**, 486–499.

Regge, T., 1958. Symmetry properties of Clebsch-Gordan's coefficients, *Nuovo Cimento*, **10**, 544–545.

Reid, H. F., 1910. *The Mechanics of the Earthquake, Report of the California State Earthquake Investigation Commission*, volume 11, Carnegie Institution of Washington Publication No. 87, Washington, D.C.

Resovsky, J. S. & Ritzwoller, M. H., 1995. Constraining odd-degree Earth structure with coupled free-oscillation overtones, *Geophys. Res. Lett.*, **4**, 372–447.

Resovsky, J. S. & Ritzwoller, M. H., 1998. New and refined constraints on three-dimensional Earth structure from normal modes below 3 mHz, *J. Geophys. Res.*, **103**, 783–810.

Revenaugh, J. & Jordan, T. H., 1991. Mantle layering from ScS reverberations—1. Waveform inversion of zeroth-order reverberations, *J. Geophys. Res.*, **96**, 19749–19762.

Richards, P. G., 1971. An elasticity theorem for heterogeneous media, with an example of body wave dispersion in the Earth, *Geophys. J. Roy. Astron. Soc.*, **22**, 453–472.

Richards, P. G., 1974. Weakly coupled potentials for high-frequency elastic waves in continuously stratified media, *Bull. Seismol. Soc. Am.*, **64**, 1575–1588.

Riedesel, M. A., Agnew, D., Berger, J. & Gilbert, F., 1980. Stacking for the frequencies and Qs of ${}_0\mathrm{S}_0$ and ${}_1\mathrm{S}_0$, *Geophys. J. Roy. Astron. Soc.*, **62**, 457–471.

Ritzwoller, M. H. & Lavely, E. M., 1995. Three-dimensional models of the Earth's mantle, *Rev. Geophys.*, **33**, 1–66.

Ritzwoller, M. H., Masters, G. & Gilbert, F., 1986. Observations of anomalous splitting and their interpretation in terms of aspherical structure, *J. Geophys. Res.*, **91**, 10203–10228.

Ritzwoller, M. H., Masters, G. & Gilbert, F., 1988. Constraining aspherical structure with low-degree interaction coefficients: Application to uncoupled multiplets, *J. Geophys. Res.*, **93**, 6369–6396.

Roberts, P. H. & Ursell, H. D., 1960. Random walk on a sphere and on a Riemannian manifold, *Phil. Trans. Roy. Soc. Lond., Ser. A*, **252**, 317–356.

Robin, L., 1958. *Fonctions Sphériques de Legendre et Fonctions Sphéroïdales*, Gauthier-Villars, Paris.

Rodi, W. L., Glover, P., Li, M. C. & Alexander, S. S., 1975. A fast, accurate method for computing group-velocity partial derivatives for Rayleigh and Love modes, *Bull. Seismol. Soc. Am.*, **65**, 1105–1114.

Romanowicz, B. A., 1987. Multiplet-multiplet coupling due to lateral heterogeneity: Asymptotic effects on the amplitude and frequency of the Earth's normal modes, *Geophys. J. Roy. Astron. Soc.*, **90**, 75–100.

Romanowicz, B. A., 1990. The upper mantle degree 2: Constraints and inferences from global mantle wave attenuation measurements, *J. Geophys. Res.*, **95**, 11051–11071.

Romanowicz, B. A., 1991. Seismic tomography of the Earth's mantle, *Ann. Rev. Earth Planet. Sci.*, **19**, 77–99.

Romanowicz, B. A., 1995. A global tomographic model of shear attenuation in the upper mantle, *J. Geophys. Res.*, **100**, 12375–12394.

Romanowicz, B. A. & Lambeck, K., 1977. Mass and moment of inertia of the Earth, *Phys. Earth Planet. Int.*, **15**, P1–P4.

Romanowicz, B. A. & Roult, G., 1986. First-order asymptotics for the eigenfrequencies of the Earth and application to the retrieval of large-scale lateral variations of structure, *Geophys. J. R. Astron. Soc.*, **87**, 209–239.

Rotenberg, M., Bivins, R., Metropolis, N. & Wooten Jr., J. K., 1959. *The 3-j and 6-j symbols*, The Technology Press, Massachusetts Institute of Technology, Cambridge, Massachusetts.

Roult, G., Romanowicz, B. A. & Montagner, J. P., 1990. 3-D upper mantle shear velocity and attenuation from fundamental mode free oscillation data, *Geophys. J. Roy. Astron. Soc.*, **101**, 61–80.

Russakoff, D., Ekström, G. & Tromp, J., 1998. A new analysis of the great 1970 Colombia earthquake and its isotropic component, *J. Geophys. Res.*, **102**, 20423–20434.

Sailor, R. V. & Dziewonski, A. M., 1978. Measurements and interpretation of normal mode attenuation, *Geophys. J. Roy. Astron. Soc.*, **53**, 559–581.

Saito, M., 1967. Excitation of free oscillations and surface waves by a point source in a vertically heterogeneous Earth, *J. Geophys. Res.*, **72**, 3689–3699.

Sambridge, M. S., 1990. Non-linear arrival time inversion: Constraining velocity anomalies by seeking smooth models in 3-D, *Geophys. J. Int.*, **102**, 653–677.

Satake, K., 1988. Effects of bathymetry on tsunami propagation: Application of ray tracing to tsunamis, *Pure Appl. Geophys.*, **126**, 27–36.

Sâto, Y., 1955. Analysis of dispersed surface waves by means of Fourier transform, *Bull. Earthquake Res. Inst. Tokyo Univ.*, **33**, 33–47.

Sâto, Y., 1958. Attenuation, dispersion and the wave guide of the G wave, *Bull. Seismol. Soc. Am.*, **48**, 231–251.

Savage, J. D., 1969. Steketee's paradox, *Bull. Seismol. Soc. Am.*, **59**, 381–384.

Schiff, L. I., 1968. *Quantum Mechanics*, McGraw-Hill, New York.

Scholte, J. G., 1947. The range of existence of Rayleigh and Stoneley waves, *Mon. Not. Roy. Astron. Soc., Geophys. Suppl.*, **5**, 120–126.

Schulten, K. & Gordon, R. G., 1975a. Exact recursive evaluation of $3j$ and $6j$ coefficients for quantum-mechanical coupling of angular momenta, *J. Math. Phys.*, **16**, 1961–1970.

Schulten, K. & Gordon, R. G., 1975b. Semiclassical approximations to $3j$ and $6j$ coefficients for quantum-mechanical coupling of angular momenta, *J. Math. Phys.*, **16**, 1971–1988.

Schulten, K. & Gordon, R. G., 1976. Recursive evaluation of $3j$ and $6j$ coefficients, *Comp. Phys. Comm.*, **11**, 269–278.

Seidelmann, P. K., 1982. 1980 IAU theory of nutation: The final report of the IAU working group on nutation, *Celest. Mech.*, **27**, 79–106.

Sezawa, K., 1927. Dispersion of elastic waves propagated on the surface of stratified bodies and on curved surfaces, *Bull. Earthquake Res. Inst. Tokyo Univ.*, **3**, 1–18.

Shearer, P. M., 1991. Imaging global body-wave phases by stacking long-period seismograms, *J. Geophys. Res.*, **96**, 20353–20364.

Shearer, P. M., 1994a. Imaging Earth's seismic response at long periods, *EOS, Trans. Am. Geophys. Un.*, **75**, 449–451.

Shearer, P. M., 1994b. Constraints on inner core anisotropy from ISC PKP(DF) data, *J. Geophys. Res.*, **99**, 19647–19660.

Shibata, N., Suda, N. & Fukao, Y., 1990. The matrix element for a transversely isotropic Earth model, *Geophys. J. Int.*, **100**, 315–318.

Silver, P. G. & Jordan, T. H., 1981. Fundamental spheroidal mode observations of aspherical heterogeneity, *Geophys. J. Roy. Astron. Soc.*, **64**, 605–634.

Silver, P. G. & Jordan, T. H., 1982. Optimal estimation of scalar seismic moment, *Geophys. J. Roy. Astron. Soc.*, **70**, 755–787.

Silver, P. G. & Jordan, T. H., 1983. Total-moment spectra of fourteen large earthquakes, *J. Geophys. Res.*, **88**, 3273–3293.

Sipkin, S. A. & Jordan, T. H., 1979. Frequency dependence of Q_{ScS}, *Bull. Seismol. Soc. Am.*, **69**, 1055–1079.

Sipkin, S. A. & Jordan, T. H., 1980. Regional variation of Q_{ScS}, *Bull. Seismol. Soc. Am.*, **70**, 1071–1102.

Slichter, L. B., 1961. The fundamental free mode of the Earth's inner core, *Proc. Nat. Acad. Sci. USA*, **47**, 186–190.

Slichter, L. B., 1967. Spherical oscillations of the Earth, *Geophys. J. Roy. Astron. Soc.*, **14**, 171–177.

Smirnov, V. I., 1964. *A Course in Higher Mathematics*, volume 4, Pergamon, New York.

Smith, B., Boyle, J., Dongarra, J., Garbow, B., Ikebe, Y., Klema, V. & Moler, C., 1976. *Matrix Eigensystem Routines—EISPACK Guide*, Lecture Notes in Computer Science, Springer-Verlag, New York.

Smith, M. F. & Masters, G., 1989a. Aspherical structure constraints from free oscillation frequency and attenuation measurements, *J. Geophys. Res.*, **94**, 1953–1976.

Smith, M. F. & Masters, G., 1989b. The effect of Coriolis coupling of free oscillation multiplets on the determination of aspherical Earth structure, *Geophys. Res. Lett.*, **16**, 263–266.

Smith, M. L., 1976. Translational inner core oscillations of a rotating, slightly elliptical Earth, *J. Geophys. Res.*, **81**, 3055–3065.

Smith, M. L., 1977. Wobble and nutation of the Earth, *Geophys. J. Roy. Astron. Soc.*, **50**, 103–140.

Smith, M. L. & Dahlen, F. A., 1981. The period and Q of the Chandler wobble, *Geophys. J. Roy. Astron. Soc.*, **64**, 223–281.

Smith, S. W., 1966. Free oscillations excited by the Alaskan earthquake, *J. Geophys. Res.*, **71**, 1183–1193.

Snieder, R. & Chapman, C., 1998. The reciprocity properties of geometrical spreading, *Geophys. J. Int.*, **132**, 89–95.

Snieder, R. & Nolet, G., 1987. Linearized scattering of surface waves on a spherical Earth, *J. Geophys.*, **61**, 55–63.

Song, X. D. & Helmberger, D. V., 1995. Depth dependence of anisotropy of Earth's inner core, *J. Geophys. Res.*, **100**, 9805–9816.

Stacey, F. D., 1992. *Physics of the Earth*, Brookfield Press, Brisbane.

Stein, S. & Geller, R. J., 1978. Attenuation measurements of split normal modes for the 1960 Chilean and 1964 Alaskan earthquakes, *Bull. Seismol. Soc. Am.*, **68**, 1595–1611.

Steketee, J. A., 1958. Some geophysical applications of the elasticity theory of dislocations, *Can. J. Phys.*, **36**, 1168–1197.

Stevenson, D. J., 1987. Limits on lateral density and velocity variations in the Earth's outer core, *Geophys. J. R. Astron. Soc.*, **88**, 311–319.

Stoneley, R., 1924. Elastic waves at the surface of separation of two solids, *Proc. Roy. Soc. Lond., Ser. A*, **106**, 416–428.

Stoneley, R., 1925. Dispersion of surface waves, *Mon. Not. Roy. Astron. Soc., Geophys. Suppl.*, **1**, 280–282.

Stoneley, R., 1926a. The effect of the ocean on Rayleigh waves, *Mon. Not. Roy. Astron. Soc., Geophys. Suppl.*, **1**, 349–356.

Stoneley, R., 1926b. The elastic yielding of the Earth, *Mon. Not. Roy. Astron. Soc., Geophys. Suppl.*, **1**, 356–359.

Stoneley, R., 1928. A Rayleigh wave problem, *Proc. Leeds Phil. Lit. Soc. (Sci. Sect.)*, **1**, 217–225.

Stoneley, R., 1949. The seismological implications of aeolotropy in continental structure, *Mon. Not. Roy. Astron. Soc., Geophys. Suppl.*, **5**, 343–353.

Stoneley, R., 1961. The oscillations of the Earth. In Ahrens, L. H., Press, F., Rankama, K. & Runcorn, S. K., editors, *Physics and Chemistry of the Earth*, volume 4, pages 239–250, Pergamon Press, New York.

Stoneley, R. & Tillotson, E., 1928. The effect of a double surface layer on Love waves, *Mon. Not. Roy. Astron. Soc., Geophys. Suppl.*, **1**, 521–587.

Stratton, J. A., 1941. *Electromagnetic Theory*, McGraw-Hill, New York.

Stutzmann, E. & Montagner, J. P., 1993. An inverse technique for retrieving higher mode phase velocity and mantle structure, *Geophys. J. Int.*, **113**, 669–683.

Stutzmann, E. & Montagner, J. P., 1994. Tomography of the transition zone from the inversion of higher mode surface waves, *Phys. Earth Planet. Inter.*, **86**, 99–115.

Su, W.-J., Woodward, R. L. & Dziewonski, A. M., 1994. Degree 12 model of shear velocity heterogeneity in the mantle, *J. Geophys. Res.*, **99**, 6945–6981.

Takeuchi, H., 1950. On the earth tide of the compressible Earth of variable density and elasticity, *Trans. Am. Geophys. Un.*, **31**, 651–689.

Takeuchi, H., 1959. Torsional oscillations of the Earth and some related problems, *Geophys. J. Roy. Astron. Soc.*, **2**, 89–100.

Takeuchi, H., Dorman, J. & Saito, M., 1964. Partial derivatives of surface wave phase velocity with respect to physical parameter changes within the Earth, *J. Geophys. Res.*, **69**, 3429–3441.

Takeuchi, H. & Saito, M., 1972. Seismic surface waves. In Bolt, B. A., editor, *Seismology: Surface Waves and Free Oscillations, Methods in Computational Physics*, volume 11, pages 217–295, Academic Press, New York.

Takeuchi, N., Geller, R. J. & Cummins, P. R., 1996. Highly accurate complete P-SV synthetic seismograms using modified DSM operators, *Geophys. Res. Lett.*, **23**, 1175–1178.

Tams, E., 1921. Über die Fortpflanzungsgeschwindigkeit der seismischen Oberflächenwellen längs kontinentaler und ozeanischer Wege, *Centralblatt Mineral. Geol. Paläntol.*, **2–3**, 44–52, 75–83.

Tanimoto, T., 1986. Free oscillations of a slightly anisotropic Earth, *Geophys. J. Roy. Astron. Soc.*, **87**, 493–517.

Tanimoto, T., 1989. Splitting of normal modes and travel time anomalies due to the magnetic field of the Earth, *J. Geophys. Res.*, **94**, 3030–3036.

Tarantola, A., 1987. *Inverse Problem Theory*, Elsevier, Amsterdam.

Thomson, C. J. & Chapman, C. H., 1985. An introduction to Maslov's asymptotic method, *Geophys. J. Roy. Astron. Soc.*, **83**, 143–168.

Thomson, W., 1863a. On the rigidity of the Earth, *Phil. Trans. Roy. Soc. Lond.*, **153**, 573–582.

Thomson, W., 1863b. Dynamical problems regarding elastic spheroidal shells and spheroids of incompressible liquid, *Phil. Trans. Roy. Soc. Lond.*, **153**, 583–616.

Thomson, W. & Tait, P. G., 1883. *Treatise on Natural Philosophy*, Cambridge University Press, Cambridge.

Toksöz, M. N. & Anderson, D. L., 1966. Phase velocities of long-period surface waves and structure of the upper mantle—1. Great-circle Love and Rayleigh wave data, *J. Geophys. Res.*, **71**, 1649–1658.

Toksöz, M. N. & Ben-Menahem, A., 1963. Velocities of mantle Love and Rayleigh waves over multiple paths, *Bull. Seismol. Soc. Am.*, **53**, 741–764.

Tolstoy, I., 1973. *Wave Propagation*, McGraw-Hill, New York.

Trampert, J. & Woodhouse, J. H., 1995. Global phase velocity maps of Love and Rayleigh waves between 40 and 150 seconds, *Geophys. J. Int.*, **122**, 675–690.

Trampert, J. & Woodhouse, J. H., 1996. High resolution global phase velocity distribution, *Geophys. Res. Lett.*, **23**, 21–24.

Tromp, J., 1993. Support for anisotropy of the Earth's inner core, *Nature*, **366**, 678–681.

Tromp, J., 1995. Normal-mode splitting due to inner-core anisotropy, *Geophys. J. Int.*, **121**, 963–968.

Tromp, J. & Dahlen, F. A., 1990a. Summation of the Born series for the normal modes of the Earth, *Geophys. J. Int.*, **100**, 527–533.

Tromp, J. & Dahlen, F. A., 1990b. Free oscillations of a spherical anelastic Earth, *Geophys. J. Int.*, **103**, 707–723.

Tromp, J. & Dahlen, F. A., 1992a. Variational principles for surface wave propagation on a laterally heterogeneous Earth—I. Time-domain JWKB theory, *Geophys. J. Int.*, **109**, 581–598.

Tromp, J. & Dahlen, F. A., 1992b. Variational principles for surface wave propagation on a laterally heterogeneous Earth—II. Frequency-domain JWKB theory, *Geophys. J. Int.*, **109**, 599–619.

Tromp, J. & Dahlen, F. A., 1993. Maslov theory for surface wave propagation on a laterally heterogeneous Earth, *Geophys. J. Int.*, **115**, 512–528.

Tsuboi, S. & Geller, R. J., 1987. Partial derivatives of synthetic seismograms for a laterally heterogeneous Earth, *Geophys. Res. Lett.*, **14**, 832–835.

Tsuboi, S. & Um, J., 1993. Anomalous amplification of the Earth's normal modes near the epicenter due to lateral heterogeneity, *Geophys. Res. Lett.*, **20**, 2379–2382.

Um, J. & Dahlen, F. A., 1992. Normal mode multiplet coupling on an aspherical, anelastic Earth, *Geophys. J. Int.*, **111**, 11–31.

Um, J., Dahlen, F. A. & Park, J., 1991. Normal mode multiplet coupling along a dispersion branch, *Geophys. J. Int.*, **106**, 111–135.

Unno, W., Osaki, Y., Ando, H., Saio, H. & Shibahashi, H., 1989. *Nonradial Oscillations of Stars*, University of Tokyo Press, Tokyo.

Valette, B., 1986. About the influence of pre-stress upon the adiabatic perturbations of the Earth, *Geophys. J. Roy. Astron. Soc.*, **85**, 179–208.

Van der Hilst, R. D., Widiyantoro, S. & Engdahl, E. R., 1996. Evidence for deep mantle circulation from global tomography, *Nature*, **386**, 578–584.

Van der Lee, S., 1996. The Earth's upper mantle: Its structure beneath North America and the 660 km discontinuity beneath Northern Europe, Ph.D. thesis, Princeton University.

Van der Lee, S. & Nolet, G., 1997. Upper mantle S velocity structure of North America, *J. Geophys. Res.*, **102**, 22815–22838.

Van Heijst, H. J. & Woodhouse, J. H., 1997. Measuring surface-wave overtone phase velocity using a mode-branch stripping technique, *Geophys. J. Int.*, **131**, 209–230.

Varshalovich, D. A., Moskalev, A. N. & Khersonskii, V. K., 1988. *Quantum Theory of Angular Momentum*, World Scientific, London.

Vermeersen, L. L. A. & Vlaar, N. J., 1991. The gravito-elastodynamics of a pre-stressed elastic Earth, *Geophys. J. Roy. Astron. Soc.*, **104**, 555–563.

Vidale, J. E., Goes, S. & Richards, P. G., 1995. Near-field deformation seen on distant broadband seismograms, *Geophys. Res. Lett.*, **22**, 1–4.

Wahr, J. M., 1981a. A normal mode expansion for the forced response of a rotating Earth, *Geophys. J. Roy. Astron. Soc.*, **64**, 651–675.

Wahr, J. M., 1981b. Body tides on an elliptical, rotating, elastic and oceanless Earth, *Geophys. J. Roy. Astron. Soc.*, **64**, 677–703.

Wahr, J. M., 1981c. The forced nutations of an elliptical, rotating, elastic and oceanless Earth, *Geophys. J. Roy. Astron. Soc.*, **64**, 705–727.

Wahr, J. & Bergen, Z., 1986. The effects of mantle anelasticity on nutations, earth tides and tidal variations in rotation rate, *Geophys. J. Roy. Astron. Soc.*, **87**, 633–668.

Wahr, J. & de Vries, D., 1989. The possibility of lateral structure inside the core and its implications for nutation and earth tide observations, *Geophys. J. Int.*, **99**, 511–519.

Wald, D. J. & Heaton, T., 1994. Spatial and temporal distribution of slip for the 1992 Landers, California earthquake, *Bull. Seismol. Soc. Am.*, **84**, 668–691.

Wang, Z. & Dahlen, F. A., 1994. JWKB surface-wave seismograms on a laterally heterogeneous Earth, *Geophys. J. Int.*, **119**, 381–401.

Wang, Z. & Dahlen, F. A., 1995. Validity of surface-wave ray theory on a laterally heterogeneous Earth, *Geophys. J. Int.*, **123**, 757–773.

Wang, Z., Dahlen, F. A. & Tromp, J., 1993. Surface-wave caustics, *Geophys. J. Int.*, **114**, 311–324.

Wang, Z., Tromp, J. & Ekström, G., 1998. Global and regional surface-wave inversions: A spherical-spline parameterization, *Geophys. Res. Lett.*, **25**, 207–210.

Ward, S. N., 1980. Relationships between tsunami generation and an earthquake source, *J. Phys. Earth*, **28**, 441–474.

Watada, S., 1995. Part I: Near-source acoustic coupling between the atmosphere and the solid Earth during volcanic eruptions, Part II: Nearfield normal mode amplitude anomalies of the Landers earthquake, Ph.D. thesis, California Institute of Technology, Pasadena.

Watada, S., Kanamori, H. & Anderson, D. L., 1993. An analysis of nearfield normal mode amplitude anomalies of the Landers earthquake, *Geophys. Res. Lett.*, **20**, 2611–2614.

Whitham, G. B., 1965. A general approach to linear and non-linear dispersive waves using a Lagrangian, *J. Fluid Mech.*, **22**, 273–283.

Whitham, G. B., 1974. *Linear and Non-Linear Dispersive Waves*, Wiley, New York.

Widmer, R., 1991. The large-scale structure of the deep Earth as constrained by free oscillation observations, Ph.D. thesis, University of California, San Diego.

Widmer, R., Masters, G. & Gilbert, F., 1991. Spherically symmetric attenuation within the Earth from normal mode

Widmer, R., Masters, G. & Gilbert, F., 1992. Observably split multiplets—Data analysis and interpretation in terms of large-scale aspherical structure, *Geophys. J. Int.*, **111**, 559–576. data, *Geophys. J. Int.*, **104**, 541–553.

Widmer, R. & Zürn, W., 1992. Bichromatic excitation of long-period Rayleigh and air waves by the Mount Pinatubo and El Chichón volcanic eruptions, *Geophys. Res. Lett.*, **19**, 765–768.

Widmer, R., Zürn, W. & Masters, G., 1992. Observation of low-order toroidal modes from the 1989 Macquarie Rise event, *Geophys. J. Int.*, **111**, 226–236.

Wiggins, R. A., 1972. The general linear inverse problem: Implications of surface waves and free oscillations for Earth structure, *Rev. Geophys. Space Phys.*, **10**, 251–285.

Wigner, E. P., 1959. *Group Theory and its Application to the Quantum Mechanics of Atomic Spectra*, Academic Press, New York.

Williams, J. G., 1994. Contributions to the Earth's obliquity rate, precession and nutation, *Astron. J.*, **108**, 711–724.

Willmore, T. J., 1959. *An Introduction to Differential Geometry*, Oxford University Press, Oxford.

Wolf, A. A., 1969. Rotation operators, *Am. J. Phys.*, **37**, 531–536.

Wong, Y. K., 1989. Upper mantle heterogeneity from phase and amplitude data of mantle waves, Ph.D. thesis, Harvard University.

Woodhouse, J. H., 1974. Surface waves in a laterally varying structure, *Geophys. J. Roy. Astron. Soc.*, **37**, 461–490.

Woodhouse, J. H., 1976. On Rayleigh's principle, *Geophys. J. Roy. Astron. Soc.*, **46**, 11–22.

Woodhouse, J. H., 1978. Asymptotic results for elastodynamic propagator matrices in plane-stratified and spherically-stratified Earth models, *Geophys. J. Roy. Astron. Soc.*, **54**, 263–280.

Woodhouse, J. H., 1980. The coupling and attenuation of nearly resonant multiplets in the Earth's free oscillation spectrum, *Geophys. J. Roy. Astron. Soc.*, **61**, 261–283.

Woodhouse, J. H., 1981a. The excitation of long-period seismic waves by a source spanning a structural discontinuity, *Geophys. Res. Lett.*, **8**, 1129–1131.

Woodhouse, J. H., 1981b. A note on the calculation of travel times in a transversely isotropic Earth model, *Phys. Earth Planet. Int.*, **25**, 357–359.

Woodhouse, J. H., 1983. The joint inversion of seismic waveforms for lateral heterogeneity in Earth structure and earthquake source parameters. In Kanamori, H. & Boschi, E., editors, *Fisica dell'interno della Terra, Rendiconti della Scuola Internazionale di Fisica "Enrico Fermi"*, Varenna, Italy, course LXXXV, pages 366–397, North-Holland, Amsterdam.

Woodhouse, J. H., 1988. The calculation of the eigenfrequencies and eigenfunctions of the free oscillations of the Earth and Sun. In Doornbos, D. J., editor, *Seismological Algorithms: Computational Methods and Computer Programs*, pages 321–370, Academic Press, New York.

Woodhouse, J. H. & Dahlen, F. A., 1978. The effect of a general aspherical perturbation on the free oscillations of the Earth, *Geophys. J. Roy. Astron. Soc.*, **53**, 335–354.

Woodhouse, J. H. & Dziewonski, A. M., 1984. Mapping the upper mantle: Three-dimensional modeling of Earth structure by inversion of seismic waveforms, *J. Geophys. Res.*, **89**, 5953–5986.

Woodhouse, J. H., Giardini, D. & Li, X.-D., 1986. Evidence for inner-core anisotropy from splitting in free oscillation data, *Geophys. Res. Lett.*, **13**, 1549–1552.

Woodhouse, J. H. & Girnius, T. P., 1982. Surface waves and free oscillations in a regionalized Earth model, *Geophys. J. Roy. Astron. Soc.*, **68**, 653–673.

Woodhouse, J. H. & Wong, Y. K., 1986. Amplitude, phase and path anomalies of mantle waves, *Geophys. J. Roy. Astron. Soc.*, **87**, 753–773.

Woods, M. T. & Okal, E. A., 1987. Effect of variable bathymetry on the amplitude of teleseismic tsunamis: A ray-tracing experiment, *Geophys. Res. Lett.*, **14**, 765–768.

Woodward, R. L. & Masters, G., 1991. Upper mantle structure from long-period differential travel times and free oscillation data, *Geophys. J. Int.*, **109**, 275–293.

Yomogida, K., 1985. Gaussian beams for surface waves in laterally slowly varying media, *Geophys. J. Roy. Astron. Soc.*, **82**, 511–533.

Yomogida, K. & Aki, K., 1985. Waveform synthesis of surface waves in a laterally heterogeneous Earth by the Gaussian beam method, *J. Geophys. Res.*, **90**, 7655–7688.

Yuen, D. A. & Peltier, W. R., 1982. Normal modes of the viscoelastic Earth, *Geophys. J. Roy. Astron. Soc.*, **69**, 495–526.

Yuen, P. C., Weaver, P. F., Suzuki, R. K. & Furumoto, A. S., 1969. Continuous, traveling coupling between seismic waves and the ionosphere evident in May 1968 Japan earthquake data, *J. Geophys. Res.*, **74**, 2256–2264.

Zener, C., 1948. *Elasticity and Anelasticity of Metals*, University of Chicago Press, Chicago.

Zhang, Y.-S. & Lay, T., 1996. Global surface wave phase velocity variations, *J. Geophys. Res.*, **101**, 8415–8436.

Zhang, Y.-S. & Tanimoto, T., 1993. High-resolution global upper mantle structure and plate tectonics, *J. Geophys. Res.*, **98**, 9793–9823.

Zhao, L. & Dahlen, F. A., 1993. Asymptotic eigenfrequencies of the Earth's normal modes, *Geophys. J. Int.*, **115**, 729–758.

Zhao, L. & Dahlen, F. A., 1995a. Asymptotic normal modes of the Earth—Part II. Eigenfunctions, *Geophys. J. Int.*, **121**, 585–626.

Zhao, L. & Dahlen, F. A., 1995b. Asymptotic normal modes of the Earth—Part III. Fréchet kernel and group velocity, *Geophys. J. Int.*, **122**, 299–325.

Zhao, L. & Dahlen, F. A., 1996. Mode-sum to ray-sum transformation in a spherical and aspherical Earth, *Geophys. J. Int.*, **126**, 389–412.

Zhao, L. & Jordan, T. H., 1998. Sensitivity of frequency-dependent travel times to laterally heterogeneous, anisotropic Earth structure, *Geophys. J. Int.*, **133**, 683–704.

Zharkov, V. N., 1978. *Physics of Planetary Interiors*, Pachart Publishing House, Tucson, Arizona.

Zharkov, V. N. & Lyubimov, V. M., 1970a. Torsional oscillations of a spherically asymmetrical model of the Earth, *Izv. Bull. Akad. Sci. USSR., Phys. Solid Earth*, **2**, 71–76.

Zharkov, V. N. & Lyubimov, V. M., 1970b. Theory of spheroidal vibrations for a spherically asymmetrical model of the Earth, *Izv., Bull. Akad. Sci. USSR, Phys. Solid Earth*, **10**, 613–618.

Zielhuis, A. & Nolet, G., 1994. Shear wave velocity variations in the upper mantle beneath central Europe, *Geophys. J. Int.*, **117**, 695–715.

Zschau, J., 1978. Tidal friction in the solid Earth: Loading tides versus body tides. In Brosche, P. & Sündermann, J., editors, *Tidal Friction and the Earth's Rotation*, pages 62–94, Springer-Verlag, Berlin.

Zschau, J., 1986. Tidal friction in the solid Earth: Constraints from the Chandler wobble period. In Anderson, A. J. & Cazenave, A., editors, *Space Geodesy and Geodynamics*, pages 315–344, Academic Press, London.

Zürn, W. & Widmer, R., 1996. World-wide observation of bichromatic long-period Rayleigh waves excited during the June 15, 1991 eruption of Mount Pinatubo. In Newhall, C. G. & Punongbayan, R. S., editors, *Fire and Mud, Eruptions of Mount Pinatubo, Philippines*, pages 615–624, Philippine Institute of Volcanology and Seismology, Quezon City and University of Washington Press, Seattle.

Index

www.ingramcontent.com/pod-product-compliance
Ingram Content Group UK Ltd.
Pitfield, Milton Keynes, MK11 3LW, UK
UKHW020845050325
455862UK00012B/471

9 780691 001241